D1213639

Interrelationships
of Fishes

Interrelationships of Fishes

Edited by

Melanie L. J. Stiassny

Department of Ichthyology
American Museum of Natural History
New York, New York

Lynne R. Parenti

Division of Fishes
National Museum of Natural History
Smithsonian Institution
Washington, D.C.

G. David Johnson

Division of Fishes
National Museum of Natural History
Smithsonian Institution
Washington, D.C.

Academic Press

San Diego   London   Boston   New York   Sydney   Tokyo   Toronto

Front cover photograph: Insert photographs are cleared and stained specimens of *Salangichthys* (top), *Horaichthys setnai* (middle), and the caudal skeleton of *Parasadis* (bottom). Background photograph courtesy of Image Club Graphics.

This book is printed on acid-free paper. ∞

Copyright © 1996 by ACADEMIC PRESS

All Rights Reserved.
No part of this publication may be reproduced or transmitted in any form or by any means, electronic or mechanical, including photocopy, recording, or any information storage and retrieval system, without permission in writing from the publisher.

Academic Press, Inc.
525 B Street, Suite 1900, San Diego, California 92101-4495, USA
http://www.apnet.com

Academic Press Limited
24-28 Oval Road, London NW1 7DX, UK
http://www.hbuk.co.uk/ap/

Library of Congress Cataloging-in-Publication Data

Interrelationships of fishes / edited by Melanie L. J. Stiassny, Lynne
    R. Parenti, G. David Johnson.
            p.   cm.
    Includes index.
    ISBN 0-12-670950-5
    1. Fishes--Phylogeny.   I. Stiassny, Melanie L. J.   II. Parenti,
Lynne R.   III. Johnson, G. David.
    QL618.2.I59   1996
    597'.038--dc20   96-28224
                                                      CIP

PRINTED IN THE UNITED STATES OF AMERICA
96 97 98 99 00 01 EB 9 8 7 6 5 4 3 2 1

Contents

CHAPTER

11

Interrelationships of Ostariophysan Fishes (Teleostei)

Sara V. Fink and William L. Fink

CHAPTER

12

Relationships of Lower Euteleostean Fishes

G. David Johnson and Colin Patterson

CHAPTER

13

Interrelationships of Stomiiform Fishes

Antony S. Harold and Stanley H. Weitzman

CHAPTER

14

Interrelationships of Aulopiformes

Carole C. Baldwin and G. David Johnson

Contributors

Numbers in parentheses indicate the pages on which the authors' contributions begin.

Per Erik Ahlberg (445) Department of Palaeontology, The Natural History Museum, London SW7 5BD, United Kingdom

Carole C. Baldwin (355) Department of Vertebrate Zoology, National Museum of Natural History, Smithsonian Institution, Washington, D.C. 20560

William E. Bemis (85) Department of Biology, University of Massachusetts, Amherst, Massachusetts 01003

Richard Cloutier (445) Université des Sciences et Technologies de Lille, Sciences de la Terre, 59655 Villeneuve d'Ascq., France

Marcelo R. de Carvalho (35) Department of Herpetology and Ichthyology, American Museum of Natural History, New York, New York 10024, and Graduate School and University Center of The City University of New York, New York, New York

Mário C. C. de Pinna (147) Departamento de Zoologia, Instituto de Biociencias, Universidade de Sao Paulo, Sao Paulo-SP 05422-970, Brazil

Katherine A. Dunn (63) Department of Wildlife and Fisheries Sciences, Texas A&M University, College Station, Texas 77843

Sara V. Fink (209) Museum of Zoology, The University of Michigan, Ann Arbor, Michigan 48105

William L. Fink (209) Museum of Zoology, The University of Michigan, Ann Arbor, Michigan 48105, and Department of Biology, The University of Michigan, Ann Arbor, Michigan 48105

P. L. Forey (175) Department of Palaeontology, The Natural History Museum, London SW7 5BD, United Kingdom

Brian G. Gardiner (1, 117) Division of Life Sciences, Department of Biological Sciences, King's College London, London W8 7AH, United Kingdom

Lance Grande (85) Department of Geology, Field Museum of Natural History, Chicago, Illinois 60605

Li Guo-Qing (163) Department of Biological Sciences and Laboratory for Vertebrate Paleontology, University of Alberta, Edmonton, Alberta T6G 2E9, Canada, and Institute of Vertebrate Paleontology and Paleoanthroplogy, Academia Sinica, Beijing 100044, People's Republic of China

Antony S. Harold (333) Department of Ichthyology and Herpetology, Royal Ontario Museum, Toronto, Ontario M5S 2C6, Canada

G. David Johnson (251, 355) Division of Fishes, National Museum of Natural History, Smithsonian Institution, Washington, D.C. 20560

G. Lecointre (193) Laboratoire d'Ichtyologie Générale et Appliqué, Muséum National d'Histoire Naturelle, 75231 Paris Cedex 05, France

D. T. J. Littlewood (117, 175) Department of Palaeontology, The Natural History Museum, London SW7 5BD, United Kingdom

John G. Maisey (117) Department of Vertebrate Paleontology, American Museum of Natural History, New York, New York 10024

John D. McEachran (63) Department of Wildlife and Fisheries Sciences, Texas A&M University, College Station, Texas 77843

A. Meyer (175) Department of Biology, State University of New York at Stony Brook, Long Island, New York 11794

Tsutomu Miyake (63) Department of Biology, University of Dalhousie, Halifax, Nova Scotia B3H 411, Canada

G. Nelson (193) Department of Ichthyology, American Museum of Natural History, New York. New York 10024, and School of Botany, The University of Melbourne, Victoria 3052, Australia

Lynne R. Parenti (427) Division of Fishes, National Museum of Natural History, Smithsonian Institution, Washington, D.C. 20560

Colin Patterson (251) Department of Paleontology, Natural History Museum, London SW7 5BD, United Kingdom

P. Ritchie (175) Department of Biology, State University of New York at Stony Brook, Long Island, New York 11794

Bobb Schaeffer (1) Curator Emeritus, American Museum of Natural History, New York, New York 10024

Shigeru Shirai (9) Marine Ecology Section, Seikai National Fisheries Research Institute, Nagasaki 850, Japan

Jiakun Song (427) Division of Fishes, National Museum of Natural History, Smithsonian Institution, Washington, D.C. 20560, and Department of Zoology, University of Maryland, College Park, Maryland 20742

Melanie L. J. Stiassny (405) Department of Ichthyology, American Museum of Natural History, New York, New York 10024

Stanley H. Weitzman (333) Division of Fishes, National Museum of Natural History, Smithsonian Institution, Washington, D.C. 20560

Mark V. H. Wilson (163) Department of Biological Sciences and Laboratory for Vertebrate Paleontology, University of Alberta, Edmonton, Alberta T6G 2E9, Canada

Preface

As editors of this volume we are well aware that the first *Interrelationships of Fishes* presents us with what must surely be considered a "hard act to follow," and perhaps not surprisingly we initially greeted the invitation to produce a "new" *Interrelationships* with some trepidation. However, after a generally enthusiastic and encouraging response from our colleagues when we mooted the idea, we soon became convinced that the time was right for a reappraisal of the "state of the art" of fish systematics.

The first *Interrelationships*, published almost a quarter of a century ago, was one of the most influential books in the modern literature of ichthyology. Looking back, it now seems clear that its profound influence was due in large part to the volume's essentially phylogenetic orientation. With that methodological framework, the first *Interrelationships* provided the community with the outlines of a model for more rigorous study of the relationships of fishes, and as a result played a pivotal role in ushering-in the "modern age" of systematic ichthyology. As Humphry Greenwood, Roger Miles, and Colin Patterson noted in their preface, "Readers of the volume may be struck by similarities in the way that most of the contributors have approached the problem of investigating and describing relationships. These methods, more precise and explicit than those traditionally used, are due to the influence of Hennig, not to editorial pressure" (1973, ix). The "influence of Hennig" and the many refinements of phylogenetic methodology and techniques of data analysis continue to invigorate the science of systematics, and in ichthyology, as in other disciplines, the result is an active, ongoing process of evaluation and reestimation of historical relationships.

Since the publication of the first *Interrelationships* in 1973, there have been many revisions and reappraisals, and in some ways the present volume is a summary of where these new inquiries have taken us.

The first volume had as its original objective "a survey of the relationships between and within all the major groups of gnathostome fishes, living and extinct" (1973, IX). With this new volume our aim has been, as far as possible, to present a broad coverage of the phylogenetic systematics of fishes that updates the 1973 volume as well as complements subsequent publications. Since 1973 a number of influential volumes have been published on the systematics and biology of higher fish taxa or pivotal groups, such as lungfishes (Bemis *et al.*, 1987), coelacanths (McCosker and Lagios, 1979; Musick *et al.*, 1991), placoderms (Denison, 1978), acanthodians (Denison, 1979), chondrichthyians (Cappetta, 1987), gadiforms (Cohen, 1989), and assorted percomorphs and their allies (Johnson and Anderson, 1993). Another compendium of note in this context is the so-called "Ahlstrom Symposium" (Moser *et al.*, 1984), which approached the question of teleostean interrelationships from an explicitly ontogenetic perspective. Most recently, three volumes concerned primarily with early vertebrate systematics and evolution have appeared (Chang *et al.*, 1991; Arsenault *et al.*, 1995; Lelièvre *et al.*, 1995). It is worthy of note that the phylogenetic analyses in the current volume, as with most of these others cited, utilize characters principally of the skeleton and soft anatomy of fishes. This strongly morphological emphasis reflects our view that these systems contain a vast reservoir of untapped information for phylogenetic systematics and, it is hoped, serves also to underline the enduring role of morphological data in contemporary systematic studies. Given our explicitly morphological orientation, we believe that this new volume also nicely complements the upcoming volume on the molecular systematics of teleost fishes (Kocher and Stepien, submitted for publication).

In October 1993, at the age of 60, Colin Patterson officially retired from the Natural History Museum, London, where he continues his research with undiminished fervor and productivity. It is with much satisfaction that we mark that milestone by dedicating this volume to him in celebration of his remarkable and enduring influence on the field of fish systematics. One can only imagine what the state of the art in this field would be today had it been denied the intellect and published works of Colin Patterson. Clearly, it would have been profoundly different, and our knowledge of the anatomy and history of fishes would be considerably less advanced. Like few before him, he has bridged the gap between paleontology and neontology, and in recent years has done the same with morphology and molecules. While Colin has always remained at the forefront of systematic theory and methodology, he is first and foremost a consummate comparative morphologist who retains a deep and abiding interest in, and respect for, the primary data of our discipline. The search for pattern in the anatomy of fishes remains the driving passion in his work, and it is this descriptive aspect that he believes will be his greatest legacy—as he has often been heard to say: "... get the anatomy wrong and the rest is meaningless." Regardless of current fashion in theory or level of analytical sophistication, to know that the anatomy was done properly, with care, accuracy, and critical insight, is to know that it will last and be relied upon by generations to come.

While there can be no doubt that the computational and analytical advances of recent years have brought a new rigor to systematics, they have sometimes also been accompanied by a misplaced notion of "objectivity" that manifests as a shift from knowledge of organisms and their characters to a preoccupation with the intricacies of computer-generated analysis. Two, often conflated, trends are evident—in one, the computational analysis is a block box "into which data are fed and out of which 'The Tree' springs"; in the other, the elaboration and sophistication of analysis are somehow seen to substitute for a solidly worked and comprehensive character base. Both are misguided. It is ironic that a result that may be both computationally accurate and robust may not necessarily approximate the underlying phylogeny. Ultimately, the results of any analysis are as meaningful, or as meaningless, as the data on which they are based. We believe that it is imperative that we do not lose sight of our organisms, and while, of course, solid and well-conceived analyses are essential, the enduring core of our contribution will ultimately be judged by the richness of our observations.

We are grateful to the following individuals who reviewed one or more chapters prior to publication: Marc Allard, Carole C. Baldwin, Robert L. Boord, Marcelo R. de Carvalho, Mario C. C. de Pinna, Dominique Didier, William L. Fink, Brian G. Gardiner, Anthony C. Gill, Lance Grande, P. Humphry Greenwood, Robert K. Johnson, Tom Kocher, Nathan R. Lovejoy, John G. Maisey, Richard L. Mayden, Amy R. McCune, John D. McEachran, John Morrissey, Randall D. Mooi, Jon A. Moore, Gavin Naylor, Colin Patterson, John R. Paxton, Hans-Peter Schultze, David G. Smith, Victor G. Springer, Richard P. Vari, Ed O. Wiley, and Richard Winterbottom.

For their help with proofreading and numerous other tasks we are particularly grateful to Lita Elvers, Monica Toledo-Piza, Marcelo de Carvalho, and Carole C. Baldwin. Our thanks also to Joseph S. Nelson and Mark V. H. Wilson, who graciously hosted the 75th annual meeting of the American Society of Ichthyologists and Herpetologists where a number of contributors to this volume presented their preliminary findings in a symposium on the Interrelationships of Fishes. Our editor, Charles R. Crumly, at Academic Press initiated this project and we are grateful to him for entrusting to us the task of bringing it to fruition. And last, but by no means least, our thanks to all of the contributors to this volume—it has been something of a "long haul" to get here and the perseverance and patience of all concerned are much appreciated.

Melanie L. J. Stiassny
Lynne R. Parenti
G. David Johnson

Arsenault, M., Lelièvre, H., and Janvier, P. (eds.) (1995). VIIth International Symposium Studies of Early Vertebrates. Bulletin du Musèum National d'Histoire Naturelle, Paris.

Bemis, W. E., Burggren, W. W., and Kemp, N. E. (eds.) (1987). The biology and evolution of lungfishes. A. R. Liss, New York.

Cappetta, H. (1987). Chondrichthyes II, Mesozoic and Cenozoic Elasmobranchii. *In* "Handbook of Paleoichthyology, 3B" (H.-P. Schultze, ed.), Stuttgart, Gustav Fischer Verlag.

Chang, M.-M., Y.-H., Lin, and Zhang, G.-R. (eds.) (1991). Early vertebrates and related problems of evolutionary biology. Science Press, Beijing.

Cohen, D. M. (ed.) (1989). Papers on the systematics of gadiform fishes. No. 32. Science Series, The Natural History Museum of Los Angeles County.

Denison, R. H. (1978). Placodermi. *In* "Handbook of Paleoichthyology, 2" (H.-P. Schultze, ed.), Stuttgart, Gustav Fischer Verlag.

Denison, R. H. (1979). Acanthodii. *In* "Handbook of Paleoichthyology, 5" (H.-P. Schultze, ed.), Stuttgart, Gustav Fischer Verlag.

Johnson, G. D., and Anderson, Jr., W. D. (eds.) (1993). Proceedings of the symposium on phylogeny of Percomorpha. *Bull. Marine Science*, **52(1),** 1–626.

Lelièvre, H., Wenz, S., Blieck, A., and Cloutier, R. (eds.) (1995). Premiers Vertébrés et Vertébrés Inférieurs. Geobios, Mémoire spécial no. 19.

McCosker, J. E., and Lagios, M. D. (eds.) (1979). The biology and physiology of the living coelacanth. Occasional papers of the California Academy of Sciences, No. 134.

Moser, H. G., Richards, W. J., Cohen, D. M., Fahay, M. P., Kendall, Jr., A. W., and Richardson, S. L. (eds.) (1984). Ontogeny and systematics of Fishes. Spec. Publ. 1, American Society of Ichthyologists and Herpetologists.

Musick, J. A., Bruton, M. N., and Balon, E. K. (eds). (1991). The biology of *Latimeria chalumnae* and evolution of coelacanths. Environmental Biology of Fishes, 32.

Colin Patterson, October 1995

1

An Annotated Bibliography of the Work of Colin Patterson

BOBB SCHAEFFER
Curator Emeritus
American Museum of Natural History
New York, New York, 10024

BRIAN G. GARDINER
Division of Life Sciences
King's College London, Atkins Building
Campden Hill Road
London W8 7AH, England

A perusal of Colin Patterson's publications indicates that they can be somewhat arbitrarily divided into several categories or topics such as comparative morphology, theory and practice of cladistics as related to systematics, phylogeny as related to homology and ontogeny, paleobiogeography, molecular systematics, and evolutionary theory. But the central theme underlying this broad spectrum is an abiding interest in the history of life and in the modes and techniques available to interpret it. Several of Patterson's early papers concentrated on interpreting the morphology of the skull and tail in both fossil and Recent actinopterygians. With the onset of the cladistic revolution in the late sixties, his research took on new meaning as he enthusiastically embraced phylogenetic systematics.

Patterson was born in London on 13 October 1933. He was educated at Tonbridge School and Imperial College from which he graduated in 1957 with 1st Class Honors in parasitology (at that time the college had only two programs in zoology: parasitology and entomology). At the end of his second undergraduate year, persuaded that fossils (particularly vertebrates) would reveal the true course of evolution, he applied for a vacation studentship at the Natural History Museum in London. And there in the summer of 1956 he got his first exposure to fossil fishes and met Errol White.

Between 1957 and 1962 Patterson was a lecturer in biology at Guy's Hospital Medical School, London.

During this interval he also started a detailed study of teleosts in the Natural History Museum (for his Ph.D.) and was attracted to the large collection of fishes from the Cretaceous English Chalk. With the approval and encouragement of Keeper Errol White, he employed the new acid technique to expose the largely unknown details of the acanthopterygian and ctenothrissiform teleosts from the English Chalk. The description of these fishes and their extant relatives, along with comments on related taxa from other areas, resulted in a new classification of the beryciforms and a consideration of acanthopterygian ancestry. Patterson concluded that the perciforms as then classified were polyphyletic and that several lineages independently attained the beryciform level. Well illustrated by his own hand, this work was accepted as his Ph.D. thesis and was published by the Royal Society in 1964 (2).

In 1962 Patterson joined the permanent staff of the museum, as scientific officer in charge of the fossil fish collection, and rapidly became the epitome of a conscientious employee engaging in both research and museum display.

The need for museum displays temporarily channeled his early research away from his favorite teleosts into the phylogeny of chondrichthyans, particularly chimaeroids (3,11) and Wealden sharks (4). Discussion in 1966 of the diverse dentitions of the British Wealden hybodont sharks was no doubt prompted

1

Copyright © 1996 by Academic Press, Inc.
All rights of reproduction in any form reserved.

by the abundance of the sharks in the collections of the Natural History Museum. The needs of identification are reflected in his much later paper on chimaeroid tooth plates (67).

During 1967 and 1968 he published six papers, several with Greenwood, on berycoids and osteoglossomorphs. Increasing interest in the systematic potential of the teleost caudal fin skeleton, as demonstrated by Monod's caudal atlas, and further expounded by Nybelin in 1963, led to a detailed analysis of the caudal skeleton in the early Jurassic pholidophoroids (10), in which his recognition of skeletal elements called urodermals and uroneurals gave the first insight into modifications in the dermal–endochondral skeleton and allowed him to consider the pholidophorids as generalized teleosts. In the same year he discussed the complicated caudal fin morphology of Mesozoic acanthopterygian fishes (12).

In the late 1960s, Rosen discovered in Patterson a colleague with similar interests and a wealth of knowledge about extinct as well as living teleosts. They published their first paper together in 1969 (13) on paracanthopterygians, including discussion of fossil taxa. They analyzed the caudal skeleton and the jaw musculature to elucidate the interrelationships of numerous proposed monophyletic groups, which together they regarded as the sister-group of the acanthopterygians. A second joint contribution (22) involved a critical study of the Mesozoic ichthyodectiform fishes, a cladistic analysis (pre-PAUP) of the major groups of teleosts, a "framework" of neopterygian interrelationships of fossil and recent teleosteans, and finally a commentary on the classification of fossils in relation to Mesozoic and Recent teleosts. In this last section, they considered the problem of classifying fossils in relation to extant organisms. With the rise of cladistic analysis, this problem became more controversial. Patterson and Rosen offered a novel but now rarely used solution—the fossil taxa "sequenced in a classification . . . should be called *plesions* . . . and may be inserted at any level in a classification without altering the rank or name of any other group."

Several outstanding papers appeared in the seventies that remain important sources of information for neopterygian systematic analysis. One on the interrelationships of the holosteans (17) critically reviewed the status of many non-teleost actinopterygians and set the stage for the characterization of the teleosts, which he showed included the pholidophorids. In a monumental study of the actinopterygian braincase (20), based to a large extent on acid-prepared fossil specimens, and including comparison with various extant taxa, he demonstrated the importance of neurocranial ossification patterns in distinguishing major

actinopterygian groups from the lower actinopterygians ("palaeonisciforms") to the generalized teleosts. As one reviewer said, "Although there are no practical limits to comparative studies, and further development may be confidently predicted, it is fair to conclude that in a hundred years time Patterson's work will stand much as it does now, a milestone in comparative paleontology." And last but not least Patterson reviewed the various types of vertebrate bone or the exoskeleton *versus* the endoskeleton (23).

Observations about the meaning and practice of cladistics are discussed or implied in most of Patterson's writings since the middle 1970s. He, like a few older survivors, attended Brundin's lecture on phylogenetic principles during the 1967 Nobel Symposium on current problems in lower vertebrate phylogeny. Following Brundin, Nelson and Patterson carried the message to New York and London (24,25,28,30,58). In spite of opposition, it spread rapidly thereafter. Colin's ideas about cladistics are tied to his thinking about homology (38,54), phylogeny, ontogeny (40), the role of fossils in deciphering evolutionary relationships, taxonomy, classification (34,35), perceived patterns of evolution, and molecular systematics (48). These contributions cannot be properly summarized here, so we shall mention only some of his most salient observations. A short paper on cladistics as "a method of systematics" (30) explains succinctly this strategy in relation to the controversy surrounding it at that time. He noted that "the cladistic method forces systematists to be explicit about the groups they recognize, and the characters of those groups—that is, the characters they regard as homologies." Authoritarianism no longer rules?

Based on his studies of Triassic and Jurassic neopterygians, Patterson (24) provided an original approach to teleost phylogeny. By analyzing 52 characters in 18 taxa, he produced a cladogram that, in effect, represents a theory of teleost relationships keyed to a time axis. In his words, "the main function (of the cladogram) is not to suggest ancestor–descendant sequences, but to give an indication of the sequence in which the various teleost characters were acquired, with an approximate time scale." The cladogram is rooted in the halecomorphs (*Amia*) and ends with the euteleosts. The characters at certain basal nodes indicate polymorphism and a few questionable interpretations due to poor preservation. Without further comment, we recommend this contribution for its discussion of problems related to paraphyly, absent characters, convergence, and others that may influence the selection of extinct taxa. In any case, this paper represents an effort to consider change through time in some detail and within a cladistic framework.

Regarding the significance of fossils in determining evolutionary relationships, he has expressed his views in several papers (28,29,34,35). He has stated that fossils alone cannot determine evolutionary relationships. This must be first accomplished by analyzing living taxa. Fossils may influence polarity decisions and, on occasion, support a relationship decision arrived at through neontology. Now that the focus in paleontological research has moved away from a search for ancestors, fossils may be used to corroborate and elaborate theories based initially on extant forms. Patterson admits that his conclusions regarding teleost relationships have been influenced by his prior knowledge of fossil teleosts and halecomorphs (24), but he notes that fossils have rarely overturned theories based on the relationships of Recent groups.

For a highly diversified supergroup such as the teleosts, Patterson noted (36) that the assignment of a fossil taxon to a Recent group must be based on shared synapomorphies and that both extinct and extant groups must be monophyletic. He has proposed that fossils, sometimes poorly preserved, do not always permit recognition of synapomorphies. Such nominal taxa should be regarded as *incertae sedis*. He suggests two standards against which the status of a presumably related fossil may be checked. One, as noted, involves the relationships of Recent taxa (based on morphology), and the second involves an independent check of relationships through molecular analysis of the Recent forms.

A description of Upper Jurassic fishes from the western United States with Schaeffer (44) produced some enigmatic problems in classification. Of the seven taxa in this assemblage, four are known from Europe and elsewhere (†*Hybodus,* †*Ischyodus,* †*Lepidotes,* and †*Caturus*). Three are new, monotypic genera. Each of the latter has an uncertain affinity at some taxonomic level. †*Hulettia* can be only assigned to the Halecostomi *incertae sedis,* †*Occithrissops* to the Ichthyodectiformes *incertae sedis,* and †*Todiltia* to the Teleostei *incertae sedis.* As Patterson has noted several times, cladistic analysis may lead to uncertainty regarding taxonomic relationships, which must lead to the use of *incertae sedis,* as opposed to traditional methodology, which in turn can only lead to "spurious certainty."

Patterson (29) presented a succinct historical review of opinions on tetrapod origins around the time when paleontologists were beginning to think about phylogeny reconstruction in terms of cladistic strategy. He noted that the osteolepiform rhipidistians as well as the Paleozoic amphibians should be regarded as polyphyletic groups and that this situation had led to obvious disagreement regarding the interpretation of the

sarcopterygian fish-tetrapod interrelationship. The following year, Rosen, Forey, Gardiner, and Patterson (32) published a controversial paper, which was perhaps the first attempt to utilize the concepts of phylogenetics as the way to overcome the "sadly tenuous" and "wishful thinking" that characterized many previous papers on this topic. A broad spectrum of morphological comparison and interpretation led them to the conclusion that tetrapods are the sister-group of lungfishes and that coelacanths are the sister-group of these two.

The definition and significance of homology in the cladistic sense have played an important role in Patterson's thinking about monophyly and natural groups. He has emphasized that homology and synapomorphy are identical, and he recognizes three tests for homology and synapomorphy. The most important is congruence (no conflict in homology recognition). The others are conjunction (obviously two homologues may not occur in the same organism) and similarity in ontogeny, topography, and tissue composition. Annotation of Patterson's (38,54) papers on homology, because of complexity and comprehensiveness, is beyond further comment here—except to note one of his final statements, "that the role of homology in phylogenetic reconstruction is limited to the production of cladograms, or classification (54)."

Several more general papers need comment. One on the relationship between phylogeny and ontogeny (40) includes a particularly cogent remark that clarifies the role of the ontogeny in an evolutionary context: "... since ontogenetic transformation is consistently observed to be in one direction, and never in reverse, we have direct evidence of transformation, and may rate the untransformed state as more general (primitive) than the transformed state (derived)." Also, "without the transformations of ontogeny, which by themselves define nested sets (outgroup + ingroup → ingroup), systematics would be impossible...."

Another (55) is a documented commentary on the ways that taxonomic method influences our perception of evolutionary patterns. The point of the commentary, with A. B. Smith, is that arbitrarily interpreted taxa will not lead to valid paleobiological observations. As Patterson has stated numerous times, "arbitrary taxa produce spurious results." These observations pertain to nonmonophyletic groups and attempts at dealing with rates and periodicity of extinction. Examples are drawn mostly from families and genera of echinoderms and fishes.

Current investigation in historical biogeography involves cladistic analysis of fossil organisms (when appropriate) and, of course, extant ones. It also requires an understanding of continental configurations and relationships during the time period under consideration. In Patterson's papers on this topic (16,31,33,41) he has stressed that the fossil record may be uninformative because the remains of extinct organisms frequently offer too few characters to permit a meaningful hypothesis of relationship. Vicariance biogeography has a limited meaning because an understanding of the interrelationships of the organisms comprising a biota is usually fragmented through tectonic and climatic change. Hence, as Patterson (31) has noted, research in historical biogeography requires the integration of cladistic information for all pertinent groups of plants and animals. Still another controversial topic involves the evidence for an earth of constant diameter *versus* one that has expanded to its present diameter during the last 180–200 million years. This problem has recently been discussed by Patterson and Owen (63) in reference to the apparent absence of endemicity in the Jurassic–Recent fishes of India which should be expected in a continent isolated for 100 million years. Their opinion is that India came into contact with Asia in the early Paleocene. The jury remains out on this one.

A major reorientation, or better, a supplement to Patterson's vision of how to deal with phylogeny, is evident in his contributions to symposium volumes primarily concerned with molecular phylogenetics (48,58,86). In both he returned to the fundamentals of cladistic theory, particularly as they are shared (or not shared) by the morphological and molecular approaches. These include bifurcating trees (dichotomous diagrams) along with similarities and differences in the meaning of homology. Both share "a range of parsimony-based packages" (58, p. 473). Morphologists have emphasized the search for synapomorphies and data (characters), while molecular systematists have necessarily concentrated on the recovery of meaningful molecular data and the ways of analyzing it. At the same time (48) he also outlined and classified the many problems associated with molecular homology.

A central problem is to reconcile an overall phylogenetic pattern based on morphological and molecular sequence data (75). In the morphological mode, character selection, shrewd analysis, and evaluation of results in terms of parsimony and congruence between data sets are essential. However, methods of sampling molecular sequence data are still controversial because of their noise and contradictory signals. As Patterson notes, this is evident from analysis of eukaryotic molecular data which shows conflict between distance analysis and parsimony (58). Patterson rightly concludes that reconciliation between morphological and molecular data remains a primary objective in systematics. The debate has since been formalized

in the issues of "total evidence" *versus* "consensus" (see Kluge and Wolfe, *Cladistics*: 1994). In a subsequent paper with Williams and Humphries (75) the issue of molecules *versus* morphology was again addressed but this time in both green plants and metazoans. The authors concluded that consensus was difficult due to problems of sampling, i.e., different molecules sampled different taxa, and molecules generally were undersampled relative to morphological studies. Although the molecular surpass the morphological in quantity, these data are one-dimensional, while the morphological are supported by transformation during ontogeny plus support from the fossil record. Although potential for the resolution of phylogenetic problems by molecular analysis seems evident, further diversification of both morphological and molecular sampling will surely improve resolution. Parenthetically, the one-dimensionality of molecular analysis may lead to a phylogeny, but only the morphological approach will provide meaning to the phylogeny in organismic terms (85).

In conclusion we wish to mention Patterson's perspicacity; this has proved to be a veritable magnet attracting a steady stream of visitors including zoologists, paleontologists, molecular biologists, science historians, philosophers, editors, and radio and television interviewers from all over the world to the fossil fish section at London's Natural History Museum. In the late 1980s, when the fate of taxonomy everywhere and the whole systematic future of the Natural History Museum was in question, Patterson was invited to give evidence before the House of Lords subcommittee on systematic biology research. His evidence helped convince the government that the Natural History Museum is an original, unparalleled systematic institution which should be protected and encouraged.

Finally, Patterson was elected to the Royal Society in March 1993, a long overdue recognition for his outstanding originality and for years of service to biological science including a 7-year sojourn as editorial secretary and zoological editor of the Linnean Society.

Bibliography

1. Griffith, J., and Patterson, C. (1963). The structure and relationships of the Jurassic fish *Ichthyokentema purbeckensis*. *Bull. Br. Mus. (Nat. Hist.), Geol.* **8,** 1–43.

2. Patterson, C. (1964). A review of Mesozoic acanthopterygian fishes, with special reference to those of the English Chalk. *Philos. Trans. R. Soc. London, Ser. B* **247,** 213–482.

3. Patterson, C. (1965). The phylogeny of the chimaeroids. *Philos. Trans. R. Soc. London, Ser. B* **249,** 101–219.

4. Patterson, C. (1966). British Wealden sharks. *Bull. Br. Mus. (Nat. Hist.), Geol.* **11,** 281–350.

5. Patterson, C. (1967). New Cretaceous berycoid fishes from the Lebanon. *Bull. Br. Mus. (Nat. Hist.), Geol.* **14**, 67–109.

6. Patterson, C. (1967). A second specimen of the Cretaceous teleost *Protobrama* and the relationships of the suborder Tselfatioidei. *Ark. Zool.* [2] **19**, 215–234.

7. Greenwood, P.H., and Patterson, C. (1967). A fossil osteoglossoid fish from Tanzania (E. Africa). *J. Linn. Soc. London, Zool.* **47**, 211–223.

8. Patterson, C. (1967). Teleostei; Elasmobranchii. *In* "The Fossil Record" (W. B. Harland *et al.*, eds.), pp. 654–683. Geological Society, London.

9. Patterson, C. (1967). Are the teleosts a polyphyletic group? *Colloq. Int. C.N.R.S.* **163**, 93–109.

10. Patterson, C. (1968). The caudal skeleton in Lower Liassic pholidophorid fishes. *Bull. Br. Mus. (Nat. Hist.), Geol.* **16**, 201–239.

11. Patterson, C. (1968). *Menaspis* and the bradyodonts. *In* "Current Problems of Lower Vertebrate Phylogeny" (T. Ørvig, ed.), Nobel Symp. 4, pp. 171–205. Almqvist & Wiksell, Stockholm.

12. Patterson, C. (1968). The caudal skeleton in Mesozoic acanthopterygian fishes. *Bull. Br. Mus. (Nat. Hist.), Geol.* **17**, 47–102.

13. Rosen, D. E., and Patterson, C. (1969). The structure and relationships of the paracanthopterygian fishes. *Bull. Am. Mus. Nat. Hist.* **141**, 357–474.

14. Patterson, C. (1970). A clupeomorph fish from the Gault (Lower Cretaceous). *Zool. J. Linn. Soc.* **49**, 161–182.

15. Patterson, C. (1970). Two Upper Cretaceous salmoniform fishes from the Lebanon. *Bull. Br. Mus. (Nat. Hist.), Geol.* **19**, 205–296.

16. Cressey, R. and Patterson, C. (1973). Fossil parasitic copepods from a Lower Cretaceous fish. *Science* **180**, 1283–1285.

17. Patterson, C. (1973). Interrelationships of holosteans. *In* "Interrelationships of Fishes" (P. H. Greenwood, R. S. Miles and C. Patterson, eds.), pp. 233–305. Academic Press, London.

18. Patterson, C. (1974). Atheriniformes. "Encyclopaedia Britannica," 15th ed., Vol. 2, pp. 269–274. Encyclopedia Britannica, London.

19. Patterson, C. (1974). Elopiformes. "Encyclopaedia Britannica," 15th ed., Vol. 6, pp. 729–731. Encyclopaedia Britannica, London.

20. Patterson, C. (1975). The braincase of pholidophorid and leptolepid fishes, with a review of the actinopterygian braincase. *Philos. Trans. R. Soc. London, Ser. B* **269**, 275–579.

21. Patterson, C. (1975). The distribution of Mesozoic freshwater fishes. *Mém. Mus. Nat. Hist. Nat., Ser. A* **88**, 156–173.

22. Patterson, C., and Rosen, D. E. (1977). Review of ichthyodectiform and other Mesozoic teleost fishes and the theory and practice of classifying fossils. *Bull. Am. Mus. Nat. Hist.* **158**, 81–172.

23. Patterson, C. (1977). Cartilage bones, dermal bones and membrane bones, or the exoskeleton versus the endoskeleton. *In* "Problems in Vertebrate Evolution" (S. M. Andrews, R. S. Miles, and A. L. Panchen, eds.), pp. 77–121. Academic Press, London.

24. Patterson, C. (1977). The contribution of paleontology to teleostean phylogeny. *In* "Major Patterns in Vertebrate Evolution" (M. K. Hecht, P. C. Goody, and B. M. Hecht, eds.), pp. 579–643. Plenum, New York.

25. Patterson, C. (1978). Verifiability in systematics. *Syst. Zool.* **27**, 218–222.

26. Patterson, C. 1978. "Evolution." British Museum (Natural History) and Routledge & Kegan Paul, London; Cornell University Press; Ithaca, NY; University of Queensland Press, Brisbane (subsequent editions in Danish, Dutch and Japanese).

27. Patterson, C. (1978). Arthropods and ancestors. *Antenna* **2**, 99–103.

28. Nelson, G. J., Patterson, C., and Rosen, D. E. (1979). Foreword. *In* "Phylogenetic Systematics" (W. Hennig, ed.), 2nd ed., pp. viii–xiii. University of Illinois Press, Urbana.

29. Patterson, C. (1980). Origin of tetrapods: Historical introduction to the problem. *In* "The Terrestrial Environment and the Origin of Land Vertebrates" (A. L. Panchen, ed.), pp. 159–175. Academic Press, London.

30. Patterson, C. (1980). Cladistics. *Biologist* **27**, 234–240 [reprinted in "Evolution Now" (J. Maynard Smith, ed.), pp. 110–120. Macmillan, London, 1982].

31. Patterson, C. (1981). Methods of paleobiogeography. *In* "Vicariance Biogeography: A Critique" (G. Nelson and D. E. Rosen, eds.), pp. 446–500. Columbia University Press, New York.

32. Rosen, D. E., Forey, P. L., Gardiner, B. G., and Patterson, C. (1981). Lungfishes, tetrapods, paleontology, and plesiomorphy. *Bull. Am. Mus. Nat. Hist.* **167**, 159–276.

33. Patterson, C. (1981). The development of the North American fish fauna—a problem of historical biogeography. *In* "The Evolving Biosphere" (P. L. Forey, ed.), pp. 265–281. Br. Mus. (Nat. Hist.), London.

34. Patterson, C. (1981). Agassiz, Darwin, Huxley, and the fossil record of teleost fishes. *Bull. Br. Mus. (Nat. Hist.), Geol.* **35**, 213–224.

35. Patterson, C. (1981). Significance of fossils in determining evolutionary relationships. *Annu. Rev. Ecol. Syst.* **12**, 195–223.

36. Patterson, C. (1982). Cladistics and classification. *New Sci.* **94**, 303–306 [reprinted in "Darwin up to Date" (J. Cherfas, ed.), pp. 35–39. New Science Publications, London, 1982].

37. Patterson, C. (1982). Morphology and interrelationships of primitive actinopterygian fishes. *Am. Zool.* **22**, 241–259.

38. Patterson, C. (1982). Morphological characters and homology. *In* "Problems of Phylogenetic Reconstruction" (K. A. Joysey and A. E. Friday, eds.), pp. 21–74. Academic Press, London.

39. Patterson, C. (1982). Classes and cladists or individuals and evolution. *Syst. Zool.* **31**, 284–286.

40. Patterson, C. (1983). How does ontogeny differ from phylogeny? *In* "Development and Evolution" (B. C. Goodwin, N. Holder, and C. C. Wylie, eds.), pp. 1–31. Cambridge University Press, Cambridge, UK.

41. Patterson, C. (1983). Aims and methods in biogeography. *In* "Evolution, Time and Space: The Emergence of the Biosphere" (R. W. Sims, J. S. Price, and P. E. S. Whalley, eds.), pp. 1–28. Academic Press, London.

42. Patterson, C. (1984). *Chanoides*, a marine Eocene otophysan fish. *J. Vertebr. Paleontol.* **4**, 430–456.

43. Patterson, C. (1984). Family Chanidae and other teleostean fishes as living fossils. *In* "Living Fossils" (N. Eldredge and S. M. Stanley, eds.), pp. 132–139. Springer-Verlag, New York.

44. Schaeffer, B., and Patterson, C. (1984). Jurassic fishes from the Western United States, with comments on Jurassic fish distribution. *Am. Mus. Novit.* **2796**, 1–86.

45. Schaeffer, B., and Patterson, C. (1985). Comments on western hemisphere Jurassic fishes. *Ameghiniana* **21**, 332–334.

46. Forey, P. L., Monod, O. and Patterson, C. (1985). Fishes from the Akkuyu Formation (Tithonian), Western Taurus, Turkey. *Geobios* **18**, 195–201.

47. Patterson, C. (1986). Cladistics. *In* "A Dictionary of Birds" (B. Campbell and E. Lack, eds.), p. 88. T. & A. D. Poyser, Calton.

48. Patterson, C. (1987). Introduction. *In* "Molecules and Morphology in Evolution: Conflict or Compromise?" (C. Patterson, ed.), pp. 1–22. Cambridge University Press, Cambridge, UK.

49. Patterson, C. (1987). Evolution: neo-Darwinian theory. *In* "The Oxford Companion to the Mind" (R. L. Gregory, ed.), pp. 234–244. Oxford University Press, Oxford.

50. Patterson, C. (1987). Otoliths come of age. *J. Vertebr. Paleontol.* **7**, 346–348.

51. Patterson, C. and Longbottom, A. E. (1987). Fishes. *In* "Fossils

of the Chalk'' (A. B. Smith, ed.), pp. 238–265. Palaeontological Association, London.

52. Patterson, C., and Smith, A. B. (1987). Is the periodicity of extinctions a taxonomic artefact? *Nature (London)* **330**, 248–252.

53. Patterson, C. (1988). The impact of evolutionary theories on systematics. *In* ''Prospects in Systematics'' (D. L. Hawksworth, ed.), pp. 59–91. Oxford University Press (Clarendon), Oxford.

54. Patterson, C. (1988). Homology in classical and molecular biology. *Mol. Biol. Evol.* **5**, 603–625.

55. Smith, A. B. and Patterson, C. (1988). The influence of taxonomic method on the perception of patterns of evolution. *Evol. Biol.* **23**, 127–216.

56. Patterson, C. and Rosen, D. E. (1989). The Paracanthopterygii revisited: Order and disorder. *Sci. Ser., Nat. Hist. Mus. Los Angeles Cty.* **32**, 5–36.

57. Patterson, C., and Smith, A. B. (1989). Periodicity in extinction: The role of systematics. *Ecology* **70**, 802–811.

58. Patterson, C. (1989). Phylogenetic relations of major groups: Conclusions and prospects. *In* ''The Hierarchy of Life'' (B. Fernholm, K. Bremer, and H. Jörnvall, eds.), Nobel Symp. 70, pp. 471–488. Excerpta Medica, Amsterdam.

59. Patterson, C. and Longbottom, A. E. (1989). An Eocene amiid fish from Mali, West Africa. *Copeia*, pp. 827–836.

60. Patterson, C. (1990). Erik Helge Osvald Stensiö. *Biogra. Mem. Fellows R. Soc.* **35**, 363–380.

61. Patterson, C. (1990). Metazoan phylogeny: Reassessing relationships. *Nature (London)* **344**, 199–200.

62. Rosen, D. E., and Patterson, C. (1990). On Müller's and Cuvier's concepts of pharyngognath and labyrinth fishes and the classification of percomorph fishes, with an atlas of percomorph dorsal gill arches. *Am. Mus. Novit.* **2983**, 1–57.

63. Patterson, C., and Owen, H. G. (1991). Indian isolation or contact? A response to Briggs. *Syst. Zool.* **40**, 96–100.

64. Forey, P. L., Gardiner, B. G. and Patterson, C. (1991). The lungfish, the coelacanth, and the cow revisited. *In* ''Origins of the Higher Groups of Tetrapods'' (H.-P. Schultze and L. Trueb, eds.), pp. 145–172. Cornell University Press (Comstock), Ithaca, NY.

65. Tyler, J. C., and Patterson, C. (1991). The skull of the Eocene *Triodon antiquus* (Triodontidae, Tetraodontiformes): Similar to that of the Recent threetooth pufferfish *T.macropterus. Proc. Biol. Soc. Wash.* **104**, 878–891.

66. Lewy, Z., Milner, A. C., and Patterson, C. (1992). Remarkably preserved natural endocranial casts of pterosaur and fish from the Late Cretaceous of Israel. *Geol. Surv. Isr., Curr. Res.* **7**, 31–35.

67. Patterson, C. (1992). Interpretation of the toothplates of chimaeroid fishes. *Zool. J. Linn. Soc.* **106**, 33–61.

68. Patterson, C. (1992). Supernumerary median fin-rays in teleostean fishes. *Zool. J. Linn. Soc.* **106**, 147–161.

69. Patterson, C. (1993). An overview of the early fossil record of acanthomorph fishes. *Bull. Mar. Sci.* **52**, 29–59.

70. Patterson, C. (1993). Lampridiforms or lampriforms, Lamprididae or Lampridae? *Bull. Mar. Sci.* **52**, 168–169.

71. Johnson, G. D., and Patterson, C. (1993). Percomorph phylogeny: A survey of acanthomorphs and a new proposal. *Bull. Mar. Sci.* **52**, 554–626.

72. Patterson, C. (1993). Teleostei. *In* ''The Fossil Record'' (M. J. Benton, ed.), Vol. 2, pp. 619–654. Chapman & Hall, London.

73. Nelson, G. J., and Patterson, C. (1993). Cladistics, sociology and success: A comment on Donoghue's critique of David Hull. *Biol. Philos.* **8**, 441–443.

74. Patterson, C. (1993). Bird or dinosaur? *Nature (London)* **365**, 21–22.

75. Patterson, C., Williams, D. M., and Humphries, C. J. (1993). Congruence between molecular and morphological phylogenies. *Annu. Rev. Ecol. Syst.* **24**, 153–188.

76. Patterson, C. (1993). Naming names. *Nature (London)* **366**, 518.

77. Duffin, C. J., and Patterson, C. (1994). I pesci fossili di Osteno: Une nuova finestra sulla vita del Giurassico inferiore. *Paleocronache 2* **2**, 18–38.

78. Patterson, C. (1994). Null or minimal models. *In* ''Models in Phylogeny Reconstruction'' (R. Scotland, D. J. Siebert, and D. M. Williams, eds.), Syst. Assoc. Spe. Vol. 52, pp. 173–192. Oxford University Press, Oxford.

79. Patterson, C. (1994). Bony fishes. *In* ''Major Features of Vertebrate Evolution'' (D. R. Prothero and R. M. Schoch, eds.), Short Courses Paleontol., No. 7, pp. 57–84. Paleontological Society, University of Tennessee, Knoxville.

80. Patterson, C., and Johnson, G. D. (1995). The intermuscular bones and ligaments of teleostean fishes. *Smithson. Contrib. Zool.* **559**, 1–85.

81. Patterson, C. (1996). Comments on Mabee's ''Empirical rejection of the ontogenetic polarity criterion.'' *Cladistics* (to be published).

82. Johnson, G. D., and Patterson, C. (this volume, Chapter 13).

83. Patterson, C. (1996). ''Evolution,'' 2nd ed., British Museum (Natural History), London.

84. Patterson, C., and Greenwood, P. H. (1967). ''Fossil Vertebrates.'' Academic Press, London.

85. Greenwood, P. H., Miles, R. S., and Patterson, C., eds. (1973). ''Interrelationships of Fishes.'' Academic Press, London.

86. Patterson, C. (1987). ''Molecules and Morphology in Evolution: Conflict or Compromise?'' Cambridge University Press, Cambridge, UK.

Reviews, Obituaries, Abstracts

1968a. Fossil fishes. Review of ''Fundamentals of Palaeontology. Vol. 11: Agnatha, Pisces (D. V. Obruchev, ed.), 1967. *Nature (London)* **220**, 514.

1968b. Fishes from Maikop. Review of ''Bony Fishes of the Maikop Deposits of the Caucasus'' (P. G. Danil'chenko), 1968. *Nature (London)* **220**, 936.

1968c. Eigil Nielsen (obituary). *The Times*, December 1968.

1971. Review of ''Les poissons fossiles du Monte Bolca classés jusqu'ici dans les familles des Carangidae, Menidae, Ephippidae, Scatophagidae'' (J. Blot), 1969. *Copeia*, pp. 187–189.

1972. Teleosts. Review of ''Functional Morphology and Classification of Teleostean Fishes'' (W. A. Gosline), 1971. *Science* **176**, 399.

1978a. Review of ''The Structure of Scientific Theories'' (F. Suppe, ed.), 1978. *New Sci.* **79**, 489.

1978b. Review of ''Paleontology and Plate Tectonics'' (R. M. West, ed.), 1977. *Palaeontol. Assoc. Circ.* **93**, 15–17 (with H. G. Owen).

1978c. Review of ''Evolution of Living Organisms'' (P.-P. Grassé), 1978. *New Sci.* **79**, 694.

1979a. The salmon, the lungfish and the cow: A reply. Nature (London) **277**, 1575–1576 (with B. G. Gardiner, P. Janvier, P. L. Forey, P. H. Greenwood, R. S. Miles, and R. P. S. Jefferies).

1979b. Review of ''Life on Earth'' (BBC television series). *New Sci.* **81**, 193.

1979c. The roots of human nature. Review of ''Beast and Man'' (M. Midgley), 1979. *New Sci.* **81**, 883.

1980a. Review of ''Arthropod Phylogeny with Special reference to Insects'' (H. B. Boudreaux), 1979. *Biologist* **27**, 56.

1980b. Review of ''Macroevolution: Pattern and Process'' (S. M. Stanley), 1979. *Palaeontol. Assoc. Circ.* **101**, 6–7 (reprinted in *Canadian Association of Palynologists Newsletter*, November 1980).

1980c. Phylogenies and fossils. Reviews of "Phylogenetic Analysis and Paleontology" (J. Cracraft and N. Eldredge, eds.), 1979; and "Models and Methodologies in Evolutionary Theory" (J. H. Schwartz and H. B. Rollins, eds.), 1979. *Syst. Zool.* **29**, 216–219.

1980d. Museum pieces (correspondence). *Nature (London)* **288**, 430.

1981a. Biogeography: In search of principles. *Nature (London)* **291**, 612–613.

1981b. Darwin's survival (correspondence). *Nature (London)* **290**, 82.

1981c. Vertebrate morphology. Review of "Basic Structure and Evolution of Vertebrates" (E. Jarvik), 1980. *Science* **214**, 431–432.

1983a. Darwinian delicacies. Review of "Hen's Teeth and Horse's Toes: Further Reflections in Natural History" (S. J. Gould), 1983. *Nature (London)* **302**, 777.

1983b. Review of "Biochemical Aspects of Evolutionary Biology" (M. H. Nitecki, ed.), 1982. *Biologist* **30**, 118.

1983c. A leaf out of the biologists' book. Review of "Organisational Systematics" (B. McKelvey), 1983. *New Sci.* **100**, 363.

1984a. George Eric Howard Foxon (1908–1982) (obituary). *Linnean* **1**(1), 24–25.

1984b. Frederick Charles Stinton (1916–1982) (obituary). *Linnean* **1**(3), 25.

1984c. Preformation to punctuation. Review of "Evolution: The History of an Idea" (P. J. Bowler), 1984. *Nature (London)* **311**, 587.

1985. The trouble with book making. Review of "Oxford Surveys in Evolutionary Biology, Vol. 1" (R. Dawkins and M. Ridley, eds.), 1985. *Nature (London)* **316**, 492.

1986. Star on the make? Review of "The Nemesis Affair: A Story of the Death of Dinosaurs and the Ways of Science" (D. M. Raup), 1986. *Nature (London)* **322**, 693.

1987a. Review of "Museum Collections: Their Roles and Future in Biological Research" (E. H. Miller, ed.), 1985. *Am. Sci.* **75**, 95.

1987b. In the limelight. Review of "Ancestors: The Hard Evidence" (E. Delson, ed.), 1985. *Paleobiology* **13**, 253–256.

1987c. Harry Ashley Toombs, 1909–1987 (obituary). *London Nat.* **66**, 191–193.

1988a. Capital letters. Review of "Simple Curiosity: Letters from George Gaylord Simpson to his Family" (L. F. LaPorte, ed.), 1987. *Nature (London)* **332**, 317.

1988b. Creationism and common sense (correspondence). *Nature (London)* **332**, 580.

1988c. Review of "Biological Metaphor and Cladistic Classification. An Interdisciplinary Perspective" (H. R. Hoenigswald and L. F. Wiener, eds.), 1987. *Palaeontol. Assoc. Circ.* **132**, 9–10.

1989a. Puncturing punctuation. Review of "Genetics, Paleontology and Macroevolution" (J. Levinton), 1988. *Trends Ecol. Evol.* **4**, 89–90.

1989b. Review of "The Cuvier-Geoffroy Debate: French Biology in the Decades before Darwinism" (T. A. Appel), 1987. *Palaeontol. Assoc. Newsl.* **5**, 36–37.

1990a. An overview of the early fossil record of Acanthomorpha. *Abstr., 70th Annu. Meet. Am. Soc. Ichthyol. Herpetol.*, p. 139.

1990b. Vote of no confidence (correspondence). *Nature (London)* **347**, 419 (with R. P. S Jefferies, K. Sattler, P. Wheatcroft, J. Clutton-Brock, C. J. Humphries, C. R. Hill, and P. H. Greenwood).

1991a. Beverly Halstead (obituary). *The Independent*, May 3, 1991, p. 22.

1991b. Animated nature. Review of "History of the Natural World" (O. Goldsmith, ed.), 1990. *Nature (London)* **349**, 115–116.

1991c. Review of "Fish Evolution and Systematics: Evidence from Spermatozoa" (B. G. M. Jamieson), 1991. *Biologist* **38**, 197.

1992a. Bulldoggish persuasion. Review of "Thomas Henry Huxley: Communicating for Science" (J. V. Jensen), 1991. *Nature (London)* **355**, 782.

1992b. The meaning of fossils, 1992. *Abstr., 5th North Am. Paleontol. Conv.*, p. 231.

1992c. Review of "Reading the Shape of Nature: Comparative Zoology at the Agassiz Museum" (M. P. Winsor), 1991. *J. Nat. Hist.* **26**, 691.

1992d. Review of "The Individual in Darwin's World. The Second Edinburgh Medal Address" (S. J. Gould), 1990. *Palaeontol. Assoc. Newsl.* **15**, 22.

1992e. Review of "Classification of Fossil and Recent Organisms," (N. Schmidt-Kittler and R. Willmann, eds.), 1991. *J. Nat. Hist.* **26**, 1115–1116.

1993a. Review of evidence on ostariophysan interrelationships. *Abstr., 73rd Annu. Meet. Am. Soc. Ichthyol. Herpetol.*, p. 244 (with S. V. Fink and W. L. Fink).

1993b. The intermuscular bones and ligaments of teleostean fishes: I. Structure. *Abstr., 73rd Annu. Meet. Am. Soc. Ichthyol. Herpetol.*, p. 245 (with G. D. Johnson).

1993c. The intermuscular bones and ligaments of teleostean fishes: II. Systematic consequences. *Abstr., 73rd Annu. Meet. Am. Soc. Ichthyol. Herpetol.*, p. 245 (with G. D. Johnson).

1994. Archetypes and ancestors. Review of "Richard Owen: Victorian Naturalist" (N. A. Rupke), 1994. *Nature (London)* **368**, 375.

1995. Interrelationships of lower Euteleostei. *Abstr., 75th Annu. Meet. Am. Soc. Ichthyol. Herpetol.*, p. 124 (with G. D. Johnson).

2

Phylogenetic Interrelationships of Neoselachians (Chondrichthyes: Euselachii)

SHIGERU SHIRAI

Marine Ecology Section
Seikai National Fisheries Research Institute
Nagasaki 850 Japan

I. Introduction

When studying the phylogeny of a certain group, we have to answer the following questions: where are the members of this group located among related taxa, and how are these members interrelated within this group? To the first question in elasmobranch fishes, Compagno (1973) gave an explicit answer with a euselachian concept. His Euselachii is composed of all Recent and hybodont–ctenacanth fossil sharks that were abundant during the Paleozoic and Mesozoic strata. He further suggested that living sharks and rays have evolved from a common ancestor within these fossil taxa [= Schaeffer's (1967) "hybodont level"]. To the second, Compagno divided the Recent taxa into four superorders, Squalomorphii, Batoidea, Squatinomorphii, and Galeomorphii, and hypothesized their interrelationships. Zangerl (1973) reached similar conclusions to Compagno's (1973) euselachian hypothesis using the term "phalacanthous design." These studies, published in the first *Interrelationships of Fishes*, set the stage for further elucidation of the interrelationships of elasmobranch fishes.

During the past two decades, various insights regarding the phylogeny of elasmobranchs have been proposed that are closely in accordance with the generalization of cladistic methodology (Hennig, 1966; Wiley, 1981). Therefore, our understanding of elasmobranch evolution has recently undergone remarkable

changes. To detail these changes is the first aim of this chapter. Here I also propose the most probable hypothesis of higher-level interrelationships based on information from neoselachian morphology.

II. Compagno's Hypothesis and Subsequent Studies

A. Euselachian and Neoselachian Concept

Based on the following morphological features shared by most living forms, Compagno (1973) suggested a common ancestry for living sharks and rays: (1) synchronomorial type of dermal denticles, (2) fused coracoid and puboischiadic bars, (3) one or two segments (mixipterygial cartilage) between the elongate metapterygium and the clasper shaft, (4) a tribasal pectoral fin with a reduced metapterygial axis, (5) aplesodic pectoral fins, (6) a short otic region of the neurocranium, (7) an incomplete postorbital wall, (8) a well-calcified vertebral centrum, and (9) a nonlunate caudal fin. He considered that these characters also suggested the placement of living forms in a more inclusive group, the "Euselachii," because many of these features are also found in several groups of Mesozoic and Paleozoic sharks. Such a definition of two groups seems rather vague, but we must recall that Compagno (1973) followed the method of evolu-

Copyright © 1996 by Academic Press, Inc.
All rights of reproduction in any form reserved.

tionary systematics and/or phenetics. The outline of the Euselachii was explained as follows: the Euselachii "also includes hybodonts and other fossil elasmobranchs related to living forms [ctenacanths]" (1973, p. 57; the expression in brackets was used on p. 29); and "the Euselachii is in opposition to the fossil elasmobranch groups that are not in the direct lineage of living forms (cladoselachians, cladodonts, xenacanths, and edestoids)" (1973, p. 29). However, despite these explicit statements, the meaning of the term "Paleozoic sharks" (especially "ctenacanths") was considered ambiguous in the early 1970s. In the succeeding pages, based on the monophyly of the euselachian assemblage, Compagno (1973) criticized counter hypotheses which stated that living forms are composed of two distinct lineages originating from placoderms (Holmgren, 1942; sharks + rays) or Paleozoic chondrichthyans (Glickman, 1964, 1967; osteodonts + orthodonts). He also rejected the analysis of Blot (1969) who, following Schaeffer (1967), excluded some of the living sharks from the clade of Recent taxa.

Zangerl (1973) stated that all living forms and the Mesozoic fossil taxa originated from the "phalacanthous design," one of six major divisions of Paleozoic sharks. The phalacanthous design, represented by the so-called ctenacanths (e.g., †*Ctenacanthus*, †*Goodrichthys*, and †*Tristychius*), is characterized by a structure of the dorsal fin spines and the pectoral fin skeleton, similar to that of Recent forms. Maisey (1975) postulated that dorsal fin spines with an outer coat of orthodentine have appeared once in chondrichthyan evolution. Because of this type of dorsal fin spine and other characters (e.g., the tribasal pectoral fin), the Recent forms, hybodonts, and ctenacanths form a monophyletic group under the category of the phalacanthous sharks. Many living sharks and rays without a dorsal fin spine ("anacanthous sharks") were regarded to have lost it secondarily. The framework of the euselachian concept was established through these studies.

There were various views about the placement of living sharks and rays in the Euselachii. Most previous works had proposed close relationships between hybodonts and some living sharks (*Heterodontus*, *Chlamydoselachus*, and/or hexanchoids) (e.g., Hasse, 1879–1885; Brown, 1900; White, 1937; Schaeffer, 1967). Compagno rejected these proposals because these "primitive" living sharks have more affinity with the remaining living forms than the Mesozoic sharks. This forms an important basis for the definition of his Galeomorphii and Squalomorphii (see below). However, Compagno's discussion was provisional because the relationships between the living forms and well-known hybodonts were very uncertain.

Maisey (1975) proposed three orders for the phalacanthous sharks of Zangerl (1973), and one of these comprises all living sharks and rays separate from all hybodonts and ctenacanths. This order was named the Euselachiformes following the usage of Moy–Thomas (1939). However, this is a little confusing given the Euselachii of Regan (1906) and Compagno (1973), so here I follow the latter terminology. Based mainly on similarities of dorsal fin spine morphology, Maisey (1975) suggested that the living forms have an origin among ctenacanth sharks rather than a hybodont–living form relationship.

Maisey's (1975) concepts (the phalacanthous lineage and the ctenacanth ancestor for living forms) were reviewed by Compagno (1977). Compagno (1977) reconfirmed his hypothesis of 1973 by cladistic analysis. In this work, he described a common ancestor of living forms and informally called this ancestor and all its descendants "neoselachians." The limit of this group remained ambiguous, and he explained that it includes "ordinal groups of living sharks and rays and certain Mesozoic sharks, including palaeospinacids, and possibly orthacodonts and anacoracids" (1977, p. 303). Compagno listed 10 derived characters for neoselachians, some of which are also seen in the hybodonts (or a part of them), such as a nonlunate caudal fin, a short metapterygial axis of the pectoral fin, fused puboischiadic bar, and the aplesodic pectoral fins which were regarded as convergences. Since then, most ichthyologists and paleontologists have agreed with the monophyly of living forms including some Mesozoic and Cenozoic fossils, apart from preferences of the term "Neoselachii" (e.g., Reif, 1977; Schaeffer, 1981; Zangerl, 1981; Young, 1982; Maisey, 1982, 1984a,b; Thies, 1983; Cappetta, 1987; Gaudin, 1991). These authors gave the following characters as evidence for the neoselachians: a long pelvic metapterygium, only a few mixipterygial cartilages, calcified vertebrae with notochordal constriction, and modern tooth and dermal denticle structures.

Nevertheless, the origin of neoselachians is still under discussion. Dick (1978) questioned Maisey's (1975) view because it was based on a single plesiomorphic character. Young (1982) revived the hypothesis of the hybodont–neoselachian relationship, in which he placed the living forms sister to three representatives of the Mesozoic and Paleozoic hybodonts (†*Hybodus*, †*Tristychius*, and †*Onychoselache* known from almost complete articulated specimens). This hypothesis requires a longer age for living forms (from the early Carboniferous) than provided by the fossil record (from the Triassic, or at best the late Paleozoic; Thies and Reif, 1985); this time disparity has been discussed by various authors (Dick, 1978; Maisey,

1984b; Compagno, 1988). Maisey (1984a,b) changed his views from his earlier work and suggested that a sister-group of living forms was restricted to the Mesozoic hybodonts represented by the genus *Hybodus*. Recently Gaudin (1991) examined the phylogeny of fossil chondrichthyans using cladistic methodology and supported Maisey's more recent view on the origin of living forms.

B. Interrelationships of Neoselachians

Compagno (1973) divided all the living elasmobranchs into four coordinate groups, i.e., Squalomorphii, Squatinomorphii, Batoidea, and Galeomorphii. Although he recognized striking differences between living sharks and rays, he explained that their ranking in two equal groups, adopted by many previous workers, "places undue emphasis on the unique characters of the rays while obscuring the many differences between the various shark groups" (1973, p. 21). Compagno also stated that recognition of these four superorders "serves to balance the uniqueness of the rays against the distinctness of the major shark groups from each other and from the rays" (1973, p. 21). Compagno followed Holmgren (1941) in allocating *Chlamydoselachus* and hexanchoids to the Squalomorphii and *Heterodontus* to the Galeomorphii. However, concerning the interrelationships among the four superorders, he failed to reach a clear conclusion but summarized his speculation in a phylogenetic tree (Compagno, 1973, fig. 5), in which each of the four superorders were evolved independently from ancestors of the hybodont level.

Subsequently Compagno defended the majority of his 1973 hypotheses by the cladistic method (e.g., Compagno, 1977, 1988, 1990). In these studies, he remained uncertain about higher neoselachian phylogeny. Compagno (1977) summarized his ideas of the higher-level interrelationships into six alternatives (Figs. 1c–1h), which included a sister-relation between the batoids and pristiophorids and the independence of a group composed of *Heterodontus* and orectoloboids from the other galeomorphs. Compagno sometimes ignored the monophyly of his Squalomorphii or Galeomorphii. His discussion suggested that his four superorders are phenetic assemblages rather than phylogenetic ones. Compagno (1988) noted that a classification also reflecting morphological diversity is more useful than a simple cladogram; he hypothesized familial or suprafamilial groups on a phenetic basis and then considered the cladistic relationships among these groups.

Maisey (1980) reviewed the jaw suspension of elasmobranchs and made an important proposal that con-

tradicted the traditional hypothesis of the hyostylic and amphistylic modes. He suggested living sharks were divided into two groups based on the palatoquadrate–basicranium articulation. One of these groups, the "orbitostylic" group, shares the orbital articulation he redefined. In this type, the palatoquadrate bears a developed ascending process (called the "orbital process") situated behind the optic nerve foramen of the neurocranium. The ascending process of *Heterodontus* and galeoids is in front of this foramen and could be analogous to the orbital process of the orbitostylic taxon on an ontogenetic basis (Holmgren, 1940; Jollie, 1971). Maisey (1980) discussed the phylogenetic intrarelationships of the orbitostylic sharks, but was inconclusive about the phylogenetic position of pristiophorids. He was also uncertain concerning the origin of the orbitostylic group. Maisey (1984a, 1985a) was also uncertain concerning the origin of the orbitostylic group (Fig. 1i).

Thies (1983) proposed a phylogenetic scheme for Mesozoic neoselachian taxa chiefly based on dental morphology (Fig. 1j). He placed living *Chlamydoselachus* and hexanchoids with palaeospinacids and orthacodonts as a primitive sister-taxon of the remaining neoselachians. In the latter group, Thies recognized three phylogenetic stems, i.e., batoids, galeoids + *Squatina*, and *Heterodontus* + squalomorphs (except *Chlamydoselachus* and hexanchoids). Thies (1983) estimated that the early differentiation of neoselachians occurred in the Triassic, although only several records of Triassic neoselachians are available.

Shirai (1992b) proposed the hypnosqualean group composed of *Squatina*, pristiophorids, and all batoids based on five synapomorphies. He placed the pristiophorids as sister to the batoids based on five derived characters. The hypnosqualeans were further grouped with "squaloids," *Chlamydoselachus*, and hexanchoids, and these composed the Squalea (Shirai, 1992c).

According to recent paleontological studies, the position of fossil neoselachians has been gradually clarified, although this is not the primary purpose here. Thies and Reif (1985) placed representative Mesozoic neoselachians in Compagno's (1973) system. In their list (Thies and Reif, 1985, table 1), several supraspecific groups were treated as superorders *incertae sedis*. Of these, palaeospinacids (†*Palaeospinax* and †*Synechodus*) recently located in a particular lineage among neoselachians, i.e., in the lineage reaching to *Heterodontus* or galeomorph sharks (Reif, 1974; Maisey, 1985a) or near hexanchiforms with orthacodonts (Thies, 1983; Duffin and Ward, 1993).

The status of Compagno's (1973) superorders is reviewed below.

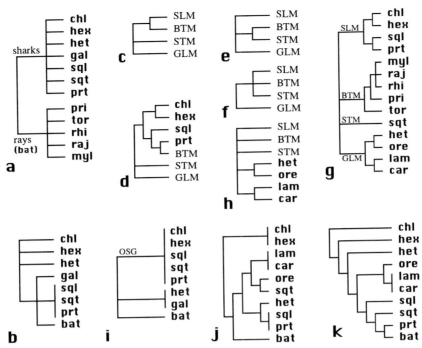

FIGURE 1 Summary of topologies translated from previous hypotheses into cladogram style. Taxa of each cladogram have been adjusted to the subordinal level. Citations
are: (a) Bigelow and Schroeder (1948, 1953); (b) Schaeffer (1967); (c–h) Compagno
(1977), where these topologies show his alternatives 1 to 5 in order, and his alternative
5 includes the two possibilities expressed in (h) and (g); (i) Maisey (1980, 1984a); (j)
Thies (1983) and Thies and Reif (1985); (k) G. Dingerkus (unpublished, cited by Seret,
1986). Abbreviations: bat, batoids; BTM, Batomorphii; car, carcharhinoids; chl, chla-
mydoselachoids; gal, galeoids (orectoloboids + lamnoids + carcharhinoids); GLM,
Galeomorphii; het, heterodontoids; hex, hexanchoids; lam, lamnoids; myl, myliobatoids; ore, orectoloboids; OSG, orbitostylic group; pri, pristoids; prt, pristiophoroids;
raj, rajoids; rhi, rhinobatoids; SLM, Squalomorphii; sql, "squaloids"; sqt, squatinoids;
STM, Squatinomorphii; tor, torpedinoids.

Squalomorphii—This superorder was composed of
three orders, the Squaliformes (= "squaloids" sensu
stricto: Squaloidea of Bigelow and Schroeder, 1957),
Pristiophoriformes, and Hexanchiformes (*Chlamydoselachus* and hexanchoids). Compagno (1973) agreed
with Holmgren's (1941) treatment of Hexanchiformes
which recognized similarities between hexanchoids
and "squaloid" sharks. According to Compagno
(1973) these three orders have an unresolved relationship with squaliforms "intermediate" between hexanchiforms and pristiophoriforms, although the two latter groups diverge more from each other than either
does from the squaliforms. Compagno (1977) listed
synapomorphies for the Squalomorphii and the
squaliform–pristiophoriform sister-relation (Fig. 1g).
Maisey (1980) stated that there was no evidence
for "squaloid" monophyly and left the position of
pristiophorids open. Shirai (1992b,c) recognized the

paraphyly of "squaloids," which form successive
sister-groups of the hypnosqualeans. Further evidence for Compagno's Squalomorphii has not
been found.

Batoidea—Living batoids form a coherent group
based on many reliable synapomorphies (e.g., Compagno, 1973, 1977; Heemstra and Smith, 1980; Maisey,
1984a; Nishida, 1990), but the interrelationships
within this group remain ambiguous. Since Regan
(1906), many authors followed a dichotomous classification based on the distinctiveness of the torpedinoids. Compagno (1973) criticized such an approach
because "it places undue emphasis on the specializations of the Torpediniformes and masks the distinctness of the Pristiformes and Myliobatiformes" (1973,
p. 40). Following Bigelow and Schroeder's (1953) system, Compagno (1973) proposed four orders, i.e.,
Rajiformes (Rhinobatoidei + Rajoidei), Pristiformes,

Torpediniformes, and Myliobatiformes. Compagno considered that ancestral features of batoids were retained in rhinobatoids, from which other taxa could have been derived.

Compagno (1977) reconsidered batoid interrelationships (Fig. 1g): the torpedinoids were sister to the other batoids because only torpedinoids (but not all) retained a sharklike feature, the absence of the hyomandibular–ceratohyal connection. Nevertheless, differentiation of the higher batoid orders remained unresolved. Heemstra and Smith (1980) considered pristoids sister to the other batoids based on three synapomorphies of the nonpristoid batoids. Maisey (1984a) supported Compagno's (1977) view because (1) Jurassic "rhinobatoids" (†*Belemnobatis* and †*Spathobatis*) have more derived hyoid arrangements than torpedinoids, and (2) these fossil rhinobatoids and *Torpedo* share sharklike basibranchial copula, in which all hypobranchials are articulated with the cardiobranchial (basibranchial plate). Nishida (1990) proposed an alternative hypothesis as follows: (1) pristoids are the most plesiomorphic rays; (2) rhinobatoids are artificial, not united by any synapomorphies; (3) torpedinoids are a plesiomorphic sister-group of myliobatidoids, rajoids, and rhinobatoids (excluding *Rhina* and *Rhynchobatus*); and (4) rajoid–myliobatidoid sister-relation is probable.

Squatinomorphii—This superorder, created by Compagno (1973), is composed of a single genus *Squatina*. Because of its raylike morphology, the phylogenetic position of this shark has been questioned since the last century. Regan (1906) considered that *Squatina* was a typical member of his squaloid sharks and that similarities to batoids were due to a common ancestor of *Squatina* and batoids; this consideration seems to have been a logical basis of the dichotomy of sharks and rays (e.g., Garman, 1913; Bigelow and Schroeder, 1948, 1953). Iselstöger (1937) and Holmgren (1941), and even Thies (1983), recognized close relations between *Squatina* and a part of galeoid sharks (orectolobids). Schaeffer (1967) suggested that *Squatina* could have arisen from a hybodont, galeoid, or squaloid stock mainly because of its mosaic distribution of characters. Compagno (1973) agreed with these previous opinions, and formalized them as a new theory. He noted that "the Squatinomorphii combines a blend of batoid, squalomorph, and a few galeoid characters with many unique features, suggesting that the Squatinomorphii are an offshoot from the euselachian base of the three other groups" (1973, p. 46). Compagno (1977) advocated this view with two other possibilities, the sister-relation with batoids or batoids + squalomorphs (Figs. 1e and 1f).

Maisey (1980) placed *Squatina* at the most plesio-

morphic sister-group of the remaining orbitostylic sharks. Shirai (1992b) questioned the evidence for squaloid–*Squatina* and galeoid–*Squatina* relations on the basis of their ambiguous synapomorphies and found that batoids and *Squatina* are not monophyletic unless pristiophorids are included. The placement of *Squatina* in the hypnosqualeans was later corroborated in a cladistic analysis (Shirai, 1992c).

Galeomorphii—Compagno (1973) placed *Heterodontus* in the traditional assemblage of galeoid sharks (orectoloboids, carcharhinoids, and lamnoids) mostly based on the strong similarities between *Heterodontus* and orectoloboids reported by Holmgren (1941) and Applegate (1972). He proposed the name Galeomorphii for this assemblage and subsequently supported its phylogenetic significance with a list of synapomorphies (Compagno, 1977, 1988). However, Compagno also pointed out the necessity of additional work to clarify galeomorph monophyly. This is probably because Compagno (1977, 1988) also retained the possibility of the independent origin of heterodontoid–orectoloboid lineage from the neoselachian base (Fig. 1h). Maisey (1984a, 1985a) suggested that a galeoid lineage can be hypothesized when similarities between the palaeospinacid †*Synechodus* and *Heterodontus* neurocrania are considered; the latter two genera were considered to be successive sister-groups of galeoids. Shirai (1992c) also recognized the galeoid assemblage based on comparisons in hard parts among living and fossil forms.

The monophyly of each galeomorph order is defended in Compagno (1973, 1977, 1988, 1990). This is not in accordance with previous classifications. His Orectolobiformes, first proposed in Regan's (1906) Orectolobidae, has not been recognized as monophyletic unless lamnoids and/or carcharhinoids (or a part of them) are included (White, 1937; Applegate, 1972; Thies, 1983; Maisey, 1984a). Contrary to these views, Compagno (1988) indicated that due to the large number of common features in recent orectoloboids (the "orectoloboid suite"), it was improbable to assume the secondary loss of the orectoloboid suite in the lineage(s) of lamnoids and/or carcharhinoids. The monophyly of the Lamniformes is also his proposal (Compagno, 1973) based on dentition and many other anatomical features. Lamnoid sharks have often been regarded as having a close relationship with advanced carcharhinoids because of their superficial similarities (i.e., body proportions, fin positions, features of the fin skeleton, and tooth morphology) (White, 1937; Maisey, 1984a). Compagno (1988) explained these characters were symplesiomorphies or convergently acquired and hypothesized the monophyly of the Lamniformes and Carcharhiniformes.

III. Character Analysis

A. Systematic Method

The primary goal of this study is to reconstruct a theory of neoselachian interrelationships based on morphological information. To do this, I have used all codable discrete characters representing the current state of knowledge of neoselachian morphology. These were extracted mainly from the following sources: Schaeffer (1967), Compagno (1973, 1977, 1984a,b, 1988, 1990), Nakaya (1975), Heemstra and Smith (1980), Maisey (1980, 1984a,b, 1985b), Thies (1983, 1987), Thies and Reif (1985), Cappetta (1987), Nishida (1990), and Shirai (1992a,b,c). All morphological changes were summarized in binary and multistate transformation series (TS); multistate TS were ordered when character relationships could be hypothesized morphologically. Characters restricted to a single terminal taxon (autapomorphies) were not included in the analysis. Terminology for skeletal, muscular, and external features follows that of Shirai (1992c).

Terminal taxa, as detailed in the data matrix (Appendix 1) for the analyses, were supraspecific groups ranked at the genus or family level. These were selected from the above publications. Character polarization follows the outgroup method (Maddison *et al.*, 1984). As noted in Section II, studies have revealed living sharks and rays to be monophyletic, forming the neoselachian group. The sister-group of the neoselachians is still controversial, but I agree with Zangerl's (1973) and Maisey's (1975) considerations, in which neoselachians were suggested to have originated from the Mesozoic phalacanthous sharks characterized by dorsal fin spine morphology. Here I treat the fossil phalacanthous groups represented by the hybodonts and ctenacanths as a single outgroup, although they possibly are not monophyletic (Maisey, 1984b).

I have used PAUP (version 3.1; Swofford, 1993) on a Macintosh computer for all analyses reported here. The random-addition-sequence option of PAUP for the stepwise addition of taxa, with 100 replicates per search, and the MULPARS option to save all minimum-length trees have been used. All analyses were carried out with the heuristic search algorithm of PAUP, because the large number of taxa and complex data precluded the practical use of the branch-and-bound and exhaustive algorithms. To infer ancestral conditions at internal nodes I included only conditions that are consistent with all three character optimization methods available in PAUP (ACCTRAN, DELTRAN, and MIN F). To compare the results with various competing hypotheses, I applied the constraint option. This can search for the minimum number of additional steps (evolutionary events) required for these hypotheses.

B. Characters

For the analyses of the morphological data assembled here, 105 binary and multistate TS were compiled. Brief descriptions follow. Alternative character states ("State" in the following description) are preceded by the code number in brackets. The condition of the outgroup (OG) is added. Numerals in braces, e.g., {1}, indicate the character numbers used in Shirai (1992c). Character numbers 1–105 correspond to the columns of the data matrix.

Neurocranium

1. Rostral process: [0] absent; [1] composed of ventral rostral rod or its derivative (trough-like rostrum); [2] composed of lateral and ventral rostral rods to form a tripodal rostrum [ordered, OG = 1] {9}. Homology of the rostrum follows Holmgren (1940, 1941). Orectoloboids have State 1, although they are often regarded as having secondarily lost the lateral rostral rod (for detailed discussion, see Maisey, 1984a).

2. "Anterior process" of neurocranium: [0] absent; [1] present [OG = 0]. Compiled from Nishida (1990); State 1 is restricted to mobuline species.

3. Precerebral fossa: [0] absent; [1] present as a circular or ovoid concavity; [2] extending anteriorly and roofed to form a tube [ordered, OG = 1] {14}. Homology of the rostrum follows Holmgren (1940, 1941). Myliobatidoids lack the precerebral fossa [0], because the cranial cavity is expanded forward to the anterior terminus of the neurocranium, where the prefrontal fontanelle opens in other batoids (Nishida, 1990).

4. Nasal capsule: [0] not expanded; [1] expanded ventrolaterally [OG = 0]. Compiled from Nishida (1990).

5. Nasal capsule: [0] almost attached to orbit; [1] well separate from orbit [OG = 0] {2}. *Heterodontus* and orectoloboids possess a prolonged interorbitonasal region, which forms a pedicel of the "trumpet-shaped nasal capsule" (Compagno, 1973).

6. Ethmoidal canal: [0] absent; [1] present, transmitting the ethmoidal nerve to the ventral or ventrolateral side; [2] present with an exit(s) at the dorsal side of the nasal capsule [unordered, OG = 0] {25–28}. State 2 corresponds to characters 26, 27, and 28 of Shirai (1992c).

7. Subnasal fenestrae: [0] absent; [1] present [OG = 0] {15}.

8. Supraorbital crest: [0] absent; [1] present, moderately developed; [2] expanded laterally [ordered, OG = 1]. Galeoids exhibit considerable variation in the development of the supraorbital crest. Compiled from Compagno (1988, 1990).

9. Superficial ophthalmic nerve passing through: [0] the foramen prooticum with the remaining branches of the trigeminal and facial nerves (excluding the hyomandibularis); [1] a separate foramen [OG = 0] {5}. Compiled from Maisey (1985a) and Compagno (1988). Although the absence of a separate foramen for the hyomandibular nerve has been considered another galeoid feature, some exceptions are observed (Compagno, 1988).

10. Ectethmoid process: [0] present as the antorbital process; [1] absent or indistinct if present; [2] present as dense connective tissue; [3] present as a chondrified antorbital cartilage; [4] present with an antler–like anterior expansion [ordered, OG = 1] {33–35}. Definition of the ectethmoid process follows Holmgren (1941) and Carvalho and Maisey (1996). "Squaloid" sharks, which have a weak ectethmoid process as a "membranous tissue" according to Holmgren (1941), are here categorized in State 1. Refer to Shirai (1992c, Plates 1–12) for variations of their cranial postnasal wall.

11. Craniopalatine articulation: [0] with the ethmoidal articulation; [1] present in "orbitostylic" mode; [2] absent [unordered, OG = 0] {43}. Compiled from Maisey (1980), who categorized the craniopalatine articulation into two distinct types between living and fossil sharks. *Chlamydoselachus, Echinorhinus,* and *Pseudocarcharias,* which were pointed out as exceptions to this theory by Compagno (1988), can be distinguished also by the relation between the ascending process of the palatoquadrate and the optic nerve (Shirai, 1992c).

12. Suborbital shelf: [0] absent; [1] present; [2] present with prominent lateral wing behind orbital notches [ordered, OG = 1]. Definition of the "suborbital shelf" and "basitrabecular process" follows Shirai (1992b,c), who suggested these are different structures based on ontogenetic evidence (Holmgren, 1940). Here I treat them as separate characters, agreeing with Carvalho and Maisey (1996). State 2 is compiled from Compagno (1990).

13. Basitrabecular process: [0] absent; [1] present [OG = 0] {44}. See TS–12.

14. Postorbital articulation: [0] absent; [1] disengaging when jaws are protracted; [2] present with an articular facet [unordered, OG = 0] {32}.

15. Postorbital fenestra: [0] absent; [1] present [OG = 0] {31}. This perforation is restricted in somniosid genera (Shirai, 1992c, plate 6 and cover plate).

16. Sphenopterotic ridge: [0] present; [1] absent [OG = 0] {50}. Definition of this character follows Shirai (1992c).

17. Hyomandibular fossa: [0] separate from orbit; [1] immediately behind orbit [OG = 0] {1}. State 1 corresponds to the "shortened auditory capsule" of galeomorphs (Compagno, 1973, 1988).

18. Hyomandibular fossa composed of: [0] a single ovoid or groovelike concavity, or two inconspicuous concavities, positioned vertically or slightly obliquely; [1] an obvious dual concavity divided horizontally [OG = 0] {54}.

19. Depression below the hyomandibular fossa: [0] absent; [1] present [OG = 0] {51}.

20. Basioccipital fovea: [0] present; [1] absent [OG = 0] {52}. The basioccipital fovea (Shirai, 1992b) is an unpaired concavity just below the foramen magnum, filled with the occipital hemicentrum for the craniovertebral conjunction. It may be homologous with the occipital cotylus (Maisey, 1982).

21. Occipital hemicentrum: [0] absent; [1] present [OG = 0] {53}.

22. Obliquus inferior β: [0] absent; [1] present [OG = ?] {55}.

23. Rectus externus arising from: [0] the eye stalk or interorbital floor; [1] own deep fossa [ordered, OG = 0] {42}. The fossa for the rectus externus (State 1), just in front of the foramen prooticum, restricted in etmopterids and euprotomicrines (Shirai and Okamura, 1992, fig. 1).

Visceral Arches (Skeleton)

24. Right and left halves of palatoquadrate and mandibular cartilages: [0] articulated with the antimere; [1] fused with the antimere [OG = 0]. Symphysial fusion of these cartilages (State 1) is seen in *Aetomylaeus* and mobuline rays (Nishida, 1990).

25. Wing-like process of mandibular cartilage: [0] absent; [1] present [OG = 0]. Compiled from Nishida (1990). State 1 is restricted in myliobatidid rays.

26. Hyoid arch: [0] reduced, nonsuspensory, having no insertion of the dorsal constrictor muscle; [1] massive, holding the mandibular arch from behind; [2] composed of reduced ventral parts and developed hyomandibula, the latter suspending the lower jaw directly; [3] similar to State 2, but the articulation between the hyomandibula and mandible is interrupted by a ligament [ordered, OG = 1] {72 –73}. State 3, seen in urolophids, dasyatids, and myliobatidids (excluding *Mobula* and *Manta*), is compiled from Nishida (1990).

27. Pseudohyoid bar: [0] absent; [1] present; [2] present with a fusion between the ventral bar and ceratobranchial 1 [ordered, OG = 0] {74}. State 2 is introduced by Nishida (1990) as the synapomorphy of myliobatidoids.

28. Extrabranchial cartilage on hyoid arch: [0] absent; [1] only the dorsal element present; [2] both dorsal and ventral elements present [ordered, OG = ?] {78, 79}.

29. Ceratobranchial 5: [0] free from the shoulder girdle; [1] firmly attached to the anterior margin of the coracoid [OG = 0] {86}.

30. Sixth branchial unit (hexanchoid type): [0] absent; [1] present with a unique paired basibranchial cartilage [OG = 0] {80}. Definition of the extra branchial arches follow Shirai (1992a,c); morphological information suggests that the sixth branchial unit of hexanchoids is analogous to that of *Chlamydoselachus*.

31. Seventh branchial unit: [0] absent; [1] present [OG = 0] {81}.

32. Anterior separate basibranchial series: [0] present; [1] absent [OG = 0] {90}. Distribution of this character among galeomorphs follows Luther (1909) and Dingerkus (1986).

33. Hypobranchial 2: [0] paired; [1] fused with the antimere forming a transverse bar [OG = 0] {91}. State 1 is known as a "hypobranchial bar" in dalatiids (excluding *Isistius*), *Oxynotus*, and *Somniosus* (Shirai, 1992c).

34. Dorsal ends of branchial arches 4 and 5: [0] separated each other; [1] not fused but attached; [2] fused to make a single plate (gill pickaxe) [ordered, OG = 0]. Most living neoselachians have the gill pickaxe (State 2) except some batoids; their pharyngobranchials are often modified considerably, and cannot simply be compared with that of sharklike forms. The "incomplete" gill pickaxe (State 1) is seen in *Heterodontus*, hexanchoids, and *Trigonognathus*, and the dorsal ends of arches 4–5 (posteriormost and penultimate arches) are separate in *Chlamydoselachus* (Shirai, 1992c, Plates 29–32).

35. Pharyngobranchial blade: [0] absent; [1] present [OG = 0] {10}. The pharyngobranchial blade is a proximoexternal expansion of the pharyngobranchial from which the arcualis dorsalis arises. Shirai (1992b,c) concluded that the absence of it (State 0) is seen in nongaleomorph lineages as a derived condition. The polarity of this character, as criticized by Carvalho and Maisey (1996), seems to have been based on ambiguous grounds. According to their observations, the representative hybodontids, †*Egertonodus fraasi* and †*Hybodus hauffianus*, and Permian

xenacanths, show no indication of the pharyngobranchial blade. Although there are only rare examples with preserved pharyngobranchials, I follow Carvalho and Maisey (1996), and the pharyngobranchial blade in galeomorphs (State 1) is regarded as derived.

Visceral Arches (Musculature)

36. Adductor mandibulae divided into: [0] three or more subdivisions; [1] ordinal dorsal and ventral blocks; [2] *lateralis i* and *ii* [unordered, OG = ?] {94}. State 0 seen in hemiscylliids and rhincodontids is compiled from Moss (1977) and Compagno (1988).

37. Adductor mandibulae medialis: [0] absent; [1] present [OG = ?] {93}.

38. Suborbitalis: [0] absent; [1] single; [2] with two divided heads [unordered, OG = ?] {98}. State 2 is compiled from Compagno (1988). The carcharhinoids uniquely have an additional slip of the suborbitalis, which inserts on the palatoquadrate via a long tendon (Moss, 1972). Compagno (1988) reported that this muscle slip occurs in all the carcharhinoids he examined.

39. Suborbitalis arising from: [0] the suborbital surface; [1] the interorbital wall; [2] the lateral surface of the nasal capsule; [3] the supraorbital surface; [4] the antorbital process [unordered, OG = ?] {95, 97}. State 3 is compiled from Compagno (1988).

40. Suborbitalis inserted on: [0] the adductor mandibulae and/or subcutaneous tissue near the mouth corner; [1] the mandibula directly; [2] the mandibula by a tough tendon [unordered, OG = ?] {96}. State 2 is shown in pristiophorids and most batoids (Nishida, 1990; Miyake and McEachran, 1991; Shirai, 1992c).

41. Adductor m. superficialis: [0] absent; [1] present [OG = ?] {103}.

42. Adductor m. superficialis inserted on: [0] subcutaneous tissue around the eye; [1] also onto the antorbital process; [2] the nasal capsule; [3] hyomandibula [unordered, OG = ?] {99–102}. State 3 is seen in *Heterodontus* (Shirai, 1992c, plate 39).

43. Nictitating lower eyelid: [0] absent; [1] "rudimentary-type"; [2] secondary lower eyelid developed; [3] retractor and depressor reduced [ordered, OG = ?]. Carcharhinoid and some orectoloboid sharks (parascylliids, orectolobids, and brachaelurids) show various forms of the nictitating membrane. Compiled from Compagno (1970, 1988).

44. Levator palatoquadrati: [0] also inserted on the spiracle; [1] separate from the spiracularis [OG = ?] {105}. Compiled from Compagno (1988) and Shirai (1992c).

45. Constrictor hyoideus dorsalis inserted onto: [0] the palatoquadrate only; [1] the hyomandibula and

palatoquadate; [2] the hyomandibula [unordered, OG = ?] {111, 112}.

46. Levator rostri: [0] absent; [1] present [OG = ?] {116}.

47. Depressor rostri: [0] absent; [1] present [OG = ?] {117}.

48. Arcualis dorsalis: [0] composed of a single slip; [1] divided into anterior and posterior slips ("double headed" condition) [OG = ?] {118}.

49. Subspinalis: [0] present; [1] absent [OG = ?] {120}.

50. Interpharyngobranchialis: [0] present; [1] absent [OG = ?] {122}.

51. Geniocoracoideus arising from: [0] the fascia of the rectus cervicis; [1] the coracoid or pericardial membrane [OG = ?] {123, 124}. State 1 comprises Shirai's (1992c) separate conditions (characters b and c of TS 2.8A) for the global comparison among living forms.

52. Coracohyoideus: [0] present; [1] absent [OG = ?]. Compiled from Nishida (1990).

53. Coracohyoideus: [0] attached to the coraco-arcualis to form massive rectus cervicis; [1] separate from the coracoarcualis, arising from the fascia on the coracohyomandibularis [OG = ?] {128}.

54. Coracohyoideus inserted on: [0] the basal part of the hyoid arch; [1] the antimere via raphe [OG = ?]. Compiled from Nishida (1990).

55. Coracobranchialis 1 arising from: [0] the fascia of the rectus cervicis; [1] the coracoid bar [OG = ?]. Compiled from Nishida (1990). State 1 is observed as a developed coracohyomandibularis in narkine and narcinine rays.

56. Coracobranchialis 1 inserted on: [0] the basihyal and/or hyoid cartilage; [1] the hyomandibula via a long tendon, termed the "coracohyomandibularis" [OG = ?] {130}.

Girdles and Paired Fins

57. Symphysial fusion of right and left halves of coracoid cartilages: [0] articulated; [1] fused incompletely; [2] fused to form a transverse bar [ordered, OG = 0]. The coracoids are separate in outgroup taxa and hexanchoids (State 0), as opposed to being a U–shaped coracoid bar in most living forms (State 2). State 1 is seen in *Chlamydoselachus* and *Aculeola*, where coracoid cartilages are prolonged anteriorly as in hexanchoids but have a characteristic wedge–shaped notch at the posterior surface of the symphysial portion (Shirai, 1992c, plate 44).

58. Scapular process: [0] separate from the vertebrae; [1] attached to the vertebrae; [2] fused to the synarcual; [3] articulated with the synarcual by a ball

and socket articulation [unordered, OG = 0] {135}. States 2 and 3 are compiled from Compagno (1973, 1977), Heemstra and Smith (1980), and Nishida (1990).

59. Articular process of the coracoid for the pectoral basals: [0] single, irregular-shaped condyle oriented vertically or obliquely; [1] two distinct condyles, one for the pectoral metapterygium; [2] three (or more) condyles including one for the propterygium [ordered, OG = 0] {136, 137}.

60. Pectoral propterygium and mesopterygium: [0] separated; [1] fused [OG = 0] {138}.

61. Pectoral mesopterygium and metapterygium: [0] separated; [1] fused [OG = 0] {139}.

62. Pectoral mesopterygium and metapterygium: [0] not expanded distally; [1] expanded distally and curved in opposite direction to make an interspace between them [OG = 0]. Although Compagno (1988) noted that the two basals in *Heterodontus* and orectoloboids are elongate and expanded distally, *Heterodontus* is more similar to some "squaloid" sharks than to orectoloboids in the form of these skeletal elements (e.g., Daniel, 1934, fig. 82; Shirai, 1992c, plate 48). Orectoloboids, here grouped together, are distinguished from the other taxa by a characteristic interspace between mesopterygium and metapterygium (see Dingerkus, 1986, fig. 6).

63. Pectoral propterygium: [0] with radials; [1] reduced with no radials [OG = 0] {141}.

64. Pectoral propterygium: [0] directed posterolaterally; [1] extending anteriorly, terminated behind the nasal capsule; [2] reached to the nasal capsule; [3] articulated with the antorbital cartilage; [4] well extended anteriorly and supporting the snout [ordered, OG = 0] {142, 143}. States 2, 3, and 4 are seen among batoids, compiled from Nishida (1990).

65. Pectoral metapterygium: [0] simple; [1] segmented [OG = 0] {144}.

66. Pectoral fin radials: [0] aplesodic; [1] plesodic [OG = 0]. Compiled from Compagno (1988) and Nishida (1990).

67. Neighboring radials of the pectoral fin: [0] separate; [1] interlocked [OG = 0]. Compiled from Nishida (1990).

68. Pelvic fin: [0] aplesodic; [1] plesodic [OG = ?]. Compiled from Heemstra and Smith (1980) and Nishida (1990).

69. Ventral marginal cartilage: [0] not elongated; [1] well elongated [OG = 0]. Compiled from White (1937) and Compagno (1988).

70. Accessory terminal (T–3) cartilage: [0] spinous; [1] not spinous; [2] modified into the external mesorhipidion [ordered, OG = ?]. Compiled from Compagno (1988).

71. Pectoral electric organs: [0] absent; [1] present [OG = ?].

Axial Skeleton and Unpaired Fins

72. First (cervical or cervicothoracic) synarcual: [0] absent; [1] present [OG = 0] {150}.

73. Second (thoracolumbar) synarcual: [0] absent; [1] present [OG = 0]. Compiled from Compagno (1973) and Nishida (1990).

74. Vertebral ribs: [0] present; [1] absent [OG = 0]. Compiled from Nishida (1990).

75. Primary calcification of vertebrae: [0] reduced or absent; [1] restricted terminally; [2] developed [ordered, OG = 0] {151, 152}.

76. Secondary calcification of vertebrae: [0] absent or poorly developed; [1] with endochordal radii radiating from the notochordal sheath; [2] with developed solid intermedialia and diagonal calcified lamellae; [3] compact mass [unordered, OG = 0] {155}. States 1 and 2 are compiled from Compagno (1988).

77. Supraneurals: [0] absent or poorly developed; [1] enlarged at least in front of the second dorsal fin; [2] also enlarged in the abdominal region [ordered, OG = 0] {156, 157}.

78. Hemal arch: [0] not arched at anterior precaudal tail vertebrae; [1] almost complete in the entire region of the precaudal tail [OG = 0] {159}.

79. Hemal processes at the precaudal tail: [0] shorter than that of the lower lobe of the caudal fin; [1] elongate and equal to those of the caudal fin [OG = 0] {160}.

80. Dorsal fin: [0] dual, developed; [1] single, developed (*Chlamydoselachus*); [2] single, developed (hexanchoids); [3] reduced or absent [unordered, OG = 0] {179, 180}. State 3 is compiled from Nishida (1990).

81. Dorsal fin skeleton composed of: [0] a triangular or rectangular basal cartilage and radials; [1] an elongate basal with radials; [2] radials only (no basal) [unordered, OG = 0] {162–165}. Shirai (1992c) recognized six patterns of the dorsal fin skeletons. In this paper, I recategorize them into three characters based mainly upon the presence or absence of the basal cartilage.

82. Dorsal radials: [0] on the basal cartilage; [1] independent in part from the basal [OG = 0] {166}. This TS is compared among species with the basal cartilage of the dorsal fin.

83. First dorsal fin radials: [0] aplesodic; [1] semiplesodic; [2] plesodic [unordered, OG = 0]. Compiled from Compagno (1990) and Nishida (1990).

84. Anal fin: [0] present; [1] absent [OG = 0] {167}.

85. Anal fin skeleton composed of: [0] the basal cartilage and radials; [1] segmented radials only [OG = 0] {8}.

86. Caudal fin: [0] with a developed lower lobe to make a "lunate" form; [1] heterocercal; [2] reduced to the plesodic or tail folds; [3] without any tail folds or finlet [ordered, OG = 1]. Compiled from Compagno (1990) and Nishida (1990).

87. Perihypochordal cartilages: [0] not elongated; [1] well elongated anteriorly [OG = 0] {161}. This series of cartilages (State 1) is seen in somniosid genera (Shirai, 1992c, plate 53).

88. Inclinator dorsalis arising from: [0] dorsal surface of body; [1] vertebrae [OG = ?] {171}.

89. Ventral bundle of body muscle: [0] reduced; [1] well developed in the precaudal region [OG = ?] {168}.

90. Flexor caudalis: [0] restricted within the caudal fin; [1] advanced at least up to the lower caudal origin; [2] extending anteriorly beyond the lower caudal origin [ordered, OG = ?] {169, 170}.

External Features

91. Cephalic lobes: [0] absent; [1] present; [2] present as the cephalic fin [ordered, OG = 0]. Compiled from Nishida (1990).

92. Nasoral groove: [0] present with an expanded lobe; [1] absent [OG = ?]. The nasoral groove is present in *Heterodontus*, orectoloboids, some scyliorhinid species, triakid *Scylliogaleus*, *Squatina*, and rajoids except pristids and most "rhinobatoids" (e.g., Compagno, 1984a,b; Last and Stevens, 1994).

93. Mesonarial flap: [0] present; [1] absent [OG = ?] {16}.

94. Nasal barbel innervated by a branch of the ophthlamicus superficialis: [0] absent; [1] present [OG = ?]. This type of the nasal barbel is seen in orectoloboids, not homologous with that of some squalids (Shirai, 1992c, fig. 5.24).

95. Circumnarial folds: [0] absent; [1] present [OG = ?]. Compiled from Compagno (1973, 1988).

96. Lower jaw teeth arranged: [0] diagonally; [1] linearly along the jaw margin without an imbrication; [2] linearly, overlapping to form a continuous cutting edge [ordered, OG = 0] {66, 68}.

97. Upper jaw teeth arranged: [0] diagonally; [1] linearly along jaw margin and if overlapping not forming a complete continuous cutting edge; [2] linearly with complete imbrication [ordered, OG = 0] {67, 69, 70}. Shirai's (1992c) characters 67 and 69 are recategorized into a single character (State 1).

98. Aprons of jaw teeth: [0] absent; [1] present at the labial and lingual sides [OG = 0]. Compiled from Thies and Reif (1985).

99. Upper eyelid: [0] present; [1] absent [OG = ?] {173}.

100. Spiracle valve: [0] absent; [1] present [OG = ?]. In "squaloids" (excluding *Echinorhinus*),

pristiophorids, and batoids, the spiracle with a valvular structure probably with a respiratory function.

101. Lateral sensory canal of trunk: [0] grooved; [1] closed [OG = 0] {4}.

102. Luminous organs: [0] absent; [1] present on the ventrolateral surface of body; [2] well developed, forming characteristic dark markings [ordered, OG = ?] {177, 178}.

103. Precaudal pit: [0] absent; [1] present at the origin of the upper caudal lobe [OG = ?] {181}.

104. Precaudal keel: [0] absent; [1] present [OG = ?] {182}.

105. Intestinal valve type: [0] spiral; [1] ring; [2] scroll [unordered, OG = ?]. Three basic types of the intestinal valves (Compagno, 1973, 1988) are estimated here.

C. Results

Despite the exhaustive studies by Compagno (1973, 1977, 1988, 1990), encodable characters for galeomorph sharks are considerably fewer than those for squalean sharks and rays. This unfortunately biases the data matrix in favor of squaleans in both numbers of TS and terminal taxa. The present data matrix has this property, but nonetheless I believe that the results of this study form the basis for an alternative view of the phylogenetic interrelationships of living neoselachians.

Parsimony analyses resulted in 48 minimum-length trees of 261 steps (CI = 0.59, RI = 0.92). These trees were summarized in the strict consensus diagram presented as Fig. 2. Topological variations among the 48 original trees are reflected in five unresolved polychotomies in the consensus tree. Apart from the ones that appeared terminally (Clades 4, 23, 39, and 40 in Fig. 2), the remaining one polychotomy (Clade 20) is commented on briefly here. The variations in this polychotomy, represented by three patterns, are related to the position of centrophorids (*Centrophorus* + *Deania*). One of these three resolutions indicated that centrophorids and dalatiiforms (Clade 21) are paraphyletic groups to squalids + hypnosqualeans (Clade 28) [(Dalatiiformes, (Centrophoridae, (Squalidae, hypnosqualeans)))], whereas the second pattern placed centrophorids + dalatiiforms as the sister-taxon to squalids + hypnosqualeans [((Dalatiiformes, Centrophoridae), (Squalidae, hypnosqualeans))]. The third pattern placed centrophorids as the primitive sister-taxon to dalatiiforms + squalids + hypnosqualeans [(Centrophoridae, (Dalatiiformes, (Squalidae, hypnosqualeans)))]. These three topologies, also seen in my previous research (Shirai, 1992c), are caused by the rather generalized

conditions of centrophorids among the higher squaleans (Clade 20). Therefore, the trichotomy in Clade 20 is adequate for the phylogenetic speculation presented here.

The unambiguous characters of each internal node are summarized in Appendix 2. My results (the strict consensus tree) support Compagno's Galeomorphii (Clade 1) with five unambiguous synapomorphies (17, 35, 57, 75, and 101). Evidence for his four ordinal clades, i.e., Heterodontiformes, Orectolobiformes, Lamniformes, and Carcharhiniformes, were also found here, but the present results support a group of galeoids (orectoloboids + lamnoids + carcharhinoids) and not a sister-relation between heterodontoids and orectoloboids (but see Discussion). Opposing this clade is a large sister-group (Clade 15) corresponding to the Squalea (Shirai, 1992c) that includes all non-galeomorph sharks and rays; the hypnosqualean assemblage (Shirai, 1992b; Clade 29) is also corroborated by the reanalyzed synapomorphies (6, 7, 20, 59, 88, 90, 96–1, and 96–2). As in Shirai (1992c), *Chlamydoselachus* and hexanchoids are paraphyletic within the squalean clade, and "squaloid" sharks are divided into four paraphyletic groups. The batoids are the most derived form of the squaleans, and the outline of their intrarelationships agree with Nishida's (1990) conclusion.

The constraint analyses were performed as 18 independent tests, shown in Table 1 with their results. The hypotheses competing with the most parsimonious trees of the present study are the following: (a) shark–ray dichotomy, (b) Schaeffer (1967), (c) Compagno (1973, 1977), (d) Maisey (1980, 1984a), (e) Thies (1983), (f) Dingerkus's unpublished tree, and (g) subsets including a particular clade(s). Searches constrained to keep previous hypotheses of a–f required considerable extra steps except for Dingerkus's tree (Test 13). In the tests for a particular grouping, a clade of heterodontoids + orectoloboids (Test 14), Hexanchiformes of Compagno (*Chlamydoselachus* + hexanchoids) (Test 15), all living taxa excluding *Chlamydoselachus* and hexanchoids (Test 16), and traditional "squaloids" (Test 18), needed only two additional steps each.

IV. Discussion

A. Higher-Level Phylogeny

The resulting tree (Fig. 2) is largely in accordance with the phylogeny of the Squalea including the hypothesized outgroup interrelationships (Shirai, 1992c, figs. 4.1, 6.1). This phylogenetic inference is based on the hypnosqualean clade (*Squatina* + pristiophorids +

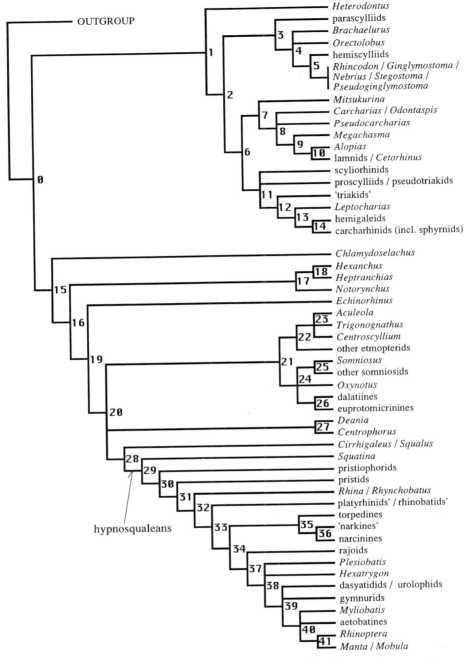

FIGURE 2 Strict consensus of 48 minimum-length trees resulting from maximum parsimony analysis. The ingroup (0) and all interior (1–41) nodes are labeled, and the apomorphies supporting each interior node are given in Appendix 2.

batoids; Shirai, 1992b) and differs from most previous hypotheses in the position of batoids. The latter placed batoids separate from the shark group or in parallel with some shark groups; recent hypotheses of Compagno (1973, 1977) and Maisey (1984a) are also included here. According to the present constrained analyses, these previous hypotheses are all substantially less parsimonious than my own, when optimized on my data matrix (Tests 1–13). A phylogenetic tree of G. Dingerkus's unpublished manuscript re-

TABLE 1 **Summary of Constraint Analysis**

Test	Constraint	Length (No. of trees)	CI (RI)	Remarks
	None (present study)	261 (48)	0.594 (0.917)	
	Shark–Ray dichotomy			
1	Basal dichotomy only	280 (192)	0.554 (0.903)	
2	Bigelow and Schroeder's (1948, 1953): suborder-level	281 (48)	0.552 (0.902)	Fig. 1a
	Schaeffer (1967)			
3	Galeoids + squaloids + batoids	280 (32)	0.554 (0.903)	Fig. 1b
	Compagno (1973, 1977)			
4	Alternative 1	281 (48)	0.552 (0.902)	Fig. 1c
5	Alternative 2	280 (48)	0.554 (0.903)	Fig. 1d
6	Alternative 3	280 (48)	0.554 (0.903)	Fig. 1e
7	Alternative 4	277 (48)	0.560 (0.905)	Fig. 1f
8	Alternative 5	280 (48)	0.554 (0.903)	Fig. 1g
9	Alternative 5'	285 (64)	0.544 (0.899)	Fig. 1h
10	Order-Level (alternative 5)	290 (864)	0.534 (0.895)	Fig. 1g
	After 1973			
11	Maisey (1980, 1984a)	281 (48)	0.552 (0.902)	Fig. 1i
12	Thies (1983), Thies and Reif (1985)	309 (48)	0.502 (0.880)	Fig. 1j
13	G. Dingerkus (cited by Seret, 1986)	269 (16)	0.576 (0.911)	Fig. 1k
	Subsets including particular clades			
14	Heterodontoids + orectoloboids	263 (95)	0.589 (0.916)	Compagno (1973)
15	Hexanchiformes	263 (48)	0.589 (0.916)	Compagno (1973)
16	All living taxa excluding *Chlamydoselachus* and hexanchoids	263 (48)	0.589 (0.916)	Thies (1983)
17	Squaliformes	280 (48)	0.554 (0.903)	Compagno (1973)
18	"Squaloids"	263 (48)	0.589 (0.916)	

quired only eight additional steps and included hypnosqualeans (Test 13). This consequently suggests that it is unreasonable to make an alternate hypothesis ignoring the hypnosqualean assemblage.

Several proposals have been presented that have allied the batoids with *Squatina* and/or pristiophorids (e.g., Hasse, 1879–1885; Hoffmann, 1913). Opposed to these proposals, Maisey (1980) stated that he did not find the "evidence for close affinity between batoids and these shark groups [*orbitostylic elasmobranchs* in original] to be very compelling" (p. 10). The batoids indeed differ from shark taxa morphologically, and their extreme condition of the jaw suspension with no craniopalatine articulation is one of the characteristic examples. The hypnosqualean hypothesis therefore demands the secondary loss of the "orbitostylic" jaw articulation, which must have been acquired among one of the neoselachian clades (Maisey, 1980).

However, the feeding apparatus of batoids is totally different from that of sharks, not only in the presence or absence of the articulation. The mandibular arch of sharks is supported by the whole hyoid arch posteromedially, and its protrusion is related to the rotat-

ing of the hyoid arch posteroventrally. Accordingly sharks generally have a well-developed ventral hyoid unit (basihyal and hyoid cartilages) and related muscles (coracohyoideus and coracobranchialis 1). In batoids, the hyomandibula directly suspends the whole mandibular arch and makes the jaws protrude by its swinglike motion ventrally. These two basic modes of the hyoid function can be seen in the differences of the hyomandibular fossa (TS 18). The basihyal and related structures of batoids are very reduced or absent (Nishida, 1990; Miyake and McEachran, 1991), and the coracobranchialis 1 has a different insertion point onto the distal end of the hyomandibula to work as a massive coracohyomandibularis (TS 56) (Shirai, 1992b; Miyake *et al.*, 1992). Although they have a loose, but distinct, "orbitostylic" mode of the articulation, pristiophorids also possess these batoid features (but no pseudohyoid bar) (Shirai, 1992b,c). From these features and other synapomorphies among hypnosqualeans, the idea that batoids are highly modified squalean sharks is not so extraordinary when compared with the alternative theories concerning the higher-level phylogeny of euselachians.

The hypnosqualean group is in accordance with Hasse's (1879–1885) "Tectospondyli" and similar to the hypothetic lineage of White (1937). This clade is also shown in the tree of G. Dingerkus (Seret, 1986). Although Dingerkus's study was also cited favorably by Nelson (1994), it has not been formally published (personal communication with B. Seret and M. Carvalho). After Shirai (1992b,c), Deets (1994) supported the interrelationships of "squaloid" sharks–*Squatina*–pristiophorids–batoids from a phylogenetic study of the parasitic copepod genus *Eudactylina*. Carvalho and Maisey (1996) reconfirmed the hypnosqualean lineage and formally placed them in the superorder Hypnosqualea. They also included †*Protospinax*, a well-preserved neoselachian fossil from the Late Jurassic, in this superorder. Consequently, the support for the hypnosqualean lineage appears to settle the controversy about batoid phylogeny.

The Squalea (Shirai, 1992c), including hypnosqualeans, is composed of Compagno's Squaliformes, Squatiniformes, and Batoidea. In the present analysis, this assemblage is present in all the most parsimonious trees. Of the synapomorphies for squaleans, we have considered previously the "orbitostylic" mode of the craniopalatine articulation (Maisey, 1980). As noted, this type of articulation seems to have been lost secondarily in batoids. This character (TS 11) therefore forms a good basis for squalean monophyly together with the loss of the suborbital shelf (TS 12).

Notwithstanding such morphological evidence, the squalean hypothesis still seems to be on weak grounds. *Chlamydoselachus* and hexanchoid genera (and/or *Echinorhinus*) retain many ancestorlike conditions, and their existence makes the basal squaleans obscure. Some of their features indicate unresolved reversal and parallel steps: e.g., the incomplete gill pickaxe (TS 34) and coracoid bar (TS 57), no radials on the pectoral propterygium (TS 63), poorly developed primary calcification of vertebrae (TS 76), and a single dorsal fin (TS 81). In a case that *Chlamydoselachus* and hexanchoids are moved from the Squalea to the most primitive sister-group of all other neoselachians like the phylogenetic tree (Fig. 1j), only two additional steps are needed (Test 16). Nevertheless, this is not a positive reason to contradict the squalean clade. These "primitive" representatives may still be key to finding more convincing neoselachian phylogenies.

The galeomorph assemblage is also detected in all the most parsimonious trees of the present analysis. In a series of studies, Compagno was inconclusive concerning galeomorph monophyly, although this group forms an important part of his hypothesis. Compagno (1977, 1988) indicated five galeomorph synapomorphies, and the present study proposed five different apomorphic features for them (including four new characters), two of which are unique to galeomorphs (TS 17 and 35). However, the present character analysis is biased in favor of squaleans, and further research is needed to reconfirm galeomorph monophyly. In particular, heterodontoid–orectoloboid interrelationships will probably hold a key to the solution, because they may occupy the basal part of galeomorph phylogeny. Based on the transferrin immunological method, Davies *et al.* (1987) regarded *Heterodontus* as a galeomorph shark, because the separation between *Heterodontus* and scyliorhinid species showed a similar level to those found in two scyliorhinid species. Such information from a nonmorphological viewpoint is expected to be particularly helpful with this group.

B. Galea

As formalized in the next section, I propose a term "Galea" for Compagno's (1973) galeomorphs. Within the galeans, the present analyses suggest the following successive sister-relations in the higher-level interrelationships, i.e., (heterodontoids, (orectoloboids, (lamnoids, carcharhinoids))). Compagno (1973, 1977, 1988) preferred the sister-group relations between heterodontoids + orectoloboids and lamnoids + carcharhinoids to the galeoid hypothesis (orectoloboids + lamnoids + carcharhinoids). The "orectoloboid suite" of Compagno (1988), evidence for the heterodontoid–orectoloboid monophyly, is summarized as follows: (1) deep anteroposteriorly oriented nasoral grooves (TS 91 of the present analysis), (2) circumnarial folds and grooves (TS 94), (3) short broadly arched or nearly straight mouth entirely in front of eyes, (4) labial furrows and cartilages displaced far anterior on jaws, (5) trumpet–shaped nasal capsules (TS 5), (6) low, broadened medial ascending processes with a sliding articulation in specialized ethmopalatine grooves on the cranium (TS 5), (7) distally expanded and elongated pectoral mesopterygium and metapterygium (TS 62), (8) vertical suborbitalis muscle (TS 40), and (9) no anterior notches on the adductor mandibulae muscles. This study analyzes six of these. (Compagno's characters 3, 4, and 9 were not used because of the difficulty in undertaking comparisions among all neoselachians.) As a reason for rejecting the galeoid hypothesis, Compagno (1988) suggested that it was improbable to have secondarily lost the orectoloboid suite (represented by the above characters) in the lineage to lamnoids and carcharhinoids. The present analyses, however, confirm the galeoid assemblage based on several characters (TS 3, 9, 34, 69, 81, and 85) despite the recognition that the "improbable" secondary loss has occurred among galeoids. However, the search constrained to keep a clade of heterodontoids + orectoloboids (Test 14) requires only two additional steps. To find a more defendable higher-level hypothesis of interrelationships, further investigations are needed concerning the homology of similar features among all galean sharks.

The sister–group relationship between lamnoids and carcharhinoids is based on the tripodal rostrum (TS 1); four reversals of the "orectoloboid suite" (TS 5, 40, 91, and 97) should have been acquired here. Compagno (1988) proposed three other synapomorphies for them (preoral snout elongated; labially expanded bilobed tooth roots; and clasper siphons greatly expanded into the abdominal wall anterior to the pelvic bases); these characters are not analyzed here because of uncertainty of their homology.

Interrelationships within the ordinal groups of galeoids remain uncertain. Based on recent research, the present situation for each group is described below.

Orectolobiformes—Dingerkus (1986) and Compagno (1988) proposed contrasting hypotheses of orectoloboid interrelationships through cladistic analyses. Although they used restricted and very different character sets, their conclusions agreed in two points. One is the sister-relation between Orectolobidae and Brachaeluridae. Although not found in my tree, this relationship is probable because of some characteristic features (e.g., symphysial groove of the lower labial and a ridge around the spiracle rim; Compagno, 1988). The other is that *Rhincodon* forms an advanced lineage with Ginglymostomatidae, *Stegostoma*, and *Pseudoginglymostoma*. The present study agrees with Compagno (1988) in that parascylliids are the sister-group of the remaining orectoloboids. The extinct taxa of orectoloboids are separate from each other in morphology, probably affected by the long independent history of their evolution (Cappetta, 1987). Further studies of these extinct taxa will probably throw much light on the basal radiation of the galean lineages.

Lamniformes—As the first to recognize the lamnoid assemblage, Compagno (1973) proposed six families following Bigelow and Schroeder (1948) with a tentative elevation of *Pseudocarcharias* to family level. The megamouth shark (*Megachasma*) discovered in 1976 was first considered to occupy a new family and was considered a primitive sister-group to the other lamnoids (Taylor *et al.*, 1983). Maisey (1985b) expanded the Cetorhinidae to include *Megachasma* based on their similarities in jaw suspension and dental arrays. Maisey also suggested that his Cetorhinidae is the

sister-group of lamnids + *Alopias* because of the plesodic radials of the pectoral fin. Compagno (1990) disagreed with Maisey on phenetic and cladistic evidence and proposed a sister-group relationship between *Megachasma* and lamnids + *Cetorhinus* + *Alopias*. The present analysis supports Compagno's view and also the view that *Mitsukurina* is the sister-group of the others. Because of the enormous range of modifications among fossil lamnoids (e.g., Cappetta, 1987), we cannot understand their phylogeny correctly unless extinct lineages are included.

Carcharhiniformes—Compagno (1973) agreed with White (1936, 1937) and subsequent studies, that scyliorhinids have evolved into carcharhinids and sphyrnids through "triakid" forms. Carcharhinoid sharks contain many convergences and reversals, and even the general framework of their phylogeny is ambiguous (Compagno, 1973, 1988; Nakaya, 1975; Maisey, 1984a). The present analysis does not contribute much to the resolution of this problem. *Leptocharias* is placed as the sister-taxon to the hemigaleids + carcharhinids in Fig. 2, but support for this resolution is weak. Compagno's (1988) detailed study explained carcharhinoid interrelationships as follows. Carcharhiniformes is clearly defined, but it is difficult to analyze their modifications phylogenetically, because some nonmonophyletic higher taxa, for example, "triakids" and carcharhinids, must be postulated. The higher carcharhinoids comprising hemigaleids, carcharhinids, and sphyrnids are the most derived, forming a single phyletic unit. This is recognized here, but Compagno designated the triakid Galeorhinini (*Galeorhinus* + *Hypogaleus*) as the sister-group of this unit. The relationships within other "lower" assemblages are poorly resolved, and Compagno considered five alternatives. Assuming the carcharhinoid–lamnoid sister-group relationship, the proscylliid (*Proscyllium*, *Eridacnis*, and *Ctenacis*) or triakid form is regarded as the most primitive, and the scyliorhinids are a true clade derived from proscylliid ancestors or their close relatives contrary to the previous view. Regardless, *Gollum* is close to *Pseudotriakis*, but their position remains uncertain.

Several recent studies, e.g., Compagno (1973, 1988), Springer (1979), and Garrick (1982, 1985), recognized the limit to understanding carcharhinoid phylogeny based on morphology. Applications of nonmorphological studies are useful especially in this group. Based on allozyme electrophoretic analyses, Naylor (1990) and Lavery (1992) agreed that Carcharhinidae and *Carcharhinus* are paraphyletic, both stating that their molecular data is incongruent with previous systematic hypotheses. Parasite–host data suggest a close relationship between Carcharhinidae and Sphyrnidae, and between all these together and "triakids" (Caira, 1985; Deets, 1987, 1994).

C. Squalea

The interrelationships among squalean ordinal groups are in accordance with Shirai (1992c). The paraphyly of the *Chlamydoselachus*–hexanchoid group was evidenced by three unambiguous characters (TS 48, 78, and 96) in Clade 16. Many previous works united these two groups in a single higher taxon (e.g., Compagno, 1973, 1984a; Maisey and Wolfram, 1984; Maisey, 1986; Nelson, 1994). I have already stressed the weaknesses of the combination of these two groups, as some of their similarities are probably caused by nonevolutionary processes. In particular, their single dorsal fin and sixth branchial unit were regarded as the only characters for their monophyly (Maisey, 1986), whereas the homology of these characters is fairly problematic (Shirai, 1992a,c). Assuming one keeps a clade of *Chlamydoselachus* + hexanchoids in the constraint analyses (Test 14), only two additional steps were required along with the loss of Clades 15 and 16. However, I see no reason to prefer the less parsimonious arrangement to the present result, although this problem is still controversial.

The monophyly of "squaloid" sharks was initially questioned by Shirai (1992c). This taxon has been an accepted systematic unit before Maisey (1980) found no morphological evidence for it. The genus *Echinorhinus* is exceptionally distinct from the others, as suggested by its morphology (Compagno, 1973, 1977; Pfeil, 1983; Herman *et al.*, 1989) and molecular analyses (Bernard and Powers, 1992). The latter authors proposed the separate position for this genus within the "squaloids" or on the outside of this group. My previous hypothesis further suggested that even "squaloids" excluding *Echinorhinus* were not monophyletic because some characteristic features are uniquely found in hypnosqualeans and revised "squalid" genera (*Squalus* and *Cirrhigaleus*). Extant "squaloid" sharks must be an "offspring" of an ancestral group in common with the hypnosqualeans, and they seem to have preserved the ancestral conditions that make it difficult to assume their probable pathways (Shirai, 1992c). Although the "squaloid" constrained analysis (Test 18) required only two additional steps, this resolution did not keep a squalean clade but required that hexanchoids and *Chlamydoselachus* are successive sister-groups of a clade for the remaining living neoselachians. The paraphyletic res-

olution of the "squaloid" group (Shirai, 1992c) is therefore reasonable.

The interrelationships of the following two orders (Dalatiiformes and Rajiformes) deserve comment because one is slightly different from that described by Shirai (1992c) and the other was not covered in my previous work.

Dalatiiformes—Shirai (1992c) placed *Oxynotus* under the Dalatiidae, whereas the present analysis indicates that this genus is a trichotomous sister-group with somniosids and other dalatiids (*Dalatias, Isistius, Euprotomicrus,* and *Squaliolus*). The former treatment depended on the similarity of the reduced internasal space in *Oxynotus* and other dalatiid genera. This character is, however, inadequate for the comparative study among all neoselachians because of an ambiguity in definition (Carvalho and Maisey, 1996). Although the probable theory of the dalatiiform phylogeny is still an interesting subject, *Oxynotus* must occupy a position near the somniosids or dalatiids and is not a separate group from the other "squaloids" as proposed by Bigelow and Schroeder (1957) and Compagno (1984a).

Rajiformes—The batoid phylogeny of this analysis is close to Nishida (1990) in the following points: (1) pristids are the most plesiomorphic batoids, (2) "rhinobatoids" are not monophyletic and are composed of two or more different phyletic lineages, (3) rajoids are the sister-group to myliobatidoids, and (4) *Hexatrygon* is a myliobatidoid and does not comprise an independent monotypic higher taxon. The traditional idea that torpedinoids are the sister-group of all other batoids (e.g., Regan, 1906; Compagno, 1977; Maisey, 1984a) is not corroborated because of a lack of morphological evidence. Since the pristiophorid–batoid monophyly is fairly probable, several pristiophorid-like conditions in pristids and a part of "rhinobatoids" (e.g., an elongate and roofed precerebral fossa, the presence of an oblique inferior β muscle, and a small pectoral fin with a short propterygium not reaching the antorbital cartilage) indicate that these ray groups should occupy plesiomorphic positions. Further phylogenetic studies are needed to further elucidate "rhinobatoids" and torpedinoids in the higher–level interrelationships of batoids.

Contrary to the probability of their monophyly, the interrelationships within torpedinoids are not at all clear. Counter to the traditional subdivision based on the number of dorsal fins (Fowler, 1941; Bigelow and Schroeder, 1953), Compagno (1973) recognized two phyletic groups, Torpedinoidea (composed of *Torpedo* and *Hypnos*) and Narcinoidea, based on the form of the antorbital cartilages and presence or absence of the ceratohyal and labial cartilages. The Narcinoidea

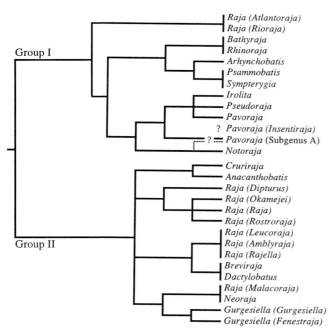

Group I

Raja (Atlantoraja)
Raja (Rioraja)
Bathyraja
Rhinoraja
Arhynchobatis
Psammobatis
Sympterygia
Irolita
Pseudoraja
Pavoraja
? Pavoraja (Insentiraja)
? Pavoraja (Subgenus A)
Notoraja
Cruriraja
Anacanthobatis
Raja (Dipturus)
Raja (Okamejei)
Raja (Raja)
Raja (Rostroraja)
Raja (Leucoraja)
Raja (Amblyraja)
Raja (Rajella)
Breviraja
Dactylobatus
Raja (Malacoraja)
Neoraja
Gurgesiella (Gurgesiella)
Gurgesiella (Fenestraja)

Group II

FIGURE 3 Interrelationships of rajoids according to McEachran and Miyake (1990), with subsequent notes of Yearsley and Last (1992) and McEachran and Last (1994). "Subgenus A" (McEachran, 1984), is said to comprise two nominal (*Bathyraja asperula* and *B. spinifera*) and several undescribed species. "Groups I and II" are also informal taxa at the family level (McEachran and Miyake, 1990).

The phylogeny of the Myliobatidoidei was clarified by the contribution of Nishida (1990). The present analyses agree with his conclusions in the points that *Plesiobatis* and *Hexatrygon* are the plesiomorphic sister-groups of the remains, and that mobulid and rhinopterid rays form an advanced lineage with *Myliobatis* and aetobatids. The position of the gymnurids is unresolved here, but Nishida concluded that they are the sister-group of *Myliobatis* + aetobatids + mobulids + rhinopterids. According to him, *Potamotrygon* should be allied with *Taeniura*, *Dasyatis*, and *Himantura*, rather than forming a monotypic family Potamotrygonidae (cf. Bigelow and Schroeder, 1953; Compagno, 1973).

V. Classification

Although a formal classification is perhaps premature given our current level of knowledge, it seems appropriate to propose a tentative framework based on the results of this study (Fig. 2). This new classification of neoselachians is shown in Appendix 3. The present analysis cannot resolve the interrelationships among some family groups of galeans and batoids, and several paraphyletic or polyphyletic taxa still persist (indicated in double quotation marks). A question mark preceding a generic name indicates that its position is uncertain.

The ranking follows the rules and conventional methods of phylogenetic classification outlined by Hennig (1966), Nelson (1972), and Wiley (1981). I also take into account Shirai (1992c), where two lineages with the same duration are ranked in the same category. Practically, the first appearance in the stratigraphical record of a fossil with one or more characteristics of a clade defines the latest divergence time of that clade. For each pair of sister-taxa I took the earlier first appearance to date the time of sister-group divergence and hence to date the node on the resulting tree. This provides estimates of the absolute time represented by branches on the tree (Fig. 4). Information concerning the fossil record is based mainly on Herman (1977), Thies (1983), and Cappetta (1987). However, the result of this procedure is not enough to fully describe neoselachian evolution because of the incompleteness of the fossil record and therefore can only be used for producing a rough framework of the classification. As a rule, sister-groups diverging in the Jurassic and having existed to date are ranked as an order. In the same way, a subordinal rank is given for a node in the Lower Cretaceous and a familial rank for one in the Upper Cretaceous, and additional subcategories are given as necessary.

were subdivided into two families (Narcinidae and Narkidae) comprised of several genera each. Of these, the monophyly of the Narcinidae was corroborated by Fechhelm and McEachran (1984) and Nishida (1990), but no evidence was found for other aspects of torpedinoid intrarelationships. On tooth forms, Cappetta (1988) stated that species of *Narcine* were near the common ancestor of this order.

The present analysis does not examine the rajoid relationships. Despite several comparative studies (e.g., Ishiyama, 1958; Bigelow and Schroeder, 1953; Stehmann, 1970; Hulley, 1972; McEachran, 1984), higher rajoid classification is still confused. The most comprehensive discussion was given by McEachran and Miyake (1990). They analyzed rajoid phylogeny using a strict cladistic method and reached a new hypothesis (Fig. 3), which was said to show poor confidence because of too many parallelisms and reversals. According to their cladogram: (1) rajoids are composed of two novel familial taxa, "Groups I and II" (not formalized); (2) Arhynchobatidae, Anacanthobatidae, Crurirajidae, Pseudorajidae, and Gurgesiellidae are not independent lineages each; and (3) the genus *Raja* should be highly polyphyletic, and thus the Rajidae *sensu stricto* is also rejected.

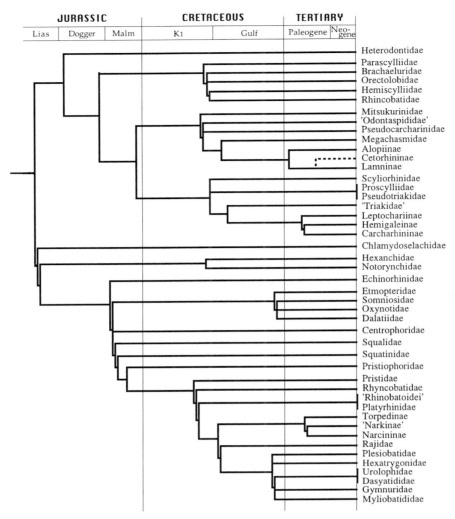

FIGURE 4 Calibrated phylogenetic tree derived using the fossil record. See text for detail.

Two neoselachian stems, galeans and squaleans, are ranked as superorders according to Compagno (1973) and Shirai (1992c). Here I propose the name Galea for a clade composed of the heterodontoids and galeoids, although this is equal to the "Galeomorphii" of Compagno (1973) in composition. The "Galea," the galeoid sharks according to White (1937), was an opposing concept to her "Squalea." This opposition is in concert with the result of this study. Moreover, the Galeomorphii was sometimes used on the phenetic basis by Compagno himself (see Section II), and therefore, such a name may be unsuitable for the present hypothesis. The ranking within each superorder is described below.

Galea—The Galea comprises four major lineages, the heterodontoids, orectoloboids, lamnoids, and carcharhinoids, and each must be treated equally. Each

lineage is considered to have existed at the end of the Jurassic, and thus ordinal rank is assigned to them. The suprafamilial category within each of them is not created because of uncertainties of their interrelationships. The Orectolobiformes includes five families; the Rhincodontidae agrees with Dingerkus (1986) and is composed of *Ginglymostoma*, *Nebrius*, *Pseudoginglymostoma*, *Rhincodon*, and *Stegostoma*, because their differentiations cannot be traced back to the Cretaceous. In the Lamniformes, *Carcharias* and *Odontaspis* are tentatively placed in a family "Odontaspididae," although Compagno (1988) indicated that it is paraphyletic. *Alopias*, *Cetorhinus*, and "lamnid" genera are assembled in a single family, Alopiidae, and three subfamilies are assigned for them. The Carcharhiniformes basically follow Compagno (1988) because their interrelationships cannot be detailed based on the present analysis. Here I tentatively recognize six lineages treated equally: Scyliorhinidae, Proscylliidae, Pseudotriakidae, "Triakidae," Leptochariidae, and Carcharhinidae. Of these, the Proscylliidae and Pseudotriakidae are not separate in my result; the relationships of *Leptocharias* have been problematic, and its familial allocation follows Compagno (1988).

Squalea—The taxonomy of squalean sharks follows Shirai (1992c) except for the Oxynotidae for *Oxynotus*. In batoids (= Rajiformes), four suborders are recognized for groups originated during the Lower Cretaceous, i.e., Pristoidei, Rhynchobatoidei, "Rhinobatoidei," and Rajoidei. The "Rhinobatoidei" tentatively recognized here includes the paraphyletic "rhinobatoid" groups (except rhynchobatids) because their phylogenetic position remains fairly ambiguous. In the Rajoidei, three superfamilies are assigned to three distinct forms, Torpedinoidea, Rajoidea, and Myliobatidoidea. In the Torpedinoidea, *Hypnos* is placed in the Torpedininae, and Compagno's Narkidae is recognized as the Narkinae, although its monophyly is ambiguous. In Myliobatidoidea, the family Myliobatididae is assigned to a group of *Myliobatis* + aetobatids + rhinopterids + mobulids agreeing with Nishida (1990).

VI. Summary

Since Compagno's paper was published in the first edition of the *Interrelationships of Fishes* (1973), studies of comparative anatomy, paleontology, and molecular biology have further elucidated neoselachian evolution (= modern sharks and rays). Compagno's hypothesis and subsequent studies are reviewed to examine how our knowledge has changed over the past two decades. This study presents an analysis of 105 independently assembled transformation series, representing the current state of knowledge of neoselachian morphology. All of the most parsimonious resolutions support two main lineages, the Galea (formalized here) and Squalea. This analysis also reveals that the previous hypothesis including the shark–ray dichotomy, Compagno's four superorders, and Maisey's "orbitostylic" group can only be supported by adding multiple ad hoc proposals of evolutionary events. The traditional "galeoid" idea is supported, despite the striking similarities between *Heterodontus* and orecoloboids. Within the Squalea, the hypnosqualeans are reconfirmed as monophyletic, sharing a common ancestor with two to four lineages of "squaloid" sharks. *Chlamydoselachus* and hexanchoids are successive sister-groups of the remaining squaleans, although these relationships are

not well corroborated. A classification, in which living neoselachians comprise two superorders and 13 orders, is proposed.

Acknowledgments

I respectfully dedicate this work to Yoshiaki Tominaga, who died in the spring of 1994 while I was preparing this manuscript. Here I express my deepest sadness at his untimely death and offer my condolences to his family. Teruya Uyeno and Kazuo Sakamoto supported me during this study, and without their encouragement and cooperation this paper could not have been completed. I thank M. L. J. Stiassny, L. R. Parenti, and G. D. Johnson for inviting me to present this contribution. Marcelo R. de Carvalho kindly read my careless manuscript and gave me many suggestions for this study, as did an anonymous reviewer. The following colleagues kindly gave me valuable information for this study: H. Cappetta, F. Cigala–Fulgosi, H. Endo, M. Goto, T. S. Goto, J. Herman, H. Ishihara, J. G. Maisey, K. Matsuura, M. Miya, and K. Nakaya. I am most grateful also to Chris Norman and Rumi S. for their help. In part this research was generously supported by the Fujiwara Natural History Foundation in 1994–1996.

References

Applegate, S. P. (1972). A revision of the higher taxa of orectolobids. *J. Mar. Biol. Assoc. India* **14**, 743–751.

Bernard, G., and Powers, D. A. (1992). Molecular phylogeny of the prickly shark, *Echinorhinus cookei*, based on a nuclear (18S rRNA) and a mitochondrial (Cytochrome b) gene. *Mol. Phylogenet. Evol.* **1**, 161–167.

Bigelow, H. B., and Schroeder, W. C. (1948). Sharks. *In* "Fishes of the Western North Atlantic" (J. Tee–Van, C. M. Breder, S. F. Hildebrand, A. E. Parr, and W. C. Schroeder, eds.), Part 1, pp. 59–576. Yale University, New Haven, CT.

Bigelow, H. B., and Schroeder, W. C. (1953). Sawfishes, guitarfishes, skates and rays. *In* "Fishes of the Western North Atlantic" (J. Tee–Van, C. M. Breder, S. F. Hildebrand, A. E. Parr, and W. C. Schroeder, eds.), Part 2, pp. 1–514. Yale, University, New Haven, CT.

Bigelow, H. B., and Schroeder, W. C. (1957). A study of the sharks of the suborder Squaloidea. *Bull. Mus. Comp. Zool.* **117**, 1–150.

Blot, J. (1969). Holocéphales et élasmobranches. Systématique. *In* "Traité de Paléontologie" (J. Piveteau, ed.), Vol. 4, Part 2, pp. 702–776. Masson, Paris.

Brown, C. (1900). Ueber das Genus *Hybodus* und seine systematische Stellung. *Palaeontographica* **44**, 149–174.

Caira, J. N. (1985). An emendation of the generic diagnosis of *Phoreiobothrium* Linton, 1889 (Tetraphyllidea: Onchobothriidae) with detailed description of bothridia and hooks. *Can. J. Zool.* **63**, 1199–1206.

Cappetta, H. (1987). "Handbook of Paleoichthyology. Volume 3B. Chondrichthyes II. Mesozoic and Cenozoic Elasmobranchii." Fischer, Stuttgart.

Cappetta, H. (1988). Les Torpédiniformes (Neoselachii, Batomorphii) des phosphates du Maroc. Observations sur la denture des genres actuels. *Tertiary Res.* **10**, 21–52.

Carvalho, M. R., and Maisey, J.G. (1996). Phylogenetic relationships of the Late Jurassic shark *Protospinax* Woodwad, 1919 (Chondrichthyes: Elasmobranchii). *In* "Mesozoic Fishes: Systematics and Paleoecology" (G. Arratia and G. Viohl, eds.), pp. 9–46. Verlag Dr. Friedrich Pfiel, Munich.

Compagno, L. J. V. (1970). Systematics of the genus *Hemitriakis*

(Selachii: Carcharhinidae), and related genera. *Proc. Calif. Acad. Sci.* **33**, 63–98.

Compagno, L. J. V. (1973). Interrelationships of living elasmobranchs. *Zool. J. Linn. Soc.* **53**, Suppl. 1, 15–61.

Compagno, L. J. V. (1977). Phyletic relationships of living sharks and rays. *Am. Zool.* **17**, 303–322.

Compagno, L. J. V. (1984a). FAO species catalogue. Vol. 4. Sharks of the world. An annotated and illustrated catalogue of sharks species known to date. Part 1. Hexanchiformes to Lamniformes. *FAO Fish. Synop.* **125** (4, Part 1), 1–249.

Compagno, L. J. V. (1984b). FAO species catalogue. Vol. 4. Sharks of the world. An annotated and illustrated catalogue of sharks species known to date. Part 1. Carcharhiniformes. *FAO Fish. Synop.* **125** (4, Part 2), 251–655.

Compagno, L. J. V. (1988). "Sharks of the order Carcharhiniformes." Princeton University Press, Princeton, NJ.

Compagno, L. J. V. (1990). Relationships of the megamouth shark, *Megachasma pelagios* (Lamniformes; Megachasmidae), with comments on its feeding habits. *In* "Elasmobranchs as Living Resources: Advances in the Biology, Ecology, Systematics, and the Status of the Fisheries" (H. L. Pratt, Jr., S. H. Gruber, and T. Taniuchi, eds.), NOAA Tech. Rep. NMFS 90, pp. 357–379. U.S. Department of Commerce, Washington, DC.

Daniel, J. F. (1934). "The Elasmobranch Fishes," 3rd ed. University of California Press, Berkeley.

Davies, E. H., Lawson, R., Burch, S. J., and Hanson, J. E. (1987). Evolutionary relationships of a "primitive" shark (*Heterodontus*) assessed by micro–complement fixation of serum transferrin. *J. Mol. Evol.* **25**, 74–80.

Deets, G. B. (1987). Phylogenetic analysis and revision of *Kroyerina* Wilson, 1932 (Siphonostomatoida: Kroyeriidae), copepods parasitic on chondrichthyans, with descriptions of four new species and the erection of a new genus, *Prokroyeria*. *Can. J. Zool.* **65**, 2121–2148.

Deets, G. B. (1994). Copepod-chondrichthyan coevolution: a cladistic consideration. Unpublished Ph.D. Dissertation, University of British Columbia, Victoria.

Dick, J. R. F. (1978). On the Carboniferous shark *Tristychius arcuatus* Agassiz from Scotland. *Trans. J. R. Soc. Edinburgh* **70**, 63–109.

Dingerkus, G. (1986). Interrelationships of orectolobiform sharks (Chondrichthyes: Selachii). *In* "Indo–Pacific Fish Biology: Proceedings of the Second International Conference on Indo–Pacific Fishes" (T. Uyeno, R. Arai, T. Taniuchi, and K. Matsuura, eds.), pp. 227–245. Ichthyological Society of Japan, Tokyo.

Duffin, C. J., and Ward, D. J. (1993). The early Jurassic palaeospinacid sharks of Lyme Regis, southern England. *In* "Elasmobranches et Stratigrahie" (J. Herman and H. Van Waes, eds.), Prof. Pap., 1993/6, pp. 53–101. Service Geologique de Belgique.

Fechhelm, J. D., and McEachran, J. D. (1984). A revision of the electric ray genus *Diplobatis*, with notes on the interrelationships of Narcinidae (Chondrichthyes, Torpediniformes). *Bull. Fla. State. Mus., Biol. Sci.* **29**, 171–209.

Fowler, H. W. (1941). The fishes of the groups Elasmobranchii, Holocephali, Isospondyli, and Ostariophysi obtained by the United States Bureau of Fisheries Steamer "Albatross" in 1907 to 1910, chiefly in the Philippine Islands and adjacent seas. *Bull.—U.S. Natl. Mus.* **100** (13), 1–879.

Garman, S. (1913). The Plagiostomia. *Mem. Mus. Comp. Zool. Harv.* **36**, 1–515.

Garrick, J. A. F. (1982). Sharks of the Genus *Carcharhinus*. *NOAA Tech. Rep., NMFS Circ.* **445**.

Garrick, J. A. F. (1985). Additions to a revision of the Shark Genus *Carcharhinus*: Synonymy of *Aprionodon* and *Hypoprion*, and description of a new species of *Carcharhinus*. *NOAA Tech. Rep. NMFS* **34**.

Gaudin, T. J. (1991). A re-examination of elasmobranch monophyly and chondrichthyan phylogeny. *Neues Jahrb. Geol. Paläeontol. Abh.* **182**, 133–160.

Glickman, L. S. (1964). "Akuly Paleogena i ikh Statigraficheskoe Znachenie." Akademia Nauk, SSSR.

Glickman, L. S. (1967). Subclass Elasmobranchii (sharks). *In* "Fundamentals of Paleontology," (Y. A. Orlov and D. V. Obruchev, eds.), Vol. 2, pp. 292–352. Israel Program Sci. Transl., Jerusalem.

Hasse, J. C. F. (1879–1885). "Das natürliche System der Elasmobranchier auf grundlage des Baues und der Entwicklung ihrer Wirbelsäule." Fischer, Jena.

Heemstra, P. C., and Smith, B. G. (1980). Hexatrygonidae, a new family of stingrays (Myliobatiformes: Batoidea) from South Africa, with comments on the classification of batoid fishes. *Ichthyol. Bull. J. L. B. Smith Inst. Ichthyol.* **43**, 1–17.

Hennig, W. (1966). "Phylogenetic Systematics." University of Illinois Press, Urbana.

Herman, J. (1977). Les sélaciens des terrains néocrétacés et paléocènes de Belgique et des contrées limitrophes. Eléments d'une biostratigraphique inter-continentale. *Mém. Exp. Cartes Géol. Minér. Belg., Sér. Géol.* **15**, 1–450.

Herman J., Hovestadt-Euler, M., and Hovestadt, D. C. (1989). Contributions to the study of the comparative morphology of teeth and other relevant ichthyodorulites in living supraspecific taxa of chondrichtyan fishes. Part A: Selachii. No. 3: Order: Squaliformes. Families: Echinorhinidae, Oxynotidae and Squalidae. *Bull. Inst. R. Sci. Nat. Belg., Biol.* **59**, 101–157.

Hoffmann, L. (1913). Zur Kenntnis des Neurocraniums der Pristiden und Pristiophoriden. *Zool. Jahrb., Abt. Anat. Ontog. Tiere* **33**, 239–360.

Holmgren, N. (1940). Studies on the head in fishes. Embryological, morphological, and phylogenetical researches. Part I: Development of the skull in sharks and rays. *Acta Zool. (Stockholm)* **21**, 51–267.

Holmgren, N. (1941). Studies on the head in fishes. Embryological, morphological, and phylogenetical researches. Part II: Comparative anatomy of the adult selachian skull, with remarks on the dorsal fins in sharks. *Acta Zool. (Stockholm)* **22**, 1–100.

Holmgren, N. (1942). Studies on the head in fishes. Embryological, morphological, and phylogenetical researches. Part III: The phylogeny of elasmobranch fishes. *Acta Zool. (Stockholm)* **23**, 129–261.

Hulley, P. A. (1972). The origin, interrelationship and distribution of southern African Rajidae (Chondrichthyes, Batoidei). *Ann. S. Afr. Mus.* **60**, 1–103.

Iselstöger, H. (1937). Das Neurocranium von *Rhina squatina* und einige Bemerkungen über ihre systematische Stellung. *Zool. Jahrb., Abt. Anat. Ontog. Tiere* **62**, 349–394.

Ishiyama, R. (1958). Studies on the rajid fishes (Rajidae) found in the waters around Japan. *J. Shimonoseki Coll. Fish.* **7**, 1–394.

Jollie, M. (1971). Some developmental aspects of the head skeleton of the 35–37 mm *Squalus acanthias* foetus. *J. Morphol.* **133**, 17–40.

Last, P. R., and Stevens, J. D. (1994). "Sharks and Rays of Australia." CSIRO, Australia.

Lavery, S. (1992). Electorophoretic analysis of phylogenetic relationships among Australian carcharhinid sharks. *Aust. J. Mar. Freshwater Res.* **43**, 97–108.

Luther, A. F. (1909). Beiträge zur Kenntnis von Muskulatur und Skelett des Kopfes des Haies *Stegostoma tigrinum* Gm. und der Holocephalen mit einem Anhang über die Nasenrinne. *Acta Soc. Sci. Fenn.* **37**, 1–60.

Maddison, W. P., Donoghue, M. J., and Maddison, D. R. (1984). Outgroup analysis and parsimony. *Syst. Zool.* **33**, 83–103.

Maisey, J. G. (1975). The interrelationships of phalacanthous sela-chians. *Neues Jahrb. Geol. Palaeontol.* **9**, 553–567.

Maisey, J. G. (1980). An evaluation of jaw suspension in sharks. *Am. Mus. Novit.* **2706**, 1–17.

Maisey, J. G. (1982). The anatomy and interrelationships of Meso-zoic hybodont sharks. *Am. Mus. Novit.* **2724**, 1–48.

Maisey, J. G. (1984a). Higher elasmobranch phylogeny and biostra-tigraphy. *Zool. J. Linn. Soc.* **82**, 33–54.

Maisey, J. G. (1984b). Chondrichthyan phylogeny: A look at the evidence. *J. Vertebr. Paleontol.* **4**, 359–371.

Maisey, J. G. (1985a). Cranial morphology of the fossil elasmo-branch *Synechodus dubrisiensis. Am. Mus. Novit.* **2804**, 1–28.

Maisey, J. G. (1985b). Relationships of the megamouth shark, *Mega-chasma. Copeia*, pp. 228–231.

Maisey, J. G. (1986). The Upper Jurassic hexanchoid elasmobranch *Notidanoides* n. g. *Neues Jahrb. Geol. Paläeontol. Abh.* **172**, 83–106.

Maisey, J. G., and Wolfram, K. E. (1984). *Notidanus. In* "Living Fossils" (N. Eldredge and S. Stanley, eds.). pp. 170–180. Springer-Verlag, Berlin.

McEachran, J. D. (1984). Anatomical investigations of the New Zealand skates *Bathyraja asperula* and *B. spinifera,* with an evalua-tion of their calssification within the Rajoidei (Chondrichthyes). *Copeia*, pp. 45–58.

McEachran, J. D., and Last, P. R. (1994). New species of skate, *Notoraja ochroderma,* from off Queensland, Australia, with com-ments on the taxonomic limits of *Notoraja* (Chondrichthyes, Rajoidei). *Copeia*, pp. 413–421.

McEachran, J. D., and Miyake, T. (1990). Phylogenetic interrelation-ships of skates: A working hypothesis (Chondrichthyes, Rajoidei). *In* "Elasmobranchs as Living Resources: Advances in the Biology, Ecology, Systematics, and the Status of the Fisher-ies" (H. L. Pratt, Jr., S. H. Gruber, and T. Taniuchi, eds.), NOAA Tech. Rep. NMFS 90, pp. 285–304. U.S. Department of Commerce, Washington, DC.

Miyake, T., and McEachran, J. D. (1991). The morphology and evolution of the ventral gill arch skeleton in batoid fishes (Chon-drichthyes: Batoidea). *Zool. J. Linn. Soc.* **102**, 75–100.

Miyake, T., McEachran, J. D., and Hall, B. K. (1992). Edgeworth's legacy of cranial muscle development with an analysis of mus-cles in the ventral gill arch region of batoid fishes (Chondrich-thyes: Batoidea). *J. Morphol.* **212**, 213–256.

Moss, S. A. (1972). The feeding mechanism of sharks of the family Carcharhinidae. *J. Zool.* **167**, 423–436.

Moss, S. A. (1977). Feeding mechanisms in sharks. *Am. Zool.* **17**, 355–364.

Moy-Thomas, J. A. (1939). The early evolution and relationships of the elasmobranchs. *Biol. Rev. Cambridge Philos. Soc.* **14**, 1–26.

Nakaya, K. (1975). Taxonomy, comparative anatomy and phylog-eny of Japanese catsharks, Scyliorhinidae. *Mem. Fac. Fish., Hok-kaido Univ.* **23**, 1–94.

Naylor, G. J. P. (1990). The phylogenetic relationships of carcha-rhiniform sharks inferred from electorophoretic data. Unpub-lished Ph.D. Dissertation, University of Maryland, College Park.

Nelson, G. (1972). Phylogenetic relationships and classification. *Syst. Zool.* **21**, 227–231.

Nelson, J. S. (1994). "Fishes of the World," 3rd ed. Wiley, New York.

Nishida, K. (1990). Phylogeny of the suborder Myliobatidoidei. *Mem. Fac. Fish., Hokkaido Univ.* **37**, 1–108.

Pfeil, F. H. (1983). Zahnmorphologische Untersuchungen an rezen-ten und fossilen Haien der Ordnungen Chlamydoselachiformes und Echinorhiniformes. *Paleoichthyologica* **1**, 1–315.

Regan, C. T. (1906). A classification of the selachian fishes. *Proc. Zool. Soc. London* pp. 722–758.

Reif, W.-E. (1974). Morphogenes und Musterbildung des Hant-zahnchen–Skelettes von *Heterodontus. Lethaia* **7**, 25–42.

Reif, W.-E. (1977). Tooth enameroid as a taxonomic criterion: 1. A new euselachian shark from the Rhaetic–Liassic boundary. *Neues Jahrb. Geol. Palaeontol., Monatsh.*, pp. 565–576.

Schaeffer, B. (1967). Comments on elasmobranch evolution. *In* "Sharks, Skates, and Rays" (P. W. Gilbert, R. F. Mathewson, and D. P. Rall, eds.), pp. 3–35. Johns Hopkins University Press, Baltimore.

Schaeffer, B. (1981). The xenacanth shark neurocranium, with com-ments on elasmobranch monophyly. *Bull. Am. Mus. Nat. Hist.* **169**, 1–66.

Seret, B. (1986). Classification et phylogénèse des Chondrichtyens. *Océanis* **12**, 161–180.

Shirai, S. (1992a). Identity of extra branchial arches of Hexanch-iformes (Pisces, Elasmobranchii). *Bull. Fac. Fish., Hokkaido Univ.* **43**, 24–32.

Shirai, S. (1992b). Phylogenetic relationships of the angel sharks, with comments on elasmobranch phylogeny (Chondrichthyes, Squatinidae). *Copeia*, pp. 505–518.

Shirai, S. (1992c). "Squalean Phylogeny: A New Framework of 'Squaloid' Sharks and Related Taxa." Hokkaido University Press, Sapporo.

Shirai, S., and Okamura, O. (1992). Anatomy of *Trigonognathus kabeyai,* with comments on feeding mechanism and phyloge-netic relationships (Elasmobranchii, Squalidae). *Jpn. J. Ichthyol.* **39**, 139–150.

Springer, S. (1979). A revision of the Catsharks, Family Scyliorhin-idae. *NOAA Tech. Rep., NMFS Circ.* **422**.

Stehmann, M. (1970). Vergleichend Morphologische und Anato-mische Untersuchungen zur Neuordnung der Systematik der nordostatlantischen Rajidae (Chondrichthyes, Batoidei). *Arch. Fischerei Wiss.* **21**, 73–164.

Swofford, D. L. (1993). "Phylogenetic Analysis Using Parsimony (PAUP), Version 3.1." Smithsonian Institution, Washington, DC.

Taylor, L. R., Compagno, L. J. V., and Struhsaker, P. J. (1983). Megamouth—a new species, genus, and family of lamnoid shark (*Megachasma pelagios,* family Megachasmidae) from the Hawaiian Islands. *Proc. Calif. Acad. Sci.* **43**, 87–110.

Thies, D. (1983). Jurazeitliche Neoselachier aus Deutschlan und S-England. *CFS, Cour. Forschungs inst. Senckenberg* **58**, 1–116.

Thies, D. (1987). Comments on hexanchiform phylogeny (Pisces, Neoselachii). *Z. Zool. Syst. Evolutionsforsch.* **25**, 188–204.

Thies, D. and Reif, W.-E. (1985). Phylogeny and evolutionary ecol-ogy of Mesozoic Neoselachii. *Neues Jahrb. Geol. Palaeontol. Abh.* **169**, 333–361.

White, E. G. (1936). Some transitional elasmobranchs connecting the Catuloidea with the Carcharhinoidea. *Am. Mus. Novit.* **879**, 1–22.

White, E. G. (1937). Interrelationships of the elasmobranchs with a key to the order Galea. *Bull. Am. Mus. Nat. Hist.* **74**, 25–138.

Wiley, E. O. (1981). "Phylogenetics." The Theory and Practice of Phylogenetic Systematics." Wiley (Interscience), New York.

Yearsley, G. K., and Last, P. R. (1992). *Pavoraja (Insentiraja) laxipella,* a new subgenus and species of skate (Chondrichthyes: Rajoidei) from the western Pacific. *Copeia*, pp. 839–850.

Young, G. C. (1982). Devonian sharks from south-eastern Australia and Antarctica. *Palaeontology* **25**, 817–843.

Zangerl, R. (1973). Interrelationships of early chondrichthyans. *Zool. J. Linn. Soc.* **53**, Suppl. 1, 1–14.

Zangerl, R. (1981). "Handbook of Paleoichthyology," Vol. 3A. Fischer, Stuttgart.

Appendix 1

Data Matrix of 105 Transformation Series for 53 Terminal Taxa

No.	Terminal taxa	10	20	30	40	50	60	70	80	90	100	
1	OUTGROUP	1010000101	0100000000	0?00010?00	0000??????	??????????	????0?0000	0000000?0?	?000000000	0000010???	0?????00??	0????
2	*Heterodontus*	0010100101	0100001000	1000010100	0001110121	1300200000	0000002000	0000000000	0000200000	0000010000	0100100100	10000
3	parascylliids	1000100011	0100001000	1000010100	0002110121	0000200000	0000002000	0100000010	0000200000	2?00110000	0101000100	10000
4	*Brachaelurus*	1000100111	0100001000	1000010100	0002110121	0000200000	0000002000	0100000010	0000210000	2?00110000	0101100100	10001
5	*Orectolobus*	1000100111	0100001000	1000010100	0102110121	0000200000	0000002000	0100000010	0000210000	2?00110000	0101100100	10001
6	hemiscylliids	1000100111	0100001000	1000010100	0002100131	0000200000	0000002000	0100000010	0000210000	2?00110000	0101100100	10001
7	rhincodontids	1000100211	0100001000	1000010100	0002100131	0000200000	0000002000	0100000010	0000210000	2?00110000	0101000100	10001
8	*Mitsukurina*	2000000011	0100001000	1000010100	0002110120	0000200000	0000002000	0000000010	0000210000	2?00110000	0000000000	10001
9	*Carcharias/Odontaspis*	2000000111	0100001000	1000010100	0002110120	0000200000	0000002000	0000000010	0000210000	2?00110000	0000000000	10101
10	*Pseudocarcharias*	2000000111	0101001000	1000010100	0002110120	0000200000	0000002000	0000000010	0000210000	2?00110000	0000000000	10111
11	*Megachasma*	2000000111	0100001000	1000010100	0002110120	0000200000	0000002000	0000010010	0000210000	2?00110000	0000000000	10101
12	*Alopias*	2000000111	0100001000	1000010100	0002110120	0000200000	0000002000	0000010010	0000210000	2210110000	0000000000	10101
13	*Cetorhinus*/lamnids	2000000111	0200001000	1000010100	0002110120	0000200000	0000002000	0000010010	0000210000	2210100000	0000000000	10111
14	scyliorhinids	2000000?11	0100001000	1000010100	0002110220	0210200000	0000002000	0000000011	0000200000	2?00110000	0?00000000	10000
15	proscylliids/pseudotriakids	2000000111	0100001000	1000010100	0102110220	0210200000	0000002000	0000000011	0000200000	2?00110000	0000000000	10000
16	*Leptocharias*	2000000111	0100001000	1000010100	0002110220	0220200000	0000002000	0000000011	0000220000	2?00110000	0000000000	10000
17	triakids'	2000000111	0100001000	1000010100	0002110220	0220200000	0000002000	0000000011	0000220000	2?00110000	0000000000	10000
18	hemigaleids	2000000011	0100001000	1000010100	0002110220	0230200000	0000002000	0000010012	0000220000	2?00110000	0000000000	10100
19	carcharhinids (incl. sphyrnids)	2000000011	0100001000	1000010100	0002110220	0230200000	0000002000	0000010012	0000220000	2?00110000	0000000000	10102
20	*Chlamydoselachus*	1010011100	1011000000	1000010000	0000010140	0700200000	0000001000	0010000000	0000100001	2?00110000	0000000000	00000
21	*Hexanchus*	1010011100	1012000010	0000000101	0001010140	1100000100	0000000000	0010100000	0000000102	1000010000	0000010000	10000
22	*Notorynchus*	1010000100	1012000000	0000000101	1001010140	1100000100	0000000000	0010100000	0000000102	1000010000	0000010000	00000
23	*Heptranchias*	1010011100	1012000010	0000000101	1001010140	1100000100	0000002001	0010100000	0000000102	1000010000	0000010000	10000
24	*Echinorhinus*	1010010100	1010000000	0000010200	0002010140	1000200100	1000001000	1000000000	0000000100	0001?10000	0010011100	00000
25	*Aculeola*	1010011101	1011010000	1000010200	0002010110	1000100100	1000001000	0000000000	0000000100	0001?10000	0010000101	12000

No.	Taxon	Character states
26	*Trigonognathus*	1010011101 1010010000 1010010200 0101010?? 1000100000 1000002001 0000000000 0000200100 0001?10000 0010000101 12000
27	*Centroscyllium*	1010011101 1010010000 1010010200 0002010110 1000100100 1000002000 0000000000 0000200100 0001?10000 0010000101 12000
28	other etmopterids	1010011101 1010010000 1010010200 0002010110 1000100100 1000002000 0000000000 0000200100 0001?10000 0010020101 12000
29	*Somniosus*	1010011101 1010100000 1000010200 0012010111 1000100100 1000002001 0000000000 0000000100 0001?11000 0010020101 10000
30	other somniosids	1010011101 1010100000 1000010200 0012010111 1000100100 1000002001 0000000000 0000200100 0001?11000 0010020101 11000
31	*Oxynotus*	1010010101 1010000000 1000010200 0012010111 1000100100 1000002001 0000000000 0000200100 0001?10000 0010020101 11000
32	dalatiines	1010011101 1010000000 1000010200 0012010111 1000100111 1000002001 1000000000 0000200100 0001?10000 0010020101 11000
33	euprotomicrinines	101001p101 1010000000 1010010200 0012010111 1000100100 1000002001 1000000000 0000200100 p001?10000 0010020101 11000
34	*Deania*	1010011101 1010000000 1000010200 0102010110 1000100100 0000002000 1000000000 0000200100 0101?10000 0010021101 10000
35	*Centrophorus*	1010011101 1010000000 1000010200 0002010110 1000100100 0000002000 1000000000 0000200100 0101?10000 0010021101 10000
36	*Cirrhigaleus/Squalus*	1010011101 1010000000 1000010200 0002010110 1000100100 0000002000 0000000000 0000201110 0001?10001 0010022101 10110
37	*Squatina*	1010020101 1010000001 1000010200 0002010?? 1000200100 ?000002010 0001000000 0000201110 0001?10112 0010000100 10010
38	pristiophorids	1020020102 1010000101 0100010200 0002010102 1000100100 0000012020 0000000000 0000022110 0001?10102 0110000100 10010
39	pristids	1020020103 2000000101 0100021210 0102021102 1201211011 0010012120 0001010100 0100232110 0001?10112 0010000101 10010
40	*Rhina/Rhynchobatus*	1020020103 2000000101 0100021210 0102021102 1201211011 ?010012220 0001010100 0100232110 0001?10112 0010000111 10010
41	torpedines	1011000004 2000000101 0000021210 0102021102 0?01211111 01??112120 0002010100 1100232110 0001?10112 0010000111 10010
42	narkines	1011000004 2000000101 0000021210 0102021102 0?01211?11 01??112120 0002010100 1100232110 0001?10112 0110000111 10010
43	narcinines	1011000004 2000000101 0000021210 0100021102 0?01211111 01??112120 0002010100 1100232110 0001?10112 0110000111 10010
44	platyrhinids/'rhinobatids'	1010020103 2000000101 0100021210 0102021102 1201211011 0010012220 0003010100 0100232110 0001?10112 0110000111 10010
45	rajoids	1010020103 2000000101 0000021010 0102021102 1201211?11 0010012220 0003010100 0100232110 0021?20112 0010000111 10010
46	*Plesiobatis*	0000000103 2000000101 0000022010 0100021102 0?01201?11 0010012320 0004010100 0100232123 ?2?1?20112 0110000111 10000
47	*Hexatrygon*	0000000103 2000000101 0000022010 0100021112 0?01201?11 0010012320 0004010100 0111230123 ?2?1?20112 0110000111 10000
48	dasyatidids/urolophids	0001000103 2000000101 0000032010 0102021112 0?01201?11 0010012320 0004010100 0111230123 ?2?1?20112 0110000111 10000
49	gymnurids	0001000103 2000000101 0000022010 0102021112 0?01201?11 0010012320 0004010100 0111230123 ?2?1?30112 0110000111 10000
50	*Myliobatis*	0001000103 2000000101 0000132010 0102021112 0?01201?11 0011012320 0004011100 0111230123 0021?30112 0110000111 10000
51	aetobatines	0001000103 2000000101 000p132010 0102021112 0?01201?11 0011012320 0004011100 0111230123 0021?30112 1110000111 10000
52	*Rhinoptera*	0101000103 2000000101 0001132010 0102021112 0?01201?11 0011012320 0004010100 0111230123 0021?30112 1110000111 10000
53	*Manta/Mobula*	0101000103 2000000101 0001122010 0102021112 0?01201?11 0011012320 0004010100 0111230123 0021?30112 2110000111 10000

Note: Codes "p" and "?" denote polymorphic and unknown (or inapplicable), respectively.

Appendix 2

The following are the unambiguous synapomorphies of the root (Clade 0) and each internal node (Clades 1–41) in Fig. 2. Character numbers in bold indicate that an exchange of the character has occurred once within the neoselachians; a character modification is from State 0 to State 1, unless an additional note is in parentheses, e.g., "(1 to 0)." A letter "r" or "p" added to a character number denotes a reversal or parallel of acquisition, respectively.

Clade 0   57
1   **17,35,**57p (1 to 2), 75p (1 to 2), 101p
2   3p (1 to 0), **9,**34p (1 to 2), **69,**81p (0 to 2), 85p
3   44p,**62,94**
4   76p,95p,105p
5   **36** (1 to 0), 39 (2 to 3)
6   1 (1 to 2)
7   76p,105p
8   103p
9   66p
10   **83**
11   **38** (1 to 2), **43,70**
12   **43** (1 to 2), 76 (0 to 2)
13   8p (1 to 0)
14   **43** (2 to 3), 66p, **70** (1 to 2), 103p
15   10 (1 to 0), **11,12** (1 to 0), 13
16   48,78,96
17   14 (0 to 2), 26 (1 to 0), **30,42,45** (2 to 0), 57r (1 to 0), **65,80** (0 to 2), 81
18   7p,**19,**101p
19   28 (1 to 2), 34p (1 to 2), 57p (1 to 2), **84,93,**98p
20   7p,10r,39 (4 to 1), 45 (2 to 1), 75 (0 to 2), 96 (1 to 2), 100,101p
21   **51,102**
22   **16,**23p,102 (1 to 2)
23   96rp (2 to 1), 96rp (1 to 0)
24   **33,40,**60p
25   **15,87**
26   61p
27   61p,**82**
28   39 (1 to 0), 77,**79,90,**104p
29   6 (1 to 2), 7rp (1 to 0), **20,59,88,90** (1 to 2), 96rp (2 to 1), 96rp (1 to 0)
30   3 (1 to 2), 10 (1 to 2), **18,** 21r (1 to 0), 22,32p,**36** (1 to 2), **56,59** (1 to 2), 77 (1 to 2)
31   10 (2 to 3), **11** (1 to 2), 13r (1 to 0), 26 (1 to 2), **27,29,37,42** (0 to 2), 44p,46,47,48r (1 to 0), 49p,50p,**53,72,**76 (0 to 3), **99**
32   58,66p,**68**
33   3r (2 to 1), 64 (1 to 2)
34   22r (1 to 0), 92p

35   4p, 8p (1 to 0), 10 (3 to 4), **52,** 58r (2 to 1), **71**
36   **55**
37   28r (2 to 1), 28rp (1 to 0), **80** (0 to 3), **83** (0 to 2), 86 (1 to 2)
38   1 (1 to 0), 3p (1 to 0), **27** (1 to 2), 39r,46r (1 to 0), 58 (2 to 3), 64 (3 to 4), **73,74,**104r (1 to 0)
39   4p
40   **25,54**
41   **2,24,91**

Appendix 3

Classification based on a revised diphyetic hypothesis (Galea–Squalea). Higher taxa in double quotation marks indicate monophyly is ambiguous. A genus name preceded by "?" indicates its uncertain position.

Superorder Galea
  Order Heterodontiformes
    Family Heterodontidae
      Heterodontus
  Order Orectolobiformes
    Family Parascylliidae
      Cirrhoscyllium
      Parascyllium
    Family Brachaeluridae
      Brachaelurus
      Heteroscyllium
    Family Orectolobidae
      Eucrossorhinus
      Orectolobus
      Sutorectus
    Family Hemiscylliidae
      Chiloscyllium
      Hemiscyllium
    Family Rhincodontidae
      Ginglymostoma
      Nebrius
      Pseudoginglymostoma
      Rhincodon
      Stegostoma
  Order Lamniformes
    Family Mitsukurinidae
      Mitsukurina
    Family "Odontaspididae"
      Carcharias
      Odontaspis
    Family Pseudocarcharinidae
      Pseudocarcharias
    Family Megachasmidae
      Megachasma
    Family Alopiidae
      Subfamily Alopiinae

Alopias
Subfamily Cetorhininae
　Cetorhinus
Subfamily Lamninae
　Carcharodon
　Isurus
　Lamna
Order Carcharhiniformes
　Family Scyliorhinidae
　　Apristurus
　　Asymbolus
　　Atelomycterus
　　Aulohalaelurus
　　Cephaloscyllium
　　Cephalurus
　　Galeus
　　Halaelurus
　　Haploblepharus
　　Holohalaelurus
　　Parmaturus
　　Pentanchus
　　Poroderma
　　Schroederichthys
　　Scyliorhinus
　Family Proscylliidae
　　Ctenacis
　　Eridacnis
　　Proscyllium
　Family Pseudotriakidae
　　Gollum
　　Pseudotriakis
　Family "Triakidae"
　　Furgaleus
　　Galeorhinus
　　Gogolia
　　Hemitriakis
　　Hypogaleus
　　Iago
　　Mustelus
　　Scylliogaleus
　　Triakis
　Family Leptochariidae
　　Leptocharias
　Family Carcharhinidae
　　Subfamily Hemigaleinae
　　　Chaenogaleus
　　　Hemigaleus
　　　Hemipristis
　　　Paragaleus
　　Subfamily Carcharhininae
　　　Carcharhinus
　　　Eusphyra
　　　Galeocerdo
　　　Glyphis

Isogomphodon
Lamiopsis
Loxodon
Nasolamia
Negaprion
Prionace
Rhizoprionodon
Scoliodon
Sphyrna
Triaenodon
Superorder Squalea
　Order Chlamydoselachiformes
　　Family Chlamydoselachidae
　　　Chlamydoselachus
　Order Hexanchiformes
　　Family Hexanchidae
　　　Heptranchias
　　　Hexanchus
　　Family Notorynchidae
　　　Notorynchus
　Order Echinorhiniformes
　　Family Echinorhinidae
　　　Echinorhinus
　Order Dalatiiformes
　　Family Etmopteridae
　　　Aculeola
　　　Centroscyllium
　　　Etmopterus
　　　Miroscyllium
　　　Trigonognathus
　　Family Somniosidae
　　　Centroscymnus
　　　Mollisquama
　　　Scymnodalatias
　　　Scymnodon
　　　Somniosus
　　　Zameus
　　Family Oxynotidae
　　　Oxynotus
　　Family Dalatiidae
　　　?*Euprotomicroides*
　　　Euprotomicrus
　　　?*Heteroscymnoides*
　　　Isistius
　　　Dalatias
　　　Squaliolus
　Order Centrophoriformes
　　Family Centrophoridae
　　　Centrophorus
　　　Deania
　Order Squaliformes
　　Family Squalidae
　　　Cirrhigaleus
　　　Squalus

Order Squatiniformes
  Family Squatinidae
    Squatina
Order Pristiophoriformes
  Family Pristiophoridae
    Pliotrema
    Pristiophorus
Order Rajiformes
  Suborder Pristoidei
    Family Pristidae
      Anoxypristis
      Pristis
  Suborder Rhynchobatoidei
    Family Rhynchobatidae
      Rhina
      Rhynchobatus
  Suborder 'Rhinobatoidei'
    Family 'Rhinobatidae'
      Aptychotrema
      Rhinobatos
      Trygonorrhina
      Zapteryx
    Family Platyrhinidae
      Platyrhina
      Platyrhinoidis
      Zanobatus
  Suborder Rajoidei
    Superfamily Torpedinoidea
      Family Torpedinidae
        Subfamily Torpedininae
          Hypnos
          Torpedo
        Subfamily 'Narkinae'
          Heteronarce
          Narke
          Temera
          Typhlonarke
        Subfamily Narcininae
          Benthobatis

          Diplobatis
          Discopyge
          Narcine
    Subfamily Rajoidea
      Family Rajidae
      (see McEachran and Miyake,
      1990)
    Subfamily Myliobatidoidea
      Family Plesiobatididae
        Plesiobatis
      Family Hexatrygonidae
        Hextrygon
      Family Urolophidae
        Urolophus
        Urotrygon
      Family Dasyatididae
        Dasyatis
        Himantura
        ?*Hypolophus*
        ?*Paratrygon*
        ?*Plesiotrygon*
        Potamotrygon
        Taeniura
        ?*Urogymnus*
        ?*Urolophoides*
      Family Gymnuridae
        Aetoplatea
        Gymnura
      Family Myliobatididae
        Subfamily Myliobatidinae
          Myliobatis
        Subfamily Aetobatinae
          Aetobatus
          Aetomylaeus
          ?*Pteromylaeus*
        Subfamily Mobulinae
          Manta
          Mobula
          Rhinoptera

Higher-Level Elasmobranch Phylogeny, Basal Squaleans, and Paraphyly

MARCELO R. DE CARVALHO

Department of Herpetology and Ichthyology
American Museum of Natural History
New York, New York 10024;
and
Graduate School and University Center
The City University of New York
New York, New York 10036

Investigations of elasmobranch anatomy have led to the development of several controversial classifications

Dunn and Morrissey (1995, p. 526).

I. Introduction

The first volume of *Interrelationships of Fishes* (1973) contained a chapter by Compagno that presented what quickly became a widely adopted classification for Recent elasmobranchs (sharks, skates, and rays). Compagno's paper contains a wealth of anatomical information and served as a landmark for the understanding of elasmobranch phylogeny, even though diagnoses of the major groups were maintained on a phenetic basis. Compagno (1977) attempted to rectify this problem by presenting what he considered diagnostic characters for his major groups. The recognized 12 orders of Recent elasmobranchs were grouped into four superorders as in 1973 (Fig. 1): the Galeomorphii, Squalomorphii, Squatinomorphii, and Batoidea. Again, the relationships among the superorders were left unresolved. Compagno's (1973, 1977) orders were widely accepted by many researchers (Compagno,

1984; Springer and Gold, 1989; Eschmeyer, 1990; Pratt *et al.*, 1990) even though some groups, e.g., the Squaliformes, seem to lack defining features.

The "orbitostylic" hypothesis of Maisey (1980) was the first attempt to relate at least some of Compagno's superorders. The Squatinomorphii (containing the angel-shark *Squatina*) was considered the sister-group to the Squalomorphii (which included the Hexanchiformes, or six- and seven-gilled sharks, and Pristiophoriformes, the sawsharks). The relationships of batoids (comprising all the rays and skates) and galeomorphs were left unresolved. Maisey (1984a) again adopted this threefold division of Recent elasmobranchs ("orbitostylic" sharks, galeomorphs, and batoids), without further elaborating on their phylogenetic placement.

The first hypothesis to relate all of Compagno's superorders was that of Shirai (1992c). In that work Shirai defended a sister-group relationship between Pristiophoriformes and batoids, which together form the sister-group to *Squatina*. This scheme became known as the "hypnosqualean hypothesis" and was formalized as the superorder Hypnosqualea by Carvalho and Maisey (1996). The "squaloid" sharks (dog-

Copyright © 1996 by Academic Press, Inc.
All rights of reproduction in any form reserved.

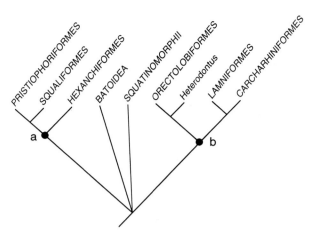

FIGURE 1 The supraordinal elasmobranch groups of Compagno (1973, 1977). *a*, Squalomorphii; *b*, Galeomorphii.

fish sharks; Squaliformes of Compagno, 1973, 1977, 1984; Squaloidea of Bigelow and Schroeder, 1948) were considered paraphyletic, forming successive sister-groups to hypnosqualeans. Hexanchiform sharks were also deemed paraphyletic with hexanchoids related more closely to all other squaleans than to *Chlamydoselachus* (Shirai, 1992c; Fig. 2).

Perhaps one of the most significant advances in the "cladistic revolution" is the concept of paraphyly (Nelson, 1989a,b; Farris, 1991). From this perspective, recent contributions in the systematics of elasmobranchs may best be understood by reviewing those papers in which long-recognized taxa have succumbed under the scrutiny of phylogenetic analyses

and parsimony algorithms. Shirai's (1992a,c) papers concerning the higher-level phylogeny of Compagno's (1973, 1977) superorders are a case in point. In those papers, Shirai has argued extensively that the recognition of a monophyletic Hexanchiformes and Squaliformes (as traditionally perceived, e.g., Compagno, 1984) is unsubstantiated by character evidence. To further investigate these claims, I analyzed a matrix containing 67 morphological features which includes characters incongruent with some of Shirai's components. My analysis reports on the paraphyly of two of Shirai's own higher-level clades (his unnamed hexanchoids + ("squaloids" + hypnosqualeans) and an unnamed Squalidae + Hypnosqualea) and supports the monophyly of six- and seven-gilled sharks (Hexanchiformes), as well as a slightly modified Squaliformes. The classification presented here is close to a compromise between Compagno's and Shirai's conclusions and stems from the inclusion of characters not considered by Shirai as well as from differences in the interpretation and coding of many features.

II. Methods

The phylogenetic analysis that follows was based on characters described by previous authors, including Regan (1906), Garman (1913), White (1937), Holmgren (1940, 1941, 1942), Compagno (1973, 1977, 1988), and Maisey (1984a,b, 1985, 1986a,b). Many of these characters have been compiled and summarized by Shirai (1992c), whose analysis included new characters and a data matrix. Specimens examined to substantiate these characters are listed in Appendix 1.

Shirai (1992c) presented a detailed and rigorous phylogenetic analysis of many Recent elasmobranch taxa representing all higher-level groups. The matrix presented in Appendix 3, which is the matrix adopted in this chapter, is based in part on the matrix presented by Shirai (1992c). Because the interrelationships of skates and rays are not considered here (for reviews, see Nishida, 1990; Lovejoy, 1996; McEachran *et al.*, this volume; Shirai, this volume), all ray and skate (batoid) groups are represented as a single terminal (Rajiformes). Justifications for the coding of characters are discussed in the character section.

The single outgroup represents a hypothetical ancestor based on the phylogenies of Schaeffer and Williams (1977), Schaeffer (1981), Maisey (1984a,b, 1986b, 1989a), Young (1982), and Gaudin (1991). These authors have advanced a well-corroborated hypothesis (summarized in Fig. 3) establishing a sequence of fossil taxa that form successive sister-groups to Recent sharks and rays, facilitating outgroup comparison

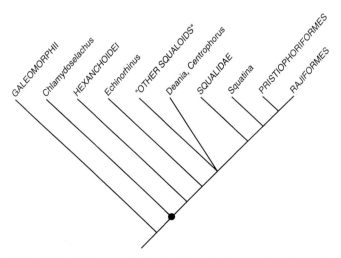

FIGURE 2 Phylogenetic relationships of Recent elasmobranchs advocated by Shirai (1992c). The marked node delimits Squalea. Note the paraphyly of Hexanchiformes and Squaliformes (as defined by Compagno, 1973, 1977, 1984).

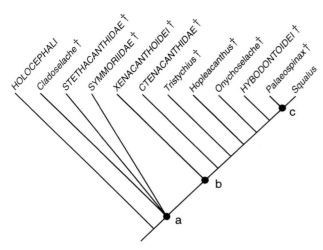

FIGURE 3 Relationships among the major Paleozoic and Mesozoic elasmobranch groups, modified from Gaudin (1991) and demonstrating the limits of the Euselachii and Neoselachii as adopted here (only a few representative taxa are included). *Squalus* represents all Recent elasmobranchs. *a*, Elasmobranchii; *b*, Euselachii; *c*, Neoselachii. Daggers indicate fossil taxa. The Hybodontoidei is delimited as in Maisey (1989a). The Neoselachii probably includes other taxa (such as *Synechodus*, orthacodontids, and anacoracids).

(Shirai, 1992c; Carvalho and Maisey, 1996). Although there is disagreement as to the placement of certain fossil taxa (primarily the Paleozoic †*Cladoselache* and symmoriids, cf. Zangerl, 1981; Lund, 1985), this does not affect the optimization of the characters discussed in this chapter.

A list of characters present in the matrix (Appendix 3) is detailed in Appendix 2, which contains information on the different states of each character. All 16 multistate characters were analyzed as nonadditive (unordered). Autapomorphies present in the tree have homoplastic distributions—no single entries were included in the matrix (Carpenter, 1988; cf. Yeates, 1992), except when the autapomorphy represents part of a multistate series or is coded as present in one taxon and as a "?" in another. Many taxa presented have numerous defining characters, but only those with potential for grouping (synapomorphy) are included.

Parsimony analysis was undertaken using version 1.5 of Hennig86 (Farris, 1988). Although there is much homoplasy required by the resulting tree, an exact tree construction procedure was implemented (ie* algorithm). The resulting single most-parsimonious tree is summarized in Fig. 4 (ordinal level relationships) and shown in detail in Fig. 5. Tree diagnostics and optimizations were performed using the Dos Equis function of Hennig86 and the program Clados (version 1.1; Nixon, 1992). The results obtained with Hennig86 were confirmed using PAUP (version 3.1.1;

Swofford, 1993), which produced the same number of equally most-parsimonious trees of the same length. When conflict in optimization occurred, characters were placed at the most basal node possible, favoring reversals over independent gains. This was done to maximize initial putative homology propositions and to give characters a greater generality (for a review, see de Pinna, 1991). Alternative optimizations are discussed under the appropriate character and were diagnosed with the aid of MacClade (version 3.01; Maddison and Maddison, 1992) and Clados (version 1.1; Nixon, 1992) and by hand.

III. Recent Elasmobranchs

Recent elasmobranchs, as defined below, are monophyletic and are divided into two large monophyletic groups, the Galeomorphii and Squalea. Recent elasmobranchs are not equivalent to Neoselachii (Fig. 3), which is a more inclusive taxon. Neoselachian monophyly is discussed in Compagno (1973, 1977), Schaeffer and Williams (1977), Reif (1977), Schaeffer (1981), Maisey (1984a,b, 1986b, 1994), Thies and Reif (1985), Cappetta (1987), Gaudin (1991), and references cited therein. The definition of Neoselachii (Euselachii of Moy-Thomas, 1939; Reif, 1977; Euselachiformes of Maisey, 1975) varies and may be restricted to Recent

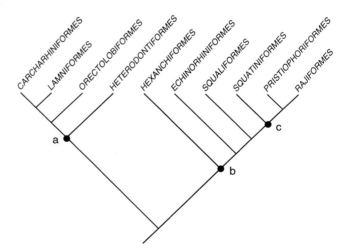

FIGURE 4 Ordinal relationships of the Recent Elasmobranchii. This tree is derived from the phylogenetic analysis discussed in this chapter (see Methods section for details). Characters optimized at the root are the following (numbers as in Appendixes 2 and 3, states of a multi-state series are in parentheses, and characters with more than one equally most-parsimonious optimization are indicated with an "!"): 19(1), 28!, 37!, and 59!. *a*, Galeomorphii; *b*, Squalea; *c*, Hypnosqualea (characters for these nodes are listed in Fig. 5).

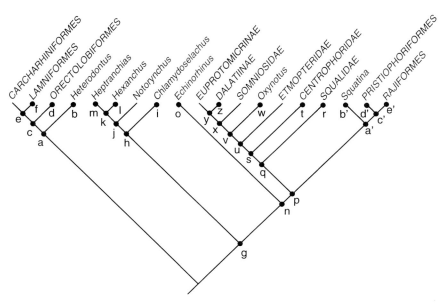

FIGURE 5 Relationships within Recent elasmobranchs, depicting interrelationships within Hexanchiformes and Squaliformes. Character numbers are as in Appendices 2 and 3. Refer to the text for a discussion of many characters listed here (*cf.* Shirai, 1992c). Characters at the root are listed in Fig. 4. Characters for nodes are as follows (reversals have "-" before the character, independent occurrences are denoted with an asterisk, states of a multi-state series are indicated in parentheses, and characters with multiple equally most-parsimonious optimizations show an "!" after the character number): node *a* (Galeomorphii): 2*, 4!, 13, 26(5)!, 27(1)*!, 55*!, 67; node *b* (Heterodontea): 1(2), 57; node *c* (Galeoidea): 7, 24*, -28*!, 43, 49(2)*, 51*; node *d* (Orectolobiformes): 30*; node *e* ("higher galeoids"): 1(1), -4!, 26(2)!, -27(0)!, -55!, -59*!; node *f* (Lamniformes): 64; node *g* (Squalea): 8(1)!, 9, 10, 11, 26(4)!, 32!, 44!, 47!, 49(1)*; node *h* (Hexanchiformes): 12(1)!, 20, -37, 40, 44(1)!, 45, 50, -59!; node *i* (Chlamydoselachoidea): -19(0), -28!, -32*!, -47!, 49(2), 51*, 65*; node *j* (Hexanchoidea): 12(2)!, 17, 21!, 29(1), 31(1), 42, 58(1); node *k* (Hexanchidae): 2*, 5(1)*, 6*, 15; node *l*: -21!; node *m*: 65*; node *n* (Echinorhinoidea + Squaloidea + Hypnosqualea): 5(1)*, 19(2), 24, 56, 66; node *o* (Echinorhiniformes): 17*, 29(3), 39(3)*, 58(1)*; node *p* (Squaloidea + Hypnosqualea): 2*, -8(0)!, -26(0)!, -44!, 46(1)!, 48!, 60!, 63!; node *q* (Squaliformes): 6*, 31(2)*, -49(0), 58(2); node *r* (Squaloidei): 54(1), 62; node *s* (Dalatioidei): 26(1)!, -46(0)!, -48!, -63!; node *t*: 39(2); node *u* (unnamed): 35, 61; node *v* (unnamed): 23, 27(1)!, 39(1); node *w* (Oxynotidae): -6; node *x* (unnamed): 49(1)*; node *y* (Dalatiidae): 39(3)*; node *z*: 33*, 34*; node *a'* (Hypnosqualea): 5(2), 16, 38(1)!, 41!, 52, 53!, 54(2); node *b'* (Squatiniformes): 55*!, -60!; node *c'* (Pristiorajea): 3, 8(2)!, 14, 17*, 18, 22, 25, 27(2)!, 36, 38(2)!, 46(2)!; node *d'*: 31(2)*, -41!, -53!; node *e'*: 8(3)!, -9, -10, -11, 29(2), 30*, -32*!, 33*, 34*.

elasmobranchs (which include only fossil taxa that are phylogenetically placed well within Recent groups, e.g., numerous Cenozoic forms known from isolated teeth; Maisey, 1986b), or the definition may include fossil taxa that fall just outside Recent groups (e.g., †*Palaeospinax*, †*Synechodus*?, †*Acronemus*?, †*Sphenodus*?, and anacoracids),[1] which is the usage here (Compagno, 1988). Maisey (1986b) mentions only two characters that unite Recent forms to the exclusion of *Palaeospinax*, and both characters are unavailable in fossils (both biochemical; but see Maisey, 1984b).

Therefore, it is quite apparent that discussions concerning the monophyly and higher-level interrelationships of living elasmobranchs should not be based solely on living forms but must take the above fossil taxa into consideration (more on this issue below).

Parsimony analysis results in a single most-parsimonious tree (L = 132 steps, CI = 0.67, RI = 0.80) depicted in Fig. 5, and this is the tree adopted for the character section. My results conflict with those of Shirai (1992c) but also agree in some fundamental aspects (cf. Fig. 2). Characters that are discussed here are numbered as in Appendix 2. The terms "component," "node," and "clade" are used interchangeably and refer to a particular monophyletic group. Complete diagnoses of all nodes are provided in the legend

---

[1] Only the first mention of a fossil taxon is indicated by †.

of Fig. 5. The consistency index (CI) for each character is given in parenthesis following the character number.

Recent elasmobranchs are defined by four characters, all of which are reversed at less inclusive levels in the phylogeny:

Extrabranchial cartilages on hyomandibula (epihyal) only (character 19) [CI = 0.66]. This character is unknown in immediate outgroups (perhaps due to preservation), and because it is only present in galeomorphs and hexanchoids it can be considered a synapomorphy at this level. The alternative view is to consider the plesiomorphic condition for Recent taxa as the absence of extrabranchial cartilages on the hyomandibula (the state in *Chlamydoselachus*, rajoids, and myliobatidoids), in which case this feature would characterize galeomorphs and hexanchoids independently.

Adductor mandibulae superficialis present (character 28) [CI = 0.5]. This character has an interesting distribution, occurring in *Heterodontus* and squaleans except *Chlamydoselachus*, torpedinoid, and myliobatidoid rays. It illustrates that characters presently included in the matrix shared by *Heterodontus* and squaleans can be considered synapomorphies of Recent elasmobranchs, and therefore loss of any such character would define galeomorphs minus *Heterodontus* (= Galeoidea).

Right and left coracoids fused (character 37) [CI = 0.5]. This character is reversed to the primitive condition in hexanchiforms. Gaudin (1991) considers the single coracoid a defining feature of a monophyletic group including †*Hybodus*, *Palaeospinax*, and Recent elasmobranchs, but there is little support for this. Nothing is available in descriptions of the Lower Jurassic *Palaeospinax* to suggest that a single coracoid element was present (Dean, 1909; Schaeffer, 1967; Maisey, 1977; Cappetta, 1987) nor in the specimen examined (the type species, †*P. priscus* AMNH VP7085). Maisey (1989a) reconstructs the Upper Pennsylvanian hybodont †*Hamiltonichthys* with separate coracoid bars, and his descriptions of more "modern" hybodonts do not give the impression that they differ from *Hamiltonichthys* in this respect (Maisey, 1982, 1986c, 1987, personal communication). The Cretaceous *Synechodus* has unfused coracoids, which lends support for its placement as a primitive neoselachian along with *Palaeospinax* and not as a basal galeomorph as advocated by Maisey, although there is evidence for its galeomorph affinities as well (Maisey, 1985). Although the Carboniferous symmoriid †*Denaea* is credited with possibly showing a single coracoid unit (Zangerl, 1981), it is more reasonable to assume that it lacked this charac-

ter. Recent chimaeroids also have a fused coracoid (Didier, 1995), but this is homoplastic as basal euselachians have unfused coracoids. Compagno (1973) regarded the single coracoid as a character of living forms only, and this is the usage here, which contradicts the interpretation of Shirai (1992c) and Carvalho and Maisey (1996), who did not support a monophyletic Hexanchiformes, a prerequisite for this optimization. The discovery of a single, ventrally fused shoulder girdle in *Palaeospinax* would not require extra steps in this character in the phylogeny adopted here.

Teeth with "aprons" (character 59) [CI = 0.33]. This character can alternatively be considered a synapomorphy of Galeomorphii (lost in "higher" galeoids) and of nonhexanchiform squaleans independently.

A. Galeomorphii

The monophyly of this supraordinal clade is highly corroborated by seven characters (Fig. 4, and node *a* of Fig. 5), two devoid of homoplasy. The findings of this analysis (and that of Shirai, 1992c) support a twofold division within the Galeomorphii, the superorders Heterodontea and Galeoidea (cf. Compagno, 1973, 1977, 1988). Shirai's (1992c) characterization of this group differs from that given below in that he considers the pharyngobranchial blade to be primitive at this level. His optimization of the insertion of the suborbitalis muscle (character 27) also differs. Of Compagno's (1988, p. 382) five galeomorph synapomorphies, only "shortened otic capsules" is considered below (character 13, differently stated).

Closed lateral line canal (character 2) [CI = 0.33]. Among Recent elasmobranchs the lateral sensory canal is open only in *Chlamydoselachus* (Allis, 1923), *Notorynchus* (Daniel, 1934), and *Echinorhinus* (Garrick, 1960). Daniel (1934) credits "hexanchoids" as possessing an open lateral line canal, but this was contradicted by the specimens examined here. The condition in *Heptranchias* is coded here as in Shirai (1992c). Recent chimaeroids (Didier, 1995) and fossil outgroups (e.g., hybodonts, particularly †*Egertonodus basanus*, Maisey, 1983; many Paleozoic taxa, Zangerl, 1981; cf. Lund, 1985) have the open condition as well, which is why the apomorphic expression is hypothesized as a defining feature here. Other neoselachians such as *Palaeospinax* may have the closed condition, which would then alter the optimization presently adopted, but data for this is lacking (unless *Palaeospinax* is erroneously considered a galeomorph, as did Glickman, 1967; Cappetta, 1987; Cappetta *et al.*, 1993). A minimum of three steps is required.

Ethmoidal region of neurocranium downcurved (character 4) [CI = 0.5]. Shirai's (1992c, p. 20) characterization of this feature is slightly different, although it implies the same attribute. This character is stated here as in Maisey (1985, p. 18, character 16). The downcurved ethmoidal area is present only in *Heterodontus* and orectolobiforms (and secondarily in some rajiforms, according to Maisey, 1985, p. 7), but it can be optimally interpreted as a galeomorph character. Compagno's (1988, p. 382) "sliding articulation in specialized ethmopalatine grooves on the cranium" used in support of a *Heterodontus* + Orectolobiformes clade corresponds to the downward curvature between the ethmoidal and orbital areas. This character illustrates that any character previously used to support a monophyletic *Heterodontus* + Orectolobiformes can be optimized as a defining feature of Galeomorphii due to the placement of orectolobiforms as the sister-group to "higher" galeomorphs on the phylogeny of Figs. 4 and 5 (other characters also behaved in this fashion).

The Mesozoic †*Synechodus dubrisiensis* and †*Sphenodus nitidus* (= †*S. macer?*) also have this feature (Maisey, 1985). If these taxa are regarded as basal neoselachians, the ethmoidal downcurvature must then be primitive at this level with the loss of such a condition also characterizing squaleans (as well as lamniforms + carcharhiniforms). Because the precise phylogenetic position of *Synechodus*, and especially *Sphenodus*, are still unresolved, this character is interpreted as a galeomorph synapomorphy (Maisey, 1985).

Hyomandibular fossa located anteriorly in the otic region (character 13) [CI = 1.0]. This character corresponds to Compagno's (1988, p. 382) "shortened otic capsules" and is used here following Maisey (1985, p. 18, character 18). Paleozoic sharks (e.g., †*Xenacanthus* and †*Tamiobatis*) and Mesozoic hybodonts have the hyomandibular facet located far from the post-orbital processes, as do squalean elasmobranchs (Maisey, 1985). The condition in galeomorphs, wherein the hyomandibular fossa abuts the orbit, is therefore derived (also in *Synechodus*: Maisey, 1985).

Suborbitalis muscle originating on upper preorbital wall (character 26) [CI = 0.8]. This character is present only in *Heterodontus*, parascylliids, and orectolobids (Shirai, 1992c), but due to their position on the tree it can be considered a galeomorph synapomorphy. Hemiscylliidae and Rhincodontidae possess this muscle (= levator labii superioris or preorbitalis of other authors) originating on the supraorbital surface, while all remaining galeomorphs show the suborbitalis arising from the posterior aspect of the nasal capsule. Orectolobiforms were scored with the same state as

in *Heterodontus* (i.e., state 4), however, because this is believed to be the primitive condition within the order [Compagno, 1988; note that the phylogeny of Dingerkus (1986) implies a different coding]. The outgroup condition is unknown, and the discovery of this character in a related neoselachian, albeit highly improbable, may change this interpretation. The origin of the suborbitalis from either the upper preorbital wall (state 4) or from the ectethmoid process (state 3) can be interpreted as a defining feature of Recent elasmobranchs, as the placement of either of these conditions at the base of the tree (ingroup node) requires no extra steps. The state present in basal squaleans (state 3) is not considered diagnostic for all extant elasmobranchs, however, because it is only present in taxa endowed with an ectethmoid process (which is absent, so far as known, from *Palaeospinax* and other fossil neoselachians). State 3 is left as a squalean synapomorphy (transformed again for Squaloidea + Hypnosqualea, see below). State 4 was also not considered a synapomorphy of Recent elasmobranchs (which would imply two independent transformations, with state 4 at the root of the tree: 4 → 2 [within galeomorphs] and 4 → 3 → 0 → 1 [within squaleans]). The placement of this character state as a galeomorph feature is maintained because those taxa with it show corresponding modifications in neurocranial architecture (Maisey, 1985) not present in stem-group neoselachians. Although Maisey (1985, p. 11) noted similarities in the ethmo-orbital regions of *Synechodus*, *Heterodontus*, and orectolobiforms, he concluded that they do not support the presence of a suborbitalis originating from the upper preorbital surface in *Synechodus* because its postorbital articulation presents constraints for suction feeding.

Suborbitalis muscle inserting directly on mandibula anterior to adductor mandibulae (character 27) [CI = 0.5]. As described by Shirai (1992c, p. 64), this derived condition occurs only in *Heterodontus* and basal orectolobiforms (coding for these taxa follows the same reasoning as character 26). Homoplasy is required as "higher galeoids" show the plesiomorphic condition that is also present in squaleans.

Nasoral groove present (character 55) [CI = 0.5]. Compiled from Compagno (1988, p. 382), this character has the same distribution as character 4 (and, within galeomorphs, of characters 26 and 27). *Squatina* also has, independently, the derived state of this character (Maisey, 1984a).

Pharyngobranchial blade present (character 67) [CI = 1.0]. The pharyngobranchials of many Recent elasmobranchs present ventrolateral projections of varying degrees, originating from the proximal terminus (site of articulation with epibranchials). The pharyngobranchial blade is one such projection in which the arcualis dorsalis and interpharyngobranchialis muscles originate. Shirai (1992a,c) regards the loss of this feature as a squalean synapomorphy in the belief that fossil euselachians also possess the pharyngobranchial blade. However, Carvalho and Maisey (1996) argue that there is only evidence for the presence of this character in galeomorphs, and this is the interpretation followed here. Because the proximal terminus of the pharyngobranchial may present various projections (e.g., double headed articulations, the "joint for pharyngobranchial and succeeding epibranchial" of Shirai, 1992a,c), this character needs confirmation from the dissection of additional taxa, as knowledge of the dorsal gill arch musculature is necessary in discerning the correct structure. *Heterodontus francisci* clearly presents the pharyngobranchial blade (observed in AMNH 96795), as does *Heterodontus zebra* (Shirai, 1992c). Carvalho and Maisey (1995) present data concerning the shape of the pharyngobranchial in other galeomorph and squalean taxa.

Comments. The monophyly and interrelationships of galeomorphs is discussed at length by Compagno (1973, 1977, 1988), who also reviews the earlier literature. The preceding definition for this group differs from that of Compagno (1988). The pseudosiphon could not presently be evaluated in light of the cluttered terminology surrounding mixopterygial components, but its distribution solely within galeomorphs, as advocated by Compagno (1988), is possible. The "rostrum not trough-shaped, possibly reduced" (another galeomorph character of Compagno, 1988) is coded differently here, where a rostrum representing a trabecular outgrowth (presumably of the trabecula communis) is primitive for neoselachians in general. The tripodal rostrum of carcharhiniforms and lamniforms (possibly all formed as further extensions and chondrification of the lamina orbitonasalis) is therefore derived (see below). A rostrum that is not "trough-shaped" is present in many squalean taxa (e.g., *Dalatias, Squaliolus, Centroscymnus,* personal observation; *Isistius, Zameus,* Shirai, 1992c). In summary, within galeomorphs there appears to be two conditions that are derived: the tripodal rostrum and the secondary loss of the rostrum (*Heterodontus*). The rostral process of orectolobiforms is primitive in that it is formed as a projection from the trabecula (although the specific shape may be derived; Dingerkus, 1986).

The "loss of the lateral commissure" also mentioned by Compagno (1988) occurs outside of galeomorphs as well, although it is retained (secondarily?) by some squaleans (*Squatina,* basal rajiforms, and some "squaloids"; Holmgren, 1941; Maisey, 1984a,b; Shirai, 1992c). The lateral commissure was not included here because its precise distribution is unclear. Compagno's (1988) "reduced post-orbital processes" is probably primitive at this level, as basal squaleans also have reduced postorbital processes compared to hybodonts and Paleozoic xenacanths. This character may not be distinct from the reduced lateral commissure.

The patterns formed by secondary calcifications of the vertebral centra, although not included, have been given much weight by previous authors. These patterns are difficult to interpret phylogenetically in Recent sharks. The monograph of Hasse (1879–1885) has influenced many workers who have used secondary calcification patterns in their classifications (most notably White, 1937). Coding this character proved extremely difficult as generalizations regarding the specific patterns of calcification have many exceptions. Furthermore, there seems to be a correlation, albeit tenuous, between habitus and the specific pattern formed (Applegate, 1967; Compagno, 1988; J. G. Maisey, personal communication). Squaleans generally do not have secondary calcification patterns comparable to most galeomorphs, which may have calcifications formed either as "solid" intermedialia or radii that originate from the outer zone cartilage of the centrum (Compagno, 1988). This is what has been termed "radial asterospondylic," even though Hasse's terminology has fallen from use by many authors. Some squaleans show conspicuous patterns, such as *Squatina* (the "squatinoid" type of Applegate, 1967) which is similar to the vertebral centra of *Cetorhinus* (Compagno, 1988). It seems that the patterns derived from secondary calcifications (those stemming from the intermedialia) may potentially define carcharhinoids or subgroups therein (e.g., the "Maltese cross" type), or even galeomorphs, but these are difficult to delimit with any precision. This has been the attitude of some recent reviewers (Applegate, 1967; Compagno, 1977, 1988; cf. Maisey, 1984a) and is reluctantly followed here. Shirai (1992c) correctly codes terminal squalean taxa (squatinoids, pristiophoroids, and rajiforms) with derived conditions, but only as different autapomorphies, therefore crediting this character with no potential for grouping.

Superorder Heterodontoidea: Heterodontiformes (Node *b* in Fig. 5)

This order contains the sole genus *Heterodontus.* Along with its many autapomorphies, *Heterodontus* is characterized by the following features:

Absence of rostral cartilages (character 1) [CI = 1.0]. Also shared with sting-rays, this condition in myliobatidoids is distinct due to the presence of the anterior

A   **B**

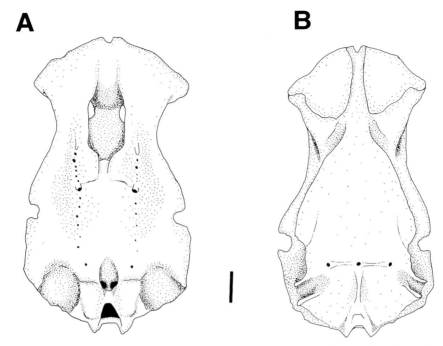

FIGURE 6   Dorsal (A) and ventral (B) view of the chondrocranium of *Heterodontus francisci* (AMNH 217862), anterior to top. Scale bar = 1 cm.

section of the rostral cartilage at least in the juvenile of some forms (Miyake *et al.*, 1992, as "rostral appendix"; McEachran *et al.*, this volume). Certain skates also lack a distinct rostrum (e.g., *Pavoraja* and *Pseudoraja*; McEachran *et al.*, this volume). This character is complex; attempts to code it over a wide range of taxa become increasingly difficult as more specializations are described (Miyake *et al.*, 1992). A distinct rostral process is not present in *Heterodontus* (Fig. 6) (Daniel, 1934; Compagno, 1977).

Circumnarial grooves and folds present (character 57) [CI = 1.0]. This character is present in orectolobids and hemiscylliids, but orectolobiforms were coded with a "?." More homoplasy is required in a more inclusive matrix.

Heterodontus plesiomorphically retains a dorsal fin endoskeleton composed of a basal cartilage and finspine (Compagno, 1977), an adductor mandibulae superficialis muscle, and unfused posteriormost dorsal gill arch elements ("gill pickaxe:" Daniel, 1934; further discussion follows).

Superorder Galeoidea: Orectolobiformes + Lamniformes + Carcharhiniformes (Node *c* in Fig. 5)

Six features characterize this component (Fig. 5). Shirai (1992c) interprets differently both the "gill pickaxe" and the adductor mandibulae superficialis muscle, here used as galeoid characters. This node is not present in Compagno (1988).

Separate foramen for superficial ophthalmic nerve (character 7) [CI = 1.0]. This character is described in detail by Goodrich (1986), Holmgren (1940, 1941), Maisey (1985), and Compagno (1988). Although not included in my analysis, the absence of a prefacial commissure (separating the hyomandibular foramen from the prootic foramen or orbital fissure) in orectolobiforms and carcharhiniforms (Compagno, 1988) is correlated with this character.

Complete fusion of posterior dorsal gill arch elements ("gill pickaxe") (character 24) [CI = 0.5]. The fusion of pharyngobranchial 5 to epibranchial 5, which in turn fuse to pharyngobranchial 4, forming a wishbone shaped structure, (Fig. 7) was used as an elasmobranch synapomorphy by previous authors (Daniel, 1934). This fusion, termed "gill pickaxe" by Shirai (1992b,c), is incomplete in *Chlamydoselachus* (Allis, 1923), hexanchoids (Shirai, 1992c), *Heterodontus* (Daniel, 1934), and, according to Shirai (1992c), in *Trigonognathus*, *Narcine*, and *Plesiobatis*. Shirai (1992c) chose to exclude this character on the basis that outgroup information is lacking or ambiguous. Some of the taxa listed above have these elements contacting each other, and partial fusion does occur, forming different patterns. In *Heptranchias* the last two pharyngobranchials appear to be fused together, and these articulate with, or contact, the last epibranchial. In *Chlamydoselachus* the last pharyngobranchial and epibranchial seem to be united, and these contact the penultimate pharyngobranchial (Allis, 1923). According to Shirai

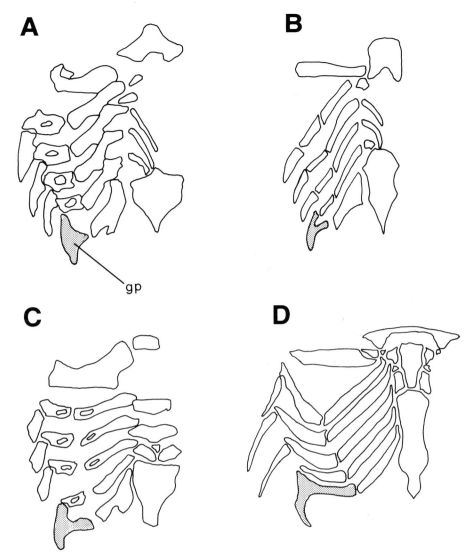

FIGURE 7 Dorsal view of gill and hyoid arches, depicting the gill "pickaxe" (gp) in representative galeomorph and squalean taxa (anterior to top). The hyomandibula is not shown. Dorsal elements pulled to left (right side not shown). The gill "pickaxe" is stippled. (A) *Hemiscyllium ocellatum* (AMNH 38151); (B) *Pseudocarcharias kamoharai* (AMNH 71984); (C) *Dalatias licha* (AMNH 19446); (D) *Pristis pristis* (AMNH 44011).

(1992c), the condition in *Hexanchus* seems similar to that of *Heptranchias*, although these sharks have a different number of gill arches. *Heterodontus* parallels *Chlamydoselachus* in that the fusion (of whatever extent) is between the last pharyngobranchial and epibranchial, which in turn contacts pharyngobranchial 4 (Daniel, 1934). Hamdy (1961) provides ontogenetic evidence that for various rajiform species the fusion between the last pharyngobranchial and epibranchial takes place before the addition (fusion) of the penultimate pharyngobranchial. This sequence was observed in embryos of *Sphyrna lewini* (AMNH 12464) by G. Dingerkus (unpublished). If this is a generalized occurrence, then it is plausible to hypothesize that the

condition in hexanchoids is secondary to that of, e.g., *Chlamydoselachus*, and would represent an additional character of hexanchoids. Fossil euselachian taxa tend to have poorly preserved dorsal gill arch elements (Zangerl, 1981; Maisey, 1982, 1989b). However, pharyngobranchial cartilages are preserved in the only known hybodont specimen from Solnhofen (*Egertonodus fraasi*, Brown, 1900; Maisey, 1986c) and in Permian xenacanths from Germany (Fritsch, 1889, 1895; Jaekel 1895, 1906; Koken, 1889). Not much anatomical detail of these structures is available in these specimens, but it is possible to confirm that the posterior pharyngobranchials were not fused into a larger structure. Many Recent chimaeriforms, however, have the last three pharyngobranchials and the last two epibranchials fused into a robust plate (Didier, 1995), and there is evidence that this fusion also occurs in fossil holocephalans and paraselachians (E. Grogan and R. Lund, personal communication). Two steps are required by this character.

Adductor mandibulae superficialis absent (character 28) [CI = 0.5]. This character shows considerable homoplasy as *Chlamydoselachus*, torpedinoid, and myliobatidoid rays also lack the adductor mandibulae superficialis (Shirai, 1992c).

Elongated ventral marginal cartilage of clasper (character 43) [CI = 1.0]. White (1937) used this character in support of her "Galea" (= Galeoidea), but Compagno (1977, 1988) has given it little importance, noting that it may not be universally distributed within galeoids (*Parascyllium* has relatively short marginals: Compagno, 1977) and that elongated marginals also occur in rays. The basal rajiform condition is coded here as not having elongated marginals, however, and this feature is therefore free of homoplasy.

Dorsal fin endoskeleton composed of radial elements only (character 49) [CI = 0.4]. Among Recent taxa, this condition occurs in *Chlamydoselachus*. Many Paleozoic elasmobranchs, such as the symmoriids (e.g., †*Cobelodus*, †*Symmorium*, and *Denaea*), have a dorsal fin with only endoskeletal radials (Zangerl, 1981), but galeoids have developed them independently, as the condition in more closely related euselachians is a basal cartilage with a finspine.

Anal fin endoskeleton composed of radials only (character 51) [CI = 0.5]. The anal fin is absent from the majority of Paleozoic elasmobranchs (Zangerl, 1981), being present most notably in xenacanthids (sometimes as a conspicuous "double-anal" fin). Some ctenacanthids (e.g., †*Goodrichthys*) are reconstructed with an anal fin composed solely of radials, but this is to some extent questionable. Even so, the state in galeoids, intriguingly similar in *Chlamydoselachus*, is independent as other Mesozoic euselachians have an anal fin with basal cartilages and radials.

Orectolobiformes (Node *d* in Fig. 5)

This order has been divided into as many as eight families (Compagno, 1973). Both Dingerkus (1986) and Compagno (1988) present phylogenies for the order and discuss characters unique to it. Additionally, orectolobiforms have a levator palatoquadrati not separated from the spiracularis muscle (character 30; CI = 0.5), a feature also present in rajiforms.

Lamniformes + Carcharhiniformes (Node *e* in Fig. 5)

This node contains the "higher" galeoids of previous authors. Six characters group lamniforms with carcharhiniforms, with almost all showing some degree of homoplasy. These include ethmoidal region of neurocranium not downcurved (character 4, CI = 0.5), suborbitalis originating on posterior aspect of nasal capsule (character 26, 0.8), suborbitalis inserting on adductor mandibulae and/or dermal tissues around mouth corner (character 27, 0.5), loss of nasoral groove (character 55, 0.5), jaw teeth without aprons (character 59, 0.33), and tripodal rostrum (character 1, 1.0). Characters 4, 26, 27, and 55 are all mentioned above under Galeomorphii, character 59 under Recent Elasmobranchs.

Lamniformes (Node *f* in Fig. 5)

Characters defining this order can be found in Compagno (1973, 1977, 1984, 1990) and Cappetta (1987). Relationships within lamniforms were most recently analyzed by Compagno (1990), but there is evidence supporting a node that contradicts his results. The node *Alopias* + ((*Mitsukurina* + *Pseudocarcharias*) + (*Carcharias* + *Odontaspis*)) is defended by the presence of lateral rostral fenestrae, dental bullae, and a rostral apex. These characters and some others that define groups within this node are still under scrutiny and it would be premature to include them here (along with other features, they are part of an ongoing project; U. L. Gomes *et al.*, unpublished). Lamniforms are characterized by one feature that was coded "?" in orectolobiforms (intestinal valve of "ring" type; character 64, CI = 1.0].

Carcharhiniformes

Compagno (1988) reviews many aspects of the biology and phylogeny of this order, and provides a taxonomic revision as well. No further characters defining carcharhiniforms were discovered in this analysis.

B. Squalea

This supraordinal taxon, presented by Shirai (1992c), includes three of Compagno's (1973, 1977) superorders, i.e., the Squalomorphii, Squatinomorphii, and Batoidea. Maisey's "orbitostylic" group was identical to the Squalea excluding the Rajiformes. Other authors have defended a higher-level group similar in composition, but at least one component was always excluded from their classifications (e.g., hexanchiforms; Goodrich, 1986; Glickman, 1967). Shirai's (1992c) analysis provided the necessary evidence of a monophyletic Squalomorphii + Squatinomorphii + Batoidea. The Squalea (node *g* of Fig. 5) is composed of four superorders, the Notidanoidea, Echinorhinoidea, Squaloidea, and Hypnosqualea, and is defended by nine characters.

Ectethmoid process present (character 8) [CI = 0.75].

Considerable confusion exists regarding what is termed the "ectethmoid" process. Posteriorly directed projections from the postnasal wall area are common features in many elasmobranch groups. There is more than one "type" of projection, however, and terminology must be coherently applied to correctly interpret their phylogenetic significance (Carvalho and Maisey, 1996). The ectethmoid process proper is defined as the ventrally projecting process originating on the posterior aspect of the nasal capsule and lying beneath the orbitonasal canal (which connects the nasal and orbital venous sinuses transmitting the orbitonasal vein). As such, it is restricted in distribution, occurring in *Chlamydoselachus*, hexanchoids, and *Echinorhinus*. *Centroscymnus* has an "ectethmoid" process similar to *Echinorhinus*, projecting from the posterior wall of the nasal capsule but less pronounced (Compagno, 1973). The "antorbital" cartilage of rajiforms and the unchondrified equivalent of pristiophorids has been considered the putative homolog of the ectethmoid process (Holmgren, 1941; Shirai, 1992c; Carvalho and Maisey, 1996). These structures are therefore regarded here in the same multistate series. The "membranous antorbital cartilage" of squaliforms (as here defined), referred to by Holmgren (1941), would further fill in the gap between taxa, but this structure has not been confirmed or properly described. Many elasmobranch taxa possess a dorsal posterior projection that overlies the orbitonasal canal and has been interpreted as the preorbital process (Compagno, 1977). This process is widespread among many Palaeozoic and Mesozoic taxa and has been regarded as a synapomorphy at a higher level (Maisey, 1984b, 1986b, as "ectethmoid" process). The ethmopalatine process of *Egertonodus basanus* lies ventral to the preorbital process but is very distinct from the ectethmoid process of Recent taxa and does not bear the same topographic relationship to the orbitonasal canal (Maisey, 1983). Four steps are minimally required by the tree, one of which is autapomorphic (for rajiforms).

Orbital articulation present (character 9) [CI = 0.5]. This character is used here as outlined by Maisey (1980) and followed by Shirai (1992a,c) and Carvalho and Maisey (1996; cf. Compagno, 1988).

Suborbital shelf absent (character 10) [CI = 0.5]. Usage follows Carvalho and Maisey (1996) who argue that the suborbital shelf of galeomorphs does not transform into the basitrabecular process of squaleans (see below), but is instead lost at this node.

Basitrabecular process present (character 11) [CI = 0.5]. This character was interpreted by Shirai (1992c; Carvalho and Maisey, 1996) to be a separate feature from the suborbital shelf on the basis of its topographic relationships and development (Holmgren, 1941).

Suborbitalis muscle originating on ectethmoid process (character 26) [CI = 0.8]. This character is strongly correlated with the presence of the ectethmoid process and has an almost identical distribution (this muscle originates from the suborbital surface in hypnosqualeans). See discussion under Galeomorphii above.

Arcualis dorsalis originating from two muscle slips (character 32) [CI = 0.33]. This muscle originates on the pharyngobranchials (with one or two heads) and inserts on the epibranchials. This character shows considerable incongruence, requiring three steps; it is lost in *Chlamydoselachus* and rajiforms (in electric rays the arcualis dorsalis originates from two muscle slips, and in *Trigonognathus* the primitive condition is again present, thus requiring two extra steps, but these taxa are not scored here; Shirai, 1992c).

Notochordal constriction reduced (character 44) [CI = 0.5]. The notochord is constricted by its expanding sheath early in development (Goodrich, 1986). The sheath expands and chondrifies between the membrana elastica externa and interna due to the invasion of matrix-depositing cells at the bases of the dorsal and ventral arcualia (the membranae become ruptured and eventually disappear into the "chordal" centra, in contrast to the "perichordal" centra of actinopterygians and tetrapods, according to Goodrich, 1986). The designation "primary calcification" was used by Shirai (1992c) to distinguish this process from the "secondary" calcifications that later originate from the intermedialia [Shirai's terms reflect Maisey's (1985, p. 17) characters 7 and 8, respectively].

According to Goodrich (1986), *Chlamydoselachus* maintains an unconstricted notochord in the trunk region. This is in contrast to the plesiomorphic condition seen in the vast majority of extant elasmobranchs (along with *Palaeospinax* and *Synechodus*; Maisey, 1977, 1984a,b, 1985), in which the notochord is constricted along the entire vertebral column (except intervertebrally in some taxa). Hexanchoids, *Echinorhinus*, and certain squaliforms show the more extreme condition of having an unconstricted notochord along practically the entire column (Shirai, 1992c). However, the degree to which this is different from the partial constriction in *Chlamydoselachus* is difficult to determine. Moreover, partial constriction of the notochord has been reported in *Heptranchias* and possibly *Hexanchus* (Regan, 1906; Schaeffer, 1967; Compagno, 1977). These taxa were therefore coded equally here. A reversal to the plesiomorphic state further defines Squaloidea + Hypnosqualea, even though some isolated squaliform taxa lack notochordal constriction as well (e.g., some *Somniosus* species and *Aculeola*; Ridewood, 1921; Shirai, 1992c), requiring further homoplasy in a generic-level analysis.

Complete hemal arches in precaudal tail region (character 47) [CI = 0.5]. A "complete" hemal arch is one in which the basiventral processes unite ventrally at the midline. The interpretation of this character heavily depends on the condition of fossil outgroups, but little information is available. Zangerl (1981, p. 27) mentions "hemal arches" in the caudal peduncle of some Paleozoic forms of which adequate skeletons are available, noting that they are "paired, fairly delicate rods that tend to fuse distally; near the base of the tail fin they usually are considerably enhanced . . . ." The questions to be answered are whether the hemal elements form true arches and where along the vertebral column the arches begin (if in the beginning of the precaudal tail region, above the anal fin, etc.). Hemal elements from the caudal origin are known from some xenacanths (primarily †*Orthacanthus*; J. G. Maisey, personal communication), ctenacanths, †*Tristychius* (Dick, 1978), †*Onychoselache* (Dick and Maisey, 1980), and possibly symmoriids, among others. Precaudal hemal elements, however, are more scarce. More importantly, the condition of these elements in forms closer to Recent elasmobranchs is also poorly known. The oldest "complete" hybodontoid, *Hamiltonichthys* (Maisey, 1989a), has been restored with precaudal hemal arches similar to those of chimaeroids (Patterson, 1965), galeomorphs, and *Chlamydoselachus*, i.e., true arches form only from the anal fin region and not from the second dorsal fin. This is believed to be the case for other hybodontoids as well (the Triassic †*Lissodus cassangensis*; Maisey, 1990, personal communication). The Jurassic *Palaeospinax* might have been similar, although no basiventral processes are apparent in the specimen examined (AMNH VP7085) or from Dean's (1909) account of the same specimen of *P. priscus*. The lack of good preservation in the precaudal region of many of the taxa above might be an indication that the basiventral processes were not very developed or calcified and possibly not forming arches at all. Accordingly, the formation of true hemal arches in the precaudal tail region of squaleans is considered derived.

Dorsal fin endoskeleton composed of basal cartilage without a finspine (character 49) [CI = 0.4]. The plesiomorphic condition of this character is found in Recent as well as in a great number of fossil groups (Maisey, 1982). Although homoplastic (five steps are required), the loss of a finspine characterizes squaleans, with a reversal to the primitive state occurring for Squaliformes. A reversal to the squalean condition occurs for higher squaliforms (Somniosidae + Dalatiidae).

Comments. The definition of Squalea provided above differs from that of Shirai (1992c). Shirai (1992c) regarded the loss of the pharyngobranchial blade as a squalean synapomorphy, here considered to be a galeomorph feature. One character derived for hexan-

chiforms was also considered a squalean character (absence of pectoral propterygial radials) because Shirai did not recognize a monophyletic Hexanchiformes. Shirai's (1992c) "nasal capsules separated from each other by a wide internasal space" has been disputed by Carvalho and Maisey (1996) who found much variation, at least in relation to how this character is presently defined, and it is accordingly not incorporated here.

Superorder Notidanoidea: Hexanchiformes (Node *h* in Fig. 5)

This node was not present in the classification of Shirai (1992c), and is characterized by seven features:

Postorbital articulation present but without complete articular facet ("nonsuspensory postorbital processes") (character 12) [CI = 1.0]. This character is best described as "postorbital processes and upper jaws connected by loose connective tissue, but the upper jaws are relatively mobile and the postorbital articulations readily disarticulate when the jaws drop" (Compagno, 1977, p. 308). This feature occurs in *Chlamydoselachus*, *Pseudocarcharias*, and *Aculeola* among Recent taxa (Maisey, 1980; Shirai, 1992c), but it is equally parsimonious to consider it a hexanchiform synapomorphy. The articulation in *Pseudocarcharias* is not adequately described (Compagno, 1973, 1977) and may differ from that of *Chlamydoselachus* in some respects, but both taxa were coded equally here because their condition is distinct from the "complete" postorbital articulation of hexanchoids (Carvalho and Maisey, 1996). Although analyzed as one multistate character, the complete articulation of hexanchoids is far more restrictive to jaw movement and protusability, reminiscent of many Paleozoic sharks (e.g., *Xenacanthus*; Maisey, 1980; Schaeffer, 1981). Alternatively, if this character is coded in binary form (i.e., as two separate characters), its presence in *Chlamydoselachus* is autapomorphic and cannot be used as a hexanchiform synapomorphy.

Sixth branchial unit present (character 20) [CI = 1.0]. This character has long been recognized as a defining feature of hexanchiforms but has been severely criticized by Thies (1987) and Shirai (1992b). The arguments presented by Thies amount to the naive belief that if a character occurs outside of the group in question (i.e., is homoplastic), then it cannot be a reliable indicator of monophyly. Accordingly, Thies (1987) discards the extra branchial arch as a hexanchiform attribute because a sawshark (*Pliotrema*) and a myliobatidoid ray (*Hexatrygon*) also possess six branchial arches (the criticisms of Thies regarding other features homologous for hexanchiforms are discussed below). Shirai (1992b) disagrees with this character on the basis of morphological dissimilarities in the extra branchial arches of hexanchoids and *Chlamydoselachus*. His detailed work demonstrates that a paired basibranchial is present in the antipenultimate arch of *Hexanchus*, *Notorynchus*, and *Heptranchias* but absent in the corresponding arch of *Chlamydoselachus*. Hexanchoids also present the derived condition of having the arcualis dorsalis muscle originate from two distinct muscle slips on the pharyngobranchials; whereas, according to Shirai (1992b), *Chlamydoselachus* has a single head of origin for this muscle. These taxa also exhibit different patterns in the dorsal aspect of the posteriormost arches. Shirai (1992b,c) believes these three distinctions are enough to render this character unreliable. In my view, these features simply highlight the fact that hexanchoids and *Chlamydoselachus* are indeed distinct, but this does not negate their monophyly. The differences described by Shirai amount to further synapomorphies of hexanchoids (the distinct pattern of the posteriormost dorsal elements, perhaps, and the paired basibranchials of the antipenultimate arch) or features defining a more inclusive group (double-headed origin of arcualis dorsalis). Basically, Shirai's argument is that because the extra arches of *Chlamydoselachus* and hexanchoids are not *identical*, they cannot be scored the same in a matrix. It is as if there can be no minor distinctions in characters hypothesized to be homologous, but this would render most characters in any matrix unusable. There is much variation in the shape of the jaws in elasmobranchs, but I doubt that anyone would consider jaws nonhomologous. In any event, many earlier authors used this character as a hexanchiform feature (Regan, 1906; White, 1937; Berg, 1947), and this is the usage followed here.

Scapulocoracoids articulating (unfused) ventrally (character 37) [CI = 0.5]. This character was discussed above under Recent Elasmobranchs.

Pectoral propterygium not contacting radials (character 40) [CI = 1.0]. This character is free of homoplasy. Contact between the propterygium and radial elements is apparently a widespread condition in fossil sharks endowed with a tribasal pectoral fin, most notably Paleozoic ctenacanths (Zangerl, 1973, 1981) and hybodonts (Maisey, 1982) including the Pennsylvanian hybodont *Hamiltonichthys* (Maisey, 1989a). This is also the case in Recent chimaeroids (Patterson, 1965; Didier, 1995), although only two basal cartilages are present. However, Thies (1987) claims that radial elements are indeed articulating with the propterygium in *Notorynchus*, based on the description given by Daniel (1934, pp. 50, 78). Examination of the text cited by Thies, however, does not lead to that conclusion, and neither does Daniel's illustration of the pectoral endoskeleton of that species. The only other evidence cited

by Thies reporting contact between the propterygium and radials in a hexanchiform (Gegenbaur, 1865) could not be examined. However, the species on which Gegenbaur apparently based his claim is *Hexanchus griseus* (Thies, 1987), which is figured by Shirai (1992c, pl. 48B) as clearly demonstrating an anterior mesopterygial expansion that does not permit contact between the propterygium and radial cartilages. The derived condition has been convincingly demonstrated in *Chlamydoselachus* and *Notorynchus* as well (Compagno, 1977; Shirai, 1992c). Partial dissection of an embryo *Chlamydoselachus* specimen (AMNH 59327, 317mm TL) revealed the mesopterygial anterior expansion figured by these authors. Regan (1906), White (1936a, 1937), Berg (1947), Compagno (1977), and Shirai (1992c) agreed that the hexanchiform condition is derived (or "diagnostic" for the former two authors), and this is the usage adopted here.

Only one dorsal fin present (character 50) [CI = 1.0]. The first usage of this character was by Rafinesque (1810, for *Hexanchus* and *Heptranchias*; Gill, 1861). Among Recent elasmobranchs, a single dorsal fin is also present in a variety of rajiform taxa (Nishida, 1990), but the plesiomorphic condition for the rays is to have both dorsal fins present. The scyliorhinid *Pentanchus*, known only from the holotype, also has only one dorsal fin, but this specimen may be anomalous as the species is otherwise identical to *Apristurus herklotsi* (Compagno, 1984). This character is free of homoplasy, but would certainly show incongruence in a phylogeny with more terminal rajiform groups included. Among Paleozoic taxa, the long dorsal fin of xenacanths (extending from just posterior to the cranium to the caudal fin, with which it can be confluent) cannot be considered the equivalent of the single dorsal in any other group. A single dorsal fin is seemingly common in many Paleozoic sharks, however. Symmoriid sharks have a single dorsal that is positioned slightly anterior to the pelvic fins and with an endoskeleton composed of radial cartilages only, not unlike the dorsals of galeomorphs (Zangerl, 1981). Sharks of the extinct order †Eugeneodontida also possessed a single dorsal fin (Zangerl, 1981; the monophyly of this order is uncertain). These occurrences are independent from the state present in six- and seven-gilled sharks, irrespective of the phylogeny (of the aforementioned Paleozoic taxa) adopted. As with the extra gill arches, Thies (1987) disregards this character as a synapomorphy of hexanchiforms due to its occurrence in other taxa. Carvalho and Maisey (1996) placed this character as a squalean feature, with a reversal to the primitive condition occurring for the node uniting *Echinorhinus* and all other squaleans, because Hexanchiformes was not considered monophyletic. However, Compagno (1973, 1977, 1984), Maisey and Wolfram (1984), and Maisey (1986a) used the single dorsal fin as a defining feature of hexanchiforms, along with numerous previous authors, and this is corroborated by this analysis.

Teeth devoid of labial and lingual "aprons" (character 59) [CI = 0.33]. Three steps are needed to accommodate this character. The "apron" is an expansion from the crown that covers the root from both labial and lingual view [Thies and Reif, 1985; Cappetta's (1987) "uvula" corresponds to the lingual apron], but it is absent in hexanchiforms and higher galeomorphs (lamniforms and carcharhiniforms; see above). The distribution of this character needs further study, but it is accepted here as described by Thies and Reif (1985).

Additional (fourth) column of heart valves present (character 45) [CI = 1.0]. Both Garman (1913) and White (1936b, 1937) have described and figured the valves of the conus arteriosus in a large number of taxa. White's comments directly support the view that the extra (fourth) column ("series") of valves present in the conus of hexanchiforms (her Hexanchea) is derived relative to the fewer number (three) of other Recent chondrichthyans. White emphasized that the high number of rows is also derived in these sharks, even though three other species of sharks (a carcharinoid and two squaliforms) and many rays were described as having four rows as well. Only the additional column was included here, but the additional rows are probably derived for hexanchiforms as White suggested. As with the lack of propterygial radials (character 40 above), this feature is free of homoplasy.

Suborder Chlamydoselachoidei (*Chlamydoselachus anguineus*) (Node *i* in Fig. 5)

Many characters define this taxon (Compagno, 1984), but those used here also occur in other taxa (listed in the legend to Fig. 5, cf. Appendix 2).

Suborder Hexanchoidei (Node *j* in Fig. 5)

This taxon is composed of the three remaining hexanchiform genera (*Notorynchus*, *Hexanchus*, and *Heptranchias*), all of which are placed in the same family following Compagno (1984; cf. Shirai, 1992c). This clade is defined by seven characters.

Complete postorbital articulation (character 12) [CI = 1.0]. If interpreted as one multistate character (as done here), then the postorbital suspension has been lost prior to the node uniting hybodonts and neoselachians, as the postorbital articulation is present in many Paleozoic taxa. It is therefore reacquired here and not lost twice (and retained by hexanchoids) as advocated by Maisey (once for hybodonts and once for neoselachians; Maisey, 1980, p. 7).

Loss of the occipital hemicentrum (character 17) [CI = 0.33]. The presence of this feature defines Neoselachii, not only extant elasmobranchs, and requires three steps: loss at nodes *j*, *o*, and *c'* independently (Fig. 5). The presence of complete calcified vertebrae in fossil hexanchoids (Maisey, 1986a; Thies, 1987) does not change this optimization, as no extra steps are required when fossil hexanchoids (for which this character is known, e.g., †*Notidanoides* and †*"Hexanchus" gracilis*) are added, as long as they remain basal to living hexanchoids, as is currently believed (Maisey, 1986a; Thies, 1987). If an alternative optimization is adopted, e.g., the independent gain of the occipital hemicentrum for Galeomorphii, *Chlamydoselachus*, and node *p* (Fig. 5), then an extra step is required if fossil hexanchoids are included. The presence of the hemicentrum in these fossil taxa, however, is still questionable. Because the occipital hemicentrum is present in the basal neoselachian *Palaeospinax* (late Jurassic; Maisey, 1977), the DELTRAN optimization is less parsimonious (more steps are also required to account for its absence in pristiophoroids + rajiforms in this scenario). The fossil genera †*Protospinax* (late Jurassic; Carvalho and Maisey, 1996) and *Synechodus* (late Cretaceous; Maisey, 1985) also have this character, but they are systematically resolved as a basal hypnosqualean and galeomorph (?), respectively (the relationships of the latter are still problematic; it may be a basal neoselachian as suggested by previous authors; M. R. de Carvalho and J. G. Maisey, unpublished).

Seventh branchial unit present (character 21) [CI = 0.5]. This character is lost in *Hexanchus*.

Adductor mandibulae superficialis inserting on ectethmoid process (character 29) [CI = 1.0]. This character is adopted from Shirai (1992c), who considers the craniomandibular muscle of *Heterodontus* to be homologous (not coded as such here). Shirai (1992c) presents many different patterns associated with the origin and insertion of this muscle, which is here divided into two characters (28 and 29). Many of Shirai's (1992c) different conditions amount to autapomorphies (states 2 and 3 here, for rajiforms and *Echinorhinus*, respectively).

Constrictor hyoideus dorsalis inserting on palatoquadrate (character 31) [CI = 0.66]. Although there is homoplasy in this character, the state present in hexanchoids does not occur in other taxa.

Pectoral metapterygium contacts short proximal segment (character 42) [CI = 1.0]. The "short proximal segment" refers to the anterior portion of the metapterygium, which is divided into two sections in hexanchoids (Shirai, 1992c).

Linearly arranged tooth rows (character 58) [CI = 0.66].

This character is coded as in Shirai (1992c), with diagonally arranged tooth rows representing the primitive condition (seen in fossil outgroups, galeomorphs, and, although somewhat less distinctly, in *Chlamydoselachus* as well).

Subfamily Hexanchinae (Node *j* in Fig. 5)

This node contains *Hexanchus* and *Heptranchias* and is defined by four characters. The closed lateral line canal (character 2, CI = 0.33) is discussed under Galeomorphii. The condition in *Chlamydoselachus* and *Notorynchus* (open lateral line canal) is a primitive feature, which renders the closed lateral line a *de novo* appearance for hexanchids. The ethmoidal canal is present but does not open on the dorsal surface of nasal capsules (character 5, CI = 0.66). This character is differently optimized by Shirai (1992c; see also Carvalho and Maisey, 1996, for further description). The subnasal fenetra is present (character 6, CI = 0.33) and is characterized following Shirai (1992c). This character also defines Squaliformes (lost for *Oxynotus*, see Fig. 5). The presence of a concavity ventral to the hyomandibular fossa (character 15, CI = 1.0), described by Holmgren (1941), also defines this component.

Superorders Echinorhinoidea + Squaloidea + Hypnosqualea (Node *n* in Fig. 5)

Echinorhinus, all remaining "squaloids," *Squatina*, pristiophorids, and all ray groups are defined by five characters: an ethmoidal canal present but not opening on dorsal surface of the nasal capsule [character 5, CI = 0.66; see Shirai (1992c) for a description of its alternative states], extrabranchial cartilages present on entire hyoid arch (character 19, CI = 1.0; discussed above under Recent Elasmobranchs), complete fusion of posterior dorsal gill arch elements (character 24, CI = 0.5; see discussion above under Galeoidea), absence of mesonarial flap (character 56, CI = 1.0), and absence of the anal fin (character 66, CI = 1.0). The anal fin is primitively present in neoselachians, and its absence here is derived (Compagno, 1984). Many Paleozoic sharks lack an anal fin, and this character has been used as a defining feature of a monophyletic group including *Cladoselache*, stethacanthids, and symmoriids (Lund, 1985). As with the single dorsal fin discussed previously, the lack of the anal fin is derived at this node irrespective of the relationships of the Paleozoic taxa that also lack this character.

Superorder Echinorhinoidea (Order Echinorhiniformes: *Echinorhinus*) (Node *o* in Fig. 5)

Along with its many diagnostic characters (Compagno, 1984), this genus is further characterized by the loss of the occipital hemicentrum (character 17, CI = 0.33; discussed above under Recent Elasmobranchs), adductor mandibulae superficialis inserting on mandibula (character 29, CI = 1.0), all three basal cartilages of pectoral fin fused (character 39, CI = 0.75; Shirai, 1992c), and linearly arranged tooth rows (character 58, CI = 0.66), which are also present (homoplastically) in hexanchoids.

Echinorhinus primitively retains an open lateral-line canal (Garrick, 1960). Placing *Echinorhinus* in an order of its own was advocated by Pfeil (1983) and further supported by Herman *et al.* (1989) on the basis of its distinctive teeth. The latter authors also mention the similarities between *Echinorhinus* and hexanchiforms (again on the basis of teeth morphology), a notion not entirely without support. Carvalho and Maisey (1996) found that the coding given to *Echinorhinus* for characters related to the ectethmoid process and suborbitalis muscle had an impact on hexanchiform monophyly, even though *Echinorhinus* and Hexanchiformes were not supported as monophyletic in any of the shortest trees they obtained. Molecular evidence also seems to support distinct ordinal placement for *Echinorhinus* (Bernardi and Powers, 1992).

Superorders Squaloidea + Hypnosqualea (Node *p* in Fig. 5)

Eight characters support this node, five of which were not considered by Shirai (1992c), who characterized this group with features that are differently optimized here.

Closed lateral line canal (character 2) [CI = 0.33]. This character is discussed above under Galeomorphii.

Ectethmoid process lost (character 8) [CI = 0.5]. The discovery of the "membranous ectethmoid process" mentioned by Holmgren (1940) would fill in this "gap" in the tree, insofar as the ectethmoid process is reacquired higher in the phylogeny by pristiophoroids + rajiforms (as the "antorbital cartilage").

Suborbitalis muscle originating from suborbital surface (character 26) [CI = 0.8]. Different states of the origin of the suborbitalis were described by Shirai (1992c) and are further discussed under Galeomorphii.

Notochord constricted along entire vertebral column (character 44) [CI = 0.5]. See discussion under Squalea above.

Enlarged supraneurals preceding second dorsal fin (character 46) [CI = 0.66]. This multistate character shows homoplasy in that the condition in dalatioids (see below) is a reversal to the state present in galeomorphs and basal squaleans (lack of obvious supraneurals). In squaloids, enlarged supraneurals are present anterior to the second dorsal fin, as in *Squatina* (Shirai, 1992c). The most apomorphic state of this character is seen in pristiophorids and rays (at least basally), where well-developed supraneurals are present in precaudal tail and anterior to caudal fin. This character was used as evidence of a Squalidae + Hypnosqualea clade (Shirai, 1992c), but this is refuted here. This character can be used as a synapomorphy at this level or as independently acquired for squaloids and *Squatina*.

Precaudal hemal processes as elongate as lower caudal skeleton (character 48) [CI = 0.5]. This character refers to the uniquely elongated hemal processes that fully partition the hypaxial musculature, contrasting to the primitive state in which the hemal processes are less expanded ventrally (Shirai, 1992c). This character is difficult to detect in fossil outgroups and Paleozoic taxa (Zangerl, 1981; Maisey, 1983).

Spiracle valve present (character 60) [CI = 0.5]. The coding of this character is after S. Shirai (personal communication).

Longitudinal precaudal keel present (character 63) [CI = 0.5]. According to Shirai (1992c), the derived condition of this character has the same distribution as the expanded supraneurals and elongated precaudal hemal processes discussed above, but this still was not enough to support a monophyletic Squalidae + Hypnosqualea, as defended by Shirai (1992c).

Superorder Squaloidea: Squaliformes (Node *q* in Fig. 5)

This component is not the equivalent of Squaliformes of Compagno (1984) and is novel to this analysis. The order is divided into two suborders, the Squaloidei and Dalatioidei. Four characters support its monophyly, but all exhibit homoplasy. Three of these are described in Shirai (1992c), even though their optimization differs here (subnasal fenestra present, character 6, CI = 0.33; constrictor hyoideus dorsalis inserting partly on hyomandibula and partly on palatoquadrate, character 31, CI = 0.66; linearly arranged tooth rows with teeth overlapping each other, character 58, CI = 0.66).

Dorsal fin endoskeleton composed of basal cartilage and finspine (character 49) [CI = 0.4]. Although this character is traditionally and correctly regarded as a primitive neoselachian feature, its placement here is a synapomorphy of squaliforms. The finspine appears *de novo* at this node and basal squaleans (hexanchiforms and *Echinorhinus*) do not have a finspine. Further homoplasy is required as the clade Somniosidae + Dalatiidae lacks a finspine. Five steps are therefore minimally imposed by this character. A finspine appears in fossil "rhinobatoid" taxa (e.g., †*Belemnobatis* and †*Spathobatis*, de Saint-Seine, 1949; †*Breviacanthus*, Maisey, 1976; Cappetta, 1987) and in the basal hyp-

nosqualean *Protospinax* (Carvalho and Maisey, 1996), therefore necessitating more homoplasy in a more inclusive matrix.

Suborder Squaloidei (Node *r* in Fig. 5)

This suborder is defined by two characters described in Shirai (1992c): a flexor caudalis muscle which extends to the origin of caudal fin (character 54, CI = 1.0) and the presence of an upper precaudal pit (character 62, CI = 1.0).

Suborder Dalatioidei: Centrophoridae + Etmopteridae + Oxynotidae + Somniosidae + Dalatiidae (Node *s* in Fig. 5)

This node is supported by a suborbitalis muscle originating from the interorbital wall (character 26, CI = 0.8), and the reversal to the primitive condition for characters 46 (CI = 0.66), 48 (CI = 0.5), and 63 (CI = 0.5) (discussed above under Squaloidea + Hypnosqualea). The (unnamed) node Etmopteridae + Oxynotidae + Somniosidae + Dalatiidae and the rest of the components within this group (nodes *t* through *z* of Fig. 5) correspond to Shirai's (1992c) Dalatiiformes, and will not be discussed further (diagnosed in the legend to Fig. 5).

Superorder Hypnosqualea: *Squatina* + Pristiophoriformes + Rajiformes (Node *a'* in Fig. 5)

This component, considered a superorder by Carvalho and Maisey (1995), is defined here by seven features, all of which are present in the definition of this group given by Shirai (1992c). Three are from myology (character 52, CI = 0.5; character 53, CI = 1.0; and character 54, CI = 1.0). The anterior expansion of the pectoral fin (elongated propterygium, character 41, CI = 0.5) is lost for pristiophoroids (along with character 53).

Shirai (1992c) defines this group, novel in his earlier account (1992a), with additional characters not optimized in the same fashion here.

Squatiniformes: *Squatina* (Node *b'* in Fig. 5)

Squatina is defined by many autapomorphies (Compagno, 1984) as well as two homoplastic features: presence of nasoral groove (character 55, CI = 0.33) and loss of the spiracle valve (character 60, CI = 0.5).

Pristiophoriformes + Rajiformes: Pristiorajea (Node *c'* in Fig. 5)

This node, and those contained within it, are not further elaborated upon, except to note that the characters listed in Fig. 5 are those that occur homoplastically in other groups or are states in multistate series.

Shirai (1992c; Carvalho and Maisey, 1996) and references he cites provide explanations of some of those characters (see legend to Fig. 5). Other characters defining nodes *d'* and *e'* (Fig. 5) can be found in Compagno (1973, 1977), Nishida (1990), and McEachran *et al.* (this volume).

IV. Discussion

A. Basal Squaleans: Hexanchiform Monophyly, and Autapomorphy

Six- and seven-gilled sharks (cow sharks) have traditionally been viewed as a primitive but natural group. Under one name or another, the group was included in the classifications of many earlier authors, such as Regan (1906), Goodrich (1909, 1986), Garman (1913), White (1936a, 1937), and Berg (1947). The first author to imply a possible close relationship between *Chlamydoselachus* and hexanchoids was Garman (1885; he described *Chlamydoselachus* the previous year), and this was followed by Gill (1893; cited in Compagno, 1984) and Jordan and Evermann (1896). Hexanchiform monophyly is entrenched in many volumes dealing with the identification and/or biology of elasmobranchs in general (e.g., Compagno, 1984; Smith and Heemstra, 1986; Cappetta, 1987; Springer and Gold, 1989; Eschmeyer, 1990; J.S. Nelson, 1994), a notion that stems more recently from Compagno (1973, 1977) and is supported by Maisey and Wolfram (1984) and Maisey (1986a). In addition, Ida *et al.* (1986) present karyological evidence in support of hexanchiform monophyly. The most outspoken opponents to this view are Thies (1987) and Shirai (1992b,c).

Compagno (1973) offers an extensive array of characters that occur in *Chlamydoselachus* and hexanchoids; however, many represent characters primitive for elasmobranchs in general or for other higher-level groups (e.g., lack of calcification in the vertebral centra and unconstricted notochord, presence of ectethmoid processes, and a trough-like rostrum) or characters vaguely defined or spurious (e.g., "eyes laterally situated on the head" and "head moderately depressed or conical but not greatly flattened"). Compagno (1977) improved upon this by explicitly listing characters that he believed to be derived for hexanchiforms, e.g., extra gill arches, lack of lateral commissures on neurocranium, long ectethmoid processes on nasal capsules, no finspines, a single dorsal fin (presumed to be the second dorsal fin), and propterygium not contacting any radial cartilages in the pectoral fin. Maisey (1986a) reduced this list, considering only the single dorsal fin and extra gill clefts as defining features (also

Maisey and Wolfram, 1984). Along with these two characters, the present analysis identifies five additional synapomorphies for the order.

The criticisms of Thies (1987) and Shirai (1992b,c) have been partially considered above (under Notidanoidea). In his 1992 book, Shirai (1992c) coded the sixth gill arch of hexanchoids and *Chlamydoselachus* as an *independent* character, therefore eliminating any possibility of it becoming a synapomorphy of hexanchiforms (as was also true of the coding given to the single dorsal fin; cf. Carvalho and Maisey, 1996). This coding derives from an a priori decision to consider the extra gill arch and single dorsal fin to be nonhomologous in hexanchoids and *Chlamydoselachus* and thereby effectively prohibiting these characters from being homologous *before* any reference to a phylogeny. The determination of homology for the extra gill arch was based on Shirai (1992b), a paper that does not present or make reference to a tree. Shirai's determination of homology was therefore a conclusion from observation and not from a phylogenetic analysis (see also Shirai, this volume). As discussed above (Notidanoidea, character 20), the distinctions reported in the arches of *Chlamydoselachus* and hexanchoids by Shirai further characterize hexanchoids, or define even more inclusive taxa, but do not affect the monophyly of Hexanchiformes. To further substantiate this, I experimentally coded *Chlamydoselachus* exactly as Shirai did for the extra gill arch (i.e., as not having it). Hexanchiformes was still monophyletic in the single tree obtained. The only way to decide if the extra gill arch is indeed independent is by congruence, not a direct conclusion from observation.

When in doubt if a feature is indeed homologous, one should consider coding that character (believed to be independent but in which there is substantial doubt) equally for those taxa presenting it (as long as this is reasonable for a particular character in question and there are not a great number of characters receiving this same treatment). In other words, if nonhomology is believed to be likely, phylogenetic analysis should not be conducted in a manner that presumes nonhomology but in one that *tests* for it (Carpenter, 1994). I doubt that anyone seriously believes that there is no basis for coding the extra gill arch and single dorsal fin equally for *Chlamydoselachus* and hexanchoids. If the result falsifies a monophyletic Hexanchiformes, then this is the necessary evidence that these characters are distinct (depending on the final topology, as different optimizations may still allow for their single origin).

The arguments presented by Thies (1987) in relation to the single dorsal fin and extra gill arch are reminiscent of compatibility (clique) analysis, wherein characters not congruent with the largest subset of mutually compatible characters (the clique) are discarded. In other words, he advocates that incongruent (homoplastic) characters are not accurate indicators of relationships and should be eliminated, once again implying that homoplasy can be determined prior to a phylogenetic analysis (cf. Farris, 1983). Any textbook on vertebrates correctly cites homeothermy as a mammalian character (J. Z. Young, 1981), even though birds are homeothermic as well.[2] A character that is homoplastic is also diagnostic (see below). Thies' (1987) rejection of the mesopterygial forward expansion which prevents the propterygium from contacting the radials was discussed above (Notidanoidea, character 45) and shown to be unfounded.

Other authors that have cast doubt on hexanchiform monophyly include Bigelow and Schroeder (1948), Applegate (1974), Bass et al. (1975), and Herman et al. (1993, 1994a). Bigelow and Schroeder (1948) chose to leave Chlamydoselachus in a separate suborder (Chlamydoselachoidea) due to differences in its suspensorium and vertebral column when compared to the "notidanoids" (their hexanchoids). However, their decision is symptomatic of a more encompassing issue, that of being misled by data that have no real bearing on relationships (i.e., autapomorphy). Chlamydoselachus and hexanchoids differ in many attributes, no doubt, and this has been used as an indication of their nonmonophyly, but many of these attributes also serve to distinguish these taxa from most or even all other elasmobranchs. The suspensorium of hexanchoids, just mentioned, is an example, as is the paired basibranchial of the antipenultimate arch, used by Shirai (1992b) as evidence for the nonmonophyly of hexanchiforms. Both these characters are nothing but defining features of hexanchoids that do not help to clarify their relationships to other elasmobranchs.[3] Herman et al. (1993, 1994a) recommend placing Chlamydoselachus in an order of its own because of distinctions in tooth structure (vascularization of the root) with hexanchoids. Even though they notice similarities in these features that suggest (to

them) a possible close relationship between Chlamydoselachus and orectolobiforms, the great distinctness from hexanchoid tooth roots was more critical in their judgement.

Autapomorphy has, in the past, misled many authors constructing classifications for elasmobranchs. For example, "sharks" were considered monophyletic without the rays due to the large number of derived features of the latter group (Müller and Henle, 1838–1841; Duméril, 1865; Günther, 1870; Regan, 1906; Garman, 1913; this dichotomy can still be seen in modern publications, even with much evidence to the contrary; Herman et al., 1987, 1994b; Dunn and Morrissey, 1995). This tradition was begun by Linnaeus (1758) who established a single genus for sharks and one for rays. Compagno (1973, p. 21) was well aware of this tendency, noting that the "ranking of modern elasmobranchs in two equal groups of sharks and rays places undue emphasis on unique characters of the rays while obscuring the many differences between the various shark groups." However, Compagno's (1973, 1977) interpretation of the phylogenetic placement of Squatina was no different, resulting from the many distinctions between angel sharks and all other elasmobranchs. Consequently, Squatina was left in a superorder of its own, equal in rank to the three other superorders containing all other sharks and rays, and its relationships remained elusive. Another example can be seen in a tree attributed to Dingerkus (Séret, 1986), in which he considered Chlamydoselachus to be the sister-group to all remaining elasmobranchs. This hypothesis was heavily influenced by the peculiarities of Chlamydoselachus' ("cladodont") tooth morphology (an autapomorphy rather than the primitive condition as is seen in the †Cladodontida of Lund, 1985). The relationships of Chlamydoselachus have been seen as problematic since its description, due to similarities shared with "cladodonts" (Garman, 1885; summarized in Gudger and Smith, 1933), even though these similarities are independently derived and not primitively retained by Chlamydoselachus or indicative of a monophyletic group comprised of the frilled shark and "cladodonts."

---

[2]This example does not hold if one accepts the Haematothermia of Gardiner (1982, 1993), but examples of homoplastic characters are abundant.

[3]The postorbital articulation is herein interpreted as a multistate character in which the condition of the frilled shark (state 1) is intermediate between the state in hexanchoids (2) and the primitive absence of the character (0), thereby implying a possible relationship between the hexanchoid condition and the incomplete postorbital connection. However, coding the postorbital articulation in hexanchoids as a character separate from the incomplete articulation in Chlamydoselachus does not affect their phylogenetic placement.

B. Homoplastic but Diagnostic

Intriguing character distributions suggested an affinity, albeit falsified, between Heterodontus and most squaleans (character 28), hexanchiforms and "higher" galeoids (character 59), Chlamydoselachus and galeoids (galeomorphs minus Heterodontus; characters 49 and 51), orectolobiforms and rajiforms (character 30), and rajiforms and dalatiines (characters 33 and 34), among other interesting combinations. Interestingly, many of these are characters of myology.

Similar to the views of Shirai (1992b,c) and Thies (1987), G. Dingerkus (unpublished) rejects hexanchiform monophyly because both the extra gill arch and single dorsal fin are "distinct." However, Hexanchiform monophyly is corroborated by more than these two characters. In any case, it is quite odd that *Chlamydoselachus* and hexanchoids share *two* features that require equal coding in a matrix but were rejected as such beforehand by these authors, thereby leaving no chance for the discovery of a monophyletic Hexanchiformes. Rejection of these characters (single dorsal fin and extra gill arch) because of their homoplastic distributions would require the rejection of a great many other features as well and derives from the misguided belief that homoplasy is a rare phenomenon. Yet this is an expectation falsified by many published phylogenies created with large data matrices (Sanderson and Donoghue, 1989).

Because phylogenetic analysis consists of minimizing homoplasy, "bad" data are not those with elevated levels of homoplasy per se, but those in which homoplasy is least minimizable (Goloboff, 1991). In other words, as long as there is resolution (congruence), then the (high) levels of homoplasy are just that, and not an indication of "bad" data. The resulting single most parsimonious tree indicates that the data are, in fact, *good*. Furthermore, because there is much discordance (between characters and the tree), the discovery of additional characters that are congruent with homoplastic features may change tree topology in their favor (in which case the "homoplasy" becomes synapomorphy). Discarding these characters a priori prohibits this.

For the resulting phylogeny (Fig. 5), almost half the characters (see Appendices 2 and 3) have homoplastic distributions (either as reversals or independent gains). This was also the case in Carvalho and Maisey (1996) and Shirai (1992c), and similarly high-levels of homoplasy have been reported in other studies dealing with elasmobranch taxa (e.g., McEachran and Miyake, 1990). In Fig. 5 almost every node contains characters that occur homoplastically elsewhere on the tree. This situation, clearly not restricted to (or uncommon in) chondrichthyans, makes careful character coding crucial, as the support for nodes is very sensitive to changes in coding strategies. Some of the nodes are supported only by homoplastic characters, and only very few nodes are supported by perfectly congruent characters (Fig. 5).

The effects of homoplasy may be elusive, as evidenced by the otherwise excellent work of Nishida (1990) on the higher-level relationships of rays. Nishida's data were not subjected by him to a parsimony algorithm, even though he provides a character matrix. Reanalysis using Hennig86 (mh* and bb* options

in conjunction) produces a strict consensus topology radically different from the tree provided by Nishida (1990). This indicates that his data are more complex than otherwise thought (Nishida's tree is not among the equally most parsimonious trees obtained, either).

Compagno (1973, 1977, 1988) has extensively reviewed the relationships within galeomorphs, bringing much anatomical data into consideration. In those papers, he defended a sister-group relationship between *Heterodontus* and orectolobiforms, but without much conviction, as evidenced by his last review (Compagno, 1988). The characters incorporated in the present analysis do not support a *Heterodontus* + Orectolobiformes clade, and they also reject Compagno's "second" choice phylogeny, in which orectolobiforms and carcharhiniforms are sister-groups. Characters shared by *Heterodontus* and orectolobiforms were included in my analysis but are best optimized as defining features of Galeomorphii (with the secondary loss further supporting lamniforms + carcharhiniforms). However, it is unclear if the inclusion of additional characters putatively derived for *Heterodontus* and orectolobiforms will render this a monophyletic group. These taxa share the derived condition of characters 4 (downcurved ethmoidal area), 27 (insertion of the suborbitalis muscle anterior to adductor mandibulae), and 55 (presence of nasoral groove; also in *Squatina*). Because there is homoplasy in these characters (Fig. 5), it is possible that perhaps one more character shared by these taxa will be enough to render them monophyletic. This situation occurs in many other branches of the tree and is a function of the high number of discordant characters.

As is now quite obvious, high levels of homoplasy render the use of parsimony algorithms a mandatory procedure when attempting to resolve the phylogenetic relationships of elasmobranchs. This kind of computational analysis came relatively late to studies of chondrichthyan systematics, however, and Shirai (1992c) is definitively among the first to implement computer-based tree searches. The tardy application of these more rigorous systematic methods (with or without parsimony algorithms), which are routine in the study of many groups, may well be the result of the fact that there are few practicing elasmobranch systematists and the effect of a virtually unchallenged acceptance of Compagno's classifications by most elasmobranch researchers.

C. Fossil Neoselachians

Fossil taxa are important when addressing relationships among basal Recent elasmobranchs. As used here, Neoselachii includes at least *Palaeospinax* (and possibly *Synechodus*), a taxon important for the optimization of characters previously restricted to living forms or their (mainly Tertiary) immediate relatives. Alternatively, if living chimaeroids are used as an outgroup, in an analysis of only Recent taxa, the information contained in Paleozoic and Mesozoic sharks would be ignored, thereby changing the interpretation of many features [e.g., fusion of posteriormost dorsal gill arch elements (gill pickaxe), fused coracoids, etc.]. Because the objective of phylogenetic analysis is to determine the origin (and therefore distribution) of a character on a tree (i.e., relationships), data from fossil elasmobranchs must be included at this level; otherwise, the origin of a specific feature may be erroneously hypothesized to be at a less inclusive node.

The present analysis identified four characters in support of a group of "modern level elasmobranchs," less inclusive than Neoselachii, composed of Recent elasmobranchs and fossil taxa placed within extant groups (represented by *Squalus* in Fig. 3). This distinction was present in the scheme put forth by Moy-Thomas (1939) and followed to a varying degree by subsequent authors (Arambourg and Bertin, 1958; Compagno, 1973, 1977; Schaeffer, 1981; Thies, 1983; Maisey, 1982, 1983, 1984a,b, 1986b). However, a more complete understanding of the anatomy of *Palaeospinax* (Schaeffer, 1967; Reif, 1974; Maisey, 1977) and *Synechodus* (Schweizer, 1964; Reif, 1973) has led to the belief that "so far, no convincing higher elasmobranch synapomorphies have been established that would exclude these fossils from a taxon consisting only of extant elasmobranchs" (Maisey, 1984a, p. 36). Of the four characters proposed here to unite "modern elasmobranchs," one concerns muscles (character 28), and another cartilage that is at best very weakly calcified (character 19, extrabranchial cartilages on hyoid arch). Because this level of the phylogeny seems more subjected to the vagaries of preservation, Maisey's statement is to some degree still applicable in spite of these characters. The biochemical characters cited in Maisey (1986b; not included in my matrix) also are unavailable in stem-group neoselachians. Accordingly, Shirai (1992c, p. 14) does not recognize synapomorphies of Recent elasmobranchs, which he accepted as "traditional," even though his analysis contained characters that are optimized as synapomorphies of that group here.

A paradoxical aspect inherent in dealing with fossil and extant taxa together in the same analysis is the optimization of characters that are not preserved in the fossils, such as myology and other "soft" anatomy, and coloration, etc. The matrix in Appendix 3 includes many myological features. When these characters involve transformations occurring at the base of the tree (i.e., at or close to the boundary that separates extant elasmobranchs from stem-group neoselachians), different options for their optimization are ambiguous. For example, consider character 26. This character (origin of suborbitalis) has five states, of which two (states 3 and 4) can be equally optimized at the root. There is really no basis for deciding which state best defines the ingroup (Recent elasmobranchs). The fossil outgroup taxa certainly do not help here, but unquestionably *Palaeospinax* must have had some state of this character. Because this information is unavailable, an extra step is counted for the tree, due to the placement of both states at the nodes where they occur unequivocally (node *a* for state 4, node *g* for state 3), even though the actual placement of one of these states lies at a more inclusive node. In comparison, Hennig86 saved a step by placing one of the states at the outgroup node, thus assuming the "?" entry in the outgroup to be "1," and thereby pushing the character farther down the tree.

The studies of Reif (1973, 1974, 1977, 1978) concerning the fine structure of tooth enameloid has led to more characters being proposed as defining features of Neoselachii, notably triple-layered enameloid (Maisey, 1984a, 1986b). Reif's (1974) own study of *Palaeospinax*, however, does not reveal three layers, and the third and innermost layer (haphazardly fibered or "tangled" enameloid) apparently is absent. *Synechodus* is described as having all three layers (Reif, 1973; cf. Maisey, 1985), which might place it closer to Recent elasmobranchs than *Palaeospinax*, as suggested, on the basis of other evidence, by Maisey (1985). The presence of "shiny-layer" and "parallel-fibered" enameloid ("tooth type β") are uncontested neoselachian characters, even though homoplasy is required, as rajiforms apparently lack specialized enameloid layers (Reif, 1973, 1977). Curiously, Reif (1977) downplays the importance of haphazardly fibered enameloid, stating that it is of "little diagnostic value" (1977, p. 570), even though it is absent from the hybodont and ctenacanth fossils he examined. Confusingly, in a later study, Reif (1978) credits all neoselachians (his Euselachii) as having triple-layered enameloid. However stated, the enameloid ultrastructure is specialized for neoselachians, even though the condition of the third layer in *Palaeospinax* is unclear.

Fossil neoselachians are known for the most part from isolated teeth and incomplete skeletal remains (an exception is the neurocranium of *Synechodus* described by Maisey, 1985). The scarce anatomical data available leaves the position of these taxa open to some speculation, although they are without a doubt closer to Recent elasmobranchs than to other euselachians (Maisey, 1986b; Fig. 3). Cappetta (1987, p. 128) and Cappetta *et al.* (1993) leave the †Palaeospinacidae (in which they include *Synechodus*) as "Galeomorphii *incertae ordinis.*" Maisey (1985) considered two alternatives regarding the phylogenetic placement of *Synechodus*, either as the sister-group to all Recent elasmobranchs or as the most basal galeomorph. *Palaeospinax* and *Synechodus* are known to possess characters diagnostic of "higher" or "modern" elasmobranchs (e.g., occipital hemicentrum, primary calcification of notochord, and specialized enameloid layering in teeth). However, only *Synechodus* has been shown to share similarities with galeomorphs that are possibly derived (Maisey, 1985), even though this remains unresolved. Hence, *Palaeospinax* is interpreted as a basal neoselachian in this study (cf. Schaeffer, 1967), a situation that may change once more information is available.

Other contenders for neoselachian membership are the orthacodontids and anacoracids, which are known almost exclusively from isolated teeth (the exception is *Sphenodus nitidus* [= *S. macer*?] from Solnhofen; de Beaumont, 1960). In recent reviews, these taxa have been considered to be hexanchoids and lamniforms, respectively (Cappetta, 1987; Cappetta *et al.*, 1993), but it is unclear if the similarities on which these assignments are based are in fact derived, primitive, or homoplastic. These taxa are perhaps best considered *incertae sedis* at present.

V. Conclusions

The classification resulting from the current analysis is intermediate when compared to the previous classifications of Compagno (1973, 1977, 1984) and Shirai (1992c). Shirai's Squalea is corroborated, but his components (hexanchoids + (*Echinorhinus* + (Dalatiiformes + (Squalidae + Hypnosqualea)))) and (Squalidae + Hypnosqualea) are not. Instead, Hexanchiformes and a slightly modified Squaliformes (excluding *Echinorhinus*) are defended as monophyletic. Recent elasmobranchs are divided into 10 monophyletic orders. The relationships among the four superorders of Compagno are fully resolved, corroborating further the work of Shirai at this level. The following hierarchical summary of ordinal groups includes only living forms, although, as mentioned, some fossil groups are well corroborated systematically:

Class Chondrichthyes
    Subclass Holocephali
    Subclass Elasmobranchii
        Infraclass Neoselachii

        Division Galeomorphii
            Superorder Heterodontoidea
                Order Heterodontiformes
            Superorder Galeoidea
                Order Orectolobiformes
                Order Lamniformes
                Order Carcharhiniformes
        Division Squalea
            Superorder Notidanoidea
                Order Hexanchiformes
                    Suborder Chlamydoselachoidei
                    Suborder Hexanchoidei
            Superorder Echinorhinoidea
                Order Echinorhiniformes
            Superorder Squaloidea
                Order Squaliformes
                    Suborder Squaloidei
                    Suborder Dalatioidei
            Superorder Hypnosqualea
                Order Squatiniformes
                Order Pristiophoriformes
                Order Rajiformes

This paper presents a broad overview of recent developments in elasmobranch phylogeny. As such, much emphasis has been placed on the work of Shirai (1992c), who has made substantial progress in creating a more rigorous and modern footing for the systematics of certain shark groups and especially in clarifying how these groups are related to the rays. Particular to Shirai's contributions is the demise of a taxon variously denoted "Squaliformes" or "Squaloidei," which, although lacking defining characters, was entrenched in the literature. The major problem in maintaining a monophyletic Squaliformes, however, was more an outcome of the cryptic effects of homoplasy than the "right" character(s) never having been found by the authors supporting it. The substantial amount of incongruence exhibited by any phylogeny constructed with the taxa in question is capable of misleading any systematist not equipped with parsimony algorithms. The matrix used here contains characters that were used by previous authors, of course, but the novel placement of some of these characters derives from a fuller exploration of their interactions, which is only possible with the application of computer-based parsimony procedures.

VI. Summary

Recent higher-level elasmobranch taxa are subjected to a cladistic analysis to clarify the interrelationships of Compagno's (1973) superorders, to test the validity of Shirai's (1992c) Squalea and the reported nonmonophyly of Hexanchiformes, and to review the long suspected notion that Squaliformes is also paraphyletic. A matrix containing 20 terminal taxa and 67 morphological characters, compiled from numerous published accounts and confirmed on specimens, was analyzed using computer-based parsimony procedures. The resulting single most parsimonious tree (L = 132 steps, CI = 0.67, RI = 0.80), further supports the Squalea, but two of Shirai's higher-level nodes are falsified. Hexanchiformes is unambiguously defended as monophyletic by seven morphological features (including the single dorsal fin, extra gill arch, and extra column of heart valves in the conus arteriosus). The Squaliformes is also monophyletic, but only if *Echinorhinus* is excluded. *Echinorhinus* is placed in its own order and is the sister-group to a node composed of Squaliformes + Hypnosqualea. Resolution within galeomorphs indicates that *Heterodontus* is the sister-group to an (Orectolobiformes + (Lamniformes + Carcharhiniformes)) clade. A proposed sister-group relationship between *Heterodontus* and Orectolobiformes is therefore falsified, but due to the high amount of discordance this node may eventually be supported if further characters, possibly homologous in *Heterodontus* and orectolobiforms, are included (e.g., the space between meso- and metapterygium, not considered here due to lack of data). Certain fossil taxa, especially the Jurassic *Palaeospinax*, play an important role in the optimization of some characters at the base of the phylogeny of living taxa. The high degree of misfit or homoplasy, the misleading effects of autapomorphy, and the tardy application of more rigorous phylogenetic procedures have inhibited progress in elasmobranch systematics when compared to bony fishes. Recent elasmobranchs are divided into two higher divisions, the Galeomorphii and Squalea, with 10 included monophyletic orders. The interrelationships of Compagno's four superorders, a problem that has persisted with little notice since 1973, is resolved here in accordance with Shirai's work concerning these higher levels. The resulting classification is intermediate between the schemes of Compagno and Shirai.

Acknowledgments

I thank Melanie L. J. Stiassny, G. David Johnson, and Lynne R. Parenti for inviting me to contribute to this volume. For discussing topics related to this paper, thanks are due to Richard Lund, Eileen Grogan, Gary Nelson, Joel Radding, Melanie L. J. Stiassny, Shigeru Shirai, and especially Monica Toledo-Piza. I benefitted a great deal from conversations with John G. Maisey concerning elasmobranch anatomy and phylogeny. For commenting on earlier drafts of this paper, I thank John D. McEachran, John F. Morrissey, and Dominique Didier. I especially acknowledge Shigeru Shirai for tolerating my attention to his studies. This chapter came to fruition thanks to the support and help of the staff of the Department of Herpetology and Ichthyology (AMNH), Melanie L. J. Stiassny, Monica Toledo-Piza, and especially Mary S. Andriani (without whom this paper would probably never have been completed).

The researchers of the Laboratório de Elasmobrânquios of the Universidade do Estado do Rio de Janeiro (Brazil), particularly Ulisses L. Gomes, Andrea Siqueira, and Alessandra Marques, have given me much needed encouragement over the years. Financial support was provided by the Conselho Nacional de Desenvolvimento Científico e Tecnológico (CNPq) of the Brazilian Federal Government, a Dissertation Grant from The City University of New York, and a Samuel Gruber Award from the American Elasmobranch Society (AES). A synopsis of this chapter was presented at the 75th meeting of the American Society of Ichthyologists and Herpetologists in Edmonton. The AES, in particular Sanford A. Moss, John F. Morrissey, and Gregor M. Cailliet, provided much incentive. Funding to attend that meeting was provided by a grant from the Donn Rosen Fund of the Department of Herpetology and Ichthyology of the AMNH and a Travel Award from the AES.

References

Allis, E. P. (1923). The cranial anatomy of *Chlamydoselachus anguineus*. *Acta Zool.* **4**, 123–221.

Applegate, S. P. (1967). A survey of shark hard parts. *In* "Sharks, Skates, and Rays" (P. W. Gilbert, R. F. Mathewson, and D. P. Rall, eds.), pp. 37–69. Johns Hopkins University Press, Baltimore.

Applegate, S. P. (1974). A revision of the higher taxa of orectolobids. *J. Mar. Biol. Assoc. India* **14**, 743–751.

Arambourg, C., and Bertin, L. (1958). Classe des Chondrichthyens (Chondrichthyes). *In* "Traité de Zoologie" (P.-P. Grassé, ed.), Vol. 3, pp. 2010–2067. Masson, Paris.

Bass, A. J., D'Aubrey, J. D., and Kistnasamy, N. (1975). "Sharks of the East Coast of Southern Africa. V. The Families Hexanchidae, Chlamydoselachidae, Heterodontidae, Pristiophoridae and Squatinidae," Invest. Rep. No. 43. Oceanographic Research Institute, Durban.

Berg, L. S. (1947). "Classification of Fishes, both Recent and Fossil" (transl.), pp. 87–517. Edwards, Ann Arbor, MI.

Bernardi, G., and Powers, B. A. (1992). Molecular phylogeny of the prickly shark, *Echinorhinus cookei*, based on a nuclear (18SrRNA) and a mitochondrial (cytochrome b) gene. *Mol. Phylogenet. Evol.* **1**, 161–167.

Bigelow, H. B., and Schroeder, W. C., eds. (1948). "Fishes of the Western North Atlantic," Sears Found. Mar. Res., Mem. No. 1, Part 3, pp. 53–576. Yale University, New Haven, CT.

Brown, C. (1900). Ueber das Genus *Hybodus* und seine systematische Stellung. *Paleontographica* **46**, 149–174.

Cappetta, H. (1987). Chondrichthyes II. Mesozoic and Cenozoic Elasmobranchii. *In* "Handbook of Paleoichthyology" (H. P. Schultze, ed.). Vol. 3B. 193 pp. Fischer, Stuttgart.

Cappetta, H., Duffin, C., and Zidek, J. (1993). Condrichthyes. *In* "The Fossil Record" (M. J. Benton, ed.), Vol. 2, pp. 593–609. Chapman & Hall, London.

Carpenter, J. M. (1988). Choosing among multiple equally parsimonious cladograms. *Cladistics* **4**, 291–296.

Carpenter, J. M. (1994). Successive weighting, reliability and evidence. *Cladistics* **2**, 215–220.

Compagno, L. J. V. (1973). Interrelationships of living elasmobranchs. *Zool. J. Linn. Soc.* **53**, 15–61.

Compagno, L. J. V. (1977). Phyletic relationships of living sharks and rays. *Am. Zool.* **17**, 303–322.

Compagno, L. J. V. (1984). FAO species catalogue. Vol. 4. Sharks of the World. An annotated and illustrated catalogue of shark species known to date. Part 1. Hexanchiformes to Lamniformes. *FAO Fish. Synop.* **125**(4), viii–249.

Compagno, L. J. V. (1988). "Sharks of the Order Carcharhiniformes." Princeton University Press, Princeton, NJ.

Compagno, L. J. V. (1990). Relationships of the megamouth shark, *Megachasma pelagios* (Lamniformes: Megachasmidae), with comments on its feeding habits. *In* "Elasmobranchs as Living Resources: Advances in the Biology, Ecology, Systematics, and the Status of the Fisheries" (H. L. Pratt, Jr., S. H. Gruber, and T. Taniuchi, eds.), NOAA Tech. Rep. NMFS, **90**, 357–379. U.S. Department of Commerce, Washington, DC.

Daniel, J. F. (1934). "The Elasmobranch Fishes," 3rd ed. University of California, Berkeley.

Dean, B. (1909). Studies on fossil fishes (sharks, chimaeroids and arthrodires). *Mem. Am. Mus. Nat. Hist.* **9**, 209–287.

de Beaumont, G. (1960). Contribuition á l'étude des genres *Orthacodus* Woodw. et *Notidanus* Cuv. (Selachii). *Schweiz. Palaeontol. Abh.* **77**, 1–46.

de Carvalho, M. R., and Maisey, J. G. (1996). Phylogenetic relationships of the Late Jurassic shark *Protospinax* Woodward 1919 (Chondrichthyes: Elasmobranchii). *In* "Mesozoic Fishes: Systematics and Ecology" (G. Arratia, and G. Viohl, eds.), pp. 9–46. Verlag Dr. Friedrich Pfiel, Munich.

de Pinna, M. C. C. (1991). Concepts and tests of homology in the cladistic paradigm. *Cladistics* **7**, 367–394.

De Saint-Seine, P. (1949). Les poissons des calcaires lithographiques de Cerin (Ain) Nlles. Arch. Mus. Hist. Nat. Lyon **1**(II), 1–357.

Dick, J. R. F. (1978). On the carboniferous shark *Tristychius arcuatus* Agassiz from Scotland. *Trans.—R. Soc. Edinburgh* **70**, 63–109.

Dick, J. R. F., and Maisey, J. G. (1980). The Scottish lower Carboniferous shark *Onychoselache traquairi*. *Palaeontology (23)* 2, 363–374.

Didier, D. A. (1995). Phylogenetic Systematics of extant chimaeroid fishes (Holocephali, Chimaeroidei). *Am. Mus. Novit.* **3119**, 1–86.

Dingerkus, G. (1986). Interrelationships of orectolobiform sharks (Chondrichthyes: Selachii). *In* "Indo-Pacific Fish Biology: Proceedings of the Second International Conference on Indo-Pacific Fishes" (T. Uyeno, R. Arai, T. Taniuchi, and K. Matsuura, eds.), pp. 227–245. Ichthyological Society of Japan, Tokyo.

Duméril, A. (1865). "Histoire Naturelle des Poissons ou Ichthiologie Général," Vol. 1. Libraire Encyclopédique de Roret, Paris.

Dunn, K. A., and Morrissey, J. F. (1995). Molecular phylogeny of elasmobranchs. *Copeia* 3, 526–531.

Eschmeyer, W. (1990). "Catalog of Genera of Recent Fishes." California Academy of Sciences, San Francisco.

Farris, J. S. (1983). The logical basis of phylogenetic analysis. *In* "Advances in Cladistics" (N. I. Platnick and V. A. Funk, eds.), pp. 7–36. Columbia University Press, New York.

Farris, J. S. (1988). "Hennig86, Version 1.5," Program and documentation. Farris, Port Jefferson Station, NY.

Farris, J. S. (1991). Hennig defined paraphyly. *Cladistics* **7**, 297–304.

Fritsch, A. (1889). "Fauna der Gaskohle und der Kalksteine der Permformation Böhmens," Vol. 2. Prague.

Fritsch, A. (1895). "Fauna der Gaskohle und der Kalksteine der Permformation Böhmens," Vol. 3. Prague.

Gardiner, B. (1982). Tetrapod classification. *Zool. J. Linn. Soc.* **74**, 207–232.

Gardiner, B. (1993). Haematothermia: Warm-blooded amnoites. *Cladistics* **9**, 369–395.

Garman, S. (1885). *Chlamydoselachus anguineus* Garman—a living species of cladodont shark. *Bull. Mus. Comp. Zool.* **12**(1), 1–35.

Garman, S. (1913). The Plagiostomia. *Mem. Mus. Comp. Zool. Harv.* **36**(2), 1–515.

Garrick, J. A. F. (1960). Studies on New Zealand Elasmobranchii. Part 10. The genus *Echinorhinus*, with an account of a second species *E. cookei* Pietchmann, 1928. *Trans. R. Soc. N.Z.* **88**(1), 105–117.

Gaudin, T. J. (1991). A re-examination of elasmobranch monophyly and chondrichthyan phylogeny. *Neues Jahrb. Geol. Paleontol. Abh.* **182**(2), 133–160.

Gegenbaur, C. (1865). "Untersuchungen zur vergleichenden Anatomie der Wirbelthiere," Part 2. Engelmann, Leipzig.

Gill, T. (1861). Analytical synopsis of the order Squali; and revision of the nomenclature of the genera. *Ann. Lyceum Nat. Hist. N.Y.* **8**, 1–47.

Gill, T. (1893). Families and subfamilies of fishes. *Mem. Natl. Acad. Sci.* **6**(6), 125–138.

Glickman, L. S. (1967). Subclass Elasmobranchii (sharks). *In* "Fundamentals of Paleontology" (Y. A. Orlov and D. B. Obruchev, eds.), Vol. 2, pp. 292–352. Israel Program Sci. Transl., Jerusalem.

Goloboff, P. (1991). Random data, homoplasy and information. *Cladistics* **4**, 395–406.

Goodrich, E. S. (1909). Vertebrata Craniata. First fascicle: Cyclostomes and fishes. *In* "A Treatise on Zoology" (R. Lankester, ed.), Vol. 9, pp. 1–518. Black, London.

Goodrich, E. S. (1986). "Studies on the Structure and Development of Vertebrates" (reprint edition). University of Chicago Press, Chicago.

Gudger, E. W., and Smith, B. G. (1933). "The Natural History of the Frilled Shark, *Chlamydoselachus anguineus*," The Bashford Dean Memorial Volume: Archaic Fishes, Artic. 5, pp. 245–319. Am. Mus. Nat. Hist., New York.

Günther, A. (1870). "Catalog of the Fishes in the British Museum," Vol. 8. Br. Mus. (Nat. Hist.), London.

Hamdy, A. R. (1961). The development of the branchial arches of *Rhinobatus halavi*. *Publ. Mar. Biol. Stn. Ghardaqa, Red Sea* **11**, 205–213.

Hasse, J. C. F. (1879–1885). "Das Naturliche Syatem der Elasmobranchier auf Grundlage des Baues und der Entwicklung ihrer Wirbelsaule." Jena.

Herman, J., Hovestadt-Euler, M., and Hovestadt, D. C. (1987). Contributions to the study of comparative morphology of teeth and other relevant ichthyodorulites in living supraspecific taxa of chondrichthyan fishes. Part A: Selachii. No. 1: Order Hexanchiformes—Family: Hexanchidae. Commissural teeth. *Bull. Inst. R. Sci. Nat. Belg., Biol.* **57**, 43–56.

Herman, J., Hovestadt-Euler, M., and Hovestadt, D. C. (1989). Contributions to the study of the comparative morphology of teeth and other relevant ichthyodorulites in living supraspecific taxa of chondrichthyan fishes. Part A: Selachii. No. 3: Order Squaliformes Families: Echinorhinidae, Oxynotidae and Squalidae. *Bull. Inst. R. Sci. Nat. Belg., Biol.* **59**, 101–157.

Herman, J., Hovestadt-Euler, M., and Hovestadt, D. C. (1993). Contributions to the study of the comparative morphology of teeth and other relevant ichthyodorulites in living supraspecific taxa of chondrichthyan fishes. Part A: Selachii. No. 1b: Order: Hexanchiformes—Family Chlamydoselachidae. *Bull. Inst. R. Sci. Nat. Belg., Biol.* **63**, 185–256.

Herman, J., Hovestadt-Euler, M., and Hovestadt, D. C. (1994a). Contributions to the study of the comparative morphology of teeth and other relevant ichthyodorulites in living supraspecific taxa of chondrichthyan fishes. Addendum to Part A, No.1: Order: Hexanchiformes—Family Hexanchidae. Odontological results supporting the validity of *Hexanchus vitulus* Springer & Waller, 1969 as the third species of the genus *Hexanchus* Rafinesque, 1810, and suggesting intrafamilial reordering of the Hexanchidae. *Bull. Inst. R. Sci. Nat. Belg., Biol.* **64**, 147–163.

Herman, J., Hovestadt-Euler, M., and Hovestadt, D. C. (1994b). Contributions to the study of the comparative morphology of teeth and other relevant ichthyodorulites in living supraspecific taxa of chondrichthyan fishes. Part B: Batomorphii No. 1a: Order Rajiformes—Suborder Rajoidei—Family: Rajidae. *Bull Inst. R. Sci. Nat. Belg., Biol.* **64**, 165–207.

Holmgren, N. (1940). Studies of the head in fishes. Part 1. Development of the skull in sharks and rays. *Acta Zool. (Stockholm)* **21**, 51–257.

Holmgren, N. (1941). Studies of the head in fishes. Part 2. Comparative anatomy of the adult selachian skull with remarks on the dorsal fins in sharks. *Acta Zool. (Stockholm)* **22**, 1–100.

Holmgren, N. (1942). Studies on the head in fishes. Part 3. The phylogeny of elasmobranch fishes. *Acta Zool. (Stockholm)* **23**, 129–262.

Ida, H., Asahida, T., and Yano, K. (1986). Karyotypes of two sharks, *Chlamydoselachus anguineus* and *Heterodontus japonicus*, and their systematic implications. *In* "Indo-Pacific Fish Biology: Proceedings of the Second International Conference on Indo-Pacific Fishes" (T. Uyeno, R. Arai, T. Taniuchi, and K. Matsuura, eds.), pp. 158–163. Ichthyological Society of Japan, Tokyo.

Jaekel, O. M. J. (1895). Ueber die Organisation der Pleuracanthiden. *Sitzungsber. Ges. Naturforch. Freunde Berlin*, pp. 69–85.

Jaekel, O. M. J. (1906). Neue Rekonstruktionen von *Pleuracanthus sessilis* und von *Polyacrodus (Hybodus) hauffianus*. *Sitzungsber. Ges. Naturforsch. Freunde Berlin*, pp. 155–159.

Jordan, D. S., and Evermann, B. W. (1896). The fishes of north and middle America. *Bull.—U.S. Natl. Mus.* **47**(1), 1–1240.

Koken, E. (1889). Ueber *Pleuracanthus* Ag. oder *Xenacanthus*. *Sitzungsber. Ges. Naturforsch. Fruende Berlin*, pp. 77–94.

Linnaeus, C. (1758). "Systema Naturae. Vol. 1. Regnum Animale." Holmiae.

Lovejoy, N. (1996). Systematics of myliobatoid elasmobranchs: with emphasis on the phylogeny and historical biogeography of Neotropical freshwater stingrays (Potamotrygonidae: Rajiformes). *Zool. J. Linn. Soc.* (in press).

Lund, R. (1985). Stethacanthid elasmobranch remains from the Bear Gulch Limestone (Namurian E2b) of Montana. *Am. Mus. Novit.* **2828**, 1–24.

Maddison, W. P., and Maddison, D. R. (1992). "MacClade: Analysis of Phylogeny and Character Evolution. Version 3.04." Sinauer Assoc., Sunderland, MA.

Maisey, J. G. (1975). The interrelationships of phalacanthous selachians. *Neues Jahrb. Geol. Palaeontol. Monatsh.*, pp. 553–567.

Maisey, J. G. (1976). The middle Jurassic selachian fish *Breviacanthus* nov. gen. *Neues Jahrb. Geol. Palaeontol. Monatsh.* 432–438.

Maisey, J. G. (1977). The fossil selachian fish *Palaeospinax* Egerton 1872 and *Nemacanthus* Agassiz 1837. *Zool. J. Linn. Soc.* **60**, 259–273.

Maisey, J. G. (1980). An evaluation of jaw suspension in sharks. *Am. Mus. Novit.* **2706**, 1–17.

Maisey, J. G. (1982). The anatomy and interrelationships of Mesozoic hybodont sharks. *Am. Mus. Novit.* **2724**, 1–48.

Maisey, J. G. (1983). Cranial anatomy of *Hybodus basanus* Egerton from the Lower Cretaceous of England. *Am. Mus. Novit.* **2758**, 1–64.

Maisey, J. G. (1984a). Higher elasmobranch phylogeny and biostratigraphy. *Zool. J. Linn. Soc.* **82**, 33–54.

Maisey, J. G. (1984b). Chondrichthyan phylogeny: A look at the evidence. *J. Vertebr. Paleontol.* **4**, 359–371.

Maisey, J. G. (1985). Cranial morphology of the fossil elasmobranch *Synechodus dubrisiensis*. *Am. Mus. Novit.* **2804**, 1–28.

Maisey, J. G. (1986a). The Upper Jurassic hexanchoid elasmobranch *Notidanoides* n.g. *Neues Jahrb. Palaeontol. Geol. Abh.* **172**, 83–106.

Maisey, J. G. (1986b). Heads and tails: a chordate phylogeny. *Cladistics* **2**, 201–256.

Maisey, J. G. (1986c). Anatomical revision of the fossil shark *Hybodus fraasi* (Chondrichthyes: Elasmobranchii). *Am. Mus. Novit.* **2857**, 1–16.

Maisey, J. G. (1987). Cranial anatomy of the Lower Jurassic shark *Hybodus reticulatus* (Chondrichthyes: Elasmobranchii), with comments on hybodontid systematics. *Am. Mus. Novit.* **2878**, 1–39.

Maisey, J. G. (1989a). *Hamiltonichthys mapesi*, g. & sp. nov. (Chondrichthyes; Elasmobranchii), from the Upper Pennsylvanian of Kansas. *Am. Mus. Novit.* **2931**, 1–42.

Maisey, J. G. (1989b). Visceral skeleton and musculature of a late Devonian shark. *J. Vertebr. Paleontol.* **2**, 174–190.

Maisey, J. G. (1990). Selachii. *In* "Triassic Fishes from the Cassange Depression (R.P. de Angola)" (M. T. Antunes, ed.), Clencias da Terra, Número Especial, pp. 16–19, Universidad Nova de Lisboa, Lisbon.

Maisey, J. G. (1994). Ganthostomes (jawed vertebrates). *In* "Major Features of Vertebrate Evolution" (D. R. Prothero and R. M. Schoch, eds.), Short Courses Paleontol., No. 7, pp. 38–56. Paleontological Society, University of Tennessee, Knoxville.

Maisey, J. G., and Wolfram, K. (1984). *Notidanus*. *In* "Living Fossils" (N. Eldredge and S. Stanley, eds.), pp. 170–180. Springer-Verlag, New York.

McEachran, J. D., and Miyake, T. (1990). Phylogenetic interrelationships of skates: A working hypothesis (Chondrichthyes, Rajoidei). *In* "Elasmobranchs as Living Resources: Advances in the Biology, Ecology, Systematics, and the Status of the Fisheries" (H. L. Pratt, S. H. Gruber, and T. Taniuchi, eds.), NOAA Tech. Rep. NMFS 90, pp. 285–304. U.S. Department of Commerce, Washington, DC.

Miyake, T., McEachran, J. D., Walton, P. J., and Hall, B. K. (1992). Development and morphology of rostral cartilages in batoid fishes (Chondrichthyes: Batoidea), with comments on homology within vertebrates. *Biol. J. Linn. Soc.* **46**, 259–298.

Moy-Thomas, J. A. (1939). The early revolution and relationships of the elasmobranchs. *Biol. Rev. Cambridge Philos. Soc.* **14**, 1–26.

Müller, J., and Henle, F. G. J. (1838–1841). "Systematische Beschreibung der Plagiostomen." Veit, Berlin.

Nelson, G. J. (1989a). Cladistics and evolutionary models. *Cladistics* **5**, 275–289.

Nelson, G. J. (1989b). Phylogeny of major fish groups. *In* "The Hierarchy of Life" (B. Fernholm, K. Bremer, and H. Jörnvall, eds.), pp. 325–336. Elsevier, Amsterdam.

Nelson, J. S. (1994). "Fishes of the World," 3rd ed., Wiley, New York.

Nishida, K. (1990). Phylogeny of the suborder Myliobatidoidei. *Mem. Fac. Fish., Hokkaido Univ.* **37**, 1–108.

Nixon, K. (1992). "Clados Version 1.07." Program and reference. Nixon, Ithaca, NY.

Patterson, C. (1965). The phylogeny of the chimaeroids. *Philos. Trans. R. Soc. London, Ser. B* **249**(757), 101–219.

Pfeil, F. H. (1983). Zahnmorphologische Untersuchungen in rezenten und fossilen Haien der Ordnungen Chlamydoselachiformes und Echinorhiniformes. *Palaeoichthyologica* **1**, 1–315.

Pratt, H. L., Gruber, S., and Taniuchi, T., eds. (1990). "Elasmobranchs as Living Resources: Advances in the Biology, Ecology, Systematics and the Status of the Fisheries," NOAA Tech. Rep. NMFS 90. U.S. Department of Commerce, Washington, DC.

Rafinesque, C. S. (1810). "Caraterri di alcuni nuovi generi e nuove specie di animali e piante della Sicilia." Palermo.

Regan, C. T. (1906). A classification of selachian fishes. *Proc. Zool. Soc. London*, pp. 722–758.

Reif, W.-E. (1973). Morphologie und ultrastruktur des Hai-"Shmelzes." *Zool. Scri.* **2**, 231–250.

Reif, W.-E. (1974). *Metopacanthus* sp. (Holocephali) and *Palaeospinax egertoni* S. Woodward (Selachii) aus dem unteren Toarcium von Holzmaden. *Stuttg. Beitr. Natuurkd., Ser. B* **10**, 1–9.

Reif, W.-E. (1977). Tooth enameloid as a taxonomic criterion. 1. A new euselachian shark from the Rhaetic-Liassic boundary. *Neues Jahrb. Geol. Palaeontol., Monatsh.* **H9**, 565–576.

Reif, W.-E. (1978). Tooth enameloid as a taxonomic criterion 2. Is "*Dalatias*" *barnstonensis* Sykes, 1971 (Triassic, England) a squalomorphic shark? *Neues Jahrb. Geol. Palaeontol., Monatsh.* **H1**, 42–58.

Ridewood, W. C. (1921). On the calcification of the vertebral centra in sharks and rays. *Philos. Trans. R. Soc. London, Ser. B* **210**, 311–407.

Sanderson, M., and Donoghue, M. (1989). Patterns of variation in levels of homoplasy. *Evolution (Lawrence, Kans.)* **43**, 1781–1795.

Schaeffer, B. (1967). Comments on elasmobranch evolution. *In* "Sharks, Skates, and Rays." (P. W. Gilbert, R. F. Matthewson, and D. P. Rall, eds.), pp. 3–35. Johns Hopkins University Press, Baltimore.

Schaeffer, B. (1981). The xenacanth shark neurocranium, with comments on elasmobranch monophyly. *Bull. Am. Mus. Nat. Hist.* **169**(1), 3–66.

Schaeffer, B., and Williams, M. (1977). Relationships of fossil and living elasmobranchs. *Am. Zool.* **17**(2), 293–302.

Schweizer, R. (1964). Die Elasmobranchier und Holocephalen aus den Nusplinger Plattenkalken. *Palaeontographica, Abt. A* **123**, 58–110.

Seret, B. (1986). Classification et phylogénèse des Chondrichthyens. *Océnis* **12**, 161–180.

Shirai, S. (1992a). Phylogenetic relationships of the angel sharks, with comments on elasmobranch phylogeny (Chondrichthyes, Squatinidae). *Copeia* **2**, 505–518.

Shirai, S. (1992b). Identity of extra branchial arches of Hexanchiformes (Pisces, Elasmobranchii). *Bull. Fac. Fish., Hokkaido Univ.* **43**, 24–32.

Shirai, S. (1992c). "Squalean Phylogeny: A New Framework of 'Squaloid' Sharks and Related Taxa." Hokkaido University Press, Sapporo.

Smith, M. M., and Heemstra, P. (eds.) (1986). "Smith's Sea Fishes." Macmillan South Africa, Johannesburg.

Springer, V. G., and Gold, J. (1989). "Sharks in Question." Smithsonian Institution, Washington, DC.

Swofford, D. (1993). "Phylogenetic Analysis Using Parsimony, Version 3.1.1," Program and documentation. Smithsonian Institution, Washington, DC.

Thies, D. (1983). Jurazeitliche Neoselachier aus Deutschland und S. England. *CFS, Courier Forschungsinst. Senkenenberg* **58**, 1–116.

Thies, D. (1987). Comments on hexanchiform phylogeny (Pisces, Neoselachii). *Z. Zool. Syst. Evolutions forsch.* **25**, 188–204.

Thies, D. and Reif, W.-E. (1985). Phylogeny and evolutionary ecology of Mesozoic Neoselachii. *Neues Jahrb. Geol. Palaeontol. Abh.* **3**, 333–361.

White, E. G. (1936a). A classification and phylogeny of the elasmobranch fishes. *Am. Mus. Novit.* **837**, 1–16.

White, E. G. (1936b). The heart valves of the elasmobranch fishes. *Am. Mus. Novit.* **838**, 1–21.

White, E. G. (1937). Interrelationships of the elasmobranchs with a key to the order Galea. *Bull. Am. Mus. Nat. Hist.* **(74)** 2, 25–138.

Yeates, D. (1992). Why remove autapomorphies? *Cladistics* **8**, 387–389.

Young, G. C. (1982). Devonian sharks from south-eastern Australia and Antarctica. *Paleontology* **25**, 817–843.

Young, J. Z. (1981). "The Life of Vetebrates," 3rd ed. Oxford University Press (Clarendon), Oxford.

Zangerl, R. (1973). Interrelationships of early chondrichthyans. *In* "Interrelationships of Fishes" (H. P. Greenwood, R. Miles, and C. Patterson, eds.), pp. 1–14. Academic Press, London.

Zangerl, R. (1981). Chondrichthyes I. Paleozoic Elasmobranchii. *In* "Handbook of Paleoichthyology" (H. P. Schultze, ed.), Vol. 3A. Fischer, New York. 115 pp.

Appendix 1

A list of Recent elasmobranch specimens used in this study follows. Specimens are housed in the Department of Herpetology and Ichthyology, AMNH (except for the Jurassic *Palaeospinax priscus*, which is in the Department of Vertebrate Paleontology, AMNH VP7085). CS, cleared and stained preparations; SW, wet skeletons; AMNH, American Museum of Natural History.

Galeomorphii

Carcharhinus leucas, AMNH 55610 (CS); *Galeocerdo cuvieri*, AMNH 55042 (CS); *Prionace glauca*, AMNH 49517 (CS), AMNH 32701 (CS); *Scyliorhinus retifer*, AMNH 36777 (CS); *Scyliorhinus canicula*, AMNH 4120 (CS); *Mustelus cf. dorsalis*, AMNH 5579 (CS); *Mustelus laevis*, AMNH 4138 (CS); *Pseudotriakis microdon*, AMNH 71983 (CS); *Sphyrna lewini*, AMNH 12464 (CS), embryo; *Sphyrna tiburo*, AMNH 53030 (CS), gill and hyoid arches only; *Pseudocarcarcharias kamoharai*, AMNH 71984 (CS); *Cetorhinus maximus*, AMNH 38141 (SW), 57264 (SW); *Chiloscyllium punctatum*, AMNH 44013 (CS), gill and hyoid arches only; *Ginglymostoma cirratum*, AMNH uncatalogued (CS); *Hemiscyllium ocellatum*, AMNH 38151 (CS); *Heterodontus francisci*, AMNH 96795 (SW), AMNH 217862 (SW).

Squalea

Chlamydoselachus anguineus, AMNH 13814, 13815, 38149, 59327; *Heptranchias perlo*, AMNH (217861); *Hexanchus griseus*, AMNH 38185, 53092, 78170, 78172, 97382, 98254; *Hexanchus vitulus*, AMNH 33474 (SW); *Notorynchus cepedianus*, AMNH 55711, 55712, 55748; *Centroscymnus coelolepis*, AMNH 78237 (SW), chondrocranium only; *Centroscymnus crepidator*, AMNH 55521 (CS); *Etmopterus princeps*, AMNH 78343 (SW), chondrocranium only; *Dalatias licha*, AMNH 19446 (CS); *Squaliolus laticaudus*, AMNH 15731 (CS); *Squalus acanthias*, AMNH 38136 (2 specimens) (CS), 55513 (SW), chondrocranium only; *Pliotrema warreni*, AMNH 44012 (CS), gill and hyoid arches only; *Pristiophorus cirratus*, AMNH 30169 (CS); *Squatina californica*, AMNH 55686 (CS); *Anoxypristis cuspidata*, AMNH 3268 (CS); *Pristis pristis*, AMNH 44011 (CS), gill arches only; *Pristis zijsron*, AMNH 44048 (CS), gill arches only; *Platyrhina sinensis*, AMNH 44055 (CS); *Rhinobatos percellens*, AMNH 55622 (CS); *Narcine brasiliensis*, AMNH 2488 (CS); *Torpedo torpedo*, AMNH 4128 (CS).

Holocephali

Callorhynchus capensis, AMNH 36943 (CS); *Hydrolagus colliei*, AMNH 40801 (CS), AMNH 44129 (2 specimens) (CS); *Hydrolagus alberti*, AMNH 39108 (CS); *Chimaera monstrosa*, AMNH 55040 (CS); *Harriotta raleighana*, AMNH 55039 (CS).

Appendix 2

A list of characters and their states coded for the matrix in Appendix 3 follows:

1. Rostral process ("rostrum") a single (ventral) projection, formed solely by the trabecula (0); tripodal rostral cartilages (1); absence of rostral process (2).
2. Open lateral line canal (0); closed lateral line canal (1).
3. Precerebral fossa absent or uncovered (0); precerebral fossa roofed (1).
4. Ethmoidal region of neurocranium not down-curved ethmoidal area (1).
5. Ethmoidal canal absent (0); present (1); present with exit on dorsal surface of nasal capsules (2).
6. Subnasal fenestra absent (0); present (1).
7. Superficial ophthalmic nerve exists neurocranium through prootic foramen (0); separate foramen for superficial ophthalmic nerve (1).
8. Ectethmoid process absent (0); present (as in Carvalho and Maisey, 1996) (1); present as "unchondrified antorbital cartilage" (2); antorbital cartilage (3).
9. Orbital articulation absent (0); present (1).
10. Suborbital shelf present (0); absent (1).
11. Basitrabecular process absent (0); present (1).
12. Post-orbital articulation absent (0); present but without complete articular facet (1); present with complete articular facet (2).
13. Hyomandibular fossa in the posterior part of otic region (0); hyomandibular fossa anteriorly situated in otic region (1).
14. Hyomandibular fossa not composed of two horizontally adjacent concavities (0); articular fossa for hyomandibula composed of two horizontally arranged concavities (1).
15. Concavity ventral to hyomandibular fossa absent (0); present (1).
16. Basioccipital fovea present (0); absent (1).
17. Occipital hemicentrum present (0); absent (1).
18. Obliquus inferior β absent (0); present (1).
19. Absence of extrabranchial cartilages on hyoid arch (0); extrabranchials on dorsal side only (1); extrabranchials present on entire arch (2).
20. Sixth branchial unit absent (0); present (1).
21. Seventh branchial unit absent (0); present (1).
22. Anterior basibranchials present (0); absent (1).
23. Hypobranchial bar absent (0); present (1).
24. Posteriormost elements of dorsal gill arches not completely fused (0); complete fusion of posterior dorsal gill arch elements ("gill pickaxe") (1).
25. Adductor mandibulae composed of dorsal and ventral components (0); composed of "lateralis i and ii" (1).
26. Suborbitalis muscle originating on suborbital surface (0); originating on interorbital wall (1); originating on posterior aspect of nasal capsule (2); originating on ectethmoid process (3); originating on upper preorbital wall (4).
27. Suborbitalis inserting on adductor mandibulae and/or dermal tissues around mouth corner (0); suborbitalis inserting directly on mandibula anterior to adductor mandibulae (1); suborbitalis inserting on mandibula via elongate tendon (2).
28. Adductor mandibulae superficialis absent (0); present (1).
29. Adductor mandibulae superficialis inserting on subcutaneous tissue around the eye opening (0); inserting on ectethmoid process (1); inserting on posterior aspect of nasal capsule (2); inserting on mandibula (3).
30. Levator palatoquadrati muscle not separated from spiracularis (0); separated (1).
31. Constrictor hyoideus dorsalis inserting on hyomandibula (0); inserting on palatoquadrate (1); inserting partly on both (2).
32. Arcualis dorsalis muscle with single origin (0); originating from two heads (1).
33. Subspinalis muscle absent (0); present (1).
34. Interpharyngobranchialis muscles present (0); absent (1).
35. Genio–coracoideus originating on fascia covering coraco–arcualis (rectus cervicis) (0); genio–coracoideus originating from coracoid or pericardial membrane (1).
36. Coraco-hyomandibularis muscle absent (0); present and originating on the coraco-arcualis muscle (1).
37. Right and left coracoids separate (0); fused (1).
38. Single facet for the articulation of pectoral basals (0); separate articular condyle for pectoral metapterygium present (1); all three basals with separate condyles for articulation with scapulocoracoid (2).
39. Tribasal pectoral fin (0); propterygium and mesopterygium fused (1); mesopterygium and

metapterygium fused (2); all three basals fused (3).
40. Pectoral propterygium articulating with radials (0); propterygium not contacting radials (1).
41. Pectoral propterygium not elongated anteriorly (0); propterygium extended forward (1).
42. Pectoral metapterygium without short proximal segment (0); metapterygium contacts short proximal segment (1).
43. Ventral marginal cartilage of clasper not elongated (0); elongated (1).
44. Notochord constricted along entire vertebral column (0); constricted only anteriorly and posteriorly, or primarily absent (1).
45. Three columns of heart valves (0); four columns (1).
46. Supraneurals not enlarged anterior to second dorsal fin (0); enlarged supraneurals preceding second dorsal fin (1); enlarged supraneurals also in abdominal and precaudal regions (2).
47. Hemal spines not forming complete arches in precaudal region (0); complete hemal arches in precaudal region (1).
48. Precaudal hemal process not elongated (0); elongated similar to lower caudal skeleton (1).
49. Dorsal fin endoskeleton composed of basal cartilage with a finspine (0); dorsal fin endoskeleton with basal cartilage and no finspine (1); dorsal endoskeleton composed of radials only (no finspine) (2).
50. Two dorsal fins (0); one dorsal fin (1).
51. Anal fin skeleton composed of basal cartilage and radials (0); composed of radials only (1).
52. Inclinator dorsalis muscle originating on epaxial musculature (0); originating from vertebrae (1).
53. Ventral tail muscle bundle not developed (0); well-developed (1).
54. Flexor caudalis muscle restricted to hemal processes of caudal fin (0); flexor caudalis extending to origin of caudal fin (1); flexor caudalis extending anterior to origin of caudal fin (2).
55. Nasoral groove absent (0); present (1).
56. Mesonarial flap present (0); absent (1).
57. Circumnarial folds and grooves absent (0); present (1).
58. Diagonal tooth rows (0); linearly arranged tooth rows (1); linearly arranged rows with teeth overlapping each other (2).
59. Teeth devoid of "aprons" (0); with "aprons" (1).
60. Spiracle valve absent (0); present (1).
61. Luminous organs absent from body surface (0); present (1).
62. Upper precaudal pit absent (0); present (1).
63. Longitudinal precaudal keel absent (0); present (1).

64. Intestinal spiral valve (0); intestinal ring valve (1).
65. Basihyal elongated (0); basihyal as broad as long (1).
66. Anal fin present (0); absent (1).
67. Pharyngobranchial blade absent (0); present (1).

Appendix 3

Matrix of characters and taxa derived from Appendix 2. This matrix was used for the phylogenetic analyses presented in this paper.

Outgroup
0000000000 0000000??0 0000?????? ??????0000 0000?00000 0??????00? ?????00

Heterodontus
2101000000 0010000010 0000041100 0000001000 0000000000 0000101010 0000001

Orectolobiformes
0101001000 0010000010 0001?410?1 0000001000 0010000020 100010?010 000?001

Lamniformes
1100001000 0010000010 0001020 0?0 0000001000 0010000020 1000000000 0??1001

Carcharhiniformes
1100001000 0010000010 0001020 0?0 0000001000 0010000020 1000000000 0000001

Chlamydoselachus
0000000111 1100000001 00000300?0 0000000001 0001100021 1000000000 0000100

Hexanchus
0100110111 1200101011 0000030110 1100000001 0101101011 0000000100 0000000

Notorynchus
0000000111 1200001011 1000030110 1100000001 0101101011 0000000100 0000000

Heptranchias
0100110111 1200101011 1000030110 1100000001 0101101011 0000000100 0000100

Echinorhinus
0000100111 1000001020 0001030130 0100001030 0001001010 ?000010110 0000010

Etmopteridae
0100110011 1000000020 0001010100 2100101000 0000001000 ?000010211 1000010

Somniosidae
0100110011 1000000020 0011011100 2100101010 0000001010 ?000010211 1000010

Oxynotus
0100100011 1000000020 0011011100 2100101010 0000001000 ?000010211
1000010

Dalatiinae
0100110011 1000000020 0011011100 2111101030 0000001010 ?000010211
1000010

Euprotomicrininae
0100110011 1000000020 0011011100 2100101030 0000001010 ?000010211
1000010

Centrophoridae
0100110011 1000000020 0001010100 2100001020 0000001000 ?000010211
0000010

Squalidae
0100110011 1000000020 0001000100 2100001000 0000011100 ?001010211
0110010

Squatina
0100200011 1000010020 00010??100 0100?01100 1000011110 ?112110010
0010010

Pristiophoridae
0110200211 1001011120 0101102100 2100011200 0000021110 ?102010011
0010010

Rajiformes
0110200300 0001011120 0101102121 0011011200 1000021110 ?112010011
0010010

4

Interrelationships of the Batoid Fishes (Chondrichthyes: Batoidea)

JOHN D. McEACHRAN *and*
KATHERINE A. DUNN

Department of Wildlife and Fisheries Sciences
Texas A&M University
College Station, Texas 77843

TSUTOMU MIYAKE

Department of Biology
University of Dalhousie
Halifax, Nova Scotia, B3H 411, Canada

I. Introduction

In his review Compagno (1973) divided living elasmobranchs (Euselachii) into four superordinal taxa, suggested that these taxa constitute a monophyletic group or a monophyletic group with the inclusion of Mesozoic hybodontoid and ctenacanthoid sharks, and provided a wealth of anatomical data that greatly stimulated subsequent research on this taxon. One of his four superorders, Batoidea, comprises the torpediniforms (electric rays), pristiforms (sawfishes), rhinobatoids (guitarfishes), rajoids (skates), and myliobatiforms (stingrays). He suggested that this taxon is most closely related to either Squatinomorphii or to Squalomorphii of his three other major groups of euselachians. He based his inference on similarities and differences among the four superorders rather than on shared derived character states. Compagno (1973) divided the batoids into four orders (Torpediniformes, Pristiformes, Rajiformes, and Myliobatiformes) and suggested that a subgroup of the rajiforms, rhinobatoids, is ancestral to the other orders, although he stated that the relationships of the torpediniforms are rather obscure. He provided anatomical descriptions for the four orders and their suborders and families and used these character states to suggest phylogenetic relationships within the order and suborders.

However, relationships were based on a combination of derived and primitive character states.

In 1977, Compagno updated his phylogenetic hypothesis with an expanded character base, new fossil shark information, and an attempt at cladistic analysis. Composition of the four superorders remained the same, but he added a number of Mesozoic and Paleozoic sharks to his Euselachii (= Neoselachii). Little progress was made on elucidating the interrelationships of the four major taxa but he did revise his hypothesis of batoid interrelationships. Based on the retention of primitive gill arch structure (retention of the ceratohyal–hyomandibular connection in narkids) he considered the torpediniforms to be sister to the remaining taxa. Also, based on the retention of primitive pectoral girdle and vertebral structure (lack of direct union of the scapulocoracoid with the vertebral column and shortness of the synarchual in *Pristis*) he considered the pristiforms to be sister to rajiforms and myliobatiforms. However, both of these suppositions were based on the retention of primitive character states rather than the acquisition of derived character states.

Heemstra and Smith (1980) proposed a phylogenetic hypothesis for batoids in their description of a new stingray taxon (Hexatrygonidae). They considered pristiforms to be sister to the remainder of the batoids and torpediniforms sister to rajiforms and

Copyright © 1996 by Academic Press, Inc.
All rights of reproduction in any form reserved.

myliobatiforms. Maisey (1984) disagreed with Heemstra and Smith's placement of pristiforms and provisionally accepted Compagno's (1977) hypothesis. He noted that the hypobranchial skeleton of *Pristis* (Pristiformes) is considerably more specialized than that of primitive Jurassic batoids (†*Spathobatis* and †*Belemnobatis*), *Pristiophorus* (putative sister to batoids), and *Torpedo* (Torpediniformes). In these taxa all of the hypobranchial cartilages articulate with the basibranchial copula, whereas in *Pristis* and remaining batoids, other than *Torpedo* and rhinobatoids, the hypobranchial cartilages do not individually articulate with the basibranchial plate (Miyake and McEachran, 1991). Maisey (1984) also suggested that pristiophoroids and squatinoids are likely sister groups of batoids but he did not formally propose this in his cladogram.

Nishida (1990) constructed a cladogram for batoids to polarize character states within myliobatoids. According to his analysis, pristiforms are sister to the remaining batoids, rhinobatoids are polyphyletic, with *Rhynchobatus* and *Rhina* sister to the remaining taxa, and torpedinoids are sister to the remaining rhinobatoids, rajoids, and myliobatoids. Within the clade containing rhinobatoids (other than *Rhynchobatus* and *Rhina*), rajoids, and myliobatoids, the rhinobatoid genera, *Rhinobatos* and *Aptychotrema*, form a trichotomy with the remainder of the taxa. The rhinobatoid genera *Zapteryx*, *Trygonorrhina*, and *Platyrhina* + *Platyrhinoidis* form a polytomy with a clade comprised of rajoids and myliobatiforms. In Nishida's cladistic analysis of myliobatiforms, *Plesiobatis* and *Hexatrygon* form a trichotomy with the remaining myliobatiforms, which in turn form two basal clades. One of these basal clades includes *Urolophus* + *Urotrygon* as sister to an unresolved trichotomy of *Potamotrygon*, *Taeniura*, and *Dasyatis* + *Himantura*. The other basal clade includes *Gymnura* as sister to the clade of pelagic stingrays (*Myliobatis*, *Aetobatus*, *Aetomylaeus*, *Rhinoptera*, *Mobula*, and *Manta*).

Lovejoy (in press) proposed a phylogenetic hypothesis for myliobatiforms that further resolved their relationships. According to his analysis, *Plesiobatis* and *Hexatrygon* form a trichotomy with the remaining myliobatiforms; Urolophidae are paraphyletic, with *Urolophus* sister to the remainder of the taxon; *Taeniura*, Amphi-American *Himantura*, and Potamotrygonidae form a clade that is sister to the remaining myliobatiforms; and *Dasyatis* and Indo-Pacific *Himantura* form a trichotomy with a clade comprising *Gymnura*, *Myliobatis*, *Aetobatus*, *Rhinoptera*, and *Mobula*.

From this brief survey it is evident that considerable progress has been made in elucidating the interrelationships within the batoids since Compagno's (1973) review of the euselachians. However, the interrela-

tionships of several of the taxa remain ambiguous. Both torpediniforms and pristiforms have been proposed as sister to the remaining batoids. Rhinobatoids are considered to be polyphyletic, but there is disagreement as to the composition of the taxon. There is a lack of consensus regarding the sister to rajoids, myliobatiforms, and potamotygonids. Compagno (1973) complained that the lack of resolution of the interrelationships was largely due to the lack of detailed anatomical data for a wide range of batoid taxa. However, as a result of several detailed comparative anatomical studies over the last 2 decades (Chu and Wen, 1980; Rosa, 1985; Rosa *et al.*, 1988; Miyake, 1988; Nishida, 1990; Miyake and McEachran, 1991; Miyake *et al.*, 1992a,b; Lovejoy, in press) and phylogenetic studies of taxa more basal to the batoids (Maisey, 1984; Shirai, 1992a,b), the relevant anatomical states are fairly well known and the outgroups of the batoids have been better elucidated. The exception is that the rhinobatoid taxa remain poorly known. The purpose of the this study is to survey anatomical variation of the majority of rhinobatoid genera and to further elucidate the interrelationships of the batoids.

II. Methods and Materials

Representatives of most of the genera of batoids were examined for this study in addition to representatives of the first two outgroups, Pristiophoridae and Squatinidae (see Appendix 1). Rajidae was treated as a single OTU because it is a highly corroborated monophyletic group (McEachran, 1984; McEachran and Miyake, 1990; Jacob *et al.*, 1994; McEachran and Konstantinou, 1996). Shirai (1992a,b) was followed in the selection of outgroups. Specimens were cleared and stained according to Dingerkus and Uhler (1977) and/or radiographed or dissected to reveal aspects of the skeletal and muscular systems. Anatomical terminology followed Miyake (1988). Data were also analyzed from several studies (Chu and Wen, 1980; Rosa, 1985; Rosa *et al.*, 1988; Miyake, 1988; Nishida, 1990; Miyake and McEachran, 1991; Miyake *et al.*, 1992a,b; Shirai, 1992a,b; Lovejoy, in press). In nearly all cases, observations based on literature sources were verified with independent observations. In developing the character matrix, the character matrices of Nishida (1990) and Lovejoy (in press) were scrutinized, and when appropriate their characters were added to those discovered in this study. However, their autapomorphic characters and characters that were ambiguous or vague were excluded from the analysis. According to Carpenter (1988) autapomorphies inflate consistency indexes and thus should not be included in phylogenetic analyses. Justification for excluding ambiguous and vague characters is provided in the discussion of characters. Nearly all of the characters used were binary; the exceptional three to five character states were run unordered to reduce subjectivity of the analysis. The character matrix (Appendix 2) was analyzed using parsimony via PAUP version 3.1.1 (Swofford, 1993). The heuristic search option was used because of the large number of taxa. Starting trees were obtained via 10 replications of a random stepwise addition. The branch swapping algorithm was tree bisection–reconstruction (TBR). Strength of nodes was analyzed using Bremer Decay indexes (Bremer, 1994), and a total support index was obtained for the entire tree. The total support index provides a measure of tree stability in terms of support resolution. The data were then reweighted by successive approximation (Farris, 1969; Carpenter, 1994) using the retention index values to attempt to choose among equally parsimonious solutions.

For the analysis, the major taxa of batoids (torpediniforms, pristiforms, rajiforms, and myliobatiforms) are treated as orders and the two major taxa of rajiforms (rhinobatoids and rajoids) are treated as suborders following Compagno (1973). *Rhynchobatus* and *Rhina* are removed from the rhinobatoids and treated as *incertae sedis*, equivalent to the major taxa following Nishida (1990). The taxon *Himantura* is applied exclusively to the two amphi-American species, and the Indo-Pacific species are lumped with *Dasyatis* following Lovejoy (in press). Characters and character states are listed in Appendix 3.

III. Results

Batoidea are a highly corroborated monophyletic group based on the following synapomorphies: (1) Cornea is attached directly to the eye; upper eyelid is absent (Compagno, 1973; Maisey, 1984; Shirai, 1992b). (2) Palatoquadrate lacks an articulation with the neurocranium (Compagno, 1973; Maisey, 1984; Shirai, 1992b). (3) Pseudohyal is present in the hypobranchial skeleton, and it partially or totally replaces the ceratohyal (Compagno, 1973; Maisey, 1984; Miyake and McEachran, 1991; Shirai, 1992b). (4) Last ceratobranchial cartilage of the hypobranchial skeleton articulates with the medial surface of the scapulocoracoid (Maisey, 1984; Nishida, 1990; Shirai, 1992b). (5) Cervical vertebral centra, just posterior to neurocranium, are fused into a tube (synarchual) (Compagno, 1973; Maisey, 1984; Shirai, 1992b). (6). Suprascapulae are joined over the vertebral column and articulate with the vertebral column or synarchual (Compagno, 1973;

Maisey, 1984; Shirai, 1992b). (7) Antorbital cartilage directly or indirectly joins the propterygium of the pectoral girdle to the nasal capsule of the neurocranium (Holmgren, 1940; Compagno, 1973; Maisey, 1984; Shirai, 1992b). (8) Levator and depressor rostri muscles of the hyoid muscle plate are present but the levator rostri is lost secondarily in the myliobatiforms (Edgeworth, 1935; Miyake 1988; Miyake *et al.*, 1992a,b; Shirai, 1992b). Maisey (1984) included two other synapomorphies for batoids: pectoral fins expanded, with anterior lobes continuous and antorbital cartilages present. However, these characters were not judged to be independent of character 7 above. Shirai (1992b) provided several additional characters, but some of these, like character 8, are not universally present within batoids.

Within batoids a number of character complexes exhibit variation that are thought to provide phylogenetic information. These systems are presented in the following order: external morphological structures, lateral line structures, skeletal structures, and cephalic and branchial musculature. Within each complex, characters are described along the anterior–posterior body axis. Character numbers, followed by character states, are included parenthetically after the description of each character state.

A. External Morphological Structures

Pelagic stingrays (*Myliobatis, Aetobatus, Aetomylaeus, Rhinoptera, Mobula,* and *Manta*) are unique among the batoids and outgroups in possessing single or paired cephalic lobes anterior to the neurocranium that are supported by the pectoral girdle and, in several taxa, by elements of the rostral cartilage (Bigelow and Schroeder, 1953; Nishida, 1990; Miyake *et al.*, 1992b). In *Aetobatus, Aetomylaeus, Rhinoptera, Mobula,* and *Manta* the anterior extension of the pectoral fin rays is separated from the remainder of the fin rays. The cephalic lobes or prehensile cephalic fins are paired and laterally located in *Rhinoptera, Mobula,* and *Manta*. The single continuous lobe (*Myliobatis*) (9,1), the single discontinuous lobe (*Aetobatus* and *Aetomylaeus*) (9,2), and the paired discontinuous lobes or fins (*Rhinoptera, Mobula,* and *Manta*) (9,3) are considered separate derived states.

The anterior nasal lobe varies from a small process that extends partially across the naris to a nasal curtain formed by the expansion and fusion of the right and left lobes across a broad-to-narrow internarial space. In the outgroup *Squatina* the anterior lobe is specialized as a series of barbels that extend in front of the terminal mouth. In *Pristiophorus* and the squaliforms [third outgroup according to Shirai (1992a)] the ante-

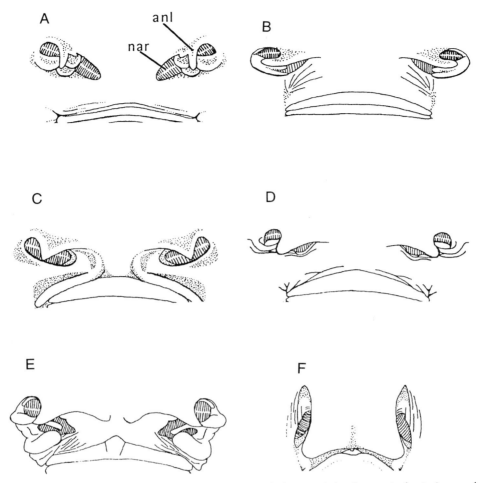

FIGURE 1 Ventral aspect of head showing nares and shape and development of anterior nasal lobe. (a) *Rhinobatos percellens* MCZ 40025; (b) *Zapteryx zyster* TCWC uncataloged; (c) *Platyrhina sinensis* CAS 15919; (d) *Platyrhinoidis triseriata* CAS 31248; (e) *Zanobatus schoenleinii* USNM 40025; (f) *Dasyatis sabina* TCWC 5824.01. Abbreviations: anl = anterior lobe, nar = nare.

rior lobe is poorly developed, extends only partially across the naris, and does not completely cover the medial aspect of the naris. The lobe is also poorly developed in pristiforms, *Rhynchobatus*, *Rhina*, and some rhinobatoids (*Aptychotrema* and *Rhinobatos*) (Fig. 1a). The anterior lobe is expanded medially to cover the medial half of the naris in all but one of the remaining rhinobatoid genera (*Zapteryx*, *Platyrhina*, *Platyrhinoidis*, and *Zanobatus*) (Fig. 1b, c, d, e). In the remaining rhinobatoid genus (*Trygonorrhina*) and the other batoids the anterior lobe is expanded and connected with its antimere across a moderately broad-to-narrow internarial space to form a nasal curtain (Fig. 1f). In all these cases, except the myliobatiform genera *Plesiobatis* and *Hexatrygon*, the nasal curtain extends to or near the mouth. The medial expansion of the naris (*Zapteryx*, *Platyrhina*, *Platyrhinoidis*, and *Zanobatus*) (10,1), the nasal curtain that falls short of the mouth

(*Plesiobatis* and *Hexatrygon*) (10,2), and the nasal curtain that reaches to or near the mouth (torpediniforms, *Trygonorrhina*, rajoids, and most myliobatiforms) (10,3) are considered separate derived character states of the anterior nasal lobe. Nishida (1990) interpreted this character differently. He considered the completely united nasal curtain that reaches the mouth to be derived and all other states (nasal curtain absent, incompletely united, or failing to reach the mouth) as the primitive state; he limited the derived state to *Trygonorrhina* and the myliobatiforms exclusive of *Hexatrygon* and *Plesiobatis*.

Urobatis and *Urotrygon* possess a tentacle on the inner margin of the spiracle during embryonic development (Miyake, 1988). This structure does not occur in any of the other batoids or in the outgroups and thus is considered a derived state (Miyake, 1988; Lovejoy, in press) (11,1).

The caudal fin of batoids varies from sharklike to absent. The outgroups possess a well developed caudal fin supported by radials and ceratotrichia. Lack of a caudal fin supported by radials is considered to be the derived state (12,1) and occurs in all myliobatiforms with the exception of *Plesiobatis*, *Hexatrygon*, *Urolophus*, *Urobatis*, and *Urotrygon*. Lovejoy (in press) also considered the presence of a dorsal fin a derived character for an assemblage of stingrays (*Aetoplatea*, *Myliobatis*, *Aetobatus*, *Aetomylaeus*, *Rhinoptera*, *Mobula*, and *Manta*). However, because dorsal fins are present in the outgroups, in the other major taxa of batoids, and are variably present in *Urolophus*, it is difficult to code this character state. If *Urolophus* with a dorsal fin (often classified as *Trygonoptera*) is sister to the remaining myliobatiforms, presence of a dorsal fin would be a plesiomorphic state.

Serrated tail spines occur exclusively in myliobatiforms and are missing in only three species within genera that also include species with spines (Nishida, 1990). Presence of the spine is considered a derived character state for myliobatiforms (13,1), although Nishida (1990) considered the character ambiguous.

Nishida (1990) considered the position of the posterior corner of the pectoral fin in relation to the origin of the pelvic fin and presence or absence of a bilobed caudal fin as characters distinguishing pristiforms, *Rhynchobatus*, and *Rhina* from the remainder of the batoids. Extension of the posterior corner of the pectoral fin to the origin of the pelvic fin was considered a derived state. However, this character was judged to be ambiguous in the present study because the posterior extension of the pectoral fin overlaps the pelvic fin origin in some squatinid species, and this character is variable in torpediniforms, rhinobatoids, and rajoids. Nishida (1990) considered the absence of a bilobed caudal fin a derived character state, but, again, this character state was judged to be ambiguous in our study because bilobed caudal fins are variably developed in squaliforms (third outgroup), absent in pristiophorids, and present in one of the two genera of pristiforms (*Anoxypristis*). Squatinids have a bilobed tail of sorts, but the tail of this taxon is unique among the neoselachians in being hypocercal. An argument could be made for considering the bilobed tail of pristiforms, *Rhynchobatus*, and *Rhina* a derived state.

B. Ventral Lateral Line Canals

Batoids exhibit considerable variation in their cephalic and trunk lateral line patterns. Torpediniforms entirely lack a cephalic lateral line system on their ventral surface but it is present in the remaining neoselachians (Chu and Wen, 1980). Absence of the ven-

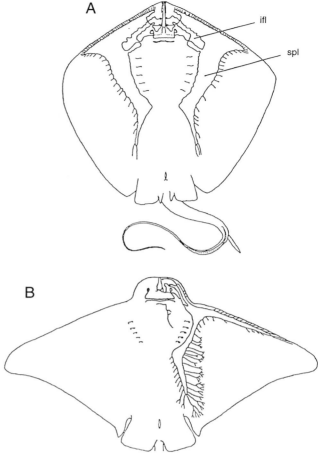

FIGURE 2 Ventral aspect of disc showing lateral line canals. (a) *Dasyatis brevis*; (b) *Myliobatis aquila*. Abbreviations: ifl = infraorbital loop, spl = subpleural loop. After Garman (1888).

tral lateral line in torpediniforms is considered a derived character state (14,1).

The infraorbital loop, formed by the suborbital and infraorbital canals and located lateral to the mouth and anterior to the first gill arch (Fig. 2a), is found only in myliobatiforms (Chu and Wen, 1980) and is considered a derived character state (15,1). The state of this character is unknown for *Hexatrygon* (15,?). On the ventral surface of the disc the infraorbital canal gives rise to the hyomandibular canal that forms variously sized and shaped subpleural loops in batoids (Garman, 1888; Chu and Wen, 1980). The subpleural loop occupies much of the lateral disc, lateral and posterior to the gill slits in myliobatiforms. In *Dasyatis* and *Gymnura* the subpleural loop closely follows the anterior margin of the disc before abruptly reversing direction and running disto-medially and lateral to the gill slits (Chu and Wen, 1980; Lovejoy, in press) (Fig. 2a). In pelagic myliobatiforms (*Myliobatis*, *Aetoba-*

tus, Rhinoptera, and *Mobula*) the reverse in direction of the subpleural loop near the lateral extreme of the disc is even more extreme and the posterior section of the canal runs parallel to the anterior section to near the level of the gill slits (Chu and Wen, 1980; Lovejoy, in press) (Fig. 2b). The character state is unknown for *Hexatrygon*. The conditions in *Dasyatis* and *Gymnura* (16,1) and in the pelagic myliobatiforms (16,2) are treated as separate derived states following Lovejoy (in press). The subpleural component of the hyomandibular canal bears lateral tubules in myliobatiforms and these are dichotomously branched in *Urobatis* and *Urotrygon* but not in other myliobatiforms (Garman, 1888; Lovejoy, in press). The condition in *Urobatis* and *Urotrygon* is considered a derived character state (17,1) following Lovejoy (in press).

The cephalic lateral line forms an abdominal canal on the coracoid bar in *Rhynchobatus, Rhinobatos, Trygonorrhina, Zapteryx, Platyrhinoidis, Platyrhina,* and *Zanobatus* (Garman, 1888; Chu and Wen, 1980; this study) (Fig. 3). Chu and Wen did not illustrate an abdominal canal for *Platyrhina sinensis* but one was found for this species in our study. This canal also occurs in several rajoid taxa [*Raja (Dipturus)* and related taxa]. Abdominal canals do not occur in other batoids or in the outgroups, thus they are considered a derived character state (18,1). Rajoids are considered polymorphic for this state (18,01).

Batoids possess scapular canals that are derived from the trunk lateral line canal, and in myliobatiforms the scapular canals form scapular loops on the dorsal surface of the scapular region (Garman, 1888; Chu and Wen, 1980; Lovejoy, in press) (Fig. 4). According to Garman (1888) the scapular loop is absent in one of the two species of *Myliobatis* that he examined and in *Aetobatus*. But according to Chu and Wen (1980) they are present in both *Myliobatis* and *Aetobatus*. Scapular loops are considered to be a derived character state (19,1). Character state is unknown for *Hexatrygon* (19,?).

C. Skeletal Structures

The rostral cartilage varies considerably in batoids (Miyake *et al.*, 1992b). In *Platyrhina* and *Platyrhinoidis* it is abbreviated and fails to reach the tip of the snout (Miyake *et al.*, 1992b) (Fig. 5) and this condition is unique for the batoids and outgroups (20,1). The rostral cartilage is either vestigial or completely lacking in *Zanobatus* and myliobatiforms (Fig. 6) (Compagno, 1973, 1977; Nishida, 1990; Miyake *et al.*, 1992b) and this is considered to be an additional derived state (20,2). The distal section of the rostral cartilage is occasionally present in a vestigial state at the tip of the

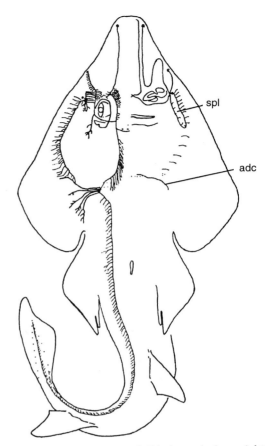

FIGURE 3 Lateral line canals of *Rhinobatos planiceps;* right side of figure is ventral aspect of body, and left side of figure is dorsal aspect. Abbreviations: spl = subpleural loop, adc = abdominal canal. After Garman (1888).

snout between the segmented propterygia in embryos and adults of *Urobatis, Urotrygon, Dasyatis, Potamotrygon,* and *Plesiotrygon* (Rosa, 1985; Rosa *et al.*, 1988; Miyake, 1988; Miyake *et al.*, 1992b). Miyake *et al.* (1992b) incorrectly identified this structure as a rostral appendix.

Each side of the tip of the rostral cartilage is ornamented with a thin sheet of cartilage (rostral appendix) in *Rhynchobatus, Rhina, Rhinobatos, Zapteryx, Trygonorrhina,* and Rajidae (Fig. 7). The appendix, formed *de novo* on either side of the rostral cartilage, is carried forward with anterior growth of the rostral cartilage (Holmgren, 1940; Miyake *et al.*, 1992a) and is considered a derived character state (21,1). Both *Platyrhina* and *Platyrhinoidis* have structures attached to the lateral aspects of their abbreviated rostral cartilage (Fig. 5), but the origin and homology of these structures are uncertain; these species are coded as unknown for this state (21,?).

In *Rhinoptera, Mobula,* and *Manta* dorsolateral com-

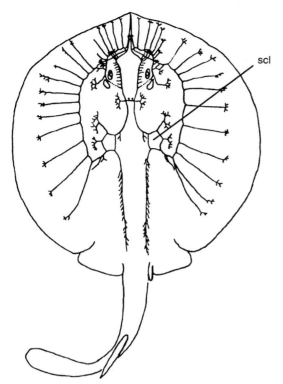

FIGURE 4 Dorsal aspect of *Urobatis halleri* showing lateral line canals. Abbreviation: scl = scapular loop. After Garman (1888).

ponents of the nasal capsule (the lamina orbitonasalis component) form a pair of anterior projections on the dorsomedial area of the nasal capsules (Miyake, *et al.*, 1992b). These structures, which support the cephalic

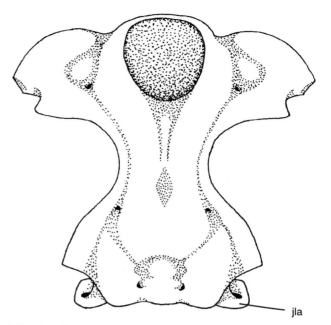

FIGURE 6 Dorsal view of neurocranium of *Zanobatus schoenleinii* USNM 222120. Abbreviation: jla = jugal arch.

lobes or fins, are unique to these three taxa and are considered a derived state (22,1) following Nishida (1990) and Lovejoy (in press).

The nasal capsules are laterally expanded in the outgroups, pristiforms, *Rhynchobatus*, *Rhina*, rhinobatoids, and rajoids and expanded ventrolaterally in torpediniforms and myliobatiforms, except for *Hexatrygon* (Miyake, 1988; Nishida, 1990; Lovejoy, in

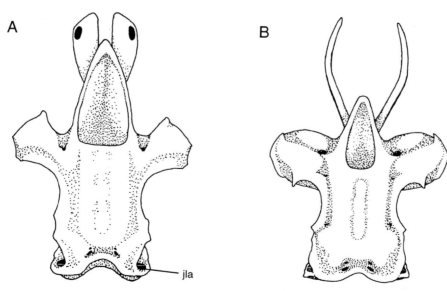

FIGURE 5 Dorsal view of neurocranium. (a) *Platyrhinoidis triseriata* CAS 31248; (b) *Platyrhina sinensis* CAS 15919. Abbreviation: jla = jugal arch.

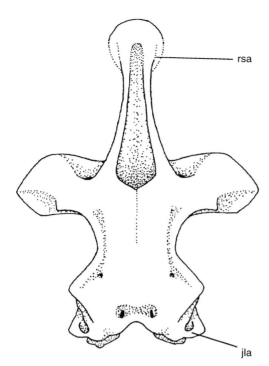

FIGURE 7 Dorsal view of neurocranium of *Trygonorrhina fasciata* MCZ 982S. Abbreviations: jla = jugal arch, rsa = rostral appendix.

press). Nishida (1990) stated that the nasal capsules are laterally expanded in *Plesiobatis*, but the specimen of this taxon examined by Miyake (1988) does have ventrolaterally expanded nasal capsules. Ventrolateral expansion of the nasal capsules is regarded as the derived state (23,1) following Nishida (1990) and Lovejoy (in press).

A basal angle on the ventral surface of the neurocranium is present in all shark-like squaleans, including the outgroups, and supports the orbital articulation of the palatoquadrate (Shirai, 1992b). The process is also present in pristiforms but absent in the remaining batoids. The absence of the process is considered the derived state (24,1) following Heemstra and Smith (1980) and Nishida (1990).

Rhinoptera, Mobula, and *Manta* lack a preorbital process at the anterior end of the supraorbital crest, and this state is regarded as derived (25,1) following Nishida (1990). According to Nishida, the preorbital crest is present in all batoids with the exception of pristiforms, *Temera* within torpediniforms, and the above three genera of myliobatiforms. However, pristiforms possess a poorly developed preorbital process (Garman, 1913; Hoffmann, 1913; Compagno, 1977). The state in narkids is coded as polymorphic (25,01) because the process is present in *Heteronarke, Narke,* and

Typhlonarke but absent in *Temera*. Lovejoy (in press) did not consider the preorbital process to be a character distinct from the anterior projections on the nasal capsules (22,1); however, the anterior projections and the preorbital processes are formed by different components of the lamina orbitonasalis (Miyake, *et al.,* 1992b) and are thus not necessarily correlated.

Torpediniforms lack a supraorbital crest that runs along the dorsal margin of the orbit in the other batoids and in the outgroups (Miyake, 1988; Nishida, 1990; Shirai, 1992b). Absence of the crest in torpediniforms is considered a derived character state (Nishida, 1990) (26,1).

The anterior preorbital foramen, through which the superficial ophthalmic nerve passes, is located on the dorsal aspect of the nasal capsule in the outgroups and in most batoids but it opens on the anterior aspect of the nasal capsule in pelagic myliobatiforms (Nishida, 1990). The superficial ophthalmic nerve passes above the nasal capsule in torpediniforms, possibly because this taxon lacks a supraorbital crest (Shirai, 1992b). The condition in pelagic stingrays is considered the derived state (27,1) following Nishida (1990), and the condition in torpediniforms is considered uncertain (27,?).

The postorbital process is present in the outgroups and batoids, with the exception of torpediniforms (28,1) (Compagno, 1973, 1977; Miyake, 1988). Absence of the process is considered a derived character state (Nishida, 1990). In myliobatiforms the postorbital process is very broad and shelf-like (28,2), but in the outgroups and other batoids it is rather narrow and triangular. The state in myliobatiforms is considered a separate derived state following Nishida (1990). The postorbital process is located in the otic region of the neurocranium in the outgroups and in batoids with the exception of *Gymnura, Aetoplatea,* and pelagic myliobatiforms (28,3). In these taxa the postorbital process extends nearly to the orbital region and this state is considered derived following Nishida (1990). The postorbital process is separated from a triangular process of the supraorbital crest in the outgroups and the majority of batoids (Compagno, 1977; Miyake, 1988; Nishida, 1990; Lovejoy, in press), but in *Plesiobatis, Urolophus, Aetobatus, Aetomylaeus, Rhinoptera,* and *Mobula* the postorbital and triangular processes are fused distally with the groove between them represented by a foramen (29,1) (Miyake, 1988; Nishida, 1990). Distal fusion of the processes is considered derived. The state in torpediniforms is considered uncertain (29,?). In myliobatids, rhinopterids, and mobulids the lateral margin of the postorbital process is prolonged and projects ventrolaterally forming a cylindrical protuberance (30,1), and following Nishida (1990) this state is considered derived. Again the state in torpediniforms is considered uncertain (30,?)

In pristiforms, *Rhynchobatus, Rhina,* rhinobatoids, and rajoids the lateral otic process forms an arch (jugal arch) between the hyomandibular facet and the posterior section of the otic capsule for passage of the jugular vein (31,1). The jugal arch is considered a derived state (Miyake, 1988; Nishida, 1990).

In *Aetobatus, Aetomylaeus, Rhinoptera, Mobula,* and *Manta* the antimeres of the upper and lower jaws are fused (32,1), unlike the case in the other batoids and outgroups (Garman, 1913; Nishida, 1990; Lovejoy, in press). Fusion of the antimeres is considered a derived character state. Fusion of the jaw antimeres is variable in *Myliobatis* (32,01) (Lovejoy, in press), and this state is coded as polymorphic. Meckel's cartilage is expanded and thickened near the symphysis in *Myliobatis, Aetobatus, Aetomylaeus,* and *Rhinoptera* (33,1) but not in the other batoids and outgroups. The expansion and thickening are considered a derived character state following Nishida. These same taxa possess grinding, pavement-like tooth plates in the jaws rather than small teeth arranged in diagonal rows as in other batoids. Nishida (1990) and Lovejoy (in press) considered the pavement-like teeth derived, but this state may be correlated with the massive nature of the jaws in these taxa and thus is not considered a separate character. Meckel's cartilage in *Myliobatis, Aetobatus, Aetomylaeus, Rhinoptera, Mobula,* and *Manta* has a pair of posteriorly expanded, winglike processes (34,1). These processes are unique to these taxa and, following Nishida (1990) are considered a derived state.

Nishida (1990) considered a small, barlike accessory hyomandibular cartilage supported by connective tissue and found in some species of *Urolophus* and *Dasyatis* to be a derived state. However, Lovejoy (in press) noted that the scattered distribution of this structure among various stingray taxa suggested that it might have more than one origin, and thus it is not considered.

The medial portion of the hyomandibula is longitudinally expanded and spans the entire length of the otico–occipital region of the neurocranium in Torpedinidae and Hypnidae (35,1) but not in the other batoids or outgroups (Garman, 1913; Compagno, 1973; Miyake, 1988). The expansion of the hyomandibula is considered a derived character state. In *Zanobatus, Plesiobatis, Urolophus, Urobatis, Urotrygon, Dasyatis, Himantura, Potamotrygon, Taeniura, Myliobatis, Aetobatus,* and *Rhinoptera* a long ligament joins the distal tip of the hyomandibular and the meckelian cartilage (36,1) (Nishida, 1990) (Fig. 8a). This ligament is absent in

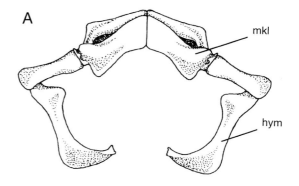

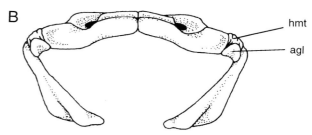

FIGURE 8   Ventral view of hyomandibula and mandibular arch. (a) *Taeniura lymma* TCWC 5278.01; (b) *Zanobatus schoenleinii* USNM 222120. Abbreviations: agl = angular cartilage, hmt = hyomandibulo-meckelian tendon, hym = hyomandibula, mkl = Meckel's arch.

Hexatrygon, Gymnura, Aetoplatea, Aetomylaeus, Mobula and *Manta,* torpediniforms, *Rhynchobatus, Rhina,* most rhinobatoids, and rajoids. Because it is absent in the outgroups, presence of the ligament is considered a derived state. *Zanobatus,* Amphi-American *Himantura,* and potamotrygonids have one or more cartilages embedded in the ligament (Fig. 8b), and this cartilage or one of these cartilages articulates with both the hyomandibula and Meckel's cartilage and is termed the angular cartilage (Holmgren, 1943; Garman, 1913; Rosa, 1985; Rosa *et al.,* 1988; Nishida, 1990; Lovejoy, in press). The cartilage in *Zanobatus* is broad and triangular and occupies the posterior section of the tendon. In *Potamotrygon* two cartilages lie in parallel in the tendon (Garman, 1913; Lovejoy, in press). *Plesiotrygon* has a single spool-shaped cartilage and the amphi-American *Himantura* have several cartilages embedded in the ligament (Lovejoy, in press). As defined by Lovejoy, the angular cartilage is located anteriorly in the tendon and articulates with both the hyomandibula and Meckel's cartilage; thus the cartilage of *Zanobatus* may not be homologous with those of the other taxa. The cartilage of *Zanobatus* (37,1) and the other taxa (37,2) are treated as separate derived states. *Urolophus, Myliobatis, Aetobatus, Rhinoptera,* and *Mobula* also possess a cartilage between the hyomandibula and Meckel's cartilage (38,1) (Garman, 1913; Lovejoy, in press), but this cartilage is located differently and is not judged homologous with either of the above angular cartilages (Lovejoy, in press). The fact that *Mobula* possesses the cartilage but not the elongated ligament suggests that Lovejoy is correct in considering this cartilage distinct from the angular cartilage. Possession of this cartilage is considered a separate derived character state.

The basihyal cartilage is slightly arched and fused to the first hypobranchial cartilages in the outgroups, pristiforms, *Rhynchobatus, Rhina,* rhinobatoids, and rajoids and in several myliobatiforms (*Plesiobatis, Hexatrygon, Urolophus, Gymnura and Aetoplatea*) (Miyake, 1988; Miyake and McEachran, 1991; Nishida, 1990; Lovejoy, in press). The basihyal is segmented in *Urobatis, Dasyatis, Taeniura, Himantura,* and Potamotrygonidae (39,1) (Miyake, 1988; Nishida, 1990; Miyake and McEachran, 1991; Lovejoy, in press). The cartilage is absent in torpediniforms and *Urotrygon* (39,2) (Miyake, 1988; Miyake and McEachran, 1991); the basihyal and first hypobranchial cartilages are absent in *Aetobatus, Aetomylaeus, Rhinoptera, Mobula,* and *Manta* (39,3) (Miyake, 1988; Nishida, 1990; Miyake and McEachran, 1991; Lovejoy, in press). The first basibranchial cartilage is present in *Myliobatis tobijei* Bleeker according to Nishida (1990) but is absent in *M. pervianus* Garman according to Miyake and McEachran (1991). Segmentation of the basihyal, absence of the basihyal, and absence of the basihyal and first basibranch are considered separate derived characters. The character state for *Myliobatis* is coded as polymorphic (39,12).

The ceratohyal cartilage articulates with the basihyal and hyomandibula in the outgroups and narkids but the ceratohyal is partially or totally replaced by the pseudohyal in the remaining torpediniforms and in all other batoids (40,1) (Compagno, 1973; Maisey, 1984; Miyake, 1988; Miyake and McEachran, 1991). Thus the condition in non-torpedinid batoids is considered derived. Nishida (1990) used fusion patterns of the ceratobranchs as derived character states. However, Miyake and McEachran (1991) and Lovejoy (1996) demonstrated that fusion patterns are variable within genera, and thus the fusion patterns constitute an ambiguous character.

The suprascapular cartilage of the scapulocoracoid articulates with the vertebral column and fuses with its antimere in all batoids. In torpediniforms and pristiforms the synarchual (fused cervical vertebrae) is short and terminates anterior to the suprascapula. The suprascapula thus fuses with its antimere and articulates with the vertebrae posterior to the synarchual. In other batoids the synarchual extends to the level of the shoulder girdle, and the suprascapula fuses with the synarchual (41,1). Because the synarchual is thought to have been very short in the Jurassic batoids (†*Spathobatis* and †*Belemnobatis*) (Maisey, 1984), the condition in torpediniforms and pristiforms is judged to be primitive, and the condition in the remaining batoids derived.

Myliobatiforms are unique among the outgroups and remainder of the batoids in possessing a ball-and-socket articulation between the scapular process of the shoulder girdle and the first synarchual (42,1) (Compagno, 1973, 1977; Heemstra and Smith, 1980), in possessing a second synarchual that is generally separated from the first by several intersynarchual vertebrae (43,1) (Compagno, 1973, 1977; Heemstra and Smith, 1980; Nishida, 1990), and in lacking ribs (44,1) (Nishida, 1990). These three states are considered derived following Nishida (1990) and Lovejoy (in press). Lovejoy polarized several character states of the first synarchual; however, these were not included in the present study because comparable data are lacking for a large number of the taxa.

The scapular process of the scapulocoracoid is short and straight in the outgroups and in all batoids except for torpediniforms. In torpediniforms it is long and posteriorly displaced (45,1), and this state is considered derived. The scapular process of the majority of myliobatiforms (*Urobatis, Urotrygon, Taeniura, Himantura, Dasyatis, Myliobatis, Aetobatus, Aetomylaeus, Rhinoptera, Mobula,* and *Manta*) has a fossa or foramen (46,1) (Rosa, 1985; Miyake, 1988; Nishida, 1990; Lovejoy, in press). This structure is lacking in the outgroups, torpediniforms, pristiforms, *Rhynchobatus, Rhina,* rhinobatoids, rajoids, and several myliobatiforms (*Hexatrygon, Plesiobatis, Urolophus,* potamotrygonids, and *Gymnura*). Presence of the fossa or foramen is considered a derived state following Nishida (1990) and Lovejoy (in press).

The scapulocoracoid of the majority of batoids has three horizontally arranged condyles along the lateral aspect, one each for the propterygium, mesopterygium, and metapterygium of the pectoral girdle (47,1). *Squatina* has two condyles that are diagonally oriented, a bilobed condyle for the propterygium and mesopterygium, and a single condyle for the metapterygium (Shirai, 1992a,b). *Pristiophorus* has three diagonally oriented condyles, one for each of the pterygia (Shirai, 1992a,b). In torpediniforms there are three condyles along the anterolateral aspect of the scapulocoracoid, and the condyles are not horizontally arranged (Miyake, 1988; Nishida, 1990; Shirai, 1992b; present study). In narkids the condyles are diagonally arranged, with the procondyle high and the metacondyle low on the scapulocoracoid, and the mesocon-

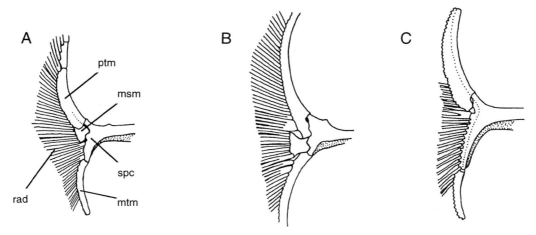

FIGURE 9 Dorsal view of left shoulder girdle traced from radiograph. (a) *Trygonorrhina fasciata* MCZ 982S; (b) *Dasyatis* sp. TCWC uncataloged; (c) *Zanobatus schoenleinii* USNM 222120. Abbreviations: msm = mesopterygium, mtm = metapterygium, ptm = propterygium, rad = radial cartilages, spc = scapulocoracoid.

dyle at mid height (Miyake, 1988; Nishida, 1990). In torpedinids and narcinids the procondyle is considerably larger and elevated with respect to the other condyles, and the metacondyle is level with or slightly elevated with respect to the mesocondyle (Miyake, 1988; Shirai, 1992b; present study). Compagno (1977) and Miyake (1988) illustrated the condyles as oriented horizontally in the narcinids (*Narcine* and *Benthobatis*, respectively), but these observations could not be verified in this study. The character state for hypnids is uncertain (47,?). The horizontal orientation of the condyles is considered the derived state. The mesocondyle is about equidistant between the procondyle and metacondyle in *Pristiophorus*, torpediniforms, pristiforms, *Rhynchobatus, Rhina, Platyrhina,* and *Platyrhinoidis*. In *Rhinobatos, Zapteryx, Trygonorrhina,* and rajoids, the scapulocoracoid is distinctly elongated between the mesocondyle and the metacondyle, and several to many radials articulate directly with the scapulocoracoid between the mesopterygium and the metapterygium (48,1) (Fig. 9a). In most myliobatiforms the scapulocoracoid is distinctly elongated between procondyle and mesocondyle (Miyake, 1988), and all of the radials articulate with one of the three pterygia (48,2) (Fig. 9b). In several taxa (*Zanobatus, Gymnura,* and pelagic myliobatiforms) the scapulocoracoid is greatly elongated, but either the mesopterygium is lacking, several mesopterygia independently articulate with the scapulocoracoid, or the mesopterygium is confined to the posterior aspect of the scapulocoracoid and a number of radials proximal to the mesopterygium directly articulate with the scapulocoracoid (48,3) (Nishida, 1990) (Fig. 9c). In the last

of the three types of elongation, the mesocondyle is replaced with a ridge spanning the distance between the procondyle and the metacondyle. These three types of elongation of the scapulocoracoid are considered separate derived states.

The antorbital cartilage directly joins the propterygium of the pectoral girdle to the nasal capsule of the neurocranium in all batoids except torpediniforms, pristiforms, *Rhynchobatus,* and *Rhina* (49,1). In the latter taxa the propterygium does not extend to the antorbital cartilage (Miyake, 1988; Nishida, 1990). The direct connection between the antorbital cartilage and the propterygium is considered a derived state following Nishida (1990). Nishida, however, distinguished between pectoral propterygia and radials extending to the nasal capsule (a character state shared among all batoids exclusive of pristiforms, *Rhynchobatus,* and *Rhina*) and lateral articulation between antorbital cartilage and pectoral propterygium whereby the antorbital cartilage provides direct, lateral support for the pectoral propterygia. However, the former state may not be homologous in the torpediniforms and the remainder of the batoids, exclusive of pristiforms, *Rhynchobatus,* and *Rhina* because of the unique nature of the antorbital cartilage in torpediniforms. The antorbital cartilages of torpediniforms are unique among the batoids in being anteriorly expanded and fan- or antler-like (50,1), and this condition is considered to be derived following Compagno (1977) and Nishida (1990).

The distal section of the propterygium extends beyond the procondyle of the scapulocoracoid and articulates with the scapulocoracoid between the pro-

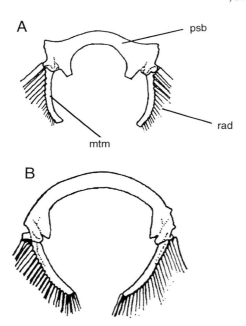

FIGURE 10 Dorsal view of pelvic girdle traced from radiograph. (a) *Zanobatus schoenleinii* USNM 222120; (b) *Dasyatis* sp. TCWC uncataloged. Abbreviations: mtm = metapterygium, psb = puboischiac bar, rad = radial cartilages.

condyle and mesocondyle in *Zanobatus* and myliobatiforms (51,1) (Fig. 9b,c) (Miyake, 1988). This condition does not occur in the outgroups and other batoids and is considered to be derived. In several myliobatiforms (*Gymnura, Aetoplatea, Myliobatis, Aetobatus,* and *Aetomylaeus)* some of the pectoral radials supported by the propterygium are expanded distally and articulate with the surface of adjacent radials (52,1). This state is unique to these taxa and is considered derived (Nishida, 1990). The pectoral and pelvic fins are aplesodic (radials are replaced by ceratotrichia near the margins of the fins) in the outgroups and pristiforms. Both fins are plesodic (radials extend to the margins of the fins, and ceratotrichia are reduced or absent) in the remaining batoids (53,1), and this state is considered derived following Heemstra and Smith (1980) and Nishida (1990). Nishida considered the condition of the pectoral and pelvic fins to be derived states for separate characters. However, this does not appear to be justified because the structure of the paired fins is very likely to be correlated.

The puboischiadic bar of the pelvic girdle is rather broad, platelike, and straight to slightly arched in the outgroups, torpediniforms, pristiforms, most rhinobatoids, and rajoids. In one rhinobatoid, *Zanobatus,* and myliobatiforms it is rather narrow and strongly arched (54,1) (Fig. 10a,b), and this condition is considered derived. Lovejoy (in press) applied a more lim-

ited version of this state, strongly arched puboischiadic bar, exclusively to *Aetobatus, Rhinoptera,* and *Mobula.* However, the bar is only slightly less arched in *Gymnura, Aetoplatea,* and *Myliobatis.* The shape of the bar is unique in *Zanobatus* and myliobatiforms, but assessing the degree of arching within these taxa is somewhat subjective. In *Platyrhina* and *Platyrhinoidis* the puboischiadic bar is unique in possessing a pair of triangular processes on the posterior margin (55,1), and these processes are considered a derived state following Nishida (1990).

The tail vertebrae are diplospondylous from about the pelvic girdle to the tip in the outgroups, torpediniforms, pristiforms, *Rhynchobatus, Rhina,* rhinobatoids, rajoids, and in several myliobatiforms (*Hexatrygon, Plesiobatis, Urolophus, Urobatis,* and *Urotrygon)* (Nishida,1990; Lovejoy, in press). In the remaining myliobatiforms the tail vertebrae posterior to the serrated spine are fused into a tube (56,1), and this state, following Lovejoy (in press), is considered derived.

D. Cephalic and Branchial Musculature

The muscles of the head and branchial skeleton are more complex than those of sharks and display considerable variation within the batoids. The cranial muscle, ethmoideo-parethmoidalis, originates on the posterolateral aspect of the nasal capsule, runs posteriorly along the medial aspect of the propterygium, and inserts on the inner surface of the propterygium and antorbital cartilage (Edgeworth, 1935; Miyake, 1988; Nishida, 1990). This muscle is found in all batoids except torpediniforms and outgroups, and it is considered a derived state (57,1) following Miyake (1988).

The mandibular plate muscle, intermandibularis, is present in the outgroups and torpediniforms; its absence in the other batoids is considered a derived character state (58,1). In sharks it is a sheetlike muscle that originates in the facia along the midventral line and inserts on the ventral posterior edge of Meckel's cartilage (Shirai, 1992b). In torpediniforms it is a narrow band of muscle that originates on the hyomandibula and inserts on the posterior margin of Meckel's cartilage (58,2) (Miyake *et al.,* 1992a), and this condition of the muscle is considered a separate derived state. In Narcinidae and Narkidae there is a ligamentous sling at the symphysis of Meckel's cartilage that supports the intermandibularis, coracomandibularis, and depressor mandibularis muscles (59,1), and this condition is considered derived (Miyake, 1988; Miyake *et al.,* 1992a). Nishida (1990) identified the depressor mandibularis as the intermandibularis posterior in myliobatiforms and stated that in pelagic myliobatiforms it runs along the mandibular cartilage under the winglike process and joins its antimere near the symphysis. According to Miyake *et al.* (1992a) the depressor mandibularis does not exist as an independent muscle in pelagic myliobatiforms. Whether the muscle exists in a derived state or is absent in pelagic myliobatiforms, the state in these taxa is considered derived (60,1). The depressor mandibularis is absent in *Hexatrygon*, and this state is similarly coded (60,1).

In the outgroups, the spiracularis (mandibular plate muscle) originates on the otic region of the neurocranium, extends along the prespiracular wall, and inserts on the palatoquadrate or the hyomandibula (Miyake, 1988; Shirai, 1992a,b). In torpediniforms a separate bundle extends from the ventral margin of the main muscle mass and enters the dorsal oral membrane underlying the neurocranium (61,1) (Miyake, 1988; Miyake *et al.*, 1992a). In *Plesiobatis, Urolophus, Urobatis, Urotrygon,* and *Dasyatis* the muscle splits into lateral and medial bundles (Miyake, 1988; Miyake *et al.*, 1992a). The medial bundle inserts onto the posterior surface of Meckel's cartilage and the lateral bundle inserts onto the side of the dorsal edge of the hyomandibula (61,2). In *Taeniura, Himantura, Potamotrygon,* and *Plesiotrygon* the muscle extends beyond the hyomandibula and Meckel's cartilage (61,3) (Miyake, 1988; Miyake *et al.*, 1992a; Lovejoy, in press). In *Taeniura* the medial component of the muscle has fibers that insert near the midline just dorsal to the coracomandibularis muscle, but in *Potamotrygon, Plesiotrygon,* and *Himantura,* the medial component of the muscle has fibers that insert near the midline ventral to the coracomandibularis muscle (Lovejoy, in press). The spiracularis is undivided in *Myliobatis* and inserts on the hyomandibula (61,0); however, in *Rhinoptera,* it is subdivided proximally and inserts separately onto the palatoquadrate and the hyomandibula (61,4) (Miyake, 1988). The states in *Hexatrygon, Gymnura, Aetobatus, Mobula,* and *Manta* are uncertain. The states in torpediniforms; *Plesiobatis, Urolophus, Urobatis, Urotrygon,* and *Dasyatis; Taeniura, Potamotrygon, Plesiotrygon,* and *Himantura;* and *Rhinoptera* are considered separate derived states. The states in *Taeniura, Himantura, Potamotrygon,* and *Plesiotrygon,* although differing in their insertion with respect to the coracomandibularis, are considered to be the same state pending further research.

Some of the branchial musculature of torpediniforms is converted into electric organs (62,1), and this state is unique and derived for the taxon. The coracobranchialis, of the branchial muscle plate, consists of three, four, or five components that insert on the second, third, fourth, and possibly fifth hypobranchs in the outgroups and all batoids with the exception

of narcinid and narkid torpediniform rays (63,1) (Miyake, 1988; Miyake *et al.*, 1992a), and the state in narkids is considered derived.

The coracohyomandibularis, of the hypobranchial muscle plate, has a single origin on the fascia that is continuous with the pericardial membrane in the outgroups and in most batoids, with the exception of narkid torpediniform rays and myliobatiforms. In narkids it has separate origins on the fascia supporting the insertion of the coracoarcualis and on the pericardial membrane (64,1). In myliobatiforms, the muscle has separate origins on the anterior portion of the ventral gill arch region and on the pericardial membrane (64,2). The state is uncertain in *Hexatrygon* (64,?). The double origins of the coracohyomandibularis in narkids and myliobatiforms are considered separate derived states. The coracohyoideus, of the hypobranchial muscle plate, in the outgroups, *Rhynchobatus, Rhina,* some rhinobatoids (*Aptychotrema, Rhinobatos,* and *Trygonorrhina*), and rajoids runs parallel to the longitudinal body axis, originates medial to the branchial cavity, and inserts on the ventral lateral aspect of the basihyal or on the first basibranch (Nishida, 1990; Miyake *et al.*, 1992a) (Fig. 11a). The muscle is lacking in torpediniforms (65,1) (Miyake, 1988; Miyake *et al.*, 1992a). In pristiforms it runs parallel to the body axis but is very short and originates in the wall of the first gill slit (65,2) (Miyake *et al.*, 1992a). In platyrhinids (*Platyrhina, Platyrhinoidis* and *Zanobatus*) and benthic myliobatiforms, the coracohyoideus is diagonally oriented, originates on the wall of the first two gill slits, and inserts on the posteromedial aspect of the basihyal or first basibranch (65,3) (Fig. 11b). The state is uncertain for *Hexatrygon* (65,?). In pelagic myliobatoids the muscle is similarly oriented as in benthic myliobatiforms, but each muscle fuses with its antimere by means of a raphe near its insertion on the first hypobranch (65,4) (Nishida, 1990). The states in torpediniforms, pristiforms, platyrhinids and benthic myliobatiforms, and pelagic myliobatoids are considered to be separate derived states.

E. *Phylogenetic Analysis*

The PAUP 3.1.1 analysis of the data matrix (Appendix 2) of 30 taxa (including two outgroups) and 65 characters produced 22 equally parsimonious trees of 115 steps with a consistency index of 0.783, retention index of 0.932, and homoplasy index of 0.252. The strict consensus tree (Fig. 12) places torpediniforms as sister to the remaining batoids and pristiforms as sister to *Rhina, Rhynchobatus,* "rhinobatoids," rajoids, and myliobatiforms. Of the currently recognized higher taxa of batoids, torpediniforms, pristiforms,

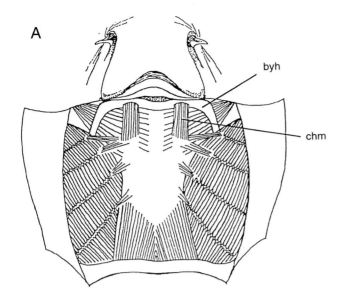

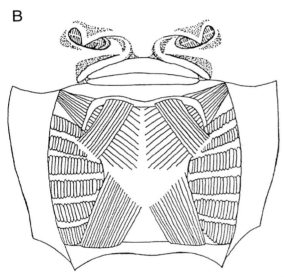

FIGURE 11 Hypobranchial muscles. (a) *Raja miraletus* TCWC 6451.01; (b) *Platyrhina sinensis* CAS 15919. Abbreviations: byh = basihyal, chm = coracohyoideus muscle.

rajoids, and myliobatiforms are monophyletic, although pristiforms and rajoids were treated as terminal nodes and thus were not tested in the analysis. Within torpediniforms, torpedinids are sister to hypnids, and narcinids are sister to narkids. *Rhina* are sister to *Rhynchobatus*, "rhinobatoids," rajoids, and myliobatiforms, and *Rhynchobatus* are sister to "rhinobatoids," rajoids, and myliobatoids. "Rhinobatoids" are polyphyletic, with *Rhinobatos, Zapteryx, Trygonorrhina*, and rajoids forming a polytomy with *Platyrhina + Platyrhinoidis* and *Zanobatus* + myliobatiforms.

Within myliobatiforms, *Hexatrygon* is sister to the remainder of the clade, and *Plesiobatis* and *Urolophus* form a trichotomy with the remaining myliobatiforms. *Urobatis* and *Urotrygon* are sister taxa and together are sister to the remaining myliobatiforms. Urolophidae are polyphyletic and Dasyatidae are paraphyletic, with *Taeniura* sister to *Himantura* + potamotrygonids and *Dasyatis* sister to *Gymnura* + the pelagic myliobatiforms.

The plots of the Bremer Decay indexes and unambiguous character states indicate high support for tree stability in terms of supported resolution. The total support index (ratio of total branch support over the tree divided by the length of the most parsimonious trees) (Bremer, 1994) is 61/115 or 0.53. Support was strong both for Bremer Decay indexes and unambiguous character states for the torpediniform branch, *Rhina*, and the remainder of the batoids branch; *Zanobatus* and the myliobatiform branch; *Hexatrygon* and the remaining myliobatiform branch; gymnurid and pelagic myliobatiform branch; and pelagic myliobatiform branch. These branches were supported by Bremer Decay indexes of 2 or greater and by 2 to 11 unambiguous character states. The remaining branches were supported by Bremer Decay indexes of 1, and these branches were poorly supported by unambiguous characters and/or supported by ambiguous characters. The lack of unambiguous characters on several branches is due to polytomies preceding or following the branch; changes at these branches depend on how the polytomies might be resolved.

Successive approximation character weighting (Farris, 1969; Carpenter, 1988), based on the retention index, resulted in five equally parsimonious solutions. The strict consensus tree of the five equally parsimonious solutions (Fig. 13) unites *Rhinobatos, Zapteryx, Trygonorrhina*, and Rajoidei into a clade, whereas in the strict consensus tree (Fig. 12) they formed a polytomy with (*Platyrhina + Platyrhinoidis*) + (*Zanobatus* + myliobatiforms). In addition, the successive approximation strict consensus tree of the five equally parsimonious solutions placed *Pleisobatis* as sister to *Urolophus* and the remainder of the myliobatoids exclusive of *Hexatrygon*, whereas in the strict consensus tree they formed a polytomy as well.

IV. Discussion

Considerable progress has been made in resolving the interrelationships of batoids since Compagno's (1973) review of the euselachians. However, several areas of conflict and lack of resolution remain. Heemstra and Smith (1980) and Nishida (1990) found pristi-

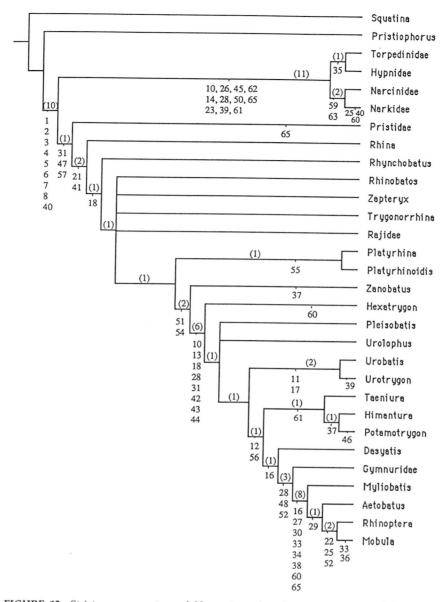

FIGURE 12 Strict consensus tree of 22 most parsimonious trees generated from 65 characters of 28 batoid taxa and two outgroups. Bremer Decay indexes are given above branches in parentheses, and unambiguous characters are given below branches.

forms to be sister to the remaining batoids; however, in this study torpedinoids are sister to the remaining batoids. The disparity in these findings is probably due both to the uniqueness of the two taxa and to differences in the characters utilized. Both taxa are defined by a large number of autapomorphies and retain a number of plesiomorphies but share few synapomorphies with other batoid taxa. Heemstra and Smith (1980) used the absence of a basal angle on the ventral surface of the neurocranium, absence of ceratotrichia in the pectoral and pelvic fins, and de-

gree of forward extension of the propterygium of the pectoral fin in defining the relationship between torpediniforms and pristiforms, and the other batoids. They claimed that torpediniforms had the derived state, and pristiforms the primitive state, for these three characters. Nishida (1990) used the same characters but treated the absence of ceratotrichia in the pectoral and pelvic fins as derived states of separate characters and coded the pristiforms, *Rhynchobatus*, and *Rhina* as sharing the plesiomorphic absence of full extension of the propterygium. In this study, ab-

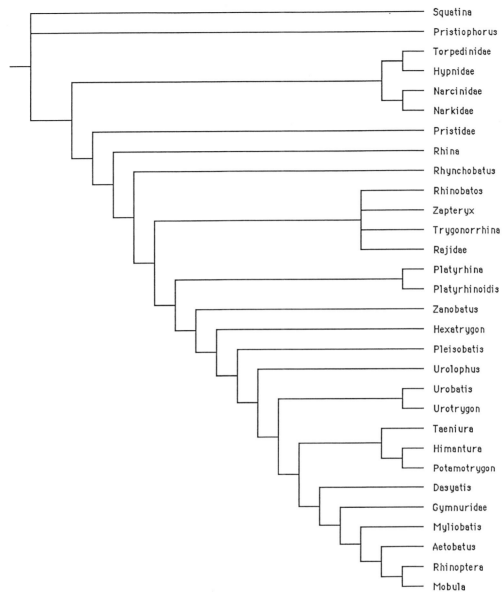

FIGURE 13 Strict consensus tree of five equally parsimonious trees resulting from successive approximation character weighting based on retention index.

sence of the basal angle (24,1) and absence of cerato-trichia (53,1) were used, but the absence of cerato-trichia in the pectoral and pelvic fins was considered the derived state of a single character. The forward extension of the propterygia was not used because the state in torpediniforms is thought to be distinct from that in other batoids and possibly correlated with other characters utilized in the analysis, e.g., structure of the antorbital cartilage and/or development of electric organs in torpediniforms. In torpediniforms the propterygium reaches to the level of the nasal capsule

but is widely separated from the nasal capsule, with the antorbital cartilage anterolaterally oriented. In rhinobatoids, rajoids, and myliobatiforms the prop-terygium is very close to the nasal capsule and directly articulates with the antorbital cartilage, and the antor-bital cartilage is lateral to the nasal capsule and pos-terolaterally oriented. This study also utilized the presence of the jugal arch (31,1), orientation of the three condyles of the scapulocoracoid (47,1), and the presence of the ethmoideo-parethmoidalis muscle (57,1) in defining the relationship between torpediniforms and pristiforms and the other batoids. Pristiforms have the derived state and torpediniforms have the primitive state for these three characters. The jugal arch, formed by the lateral otic process between the hyomandibular facet and the posterior section of the otic capsule, is unique to pristiforms, *Rhynchobatus*, *Rhina*, rhinobatoids, and rajoids. The three condyles of the scapulocoracoid are horizontal in all batoids, with the exception of torpediniforms, and the condition in torpediniforms is very similar to that in the first outgroup, Pristiophoridae. The scapular process of the scapulocoracoid is uniquely shaped in torpediniforms, but the surface of the scapulocoracoid that articulates with the three pterygia of the pectoral fin is very similar to that of the outgroups. The ethmoideoparethmoidalis is unique to batoids exclusive of torpediniforms. In addition to these character states, narkids lack a synapomorphy of all other batoids, loss of an articulation between the hyomandibula and ceratohyal (43,0). Batoids, with the exception of narkids, have a reduced or no ceratohyal. The fact that narkids possess the plesiomorphic condition suggests that this character state is a reversal or that loss of the ceratohyal occurred in parallel in torpediniforms and the remaining batoids because torpediniforms are a highly corroborated monophyletic group. Thus one of the two possibilities implies torpediniforms are sister to the remaining batoids, a hypothesis supported by the parsimony analysis. However, the Bremer Decay index does not offer strong support for the branch including the pristiforms and remaining batoids, suggesting that the relationships among torpediniforms, pristiforms, and the remaining batoids need to be investigated further.

Nishida (1990) revealed rhinobatoids to be polyphyletic, with *Rhynchobatus* + *Rhina* sister to the other batoids exclusive of pristiforms. His study left the relationships among the other rhinobatoids largely unresolved, with the exception of *Rhinobatos* and *Aptychotrema* forming a trichotomy with the remaining rhinobatoids, rajoids, and myliobatiforms, and *Platyrhinoidis* and *Platyrhina* comprising a sister relationship. However, when the data for his phylogeny of rajiforms (= Batoidea) were reanalyzed using PAUP 3.1.1 (Swofford, 1993), 14 most parsimonious trees were found. In the strict consensus tree *Rhynchobatus*, *Rhina*, *Platyrhinoidis*, *Trygonorrhina*, and *Zapteryx* form a polytomy with a clade comprising torpediniforms, *Rhinobatos*, *Aptychotrema*, *Platyrhina*, and rajoids + myliobatiforms. In this study, *Rhina* are sister to *Rhynchobatus*, "rhinobatoids," rajoids, and myliobatiforms, and *Rhynchobatus* are sister to "rhinobatoids," rajoids, and myliobatoids. *Rhinobatos*, *Zapteryx*, *Trygonorrhina*, rajoids, and *Platyrhinoidis* + *Platyrhina* form a polyt-

omy with *Zanobatus* + myliobatiforms. Thus the "rhinobatoids" are truly polyphyletic with subunit taxa sister to both rajoids and myliobatiforms.

According to Nishida, myliobatiforms are divided into two clades, one comprising *Gymnura* and *Aetoplatea*, and pelagic myliobatiforms, and the other comprising the remaining myliobatiforms. Within the latter clade, potamotrygonids form a trichotomy with *Taeniura* and *Dasyatis* + *Himantura*. However, in reanalyzing Nishida's data using PAUP 3.1.1 (Swofford, 1993) the latter clade breaks down. *Hexatrygon* are sister to a polytomy consisting of *Plesiobatus*, *Urolophus*, *Urotrygon*, *Potamotrygon*, and *Taeniura*, that in turn is sister to a clade comprising the remaining myliobatiforms. According to Lovejoy (in press) *Taeniura* and *Himantura* + Potamotrygonidae form a clade that is sister to a clade comprising *Dasyatis*, *Gymnura*, and pelagic myliobatiforms. The differences in resolution between these studies are due to differences in taxa analyzed, characters used, and coding character states. Lovejoy included three taxa that were not included in Nishida's study (amphi-American *Himantura* and two additional genera of potamotrygonids). The amphi-American *Himantura* possess angular cartilages that define the sister-group relationship between *Himantura* and potamotrygonids (Lovejoy, in press). Lovejoy utilized a number of lateral line characters that had not been included in previous studies, and these characters helped elucidate the relationships within the myliobatiforms. Nishida (1990) coded the loss of the basihyal and first basibranch (his hypohyal) as a derived character state. Lovejoy (in press) coded the segmentation of the basihyal and loss of the basihyal as separate derived states. This study corroborates the findings of Lovejoy.

The phylogenetic hypothesis generated in this study is not fully resolved and suggests that research is needed to further elucidate the relationships among "rhinobatoid" taxa and rajoids, and benthic myliobatiforms. The resulting successive approximation character weighting analysis (Fig. 13) that weighs characters *a posteriori* according to their retention index supports a clade of *Rhinobatos*, *Zapteryx*, *Trygonorrhina*, and rajoids, and places *Plesiobatis* as sister to *Urolophus* and myliobatiforms, exclusive of *Hexatrygon*. *Rhinobatos*, *Zapteryx*, *Trygonorrhina*, and rajoids share a unique character state: a scapulocoracoid that is expanded between the mesocondyle and metacondyle with radials that directly articulate with the scapulocoracoid between the mesopterygium and metacondyle (48,1). *Plesiobatis* shares two derived character states with the remainder of the myliobatiforms, exclusive of *Hexatrygon*: an infraorbital loop of the lateral line canal (15,1) and ventrolateral expansion of the nasal capsules (24,1). However, the structure of the lateral line canals is unknown for *Hexatrygon*, and ventrolateral expansion of the nasal capsules also occurs in torpediniforms. Further research is needed to test these hypotheses. Additionally, some of the branches on the cladogram are supported by ambiguous character states, e.g., the clade comprising *Rhinobatos, Zapteryx, Trygonorrhina*, rajoids, and the remaining batoids; the clade comprising *Platyrhina + Platyrhinoidis* and the remainder of the batoids; the clade comprising *Plesiobatis* and the remaining myliobatiforms; and the clade comprising *Urobatis + Urotrygon* and the remaining myliobatiforms. The lack of unambiguous character support is due to polytomies on either side of the node of these branches that make determination of ancestral character states difficult. Further resolution of the polytomies thus may increase the branch support.

Although not fully resolved, the phylogenetic hypotheses proposed here suggest that several changes in batoid classification are needed. Consequently, *Rhina* and *Rhynchobatus* are given ordinal status pending further research. *Rhinobatos, Zapteryx*, and *Trygonorrhina* are regarded as *incertae sedis* and placed between Rhynchobatiformes and Rajiformes pending further research. Platyrhinidae is limited to *Platyrhinoidis* and *Platyrhina* and placed with *Zanobatus* in Myliobatiformes. *Zanobatus* is ranked as a family. *Plesiobatis* and *Urolophus* are classified as *incertae sedis* within Myliobatiformes pending further research. *Urobatis* and *Urotrygon* are ranked as a family in Dasyatoidea. Potamotrygonidae includes *Taeniura*, Amphi-America *Himantura, Potamotrygon, Plesiotrygon*, and *Paratrygon*. Dasyatidae includes *Dasyatis, Gymnura, Myliobatis, Aetobatus, Mobula*, and *Manta*.

Batoid classification based on the parsimony analysis is as follows:

Class Chondrichthyes
   Subclass Neoselachii
     Cohort Squalea
      Superorder Batoidea
       Order Torpediniformes
        Suborder Torpedinoidei
         Family Torpedinidae Bonaparte 1838: *Torpedo* Duméril 1806
         Family Hypnidae Gill 1862: *Hypnos* Duméril 1852
        Suborder Narcinoidei
         Family Narcinidae Gill 1862: *Benthobatis* Alcock 1898; *Diplobatis* Bigelow and Schroeder 1948; *Discopyge* Heckel 1846; *Narcine* Henle 1834

         Family Narkidae Fowler 1934: *Heteronarce* Regan 1921; *Narke* Kaup 1826; *Temera* Gray 1831; *Typhlonarke* Waite 1909
       Order Pristiformes
        Family Pristidae Bonaparte 1838: *Anoxypristis* White and Moy-Thomas 1941; *Pristis* Latham 1794
       Order Rhiniformes
        Family Rhinidae Gray 1851: *Rhina* Bloch and Schneider 1801
       Order Rhynchobatiformes
        Family Rhynchobatidae Garman 1913: *Rhynchobatus* Müller and Henle 1837
          incertae sedis Rhinobatos Link 1790
          incertae sedis Aptychotrema Norman 1926
          incertae sedis Zapteryx Jordan and Gilbert 1880
          incertae sedis Trygonorrhina Müller and Henle 1838
       Order Rajiformes
        Rajidae Bonaparte 1831
         Subfamily Rajinae: *Amblyraja* Malm 1877; *Anacanthobatis* von Bonde and Swart 1923; *Breviraja* Bigelow and Schroeder 1948; *Cruriraja* Bigelow and Schroeder 1948; *Dactylobatus* Bean and Wead 1909; *Dipturus* Rafinesque 1810; *Fenestraja* McEachran and Compagno 1982; *Gurgesiella* de Buen 1959; *Leucoraja* Malm 1877; *Malacoraja* Stehmann 1970; *Neoraja* McEachran and Compagno 1982; *Okamejei* Ishiyama 1958; *Raja* Linnaeus 1758; *Rajella* Stehmann 1970; *Rostroraja* Hulley 1972
         Subfamily Arhynchobatinae: *Atlantoraja* Menni 1972; *Arhynchobatis* Waite 1909; *Bathyraja* Ishiyama 1958; *Irolita* Whitley 1931; *Notoraja* Ishiyama 1958; *Pavoraja* Whitley 1939; *Psammobatis* Günther 1870; *Pseudoraja* Bigelow and Schroeder 1954; *Rhinoraja* Ishiyama 1952; *Rioraja* Whitley 1939; *Sympterygia* Müller and Henle 1837
       Order Myliobatiformes
        Suborder Platyrhinoidei
         Family Platyrhinidae Jordan 1923:

Platyrhina Müller & Henle 1838;
Platyrhinoidis Garman 1881
Suborder Zanobatoidei
    Family Zanobatidae: *Zanobatus* Garman 1913
Suborder Myliobatoidei
    Superfamily Hexatrygonoidea Heemstra and Smith 1980
    Family Hexatrygonidae Heemstra and Smith 1980: *Hexatrygon* Heemstra and Smith 1980
      insertae sedis Plesiobatis Nishida 1990
      insertae sedis Urolophus Müller and Henle 1837
    Superfamily Dasyatoidea
    Family Urotrygonidae: *Urobatis* Garman 1913; *Urotrygon* Gill 1863
    Family Potamotrygonidae Garman 1913: *Taeniura* Müller and Henle 1837; *Himantura* Müller & Henle 1837 (in part: amphi-American species); *Paratrygon* Duméril 1865; *Plesiotrygon* Rosa, Castello, and Thorson 1987; *Potamotrygon* Garman 1877
    Family Dasyatidae Rafinesque 1810
      Subfamily Dasyatinae: *Dasyatis* Rafinesque 1810; *Pastinachus* Rüppell 1828; *Urogymnus* Müller & Henle 1837; *Urolophoides* Lindberg 1930; *Himantura* Müller and Henle 1837 (in part: Indo-West Pacific species)
      Subfamily Gymnurinae Fowler 1934: *Aetoplatea* Valenciennes 1841; *Gymnura* van Hasselt 1823
      Subfamily Myliobatinae Bonaparte 1838: *Aetobatus* Blainville 1816; *Aetomylaeus* Garman 1908; *Myliobatis* Cuvier 1816; *Pteromylaeus* Garman 1913; *Rhinoptera* Cuvier 1829; *Mobula* Rafinesque 1810; *Manta* Bancroft 1829

V. Summary

Considerable progress in elucidating the interrelationships within the batoids has been made since Compagno's (1973) review of the euselachians. However, interrelationships of several of the taxa remain ambiguous, in part, because of lack of morphological

information on rhinobatoids. Survey of the majority of rhinobatoid genera eliminated some ambiguities allowing further resolution of batoid interrelationships. A parsimony analysis revealed that torpediniforms are sister to the remaining batoids; pristiforms are sister to the rhinobatoids, rajoids, and myliobatiforms; and rhinobatoids are polyphyletic. *Rhina* are sister to *Rhynchobatus* + the remaining rhinobatoids, rajoids, and myliobatiforms. *Rhinobatos, Zapteryx, Trygonorrhina,* and rajoids form a polytomy with *Platyrhina* + *Platyrhinoidis* and *Zanobatus* + myliobatiforms. Within myliobatiforms, *Hexatrygon* are sister to the remainder of the clade, and *Plesiobatis* and *Urolophus* form a trichotomy with remaining myliobatiforms. *Urobatis* and *Urotrygon* are sister taxa and together are sister to the remaining myliobatiforms. The remaining myliobatiforms form two completely resolved clades, one with *Dasyatis* sister to *Gymnura* and the pelagic myliobatiforms and the other with *Taeniura* sister to *Himantura* + Potamotrygonidae.

Acknowledgments

Nathan Lovejoy kindly permitted us to cite his manuscript (in press) on the interrelationships of the myliobatoids, and for this he is sincerely thanked. Carolyn J. Rose prepared the anatomical illustrations. Curators at the following institutions kindly loaned us batoid specimens utilized in this study: Bernice P. Bishop Museum; California Academy of Sciences; J.L.B. Smith Institute of Ichthyology; Kyoto University, Department of Fisheries, Faculty of Agriculture; Los Angeles County Museum; Museum of Comparative Zoology; National Museum of Natural History, Smithsonian Institution; and Scripps Institution of Oceanography. Colin Patterson, through his important contributions to systematic ichthyology, was a major inspiration for this study. The study was supported in part by grants from the National Science Foundation to JDM (DEB82–04661 and BSR87–00292). Travel funds to visit the MCZ were provided by the Ernst Mayr Grant Fund to JDM and TM. Part of this work was carried out in the Center for Biosystematics and Biodiversity, a facility funded, in part, by the National Science Foundation (Award DIR-8907006). This paper represents Contribution No. 49 of the Center for Biosystematics and Biodiversity at Texas A. & M. University.

References

Bigelow, H. B., and Schroeder, W. C., eds. (1953). "Fishes of the Western North Atlantic. Sawfishes, Guitarfishes, Skates and Rays, and Chimaeroids," Sears Found. Mar. Res., Mem. No. 1, Part 2. Yale University, New Haven, CT.

Bremer, K. (1994). Branch support and tree stability. *Cladistics* **10**, 295–304.

Carpenter, J. M. (1988). Choosing among multiple equally parsimonious cladograms. *Cladistics* **4**, 291–296.

Carpenter, J. M. (1994). Successive weighting, reliability and evidence. *Cladistics* **2**, 215–220.

Chu, Y. T., and Wen, M. C. (1980). "A Study of the Lateral-line Canals System and that of Lorenzini Ampullae and Tubules of Elasmobranchiate Fishes of China," Monograph of Fishes of

China, 2. Shanghai Science and Technology Press, Shanghai, Peoples Republic of China.

Compagno, L. J. V. (1973). Interrelationships of elasmobranchs. In "Interrelationships of Fishes" (P. H. Greenwood, R. S. Miles, and C. Patterson, eds.). pp. 15–61. Academic Press, New York.

Compagno, L. J. V. (1977). Phyletic relationships of living sharks and rays. Am. Zool. 17, 303–322.

Dingerkus, G., and Ulher, L. D. (1977). Enzyme clearing of alcian blue stained whole small vertebrates for demonstration of carti-lage. Stain Technol. 52, 229–232.

Edgeworth, F. H. (1935). "The Cranial Muscles of Vertebrates." Cambridge University Press, Cambridge, UK.

Farris, J. S. (1969). A successive approximation approach to charac-ter weighting. Syst. Zool. 18, 374–385.

Garman, S. (1888). On the lateral line canal system of the Selachia and Holocephala. Bull. Mus. Comp. Zool. 6, 167–172.

Garman, S. (1913). The Plagiostomia. Mem. Mus. Comp. Zool. Harv. 36, 1–515.

Heemstra, P. C., and Smith, M. M. (1980). Hexatrygonidae, a new family of stingrays (Myliobatiformes: Batoidea) from South Af-rica, with comments on the classification of batoid fishes. Ich-thyol. Bull. J.L.B. Smith Inst. Ichthyol. 43, 1–17.

Hoffmann, L. (1913). Zur Kenntnis des Neurocraniums der Pristi-den und Pristiophoriden. Zool. Jahrb., Abt. Anat. Ontog. Tiere 33, 239–360.

Holmgren, N. (1940). Studies on the head of fishes. Embryological, morphological, and phylogenetical researches. Part I: Develop-ment of the skull in sharks and rays. Acta Zool. (Stockholm) 21, 51–267.

Holmgren, N. (1943). Studies on the head of fishes. An embryologi-cal, morphological and phylogenetic study. Acta Zool. (Stock-holm) 24, 1–188.

Jacob, B. A., McEachran, J. D., and Lyons, P. L. (1994). Electric organs in skates: Variation and phylogenetic significance (Chon-drichthyes: Rajoidei). J. Morphol. 221, 45–63.

Leviton, A. E., Gibbs, R. H., Jr., Heal, E., and Dawson, C. E. (1985). Standards in herpetology and Ichthyology: Part 1. Standard symbolic codes for institutional resource collections in herpetol-ogy and ichthyology. Copeia pp. 802–832.

Lovejoy, N. R. (1996). Systematics of myliobatoid elasmobranchs: With emphasis on the phylogeny and historical biogeography of neotropical freshwater stingrays (Potamotrygonidae: Raji-formes). Zool. J. Linn. Soc. (in press).

Maisey, J. G. (1984). Higher elasmobranch phylogeny and biostrati-graphy. Zool. J. Linn. Soc. 82, 33–54.

McEachran, J. D. (1984). Anatomical investigations of the New Zealand skates Bathyraja asperula and B. spinifera, with an evalua-tion of their classification within the Rajoidei (Chondrichthyes). Copeia, pp. 45–58.

McEachran, J. D., and Konstantinou, H. (1996). Survey of the varia-tion in alar and malar thorns in skates: phylogenetic implications (Chondrichthyes: Rajoidei). J. Morphol. 228, 165–178.

McEachran, J. D., and Miyake, T. (1990). Phylogenetic relationships of skates: A working hypothesis (Chondrichthyes: Rajoidei). In "Elasmobranchs as Living Resources: Advances in the Biology, Ecology, Systematics, and the Status of the Fisheries" (H. L. Pratt, Jr., T. Taniuchi, and S. H. Gruber, eds.), NOAA Tech Rep. NMFS 90, pp. 285–304. U.S. Department of Commerce, Washington, DC.

Miyake, T. (1988). The systematics of the stingray genus Urotrygon with comments on the interrelationships within Urolophidae (Chondrichthyes, Myliobatiformes). Ph.D. Dissertation, Texas A&M University, College Station.

Miyake, T., and McEachran, J. D. (1991). The morphology and

evolution of the ventral gill arch skeleton in batoid fishes (Chon-drichthyes: Batoidea). Zool. J. Linn. Soc. 102, 75–100.

Miyake, T., McEachran, J. D., and Hall, B. K. (1992a). Edgeworth's legacy of cranial development with an analysis of muscles in the ventral gill arch region of batoid fishes (Chondrichthyes: Batoidea). J. Morphol. 212, 213–256.

Miyake, T., McEachran, J. D., Walton, P. J., and Hall, B. K. (1992b). Development and morphology of rostral cartilages in batoid fishes (Chondrichthyes: Batoidea), with comments on homology within vertebrates. Biol. J. Linn. Soc. 46, 259–298.

Nishida, K. (1990). Phylogeny of the suborder Myliobatoidei. Mem. Fac. Fish., Hokkaido Univ. 37, 1–108.

Rosa, R. S. (1985). A systematic revision of the South American freshwater stingrays (Chondrichthyes: Potamotrygonidae). Ph.D. Dissertation, College of William and Mary, Williams-burg, VA.

Rosa, R .S., Castello, H. P., and Thorson, T. B. (1988). Plesiotrygon iwamae, a new genus and species of neotropical freshwater sting-ray (Chondrichthyes: Potamotrygonidae). Copeia, pp. 447–458.

Shirai, S. (1992a). Phylogenetic relationships of the angel sharks, with comments on elasmobranch phylogeny (Chondrichthyes, Squatinidae). Copeia, pp. 505–518.

Shirai, S. (1992b). "Squalean Phylogeny, a New Framework of "Squaloid" Sharks and Related Taxa." pp. 1–151. Hokkaido University Press, Sapporo.

Swofford, D. L. (1993). "PAUP: Phylogenetic Analysis Using Parsi-mony, Version 3.1.1." Smithsonian Institution, Washington, DC.

Appendix 1

Specimens Examined

The specimens examined are housed at the following museums and institutions: BPBM: Bernice P. Bishop Museum, Honolulu; CAS: California Academy of Sciences, San Francisco; FAKU: Kyoto University, Department of Fisheries, Faculty of Agriculture, Kyoto; LACM: Los Angeles County Museum, Los Angeles; MCZ: Museum of Comparative Zoology, Cambridge; RUSI: J.L.B. Smith Institute of Ichthyology, Grahams-town; SIO: Scripps Institution of Oceanography, La Jolla; TCWC: Texas Cooperative Wildlife Collection, College Station; USNM: National Museum of Natural History, Smithsonian Institution, Washington, D.C.

Acronyms of Institutions follow Leviton et al. (1985).

Squatina dumeril (TCWC 4214.02, 275 mm TL); *Pristi-ophorus japonicus* (MCZ 42152, 558 mm TL); *Torpedo californica* (MCZ 43, 334 mm TL); *Torpedo marmorata* (MCZ 42, 270 mm TL); *Torpedo tremens* (TCWC uncata-loged); *Hypnos subnigrum* (MCZ 38602, 282 mm TL); *Narke japonica* (MCZ 1339, 270 mm TL); *Typhlonarke aysoni* (FAKU 46477, 317 mm TL, FAKU 47178, 306 mm TL); *Narcine brasiliensis* (TCWC 2923.01, 356 mm TL; TCWC uncataloged, 235 mm TL; TCWC uncata-loged, 34 mm TL; TCWC uncataloged, 36 mm TL; TCWC uncataloged, 53 mm TL); *Discopyge tschudii* (FAKU 105040, 456 mm TL; FAKU 105043, 394 mm TL); *Benthobatis marcida* (TCWC 443.01, 172 mm TL;

TCWC 1903.01, 137 mm TL); *Diplobatis pictus* (TCWC 1900.01, 119 mm TL; TCWC 1909.01, 99 mm TL; TCWC 5291.01, 119 mm TL); *Pristis pectinatus* (MCZ 36960, 1,028 mm TL); *Rhina ancylostoma* (TCWC uncataloged); *Rhynchobatus djiddensis* (MCZ 806, 490 mm TL); *Rhinobatos percellens* (MCZ 40025, 415 mm TL); *Rhinobatos planiceps* (TCWC uncataloged); *Zapteryx exasperata* (MCZ 833S, 198, 210 mm TL); *Zapteryx zyster* (TCWC uncataloged); *Trygonorrhina fasciata* (MCZ 982S, 256 mm TL); *Raja miraletus* (TCWC 6451, 01 mm TL); *Platyrhina sinensis* (CAS 15919, 353 mm TL); *Platyrhinoidis triseriata* (CAS 31248, 312 mm TL, 423 mm TL); *Zanobatus schoenleinii* (USNM 222120, 300 mm TL; TCWC uncataloged, 367 mm TL); *Plesiobatis daviesi* (BPBM 24578, 481 mm TL; RUSI 7861, 1,717 mm TL; TCWC uncataloged, 479 mm TL); *Urobatis halleri* (SIO uncataloged, 381 mm TL); *Urobatis jamaicensis* (TCWC 815.01, 285 mm TL); *Urotrygon aspidura* (CAS 51834, 284 mm TL; CAS 51835, 250 mm TL); *Urotrygon chilensis* (LACM 7013, 352 mm TL; USNM 29542, 300 mm TL); *Urotrygon microphthalmum* (USNM 222692, 244 mm TL); *Urotrygon venezuelae* (USNM 121966, 256 mm TL); *Dasyatis* sp. (TCWC uncataloged); *Dasyatis americana* (TCWC 2749.01, 1,607 mm TL; TCWC 5820.01, 614 mm TL); *Dasyatis sabina* (TCWC 5824.01); *Dasyatis violacea* (MCZ 57667, 958 mm TL); *Taeniura lymma* (TCWC 5278.01); *Potamotrygon constellata* (MCZ 295S); *Paratrygon aiereba* (MCZ 606S, 522 mm TL); *Gymnura micrura* (TCWC 642.08, 292 mm TL; TCWC uncataloged, 203 mm TL); *Myliobatis californicus* (MCZ 395, 500 mm TL); *Myliobatis goodei* (TCWC 3699, 725 mm TL); *Aetobatus narinari* (MCZ 1400, 950 mm TL); *Rhinoptera bonasus* (TCWC 4423.01, 604 mm TL); *Mobula hypostoma* (MCZ 36406, 490 mm disc width)

Appendix 2   Data Matrix

Squatina	0000000000	0000000 0 00	0000 0 00000	0 0 00000000	0000000000	0000000000	00000
Pristiophorus	0000000000	0000000 0 00	0000 0 00000	0 0 00000000	0000000000	0000000000	00000
Torpedinidae	1111111103	0001000 0 00	0011 0 1?1??	0 0 00100021	0000100001	0010000200	11001
Hypnidae	1111111103	0001000 0 00	0011 0 1?1??	0 0 00100021	000010?001	0010000200	11001
Narcinidae	1111111103	0001000 0 00	0011 0 1?1??	0 0 00000021	0000100001	0010000210	11101
Narkidae	1111111103	0001000 0 00	0011&1?1??	0 0 00000020	0000100001	0010000210	11111
Pristidae	1111111100	0000000 0 00	0000 0 00000	1 0 00000001	0000001000	0000001100	00002
Rhynchobatus	1111111100	0000000 1 00	1001 0 00000	1 0 00000001	1000001000	0010001100	00000
Rhina	1111111100	0000000 0 00	1001 0 00000	1 0 00000001	1000001000	0010001100	00000
Rhinobatos	1111111100	0000000 1 00	1001 0 00000	1 0 00000001	1000001110	0010001100	00000
Zapteryx	1111111101	0000000 1 00	1001 0 00000	1 0 00000001	1000001110	0010001100	00000
Trygonorrhina	1111111103	0000000 1 00	1001 0 00000	1 0 00000001	1000001110	0010001100	00000
Platyrhina	1111111101	0000000 1 01	?001 0 00000	1 0 00000001	1000001010	0010101100	00003
Platyrhinoidis	1111111101	0000000 1 01	?001 0 00000	1 0 00000001	1000001010	0010101100	00003
Zanobatus	1111111101	0000000 1 02	0001 0 00000	1 0 00011001	1000001310	1011001100	00003
Rajidae	1111111103	0000000&00	1001 0 00000	1 0 00000001	1000001110	0010001100	00000
Hexatrygon	1111111102	0010??0 0 ?2	0001 0 00200	0 0 00000001	1111001210	1011001101	?00??
Pleisobatis	1111111102	0010100 0 12	0011 0 00210	0 0 00010001	1111001210	1011001100	20023
Urolophus	1111111103	0010100 0 12	0011 0 00210	0 0 00010101	1111001210	1011001100	20023
Urobatis	1111111103	1010101 0 12	0011 0 00200	0 0 00010011	1111011210	1011001100	20023
Urotrygon	1111111103	1010101 0 12	0011 0 00200	0 0 00010021	1111011210	1011001100	20023
Dasyatis	1111111103	0110110 0 12	0011 0 00200	0 0 00010011	1111011210	1011011100	20023
Taeniura	1111111103	0110100 0 12	0011 0 00200	0 0 00010011	1111011210	1011011100	30023
Himantura	1111111103	0110100 0 12	0011 0 00200	0 0 00012011	1111011210	1011011100	30023
Potamotrygon	1111111103	0110100 0 12	0011 0 00200	0 0 00012011	1111001210	1011011100	30023
Gymnuridae	1111111103	0110110 0 12	0011 0 00300	0 0 00000001	1111001310	1111011100	?0023
Myliobatis	1111111113	0110120 0 12	0011 0 01301	0&110101$1	1111011310	1111011101	00024
Aetobatus	1111111123	0110120 0 12	0011 0 01311	0 1 11010131	1111011310	1111011101	?0024
Rhinoptera	1111111133	0110120 0 12	0111 1 01311	0 1 11010131	1111011310	1011011101	40024
Mobula	1111111133	0110120 0 12	0111 1 01311	0 1 01000131	1111011310	1011011101	?0024

Note: ?, unknown; &, (0,1); $, (1,2).

Appendix 3

List of Characters and Character States

1. Upper eyelid: 0 = present, 1 = absent. 2. Palatoquadrate: 0 = articulates with cranium, 1 = does not articulate with cranium. 3. Pseudohyal: 0 = absent, 1 = present. 4. Last ceratobranch: 0 = independent of scapulocoracoid, 1 = articulates with scapulocoracoid. 5. Synarchual: 0 = absent, 1 = present. 6. Suprascapulae: 0 = free of vertebral column, 1 = articulate with vertebral column. 7. Antorbital cartilage: 0 = free of propterygium, 1 = joins propterygium and nasal capsule. 8. Levator and depressor rostri muscles: 0 = absent, 1 = present. 9. Cephalic lobes(s) or fins: 0 = absent, 1 = single and continuous, 2 = single and discontinuous, 3 = paired and discontinuous. 10. Anterior nasal lobe: 0 = poorly developed, 1 = medially expanded, 2 = forms nasal curtain separated from mouth, 3 = forms nasal curtain close to mouth. 11. Spiracular tentacle: 0 = absent, 1 = present. 12. Caudal fin with radial cartilages: 0 = present, 1 = absent. 13. Serrated tail spine(s): 0 = absent, 1 = present. 14. Cephalic lateral line on ventral surface: 0 = present, 1 = absent. 15. Infraorbital loop of suborbital and infraorbital canals: 0 = absent, 1 = present. 16. Subpleural loop of hyomandibular canal: 0 = broadly rounded, 1 = abruptly reversing, 2 = abruptly reversing and anterior and posterior sections nearly parallel. 17. Lateral tubules of subplural loop: 0 = unbranched, 1 = dichotomously branched. 18. Abdominal canal on coracoid bar: 0 = absent, 1 = present. 19. Scapular loops of scapular canals: 0 = absent, 1 = present. 20. Rostral cartilage: 0 = complete, 1 = abbreviated, 2 = completely absent. 21. Rostral appendix: 0 = absent, 1 = present. 22. Dorsolateral components of nasal capsules: 0 = absent, 1 = present. 23. Nasal capsules: 0 = laterally expanded, 1 = ventrolaterally expanded. 24. Basal angle of neurocranium. 0 = present, 1 = absent. 25. Preorbital process: 0 = present, 1 = absent. 26. Supraorbital crest: 0 = present, 1 = absent. 27. Anterior preorbital foramen: 0 = dorsally located, 1 = anteriorly located. 28. Postorbital process: 0 = narrow and in otic region, 1 = absent, 2 = broad and in otic region, 3 = broad and in orbital region. 29. Postorbital process: 0 = separate from triangular process, 1 = fused with triangular process. 30. Postorbital process: 0 = projects laterally, 1 = projects ventrolaterally. 31. Jugal arch: 0 = absent, 1 = present. 32. Antimeres of upper and lower jaws: 0 = separate, 1 = fused. 33. Meckel's cartilage = not expanded medially, 1 = expanded medially. 34. Winglike processes of Meckel's cartilage: 0 = absent, 1 = present. 35. Medial part of hyomandibular: 0 = narrow, 1 = expanded. 36. Ligament between hyomandibula and Meckel's cartilage: 0 = absent, 1 = present. 37. Angular cartilage: 0 = absent, 1 = located anteriorly, 2 = located posteriorly. 38. Small cartilage(s) between hyomandibula and Meckel's cartilage: 0 = absent, 1 = present. 39. Basihyal and first hypobranchial cartilages: 0 = both present, 1 = former segmented, 2 = former absent, 3 = both absent. 40. Ceratohyal: 0 = fully developed, 1 = reduced or absent. 41. Suprascapula: 0 = articulated with vertebra, 1 = fused with synarchual. 42. Ball-and-socket articulation between scapular process and synarchual: 0 = absent, 1 = present. 43. Second synarchual: 0 = absent, 1 = present. 44. Ribs: 0 = present, 1 = absent. 45. Scapular process: 0 = long, 1 = short. 46. Scapular process: 0 = without fossa, 1 = with fossa. 47. Scapulocoracoid condyles: 0 = not horizontal, 1 = horizontal. 48. Mesocondyle: 0 = equidistant, 1 = closer to procondyle, 2 = closer to metacondyle, 3 = lacking, segmented or posteriorly located. 49. Antorbital cartilage: 0 = indirectly joins propterygium and nasal capsule, 1 = directly joins cartilages. 50. Antorbital cartilages: 0 = not anteriorly expanded, 1 = anteriorly expanded. 51. Distal section of propterygium: 0 = extends to procondyle, 1 = expends posterior to procondyle. 52. Pectoral radials: 0 = not expanded distally, 1 = expanded distally. 53. Fins: 0 = aplesodic, 1 = plesodic. 54. Puboischiadic bar: 0 = platelike, 1 = narrow and arched. 55. Puboischiadic bar: 0 = without triangular processes, 1 = with triangular processes. 56. Tail vertebrae: diplospondylous, 1 = fused into tube. 57. Ethmoideo-parethmoidalis: 0 = absent, 1 = present. 58. Intermandibularis muscle: 0 = present, 1 = absent, 2 = present but specialized. 59. Ligamentous sling on Meckel's cartilage: 0 = absent, 1 = present. 60. Depressor mandibularis muscle: 0 = present, 1 = absent. 61. Spiracularis muscle: 0 = undivided, 1 = divided and one bundle enters dorsal oral membrane, 2 = separate bundles insert on Meckel's cartilage and on hyomandibula, 3 = separate bundles extend beyond Meckel's cartilage and hyomandibula, 4 = one of two bundles inserts onto palatoquadrate. 62. Electric organs derived from branchial muscles: 0 = absent, 1 = present. 63. Coracobranchialis muscle: 0 = consists of three to five components, 1 = consists of single component. 64. Coracohyomandibularis muscle: 0 = single origin on facia continuous with pericardial membrane, 1 = separate origins on facia supporting coracoarcualis and pericardial membrane, 2 = separate origin on anterior portion of ventral gill arch region and on pericardial membrane. 65. Coracohyoideus muscle: 0 = parallel to body axis, 1 = absent, 2 = parallel to body axis but very short, 3 = diagonal to body axis, 4 = diagonal to body axis and fused to antimere.

Interrelationships of Acipenseriformes, with Comments on "Chondrostei"

Lance Grande
Department of Geology
Field Museum of Natural History
Chicago, Illinois 60605

William E. Bemis
Department of Biology
University of Massachusetts
Amherst, Massachusetts 01003

I. Introduction

In the terminology of Gardiner (1993), "lower actinopterygian" fishes are all ray-finned fishes other than Ginglymodi (gars and their close relatives), Halecomorphi (bowfins and their close relatives), and teleosts. Those three "higher" actinopterygian groups are collectively referred to as Neopterygii, a monophyletic group. Lower actinopterygians form a non-monophyletic group of convenience, which until recently was commonly referred to as "Chondrostei" (e.g., Berg, 1940; Romer, 1966; Schaeffer, 1973) and is still occasionally referred to by that name (e.g., Nelson, 1994). Lower actinopterygians include only two extant lineages, the Polypteriformes (bichirs and reedfishes) and Acipenseriformes (sturgeons and paddlefishes). However, lower actinopterygians also include about 270 fossil genera in about 60 nominal families (Gardiner, 1993).

Lower actinopterygians remain one of the great frontiers in paleoichthyology. That is to say, we still know relatively little about most of these fishes, especially in terms of their comparative anatomy and phylogenetic relationships, and there is much misinformation entrenched in the current descriptive literature. Today, a wealth of new, well-preserved fossil material, and much previously described older material, exists in museum collections. Most of this material needs fine preparation using modern techniques so it can be accurately described (or redescribed in many cases) in a broad comparative manner together with living and other well-preserved fossil species. Without such new updated descriptive work, comprehensive phylogenetic analyses of major extinct lower actinopterygian groups (e.g., paleonisciforms) is premature. Based on our experience, most of the existing descriptive literature is insufficient for such analyses. Therefore, we will focus primarily on Acipenseriformes in this paper. Because Acipenseriformes contains living species, it is an ideal group with which to make a fresh start in the study of lower actinopterygians. Acipenseriformes includes most of the living species of lower actinopterygians (about 25 species versus 12 species of Polypteriformes), has a good and interesting fossil record, and is the subject of renewed descriptive and phylogenetic studies (e.g., Grande and Bemis, 1991; Findeis, 1991, 1993, 1997; Birstein *et al.*, 1997; Bemis *et al.*, 1997). Here we review phylogenetic interpretations about Acipenseriformes including an important analysis by Jin (1995). We also add new descriptive and phylogenetic information on newly discovered important Chinese material, primarily the extinct acipenseriform family, †Peipiaosteidae, based on material that was unavailable

Copyright © 1996 by Academic Press, Inc.
All rights of reproduction in any form reserved.

to us for our earlier papers on acipenseriform phylogeny.

Acipenseriformes is an ancient group known to be at least as old as early Jurassic (approximately 200 million years before present). Their known geographic range is Holarctic. Acipenseriforms are generally included in studies of basal osteichthyan relationships, and living members of the order are often referred to as "living fossils" (e.g., Gardiner, 1984; Grande and Bemis, 1991). In spite of their putative importance, little comprehensive phylogenetic work has been done on Acipenseriformes. Even the monophyly of the largest and most wide-ranging Recent genus (*Acipenser*) is still in question (Findeis, 1997). Of the two extant acipenseriform families, only Polyodontidae (paddlefishes) has undergone comprehensive phylogenetic study (Grande and Bemis, 1991). Since that paper, the discovery of another new paddlefish (†*Protopsephurus* Lu, 1994) indicates that Polyodontidae originated at least by the Lower Cretaceous or Upper Jurassic (see following discussion). Thus we can predict the eventual discovery of late Jurassic or early Cretaceous fossils of Acipenseridae.

Two of the new specimens of *Peipiaosteus* illustrated in this paper are deposited at the Field Museum of Natural History, Chicago, Illinois (FMNH). Other materials discussed are in the collection of the Institute of Vertebrate Paleontology and Paleoanthropology, Beijing, China (IVPP); the Kitakyushu Museum and Institute of Natural History, Kitakyushu, Japan (KMNH); the Paleontological Institute, Russian Academy of Sciences, Moscow (PIN); and the University of Michigan Museum of Paleontology, Ann Arbor, Michigan (UMMP). A dagger (†) precedes the name of an exclusively fossil (i.e., extinct) taxon.

II. Historical Overview of Acipenseriformes

"Acipenseriformes" was first used by Berg (1940) as an ordinal-level name for sturgeons (Acipenseridae) and their close relatives. Berg (1948a,b) and Vladykov and Greeley (1963) summarized much information concerning these fishes. The common or Baltic sturgeon, *Acipenser sturio*, was first described by Linnaeus in 1758, who regarded it as a species of shark. This is not surprising because in Linnaeus' time, gnathostome fishes were all classified either as Chondropterygii (= chondrichthyans) or Ossei (= osteichthyans). Sturgeons were the only nonteleostean osteichthyans known to pre-Linnaean ichthyologists (Patterson, 1982). Agassiz (1833–1944) was the first to distinguish sturgeons from sharks by placing them in his Ganoidei.

Some time after Linnaeus' description of *Acipenser*, the first paddlefish (Polyodontidae) was mentioned (Maduit, 1774) and eventually described by Walbaum (1792). As with *Acipenser*, this species was first classified as a shark (*Squalus spathula*). This species was later placed in a new genus, *Polyodon*, by Lacépède (1797).

By the late 1830s, fossil lungfishes and Recent polypterids had been described and added to Agassiz's Ganoidei, and this group was increasingly recognized as unnatural. Müller (1844), noting that the group as now constructed had no diagnostic characters, erected Chondrostei to contain sturgeons and paddlefishes and exclude the other members of Ganoidei. Cope (1887) renamed the Chondrostei as Podopterygia, but his name was never widely accepted or used. During the latter half of the 19th century and continuing through the 1960s, many unrelated fossil taxa were incorporated into Chondrostei, fundamentally altering Müller's (1844) original concept of the group. The resulting grade group "Chondrostei" has confused generations of vertebrate biologists (e.g., Romer, 1966) and no doubt frustrated basic research on many taxa traditionally included within it. Thus, one objective of this chapter is to change the legacy of confusion about "Chondrostei" by focusing on well-preserved, well-prepared specimens and reexamining the definable, monophyletic groups within "Chondrostei," in this case, the Acipenseriformes.

In our review of the biology of Acipenseriformes, Bemis *et al.* (1997) consolidated the phylogenetic analysis of Grande and Bemis (1991) for Acipenseriformes and Polyodontidae with those of Findeis (1997) for Acipenseridae and produced a summary cladogram complete with some additional and revised characters. A summary of that analysis (Fig. 1) serves as the starting point for the new work reported in this paper. Figure 1 shows that among the taxa we surveyed, †*Birgeria* (Triassic of eastern Greenland) is the sister taxon of Acipenseriformes, a conclusion that we based on three osteological synapomorphies (characters A1–A3, Fig. 1; see comments where this node is discussed in Phylogeny of Acipenseriformes). As the detailed osteology of additional lower actinopterygians (such as †*Errolichthys*) becomes better known, they should be included in analyses. As an outgroup for examining Acipenseriformes, †*Birgeria* is nearly ideal, thanks to the detailed osteological description provided by Nielsen (1949).

Several authors examined evidence that confirms monophyly of Acipenseriformes (e.g., Jollie, 1980; Patterson, 1982; Gardiner and Schaeffer, 1989; Grande and Bemis, 1991). Our review (Bemis *et al.*, 1997) reported and discussed eight osteological synapomorphies of Acipenseriformes (Node B, Fig. 1;

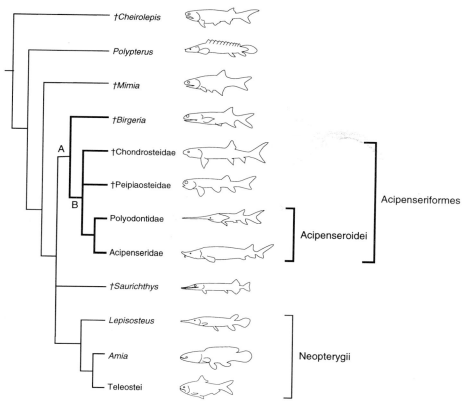

FIGURE 1 Phylogeny of certain lower actinopterygians from Bemis *et al.* (1997). Heavy lines indicate Acipenseriformes and an outgroup taxon, †*Birgeria*, used to polarize characters. This diagram explains basic nomenclature for the taxa included in this paper and is the starting point for the new analysis presented here. See Fig. 11 for summary of the phylogenetic conclusions from this paper.

these features are summarized in Phylogeny of Acipenseriformes). Thus Acipenseriformes is almost certainly monophyletic.

Within the order Acipenseriformes, we recognize four families (Fig. 1). These are the following: †Chondrosteidae Traquair 1877 (Lower Jurassic deposits of England and Germany), †Peipiaosteidae Liu and Zhou 1965 (Upper Jurassic to Lower Cretaceous of central Asia and northeastern China; see Jin, 1995: fig. 1, for map of localities), Polyodontidae Bonaparte 1838 (Upper Jurassic or Lower Cretaceous of Peipiao-Liaoning Province, China, to Recent of China and North America), and Acipenseridae Bonaparte 1831 (Cretaceous to Recent of North America, Europe, and Asia). Bemis *et al.* (1997) reaffirmed the validity of Acipenseroidei, a clade containing Polyodontidae and Acipenseridae (Fig. 1). This particular composition of Acipenseroidei was first proposed by Grande and Bemis (1991) but challenged in a subsequent paper by Zhou (1992). The many problems with Zhou's (1992) analysis are discussed in detail in Bemis *et al.* (1997).

In particular, we note that Zhou's (1992) "new" phylogenetic hypothesis is the same as the one proposed by Gardiner (1984) with the addition of †*Peipiaosteus*. Zhou (1992) did not examine most of the taxa included in this scheme, nor did the scheme include the new findings of Grande and Bemis (1991). Over and above these problems, our reanalysis of Zhou's (1992) own data showed that a tree recognizing our Acipenseroidei was two steps shorter than the one he proposed (CI: 1.00 versus CI: 0.78; see Bemis *et al.*, 1997, for analysis and further information). Thus, Acipenseroidei, in the sense defined by Grande and Bemis (1991), survives Zhou's (1992) challenge.

As shown in Fig. 1, Bemis *et al.* (1997) left relationships among †Peipiaosteidae, †Chondrosteidae, and Acipenseroidei an unresolved trichotomy, primarily because of inadequate knowledge and conflicting information about the osteology of any member of †Peipiaosteidae. A practical objective of this chapter is to carry the analysis of acipenseriform relationships one step further by providing a new, detailed study of

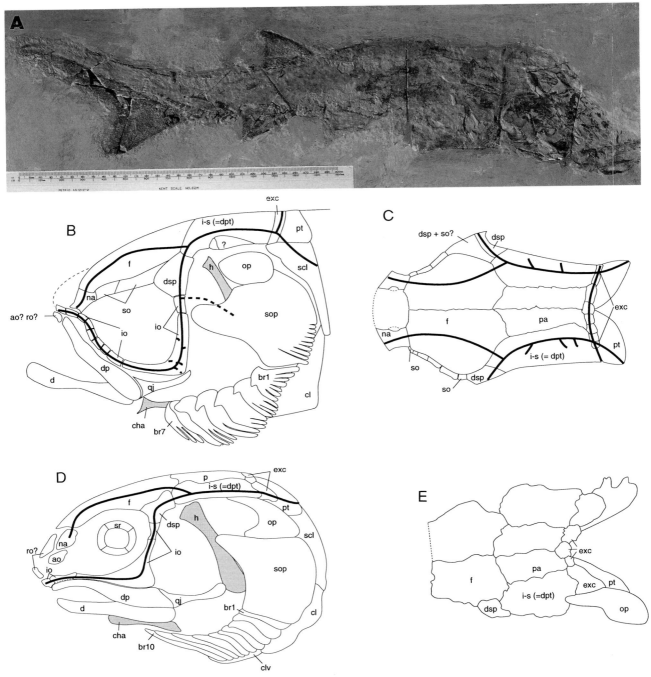

FIGURE 2 Cranial osteology of †*Stichopterus* and †*Chondrosteus*. (A) Photograph of a specimen referred to †*Stichopterus popovi*, anterior facing right (uncataloged PIN specimen on touring exhibit at Queensland Museum, Australia). (B) and (C) †*Stichopterus reissi* redrawn and relabeled after Jakovlev (1977, figs. 6 and 7), anterior facing left. The peculiar opercular and branchiostegals of †*Stichopterus* closely resemble those of †*Peipiaosteus*. We disagree with certain interpretations of these figures by Jakovlev, e.g., his premaxilla is probably an antorbital or rostral bone, and his anterior infraorbitals may be rostral bones, but we cannot yet treat all of these discrepancies in detail because we have not yet studied material of †*Stichopterus* first hand. (D) Lateral view of the skull of †*Chondrosteus acipenseroides*, redrawn from Gardiner and Schaeffer (1989, fig. 21), anterior facing left. (E) Dorsal view of the skull of †*Chondrosteus acipenseroides*; portion of skull roof redrawn from Traquair (1887, fig. 1), anterior facing left. Elements are relabeled to reflect our terminology (see text for explanation). We note two important differences between our interpretation of †*Chondrosteus* and that of Gardiner and Schaeffer (1989). The first concerns the element we interpret as a

†*Peipiaosteus pani*, which allows us to resolve that trichotomy. We also add information about the paddlefish †*Protopsephurus* and comment on †*Stichopterus*, and †*Spherosteus*, as well as an acipenseriform from the Upper Jurassic–Lower Cretaceous of China, †*Yanosteus longidorsalis* Jin *et al.* (1995). We include a revised tree and character scheme summarizing our information and interpretations.

III. Descriptive Osteology of †*Peipiaosteus Pani* Liu and Zhou, 1965

In our scheme, †Peipiaosteidae includes four genera, †*Peipiaosteus*, †*Stichopterus*, †*Spherosteus*, and †*Yanosteus*. Here, we redescribe the osteology of †*Peipiaosteus pani* Liu and Zhou (1965), which is the type and best-known species of the family †Peipiaosteidae. In Grande and Bemis (1991, p. 116) the available descriptive data were too limited to allow inclusion of †Peipiaosteidae. Later, on the basis of Zhou's (1992) study, Bemis *et al.* (1996) concluded that †Peipiaosteidae does belong to Acipenseriformes, yet the available descriptive information was still too limited to allow detailed comparative study. Our redescription is based on three specimens and organized following our standard format for systematic osteology.

Genus †*Peipiaosteus* Liu and Zhou 1965

Type species—†*Peipiaosteus pani* Liu and Zhou 1965. The genus is monotypic. A second species of †*Peipiaosteus*, †*P. fengnengensis* Bai 1983, was described, but later synonymized with †*P. pani* by Zhou (1992).

Generic diagnosis—†*Peipiaosteus* differs from other genera of †Peipiaosteidae by the following features: (1) The presence of a short dorsal fin (differentiating it from †*Yanosteus* and †*Sperosteus*) and (2) only a single pair of extrascapular bones (versus several pairs in †*Stichopterus*).

Remarks—Another genus usually included in †Peipiaosteidae and usually considered to be very close to †*Peipiaosteus* is †*Stichopterus*, which is known from the Upper Jurassic–Lower Cretaceous of Siberia and Mongolia (review in Jin, 1995). †*Stichopterus woodwardi* Reis (1910) is from the Turgen Formation of Transbaikalia (Siberia). †*Stichopterus popovi* Jakovlev (1986) is from the Gurvaneren Formation of Mongolia. A second Mongolian species, possibly from the Shinkhundukian of Mongolia, is known as †*S. reissi*, but authorship of this species is in question (see Jakovlev, 1977, and Jin, 1995, for discussion). Very little is known about the anatomy of †*Stichopterus*, and the available information is confusing. A photograph of a specimen referred to †*Stichopterus popovi* is shown in Figure 2A. Diagrams of cranial bones from specimens referred to †*Stichopterus reissi* (redrawn from Jakovlev, 1977, figs. 6 and 7) are shown in our Figures 2B and 2C. We relabeled the diagrams in Figures 2B and 2C to reflect our osteological nomenclature and to reinterpret some features. For example, Jakovlev (1977) considered that a premaxilla was present in †*Stichopterus*, but we interpret the element he labeled as premaxilla to be either a portion of the antorbital bone or a rostral canal ossification (Fig. 2B, ao? ro?). Many of the small ossifications in the rostral region of acipenseriforms are difficult to interpret. It is highly unlikely, however, that a premaxilla is present in †*Stichopterus*, for no other acipenseriform has this element, and there are no diagnostic features of the premaxilla that would distinguish it from other dermal elements of the rostral region. A second reinterpretation included in Figure 2B concerns the element traditionally considered to be a maxilla. Findeis (1991) and Bemis *et al.* (1997) recently argued that the maxilla is absent in acipenseriforms and that the bone traditionally interpreted as a maxilla is actually a dermopalatine (Fig. 2B, dp [= mx]; for additional comments, see description of the jaws of †*Peipiaosteus pani*, below).

We cannot provide a meaningful differential diagnosis between †*Peipiaosteus* and †*Stichopterus* because we have not been able to study †*Stichopterus*, but based on Jakovlev's (1977) drawings, the two genera are very closely related and possibly synonymous. If they are synonymous, then the name †*Stichopterus* Reis 1910 will have priority. The proposed differen-

---

quadratojugal (qj), which they labeled as a preopercle; our interpretation is consistent with the condition in other Acipenseriformes, none of which has a preopercle. The second concerns the lack of contact between the dermopalatine (dp [= mx]) and nearby infraorbital bones (io). The diagram in Gardiner and Schaeffer (1989) does not indicate any space between these elements but seems instead to imply that they were suturally connected. There is no evidence for this in the specimens of †*Chondrosteus* that we have seen, nor does such a connection between the upper jaw and the infraorbital bones or other elements of the cheek occur in any other acipenseriform, which is why we have drawn dp (= mx) and io with a space between them.

tial diagnosis of the genera provided by Zhou (1992, p. 100) reads:

Peipiaosteus [is] distinguishable from *Stichopterus* mainly by: the two supraorbitals present in the latter are absent in the former; the infraorbitals are narrower in *Peipiaosteus* than in *Stichopterus*; the endopterygoid and the palate found in *Stichopterus* have disappeared in *Peipiaosteus*; the dorsal fin is longer in *Stichopterus* than in *Peipiaosteus*; in caudal fin the rhomboid scales of the upper lobe remain in *Stichopterus* but totally lost in *Peipiaosteus*. From above comparison it is reasonable to regard *Peipiaosteus* and *Stichopterus* as two sister genera, and the former is more derived than the latter.

Most of these supposedly diagnostic differences are problematic, possibly because described material of †*Stichopterus* is from large adult specimens, presumably at more advanced developmental stages than most or all of the described material for †*Peipiaosteus*. Thus, several "diagnostic differences" may be instead ontogenetic differences. For example, the narrower infraorbitals, and lack of rhomboid scales on the caudal fin of †*Peipiaosteus* (as opposed to slightly better developed infraorbitals and a few poorly developed rhomboid scales on the caudal fin of †*Stichopterus*), could reflect ontogenetic differences. Both smaller †*Stichopterus* and larger †*Peipiaosteus* specimens need to be included in future comparisons. Zhou's (1992) report of †*Stichopterus* having a longer dorsal fin than †*Peipiaosteus* seems incorrect or at least unsupported by available meristic data. Finally, the "endopterygoid and palate" differences mentioned by Zhou (1992) are unclear and need further study and diagrammatic comparisons. One feature that appears to differentiate †*Stichopterus* from †*Peipiaosteus* is a difference in the number of extrascapular elements: †*Stichopterus* is reported to have as many as seven extrascapular bones (labeled *exc*, Fig. 2C), whereas †*Peipiaosteus* has only two lateral extrascapulars. This region of the skull roof, however, is notoriously variable in living acipenseriforms (e.g., Findeis, 1997, fig. 5), so this apparent generic difference should be checked and illustrated for a large series of specimens.

The genus †*Yanosteus* Jin *et al.* 1995, is readily distinguishable from †*Peipiaosteus*, although the two genera are stratigraphically concurrent. †*Yanosteus* and the very poorly known genus †*Spherosteus* appear to be unique among acipenseriforms in having an elongate dorsal fin (Jin *et al.*, 1995; fig.1 for †*Yanosteus* and Jakovlev, 1977, pl. 13 for †*Spherosteus*). Jin *et al.* (1995) also note the "strong resemblance" of †*Yanosteus* to †*Peipiaosteus* and †*Chondrosteus*.

Species †*Peipiaosteus pani* Liu and Zhou 1965

Holotype—IVPP V3049.1, a nearly complete skeleton with an estimated total length of about 210 mm; illustrated in Liu and Zhou (1965, fig. 1).

Distribution (locality and age)—Upper Jurassic or Lower Cretaceous deposits of Liaoning and Hebei provinces, northeastern China (exact age of deposits is currently controversial; Jin, 1995).

Habitat—Specimens are preserved mostly as impressions in a buff to light gray, finely laminated claystone, thought to have been deposited in fresh water. Specimens are commonly associated with †*Lycoptera* (Hiodontoidea and Osteoglossomorpha), a common Asiatic teleost from the Upper Jurassic and Lower Cretaceous. †*Peipiaosteus pani* is also fairly common where it occurs (e.g., Liu and Zhou, 1965, p. 244), but until the 1990s there were few if any specimens in collections outside of China.

Material examined—Three nearly complete skeletons from the type locality. These are FMNH PF14370 (184 mm total length), FMNH PF14371 (estimated 320 mm total length), and KMNH VP100,239 (189 mm total length). We prepared specimens for study by immersing them in 30% technical grade HCl to etch away remaining fragments of bone. The specimens were then rinsed in water for one day and dried; black liquid latex rubber was applied to the fine impressions in the claystone. The resulting latex peels provided finely detailed casts of the skeleton.

Emended diagnosis—As for genus (monotypic).

Etymology—*Peipiao*, after the type locality. *Osteus*, bone; from the Greek.

Comments—Despite three previous descriptions of †*Peipiaosteus* (Liu and Zhou, 1965; Bai, 1983; Zhou, 1992), this taxon needed redescription. Based on the fully prepared material described here, we can add new information to what is known about †*Peipiaosteus pani* (the type and probably the only valid species) and present some different interpretations of previously described morphological features.

Anatomical Description

Measurements—Table 1 provides selected measurements of our new material and the holotype specimen described by Liu and Zhou (1965).

The largest specimen observed by us was about 600 mm total length in a private collection in Germany. Unfortunately, we were unable to borrow the specimen for preparation and analysis. The three specimens in hand are well-preserved, but larger specimens should be examined for elements that may ossify in later ontogenetic stages. The largest reported specimen is 900 mm TL (Zhou, 1992, p. 99). We believe the restoration in Zhou (1992, fig. 1) combined the head of a young individual with the postcranial skeleton of a much older individual. Compared to Zhou's restoration, our larger specimen (Fig. 3B) has a more fully ossified skull (e.g., with better developed sutural connections between dermal bones), yet it has an ear-

TABLE 1 Measurements of Three Specimens of †*Peipiaosteus*

Specimen number (total length)[a]	Head length (as % of total length)	Mid-eye to anterior tip of rostrum (as % of total length)	Pre-dorsal length (as % of total length)	Pre-pelvic length (as % of total length)	Pre-anal length (as % of total length)	Caudal length (as % of total length)
FMNH PF14370 (184 mm)	31 mm (17)	? (?)	100 mm (50)	78 mm (42)	105 mm (57)	48 mm (26)
FMNH PF14371 (est. 320 mm)	52 mm (16)	13 mm (4)	175 mm (55)	143 mm (45)	184 mm (58)	est. 85 mm (27)
IVPP 3041.1 (est. 215 mm)	35 mm (16)	? (?)	120 mm (56)	96 mm (45)	130 mm (60)	? (?)

[a]Measurements for two nearly complete skeletons of †*Peipiaosteus pani* described here, compared with measurements of holotype. Caudal length is measured along horizontal axis of fish from anterior tip of dorsal caudal fulcrum to below posterior tip of upper caudal lobe. We did not see the actual specimen of the holotype, and our measurements are taken from a photograph which is supposedly at life size (Liu and Zhou, 1965, pl. 1, fig. 1). Consequently, the individual measurements for the holotype may be slightly inaccurate, but the proportional percentages should be accurate.

FIGURE 3 †*Peipiaosteus pani* Liu and Zhou 1965, from Upper Jurassic–Lower Cretaceous freshwater deposits of Peipiao, Liaoning province, China. Two specimens: one (A) before acid preparation and another (B) after acid preparation. (A) Small specimen with skin pigmentation pattern preserved (FMNH PF14370; 184 mm total length). Anterior part of body in dorsal view, caudal region in lateral view (body is twisted 90° between pelvic region and caudal fin). Latex peels and drawings of prepared specimen shown in Figs. 4, 5, and 9. (B) Larger specimen (FMNH PF14371; estimated 320 mm total length) in lateral view, posterior end of upper caudal lobe missing. Latex peels and drawings of prepared specimen shown in Figs. 7 and 8.

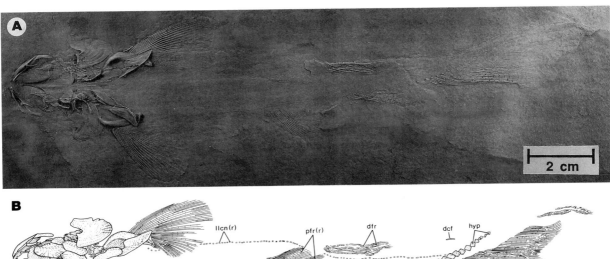

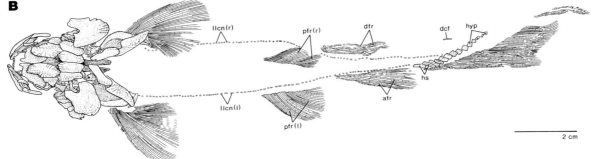

FIGURE 4 †*Peipiaosteus pani*. Specimen from Figure 3A after acid preparation (FMNH PF41370, 184 mm total length). Close-up of head region is illustrated in Fig. 5 and close-up of caudal region in Fig. 9. (A) Latex peel of specimen lightly dusted with ammonium chloride. (B) Line drawing of A. Abbreviations in Appendix 1.

lier developmental stage of the postcranial skeleton (e.g., the posterior vertebral elements and posterior-most proximal and middle radials of median fins were not yet mineralized).

Bones of the skull—The head region is preserved in three different orientations in our specimens. In the smaller individuals (FMNH PF14370; 184 mm total length and KMNH VP100,239; 189 mm total length), the head is preserved in dorsal view (Figs. 3A, 4, and 5) and ventral view (Fig. 6). The head of the larger individual (FMNH PF14371; estimated 320 mm total length) is preserved in lateral view (Figs. 3B, 7, and 8). Our interpretation of the skull and shoulder girdle is very different from that of Zhou (1992). We believe this is largely due to our better-preserved material representing larger, better-developed individuals than those studied by Zhou (1992).

The skull roof of our specimens of †*Peipiaosteus* is poorly ossified. The posteriormost dorsal bones of the cranial midline are the paired parietal bones (*pa*, Figs. 5 and 8). There is no trace of a sensory canal in or above these bones. In our smaller specimen (Fig. 5), as in an even younger individual illustrated by Zhou (1992, fig. 3), the parietal bones are in contact medially, but the frontals (*fr*, Fig. 5) are not, and the medial

edges of the frontals are relatively smooth. The lack of medial contact between the frontal bones is probably a developmental feature because the medial edges of the frontal bones in our larger specimen show deep interdigitating sutures (Fig. 8). The frontal canal traverses the length of this bone (Figs. 5 and 8).

Lateral to each parietal bone in †*Peipiaosteus* is a bone usually termed dermopterotic in neopterygians (e.g., see Patterson, 1973, fig. 3, for *Lepisosteus*, *Amia* and *Elops*) and intertemporo-supratemporal in many non-neopterygian actinopterygians (e.g., Gardiner and Schaeffer, 1989; Grande and Bemis, 1991). Here, we discuss evidence for the homology of this bone in actinopterygians. Although the term intertemporo-supratemporal may be more reflective of the phylogenetic origin of the bone, dermopterotic is much more widely used in anatomical studies of actinopterygians. This bone, *is* (= *dpt*) in Figs. 5 and 8, is identified by the presence of the supratemporal sensory canal which bifurcates in the anterior part of the bone to connect medially to the supraorbital canal and laterally to the infraorbital canal. In polyodontids, where the bony tubes of the sensory canals lie superficial to the dermal bones, the dermopterotic is identified topologically (i.e., it is the bone or bones lying beneath

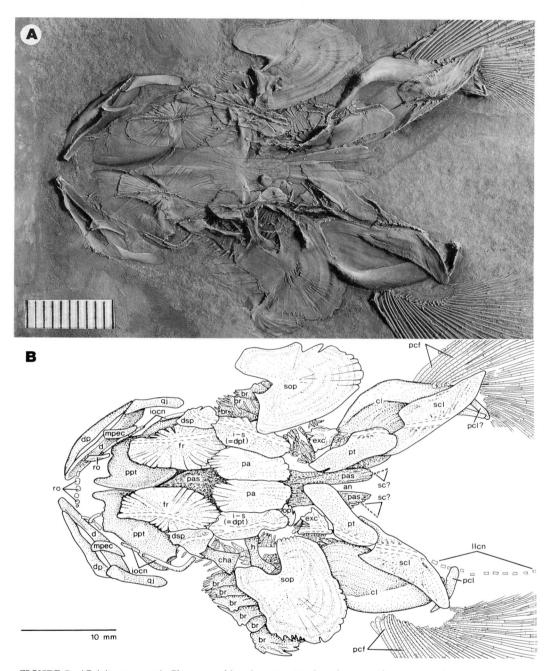

FIGURE 5 †*Peipiaosteus pani*. Close-up of head region (in dorsal view) of specimen shown in Figs. 3A and 4. (A) Latex peel of specimen lightly dusted with ammonium chloride. (B) Line drawing of A. Abbreviations in Appendix 1.

the anterior bifurcation of the supratemporal sensory canal and lateral to the parietal; see element labeled *i-s* in Grande and Bemis, 1991, figs. 13 and 6). The term *intertemporo-supratemporal* (e.g., Holmgren and Stensïo, 1936; Jollie, 1980; Grande and Bemis, 1991) reflects the notion that the bone is phylogenetically (and in some observable cases, ontogenetically) derived from autogenous intertemporal and supratem-

poral bones. Phylogenetic evidence for this is based partly on the observation that the bone "occupies the area primitively covered by the intertemporal and supratemporal" (Gardiner and Schaeffer, 1989, p. 148). In many extinct stem-actinopterygians, such as †*Cheirolepis*, †*Mimia*, †*Howqualepis*, †*Moythomasia*, and †*Tegeolepis*, separate intertemporal and supratemporal bones occur (Gardiner and Schaeffer, 1989, fig. 2). In adult

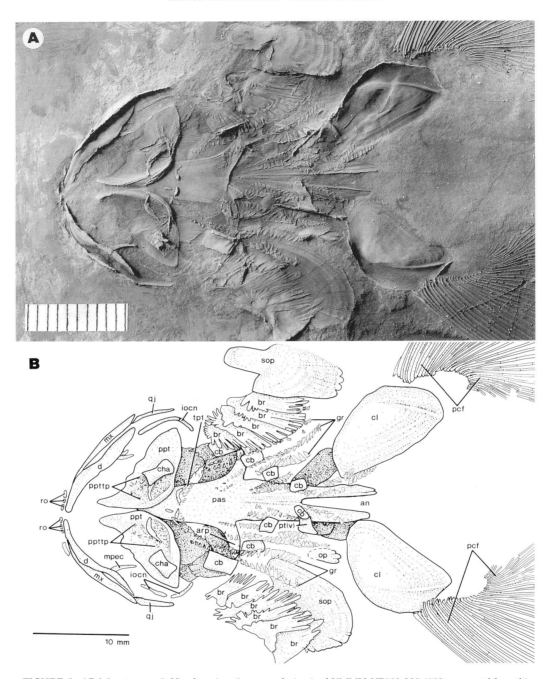

FIGURE 6 †*Peipiaosteus pani*. Head region (in ventral view) of KMNH VP100,239 (189 mm total length). (A) Latex peel of specimen lightly dusted with ammonium chloride. (B) Line drawing of A. Abbreviations in Appendix 1.

Polypterus, however, (a living stem-actinopterygian) only a single bone is present. Most other actinopterygians also have only a single bone in this region. An exception is the individual variation reported in *Polyodon spathula* (Grande and Bemis, 1991), where sometimes there is a single element, but often there are two. The case for fusion of two elements to form a single bone is supported by developmental studies by Pehrson (1947, pp. 438–440, for *Polypterus*; 1944, p. 37, for *Acipenser*), Jollie (1980, p. 245 for *Polyodon*), and Pehrson (1922, p. 54; 1940, p. 14 for *Amia*). Gardiner and Schaeffer (1989, pp. 148–150) theorize, based on phylogenetic evidence, that the fusion of separate intertemporal and supratemporal bones "took place in-

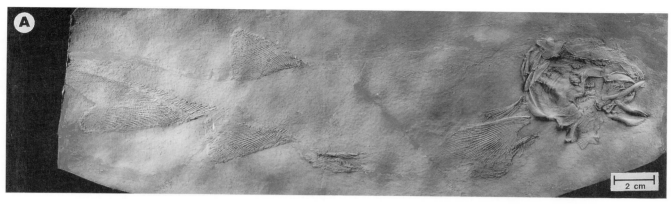

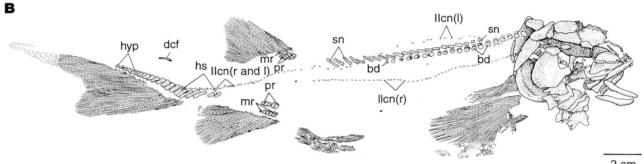

FIGURE 7 †*Peipiaosteus pani.* Specimen shown in Figure 3B after acid preparation (FMNH PF14371, estimated 320 mm total length). Close-up of head region is illustrated in Figure 8. (A) Latex peel of specimen lightly dusted with ammonium chloride. (B) Line drawing of A.

dependently within the unknown *Polypterus* lineage ... and once more in advanced lower actinopterygians [presumably a group containing more advanced lower actinopterygians plus neopterygians]."

Anterior to the dermopterotic and lateral to the frontal in †*Peipiaosteus* is the dermosphenotic (dsp, Figs. 5 and 8). This element is the last bone in the infraorbital series, and it starts out as an autogenous canal bone, as in other actinopterygians (e.g., see illustration of a younger juvenile specimen of †*Peipiaosteus* in Zhou, 1992, fig. 3). As this bone develops, it enlarges and incorporates itself into the skull roof with laminar bone and interdigitating sutural connections to the frontal and dermopterotic bones (Fig. 8). The other infraorbital elements of †*Peipiaosteus* remain as small autogenous ossified tubular canal bones surrounding the lower part of the orbit (iocn, Figs. 5, 6, and 8).

Anterior to the frontals, the supraorbital canal continues forward in a series of ossified tubes (sobn, Fig. 8) that connect to the nasals (na, Fig. 8) which are also little more than ossified canal tubes. These ossifications of the supraorbital canal are not identifiable in our smaller specimen. Based on what happens to canal bones in living acipenseriforms, they would prob-

ably be disproportionately larger in larger individuals. Lateral to the nasals are small Y-shaped canal tubes thought here (and in Zhou, 1992, fig. 2) to be antorbital bones (ao, Fig. 8). A series of short canal tubes termed the rostrals (ro, Figs. 5 and 6) form commissures between the right and left sides of the lateral line canal systems anterior to the nasals.

Zhou (1992) provided the first detailed description of the head region of †*Peipiaosteus.* Our interpretation of the posterior region of the skull (and upper shoulder girdle) is markedly different from his, based on our apparently better preserved material. What Zhou (1992, Fig. 3) identified as a posttemporal, we identify as an extrascapular (exc, Figs. 5 and 8), sometimes also referred to as the supratemporal by various authors (e.g., Patterson, 1973). This bone cannot be a posttemporal, in our opinion, because it contains the supratemporal commissural canal (the canal branching medially in the exc bones of Figs. 5 and 8), a feature typically used to identify extrascapular bones. Also, we identify another canal-bearing bone between the extrascapular and the supracleithrum that more clearly resembles the posttemporal of other acipenseriforms (see discussion of posttemporal). Only two laterally placed extrascapular bones are

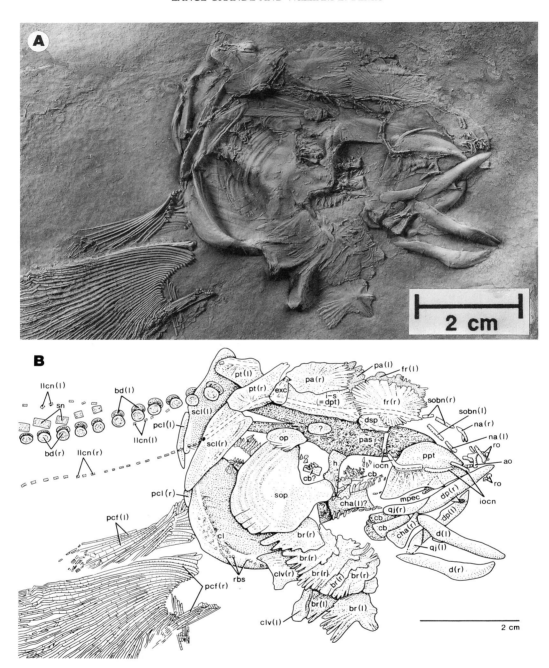

FIGURE 8 †*Peipiaosteus pani*. Head region (in lateral view except for *pas* which is in dorsal view) from specimen shown in Figures 3B and 7. Some of the round based scales of shoulder girdle region (rbs) enlarged in Figure 9C. (A) Latex peel of specimen lightly dusted with ammonium chloride. (B) Line drawing of A. Abbreviations in Appendix 1.

present in our material. We found no trace of any extrascapular elements between the two such as exist in †*Stichopterus* (*exc*, Figs. 2B and C) and †*Chondrosteus* (*exc*, Figs. 2D and E), †*Strongylosteus hindenburgi* (see Jaekel, 1929, pl. XI, fig. 1), Polyodontidae (e.g., see Grande and Bemis, 1991, figs. 7 and 8), and Acipenser-idae (e.g., see Jollie, 1980, fig. 2, for *Acipenser ruthenus* and Findeis, 1997, fig. 5, for other acipenserids). The homologies of canal-bearing bones in the occipital region of acipenserids are somewhat more complicated. The lateral extrascapular usually can be fused to the posttemporal, but variation exists, and according to

Jollie (1980, p. 236) in *Acipenser sturio* "some specimens have a separate lateral extrascapular intercalated between the intertemporo-supratemporal and the post-temporal" (also see Findeis, 1997, fig. 5).

Posterior to the extrascapulars are the long post-temporal bones (*pt*, Figs. 5 and 8). Zhou (1992, figs. 1 and 3) drew the posttemporal and supracleithrum as a single bone, which he identified as a supracleithrum. Our specimens clearly show two elements, in agreement with the earlier description by Liu and Zhou (1965). The posttemporal of †*Peipiaosteus* is similar to that of †chondrosteids and polyodontids in that it has an anteromedially pointing arm that articulates with the posterior edge of the parietals (covered by the median extrascapular elements in †chondrosteids and polyodontids). In †*Peipiaosteus* and †chondrosteids, where the extrascapulars are more than ossified canal tubes and are sutured into the skull roof, the posttemporals do not contact the dermopterotic. In polyodontids the extrascapulars are poorly developed, often little more than ossified canal tubes which form above the other posterior dermal bones of the skull roof. The posttemporals of polyodontids are also greatly enlarged, with prominent ventral wings (e.g., Grande and Bemis, 1991, fig. 10). Consequently, in polyodontids the posttemporal does contact the dermopterotic. In acipenserids the condition is variable because the extrascapular usually fuses into the posttemporal (see previous discussion of extrascapulars above).

The posttemporal of †*Peipiaosteus* articulates with the supracleithrum (*scl*, Figs. 5 and 8). The main trunk sensory canal, which runs through the posttemporal to the supracleithrum, connects posteriorly to the trunk lateral line system (*llcn*, Figs. 4–8).

There is no trace of any median nuchal plates in our material of †*Peipiaosteus*. Nuchal plates of acipenseriforms are problematic, and the literature is confusing because these bones were originally defined by the condition in sturgeons. Among acipenseriforms, sturgeons uniquely possess five rows of bony scutes armoring the body, and the most antero-dorsal of these scutes becomes fully incorporated into the skull roof during ontogeny (e.g., Findeis, 1997, *ds1*, fig. 5). This antero-dorsal scute is typically identified as a median nuchal plate (e.g., Jollie, 1980, M, fig. 2). Grande and Bemis (1991, *nu?*, fig. 66) considered that a median nuchal might be present in †*Paleopsephurus*, although this area of the skull of the type and only known specimen is poorly preserved, and no other scutes are present in any paddlefish. Lu (1994, *nu*, fig.1B) reported a median nuchal plate in the polyodontid †*Protopsephurus*; again, unfortunately, this region is questionably preserved and the element in question could prove to be a median extrascapular.

†Chondrosteids have a median bone in the posterior part of the skull (e.g., *exc*, Fig. 2E), but the element is a small bone that does not resemble the scutes of sturgeons. Based on ontogenetic studies of the sturgeon skeleton reported by Jollie (1980, E, fig. 2) and comparative studies by Findeis (1997, *excm*, fig. 5), the median element in †*Chondrosteus* seems likely to be a median extrascapular. This interpretation also appears to be consistent with the condition in †*Stichopterus* (*exc*, Fig. 2C).

The parasphenoid of †*Peipiaosteus pani* (*pas*, Figs. 5, 6, and 8) is incompletely visible in our specimens, but enough of it shows to indicate a deep aortic notch (*an*, Figs. 5 and 6) and lack of an ossified basioccipital as in other acipenseriforms. One of our specimens also shows well-developed ascending rami pointing nearly perpendicular to the sagittal plane (*arp*, Fig. 6). Liu and Zhou (1965, pl. 1, fig. 3, and text-fig. 2) illustrated a complete isolated parasphenoid from †*Peipiaosteus pani* showing that the element is relatively wide throughout most of its length but comes to a fine median point anteriorly.

Sensory canals—Sensory canals are indicated by two types of preservation in our specimens (e.g., Figs. 5, 6, and 8). The first type is in the form of simple ossified tubes, which presumably enclosed the sensory canal as in living sturgeons and paddlefishes. These include the infraorbital bones other than the dermosphenotic (*iocn*), supraorbital canal ossifications (*sobn*), nasals (*na*), antorbitals (*ao*), rostrals (*ro*), and trunk lateral line tubes (*llcn*). The other type of preservation of sensory canals is in the form of clear canals (either open or only thinly covered with bone) along the surface of the better developed dermal bones (indicated in our drawings by dashed lines). Sensory canals with this type of preservation occur in the frontals, dermosphenotics, dermopterotics, extrascapulars, posttemporals, and supracleithra. The paths of the sensory canals are best illustrated in Figs. 5 and 8. There is no ossified mandibular canal in †*Peipiaosteus* (a reduced or absent mandibular canal is a synapomorphy of Acipenseriformes; character B6, Table 2).

Jaws and suspensorium—The large, terminal mouth of †*Peipiaosteus* more closely resembles that of polyodontids and †chondrosteids than of acipenserids, which typically have smaller, subterminal mouths. Based on Findeis (1991; also see discussion in Bemis *et al.*, 1997), we consider that the bone traditionally interpreted as a maxilla in Acipenseriformes is actually a dermopalatine and that the maxilla is absent in all members of the order. As in acipenserids, †chondrosteids, and primitive polyodontids (see Grande and Bemis, 1991), the upper jaw of †*Peipiaosteus* includes two major bones, an expansive palato-

TABLE 2   **Data Matrix for Cladogram in Figure 11**

Characters (coded to text descriptions and Figure 11)

Species	A1	A2	A3	B1	B2	B3	B4	B5	B6	B7	B8	C1	C2	C3	D1	D2	E1	E2	E3	F1	F2	G1	G2	H1	H2	H3	H4	H5	I1	I2	I3	I4	K1	K2	M1	M2	N1	N2
†*Birgeria groenlandica*[a]	1	1	1	0	0	?	?	0	0	0	0	0	0	0	0	0	0	?	0	0	0	0	0	0	0	0	0	?	0	0	0	0	0	0	0	?	0	0
†*Chondrosteus acipenseroides*[b]	1	1	1	1	0	1	?	1	1	1	1	1	1	1	0	0	0	0	0	0	?	0	0	0	0	0	0	0	0	0	0	0	0	0	0	?	0	0
†*Strongylosteus hindenburgi*[c]	1	1	1	1	1	1	?	1	1	1	1	1	1	1	0	0	0	0	0	0	0	0	0	0	0	0	0	?	0	0	0	0	0	0	0	?	0	0
†*Peipiaosteus pani*	1	1	1	1	1	1	1	1	1	1	1	0	0	1	0	1	1	1	1	0	0	0	0	0	0	0	0	0	0	0	0	0	0	0	0	?	0	0
†*Stichopterus reissi*[d]	1	?	1	1	?	?	?	1	1	1	1	0	0	0	1	1	1	1	1	1	0	0	1	0	0	0	0	0	0	0	0	0	0	0	0	?	0	0
†*Spherosteus scharovi*[d,e]	?	?	1	?	?	?	?	1	?	1	1	?	0	0	1	1	1	?	1	1	0	0	1	0	0	0	?	?	?	0	?	?	0	0	?	?	0	0
†*Yanosteus longidorsalis*[e,f]	1	1	1	1	?	1	?	1	1	1	1	0	0	1	1	1	1	1	1	1	1	0	1	0	0	0	?	0	0	0	0	0	0	0	0	?	0	0
†*Protopsephurus liui*	1	1	1	1	?	1	?	1	1	1	1	0	0	1	1	1	1	1	0	1	0	1	?	1	0	0	?	0	0	0	0	0	0	1	0	?	0	0
†*Paleopsephurus wilsoni*	1	?	1	1	?	1	?	1	1	1	1	0	?	1	1	1	0	0	0	0	?	1	0	1	0	0	0	0	0	1	1	1	1	1	0	?	?	0
Psephurus gladius	1	1	1	1	1	1	1	1	1	1	1	0	?	1	1	1	0	0	0	0	0	1	0	1	0	0	0	0	1	1	1	1	1	1	1	0	0	0
†*Crossopholis magnicaudatus*	1	1	1	1	1	1	1	1	1	1	1	0	0	1	1	1	0	0	0	0	0	1	0	1	0	0	0	0	0	0	0	0	1	1	1	1	0	0
Polyodon spathula	1	1	1	1	1	1	1	1	1	1	1	0	0	1	1	1	0	0	0	0	0	1	0	1	0	0	0	0	0	0	0	0	1	1	1	1	0	0
Huso huso	1	1	1	1	1	1	1	1	1	1	1	0	0	1	1	1	0	0	0	0	0	1	0	1	1	1	1	1	0	0	0	0	0	0	0	0	0	0
Acipenser brevirostrum	1	1	1	1	1	1	1	1	1	1	1	0	0	1	1	1	0	0	0	0	0	1	0	1	1	1	1	1	0	0	0	0	0	0	0	0	0	0
Acipenser oxyrhyncus	1	1	1	1	1	1	1	1	1	1	1	0	0	1	1	1	0	0	0	0	0	1	0	1	1	1	1	1	0	0	0	0	0	0	0	0	0	0
Acipenser transmontanus	1	1	1	1	1	1	1	1	1	1	1	0	0	1	1	1	0	0	0	0	0	1	0	1	1	1	1	1	0	0	0	0	0	0	0	0	0	0
Scaphirhynchus platorynchus	1	1	1	1	1	1	1	1	1	1	1	0	0	1	1	1	0	0	0	0	0	1	0	1	1	1	1	1	0	0	0	0	0	0	0	0	0	0
†*Protoscaphirhynchus squamosus*	1	?	?	?	?	?	?	1	?	?	?	0	?	?	1	1	0	?	0	0	0	?	?	??	?	?	?	?	0	0	0	0	?	0	?	?	?	0
Pseudoscaphirhynchus kaufmanni	1	1	1	1	1	1	1	1	1	1	1	0	0	1	1	1	0	0	0	0	0	1	0	1	1	1	1	1	0	0	0	0	0	0	0	0	0	0

Characters (coded to text descriptions and Figure 11)

Species	O1	O2	O3	O4	P4	Q1	Q2	Q3	Q4	Q5	Q6	R1	R3	R4	R5	R6	R7	R8	S1	S2	T1	T3	T4	T5	T6	V1	V2	V3	W1	W2	W3	X1	X2
†*Birgeria groenlandica*[a]	0	0	0	n	0	0	?	?	?	1	1	0	0	0	0	0	0	0	?	?	0	n	0	0	0	0	?	0	0	?	?	?	n
†*Chondrosteus acipenseroides*[b]	0	0	?	n	0	0	?	?	?	?	0	0	0	0	0	0	0	0	?	?	0	n	0	0	0	0	?	0	0	?	?	?	0
†*Strongylosteus hindenburgi*[c]	0	0	0	n	0	0	?	?	?	?	0	0	0	0	0	0	0	0	?	?	0	n	0	0	0	0	?	0	0	?	?	0	0
†*Peipiaosteus pani*	0	0	0	n	0	0	0	?	0	?	0	0	0	0	0	0	0	0	?	?	0	n	0	0	0	0	0	0	0	0	0	0	0
†*Stichopterus reissi*[d]	0	0	0	n	0	0	?	0	?	0	0	0	0	0	0	0	0	0	?	?	0	n	0	0	0	0	?	0	0	0	0	0	1
†*Spherosteus scharovi*[d,e]	?	0	?	n	0	?	?	?	?	?	0	0	0	?	0	0	0	0	?	?	0	n	?	?	0	0	?	0	0	?	?	?	?
†*Yanosteus longidorsalis*[e,f]	0	0	0	n	0	?	?	?	?	?	0	0	0	?	?	0	0	0	?	?	0	n	?	?	0	0	?	0	0	?	?	0	1
†*Protopsephurus liui*	0	0	0	n	0	?	?	0	?	0	0	0	0	?	0	0	0	0	?	?	0	0	?	0	0	0	?	0	0	0	0	0	0
†*Paleopsephurus wilsoni*	0	?	?	0	0	?	?	?	0	0	0	0	0	?	?	0	0	0	?	?	0	?	?	?	0	0	?	0	0	0	0	0	0
Psephurus gladius	0	0	0	0	0	0	0	0	0	0	0	0	0	0	0	0	0	0	0	0	0	0	0	0	0	0	0	0	0	0	0	0	0
†*Crossopholis magnicaudatus*	1	1	1	1	1	1	1	1	1	0	0	1	1	1	1	1	1	0	0	?	1	1	0	0	0	0	0	0	0	0	0	0	0
Polyodon spathula	1	1	1	1	1	1	1	1	1	0	1	1	1	1	1	1	1	0	?	0	1	1	0	0	0	0	0	0	0	0	0	0	0
Huso huso	0	0	0	n	0	0	0	0	0	0	0	0	0	0	0	0	0	0	0	0	0	0	0	0	0	0	0	0	0	0	0	0	0
Acipenser brevirostrum	0	0	0	n	0	0	0	0	0	0	0	1	1	1	1	1	1	1	0	?	1	1	0	0	0	0	0	0	0	0	0	0	0
Acipenser oxyrhyncus	0	0	0	n	0	0	0	0	0	0	0	1	1	1	1	1	1	1	0	?	1	1	0	0	0	0	0	0	0	0	0	0	0
Acipenser transmontanus	0	0	0	n	0	0	0	0	0	0	0	1	1	1	1	1	1	1	0	0	1	1	0	0	0	0	0	0	0	0	0	0	0
Scaphirhynchus platorynchus	0	0	n	n	0	0	0	0	0	0	0	1	1	1	1	1	1	1	0	0	1	1	1	1	1	1	1	1	1	1	1	0	1
†*Protoscaphirhynchus squamosus*	?	?	?	?	?	?	?	?	?	?	?	1	1	1	1	1	1	1	?	?	1	?	?	?	1	1	?	?	?	1	?	?	1
Pseudoscaphirhynchus kaufmanni	0	0	n	n	0	0	0	0	0	0	0	1	1	1	1	1	1	1	0	0	1	1	1	1	1	1	1	1	0	1	1	1	1

Note. Abbreviations: 0, absent; 1, present; ?, condition unknown; n, not applicable.

[a]Nielsen (1949)
[b]Traquair (1877)
[c]Hennig (1925)
[d]Jakovlev (1977)
[e]Jin et al. (1995)
[f]Jin (1995)
[g]The bone identified as an opercle in the original description is actually a subopercle.

pterygoid (*ppt*, Figs. 5, 6, and 8), and the long, narrow dermopalatine [*dp* (= *mx*), Figs. 5 and 8] joined posteriorly by an elongate quadratojugal (*qj*, Figs. 5 and 8). Associated with the ventral (oral) surface of each palatopterygoid is a patch of tiny, pointed teeth (*ppttp*, Fig. 6). The tooth patches were evidently anchored in soft tissues of the palate rather than in the palatopterygoid itself because none of the teeth appear to be socketed into the bone. There is also a small element which appears to be an autogenous ectopterygoid bone (*mpec*, Figs. 5 and 8). We found no trace of a mineralized autopalatine, possibly because our individuals were small. Typically, the autopalatine ossifies only in large individuals of many acipenseriform species (e.g., Grande and Bemis, 1991, p. 23).

The lower jaw of †*Peipiaosteus pani* consists only of the dentary in our specimens (*d*, Figs. 5 and 8). The right and left dentary bones do not appear to be tightly sutured at the symphysis. There was no indication of ossified prearticular or mentomeckelian bones in our specimens, but this could be due to their small size. These elements are also unossified in small acipenseriforms (e.g., Grande and Bemis, 1991, figs. 78, L and M).

The jaws of our specimens of †*Peipiaosteus* lack teeth.

The only apparent bone of the suspensorium is the hyomandibula (*h*, Figs. 5 and 8). As in other acipenseriforms, this bone is hourglass-shaped, but it appears relatively smaller in size than in large adults of other acipenseriform species. This may further indicate that we do not have fully developed material.

Opercular series—The opercular series of †*Peipiaosteus pani* retains the opercle (*op*, Figs. 5, 6, and 8), although this element is clearly reduced in size from that of non-acipenseriforms. Among acipenseriforms, an opercle is present also in †Chondrosteidae and †*Stichopterus*, where it is also reduced in relative size from that of other actinopterygians (e.g., Figs. 2B and D) and entirely lost in Polyodontidae and Acipenseridae. The posterior edges of the opercular and branchiostegal elements of †*Peipiaosteus* have numerous clefts and rod-like projections.

The subopercle (*sop*, Figs. 5 and 8) is well-developed and broadly expansive, with a well-developed anterior arm, as in †chondrosteids (*sop*, Fig. 2D) and acipenserids (e.g., Findeis, 1997, fig. 7). In polyodontids, the subopercle has a well-developed anterior arm, but the bone is less broad and has a fringe of rod-like structures along the posterior edge (e.g., Grande and Bemis, 1991, figs. 9, 34, 54, and 68).

We find 6 branchiostegal elements on each side in our material of †*Peipiaosteus pani* (*br*, Figs. 5, 6, and 8). They are more numerous in †*Peipiaosteus* than in Polyodontidae (where the number has been reduced

to 1 in all except †*Protopsephurus*) and Acipenseridae (where the number is usually 2, but ranges from 1 to 3) and less numerous than in †Chondrosteidae, where they number 8 to 10 (Fig. 2D; also see Grande and Bemis, 1991, p. 108). Jakovlev's (1977) drawing of †*Stichopterus* shows 7 branchiostegals (1977, fig. 2B). Observations by Findeis (1993) and our new information about the branchiostegal series of †*Peipiaosteus* cause us to revise one of our previous theories about the origin of the branchiostegal element of Polyodontidae. Based mostly on the shape of the bone in polyodontids, we speculated (Grande and Bemis, 1991, p. 5) that the bone was made up of numerous branchiostegal rays fused anteriorly. But in †*Peipiaosteus*, each of the branchiostegals looks much like the single branchiostegal element in polyodontids (e.g., the lower elements figured in Grande and Bemis, 1991, fig. 15). Therefore, we now believe that the single branchiostegal element in polyodontids resulted from loss of other elements rather than fusion.

Gill arches—Among our three specimens of †*Peipiaosteus pani*, the ossified gill arch elements were best preserved in the specimen illustrated in Fig. 6. The specimen has four well-preserved ceratobranchials on each side (*cb*) and two rows of short, tooth-shaped gill rakers bordering what was probably the position of the cartilaginous portion of many of the branchial arches (*gr*, and unlabelled in Fig. 6). The numerous gill rakers are similar in length to those found in all other living and fossil acipenseriforms except for *Polyodon* (which has long, comb-like gill rakers for filter-feeding; Grande and Bemis, 1991, fig. 26) and *Scaphirhynchus* (which has branched gill rakers for feeding on soft-bodied benthic invertebrates; Findeis, 1996: fig. 25). The anterior ceratohyal bones are also preserved (*cha*, Figs. 6 and 8). There also appear to be two elongate tooth patches of tiny, sharply pointed teeth (pointing back towards the throat region) that may represent newly forming tooth plates of the first branchial arch (*tp1*, Fig. 6). Tooth plates are often found in this position in living Acipenseriformes (e.g., see Nelson, 1969, pl. 79, fig. 1 for *Polyodon*).

Vertebral column—As in other acipenseriforms, †*Peipiaosteus* lacks ossified centra. The only observed vertebral elements in our specimens are basidorsal ossifications (*bd*, Fig. 7), supraneurals (*sn*, Fig. 7), and hemal spines (*hs*, Figs. 4, 7, and 8). In our smaller specimen (Fig. 4), only the hemal spines are ossified. Larger specimens of †*Peipiaosteus* might demonstrate basiventral ossifications, such as are seen on very large individuals of other acipenseriforms (e.g., Grande and Bemis, 1991, fig. 53).

The vertebral elements are less completely developed and less numerous in our larger specimen (estimated 320 mm total length) than in Zhou's (1992, fig.

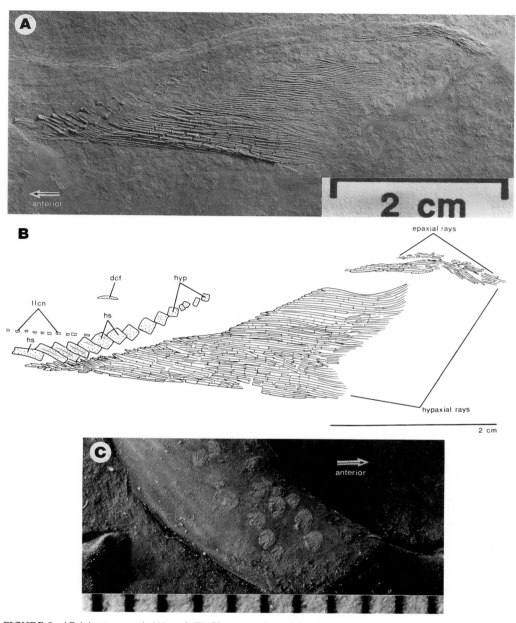

FIGURE 9 †*Peipiaosteus pani*. (A) and (B) Close-up of caudal region from specimen in Figure 4; anterior facing left. (C) Close-up of round based scales in shoulder girdle region of specimen in Figure 7; anterior facing right. Scale in millimeters.

1) restoration. This difference is probably because he used larger specimens as a model for his postcranial restoration (he reports having specimens up to 900 mm total length).

Caudal fin and supports—The caudal fin of †*Peipiaosteus* is asymmetrical (heterocercal), with the upper lobe being much longer than the lower lobe (Figs. 4 and 7). An asymmetrical tail is present in most acipenseriforms (other than *Polyodon* and †*Crossopholis*) and many other lower actinopterygians. The upper lobe of the caudal fin has both epaxial and hypaxial rays (Fig. 9).

The ossifications of the caudal skeleton of †*Peipiaosteus* appear to be reduced in comparison to other acipenseriforms. The single small, thin dorsal caudal fulcrum (*dcf*, Figs. 4, 7 and 9) contrasts with the numerous, more massively developed dorsal caudal fulcra present in all other acipenseriforms (Grande and Bemis, 1991). The caudal fin "fulcra" reported by Zhou (1992, p. 96) appear to be possibly anterior epaxial caudal fin rays (e.g., see his plate 1, Fig. 3). If these are caudal fulcra, then they are extremely tiny and much less well-developed than in any other known acipenseriform. Zhou (1992) did not report the one larger fulcrum unambiguously present in both of our specimens (e.g., Figs. 4, 7, and 9).

Dorsal and anal fins and supports—The many rays of the dorsal and anal fins are supported by series of ossified middle and proximal radials (*pr* and *mr*, Fig. 7). In the dorsal fin of our larger specimen, we counted one tiny procurrent ray followed by 34 segmented rays. Dorsal fin rays were uncountable in our smaller specimen. In the anal fin we counted 2 procurrent rays plus 32 rays in our larger specimen and 2 procurrent plus 28 segmented rays in our smaller specimen. The series of proximal and middle radials are incompletely ossified even in our larger specimen, in contrast to Zhou's (1992) reconstruction, which we think must be based on much larger specimens (Zhou, 1992, fig. 1). As in all acipenseriforms, the dorsal fin is slightly anterior in position to the anal fin but well posterior to the pelvic fins.

Rhomboid scales are absent from the lateral surfaces of the upper caudal lobe of our †*Peipiaosteus* specimens. Such scales are present in all other acipenseriforms and many other lower actinopterygians (Fig. 10). Even in large specimens of †*Peipiaosteus* reported by other workers (e.g., Liu and Zhou, 1965), the rhomboid caudal scales are absent. This contrasts with the condition reported for †*Stichopterus*, in which at least some rhomboid scales are said to be present (Zhou, 1992). The reduction of dorsal caudal fin fulcra and the loss of rhomboid caudal scales thus appear to be derived features of †*Peipiaosteus*.

Pectoral and pelvic fins and girdles—The ossified part of the pectoral fin girdle in †*Peipiaosteus* contains a series of bones. Articulating with the medial surface of the posttemporal is the supracleithrum (*scl*, Figs. 5 and 8). A sensory canal runs along its surface, connecting anteriorly to the posttemporal canal and posteriorly to the trunk lateral line.

Laterally, the ventral part of the supracleithrum overlies the postcleithrum (*pcl*, Figs. 5 and 8) and the cleithrum (*cl*, Figs. 5 and 8). The cleithrum is a well-developed bone including a narrow lateral region in contact with the skin on the lateral surface of the head and a broad, medially pointing sheet of smooth bone with prominent growth lines, presumably forming the posterior wall of the opercular chamber. The medially directed portion of the clavicle (*clv*, Fig. 8) also forms part of this posterior wall. This well-developed medially directed wall formed by the cleithrum and clavicle is also present in all Acipenseridae. The wall is greatly reduced in polyodontids (e.g., Grande and Bemis, 1991, figs. 20, 42, and 61). The medially directed portions of these bones are greatly developed in Acipenseridae and were considered synapomorphic for that family by Bemis *et al.* (1997, character 31) and Findeis

(1993, 1996). There is no trace of an interclavicle bone in our specimens. Covering the posteroventral region of the cleithrum is a series of small round-based scales, described below.

The pectoral radials of our specimens of †*Peipiaosteus* were not ossified and so were not preserved. The fin contains numerous rays (32 on each of our specimens). There is no trace of a pelvic spine such as the type found in Acipenseridae.

The pelvic fin in our specimens also had numerous rays (30 on the smaller specimen; too numerous to count on the larger). No trace of the pelvic fin girdle was preserved on our specimens.

Scales—†*Peipiaosteus* has two types of scales. One is the "round-based scales of the shoulder girdle," which possess three posteriorly directed tips (*rbs*, Figs. 8 and 9C). This morphology closely resembles that discussed for polyodontids by Grande and Bemis (1991, p. 37). These scales were also illustrated in the three previous descriptions of †*Peipiaosteus* (Liu and Zhou, 1965; Bai, 1983; Zhou, 1992). The presence (or absence) of such scales in †chondrosteids is yet unknown. Findeis (1997, fig. 14) reported a modified form of these scales in basal acipenserids. In particular, the scales of the shoulder girdle of *Huso* and *Acipenser* are elongate and have as many as five posteriorly directed tips. Scaphirhynchines (*Scaphirhynchus* and *Pseudoscaphirhynchus*) lack any scales on the shoulder girdle.

The trunk scale is the second type of scale occurring in our specimens of †*Peipiaosteus*. These occur as a row of small ossified tubes on each side of the body which apparently formed around the trunk lateral lines (*llcn*, Figs. 4–8).

No trunk scutes (such as those found in acipenserids) occur in †*Peipiaosteus*.

IV. Phylogeny of Acipenseriformes

In this section, we review taxa and characters in support of the cladogram presented in Fig. 11. In many cases, we are aware of additional characters proposed for various nodes in the cladogram. We avoid multistate characters, though it might be possible to construct them in some cases. We present only those characters that we can score confidently in most of the taxa or which ought to be readily scorable if additional fossil material is discovered and prepared.

Series Chondrostei *sensu* Patterson, 1982 (Node A, Fig. 11)

To make Chondrostei monophyletic, we exclude many taxa traditionally included within it, focusing

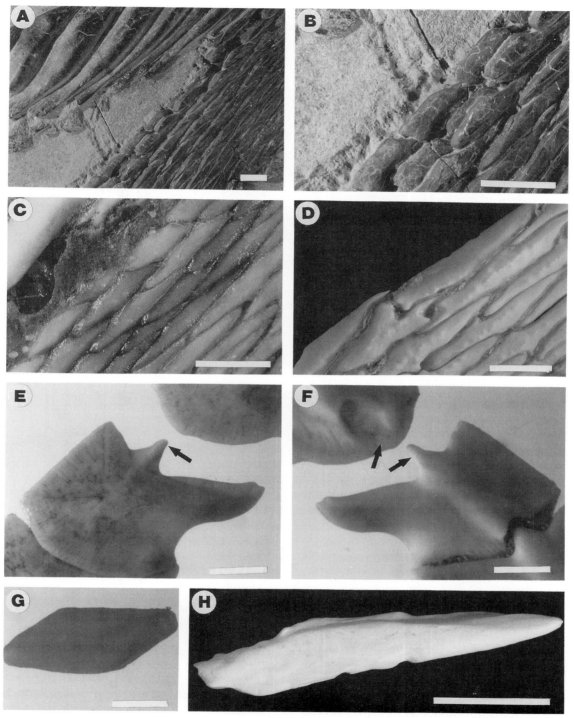

FIGURE 10 Morphology of rhombic caudal scales in selected Acipenseriformes, with comparisons to *Polypterus*. Dorsal to top of page; anterior to left for A–D and F–H. (A) Caudal scales of †*Chondrosteus acipenseroides*; scale bar = 10 mm. (B) Close-up of specimen in A, showing absence of peg-and-socket articulation between scales; scale bar = 5 mm. (C) Caudal scales of *Polyodon spathula*; scale bar = 5 mm. (D) Caudal scales of *Acipenser oxyrhynchus* (also see H); scale bar = 10 mm. (E) and (F) *Polypterus ornatapinnis* flank scales showing peg-and-socket articulation (arrows) between scales in external (E) and internal (F) views; scale bar = 2 mm. (G) *Polypterus ornatapinnis* caudal scale showing absence of peg-and-socket articulation between scales; scale bar = 2 mm. (H) Close-up of scale from specimen in D, showing absence of peg-and-socket articulation; scale bar = 10 mm.

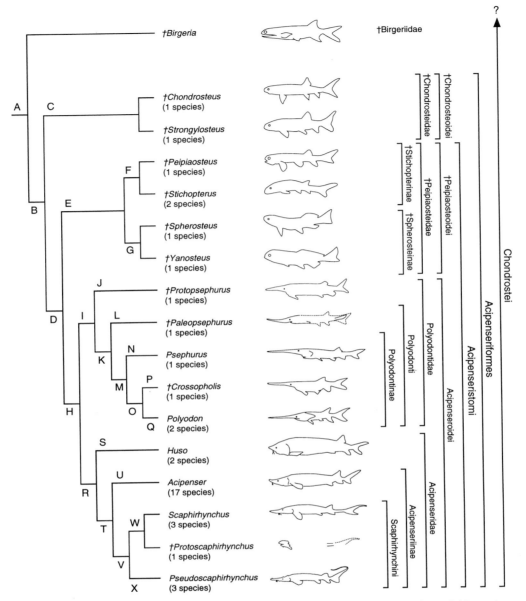

FIGURE 11 Cladogram of Acipenseriformes. Characters at the individual nodes (A through X) are given in the text. See Nomenclatural Recommendations section in the text for full nomenclature and authorship of names.

our attention on a working definition that is synapomorphy based. Many additional taxa may eventually meet these criteria. This chapter explains but does not focus on the following synapomorphies at Node A:

(A1) Reduction of the opercle. See Bemis *et al.* (1997) for definition and discussion of this character. The opercle is further reduced and eventually lost within Acipenseriformes (character H 1 below).

(A2) Elongate posterior extension of the parasphenoid.

This diagnostic feature has important functional anatomical correlates discussed by Bemis *et al.* (1997).

(A3) Body scaling reduced to tiny isolated elements or absent. This character refers to the loss of integrated, sheet-like squamation as exemplified by *Polypterus*. See Bemis *et al.* (1997) for definition and discussion.

Comments—Bemis *et al.* (1997) considered that among the taxa that they surveyed, †*Birgeria* is the sister taxon of Acipenseriformes. Our interpretation builds upon analyses by Rieppel (1992), who refuted the sister group relationship of †*Saurichthys* and Acipenseriformes, and contradicts the placement of †*Birgeria* proposed by Gardiner and Schaeffer (1989, p. 177 and table 1, fig. 12). As a general observation, some problems stem from their assumption that *Acipenser* is "representative" of the group (Gardiner and Schaeffer, 1989, p. 161, caption to fig. 11). Acipenserids are in fact very derived for many of the skeletal characters they consider, and to build a complete image of the group's characteristics it is necessary to include basal taxa, such as †*Peipiaosteus* and †*Chondrosteus*, as well as polyodontids. Specifically, characters at "Q" and "R" in their Figure 12 (Gardiner and Schaeffer, 1989, p. 163) warrant comment here. For "Q" Gardiner and Schaeffer (1989) proposed "numerous irregular anamestic supraorbital bones between nasal and keystone-shaped dermosphenotic." We consider this character to be problematic because a series of anamestic supraorbital bones is reported in †*Stichopterus* (Figs. 2B and 2C). Supraorbitals may also be present in polyodontids (Grande and Bemis, 1991, suo?, fig. 6) and acipenserids (see discussion of character R 3 below), although in neither case are the bones numerous or anamestic. In keeping with the general loss of cheek bones in Acipenseriformes, it seems to us likely that their ancestor had supraorbital bones and that they have been lost within the group. At their "R," Gardiner and Schaeffer (1989) proposed two characters: "premaxilla and antorbital bones present" and "neural spines in caudal region median." The premaxilla is a primitive character of Actinopterygii that we interpret as lost in Acipenseriformes (character B 8, below). Also, we found evidence in †*Peipiaosteus* of a Y-shaped canal bone that we interpret as an antorbital (*ao*, Fig. 8). Thus the first character appears to be shared by Acipenseriformes. In so far as we understand their second character at node R, it is unscorable in Acipenseriformes because caudal neural spines are not present. Thus we stand by our placement of †*Birgeria* as the sister taxon of Acipenseriformes.

Order Acipenseriformes (Node B, Fig. 11)

Characters available to support this node are discussed and illustrated by Grande and Bemis (1991) and Bemis *et al.* (1997).

Diagnosis—A group of Chondrostei that differs from all other groups by the following characters:

(B1) Palatoquadrates with an anterior symphysis. This is a traditionally recognized character often cited as evidence of acipenseriform monophyly. See Grande and Bemis (1991, p. 106, character 1) and Bemis *et al.* (1997, character 4).

(B2) Palatoquadrate with broad autopalatine portion, *palatoquadrate bridge, and quadrate flange.* This character may be correlated with character B 1. See Bemis *et al.* (1997, character 5).

(B3) Presence of triradiate quadratojugal bone. This character is lost in the genus *Polyodon* (character Q 2, below). See Grande and Bemis (1991, p. 106, character 6) and Bemis *et al.* (1997, character 6).

(B4) Gill-arch dentition confined to first two hypobranchials and upper part of first arch. See Grande and Bemis (1991, p. 106, character 2) and Bemis *et al.* (1997, character 7).

(B5) Subopercle possesses an anterior process. This character was recognized as a synapomorphy of Acipenseriformes by Findeis (1993) and Bemis *et al.* (1997, character 8). Formerly, Grande and Bemis (1991, character 17) had included portions of this character in the definition of a synapomorphy of Polyodontidae.

(B6) Preopercular canal in a series of ossicles, mandibular canal short or absent. See Grande and Bemis (1991, p. 106, character 4) and Bemis *et al.* (1997, character 9). Characters B 6 and B 7 appear to be related to the reduction and loss of dermal bone in the cranium, particularly in the cheek region, which occurs in Acipenseriformes. This phenomenon may be linked to paedomorphosis (see discussion in Bemis *et al.*, 1997). We interpret the lack of a preopercular bone and the absence of pores in the mandible as indicative of this character in fossil taxa.

(B7) Infraorbital sensory canal in a series of ossicles. See Bemis *et al.* (1997, character 10).

(B8) Loss of premaxillary and maxillary bones. See Bemis *et al.* (1997, character 11). The element historically regarded as the maxilla of Acipenseriformes is better interpreted as a dermopalatine. See Findeis (1991, 1993) and Bemis *et al.* (1997) for discussion. Based on the similarity in the shape of the dermopalatine and its lack of sutural contact with the infraorbital and other bones of the cheek, we interpret it as homologous in all Acipenseriformes.

Comments—Monophyly of Acipenseriformes is accepted by most recent authors (e.g., Gardiner and Schaeffer, 1989).

Suborder †Chondrosteoidei, Family †Chondrosteidae (Node C, Fig. 11)

Diagnosis—A group within Acipenseriformes that differs from all others by the following:

(C1) Anterior part of palatopterygoid club-shaped. This feature is best understood by comparing Woodward (1895, mpt, fig. 6) with Grande and Bemis (1991, fig. 79). We (Bemis *et al.*, 1997, character 12) cited this feature as an autapomorphy of †Chondrosteidae.

(C2) Complete loss of trunk scalation. This is a well-

known character (e.g., Liu and Zhou, 1965). Even the lateral line ossifications found in †*Birgeria*, †Peipiaosteidae and Acipenseroidei appear to be absent in †Chondrosteidae. There is some doubt, however, whether any chondrosteid has ever been prepared sufficiently well to reveal whether these elements are present.

(C3) Extreme reduction or loss of ossification of vertebral and supraneural elements. This is a newly reported character. Based on available descriptions of even very large specimens (e.g., †*Strongylosteus hindenburgi*; see Hennig, 1925), the vertebral elements of †Chondrosteidae are greatly reduced. This is unique among Acipenseriformes.

Comments—The family †Chondrosteidae includes two genera (†*Chondrosteus* and †*Strongylosteus*). We cannot yet provide meaningful generic diagnoses for these taxa; we include both in our summary cladogram primarily for completeness. †*Gyrosteus* is usually considered to be a member of the family †Chondrosteidae (e.g., Woodward, 1891; Grande and Bemis, 1991), but we now consider it too fragmentary to evaluate here given that most other genera of Acipenseriformes are known from more complete skeletons. †Chondrosteidae warrants renewed attention and descriptive study, especially in light of the recent emphasis on all other families of the order Acipenseriformes.

Division Acipenseristomi, new name (Node D, Fig. 11)
Diagnosis—A group of Acipenseriformes that differs from all others by the following:

(D1) Loss of sclerotic rings. This is the first definition of this character of which we are aware. Sclerotic rings are well-developed in non-acipenseriform outgroups (e.g., †*Birgeria*) and †Chondrosteidae (*sr*, Fig. 2D) but absent in all members of Acipenseristomi.

(D2) Reduction in number of branchiostegal rays to fewer than 7. This character was also noted by Jin (1995, 18, node B). In non-Acipenseriformes, such as †*Birgeria*, about 14 branchiostegal rays are present. In †Chondrosteidae, 8–10 branchiostegal rays are present, and we judge this to approximate the original condition for Acipenseriformes. A further reduction in the number of branchiostegal rays characterizes Acipenseroidei (character H 2, below).

Comments—Jin (1995) proposed two synapomorphies in support of a clade comparable to our Acipenseristomi. One of these is our character D 2. The other is the loss of peg-and-socket articulation on the rhombic caudal scales. Gardiner and Schaeffer (1989, p. 176) noted that the two dorsalmost rows of caudal scales of †*Chondrosteus* have peg-and-socket articulations. This seemed odd to us. To investigate this, we surveyed

caudal scales of †*Chondrosteus*, *Polyodon*, and *Acipenser* and were unable to detect any significant differences in the shapes of the caudal rhombic scales (Figs. 10A, C, and D). We found no peg-and-socket articulations in a carefully prepared †*Chondrosteus* (Fig. 10B), as in *Acipenser oxyrhynchus* (Fig. 10H). Next, we examined the peg-and-socket articulation between trunk scales of *Polypterus ornatapinnis* (Fig. 10E and F). Interestingly, peg-and-socket articulations are absent in scales from the caudal region of the same specimen (Fig. 10G). It seems that the absence of peg-and-socket articulation between rhombic caudal scales cannot be used to diagnose our Acipenseristomi but may instead be a character at a much higher level within Actinopterygii.

Suborder †Peipiaosteoidei, new name, Family †Peipiaosteidae (Node E, Fig. 11)
Diagnosis—A group within Acipenseristomi that differs from all others by the following characters:

(E1) Extreme reduction (e.g., †Stichopterus) or loss (e.g., †Peipiaosteus) of the rhomboid scales on the upper lobe of the caudal fin.

(E2) The absence of an interclavicle. This bone is also lost in *Polyodon*, but it is present in more primitive polyodontids.

(E3) The peculiar branched shapes of the bones of the opercular and branchiostegal series. Jin et al. (1995) described the posterior margin of the subopurculum of †*Yanosteus* as "ctenoid," an excellent descriptor of its shape.

Comments—In addition to †*Peipiaosteus*, this family also contains †*Stichopterus* Reis 1910 from the Lower Cretaceous of Transbaikal (Siberia) and western Mongolia (see Jin, 1995, for comments on stratigraphy and distribution). The genus †*Yanosteus* (Jin et al. 1995) from the Upper Jurassic-Lower Cretaceous Yixian Formation in the northern Hebei Province of China is also assigned to this family. A fourth genus from the Upper Jurassic of Kazakhstan, †*Spherosteus* (Jakovlev 1977), is as yet poorly described, although Jin (1995) considered it similar to †*Yanosteus*, chiefly in its possession of an elongate dorsal fin.

Jin (1995) proposed "rostral bones reduced to tube-like ossicles" as a character of Peipiaosteidae, but the polarity of this change is not clear. The rostral bones present in †*Chondrosteus* (Fig. 2D) appear to resemble those we found in †*Peipiaosteus* (above). Moreover, there appears to be an issue of homology here: if these tube-like ossicles are in fact canal bones, then they are fundamentally different from the flat rostral bones that characterize Acipenseroidei (character H 4, below). Rostral bones of Acipenseroidei are developmentally and structurally separate from the "tube-like ossicles" associated with the rostral portion of the infraorbital sensory canal. This question needs more study.

Subfamily †Stichopterinae, new name (Node F, Fig. 11)
   Diagnosis—A group within †Peipiaosteidae that differs from all others by the following:

(F1) Reduction of dorsal caudal fulcra to a single element. In †*Yanosteus* (Jin *et al.*, 1995) and †*Spherosteus* (Jakovlev, 1977, pl. 13) several dorsal caudal fulcra are present, whereas in †*Peipiaosteus* (described above) and †*Stichopterus* (at least based on available photographs of †*Stichopterus)*, only a single element is present.

(F2) Loss of ventral caudal fulcrum or reduction of it to a bone too small to be identified in existing material. There is a well-developed, ventral caudal fulcrum in all acipenserids, polyodontids, and †chondrosteids, but this element appears to be absent (or at least highly reduced) in †Stichopterinae.

   Comments—Additional remarks about the genera †*Peipiaosteus* and †*Stichopterus* are provided in the descriptive osteology section above.

Subfamily †Spherosteinae, new name (Node G, Fig. 11)
   Diagnosis—A group within †Peipiaosteidae that differs from all others by the following:

(G1) Elongate dorsal fin. This character was noted by Jin *et al.* (1995) and Jin (1995) in describing the probable relationship between two Mesozoic genera from Asia, †*Spherosteus* and †*Yanosteus.* Few details of fin morphology in these taxa are available.

   Comments—†Spherosteinae is much better known since the description of †*Yanosteus* by Jin *et al.* (1995).

Suborder Acipenseroidei (Node H, Fig. 11)
   This group was first recognized and defined by Grande and Bemis (1991). It appears to be a well-supported clade.
   Diagnosis—A group within Acipenseristomi that differs from all others by the following:

(H1) Loss of opercle. This character was defined by Grande and Bemis (1991: 108, character 7) and further commented on by Bemis *et al.* (1997, character 14).

(H2) Further reduction in number of branchiostegals supporting gill cover. This character was defined by Grande and Bemis (1991, p. 108, character 8) and further commented on by Bemis *et al.* (1997, character 15). In the context of character D 2 (above) we note that fewer than four branchiostegal rays are present in all fossil and living Acipenseroidei.

(H3) Endocranium with extensive rostrum. This char-

acter was defined by Grande and Bemis (1991, p. 108, character 9). The historic and often used reconstruction of †*Chondrosteus* (Woodward, 1891, fig. 12) greatly exaggerates the outline of the snout, presumably because it is based on that of *Acipenser.* In fact, there is no evidence of such an elongate rostrum in either of the well-preserved genera of †Chondrosteidae.

(H4) Dorsal and ventral rostral bones. This character as originally formulated by Grande and Bemis (1991, p. 108, character 10) was expanded by Bemis *et al.* (1997, character 17).

(H5) Ventral process of posttemporal bone. This character was defined by Bemis *et al.* (1997, character 18).

   Comments—Zhou's (1992) challenge to Acipenseroidei failed; see Historical Overview of Acipenseriformes, above.

Family Polyodontidae (Node I, Fig. 11)
   The paddlefish family, Polyodontidae Bonaparte 1838, was reexamined in detail by Grande and Bemis (1991); additional interpretations incorporating the new genus †*Protopsephurus* (Lu 1994) based on the literature were discussed by Bemis *et al.* (1997). Since then, we have seen two nearly complete specimens of †*Protopsephurus* at KMNH. We comment only briefly on them here, and a description of these specimens will be forthcoming (Fan Jin, personal communication). The above papers and the newly discovered specimens are the basis for our present analysis.
   Diagnosis—A group within Acipenseristomi that differs from all others by the following:

(I1) A series of very elongate dorsal median rostral and ventral median rostral bones with cylindrical cross-sections. These are remarkably conservative across Polyodontidae and Acipenseridae. This character was described and figured by Grande and Bemis (1991, p. 108, character 16) and modified by Bemis *et al.* (1997, character 20).

(I2) A peculiarly shaped subopercle with a posteriorly pointing fan of rod-like ossifications. This character was described and figured by Grande and Bemis (1991, p. 108, character 17) and discussed by Bemis *et al.* (1997, character 21). Although all acipenseriforms have an anteriorly pointing arm, non-polyodontids lack the fan of posteriorly pointing projections.

(I3) Development of large elongate anterior and posterior divisions of the fenestra longitudinalis of the skull roof. This character was described and figured by Grande and Bemis (1991, p. 109, character 18) and discussed in Bemis *et al.* (1997, character 22). Jin (1995, caption to fig. 2) states that this feature is present in †*Protopsephurus.*

(I4) Posttemporal with long anterolateral arm suturing

into the dermosphenotic. This character was described and figured by Grande and Bemis (1991, p. 109, character 19) and discussed by Bemis *et al.* (1997, character 23).

Comments—We have examined specimens of all polyodontids, including †*Protopsephurus.* A new description of †*Protopsephurus,* based on the complete specimens, is still needed. Gardiner (1984) proposed a diphyletic origin of paddlefishes within Acipenseriformes based on the purported absence of stellate bones and other problematic characters in †*Paleopsephurus.* The diphyletic origin of paddlefishes was refuted by Grande and Bemis (1991, pp. 98–106). The trunk scale morphology is complex and extremely similar among adult-sized specimens of †*Crossopholis,* †*Paleopsephurus,* and †*Protopsephurus* (e.g., see Grande and Bemis, 1991, fig. 63E for †*Crossopholis*) which might mean that this is an additional polyodontid synapomorphy, primitively derived for the family and further modified in *Polyodon* (as were so many other features) where trunk scales are much reduced in number and degree of ossification. The trunk scales of large adult *Psephurus* (e.g., 2 meters or more) are unknown, but the poorly ossified trunk scales of small individuals (e.g., Grande and Bemis, 1991, fig. 45) somewhat resemble those of an early ontogenetic stage in †*Crossopholis* (see Grande and Bemis, 1991, fig. 63F) and they are numerous.

Subfamily †Protopsephurinae, new name, Genus †*Protopsephurus* (Node J, Fig. 11)

Diagnosis—This monotypic subfamily is currently diagnosed within Polyodontidae by the following symplesiomorphies: (*A*) lack of the numerous stellate bones that make up the lateral supports of the paddle in all other polyodontid species, (*B*) more than a single branchiostegal ray (vs only one branchiostegal ray in all other paddlefishes), and (*C*) relatively shorter head length than any other polyodontid (i.e., smaller head-length-to-total-length ratio). There appear to also be certain meristic features that are synapomorphic for †*Protopsephurus,* but these will not be described here. Other descriptive information can be found in Lu (1994) based on partial specimens, and the nearly complete skeletons at KMNH will be described in detail in a future paper.

Series Polyodonti, new name (Node K, Fig. 11)

Diagnosis—A group within Polyodontidae that differs from all others by the following:

(*K1*) *Single branchiostegal ray.* This character was defined by Grande and Bemis (1991, p. 109, character 20) at the level of Polyodontidae; it appears that †*Protopsephurus* has two or three branchiostegal rays (Lu,

1994, and preliminary examination of KMNH specimens here). Whether this decrease resulted by fusion or loss of several branchiostegal rays (as proposed by Findeis, 1993), it seems likely to be a loss based on the branchiostegals of †*Peipiaosteus* (see previous description).

(*K2*) *Many small, stellate bones making up the lateral supports for the paddle.* This character was described and figured by Grande and Bemis (1991, p. 108, character 15). Lu (1994) was uncertain about the presence of stellate bones in the genus †*Protopsephurus* based on his material. Based on our examination of two well-preserved specimens of †*Protopsephurus* (uncataloged specimens at KMNH) we confirm what Lu (1994) implied: there are few if any stellate bones on either of the two well-preserved KMNH specimens we examined. Numerous stellate bones are present in all paddlefish species except for †*Protopsephurus.*

Comments—The head appears to be proportionately longer in Polyodonti than it is in †*Protopsephurus* and other Acipenseriformes. Thus head length is another potential character for node K.

Subfamily †Paleopsephurinae, Genus †*Paleopsephurus* (Node L, Fig. 11)

Diagnosis—Grande and Bemis (1991, p. 109) reported no autapomorphies at this node, and no additional information has become available since then.

Comments—MacAlpin (1941a,b, 1947) described †*Paleopsephurus;* additional information and interpretations are provided by Grande and Bemis (1991). More specimens are needed for a complete description.

Subfamily Polyodontinae (Node M, Fig. 11)

Diagnosis—A group within Polyodonti that differs from all others by the following:

(*M1*) *Loss of median elements at the rear of the skull roof.* Here we modify this character from its original description in Grande and Bemis (1991, p. 109, character 22). We base our new interpretation on information about extrascapular bones in other acipenseriforms discussed in the description of †*Peipiaosteus pani* above. In particular, we noted above that nuchal plates were probably never present in the common ancestor of Acipenseriformes, let alone Acipenseristomi (Fig. 11). Thus, we now consider the median elements lost at Node M to have been median extrascapulars.

(*M2*) *Loss of ectopterygoid process of the palatopterygoid contacting the dermopalatine (= maxilla of Grande and Bemis, 1991).* This character is defined by Grande and Bemis (1991, p. 109, character 23).

Comments—This group contains *Psephurus*, †*Crossopholis*, and *Polyodon*.

Genus *Psephurus* (Node N, Fig. 11)

Diagnosis—A group within Polyodontinae that differs from all others by the following:

(N1) Fusion of the first two to five proximal radials of the anal fin. This character is defined by Grande and Bemis (1991, p. 109, character 24). This is a large laterally compressed spherical element in our largest *Psephurus* specimen.

(N2) Dorsal caudal fulcra reduced in number and, in adults, much more heavily ossified than in any other acipenseriform. This character is defined by Grande and Bemis (1991, p. 109, character 25).

Comments—The single living species of *Psephurus*, *P. gladius*, occurs in the Yangtze River of China. Based on the examination of the KMNH specimen of †*Protopsephurus*, we can verify that this extinct genus also lacks characters N 1 and N 2.

Tribe Polyodontini (Node O, Fig. 11)

Diagnosis—A group within Polyodontinae that differs from all others by the following:

(O1) Extreme elongation of supracleithrum. This character was defined by Grande and Bemis (1991, p. 109, character 26). Now, on the basis of additional comparisons, we refine our definition to refer to the elongate, narrow shape of the supracleithrum because several other acipenseriforms (e.g., †*Stichopterus*) could be said to have elongate supracleithra.

(O2) Lateral expansion of stellate bone sheets on posterior half of rostrum. This character is defined by Grande and Bemis (1991, p. 110, character 27).

(O3) Prearticular developed into a long splintlike bone about half the length of the lower jaw. This character is defined by Grande and Bemis (1991, p. 111, character 28).

(O4) Reduction in size of the middle division of the fenestra longitudinalis. This character is defined by Grande and Bemis (1991, p. 111, character 29). This character is not applicable for non-polyodontids because they do not have a fenestra longitudinalis.

Comments—This group existed in North America east of the present continental divide since at least the Paleocene.

Genus *Crossopholis* (Node P, Fig. 11)

Diagnosis—A group within Tribe Polyodontini that differs from all others by the following:

(P1) Presence of a pronounced anterior projection of the parasphenoid separating the vomers along most of their length. This character is defined by Grande and Bemis (1991, p. 111, character 30).

Comments—†*Crossopholis magnicaudatus* is known from the Green River Formation of Wyoming. Although the species is rare, many complete skeletons were reported by Grande and Bemis (1991).

Genus *Polyodon* (Node Q, Fig. 11)

Diagnosis—A group within Tribe Polyodontini that differs from all others by the following:

(Q1) Gill rakers in adults extremely numerous, long and thin, modified for filter-feeding. This character is defined and illustrated by Grande and Bemis (1991, p. 111, character 31).

(Q2) Loss of quadratojugal. This character is defined by Grande and Bemis (1991, p. 111, character 32).

(Q3) Ossified elements of gill arches compressed into wafer-thin plates. This character is defined by Grande and Bemis (1991, p. 111, character 34).

(Q4) Upper jaw symphysis firmly attached to braincase and loss of jaw protrusibility. This character is defined by Grande and Bemis (1991, p. 112, character 35).

(Q5) Loss of ectopterygoid process of dermopalatine (= maxilla of Grande and Bemis, 1991). This character is defined by Grande and Bemis (1991, p. 112, character 36).

(Q6) Very large relative lengths of upper and lower jaws and correspondingly large gape of mouth. This character is defined by Grande and Bemis (1991, p. 112, character 37). †*Birgeria* also has a large gape, but we see the condition in *Polyodon* as clearly derived, based on the distribution of other characters.

Comments—The genus *Polyodon* includes one extant species (*P. spathula*) and one fossil species (†*P. tuberculata*) from the Cretaceous of North America. The interclavicle is also lost in *Polyodon* (Grande and Bemis, 1991, p. 111, character 33), a convergence with †Peipiaosteidae (see character E2).

Family Acipenseridae (Node R, Fig. 11)

Osteology and systematics of the sturgeon family, Acipenseridae Bonaparte 1831, have been reexamined in several recent studies particularly those of Findeis (1991, 1993, 1997; also see Grande and Bemis, 1991; Bemis *et al.* 1997). Acipenseridae, as used here, contains four extant genera and one extinct genus, †*Protoscaphirhynchus*, although this genus may not warrant generic status separate from *Scaphirhynchus*. Other fossil material has been described, but it is either fragmentary material of doubtful phylogenetic significance, or it has since been removed from the family. Intrafamilial relationships, discussed by Findeis (1997), are the basis for many characters incorporated here. Artyukhin (1995) and Birstein *et al.* (1997) have also examined relationships within the family Acipenseridae using molecular and karyological evidence as well as morphology. There is much doubt among workers concerning the traditional division of the family; here we follow Findeis' (1997) interpretation in dividing Acipenseridae into two subfamilies, Husinae and Acipenserinae.

Diagnosis—A group within Acipenseroidei that differs from all others by the following:

(R1) The presence of five longitudinal rows of bony scutes along the trunk. This traditional character for acipenserids has recently been reviewed and illustrated (Grande and Bemis, 1991, p. 109, character 11 and figs. 78–79; Findeis, 1997, character 1).

(R2) The presence of a long, stout fin spine along the leading edge of the pectoral fin. This traditional character for acipenserids has recently been reviewed and illustrated (Grande and Bemis, 1991, p. 109, character 13 and figs. 78–79; Findeis, 1997, character 2, fig. 2).

(R3) The presence of a platelike supraorbital with a preorbital descending process. This character is unfortunately subject to different terminology. Jollie (1980, p. 235) called it a supraorbital. Grande and Bemis (1991, suo?, fig. 6A) identified a bone above the eye of *Polyodon* that they tentatively identified as a supraorbital. Findeis (1997, character 4, fig. 4) termed this bone an antorbital, and Bemis *et al.* (1997, character 27) maintained that terminology. Henceforth we will refer to this element as a supraorbital, for this more closely agrees with our current conception of the homologies of this element. In any event, the condition in question is certainly unique to acipenserids.

(R4) Rostral canals curve lateral to barbels. This character is defined and illustrated by Findeis (1996, character 6, fig. 6). In fossils, we score this character by looking for a characteristic U-shaped lateral bend in the rostral part of the infraorbital canal. It is one of the most striking specializations of the sensory canals in any lower actinopterygian, and it is certainly unique to Acipenseridae.

(R5) The supracleithrum is tightly joined to and forms part of the dermal skull roof. This condition is achieved by a more or less elongate anterior process of the supracleithrum. The character is defined and illustrated by Findeis (1997, character 9, fig. 8). In keeping with the hypertrophy of the posterior part of the skull roof of acipenserids, the pectoral girdle is fully integrated and tightly linked to the skull.

(R6) The presence of a medial opercular wall formed by medial extensions of the cleithrum and clavicle. This character is defined and illustrated by Findeis (1996, character 10, fig. 8). Although many fishes possess a branchial lamina formed by extensions of the pectoral girdle, the condition in acipenserids is extreme and distinctive within lower actinopterygians.

(R7) Presence of a cardiac shield. This character is defined and illustrated by Findeis (1997, character 11, fig. 8). This structure is formed by the right and left clavicles, which suture tightly together in the ventral midline. It encases the pericardial cavity.

(R8) The presence of a cleithral process limiting the forward mobility of the fin spine. This character is defined and illustrated by Findeis (1997, character 12, fig. 8). This character appears to be linked to the presence of the fin spine (character R 2 in this chapter).

Comments—Many additional putative characters of Acipenseridae are reported and discussed by Findeis (1997). Because there is little doubt of the monophyly of the family, and because we do not know how to score some of Findeis' (1997) characters in other Acipenseriformes, we omit discussion of them here. Many of these characters appear to be outstanding diagnostic features of Acipenseridae, but those involving cartilaginous features are difficult to survey fully in other Acipenseriformes without making new original studies. This should certainly be a component of future work on the phylogenetics of Acipenseriformes.

We do wish to make one correction to Findeis (1997) and Bemis *et al.* (1997) based on the work reported in this chapter. Findeis (1997, character 5) considered that the presence of a median extrascapular bone was synapomorphic for Acipenseridae. This character was also expressed as relating to the commissure of occipital canals in a median extrascapular bone (Bemis *et al.*, 1997, character 28). Upon further evaluation, we think this character is not a synapomorphy of acipenserids, for it appears that median extrascapular bones may be present in both †peipiaosteids (*exc*, Fig. 2c) and †chondrosteids (*exc*, Fig. 2E). Where median extrascapular bones are present, any commissure would presumably occur in these bones.

Subfamily Husinae (*sensu* Findeis, 1993), genus *Huso* (Node S, Fig. 11)

Diagnosis—A group within Acipenseridae that differs from all others by the following:

(S1) Basitrabecular processes form flattened shelves. This character is defined and illustrated by Findeis (1997, character 25, fig. 15a).

(S2) No palatoquadrate–interhyal joint. This traditional character of *Huso* is defined and illustrated by Findeis (1997, character 26, fig. 16).

Comments—The genus *Huso* includes two species, *H. huso* and *H. dauricus*. Species of *Huso* occur in the

Adriatic, Black, Caspian, and Okhotsk Seas, and in the Amur River Basin.

Subfamily Acipenserinae (*sensu* Findeis, 1993) (Node T, Fig. 11).

Diagnosis—A group within Acipenseridae that differs from all others by the following:

(T1) Dorsal rostral series forms an expanded shield. This character is defined and illustrated by Findeis (1997, character 28, fig.18).

(T2) The presence of border rostral bones along the lateral margin of the snout region. This character is defined and illustrated by Findeis (1997, character 29, fig. 18).

(T3) The first ventral rostral bone is single. This character is defined and illustrated by Findeis (1997, character 30, fig. 19). This character is not applicable to non-acipenseroids (e.g., outside of Node H, Fig. 11) because they lack ventral rostral bones.

(T4) The presence of pineal bones. This character is defined and illustrated by Findeis (1997, character 32, fig. 18).

(T5) The lower jaw is straight rather than curved toward symphysis. The lower jaws of Acipenserinae lack temporal processes, which are present in all other Acipenseriformes. This character is defined and illustrated by Findeis (1997, character 33, fig. 21).

(T6) The dermopalatine (= maxilla of Grande and Bemis, 1991) has a bladelike ventral expansion on the anterior ventral edge. This striking character is defined and illustrated by Findeis (1997, character 34, fig. 22). It is associated with the unusual feeding system of Acipenserinae, in which food is processed by crushing between the palate and tongue.

Comments—Findeis (1997) identified several additional putative synapomorphies of Acipenserinae. Subfamily Acipenserinae is divided into two tribes, Acipenserini and Scaphirhynchini, discussed next.

Tribe Acipenserini and Genus *Acipenser* (Node U, Fig. 11)

Diagnosis—Osteological synapomorphies of this group are unknown.

Comments—Acipenserini and the genus *Acipenser* need much phylogenetic study. *Acipenser* is probably not monophyletic (Findeis, 1993, 1997), and we present no phylogenetic characters diagnosing this group. New interpretations of *Acipenser*, based on molecular and karyological data as well as more traditional anatomical characters, are forthcoming (e.g., Birstein *et al.*, 1997). Some of these studies group *Huso* within the genus *Acipenser*. Fossils assigned to *Acipenser* occur in widespread localities in Asia and western North America, although most of the material is fragmentary. The 17 nominal Recent species included

in this genus are mostly endemic to Europe and Asia, but 5 species are endemic to North America. Most of these inhabit both fresh and salt water.

Tribe Scaphirhynchini (Node V, Fig. 11)

Diagnosis—A group within Acipenserinae that differs from all others by the following:

(V1) The presence of ampullary bones. This character is defined and illustrated by Findeis (1997, character 40, fig. 18). These bones are associated with the ampullary electroreceptive organs and form later in development than adjacent rostral bones.

(V2) Loss of prearticular bones. This character is defined by Findeis (1997, character 48).

(V3) The presence of a long, well-developed caudal fin filament. This traditionally known character is defined and illustrated by Findeis (1997, character 53, fig. 23).

Comments—Findeis (1997) noted many other characters at Node V. These include some features, such as central spines on the dermal bones of skull roof and shoulder girdle, which occur in most but not all species of Scaphirhynchini. The caudal fin filament, or cercus, is distinctive at this node but unfortunately cannot be scored in fossils. Some of Findeis' other characters are cartilaginous or incompletely known in other Acipenseriformes. Scaphirhynchini is divided into two sections.

Section one of Scaphirhynchini (Node W, Fig. 11)

Diagnosis—A group within tribe Scaphirhynchini that differs from all others by the following:

(W1) Caudal peduncle and preanal area entirely encased in an armor of scutes. This traditional character is defined by Findeis (1997, character 65; see Grande and Bemis, 1991, fig. 78 for photograph). This character is present in †*Protoscaphirhynchus*.

(W2) Caudal peduncle is flat and elongate. This traditional character is defined by Findeis (1997, character 64; see Grande and Bemis, 1991, fig. 78 for photograph). This character is present in †*Protoscaphirhynchus*.

(W3) Gill rakers are crenelated. This traditionally known character is defined and figured by Findeis (1997, character 63 and fig. 25). The condition in †*Protoscaphirhynchus* is unknown. Gill rakers of all other Acipenseriformes, except for *Polyodon* and Section two of Scaphirhynchini (*Pseudoscaphirhynchus*), are stubby and triangular. We cannot fully assess the polarity of change in gill raker morphology within Scaphirhynchini (i.e., whether the different conditions in Section one and Section two were independently derived from a common pattern, or whether one was derived from the other; see character X 1 in this chapter).

Nevertheless, the difference in gill raker morphology between Sections one and two is diagnostic.

Comments—Section one of Scaphirhynchini contains the extant genus *Scaphirhynchus* and the fossil genus †*Protoscaphirhynchus*. Findeis (1997) reports several additional synapomorphies for this node. The genus *Scaphirhynchus*, sometimes referred to as the shovelnose or river sturgeons, contains three nominal species, all restricted to fresh, occasionally brackish waters of North America (Findeis, 1997; Bemis *et al.*, 1997). The monotypic fossil genus †*Protoscaphirhynchus* Wilimovsky 1956, is from Upper Cretaceous freshwater deposits of the Hell Creek beds, Montana. Wilimovsky (1956), in his original description, noted the similarity of this fossil species to the extant *Scaphirhynchus*. Gardiner (1984) later removed this species from Acipenseriformes for several reasons. Gardiner (1984a, 152) stated: "Wilimovsky (1956) considered [†*Protoscaphirhynchus*] an undoubted member of the Acipenseridae, but the genus is completely armored with ganoine-covered scales and with small post-temporals, frontals, and parietals quite unlike any living or fossil acipenseriform. On this evidence *Protoscaphirhynchus* is certainly not a member of the Acipenseriformes."

After studying the only available specimen (UMMP 22210), we disagree with Gardiner's (1984) interpretation and conclude that †*Protoscaphirhynchus* is closely related to *Scaphirhynchus*. The "ganoine-covered scales" that cover the caudal region of †*Protopsepherus* we consider to be homologous to those of other species of *Scaphirhynchus* (see Findeis, 1997).

Section two of Scaphirhynchini (Node X, Fig. 11)
Diagnosis—A group within tribe Scaphirhynchini that differs from all others by the following:

(X1) *Presence of "pronged" gill rakers.* This traditional character of *Pseudoscaphirhynchus* was critically defined and illustrated by Findeis (1997, character 58, fig. 25). Gill rakers of all other Acipenseriformes, except for *Polyodon* and Section one of Scaphirhynchini (*Scaphirhynchus* + †*Protoscaphirhynchus*), are stubby and triangular. We cannot fully assess the polarity of change in gill raker morphology within Scaphirhynchini (i.e., whether the different conditions in Sections one and two were independently derived from a common pattern or whether one was derived from the other; see character W3, discussed previously). Nevertheless, the difference in gill raker morphology between Sections one and two is diagnostic.

(X2) Lateral extrascapular bones bear the triradiation of the trunk, occipital, and supratemporal canals. This character was defined and illustrated by Findeis

(1996, character 56, fig. 5). Among Acipenseriformes, it only occurs in *Pseudoscaphirhynchus*. However, a similar condition seems to occur in †*Peipiaosteus* and †*Yanosteus*.

Comments—Section two of Scaphirhynchini contains only the extant genus *Pseudoscaphirhynchus* from the Aral Sea basin in central Asia. Two of the three described species appear to be extinct (Birstein, 1993), and populations of *P. kaufmanni* appear to be much smaller than historically reported. Findeis (1997) defined several additional characters applicable to Node X (Fig. 11). Many of these appear to be good diagnostic features for the group.

V. Nomenclatural Recommendations

The cladogram in Fig. 11 includes some new and redefined nomenclature. A summary follows:

Series Chondrostei *sensu* Patterson, 1982 and Grande and Bemis, 1991

    Family †Birgeriidae *sensu* Nielsen 1949
        Genus †*Birgeria* Stensiö 1919
    Order Acipenseriformes Berg 1940
    Suborder †Chondrosteoidei *sensu* Grande and Bemis 1991
        Family †Chondrosteidae Egerton 1858
            Genus †*Chondrosteus* Agassiz 1833–1844
            Genus †*Strongylosteus* Agassiz 1833–1844
    Division Acipenseristomi—new name
    Suborder †Peipiaosteoidei—new name
    Family †Peipiaosteidae Liu and Zhou 1965
        Subfamily †Stichopterinae—new name
            Genus †*Stichopterus* Reis 1910
            Genus †*Peipiaosteus* Liu and Zhou 1965
        Subfamily †Spherosteinae—new name
            Genus †*Spherosteus* Jakovlev 1977
            Genus †*Yanosteus* Jin, Tian, Yang, and Deng 1995
    Suborder Acipenseroidei *sensu* Grande and Bemis 1991
        Family Polyodontidae Bonaparte 1838
            Subfamily †Protopsephurinae—new name
                Genus †*Protopsephurus* Lu 1994

Series Polyodonti—new name
  Subfamily †Paleopsephurinae
  Grande and Bemis 1991
    Tribe †Paleopsephurini Grande
    and Bemis 1991
      Genus †*Paleopsephurus* MacAl-
      pin 1941a
  Subfamily Polyodontinae Grande
  and Bemis 1991
    Tribe Psephurini Grande and
    Bemis 1991
      Genus *Psephurus* Günther
      1873
    Tribe Polyodontini Grande and
    Bemis 1991
      Genus †*Crossopholis* Cope 1883
      Genus *Polyodon* Lacepede 1797
Family Acipenseridae Linnaeus 1758
  Subfamily Husinae Findeis 1993
    Genus *Huso* Brandt 1869
  Subfamily Acipenserinae *sensu* Fin-
  deis 1993
    Tribe Acipenserini Findeis 1993
      Genus *Acipenser* Linnaeus
      1758
    Tribe Scaphirhychini Bonaparte
    1846
      Section one—new section
        Genus *Scaphirhynchus* Heckel
        1836
        Genus †*Protoscaphirhynchus*
        Wilimovsky 1956
      Section two—new section
        Genus *Pseudoscaphirhynchus*
        Nikolskii 1900

VI. New Approaches to Phylogenetics of "Chondrostei" and Lower Actinopterygians

The cladogram in Figure 11 differs from the cladogram presented by Grande and Bemis (1991, fig. 76B) in its greater resolution of Acipenseridae, *sensu* Findeis, 1993, 1997) †Peipiaosteoidei, and in the inclusion of †Birgeriidae. Even with this additional taxon, our concept of Chondrostei includes many fewer taxa than have been included elsewhere (e.g., Gardiner, 1967; Carroll, 1987; Nelson, 1994), but we find insufficient evidence to include more taxa in Chondrostei without making that group non-monophyletic. This is not unlike the case of lungfishes and tetrapod origins, where Rosen *et al.* (1981) "cleaned the slate" and provoked a decade of productive research. We follow Patterson (1982) in doing the same for "Chondrostei."

We acknowledge the monumental efforts of Gardiner and Schaeffer (1989) to synthesize data on lower actinopterygians from previous descriptive studies and analyze it with PAUP. Unfortunately, as they point out (Gardiner and Schaeffer, 1989, p. 137), much or most of the skeleton is yet unknown for the vast majority of lower actinopterygians. Combining this informational void with published misinformation derived from misinterpretation of poorly prepared specimens presents a significant problem. No matter how sophisticated a phylogenetic analytical program we use, the meaningfulness of the resulting phylogeny depends on the quality of the data matrix. The real frontier in lower actinopterygian paleontological research is to study the abundance of newly discovered, well-preserved fossil (and Recent) material using modern preparation techniques to extract a high quality of anatomical information and then to use modern illustration and publishing techniques to unambiguously document that information. To allow reliable interpretation, these materials must be examined in broadly based comparative studies. The use of morphology to investigate phylogeny is under fire by some (e.g., Ho, 1988); however, we believe that comparative morphological research is resurgent, particularly for groups such as the lower actinopterygians. By emphasizing the strengths of comparative morphological analyses and integrating studies of fossil and living species, we hope to demonstrate the essential role of morphological data for reconstructing phylogenies.

Acknowledgments

We dedicate this paper to Colin Patterson, who has been a great inspiration to us both. Much of the first author's early interest in systematic ichthyology was the result of encouragement from Colin and from reading his groundbreaking papers in ichthyology, systematics, evolution, and developmental biology. Colin set new standards for work on fossil fishes, with a clear and logical approach to each of his studies. He will probably cringe at the dedications contained within this book, because he never seemed to desire the limelight, even though he deserves it as much as anyone working in systematics and evolutionary biology today. Like it or not, Colin (to paraphrase Budweiser) this volume's for you.

Thanks also to Eric K. Findeis, E. O. Wiley, and Amy McCune for reading and commenting on an earlier draft of this manuscript. Ralph Molnar (Queensland Museum) kindly sent us the photograph of †*Stichopterus popovi*. We especially thank Yoshitaka Yabumoto (Kitakyusha Museum of Natural History) for allowing us to study specimens in his care. Final inking and labeling for most of the stipple drawings was completed primarily by Lori Grove. Expert photographic assistance was provided by John Weinstein. Our research program on non-teleostean actinopterygians is supported by NSF (DEB 8806539, 9119561 and 9220938) and the Tontogany Creek Fund.

References

Agassiz, L. (1833–1844). "Recherches sur les Poissons Fossiles." Petitpierre Imprimiere, Neuchâtel.

Artyukhin, E. N. (1995). On biogeography and relationships within the Genus *Acipenser*. *Sturgeon Q.* **3**(2), 6–8.

Bai, Y. (1983). A new species of the Beipiao sturgeon from Fengning County, Hebei Province: Gu Jizhui Dongue yu Gu Renlei. *Vertebr. PalAsiat.* **21**(4), 341–346 (in Chinese with an English summary).

Bemis, W. E., Findeis, E. K., and Grande, L. (1997). An overview of Acipenseriformes. *Environ. Biol. Fishes.* **47(1)**

Berg, L. S. (1940). "Classification of Fishes Both Recent and Fossil." Thai National Documentation Center, Bangkok (English translation of Russian original, 1965).

Berg, L. S. (1948a). "Freshwater Fishes of the USSR and Adjacent Countries." 4th ed., Vol. 1. Israel Program for Scientific Translation, Jerusalem (English translation of Russian original, 1962).

Berg, L. S. (1948b). On the position of the Acipenseriformes in the system of fishes. *Tr. Zool. Inst., Akad. Nauk SSSR* **7**, 7–57.

Birstein, V. J. (1993). Sturgeons and paddlefishes: Threatened fishes in need of conservation. *Conserv. Biol.* **7**, 773–787.

Birstein, V. J., Hammer, R., and DeSalle, R. (1997). Phylogeny of the Acipenseriformes: Cytogenetical and molecular approaches. *Environ. Biol. Fishes.* **47(1)**

Bonaparte, C. L. (1831). Saggio di una distribuzione metodico delgi animali vertebrati. *G. Arcadico di Sci., Lett. Arti* **49**, 1–77.

Bonaparte, C. L. (1838). Selachorum tabula analytica. *Nouv. Ann. Sci. Nat.* **2**, 195–214.

Bonaparte, C. L. (1846). "Catalogo metodico dei pesci Europi." Napoli.

Brandt, J. F. (1869). Einige worte uber die europaich-asiatischen Storarten (Sturionides). *Mélanges Biol.* **7**, 110–116.

Carroll, R. L. (1987). "Vertebrate Paleontology and Evolution." Freeman, New York.

Cope, E. D. (1883). A new chondrostean from the Eocene. *Am. Nat.* **17**, 1152–1153.

Cope, E. D. (1887). Zittel's manual of palaeontology. *Am. Nat.* **21**, 1014–1019 (a review of the above work of Professor Zittel, and containing a classification of the teleostomous fishes).

Egerton, P. de M. G. (1858). On *Chondrosteus*, an extinct genus of the Sturionidae, found in the Lias Formation at Lyme Regis. *Philos. Trans. R. Soc. London* **148**, 871–885.

Findeis, E. K. (1991). Ontogeny and homology of the "maxilla" of sturgeons: Evolution of mobile jaw elements in early actinopterygian fishes. *Am. Zool.* **31**, 8A (abstr.).

Findeis, E. K. (1993). Skeletal anatomy of the North American shovelnose sturgeon *Scaphirhynchus platorynchus* (Rafinesque 1820) with comparisons to other Acipenseriformes. Doctoral Dissertation, University of Massachusetts, Amherst.

Findeis, E. K. (1997). Comparative osteology and relationships of Recent sturgeons (Acipenseridae). *Environ. Biol. Fishes.* **47(1)**

Gardiner, B. G. (1967). Further notes on paleoniscids with a classification of the Chondrostei. *Bull. Br. Mus. (Nat. Hist.), Geol.* **8**, 255–325.

Gardiner, B. G. (1984). Sturgeons as living fossils. *In* "Living Fossils." (N. Eldredge and S. M. Stanley, eds.), pp. 148–152. Springer-Verlag, New York.

Gardiner, B. G. (1993). Basal Actinopterygians. *In* "The Fossil Record" (M. J. Benton, ed.) Vol. 2. Chapman & Hall, London.

Gardiner, B. G., and Schaeffer, B. (1989). Interrelationships of lower actinopterygian fishes. *Zool. J. Linn. Soc.* **97**, 135–187.

Grande, L., and Bemis, W. E. (1991). Osteology and phylogenetic relationships of fossil and Recent paddlefishes (Polyodontidae) with comments on the interrelationships of Acipenseriformes. *J. Vertebr. Paleontol.* **11**, Suppl. 1, 1–121.

Günther, A. (1873). *Psephurus gladius. Ann. Mag. Nat. Hist.* [4] **12**, 250.

Heckel, J. J. (1836). *Scaphirhynchus*, eine neue fischgattung aus der ordnung der Chondropterygier mit freien kiemen. *Ann. Wien. Mus.* **1**, 68–78.

Hennig, E. (1925). *Chondrosteus hindenburgi* Pomp., ein "Stor" des württemberg Oelschiefers (Lias). *Paleontographica Stuttgart* **67**, 115–133.

Ho, M.-W. (1988). How rational can rational morphology be? A post-Darwinian rational taxonomy based on a structuralism of process. *Rev. Biol.* **81**, 11–55.

Holmgren, N., and Stensiö, E. (1936). Kranium und Visceralskelett der Acranier. Cyclostomen und Fische. *In* "Handbuch der vergleichenden Anatomie der Wirbeltiere." (L. Bolk, E. Göppert, E. Kallius, and W. Lubosch, eds.), Vol. 4, pp. 233–500. Urban & Schwarzenberg, Berlin.

Jaekel, O. (1929). Die Morphogenie der ältesten Wirbeltiere. *Monogr. Geol. Palaeontol.* **1**(3), 1–198.

Jakovlev, V. N. (1977). Phylogenesis of acipenseriforms. *In* "Essays on Phylogeny and Systematics of Fossil Fishes and Agnathans" (V. V. Menner, ed.), pp. 116–143. USSR Academy of Sciences, Moscow (in Russian).

Jakovlev, V. N. (1986). Fish. *Transactions* **28**, 178–181.

Jin, F. (1995). Late Mesozoic Acipenseriformes (Osteichthyes: Actinopterygii) in Central Asia and their biogeographic implications. *In* "Sixth Symposium on Mesozoic Terrestrial Ecosystems and Biota, Short Papers" (A. Sun and Y. Wang, eds.), pp. 15–21. China Ocean Press, Beijing.

Jin, F., Tian, Y.-P., Yang, Y.-S., and Deng, S. Y. (1995). An early fossil sturgeon (Acipenseriformes, Peipiaosteidae) from Fenging of Hebei, China. *Vertebr. PalAsiat.* **33**, 1–18.

Jollie, M. (1980). Development of the head and pectoral girdle skeleton and scales in *Acipenser. Copeia*, pp. 226–249.

Lacépède, B. G. E. (1797). Sur la *Polyodon* Fueille (*Spatularia folium*). 1. *Bull. Sci. Soci. Philomatique Paris* **2**, 49.

Linnaeus, C. (1758). "Systema Naturae," Editio X. [Systema naturae per regna tria naturae, secundum classes, ordines, genera, species, cum characteribus, differentiis, synonymis, locis. Tomus I. Editio decima, reformata.] Holmiae.

Liu, H.-T., and Zhou, J. J. (1965). A new sturgeon from the upper Jurassic of Liaoning, north China. *Vertebr. PalAsiat.* **9**, 237–248.

Lu, L. (1994). A new paddlefish from the Upper Jurassic of northeast China. *Vertebr. PalAsiat.* **32**(2), 32.

MacAlpin, A. (1941a). †*Paleopsephurus wilsoni*, a new polyodontid fish from the Upper Cretaceous of Montana, with a discussion of allied fish, living and fossil. *Bull. Geol. Soc. Am.* **52**, 1989 (abstr.).

MacAlpin, A. (1941b). †*Paleopsephurus wilsoni*, a new polyodontid fish from the Upper Cretaceous of Montana, with a discussion of allied fish, living and fossil. Doctoral Dissertation, University of Michigan, Ann Arbor.

MacAlpin, A. (1947). †*Paleopsephurus wilsoni*, a new polyodontid fish from the Upper Cretaceous of Montana, with a discussion of allied fish, living and fossil. *Contrib. Mus. Geol. (Paleontol.), Univ. Mich.* **6**, 167–234.

Maduit, M. (1774). *J. Phys.* **4**, 384–400 (typographical error in original article numbers first page as p. 284).

Müller, J. (1844). On the structure and characters of the Ganoidei, and on the natural classification of fish. *Sci. Mem.* **4**, 499–542. 1846 (English translation of Müller, 1844).

Nelson, G. (1969). Gill arches and the phylogeny of fishes with notes on the classification of vertebrates. *Bull. Am. Mus. Nat. Hist.* **141**, 475–552.

Nelson, J. S. (1994). "Fishes of the World," 3rd ed. Wiley, New York

Nielsen, E. (1949). Studies on Triassic fishes from East Greenland. II. *Australosomus* and *Birgeria. Palaeozool. Groenl.* **3**, 1–309.

Nikolskii, A. M. (1900). *Pseudoscaphirhynchus rossikowi*, n. gen. et spec. *Ann. Mus. Imp. Sci. St. Petersburg* **4**, 257–260 (text in Russian).

Patterson, C. (1973). Interrelationships of holosteans. *Zool. J. Linn. Soc.* **53**, Suppl. 1, 233–306.

Patterson, C. (1982). Morphology and interrelationships of primitive actinopterygian fishes. *Am. Zool.* **22**, 241–259.

Pehrson, T. (1922). Some points in the cranial development of teleostomian fishes. *Acta Zool.* (Stockholm) **3**, 1–63.

Pehrson, T. (1940). The development of dermal bones in the skull of *Amia calva. Acta Zool.* (Stockholm) **21**, 1–50.

Pehrson, T. (1944). Some observations on the development and morphology of the dermal bones in the skull of *Acipenser* and *Polyodon. Acta Zool.* (Stockholm) **25**, 27–48.

Pehrson, T. (1947). Some new interpretations of the skull in *Polypterus. Acta Zool.* (Stockholm) **28**, 399–455.

Reis, O. M. (1910). "Die Binnenfauna der Fischschiefer in Transbaikalien. Recherches géologiques et minères le long du chemin de fer Sibérie," Book 29. St. Petersburg.

Rieppel, O. (1992). A new species of the genus *Saurichthys* (Pisces: Actinopterygii) from the Middle Triassic of Monte San Giorgio (Switzerland), with comments on the phylogenetic interrelationships of the genus. *Palaeontographica, Abt. A* **221**, 63–94.

Romer, A. S. (1966). "Vertebrate Paleontology," 3rd ed. University of Chicago Press, Chicago.

Rosen, D. E., Forey, P. L., Gardiner, B. G., and Patterson, C. (1981). Lungfishes, tetrapods, paleontology and plesiomorphy. *Bull. Am. Mus. Nat. Hist.* **167**, 159–276.

Schaeffer, B. (1973). Interrelationships of chondrosteans. In "Interrelationships of Fishes" (P. H. Greenwood, R. S. Miles, and C. Patterson, eds.), pp. 207–226. Academic Press, London.

Stensiö, E. A. (1919). "Einige Bemerkungen über die systematische Stellung von *Saurichthys mougeoti* Agassiz," Vol. 1. Senckenbergiana, Frankfurt an Min.

Traquair, R. H. (1877). The ganoid fishes of the British Carboniferous Formations. Part I. Monographs. *Palaeontogr. Soc.* **31**, 1–60.

Traquair, R. H. (1887). Notes on *Chondrosteus acipenseroides. Geol. Mag.* **4**, 248–257.

Vladykov, V., and Greeley, J. R. (1963). Order Acipenseroidei. In "Fishes of the Western Northern Atlantic" (H. B. Bigelow and W. C. Schroeder, eds.), Sears Found. Mar. Res., Mem. No. 1, pp. 24–60. Yale University, New Haven, CT.

Walbaum, J. (1792). [part 3 of] Petri Artedi renovati, pars i et ii [and iii–v], i.e., Bibliotheca et philosophia ichthyologica, cura Iohannis Iulii Walbaumii edidit. 5 parts in 3 vols. Grypeswaldiae [1788] 1789 [-1793], 4 pls. Part 3 is subtitled "Petri Artedi Sueci Genera piscium in quibus systema totum ichthyologiae proponitur cum classibus, ordinibus, generum characteribus, specierum differentiis, observationibus plurimis. Redactis speciebus 242 ad genera 52. Ichthyologiae, pars iii. Emendata et aucta a Johanne Julio Walbaum."

Wilimovsky, N. J. (1956). *Protoscaphirhynchus squamosus*, a new sturgeon from the upper Cretaceous of Montana. *J. Paleontol.* **30**, 1205–1208.

Woodward, A. S. (1891). On the paleontology of sturgeons. *Proc. Geol. Assoc.* **11**, 24–44.

Woodward, A. S. (1895). "Catalogue of the Fossil Fishes in the British Museum (Natural History)," Part III. Br. Mus. (Nat. Hist.), London.

Zhou, Z. (1992). Review on *Peipiaosteus* based on new material of *P. pani. Vertebr. PalAsiat.* **30**, 85–101 (in Chinese with English summary).

Appendix 1

Anatomical Abbreviations

Unless otherwise stated, osteological terminology follows that of Grande and Bemis (1991). The use of (r) or (l) after an abbreviation indicates right- or left-side bones for paired elements:

afr, anal fin; *an*, aortic notch; *ao*, antorbital; *arp*, ascending ramus of parasphenoid; *bd*, ossification of the basidorsal element of the vertebral column; *br*, branchiostegal; *cb*, ceratobranchial; *cha*, anterior ceratohyal; *cl*, cleithrum; *clv*, clavicle; *d*, dentary; *dcf*, dorsal caudal fulcrum; *dfr*, dorsal fin; *dp* (= *mx*), dermopalatine [= maxilla of Grande and Bemis, (1991); see Findeis 1991 and Bemis *et al.* (1997) for discussion of homologies]; *dr*, distal radial of dorsal or anal fin; *dsp*, dermosphenotic; *exc*, extrascapular; *fr*, frontal; *gr*, gill rakers; *h*, hyomandibula; *hs*, hemal spine or arch; *hyp*, hypural; *iocn*, infraorbital canal ossifications; *is* (= *dpt*), dermopterotic (= intertemporo-supratemporal of Grande and Bemis, 1991; see discussion in this chapter explaining our interpretation of the homologies of this bone); *llcn*, lateral line canal ossifications; *mpec*, ectopterygoid bone?; *mr*, middle radial of dorsal or anal fin; *na*, nasal; *op*, opercle; *pa*, parietal; *pas*, parasphenoid; *pcf*, pectoral fin; *pcl*, postcleithrum; *pfr*, pelvic fin; *pt*, posttemporal; *pt(v)*, ventral process of posttemporal; *ppt*, palatopterygoid; *ppttp*, palatopterygoid tooth patch; *qj*, quadratojugal; *rbs*, round based scales of shoulder girdle region; *ro*, rostral canal ossifications; *sc?*, scales?; *scl*, supracleithrum; *so*, supraorbital bones; *sobn*, ossifications of the supraorbital sensory canal; *sop*, subopercle; *sr*, sclerotic ring; *tp1*, possible tooth plates of the first branchial arch.

6

Interrelationships of Basal Neopterygians

BRIAN G. GARDINER
Department of Life Sciences
King's College, University of London,
London, England

JOHN G. MAISEY
Department of Vertebrate Paleontology
American Museum of Natural History
New York, New York 10024

D. TIM J. LITTLEWOOD
Department of Paleontology
The Natural History Museum
London SW75BD, United Kingdom

I. Introduction*

Although the name Holostei was originally used by Müller (1844) for a group containing *Polypterus* and *Lepisosteus*, the addition of *Amia* (Müller, 1846) and the subsequent removal of *Polypterus* (Huxley, 1861) established a grouping which remained in use for the next 100 years.

Working on fossil evidence (focusing on the dermal bones of the skull) and using ancestor-descendant sequences, Gardiner concluded (1960, 1967), as had Saint-Seine (1949) before him, that the teleosts and *Amia* share a common ancestry in the parasemionotids, while *Lepisosteus* is descended from the Semionotidae (see also Goodrich, 1930, p. 293).

With the advent of cladistic methodology Nelson (1969a,b) first revived the Holostei as a clade and attempted to show that among extant taxa *Amia* and *Lepisosteus* are sister-groups. He was apparently supported by Jessen's (1972) work on the pectoral fin. Meanwhile Patterson (1973) had been working for several years on actinopterygian braincases and had already concluded that the teleosts were monophyletic (see Patterson, 1967). He subsequently decided to reexamine the evidence for the interrelationships of *Amia*, *Lepisosteus*, and teleosts and review the arrangement of those fossil groups usually placed in the Holostei. Patterson (1973) concluded that *Amia* is more closely related to the teleosts than to *Lepisosteus* and that the basic character of the Halecomorphi (a supraordinal taxon containing *Amia* and its fossil relatives, coordinate to Teleostei) is a joint between the symplectic and the lower jaw as in *Amia*. An amiid-like jaw joint with symplectic support was also found by Patterson (1973) in the extinct caturids and parasemionotids, and he proposed that these taxa plus amiids form a monophyletic group. Gardiner (1993) also united amiids and caturids but excluded parasemionotids and included another extinct group, the ophiopsids. Patterson (1994) subsequently retained parasemionotids on the basis of similarities in the jaw joint and braincase but was noncommittal about ophiopsids.

In his analysis Patterson (1973) also noted that the hypothesis that *Lepisosteus* is more closely related to the teleosts than to *Amia* had never been suggested. Ironically the first paper to challenge Patterson's conclusion was by Olsen (1984) who, in describing the jaw articulation of the parasemionotid *Watsonulus eugnathoides*, drew attention to Nielsen's (1942, 1949) descriptions of the jaw articulations in the palaeoniscids *Pteronisculus*, *Birgeria*, and *Boreosomus*. He used this, together with the congruent characters of preopercular shape and presence–absence of clavicles, to suggest that the compound jaw joint involving the symplectic and lower jaw as seen in *Watsonulus*, *Amia*, and stem-group neopterygians is primitive, whereas the quadrate-articular joint of gars and teleosts is de-

---

*All taxa cited in this study are fossil with the exception of the following extant genera: *Amia*, *Arapaima*, *Elops*, *Gasterosteus*, *Hiodon*, *Hoplias*, *Lepisosteus*, *Lophius*, *Megalops*, *Onchorhynchus*, *Polypterus*, *Polyodon*, *Psephurus*, and *Salmo*.

Copyright © 1996 by Academic Press, Inc.
All rights of reproduction in any form reserved.

rived. Thus gars and teleosts are more closely related to each other than either is to *Amia* in Olsen's 1984 cladogram.

Then in 1991 Olsen and McCune, in a study of *Semionotus*, concluded that the Semionotiformes, including lepisosteids, macrosemiids, and semionotids, constitute a monophyletic group which is either the sister group of teleosts (sharing with them a symplectic removed from the jaw joint and ending blindly on the quadrate) or of *Amia* and *Caturus*. This second possibility, which revives the clade Holostei, is based on the presumed congruent characters of greatly elongated nasal processes, loss of opisthotic, ossification pattern of the endoskeletal pectoral girdle, and the presence of equal numbers of fin rays and supports in the dorsal and anal fins. However, in justification of either grouping, Olsen and McCune (1991, p. 287) reiterate the hypothesis that the double articulation in halecomorphs is primitive, pointing out that Véran (1988) had recently demonstrated that the symplectic has a similar contact with the mandible in a variety of forms usually grouped in the Palaeonisciformes, suggesting that the joint between the symplectic and the lower jaw is primitive for actinopterygians as postulated by Olsen (1984). Meanwhile, Jollie (1984) had reviewed the evidence from the Recent forms and also concluded that the Holostei are monophyletic. From their studies of *Watsonulus* and *Semionotus* Olsen and McCune (1991, p. 288) concluded that "at this time" it did not seem possible to resolve neopterygian relationships and that molecular data and the study of more fossil taxa were required. They considered fossil taxa "especially critical in groups like the Neopterygii where much of the morphological diversity can only be found in fossils."

The basis for Olsen's (1984) original interpretation was the fact that the parasemionotid *Watsonulus* has two primitive characters absent in other neopterygians: unreduced clavicles and a preoperculum with a broad dorsal limb. His assumption that reduction of the clavicles and the dorsal portion of the preopercular occurred once led Olsen to the conclusion that the derived condition of those features in other neopterygians make them monophyletic, with *Watsonulus* as their sister-group. It follows from this scheme of relationships that the symplectic–lower jaw articulation shared by *Watsonulus* and some neopterygians (caturids and amiids) is either primitive or independently derived. Although reduced, the clavicle in *Lepidotes* (see Natural History Museum , London [NHM] specimen P. 34511) still caps the end of the cleithrum while a preoperculum with a broad dorsal limb is present in the pycnodont *Brembodus* (Tintori, 1980).

Olsen and McCune's (1991) reason for considering the double joint as primitive was based primarily on Olsen's (1984) analysis of *Watsonulus*, which included a reinterpretation of Nielsen (1942, 1949), though they cited as additional evidence the work of Véran (1988). Véran[1] found that the short, broad symplectic of palaeoniscoids (e.g., *Pygopterus*, *Pteronisculus*, and *Acrorhabdus*) and ptycholepiforms (*Ptycholepis* and *Boreosomus*) lies *in series* between the tip of the hyomandibular and the interhyal and articulates anteriorly with both the quadrate and lower jaw. In neopterygians, the relations of the symplectic and interhyal are different since the latter lies in parallel (rather than in series) with the symplectic and articulates with the posterior part of the broad distal end of the hyomandibular. The issue with regard to the jaw joint is whether palaeoniscoids and ptycholepiforms have a true articulation (a joint) between the symplectic and lower jaw, homologous with that in parasemionotids and halecomorphs (amiids, caturids and ionoscopids). In our view, such a joint could exist only if the area of contact between the symplectic and articular lacks perichondral bone. In *Pteronisculus* this is not so (Nielsen, 1942, p. 175; Véran, 1988, pl. 6D). In *Boreosomus* and *Acrorhabdus* Véran (1988, figs. 3–5) drew the surface of the symplectic that contacts the lower jaw, and the hind end of the lower jaw in *Boreosomus*, with a type of shading similar to that which she employed for cartilage-capped surfaces, but she did not show or describe any interruption of the perichondral bone cover. In *Pygopterus* her drawing (Véran, 1988, fig. 7) indicated such an interruption on the symplectic but not on the articular, and her stereopair of the specimen (Véran, 1988, pl. 5C) is ambiguous. As yet, there is no demonstrable "joint" (opposed surfaces which lack perichondral bone and were cartilage-covered in life) between the symplectic and lower jaw in palaeoniscoids or ptycholepiforms.

In *Perleidus*, Véran found a symplectic in only one specimen, and we have not yet seen one. Although the relation between the symplectic and lower jaw is not evident in Véran's specimen, the symplectic is clearly short and of the same caliber as the distal end of the hyomandibular; this short, broad symplectic is therefore more like those of palaeoniscoids and ptycholepiforms than the elongate, tapering symplectic of neopterygians. The acid-prepared *Perleidus* available to us show that the lower jaw had a single, transversely orientated articular cotylus, which appears to match the broad quadrate condyle in size and shape, leaving no room for a symplectic component in the jaw

---

[1]For these comments and the resume of Véran (1988) we are indebted to our friend and colleague Colin Patterson.

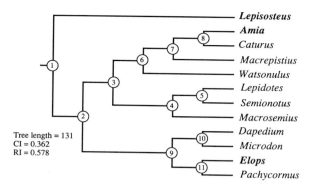

FIGURE 1 Phylogeny of basal neopterygians. The numbered nodes are supported by the following characters (see Appendix 1). Node 2: 6(0–2), 8(0–1), 9(0–2), 21(0–1), 22(0–3), 23(0–1), 24(0–2), 27(0–1), 31(0–1), 32(0–1). Node 3: 17(0–2), 18(0–1), 19(0–2), 33(0–1). Node 4: 4(0–3), 5(0–3), 28(0–1), 29(0–1), 37(0–2). Node 5: 2(0–2), 24(2–1), 36(0–1). Node 6: 1(2–1), 11(2–0), 16(0–1), 18(1–2), 22(3–2), 25(1–2), 26(1–2). Node 7: 4(0–1), 5(0–2), 10(0–1), 17(1–0), 25(2–3), 30(0–1), 34(0–1), 37(0–3). Node 8: 1(1–2), 2(0–2). Node 9: 3(1–2), 9(2–3), 13(1–2), 14(0–1), 20(2–1), 37(0–4). Node 10: 11(2–1), 12(1–2), 21(1–2), 29(0–2), 30(0–3), 36(0–2). Node 11: 5(0–1), 7(1–3), 10(0–1), 19(0–4), 20(1–3), 22(3–1), 34(0–1), 35(0–1).

articulation (see also Schaeffer and Patterson, 1984; Patterson, 1994).

There is valid controversy whether fossils can overturn theories of relationship or whether they can change decisions on homology (and enable us to demonstrate characters which have been acquired in parallel) (Gauthier *et al.*, 1988; Gardiner, 1994). In this chapter we have treated various neopterygian fossils like any other terminal taxon; in our analysis we have included only those characters found in both fossil and extant forms. A parsimony analysis of the resultant data matrix (Appendix 1) indicates that the Holostei are a paraphyletic group with *Amia* more closely related to teleosts than to *Lepisosteus*. Macrosemiidae and Semionotidae are stem-group halecomorphs, while Dapediidae and Pycnodontiformes are immediate sister-groups and stem-group teleosts (Fig. 1). Parasemionotidae, Ionoscopidae, and Caturidae are halecomorphs.

II. Fossils Included in Analysis

Many fossil halecomorph fishes have been described since Patterson's (1973) review of holosteans, but their relationships to *Amia* and to each other remain somewhat vague. Several alternative schemes of interrelationship for fossil amiids have appeared (e.g., Boreske, 1974; Taverne, 1974; Wenz, 1977, 1979; Chalifa and Tchernov, 1982; Bryant, 1987; Maisey, 1991a), but these have all been weakly supported phy-

logenetic hypotheses including few of the existing taxa and often omitting critical data from *Amia calva* (Grande, 1995). There has also been little discussion of higher level relationships among non-amiid fishes possessing an *Amia*-like jaw joint (e.g., Bartram, 1975; Maisey, 1991b; Lambers, 1992).

For our phylogenetic analysis the selected fossils were chosen mainly because they are represented by well-preserved specimens (especially where the cranial anatomy is concerned) that are either clearly described in the literature or are available to us in collections. Most of the data for amiids, caturids, ionoscopids, and ophiopsids are taken from Maisey (1996); other data, principally for semionotids and macrosemiids, were contributed by B.G.G. Much of the material examined in this study is housed in The Natural History Museum, London (NHM) and the American Museum of Natural History, New York (AMNH).

Calamopleurus (also known in the literature as *Enneles*; see Maisey (1991a) for its nomenclatural history) is an amiid that is common in the Albian (lower Cretaceous) Santana Formation of Brazil. Information was obtained from several beautiful acid-prepared AMNH specimens. *Sinamia* and *Ikechoamia* are also regarded as closely related to amiids and are from the Early Cretaceous of China; all the information included here regarding the Chinese fossils was obtained from the literature (Stensiö, 1935; Chang and Hong, 1980; Liu and Su, 1983).

Watsonulus is a parasemionotid from the lower Triassic of Madagascar, and we have supplemented Olsen's (1984) description of its cranial anatomy with information from an undetermined parasemionotid in the NHM (see Patterson, 1973, fig. 23; 1975, fig. 98).

Caturus is a well-known fossil fish that includes several species groups. Character information was obtained both from acid-prepared AMNH specimens of *Caturus* from the Kimmeridgian Lithographic stone of Bavaria and from numerous NHM specimens from various localities.

Ionoscopus is a problematic taxon; the type species is *I. petraroiae* from the Cretaceous of Italy, but it is in need of revision. The best known fossils referred to *Ionoscopus* are from the Kimmeridgian of Bavaria and Cerin. Character information for *Ionoscopus* was obtained from Saint-Seine (1949), an unpublished thesis by Steutzer (1972), and acid-prepared specimens such as AMNH 9700 and a magnificent skull in London (NHM 37795A; Maisey, in preparation). *Oshunia* was described on the basis of a single specimen from the Albian of Brazil by Wenz and Kellner (1986) who regarded it as an ionoscopid. Additional specimens of *Oshunia* in the AMNH collection have yielded a great deal more anatomical information, particularly AMNH 12793 (Maisey, 1991b, in prep.).

Ophiopsis is best known from the descriptions by Bartram (1975) of some Kimmeridgian material from Solnhofen, and that work provided much of the character information included here, supplemented by data from some Solnhofen specimens in the AMNH collection that have been prepared in acid.

Macrepistius is known only from the holotype (an almost complete fish) and an isolated braincase (Schaeffer, 1960, 1971). This fossil was originally included with caturids although for no particular reason. Further preparation of the holotype in the AMNH has revealed new information about this fish (Maisey, 1996), reinforcing Bartram's (1975) opinion that it is closely related to *Ophiopsis*. Ophiopsids were excluded from Patterson's (1973) review of holosteans because at that time Alan Bartram was working on both them and macrosemiids. Since that time, new data for ophiopsids were provided by Bartram (1975) and Applegate (1988), both of whom presented useful diagnoses for the family. The jaw joint is still unknown in *Ophiopsis*, but we have found an *Amia*-like symplectic which articulates with the quadrate in *Macrepistius*. Gardiner (1993) listed ophiopsids within halecomorphs, and several of the features discussed in this chapter support that placement.

Other fossils previously identified as caturids include *Brachyichthys* and *Heterolepidotus*. These taxa are usually placed in synonymy (e.g., Woodward, 1895; Gardiner, 1960), and in many respects they are similar, but we have detected some differences in rostral and parietal morphology between AMNH specimens of the Kimmeridgian *Brachyichthys typicus* and the Sinemurian *Heterolepidotus latus*, and consequently we prefer to distinguish them at the generic level.

Reference will be made below to some fossil braincases from the Bathonian of England, first described by Rayner (1948) under the name *Aspidorhynchus*. Patterson (1973) redescribed them and pointed out that they were not from an aspidorhynchid; instead, he regarded them as caturid braincases. They are remarkably similar (both in general appearance and in many details) to the braincase of *Oshunia* (Maisey, 1991b, in prep.), but when they were included in phylogenetic analysis their position was unresolved. Other poorly known taxa discussed below include the "caturids" *Furo* and *Osteorachis*. As with the "*Aspidorhynchus*" braincases, inclusion of these incompletely known taxa in our analysis was cladistically uninformative, and they were finally omitted from the phylogenetic analysis.

Other fossils included in this analysis (by B.G.G.)

are *Semionotus*, *Lepidotes*, *Macrosemius*, and the pycnodontid, *Neoproscinetes*.

III. Anatomical Features

Symplectic involvement in jaw joint: An *Amia*-like symplectic articulation with the articular occurs in *Caturus* and *Furo* (Patterson, 1973; Bartram, 1977) and *Oshunia* and *Ionoscopus* (Wenz and Kellner, 1986; Lambers, 1992) and in *Macrepistius*, *Heterolepidotus*, and *Brachyichthys* (J.G.M., personal observation). There is also a symplectic contribution to the jaw joint in *Watsonulus* (Olsen, 1984; see also Patterson, 1973, fig. 23). The condition of the symplectic in *Ophiopsis* remains elusive, and we have been unable to observe it in acid prepared specimens in the AMNH collection, although we have found an *Amia*-like symplectic in *Macrepistius*, which may be closely related to *Ophiopsis* (Bartram, 1975; see below).

Symplectic involvement in the jaw joint is widely accepted as the halecomorph synapomorphy (Patterson, 1973, 1994; Bartram, 1975, 1977; Wiley, 1976; Lauder and Liem, 1983; Gardiner, 1984; Schultze and Wiley, 1984; Wiley and Schultze, 1984; Wenz and Kellner, 1986; Maisey, 1991a). It occurs in parasemionotids, which otherwise lack amiiform characters discussed here (Appendix 2). The placement of *Watsonulus* and other "parasemionotids" within neopterygian fishes is controversial, but Patterson's (1994) view that the amiid jaw joint represents a derived character within the neopterygian clade has not been refuted. Thus although the jaw joint of pycnodonts involves the symplectic it is only superficially similar to the double joint of halecomorphs; this study, however, suggests that a monophyletic clade of halecomorph fishes is recognizable on the basis of a number of characters other than the jaw.

In *Pholidophorus* and *Dapedium* the symplectic does not articulate with the lower jaw as in *Amia* (Patterson, 1973, fig. 26; Thies, 1988). According to Brito (1988, p. 822), the symplectic of the aspidorhynchid *Vinctifer* lies behind the quadrate, and both bones articulate with the lower jaw. In specimens of *Vinctifer* we have examined, the symplectic does not reach the lower jaw; furthermore it is in contact with the quadrate (Maisey, 1991c). The symplectic in aspidorhynchids is not attached to the preopercular but is separated from it, teleost fashion, by the quadrate. The symplectic in the specimen illustrated by Brito (1988) seems to be displaced when compared with material figured by Maisey (1991c). In *Vinctifer* the surface of the quadrate is flat where it is pressed against the preopercular.

The symplectic is not flattened against the preopercular except at its base; farther dorsally it is overlain by the quadrate. We conclude from this that *Vinctifer* did not have an *Amia*-like jaw joint, as suggested by Brito (1988); instead it had a primitive teleost-like joint in which the non-suspensory symplectic was separated by the quadrate from the preopercular.

Symplectic bound to preopercular: In *Amia* the symplectic has membranous bone outgrowths which bind it to the inner face of the preopercular (Allis, 1898). We have found similar outgrowths in *Calamopleurus* (and the pycnodont *Neoproscinetes*; see page 134) and Patterson (1973, p. 281) noted a similar development in "some *Caturus* species," but they apparently are absent in others (Lambers, 1992, p. 169). In other non-amiid halecomorphs (e.g., *Ionoscopus*, *Oshunia* and *Macrepistius*) the symplectic lacks membranous outgrowths (Maisey, in prep.). These are also absent in *Watsonulus* (Olsen, 1984). The symplectic is not attached to the preopercular in aspidorhynchids.

Interhyal: Primitively there is a single element between the hyomandibula and ceratohyal in osteichthyans: the interhyal (= stylohyal). This is the condition in *Acanthodes, Polypterus, Mimia, Moythomasia, Eusthenopteron,* actinistians, and porolepids. Additionally in actinistians there is a third element lying lateral to the hyoid arch which in *Laugia* is completely perichondrally ossified. This element is usually called a symplectic.

The ceratohyal in stem-group neopterygians (viz. *Pteronisculus, Boreosomus, Pygopterus, Acrorhabdus* and *Ptycholepis*) as well as all other neopterygians, consists of two bones in one cartilage. In chondrosteans the ceratohyal is comprised of two cartilages. The ceratohyal of *Eusthenopteron* is also made up of two ossifications.

In our opinion the bone described as an interhyal in *Pteronisculus, Boreosomus,* and *Pycholepis* by Véran (1981, 1988) is better interpreted as a posterior ceratohyal, while that of *Acrorhabdus* as an articular. Thus, according to this interpretation, the symplectic of both Nielsen (1942, 1949) and Véran (1981, 1988) is equivalent to the interhyal of *Polypterus*. This means that the symplectic (which in ontogeny is derived from the ventral end of the hyomandibula) is a synapomorphy of neopterygians.

If we accept that the symplectic is a synapomorphy of neopterygians, or that the relations of the symplectic and interhyal are different in neopterygians and palaeoniscoids, then it follows that a polarity decision on the neopterygian jaw joint as to which condition is the more primitive can only be made by testing against trees and/or phylogenies.

Separate articulars (Bridge's ossicles): The condition in *Amia*, with two separate articulars (Bridge's ossicles "c" and "d" of Allis, 1897), is probably a derived condition (Patterson, 1973). The same condition is found in *Calamopleurus* (e.g., AMNH 11829) but has not been described in other amiids. Stensiö (1935) did not describe the articular in *Sinamia*, but according to his illustration (Stensiö, 1935, pl. VI, fig. 1) there is a single large articular. In *Amia* and *Calamopleurus* the quadrate articulates with the angular and both articulars, and the symplectic articulates with the posterior face of the posteriormost articular (Bridge's ossicle "d"). Separate articulars (Bridge's ossicles) are known in only living and certain fossil amiids.

There is a single articular bone in *Oshunia* (AMNH 12000), *Macrepistius* (AMNH 2435), and *Brachyichthys* (AMNH 7741). Lambers (1992, p. 165) found the same condition in *Ionoscopus*. This is also the condition in *Caturus* (Lambers, 1992, p. 165), *Furo*, and *Heterolepidotus* (Patterson, 1973). *Neoproscinetes* also has a single articular. In all these fossils the symplectic articulates with the posterior face of the articular, and the quadrate meets the articular farther dorsally.

Quadrate and quadratojugal: Patterson (1973) proposed that *Amia* and teleosts have lost the quadratojugal as an independent element. A splintlike dorsal projection from the posterior margin of the quadrate in teleosts usually is identified as a quadratojugal (for references, see Lambers, 1992). A small separate bone, also identified as a quadratojugal, occurs in gars (Wiley, 1976; Arratia and Schultze, 1991). Lambers (1992, p. 170) concluded that the teleost process is a new structure, not a quadratojugal. Developmental studies of teleosts have failed to demonstrate that the supposed quadratojugal is ever an independent element (Arratia and Schultze, 1991). Nevertheless a free, splint-like quadratojugal is present in *Dapedium* and some pycnodonts.

A teleost-like process on the quadrate is absent in *Amia*, but there is a flange of bone at its posterior margin (Allis, 1897). In many non-amiid halecomorphs there is a small flange of bone on the posterior or posterolateral margin of the quadrate, for example, in *Caturus, Furo,* and taxa classified as parasemionotids (Patterson, 1973; Bartram, 1977; Olsen, 1984). A similar flange occurs in *Oshunia* (Maisey, in prep.), and in *Ionoscopus* the corresponding flange is extended into a short process (Lambers, 1992, fig. 33).

Variation in the size and extent of the process is reported in aspidorhynchids; there is a small flattened shelf or flange in *Vinctifer* and *Aspidorhynchus* like that at the base of the process in *Pholidophorus germanicus* (Patterson, 1973, fig. 7). The flattened area of the flange is pressed against the mesial surface of the preopercular, in part separating it from the symplectic. In *Belonostomus* there is a relatively large posteroventral process (Taverne, 1981; Brito, 1988; Maisey, 1991c; Lambers, 1992). Thus although quadrate morphology is similar in primitive teleosts and halecomorphs, the presence of a separate quadratojugal in the presumed stem-group teleosts *Dapedium* and *Macromesodon* suggests that the process has been lost in more derived halecomorphs (Patterson, 1973).

Coronoids and coronoid process: Lambers (1992) proposed that the presence of only two coronoids (plus the prearticular) is a synapomorphy of caturids in his restricted sense. There are more than two coronoids in many palaeoniscids and *Amia*, but there are only two (plus the prearticular) in *Polypterus*, *Ospia*, and *Lepisosteus* (Gardiner, 1984, p. 332), casting doubt on the significance of Lambers' (1992) character. Coronoids are present in pholidophorids but are absent in higher teleosts.

In *Ionoscopus* and *Oshunia* there is a massive, vertical coronoid process on the lower jaw, involving the angular and prearticular, with relatively little contribution by the surangular. *Polypterus* and *Neoproscinetes* have a coronoid process formed exclusively by the prearticular. In coelacanths the process is formed by the coronoid series. Although a prearticular is certainly present in teleosts, and a small splint of the prearticular extends into the coronoid process in *Pholidophorus germanicus*, extensive prearticular involvement in the coronoid process apparently is confined to polypterids, pycnodontiforms, and ionoscopids (there is also a minor contribution in *Caturus*). Where a coronoid process is present in teleosts it is formed mainly by the dentary and angular, primitively accompanied by a small autocoronoid (e.g., *Pholidophorus germanicus*) and surangular (e.g., *Pholidophorus germanicus* and *Proleptolepis* sp.; Patterson, 1973; fig. 7; Patterson and Rosen, 1977, fig. 32A; absent in modern teleosts).

Parasphenoid ascending process: In *Oshunia* and *Ionoscopus* the ascending process of the parasphenoid extends dorsally to meet the lateral face of the sphenotic (Maisey, in prep.). The ascending process also meets the sphenotic in *Amia*, *Calamopleurus*, and *Sinamia*, as well as in Rayner's (1948) "*Aspidorhynchus*" braincases and *Macrepistius*. Lambers (1992, fig. 14) shows the same condition in *Caturus furcatus*, and Gardiner (1960) found it in *Caturus chirotes*. In crushed braincases of *Ophiopsis* (e.g., AMNH 19237, 19238) there is an elongate ascending process that may have contacted the sphenotic, but better material is needed to confirm this. In gars, pholidophorids, leptolepids, and *Perleidus* the ascending process fails to reach the

sphenotic, but it does so in *Dapedium* (Patterson, 1975, fig. 112) and *Neoproscinetes*. Among modern teleosts this contact occurs in only a few clupeocephalan fishes (e.g., *Gasterosteus*, *Lophius*, and some eels; Patterson, 1975). The condition in *Watsonulus* is uncertain; however, in the undetermined parasemionotid from eastern Greenland the ascending process meets the sphenotic (see Patterson, 1975).

The Sphenotic: The sphenotic of neopterygians usually ossifies in the postorbital process and has a complete covering of dermal bones (including the dermosphenotic and the anterior end of the dermopterotic). Primitively the lower opening of the spiracular canal lies at the junction of the sphenotic and prootic, and the canal passes medial to the sphenotic in most neopterygians (other than teleosts; Patterson, 1975) while the anterior part of the dermopterotic lamina passes down into the upper opening of the spiracular canal.

Olsen (1984, p. 485) and Olsen and McCune (1991, p. 275) concluded that the sphenotic participated in the skull roof in a variety of neopterygians including *Watsonulus*, *Semionotus*, gars, *Ophiopsis*-like caturids, and *Amia*; and that in *Watsonulus* the sphenotic was "covered by dermal ornamentation characteristic of the rest of the skull roof" (Olsen, 1984). More recently Lambers (1992, p. 157) has suggested that an exposed sphenotic may be primitive for neopterygians.

In very large (old) gars there is occasionally a small "supernumerary" dermal ossification associated with the dorsolateral extremity of the sphenotic. However, in our estimation the participation of the sphenotic in the skull roof is limited to halecomorphs. Primitively as in parasemionotids (see also *Calamopleurus*) the dermosphenotic (discussed below in more detail) is incorporated into the roof of the skull, fitting into a notch or recess in the anterior margin of the sphenotic. In more cladistically derived halecomorphs the dermosphenotic is elaborate and plastered against the anterior face of the sphenotic, allowing the latter to project posteriorly between it and the dermopterotic (see Patterson, 1975, figs. 99 and 101; Bartram, 1975, fig. 2; Maisey, 1991c, p. 159). Such a condition is considered derived and a synapomorphy of ionoscopids, caturids, and amiids. Proof of the participation of the sphenotic in the skull roof, and that in life it was covered by skin, is provided by the presence of dermal ornament (sometimes lacking ganoine) in at least "*Aspidorhynchus*" (Patterson, 1975, figs. 99 and 100), *Ophiopsis*, *Osteorachis*, *Furo*, and *Heterolepidotus*.

Anterior extent of parietals: *Ophiopsis* and *Macrepistius* have parietals and dermopterotics of similar length, mainly as a result of a shortening of the dermopterotics. Primitively the dermopterotics extend much farther anteriorly than the parietals, flanking the posterior part of the frontal, as in primitive teleosts (including *Pholidophorus*), caturids, ionoscopids, and amiids (although in Recent *Amia* the anterior extent of the dermopterotic is variable and can be quite short). The dermopterotic of *Brachyichthys* in AMNH 7741 is short, but the full extent of the parietal cannot be seen. In *Dapedium* the relative extent of these bones is unknown because the skull has a continuous roof formed (presumably) by the parietals, frontals, and dermopterotics (Gardiner, 1960; Thies, 1988).

Absence of pterotic: Paired pterotics are present in teleosts including pholidophorids and leptolepids, and they are inferred to be present in *Dapedium* (Patterson, 1975). In gars and *Lepidotes* the "epiotics" may represent pterotics which have grown backward (Patterson, 1975, fig. 114). The absence of a pterotic bone in the skull of *Amia* is a presumed loss (Patterson, 1973, p. 256). A pterotic has been reported in the fossil amiid *Calamopleurus* (= "*Enneles*"; Taverne, 1974; Maisey, 1991a, p. 147), but the report is erroneous as the bone in question is actually a median unpaired endochondral ossification beneath the skull roof (supraotic; see following discussion). Pterotics are still unknown in any amiid, although they are present in extinct non-amiid halecomorphs such as ionoscopids. Pterotics are supposedly absent in caturids (Patterson, 1975; Schultze and Wiley, 1984; Lambers, 1992).

The absence of the pterotic in fossils can be difficult to determine because the bone can be extremely small and hidden within the posttemporal fossa. In some cases it may not be absent but merely elusive (and "pterotic elusive" is hardly an apomorphic character!). For example, there is a small pterotic in *Oshunia*, but it is located so deep within the posttemporal fossa that it can be seen only by looking straight down from the occiput in well-preserved acid-prepared fossils (e.g., AMNH 12793). The pterotic is not at all apparent externally, and from a cursory examination of the braincase one might easily, yet incorrectly, assume it was absent. A very small pterotic, wedged between the epioccipital and dermopterotic within the posttemporal fossa, is also present in *Ionoscopus cyprinoides*. Much of the inner wall of this fossa is formed by the epioccipital in both *Oshunia* and *Ionoscopus* (Maisey, in prep.). In *Oshunia* the suture between the epioccipital and pterotic (representing the dorsolateral part of the fissura otico-occipitalis) is located on its medial wall, about half way into the posttemporal fossa. In *Ionoscopus*, *Macrepistius*, and "*Aspidorhynchus*" the pterotic contributes more to the inner wall of this fossa than in *Oshunia*.

The pterotic therefore is primitively present in halecomorphs, and its absence is a potential synapomorphy of amiids and *Caturus*. The small size and internal

position of the pterotic in *Oshunia* and *Ionoscopus* mean that statements regarding its absence in other fossil fishes (such as caturids) should be treated cautiously unless the braincase is known in detail. The condition of the pterotic in *Ophiopsis* is unknown. The pterotic is absent in *Mesturus* and *Neoproscinetes*.

Supraotic bone: The supraotic is a median dorsal endochondral bone in the otic region of the braincase. Patterson (1975) first drew attention to this bone in the isolated braincases described previously by Rayner and identified as *Aspidorhynchus*. Rayner had identified the bone as a supraoccipital, but Patterson (1975) regarded it as a new ossification, not homologous with the teleost supraoccipital. A similar bone has now been found in *Oshunia*, *Ionoscopus*, and *Calamopleurus* (Maisey, in prep.), i.e., among both amiid and non-amiid halecomorphs. It is exposed on the skull table posteriorly in the "*Aspidorhynchus*" braincases and in *Ionoscopus* and *Oshunia*, but it is completely overlain by the parietals in *Calamopleurus*. In *Ionoscopus* and *Calamopleurus* the bone forms a roof over the labyrinth region, enclosing parts of the anterior and posterior semicircular canals and the sinus superior (in *Oshunia* the internal surface of the bone is difficult to see without destroying the braincase). In most teleosts the supraoccipital remains superficial to the labyrinth canals, but in *Arapaima gigas* it encloses the upper part of the posterior semicircular canal (J.G.M., personal observation), and C. Patterson (personal communication) has found the same condition in *Hiodon*, *Elops*, and *Megalops*. As far as we know, however, in no teleost does the anterior semicircular canal become enclosed by the supraoccipital. Part of the anterior semicircular canal is enclosed by endochondral bone in *Tetragonolepis* (Thies, 1989a, figs. 13, 14), but this could be part of the pterotic, as in many other fishes.

A median endochondral bone occurs in the otico-occipital region of *Neoproscinetes* (Maisey, in prep.). This bone does not enclose the semicircular canals to the same extent as in *Ionoscopus* or *Calamopleurus*, but there are paired depressions for parts of the anterior as well as the posterior semicircular canals. The topographic position of this bone relative to the dorsal part of the otico-occipital fissure is difficult to determine in *Neoproscinetes* because the posterior region of the braincase is poorly ossified. The "supraotic" reportedly present in a single specimen of *Lepidotes* (Patterson, 1975) is very different from the supraotic in the other fossils and has never been confirmed in other semionotids. It is still uncertain whether a supraoccipital is present in *Dapedium* (see Patterson, 1975, p. 455, for a discussion of the problem).

The different relationship of the halecomorph supraotic and teleost supraoccipital to the otico-occipital fissure could, as Patterson (1975) suggested, indicate nonhomology. Alternatively, this topographic difference could merely reflect failure of the dorsal part of the fissure to separate the otic and occipital segments completely (a commonly observed phenomenon in vertebrate embryogenesis; de Beer, 1937). In this scenario, a single dorsally-situated median ossification center could have been located either entirely within the occipital segment (as in some pholidophorids), or it could straddle both the otic and occipital segments (as in most teleosts), or it could have been confined to the otic region (as in fossil halecomorphs). The fact that in some teleosts the posterior semicircular canal is enclosed by the supraoccipital suggests that part of this bone has ossified within the wall of the otic segment and not simply invaded its space. Aside from the greater anterior extent of the halecomorph supraotic, which has come to enclose the anterior semicircular canals as well as the posterior pair, there does not seem to be much difference from the teleostean supraoccipital. Unfortunately, because *Amia* lacks a supraotic, the absence of developmental data for this bone in halecomorphs means that its homology with the teleostean supraoccipital must remain conjectural. It is nevertheless possible that the topographic position of the halecomorph supraotic, anterior to the otico-occipital fissure, represents a derived state of the supraoccipital, much like the teleostean "epiotic" which Patterson (1975) convincingly showed was really the epioccipital now located entirely within the otic segment of the teleost braincase.

Whether the supraotic is a new bone, as Patterson (1975) contended, or a special kind of supraoccipital as suggested here, both the position of this median bone entirely anterior to the otico-occipital fissure and the enclosure in bone of the anterior semicircular canal probably are apomorphic characters. The first of these features characterizes all known examples of the supraotic we have seen, but the second is known with certainty only in *Calamopleurus* and *Ionoscopus*. Although there are impressions of parts of the anterior semicircular canals in the median bone of *Neoproscinetes*, they are not enclosed by bone in the same way as in *Calamopleurus* and *Ionoscopus*.

The absence of a supraotic in *Amia* is likely to be derived, in view of its presence in the fossil amiid *Calamopleurus*. Having the supraotic exposed in the posterior skull table, as in ionoscopids, is regarded here as a primitive condition (like the supraoccipital in teleosts); the completely internal supraotic in *Calamopleurus* may represent a derived condition within amiids. The braincase is poorly known in *Ophiopsis*, and it is unknown whether a supraotic was present. A supraotic has not been reported in *Macrepistius* or

caturids, but it has not been determined whether this elusive bone is absent or internal in these fossils.

Extent and form of intercalar: Where known, the halecomorph intercalar is of an "amiid type" *sensu* Patterson (1973), extending over the lateral wall of the saccular chamber, below the jugular vein. In non-amiid halecomorphs (including ionoscopids and caturids) there is primitively an endochondral component to the intercalar, but *Calamopleurus* has a membranous intercalar as in *Amia*, suggesting that the amiid condition is derived in lacking an endochondral component. *Neoproscinetes* (along with *Mesturus* and many other pycnodontiforms) lacks an intercalar.

Patterson (1973) noted that in *Macrepistius* the intercalar meets a flange from the posterior process of the parasphenoid. In teleosts, gars, *Amia*, and *Watsonulus* the intercalar does not extend to meet the parasphenoid, but there is contact between these bones in the braincases of *Oshunia*, *Ionoscopus*, "*Aspidorhynchus*," *Heterolepidotus*, *Caturus*, *Calamopleurus*, perhaps *Sinamia* (Stensiö, 1935, p. 15), and apparently *Ophiopsis* (e.g., AMNH 19238). This feature may represent a synapomorphy of a large group of halecomorph fishes. In *Dapedium* the extent of the intercalar is unclear; Patterson (1975, p. 455, fig. 113) inferred from the shape of the braincase that an endochondral intercalar was present, but its limits cannot be determined because no sutures are preserved.

Extent of orbitosphenoid and basisphenoid: The orbitosphenoid is well-developed in *Ionoscopus*, *Oshunia*, "*Aspidorhynchus*," *Heterolepidotus*, *Macrepistius*, and *Watsonulus*. It is also extensive in *Sinamia* and *Calamopleurus*. As far as is known all these taxa also have an ossified basisphenoid. In *Amia* the orbitosphenoid is comparatively small, and we did not see a basisphenoid, although L. Grande (personal communication) reports one in *Amia*. There is a basisphenoid and a well-developed orbitosphenoid in *Mesturus* (Tübingen: No. 1261); in *Neoproscinetes* both these bones are absent.

Extent of lateral ethmoid: In *Calamopleurus* and *Amia* the pre-ethmoid bone makes contact with the vomer, but the lateral ethmoid bone does not. The latter also fails to contact the parasphenoid ventrally and the orbitosphenoid posteriorly, although they are connected by cartilage. More extensive ossification of the ethmoid cartilage, giving rise to contact between the lateral ethmoid and parasphenoid, is undoubtedly a primitive condition. This occurs in *Sinamia* (Stensiö, 1935, fig. 2), *Lepidotes* (Stensiö, 1932, fig. 80), *Ionoscopus* and *Oshunia* (Maisey, in prep.), and apparently (although the sutures are obliterated) *Caturus*, *Macrepistius*, *Watsonulus*, and *Heterolepidotus* (Gardiner, 1960, fig. 36; Schaeffer, 1971, figs. 3, 7; Olsen, 1984, fig. 8; Patterson, 1975, figs. 102, 103). In a badly crushed acid-prepared braincase of *Ophiopsis* (AMNH 19237) the corresponding area of the lateral ethmoid also connects with the anterior part of the parasphenoid. In *Neoproscinetes* and other pycnodontiforms (e.g., *Coelodus*; Wenz, 1989, fig. 1) the lateral ethmoid is an extensive bone that forms a large part of the interorbital septum anteriorly. The lateral ethmoid meets the vomer and parasphenoid ventrally but the orbitosphenoid is often unossified. The lateral ethmoid is similarly sheathed by the vomer in *Dapedium* and meets the parasphenoid ventrally. Reduction in ossification of the lateral ethmoid, so that the bone no longer makes contact with the parasphenoid, is regarded as a derived condition that arose within amiids.

Antorbital and X-bone morphology: Schaeffer (1960, figs. 2, 8, 9) found a narrow bone between the frontal and first infraorbital in *Macrepistius*, *Furo*, and *Heterolepidotus* but was uncertain as to its identity and designated it "X." Applegate (1988, fig. 5) identified a similar X-bone in the ophiopsid *Teoichthys*, although Bartram (1975) did not identify it in *Ophiopsis*. We find no evidence that the bone designated "X" in *Macrepistius* and *Heterolepidotus* is anything other than the antorbital crushed down on top of the lateral ethmoid. The antorbital in *Oshunia* is an elaborate bone, shaped like a hook, extending from beneath the nostrils back to the orbit (Maisey, 1991c). Its posterior part overlies the lateral ethmoid, and anteriorly it curves under the nasal as in *Ophiopsis* (Bartram, 1975, fig. 3). In *Ophiopsis*, however, the antorbital does not reach the orbit (Bartram, 1975, fig. 5). The anterior part of the X-bone identified by Schaeffer (1960, fig. 2B) in *Macrepistius* is reinterpreted here as the lateral ethmoid, and the posterior part (reaching the orbit) is reinterpreted as the antorbital. Conversely, the X-bone identified by Applegate (1988, figs. 4, 5) in *Teoichthys* is probably the posterior part of the lateral ethmoid, behind and below the antorbital. Interpretation of the antorbital and lateral ethmoid in these fossils is difficult because the two bones are crushed together.

Opisthotic: *Oshunia* retains an opisthotic, as do other non-amiid halecomorphs, but *Caturus*, *Calamopleurus*, and *Sinamia* resemble *Amia* in lacking this bone. A small opisthotic is present in pholidophorids, but it is absent in *Lepidotes*, leptolepids, and more advanced teleosts (Patterson, 1975; Gardiner, 1984, p. 212), as well as in pycnodontiforms (e.g., *Mesturus* and *Neoproscinetes*).

Dermosphenotic: In *Amia* and most fossil amiids the dermosphenotic is incorporated into the skull roof and is bound to the sphenotic by an anteroventral flange (Patterson, 1973, p. 262). In *Amia* this flange does not pass into the orbit, nor does it contain the infraorbital canal. This appears to be a general condition among amiids.

In *Calamopleurus* the dermosphenotic is unusual for amiids in that it is not firmly sutured to the skull roof (Da Silva Santos, 1960), and it lacks a descending lamina. It is nevertheless incorporated into the roof in that it occupies a distinct recessed area in the sphenotic and frontal. In this position the dermosphenotic cannot be regarded as free from the skull roof as in teleosts. In many non-amiid halecomorphs the dermosphenotic is also fused into the skull roof, suggesting that the condition in *Calamopleurus* represents an autapomorphy. The infraorbital canal within the dermosphenotic of *Calamopleurus* is positioned as in *Amia*.

According to Patterson (1973, p. 279), the dermosphenotic in *Heterolepidotus*, *Caturus heterurus*, "*Aspidorhynchus*," and *Macrepistius* wraps around the anterior surface of the sphenotic "in just the same way as the dermosphenotic of *Amia*....*" There is, however, an important additional feature in all of these fossils that is not found in *Amia*. The dermosphenotic flange in these fossils is canal-bearing, whereas in *Amia* the infraorbital canal exits from the dermosphenotic farther dorsally. The presence of the canal in the dermosphenotic flange is regarded as a separate apomorphic character within halecomorphs.

The dermosphenotics in *Oshunia*, *Ionoscopus*, *Caturus*, *Ophiopsis*, *Macrepistius*, *Osteorachis*, *Furo*, and *Heterolepidotus* are fully sutured into the skull roof, as in *Amia*. In *Oshunia* this bone is wrapped around the anterior face of the sphenotic and has a membranous orbital lamina which extends deep into the orbit to overlap part of the pterosphenoid dorsally (AMNH 12000, 12793). The infraorbital sensory canal occupies a short, ventrally-directed tubular extension of this lamina and appears to meet the canal-bearing part of the last infraorbital deep in the orbit.

In *Ionoscopus cyprinoides* (NHM 37795A) there is a dermosphenotic descending lamina containing the infraorbital canal, similar to that of *Oshunia*. This lamina meets the sphenotic posteriorly, but then projects sharply into the orbit, with a distinct gap between its canal-bearing part and the posterior wall of the orbit.

The remarkable structure of the canal-bearing part of the dermosphenotic in *Oshunia* and *Ionoscopus* is similar to that depicted in *Ophiopsis procera* by Bartram (1975, fig. 5) and in the "*Aspidorhynchus*" braincase figured by Patterson (1975, figs. 99, 101). Among these fossils, the condition in *Ionoscopus* seems to be the most derived because the bone enclosing its infraorbital canal is free distally from the posterior wall of the orbit. The dermosphenotic infraorbital canal also enwraps the anterior surface of the sphenotic in all of the species of *Caturus* that we have been able to check including *Caturus heterurus* (Patterson, 1973), *Caturus pachyurus*, *C. giganteus*, *C. furcatus*, and *C. latipennis*, but like in *Osteorachis* and *Furo* the descending lamina does not project into the orbit.

A reexamination by J.G.M. of the two known skulls of *Macrepistius* described by Schaeffer (1960,1971) suggests first that his "supraorbital" in the skull roof is the dermosphenotic, with a lamina extending into the orbit. Second, his "dermosphenotic" is reinterpreted as an exposed part of the sphenotic, as in *Oshunia*. Third, the infraorbital canal in *Macrepistius* is contained by a bony tube within the orbital lamina of the dermosphenotic, as in *Oshunia*. In *Heterolepidotus latus* (AMNH 4691) the upper end of the last infraorbital is located deep in the orbit, as in *Oshunia*. We agree with Bartram (1975) that the element identified by Schaeffer (1960) in this specimen as the dermosphenotic is probably the sphenotic. Thus *Oshunia*, *Ionoscopus*, *Ophiopsis*, *Macrepistius*, and *Heterolepidotus* possess a similar arrangement of their infraorbital sensory canal, involving unusual modifications to their otherwise *Amia*-like dermosphenotic and to the last infraorbital.

The dermosphenotic in *Watsonulus* occupies a recess in the anterior margin of the dermopterotic (Olsen, 1984), so it overlies, but is not sutured into, the skull roof, much as in *Calamopleurus*. In primitive actinopterygians the dermosphenotic is little more than the last infraorbital (e.g., *Mimia* and *Moythomasia*; Gardiner, 1984), but even here it can rest dorsally on a ledge formed by the neurocranium. In these fishes, plus gars, teleosts, and many sarcopterygians, the dermosphenotic is loosely attached or hinged to the skull roof, and this is customarily regarded as the primitive condition. In living chondrosteans the dermosphenotic is sutured to other skull roofing bones, but the arrangement is clearly divergent from that seen in *Amia* and is regarded as an independent specialization; in the Early Jurassic *Chondrosteus* the dermosphenotic is arranged as in primitive fossil actinopterygians and is not firmly attached to the skull roof. The dermosphenotic lacks a descending lamina in palaeoniscoids, chondrosteans, gars, macrosemiids, semionotids, *Dapedium*, pycnodonts, pachycormids, pholidophorids, leptolepids, and more advanced teleosts.

Bartram (1975) suggested that the lack of an anastomosis between the supraorbital and infraorbital canals within the dermosphenotic of *Ophiopsis* and *Macrepistius* is unusual and may represent a synapomorphy of these taxa. Such a connection is absent in many fishes (*Hiodon* and *Hoplias*; Nelson, 1972; alepocepha-loids; Gosline, 1969; pholidophorids; Nybelin, 1966; *Watsonulus*, Olsen, 1984; *Mimia*, *Moythomasia*; Gardiner, 1984). The utility of this feature as an ophiopsid synapomorphy is therefore dubious.

Infraorbitals: Several features of potential phylogenetic interest are worth mentioning here. First, the infraorbital sensory canal in the last infraorbital may turn mesially to meet the descending flange of the dermosphenotic. In *Amia* this does not occur, and the infraorbital canal is entirely superficial to the orbit. In fossils where the infraorbital canal of the dermosphenotic is in part plastered against the sphenotic within the orbit, the canal-bearing upper part of the last infraorbital bone is turned inward (e.g., *Ophiopsis*, *Macrepistius*, and *Oshunia*).

The obliquely (posterodorsally) oriented lower margin of the last infraorbital is a distinctive feature in *Ophiopsis*, *Macrepistius*, *Heterolepidotus*, and *Furo* (Bartram, 1975). It also occurs in *Oshunia* and *Brachyichthys*, but it is not found in *Ionoscopus*.

The infraorbitals below the orbit are comparatively shallow in comparison with their depth in *Amia*, fossil amiids, *Caturus*, *Ionoscopus*, *Watsonulus*, and many primitive teleosts. Deeper infraorbitals occur below the eye in *Heterolepidotus*, *Osteorhachis*, *Brachyichthys*, *Oshunia*, *Ophiopsis*, *Macrepistius*, and *Caturus chirotes*.

It has been suggested that reduction in the number of infraorbitals below the orbit (to two) is a synapomorphy of caturids in a restricted sense (Lambers, 1992, p. 180), but more probably this is the primitive condition.

Supraorbitals: These are generally absent in halecomorphs but present in caturids and the amiid *Calamopleurus*, where they meet the infraorbitals. This condition seems to be derived and is also present in semionotids and macrosemiids, *Watsonulus*, caturids, ionoscopids, and ophiopsids.

Rostral morphology: The actinopterygian snout is covered primitively by a large median rostral plate (e.g., *Mimia* and *Moythomasia*: Gardiner, 1984: figs. 42, 45), and a similar plate-like rostral bone is widespread among primitive neopterygians (e.g., pholidophorids, *Dapedium*, and *Hulettia*: Patterson, 1975; Thies, 1988; Schaeffer and Patterson, 1984). By contrast in *Amia* and many extinct halecomorphs the rostral is V-shaped, with lateral horns containing the sensory commissure. The relative lengths of the rostral horns and extent of their articulation with the antorbitals may be poor characters to use because these features are variable within fossil amiids. Nevertheless, the very short rostral horns in *Ophiopsis* and *Macrepistius* are unusual, and a similar rostral bone with short lateral horns is present in *Brachyichthys*.

Reduction of the rostral bones to a narrow tube with lateral processes may unite macrosemiids and semionotids with all other halecomorphs.

Maxilla: The maxilla carries a branch of the infraorbital sensory canal in *Ophiopsis* (Bartram, 1975, fig. 5), but the condition in *Macrepistius* is unknown because the maxillae are not preserved in any specimens. In *Brachyichthys* the maxilla lacks a sensory canal.

Presence of a sensory canal on the maxilla is unusual, but Thies (1989b) found traces of a pit line on the maxilla in *Lepidotes elvensis* and suggested that it is homologous either with the anterior part of the supramaxillary line in lower vertebrates or with the infraorbital pit line in some modern teleosts.

Uppermost branchiostegal enlarged, truncated proximally: The shape and size of the uppermost branchiostegal has not been widely commented upon in fossil halecomorphs; the structure seems to be unmodified even in fossils close to *Amia* (e.g., *Pachyamia*). Enlargement of this bone in *Amia* may represent an autapomorphy or a character which arose late in amiid history.

Hypural-ural fusion: Fusion of all hypurals except the first to their respective ural centra occurs in *Amia* but not in other amiids such as *Calamopleurus*, *Vidalamia*, *Ikechaoamia*, *Urocles*, and *Amiopsis* (Maisey, 1991a) nor in non-amiid halecomorphs. This feature apparently arose late in amiid phylogeny; it does not define the amiid group as a whole, although it may help define a monophyletic group within amiids.

Hypural-fin ray arrangement: The pattern found in *Amia* seems to be a result of gradual changes, and when extended to include fossils it becomes a multistate character within amiids. There is a one-to-one hypural–fin ray relationship with 10 or more hypurals in *Amia*, *Vidalamia*, *Urocles*, and *Amiopsis*; in *Calamopleurus* and *Ikechoamia* there are only 8 or 9 hypurals. The character is absent in *Liodesmus*, caturids, and ionoscopids.

Another, related multistate character among some amiids is a one-to-one relationship of fin rays and preural hemal arches (parhypural plus five or six pre-ural hemal arches in *Amia* and *Pachyamia*, four or five in *Urocles*, three or four in *Vidalamia*, none in *Calamopleurus*, and apparently none in *Sinamia* and *Ikechoamia*).

Ural neural arches and spines: Patterson (1973) suggested that having only one or two ural neural arches, with long neural spines, may be an amiid synapomorphy. That view was weakened by the discovery of up to three ural neural arches in some specimens of *Amia* (Schultze and Arratia, 1986), although the number is generally lower. The general condition in most fossil amiids is nonetheless similar to that in *Amia*, and a small number of ural neural arches probably are derived. Long ural neural spines occur in fossil amiids (e.g., *Calamopleurus*, *Vidalamia*, *Amiopsis*, and *Urocles*; Wenz, 1971, 1977, 1979) but are otherwise unusual. In *Ikechoamia* only two or three ural centra appear to have neural arches, but they are long (Chang and Hong, 1980, fig. 3; Liu and Su, 1983, fig. 7).

The presence of short, block-like ural neural arches in *Caturus* is considered to be a derived character (Patterson, 1973; Schaeffer and Patterson, 1984). The caudal skeletons of *Ionoscopus* and *Oshunia* contain more than 13 ural centra, up to 20 hypurals, several epurals, and up to six ural neural arches. It is uncertain whether the large number of these elements is primitive or represents a derived condition perhaps uniting *Oshunia* with *Ionoscopus*, *Oligopleurus*, *Spathiurus*, and *Callopterus* (which were tentatively united on the basis of this feature by Patterson, 1973).

Vertebrae: In *Amia* the vertebrae are said to differ both from gar vertebrae (which are perichondral ossifications of dorsal and ventral arcualia) and from teleost vertebrae (in which a variety of ossification patterns occur, involving dorsal and ventral arcualia and arcocentra). Arguments of nonhomology of these vertebrae are usually couched in developmental terms, but palaeontological data have provided the most appealing (if tautological) evidence that vertebrae arose independently in amiids, gars, and teleosts.

Ossified vertebrae are present in modern gars, *Amia*, and teleosts, and in all known fossil gars, but they are absent in many fossils classified as primitive teleosts and halecomorphs. The reciprocal illumination of cladistic analysis suggests that vertebral ossification arose several times within neopterygians.

Among non-amiid halecomorphs, ossified vertebral centra occur in *Ionoscopus*, *Oshunia*, *Ophiopsis*, *Macrepistius*, and some caturids but not in parasemionotids. Ossified annular centra are absent in *Caturus*, but centra (apparently perichordal) are present in *Furo* (earlier members possess hemicaudacentra similar to those in *Osteorachis*, *Heterolepidotus*, and some caturids) and *Neorhombolepis* (Patterson, 1973, p. 281).

At least two kinds of vertebrae are distinguishable in fossil halecomorphs. One kind is the oval drumlike "amiid vertebrae," like those of *Amia*. All fossils generally regarded as amiids possess these vertebrae apart from *Liodesmus* (whose inclusion with amiids is doubtful). The vertebrae of *Ophiopsis* and *Macrepistius* also have drumlike centra. In *Amia* the preural vertebral morphology develops by elongation of the dorsal and ventral arcualia just external to the outer notochordal sheath (elastica externa) and by simultaneous cancellous bone formation between the arcualia and around the notochordal sheath (Hay, 1895; Schaeffer, 1967). The solid "drum" of the centrum consists mainly of this cancellous bone surrounding radial struts of perichordal bone.

A different vertebral pattern occurs in *Ionoscopus*, *Oshunia* (also *Callopterus*, and *Spathiurus*), and apparently the amiid *Ikechoamia*. The vertebral body lacks a cancellous bone layer around the notochordal sheath. In *Ionoscopus* and *Oshunia* each centrum has two deep lateral pits on each side, separated by a horizontal ridge of bone, and also a deep ventral pit between the facets for the dorsal and ventral arcualia. If we were to take an *Amia* vertebra and remove the cancellous bone from around its perichordal sheath, the remaining structure would resemble the vertebrae of *Ionoscopus*, *Oshunia*, and *Ikechoamia*. These centra superficially resemble those of teleosts but lack pre- and post-zygapophyses; furthermore, there is no evidence that these vertebrae consist of chordacentra and autocentra as in teleosts.

Diplospondylic caudal vertebrae are present in *Amia*, many fossil amiids, *Ophiopsis*, *Macrepistius*, and *Neorhombolepis*. In amiids and ophiopsids the abdominal vertebrae are monospondylic, but most of the caudal vertebrae are diplospondylous. Diplospondylous caudal vertebra are not present in *Ikechoamia* (Chang and Hong, 1980, fig. 3) or *Sinamia* L. Grande (personal communication). The middle part of the caudal region in *Caturus* is diplospondylic (Rosen *et al.*, 1981, p. 249, fig. 59), and diplospondylous hemicentra were also found in *Pholidophorus* and *Australosomus* (Stensiö, 1932, pls. 35–37). By contrast, the entire vertebral column is monospondylic in *Ionoscopus* and *Oshunia*. The arrangement of neural and hemal arcualia in *Neoproscinetes* is monospondylic throughout the vertebral column, and there is no indication of any additional caudal interventrals or interdorsals between the interlocking neural and hemal arches.

Scale morphology: "Amiid scales," with their distinctive pattern of thin longitudinal striae and lack of ganoine, occur in many fossil amiids but are absent in the supposed amiid *Sinamia*, which has thick ganoid scales. Thin scales with sub-parallel striae and lacking ganoine occur also in *Caturus* and ionoscopids. Moreover, amiid scales have also been found in *Eurycormus*, pholidophorids (*Pholidophoropsis*; Schultze, 1966; Patterson, 1973; *Neopholidophoropsis*; Taverne, 1981), and macrosemiids, sometimes along with ganoid scales in the same fish (Lambers, 1992, p. 179). The disjunct distribution of this distinctive scale morphology within neopterygians may reflect some underlying ontogenetic similarity that is only occasionally manifested in the adult scale.

The most striking feature of *Amia* scales is their subparallel alignment of superficial striae, unlike the concentric striation pattern of cycloid scales in teleosts. There are indications of a subparallel scale striation pattern even where ganoine is present, however, for example, in the amiid *Ikechoamia* (Chang and Hong, 1980, fig. 4). A thick ganoine layer is present in ophiopsid scales, but here too there is a strong subparallel alignment of ridges on the scale surface. Similar ridges also occur in ganoid scales of other fishes, including semionotids and various palaeoniscids, and may therefore reflect a primitive structural pattern. Viewed in this way, the concentric circuli of modern teleost scales are derived, and the subparallel arrangement of scale striae in various neopterygians may have resulted from loss of ganoine (perhaps several times) from a fundamentally more primitive scale morphotype. The disjunct distribution of the amiid scale pattern is incongruent with all established phylogenetic hypotheses of neopterygian relationships.

Basal and fringing fulcra: It is well-known that the presence of these specialized scales is primitive, and their reduction or loss is a derived condition seen especially in teleosts. These structures are greatly enlarged in semionotids, macrosemiids, and some halecomorphs, a possible synapomorphy of these taxa.

Lateral line ossicles: One reason Bartram (1975, p. 202) included *Macrepistius* in ophiopsids is because the main lateral line extends onto the caudal fin. In these taxa there is a series of ossicles running along the division between upper and lower caudal fin rays. *Teoichthys* also has this feature (Applegate, 1988; there are no characters distinguishing this genus from *Ophiopsis*). In an acid-prepared specimen of *Brachyichthys typicus* (AMNH 7741) the lateral line also extends between the caudal fin rays as a series of small ossicles, located in the same position as in ophiopsids. In actinopterygians with homocercal tails the lateral line typically terminates at or near the caudal peduncle and does not extend onto the tail fin; however, in *Amia* and many if not all amiids (Grande, 1995) the lateral line also extends onto the caudal fin. Where the body lobe extends farther into the tail, as in paddlefishes, the lateral line is correspondingly longer and extends along the body lobe almost to its tip, just under the series of rhomboid scales (*Polyodon*, *Psephurus*, *Crossopholis*; Grande and Bemis, 1991), but does not extend between the fin rays.

IV. Monophyletic Neopterygian Groups

Symplectic involvement in the jaw hinge, regarded as a basal halecomorph synapomorphy by Patterson (1973), helps define a large monophyletic

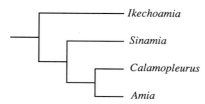

FIGURE 2 Consistently recognized clade of amiids.

clade of halecomorph fishes. Several other features (e.g., dermosphenotic with a descending lamina and is sutured into the skull roof; symplectic bound to preoperculum; parasphenoid contacts the sphenotic and the intercalar; rostral bone V-shaped, with long lateral horns) help define another clade within halecomorphs.

According to the analyses of data presented in Appendix 2, non-amiid halecomorphs (all of which are extinct) do not form a monophyletic group. This assemblage includes some groups that were never diagnosed by apomorphic characters and probably are not monophyletic (such as ionoscopids and caturids *sensu lato*), as well as other groups (such as ophiopsids and caturids *sensu* Lambers, 1992) that are more readily diagnosed. Two related problems have arisen where apomorphic characters have not been clearly identified: first, how to diagnose the group and second, which fossils to include.

A. Amiidae

This group of fishes is represented by a solitary living species, *Amia calva* (the bowfin), which is restricted to fresh waters of the eastern regions of the United States and Canada.

Fossil representatives of *Amia* and closely related fossil amiids show that although *Amia* occurred in North America from the Late Cretaceous (Maestrichtian and Campanian) onwards and is also recorded from the Early Cretaceous of South America, *Amia* was once widespread. In Europe Amiidae occurs from the Late Jurassic to Oligocene, in West Africa from the Late Cretaceous to Eocene, and in central and eastern Asia in the Paleogene (Patterson, 1994, p. 66). The interrelationships of some amiids have recently been discussed by Maisey (1991b, p. 154) and Grande (1995). Grande and Bemis are currently working on a detailed phylogenetic study of the family.

Our halecomorph analysis recognized a monophyletic group of four taxa (*Amia, Calamopleurus, Sinamia,* and *Ikechoamia*; Fig. 2). *Ikechoamia* shares one character with *Amia* (parietals and dermopterotics of equivalent length). *Sinamia* is further united with *Amia* by having

no endochondral bone in the intercalar and by the presence of drum-like "amiid" vertebrae [this is variable in *Sinamia*; one of Stensiö's specimens has it; the other has lateral fossae (L. Grande personal communication)], an ambiguous character also found in ophiopsids. There are several characters, unique among halecomorphs, that indicate a monophyletic Sinamiidae (*Sinamia* + *Ikechoamia*). These include median parietal and morphology of the extrascapular series; moreover, there is evidence that one or both of the *Ikechoamia* species are merely juvenile *Sinamia* sp. (L. Grande personal communication). *Calamopleurus* additionally shares with *Amia* two unambiguous characters (separate Bridge's ossicles; lateral ethmoid fails to reach parasphenoid) and two equivocal characters (the one-to-one hypural–fin ray relationship and the low number of ural neurals; neither feature is known in *Sinamia*).

This grouping of taxa proved to be robust and survived even in radically different hypotheses of halecomorph relationships where other parts of the phylogeny were unstable, supporting the view that amiid fishes are monophyletic even though their relationship to other halecomorphs and their position within neopterygians are still unsettled. Morphological data for fossil amiids are here strongly biased towards a few taxa. Many important characters are not known with certainty among some fossils usually identified as amiids. *Calamopleurus* is known from well-preserved, three-dimensional acid-prepared fossils, and we consequently can compare more of its anatomy with *Amia* than is possible with most other fossil amiids.

Maisey (1991b), like Patterson (1973, p. 277), considered *Liodesmus* to be the primitive sister-group of all other amiids (*contra* Schultze and Wiley, 1984). Additional support for Patterson's view has recently been provided by Lambers (1992, fig. 37) in his illustration of the tail of *Liodesmus*. Lambers (1992), however, considered the type of tail seen in *Liodesmus*, ionoscopids, caturids, and primitive teleosts, where a dorsal bunch of rays extend over several hypurals, to be derived and accordingly removed *Liodesmus* from the Amiidae. But both our phylogeny and the condition in gars show that the pattern of the tail of *Liodesmus* is primitive relative to all amiids. Since it seems highly unlikely that there has been a loss of at least three ossified ural neural arches in *Liodesmus*, and at the same time the independent acquisition of a one-to-one hypural–fin ray pattern in sinamiids and higher amiids, we support the view of Maisey (1991b) that *Liodesmus* is not an amiid.

Despite using only morphological characters common to fossils and living forms, *Amia* always emerged as the sister-group of teleosts in all analyses. How-ever, were we to add in all those derived features of soft anatomy, the case for the Halecostomi becomes overwhelming. Thus, although *Lepisosteus* and *Amia* possess dorsal retractor muscles and an upper pha-ryngeal dentition of tooth plates not fused with the pharyngobranchials (Nelson, 1969a,b), *Amia* uniquely shares with teleosts some six or more synapomor-phies. These include the structure of the heart in which there is a noncontractile bulbus arteriosus and a reduction of the conus and its contained tiers of valves (Goodrich, 1930), the working of the mouth with its levator operculi coupling (Lauder, 1980), the presence of a supracarinalis muscle in the caudal fin, and the presence of interradialis muscles between the caudal fin-rays (Lauder, 1989).

B. Caturidae

Following Lambers (1992) the main distinguishing features of caturids are the slender barlike maxilla; long, thin supramaxilla; the paired, blocklike ural neu-ral arches; and the branchiostegal ray size and number (at least 24). Within the genus *Caturus* there are several species groups which may turn out to be separate genera when properly defined. Some of these taxa may need to be removed from Caturidae as defined here.

Schaeffer and Patterson (1984, p. 40) have already indicated two caturid groups: one characterized by fused anterior hypurals (including *C. heterurus, C. smithwoodwardi,* and *C. dartoni*) and the other by a mosaic of numerous small supraorbitals (including *C. furcatus, C. porteri,* and *C. driani*). To these may be added a third group, characterized by greatly deep-ened infraorbitals and very stout premaxillary teeth (including *C. chirotes* and *C. latidens*). Similar premaxil-lary teeth, stouter than those on the dentary, are also found in *C. pachyurus.*

Greatly deepened infraorbitals below the orbit also occur in some ionoscopids (see previous discussion). In caturids the last infraorbital, though deeper than broad, always slopes backwards. This last character may be correlated with the position of the descending lamina of the dermosphenotic.

The relationships of caturids to both ionoscopids and amiids (Fig. 3) are best demonstrated by the form of the intercalar which has extensive membra-nous outgrowths over the lateral surface of the saccular chamber, below the jugular vein, and in front of the glossopharyngeal foramen in all three groups (Patterson, 1975, p. 437). Other features which point to a sister-group relationship between

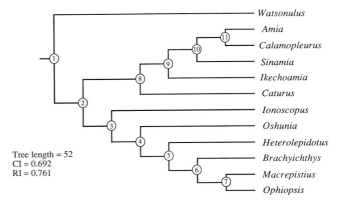

FIGURE 3 Phylogeny of halecomorphs. The numbered nodes are supported by the following characters (see Appendix 2). Node 1: 1(0–1). Node 2: 3(0–1), 4(0–1), 5(0–1), 6(0–1), 7(0–1), 22(0–1), 23(0–2). Node 3: 8(0–1), 11(0–1), 25(0–1), 33(0–1). Node 4: 26(0–1), 27(0–1), 30(0–1). Node 5: 8(1–0), 11(1–0), 21(0–1), 25(1–0). Node 6: 23(2–1). Node 7: 8(0–2), 24(0–1), 28(0–1). Node 8: 12(0–1), 13(0–1). Node 9: 8(0–2), 14(0–1), 15(0–1). Node 10: 2(0–1), 11(0–1), 29(0–1), 32(0–1). Node 11: 9(0–1), 10(0–1), 16(0–1), 17(0–1). Note: Four equally parsimonious trees were produced which varied only in the relative positions of *Heterolepidotus* and *Brachyichthys* (see Fig. 2).

caturids and amiids include the presence of serrated appendages (Bartram, 1977, fig. 50) and the loss of the basipterygoid process (missing also in *Macrepis-tius,* pachycormids, and most teleosts), the opisthotic (lost also in *Lepidotes,* leptolepids, and many tele-osts), and the pterotic.

Lambers (1992) recognized a monophyletic group comprising some but not all caturids including *Ambly-semius* (= *C. pachyurus*), *Caturus,* and possibly *Furo.* He united these taxa by the following characters: two infraorbitals below the orbit; block-like ural neural arches; and presence of only two coronoids. The last character is widespread and occurs both in gars and in primitive teleosts. Two coronoids are present in *Caturus chirotes,* while *Furo* does not possess block-like ural neural arches in our opinion. *Caturus* resemble amiids in lacking an opisthotic and pterotic. The ab-sence of an opisthotic and pterotic in *Caturus* united it with amiids in all analyses.

Finally we did not include *Osteorachis* or *Furo* in our analyses for want of information. *Furo* neverthe-less shares at least two characters with genera in-cluded in the Ionoscopidae: obliquely inclined lower margin of the last infraorbital, as in *Oshunia, Hetero-lepidotus, Ophiopsis, Brachyichthys,* and *Macrepistius,* and greatly elongated antorbital extending into the orbit margin, as in *Oshunia, Heterolepidotus,* and *Macrepistius.* We consider *Furo* to be a stem-group io-noscopid.

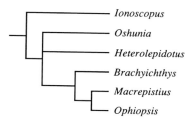

FIGURE 4 Monophyletic clade of ophiopsids plus ionoscopids (Adams consensus of trees derived from matrix in Appendix 2).

C. Ionoscopidae and Ophiopsidae

Trees generated by the halecomorph analysis consistently recognized a monophyletic group comprising *Ophiopsis*, *Macrepistius*, *Brachyichthys*, *Heterolepidotus*, *Oshunia*, and *Ionoscopus* (Fig. 4). Uncertainty in the position of *Heterolepidotus* accounts for differences between the four trees generated from the data matrix (Appendix 2).

A sister-group relationship is postulated between *Ophiopsis* and *Macrepistius* (both of which have short rostral horns and first infraorbitals with a depth more than twice their length plus drum-like amiid vertebrae). All four trees in this analysis recognized a monophyletic group comprising *Ophiopsis*, *Macrepistius*, and *Brachyichthys*, although some trees also placed *Heterolepidotus* with *Brachyichthys*. An Adams consensus tree (Adams, 1972; Wiley *et al.*, 1991) retained the relationship between *Ophiopsis*, *Macrepistius*, and *Brachyichthys* (Fig. 4). *Brachyichthys* is united with *Ophiopsis* and *Macrepistius* by three characters: lateral line ossicles in tail, rectangular rostral with short lateral "horns," and parietals equal in length to dermopterotics. Of five characters given by Bartram (1975, p. 203) uniting *Ophiopsis* and *Macrepistius*, four also are found in *Heterolepidotus*, three of the four also occur in *Furo*, one occurs in *Caturus*, and another occurs in *Osteorachis*.

Oshunia resembles ophiopsids in three characters not present in *Ionoscopus* (short maxilla, deep infraorbitals, and obliquely inclined lower margin of last infraorbital; Maisey, in prep.). However, greatly deepened infraorbitals also occur in *Caturus chirotes* and *Osteorachis*, while an obliquely lower margin of last infraorbital is seen elsewhere in *Furo*. *Ionoscopus* is united with *Oshunia*, *Brachyichthys*, *Heterolepidotus*, and ophiopsids by having the canal-bearing descending lamina of the dermosphenotic within the orbit. In *Caturus*, *Furo*, and *Osteorachis* the descending lamina is shorter and does not project into the orbit. In all four analyses, *Oshunia* was placed closer to ophiopsids than to *Ionoscopus* despite their similar coronoid pro-

cess morphology. Both taxa also have thin amiid scales, vertebrae with transverse ridges separated by deep pits, and a caudal fin with numerous slender ural neural arches (and spines) and ural centra.

D. Parasemionotidae

In both the overall phylogenetic analysis (Fig. 1, Appendix 1) and that involving just the halecomorphs (Fig. 3, Appendix 2), *Watsonulus* is resolved as the plesiomorph sister to all remaining halecomorphs.

The parasemionotids appear to be a monophyletic assemblage supported by the form of their ovoid preoperculum. The preoperculum is often notched ventrally and in *Watsonulus* (see also *Ospia* Stensiö, 1932, fig. 84) is assumed by Olsen (1984, fig. 1) to have been filled by the quadratojugal. Elsewhere the notch reveals either the underlying quadrate (*Eurycormus* Patterson, 1973, fig 14) or the underlying symplectic (*Neoproscinetes* Nursall and Maisey, 1991, p. 131).

We have reexamined the acid-prepared parasemionotid of Patterson (1973, fig. 23) as well as all the other eastern Greenland material and Madagascar nodules in the NHM. We confirm that there is a vertical flange of membrane bone projecting fore and aft on the outer surface of the quadrate and that we have been unable to find any trace of a rectangular bone (as figured by Olsen, 1984) in the preopercular notch in any specimen examined.

To our knowledge the only stem-group neopterygians in which the quadratojugal is particularly exposed are *Australosomus* and *Pteronisculus*; in these it is overlapped by both the preoperculum and maxilla, although mesially it is closely applied to the lateral face of the quadrate. Thus, like Patterson (1973), we conclude that this flange of membrane bone on the quadrate has the same topographical relations of a quadratojugal and is therefore considered as such. A smaller flange of membrane bone is similarly fused to the posterior edge of the quadrate in *Caturus* and *Furo* (Bartram, 1977, figs. 47,48). As well as resembling other halecomorphs in the form of their quadratojugal, the parasemionotids also share with them a double jaw joint in which the distal head of the symplectic articulates with the articular of the lower jaw, just posterior to the quadrate articulation. There are, however, several other synapomorphies which the parasemionotids uniquely share with the halecomorphs including the form of the dermosphenotic, maxilla, and antorbital.

The antorbital of parasemionotids has an enlarged plate-like posterior portion reaching up to the supraorbitals whereas ventrally it forms a very long, tubular rostral process (see Patterson, 1975, fig. 137).

The rostral is broad in stem-group neopterygians with a pronounced rostro–caudal extent bordered by the nasals. This type of rostral is present in *Acentrophorus*, *Hulettia*, and *Dapedium*. The rostral of pholidophorids and the rostro-dermethmoid of pachycormids is markedly broader (Patterson, 1975, p. 509). Although the median rostral is little more than a tube round the ethmoid commissure in gars, it is still quite broad spanning well over half of each premaxilla. Stem-group halecomorphs (macrosemiids and semionotids) are characterized by a much narrower, tubelike rostral beneath which the long rostral processes of the antorbital pass to overly the premaxillae. The halecomorph antorbital differs from that of basal halecostomes and from gars in that ventrally it is produced into a very long tube, whereas dorsally it remains plate-like.

The maxilla of parasemionotids is characterized by a notch or concavity in the posterior margin and by the tooth row that posteriorly comprises finer teeth and, in the region of the maxillary process, markedly stouter teeth. A similarly posteriorly indented maxilla with anteriorly enlarged teeth is also seen in ionoscopids, caturids, and amiids. Elsewhere a posteriorly notched maxilla occurs in some pholidophorids (Nybelin, 1966, figs. 9, 12), but in these fishes the notch is nearer to the dorsal margin than in halecomorphs and is therefore considered an apomorphy of pholidophorids.

The caudal fin of parasemionotids has from 22 to 25 principal rays, while the insertion of the dorsalmost fin rays which do not cross the corresponding hypurals is probably a specialization. In caturids and ionoscopids there are from 14 to 22 hypurals, and this is derived since in gars there are up to 12 and in *Dapedium*, *Hulettia*, and *Pleuropholis* (Patterson, 1973, fig. 17) there are only 8 whereas in pholidophorids there are 9–12. Further, in caturids, ionoscopids, and *Liodesmus* the most dorsal 8–10 rays (which are often recurved) narrow and converge proximally and overlap up to 14 hypurals in *Ionoscopus* and 22 in *Caturus*.

E. Macrosemiidae and Semionotidae

The extinct macrosemiids are resolved here as stem-group halecomorphs, sharing four features with them (supraorbitals meet infraorbitals, antorbital shape, rostral morphology, and greatly enlarged basal and fringing fulcra on the caudal fin). The posttemporal fossa appears to be absent (however, this can not be verified from the material available in the NHM) and there is no posttemporal process.

Most halecomorphs (as well as *Semionotus* and *Lepidotes*) possess a posttemporal with a process which projects antero–ventrally and which in halecomorphs and teleosts often articulates with the intercalar. Since there is a ventral nub of bone on the posttemporal of *Lepisosteus* the absence of such a process in *Macrosemius* could be primitive or derived. On the other hand the absence of the gular and the opisthotic in *Macrosemius* are not considered indicative of relationships. A gular is also missing in *Lepisosteus*, *Lepidotes*, *Semionotus*, pycnodonts, aspidorhynchids, and many teleosts while an opisthotic is absent in *Lepisosteus*, *Amia*, *Semionotus*, *Lepidotes*, *Microdon*, and *Elops*. As in *Lepisosteus* the absence of a supramaxilla is considered primitive. Other neopterygians in which the supramaxilla is wanting include the pycnodonts, *Acentrophorus*, and *Hulettia* (Schaeffer and Patterson, 1984).

One specialization of macrosemiids claimed by Bartram (1977, p. 144) is the possession of an efferent pseudobranchial foramen through the parasphenoid. Such a foramen has only been reported elsewhere in *Hulettia* (Schaeffer and Patterson, 1984), *Dapedium* (Patterson, 1975, fig. 112), *Neoproscinetes* (Nursall and Maisey, 1991), and teleosts. The Tübingen braincase of *Mesturus* (specimen number 1261) shows only a groove for the efferent pseudobranchial artery, while we remain unconvinced of the presence of a foramen for such an artery in the parasphenoid of *Macrosemius*.

The splint-like quadratojugal was considered by Bartram (1977, p. 148) to be fused to the quadrate in the region of the condyle in *Macrosemius* and to be free in *Propterus*. In other halecomorphs (and pachycormids) the quadratojugal is reduced to a small process on the quadrate or is absent altogether (*Amia*), whereas in *Dapedium* and stem-group teleosts (Patterson, 1973, fig. 7) the quadratojugal is free.

Both *Lepidotes* and *Semionotus* have a series of infraorbital bones which extend on the snout anterior to the circumorbital ring. This is also the case in *Lepisosteus* and macrosemiids, and although Olsen and McCune (1991, fig. 16) considered it to be a synapomorphy uniting all three groups, it is more economically interpreted as either primitive for neopterygians or as homoplasy: in *Lepisosteus* there are at least nine infraorbital bones anterior to the ring, whereas there are only two in *Lepidotes* and *Semionotus* and two or maybe three in macrosemiids. Moreover the anteriormost supraorbital also abuts the infraorbital ring in *Acentrophorus*, *Watsonulus*, *Macrepistius*, *Ophiopsis*, *Caturus*, *Calamopleurus*, and *Elops*. While this might suggest that contact between infraorbitals and supraorbitals is primitive for neopterygians, in *Megalops* the supraorbital does not meet the infraorbital but only the antorbital, as in *Dapedium* (Patterson, 1973, fig. 25), some parasemionotids (Patterson, 1975, fig. 137), *Hulettia, Pleuropholis* (Patterson, 1973, fig. 16), and *Pholidophorus* (where they almost contact) this is presumed to be derived. In stem-group neopterygians (viz. perleidids) the supraorbitals invariably contact the nasals. The premaxillae of *Lepidotes* and *Semionotus* have long nasal processes perforated by the olfactory nerve as in *Lepisosteus* and *Amia*. Long nasal processes are also found in macrosemiids, caturids, and ionoscopids, but, though notched in *Caturus*, they are never perforated. A short nasal process as in *Perleidus* (Patterson, 1975, fig. 38), parasemionotids, *Dapedium, Hulettia*, and *Acentrophorus* is believed to be primitive.

The axial skeleton of the Semionotidae is considered derived with respect to that of gars in that the last 3–5 pre-ural neural spines are median. Although Olsen and McCune (1991, p. 284) reported the presence of six median pre-ural neural spines in a specimen of *Lepisosteus*, most of the caudal preural neural spines are paired.

A distinctive feature of the Semionotidae is the enlargement of the basal and fringing fulcra on all of the fins (seen also in *Woodthorpea* and *Acentrophorus*), and this is regarded as a synapomorphy of the group, but enlargement of basal and fringing fulcra in the tail appears to be a shared derived feature uniting semionotids with both macrosemiids and halecomorphs (other than *Heterolepidotus*, which has normally developed tail fulcra much as in *Perleidus, Hulettia*, and *Dapedium*). The internal structure of the semionotid tail also appears derived with respect to gars. In gars (as in most halecomorphs, pholidophorids, and basal teleosts) a bunch of dorsal rays extend over several hypurals in characteristic fashion with the ends of the rays recurved proximally. By contrast the rays in the semionotid tail have a one-to-one relationship to the hemal spines, while the large number of preural hemal spines (13 according to Olsen and McCune, 1991, p. 283) is a rare condition. There are 17–19 principal caudal rays in *Lepidotes* and *Semionotus* of which 9 are in the lower lobe. This relatively small number of principal rays is seen elsewhere in macrosemiids (11–16), *Acentrophorus* (17), *Lepisosteus* (11), *Hulettia* (19–20; Schaeffer and Patterson, 1984, p. 34), and such stem-group teleosts as *Pleuropholis* (15) and *Ichthyokentema* (19).

Olsen and McCune (1991, p. 286) considered absence of an intercalar from the neurocranium of *Lepisosteus*, macrosemiids, and semionotids (*Lepidotes, Semionotus*) as indicative of relationship. We are uncertain as to whether an intercalar is absent in macrosemiids (the prepared material in the NHM is equivocal).

Furthermore, if we accept that the small process on the exoccipital of *Lepidotes* (to which the ventral limb of the posttemporal attaches) is topographically homologous with that on the intercalar of *Amia* and teleosts (Patterson, 1975, p. 451), then it seems likely that the intercalar has been lost at least twice: once in *Lepisosteus* and once in semionotids. An intercalar is also missing in *Neoproscinetes* and other pycnodonts.

F. Extinct Relatives of Teleosts

According to this analysis, *Dapedium* and pycnodonts are stem-group teleosts. Their position on the cladogram is mainly the result of characters they share with teleosts. These characters include the following: absence of epioccipital, premaxilla with a small nasal process, absence of clavicles, a parasphenoid with foramina for the internal carotid and efferent pseudobranchial arteries [ironically the latter is not present in *Elops* though found in many stem-group teleosts such as pholidophorids, leptolepids, and *Ichthyokentema* (Patterson, 1975)], a posttemporal fossa that communicates with the fossa bridgei (but does not do so in pholidophoids), and a median unpaired vomer.

A foramen for the internal carotid in the parasphenoid is also present in *Pachycormus, Hulettia* (Schaeffer and Patterson, 1984, p. 36), some stem-group neopterygians (e.g., *Boreosomus*), as well as teleosts. A foramen for the efferent pseudobranchial, though present in *Hulettia* and teleosts, does not occur elsewhere. A median vomer is also present in *Lepidotes, Hulettia*, and *Bobasatrania* (a stem-group neopterygian).

If pachycormids are primitive stem-teleosts (Patterson, 1973, p. 276), it is likely that both the median vomer and the foramen for the efferent pseudobranchial artery have arisen more than once within the neopterygians. In fact our phylogeny suggests that the median vomer has evolved on at least three occasions: in *Lepidotes*, in (*Dapedium* + pycnodonts), and in teleosts. The median vomer appears to have developed in relation to a crushing dentition in *Lepidotes* and (*Dapedium* + pycnodonts), whereas in *Hulettia* and generalized teleosts the dentition is unmodified. This leaves the foramen for the internal carotid artery in the parasphenoid as a possible synapomorphy of (*Dapedium* + pycnodonts) and teleosts.

In many respects *Dapedium* exemplifies the basic halecostome condition. Thus the pattern of its snout is similar to that of *Acentrophorus*, parasemionotids, *Hulettia* (Schaeffer and Patterson, 1984, p. 15), and pholidophorids. The caudal skeleton is also relatively primitive (Patterson, 1973, fig. 27) and similar to that of parasemionotids. The last four pre-ural neural spines are short and paired, the ural neural arches are paired, and *Dapedium* has seven epurals as in *Hulettia* and *Pachycormus*. The tail of *Tetragonolepis* resembles that of *Dapedium* except that the former, like *Hulettia* and pholidophorids, possesses hemicaudacentra.

The premaxillae of *Dapedium* have very short nasal processes much as in *Acentrophorus*, *Hulettia*, parasemionotids, and stem-group neopterygians. The nasal processes are somewhat longer in the dapediid *Heterostrophus* and meet in the midline basally as in *Hulettia*. The nasal processes of pycnodonts are considerably longer but unperforated. The nasal process is perforated by the olfactory nerve in *Lepisosteus*, *Amia*, semionotids. The premaxilla bears three incisiform teeth with bifid crowns much as in the stem pycnodont *Gibbodon*, while *Heterostrophus*, like some other pycnodonts, has just a pair of incisiform teeth. The corresponding dentary teeth in *Dapedium*, as in *Gibbodon* (Tintori, 1980, p. 807), also have bifid crowns. The maxilla of *Dapedium* is rounded posteriorly with a very short maxillary process and very reduced tooth row. There are just two or three teeth anteriorly in *Dapedium*, while the maxilla of *Sargodon* and *Heterostrophus*, like pycnodonts, is edentulous. The lower jaw of *Dapedium* is short and massive with a very well-developed coronoid process. The dentary has some 10 or more incisiform teeth, and the prearticular bears four rows of grinding teeth (with five cusps). The vomer also supports four rows of grinding teeth much as in *Sargodon* (Tintori, 1982, p. 422).

Although the dorsal ridge scales are small in *Dapedium*, the ventral ridge scales are somewhat larger, and the small pelvic fin on each side is inserted just above them. The ventral ridge scales of *Sargodon* have a crenulate ornament similar to that in many pycnodonts (viz. *Macromesodon* and *Neoproscinetes*).

Pycnodonts appear to have lost a considerable number of dermal bones including supraorbitals, suborbitals, supramaxilla, interoperculum, suboperculum, and gular plate, while the number of branchiostegal rays is reduced to two. Among neopterygians this reduction in branchiostegals is seen elsewhere in gars (0), *Dapedium* (4–5), *Sargodon* (4), and *Tetragonolepis* (3).

The parasphenoid of pycnodonts and *Dapedium* is edentulous, inflected downwards posteriorly, and has a long ascending process which reaches the margin of the hyomandibular facet, while posteriorly it completely embraces the aortic canal apart from a ventral fenestra for the carotid and efferent pseudobranchial arteries. This type of parasphenoid with a posterior extension below the aortic canal is unique to *Dapedium* and pycnodonts. A further similarity between pycnodonts and *Dapedium* is the form of the

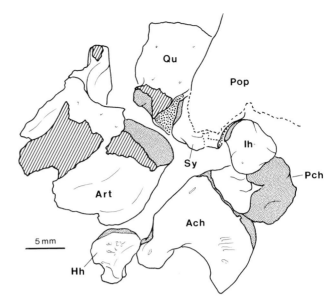

FIGURE 5 Lateral view of the jaw joint in the pycnodont *Neoproscinetes penalvai*, (AMNH 11893) from Albian, Santana Formation, Brazil. Fine stipple, unfinished cartilage; coarse stipple, concave articular surfaces of quadrate and symplectic; diagonal lines, broken cartilage; dashed line, broken margin of preopercular. Abbreviations: Ach, anterior ceratohyal; Art, Articular; Hh, hypohyal; Ih, interhyal; Pch, posterior ceratohyal; Pop, preopercular; Qu, quadrate; Sy, symplectic.

ceratohyal where the proximal bone is very short and unfinished in perichondral bone dorsally.

A unique feature of pycnodonts is the form of the jaw articulation. Though superficially similar to the double joint of halecomorphs it has its own special characteristics (Nursall and Maisey, 1991, p. 132; Maisey, in prep.). In *Neoproscinetes* the symplectic has an articular surface, lacking finished perichondral bone (Fig. 5), corresponding with an unfinished surface on the articular. These surfaces would have formed a joint similar to that in *Amia*, and the symplectic evidently played a role in supporting the lower jaw (Nursall and Maisey, 1991). The anterior surface of the symplectic is smoothly rounded and fits within a shallow concavity in the thickened posterior margin of the quadrate. This is not a true synovial joint because the adjacent surfaces consist of perichondral bone. The symplectic is bound to the internal surface of the preopercular by bony laminae (Fig. 6). There is a gutter in the lateral surface of the symplectic, roofed by the preopercular to form a canal, apparently for an external ramus of the hyomandibular nerve. The corresponding nerve in *Amia* passes lateral to the posterior part of the symplectic and mesial to the preopercular but is not confined by a canal (Allis, 1897, pl. XXIX, figs. 36, 37). The canal in *Neoproscinetes*

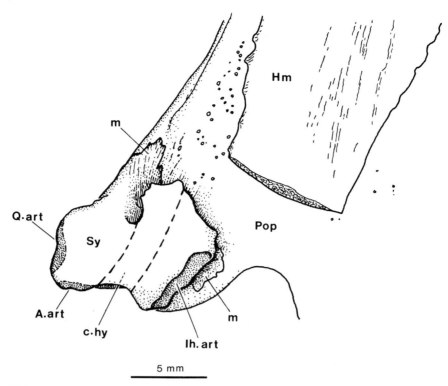

FIGURE 6 Symplectic of *Neoproscinetes penalvai* attached to the mesial surface of the preopercular (AMNH 11843). Abbreviations: Ih. art, articulation surface with interhyal; A. art, articulation surface for articular; c. hy, canal through symplectic; Hm, hyomandibular; m, membrane bone outgrowths attaching symplectic to preopercular; Pop, preopercular; Q. art, articulation surface for quadrate; Sy, symplectic.

passes dorsoventrally and emerges just behind the articulation between the symplectic and articular. Farther dorsally the hyomandibular nerve passes through a partially-enclosed gutter formed by the anterior margins of the properculum and hyomandibular. This gutter widens ventrally, reaching a maximum just above the symplectic. The hyomandibular nerve may have branched just before it entered the canal between the symplectic and preoperculum because there is also a short open groove extending from the canal across the anteromesial surface of the symplectic. That the symplectic articulated with the quadrate or propped it can be deduced from several pycnodonts in the NHM including *Palaeobalistum goedeli* (P.39231) and *Pycnodus gibbus* (P.44520).

The quadratojugal in other pycnodonts (e.g., *Macromesodon*) is splint-like and braces the quadrate condyle much as in *Lepidotes* and *Dapedium* (Patterson, 1973, fig. 26). However, on one side of one specimen of *Gyrodus hexagonus* (NHM P.4633) the quadratojugal is fused with the quadrate as in *Pholidophorus* (Patterson, 1973, fig. 7) and teleosts. In pycnodonts such as *Neoproscinetes* it has been suggested (Nursall and

Maisey, 1991, p. 132) that the posterior crest on the quadrate represents a fused quadratojugal as in halecomorphs and teleosts.

Well-preserved acid-prepared skeletons of *Neoproscinetes* also raise other questions of interpretation. For example, there is in *Neoproscinetes* a large triangular canal-bearing dermal bone in the temporal region which is sutured to the sphenotic beneath it; it is also sutured mesially to the parietal and anteriorly to the frontal. The bone contributes to the orbital margin dorsally but is separated from the orbit ventrally by the sphenotic. Some earlier interpretations (Da Silva Santos, 1970; Figueredo and Da Silva Santos, 1990) show two bones in this position, identified as an anterior dermosphenotic and a posterior dermopterotic, but in acid-prepared specimens of *Neoproscinetes* in the AMNH there is only one bone (identified as the dermopterotic by Nursall and Maisey, 1991, p. 131). The posterior margin of this bone in the AMNH specimens is smooth and not sutured to anything behind it. Most pycnodonts also have a single bone in this position, although some (e.g., *Macromesodon*, Lehman, 1966, fig. 169; *Paramesturus*, Taverne, 1981, fig.

2; *Coelodus*, Wenz, 1989, fig. 1; and *Gyrodus*, Lambers, 1992, fig. 9) have two bones, the more posterior of which is the parietal. This pattern where the parietal borders the dermopterotic posteriorly rather than laterally is unique and apparently confined to the Gyrodontidae.

In *Neoproscinetes*, the bone in question contains a sensory canal along its anterior margin, with an opening that aligns with the canal contained by the elongate dermosphenotic (or first infraorbital, according to some interpretations). This opening is the start of the infraorbital canal where it meets the supraorbital. There is another opening along a sensory canal in the posterior margin of the same bone, passing toward the parietal. This may correspond partially with the anterior part of the main lateral line canal but could also include part of the middle pit-line. A canal–pit-line in an identical position also occurs in the dermopterotic of *Dapedium*.

As far as we are aware the anastomosis of the supraorbital and infraorbital canals within the dermopterotic is a rare occurrence and is seen elsewhere only in *Dapedium* (Wenz, 1968, fig. 28) and its relatives (e.g., *Tetragonolepis*, Gardiner, 1960; *Heterostrophus*, see Royal Scottish Museum, 1905, 62.7). Elsewhere anastomosis of the infraorbital and supraorbital canals occurs in many teleosts, but here the fusion occurs within the dermosphenotic. If the bone in question in pycnodonts is a dermopterotic, anastomosis of sensory canals within it may be interpreted as a synapomorphy of pycnodonts and dapediids. An elongate dermosphenotic (= "first infraorbital"), on the other hand, defines a subset of pycnodonts.

Fringing fulcra occur on the unpaired fins of primitive pycnodonts (viz. *Brembodus*; Tintori, 1980, p. 796), and as in many halecostomes there are several caudal fin rays for each endoskeletal support.

G. Ginglymodi

Living lepisosteids or gars are confined to freshwaters in the south-eastern United States, Cuba, and Central America, while fossil representatives are known from the Cretaceous of South America and West Africa and the Paleogene of Europe and India; in North America they occur from the Late Cretaceous onward. In all our analyses gars emerge as the sister-group of the remaining neopterygians. The absence of a supramaxilla, interopercular, post-temporal fossa, and median neural spines are considered primitive.

Although Olsen and McCune (1991) reported the presence of some median neural spines in their material most of the caudal preural neural spines of gars

are paired. In halecomorphs and teleosts most of the caudal preural neural spines are median. The immobile maxilla and the absence of a posterior myodome are both considered apomorphies rather than the basal neopterygian condition since the maxilla of the fossil gar *Obaichthys* apparently bears an anterior peg, whereas stem-group neopterygians, such as perleidids and peltopleurids, all possess well-developed posterior myodomes.

V. Molecular Evidence

Normark *et al.* (1991) presented an analysis of partial sequence data of three mitochondrial genes from a variety of neopterygians. In their analysis of cytochrome oxidase I (COI), cytochrome oxidase II (COII), and cytochrome b (Cytb), based largely on amino acid translations rather than primary sequence data, *Lepisosteus* and *Amia* tended to be resolved as a monophyletic holostean clade. More recent analyses of a larger data set drawn from partial large subunit nuclear ribosomal RNA (28S rRNA) sequences of gnathostomes have yielded conflicting topologies with respect to the holosteans (Lê *et al.*, 1993; Lecointre *et al.*, 1994). This reflects both a limited data set and differences in the treatment of the data, although, on the basis of bootstrap support, the holostean monophyly was strongly supported. Where holosteans were resolved as paraphyletic in either the mitochondrial or 28S rRNA studies, *Amia* emerged as the sister-group to the teleosts.

The purpose of this section is to present new sequence data from complete small subunit ribosomal RNA (18S rRNA), to analyze the mitochondrial gene sequences rather than their amino acid equivalents, and to combine the data into mitochondrial DNA (mtDNA), nuclear DNA (nucDNA), and total molecular data sets. For consistency and to minimize the introduction of missing data we have treated this problem as a three-taxon statement employing data for a "generalized" teleost (individual or combined data from *Salmo trutta* and *Oncorhynchus mykiss*), *Lepisosteus* (individual or combined data from *L. oculatus* and *L. oseus*), and *Amia calva*. The choice of taxa was limited to previously published sequence data of which there is little concerning actinopterygians. Trees were rooted using data from either *Polyodon spathula* or *Polypterus retropinnus*. Where necessary, data from the outgroup taxa were combined.

Sources of the data are presented in Appendix 3. Details of the extraction, amplification, and sequencing of new 18S rRNA genes (from *Polyodon spathula*,

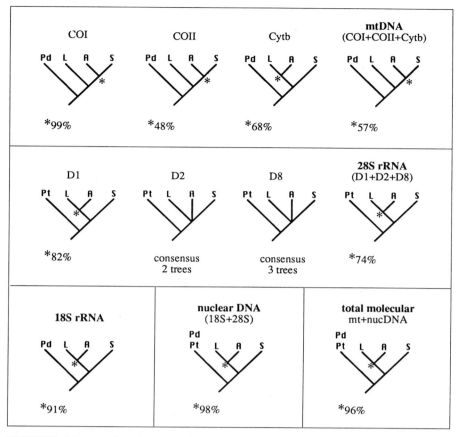

FIGURE 7 Most parsimonious trees determined by heuristic searches of the individual and combined data sets. COI = cytochrome oxidase I; COII = cytochrome oxidase II; Cytb = cytochrome b; D1, D2, and D8 = D1, D2, and D8 domains of large subunit (28S) rRNA gene; 18S = small subunit rRNA gene; Pd = *Polyodon spathula*; Pt = *Polypterus retropinnus*; A = *Amia calva*; L = individual or combined data from *Lepisosteus oculatus* and *L. osseus*; S = individual or combined data from *Salmo trutta* and *Oncorhynchus mykiss* (see text for details). Asterisks indicate percentage bootstrap support (1000 replicates) for particular nodes. Strict consensus trees are shown where more than one tree resulted.

Lepisosteus osseus, Amia calva, and *Salmo trutta*) are the same as those detailed elsewhere for the same gene (Forey *et al.,* this volume). Sequence alignment was accomplished by eye with reference to secondary structure, and insertions were treated as fifth character states. There were few ambiguous positions, none of which altered the topology of resulting trees. Only homologous sites present in all the taxa were included in the analyses, thus obviating the need to include missing data other than that brought about through sequencing. Each of the data sets was analyzed cladistically using PAUP 3.1.1 (Swofford, 1993).

Results are presented in Fig. 7 and Appendix 4. Among the cytochrome genes two solutions were apparent. Cytochrome oxidase subunits I and II and the combined mitochondrial data each suggested phylog-

enies consistent with Patterson (1973). Only cytochrome b treated alone maintained holostean monophyly. Bootstrap support was high for COI and the tree (length = 180) was 9 steps shorter than the next shortest tree. The combined mtDNA solution (length = 440 steps) was 5 steps shorter than the next shortest tree. Considerably fewer phylogenetically informative sites were available with the 28S rRNA data. Combined data from the variable domains resulted in a monophyletic holostean clade, but the tree (length = 178) was only 2 steps shorter than the next most parsimonious solution. Holostean monophyly was more convincingly suggested by the 18S rRNA gene and combined nuclear DNA data. Further, the combined molecular data strongly supports holostean monophyly through high bootstrap support (>95%)

and a tree 16 steps (more than 2% of the total tree; length = 799) longer than a paraphyletic solution.

VI. Conclusions

From the morphological analysis we conclude, like Patterson (1973, p. 299), that within living neopterygians *Amia* is the sister-group of the teleosts and that *Lepisosteus* is the sister-group of these two combined: we note, however, that molecular data suggest a different phylogenetic pattern.

As might be expected, the many unique characters of *Amia* that separate it from teleosts apparently represent several different levels of monophyly among halecomorph fishes (characters of all halecomorphs, characters of halecomorphs other than parasemionotids, characters of all amiids, characters acquired within the amiid clade, and autapomorphic characters of *Amia* itself). Philosophically, it could be argued that having more than one higher taxonomic name is a redundancy for an extant monotypic genus such as *Amia*. When confronted with all the fossils, however, it seems prudent to recognize successive taxonomic levels in a formal way; all amiids (the living crown group) are halecomorphs, but not all halecomorphs (the stem group, including many fossils) are amiids.

Patterson's (1973) view that *Amia* is the sole surviving member of a formerly diverse monophyletic halecomorph clade has survived past attempts at refutation and is supported by a suite of derived characters in addition to symplectic involvement in the jaw joint. This clade includes amiids, caturids, ionoscopids (including ophiopsids), and many plesiomorphic taxa including *Watsonulus* and other parasemionotids. Some fossils which have been loosely regarded in the past as caturids can be placed instead closer to other extinct halecomorph taxa such as ionoscopids and/or ophiopsids (e.g., *Brachyichthys* and *Heterolepidotus*). According to the results of our analysis, there is an additional branch at the base of this clade that includes macrosemiids and semionotids.

Several levels of monophyly are recognizable among halecomorph fishes. *Amia* represents the only surviving halecomorph. Another monophyletic group is recognized, containing "ionoscopids" (*Oshunia* and *Ionoscopus*), ophiopsids (including *Ophiopsis* and *Macrepistius*), and *Brachyichthys*. The amiid–caturid and ionoscopid/ophiopsid clades are regarded as sister taxa above the parasemionotid grade represented by *Watsonulus*. Further study of unplaced "caturids" such as *Osteorachis*, *Furo*, and the enigmatic "*Aspidorhynchus*" is necessary to determine where they belong relative to the two better resolved clades of haleco-

morphs. *Ionoscopus* is regarded as a primitive sister taxon to *Oshunia* and ophiopsids, but there is support from some trees that ionoscopids form a monophyletic group (although we are doubtful that it contains all the taxa included by Patterson, 1973).

We hope future research into fossil halecomorphs will resolve the phylogenetic positions of various plesiomorphic taxa traditionally placed loosely within parasemionotids and caturids. Lambers' (1992) more exclusive diagnosis of caturids helps clarify the phylogenetic relationships of at least some of these generalized taxa. Other "caturids" seem to have affinities elsewhere among halecomorphs. In certain cases this has been suspected for some time (e.g., *Macrepistius* and *Heterolepidotus* with ophiopsids; Bartram, 1975). The enigmatic "*Aspidorhynchus*" braincases described by Rayner (1948) and Patterson (1975) lack the derived features of caturids in the restricted sense of Lambers (1992), but they closely resemble the braincase of *Oshunia* in many details (Maisey, in prep.). Remaining "caturids" (e.g., *Furo*, *Osteorachis*, and *Liodesmus*) can be characterized only as primitive halecomorphs above the parasemionotid level of organization, and all parasemionotids are at present in phylogenetic limbo as plesiomorphic halecomorphs lacking any apomorphic characters of their own (although the form of the preoperculum may specify certain groups). They may not form a monophyletic group and are best regarded as stem halecomorphs in need of further phylogenetic resolution.

The evolutionary history of halecomorph fishes displays partial congruence with the geologic time scale. The earliest amiids are of Late Jurassic (Oxfordian) age. Caturids (in the restricted sense of Lambers, 1992) are restricted to the Jurassic—much less than the range (Norian to Albian) suggested by Gardiner (1993), who had a far broader concept of caturids and included fossils that are excluded here. The first "ionoscopids" are Oxfordian and may range into the Early or Late Cretaceous (depending on whether one includes *Oshunia* and various "oligopleurids" discussed by Patterson, 1973), but this seems to be an artificial assemblage and its range is therefore irrelevant. Ophiopsids (including *Macrepistius*) range from the Anisian to the Turonian. The geologic range of fossils classified as parasemionotids apparently is restricted to the Early Triassic (Scythian). Macrosemiids range from the late Triassic to the early Cretaceous, and the earliest supposed semionotid (*Acentrophorus*) is Permian. Of all these taxa only amiids survive into the Tertiary.

Amia is commonly regarded as a "living fossil" (for a discussion, see Schultze and Wiley, 1984). In this chapter we have examined characters that bear upon higher levels of relationship among halecomorph fossils. The degree to which these characters, representing successive levels of monophyly, have become superimposed by the extinction of sister taxa may help explain why *Amia* has been characterized as a living fossil. Organisms have been given this designation for a variety of reasons (Schopf, 1984; Eldredge, 1984) but are commonly viewed as isolated living taxa that evolved slowly (in the rate of acquisition and loss of observable characters) relative to their extant sister taxa. Living fossils are popularly thought of as successful in terms of temporal longevity but not in taxonomic diversity (an admittedly qualitative and unempirical attitude). Some living fossils such as *Amia* are terminal survivors of formerly extensive clades, however.

Isolation of *Amia* by extinction of taxa lower on the halecomorph clade has resulted in the loss of character information. Numerous apomorphic characters seen in *Amia* seem to define several different levels of monophyly within the clade leading to amiids but are now collapsed together (loss of resolution); *as taxa become extinct, so do phylogenetic nodes*. Other characters, not seen in *Amia* but found in fossils, may represent apomorphic traits of extinct monophyletic groups of non-amiid halecomorphs or may have appeared and disappeared along the lineage leading to amiids (e.g., the supraotic and deep infraorbitals). In any event, these features have been lost; *as taxa become extinct, so do characters*. Extinction within the halecomorph clade has dramatically reduced our ability to perceive both the depth and breadth of previous evolutionary richness because of these losses in resolution and characters. The former evolutionary success and richness of this clade can only be revealed today by teasing apart character distributions among different fossils. Palaeontology reveals that the evolutionary pedigree of *Amia* is in essence far richer (or more successful in terms of accumulation of characters and consequently in terms of taxonomic diversity) than that of, for example, *Latimeria* or *Lingula*, although these taxa share the attribute of sole survivorship in their respective clades.

We also conclude that the molecular evidence is equivocal but note that the rate of morphological change appears not to be correlated (at least in actinopterygians) with that of molecular change (Littlewood, C. Patterson, and Smith, personal communication).

To summarize the results of our morphological analyses, the Caturidae is the sister-group of the Amiidae, and the Ionoscopidae plus the Ophiopsidae combined are the sister-group of these two. The Parasemionotidae, represented by *Watsonulus*, is the sister of all of the above, and together they constitute the Ha-

lecomorphi as defined by Patterson (1973) (Fig. 3). The Macrosemiidae and Semionotidae together are stem-group halecomorphs but lack the characteristic double jaw articulation of other halecomorphs. We further conclude that the pycnodonts are the sister-group of the Dapediidae and together are stem-group teleosts (Fig. 1).

VII. Summary

Fossil and Recent holosteans have been subjected to parsimony analysis. The resultant analysis shows that the living Holostei (gars and amiids) are a paraphyletic group and that *Amia* is more closely related to teleosts than to *Lepisosteus*. Molecular data, on the other hand, suggest that *Lepisosteus* and *Amia* form a monophyletic holostean clade.

Four characters unite the Macrosemiidae and Semionotidae with other halecomorphs. Seven other characters unite the Parasemionotidae, Ionoscopidae, Ophiopsidae, Amiidae, and Caturidae as another monophyletic group of halecomorphs. *Caturus* and amiids are united by two characters, and eight characters unite a monophyletic clade (comprising ionoscopids and ophiopsids) with amiids and *Caturus*. The Dapediidae and Pycnodontiformes are united and form a sister-group to the teleosts, with which they are united by six characters.

Acknowledgments

Colin Patterson was one of J.G.M.'s Ph.D. advisors and helped shape his early paleontological career. Both J.G.M. and B.G.G. wish to thank him for all his kindness and helpful advice through the years. We also wish to pay tribute to our former Ph.D. supervisor, Kenneth Kermack of University College, London, who gave both of us (J.G.M. and B.G.G.) the opportunity of becoming vertebrate paleontologists. Our particular thanks go to Lance Grande and an anonymous reviewer for reading and commenting on this manuscript. Thanks also to Peter Forey for allowing J.G.M. and B.G.G. to communicate rapidly via his e-mail.

References

Adams, E. N. (1972). Consensus, techniques and the comparison of taxonomic trees. *Syst. Zool.* **21,** 390–397.

Allis, E. P. (1897). The cranial muscles and cranial and first spinal nerves in *Amia calva. J. Morphol.* **12,** 487–808.

Allis, E. P. (1898). On the morphology of certain bones of the cheek and snout of *Amia calva. J. Morphol.* **14,** 425–466.

Applegate, S. P. (1988). A new species of a holostean belonging to the family Ophiopsidae, *Teoichthys kallistos,* from the Cretaceous, near Tepexi de Rodriguez, Puebla. *Rev. Univ. Nac. Auton. Mex., Inst. Geol.* **2,** 200–205.

Arratia, G., and Schultze, H.-P.(1991). The palatoquadrate and its ossifications: Development and homology within osteichthyans. *J. Morphol.* **208,** 1–81.

Bartram, A. W. H. (1975). The holostean fish *Ophiopsis* Agassiz. *Zool. J. Linn. Soc.* **56,** 183–205.

Bartram, A. W. H. (1977). The Macrosemiidae, a Mesozoic family of holostean fishes. *Bull. Br. Mus. (Nat. Hist.), Geol.* **29,** 137–234.

Boreske, J. R. (1974). A review of the North American fossil amiid fishes. *Bull. Mus. Comp. Zool.* **146,** 1–87.

Brito, P. M. (1988). La structure du suspensorium de *Vinctifer*, poisson actinoptérygien mésozoique: Remarques sur les implications phylogénétiques. *Geobios* **21**(6), 819–823.

Bryant, L. J. (1987). A new genus and species of Amiidae (Holostei, Osteichthyes) from the Late Cretaceous of North America, with comments on the phylogeny of the Amiidae. *J. Vertebr. Paleontol.* **7,** 349–361.

Chalifa, Y., and Tchernov, E. (1982). *Pachyamia latimaxillaris,* new genus and species (Actinopterygii: Amiidae) from the Cenomanian of Jerusalem. *J. Vertebr. Paleontol.* **2,** 269–285.

Chang, M.-M., and Hong, Z. (1980). Discovery of *Ikechoamia* from South China. *Vertebr. PalAsiat.* **18**(2), 89–93.

Da Silva Santos, R. (1960). A posição sistemática de *Enneles* audax Jordan e Branner da Chapada do Araripe, Brasil. *Monographias Div. Geol. Miner. Bras.* **17,** 1–24.

Da Silva Santos, R. (1970). A Paleoictiofauna da Formação Sontana-Holostei: Família Girodontidae. *An. Acad. Bras. Cienc.* **42,** 445–452.

de Beer, G. R. (1937). "The Development of the Vertebrate Skull." Oxford University Press, Oxford.

Eldredge, N. (1984). Simpson's inverse: Bradytely and the phenomenon of living fossils. *In* "Living Fossils" (N. Eldredge and S. M. Stanley, eds.), Springer-Verlag, New York.

Figueredo, F. J., and Da Silva Santos, R. (1990). Sobre *Neoproscinetes penalvai* (Silva Santos, (1970) (Pisces, Pycnodontiformes) do Cretáceo Inferior da Chapada do Araripe, Nordeste do Brasil. *An. Acad. Bras. Cienc.* **60**(3), 269–282.

Gardiner, B. G. (1960). A revision of certain actinopterygian and coelacanth fishes, chiefly from the Lower Lias. *Bull. Br. Mus. (Nat. Hist), Geol.* **4,** 239–384.

Gardiner, B. G. (1967). The significance of the preopercular in actinopterygian evolution. *J. Linn. Soc. London, Zool.* **47,** 197–209.

Gardiner, B. G. (1984). The relationships of the palaeoniscid fishes, a review based on new specimens of *Mimia* and *Moythomasia* from the Upper Devonian of Western Australia. *Bull. Br. Mus. (Nat. Hist.), Geol.* **37,** 173–428.

Gardiner, B. G. (1993). Osteichthyes: Basal actinopterygians. *In* "The Fossil Record" (M. J. Benton, ed.), Vol. 2. 611–619. Chapman & Hall, London.

Gardiner, B. G. (1994). Haematothermia: Warm-blooded amniotes. *Cladistics* **9,** 369–395.

Gauthier, J. A., Kluge, A. G., and Rowe, T. (1988). Amniote phylogeny and the importance of fossils. *Cladistics* **4,** 105–209.

Goodrich, E. S. (1930). "Studies on the Structure and Development of Vertebrates." Macmillan, London.

Gosline, W. A. (1969). The morphology and systematic position of the alepocephaloid fishes. *Bull. Br. Mus. (Nat. Hist.), Zool.* **18,** 183–218.

Grande, L. (1995). Using the extant *Amia calva* to test the monophyly of Mesozoic groups of fishes. *In* "Mesozoic Fishes: Systematics and Paleoecology" (G. Arratia and G. Viohl, eds.), pp. 181–189. Verlag Dr. Friedrich Pfeil, Munich.

Grande, L., and Bemis, W. E. (1991). Osteology and relationships of fossil and Recent paddlefishes (Polyodontidae) with com-

ments on the interrelationships of Acipenseriformes. *J. Vertebr. Paleontol.* **11,** Suppl. 1, 1–121.

Hay, P. (1895). On the structure and development of the vertebral column in *Amia*. *Field Mus. Nat. Hist., Publ.* **1,** 1–54.

Huxley, T. H. (1861). Preliminary essay upon the systematic arrangement of the fishes of the Devonian epoch. *Mem. Geol. Surv. U.K. Decade* **10,** 1–40.

Jessen, H. (1972). Schultergürtel und pectoralflosse bei Actinopterygiern. *Fossils Strata* **1,** 1–101.

Jollie, M. (1984). Development of cranial and pectoral girdle bones of *Lepisosteus* with a note on scales. *Copeia*, pp. 476–502.

Lambers, P. (1992). "On the Ichthyofauna of the Solnhofen Lithographic Limestone (Upper Jurassic, Germany)." Rijksuniversiteit Gröningen, Gröningen, Germany.

Lauder, G. V. (1980). Evolution of the feeding mechanism in primitive actinopterygian fishes: A functional anatomical analysis of *Polypterus, Lepisosteus* and *Amia*. *J. Morphol.* **163,** 283–317.

Lauder, G. V. (1989). Caudal fin locomotion in ray-finned fishes: Historical and functional analyses. *Am. Zool.* **29,** 85–102.

Lauder, G. V., and Liem, K. F. (1983). The evolution and interrelationships of the actinopterygian fishes. *Bull. Mus. Comp. Zool.* **150,** 95–197.

Lê, H. L. V., Lecointre, G., and Perasso, R. (1993). A 28S rRNA-based phylogeny of the gnathostomes: First steps in the analysis of conflict and congruence with morphologically based cladograms. *Mol. Phylogenet. Evol.* **2,** 31–51.

Lecointre, G., Philippe, H., Lê, H. L. V., and Le Guyader, H. (1994). How many nucleotides are required to resolve a phylogenetic problem? The use of a new statistical method applicable to available sequences. *Mol. Phylogenet. Evol.* **3,** 292–309.

Lehman, J. P. (1966). Actinopterygii. *In* "Traité de Paléontologie" (J. Pivetau, ed.), Vol. 4, Part 3, pp. 1–242. Masson, Paris.

Liu, H.-T., and Su, D. (1983). fossil amiids (Pisces) of China and their biogeographic significance. *Acta Paleontol. Pol.* **28,** 181–194.

Maisey, J. G. (1991a). *Calamopleurus. In* "Santana Fossils: An Illustrated Atlas" (J. G. Maisey, ed.) pp. 139–156. TFH Publications, Neptune City, NJ.

Maisey, J. G. (1991b). *Oshunia. In* "Santana Fossils: An Illustrated Atlas" (J. G. Maisey, ed.) pp. 157–168. TFH Publications, Neptune City, NJ.

Maisey, J. G. (1991c). *Vinctifer. In* "Santana Fossils: An Illustrated Atlas" (J. G. Maisey, ed.) pp. 170–189. TFH Publications, Neptune City, NJ.

Maisey, J. G. In preparation.

Müller, J. (1844). Über den Bau und die Grenzen der Ganoiden, und über das natürliche System der Fische. *Ber. Akad. Wiss. Berlin,* pp. 416–422.

Müller, J. (1846). Ueber den Bau und die Grenzen der Ganoiden und über das natürliche System der Fische. *Abh. K. Akad. Wiss. Berlin, Phys.-Math. Kl.,* pp. 117–216.

Nelson, G. J. (1969a). Gill arches and the phylogeny of fishes, with notes on the classification of vertebrates. *Bull. Am. Mus. Nat. Hist.* **141,** 475–552.

Nelson, G. J. (1969b). Origin and diversification of teleostean fishes. *Ann. N.Y. Acad. Sci.* **167,** 18–30.

Nelson, G. J. (1972). Cephalic sensory canals, pitlines, and a classification of esocoid fishes, with notes on galaxiids and other teleosts. *Am. Mus. Novit.* **2492,** 1–49.

Nielsen, E. (1942). Studies on Triassic fishes from East Greenland. I. *Glaucolepis* and *Boreosomus*. *Medd. Groenl.* **138,** 1–403.

Nielsen, E. (1949). Studies on Triassic fishes from East Greenland. II. *Australosomus* and *Birgeria*. *Medd. Groenl.* **146,** 1–309.

Normark, B. B., McCune, A. R., and Harrison, R. G. (1991). Phylogenetic relationships of neopterygian fishes, inferred from mitochondrial DNA sequences. *Mol. Biol. Evol.* **8,** 819–834.

Nursall, R., and Maisey, J. G. (1991). *Neoproscinetes. In* "Santana Fossils: An Illustrated Atlas" (J. G. Maisey, ed.) pp. 124–137. TFH Publications, Neptune City, NJ.

Nybelin, O. (1966). On certain Triassic and Liassic members of the family Pholidophoridae s.str. *Bull. Br. Mus. (Nat. Hist.), Geol.* **11,** 351–432.

Olsen, P. E. (1984). The skull and pectoral girdle of the parasemionotid fish *Watsonulus eugnathoides* from the Early Triassic Sakamena Group of Madagascar, with comments on the relationships of the holostean fishes *J. Vertebr. Paleontol.* **4,** 481–499.

Olsen, P. E., and McCune, A. R. (1991). Morphology of the *Semionotus elegans* species group from the Early Jurassic part of the Newark supergroup of eastern North America, with comments on the family Semionotidae (Neopterygii). *J. Vertebr. Paleontol.* **11,** 269–292.

Patterson, C. (1967). Are the teleosts a polyphyletic group? *Colloq. Int. C.N.R.S.* **163,** 93–109.

Patterson, C. (1973). Interrelationships of holosteans. *In* "Interrelationships of Fishes" (P. H. Greenwood, R. S. Miles, and C. Patterson, eds.), pp. 233–305. Academic Press, London.

Patterson, C. (1975). The braincase of pholidophorid and leptolepid fishes with a review of the actinopterygian braincase. *Philos. Trans. R. Soc. London, Ser. B* **269,** 275–579.

Patterson, C. (1977). The contribution of paleontology to teleostean phylogeny. *In* "Major Patterns in Vertebrate Evolution" (M. K. Hecht, P. C. Goody, and B. M. Hecht, eds.), pp. 579–643. Plenum, New York.

Patterson, C. (1994). Bony fishes. *In* "Major Features of Vertebrate Evolution" (D. R. Prothero and R. M. Schoch, eds.), Short Courses in Paleontol., No. 7, pp. 57–84. Paleontological Society, University of Tennessee, Knoxville.

Patterson, C., and Rosen, D. E. (1977). Review of ichthyodectiform and other Mesozoic teleost fishes and the theory and practice of classifying fossils. *Am. Mus. Nat. Hist.* **158,** 83–172.

Rayner, D. H. (1948). The structure of certain Jurassic fishes with special reference to their neurocrania. *Philos. Trans. R. Soc. London, Ser. B* **233,** 287–345.

Rosen, D. E., Forey, P. L., Gardiner, B. G., and Patterson, C. (1981). Lungfishes, tetrapods, paleontology and plesiomorphy. *Bull. Am. Mus. Nat. Hist.* **167,** 159–276.

Saint-Seine, P. (1949). Les poissons des calcaires lithographiques de Cerin (Ain). *Nouv. Arch. Mus. Hist. Nat. Lyon* **2,** 1–357.

Schaeffer, B. (1960). The Cretaceous holostean fish *Macrepistius. Am. Mus. Novit.* **2011,** 1–18.

Schaeffer, B. (1967). Osteichthyan vertebrae. *J. Linn. Soc. London, Zool.* **47,** 185–195.

Schaeffer, B. (1971). The braincase of the holostean fish *Macrepistius,* with comments on neurocranial ossification in the Actinopterygii. *Am. Mus. Novit.* **2459,** 1–34.

Schaeffer, B., and Patterson, C. (1984). Jurassic fishes from the western USA, with comments on Jurassic fish distribution. *Am. Mus. Novit.* **2796,** 1–86.

Schopf, T. J. M. (1984). Rates of evolution and the notion of living fossils. *Ann. Rev. Ecol. Syst.* **12,** 245–292.

Schultze, H.-P. (1966). Morphologische und histologische Untersuchungen an Schuppen mesozoischer Actinopterygier (Übergang von Ganoid- zu Rundschuppen). *Neues Jahrb. Geol. Palaeontol., Abh.* **126,** 232–314.

Schultze, H.-P., and Arratia, G. (1986). Reevaluation of the caudal skeleton of actinopterygian fishes: I. *Lepisosteus* and *Amia. J. Morphol.* **190,** 215–241.

Schultze, H.-P., and Wiley, E. O. (1984). The neopterygian *Amia* as a living fossil. *In* "Living Fossils" (N. Eldredge and S. M. Stanley, eds.) pp. 153–159. Springer-Verlag, New York.

Stensiö, E. A. (1932). Triassic fishes from East Greenland collected by the Danish expeditions in 1929–1931. *Medd. Groenl.* **83**(3), 1–305.

Stensiö, E. A. (1935). *Sinamia zdanskyi,* a new amiid from the Lower Cretaceous of Shantung, China. *Paleontol. Sin.* **3,** 1–47.

Steutzer, P. H. (1972). Morphologie, Taxonomie und Phylogenie der Ionoscopidae (Actinopterygii, Pisces). Unpublished Dissertation, University of Munich, Munich, Germany.

Swofford, D. L. (1993). "PAUP: Phylogenetic Analysis Using Parsimony, Version 3.1." Smithsonian Institution, Washington, DC.

Taverne, L. (1974). Sur le première examplaire complet d'*Enneles audax* Jordan, D. S. et Branner, J. C., 1908 (Pisces, Holostei, Amiidae) du Crétacé supérieur du Brésil. *Bull. Soc. Belge Géol., Palaeontol. Hydrol.* **83,** 61–71.

Taverne, L. (1981). Les actinoptérygiens de l'Aptien Inférieur (Töck) d'Helgoland. *Mitt. Geol.-Palaeontol. Inst. Univ. Hamburg* **51,** 43–82.

Thies, D. (1988). *Dapedium pholidotum* (Agassiz, 1832)? (Pisces, Actinopterygii) aus dem Unter-Toarcium NW-Deutschlands. *Geol. Palaeontol.* **22,** 89–121.

Thies, D. (1989a). Der Hirnschädel und das Gehirn von *Tetragonolepis semicincta* Bronn 1830 (Actinopterygii, Semionotiformes). *Palaeontographica, Abt. A* **209,** 1–32.

Thies, D. (1989b). Sinneslinien bei dem Knochenfisch *Lepidotes elvensis* (Blainville 1818) (Actinopterygii, Semionotiformes) aus dem Oberlias (Unter-Toarcium) von Grimmen in der DDR. *Neues. Jahrb. Geol. Palaeontol. Monatsh.* **11,** 692–704.

Tintori, A. (1980). Two new pycnodonts (Pisces, Actinopterygii) from the Upper Triassic of Lombardy (N. Italy). *Riv. Ital. Paleontol.* **86,** 795–824.

Tintori, A. (1982). Hypsiomatic Semionotidae (Pisces, Actinopterygii) from the Upper Triassic of Lombardy (N. Italy). *Riv. Ital. Paleontol.* **88,** 417–442.

Véran, M. (1981). Apports de l'ichthyofaune éodévonienne et éotriassique du Spitzberg dans l'étude du symplectique et de l'interhyal de l'arc hyoïdien des poissons téléostomes. *C.R. Seances Acad. Sci., Sér. 2* **293,** 199–202.

Véran, M. (1988). Les éléments accesoire de l'arc hyoïdien des poissons téléostomes (acanthodiens et osteichthyens) fossiles et actuels. *Mém. Mus. Natl. Hist. Nat., Ser. C* **54,** 1–113.

Wenz, S. (1968). "Compléments à l'étude des poissons actinoptèrygiens du Jurassique français." CNRS, Paris.

Wenz, S. (1971). Anatomie et position systématique de *Vidalamia,* poisson holostéen du Jurassique supérieur du Montsech (Province de Lérida, Espagne). *Ann. Paleontol. (Vertebr.)* **57,** 42–62.

Wenz, S. (1977). Le squelette axial et l'endosquelette caudal d'*Enneles* audax, poisson amiidé du Crétacé de Ceará (Brésil). *Bull. Mus. Natl. Hist. Nat., Sci. Terre* **67,** 341–348.

Wenz, S. (1979). Squelette axial et endosquelette caudal d'*Amiopsis dolloi,* amiidé du Wealdien de Bernissart. *Bull. Mus. Natl. Hist. Nat., Sect. C* [4] **1,** 343–357.

Wenz, S. (1989). Une nouvelle espèce de *Coelodus* (Pisces, Pycnodontiformes) du Crétacé Inférieur du Montsech (Province de Lérida, Espagne): *Coelodus subdiscus* n. sp. *Geobios* **22,** 515–520.

Wenz, S., and Kellner, A. W. A. (1986). Découverte du premier Ionoscopidae (Pisces, Halecomorphi) sud-américain, *Oshunia brevis* n.g., n.sp., dans le Crétacé Inférieur de la Chapada do Araripe (nord-est du Brésil). *Bull. Mus. Natl. Hist. Nat., Sect. C* [4] **8,** 77–88.

Wiley, E. O. (1976). The phylogeny and biogeography of fossil and recent gars (Actinopterygii, Lepisosteidae). *Misc. Publ. Univ. Kan. Mus. Nat. Hist.* **64,** 1–111.

Wiley, E. O., and Schultze, H.-P. (1984). Family Lepisosteidae (gars) as living fossils. *In* "Living Fossils" (N. Eldredge and S. M. Stanley, eds.), Springer-Verlag, New York.

Wiley, E. O., Siegel-Causey, D., Brooks, D. R., and Funk, V. A. (1991). "The Compleat Cladist." University of Kansas, Lawrence.

Woodward, A. S. (1895). "Catalogue of the Fossil Fishes in the British Museum (Natural History), London," Vol. 3. Br. Mus. (Nat. Hist.), London.

Appendix 1

Neopterygian Character List

1. Opisthotic–pterotic relationship
   0. Opisthotic larger than pterotic
   1. Opisthotic and pterotic subequal
   2. Opisthotic absent
2. Pterotic
   0. Pterotic present
   1. Pterotic fused with dermopterotic
   2. Pterotic absent
3. Epioccipital
   0. Epioccipital present bordered anteriorly by cranial fissure
   1. Epioccipital extends into otic region
   2. Epioccipital absent
4. Intercalar
   0. Endochondral with minor membranous outgrowths
   1. With extensive membranous outgrowths medial to jugular (with or without endochondral core)
   2. With extensive membranous outgrowths lateral to jugular (with or without endochondral core)
   3. Intercalar lost
5. Vagal foramen
   0. Anterior to exoccipital
   1. Lateral outgrowths from intercalar enclose lateral margin
   2. Ventral outgrowths from intercalar lateral margin enclose ventral margin
   3. Enclosed by exoccipital
6. Myodome
   0. Absent
   1. Intramural, lined by endoskeletal floor
   2. With a median fenestra in floor
7. Anterior Myodome
   0. Absent
   1. Separate, paired dorsal and ventral
   2. Through orbitonasal canal
   3. Median—foramen olfaction evehens
8. Aortic canal
   0. Short, bifurcates anteriorly
   1. Short with cuplike recess (which lies behind occipital canal)
   2. Ligament originates on basioccipital
   3. Occipital artery penetrates basioccipital
9. Fossa bridgei
   0. Fossa bridgei rudimentary—no posttemporal fossa
   1. Fossa bridgei discrete—small posttemporal fossa
   2. Fossa bridgei discrete—large posttemporal fossa
   3. Posttemporal fossa communicates with fossa bridgei
10. Basipterygoid process
    0. Well-developed dermal process with rudimentary endoskeletal component
    1. Basipterygoid process absent
11. Parasphenoid teeth
    0. Present from level of ascending process forwards
    1. Teeth absent
    2. Tooth patch on posterior portion
12. Parasphenoid
    0. Parasphenoid extends beneath basioccipital to level of vagal foramen
    1. Parasphenoid reaches posterior end of basioccipital
    2. Parasphenoid embraces aortic canal: with a ventral fenestra
13. Vomer
    0. Small
    1. Sutured to parasphenoid
    2. Median
14. Parasphenoid—internal carotid
    0. Through notch
    1. Through foramen
15. Parasphenoid—efferent pseudobranchial
    0. Through notch
    1. Through foramen
16. Dermosphenotic
    0. Hinged to skull roof
    1. Bound to, or fused to, anterior margin of sphenotic
17. Supraorbitals
    0. Anterior supraorbitals meet or lie in close proximity to antorbital
    1. Supraorbitals meet infraorbitals
    2. Supraorbitals absent
18. Antorbital
    0. Antorbital platelike, with stout anterior process
    1. Antorbital with long anterior process, "C" shaped

2. Antorbital with very long anterior process and platelike posterior expansion
3. Reduced to tubular ossification

19. Rostral
    0. A deep cap on tip of snout partially or wholly separating the nasals
    1. Of moderate to narrow size
    2. Reduced to a narrow tube with lateral processes
    3. A short tube
    4. Fused to something else (viz. rostro-derm-ethmoid)
20. Premaxilla
    0. Rudimentary nasal process
    1. Small nasal process
    2. Elongate nasal process notched or perforated by olfactory nerve
    3. Divided into lateral toothed part and median dermethmoid
21. Maxilla
    0. Fixed to cheek
    1. Free, with long, curved, cylindrical medial process
    2. Free, with short, medial process
22. Maxillary shape
    0. Elongate, broad posteriorly, stretches well behind orbit
    1. Elongate, narrow, stretches well behind orbit
    2. Elongate, narrow, stretches well behind orbit, indented posteriorly
    3. Short ends anterior to orbit, rounded posteriorly
    4. Short, notched dorsally
    5. Very short, sliver of bone
23. Supramaxilla
    0. Absent
    1. Present
24. Quadratojugal
    0. Platelike, lateral to quadrate
    1. Splintlike, free along posterior border of quadrate
    2. Fused to quadrate
    3. Absent
25. Symplectic
    0. Absent
    1. On inside of quadrate—contacts preoperculum
    2. Behind quadrate—contacts lower jaw and suboperculum
    3. Behind quadrate—contacts lower jaw and is bound to preoperculum by membrane bone
26. Jaw joint
    0. Quadrate—articular, no symplectic

1. Quadrate—articular, symplectic lies on inner face of quadrate
2. Double jaw joint, quadrate + symplectic–articular

27. Interoperculum
    0. Absent
    1. Present
28. Gular
    0. Present
    1. Absent
29. Dentition
    0. Non-crushing—maxilla with full compliment of teeth
    1. Reduction in maxillary tooth row, crushing dentition on at least vomer, palatines, and prearticulars
    2. Maxillary tooth row very reduced or absent, rows of regularly arranged crushing teeth on vomer, coronoids, and prearticulars
30. Ceratohyal
    0. Proximal ceratohyal long, relatively straight—same depth posteriorly as distal element
    1. Proximal ceratohyal long, gently curved with very small, distal element
    2. Proximal ceratohyal short and deep posteriorly
    3. Proximal ceratohyal very short—open dorsally
31. Epibranchials
    0. Slender
    1. With uncinate processes
32. Neural spines
    0. Paired
    1. Median, unpaired pre-ural neural spines
33. Fulcral scales
    0. Basal and fringing fulcra present
    1. Basal and fringing fulcra greatly enlarged
    2. Fringing fulcra very reduced or absent
34. Uppermost hypaxial caudal rays
    0. Successively shorter from bottom to top
    1. A bundle of elongate fin-ray bases extending over several hypurals
    2. Dorsal and ventral fin-ray bases symmetrical
    3. Fin-ray one-to-one on hypurals
35. Uroneurals
    0. Absent
    1. Present
36. Ridge scales
    0. Absent
    1. Present along dorsal margin (with posteriorly directed spines)
    2. Present along both dorsal and ventral margins

37. Clavicle
    0. Large, caps anterior end of cleithrum
    1. Toothed plates on postbranchial lamina of cleithrum
    2. Clavicle reduced, often with a single row of denticles
    3. Serrated organ (with 12 or more ridges of denticles)
    4. Absent

Data Matrix 1 for Neopterygian Analysis

	1 2 3 4 5 6 7 8 9 0 1 2 3 4 5 6 7 8 9 0 1 2 3 4 5 6 7 8 9 0 1 2 3 4 5 6 7
Perleidus	0 0 0 0 0 1 1 0 0 0 0 0 0 0 0 0 0 0 0 0 0 0 0 0 0 0 0 0 0 0 0 0 0 0 0 0 0
Lepisosteus	2 2 1 3 3 0 0 3 0 0 2 1 1 0 0 0 1 0 1 2 0 5 0 1 1 1 0 1 0 0 0 0 0 1 0 0 1
Macrosemius	? 0 ? ? 3 2 ? ? ? 0 1 1 1 0 0 0 2 3 1 2 1 4 0 2 1 1 1 1 1 2 1 ? 1 0 0 0 2
Watsonulus	1 0 0 0 0 2 1 1 1 0 0 0 1 0 0 1 1 2 2 1 1 2 1 2 2 2 1 0 0 0 ? 1 1 0 0 0 0
Caturus	2 2 1 1 2 2 1 1 2 1 0 1 1 0 0 1 0 2 2 2 1 2 1 2 3 2 1 0 0 1 1 1 1 1 0 0 3
Amia	2 2 1 1 2 2 2 3 2 1 2 1 1 0 0 1 2 2 1 2 1 2 1 3 3 2 1 0 0 1 0 1 2 3 0 0 3
Semionotus	2 2 1 3 3 2 1 1 ? 0 2 1 1 0 0 0 1 1 ? 2 1 3 1 1 1 1 1 1 1 1 0 1 1 1 0 0 1 2
Lepidotes	2 2 1 3 3 2 ? 1 2 0 2 1 2 0 0 0 1 1 2 3 1 1 1 1 1 1 1 1 0 1 1 1 0 0 1 2
Dapedium	? 0 ? 0 ? 1 1 1 3 0 1 2 2 1 1 0 0 0 0 1 2 3 1 1 1 1 1 0 2 3 1 1 0 0 0 2 4
Microdon	2 2 2 ? 0 2 ? 1 3 0 1 2 2 1 0 0 2 3 3 1 2 3 0 2 3 2 0 1 2 3 0 1 0 0 0 2 4
Pachycormus	1 0 2 0 1 2 3 1 3 1 2 0 1 1 0 0 2 0 4 3 1 1 1 3 1 1 1 0 0 0 0 1 0 2 1 0 4
Elops	2 1 1 2 3 2 3 2 3 1 2 1 2 1 0 0 0 1 4 3 1 1 1 2 1 1 1 0 0 0 1 1 2 1 1 0 4
Macrepistius	1 0 1 1 2 2 1 2 2 1 1 1 1 0 0 1 0 2 2 ? 1 2 1 2 3 2 1 0 0 1 1 1 1 1 0 0 ?

Appendix 2

Characters used in Halecomorph Analysis

1. Symplectic–lower jaw articulation; absent (0) or present (1)
2. Parietal; paired (0) or median (1)
3. Dermosphenotic descending lamina absent (0) or present (1)
4. Symplectic; not bound (0) or bound (1) to preoperculum
5. Parasphenoid ascending process; does not contact (0) or contacts (1) sphenotic
6. Intercalar; does not contact (0) or contacts (1) parasphenoid
7. Supraotic; absent (0) or present (1)
8. Vertebral centra; absent (0); with lateral pits (1); drumlike, "amiid" (2); teleostean (3)
9. One-to-one hypural–fin ray arrangement; absent (0) or present (1)
10. Ural neural spines; many (0) or only one, two, or three (1)
11. Caudal diplospondyly; present (0) or absent (1)
12. Opisthotic; present (0) or absent (1)
13. Pterotic; present (0) or absent (1)
14. Endochondral bone in intercalar; present (0) or absent (1)
15. Parietals; shorter than dermopterotic (0) or of comparable length (1)
16. Lateral ethmoid; meets (0) or does not meet (1) parasphenoid
17. Articular; single (0) or double Bridges ossicles (1)
18. Infraorbitals below orbit; two (1) or more than two (0)
19. Ural neural arches and spines; longer than deep (0) or short, blocklike (1)
20. Coronoids; two (1) or more than two (0)
21. Lateral line ossicles between caudal fin rays; absent (0) or present (1)
22. Rostral bone; platelike (0) or V-shaped (1)
23. Lateral horns of rostral bone; absent (0), short (1), or long (2)
24. Maxilla; without (0) or with (1) branch of infraorbital sensory canal
25. Coronoid process consisting mainly of angular and prearticular (1) or not (0)
26. Maxilla; long, extending behind orbit (0) or short, not extending behind orbit (1)
27. First infraorbitals; deeper than long (1) or not (0)
28. First infraorbitals; depth more than twice length (1) or less (0); note that this character defines a subset of character 31
29. Extrascapular series; simple morphology (0) or complex (1)
30. Lower margin of last infraorbital; not inclined (0) or inclined posterodorsally (1)
31. Orbitosphenoid reduced (1) or not reduced (0)
32. Ural arches–centra; eight or more (0) reduced number (1)
33. Lamina of dermosphenotic free distally (1) or fused to sphenotic (0)

Data Matrix for Halacomorph Analysis

Pholidophorus
0000000300 0000000000 0000000000 000
Watsonulus
1000?0?00? ?000000000 0000000000 000
Amia
1011100211 0111111000 0100000000 100
Calamopleurus
1001111211 0111111000 0120000000 000
Caturus
101111?000 0110000111 0120000000 000
Ionoscopus
1011111100 1000000000 0120100000 001
Oshunia
1011111100 1000000000 0120111001 001

Ophiopsis
?01??1?20? ???0000000?? 1111?11101 001
Macrepistius
101111?20? 00?0000070 111?011101 001
Brachyichthys
1011???00? ????0000?? 1110011001 ?01
Heterolepidotus
101?01?00? ?0?000007?1 ?120011001 001
Ikechoamia
?1?????200 1???1???0? 0??0?0001? ?1?
Sinamia
?1??1172?? 1111100??? 0120000010 01?

Character 11—caudal diplospondy was omitted from the subsequent PAUP analysis (see Fig. 4).

Appendix 3

Origin of Data Used in Molecular Analyses

Gene	Species	GenBank	Reference
COI	*Amia calva*	M64885	Normark *et al.* (1991)
	Lepisosteus oculatus	M64897	Normark *et al.* (1991)
	Oncorhynchus mykiss	L29771	Zardoya *et al.* (unpublished)
	Polyodon spathula	M64908	Normark *et al.* (1991)
COII	*Amia calva*	M64886	Normark *et al.* (1991)
	Lepisosteus oculatus	M64898	Normark *et al.* (1991)
	Oncorhynchus mykiss	L29771	Zardoya *et al.* (unpublished)
	Polyodon spathula	M64909	Normark *et al.* (1991)
Cytb	*Amia calva*	M64887	Normark *et al.* (1991)
	Lepisosteus oculatus	M64899	Normark *et al.* (1991)
	Salmo trutta	M64918	Normark *et al.* (1991)
	Polyodon spathula	M64910	Normark *et al.* (1991)
D1	*Amia calva*	Z18672	Lê *et al.* (1993)
	Lepisosteus platyrhynchus	Z18680	Lê *et al.* (1993)
	Oncorhynchus mykiss	Z18683	Lê *et al.* (1993)
	Polypterus retropinnus	Z18766	Lê *et al.* (1993)
D2	*Amia calva*	Z18697	Lê *et al.* (1993)
	Lepisosteus platyrhynchus	Z18706	Lê *et al.* (1993)
	Oncorhynchus mykiss	Z18709	Lê *et al.* (1993)
	Polypterus retropinnus	Z18714	Lê *et al.* (1993)
D8	*Amia calva*	Z18726	Lê *et al.* (1993)
	Lepisosteus platyrhynchus	Z18734	Lê *et al.* (1993)
	Oncorhynchus mykiss	Z18760	Lê *et al.* (1993)
	Polypterus retropinnus	Z18741	Lê *et al.* (1993)
18S	*Amia calva*	X98836	this study
	Lepisosteus osseus	X98837	this study
	Salmo trutta	X98839	this study
	Polyodon spathula	X98838	this study

Note: Abbreviations for gene fragments are as in the text.

Appendix 4

Details of Most Parsimonious Trees Shown in Figure 7
Using Various Combinations of Molecular Data

Data	Aligned positions	Informative positions	Tree length	CI	RI
COI	440	27	180	0.711	0.593
COII	218	23	103	0.649	0.381
Cytb	306	30	149	0.674	0.464
mtDNA	964	80	440	0.636	0.382
D1	373	7	72	0.700	0.571
D2	142	6	76	0.667	0.500
D8	72	3	30	0.600	0.333
28S	587	16	178	0.667	0.500
18S	1805	23	164	0.710	0.591
nucDNA	2389	39	342	0.691	0.553
totDNA	3353	117	799	0.632	0.386

Note: CI, consistency index excluding uninformative positions; RI, retention index.

7

Teleostean Monophyly

MÁRIO C. C. DE PINNA
Departamento de Zoologia
Instituto de Biociencias
Universidade de Sao Paulo
Sao Paulo-SP 05422-970, Brazil

I. Introduction

The Teleostei is generally considered a model group for the application of cladistic methods and was among the earliest of zoological groups to be treated by modern phylogenetic reasoning. The study of teleostean relationships accelerated in the mid-1960s, slightly preceding the onset of cladistics. A series of papers from that period focused on the Teleostei and its major subgroups (e.g., Gosline, 1961, 1965; Greenwood *et al.*, 1966; Monod, 1967; Nelson, 1969a,b, 1973a,b; Patterson, 1973, 1975; Patterson and Rosen, 1977). Others dealt with relationships within subgroups or with specific regions of the teleost cladogram (e.g., Gosline, 1960; Weitzman, 1967; Winterbottom, 1974; Rosen and Patterson, 1969; Rosen, 1973, 1974), and vigorous research on teleostean phylogeny continues (e.g., Fink and Fink, 1981; Rosen and Parenti, 1981; Fink and Weitzman, 1982; Patterson, 1984; Fink, 1984; Rosen, 1985; Grande, 1985; Stiassny, 1986; Johnson, 1992; Johnson and Patterson, 1993; Arratia, 1994; Patterson and Johnson, 1995). Molecular studies dealing directly with teleostean relationships are still few (e.g., Normark *et al.*, 1991; Lê *et al.*, 1993; Lecointre, 1993; Lee *et al.*, 1995), but their number is likely to increase markedly.

As one might expect, the Teleostei has been broadly accepted as a well-corroborated monophyletic group for nearly 30 years. However, a look through some general works (e.g., Lauder and Liem, 1983; Nelson, 1994) reveals a disappointingly short list of teleostean synapomorphies. This situation results in part from the fact that characters have been proposed over a relatively long period in numerous independent publications, and these have not been critically compiled. Also contributing to this unsatisfactory situation is the existence of several poorly known fossil species near the base of the teleostean cladogram. Such taxa make it difficult to ascertain the exact level of generality of certain synapomorphies, and as a result they have tended not to be presented as teleost synapomorphies. Additionally, some characters of soft anatomy that support the monophyly of living teleosts have remained buried in the literature, perhaps mainly because of their lack of relevance in settling controversies about the position of basal fossils. Nonetheless, these "soft anatomy" characters provide important independent evidence when comparing living teleosts with *Amia*, *Lepisosteus*, chondrosteans, and *Polypterus*. Information of this sort is becoming all the more meaningful in view of the growing utilization of molecular data in phylogeny reconstruction, also available only in Recent taxa.

This chapter reviews evidence bearing on teleost monophyly and determines the limits and composition of that group.

II. A Note on Fossils and Names

The long ladder of fossil forms at the base of the teleostean cladogram (cf. Fig. 1; Patterson, 1977a, fig. 19; Patterson, 1994, fig. 8; in a simplified form, also

Copyright © 1996 by Academic Press, Inc.
All rights of reproduction in any form reserved.

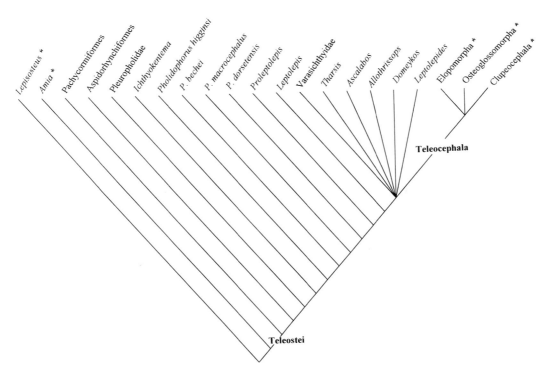

FIGURE 1 Cladogram of basal teleost relationships *Amia* and *Lepisosteus*, compiled from Patterson (1973, 1977a, 1994); Patterson and Rosen (1977) and Arratia (1991, 1995). *, indicates taxa with recent representatives. According to some hypotheses (e.g., Patterson, 1977a), *Leptolepides* may be a member of Teleocephala.

Patterson and Rosen, 1977, fig. 54; Lauder and Liem, 1983, fig. 14; Arratia and Schultze, 1987, fig. 20) shows how gradual the transition between halecomorphs and the remaining halecostomes actually is when all known taxa are taken into account. This situation poses a problem when attempting to list teleostean synapomorphies. When a character set is given as a synapomorphy list for a supraspecific taxon, the composition of the taxon must be established first. This has no phylogenetic connotation but simply refers to the level of generality at which a name will apply given a particular cladogram. Historically, the composition of the Teleostei has changed since its original conception by Müller (1844). While the inclusion of living teleosts in the group has been relatively consistent, the number of fossil taxa included has varied. Subsequent to the adoption of cladistic methods, there has been a tendency to include more fossil forms within the Teleostei. This results largely from the hypothesis that the Holostei was not a monophyletic group (Patterson, 1973, 1977a), an idea widely, but not universally, accepted (cf. Patterson, 1994).

Within a phylogenetic framework, the level at

which the name "Teleostei" is applied is arbitrary when all basal fossils are considered. The large morphological gap between Recent teleosts and living non-teleosts, such as *Amia* and *Lepisosteus*, provides an intuitive "gap" to separate teleosts from nonteleosts. However, that gap has gradually faded as forms such as pachycormiforms, aspidorhynchiforms, ichthyodectiforms, and non-monophyletic pholidophorids and leptolepids (Patterson and Rosen, 1977; Patterson, 1977a) were found to be more closely related to teleosts than to any so-called holosts. Such fossils were included among teleosts to keep them from a classificatory "paraphyletic limbo." While that tendency was fully justified, it created a problem with the definition of the name Teleostei. Awareness of this situation is not new and, for example, Regan (1923), Woodward (1942), Berg (1940), Bertin and Arambourg (1958), and Lehman (1966) considered that teleosts and holosts either could not be separated at all or could only be separated arbitrarily. Such views are historically related to the "teleostean polyphyly" tradition discussed below (see Discussion), but in part they also rely on another, independent point. The problem of determining the exact level where the name Teleostei applies stems from a general problem in biological classification.

In the case of teleosts, diverging opinions may concern two independent issues: (1) the role of fossils in phylogenetic analysis (analytical methodology) and (2) the application of taxonomic names to monophyletic groups (nomenclatural convention).

The first issue has been extensively discussed in the literature (e.g., Patterson and Rosen, 1977; Patterson, 1981; Rosen *et al.*, 1981; Gauthier *et al.*, 1988; Donoghue *et al.*, 1989). The trend in teleost systematics has always been to emphasize classifications based on living taxa, and that was already evident long before the debate in a cladistic context came to the fore. For example, Gosline (1961, p. 1) wrote,

knowledge of fossil fishes is slow in forthcoming, and the nature of the fossil record is such that many crucial data will probably never be forthcoming at all. Thus a good proportion of teleostean classification will always have to depend upon such information as can be gathered from modern fishes. In any event, the ichthyologist working on existing forms can do much to point out what fishes and what structural features need particular attention if they ever are located among fossils.

An equally pertinent statement was made by Greenwood *et al.* (1966, p. 347) a few years later when they wrote, "Paleoichthyologists who deal extensively with teleostean fossils are quite aware that the classification of living teleosts must be understood before the fossil record can be properly interpreted." Interestingly, Patterson (1977a, p. 609) wrote that their statement, ". . . rather than imperfections in the fossil record, gives the true reason why the contribution of paleontology to teleostean phylogeny had so far been insubstantial."

Few would deny that information from fossils can be important in facilitating the recognition of morphological homologues no longer evident in living taxa (Gauthier *et al.*, 1988; Donoghue *et al.*, 1989; an opinion presented earlier in Patterson, 1977a). Also, as any additional taxon (Stiassny and de Pinna, 1994), fossils may show unique combinations of character states which can have a profound impact on hypotheses of the relationships of living organisms.

The limitation of fossils, of course, is that they normally do not allow a complete character survey because of limitations in preservation. For example, most soft anatomy cannot be checked in fossil teleosts and can only be considered as evidence for the monophyly of living members of the group. Depending on the amount of missing data, the inclusion of fossils in an analysis involving living taxa may do more harm than good to the final hypothesis (Forey, 1992).

The second issue raised refers to the name Teleostei and its limits. Two of the few other taxon names whose composition is more complex than the Teleostei (as a result of the inclusion of fossil lineages) are the Mammalia and the Aves. The seemingly endless dilemma of the membership of Mammalia was recently reviewed by De Queiroz (1994), and most of the reasoning therein could be applied just as well to the Teleostei. Briefly stated, according to De Queiroz, the problem stems from a residue of essentialist thinking in taxonomy, in which the definition of a taxon is based on a description of the traits of individual organisms. De Queiroz contrasts that view with the alternative, regarded by him as nominalist, in which the definition of a taxon is stated in terms of relationships rather than organismal traits. The nominalistic approach is viewed by De Queiroz as more in tune with the phylogenetic paradigm in systematics and nomenclature. De Queiroz (1994) considers that an "apomorphy-based" delimitation of taxa (essentialistic) is inferior to the alternatives, namely "node-based" and "stem-based." Those three methods were originally discussed by Hennig (1969) and have been further elaborated upon by De Queiroz and Gauthier (1990).

The controversy with the name Aves (see Patterson, 1993a,b; Norell *et al.*, 1993) in connection with the name Dinosauria nicely illustrates the distinction between the other two alternatives (node-based and stem-based). Traditional Dinosauria is a paraphyletic group because some of the included taxa are more closely related to birds than to other dinosaurs. Two alternatives are then possible: (1) extend the name Aves downwards to include taxa previously assigned to dinosaurs (stem-based or total group) and (2) expand the name Dinosauria up to include Aves (node-based, or crown group). In both cases only monophyletic groups are recognized. In the second case, the meaning of dinosaurs is modified to include birds, so that birds become a kind of dinosaur. An equivalent move in neopterygians would be to expand the name Holostei to include the Teleostei, so that teleosts would be a kind of holost (that, of course, considers Patterson's 1973 hypothesis that the traditional Holostei is a paraphyletic group).

The history of teleostean classification clearly leans toward a stem-based definition; teleosts have seldom been seen as a kind of holost. Fossils phylogenetically located between the two have been assigned to the Teleostei, expanding the limits of the group, while the Holostei has been abandoned in formal classification. One point about which there is general agreement is that *Amia* and *Lepisosteus*, the closest living relatives of teleosts, are not teleosts. The consensus on this last issue provides a fulcral point to establish the limits of the Teleostei.

The Teleostei is here defined, in accordance with the prevailing meaning, as the largest (i.e., most inclusive) actinopterygian clade not including either the Halecomorphi (*Amia* and close relatives) and/or the Ginglymodi (*Lepisosteus* and close relatives). The definition is in the negative form, so that new fossils eventually found between halecomorphs and the most basal pachycormiformes, may be accommodated within the Teleostei. Also, the definition minimizes problems with ongoing controversies about the relative position of *Amia*-like and *Lepisosteus*-like taxa relative to each other and to teleosts (diverging views are exemplified by Jollie, 1984b; Patterson, 1973; Olsen, 1984; Olsen and McCune, 1991; Normark *et al.*, 1991; a recent summary of the controversy is presented in Patterson, 1994).

III. Characters

In view of the foregoing, the characters described in this chapter are those whose minimum level of generality is for all living teleosts, i.e., they are synapomorphies either for a group composed of all living teleosts or for a larger group including taxa down to the node immediately above the common ancestor with Halecomorphi and/or Ginglymodi. The exact level of applicability, when known, is discussed in the individual sections. Some characters occur only in fossil lineages in the basal portions of the teleostean tree and are reversed or modified beyond meaningful comparison farther up the tree. Some such characters have, on occasion, been considered as teleostean synapomorphies. However, their value as indicators of teleost monophyly depends on the cladistic structure of the basal portion of the teleost cladogram, and to date this has been investigated by relatively few authors. A quantification of all the evidence bearing on basal teleostean fossils has yet to be made, and their precise relationships cannot be considered well-established at present. Uncertainty about relationships at that level results in uncertainty about the phylogenetic significance of characters that occur only there. Also, a number of synapomorphies proposed for the base of teleostean cladogram involve fusions of bones well-documented in fossils (Patterson, 1975) but not yet adequately studied in recent forms. To be conservative, characters of that kind are not included in the presented evidence, and space constraints do not allow a discussion of them in this chapter. The reader, however, is referred to Patterson (1977a) for a summary of characters in basal fossils that are omitted from this discussion.

In the following section, "Recent teleosts" (Fig. 1) refers to the group composed of osteoglossomorphs, elopomorphs, and clupeocephalans, independent of their fossil or extant status. The formal name Teleocephala is coined for the clade and used interchangeably with Recent teleosts throughout the text. The names "pholidophoriforms" and "leptolepiforms" are used informally, since these groups are demonstrably not monophyletic (Patterson and Rosen, 1977; Patterson, 1977a).

A. Caudal Skeleton

Much of the evidence bearing on teleost monophyly and relationships comes from the caudal skeleton. The internal supports of the caudal fin have been important in characterizing teleosts since Heckel (1850), who defined the group as including fishes in which the posterior end of the notochord is roofed by uroneurals or enclosed in a terminal half-centrum. The caudal skeleton has been a source of useful systematic data ever since (e.g., Regan, 1910; Hollister, 1936). However, it was with Gosline (1960, 1961) and Nybelin (1963) that the detailed contemporary studies began, and these works set the standards that gave a renewed impulse to the study of the teleostean caudal skeleton. Major monographic works on that complex include the works of Monod (1967, 1968) and Fujita (1990). The structure of the caudal skeleton of actinopterygians has been studied recently in a series of papers by Schultze and Arratia (1986, 1988, 1989). Although the taxonomic coverage of those papers is not as wide as that of some other surveys (e.g., Monod, 1968), their level of anatomical detail and accuracy is unprecedented. In particular, the ontogenetic data presented by Schultze and Arratia unveiled a new set of morphological information relevant for understanding the phylogenetic implications of caudal-skeleton characters.

(1) Diural Caudal Skeleton

The distinction between ural and preural centra is based on the branching point of the caudal artery (Nybelin, 1963; Monod, 1967, 1968). Vertebrae posterior to that point are considered ural (starting with U1, counting posteriorly), and those anterior to it are considered preural (starting with PU1, counting anteriorly). Inferences about the homology of numerous elements in the caudal skeleton depend on that single morphological landmark. Schultze and Arratia (1989) provided some other morphological correlations indicating that the boundary between preural and ural vertebrae is indeed homologous, at least in actinopterygians with a homocercal caudal fin.

The difference between polyural and diural caudal skeletons has been a classic character separating teleosts from non-teleost actinopterygians. This difference has been the object of numerous studies of the functional significance of heterocercal and homocercal caudal fins (e.g., Lauder, 1989). The distinction is simple: polyural skeletons have three or more ural centra, while diural skeletons have only two (Nybelin, 1963; Patterson, 1973). The diural condition primitively results from an ontogenetic fusion of centra, so that the so-called first ural centrum in teleosts is in fact a compound centrum formed by the first and second ural centra, visible only in juveniles (Greenwood, 1970; Patterson and Rosen, 1977; Schultze and Arratia, 1989). In some cases, the first ural centrum is indeed a single centrum, one of the primitive centra being lost (such as in *Albula*; Schultze and Arratia, 1988). The so-called second ural centrum in diural skeletons may be formed by a variety of fusions involving up to four juvenile centra, while in some other cases it is only a single centrum (Schultze and Arratia, 1989). Regardless of the ontogenetic pathways toward the diural structure, there is little doubt that the condition in adults is homologous and derived relative to the condition in more primitive actinopterygians.

The difference between polyural and diural is associated with a host of other characteristics related to homocercal (diural) versus heterocercal (polyural) caudal fins. In diural skeletons, the posteriormost ural centrum is located anterior to the bases of the principal caudal-fin rays, while in polyural skeletons, the bases of some rays are anterior to that of some ural centra. The diural skeleton is not a character confined to teleocephalans. According to Patterson (1977a), it is a synapomorphy for †*Leptolepis coryphaenoides*, ichthyodectiforms, †*Tharsis*, and teleocephalans.

A teleostean synapomorphy previously proposed by Patterson and Rosen (1977) reads, "caudal axis turned sharply upward at level of PU1." This condition is opposed to the primitive, "gradually turned upward from PU3 to PU5." That trait is considered synonymous with the diural caudal skeleton, since the shortening associated with the diural condition necessarily results in an abrupt turn upward.

(2) Ural Neural Arches Modified as Uroneurals

Uroneurals are elongate paired bones located bilaterally along the dorsal region of the caudal skeleton. They are derived ontogenetically from the neural arches of ural centra and seem primitively to be preformed in cartilage (Schultze and Arratia, 1988, *contra* Patterson, 1968), except in acanthomorphs and the first one in clupeomorphs (other uroneurals in the latter group, however, are preformed in cartilage; Schultze and Arratia, 1989).

The presence of uroneurals is a synapomorphy for all teleosts, known fossils (pachycormiforms, aspidorhynchiforms, pholidophoriforms, leptolepiforms, ichthyodectiforms, and †*Tharsis*) included. The ural neural arches are unmodified in non-teleost actinopterygians. The value of uroneurals for diagnosing teleosts has been recognized for a long time and discussed on numerous occasions (e.g., Patterson, 1968, 1977a,b; Patterson and Rosen, 1977; Lauder and Liem, 1983; Schultze and Arratia, 1989).

In some groups, such as many clupeomorphs and ostariophysans, a uroneural is supposed to be modified into a pleurostyle (Monod, 1968). The number of uroneurals varies among teleosts, with a tendency for reduction towards the Recent members of the group (Patterson, 1968). There may be as many as seven uroneurals in fossil teleosts (e.g., †*Leptolepis coryphaenoides*; Patterson, 1968) and up to four in living taxa (*Hiodon* and *Alepocephalus*; Patterson, 1968; Schultze and Arratia, 1988), with many forms having only one or none. The phylogenetic significance of these numbers is difficult to determine because uroneurals are lacking in outgroups to the Teleostei.

(3) Division of Hypurals into Dorsal and Ventral Groups

This characteristic was noted by Monod (1968, p. 591) and explicitly presented as a character common to all teleosts. The trait has been largely ignored as a potential teleostean synapomorphy, even though its status as such seems to be valid. In all teleosts, the hypural series is clearly separated into a dorsal and a ventral group, and the caudal-fin rays are similarly separated. Such a division is absent in *Amia* and other non-teleostean actinopterygians. Even in cases where there is extensive hypural fusion, it is still usually possible to separate a dorsal and a ventral region of the compound hypural plate, due either to a notch in the corresponding line of division and/or to a hiatus in the sequence of caudal-fin rays.

(4) First Two Hypurals Supported by a Single Centrum

Hypurals one and two in non-teleost actinopterygians are supported by a separate ural centrum (one and two, respectively). In teleosts, the first two hypurals are attached to (or articulate with) the same vertebral centrum, namely ural centrum 1 (Patterson and Rosen, 1977). Schultze and Arratia (1989) have shown that the traditional first ural centrum in teleosts results from an ontogenetic fusion of two centra, which they call ural centra 1 and 2. Accordingly, the traditional ural centrum 2 is in fact ural centrum 3, or a fusion of 3 plus a varying number of more posterior centra. Consequently, the present character may also be expressed as the fusion between the first two ural centra (*sensu* Schultze and Arratia, 1989), normally without a corresponding fusion of their hypurals.

Pachycormiforms, aspidorhynchiforms, pleuropholids, and pholidophoriforms show the primitive condition for this character (Patterson, 1977a), but leptolepiforms, ichthyodectiforms, and †*Tharsis* show the derived condition along with Recent teleosts.

(5) Number of Epurals

Epurals are traditionally defined as unpaired independent ossifications located between the last preural neural spine and the dorsal margin of the caudal skeleton (Monod, 1968). Schultze and Arratia (1989) proposed that the structures called "epurals" in various actinopterygians are not necessarily homologous. They restricted the term epural to structures ontogenetically derived from detached neural spines, as seen in *Lepisosteus*, *Hiodon*, and *Elops* (Schultze and Arratia, 1989) and most other Teleocephalans. In *Amia*, by contrast, only the posteriormost element may actually be a true arcual epural (Scultze and Arratia, 1989). Since the identification of a true epural relies mostly on ontogenetic data, it is difficult to verify in fossils. Even for most Recent teleosts evidence of such ontogenetic information is still unavailable. The position of a supposed epural between neural spines may be an indication that it is not a true epural (Schultze and Arratia, 1989). However, the loss or fusion of an epural-bearing centrum (a common phenomenon; see Schultze and Arratia, 1989) without corresponding loss of its epural may result in a true epural located between neural spines. Considering the problems in identifying true epurals and the chronic scarcity of developmental data, and also that ontogeny is not necessarily the ultimate criterion of homology, we are forced to maintain the admittedly coarse-grained definition of epural. Still, even given these limitations, there seems to be some congruent information in the number of epurals in actinopterygians.

The number of epurals, *sensu lato*, in non-teleost actinopterygians is normally eight or more and is considered plesiomorphic for the group. The reduction in the number of epurals is a trend within the teleostean clade (Patterson, 1973), and several states of reduction can be recognized (Patterson, 1977a), making this a multistate character. If the states are ordered according to similarity, with the condition in outgroup actinopterygians as the plesiomorphic end of the series

(State 0), then the following derived states can be recognized:

State 1: seven or fewer—All teleosts, including the fossil Pachycormiformes, Aspidorhynchiformes, Pholidophoriformes, Pleuropholidae, Leptolepiformes, Ichthyodectiformes, and *Tharsis*.

State 2: six or fewer—All teleosts except Pachycormiformes and Aspidorhynchiformes.

State 3: five or fewer—All teleosts except Pachycormiformes, Aspidorhynchiformes, and Pleuropholidae (this state is homoplastically present in most amiids, including *Amia calva*).

State 4: four or fewer—All teleosts except Pachycormiformes, Aspidorhynchiformes, Pleuropholidae, and Pholidophoriformes.

State 5: three or fewer—All teleocephalans, plus †*Tharsis*, Ichthyodectiformes, and *Leptolepis coryphaenoides*.

(6) Seven or Fewer Hypurals

Amia, *Lepisosteus*, and other primitive actinopterygians normally have eight or more hypurals in the caudal skeleton. High hypural counts are also observed in basal fossil teleosts. In teleocephalans, the maximum number of hypurals is seven (Patterson and Rosen, 1977). Schultze and Arratia (1988) reported six to eight hypurals in adult *Hiodon*, seven being most common. In juveniles of that genus (22–38 mm SL), however, eight hypurals are always present; seven are ossified, and an eighth is cartilaginous (Schultze and Arratia, 1988, pp. 278–279). The eighth hypural is assumed to disappear during ontogeny in most individuals. This observation confirms the hypothesis that seven or fewer hypurals are derived relative to the condition of eight or more.

B. Branchial Arches and Hyoid Apparatus

(7) Presence of Two Ossified Hypohyals

In sarcopterygians and basal actinopterygians, the hyoid arch has a single hypohyal. Such is the condition in *Polypterus* (Arratia and Schultze, 1990), chondrosteans (McAllister, 1968; Grande and Bemis, 1991), lepisosteids (Wiley, 1976), and various paleonisciform fossils (Watson, 1925; Lauder, 1980; Gardiner, 1984a).

In nearly all teleosts, Recent and fossil, there are two hypohyals, commonly termed dorsal and ventral (Arratia and Schultze, 1990). The two hypohyals may be similar or different in size, and in the latter case, either the dorsal or the ventral hypohyal may be the larger. A few teleosts have a single hypohyal, a condition hypothesized to be a result of reversals. The most notable case is in osteoglossomorphs, where two hypohyals are present only in *Hiodon*. In all other osteoglossomorphs (including the fossil lycopterids) there is a single hypohyal, supposedly the dorsal one as judged from the ligamentous connection of the urohyal directly with the anterior ceratohyal (Arratia and Schultze, 1990). The condition of the hypohyals in the problematic fossil family Ostariostomidae, arguably related to osteoglossomorphs (Grande and Cavender, 1991; Li and Wilson, this volume) is unknown in detail, although at least one ossified hypohyal is present in these fish (Grande and Cavender, 1991).

A single hypohyal is also observed in some siluriforms, such as in the families Amblycipitidae, Trichomycteridae, Callichthyidae, Scoloplacidae, Loricariidae, and Astroblepidae. Other catfishes, however (including the basal Diplomystidae and Hypsidoridae), have two ossified hypohyals, indicating that a single hypohyal is a reversal within the order.

In *Amia*, there is a single hypohyal for most of the ontogenetic cycle. This single hypohyal has a hard cartilage cap on its dorsal region. Arratia and Schultze (1990) observed that in large specimens of *Amia*, a small ossification appears in the posterodorsal portion of the cartilage. Whether this ossification is actually a dorsal hypohyal or a spurious senescent calcification is at present unknown. Grande (personal communication) did not find a second hypohyal in any amiids examined, fossil or recent. If it is indeed an incipient hypohyal, then the presence of the two bones becomes a putative synapomorphy for *Amia* plus teleosts. However, even under this interpretation, the appearance of two hypohyals very early in ontogeny can still be considered a synapomorphy for teleosts. According to Patterson (1977a), two hypohyals are seen in the fossil taxa †*Tharsis*, ichthyodectiforms, and †*Leptolepis coryphaenoides*.

(8) Urohyal Formed as an Unpaired Tendon Bone

The urohyal in teleosts is a unique median structure formed as an ossification of the tendon of the sternohyoideus muscle (Arratia and Schultze, 1990). The first part to differentiate from the ligament is the dorsal urohyal crest. In later ontogenetic stages, the sternohyoideus becomes paired, and the urohyal extends into the myoseptum between the two muscles.

There is reason to believe that the teleostean urohyal is not homologous to a bone bearing the same name in other actinopterygians. Unpaired structures called "urohyals" exist in at least one fossil dipnoan (Miles, 1977; but in no extant ones, Jarvik, 1963) and in fossil and Recent coelacanths (Stensiö, 1921; Jarvik, 1963; Millot and Anthony, 1958). In all these cases,

however, the structure seems to be formed as a cartilage bone (Jarvik, 1954, 1963; Millot and Anthony, 1958; Miles, 1977).

There is some controversy regarding the structure of the peculiar urohyal in the fossil †*Pholidophorus*. According to the anatomical interpretation of Patterson (1977b), the urohyal in that genus is formed by fusion of a cartilage bone with a toothed interclavicle. If this interpretation is correct, then the pholidophorid "urohyal" would be unique to the group. However, as pointed out by Arratia and Schultze (1990), presence of an interclavicle in pholidophorids is difficult to accept phylogenetically since the bone seems to be absent in all other Neopterygii. The tooth-bearing portion in the †*Pholidophorus* urohyal is best considered as a dermal neomorph. Considering the anatomical information available, it is possible that the urohyal in that genus is formed as a tendon bone and is therefore homologous to that in other teleosts.

In *Polypterus*, there is a median Y-shaped tendon bone and a pair of elongate anterior tendon bones ligamentously connected to it (Gegenbaur, 1898; De Beer, 1937; Jarvik, 1954; Arratia and Schultze, 1990). This bone was called a urohyal by Jollie (1984a). The Y-shaped bone is connected to the sternohyoideus muscle, and anteriorly each of its arms is ligamentously connected to one elongated bone, itself attached to the hypohyal of its side. It is unclear whether any component of that complex may be considered homologous to the urohyal in teleosts. Phylogenetically, such is unlikely since all these structures are absent in clades closer to teleosts (see below). On the other hand, the bones in *Polypterus* are tendon bones, and the Y-shaped bone, in particular, is unpaired (although De Beer, 1937, stated that they were probably of paired origin). Even though morphologically there is not much reason to discard homology, such an interpretation would be unparsimonious in view of currently accepted phylogeny of actinopterygians.

There is no urohyal of any sort in known palaeonisciforms, chondrosteans, *Lepisosteus*, and *Amia*. Hammarberg (1937) reported a small cartilage in a position similar to that expected for a urohyal in juvenile *Lepisosteus*. Those observations, however, were not confirmed by Arratia and Schultze (1990) in specimens of similar, smaller, or larger size.

(9) Presence of Foramen in the Dorsal and Ventral Hypohyals for Passage of the Hyoidean Artery

In primitive actinopterygians, the afferent hyoidean artery runs along the lateral margin of the ceratohyal (Nielsen, 1942; Gardiner, 1984a) without passing through the hypohyals (a condition observed in *Amia* by Allis, 1897). In teleosts, contrastingly, the artery pierces one or both hypohyals through a foramen (Arratia and Schultze, 1990). Much variation exists in the path of the hyoidean artery through the hypohyals in teleosts (Arratia and Schultze, 1990), and more data are needed for many groups, especially for extinct basal teleosts. However, information at hand supports a preliminary hypothesis that the arterial passage through some area of the hypohyals is synapomorphic for teleosts.

(10) Independent Endoskeletal Basihyal Present

Living chondrosteans, *Amia*, and *Lepisosteus* all lack a separate basihyal, an unpaired ossification anterior to the basibranchial series. A basihyal is present in the majority of Recent teleosts (Nelson, 1969a). According to Patterson (1977a) the basihyal is absent in fossil lineages below the osteoglossomorph level and is therefore a synapomorphy for teleocephalans.

(11) Symplectic

The symplectic is a rod-shaped bone fitting into a groove between the inner face of the quadrate and its quadratojugal spine. In teleosts and "holosts" the symplectic is probably homologous. However, it is uncertain whether it is also homologous with the bone bearing the same name in other actinopterygians (Patterson, 1973). The condition in teleosts is quite distinctive when compared to that in "holosts." In *Lepisosteus*, the symplectic is located posterior to, and distant from, the quadrate. In *Amia*, it is adjacent to the quadrate, but the two bones contact without modification of their surfaces. In both holostean genera, the symplectic contacts the preopercle. However, in teleosts the symplectic does not contact the preopercle and fits into a deep groove on the inner face of the quadrate, a space supposed to be the primitive separation between the quadrate and quadratojugal (Patterson, 1973; Véran, 1988).

The fusion of the quadrate and the quadratojugal has itself been suggested as a separate diagnostic character for teleosts (Patterson, 1973; Véran, 1988). This proposition relies mainly on the idea that the spine-like posterodorsal process of the quadrate in teleosts corresponds to the primitive quadratojugal of other actinopterygians such as *Lepisosteus*, first proposed by Allis (1909) (see also Gardiner *et al.*, this volume). Actual ontogenetic fusion, however, has been poorly documented so far (cf. Patterson, 1973, pp. 248–250). Additionally, the quadratojugal is probably absent in *Amia*, at least as a separate ossification, further confusing the phylogenetic interpretation of the possible fusion.

(12) Median Tooth Plate Covering Basibranchials 1–3

The tooth plates covering the basibranchials in primitive actinopterygians are either disposed in pairs or distributed asymmetrically. In all teleosts, Recent and fossil (Nelson, 1969a), except for pachycormiforms (Patterson, 1977a), these toothplates are median. This character is at the node joining all teleosts exclusive of pachycormiforms (i.e., aspidorhynchiforms and higher). L. Grande (personal communication) finds that median basibranchial toothplates are also present in *Amia*. If this is confirmed and observed in other amiiforms, then the level of generality of this character will have to be increased.

(13) Four Pharyngobranchials

Non-teleostean actinopterygians and other teleostomes have a maximum of three pharyngobranchials in the gill-arch skeleton (Nelson, 1969a; Patterson, 1973; Grande and Bemis, 1991), and teleosts are unique in having four such elements (Patterson, 1973, 1977b). Pharyngobranchials are poorly-known in fossils because of the usually limited preservation of branchial arches in general. Four pharyngobranchials may be considered a synapomorphy for teleocephalans only.

C. Jaws

(14) Mobile Premaxilla

The premaxilla in non-teleost actinopterygians is immovably attached or ankylosed to other bones of the anterior part of the neurocranium, in particular to the rostral. In all teleosts, fossil and Recent, the premaxilla is free or nearly so, and it is capable of motion relative to surrounding bones (Regan, 1923; Patterson, 1973, 1975). Arratia and Schultze (1987) were uncertain about the mobility of the premaxilla in the fossil *Atacamichthys*. However, their illustrations (1987, figs. 2, 6 and 7) indicate that the premaxilla in that genus is not in the plesiomorphic fixed condition of primitive actinopterygians. Patterson (1977a) indicated that a mobile premaxilla characterizes all teleosts, living and fossil, except for Pachycormiformes, Aspidorhynchiformes, and Pleuropholidae.

(15) Lower Jaw Without Coronoid Bones

The coronoid bones are small, platelike dermal bones, normally bearing teeth, located medial to the marginal tooth row of the dentary. Coronoids are present in nearly all non-teleost actinopterygians, and in *Latimeria* and *Polypterus* one of the coronoid bones plays a prominent role in the structure of the jaw, forming a large coronoid process (Nelson, 1973b). Coronoid bones are absent in teleosts. Nelson (1973b) found no evidence to support possible fusion of coronoids to the teleostean dentary. Even if that were the case, however, the phylogenetic implications of the character would remain the same, as a synapomorphic fusion exclusive to teleosts. Coronoid bones are also absent, supposedly independently, in living chondrosteans.

(16) Lower Jaw Without Supraangular Bone

The supraangular is a small dermal bone located at or near the dorsal margin of the coronoid process of the lower jaw. It is not homologous to the surangular of sarcopterygians, which is a canal bone (Jarvik, 1967). *Amia* and *Lepisosteus* both have a supraangular, but *Polypterus* lacks it (Nelson, 1973b). Patterson (1977a) states that the supraangular is absent in teleocephalans, †*Tharsis*, ichthyodectiforms, and †*Leptolepis coryphaenoides*.

D. Neurocranium

(17) Posterior Myodome Extending into Basioccipital

The posterior myodome is a space in the orbitotemporal and otic regions which occurs in the skull of many teleostomes. It accommodates the posterior portion of the recti muscles of the eye, which extend posteriorly to, or beyond, the area of the basisphenoid ventrolateral to the pituitary. In *Amia*, the posterior myodome is restricted to the prootics. In teleosts, contrastingly, the myodome extends well into the basioccipital (Patterson, 1973). The well-developed myodome in *Amia* and teleosts has been considered as supporting the hypothesis that the two form a monophyletic group (Patterson, 1973, p. 254), since the structure is poorly developed or absent in *Lepisosteus*. Among more primitive actinopterygians, the myodome is absent in *Polypterus*, the fossil †*Moythomasia*, †*Cheirolepis*, and †*Mimia* (Gardiner and Schaeffer, 1989) but present in more distal "palaeoniscoids" such as †*Pteronisculus*, †*Boreosomus*, and †*Birgeria*, among others (Schaeffer and Dalquest, 1978). The myodome is absent in the chondrosteans *Acipenser* and *Polyodon* (unless it is represented by a very shallow pit, more noticeable in the latter; Gardiner and Schaeffer, 1989, p. 163).

The spotty distribution of the myodome, as well as the complex homology decisions in borderline situations, creates difficulties in assessing the relative polarities of the different conditions in various actinopterygians. These difficulties have been well-documented by Gardiner and Schaefer (1989, p. 163),

who showed that alternative interpretations of rudimentary "myodomes" can have profound impacts on hypotheses of relationship among basal actinopterygians. Nevertheless, the condition of the myodome in teleosts, extending into the basioccipital, is unique and stands as synapomorphic even in face of the uncertainty in other regions of the actinopterygian cladogram. Within teleosts, the myodome has been lost in a few groups, such as siluriforms and some gadids (Gardiner and Schaeffer, 1989). Among fossil teleosts, the basioccipital extension of the posterior myodome is present in all pholidophorids and leptolepids, ichthyodectiforms, and *Tharsis* but not in more basal lineages (Patterson, 1977a).

(18) Median Vomer

The vomer in *Amia* and *Lepisosteus* is a paired bone, as it is in dipnoans, coelacanths, and rhipidistians (if the bone called vomer in the latter three taxa is actually homologous to that in actinopterygians). A vomer is paired in living chondrosteans, and palaeoniscoids and very reduced *Polypterus* (Jollie, 1984a). In teleosts, the vomer is a median (= unpaired) structure. Exceptions within teleosts are some species of *Hiodon* and *Osmerus* (Patterson, 1975, p. 513), probably representing reversals within their respective clades. According to Patterson (1977b), a median vomer is not present in the basal teleost fossils Pleuropholidae, Aspidorhynchiformes, and Pachycormiformes, being thus a character for the group composed of *Ichthyokentema* and higher in his cladogram (Patterson, 1977b, fig. 19). However, Brito (1992, p. 149) more recently found a median vomer in the aspidorhynchiform *Vinctifer*. This indicates that the median vomer may be a character slightly more general within teleosts than so far believed.

(19) Frontals Expanded Posteriorly

Arratia and Schultze (1987, p. 9) observed that the posterior part of the frontal bone in primitive teleosts is distinctly broader than its anterior portion. The posterior widening can be gradual (e.g., †*Cladocyclus*) or abrupt (e.g., †*Atacamichthys*), but it is almost always evident. In amiiforms and most other primitive actinopterygians, the posterior part of the frontal is narrower than the anterior, or else the bone is of even width and rectangular (cf. Wenz, 1968). This derived condition applies to known "pholidophorids" and leptolepids, and probably its level is at or near the base of the Teleostei.

E. Other Characters

(20) Presence of Craniotemporal Muscle

The epaxial musculature is located dorsal to the horizontal septum in fishes. The anterior attachment of the epaxialis onto the skull is variable and is usually associated with crests and depressions in the various bones of that region (Winterbottom, 1974). Stiassny (1986) observed that in non-acanthomorph teleosts there is a well-differentiated anterior slip of epaxial fibers on each side, anteriorly attaching onto the posterodorsal margin of the dorsal limb of the movable posttemporal. That slip, which receives the name craniotemporalis (Le Danois, 1967; Stiassny, 1986) is absent in non-teleosts such as *Amia* and *Lepisosteus* (Stiassny, 1986). Within teleosts, it is also absent in acanthomorphs and most ostariophysans (Stiassny, 1986). In acanthomorphs there is often a marked epaxial encroachment on the neurocranium, and the posttemporal is firmly fixed to the neurocranium. These modifications may explain the absence of a craniotemporalis in that clade. A similar situation occurs among ostariophysans, except in characiforms, where the muscle is present (Stiassny, 1986). Both cases are probably autapomorphic reversals within teleosts.

Associated with the mobility of the posttemporal in most non-acanthomorph teleosts is the presence of a well-defined posttemporal-epioccipital ligament (Stiassny, 1986). This ligament may also be a potential teleost synapomorphy, pending more detailed studies.

(21) Presence of Two or More Postcleithra

The postcleithrum is a small dermal bone located on the posterior margin of the pectoral girdle. Basal actinopterygians have a single postcleithral element, located mesially near the cleithrum-supracleithrum limit or sometimes partly sandwiched between the two (Gottfried, 1989; Grande and Bemis, 1991). A single postcleithrum is seen in polypterids (Jollie, 1984a), lepisosteids (Jollie, 1984b), and *Amia* (Gottfried, 1989). Within chondrosteans, a single postcleithrum was reported in *Polyodon* by Gregory (1933); then it was given as absent by Jollie (1980) and Gardiner (1984b). Grande and Bemis (1991) showed beyond doubt that a single postcleithrum is actually present in all Recent polyodontids and probably in all fossil representatives of the group as well.

Teleosts have more than one postcleithral element. The usual number for teleocephalans is three (Gosline, 1980), but there is variation in both Recent and fossil taxa. Arratia (1984, 1987) has described six or seven postcleithra in the Jurassic †*Varasichthys*. More than three seems to be the condition in some other Jurassic forms such as †*Domeykos* and †*Protoclupea* (Arratia, 1994), while two are seen in †*Atacamichthys* (Arratia and Schultze, 1987). Five postcleithra are present in *Elops* (Arratia, 1984) and juvenile *Salmo* (Arratia and Schultze, 1987). Ctenosquamates (myctophiforms plus acanthomorphs; Rosen, 1973; Johnson, 1992; Sti-

assny, this volume) have two postcleithra. Because of such wide variation in the number among ingroup taxa, the exact primitive number of postcleithra for teleosts is uncertain (Gottfried, 1989). Whichever the case, the presence of more than a single postcleithrum is probably synapomorphic for the group.

Some teleostean lineages have lost a number of postcleithral elements, most notably osteoglossomorphs (usually with one postcleithrum; Taverne, 1977, 1978) and variably several clades within acanthomorphs (Gosline, 1980; Gottfried, 1989). Other groups lack postcleithra completely, such as siluriforms, engraulids, anguilliforms, and mormyrids. However, the phylogenetic position of most of these taxa is internested with that of other forms showing two or more postcleithra, clearly indicating that the reduced conditions are reversals. In some cases, the postcleithra are fused in adults but are still separated in juveniles (Johnson and Washington, 1987; Gottfried, 1989).

The situation in osteoglossomorphs deserves some comment. According to the prevailing hypothesis of teleostean relationships (Patterson and Rosen, 1977; Lauder and Liem, 1983), in which osteoglossomorphs are the sister group of all other teleocephalans, the status of two or more postcleithra as a teleostean synapomorphy depends on the knowledge of fossil taxa with two or more postcleithra (see previous discussion). However, it has also been proposed that elopomorphs are the sister group of all other recent teleosts (Arratia, 1991). If that is so, the presence of two or more postcleithra can be interpreted as a teleostean synapomorphy even in the absence of relevant fossils, provided the character is ACCTRAN-optimized on the teleostean cladogram (i.e., with a reversal to the primitive condition in osteoglossomorphs).

Postcleithral elements are poorly known in fossil teleosts because they are small and frequently obscured by larger surrounding bones. However, the several postcleithra present in a variety of basal fossil taxa (see previous discussion) indicates that the character is for a group more inclusive than that composed only of teleocephalans. However, the bone is poorly known in "pholidophorids," and its exact phylogenetic level is uncertain.

(22) Lateral Forebrain Bundle Composed of Myelinated Fibers

The lateral forebrain bundle, more technically known as fasciculus lateralis telencephali, is one of three distinct fiber systems which connect the diencephalon with the telencephalon in fishes. In polypteriforms, chondrosteans, *Amia*, and lepisosteids, the lateral bundle consists entirely of unmyelinated fibers. In teleosts, this bundle contains myelinated components (Nieuwenhuys, 1982). The state of this character in fossil teleosts is unknown.

(23) Rostrolateral Parts of Lobus Vestibulolateralis of Cerebellum Solid (Eminentiae Granulares)

The cerebellum of actinopterygian fishes is composed of three main parts: the corpus cerebelli, the valvula cerebelli, and the lobus vestibulolateralis (Nieuwenhuys, 1982). The last of these, the vestibulolateral lobe, is posteriorly continuous with the area octavolateralis, the most dorsal rhombencephalic zone. In polypteriforms, chondrosteans, and *Amia*, the rostrolateral parts of the vestibulolateral lobe are in the form of a pair of auriculae. Considering the distribution of this condition in primitive lineages, it can be considered as the primitive state within actinopterygians. In teleosts, contrastingly, the same structures are represented by solid bodies named eminentiae granulares (Nieuwenhuys, 1967, 1982). The solid condition of the vestibulolateral lobe can therefore be considered as synapomorphic for the Teleostei. It must be observed, however, that the structure of the actinopterygian brain has been studied with the necessary detail in only a few representative taxa. More data are needed in order to test the generality of this trait as a teleostean synapomorphy. Also, the condition of the character is unknown (and likely to remain so) in most fossil taxa.

It is worth pointing out two frequently overlooked cerebellum characters proposed by Nieuwenhuys (1982) for more inclusive levels of actinopterygian phylogeny. The first concerns the corpus cerebelli, which in polypteriforms and chondrosteans protrudes ventrally into the fourth ventricle. In *Amia* and teleosts, this section of the cerebellum is a dorsally evaginated structure, curving posteriorly over the fourth ventricle or extending anteriorly over the tectum and the telencephalon. This condition is apomorphic, considering the state in more basal actinopterygians, but, pending data on lepisosteids, its level of generality is as yet uncertain. A second cerebellum character is the presence of the valvula cerebelli, mentioned previously. This structure, a thin-walled body extending anteriorly into the ventricle of the midbrain, is said by Nieuwenhuys (1982, p. 297) to be "... a unique specialization present only in actinopterygians."

(24) Presence of Accessory Nasal Sacs

Nasal sacs are hypothesized to have formed from the walls of a pair of gill slits located immediately anterior to the branchial moieties of the first metamere (Bjerring, 1972). In turn, accessory nasal sacs, also called ventilation sacs, are structures that communicate with the main olfactory chamber and may be located posterior, ventral, or dorsal to it. The presence of accessory nasal sacs has been reported in a number of teleosts by Dahlgren (1908), Burne (1909), Atz (1952), and Theisen *et al.* (1991). Apparently, nasal sacs function as an aid in olfactory ventilation (Theisen *et al.*, 1991) or in nasal aspiration (Bertmar, 1969).

Chen and Arratia (1994) reported the presence of accessory sacs in *Elops* and *Hiodon*, noting also their absence in more primitive actinopterygians such as *Amia*, *Lepisosteus*, and *Polypterus*. The same authors tentatively proposed accessory nasal sacs as a synapomorphy for teleosts. It is unknown whether accessory nasal sacs are present in fossil teleosts and other actinopterygians.

(25) Four Pectoral Radials

Pectoral radials in primitive actinopterygians, either cartilaginous or ossified (Grande and Bemis, 1991), are 5 or more in number. In teleosts, the radials are rarely more than 4 (Jessen, 1972; some exceptions are some macrourids, anguillids, and batrachoidids, with 5 to 12 radials). According to Patterson, the derived condition is seen in †*Tharsis*, ichthyodectiforms, leptolepids, and †*Pholidophorus macrocephalus* but not in other pholidophoriforms, pleuropholids, aspidorhynchiforms, and pachycormiforms.

(26) Long Epineurals Along Abdominal Region

Epineurals originate as outgrowths from the neural arches, and, when present in primitive actinopterygians, they are short and restricted to the first few neural arches. In teleosts, the epineurals are long, equal in length to several vertebrae (Patterson, 1973; Patterson and Johnson, 1995). This character applies to all teleocephalans (except, of course, in cases where epineurals are absent) and at least to †*Tharsis*, ichthyodectiforms, leptolepiforms, and some pholidophoriforms (e.g., †*Pholidophorus bechei*), among fossils (Patterson and Johnson, 1995).

(27) Ossification Pattern of Epineurals

Epineurals are primitively preformed in cartilage in actinopterygians. This condition is visible in adults by their hollow structure and frequently by the cartilaginous tip of these bones (Patterson and Johnson, 1995). In all Recent teleosts, epineurals are formed of solid membrane bone, with no trace of cartilage (Emelianov, 1935; Patterson and Johnson, 1995). The solid condition is synapomorphic for teleocephalans plus a few fossils, such as †*Tharsis* and ichthyodectiforms (Patterson and Johnson, 1995). In a few other fossil taxa, such as †*Leptolepis coryphaenoides* and †*Pro-*

leptolepis, the epineurals are fully ossified and lack a cartilaginous tip, but their shaft is hollow, a condition indicative of cartilage precursor. It is possible that the obliteration of the cartilage tip is an intermediate stage between primitive epineurals and the fully ossified membrane–bone condition. If so, this character is multistate, with loss of the distal cartilage a synapomorphy for all taxa previously mentioned and total loss of cartilage synapomorphic for a less inclusive clade.

IV. Discussion and Comparison with Molecular Data

The preceding list of characters leaves little room for serious doubt regarding teleostean monophyly. Certainly as far as Recent teleosts are concerned, the notion of polyphyly of the group appears ludicrous to contemporary ichthyologists. However, until about three decades ago the situation was exactly the opposite and teleost polyphyly was the accepted paradigm in systematic ichthyology. The origin of teleosts was considered to lie independently with more than one lineage of "pholidophorid" and "leptolepid." Thus, the group was rendered polyphyletic or paraphyletic according to the systematic traditions of the time. This trend, prevailing through a good part of the 1900s, is reviewed in Gosline (1965) and Patterson (1967, 1977a). Some major representatives of the teleostean polyphyly view include Woodward (1942), Arambourg (1935, 1950), Gardiner (1960), Bardack (1965), and Greenwood *et al.* (1966). The general reasoning behind such a widespread opinion did not hinge on data but rather on systematic theory. The period was one of great confusion in the foundations of systematics. Evolutionary theory was already well-established, and it was generally accepted that classifications should somehow reflect phylogeny. However, systematic theory and methodology, in trying to accommodate evolutionary thinking, temporarily fell off the rails. The search for ancestor–descendant relationships (with its corollary of actual ancestors) was considered the central goal of systematic research. Fossils, viewed as the only direct line of evidence of phylogenetic relationship, acquired a special status in phylogeny reconstruction. All of this resulted in a definition of monophyly that gave rise to insoluble controversies about the validity of groups. The history of that period is now well-documented (e.g., Rieppel, 1988; Hull, 1988), and clearly the situation in fish systematics was no different from that of systematics in general.

The tide began to change with Gosline (1965), who pointed out that teleosts share a number of specialized traits, and that it was difficult to imagine that they had been independently acquired. Shortly thereafter, Greenwood *et al.* (1966) published their now classic paper. Although Greenwood and his co-authors appear to have subscribed to the teleostean polyphyly view (cf. Greenwood *et al.*, 1966, pp. 345, 346, 348), there is a diffuse but recurrent tendency in their paper to recognize subgroups on the basis of shared specializations. It is interesting to note that modern phylogenetic thinking was already beginning to suffuse fish systematics before the ideas of Hennig (1966) became widespread. A number of monophyletic groups were recognized on the basis of shared derived characters at a time when the ideas of Hennig were either unavailable or unknown to most ichthyologists. This prior preparation may explain why cladistics was so quickly adopted in fish systematics thereafter (Nelson, 1973a, 1989; Pietsch, 1988).

Be that as it may, teleosts were first explicitly demonstrated to be monophyletic in a modern sense by Patterson (1968), who also included in the group many fossil taxa previously considered to be "holosteans." As early as 1969, in referring to the polyphyly of teleosteans and comparing it to a similar problem in mammals, Nelson wrote that "... this polyphyletic dilemma represents hardly more than a breakdown in logic, the fallacy of which has been nicely discussed by Brundin" (1969a, p. 528). That statement came immediately after an evaluation of teleostean gill-arch features: "... from the standpoint of gill-arch structure the teleosts probably are the best defined major group of fishes, with little indication of polyphyly" (Nelson, 1969a, p. 528).

One of the problems with teleostean synapomorphies, as previously noted, is the exact level in the phylogeny where the name Teleostei will apply. If all basal teleost fossils were unknown, the evidence for teleostean monophyly (then equivalent to living teleosts) would be overwhelming, and the group would be separated by a large morphological gap from other actinopterygians. This happens because once the basal succession of adjacent sister groups is omitted, all characters spread at successive levels at the base of the cladogram would be telescoped onto a single node (Patterson, 1977a). However, basal teleostean fossils existing as they do, the synapomorphies are actually spread over several nodes.

Regardless of the details of fossil branches near the base, the monophyly of teleocephalans is an extremely well-supported hypothesis. Synapomorphies at more inclusive levels of the teleostean phylogeny evidently provide support for the monophyly of a subgroup composed of only living teleosts when a comparison is carried out with taxa outside the teleostan node. Of course, for many putative apomorphies from soft anatomy, meaningful comparisons are only possible with *Amia, Lepisosteus*, chondrosteans, and polypterids. Still, those characters withstand whatever testing is possible.

Considering the large number of synapomorphies for the group composed of Recent teleosts, it is surprising that their monophyly is not well-supported by molecular data in the few studies so far undertaken (Normark *et al.*, 1991; Lê *et al.*, 1993; see also review in Lecointre, 1993, and Lecointre and Nelson, this volume). Normark *et al.* (1991) analyzed fragments of three mitochondrial DNA sequences (cytochrome b, cytochrome oxidase I, and cytochrome oxidase II) translated into amino acid sequences, with each amino acid position treated as a character. The strict consensus of an analysis including all three sequences for 17 taxa did not find a clade equivalent to Recent teleosts. Actually, the groupings therein are incompatible with a monophyletic Teleostei, with, for example, *Megalops* closer to chondrosteans than to other teleosts, and the cichlid *Geophagus* as the sister group to all actinopterygians (*Polypterus* included). Indeed, Normark *et al.* stated, "The strict consensus tree resulting from the analysis of all three regions (fig. 1) would strike any morphologist as unacceptably bizarre" (1991, p. 825). The same authors conducted a second analysis utilizing only the cytochrome b data, this time including 25 taxa. A strict consensus of the 5360 most parsimonious trees found in that analysis left neopterygians almost totally unresolved. The only clade congruent with Teleostei came from a 50%-majority-rule consensus, but due to the nature of majority-rule consensus this is a rather circumstantial kind of support. These and other results led Normark *et al.* (1991) to suggest that in certain cases strongly supported morphology based nodes be used as topological constraints for molecular analyses.

Lê *et al.* (1993) conducted two analyses for about 500 nucleotides of 28S rRNA, one for 38 and the other for 31 species (which excluded supposedly "fast-evolving" species from the 38-species pool). Their 38-species analysis found a monophyletic Teleostei, but the bootstrap value of the teleost node (32%) was among the lowest of the tree. Their 31-taxon tree placed *Amia* as the sister group of elopomorphs plus osteoglossomorphs and thus resulted in a nonmonophyletic Teleostei.

In a recent paper, Lee *et al.* (1995) dealt obliquely with teleostean relationships on the basis of data from mitochondrial control regions. The results, however, are irrelevant to the question of teleostean monophyly because of methodological limitations inherent in the study. The study of Lee *et al.* included only one non-

teleost (*Acipenser*), which served as the outgroup and therefore delimited the root site. Under such circumstances, any result would support monophyly of the teleost ingroup.

The results of both Normark *et al.* and Lê *et al.* are clearly problematic with regard to teleostean monophyly. In no case where teleosts did not hold together can the Teleostei be rendered monophyletic by changing the root site because teleosts and non-teleosts are mixed in the unrooted tree. Leaving aside the remote possibility of actual teleostean nonmonophyly, it seems that these results strongly indicate that the particular kind of molecular data utilized is not informative for the time period during which Recent teleosts differentiated from remaining actinopterygians.

V. Summary

Teleosts are a highly corroborated monophyletic group with at least 27 known synapomorphies from various anatomical systems. The exact composition of the group is subject to debate regarding the inclusion of certain basal fossils. This uncertainty about the exact limits of the taxon results in difficulties in determining which synapomorphies are actual "teleostean synapomorphies." The Teleostei is here defined as the most inclusive actinopterygian group not including the Halecomorphi (*Amia* and relatives) and/or the Ginglymodi (*Lepisosteus* and relatives). This stem-based definition conforms with the practice of including certain basal fossils in the Teleostei and maintains the traditional exclusion of the Recent genera *Amia* and *Lepisosteus* from the group. The clade composed of all so-called Recent teleosts (here named Teleocephala; composed of osteoglossomorphs, elopomorphs, clupeomorphs, and euteleosts) is supported as monophyletic by an outstanding number of features from both hard and soft anatomy. Intriguingly, molecular data have yet to provide consistent support for teleostean monophyly.

Acknowledgments

I first must thank Colin Patterson, to whom this volume is very appropriately dedicated, for his monumental contributions to the understanding of teleostean monophyly and relationships. The timely completion of this chapter was in large measure due to the help of Monica Piza, who transformed previous e-mail versions into readable copies for review. I also thank Melanie Stiassny, Lynne Parenti, and David Johnson for their effort (and patience) in putting this volume together.

The manuscript benefited from reviews by Lance Grande and Colin Patterson.

References

Allis, E. P. (1897). The cranial muscles and cranial first spinal nerves in *Amia calva. J. Morphol.* **12**, 487–809.

Allis, E. P. (1909). The cranial anatomy of the mail-cheeked fishes. *Zoologica (Stuttgart)* **22**(2), 1–29.

Arambourg, C. (1935). Observations sur quelques poissons fossiles de l'ordre des Halécostomes et sur l'origine des Clupéidés. *C. R. Hebd. Seances Acad. Sci.* **200**, 2110–2112.

Arambourg, C. (1950). Nouvelles observations sur les Helécostomes et l'origine des Clupeidae. *C. R. Hebd. Seances Acad. Sci.* **231**, 416–418.

Arratia, G. (1981). *Varasichthys ariasi* n. gen. et sp. from the upper Jurassic of Chile (Pisces, Teleostei, Varasichthyidae, n. fam.). *Palaeontographica, Abt. A* **175**, 107–139.

Arratia, G. (1984). Some osteological features of *Varasichthys ariasi* Arratia (Pisces, Teleostei) from the Late Jurassic of Chile. *Palaeontol. Z.* **58**, 149–163.

Arratia, G. (1987). Jurassic fishes from Chile and critical comment. *In* "Bioestratigrafia de los Sistemas Regionales del Jurásico y Cretácico en América del Sur" (W. Volkheimer and E. A. Musacchio, eds.), Vol. 1, pp. 257–286. Comité Sudamericano del Jurásico y Cretácico, Mendoza.

Arratia, G. (1991). The caudal skeleton of Jurassic teleosts: A phylogenetic analysis. *In* "Early Vertebrates and Related Problems in Evolutionary Biology"(M.-M. Chang, Y.-H. Liu, and G.-R. Zhang, eds.), pp. 249–340. Science Press, Beijing.

Arratia, G. (1994). Phylogenetic and paleontographic relationships of the varasichthyid group (Teleostei) from the late Jurassic of Central and South America. *Rev. Geol. Chile* **21**(1), 119–165.

Arratia, G. (1995). Importance of specific fossils in teleostean phylogeny. *Geobios (Jodhpur, India)* **19**, 173–176.

Arratia, G. and Schultze, H.-P. (1987). A new halecostome fish (Actinopterygii, Osteichthyes) from the late Jurassic of Chile and its relationships. *Dakoterra* **3**, 1–13.

Arratia, G., and Schultze, H.-P. (1990). The urohyal: Development and homology within osteichthyans. *J. Morphol.* **203**, 247–282.

Atz, J. W. (1952). Internal nares in the teleost, *Astroscopus. Anat. Rec.* **113**, 105–115.

Bardack, D. (1965). Anatomy and evolution of chirocentrid fishes. *Univ. Kans. Paleontol. Contrib., Vertebr.* **10**, 1–88.

Berg, L. S. (1940). Classification of fishes, both recent and fossil. *Trav. Inst. Zool. Acad. Sci. Leningr.* **5**, 85–517.

Bertin, L., and Arambourg, C. (1958). Super-ordre des Téléostéens (Teleostei). *In* "Traité de Zoologie" (P.-P. Grassé, ed.), Vol. 13, pp. 2204–2500. Masson, Paris.

Bertmar, G. (1969). The vertebrate nose, remarks on its structural and functional adaptation and evolution. *Evolution (Lawrence, Kans.)* **23**, 131–152.

Bjerring, H. C. (1972). The rhinal bone and its evolutionary significance. *Zool. Scr.* **1**, 193–201.

Brito, P. M. (1992). L'endocrâne et le moulage endocrânien de *Vinctifer comptoni* (Actinopterygii, Aspidorhynchiformes) de Crétacé inférieur de Brésil. *Anna. Paléontol.* **78**, 129–157.

Burne, R. H. (1909). The anatomy of the olfactory organ of teleostean fishes. *Proc. Zool. Soc. London* **2**, 610–637.

Chen, X.-Y., and Arratia, G. (1994). Olfactory organ of Acipenseriformes and comparison with other actinopterygians: patterns of diversity. *J. Morphol.* **222**, 241–267.

Dahlgren, U. (1908). The oral opening of the nasal cavity in *Astroscopus. Science* **27**, 993–994.

De Beer, G. R. (1937). "The Development of the Vertebrate Skull." University of Chicago Press, Chicago.

De Queiroz, K. (1994). Replacement of an essentialistic perspective on taxonomic definitions as exemplified by the definition of "Mammalia." *Syst. Biol.* **43**, 497–510.

De Queiroz, K., and Gauthier, J. (1990). Phylogeny as a central principle in taxonomy: Phylogenetic definitions of taxon names. *Syst. Zool.* **39**, 307–322.

Donoghue, M., Doyle, J., Gauthier, J., Kluge, A., and Rowe, T. (1989). The importance of fossils in phylogeny reconstruction. *Annu. Rev. Ecol. Syst.* **20**, 431–460.

Emelianov, S. W. (1935). Die Morphologie der Fischrippen. *Zool. Jahrb. Abt. Anat. Ontog. Tiere* **60**, 132–262.

Fink, S. V., and Fink, W. L. (1981). Interrelationships of the ostariophysan fishes (Teleostei). *Zool. J. Linn. Soc.* **72**, 297–353.

Fink, W. L. (1984). Basal euteleosts: Relationships. *In* "Ontogeny and Systematics of Fishes" (H. G. Moser, W. J. Richards, D. M. Cohen, M. P. Fahay, A. M. Kendall, Jr., and S. L. Richardson, eds.), Spec. Publ. No. 1, pp. 202–206. American Society of Ichthyologists and Herpetologists, Lawrence, KS.

Fink, W. L., and Weitzman, S. H. (1982). Relationships of the stomiiform fishes (Teleostei), with a description of *Diplophos. Bull. Mus. Comp. Zool.* **150**(2), 31–93.

Forey, P. L. (1992). Fossils and cladistic analysis. *In* "Cladistics: A Practical Course in Systematics" (P. L. Forey, C. J. Humphries, I. L. Kitching, R. W. Scotland, D. J. Siebert, and D. M. Williams, eds.), Publ. No. 10, pp. 124–136. The Systematics Association, Oxford University Press, London.

Fujita, K. (1990). "The Caudal Skeleton of Teleostean Fishes." Tokai University Press, Tokyo.

Gardiner, B. G. (1960). A revision of certain actinopterygian and coelacanth fishes, chiefly from the lower Lias. *Bull. Br. Mus. (Nat. Hist.), Geol.* **4**, 241–384.

Gardiner, B. G. (1984a). The relationships of the palaeoniscid *Moythomasia* from the Upper Devonian of Western Australia. *Bull. Br. Mus. (Nat. Hist.), Geol.* **37**, 173–428.

Gardiner, B. G. (1984b). Sturgeons as living fossils. *In* "Living Fossils" (N. Eldgredge and S. M. Stanley, eds.), pp. 148–152. Springer-Verlag, New York.

Gardiner, B. G. and Schaeffer, B. (1989). Interrelationships of lower actinopterygian fishes. *Zool. J. Linn. Soc.* **97**, 135–187.

Gauthier, J., Kluge, A. G., and Rowe, T. (1988). Amniote phylogeny and the importance of fossils. *Cladistics* **4**, 105–209.

Gegenbaur, C. (1898). "Vergleichende Anatomie der Wirbeltiere mit Berücksichtigung der Wirbellosen," Vol. 1. Engelmann, Leipzig

Gosline, W. A. (1960). Contributions toward a classification of modern isospondylous fishes. *Bull. Br. Mus. (Nat. Hist.), Zool.* **6**(6), 325–365.

Gosline, W. A. (1961). Some osteological features of modern lower teleostean fishes. *Smithson. Misc. Collect.* **142**(3), 1–42.

Gosline, W. A. (1965). Teleostean phylogeny. *Copeia*, pp. 186–194.

Gosline, W. A. (1980). The evolution of some structural systems with reference to the interrelationships of modern lower teleostean fish groups. *Jpn. J. Ichthyol.* **27**, 1–28.

Gottfried, M. D. (1989). Homology and terminology of higher teleost postcleithral elements. *Trans. San Diego Soc. Nat. Hist.* **21**(18), 283–290.

Grande, L. (1985). Recent and fossil clupeomorph fishes, with materials for revision of the subgroups of clupeoids. *Bull. Am. Mus. Nat. Hist.* **181**, 231–372.

Grande, L., and Bemis, W. E. (1991). Osteology and phylogenetic relationships of fossil and recent paddlefishes (Polyodontidae) with comments on the interrelationships of Acipenseriformes. *J. Vertebr. Paleontol. Mem. 1* **11**, Suppl. 1, 1–121.

Greenwood, P. H. (1970). On the genus *Lycoptera* and its relationships with the family Hiodontidae (Pisces, Osteoglossomorpha). *Bull. Br. Mus. (Nat. Hist.), Zool.* **19**, 259–285.

Greenwood, P. H., Rosen, D. E., Weitzman, S. H., and Myers, G. S. (1966). Phyletic studies of teleostean fishes, with a new classification of living forms. *Bull. Am. Mus. Nat. Hist.* **131**(4), 339–456.

Gregory, W. K. (1933). Fish skulls: A study of the evolution of natural mechanisms. *Trans. Am. Philos. Soc.* **23**, 75–481.

Hammarberg, F. (1937). Zur Kentniss der ontogenetischen Entwicklung des Schädels von *Lepisosteus platystomus. Acta Zool. (Stockholm)* **18**, 209–337.

Heckel, J. J. (1850). Über das Wirbelsäulen-Ende bei Ganoiden und Teleostiern. *Sitsungsber. Akad. Wiss. Wien* **5**, 143–148.

Hennig, W. (1966). "Phylogenetic Systematics." University of Illinois Press, Urbana.

Hennig, W. (1969). "Die Stammesgeschichte der Insekten." Kramer, Frankfurt.

Hollister, G. (1936). Caudal skeleton of Bermuda shallow water fishes. I. Order Isospondyli: Elopidae, Megalopidae, Albulidae, Clupeidae, Dussumieridae, Engraulidae. *Zoologica (N.Y.)* **21**, 257–290.

Hull, D. L. (1988). "Science as a Process." University of Chicago Press, Chicago.

Jarvik, E. (1954). On the visceral skeleton in *Eusthenopteron* with a discussion of the parasphenoid and palatoquadrate in fishes. *K. Sven. Vetenskapsakad. Handl.* [4] **5**, 1–104.

Jarvik, E. (1963). The composition of the intermandibular division of the head in fishes and tetrapods and the diphyletic origin of the tetrapod tongue. *K. Sven. Vetenskapsakad. Handl.* [4] **9**, 1–74.

Jarvik, E. (1967). The homologies of frontal and parietal bones in fishes and tetrapods. *Colloq. Int. C.N.R.S.* **163**, 181–213.

Jessen, H. (1972). Schultergürtel und Pectoralflosse bei Actinopterygiern. *Fossils Strata* **1**, 1–101.

Johnson, G. D. (1992). Monophyly of the euteleostean clades—Neoteleostei, Eurypterygii, and Ctenosquamata. *Copeia* **1**, 8–25.

Johnson, G. D., and Patterson, C. (1993). Percomorph phylogeny: A survey of acanthomorphs and a new proposal. *Bull. Mar. Sci.* **52**(1), 554–626.

Johnson, G. D., and Washington, B. B. (1987). Larvae of the Moorish idol, *Zanclus cornutus*, including a comparison with other larval acanthuroids. *Bull. Mar. Sci.* **40**, 494–511.

Jollie, M. (1980). Development of head and pectoral girdle skeleton and scales in *Acipenser. Copeia*, pp. 226–249.

Jollie, M. (1984a). Development of the head and pectoral skeleton of *Polypterus* with a note on scales (Pisces, Actinopterygii). *J. Zool.* **204**, 469–507.

Jollie, M. (1984b). Development of cranial and pectoral girdle bones of *Lepisosteus* with a note on scales. *Copeia*, pp. 476–502.

Lauder, G. V. (1980). Evolution of the feeding mechanism in primitive actinopterygian fishes: A functional anatomical analysis of *Polypterus, Lepisosteus* and *Amia. J. Morphol.* **163**, 283–317.

Lauder, G. V. (1989). Caudal fin locomotion in ray-finned fishes: Historical and functional analyses. *Am. Zool.* **29**, 85–102.

Lauder, G. V. and Liem, K. F. (1983). The evolution and interrelationships of the actinopterygian fishes. *Bull. Mus. Comp. Zool.* **150**(3), 95–197.

Lê, H., L., V., Lecointre, G., and Perasso, R. (1993). A 28S rRNA-based phylogeny of the gnathostomes: First steps in the analysis of conflict and congruence with morphologically based cladograms. *Mol. Phylogenet. Evol.* **2**(1), 31–51.

Lecointre, G. (1993). Etude de l'impact de l'échantillonnage des espèces et de la longueur des séquences sur la robustesse des phylogénies moléculaires. Implication sur la phylogénie des Téléostéens. Doctoral Dissertation, Université de Paris VII, Paris.

Le Danois, Y. (1967). Quelques figures descriptives de l'anatomie de *Pantodon buchholzi* Peters. *Bull. Inst. Fondam. Afr. Noire, Ser. A* **29**, 1051–1096.

Lee, W.-J., Conroy, J., Hunting Howell, W., and Kocher, T. D. (1995). Structure and evolution of teleost mitochondrial control regions. *J. Mol. Evol.* **41**, 54–66.

Lehman, J. P. (1966). Actinopterygii. *In* "Traité de Paléontologie" (J. Piveteau, ed.), Vol. 4, Part 3, pp. 1–242. Masson, Paris.

McAllister, D. E. (1968). Evolution of branchiostegals and classification of teleostome fishes. *Bull.—Natl. Mus. Can.* **221**, 1–239.

Miles, R. (1977). Dipnoan (lungfish) skulls and the relationships of the group: A study based on new species from the Devonian of Australia. *Zool. J. Linn. Soc.* **61**, 1–328.

Millot, J., and Anthony, J. (1958). "Anatomie de *Latimeria chalumnae*. Vol. 1. Squelette, muscles et formation de soutien." CNRS, Paris.

Monod, T. (1967). Le complèxe urophore des téléostéens: Typologie et évolution (note préliminaire). *Colloq. Int. C.N.R.S.* **163**, 111–131.

Monod, T. (1968). Le complèxe urophore des poisson téléostéens. *Mem. Inst. Fondam. Afr. Noire* **81**, 1–705.

Müller, J. (1844). Ueber den Bau und die Grenzen der Ganoiden und über das natürlich System der Fische. *Ber. Akad. Wiss. Berlin*, pp. 416–422.

Nelson, G. (1969a). Gill arches and the phylogeny of fishes, with notes on the classification of vertebrates. *Bull. Am. Mus. Nat. Hist.* **141**(4), 475–552.

Nelson, G. (1969b). Origin and diversification of teleostean fishes. *Ann. N.Y. Acad. Sci.* **167**(1), 18–30.

Nelson, G. (1973a). Comments on Hennig's 'phylogenetic systematics' and its influence on ichthyology. *Syst. Zool.* **21**, 364–374.

Nelson, G. (1973b). Relationships of clupeomorphs, with remarks on the structure of the lower jaw in fishes. *Zool. J. Linn. Soc.* **53**, Suppl. 1, 333–349.

Nelson, G. (1989). Phylogeny of major fish groups. *In* "The Hierarchy of Life" (B. Fernholm, K. Bremer, and H. Jörnvall, eds.), pp. 325–336. Elsevier, Amsterdam.

Nelson, J. S. (1994). "Fishes of the World," 3rd. ed. Wiley, New York.

Nielsen, E. (1942). Studies on Triassic fishes from East Greenland. I. *Glaucolepis* and *Boreosomus. Medd. Groenl. (Palaeozool. Groenl., I)* **138**, 1–403.

Nieuwenhuys, R. (1967). Comparative anatomy of the cerebellum. *Prog. Brain Res.* 1–93.

Nieuwenhuys, R. (1982). An overview of the organization of the brain of actinopterygian fishes. *Am. Zool.* **22**, 287–310.

Norell, M., Clark, J., and Chiappe, L. (1993). Naming names. *Nature (London)* **366**, 518.

Normark, B. B., McCune, A. R., and Harrison, R. G. (1991). Phylogenetic relationships of neopterygian fishes, inferred from mitochondrial DNA sequences. *Mol. Biol. Evol.* **8**(6), 819–834.

Nybelin, O. (1963). Zur Morphologie und Terminologie des Schwanzskelettes der Actinopterygier. *Ark. Zool.* [2] **15**, 485–516.

Olsen, P. E. (1984). The skull and pectoral girdle of the parasemionotid fish *Watsonulus eugnathoides* from the early Triassic Sakamena Group of Madagascar, with comments on the relationships of the holostean fishes. *J. Vertebr. Paleontol.* **4**, 481–499.

Olsen, P. E., and McCune, A. R. (1991). Morphology of the *Semionotus elegans* species group from the early Jurassic part of the Newark Supergroup of eastern North America, with comments on the family Semionotidae (Neopterygii). *J. Vertebr. Paleontol.* **11**, 269–292.

Parenti, L. R. (1986). The phylogenetic significance of bone types in euteleost fishes. *Zool. J. Linn. Soc.* **86**, 37–51.

Patterson, C. (1967). Are the teleosts a polyphyletic group? *Colloq. Int. C.N.R.S.* **163**, 93–109.

Patterson, C. (1968). The caudal skeleton in Lower Liassic pholidophorid fishes. *Bull. Br. Mus. (Nat. Hist.), Geol.* **16**, 201–239.

Patterson, C. (1973). Interrelationships of holosteans. *In* "Interrelationships of Fishes" (P. H. Greenwood, R. S. Miles, and C. Patterson, eds.), pp. 233–305. Academic Press, London.

Patterson, C. (1975). The braincase of pholidophorid and leptolepid fishes, with a review of the actinopterygian braincase. *Philos. Trans. R. Soc. London, Ser. B* **269**, 275–579.

Patterson, C. (1977a). The contribution of paleontology to teleostean phylogeny. *In* "Major Patterns in Vertebrate Evolution" (M. K. Hecht, P. C. Goody, and B. M. Hecht, eds.), pp. 579–643. Plenum, New York.

Patterson, C. (1977b). Cartilage bone, dermal bones and membrane bones or the exoskeleton versus the endoskeleton. *Linn. Soc. Symp. Ser.* **4**, 77–121.

Patterson, C. (1981). Agassiz, Darwin, Huxley, and the fossil record of teleost fishes. *Bull. Br. Mus. (Nat. Hist.), Geol.* **35**(3), 213–224.

Patterson, C. (1984). *Chanoides*, a marine eocene otophysan fish (Teleostei: Ostariophysi). *J. Vertebr. Paleontol.* **4**(3), 430–456.

Patterson, C. (1993a). Bird or dinosaur? *Nature (London)* **365**, 21.

Patterson, C. (1993b). Reply to Norell *et al.* (1993).

Patterson, C. (1994). Bony fishes. *In* "Major Features of Vertebrate Evolution" (D. R. Prothero and R. M. Schoch, eds.), Short Courses Paleontol., No. 7, pp. 57–84. Paleontological Society, University of Tennessee, Knoxville.

Patterson, C., and Johnson, G. D. (1995). The intermuscular bones and ligaments of teleostean fishes. *Smithson. Contrib. Zool.* **559**, 1–85.

Patterson, C., and Rosen, D. E. (1977). Review of ichthyodectiform and other mesozoic teleost fishes and the theory and practice of classifying fossils. *Bull. Am. Mus. Nat. Hist.* **158**(2), 81–172.

Pietsch, T. (1988). "Tradition or Cladism, Which has Prevailed?" Program and Abstracts, p. 52. American Society of Ichthyologists and Herpetologists, Ann Arbor, MI.

Regan, C. T. (1910). The caudal fin of the Elopidae and of some other teleostean fishes. *Ann. Mag. Nat. Hist.* [8] **5**, 354–358.

Regan, C. T. (1923). The skeleton of Lepidosteus, with remarks on the origin and evolution of the lower neopterygian fishes. *Proc. Zool. Soc. London*, pp. 445–461.

Rieppel, O. (1988). "Fundamentals of Comparative Biology." Birkhäuser, Basel.

Rosen, D. E. (1973). Interrelationships of higher euteleostean fishes. *Zool. J. Linn. Soc.* **53**, Suppl. 1, 397–513.

Rosen, D. E. (1974). Phylogeny and zoogeography of salmoniform fishes, and relationships of *Lepidogalaxias salamandroides*. *Bull. Am. Mus. Nat. Hist.* **153**, 265–326.

Rosen, D. E. (1985). An essay on euteleostean classification. *Am. Mus. Novit.* **2827**, 1–57.

Rosen, D. E., and Parenti, L. R. (1981). Relationships of *Oryzias*, and the groups of atherinomorph fishes. *Am. Mus. Novit.* **2719**, 1–25.

Rosen, D. E., and Patterson, C. (1969). The structure and relationships of the paracanthopterygian fishes. *Bull. Am. Mus. Nat. Hist.* **141**(3), 357–474.

Rosen, D. E., Forey, P. L., Gardiner, B. G., and Patterson, C. (1981). Lungfishes, tetrapods, paleontology and plesiomorphy. *Bull. Am. Mus. Nat. Hist.* **141**, 357–474.

Schaeffer, B., and Dalquest, W. W. (1978). A palaeonisciform braincase from the Permian of Texas, with comments on cranial fissures and the posterior myodome. *Am. Mus. Novit.* **2658**, 1–15.

Schultze, H.-P., and Arratia, G. (1986). Reevaluation of the caudal skeleton of actinopterygian fishes: I. *Lepisosteus* and *Amia. J. Morphol.* **190**, 215–241.

Schultze, H.-P., and Arratia, G. (1988). Reevaluation of the caudal skeleton of some actinopterygian fishes: II. *Hiodon, Elops*, and *Albula. J. Morphol.* **195**, 257–303.

Schultze, H.-P., and Arratia, G. (1989). The composition of the caudal skeleton of teleosts (Actinopterygii: Osteichthyes). *Zool. J. Linn. Soc.* **97**, 189–231.

Stensiö, E. A. (1921). "Triassic Fishes from Spitzbergen," Part I. Adolf Holzhausen, Vienna.

Stiassny, M. L. J. (1986). The limits and relationships of the acanthomorph teleosts. *J. Zool. (B)* **1**, 411–460.

Stiassny, M. L. J., and de Pinna, M. C. C. (1994). Basal taxa and the role of cladistic patterns in the evaluation of conservation priorities: A view from freshwater. *In* "Systematics and Conservation Evaluation" (P. L. Forey, C. J. Humphries, and R. I. Vane-Wright, eds.), Syst. Assoc. Spec. Vol. No. 50, pp. 235–249. Oxford University Press (Clarendon), Oxford.

Taverne, L. (1977). Ostéologie, phylogénèse, et systématique des Téléostéens fossiles et actuels du super-ordre des Ostéoglossomorphes. Première partie. *Mém. Cla. Sci., Acad. R. Belg.* **42**, 1–235.

Taverne, L. (1978). Ostéologie, phylogénèse, et systématique des Téléostéens fossiles et actuels du super-ordre des Ostéoglossomorphes. Deuxieme partie. *Mém. Cla. Sci., Acad. R. Belg.* **43**, 1–168.

Theisen, B. E., Zeiske, W. L., Silver, T., Marui, and Caprio, J. (1991). Morphological and physiological studies on the olfactory organ of the striped eel catfish, *Plotosus lineatus. Mar. Biol.* **110**, 127–135.

Véran, M. (1988). Les éléments accessoires de l'arc hyoïdien des poissons téléostomes (Acanthodiens et Osteichthyens) fossiles et actuels. *Mem. Mus. Natl. Hist. Nat., Sér. C* **54**, 1–98.

Watson, D. M. S. (1925). The structure of certain palaeoniscids and the relationship of that group with other bony fish. *Proc. Zool. Soc. London*, pp. 815–870.

Weitzman, S. H. (1967). The origin of the stomiatoid fishes with comments on the classification of salmoniform fishes. *Copeia*, **3**, 507–540.

Wenz, S. (1968). "Compléments à l'étude des poissons actinoptérigiens du Jurassique française." CNRS, Paris.

Wiley, E. O. (1976). The phylogeny and biogeography of fossil and recent gars (Actinopterygii: Lepisosteidae). *Misc. Publ., Mus. Nat. Hist. Univ. Kans.* **64**, 1–111.

Winterbottom, R. (1974). A descriptive synonymy of the striated muscles of the Teleostei. *Proc. Acad. Nat. Sci. Philadelphia* **125**(12), 225–317.

Woodward, A. S. (1942). The beginning of the teleostean fishes. *Ann. Mag. Nat. Hist.* [11] **9**, 902–912.

Phylogeny of Osteoglossomorpha

LI GUO-QING

Department of Biological Sciences and
Laboratory for Vertebrate Paleontology
University of Alberta
Edmonton, Alberta T6G 2E9, Canada
and
Institute of Vertebrate Paleontology and Paleoanthropology
Academia Sinica
Beijing 100044, People's Republic of China

MARK V. H. WILSON

Department of Biological Sciences and
Laboratory for Vertebrate Paleontology
University of Alberta
Edmonton, Alberta T6G 2E9, Canada

I. Introduction

The Osteoglossomorpha were defined as a superorder of the Teleostei by Greenwood *et al.* (1966) and have been the subject of numerous subsequent systematic studies (e.g., Nelson, 1968, 1969b; Greenwood, 1973; Patterson and Rosen, 1977; Taverne, 1979; Lauder and Liem, 1983; Li, 1994a, 1996; Li and Wilson, 1994). They are interesting to both ichthyologists and paleontologists because of their anatomy, physiology, geographic distribution, and ancient fossil record. Various osteoglossomorphs display unusual specializations of anatomy, such as elongate anal and dorsal fins and peculiarities of the jaws related to feeding, and physiology and behavior, such as generation of electric fields. Exclusively freshwater fishes, at least in the modern fauna, they display interesting biogeographic distributions including apparent examples of endemism of extant suprageneric taxa (Hiodontidae in North America, and Mormyroidea and Pantodontidae in Africa), as well as circumtropical or old-world tropical distributions (Osteoglossidae and Notopteroidea).

With regard to the fossil record, Eocene fossil members of genera included in Osteoglossidae have been known since the late 1800s (Leidy, 1873). Eocene fossil Hiodontidae have been recognized since Cavender's (1966) work on †*Eohiodon*. However, the superorder was seen to be much older than Eocene when Greenwood (1970) suggested that the Asian fossil genus †*Lycoptera*, of Late Jurassic to Early Cretaceous age, might be an osteoglossomorph. Since 1970 numerous fossil genera from all major continents except Antarctica have been added to the list of potential fossil osteoglossomorphs. Examples include †*Brychaetus* Woodward (1901), †*Paralycoptera* Chang and Chou (1977), †*Laeliichthys* Da Silva Santos (1985), †*Chandlerichthys* Grande (1986), †*Yanbiania* Li (1987), and †*Ostariostoma* Schaeffer (1949) as discussed by Grande and Cavender (1991). Obviously, correct phylogenetic interpretations of these and other fossil taxa are important for understanding the historical biogeography of the superorder.

This chapter is based on more extensive studies of fossil and extant osteoglossomorphs (Li, 1994a,b, 1996; Li and Wilson, 1994, 1996), in which additional fossil taxa of osteoglossomorphs are described. We present the results of a cladistic analysis of 50 characters in 29 taxa, analyzed with the MacClade 3.04 (Maddison and Maddison, 1992) and PAUP 3.1.1 (Swofford, 1993) computer programs. Characters and their abbreviations are listed in Appendix 1, character states for each character are summarized in Fig. 1, and the

Copyright © 1996 by Academic Press, Inc.
All rights of reproduction in any form reserved.

No.	Char.	†Leptolepis	Elopoidei	†Lycoptera	†Yanbiania	†Eohiodon rosei	†E. woodruffi	†H. consteniorum	Hiodon tergisus	H. alosoides	†Tanichthys	†Laellichthys	†Sinoglossus	Heterotis	Arapaima	†Phareodus testis	†P. encaustus	†P. queenslandicus	†B. muelleri	Pantodon	†Singida	Scleropages	Osteoglossum	†Ostariostoma	Mormyroidea	Xenomystus	Notopterus	Papyrocranus	Clupeoidea	Chanos
1	Lnpu2	0	1	1	1	1	1	1	1	1	1	1	?	1	1	1	1	1	?	1	1	1	1	1	1	1	1	1	1	1
2	Ep	0	0	1	1	1	1	1	1	1	1	1	?	1	1	1	1	1	?	1	1	1	1	1	1	2	2	2	0	1
3	Smx	0	0	0	1	1	1	1	1	1	1	1	1	1	1	1	1	1	1	1	1	1	1	1	1	1	1	1	0	1
4	Vfr	0	0	2	1	1	1	1	1	1	2	2	2	2	2	2	2	2	?	2	2	2	2	2	2	2	2	2	2	0
5	So	0	0	1	1	1	1	1	1	1	1	1	1	1	1	1	1	1	1	1	1	1	1	1	1	1	1	1	0	0
6	Pbm	0	0	1	1	1	1	1	1	1	1	1	1	1	1	1	1	1	1	1	1	1	1	1	1	1	1	1	0	0
7	Cfr	0	0	1	1	1	1	1	1	1	2	2	2	2	2	2	2	?	2	2	2	2	2	2	1	2	2	2	0	0
8	Tbhb	?	0	?	?	?	?	?	1	1	?	?	?	1	1	?	?	?	?	1	?	1	1	?	1	1	1	1	0	0
9	Int	?	0	?	?	?	?	?	1	1	?	?	?	1	1	?	?	?	?	1	?	1	1	?	1	1	1	1	0	0
10	Op	0	0	2	1	1	1	1	1	1	2	2	2	2	3	3	3	3	3	3	3	3	0	0	4	4	4	0	0	0
11	Io4-5	0	0	0	1	1	1	1	1	1	0	1	1	1	1	1	1	1	1	1	1	1	1	1	1	1	1	1	0	0
12	Na	0	0	0	0	0	1	1	1	?	2	2	2	2	2	2	2	2	2	2	2	2	2	2	2	2	2	2	0	0
13	Pomc	0	0	0	0	0	0	0	0	0	0	0	0	0	1	1	1	1	1	1	1	1	1	1	1	1	1	1	0	0
14	Nu1	0	0	0	0	0	0	0	0	0	1	?	1	1	2	2	?	?	1	1	1	1	0	0	0	1	1	0	0	0
15	Io2	0	0	0	0	0	0	0	0	0	?	2	2	2	2	0	0	?	0	0	0	0	0	1	1	1	1	1	0	0
16	P-qa	0	0	0	0	0	0	0	0	0	?	1	1	1	1	1	1	1	1	1	1	1	1	0	0	0	0	0	0	0
17	Dapt	0	0	0	1	1	1	1	1	1	0	0	0	0	0	0	0	?	0	0	0	0	0	0	0	0	0	0	0	0
18	Hmh	0	0	2	1	1	2	2	2	0	0	?	0	0	0	0	0	0	0	0	0	0	0	0	0	0	0	0	0	0
19	Sdsp	0	0	1	1	1	1	1	1	0	0	0	0	0	0	0	0	0	0	0	0	0	0	0	0	0	0	0	0	0
20	CKW	0	0	0	0	0	0	0	0	0	1	1	1	1	0	0	0	0	0	0	0	0	0	0	0	0	0	0	0	0
21	Bhbb	?	0	0	0	0	0	0	0	0	?	?	?	0	0	1	1	?	?	1	?	1	1	1	?	0	0	0	0	0
22	Pena	0	0	0	0	0	0	0	0	0	?	0	0	0	0	1	1	1	1	1	1	1	1	1	0	0	0	0	0	0
23	Pfr1	0	0	0	0	0	0	0	0	0	0	0	0	0	1	1	1	1	1	1	1	1	0	0	0	0	0	0	0	0
24	Ja	0	0	0	0	0	0	0	0	0	0	0	0	0	1	2	2	2	2	2	2	0	0	0	0	0	0	0	0	0
25	Fr	0	0	0	0	0	0	0	0	0	0	0	0	0	0	2	2	2	2	0	0	0	0	0	0	0	0	0	0	0
26	Rdwo	0	0	0	0	0	0	0	0	0	0	0	0	0	1	1	1	1	0	0	0	0	0	0	0	0	0	0	0	0
27	Ses	0	0	0	0	0	0	0	0	0	0	0	0	0	1	1	1	1	0	0	0	0	0	0	0	0	0	0	0	0
28	Pop	0	0	0	0	0	0	0	0	0	0	0	0	0	1	1	1	0	0	0	0	0	0	0	0	0	0	0	0	0
29	Hm	0	0	0	0	0	0	0	0	0	?	0	0	0	1	1	1	0	0	0	0	0	0	0	0	0	0	0	0	0
30	Sio3	0	0	0	0	0	0	0	0	0	0	0	0	0	0	2	2	0	0	0	0	0	0	1	1	1	1	1	0	0
31	IOL	0	0	0	0	0	0	0	0	0	0	0	0	0	0	0	0	0	0	0	0	0	0	1	1	1	1	1	0	0
32	Ioc	0	0	0	0	0	0	0	0	0	0	0	0	0	0	0	0	0	0	0	0	0	0	1	1	1	1	1	0	0
33	Utr	?	0	?	?	?	?	?	0	0	?	?	?	0	0	?	?	?	?	?	0	?	0	0	1	1	1	1	0	0
34	Ash	0	0	0	0	1	1	1	1	1	0	0	0	0	0	0	0	0	?	0	0	0	0	0	0	0	0	0	0	0
35	Pmx	0	0	0	0	0	1	2	2	2	0	0	0	0	0	0	0	0	0	0	0	0	0	0	0	0	0	0	0	0
36	Pdpo	0	0	0	0	0	0	1	1	1	0	0	0	0	0	0	0	0	0	0	0	0	0	0	0	0	0	0	0	0
37	Mcpm	0	0	0	0	0	0	1	1	1	0	0	0	0	0	0	0	0	0	0	0	0	0	0	0	0	0	0	0	0
38	BK	0	0	0	0	0	0	0	1	1	0	0	0	0	0	0	0	0	0	0	0	0	0	0	0	0	0	0	0	0
39	Npu1	0	0	0	0	0	0	1	1	1	0	0	?	0	0	0	0	?	?	0	0	0	0	0	0	0	0	0	0	0
40	Pemx	0	0	0	0	0	0	0	0	0	0	1	1	1	0	0	0	0	0	0	0	0	0	0	0	0	0	0	0	0
41	Aio1	0	0	0	0	0	0	0	0	0	0	1	1	1	0	0	0	0	0	0	0	0	0	0	0	0	0	0	0	0
42	Bhtp	?	0	0	0	0	0	0	0	0	0	?	?	1	1	0	0	?	?	?	0	?	0	0	0	0	0	0	0	0
43	Pmop	0	0	0	0	0	0	0	0	0	0	0	0	0	0	0	0	0	0	0	0	1	1	1	0	0	0	0	0	0
44	Osh	0	0	0	0	0	0	0	0	?	0	0	0	0	0	0	0	0	0	0	0	1	1	0	0	0	0	0	0	0
45	MB	0	0	0	0	0	0	0	0	0	0	0	0	0	0	0	0	0	0	0	0	1	1	0	0	0	0	0	0	0
46	Mdc	0	0	0	0	0	0	0	0	0	0	0	0	0	0	0	0	0	0	0	0	0	0	0	1	1	1	1	0	0
47	Afcf	0	0	0	0	0	0	0	0	0	0	0	0	0	0	0	0	0	0	0	0	0	0	?	0	1	1	1	0	0
48	Fcv	0	0	0	0	0	0	0	0	0	0	0	0	0	0	0	0	0	0	0	0	0	0	0	0	0	0	0	1	1
49	Ust	0	0	0	0	0	0	0	0	0	0	0	0	0	0	0	0	0	0	0	0	0	0	0	0	0	0	0	1	1
50	Un	0	0	0	0	0	0	0	0	0	1	?	1	1	1	1	1	1	?	1	1	1	1	1	1	1	1	1	0	0

FIGURE 1 Data matrix showing the distribution of states of the 50 characters listed in Appendix 1. In total, 29 taxa are examined and included in the cladistic analysis. Abbreviations: †B., †Brychaetus; †E., †Eohiodon; H., Hiodon; †P., †Phareodus. ? = missing state.

results are summarized in Fig. 2. In the following discussion, character and state numbers are indicated in parentheses and brackets, respectively; for example, "(12[2])" means state 2 of character 12.

Comparative specimens of extant fish species related to the taxa listed in Fig. 1 are deposited in the collection of the UAMZ (Museum of Zoology, Department of Biological Sciences, University of Alberta, Edmonton, Canada) and consist of dried skeletons, specimens cleared and stained for bone with alizarin, and specimens fixed in formalin and preserved in alcohol. Fossil materials used in this study include those deposited in the collections of the AMNH (American Museum of Natural History, New York, New York), FMNH (Field Museum of Natural History, Chicago, Illinois), IVPP (Institute of Vertebrate Paleontology and Paleoanthropology, Academia Sinica, Beijing, People's Republic of China), PU (Princeton University, Princeton, New Jersey), and UALVP (Laboratory for Vertebrate Paleontology, Department of Biological Sciences, University of Alberta, Edmonton, Canada).

II. Monophyly and Extra-group Relationships of Osteoglossomorpha

A. *Synapomorphies of Osteoglossomorpha*

Patterson and Rosen (1977) believed Osteoglossomorpha to be a monophyletic group, but Shen (1993) raised doubts about its monophyly. In their original treatment, Greenwood *et al.* (1966) recognized Osteoglossomorpha based mainly on two "group characters": 1) primary bite between parasphenoid and tongue (6[1]), and (2) paired tendon bones on the second hypobranchial or second hypobranchial and basibranchial (8[1]) (also Nelson, 1969a). Our study indicates that these two group characters still support the monophyly of Osteoglossomorpha since they are not seen in other taxa. We have also checked Patterson and Rosen's (1977) four osteoglossomorph diagnostic features (18 principal caudal fin rays, a full neural spine on pu1, a large postero-ventral infraorbital bone representing the io3 and io4 of other teleosts, and intestine passing to the left of the stomach). We find that there are 18 or fewer principal caudal fin rays among osteoglossomorphs (7[1/2]). Since this range is not seen in the outgroup taxa, it is thought to be a valid character supporting the monophyly of Osteoglossomorpha. A full neural spine on pu1 (39[0]) is developed in all the taxa except *Hiodon* and is thus not useful as a synapomorphy of Osteoglossomorpha. The "large postero-ventral infraorbital bone" represents only io3, which may fuse with io2 (e.g., that in †*Phareodus testis*) but not with io4. A similar io3 is also seen in †*Anaethalion* (Arratia, 1987), †*Cladocyclus* (Maisey, 1991), Elopomorpha (Forey, 1973a,b), and *Chanos* (Gregory, 1933, also personal observation) (30[0]). It is, thus, plesiomorphic and not indicative of monophyly of Osteoglossomorpha. The condition "intestine passes to the left of the stomach" (9[1]) can be tested only in extant taxa (Nelson, 1972). Nevertheless, on available evidence it appears to be a synapomorphy of the (extant) Osteoglossomorpha.

Therefore, four of the previously defined osteoglossomorph synapomorphies still support the mono-

phyly of Osteoglossomorpha. In addition to these, our analysis yields another four:

Number of epurals decreased to one or zero—At least two epurals (2[0]) are present in all the outgroup taxa except *Chanos*, while most osteoglossomorphs have only one (2[1]). Epurals are absent in notopterids (2[2]).

Absence of supramaxilla—One or two supramaxillae are seen in primitive members of all the outgroup taxa, but supramaxillae are absent in all osteoglossomorphs (3[1]) except lycopterids.

Absence of supraorbital—A supraorbital is found in basal teleosts (Nybelin, 1974; Patterson and Rosen, 1977; Arratia, 1987), elopomorphs (Forey, 1973a,b), clupeomorphs (Grande, 1985), and even in euteleosts such as *Coregonus* and *Salmo* (personal observation; also see Gregory, 1933; Patterson, 1970), but is absent in osteoglossomorphs (5[1]).

Io4 fused with io5—In all outgroup taxa, io4 is separate from io5. They are fused with each other (11[1]) in all osteoglossomorphs except †*Laeliichthys* (Da Silva Santos, 1985) and †*Lycoptera*.

B. *Extra-group Relationships of Osteoglossomorpha as Open Question*

The extra-group relationships of the Osteoglossomorpha have also been debated. Greenwood (1973) suggested a sister-group relationship between Osteoglossomorpha and Clupeomorpha, whereas Patterson and Rosen (1977), Lauder and Liem (1983), and J. S. Nelson (1994) considered Osteoglossomorpha to be the most primitive living teleosts. Arratia (1991) argued that elopomorphs are sister to osteoglossomorphs and all other teleosts. Our study has not uncovered convincing synapomorphies for any of these alternatives; we therefore consider the extra-group relationships of the Osteoglossomorpha as still not well resolved (Fig. 2).

III. †Lycopteridae as Stem-group Osteoglossomorphs

†*Lycoptera* Müller (1847) was originally referred to the Esocidae. Woodward (1901) transferred it from Esocidae to †Leptolepidae based on the opercular elements. Cockerell (1925) named the family †Lycopteridae on features of the scales and suggested that †*Lycoptera* might be "the ancestor of the Cyprinidae and their allies." Berg (1940) grouped the family in Clupeiformes as suborder †Lycopteroidei based on the gular plate (a misidentified structure, personal obser-

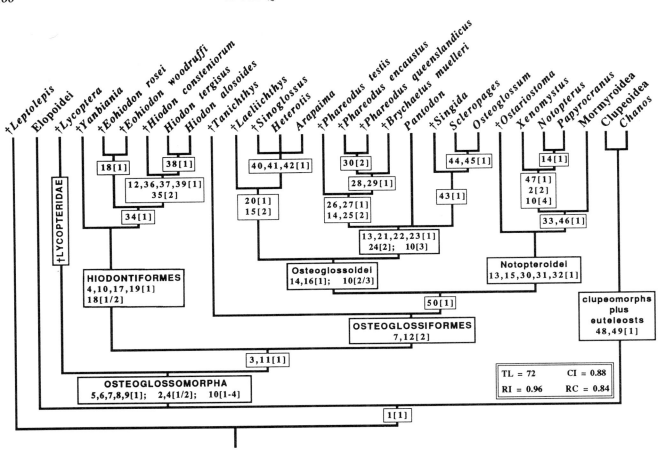

FIGURE 2 Strict consensus tree showing the phylogeny of Osteoglossomorpha. Note that †Lycopteridae are equally closely related to the Hiodontiformes *sensu stricto* and the Osteoglossiformes, and †*Ostariostoma* is a sister-clade of Notopteridae plus "Mormyroidea" *sensu* Nelson (1994). This tree also suggests either that †*Phareodus* Leidy 1873 is a paraphyletic group or that †*Brychaetus* Woodward 1901 is a junior synonym of †*Phareodus*. This cladogram was generated by PAUP 3.1.1 based on 27 shortest trees found by a branch-and-bound search from the matrix in Fig. 1. Numbers outside square brackets represent the characters listed in Appendix 1, while numbers in square brackets represent the states of characters. Abbreviations: CI, consistency index; RC, rescaled consistency index; RI, retention index; TL, tree length.

vation) and absence of the Weberian apparatus. Yakovlev (1965) noted similarities between †*Lycoptera* and *Arapaima*, but Greenwood (1970) placed †Lycopteridae in the Hiodontoidea based on the temporal fenestra, the hyoid arch, and the caudal skeleton.

We suggest rather that †Lycopteridae and Hiodontidae are not sister-taxa (Fig. 2) because the features used by Greenwood to link lycopterids with hiodontids are either shared with other Osteoglossomorpha (e.g., one epural) or are probably primitive (e.g., seven hypurals, also present in some elopomorphs; Forey, 1973b).

The phylogenetic significance of the "temporal fenestra" in lycopterids and hiodontids needs to be clarified. Ridewood (1904) suspected a possible homology between the hiodontid temporal fenestra and the clupeoid preepiotic fossa, but Greenwood (1970) consid-

ered these two structures to be more analogous than homologous. We examined these structures in lycopterids, hiodontids, and clupeoids, finding that the temporal fenestra and the preepiotic fossa are similar in position and are bounded by the same bones (anterodorsally by the parietal, ventrally by the pterotic, and posteriorly by the epiotic) in all three groups (also see Grande, 1985). We suggest, therefore, that the temporal fenestra in hiodontids and lycopterids could be homologous with the preepiotic fossa in clupeomorphs. A similar structure is also seen in *Pantodon*, *Papyrocranus*, Osmeridae, and Salmonidae (personal observation), and thus does not support a special relationship between lycopterids and hiodontids.

In our analysis, none of the 50 characters represents a unique synapomorphy of †Lycopteridae and Hiodontidae. Instead, we suggest that †Lycopteridae are stem-group (Wilson, 1992) osteoglossomorphs, sister to all extant clades of osteoglossomorphs. This relationship is supported by four synapomorphies: 2[1/2] epurals decreased to one or zero; 5[1] absence of supraorbital; 6[1] primary bite between parasphenoid and basihyal; and 7[1/2], 16 or fewer branched caudal fin rays.

IV. Phylogeny of Hiodontiformes *Sensu Stricto*

The order Hiodontiformes, erected by Taverne (1979) based on what are probably symplesiomorphies, comprised †Lycopteridae and Hiodontidae. However, our concept of Hiodontiformes is different from that of Taverne in both diagnosis (defined by derived character states) and content (containing only Hiodontidae). In recent years, hiodontids have been grouped with notopterids in suborder Notopteroidei of order Osteoglossiformes, based mainly on the swimbladder–ear connection (e.g., Lauder and Liem, 1983; Nelson, 1994), which is not a uniquely derived character state present only in Hiodontidae and Notopteridae (see Li, 1994a). This concept conflicts with the results of our analysis (see Fig. 2). If we include the three notopterid genera within the Hiodontiformes, tree length is increased greatly from 72 to 82 steps.

In contrast, our investigation supports a sister-group relationship between Hiodontiformes *sensu stricto* on the one hand and Osteoglossiformes, including Notopteroidei, on the other. This relationship is supported by two synapomorphies: 3[1] absence of supramaxilla and 11[1] fusion of io4 with io5. All the outgroup taxa except *Chanos* have two supramaxillae, and †*Lycoptera* has one. Formerly the fused io4–5 in osteoglossomorphs was considered to be homologous either with io5 (e.g., Nelson, 1969b; Kershaw, 1976) or with io4 (e.g., Greenwood and Patterson, 1967). Our comparisons with basal teleosts (Arratia, 1987; Maisey, 1991), elopomorphs (Forey, 1973b), clupeomorphs (Grande, 1985), and primitive euteleosts (Patterson, 1970) suggest that this bone is homologous with the fourth and fifth infraorbitals in the outgroup taxa. If so, a fused io4–5 is a synapomorphy for Hiodontiformes *s.s.* and Osteoglossiformes.

Hiodontiformes *s.s.* (containing only Hiodontidae) are defined by these synapomorphies:

Seven-rayed pelvic fin—We (Li and Wilson, 1994) recently discussed the seven-rayed pelvic fin in Hiodontiformes *s.s.* Basal teleosts such as †*Leptolepis*, †*Tharsis*, †*Allothrissops*, †*Cladocyclus*, elopomorphs, and primi-

tive euteleosts have more than seven pelvic rays (4[0]). †Lycopteridae, Osteoglossiformes, and clupeomorphs have six or fewer pelvic rays (4[2]). A seven-rayed pelvic fin (4[1]) is seen only in Hiodontiformes *s.s.* among osteoglossomorphs (Cavender, 1966; Wilson, 1978; Grande, 1979; Li, 1987).

Opercle an irregular parallelogram—The opercle has five general shapes: An irregularly trapezoid opercle (10[0]), usually with a round dorsal edge, is seen in basal teleosts (Nybelin, 1974; Arratia, 1987, 1996), ichthyodectiforms (Patterson and Rosen, 1977; Maisey, 1991), elopomorphs (Forey, 1973b), clupeomorphs (Grande, 1985), and lower euteleosts (personal observation); an irregularly parallelogramic opercle (10[1]) is seen in Hiodontiformes *s.s.*; an oval or kidney-shaped opercle occurs in †Lycopteridae and Heterotidinae; a subsemicircular opercle 10[3]) occurs in †*Laeliichthys* and all Osteoglossinae; and a fan-shaped opercle (10[4]) occurs only in notopterids.

Dorsal arm of posttemporal more than twice as long as ventral arm—We have seen this feature (17[1]) only in hiodontiforms. The posttemporal has a closely similar shape in all hiodontid genera (including †*Yanbiania* Li, 1987). In lycopterids and all outgroup taxa the dorsal arm may be longer than the ventral arm, but it is not twice as long (personal observation).

Hyomandibula double-headed—The hyomandibula articulates with the cranium via a single head in the outgroup taxa, but in hiodontids the double head is either two separate heads (18[1]), as in †*Eohiodon*, or two heads connected by a web of bone (18[2]), as in *Hiodon*.

Dermosphenotic triradiate—The dermosphenotic is an irregular triangle or a trapezoid in the outgroup taxa but triradiate (19[1]) in hiodontids.

V. Remarks on Osteoglossiformes

A. Monophyly of Osteoglossiformes

The remaining osteoglossomorphs in the Osteoglossiformes, with the addition of the notopterids, form a well-defined monophyletic group supported by three synapomorphies:

Fifteen or fewer branched caudal rays—All Osteoglossiformes except for some mormyrids have 15 or fewer branched caudal rays (7[2]), while Hiodontiformes have 16 branched caudal rays. Basal teleosts and primitive euteleosts have 17 branched caudal rays.

Nasal gutter-like or irregularly subrectangular—Three basic nasal shapes are seen in the 29 taxa listed in Fig. 1. A straight, tubular nasal (12[0]) is seen in basal teleosts (Patterson, 1967; Nybelin, 1974; Patterson and

Rosen, 1977; Arratia, 1987), elopomorphs (Forey, 1973b), clupeomorphs (Grande, 1985), and primitive euteleosts (Patterson, 1970; Wilson and Williams, 1991). A tubular and strongly curved nasal (12[1]) is present only in *Hiodon*. A gutter-like and irregularly subrectangular nasal (12[2]) occurs in Osteoglossiformes.

Uroneurals decreased to two or zero—Except for †*Tanichthys* Jin (1991), a decrease or fusion of uroneurals has occurred in the Osteoglossiformes. †*Singida* and †*Thaumaturus* have two uroneurals (Greenwood and Patterson, 1967; Micklich, 1992). †*Ostariostoma* has only one uroneural (Fig. 3). All other osteoglossiforms lack uroneurals. This situation was first noted by Whitehouse (1910) in *Notopterus* and later by Greenwood (1966) in *Pantodon* and in all osteoglossids. Since this state (50[1]) is not present in any of the basal teleosts, elopomorphs, or clupeomorphs and is seldom seen in more advanced teleosts (personal observation), it is considered to be an important synapomorphy of the Osteoglossiformes.

B. Three Synapomorphies of Osteoglossoidei

Osteoglossoidei traditionally contained Osteoglossidae, Pantodontidae, and †Singididae (Greenwood and Patterson, 1967). However, characters considered diagnostic of Pantodontidae by Nelson (1994) and of †Singididae are mainly autapomorphies of *Pantodon* and †*Singida*. Since these two genera share all the synapomorphies of the Osteoglossidae and Osteoglossinae, as well as one or more unique derived states with the extant *Scleropages* and *Osteoglossum*, we include them in the Osteoglossinae. Thus, the suborder Osteoglossoidei discussed here is coextensive with the Osteoglossidae. These fishes share at least three synapomorphies:

Opercle oval or subsemicircular—Among osteoglossomorphs, an oval opercle (10[2]) is seen in Heterotidinae of Osteoglossoidei, while the opercle in all Osteoglossinae is subsemicircular (10[3]) (Ridewood, 1905).

First ural centrum with one or two fully developed neural spines—Except for notopterids, all the outgroup taxa lack a fully developed neural spine on u1, whereas in Osteoglossoidei there is either one or two fully developed neural spines (14[1/2]). The similar feature seen in notopterids is here considered to be convergent.

Pterygo-quadrate area behind and below orbit completely covered by infraorbitals—All Osteoglossoidei share this state (16[1]). However, this area is more or less exposed in all the out-group taxa (personal observation).

C. Diagnosis of Notopteroidei

Because of the elimination of lycopterids and hiodontids from this suborder, and also because of the combination of mormyroids with notopterids and fossil ostariostomids in the same suborder (Fig. 2), the content of the Notopteroidei is radically different from any previously defined concept. The following five features are thought to be diagnostic of the revised Notopteroidei:

Io2 medium-sized and subrectangular—The second infraorbital has three basic states. In most taxa, it is a small, slender bone (15[0]). Heterotidinae have a large, irregular trapezoid io2 (15[2]). All the notopteroids have a subrectangular and medium-sized io2 (15[1]).

Io3 subrectangular—The third infraorbital also has three basic states in teleosts. As shown in Fig. 1, the io3 in most taxa is short, posteriorly deep, and fanlike. A subrectangular io3 (30[1]) occurs in all notopteroids except *Notopterus*.

Infraorbital ledge formed by io1 and io2—All notopteroids have an infraorbital ledge (new term) (31[1]) consisting of a prominent lateral extension of the orbital margin of io1 and io2 (e.g., †*Ostariostoma* and *Notopterus*). This structure, which differs from the "subocular shelf" traditionally used to define the inward extension of the bony lamina from infraorbitals (see Rojo, 1991), is absent in all the outgroup taxa.

Infraorbital canal on io1 to io3 in an open groove—This character state (32[1]) is shared by all Notopteroidei. A convergent condition is seen in some primitive euteleosts such as osmerids (Wilson and Williams, 1991), but it is not seen in any of the basal teleosts including elopomorphs or other osteoglossomorphs.

Utriculus completely separated from sacculus and lagena—According to Greenwood (1973) and Lauder and Liem (1983), the utriculus in all notopterids and mormyroids is completely separated from the sacculus and lagena (33[1]). Although we are not able to test this state in fossil taxa, this feature has not been reported in any other extant teleost.

D. Reassessment of the Systematic Position of †Ostariostoma

†*Ostariostoma* Schaeffer (1949) is a monotypic genus named on a single specimen from the Late Cretaceous or Paleocene of Montana. Although Schaeffer assigned this genus to the Clupeoidei, he mentioned another possible relationship with the ancestral lineage of Ostariophysi. Grande and Cavender (1991)

admitted that †*Ostariostoma* is perplexing because of both insufficient preservation and features actually preserved on the specimen. They tentatively believed that †*Ostariostoma* is closely related to Osteoglossomorpha (possibly a hiodontoid) based on presence of 16 branched caudal-fin rays, absence of supramaxilla, and possibly a hiodontid-like anal fin.

We agree with the osteoglossomorph relationships of †*Ostariostoma* but disagree with a hiodontiform relationship because it does not share the hiodontoid synapomorphies: 4[1] seven-rayed pelvic fin *vs* four-rayed in †*Ostariostoma*, 10[1] opercle an irregular parallelogram *vs* irregular trapezoid in †*Ostariostoma*, 17[1] dorsal arm of posttemporal more than twice as long as ventral arm *vs* less than twice as long in †*Ostariostoma*, 18[1/2] hyomandibula double-headed *vs* single-headed in †*Ostariostoma*, and 19[1] dermosphenotic triradiate *vs* trapezoidal in †*Ostariostoma*.

Although most of our observations of †*Ostariostoma* (Fig. 3) agree with those of Grande and Cavender (1991), we suggest: (1) the antorbital (ao) designated by Grande and Cavender is more likely to be the displaced nasal based on its gutterlike shape and its strongly curved posterior end; (2) the possible nasal (na?) labeled by Grande and Cavender is possibly the mesethmoid; (3) the large bone labeled as the lacrimal in Grande and Cavender (1991) is two bones, the io1 and io2; (4) the second infraorbital (io2) of Grande and Cavender is the third infraorbital; (5) the temporal fenestra (tf) may be the homologue of the preepiotic fossa as previously discussed; (6) the epural (epu) designated by Grande and Cavender may be the distal part of the broken neural spine of the first preural; (7) the bone labeled "?" in their paper is here thought to be the epural; and (8) we suggest that there is only one uroneural.

Our study confirms the osteoglossomorph relationship of †*Ostariostoma* (though it has some interesting convergences with certain osmeroids) because it shares with all osteoglossomorphs the following: 2[1] epurals decreased to one, 3[1] absence of supramaxilla, 5[1] absence of supraorbital, 6[1] primary bite between parasphenoid and basihyal, and 11[1] io4 fused with io5. We suggest that †*Ostariostoma* is related to Osteoglossiformes by three synapomorphies: 7[2] 15 or fewer branched caudal fin rays, 12[2] nasal gutter-like, and 50[1] uroneurals fewer than three. When we move this taxon from the Osteoglossiformes to the Hiodontiformes, tree length is increased from 72 to 81 steps. We suggest that †*Ostariostoma* forms a clade with the Notopteroidei based on four synapomorphies: 15[1] io2 rectangular and medium-sized, 30[1] io3 an irregular rectangle, 31[1] presence of infraor-

bital ledge, and 32[1] infraorbital canal on io1 to io3 an open groove.

VI. Classification of Osteoglossomorpha

In conclusion, we adopt the phylogeny of Osteoglossomorpha shown in Fig. 2 and propose the following classification (using the sequencing convention):

Superorder Osteoglossomorpha
    Family †Lycopteridae (possibly also including †*Tongxinichthys*)
        †*Lycoptera*
    Order Hiodontiformes
    Family Hiodontidae
        †*Yanbiania*
        †*Plesiolycoptera*
        †*Eohiodon*
        Hiodon
    Order Osteoglossiformes
        †*Tanichthys* (see NOTE ADDED IN PROOF)
    Suborder Osteoglossoidei
    Family Osteoglossidae
        Subfamily Osteoglossinae
            †*Phareodus* (including †*Phareoides* and †*Brychaetus*)
            Pantodon
          Unnamed genus group
            †*Singida*
            Osteoglossum
            Scleropages
        Subfamily Heterotidinae
            †*Laeliichthys*
            †*Sinoglossus*
            Heterotis
            Arapaima
    Suborder Notopteroidei
    Family †Ostariostomidae (possibly including †*Thaumaturus*)
        †*Ostariostoma*
    Family Notopteridae
        Xenomystus
        Notopterus
        Papyrocranus
    Family Mormyridae
    Family Gymnarchidae

VII. Some Implications

We suggest that the fossil and extant osteoglossomorphs represent a potentially interesting case study of the role of continental movements in the radiation

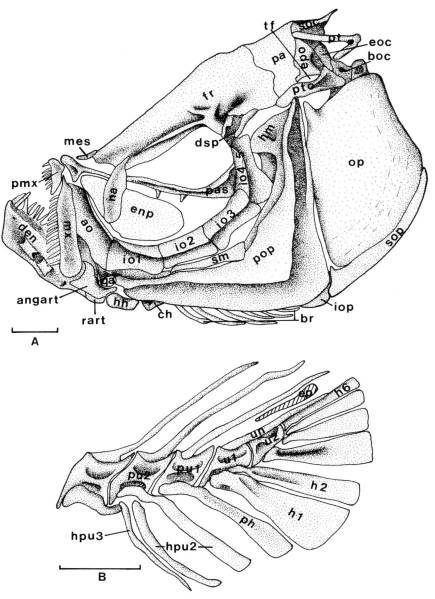

FIGURE 3 The cranial (A) and caudal (B) skeleton of †*Ostariostoma wilseyi*, drawn from the cast of PU (Princeton University) 14728. Scale bars for both A and B equal 1 mm. Abbreviations: angart, angulo-articular; ao, antorbital; boc, basioccipital; br, branchiostegals; ch, ceratohyal; hh, hypohyal; den, dentary; dsp, dermosphenotic; enp, endopterygoid; eoc, exoccipital; ep, epural; epo, epiotic; fr, frontal; h1, 1st hypural; h2, 2nd hypural; h6, 6th hypurals; hm, hyomandibula; hpu2/3, haemal spine on 2nd or 3rd preural; io1, 1st infraorbital; io2, 2nd infraorbital; io3, 3rd infraorbital; io4–5, fused 4th and 5th infraorbitals; iop, interopercle; mes, mesethmoid; mx, maxilla; na, nasal; op, opercle; pa, parietal; pas, parasphenoid; ph, parhypural; pmx, premaxilla; pop, preopercle; pt, posttemporal; pto, dermopterotic; pu1, 1st preural; pu2, 2nd preural; qu, quadrate; rart, retroarticular; sm, symplectic; soc, supraoccipital; sop, subopercle; tf, temporal fenestra (temporal fossa or pre-epiotic fossa); u1/2, 1st and 2nd urals; un, uroneural.

and distribution of freshwater organisms and of the importance of fossil taxa in the reconstruction of phylogenetic relationships of extant species. Osteoglossomorpha are a geologically ancient group of freshwater fishes (Fig. 4), old enough that their distribution should have been extensively influenced by major plate tectonic events (see Nelson, 1969b). Although the distributions of extant members seem to be consis-

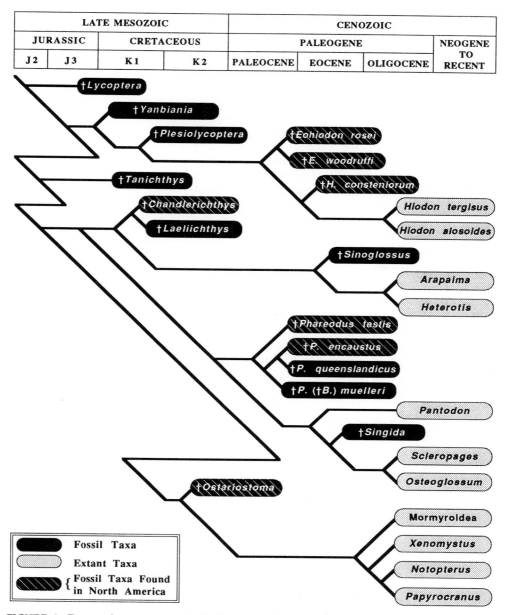

LATE MESOZOIC				CENOZOIC			
JURASSIC		CRETACEOUS		PALEOGENE			NEOGENE TO RECENT
J2	J3	K1	K2	PALEOCENE	EOCENE	OLIGOCENE	

FIGURE 4 Temporal occurrence and phylogenetic relationships of major osteoglossomorph taxa discussed in this chapter.

tent with plate boundaries, the fossil record of the group is more extensive than commonly realized. It is reasonable that members of the group have lived on all the major continental areas in the past, since most of these areas were connected during the Late Jurassic.

Our study also suggests that the Osteoglossomorpha are a well-defined clade, contrary to some expressed doubts. However, it is also true that many incompletely studied fossil taxa have been assigned to the group in the past. Therefore, it is also reasonable

that careful study of many of these will show that they are not osteoglossomorphs. We hope that we have contributed to this effort by clarifying some of the important diagnostic features of the group and its best-known members.

VIII. Summary

The Osteoglossomorpha are a geologically ancient group of Teleostei. Although their relationship to other major teleostean clades is not well-resolved, the monophyly of fossil and extant Osteoglossomorpha, including †Lycopteridae, is supported by the following: presence of fewer than two epurals, absence of supraorbital, presence of primary bite between parasphenoid and basihyal, and 16 or fewer branched caudal fin rays. In addition, all except †Lycopteridae lack a supramaxilla and have fused fourth and fifth infraorbitals suggesting that †Lycopteridae are stem-group osteoglossomorphs. Extant osteoglossomorphs share two additional synapomorphies: presence of paired tendon bones on the second hypobranchial, and intestine coiling to the left of the esophagus and stomach.

Hiodontiformes are the sister-group of Osteoglossiformes, which consist of Osteoglossoidei and Notopteroidei. Fishes in the Osteoglossoidei share at least three synapomorphies: opercle oval or subsemicircular, first ural centrum bearing one or two fully developed neural spines, and infraorbitals completely covering the pterygo-quadrate area posteroventral to orbit. The Notopteroidei, containing †Ostariostomidae, Mormyridae, Gymnarchidae, and Notopteridae, share five synapomorphies: second infraorbital more or less rectangular and moderate in size, third infraorbital subrectangular, presence of infraorbital ledge on first and second infraorbitals, infraorbital canal open on first to third infraorbitals, and utriculus completely separated from sacculus and lagena.

Potentially, osteoglossomorphs represent an interesting case study of the role of continental movements in the radiation and distribution of freshwater organisms and of the importance of fossil taxa in the reconstruction of phylogenetic relationships of extant species.

Acknowledgments

Our best thanks to Lance Grande for helpful discussions and to Grande and John Maisey for the first author's opportunity to study comparative collections. We are most grateful to Colin Patterson and the late P. Humphrey Greenwood for valuable comments and suggestions on the manuscript. Financial support was from Natural Sciences and Engineering Research Council of Canada operating grant A9180 to the second author and a Research Assistantship from the University of Alberta to the first author.

NOTE ADDED IN PROOF The generic name †Tanichthys Jin 1991 is preoccupied by Tanichthys Lin 1932 [Lingnan Sci. J. Canton, 11(3), 379–383], an extant genus of Chinese cyprinid fishes. Jin (1994) has recently published the replacement name †Tanolepis Jin 1994 [Vertebrata PalAsiatica, 32(1), 70] for the fossil osteoglossomorph.

References

Arratia, G. (1987). Anaethalion and similar teleosts (Actinopterygii, Pisces) from the Late Jurassic (Tithonian) of southern Germany and their relationships. Palaeontographica, Abt. A 200, 1–44.

Arratia, G. (1991). The caudal skeleton of Jurassic teleosts; a phylogenetic analysis. In "Early Vertebrates and Related Problems of Evolutionary Biology" (M.-M. Chang, Y.-H. Liu, and G.-R. Zhang, eds.), pp. 249–340. Science Press, Beijing.

Arratia, G. (1996). Reassessment of the phylogenetic relationships of certain Jurassic teleosts and their implications on teleostean phylogeny. In "Mesozoic Fishes: Systematics and Paleoecology" (G. Arratia and G. Viohl, eds.), pp. 219–242. Verlag Dr. Fredrich Pfeil, Munich.

Berg, L. S. (1940). Classification of fishes both recent and fossil. Tr. Zool. Inst. Leningr. 5, 87–512 (English transl., Ann Arbor, MI, 1947).

Cavender, T. (1966). Systematic position of the North American Eocene fish, "Leuciscus" rosei Hussakof. Copeia, pp. 311–320.

Chang, M.-M., and Chou, C.-C. (1977). On Late Mesozoic fossil fishes from Zhejiang Province, China. Mem. Inst. Vertebr. Paleontol. Paleoanthropol., Acad. Sin. 12, 1–59.

Cockerell, T. D. A. (1925). The affinities of the fish Lycoptera middendorffi. Bull. Am. Mus. Nat. Hist. 51, 313–317.

Da Silva Santos, R. (1985). Laeliichthys ancestralis, novo gênero e espécie de Osteoglossiformes do Aptiano da Formaç ao Areado, estado de Minas Gerais, Brasil. MME-DNPM, Geol. 27, Paleontol. Estrat. 2, 161–167.

Forey, P. L. (1973a). Relationships of elopomorphs. In "Interrelationships of Fishes" (P. H. Greenwood, R. S. Miles, and C. Patterson, eds.), pp. 351–368. Academic Press, London.

Forey, P. L. (1973b). A revision of the elopiform fishes, fossil and Recent. Bull. Br. Mus. (Nat. Hist.), Geol., Suppl. 10, 1–222.

Grande, L. (1979). Eohiodon falcatus, a new species of hiodontid (Pisces) from the late early Eocene Green River Formation of Wyoming. J. Paleontol. 53, 103–111.

Grande, L. (1985). Recent and fossil clupeomorph fishes with materials for revision of the subgroups of clupeoids. Bull. Am. Mus. Nat. Hist. 181, 231–372.

Grande, L. (1986). The first articulated freshwater teleost fish from the Cretaceous of North America. Palaeontology 29, 365–371.

Grande, L., and Cavender, T. M. (1991). Description and phylogenetic reassessment of the monotypic †Ostariostomidae (Teleostei). J. Vertebr. Paleontol. 11, 405–416.

Greenwood, P. H. (1966). The caudal fin skeleton in osteoglossoid fishes. Ann. Mag. Nat. Hist. [13] 9, 581–597.

Greenwood, P. H. (1970). On the genus Lycoptera and its relationship with the family Hiodontidae (Pisces, Osteoglossomorpha). Bull. Br. Mus. (Nat. Hist.), Zool. 19, 257–285.

Greenwood, P. H. (1973). Interrelationships of osteoglossomorphs. In "Interrelationships of Fishes" (P. H. Greenwood, R. S. Miles, and C. Patterson, eds.), pp. 307–332. Academic Press, London.

Greenwood, P. H., and Patterson, C. (1967). A fossil osteoglossoid fish from Tanzania (E. Africa). J. Linn. Soc. London, Zool. 47, 211–223.

Greenwood, P. H., Rosen, D. E., Weitzman, S. H., and Myers, G. S. (1966). Phyletic studies of teleostean fishes, with a provisional classification of living forms. Bull. Am. Mus. Nat. Hist. 131, 339–456.

Gregory, W. K. (1933). Fish skulls: A study of the evolution of natural mechanisms. Trans. Am. Philos. Soc. 23, 75–481.

Jin, F. (1991). A new genus and species of Hiodontidae from Xintai, Shandong. Vertebr. PalAsiat. 29, 46–54.

Kershaw, D. R. (1976). A structural and functional interpretation of the cranial anatomy in relation to the feeding of osteoglossoid fishes and a consideration of their phylogeny. Trans. Zool. Soc. London 33, 173–252.

Lauder, G. V., and Liem, K. F. (1983). The evolution and interrelationships of the actinopterygian fishes. Bull. Mus. Comp. Zool. 150, 95–197.

Leidy, J. (1873). Notice of remains of fishes in the Bridger Tertiary Formation of Wyoming. *Proc. Acad. Nat. Sci. Philadelphia* **25**, 97–99.

Li, G.-Q. (1987). A new genus of Hiodontidae from Luozigou Basin, east Jilin. *Vertebr. PalAsiat.* **25**, 91–107.

Li, G.-Q. (1994a). New Osteoglossomorphs (Teleostei) from the Upper Cretaceous and Lower Tertiary of North America and their phylogenetic significance. Ph.D. Thesis, University of Alberta, Canada.

Li, G.-Q. (1994b). Systematic position of the Australian fossil osteoglossid fish †*Phareodus* (= †*Phareoides*) *queenslandicus* Hills. *Mem. Queensl. Mus.* **37**(1), 287–300.

Li, G.-Q. (1996). A new species of Late Cretaceous osteoglossid (Teleostei) from the Oldman Formation of Alberta, Canada, and its phylogenetic relationships. *In* "Mesozoic Fishes: Systematics and Paleoecology" (G. Arratia and G. Viohl, eds.), pp. 285–298. Verlag Dr. Fredrich Pfeil, Munich.

Li, G.-Q., and Wilson, M. V. H. (1994). An Eocene species of *Hiodon* from Montana, its phylogenetic relationships, and the evolution of the postcranial skeleton in the Hiodontidae (Teleostei). *J. Vertebr. Paleontol.* **14**(2), 153–167.

Li, G.-Q., and Wilson, M. V. H. (1996). The discovery of Heterotidinae (Teleostei: Osteoglossidae) from the Paleocene Paskapoo Formation of Alberta, Canada. *J. Vertebr. Paleontol.* **16**(2), 198–209.

Maddison, W. P., and Maddison, D. R. (1992). "MacClade: Analysis of Phylogeny and Character Evolution, Version 3.04." Sinauer Assoc., Sunderland, MA.

Maisey, J. G. (1991). *Cladocyclus. In* "Santana Fossils: An Illustrated Atlas" (J. G. Maisey, ed.), pp. 190–207. TFH Publications, Neptune City, NJ.

Micklich, N. (1992). Ancient knights-in-armour and modern cannibals. *In* "Messel: An Insight into the History of Life and of the Earth" (S. Schaal and W. Ziegler, eds.), pp. 70–92. Oxford University Press (Clarendon), Oxford.

Müller, J. (1847). Fossile Fische. *In* "Reise in den aussersten norder und osten Sibiriens wahrend der Jahre 1843 und 1844" (A. Th. von Middendorff, ed.), Vol. 1, pp. 260–264. Kaiserlichen Akademie der Wissenschaften, St. Petersburg.

Nelson, G. J. (1968). Gill arches of teleostean fishes of the division Osteoglossomorpha. *J. Linn. Soc. London, Zool.* **47**, 261–277.

Nelson, G. J. (1969a). Gill arches and the phylogeny of fishes, with notes on the classification of vertebrates. *Bull. Am. Mus. Nat. Hist.* **141**, 475–552.

Nelson, G. J. (1969b). Infraorbital bones and their bearing on the phylogeny and geography of osteoglossomorph fishes. *Am. Mus. Novit.* **2394**, 1–37.

Nelson, G. J. (1972). Observations on the gut of the Osteoglossomorpha. *Copeia*, pp. 325–329.

Nelson, J. S. (1994). "Fishes of the World," 3rd ed. Wiley, New York.

Nybelin, O. (1974). A revision of the leptolepid fishes. *Acta Reg. Soc. Sci. Litt. Goth., Zool.* **9**, 1–202.

Patterson, C. (1967). Are the teleosts a polyphyletic group? *Colloq. Int. C.N.R.S.* **163**, 93–109.

Patterson, C. (1970). Two Upper Cretaceous salmoniform fishes from the Lebanon. *Bull. Br. Mus. (Nat. Hist.), Geol.* **19**, 205–296.

Patterson, C., and Rosen, D. E. (1977). Review of ichthyodectiform and other Mesozoic teleost fishes and the theory and practice of classifying fossils. *Bull. Am. Mus. Nat. Hist.* **158**, 81–172.

Ridewood, W. G. (1904). On the cranial osteology of the fishes of the families Mormyridae, Notopteridae, and Hyodontidae. *J. Linn. Soc. London, Zool.* **29**, 188–217.

Ridewood, W. G. (1905). On the cranial osteology of the fishes of the families Osteoglossidae, Pantodontidae, and Phractolaemidae. *J. Linn. Soc. London, Zool.* **29**, 252–282.

Rojo, A. L. (1991). "Dictionary of Evolutionary Fish Osteology." CRC Press, Boca Raton, FL.

Schaeffer, B. (1949). A teleost from the Livingston Formation of Montana. *Am. Mus. Novit.* **1427**, 1–16.

Shen, M. (1993). Fossil "osteoglossomorphs" in China and their implications on teleostean phylogeny. *Abstr., Symp. Mesozoic Fishes: Syst. Palaeoecol.*, Jura-Museum, Eichstätt, Germany.

Swofford, D. (1993). "PAUP: Phylogenetic Analysis Using Parsimony, Version 3.1.1." Smithsonian Institution, Washington, DC.

Taverne, L. (1979). Ostéologie, phylogénèse et systématique des téléosté ens fossiles et actuels du superordre des ostéoglossomorphes. Troisième Partie. *Mém. Cl. Sci., Acad. R. Belg.* **43**, Fasc. 3, 1–168.

Whitehouse, R. H. (1910). The caudal fin of the Teleostomi. *Proc. Zool. Soc. London*, pp. 590–627.

Wilson, M. V. H. (1978). *Eohiodon woodruffi* n. sp. (Teleostei, Hiodontidae) from the middle Eocene Klondike Mountain Formation near Republic, Washington. *Can. J. Earth Sci.* **15**, 679–686.

Wilson, M. V. H. (1992). Importance for phylogeny of single and multiple stem-group fossil species with examples from freshwater fishes. *Syst. Biol.* **41**, 462–470.

Wilson, M. V. H., and Williams, R. R. G. (1991). New Paleocene genus and species of smelt (Teleostei: Osmeridae) from freshwater deposits of the Paskapoo Formation, Alberta, Canada, and comments on osmerid phylogeny. *J. Vertebr. Paleontol.* **11**, 434–451.

Woodward, A. S. (1901). "Catalogue of the Fossil Fishes in the British Museum (Natural History)," Part IV. Br. Mus. (Nat. Hist.), London.

Yakovlev, V. N. (1965). Systematics of the family Lycopteridae. *Int. Geol. Rev.* **8**, 71–80 (Engl. transl.).

Appendix 1

Definitions of characters and character-states used in cladistic analysis

1. Length of neural spine on pu2 (Lnpu2): [0] shorter than npu3, [1] as long as npu3.
2. Epurals (Ep): [0] two or more, [1] one, [2] absent.
3. Supramaxilla (Smx): [0] present, [1] absent.
4. Pelvic fin rays (when present) (Vfr): [0] more than seven, [1] seven, [2] six or fewer.
5. Supraorbital (So): [0] present, [1] absent.
6. Primary bite between parasphenoid and basihyal (Pbm): [0] absent, [1] present.
7. Principal branched caudal fin rays (Cfr): [0] 17 or more, [1] 16, [2] 15 or fewer.
8. Paired tendon bones on 2nd hypobranchial (Tbhb): [0] absent, [1] present.
9. Intestine (Int): [0] coiling to right of stomach, [1] coiling to left of stomach.
10. Opercle (Op): [0] irregular trapezoid, [1] irregular parallelogram, [2] oval or kidney-shaped, [3] subsemicircular, [4] fan-shaped.
11. Fourth and 5th infraorbitals (Io4–5): [0] separate, [1] fused.
12. Nasal (Na): [0] tubular but straight, [1] tubular

and strongly curved, [2] gutter-like or irregularly subrectangular.

13. Branches of preoperculo-mandibular canal on horizontal arm of preopercle (Pomc): [0] separate, [1] connected with each other to form a horizontal groove.

14. Neural spine on u1 (Nu1): [0] absent or rudimentary, [1] one full, [2] two full.

15. Second infraorbital (Io2): [0] slender and small, [1] medium-sized and rectangular, [2] large, wide irregular trapezoid.

16. Pterygo-quadrate area behind and below orbit (P-qa): [0] not completely covered by infraorbitals, [1] completely covered by infraorbitals.

17. Dorsal arm of posttemporal (Dapt): [0] less than 1.5 times as long as ventral arm, [1] more than twice as long as ventral arm.

18. Hyomandibular heads (Hmh): [0] one, [1] two, separate, [2] two, connected.

19. Shape of dermosphenotic (Sdsp): [0] irregularly triangular or trapezoid, [1] triradiate.

20. "Cheek wall" formed by enlargement of io1 to io3 (CKW): [0] absent, [1] present.

21. Basihyal tooth plate (when present) and basibranchial tooth plate (Bhbb): [0] separate from each other, [1] fused.

22. Posterior edge of nasal (when it is gutter-like or subrectangular) (Pena): [0] straight or slightly curved, [1] strongly curved and extending backward.

23. First pectoral fin ray (Pfr1): [0] normal, [1] greatly enlarged and extremely long.

24. Angle of jaws (Ja): [0] anterior to middle vertical line of orbit, [1] between middle vertical line and posterior edge of orbit, [2] behind orbit.

25. Anterior portion of frontal (Fr): [0] narrower or slightly broader than posterior end, [1] about 1.5 times as broad as posterior end, [2] at least twice as broad as posterior end.

26. Ratio of depth to width of opercle (Rdwo): [0] less than 2, [1] greater than 2.

27. Shape of extrascapular when it is present (Ses): [0] expanded and more or less square or irregularly triangular, [1] slender and distinctly angular or branched.

28. Preopercle (Pop): [0] angular, [1] curved with no distinct horizontal arm.

29. Hyomandibula (Hm): [0] anteriorly not extended, [1] anteriorly extended to form a subtriangular anterior wing.

30. Shape of io3 (Sio3): [0] short, posteriorly deep and fanlike, [1] subrectangular, [2] long, posteriorly shallow and fanlike.

31. Infraorbital ledge formed by lateral extension of orbital margin of io1 and io2 (IOL): [0] absent, [1] present.

32. Infraorbital canal on io1 to io3 (Ioc): [0] enclosed in bony tube, [1] in open groove.

33. Utriculus (Utr): [0] attach to sacculus and lagena, [1] separate from sacculus and lagena.

34. Shape of anal fin (Ash): [0] similar in both sexes, [1] sexually dimorphic.

35. Anterior part of premaxilla (Pmx): [0] raised, [1] not raised, [2] lower than posterior part.

36. Posterodorsal projection of opercle (Pdpo): [0] absent, [1] present.

37. Mid-dorsal concavity on premaxilla (Mcpm): [0] absent, [1] present.

38. Belly keel (BK): [0] absent, [1] present.

39. Neural spine on pu1 (Npu1): [0] completely developed, [1] rudimentary, or absent.

40. Posterior end of maxilla (Pemx): [0] lying on angular, [1] lying on dentary.

41. Antorbital and first infraorbital (Aio1): [0] separate, [1] fused.

42. Tooth plate of basihyal (Bhtp): [0] present, [1] absent.

43. Posteroventral margin of opercle (Pmop): [0] rounded, [1] distinctly concave.

44. Orbitosphenoid (Osh): [0] present, [1] absent.

45. Mandibular barbels (MB): [0] absent, [1] present.

46. Mandibular (dentary) canal (Mdc): [0] enclosed in bony tube, [1] in open groove.

47. Anal fin and caudal fin (Afcf): [0] separate, [1] connected.

48. Fusion between pu1 (1st preural) and u1 (1st ural) (Fcv): [0] absent, [1] present.

49. Urostyle (formed by fusion of nspu1 and u1) (Ust): [0] absent, [1] present.

50. Number of uroneurals (Un): [0] three or more, [1] two or fewer.

9

Interrelationships of Elopomorph Fishes

P. L. FOREY *and* **D. T. J. LITTLEWOOD**
Department of Palaeontology
The Natural History Museum
Cromwell Road, London SW7 5BD, England

P. RITCHIE* *and* **A. MEYER**
Department of Biology
State University of New York at Stony Brook
Long Island, New York 11794

I. Introduction

This chapter takes as its starting point a previous publication of the same title (Forey, 1973a). That paper, arising from a history of evolutionary taxonomy, was written before molecular systematics had any marked influence on classification. It seems appropriate here to return to the same subject matter with some new approaches. In 1973 monophyly of the Elopomorpha, as proposed by Greenwood *et al.* (1966), was accepted and it continues to be accepted, but the scheme of elopomorph interrelationships proposed by Forey (1973a) has been justifiably criticized (e.g., Greenwood, 1977; Patterson and Rosen, 1977), and a number of alternative theories have been proposed (Fig. 1).

The purpose of this chapter is to evaluate the different theories of elopomorph interrelationship by examining a cladistic analysis of morphological data and by adding molecular data which has become available through the work of P.R. and A.M. (mtDNA) and D.T.J.L. (nuclear DNA). This also provides us with the opportunity to assess the relative contributions of these different types of data to a classification of elopomorphs.

Elopomorph fishes include tenpounders (*Elops*), *Megalops*, eels, the deep sea halosaurs and notacanths,

and the highly derived but poorly-known gulper eels (saccopharyngoids). These fishes were widely scattered in teleost classifications until Greenwood *et al.* (1966) united them in the superorder Elopomorpha based on the fact that representatives of all subgroups have a leptocephalus larva.[1] Some authors have disputed that the leptocephalus is a derived feature of elopomorphs, preferring instead to regard it as the generalized teleost larval condition (Harrison, 1966; Nybelin, 1971; Hulet and Robins, 1989). However, like Greenwood *et al.* (1966) and Smith (1984), we regard the leptocephalus as a specialized larva (Forey, 1973a,b) characterized by a leaf or ribbon shape, a gut which runs at the ventral mid-line to open just in front of the tail, and a body filled mostly by a mucinous pouch resorbed at metamorphosis during which there is marked shrinkage of the body and considerable change in fin positions and, especially, the anus.

Four additional elopomorph characters have been proposed (Forey, 1973a,b; Jamieson, 1991): presence

---

[1]Greenwood *et al.* (1966) hesitatingly suggested that elopomorphs and clupeomorphs might be closely related and placed them together in their Division I, which they subsequently named Cohort Taeniopaedia (in recognition of the leptocephalus larva which, however, is not seen in clupeomorphs) (Greenwood *et al.*, 1967). This action can be viewed as a reluctance to bury the category "Isospondyli," but Greenwood (1973) subsequently suggested that clupeomorphs were the sister-group of osteoglossomorphs and removed them from the Taeniopaedia, leaving elopomorphs as the only included taxon. The Taeniopaedia thus becomes an empty rank.

---

*Current address: Department of Ecology, Massey University, Private Bag 11-222, Palmerston North, New Zealand.

Copyright © 1996 by Academic Press, Inc.
All rights of reproduction in any form reserved.

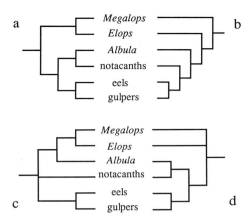

FIGURE 1 Four theories of interrelationships among elopomorph fishes. (a) Nelson (1973), Greenwood (1977); (b) Forey (1973a); (c) Greenwood *et al.* (1966); (d) Patterson and Rosen (1977).

of prenasal and rostral ossicles, presence of a pectoral splint, fusion between the angular and retroarticular bones within the lower jaw (Nelson, 1973), and sperm morphology. However, the first three of these characters are not universal among elopomorphs but are restricted to the presumed plesiomorphic members (elopiform, albuloids, and notacanthiforms).

The rostral ossicles surround the sensory canal as it passes from the bone-enclosed ethmoid commissure [the latter feature being a plesiomorphic condition in teleosts (see Nybelin, 1967)] to the infraorbital sensory canal. In albuloids the canal pierces the premaxilla for part of its traverse across the snout, and this is presumed to have resulted from the fusion of rostral ossicles with the underlying premaxilla. Those eels (e.g., *Anguilla*) that have retained the ethmoid commissure and the premaxillae (fused with the ethmoid in the adult) have no separate rostral ossicles. Most eels do not have an ethmoid commissure, but equally they have highly consolidated snouts that are usually regarded as resulting from a fusion between the mesethmoid and the vomer (the ethmovomer) and, debatably, the premaxillae. Until more extensive studies of the ontogeny of a variety of eels have been carried out, the generality of rostral ossicles and the association of the sensory canal with the premaxillae cannot be assessed. However, we do regard the presence of rostral ossicles as a putative character of elopomorphs, and it is included in our morphological data matrix (Appendix 2, character 10).

The pattern of fusion of lower jaw bones was explored by Nelson (1973) who suggested that the posterior end of the primitive teleost lower jaw comprises three separate bones: the dermal angular, endoskeletal articular, and retroarticular. Elopomorphs were re-

garded as derived in showing fusion between the retroarticular and angular while both components contribute to the jaw articulation. Most other teleosts show a fusion of the angular and articular. Nelson added two caveats to this conclusion. First, eels (anguilloids) and *Pterothrissus*, among albuloids, have a single bone at the posterior end of the lower jaw which presumably represents a completely fused angular–retroarticular–articular and, assuming that anguilloids were derived from elopoids and albuloids, he suggested that the *initial* fusion between the angular and retroarticular was a character. The validity of this character is justified by observations of Leiby (1979), who recorded a separate angulo-articular in young specimens of the ophichthid eel *Myrophis punctatus*. Second, Nelson (1973) noted that within osteoglossomorphs, which exhibit all patterns of fusion echoed in other teleost groups, *Hiodon* and *Gymnarchus* have the elopoid–*Albula* condition.

The morphology of the spermatozoa of many elopomorph taxa has been described by a number of authors and summarized by Jamieson (1991, p. 122, fig. 11.1). Jamieson (1991) recognizes five features commonly found among elopomorphs and two derived features universal to all examined members: a 9 + 0 flagellum (with nine peripheral and no central axomeres) and a division of the proximal centriole into two elongate bundles of four and five triplets forming a pseudoflagellum. Based on this brief review, it appears that the only unambiguously derived characters known to be universally present in elopomorphs are the presence of a leptocephalus larva and a particular sperm morphology.

Elops and *Megalops* have traditionally been grouped together as sister-taxa within Elopiformes. The grouping of *Elops* and *Megalops* (elopoids) is usually based on possession of plesiomorphic attributes, and Patterson and Rosen (1977) suggested equal ranking of *Elops*, *Megalops*, and the rest of the Elopomorpha (as Anguilliformes) (Fig. 1d), implying a basal trichotomy due to lack of information rather than conflicting data. The phylogeny proposed by Forey (1973a; Fig. 1b) implies that *Elops* is more closely related to anguilliforms than either is to *Megalops*. This conclusion was based on resemblances between *Megalops* and the Upper Jurassic †*Pachythrissops propterus* implying early separation of the *Megalops* lineage. This mixture of stratigraphic and phenetic argumentation was rightly criticized by Patterson and Rosen (1977), and in this chapter the two Recent taxa are kept as separate terminal taxa. *Megalops* can be recognized as a terminal taxon by virtue of the inflated intercalar surrounding the anterior diverticulum of the swimbladder (Greenwood, 1970).

Albula and *Pterothrissus* are usually grouped as Albuloidei (Greenwood *et al.*, 1966). Some of the albuloid characters are: (1) presence of a dorsal process on the ectopterygoid abutting the infraorbital, (2) unexpanded inner caudal fin rays, and (3) absence of muscle fibers attaching to the ligamentum primordium (Greenwood, 1977). Forey (1973a) proposed that albuloids are paraphyletic with *Pterothrissus* as the sister-group of notacanthiforms and anguilliforms. That theory was weakly grounded on a series of similarities (elongate snout with an inferior mouth, inturned head of the maxilla, and reduced ossification in the braincase) that are also found in *Albula*—one similarity (edentulous vomer) being a very widespread character in lower teleosts—and an alleged similarity in the sternohyoideus muscle that has subsequently been reinterpreted by Greenwood (1977). So, at best, these features might be interpreted as being albuloid–notacanthiform characters. There are, however, two seemingly derived characters shared by *Pterothrissus* and notacanthids among notacanthiforms. In both, the levator arcus palatini is divided into superficial and deep subdivisions, and the lateral line scales are deeply overlapped by the surrounding scales (Greenwood, 1977). The latter character is also shared with *Lipogenys*. *Pterothrissus* also has complete fusion between the posterior lower jaw bones (Nelson, 1973), a feature also seen in eels. The implication of these putative synapomorphies is that neither albuloids nor notacanthiforms are monophyletic.

That *Albula* is more closely related to notacanthiforms than to *Pterothrissus* has not been suggested previously, but there is one character shared by *Albula* and halosaurs among notacanthiforms (trough-like groove within the mandible housing the mandibular sensory canal which faces ventrally). The indication that the albuloids and notacanthiforms may not be monophyletic prompts our inclusion of representatives as separate terminal taxa in the morphological matrix.

The Halosauridae (Lyopomi) and Notacanthidae + *Lipogenys* (Heteromi) were grouped by Regan (1909) as the Heteromi and justified as a natural group by Marshall (1962). Marshall suggested several group characters such as projection of the snout, absence of mesocoracoid, postcleithrum, orbitosphenoid, and basisphenoid as well as the presence of a spine on the posterior margin of the maxilla (many of these characters are also seen in eels). He also described the derived structure of the swimbladder in these fishes: the swimbladder is divided into two unequal lobes, and the paired retiae and the paired gas glands are clustered at the posterior end of the pneumatic duct. Marshall pointed out that this swimbladder anatomy is also seen in eels (*Anguilla* and *Synaphobranchus* were compared), the chief difference being that, with exceptions, the retiae mirabilia of eels are bipolar whereas those of halosaurs and notacanths are unipolar. When Marshall wrote his paper the notacanth leptocephalus had not been described, although an interrogative footnote (Marshall, 1962, p. 261) predicted its discovery. This quickly followed (Mead, 1965; Harrison, 1966) and prompted Greenwood *et al.* (1966) to include them with elopomorphs. Greenwood (1977) provided the most carefully argued hypothesis of relationships among notacanthiforms in which, following a study of jaw muscles, he recognized the topology (Halosauridae (*Notacanthus* (*Lipogenys*, *Polyacanthonotus*))). Greenwood considered the Notacanthiformes to be the sister-group of Albuloidei (Fig. 1a).

Regan (1909) recognized anguilloids (Apodes) as distinct from other eel-shaped fishes by the fact that premaxillae are absent (Regan, 1909) or suturally united to the vomer and ethmoid (Regan, 1910). Later Regan (1912) provided a more thorough justification for monophyly, citing among other features restricted gill openings, reduced number of hyopalatine bones (hyomandibula, quadrate, and pterygoid remaining), elongate branchiostegals covering the branchial chamber, a single pair of upper dentigerous pharyngeals (but see Nelson, 1966), pterotic extending above prootic to contact the pterosphenoid, no basisphenoid (apparently a misidentification as it is the orbitosphenoid, not the basisphenoid, that is missing), no posttemporal, and no mesocoracoid. While these may all be derived features within teleosts, some are present in sporadic distribution across teleost subgroups. Eel synapomorphies are the following: displacement of the gill arches posteriorly so that a bony connection no longer exists with the neurocranium, expansion of the branchiostegals posterior and dorsal to the operculum, and forward extension of the pterotic to contact the pterosphenoid.

The classification of anguilloids is based largely on superficial (external) characters and currently is far from satisfactory. There have been few detailed comparative osteological studies of eels, although Nelson's (1966) study of gill arches provides a model in need of repeating with other skeletal systems. Regan's (1912) division of eels into those with separated frontals and those with fused frontals has generally been followed, and certain families appear clearly defined (Böhlke, 1989).

The Saccopharyngoidei (Lyomeri) include the deep sea gulper eels. The three families (Saccopharyngidae, Eurypharyngidae, and Monognathidae) are poorly known and little detailed osteological information is available, chiefly because specimens are rare. Nielsen and Bertelsen (1985) and Bertelsen and Nielsen (1987) provided good accounts of the external morphology of saccopharyngids and monognathids, and these have been summarized by Bertelsen et al. (1989). Tchernavin (1947a,b) provided the most complete anatomical descriptions to date showing that saccopharyngoids have a highly reduced skeleton in which it is sometimes difficult to homologize what remains with the more complete osteology of other teleosts. Indeed, Tchernavin (1946) considered the possibility that saccopharyngoids are not even bony fishes. As well as absence characters shared with eels (see data matrix and analysis) saccopharyngoids have no supraoccipital (a feature shared with some eels) or branchiostegals, and they lack the preoperculum, operculum, suboperculum, and possibly the interoperculum, as well as dermal bones in the pectoral girdle. They have peculiar papillae projecting from the lateral line organs, the median fin rays are neither segmented nor branched, the efferent arteries are joined both above and below the gill clefts (Tchernavin, 1946), and the gills are pouched with plume-like lamellae. All of these features confirm the monophyly of saccopharyngoids and demonstrate the extreme autapomorphy of members of this group.

II. Morphological Analysis

The morphological analysis is designed to evaluate the competing theories illustrated in Fig. 1. The data matrix includes representative elopomorph taxa to take account of the conflicting character distributions mentioned in the Introduction. For instance, both *Albula* and *Pterothrissus* are included instead of being lumped into a terminal Albuloidei; similarly halosaurs and notacanths are treated as separate terminals. Among the anguilloids only Anguillidae and Ophichthidae were selected for inclusion as molecular sequences for certain of the included species are available. With the available molecular data two ophichthids were sequenced, but the codings for the morphological characters were the same. Our sample of anguilloid taxa is admittedly meager, but it is one that reflects the poor state of anatomical knowledge for this group. Three osteoglossomorphs are included as potential outgroup taxa, and these match the taxa for which whole or partial molecular sequences are available. In addition *Clupea* is selected as a taxon cladistically more derived than elopomorphs. Mitochondrial genes were sequenced for two additional clupeomorphs (see below) but the morphological characters were constant in these taxa, so *Clupea*

stands as representative in the morphological analysis.

The relationships of elopomorphs to other lower teleosts is not questioned here. Elopomorphs are accepted as more derived than osteoglossomorphs (Patterson, 1977). Elopomorpha and Clupeocephala (*sensu* Patterson and Rosen, 1977) share two features: only two uroneurals extending forward beyond U2 (Patterson, 1977) and epipleural intermusculars in the abdominal and anterior caudal regions (Patterson, 1977). Patterson and Johnson (1995) identified additional derived characters shared by elopomorphs, clupeomorphs, and other lower teleosts; for example, in these taxa at least some epineurals are not fused with neural arches, and the epineurals and epipleurals are forked proximally.

Some of the difficulties in constructing a data matrix for all elopomorph fishes stem from a lack of information on anguilloid and saccopharyngoid morphology. This lack of knowledge is exacerbated by the apparently reductive nature of the skeleton of these elopomorphs. Many of the features of the head and tail skeleton of eels and saccopharyngoids appear highly simplified, but uncertainty about whether this is a result of loss, fusion, or primitive absence causes problems in coding certain characters. The problem of reductive characters in fish systematics has been addressed (Begle, 1991). There appear to be two solutions: (1) code for absence and allow parsimony to determine character evolution, or (2) if there is ontogenetic evidence of fusion or loss, code this an additional state. The latter approach is preferred, but the appropriate knowledge is rarely available for studies so far carried out on eels (but see Leiby, 1979, on *Myrophis*); it is absent for saccopharyngoids. In the wider context of teleostean phylogeny many characters describing absences might reasonably be regarded as describing losses, but we have preferred to code for absence. Fifty-six characters were coded for 13 terminal taxa. Descriptions of the characters are given in Appendix 1, and the character matrix is given in Appendix 2. In some cases we did not have information for taxa (halosaurs and saccopharyngoids in particular), so question marks are inserted. Some elopomorph characters (10, 34, 54, 55) are included. Morphological and molecular data were analyzed using PAUP version 3.1.1 (Swofford, 1993).

The morphological analysis resulted in a single most parsimonious tree (Fig. 2). The traditional higher-level relationships, with *Clupea* and elopomorphs being more derived than osteoglossomorphs, are maintained. However, some of the synapomorphies justifying this node are spurious because *Hiodon*

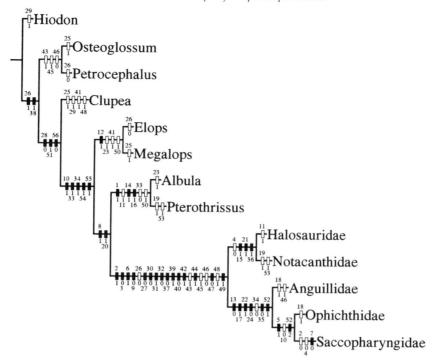

FIGURE 2 Result of morphological analysis using 56 characters described in Appendix 1. Branch-and-bound analysis using PAUP 3.1.1 (Swofford, 1993); all characters unordered. *Hiodon* used as outgroup. Length 80, C.I. = 0.705, R.I. = 0.823. Character numbers optimized as favoring reversals and plotted on tree using CLADOS; black boxes indicate synapomorphies (irrespective of whether they are an acquisition or a loss of the 1 state); open boxes indicate homoplasies. The paraphyletic status of the osteoglossomorph genera is an artifact of rooting on *Hiodon*. Exactly the same topology for ingroup taxa is obtained if all osteoglossomorphs are constrained as a monophyletic outgroup. However, the character optimizations are slightly different.

was used as the outgroup taxon. Thus characters 28 (hypobranchial 2 with ventral process) and 56 (intestine passing to the left of the stomach) would be resolved as osteoglossomorph synapomorphies if more plesiomorphic teleosts were included.

Within elopomorphs a number of observations about the tree (Fig. 2) can be made. The node linking *Elops* + *Megalops* is poorly supported by a single synapomorphy (character 12, the medial position of the posterior opening of the mandibular sensory canal). However, Patterson and Rosen (1977) pointed out that this character is widespread among stem-lineage teleosts (ichthyodectiforms and various *Leptolepis* spp.). Therefore this may be a plesiomorphic character for teleosts. We conclude that the morphological characters we have considered fail to resolve the relationships of *Elops*, *Megalops*, and higher elopomorphs (see also Patterson and Rosen, 1977; Fig. 1d). One further character that may link these taxa is the fusion of distal cartilages associated with hemal spines of PU3, PU2, parhypural, and

hypural 1 (Fujita, 1990); in teleosts, these cartilages, when present, are primitively separate from one another. Before the validity of this character may be fully assessed the condition of the distal cartilages needs to be checked in a wider range of taxa.

The node specifying a sister-group relationship between notacanthiforms and anguilliforms is well supported with numerous synapomorphies, but it is worth pointing out that many of them (characters 3, 9, 27, 30, 32, 37, 40, and 47) are resolved as losses, and some of these may be expected to occur, in sporadic fashion, within other teleost groups. Additionally, character 36 (pelvic fin webs joined in the midline) was also resolved as a synapomorphy at this node, but this resulted from the fact that illogical codings were used for anguilloids and gulpers, which lack pelvic fins. More sensibly this character is interpreted as a synapomorphy of notacanthiforms (Fig. 2). "Anguilloids" are resolved as paraphyletic; ophichthids and gulpers are more closely related to each other than either is to anguillids.

The anatomy of the leptocephalus larva is a largely untapped source of data for the discovery of interrelationships of elopomorphs. Considerable literature concerns matching larvae with adults and species recognition (see Richards, 1984; Castle, 1984; Smith, 1989, for summaries). Leptocephali of tailed elopomorphs, and of representatives of most families of notacanthiforms, anguilloids, and saccopharyngoids, are known; and leptocephali of certain higher-level taxa can be recognized (e.g., Ophichthidae; Leiby, 1984). Notacanthiform, anguilloid, and saccopharyngoid larvae are specialized relative to leptocephali of *Elops*, *Megalops*, *Albula*, and *Pterothrissus*. The latter have forked tails separated from short-based dorsal and anal fins. Notacanthiform, anguilloid, and saccopharyngoid larvae have small rounded or pointed tails that are usually greatly simplified and continuous with the dorsal and/or anal fins.

The higher number of myomeres in notacanthiform and anguilliform elopomorphs (100–200 *versus* 51–92 in forked-tailed elopomorphs; see Richards, 1984; Castle, 1984; Smith, 1989) reflects adult morphology, but there are some anguilliforms with fewer than 100 myomeres. Notacanthiform and saccopharyngoid leptocephali may be uniquely derived in having V-shaped (*versus* W-shaped) myomeres, and notacanthiforms may be unique in having vertically elongated eyes ("*Tiluropsis*" and "*Leptocephalus attenuatus*").

III. Molecular Analysis

We were able to examine sequences for the small subunit nuclear (18S) rRNA gene as well as mitochondrial 12S and 16S rRNA genes. The taxon data base for nuclear and mtDNA were different, although all taxa for which nuclear DNA was available were sequenced for mtDNA. The taxa are listed in Appendix 3.

DNA Isolation and Purification

No fresh specimens were available for analysis, and total genomic DNA was extracted from alcohol-preserved samples of white muscle (1 volume tissue to 5 volumes absolute ethanol). Prior to extraction, tissues were soaked in two to three washes of 10 mM Tris-HCl (pH 8.0), 0.1 mM EDTA and one to two washes of dH$_2$O. Approximately 0.01–0.1 g of tissue DNA was extracted either by (1) the CTAB extraction method adapted from Doyle and Doyle (1987) by M. Black, Rutgers University, New Jersey, with modifications described in Littlewood (1994), or (2) standard proteinase K/SDS dissolution and phenol-chloroform ex-

traction (Kocher *et al.*, 1989; Sambrook *et al.*, 1989). Ethanol precipitated DNA was resuspended in either dH$_2$O or TE.

Gene Amplification and Isolation

Both mitochondrial and nuclear genes were amplified using PCR (Saiki *et al.*, 1988) with previously published primers and standard protocols. Small subunit nuclear (18S) rRNA genes were cloned prior to sequencing, whereas mitochondrial (12S and 16S) rRNA genes were sequenced directly. Amplified products were gel-purified.

18S Nuclear rRNA Gene

Approximately 1800 base pairs of the nuclear 18S-like rRNA gene were amplified by "hot start" PCR (Hosta and Flick, 1991) using universal primers of Medlin *et al.* (1988) or of Embley *et al.* (1992) and reaction buffers and conditions detailed in Littlewood (1994).

12S and 16S Mitochondrial rRNA Genes

Two segments were amplified, one from the large (16S) and one from the small (12S) mitochondrial ribosomal genes, using previously published primers (Kocher *et al.*, 1989; Palumbi *et al.*, 1991) and reaction conditions (Bargelloni *et al.*, 1994). Double-stranded PCR- products were used to generate single-stranded DNA of both strands for direct sequencing, using asymmetric PCR (Gyllensten and Erlich, 1988).

Cloning of 18S rRNA Genes

PCR products were electrophoresed on 1% agarose/TAE gels, excised, and purified with QIAEX (QIAGEN Inc.). Products were cloned directly into pGEM-T (Promega) using the given protocol (see also Sambrook *et al.*, 1989). Recombinants were identified by restriction analysis of miniprep DNA. Each recombinant vector was grown in the JM109 strain of *Escherichia coli* and purified using column purification kits (Magic MiniPreps, Promega; midi- and maxi-prep columns, QIAGEN Inc.).

Sequencing

Double-stranded plasmid DNA was alkaline-denatured and sequenced using the Sanger dideoxy-sequencing method (Sanger *et al.*, 1977) (Sequenase v. 2.0, USB) with 10% dimethyl sulphoxide added to all stages of the reaction (Winship, 1989). Plasmid primers (T7 and SP6) and internal sequencing primers (listed in Littlewood and Smith, 1995) were used to sequence the 18S rRNA genes.

Single-stranded mtDNA was concentrated, desalted, and spun in columns (Millipore: Ultrafree-MC 30,000) and sequenced directly, also using Sequenase, with previously published primers (Kocher *et al.,* 1989).

In all cases both strands were sequenced. Consensus sequences for each species were assembled using AssemblyLIGN (IBI, Inc.) or ESEE (Cabot and Beckenbach, 1989). The entire length of the 18S rRNA gene was sequenced for a total of 7 taxa, and approximately 535 bp and 345 bp of 16S and 12S mitochondrial rDNA were sequenced for each of 14 taxa.

Alignment

Molecular sequences were aligned by eye and by using the computer program VSM (Christen, 1993); work proceeded from the highly conserved regions, progressively adding more divergent regions. Alignment posed virtually no problem with the 18S sequences, whereas large areas of ambiguity remained in the final alignment of the mitochondrial data even with the aid of the alignment program CLUSTAL (Higgins and Sharp, 1988). The full alignments and regions used for subsequent phylogenetic analysis are given in Appendix 4.

Phylogenetic trees were constructed using parsimony (PAUP version 3.1.1; Swofford, 1993). All characters were given equal weight, and missing bases were scored as a fifth base. Searches were made with the heuristic option for all but one data set (18S). Bootstrap replicates (1000) were carried out to establish the robustness of topologies.

Results

A total of 2750 unambiguously aligned sites were available for analysis yielding 579 variable sites (Appendix 4). The number of phylogenetically informative sites is 117, 255, and 33 from the 12S, 16S, and 18S rRNA genes, respectively. *Hiodon* was the outgroup taxon in each of the analyses. We conducted various experiments with the molecular data, including all combinations of sequences for individual genes and transversion parsimony. Only a few results are given here and these cover the spectrum from poor to good performance. By this we do not mean to imply that we ignore the tree topologies that bear no relationship to the morphological tree. But rather we acknowledge that many of the analyses, other than the combined molecular data, were poorly supported by bootstrap

values (see following discussion) and that the optimal trees were only one or a few steps shorter than suboptimal trees with different topologies (for all except the 18S analysis where exhaustive search was used, suboptimal trees were built using MacClade version 3; Maddison and Maddison, 1992). Separate analyses of the individual genes were carried out, but the results were very unsatisfactory: not only did they give unusual trees but the bootstrap support for most of the nodes was very low.

For the 18S sequences, data were available for only seven taxa (see Appendix 3). An exhaustive search found a single tree (length = 235, CI = 0.667, RI = 0.545) with the following topology: ((*Ophichthus, Echiophus*)(*Clupea*(*Elops*(*Albula,* *Megalops*)))). Bootstrap support was less than 50% for most of the nodes, except for the ophichthid eels (*Ophichthus* and *Echiophus*) which were strongly supported as a monophyletic group (98% bootstrap value).

For the 12S data (13 taxa available) four most parsimonious trees were found using heuristic search (length = 428, CI = 0.508, RI = 0.465), the strict consensus of which had the following topology: ((*Stolothrissa*(*Limnothrissa, Clupea*))((*Elops, Megalops*)((*Ophichthus, Echiophis*)(*Albula, Notacanthus*)(*Anguilla, Eurypharynx*)(*Osteoglossum, Petrocephalus*)))). Here the osteoglossomorphs (*Osteoglossum, Petrocephalus*) were grouped with three other elopomorph subgroups in a tetrachotomy to the exclusion of the clupeomorphs (*Stolothrissa, Limnothrissa,* and *Clupea*).

For the 16S data (13 taxa available) the heuristic search found two most parsimonious trees (length = 415, CI = 0.512, RI = 0.415), the strict consensus of which had the following topology: (((*Albula, Eurypharynx*)((*Elops, Megalops*)*Notacanthus*)((*Ophichthus, Echiophis*)*Anguilla*)((*Clupea, Limnothrissa*)*Stolothrissa*))(*Osteoglossum, Petrocephalus*)).

In both mitochondrial gene analyses bootstrapping strongly supported nodes for monophyletic ophichthid eels (*Ophichthus, Echiophis*), elopoids (*Elops, Megalops*), and a *Clupea* + *Limnothrissa* clade.

The mitochondrial and nuclear DNA data were added together, and for those taxa for which mitochondrial sequences were not available question marks were substituted. A single most parsimonious tree resulted, and this is shown in Fig. 3. Elopomorphs group together, and there is agreement between this and the morphological tree except that *Notacanthus* is the sister-group of *Albula* here rather than the sister-group to anguilliforms and that this clade was resolved as the most plesiomorphic. Also, *Anguilla* is resolved as the sister-group to ophichthids rather than the sister-group of ophichthids + saccopharyngoids. Bootstrap values are low for most nodes with the ex-

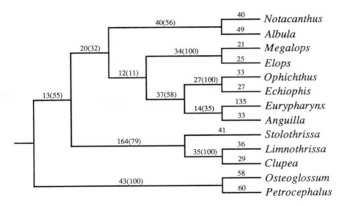

FIGURE 3 Result of combined molecular analysis using sequences from 18S (nuclear) and 12S and 16S (mitochondrial) genes. Variable site positions used are given in Appendix 4. Branch lengths are given with bootstrap values in parentheses. Heuristic search. Tree length = 1085, CI = 0.517, RI = 0.483.

ception of that supporting the monophyly of ophichthid eels, *Elops* + *Megalops*, the two internal nodes specifying the clupeomorphs, and the osteoglossomorphs. One further feature of note is the very long branch lengths leading to *Eurypharynx* and to the clupeoid fishes. Elsewhere the branch lengths are much shorter and approximately equal.

IV. Total Evidence

The combined molecular and morphological information allowed a single most parsimonious tree to be found (Fig. 4). This tree has the same topology as that found with morphology plus mtDNA, and it is congruent with the tree based on morphology alone.

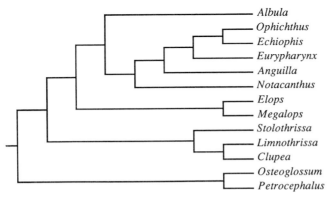

FIGURE 4 Total evidence tree from combining all molecular and morphological data. Heuristic search. Tree length = 1166, CI = 0.531, RI = 0.501.

We argue that the total evidence approach provides the best estimate of the phylogenetic relationships of elopomorphs. The molecular data do not provide a strong signal except in one important respect: resolution of *Elops* and *Megalops* as sister-groups. This signal came through in both individual and combined gene analyses.

V. Fossil Elopomorphs

The elopomorphs have a rich but patchy fossil record extending back to at least the Lower Cretaceous (Valanginian), and those taxa that can unambiguously be referred to modern clades may be used to date internal nodes. However, many elopomorph fossils have generalized features that result in their placement as *incertae sedis* at various ranks. No good evidence indicates that the inclusion of fossils will overturn the phylogenetic tree given in Fig. 4, although their consideration may call some homologies into question. The stratigraphic distribution of elopomorphs is illustrated in Fig. 5.

For the *Elops* + *Megalops* clade the fossil record extends to the Middle Cenomanian for *Elops* [represented by †*Davichthys* (Forey, 1973b)], and to the Albian for *Megalops* [represented by †*Elopoides* (Wenz, 1965)]. The generalized characters of *Elops* and *Megalops* make it difficult to assign fossils to the clade; however, there are a number of older taxa traditionally referred to the elopiforms, but these should be more correctly referred to as Elopomorpha *incertae sedis*.

Based on the structure of the caudal skeleton which has a compound neural arch associated with PU1 and U1, Patterson and Rosen (1977, p. 139) refer the Valanginian †*Anaethalion vidali* to the Elopomorpha. This character is shared with all tailed elopomorphs (a paraphyletic group). There are also some similarities between †*A. vidali* and *Megalops* in the elongate anal fin with 26 rays (22–27 in *Megalops*) and the very long first anal pterygiophore that almost contacts the vertebral centra. It is therefore possible that the dichotomy between *Elops* and *Megalops* is older than estimated by approximately 30 million years.

The albuloids (*Albula* and *Pterothrissus*) have a fossil record extending to the Middle Cenomanian where lineages leading to both of the modern genera may be recognized. Here †*Lebonichthys* bears distinct similarities with *Albula* in the possession of rounded crushing teeth, weakly ossified autopalatine, and expanded neural and hemal spines borne by PU2 and PU3 (†*L. lewisi*). †*Hajulia* appears to be a *Pterothrissus* relative showing a slightly elongated dorsal fin and a

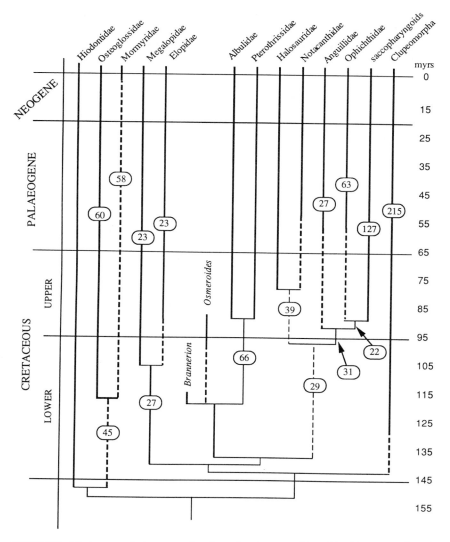

FIGURE 5 Phylogeny of elopomorphs plus other taxa considered in this paper plotted against time to show degree of agreement between the phylogeny and stratigraphy. The numbers represent nucleotide character changes from optimizing the combined molecular data set on the total evidence tree. The long branch leading to the gulper eels is probably realistic, but that leading to Clupeomorpha, represented in the data set by three taxa, is greatly overestimated because of poor taxon sampling. The fossil record of elopomorphs is briefly discussed in the text. For information on the earliest clupeomorph see Patterson (1993). Within the osteoglossomorphs the Osteoglossidae are known as far back as the Danian (65 million years). However, most classifications agree that Osteoglossidae are the sister-group of the Arapaimidae. Thus, this lineage has been drawn back to the earliest occurrence of a recognizable arapaimid (Patterson, 1993) from the Aptian (circa 120 million years). The long-range extension demanded for eels (circa 70 million years) may be artifactual since some Cretaceous eels may belong to modern families.

half-length neural spine upon PU2. The Campanian †*Istieus* is virtually identical to *Pterothrissus* (Forey, 1973b).

There are a number of genera that may be stem-group albuloids. †*Osmeroides* (Upper Albian–Upper Cenomanian) and †*Brannerion* (Aptian) show several characters of modern albuloids such as a deep subtemporal fossa which partially occludes the posttemporal fossa, a deep depression at the base of the parasphenoid ascending process, and a dorsal process on the ectopterygoid. Otherwise these genera appear to be primitive in most features relative to modern albuloids. †*Brannerion* has a parasphenoid and basibranchial dentition of rounded crushing teeth as in *Albula*, while †*Osmeroides* has a small intercalar and lacks a prootic-intercalar bridge. There is also an otolith record of an albuloid from the Hauterivian (Weiler, 1971).

True eels are known from the Middle Cenomanian as †*Anguillavus*, †*Urenchelys*, and a new genus recognized by Lu (1994), but none can be referred to any living family and may therefore be stem Anguilliformes. However, these remarks must be set in the context of our poor knowledge of the characterization and interrelationships of modern eel families, and it remains possible that some of these Cretaceous eels belong to Recent families. The Anguillidae and Ophichthidae are both known from the Ypresian (Patterson, 1993). †*Anguillavus*, among the Cenomanian eels, has pelvic fins and a separate caudal fin. These observations would suggest that characters 35 and 37 in our matrix specify a more exclusive group than "anguilloids" + saccopharyngoids as suggested by an analysis of Recent taxa only.

The fossil record of notacanthiforms is sparse. The monotypic †*Echidnocephalus* from the Campanian, regarded as a halosaur, represents the earliest record of this group. A Lutetian otolith attributed to *Notacanthus* sp. appears to be the earliest and only fossil record of the Notacanthidae (Patterson, 1993).

The saccopharyngoids may be represented by a single fossil occurrence in the Cenomanian (Patterson, 1993).

As can be seen from Fig. 5, the match between stratigraphic occurrence and phylogeny is quite good (and may be even better when eels have been studied more comprehensively). In this figure we also plot the number of nucleotide changes for the combined molecular data set. Extremely rapid rates of molecular change are seen for the lineages leading to saccopharyngoids and clupeomorphs. The long branch leading to the gulper eels is probably realistic, but that leading to the Clupeomorpha, represented in the data set by three taxa, is likely greatly overestimated because of poor taxon sampling. The rate of diversification of eels is also high. However, these generally high rates may be expected since most of the variable sites were found in the mitochondrial genes.

VI. Summary

Morphological analysis of elopomorph fishes provides a phylogeny in which internal nodes are very unevenly supported (Fig. 2). The anguilliforms clearly comprise a monophyletic group characterized by many synapomorphies of which a high percentage are reductive. A similar reductive trend also characterizes the node supporting a notacanthiform + anguilliform clade. There appears to be little support for previous hypotheses that regarded notacanthiforms and albuloids as sister-groups (Fig. 1a,d). The node linking albuloids with higher elopomorphs and the node supporting a sister-group pairing between *Elops* and *Megalops* are both poorly supported.

The molecular data have low bootstrap support except for the consistent pairing of *Elops* and *Megalops* (e.g., Fig. 3). Separate genes gave conflicting trees, but overall the mitochondrial genes were more informative. This is because they had more potentially informative sites than the nuclear gene, and there were more taxa sampled. For this reason we found the faster evolving mitochondrial genes more useful within the limits of this systematic problem.

The morphological and combined molecular analyses were complementary to the extent that nodes poorly supported by molecular data were strongly supported by morphology and *vice versa*.

The following Linnean classification is an adaptation of that given by Patterson and Rosen (1977); plesion, quotation marks, and sequencing conventions are used (see Patterson and Rosen, 1977, and Wiley, 1979, for discussions of these). True eels are designated as a paraphyletic taxon because it is strongly suspected that some will be shown to be more closely related to saccopharyngoids (e.g., in this analysis ophichthids were found to be more closely related to gulper eels than to anguillids). However, since so few eel families have been studied it seems premature to formally dissolve the taxon Anguilloidei. The classification of Notacanthiformes follows Greenwood (1977). Names after the hyphens refer to the taxa illustrated in Fig. 1.

Cohort Elopomorpha
  Order Elopiformes
      Family Elopidae—*Elops*
      Family Megalopidae—*Megalops*
  Order Albuliformes
    Plesion †*Brannerion sedis mutabilis*
    Plesion †*Osmeroides sedis mutabilis*
    Suborder Albuloidei *sedis mutabilis*
      Family Albulidae—*Albula*
      Family Pterothrissidae
  Order Notacanthiformes—notacanths
      Family Halosauridae
      Family Notacanthidae
        Subfamily Notacanthinae

Subfamily Polyacanthonotinae
    Tribe Polyacanthonotini
    Tribe Lipogenyini
Order Anguilliformes
    Suborder "Anguilloidei"—eels
        16 families listed by J. S. Nelson (1994), all
        sedis mutabilis
    Suborder Saccopharyngoidei—gulpers
        Family Saccopharyngidae *sedis mutabilis*
        Family Eurypharyngidae *sedis mutabilis*
        Family Monognathidae *sedis mutabilis*

Acknowledgments

P. L. F. has had the pleasure of working alongside Colin Patterson since the mid-1970s, sharing his company, fish skeletons, and many congenial hours. Most of all Colin has freely shared his impressive knowledge of comparative anatomy, paleoichthyology, and systematic methodology. What more could a colleague want in a working life? Just more of the same!

We thank the following individuals for help in providing teleost material for molecular analysis: Gary Nelson (American Museum of Natural History), Ed Wiley (University of Kansas), and Nigel Merrett (The Natural History Museum). One of us, D. T. J. L., was supported by NERC grant GR37960, and P. R. wishes to acknowledge support given through NSF grants BSR-9119867 and BSR-9107838.

References

Bargelloni, L., Ritchie, P. A, Patarnello, T., Battaglia, B., Lambert, D. M., and Meyer, A. (1994). Molecular evolution at subzero temperatures: Mitochondrial and nuclear phylogenies of fishes from Antarctica (Suborder Notothenioidei) and the evolution of antifreeze glycopeptides. *Mol. Biol. Evol.* **11,** 854–863.

Begle, D. P. (1991). Relationships of the osmeroid fishes and the use of reductive characters in phylogenetic analysis. *Syst. Zool.* **40,** 33–53.

Bertelsen, E., and Nielsen, J. G. (1987). The deep sea eel family Monognathidae (Pisces, Anguilliformes). *Steenstrupia* **13,** 141–198.

Bertelsen, E., Nielsen, J. G., and Smith, D. G. (1989). Suborder Saccopharyngoidei Families Saccopharyngidae, Eurypharyngidae and Monognathidae. *In* "Fishes of the Western North Atlantic" (E. Böhlke, ed.), Sears Found. Mar. Res., Mem. No. 1, Part 9, pp. 636–655. Yale University, New Haven, CT.

Bertin, L. (1936). Un nouveau genre de poissons Apodes caractérisé par l'absence de machoire supérieure. *Bull. Soc. Zool. Fr.* **61,** 533–540.

Böhlke, E. B., ed. (1989). Orders Anguilliformes. "Fishes of the Western North Atlantic," Sears Found. Mar. Res., Mem. No. 1, Part 9. Yale University, New Haven, CT.

Cabot, E. L., and Beckenbach, A. T. (1989). Simultaneous editing of multiple nucleic acid and protein sequences. *Comput. Appl. Biosci.* **5,** 233–234.

Castle, P. H. J. (1984). Notacanthiformes and Anguilliformes. *In* "Ontogeny and Systematics of Fishes" (H. G. Moser, W. J. Richards, D. M. Cohen, M. P. Fahay, A. W. Kendall, Jr., and S. L. Richardson, eds.), Spec. Publ. No. 1, Suppl. to *Copeia*, pp. 62–93. American Society of Ichthyologists and Herpetologists, Lawrence, KS.

Christen, R. (1993). "VSM16S—Database for rRNA Sequence Information." Distributed by the author, Observatoire Océanologique, F06230, Villefranche-sur-Mer, France.

Doyle, J. J., and Doyle, J. L. (1987). A rapid DNA isolation procedure for small quantities of fresh leaf tissue. *Phytochem. Bull.* **19,** 11–15.

Embley, T. M., Finlay, B. J., Thomas, R. H., and Dyall, P. L. (1992). The use of rRNA sequences and fluorescent probes to investigate the phylogenetic positions of the anaerobic ciliate *Metopus palaeoformis* and its archaeobacterial endosymbiont. *J. Gen. Microbiol.* **138,** 1479–1487.

Forey, P. L. (1973a). Relationships of elopomophs. *In* "Interrelationships of Fishes" (P. H. Greenwood, R. S. Miles, and C. Patterson, eds.), pp. 351–368. Academic Press, London.

Forey, P. L. (1973b). A revision of elopiform fishes, fossil and Recent. *Bull. Br. Mus. (Nat. Hist.), Geol., Suppl.* **10,** 1–222.

Fujita, K. (1990). "The Caudal Skeleton of Teleostean Fishes." Tokai University Press, Tokyo.

Greenwood, P. H. (1970). Skull and swimbladder connections in fishes of the family Megalopidae. *Bull. Br. Mus. (Nat. Hist.), Zool.* **19,** 119–135.

Greenwood, P. H. (1973). Interrelationships of osteoglossomorphs. *In* "Interrelationships of Fishes" (P. H. Greenwood, R. S. Miles, and C. Patterson, eds.), pp. 307–332. Academic Press, London.

Greenwood, P. H. (1977). Notes on the anatomy and classification of elopomorph fishes. *Bull. Br. Mus. (Nat. Hist.), Zool.* **32,** 65–102.

Greenwood, P. H., Rosen, D. E., Weitzman, S. H., and Myers, G. (1966). Phyletic studies of teleostean fishes, with a provisional classification of living forms. *Bull. Am. Mus. Nat. Hist.* **131,** 339–456.

Greenwood, P. H., Rosen, D. E., Myers, G., and Weitzman, S. H. (1967). Named main divisions of teleostean fishes. *Proc. Biol. Soc. Wash.* **80,** 227–228.

Gyllensten, U. B., and Erlich, H. A. (1988). Generation of single-stranded DNA by the polymerase chain reaction and its application to direct sequencing of the HLA-DQA locus. *Proc. Natl. Acad. Sci. U.S.A.* **85,** 7652–7655.

Harrison, C. M. H. (1966). On the first halosaur leptocephalus from Madeira. *Bull. Br. Mus. (Nat. Hist.), Zool.* **14,** 444–486.

Higgins, D. G., and Sharp, P. M. (1988). CLUSTAL: A package for performing multiple sequence alignments on a microcomputer. *Gene* **73,** 237–244.

Hosta, L., and Flick, P. (1991). Enhancement of specificity and yield in PCR. *U.S. Biochem. Comments* **18,** 1–5.

Hulet, W. M., and Robins, R. C. (1989). The evolutionary significance of the leptocephalus larva. *In* "Fishes of the Western North Atlantic" (D. M. Cohen, ed.), Sears Found. Mar. Res., Mem. No. 9, Part 2, pp. 669–677. Yale University, New Haven, CT.

Jamieson, B. G. M. (1991). "Fish Evolution and Systematics: Evidence from Spermatozoa." Cambridge University Press, Cambridge, UK.

Kocher, T. D., Thomas, W. K., Meyer, A., Edwards, S. V., Paabo, S., Villablanca, F. X., and Wilson, A. C. (1989). Dynamics of mitochondrial DNA evolution in animals: amplification and sequencing with conserved primers. *Proc. Natl. Acad. Sci. U.S.A.* **86,** 6196–6200.

Leiby, M. M. (1979). Morphological development of the eel *Myrophis punctatus* (Ophichthidae) from hatching to metamorphosis with emphasis on the developing head skeleton. *Bull. Mar. Sci.* **29,** 509–521.

Leiby, M. M. (1984). Ophichthidae: development and relationships. *In* "Ontogeny and Systematics of Fishes" (H. G. Moser, W. J. Richards, D. M. Cohen, M. P. Fahay, A. W. Kendall, Jr., and S. L. Richardson, eds.), Spec. Publ. No. 1, Suppl. to *Copeia*, pp.

102–108. American Society of Ichthyologists and Herpetologists, Lawrence, KS.

Littlewood, D. T. J. (1994). Molecular phylogenetics of cupped oysters based on partial 28S rRNA gene sequences. *Mol. Phylogenet. Evol.* **3**, 221–229.

Littlewood, D. T. J., and Smith, A. B. (1995). A combined morphological and molecular phylogeny for sea urchins (Echinoidea: Echinodermata). *Philos. Trans. R. Soc. London, Ser. B* **347**, 213–234.

Lu, Yi. (1994). New Upper Cretaceous teleost fishes from Namoura, a new locality in Lebanon, and the phylogeny of Aulopiformes (Euteleostei), fossil and Recent. Ph.D. Thesis, University of London.

Maddison, W. P., and Maddison, D. R. (1992). "MacClade: Analysis of Phylogeny and Character Evolution, Version 3.04." Sinauer Assoc., Sunderland, MA (Computer program and manual).

Marshall, N. B. (1962). Observations on the Heteromi, an order of teleost fishes. *Bull. Br. Mus. (Nat. Hist.), Zool.* **9**, 249–268.

Matthei, C., and Matthei, X. (1974). Spermatogenesis and spermatozoa of the elopomorpha (teleost fish). *In* The Functional Anatomy of the spermatozoon (B. A. Afzelius, ed.), pp. 211–221. Pergamon, Oxford.

McCosker, J. E. (1977). The osteology, classification, and relationships of the eel family Ophichthidae. *Proc. Calif. Acad. Sci.* [4] **41**, 1–123.

McDowell, S. B. (1973). Order Heteromi (Notacanthiformes). *In* "Fishes of the Western North Atlantic" (D. M. Cohen, ed.), Sears Found. Mar. Res., Mem. No. 1, Part 6, pp. 1–228. Yale University, New Haven, CT.

Mead, G. W. (1965). The larval form of the Heteromi (Pisces). *Breviora* **226**, 1–5.

Medlin, L., Elwood, H. J., Stickel, S., and Sogin, M. L. (1988). The characterization of enzymatically amplified 16S-like rRNA-coding regions. *Gene* **71**, 491–499.

Nelson, G. J. (1966). Gill arches of the teleostean fishes of the order Anguilliformes. *Pac. Sci.* **20**, 391–408.

Nelson, G. J. (1972). Observations on the gut of Osteoglossomorpha. *Copeia*, pp. 325–329.

Nelson, G. J. (1973). Relationships of clupeomorphs, with remarks on the structure of the lower jaw in fishes. *In* "Interrelationships of Fishes" (P. H. Greenwood, R. S. Miles, and C. Patterson, eds.), pp. 333–349. Academic Press, London.

Nelson, J. S. (1994). "Fishes of the World," 3rd ed. Wiley, New York.

Nielsen, J. G., and Bertelsen, E. (1985). The gulper-eel family Saccopharyngidae (Pisces, Anguilliformes). *Steenstrupia* **11**, 157–206.

Norman, J. R. (1926). The development of the chondrocranium of the eel (*Anguilla vulgaris*), with observations on the comparative morphology and development of the chondrocranium in bony fishes. *Philos. Trans R. Soc. London, Ser. B* **214**, 369–464.

Nybelin, O. (1967). Notes on the reduction of the sensory canal system and of the canal-bearing bones in the snout of higher actinopterygian fishes. *Ark. Zool.* **19**, 235–246.

Nybelin, O. (1971). On the caudal skeleton in *Elops* with remarks on other teleostean fishes. *Acta Reg. Soc. Sci. Litt. Goth., Zool.* **7**, 1–52.

Palumbi, S. R., Martin, A., Romano, S., McMillan, W. O., Stice, L., and Grabowski, G. (1991). "The Simple Fools Guide to PCR." University of Hawaii Press, Honolulu.

Patterson, C. (1975). The braincase of pholidophorid and leptolepid fishes, with a review of the actinopterygian braincase. *Philos. Trans R. Soc. London, Ser. B* **269**, 275–579.

Patterson, C. (1977). The contribution of palaeontology to teleostean phylogeny. *In* "Major Patterns in Vertebrate Evolution" (M. K.

Hecht, P. C. Goody, and B. M. Hecht, eds.), pp. 579–643. Plenum, New York.

Patterson, C. (1993). Osteichthyes: Teleostei. *In* "The Fossil Record 2" (M. J. Benton, ed.), pp. 621–656. Chapman & Hall, London.

Patterson, C., and Johnson, G. D. (1995). The intermuscular bones and ligaments of teleostean fishes. *Smithson. Contrib. Zool.* **559**, 1–83.

Patterson, C., and Rosen, D. E. (1977). Review of ichthyodectiform and other Mesozoic teleost fishes and the theory and practice of classifiying fossils. *Bull. Am. Mus. Nat. Hist.* **158**, 81–172.

Regan, C. T. (1909). The classification of teleostean fishes. *Ann. Mag. Nat. Hist.* [8] **3**, 75–86.

Regan, C. T. (1910). The caudal fin of the Elopidae and of some other teleostean fishes. *Ann. Mag. Nat. Hist.* [8] **5**, 354–358.

Regan, C. T. (1912). The osteology and classification of the teleostean fishes of the Order Apodes. *Ann. Mag. Nat. Hist.* [8] **10**, 377–387.

Richards, W. J. (1984). Elopiformes: development. *In* "Ontogeny and Systematics of Fishes" (H. G. Moser, W. J. Richards, D. M. Cohen, M. P. Fahay, A. W. Kendall, Jr., and S. L. Richardson, eds.), Spec. Publ. No. 1, Suppl. to *Copeia*, pp. 60–62. American Society of Ichthyologists and Herpetologists, Lawrence, KS.

Robins, C. R. (1989). The phylogenetic relationships of the anguilliform fishes. *In* "Fishes of the Western North Atlantic" (E. Böhlke, ed.), Sears Found. Mar. Res., Mem. No. 1, Part 9, pp. 9–23. Yale University, New Haven, CT.

Saiki, R. K., Gelfand, D. H., Stoffel, S., Scharf, S., Higuchi, R., Horn, R., Mullis, K. B., and Erlich, H. A. (1988). Primer-directed enzymatic amplification of DNA with a thermostable DNA-polymerase. *Science* **239**, 487–491.

Sambrook, J., Fritsch, E. F., and Maniatis, T. (1989). "Molecular Cloning: A Laboratory Manual," 2nd ed. Cold Spring Harbor Lab. Press, Cold Spring Harbor, NY.

Sanger, F., Nicklen, S., and Coulson, A. R. (1977). DNA sequencing with chain terminating inhibitors. *Proc. Natl. Acad. Sci. U.S.A.* **74**, 5463–5467.

Smith, D. G. (1984). Elopiformes, Notacanthiformes and Anguilliformes: Relationships. *In* "Ontogeny and Systematics of Fishes" (H. G. Moser, W. J. Richards, D. M. Cohen, M. P. Fahay, A. W. Kendall, Jr., and S. L. Richardson, eds.), Spec. Publ. No. 1, Suppl. to *Copeia*, pp. 94–102. American Society of Ichthyologists and Herpetologists, Lawrence, KS.

Smith, D. G. (1989). Families: Saccopharyngidae, Eurypharyngidae, and Monognathidae: Leptocephali. *In* "Fishes of the Western North Atlantic" (D. M. Cohen, ed.), Sears Found. Mar. Res., Mem. No. 9, Part 2, pp. 948–954. Yale Univesity, New Haven, CT.

Swofford, D. L. (1993). "PAUP: Phylogenetic Analysis Using Parsimony, Version 3.1.1." Smithsonian Institution, Washington, DC.

Taverne, L. (1977). Ostéologie, phylogénèse et systématique des téléostéens fossiles et actuels du super-ordre des ostéoglossomorphes. Première partie. Ostéologie des genres *Hiodon, Eohiodon, Lycoptera, Osteoglossum, Scleropages, Heterotis* et *Arapaima*. *Mém. Cl. Sci., Acad. R. Belg.* **62**, 1–235.

Tchernavin, V. V. (1946). A living bony fish which differs substantially from all living and fossil osteichthyans. *Nature (London)* **158**, 667.

Tchernavin, V. V. (1947a). Six specimens of Lyomeri in the British Museum (with notes on the skeleton of the Lyomeri). *J. Linn. Soc. London, Zool.* **41**, 287–350.

Tchernavin, V. V. (1947b). Further notes on the structure of the bony fishes of the order Lyomeri (Eurypharynx). *J. Linn. Soc. London, Zool.* **41**, 378–393.

Weiler, W. (1971). *Palealbula ventralis* n. sp. (Pisces, Clupeiformes) aus dem Neocom (Unter-Hauterive) von Engelbostei bei Hannover. *Senkenbergiana Lethaea* **52**, 1–3.

Wenz, S. (1965). Les poissons Albiens de Vallentigny (Aube). *Ann. Paléontol.* **51**, 1–24.

Wiley, E. O. (1979). An annotated Linnean hierarchy, with comments on natural taxa and competing systems. *Syst. Zool.* **28**, 308–337.

Winship, P. R. (1989). An improved method for directly sequencing PCR amplified material using dimethyl sulphoxide. *Nucleic Acids Res.* **17**, 1266.

Appendix 1

Description of Morphological Data

Within elopomorphs *Elops* and *Megalops* show the most primitive teleostean conditions with the maximum number of bones present. Notacanthiforms, and especially eels and saccopharyngoids, have far fewer braincase bones and, in the context of a wider phylogeny of teleost fishes, this is reasonably regarded as loss–fusion and is usually coded as a separate derived state. However, here simple presence (1) or absence (0) is coded.

Braincase   Character 1. Subepiotic fossa absent (0), present (1).

Character 2. Parasphenoid without high dorsal process meeting pterosphenoid (0), with dorsal process (1).

Character 3. Orbitosphenoid absent (0), present (1). The scoring for saccopharyngoids in this and other cranial characters is based on Tchernavin's (1947a) descriptions of *Eurypharynx* and *Saccopharynx*.

Character 4. Basisphenoid absent (0), present (1).

Character 5. Frontals separated by median suture (0), fused in the midline (1).

Character 6. Epaxial musculature inserting on either side of the posterior semicircular canal which is contained within the epiotic (0), inserting to rear wall of neurocranium entirely medial to posterior semicircular canal (1). In primitive teleosts the epaxial musculature inserts laterally within the roofed or unroofed posttemporal fossa and medially on the rear wall or sometimes (e.g., in albuloids) into subepiotic fossae. In eels, notacanthiforms, and saccopharyngoids the rear wall of the braincase is not excavated for muscle insertion, and in eels and notacanthiforms the epiotic extends laterally.

Character 7. Supraoccipital absent (0), present (1).

Character 8. Infraorbital branch of buccal VII not piercing the premaxilla (0), piercing the premaxilla (1). This character is related to the fact that the infraorbital sensory canal (innervated by buccal VII) may be carried over the surface of the premaxilla (sometimes in separate rostral ossicles).

Character 9. Intercalar absent (0), present (1).

Sensory Canals   Character 10. Ethmoid commissure absent or, if present, free in skin (0), bone enclosed within premaxilla or rostral ossicles (1). *Elops*, and especially *Megalops*, show the most primitive living teleost condition of the ethmoid with the commissure passing through the dermethmoid (more correctly rostrodermethmoid—see Patterson, 1975). Additionally they, with some other elopomorphs, have rostral ossicles—a derived feature. The fate of the ethmoid commissure in eels is less clear because of the high degree of bone fusion that has apparently taken place. *Anguilla* has separate, canal-bearing premaxillae which fuse with the ethmoid in the postlarval stage (Norman, 1926, p. 398). Ophichthid eels are usually stated not to have premaxillae (Robins, 1989, p. 12), but McCosker (1977, p. 18) suggests they have fused with the ethmoid and vomer based on the observation that the toothed ethmovomer is constricted, implying originally separate premaxilla and vomer components. However, Leiby's (1979) study of the ontogeny of *Myrophis punctatus* showed no separate premaxillae at any stage. Saccopharyngoids do not have premaxillae, and an ethmoid commissure is absent in both them and ophichthids.

Character 11. Mandibular sensory canal running within closed tube within lower jaw (0), running within open tube (1).

Character 12. Posterior opening of the mandibular sensory canal within the lower jaw, lateral (0), medial (1).

Character 13. Canal-bearing extrascapulars absent (0), present (1).

Palate and Jaws   Character 14. Ectopterygoid without dorsal process (0), with dorsal process (1). This process rests against the inner surface of the adjacent infraorbital. The presence of pterygoid elements in saccopharyngoids is problematic (Tchernavin, 1947a), and therefore coding for this character is scored inapplicable for this taxon.

Character 15. Large connective tissue nodule intercalated between pterygoid arch and maxilla absent (0), present (1). See Greenwood (1977) for discussion of this character.

Character 16. Fenestra within the hyomandibular and metapterygoid suture allowing deep portions of the levator arcus palatini to pass through and insert to medial surface of palate; absent (0), present (1). In some osteoglossomorphs (*Osteoglossum*, *Heterotis*, and *Arapaima*; Taverne 1977) there is an anteroventral prong on the hyomandibula which reaches towards the entopterygoid delimiting a fenestra, but muscles do not pass through to the medial surface of the palate, and it is here regarded as nonhomologous with the fenestra in some elopomorphs. Modern anguilloids and saccopharyngoids have lost the metapterygoid and therefore do not show this fenestra.

Character 17. Symplectic free (0), symplectic fused with quadrate (1). The presence of a symplectic in saccopharyngoids is questionable (see character 18).

Character 18. Hyomandibula with simple synchondrosis with quadrate (0), with interdigitate suture with quadrate (1). The primitive teleost condition is for the hyomandibula to articulate solely with the symplectic through a simple synchondrosis. The assumed derived state of this character is known to occur elsewhere (e.g., siluroids) so its use here presupposes a monophyletic Elopomorpha. The homology of elements of the hyoid arch in saccopharyngoids is not clear. Tchernavin (1947a) describes a long hyomandibula that lies in continuity with a lower element which he calls a quadrate. In this he is correct since there is a perfectly good ball-and-socket joint with the lower jaw. However, the upper synchondrosis is that expected to occur between the hyomandibula and the symplectic. It is therefore possible that the lower bone in the hyoid arch is a fused symplectic and quadrate (as in true eels).

Character 19. Levator arcus palatini undivided (0), divided (1). See Greenwood (1977) for discussion of this character.

Character 20. Premaxilla movable on ethmoid (0), firmly attached or fused–absent (1).

Character 21. Maxilla without posteriorly directed spine (0), with spine (1).

Gill Arches　　Character 22. Gill arches beneath and articulating with the neurocranium (0), displaced posteriorly free from neurocranium (1).

Character 23. Gular plate absent (0), present (1).

Character 24. Gill rakers absent (0), present (1).

Character 25. Uncinate process on epibranchial 1 absent (0), present (1).

Character 26. Uncinate process on epibranchial 2 absent (0), present (1).

Character 27. Unicinate process on epibranchial 3 absent (0), present (1).

Character 28. Hypobranchial 2 without process (0), with process (1).

Character 29. Hypobranchial 3 without process (0), with process (1).

Pectoral Girdle and Fin　　Character 30. Postcleithrum absent (0), present (1).

Character 31. Pectoral girdle attached to skull (0), free from skull (1). This character involves the absence of a posttemporal in most of the fishes considered

here. However, in halosaurs the canal bearing portion of the posttemporal persists (McDowell, 1973).

Character 32. Mesocoracoid absent (0), present (1).

Character 33. Sternohyoideus originating mainly on the cleithrum (0), sternohyoideus originating aponeurotically from hypaxial musculature (1). See Greenwood (1977) for discussion of this character.

Character 34. Pectoral splint absent (0), present (1).

Pelvic Girdle and Fin　　Character 35. Pelvic girdle and fin absent (0), present (1).

Character 36. Pelvic fin webs not joined in ventral midline (0), joined in midline (1). Recent eels and saccopharyngoids lack pelvic fins and hence this character carries an illogical coding for these taxa.

Caudal Fin　　Character 37. Distinct caudal fin absent (0), present (1). A distinct caudal fin is recognized if there is a clear gap between the fin rays of dorsal and/or anal fins supported by radials (or, in the case of notacanths by the hemal spines; Fujita, 1990) and the caudal fin rays supported by hypurals. Saccopharyngoids are quoted as lacking a recognizable caudal fin (Bertelsen *et al.*, 1989). Instead the terminal vertebra extends as a urostyle. However, the tail of *Monognathus bruuni* as figured by Bertin (1936, fig. 4) does show three fin rays associated with the terminal vertebra. Several authors have remarked that the caudal fins of notacanthiforms and, particularly, saccopharyngoids are usually broken and show signs of regeneration, so it is possible that the observations made so far on saccopharyngiform tails are not general.

Character 38. Four or more uroneurals (0), three or fewer uroneurals (1). The first condition represents the primitive teleost condition.

Character 39. Parhypural free (0), fused with PU1 (1).

Character 40. Inner caudal fin rays not expanded (0), expanded (1).

Character 41. Hypural 1 not fused with hypural 2 (0), fused (1).

Character 42. Upper hypurals remaining unfused (0), fused together (1).

Character 43. Hypural 1 remaining free from ural centrum (0), fused to ural centrum (1).

Character 44. Upper hypurals remaining free from ural centra (0), fused to ural centra (1).

Character 45. Epurals absent (0), present (1).

Character 46. Dorsal and ventral procurrent rays absent (0), present (1). Procurrent rays are those rays lying anterior to the dorsal and ventral principal rays.

Character 47. Hypural 2 remaining free from centrum (0), fused to ural centrum (1).

Character 48. PU1 free from U1 (0), fused with U1 (1).

Character 49. Cartilage plate attached to NaU1 (1), no cartilage plate. See Patterson and Rosen (1977) for discussion of this character.

Vertebral Column   Character 50. Epipleural intermuscular bones absent (0), present (1). See Patterson and Johnson (1995) for discussion of this character.

Character 51. Supraneurals absent (0), present (1).

Scales   Character 52. Main body scales present and overlapping (0), present but not imbricating (1), absent (2). Ophichthids are here scored as lacking body scales, but they do possess ossicles surrounding the lateral line (McCosker, 1977).

Character 53. Lateral line scales exposed (0), overlapped by neighboring rows (1).

Other   Character 54. Leptocephalus larva absent (0), present (1).

Character 55. Spermatozoa in which the flagellum contains two central and nine peripheral axonemes (0), flagellum with 9 + 0 axoneme arangement and the proximal centriole divided into two elongate bundles of four and five triplets structure (1) (Matthei and Matthei, 1974). See Jamieson (1991) for discussion of this character.

Character 56. Intestine passing to right of stomach (0), left of stomach (1). See Nelson (1972) for a discussion of this character.

Appendix 2

Morphological Data Set

```
1234567891111111111222222222233333333334444444444555555
          0123456789012345678901234567890123456789012345 6
```

Hiodon
00110010100010000000001001110100101000000011000010000 1
Osteoglossum
00110010100010000000001111010100101100001010100001000 1
Petrocephalus
0011001010001000000000100110101001011070101?10000100001
Elops
001100101101100000000011001001011110110100001100111001 1?
Megalops
0011001011011000000000111100101111011010000110011001?0
Albula
10110011110110100010011011001010101011000000110011001 1?
Pterothrissus
1011001111011010010011000101100101011011000000110011011 1?
Halosauridae
0100011101101010000110011001?????010111101?0??????????01?00 11?
Notacanthidae
0100011101001010001110011001?????010111101101111001101?011 ??
Anguillidae
010101110100000N11010100000000010100N0110111101101010 11?
Ophichthidae
0101111N0000000N11010100000000010?00N0110111100110102N 11?
```

saccopharyngoids
```
00001?0N000?0N?N?0?10100?????010?00N0NNNNNNN0?NN0?02N1 ??
```
*Clupea*
```
00110010100010000000000001111011017010110100001110011000 00
```
-------------------------------------------------------

*Note:* Data matrix coding 56 morphological characters described in Appendix 1: N, not applicable coding; ?, unchecked information.

## Appendix 3

Taxa used in this analysis and the molecular data available; [18S] = 18S small subunit nuclear rRNA sequence available; [mt] = 12S and 16S mitochondrial rRNA sequence available. Linnean classification from Greenwood *et al.* (1966)

Superorder Osteoglossomorpha
  Order Osteoglossiformes
    Suborder Notopteroidei
      Family Hiodontidae
        *Hiodon alosoides* (Rafinesque) [18S, mt]
    Suborder Osteoglossoidei
      Family Osteoglossidae
        *Osteoglossum ferreirai* Kanazawa [mt]
  Order Mormyriformes
    Family Mormyridae
      *Petrocephalus* sp. [mt]
Superorder Elopomorpha
  Order Elopiformes
    Suborder Elopoidei
      Family Elopidae
        *Elops hawaiiensis* Regan [18S, mt]
      Family Megalopidae
        *Megalops atlanticus* Cuvier and Valenciennes [18S, mt]
    Suborder Albuloidei
      Family Albulidae
        *Albula vulpes* (Linnaeus) [18S, mt]
  Order Notacanthiformes
    Family Notacanthidae
      *Notacanthus bonapartei* Risso [mt]
  Order Anguilliformes
    Suborder Anguilloidei
      Family Anguillidae
        *Anguilla rostrata* (Lesueur) [mt]
      Family Ophichthidae
        *Echiophis punctifer* (Kaup) [18S, mt]
        *Ophichthus rex* Böhlke and Caruso [18S, mt]
    Suborder Saccopharyngoidei
      Family Eurypharyngidae
        *Eurypharynx pelecanoides* Vaillant [mt]
Superorder Clupeomorpha
  Order Clupeiformes
    Suborder Clupeoidei
      Family Clupeidae
        *Clupea harengus* Linnaeus [18s, mt]
        *Stolothrissa tanganicae* Regan [mt]
        *Limnothrissa miodon* (Boulenger) [mt]
```

Appendix 4

Molecular Data Showing the Unambiguously Aligned Variable Positions Used

```
          |16S mtDNA.......................................................................................
                                                       1111111111111111111111111111111111111111111111
                   1111122222233333333333444444444455555555556666666666677777777778888899000001111122222233333334455556666677777777
Node               123456789012345678901234567890123456789012345678901345081234912345645678901234673701271567012345

HIODON     CGCCTGCCCAGTGATTAAATGGCCGCGGTATTTTAACCGTGCAAAGGTAGCGTAATCACTTGTCTTTTAAAGAACTCATCGGCCCAGCTTTTCCCAGTAATTGCCCAGGCGTAAACAT
ALBULA     CGCCTGCCCTGTGATTTAACGGCCGCGGTATTCTGACCGTGCAAAGGTAGCGTAATCATTTGTCTTTTAAAAAATTCATCGGCTCGACTATCCCTAATAGTTGTCCAGGCAAAAACAT
ELOPS      CGCCTGCCCCGTGATTTAACGGCCGCGGTATTTTGACCGTGCGAAGGTAGCGTAATCACTTGTCTTTTAAAGAAGTCAACGGCTCAGCCGCCTCCAGTAATTGCCCAGACACAACCAT
MEGALOP    CGCCTGCCCCGTGATTTAACGGCCGCGGTATTTTGACCGTGCCAAGGTAGCGTAATCACTTGTCTTTTAAAGAAGTCAACGGCTCAGCCGCCTCCAGTAATTGCCCAGACAACAACAT
OPHICTH    CGCCTGCCCTGTGATTTAACGGCCGCGGTATCGTAACCGTGCNAAGGTAGCGTAATCACTTNTCTTTTAAAGAATTCACCGGTTTAACTTCCCCCAGTAATTGCCCAGACAAAANAAT
ECHIOPH    CGCCTGCCCTGTGATTTAACGGCCGCGGTATCATAACCGTGCAAAGGTAGCGTAATCACTTGTCTTTTAAAGAATTCACCGGTTTAACTTCCCCCAGTAATTGCCCAGACAAAAACAT
NOTACAN    CGCCTGCCCAGTGATTAAATGGCCGCGGTATTCTGACCGTGCTAAGGTACGCAATCACTTGTCTTTTAAAGAACTCATCGGCTCACCTCTCTTCGGTAATTGCCCAGACAAACGCGT
EURYPHA    CGCCTGCCCAGTGATTTAACGGCCGCGGTACCCTAACCGTGCAAAGGTAGCATAATCACTTGTTCCTTAAAAGGTTCTCCGGTTCAGTTACTTTTAACAGTCGCCAGGGCGACACAGC
STOLOTH    NNNNNNNNNNNNNNNNNNNNNNNNNNNNNNTATTTTGACCGTGCAAAGGTAGCGCAATCACTTGTCTTTTAAAAGATTCTAAGGCTTAACCCTTTCAAGTAGTTGCCCAGGTAACAATAC
LIMNOTH    CGCCTGCCCTGTGATTTAACGGCCGCGGTATTTTAACCGTGCAAAGGTAGCGCAATCAATTGTCTTTTAAAGGATTCATAGGTTTAATTTTTTCACGTACTGCCCAGACACTTACAC
CLUPEA     NNNNNNNNNNNNNNNNNNNNNNNNNNNNNNNNNNNNNNNNNNNNNNNNNNNNNNNNNNNNNNNNNNNNNNNNNGAATTCATAGGTTTAATTTCTCCTGGTAGCTACCCAGGTGAATATAC
ANGUILL    CGCCTGCCCTGTGATTTAACGGCCGCGGTATCTGACCGTGCAAAGGTAGCGTAATCATTTGTCTTTTAAAAGATTCATAGGTTTAACTTCCCCCAGTAATTGCCCAGACAAAATACAT
OSTEOGL    AACCTGCCCAGTGATTAAATGGCCGCGGTATTTTAACCGCGCTAAGGTAGCGTAATCACTTGTCTTTTAAAGAACATACCAACCTTACTTCCTCAAGTAGTTATTCAAATCACAACAT
PETROCE    AACCTGCCCAGTGATTTAACGGCCGCGGTATTTTAACCGTGCTAAGGTAGCGTAATCACTTGTCTTTTAAAGAACATGTCGGCCCTCCTACTCCAAGTTATTGCCCAGGCACCCCCAT


                   ....................................................................................................
                   1122233333333333333333333333333333333333333333333333334444444444444444444444444444444444444445555555555555555555555
                   7900090000111111222222233366666777777788888888889999999000000001111112222222222444444555555566668999990000011134445555556
Node               843503478123578902458901256789015678901234567890036789014567890127890123456789567890123469345990234902780232029015678 90

HIODON     AAAGCGAGCAGGAACAAGCTATGTGAA-AGCCAGACACAA-------CT--GCAAGACAAA-TCCAAAAAT-GATNNCC-GATCAACGCTTATTTCCCTCAGAGAAAA-TAACGCAGC
ALBULA     AAAGCGAGCAGGGAAAGCTATGTGAA-AGCTAGTTACAG-------CT--GAAGGAA-TA-TCCATAAAT-GATCCCCTGATCAACACTTATTTCTCTTAGAGAAA-GGTAATGCAGC
ELOPS      AGAGCGAGCAGGAAACCTGCTATGTGAA-AGCCAGACACAC-------CT--GCAGGAAC-A-TCCAAAAAT-GATGGCC-GATCAACACAAGCTTCATCCGAGAA-AGTAATGCAGC
MEGALOP    AGAGCATGCAGAAATCAACTATGTGAA-AACCAGACACA-------CT--GCAGGAAA-A-TCCAAA---GACCCCCTGATCAACACTTACTCCCCTCAGAGAA-AGTAATGCAGC
OPHICTH    AATGCGAACGGGAACAAGCCGAATG-A-AACCAAATACAC-------TT--GCAAGAAA-A-TCCAAACAT-GACCCCC-GATCAATACTTATTTCCCTCCGAGAA-AGTAATGAAGC
ECHIOPH    AATGCGAACGGGAACAGACCGAATA-A-AACCAAATACAC-------TT--GTAAGAA--ATTCTAAACAT-GACCCCCTGATCAATACTTATTTCCCTCCGAGAA-AGTAATGTAGC
NOTACAN    TAAGCGATCAGGAACAGCTGTGTGAA-AACTAAGTACAA-------AT--GTAGAAAA-A-TTCAATAAT-GACCCCC-GATCAACACTTATCTCCCTCAGAGAG-AGTAATGCAGC
EURYPHA    AATATGATCAGCGACCCACTGAGCA---AGTCCGAACTACTAT---AC--TTGATCATA--A-CTC-AACG----------------A-TTATTATCTTCAAAATA-AGCGATGCAGC
STOLOTH    AAAGCGACGAGGAAAAAGCTAAGTGAA-AACCAGAGACAT-------CT--GCCAGAA-A-TCCAAACAT-GATCCCC-GATCAACACTTATCCCCCTCAGAGGG-AGTAATGCAGC
LIMNOTH    AATGCGAACGGGAACGAGCCGAGAGAA-AACCAACTACAA-------TT--GTCAAAT-G-TTCAATTAATGATCGCC-GATCAACGCTTATCCCCTCAGAGAG-AGTAATGCAGC
CLUPEA     AATGCGAGCGAGGAAAAGTCGAGAGAA-AACCAGCTACAC-------CT--GTCAAAC-A-TTCAAAAAT-GATCCCC-GATTAACACTTATCTCCCTCAGAGGA-AGTAATGCAGC
ANGUILL    AATGCGAATGGGAGAAAGCCACGAG-A-AACTAGGGACAC-------CT--GCAAGAAA-A-TCCAAA-AT-GACCCCCTGATCAACACTTATTTCCCTCAGAGAA-AGTAATGCAGC
OSTEOGL    AGAGCGAATAGGCAAAAGCTAAGAAAAGAACCACAA-------TC--GTAAGAAA-ATTCTAACAAT-GATCCC--GATCAACACTTACTTCCACCAGCGAA-AGTAACGAAAA
PETROCE    AGAGCGAATGGGCGAAAGCCAAGAAAT-CCCTAGTTAAGAGCTACAACT--ACAAGAAATA-TCTAATAAT-GATCCCCTGATCAACGCTTACTTCCCCAGGAA-AGTGGCNNNNN


                   ......................|12S mtDNA..............................................................................
                   5555555555555555555555555555 55555555555556666666666666666666666666666666666666666666666677777777777777777777777777777777
                   6666666667777777777888888 889999999999900001111122222223344444455555555667788888999999000111111123444455555666666666667777
Node               123456789012345678901234 780123456789058913579012357878012589012346892358125690123580690123689094672367802345678901459

HIODON     AGCTATTAAGGGGTTCGTTTGTTCAA--AAATATTA-TCAG-CAAGATTA-CCGAGACTGGTG-CCCGC-ACTTAGCTTAACCCTTCCACCCTCTCGCTCTACTGGAGG-ATTAATCAA
ALBULA     CGCTATTAAGGGTTCGTTTGTTCAA--AGATATTA-TCAG-CAAGGTAC-TAAGACTGACGCCTTGT-CCCTATCTCGATCTTACCACTCTTCCGCTCTATTGGAGGGAC-AGTTAT
ELOPS      CGCTATTAAGGGTTCGTTTGTTCGG--AAATATCA-TCAG-CAAGGTCAC-TAAGACTGGTG-CCCCC-ACTTAGCCTAACCCTACCACCCTTCCGCTCTACTGGAGG-ACCAATTAC
MEGALOP    CGCTATTAAGGGTTCGTTTGTTCAA--AAATATCA-TCAG-CCAGGTCAC-TAAGACTGGTG-CCCCC-ACTTAGCCTAACCCTACCACCCTTCCGCTCTACTGGAGG-ACCAATTAC
OPHICTH    CGCTATTAAGGGTTCGTTTGTTCAA--AAATACCA-CCA-ATAGAATTA-TTAAGACTGGTGCTTCAC-ACTCATCCTAATCCTTCCACTCTCTTACCCTACCAGGGGCA-TAACTAT
ECHIOPH    TGCTATTAAGGGTTCGTTTGTTCGA--AAATACCA-TCAGATAGAGCTA-ITAAGACTGGTGCTTCAC-ACTCATCCTAATCCTTCCACTCTCTTACCCTGCCAGGGGTAC-AATCAC
NOTACAN    CGCTATTAAGGGTTCGTTTGTTCAA--ACATATTA-CCAG-CAAGACTA-CTAAGACTGGTG-CTCAC-ATTTATCTTAATCTTACCACTTTTCCGCTCCCATTGGAAG-ACTAATTGT
EURYPHA    CGCTATTAAAGGTTCGTTTGTTCAATGATAT-GCA-TCAGTCAAGACTA-CTAAGTGTGGTAACTCAC-ACTCATTCTAAACCCATNTCCCTCCCACTCTATCTAGAAGAATGATCGC
STOLOTH    CGCTATTAAGGGTTCGTTTGTTCAA--AATTAAATGTCAGTCAAGATCA-CTAAAACCAGTGTCTCAC-CCTTAGCCTAATCCTACCACCTACCCG-TCTACTGGAGGAAACAAAATAA
LIMNOTH    CGCTATTAAGGGTTCGTTTGTTCAA--AACAAGTA-TCTGCCAGGGCCAC-TAAGACTGGTGCTTCGC-CCTTAGCCTAACCTCCTACTCTCCCGCCCTACTGGAGG-AGTCACTAA
CLUPEA     CGCTATTAAGGGTTCGTTTGTTCAA--AACAAATA-TCCGACAGGGCCAC-TAAGACTGGTGCTTCGC-CCTTAGCCTAACCTCCCACTCTCCCGCCCTACTGGAGGTACT-ACTAA
ANGUILL    CGCTATTAAGGGTTCGTTTGTTCAA--AAATATCA-TCAG-CAGGGTTAC-TAAGACTGGTG-CTCAC-ACTTATCCTAATCATACCATCTTCCCGCCCTGTCTGGAG-ATTAAACAC
OSTEOGL    TTCTATTAAGGGTTCGTTTGTTCAA--TCATACCA-CTAGACAGGGCATTCTACGACTGGTG-CTCCCTACTTATCTAACTACTACCGTTTACTCGCCACACTAGAGGGAAAATAGTCAA
PETROCE    NNNNNNNNNNNNNNNNNNNNNNNNNN--AAATATTA-TCA-ACAAGGCATTCCGCGACTGGTG-CCCCC-GCTTGTCCTTCCACCCACTCACTACACTGGAGG-CCCAATTAA


                   .............................................................................|18S nucDNA.......................
                                                                                        1111111111111111111111111111111111111111111111
                   7777777777777788888888888888888888888888888888888888888888899999999999999 99999990000000000111111111111111111111111111111
                   888888889999999011222222233344555555566666666777777778888880000011222222 36788890244447880000000011111111122224556666666777
Node               0134578901236788913456791447123456712345671345678901345612349784567891 46545752625682891234567123456789012452828234578012

HIODON     AGTGGTACAACCGAGGTTAGGTGGAAATGAAAGAA-ATATTACGACAACCA---GTCTGTGCAAAGAAG GTACGTAATA--TTAA-------CC----------GCCCG-TG-TC-G
ALBULA     CATGGTATGGCCGAGTATGGGTGGAAGTGGTT--C--ATATTGTAATGCT-T---GTGTGTGCAGAGAGA GTACGTAATA--TTCA-------TC----------GCCCG-TG-T---
ELOPS      TATGACACAGCTAAGGATGGGTGGAAATGAAAC-N--ATACTGTGACATA-G---GCCTGTGCAAAGGGA GTACGTAACA--TTCG-------TC----------GCCCG--TG-T---
MEGALOP    TATGACACAGCTAAGGATGGGTGGAAATGAAAC--C--ATACTGTGACATA-G---GCCTGTGCAAAGGGA GTACGTAATA--TTCA-------TC----------GCCCG-TG-T---
OPHICTH    AATGGCCCCGC-ACGGACCAGTGGAGATGAAC--C--ACACTAAGGTGTC-A---GTT-GTGCAAAGAAG GTACGTAATA--TTAA-------CTCTG-------GCCC-TTGC-CGG
ECHIOPH    AATGGC-TCCCCATGGACCGATGGAGATGAAC--C--ACACTAAGGTGTT-A---GTT-GTGCAAAGAAG GTACGTGATA--TTAA-------CTCTG-------GCCCGTTGCTCGG
NOTACAN    TATGGAATATCCGAGGATAGGTGGAAATGGTCT-C--ATATTAAGATGTT-A---G---GAGCAGAGAAA ????????????????????????????????????????????????????
EURYPHA    CATGATAAAACTAAAGTTAGGAGATGATGAAC--C--AAACTATAGTATA-AAAAT---GTACAAGAGGA ????????????????????????????????????????????????????
STOLOTH    AATGGC--ACAGAGGGCTAAGTGAAATACCA--C--ATATCACATGCACC----GCTTGTGTAAAGGAA ????????????????????????????????????????????????????
LIMNOTH    GGTGGC--AATGGAGGTCAAGTAGAAAAAGAAG--CTTAAATAAGAGTCACC----GCTTGTGCGGGGAGA ????????????????????????????????????????????????????
CLUPEA     GATGACATTCCTGAGGTCGAGTGGAAAAGAAA--CTT-ATTTAAGGCCGCC----GCCCGTGCAAAGGGA AGGGT-AGTCCC-CAACTGGCCCTCACGGGGGGCTCTAGG-C--T---
ANGUILL    TATGGTATTACCAAGGATAGATGATAATGATA--C--AAAACAAAGTG-CCA---GCT-GTGCAAAGGAA ????????????????????????????????????????????????????
OSTEOGL    AG--ACTTCGCCGAGAATAAACGGAAATGAACAAC---ACATACATAATT-A---GCTTGAGTAAATGTA ????????????????????????????????????????????????????
PETROCE    AGTGGCACAACCACGACTGGGCGGAAATGCCCAACAT-ACCAATAACGAC-A---GTTTAAGCAAAGAAA ????????????????????????????????????????????????????
```

Appendix 4 (continued)

```
                  ..........................................................................................
                  1111111111111111111111111111111111111111111111111111111111111111111111111111111112222222222222222222222222222222222
                  1112222222222222222333344444444444444444455555566666666666666666666677777778888888899999000001111122223333333333333334444
                  7780122223344468809900136677788888993589990013444556666778889999444444500115999168900455044683449000112222334667880134
Node              4637112391703665737959798934672578906156249027312304125627189135901234575723401207681245633467557701308679690585022281
--------------------------------------------------------------------------------------------------------------------------------------------
HIODON            C--CGACGCTCCCAGTTCCCTGTA--CATTTTTA-CT-ATCTTCAACAG-CTCCTCTAACGGGTACGGT--CCCCGTGCGTTAGTACA-AGCA-GGTTGGTCGGACCC-GCAATAACA
ALBULA            C--CGACGTTTCCAATTACCAGTA--CATTTTTAACTGATCTTCACCAG-CTCTTCTAACG-ATACGGT--TCTTGTGC-TTAATACA-A-CA-TACCGGACAGATTC-TCAATGACA
ELOPS             C--CGACGCTTCCAATTCCCTGTA--CATTTTTAACTGATCTTCACCAG-CTCCTCTAACGGGTACGGT--CCCTGT-C-TCAATACA-A-SA-GGTTGGACGGACCC-GCAATAGCA
MEGALOP           C--CGACGCTTTCAATTCCCTGTA--CATTTTTAACTGAACTTCACCAG-CTCTCTGACGGGTAGGGT--CCTTGT-CGTTAATACA-A-CA-GGTTGGACGGACCC-GCAATAACA
OPHICTH           C--CGAGCCTTCCAATTCCCTGTA--CATTTTTAACT-ATCTTCACCAG-CTCCTCTAACGGGTACGGT--CCCTGTCGGTTAATACA-AGCAGGGTTGGACGGACCC-GCAATAACA
ECHIOPH           C--CAACGCTTCCAATTCCCTGCA--CATTTTTAACTGATCTTCACCAG-CTCCTCTAACGGGTACGGT--CCCTGTGCGATAATACA---CAGGGTTGGACGGACCC-GCAGTAACA
NOTACAN           ?????????????????????????????????????????????????????????????????????????????????????????????????????????????????????
EURYPHA           ?????????????????????????????????????????????????????????????????????????????????????????????????????????????????????
STOLOTH           ?????????????????????????????????????????????????????????????????????????????????????????????????????????????????????
LIMNOTH           ?????????????????????????????????????????????????????????????????????????????????????????????????????????????????????
CLUPEA            TTT-G--GGCCCGTAGCCTAATTTGCTC--CAGGGTC-TTGCGTGCTGTATCTCCGCAGGAGGCGGCAGCACTGCAC-CGTTGGCCGCGA-CG-GGTTTCAAGTCTCTTGTGAGAAAG
ANGUILL           ?????????????????????????????????????????????????????????????????????????????????????????????????????????????????????
OSTEOGL           ?????????????????????????????????????????????????????????????????????????????????????????????????????????????????????
PETROCE           ?????????????????????????????????????????????????????????????????????????????????????????????????????????????????????


                  ...................................l
                  2222222222222222222222222222222222222
                  4444444444444555555566666666666666777
                  4555667777780011223155566666677788134
Node              4238252367832945478856902569037725646
--------------------------------------------------------------
HIODON            CACCCT--GG--GATCCTGAGCCGSA-GCG-GGTCT
ALBULA            GGTTCT--GCG-GGTCCTGAGCCGCGGGGACGGTTC
ELOPS             GGTCCT--GG--GATCCTGAGCCGCGGGTG-G-TCT
MEGALOP           GGTCCT--GCG-GATCCTGAGCCGCGGGTA-GGTCT
OPHICTH           GGTTCTCG---GTATCCTGTGCCGCGGGGA--GACT
ECHIOPH           GGTTCT--GG--TATCCTGAGCCGCGGGGA-GGTCT
NOTACAN           ?????????????????????????????????????
EURYPHA           ?????????????????????????????????????
STOLOTH           ?????????????????????????????????????
LIMNOTH           ?????????????????????????????????????
CLUPEA            GGTCTG--N---TTCTACCACGGCC-CC-GG-GTCT
ANGUILL           ?????????????????????????????????????
OSTEOGL           ?????????????????????????????????????
PETROCE           ?????????????????????????????????????
```

Of the original full alignment, mitochondrial data are from 16S rRNA (1–585) and 12S rRNA (586–931) genes; nuclear data are from the 18S rRNA (932–2750) gene. Deletions are marked (-); missing data are marked (?); IUPAC codes used throughout. Individual gene sequences have been deposited with EMBL/GenBank under accession numbers X98840–X98846 and X99169–99196 inclusive. Taxa used are listed in Appendix 3. Full alignment available from P.L.F.

10

Clupeomorpha, Sister-Group of Ostariophysi

G. LECOINTRE

Service Commun de Systématique Moléculaire
Laboratoire d'Ichtyologie Générale et Appliqué
Muséum National d'Histoire Naturelle
75231 Paris cedex 05, France

G. NELSON

American Museum of Natural History
New York, New York 10024
and
School of Botany
The University of Melbourne
Parkville, Victoria 3052, Australia

La nature . . . est une femme qui aime à se travestir, et dont les différents déguisements, laissant échapper tantôt une partie, tantôt une autre, donnent quelque espérance à ceux qui la suivent avec assiduité de connaître un jour toute sa personne.

D. Diderot (1754)

I. Clupeomorpha

The taxon Clupeomorpha was perceived by Linnaeus (1735). He recognized one genus, *Clupea*, with four species: the European herring, sprat, anchovy, and shad. These he placed in his order Malacopterygii in reference to fins without spines: *Pinnae osseae, quae omnes molles*. The genus, defined partly by the phrase *Venter acutus serratus*, expanded in the 10th edition of his *Systema Naturae* (1758) to include 10 species, 6 of which are today reckoned as clupeomorphs. The two added clupeomorphs are a gizzard shad and a grenadier anchovy from China, collected by his friends M. Lagerström and P. Osbeck (also added are an ostariophysan, *Gasteropelecus sternicla*, and three *nomina dubia*). He placed these 6 species, fairly representative of clupeomorph diversity even more than 2 centuries later (G. J. Nelson, 1994), in his order Abdominales in reference to the posterior position of the ventral fins: *Pinnae ventrales pone pinnas pectorales*.

Since Linnaeus, the taxon expanded to include over 80 genera and 350 species. Names for extant genera and species currently recognized were proposed by some 130 different authors. Among authors of the nineteenth century, P. Bleeker and A. Valenciennes described nearly one-half of the current genera and species as they were then known and about one-seventh of those taxa as they are known today. By 1900, more than one-half of the current taxa had been discovered.

During this period, clupeomorphs were variously arranged in one or more family, usually placed in an order Malacopterygii, Abdominales, Cycloids (scales without spines), Physostomi (persistent duct between air bladder and gut), or Isospondyli (malacopterygians, physostomes, cycloids, or abdominals with vertebrae that are all similar, i.e., not forming a Weberian apparatus that is characteristic of ostariophysans). However construed, the order was seen to include primitive teleosts without derived features.

Modern understanding of clupeomorph species and genera derives from the lifework and influence of the late P. J. P. Whitehead of the Natural History Museum (London). Among the contributions of his later years he provided the best introduction to clupeomorphs (Whitehead, 1985a) and world cata-

Copyright © 1996 by Academic Press, Inc.
All rights of reproduction in any form reserved.

logs of their species (Whitehead, 1985b; Whitehead et al., 1988). His work at the Natural History Museum was not done in isolation, for extant clupeomorphs were studied and named by most of the museum's ichthyologists (Gray, Günther, Boulenger, Regan, and Norman), as were fossil clupeomorphs by its paleontologists (Woodward, Patterson, and Forey); relationships of clupeomorphs were an abiding interest of other of his colleagues (Marshall and Greenwood). Earlier in this century, Regan (1916, 1917a,b,c, 1922) and Norman (1923) revised virtually the entire extant group as was then known, for which Norman (1957) provided a summary and synopsis. These several ichthyologists together with Whitehead and his principal student, Wongratana (1980), described about one-fourth of the current genera and species. Working primarily with the collections of the Natural History Museum, augmented by Whitehead over the years, Wongratana (1983, 1987a,b) described 30 new Indo-Pacific species, which, if proven valid, would be more clupeomorph species than described by any other person.

II. Isospondyli and Derivative Taxa

During the modern (Darwinian) period, notions of relationships conformed to the idea that one taxon (or character) is the ancestor of another, and the study of relationships was the search for ancestral taxa (or characters). Among teleosts, the basal taxon was the order Isospondyli (Cope, 1871) used by Gill (1872) in his arrangement of families of fishes, by Woodward (1895) in his catalog of fossil fishes, by Jordan (1923) in his general classification; and by Regan, Norman, and others in the Systematic Index of *The Zoological Record* (1911–1969). Other orders were seen as derivatives of Isospondyli (or the equivalent Clupeiformes of Goodrich, 1909, and Berg, 1940). Among isospondyls, relationships were never exactly perceived, but the taxon including fishes such as *Elops* and *Megalops* was seen as basal, and other taxa, such as that including clupeomorphs, were seen as derivatives. Gosline's (1960) account foreshadowed and inspired subsequent developments. He divided isospondyls into two divisions, Osteoglossi and Clupei, the latter including (as suborders) elopoids, clupeoids (with alepocephaloids), gonorhynchoids, stomiatoids, salmonoids, and esocoids. Of his division Clupei he stated it "is represented today by some primitive forms, notably the elopoids, that in most respects might stand as the ancestors of the whole division" (Gosline, 1960, p. 359).

III. Divisions I–III

Greenwood et al. (1966), who also influenced subsequent developments, dismembered the order Isospondyli, distributing isospondyls among three divisions, which soon became known as Elopomorpha (division I), Osteoglossomorpha (division II), and Euteleostei (division III) (partly after Greenwood et al., 1967). They tentatively placed Clupeomorpha in division I but suggested that the taxon "Clupeomorpha should be recognized as a distinct division of teleosts" (Greenwood et al., 1966, p. 360; also Greenwood, 1970).

The divisions of Greenwood et al. combine basal and derivative taxa of older systems. Elopomorpha combine isospondylous Elopoidei of Gosline and their derivative orders, Halosauriformes (Lyopomi), Notacanthiformes (Heteromi), Anguilliformes (Apodes), and Saccopharyngiformes (Lyomeri). Osteoglossomorpha combine isospondylous Osteoglossi of Gosline and their derivative order, Mormyriformes (Scyphophori). Euteleostei are all other teleosts save clupeomorphs: isospondylous Alepocephaloidae [sic], Gonorhynchoidei, Stomiatoidei, Salmonoidei, Esocoidei of Gosline, and all of their supposedly derivative taxa comprising all other teleosts.

Hubbs (1967, p. 41) commented that "All other teleosts—more than 90 percent—are clumped in the undefined, and I think undefinable 'Division III'" (Euteleostei). He noted that the "new scheme is justified in a long series of bare statements of major trends within 'divisions' . . . with singular avoidance of diagnostic or alternative features." Although he stated that division II (Osteoglossomorpha) "appears to me to be a natural unit," his criticism is of divisions in general: "I see little justification for these 'divisions.' It would be far more realistic and useful . . . to recognize independently a considerable number of seemingly natural taxa, and to play chess with them to our hearts' content." He gave no reason for this remarkable, even absurd, recommendation beyond that "the fossil record of teleosts is far too incomplete, too unrepresentative, and too ancient, to approach the significance of paleontological data in the classification of mammals."

Hubbs saw Elopomorpha as "an incongruous assemblage," and as "particularly naive . . . the stress laid on the common occurrence of a leptocephalus stage. In this and in other respects there was too little appraisal of convergent and adaptive characters." Subsequent study provided other characters diagnostic for Elopomorpha and Osteoglossomorpha (Nelson, 1972, 1973; Patterson and Rosen, 1977; Leiby, 1979; Smith, 1984; Mattei, 1991; Jamieson, 1991), today leaving only division III (Euteleostei) without significant empirical support (see following discussion).

The divisions were themselves soon combined to reflect a dichotomous pattern of the basal branching of teleosts. Elopomorphs, clupeomorphs, and euteleosts were combined as Elopocephala; and clupeomorphs and euteleosts, as Clupeocephala (Patterson and Rosen, 1977, based partly on lower jaw structure described by Nelson, 1973). The taxon Elopocephala in effect combines isospondylous Clupei of Gosline and all of their derivative taxa. A conceptual equivalent of the taxon Clupeocephala is not seen in earlier systems, which tend to associate clupeoids and elopoids, as does, "traditionally," division I of Greenwood *et al.* (1966, pp. 359, 393–394).

IV. Divisions of Clupeomorpha

Norman (1957) perceived Clupeomorpha as one family with four subfamilies: Chirocentrinae (wolf herrings: large teeth), Engraulinae (anchovies: prominent snout), Dussumieriinae (round herrings: no abdominal scutes), and Clupeinae (other clupeids, with abdominal scutes). His diagnosis of the family mentions reduced lateral line and intracranial penetration by the swimbladder. His synopsis divides Clupeinae into groups of genera: anal fin with more than 30 rays (*Pristigaster* and other genera), upper jaw with median notch (e.g., *Alosa* and *Dorosoma*), one supramaxillary bone (e.g., *Pellonula*), and a group of genera without such peculiarities (e.g., *Clupea*). Other authors of that time perceived Clupeomorpha as one suborder with three or four families and his groups of genera as subfamilies of Clupeidae (Svetovidov, 1952, 1963; Whitehead, 1963a): Pristigasterinae, Alosinae, Dorosomatinae, Pellonulinae, and Clupeinae, taxa still recognized today.

Notions of relationships among clupeomorphs, too, conformed to the idea of ancestral and descendant taxa. According to Chapman (1948), round herrings (Dussumieriidae) are the ancestral or basal taxon, and others are descendant, or derivative taxa. Dussumieriids have a round, rather than laterally compressed, body and lack abdominal scutes. Abdominal scutes are median, scale-like bones, usually with a posteriorly directed spine. Spines of successive scutes comprise the serrated abdominal margin noted in Linnaeus's (1735) diagnosis. Each scute has paired ascending arms, developed in the abdominal region (between pectoral and pelvic fins), the arm of each side in ligamentous connection with the distal portion of a rib.

Schaeffer (1947) promoted Woodward's (1892, 1901) idea of "doubly-armoured" clupeomorphs, fishes with dorsal as well as ventral (abdominal) scutes. Dorsal scutes were known to occur as a series of enlarged median scales, sometimes with a posteriorly directed spine, in the predorsal area of certain Cretaceous and early Tertiary fishes (†*Diplomystus* and †*Knightia*) and in certain extant clupeomorphs (*Ethmidium*, *Potamalosa*, and *Hyperlophus*). The implication is that these extant taxa are descendants of the fossil taxa, and together they form a group apart from other clupeomorphs. Ogilby's (1892) discovery of dorsal scutes in extant clupeomorphs, which led him to suggest a new genus (*Hyperlophus*), which is recognized today, prompted Woodward (1892, p. 413) to bemoan "that lamentable ignorance of extinct animals so conspicuous in a certain school of zoologists," and to regard the name *Hyperlophus* as a synonym of †*Diplomystus*. He saw the significance of Ogilby's discovery to mean that "the occurrence of *Diplomystus* at the present day in the freshwaters of Australia, is thus another interesting case of the survival of ancient types in remote places of refuge."

Whitehead (1962, 1963a,b) revised dussumieriids; described the single, and peculiarly W-shaped scute invariably present just anterior to, and partially wrapped round, the base of the pelvic fins of these fishes; and, following Chapman's notion that dussumieriids are ancestral, interpreted this single scute as primitive for Clupeomorpha: "Evolution of the pelvic scute may have preceded the evolution of the abdominal scutes, almost as it were, acting as a template" (Whitehead, 1963a; p. 743). He saw the pelvic scute as derived from a pair of bones ("pelvic scute") in *Chirocentrus*, and all of these in turn from paired pelvic splints (bones), then recently described for fishes such as *Albula* and *Megalops* (Gosline, 1961): "Pelvic scutes may have arisen, not from folded ventral scales, as seems possible in the case of the abdominal scutes, but from modified splint bones" (Whitehead, 1963a, p. 743). He saw dussumieriids as a paraphyletic series, basal to clupeids, with engraulids the sister-group of that assemblage (Whitehead, 1963b, fig. 33). He mentioned that "Jordan and Seale (1926) thought an independent development of scutes in both anchovies and herrings unlikely . . . Early engraulids may have had a w-shaped pelvic scute . . . The w-shaped pelvic scute marks a step in the evolution of scutes, not a modification, but . . . is no longer found in the Engraulidae" (Whitehead, 1963b, pp. 368–369). *Denticeps* (see following discussion) has abdominal scutes without ascending arms, and for Whitehead (1963b, p. 369) "The presence of ascending arms in the three clupeoid families . . . and the constancy of the pelvic scute, seem to indicate that the evolution of the clupeoid scutes was in some manner linked to the evolution of the pelvic scute, and that the evolution of the scutes in *Denticeps* has followed a different course . . . ."

Bardack (1965) promoted Woodward's (1902–1912, 1942) and de Saint-Seine's (1949) idea that *Chirocentrus* is descended from other Cretaceous fishes (e.g., †*Ichthyodectes* and †*Xiphactinus*) with the implication that they form a group altogether apart from clupeomorphs (cf. Patterson and Rosen, 1977, pp. 130–131). Greenwood *et al.* (1966; also Cavender, 1966) argued against this idea, giving their own assessment of the relationships of the relevant fossils (cf. Applegate, 1967). They noted Marshall's (1962, p. 265) review of the varied opinions about *Chirocentrus* and his conclusion that "clupeids and *Chirocentrus* are thus quite closely related, and in two striking respects, bauplan of swimbladder and lateral line system, they are more than suspiciously like *Denticeps*. There may even be good reason for putting the Denticipitidae, the Chirocentridae, and the Clupeidae in a division . . . of the suborder Clupeoidea." In sum, the effort of Greenwood *et al.* and the subsequent history of investigation by others refute Woodward's (1942, pp. 911–912) expectation that "no further progress can be made in studying the early evolution of the Teleosteans until large series of well-preserved fossil fishes have been discovered in Lower Cretaceous formations," an old expectation that persisted for some 20 years beyond Woodward (reflected, for example, in Bertin and Arambourg, 1958, p. 1976), but that is obsolete today, after the cladistic reform of paleontology (Nelson, 1989).

Greenwood *et al.* (1966) and Cavender (1966) emphasized characters, indicative of relationships among extant clupeomorphs, that had previously been known for one or a few species. They inferred that these are characters primitive for the entire group, and are therefore represented in all living clupeomorphs. The characters are the ear–swimbladder connection, recessus lateralis, reduced lateral line, fusion (hypural two and ural centrum one, and uroneural and preural centrum one) and separation (hypural one and ural centrum one) between bones of the caudal skeleton, and so on. Their inference is now corroborated by examination of specimens representing virtually all extant clupeomorph species and many others represented by fossils.

Central to the discussion of Greenwood *et al.* was the nearly simultaneous discovery by Clausen (1959) and Gras (1961) of a seemingly ancestral clupeomorph from African freshwater, *Denticeps clupeoides* (see Greenwood, 1965), and by Greenwood (1960) of the closely related fossil (possibly from the Oligocene), †*Palaeodenticeps tanganikae*. *Denticeps* was seen as the ancestor of all other clupeomorphs because of its complete lateral line, uroneural not fused with preural centrum one, hypural one in contact with ural centrum one, etc.: "The denticipitoid grade (apart from the loss of supramaxillae and the presence of cephalic denticles) could well be basic to that of the clupeoids. The fundamental question is, can the proto-elopomorphs and proto-denticipitoids be derived from a common stem? At present the question cannot be answered" (Greenwood *et al.*, 1966, p. 360). Greenwood later (1968, p. 271) stated that "The overall relationships of the Denticipitoidei to the Clupeoidei are probably best expressed by Hennig's concept of 'sister groups . . . .'"

Nelson (1967) found one primitive character (tooth plates unfused to basibranchial bones) in the gill arches of round herrings (*Etrumeus*), in seeming confirmation of the notions of Chapman (1948) and Whitehead, but later (Nelson, 1970a) found that one character of the gut (air bladder connects to end of stomach) convincingly relates dussumieriids (*Etrumeus*) with *Clupea* rather than with engraulids and pristigasterids, with the implication that unfused toothplates (*Etrumeus*) are secondary. Among clupeoids (*Denticeps* aside) he recognized four superfamilies, chirocentroids, engrauloids, pristigasteroids, and clupeoids, a result formally similar to the three divisions (plus clupeomorphs) of Greenwood *et al.* (1966) and not very different from Norman's (1957) synopsis (different alignments of dussumieriids and pristigasterids).

Further study soon showed that dorsal and ventral scutes are clupeomorph synapomorphies, confirming Linnaeus's early (1735) diagnosis and the assessments of Jordan and Seale (1926) and Eschmeyer (1966). As anticipated by Ogilby (1892), dorsal scutes were discovered in numerous extant clupeomorphs (Nelson, 1970b; Grande, 1982b, 1985). Contested fossils were found to comprise one or more early clade of clupeomorphs (Grande, 1982a,b, 1985), and *Chirocentrus* was found related more closely to clupeids than to engraulids and pristigasterids (Grande, 1985). These developments stemmed from study of anchovies (Engraulidae), particularly their caudal skeleton in relation to the occurrence both of abdominal and dorsal scutes in those fishes (Nelson, 1983; Grande and Nelson, 1985). The genus *Stolephorus* was found to be paraphyletic, with a branching pattern, corroborated by several characters (egg shape, sensory canals, dentition, fusion of bones of the suspensorium, gill arches, and caudal skeleton), implying scute reduction rather than development. The result for clupeomorphs, as summarized for both living and fossil forms by Grande (1985), is not a dichotomous system of their basal branching but is significant progress toward it. For Grande (1985, p. 329),

The major phylogenetic problems left among clupeomorph fishes . . . are seen here as (1) solving the interrelationships of Engrauloidea; (2) discovering the relationships of the members of Clupeinae, Alosinae and Dorosomatinae; and (3) testing the cladograms here based on the skeleton, by doing similar comprehensive studies of the muscle, internal organ, and nervous systems of clupeiform fishes.

V. Clupeomorpha and Euteleostei

As initially perceived (Greenwood *et al.*, 1966: fig. 1), the taxon Euteleostei (division III) was composed of a basal taxon (Protacanthopterygii, including some isospondyls, myctophoids, and other groups) and four derivative taxa (Ostariophysi, Atherinomorpha, Paracanthopterygii, and Acanthopterygii). Soon, some isospondyls (stomioids) and some protacanthopterygians (aulopiforms, myctophoids, and others) with their supposedly derivative taxa (Atherinomorpha, Paracanthopterygii, and Acanthopterygii) were combined as a taxon Neoteleostei (Rosen, 1973; Stiassny, 1986; Johnson, 1992). Neoteleostei, Ostariophysi, and the groups still associated as Protacanthopterygii came to be regarded as the divisions of the taxon Euteleostei. Rosen (1973, 1974) suggested a dichotomous scheme in which the taxon Euteleostei divides into Ostariophysi and Neognathi (all other euteleosteans), and the taxon Neognathi divides into Protacanthopterygii (esocoids, argentinoids, and salmonoids) and Neoteleostei (all other neognaths) (Fig. 1A). He later (1982) suggested a scheme in which the taxon Euteleostei divides into three groups: Ostariophysi, Salmoniformes (Protacanthopterygii), and Neoteleostei (Fig. 1B).

Although poorly defined, the taxon Euteleostei is provisionally considered the sister-group of Clupeomorpha (Figs. 1A–E and 1G). Patterson and Rosen (1977, p. 126) suggested three synapomorphies (Figs. 1D–E), admitting that "this definition of euteleosteans is far from satisfactory." Lauder and Liem (1983, p. 132) noted that "further evidence corroborating monophyly of this group is highly desirable." Fink (1984, p. 202) stated that "the Euteleostei . . . [are] poorly diagnosed in terms of unique traits, and most more phylogenetically advanced members lack some of the diagnostic characters." Rosen (1985) reviewed the

taxon, retained only one synapomorphy (adipose fin), and excluded esocoids (Fig. 1F). However construed, the taxon Euteleostei is not well-supported, and neither is any particular arrangement of its component taxa, as may be appreciated from summaries by J. S. Nelson (1976, 1984, 1994).

A. Ostariophysans

Rosen (1973, pp. 433–434) remarked that "Concerning the Ostariophysi, there is no evidence that they are most closely related to one or another group of euteleosteans." Fink and Weitzman (1982) placed ostariophysans within a multifurcation containing three other clades, (1) salmonids, (2) a new clade grouping argentinoids and osmeroids, and (3) neoteleosts; esocoids are the sister-group of the whole (Fig. 1C), but the authors admitted that esocoids could be included within the multifurcation. Lauder and Liem (1983, p. 132) reduced the multifurcation by "provisionally" combining salmonids and neoteleosts (neognaths); esocoids are the sister-group of the whole (Fig. 1D). Fink (1984) proposed another multifurcation of four clades: (1) ostariophysans, (2) argentinoids, (3) osmeroids, and (4) neognaths; esocoids are the sister-group of the whole (Fig. 1E). Excluding esocoids, Rosen (1985) proposed a trifurcation within euteleosteans: (1) ostariophysans, (2) argentinoids, and (3) neognaths (including osmeroids, Fig. 1F). Again excluding esocoids, Begle (1992) proposed a different trifurcation: (1) ostariophysans, (2) salmonids, and (3) Osmerae (including osmeroids and argentinoids) and neoteleosts (Fig. 1G). In short, ostariophysans are clupeocephalans (Patterson and Rosen, 1977) but not neognaths (Rosen, 1973). Their branching point is somewhere between these two nodes of the teleostean cladogram, depending on relationships of other clades in the same situation, at least clupeomorphs and esocoids, about which the same may be stated: They are clupeocephalans but not neognaths (Fig. 1H).

B. Esocoids

Rosen (1973, p. 434) left esocoids "provisionally aligned" with salmonoids within Protacanthopterygii, which were reduced through removal of certain taxa (stomioids and myctophoids) to the Neoteleostei. He later (Rosen, 1974) placed esocoids within his Salmoniformes (= reduced Protacanthopterygii), as sister-group of Salmonae (Fig. 1A). Fink and Weitzman (1982, p. 87) excluded esocoids from Salmoniformes, without finding their relationships:

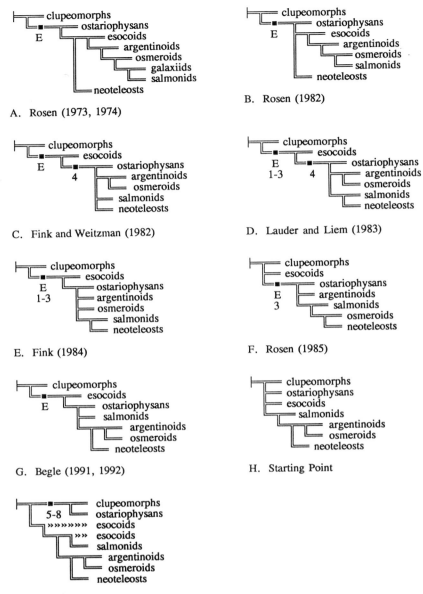

A. Rosen (1973, 1974)

B. Rosen (1982)

C. Fink and Weitzman (1982)

D. Lauder and Liem (1983)

E. Fink (1984)

F. Rosen (1985)

G. Begle (1991, 1992)

H. Starting Point

I. Molecular Data

FIGURE 1 Clupeocephalan relationships from 1973 to 1994. Abbreviation: E, Euteleostei. Characters 1–8 numbered as in text (1–3, after Patterson and Rosen, 1977).

We find no evidence to consider esocoids closely related to the other members of the group; . . . Esocoids seem to share no unique specializations with the other included taxa [Ostariophysi, Argentinoidei, Osmeroidei (including Galaxiidae), Salmonidae, Stomiiformes]; we could list Esocoids as *sedis mutabilis* at the euteleostean level or as the sister-group of all other teleosts, depending on placement of the ostariophysans (Fink and Weitzman, 1982, p. 86).

Rosen (1985) accepted Euteleostei, evidenced by one character (adipose fin), and it became one of three clupeocephalan divisions: (1) clupeomorphs, (2) esocoids, and (3) euteleosteans (Fig. 1F). Begle (1991, 1992) offered synapomorphies grouping argentinoids and osmeroids (Fig. 1G), as already suggested by Fink and Weitzman (1982, fig. 23). Given this group and Rosen's conclusions (except monophyly of Euteleostei, as evidenced by the adipose fin), there are four clupeocephalan clades (Fig. 1H): (1) clupeomorphs, (2) ostariophysans, (3) esocoids, and (4) neognaths (including osmeroids, argentinoids, and neoteleosts), among which, or as some combination of which one may suppose, is the sister-group of clupeomorphs.

VI. Molecular Data

In a parsimony tree for 38 species (190 informative sites from 28S rRNA sequences), Lê *et al.* (1993) found a clupeomorph + ostariophysan group (bootstrap value 92%). The species include two clupeids (*Sardina*, and *Clupea*) and three otophysans (*Tinca*, *Gobio*, and *Ictalurus*), with sequences that evolve slightly faster than those of other fishes, raising the possibility of long-branch attraction; however, gadiform sequences were found to evolve at the same rate, and gadiforms did not cluster with clupeomorphs + ostariophysans. The level of mutations in the data approaches saturation, and bootstrap values for some nodes (particularly deeper nodes) are weak. The sequences analyzed are probably not suitable for all nodes inferred (most nodes with high bootstrap values are terminal, grouping two species). Nevertheless, deep nodes with high bootstrap values are worth consideration (those for neopterygians, chondrichthyans, and clupeomorphs + ostariophysans) even if neighboring deep nodes are not. Lê *et al.* (1993, p. 47) concluded that if the same grouping were confirmed from wider taxonomic sampling of the two sister-groups, then "the possibility of such relationships should be seriously considered by morphologists."

Confirmation came from work of D. T. J. Littlewood, C. Patterson, and A. B. Smith (unpublished; communication by Patterson at the Fish Phylogeny Workshop, following the 13th Willi Hennig Society meeting, Copenhagen, 1994; also Patterson, 1994, p. 77), who sequenced 18S rRNA of various teleosts. Their unpublished Hennig-86 tree shows an extraordinarily high number of molecular synapomorphies shared by *Chanos* and *Clupea*. The number of taxa sampled in each of the two sister-groups (Clupeomorpha, Ostariophysi) is only one, but the ostariophysan (*Chanos*) represents the earliest branch of the group, gonorynchiforms. The two studies [Lê *et al.* with otophysans, and Littlewood *et al.* (unpublished) with gonorynchiforms], converge to the same conclusion. Without representatives of the two major ostariophysan lineages (otophysans and gonorhynchiforms) in the same molecular tree, monophyly of ostariophysans (Rosen and Greenwood, 1970; S. V. Fink and Fink, 1981) is not tested (cf. Gayet, 1986).

Müller-Schmid *et al.* (1993, p. 584) found "no indication for the existence of an euteleost infradivision" (excerpt from title) from about 200 sites of the amino-acid sequences of the ependymins of six teleost species. Four families were sampled: Clupeidae, Cyprinidae, Esocidae, and Salmonidae. The presented tree (their fig. 6, UPGMA-bootstrap) shows (Clupeidae + Cyprinidae) (Esocidae + Salmonidae), hence a clupeomorph + ostariophysan relationship, but this grouping is perhaps artifactual. With neighbor-joining, they found a different result: "Salmoniformes were significantly grouped with *C. harengus* but not with Cypriniformes (data not illustrated)." This seems only a difference in the position of the root. The problem is twofold: (1) the main result is based on a phenetic procedure (UPGMA), and (2) the authors showed a higher rate of evolution for sequences of esocids and salmonids (UPGMA is misleading in such a case). Neighbor-joining and parsimony trees are not shown. The problem with a four-taxon tree and methods yielding unrooted trees (neighbor-joining or parsimony) is that interpretation of relationships may depend on the position of the root. The authors gave no indication where they chose to root their neighbor-joining and parsimony trees, a decision crucial for their conclusions. (Were the data analyzed with neighbor-joining or parsimony, and the tree rooted on Clupeidae, one would expect a monophyletic Euteleostei.) The authors also reduced their taxonomic sample to a three-taxon problem (with *Esox* considered a salmoniform) and, with a parsimony procedure, discovered that changing the root does not change the number of steps. Because UPGMA does not require an outgroup, the clupeid + cyprinid group is possibly due to symplesiomorphy or to higher rates of evolution in esocids and salmonids. These results are inconclusive. A suitable outgroup could be added to these data for parsimony analysis.

Some studies include sequences for esocoids but none for ostariophysans or clupeomorphs. Lê *et al.* (1989) sequenced portions of the 28S rRNA for 12 fishes, including seven teleosts (*Clupea*, *Esox*, and five neoteleosts). Their tree based on a distance-matrix method shows a monophyletic Teleostei and the group (*Clupea* (*Esox* + neoteleosts)), perhaps resolving the basal trifurcation of Rosen (1985) (Fig. 1F). Problems of this study include the following: (1) no robustness indicator was given, (2) the *Esox* 28S rRNA sequence suffers from a number of undetermined nucleotides, and (3) no justification was given for discarding five species (including *Clupea*) from the data for parsimony analysis. Bernardi *et al.* (1993) analyzed a data base of 27 growth-hormone amino-acid sequences (96 informative sites) for 25 teleosts: *Anguilla*, five otophysans, *Esox*, four salmonids, and 14 acanthopterygians. In their parsimony-bootstrap tree, relationships are (*Anguilla* (otophysans ((*Esox* salmonids) tionships are (*Anguilla* (otophysans ((*Esox* salmonids) acanthopterygians))). Each node is well-supported except that of salmonids (62%) and that of salmonids + *Esox* (69%; the Salmoniformes of Rosen, 1974; see Fig. 1A). There is no clupeomorph sequence. The node for the sister-group of *Anguilla* is well-supported (100%), but that node can represent either Clupeocephala or Euteleostei (including *Esox*, as in Rosen, 1973, 1974). This study does not bear directly on the monophyly of Euteleostei. The topology obtained is that of Fig. 1A (or Fig. 1I, esocoids + salmonids) without clupeomorphs, argentinoids, osmeroids, and galaxiids.

Some studies include esocoids, ostariophysans, and clupeomorphs. That of Müller-Schmid *et al.* (1993) includes an ependymin sequence of *Esox* and describes an *Esox* + salmonid relationship, but it is unreliable (see above). In the unpublished trees of Littlewood *et al.* (see above), the clupeocephalan topology is (*Clupea* + *Chanos*) (*Esox* (*Salmo* + neoteleosteans)), in agreement with Lê *et al.* (1989) but not with Bernardi *et al.* (1993)—as in Fig. 1I (esocoids + neognaths) without argentinoids and osmeroids.

From these studies one may conclude (Fig. 1I) that esocoids are the sister-group either of salmonids (Bernardi *et al.*, 1993) or of neognaths (salmonids, osmerans, and neoteleosts) (Littlewood *et al.*, see above); and that clupeomorphs and ostariophysans are sister-groups. If so, morphological characters proposed as synapomorphies for Euteleostei are invalid, but there should be morphological synapomorphies for the clupeomorph + ostariophysan clade.

VII. Euteleostei: Doubtful Synapomorphies

A. Character 1: Nuptial Tubercles

This character, found too variable to define Euteleostei, was rejected by Rosen (1985). Some salmonids and some osmeroids have nuptial tubercles (Begle, 1991); argentinoids and esocoids do not (Patterson and Rosen, 1977; Fink, 1984; Begle, 1991). This character is not shared by all ostariophysans: it occurs in some gonorynchiforms, many cypriniforms, and few characiforms and siluroids (Fink and Fink, 1981). Fink (1984) considered that the character might be a synapomorphy of a clade uniting ostariophysans, salmonids, and osmeroids, but he found that too many other characters contradict this grouping.

B. Character 2: Stegural

In the caudal skeleton, outgrowths of laminar bone (or membranous laminae) extend anteriorly from the anterodorsal margin of uroneural one. The exact definition of this character as an euteleostean synapomorphy is "anterior membranous outgrowth of first uroneural (stegural) not joining opposite member in midline" (Patterson and Rosen, 1977). When the stegural is median (as in clupeomorphs), it is considered a synapomorphy of the clupeocephalan clade (Patterson and Rosen, 1977; Rosen, 1985; but criticized by Arratia, 1991). When there are "paired stegural outgrowths of the first uroneural" (Rosen, 1985), they are considered a synapomorphy of neognaths. The paired stegural is not a definition of the original Euteleostei, including esocoids (Patterson and Rosen, 1977). Indeed, the anterodorsal outgrowth of the first uroneural of esocoids is doubtfully homologous with the outgrowth of euteleosts (Rosen, 1985). Ostariophysans lack a stegural (Monod, 1968; Rosen and Greenwood, 1970; Fujita, 1990; M. Gayet, personal communication). In argentinoids, the supraneural lamina identified by Monod (1968, p. 304) as a stegural is doubtfully homologuous with the stegural of salmonids (Greenwood and Rosen, 1971). This lamina is linked to the underlying vertebral centrum through one or two rudiments of neural arches and is rarely fused with the anterodorsal margin of the first uroneural (Greenwood and Rosen, 1971, pp. 14, 20–21, fig. 12). According to Rosen's (1985) cladogram (Fig. 1F), the paired stegural from the first uroneural is found exclusively in neognaths.

C. Character 3: Adipose Fin

This was considered by Rosen (1985) as the only known synapomorphy of Euteleostei (excluding esocoids, which lack the adipose). The adipose fin occurs in some ostariophysans [absent from gonorynchiforms, cypriniforms, and extant gymnotoids; present in some characiforms, siluroids, and fossil gymnotoids (M. Gayet, personal communication)]. The adipose occurs in argentinoids, salmonids, osmeroids, stomiiforms, aulopiforms, and myctophoids, but not in many other neoteleosteans.

D. Character 4: No Toothplate over Fourth Basibranchial

Clupeomorphs (Nelson, 1970a, pp. 2, 7) and esocoids (Rosen, 1974, p. 274) have this plate, while ostariophysans and neognaths do not (Fink and Weitzman, 1982; Begle, 1992). However, most neognath fishes have no ossified fourth basibranchial (e.g., Rosen, 1974, p. 275).

VIII. Clupeomorpha + Ostariophysi: Possible Synapomorphies

A. Character 5: Pleurostyle

In the caudal skeleton, a paired lateral process extends posterodorsally from preural centrum one (not homologuous with the pseudurostyle of acanthopterygians: Monod, 1967). It was found by Monod (1967, 1968) only in clupeomorphs and ostariophysans (including gonorynchiforms). Fujita (1990) recorded it only in clupeomorphs and ostariophysans, in every taxon from the families he sampled except Chacidae and Plotosidae: Clupeidae, Engraulidae, Chirocentridae, Chanidae, Gonorynchidae, Cyprinidae, Cobitidae, Gasteropelecidae, Characidae, Ictaluridae, Bagridae, and Siluridae. Neither Monod (1968, pp. 128, 275, 597, 599) nor Grande (1985, p. 259) found a pleurostyle in *Denticeps*, which Fujita (1990) did not examine.

Monod (1968, pp. 275, 599, figs. 853, 856, 859) insisted that in many traits the caudal skeleton of *Denticeps* is closer to that of salmonids (type I) rather to that of clupeomorphs (type II). Grande (1985, p. 258) considered that "the general shape of the *Denticeps* uroneural, and the slight branched appearance of its posterior end, suggest that the three uroneurals have fused in denticipitoids," while "all other clupeomorphs observed have three uroneurals." He considered the pleurostyle as the fused uroneural one and preural centrum one (Grande, 1982b, pp. 7, 13, 19). If so, ancestors of *Denticeps* lacked a pleurostyle. He stated: "Although *Denticeps* has many peculiar skeletal features, it is clearly a clupeiform clupeomorph because of the presence of pterotic and prootic bullae, abdominal scutes, and a recessus lateralis. The many peculiarities of the skull and caudal skeleton are independently derived features of denticipitoids."

A pleurostyle is absent from fossil gonorynchiforms (Poyato-Ariza, 1994), and from the fossil clupeomorphs †*Armigatus* and †*Diplomystus* (Patterson and Rosen, 1977, p. 138; Grande, 1982a), but it occurs in other fossil clupeomorphs, such as †*Knightia* and †*Gosiutichthys* (Grande, 1982b). Monod (1968, p. 99) noted in †*Diplomystus brevissimus* (†*Armigatus brevissimus* of Grande, 1982a) that two uroneurals (one and two) anteriorly contact preural centrum one. He considered that fusion of uroneural two and preural centrum one could yield a pleurostyle and a typical clupeomorph caudal skeleton (in this species the caudal skeleton is otherwise of clupeomorph type, with fused hypural two and ural centrum one). Monod concluded (1968, p. 99): "Il pourra donc être très tentant de voir dans la disposition présentée par *Diplomystus* [†*Armigatus*]

un stade sinon exactement intermédiaire entre les complexes à urodermaux [uroneurals] tous libres (*Leptolepis*, *Elops*, etc.) et ceux des Clupes, du moins un état pouvant avoir immédiatement précédé celui qui caractérise ces derniers."

This interpretation of the uroneurals of *Diplomystus* is an interesting but problematic conjecture. Uroneurals one and two in †*Diplomystus longicostatus* (†*Ellimichthys longicostatus* of Grande, 1982a) do not reach preural centrum one (Patterson and Rosen, 1977), but uroneural two does so in †*D. dentatus* and †*D. birdi* (Grande, 1982a). The conjecture contradicts Monod's nomenclature: In all clupeomorph caudal skeletons of his study, independent bones called "urodermals [uroneurals] 1 and 2" are identified in the presence of a well-developed pleurostyle. More appropriate is the nomenclature of Grande and Patterson (UN2 and UN3), who supposed that the pleurostyle is the fused preural centrum one and uroneural one (rather than uroneural two).

B. Character 6: Fusion of Hypural Two and Ural Centrum One

The caudal skeleton of a great variety of teleosts in the studies of Monod (1968) and Fujita (1990) shows that extant clupeomorphs (even *Denticeps*) and otophysans all have this caudal fusion, but other extant teleosts do not. The survey of Fujita (1990, pp. 824–845) precisely shows that, except in cases of global fusion of the caudal skeleton in some fishes, fusion of these two bones is restricted to clupeomorphs and otophysans. The problem is that this fusion is not present in all anotophysans. It occurs in *Gonorynchus* (Monod, 1968, p. 199), but not in *Chanos* (Monod, 1968, p. 70; Fink and Fink, 1981, p. 340; Fujita, 1990, p. 211), *Phractolaemus* (Monod, 1968, p. 71), *Kneria*, or *Parakneria* (Lenglet, 1974; M. Gayet, personal communication). The fusion occurs in fossil clupeomorphs, †*Diplomystus* (Patterson and Rosen, 1977; Grande, 1982a), †*Ellimmichthys*, and †*Knightia* (Grande, 1982a,b); in some fossil ostariophysans, †*Ramallichthys* and †*Lusitanichthys* (Gayet, 1986); and in the fossil otophysan †*Chanoides* (Patterson, 1984); but not in †*Prochanos*, †*Rubiesichthys*, and †*Tharrhias* (Lenglet, 1974; Patterson, 1975; Wenz, 1984; M. Gayet, personal communication). The fusion is found also in some Jurassic non-clupeomorph teleosts (Arratia, 1991).

C. Character 7: Fusion of Extrascapulars and Parietals

The supratemporal sensory canal extends medially from the dermopterotic bone through one or more extrascapular bones, for example, in *Lepisosteus*, *Amia*, the fossil †*Anaethalion* (Forey, 1973, fig. 16), *Elops* and *Megalops* (Forey, 1973, figs. 16, 20, 26, 32), and the anabantid *Ctenopoma* (Daget, 1964, p. 272). When the canal passes through parietal bones, it is supposed that extrascapular and parietal bones are fused. Examples are found in Gayet (1986, pp. 13, 16–17), where the canal passes through parietals in *Chanos* and *Phractolaemus* (not in *Kneria*, which lacks parietals); the cypriniform *Beaufortia*; the fossil gonorynchiforms †*Ramallichthys* and †*Hakeliosomus* (Gayet, 1986, 1993); the fossil otophysan †*Chanoides* (Patterson, 1984, pp. 434–435), and the extant characiforms *Brycon* (Daget, 1964, p. 272), *Acestrorhynchus*, *Ctenolucius*, *Hoplias*, and *Hepsetus* (Roberts, 1969). The "supratemporal commissural sensory canal primitively passing through parietals and supraoccipitals" was considered by Grande (1985, p. 326) a synapomorphy of Clupeomorpha, an opinion based on a large survey of fossil and extant clupeomorphs and illustrated for *Dorosoma* and *Odaxothrissa*, and for the fossil clupeomorph, †*Spratticeps* (Patterson, 1970). Fusion of extrascapulars and parietals is restricted neither to ostariophysans nor to clupeomorphs but is a possible synapomorphy of a clade Clupeomorpha + Ostariophysi.

D. Character 8: Fusion of Hemal Spines and Centra Anterior to Preural Centrum Two

This character was suggested by Fink and Fink (1981, pp. 339–341) as an ostariophysan synapomorphy:

In ostariophysans, all hemal spines anterior to that of the second preural centrum are fused to the centra from a young juvenile stage ... In the primitive members of most other primitive teleostean lineages, including *Scleropages*, *Elops*, *Esox* and *Diplophos*, four or more hemal spines are autogenous in much larger juveniles. An adult *Esox* specimen has five autogenous hemal spines, and an adult *Salmo* two (dry skeletal material). Clupeomorphs have all hemal spines fused to the centra, suggesting relationships between the Clupeomorpha and the Ostariophysi. However, clupeomorphs lack the adipose fin and breeding tubercles which link ostariophysans with other members of the Euteleostei (Patterson and Rosen, 1977).

Those characters were later rejected as euteleostean synapomorphies (see previous discussion).

Fujita (1990, pp. 824–845) showed that fusion of hemal spines anterior to preural centrum two probably occurs in all clupeomorphs and ostariophysans but also in other teleostean groups. He (1990, p. 253) observed only one autogenous hemal spine in *Diplophos* (HPU2). Fishes with a derived caudal skeleton aside, fusion of the hemal spine and preural centrum three (and anterior hemal spines and centra) occurs sparsely in non-acanthomorphs: *Umbra* (but not other esocoids), *Bathylagus*, Gonostomatidae, Sternoptychidae, two genera of the Photichthyidae, *Bathypterois*, *Bathysaurus*, *Omosudis*, *Alepisaurus*, *Centrobranchus*, and *Lampanyctus*. These cases may be evaluated in terms of their frequency of occurrence and distribution among Fujita's large taxonomic sample. They involve a small minority of taxa among 39 families and 72 genera of non-acanthomorph, non-clupeomorph, and non-ostariophysan fishes sampled. All are isolated cases within their family or order (except in Gonostomatidae and Sternoptychidae). By contrast, fusions were constant in all 39 genera and 14 families of clupeomorphs and ostariophysans sampled.

IX. Clupeomorpha + Ostariophysi: Doubtful Synapomorphies

A. Character 9: Epicentral Intermuscular Bones

Clupeomorphs and gonorynchiforms have epicentral bones (Patterson and Johnson, 1995), but they are absent from most otophysans (Gayet *et al.*, 1994). They occur in gymnotoids (Meunier and Gayet, 1991, p. 225, fig. 2; Gayet *et al.*, 1994). They also occur among elopomorphs, in *Megalops* and possibly in some fossil eels (Blot, 1978, p. 18, fig. 2).

B. Character 10: Connection between Swimbladder and Ear

In many teleosts, the swimbladder connects, directly or indirectly, to the inner ear. The direct connection can be described as a diverticulum of the swimbladder extending forward to the labyrinth in some osteoglossomorphs (Notopteroidea), in all clupeomorphs, and in some holocentrids. The indirect connection is achieved in ostariophysans through perilymphatic cavities and a paired chain of bones (Weberian apparatus) between the swimbladder and the labyrinth.

In clupeomorphs, the swimbladder extends forward, bifurcating as it nears the skull (e.g., Grande, 1985, p. 253; Whitehead, 1985a, fig. 9). On each side, the diverticulum enters the skull through an opening in the exoccipital bone and terminates in two distinct vesicles, one in the prootic bone, the other in the pterotic bone. Each vesicle is surrounded by a bony bulla. The extremity of each vesicle is closely linked to a foramen closed by a fibrous membrane beyond which is the perilymph of the labyrinth of the ear. This connection strongly modifies the perilymphatic cavities, through which the vesicle transmits stimuli from the swimbladder to the utricula (Wohlfahrt, 1936; O'Connell, 1955; Denton and Blaxter, 1976; Best and Gray, 1980; Blaxter *et al.*, 1981; Blaxter and Hunter, 1982).

In otophysans, a constriction divides the swimbladder into anterior and posterior chambers, also present in gonorynchiforms (Rosen and Greenwood, 1970). The anterior chamber is linked to the posterior Weberian bones, tripus and suspensorium, through the peritoneal tunic. The four most anterior vertebrae have modified neural arches, pleural ribs, and parapophyses which form the other Weberian bones, claustrum, scaphium, intercalarium, tripus, and os suspensorium (Rosen and Greenwood, 1970). These bones, linked through ligaments, are supported by vertebral centra, which are variously fused in advanced otophysans (Roberts, 1973). In gonorynchiforms, the complete series of Weberian bones is never present. The connection between swimbladder and inner ear is indirect, and although different from that of otophysans, it, too, involves modified vertebral elements (Rosen and Greenwood, 1970; Gayet and Chardon, 1987).

Rosen and Greenwood (1970, p. 19, fig. 16) and Fink and Fink (1981) showed that in gonorynchiforms and otophysans, elements of the otophysic connection are possibly homologous. The connection between endolymphatic spaces of the inner ear and the extension of the swimbladder differ from those of clupeomorphs. Sacculae of the ears are linked through a transverse canal (ductus communicans transversus) passing under the brain. From the middle of this transverse canal, an axial endolymphatic sinus extends posteriorly, the whole filled with endolymph. Surrounding these structures, the perilymphatic space begins as an axial structure and, extending posteriorly, develops into two symmetrical atria (atria sinus imparis). Both atria pass through a median opening in the basioccipital (cavum sinus impar), beneath the foramen magnum. The cavum sinus impar is separated from the spinal cord by a shelf of bone of the exoccipitals. Behind the skull, the two atria separate as they extend backwards to each side of the first vertebra. Their extremities link to scaphium and claustrum, the two most anterior Weberian bones. These perilymphatic and endolymphatic spaces transmit vibrations from the swimbladder and Weberian bones to the labyrinth of the ear.

The anatomy of the otophysic connection differs in clupeomorphs and ostariophysans, and its different forms are not seen as homologous (Fink and Fink,

1981, p. 343). However, one may imagine the existence of a soft direct otophysic connection before the development of the indirect connection. Gayet (1986, p. 73) reasoned that

Il est très probable que ... des rapports physiologiques aient été nécessairement présents entre la vessie natatoire et le labyrinthe de l'oreille avant que ne se soit formé l'appareil de Weber. ... Une relation directe était nécessaire d'où une modification des éléments vertébraux pour transporter les informations acoustiques et hydrostatiques de la vessie à l'oreille. Une première modification a été réalisée au niveau des éléments hémaux, côtes pleurales des troisième et quatrième vertèbres, sur lesquelles la tunique péritonéale de la chambre antérieure de la vessie natatoire est attachée (Rosen and Greenwood, 1970), caractère que l'on retrouve sans exception chez tous les Ostariophysi sensu lato, quelle que soit la forme de ces troisième et quatrième côtes pleurales.

May one consider that clupeomorphs have a soft "pre-ostariophysan" otophysic connection? More exactly, may one imagine the soft otophysic connection of the last hypothetical ancestor common to clupeomorphs and ostariophysans as similar to that of clupeomorphs? Ostariophysans lack a diverticulum anterior to the swimbladder unless their perilymphatic cavities are derivatives of this diverticulum.

Ballantyne (1927) showed that in embryonic mormyrids, the swimbladder extends forward as a bifurcated diverticulum, of which the extremities come in close contact with the sacculi. During later ossification of the skull, the extremities become closed gas vesicles disconnected from the swimbladder. Vesicles are so closely linked to sacculi that they were previously thought ontogenetically derived from the ears rather than from the swimbladder.

Do these facts help to explain evolution from a direct otophysic connection (clupeomorphs) to an indirect one (ostariophysans)? The answer seems negative. First, one must explain how a structure that normally contains gas comes to contain perilymph, e.g., by hypothetical fusion between swimbladder derivates and perilymphatic structures (that has never been found before, to our knowledge). Second, perilymphatic ducts leave the skull posteriorly through the basioccipital in ostariophysans, whereas the swimbladder extensions of clupeomorphs enter the skull laterally through the exoccipital bones. Except for hypothetical possibilities, nothing can be taken as homologous (synapomorphic) in the two otophysic connections. Greenwood (1973) and Taverne (1973) separately proposed an osteoglossomorph + clupeomorph relationship based on similarities of the otophysic connection of these two groups. Patterson and Rosen (1977) refuted their comparisons (perhaps better founded but perhaps symplesiomorphic). Nevertheless, if clupeomorphs and ostariophysans are related, then, too, are their otophysic connections. If so, the exact nature of that relationship remains for the future to reveal.

X. Summary

Analysis of molecular data, even with some methodological weakness, suggests a relationship between clupeomorphs and ostariophysans. If these two groups form a clade, that clade relates in some way to esocoids and neognaths (osmerans, salmonids, and neoteleosts). Molecular data suggest that esocoids are the sister-group either of salmonids or of neognaths. If so, clupeocephalans divide into two groups, Clupeomorpha + Ostariophysi and Esocoidei + Neognathi (Fig. 1I). Four morphological characters are suggested as possible synapomorphies of a clupeomorph + ostariophysan clade: pleurostyle, fusion between hypural two and ural centrum one, fusion between hemal spines and centra anterior to preural centrum two, and fusion of extrascapulars and parietals.

Acknowledgments

We are indebted to Carl Ferraris, Mireille Gayet, Lance Grande, François Meunier, Théodore Monod, and Sylvie Wenz for information and discussion; and to Colin Patterson, to whom this volume is dedicated, for drawing attention to possible clupeomorph + ostariophysan synapomorphies during redaction of Lecointre's doctoral thesis (1993), and for review of this contribution, which to a significant degree is also Colin's.

References

Applegate, S. P. (1967). (Review of Greenwood *et al.,* 1966). *Copeia,* 693–694.

Arratia, G. (1991). The caudal skeleton of Jurassic teleosts: A phylogenetic analysis. *In* "Early Vertebrates and Related Problems of Evolutionary Biology" (M.-M. Chang, Y.-H. Liu, and G.-R. Zhang, eds.), pp. 249–340. Science Press, Beijing.

Ballantyne, F. M. (1927). Air-bladder and lungs: A contribution to morphology of the air-bladder of fish. *Trans. Roy. Soc. Edinburgh* **55,** 371–394.

Bardack, D. (1965). Anatomy and evolution of chirocentrid fishes. *Univ. Kans. Paleontol. Contrib.* **10,** 1–88.

Begle, D. P. (1991). Relationships of the osmeroid fishes and the use of reductive characters in phylogenetic analysis. *Syst. Zool.* **40,** 33–53.

Begle, D. P. (1992). Monophyly and relationships of the argentinoid fishes. *Copeia,* pp. 350–366.

Berg, L. S. (1940). Classification of fishes, both recent and fossil. *Trav. Inst. Zool. Acad. Sci. U.R.S.S.* **5,** 87–517.

Bernardi, G., d'Onofrio, G., Caccio, S., and Bernardi, G. (1993). Molecular phylogeny of bony fishes, based on the amino acid sequence of the growth hormone. *J. Mol. Evol.* **37,** 644–649.

Bertin, L., and Arambourg, C. (1958). Systématique des poissons. *In* "Traité de Zoologie" (P.-P. Grassé, ed.), Vol. 13, Fasc. 3, pp. 1967–1983. Masson, Paris.

Best, A. C. G., and Gray, J. A. B. (1980). Morphology of the utricular recess in the sprat. *J. Mar. Biol. Assoc. U.K.* **60,** 703–715.

Blaxter, J. H. S., and Hunter, J. R. (1982). The biology of clupeoid fishes. *Adv. Mar. Biol.* **20,** 1–223.

Blaxter, J. H. S., Denton, E. J., and Gray, J. A. B. (1981). The auditory bullae-swimbladder system in late stage herring larvae. *J. Mar. Biol. Assoc. U.K.* **61,** 315–326.

Blot, J. (1978). "Les Apodes fossiles du Monte Bolca," Stud. Ric. Giacimenti Terziari di Bolca, Vol. 3, Fasc. 1. Museo Civico di Storia Naturale, Verona.

Cavender, T. (1966). The caudal skeleton of the Cretaceous teleosts *Xiphactinus, Ichthyodectes,* and *Gillicus,* and its bearing of their relationship with *Chirocentrus. Occas. Pap. Mus. Zool., Univ. Mich.* **650,** 1–15.

Chapman, W. M. (1948). The osteology and relationships of the round herring *Etrumeus micropus* Temminck and Schlegel. *Proc. Calif. Acad. Sci.* [4] **26,** 25–41.

Clausen, H. S. (1959). Denticipitidae, a new family of primitive isospondylous teleosts from West African fresh-water. *Medd. Dan. Naturn. Foren.* **121,** 141–151.

Cope, E. D. (1871). Contribution to the ichthyology of the Lesser Antilles. *Trans. Am. Philos. Soc.* **14,** 445–483.

Daget, J. (1964). Le crâne des téléostéens. *Mém. Mus. Natl. Hist. Nat.,* [Ser. A] **31,** 163–342.

Denton, E. J., and Blaxter, J. H. S. (1976). The mechanical relationships between the clupeoid swimbladder, inner ear and lateral line. *J. Mar. Biol. Assoc. U.K.* **56,** 787–807.

de Saint-Seine, P. (1949). Les poissons des calcaires lithographiques de Cerin (Ain). *Nouv. Arch. Mus. Hist. Nat. Lyon* **2,** 1–357.

Diderot, D. (1994). "Oeuvres. Tome 1: Philosophie" Coll. Bouquins. Robert Laffont, Paris.

Eschmeyer, W. N. (1966). Re-evaluation of the position of the family Dussumieriidae in clupeoid phylogeny. *Forty-sixth Annu. Meet., Am. Soc. Ichthyol. Herpetol.,* Miami, FL.

Fink, S. V., and Fink, W. L. (1981). Interrelationships of the ostariophysan fishes (Teleostei). *Zool. J. Linn. Soc.* **72,** 297–353.

Fink, W. L. (1984). Basal euteleosts: Relationships. *In* "Ontogeny and Systematics of Fishes" (H. G. Moser, W. J. Richards, D. M. Cohen, M. P. Fahay, A. W. Kendall, Jr., and S. L. Richardson, eds.), Special Publ. No. 1, pp. 202–206. American Society of Ichthyologists and Herpetologists, Lawrence, KS.

Fink, W. L., and Weitzman, S. H. (1982). Relationships of the stomiiform fishes (Teleostei) with a description of *Diplophos. Bull. Mus. Comp. Zool.* **150,** 31–93.

Forey, P. L. (1973). A revision of the elopiform fishes, fossil and Recent. *Bull. Br. Mus. (Nat. Hist.), Geol., Suppl.* **10,** 1–222.

Fujita, K. (1990). "The Caudal Skeleton of Teleostean Fishes." Tokai University Press, Tokyo.

Gayet, M. (1986). *Ramallichthys* Gayet du Cénomanien inférieur marin de Ramallah (Judée), une introduction aux relations phylogénétiques des Ostariophysi. *Mém. Mus. Natl. Hist. Nat., Ser. C* **51,** 1–81.

Gayet, M. (1993). Relations phylogénétiques des Gonorhynchiformes (Ostariophysi). *Belg. J. Zool.* **123,** 165–192.

Gayet, M., and Chardon, M. (1987). Possible otophysic connections in some fossils and living ostariophysan fishes. *Proc. Congr. Eur. Ichthyol., 5th, Stockholm, 1985,* pp. 31–42.

Gayet, M., Meunier, F. J., and Kirschbaum, F. (1994). *Ellisella kirschbaumi,* Gayet et Meunier 1991, gymnotiforme fossile de Bolivie et ses relations phylogénétiques au sein des formes actuelles. *Cybium* **18,** 273–306.

Gill, T. (1872). Arrangement of the families of fishes, or Classes Pisces, Marsipobranchii, and Leptocardii. *Smithson. Misc. Collect.* **247,** 1–49.

Goodrich, E. S. (1909). Vertebrata Craniata (First Fascicle: Cyclostomes and Fishes). *In* "A Treatise on Zoology" (R. Lankester ed.), pp. 1–518. Part IX. Black, London.

Gosline, W. A. (1960). Contributions toward a classification of modern isospondylous fishes. *Bull. Br. Mus. (Nat. Hist.), Zool.* **6**, 327–365.

Gosline, W. A. (1961). Some osteological features of modern lower teleostean fishes. *Smithson. Misc. Collect.* **142**(3), 1–42.

Grande, L. (1982a). A revision of the fossil genus *Diplomystus*, with comments on the interrelationships of clupeomorph fishes. *Am. Mus. Novit.* **2728**, 1–34.

Grande, L. (1982b). A revision of the fossil genus *Knightia*, with a description of a new genus from the Green River Formation (Teleostei, Clupeidae). *Am. Mus. Novit.* **2731**, 1–22.

Grande, L. (1985). Recent and fossil clupeomorph fishes with materials for revision of the subgroups of clupeoids. *Bull. Am. Mus. Nat. Hist.* **181**, 231–372.

Grande, L., and Nelson, G. (1985). Interrelationships of fossil and Recent anchovies (Teleostei: Engrauloidea) and description of a new species from the Miocene of Cyprus. *Am. Mus. Novit.* **2826**, 1–16.

Gras, R. (1961). Contribution à l'étude des poissons du Bas-Dahomay. Description de quatre espèces nouvelles. *Bull. Mus. Natl. Hist Nat.* [2] **32**, 401–410.

Greenwood, P. H. (1960). Fossil denticipitid fishes from East Africa. *Bull. Br. Mus. (Nat. Hist.), Geol.* **5**, 1–11.

Greenwood, P. H. (1965). The status of *Acanthothrissa* Gras, 1961 (Pisces, Clupeidae). *Ann Mag. Nat. Hist.* [13] **7**, 337–338.

Greenwood, P. H. (1968). The osteology and relationships of the Denticipitidae, a family of clupeomorph fishes. *Bull. Br. Mus. (Nat. Hist.), Zool.* **16**, 215–273.

Greenwood, P. H. (1970). Skull and swimbladder connections in fishes of the family Megalopidae. *Bull. Br. Mus. (Nat. Hist.), Zool.* **19**, 119–135.

Greenwood, P. H. (1973). Interrelationships of osteoglossomorphs. *Zool. J. Linn. Soc.* **53**, Suppl. 1, 307–332.

Greenwood, P. H., and Rosen, D. E. (1971). Notes on the structure and relationships of the alepocephaloid fishes. *Am. Mus. Novit.* **2473**, 1–41.

Greenwood, P. H., Rosen, D. E., Weitzman, S. H., and Myers, G. S. (1966). Phyletic studies of teleostean fishes, with a provisional classification of living forms. *Bull. Am. Mus. Nat. Hist.* **131**, 339–456.

Greenwood, P. H., Myers, G. S., Rosen, D. E., and Weitzman, S. H. (1967). Named main divisions of teleostean fishes. *Proc. Biol. Soc. Wash.* **80**, 227–228.

Hubbs, C. L. (1967). A fresh look at teleostean evolution. *Q. Rev. Biol.* **42**, 40–41.

Jamieson, B. G. M. (1991). "Fish evolution and Systematics: Evidence from Spermatozoa." Cambridge University Press, Cambridge, UK.

Johnson, G. D. (1992). Monophyly of the euteleostean clades— Neoteleostei, Eurypterygii, and Ctenosquamata. *Copeia*, pp. 8–25.

Jordan, D. S. (1923). "A Classification of Fishes Including Families and Genera as Far as Known." Stanford University Press, Stanford, CA.

Jordan, D. S., and Seale, A. (1926). Review of the Engraulidae, with descriptions of new and rare species. *Bull. Mus. Comp. Zool.* **67**, 353–418.

Lauder, G. V., and Liem, K. F. (1983). The evolution and interrelationships of the actinopterygian fishes. *Bull. Mus. Comp. Zool.* **150**, 95–197.

Lê, H. L. V., Lecointre, G., and Perasso, R. (1993). A 28S rRNA-based phylogeny of the gnathostomes: First steps in the analysis of conflict and congruence with morphologically based cladograms. *Mol. Phylogenet. Evol.* **2**, 31–51.

Lê, H. L. V., Perasso, R., and Billard, R. (1989). Phylogénie moléculaire préliminaire des "poissons" basée sur l'analyse de sé-

quences d'ARN ribosomique 28S. *C.R. Seances Acad. Sci.* **309**, 493–498.

Lecointre, G. (1993). Etude de l'impact de l'echantillonnage des espèces et de la longueur des séquences sur la robustesse des phylogénies moléculaires: Implications sur la phylogénie des téléostéens. Thèse de Doctorat, Université Paris VII.

Leiby, M. M. (1979). Morphological development of the eel *Myrophis punctatus* (Ophichthidae) from hatching to metamorphosis, with emphasis on the developing head skeleton. *Bull. Mar. Sci.* **29**, 509–521.

Lenglet, G. (1974). Contribution à l'étude ostéologique des Kneriidae. *Ann. Soc. R. Zool. Belg.* **104**, 51–103.

Linnaeus, C. (1735). "Systema Naturae, sive Regna Tria Naturae Systematice Proposita per Classes, Ordines, Genera, & Species." Theodorum Haak, Lugduni Batavorum.

Linnaeus, C. (1758). "Systema Naturae per Regna Tria Naturae, secundum Classes, Ordines, Genera, Species, cum Characteribus, Differentiis, Synonymis, Locis. Tomus I. Editio Decima, Reformata." Laurentii Salvii, Holmiae.

Marshall, N. B. (1962). Observations on the Heteromi, an order of teleost fishes. *Bull. Br. Mus. (Nat. Hist.), Zool.* **9**, 251–270.

Mattei, X. (1991). Spermatozoon ultrastructure and its systematic implications in fishes *Can. J. Zool.* **69**, 3038–3055.

Meunier, F. J., and Gayet, M. (1991). Premier cas de morphogénèse réparatrice de l'endosquelette caudal d'un poisson gymnotiforme du Miocène supérieur bolivien. *Géobios* N.S. **13**, 223–230.

Monod, T. (1967). Le complexe urophore des téléostéens: Typologie et évolution (note préliminaire). *Colloq. Int. C.N.R.S.* **163**, 111–131.

Monod, T. (1968). Le complexe urophore des poissons téléostéens. *Mém. Inst. Fondam. Afr. Noire* **81**, 1–705.

Müller-Schmid, A., Ganss, B., Gorr, T., and Hoffmann, W. (1993). Molecular analysis of ependymins from the cerebrospinal fluid of the orders Clupeiformes and Salmoniformes: No indication for the existence of an euteleost infradivision. *J. Mol. Evol.* **36**, 578–585.

Nelson, G. J. (1967). Gill arches of teleostean fishes of the family Clupeidae. *Copeia*, pp. 389–399.

Nelson, G. J. (1970a). The hyobranchial apparatus of teleostean fishes of the families Engraulidae and Chirocentridae. *Am. Mus. Novit.* **2410**, 1–30.

Nelson, G. J. (1970b). Dorsal scutes in the Chinese gizzard shad *Clupanodon thrissa* (Linnaeus). *Jpn. J. Ichthyol.* **17**, 131–134.

Nelson, G. J. (1972). Observations on the gut of the Osteoglossomorpha. *Copeia*, pp. 325–329.

Nelson, G. J. (1973). Relationships of clupeomorphs, with remarks on the structure of the lower jaw in fishes. *Zool. J. Linn. Soc.* **53**, Suppl. 1, 333–349

Nelson, G. J. (1983). *Anchoa argentivittata*, with notes on other eastern Pacific anochovies and the Indo-Pacific genus *Encrasicholina*. *Copeia*, pp. 48–54.

Nelson, G. J. (1989). Phylogeny of major fish groups. *In* "The Hierarchy of Life: Molecules and Morphology in Phylogenetic Analysis" (B. Fernholm, K. Bremer, and H. Jörnvall, eds.), Proc. Nobel Symp. 70, pp. 325–336. Excerpta Medica, Amsterdam.

Nelson, G. J. (1994). Sardines and their allies. *In* "Encyclopaedia of Fishes" (J. R. Paxton and W. N. Eschmeyer, eds.), pp. 91–95. University of New South Wales Press, Sydney.

Nelson, J. S. (1976). "Fishes of the World." Wiley, New York.

Nelson, J. S. (1984). "Fishes of the World", 2nd ed. Wiley, New York.

Nelson, J. S. (1994). "Fishes of the World," 3rd ed. Wiley, New York.

Norman, J. R. (1923). A revision of the clupeid fishes of the genus *Ilisha* and allied genera. *Ann. Mag. Nat. Hist.* [9] **11**, 1–22.

Norman, J. R. (1957). "A Draft Synopsis of the Orders, Families and

Genera of Recent Fishes and Fish-like Vertebrates (Excluding Ostariophysi, Scleroparei, Ammodytidae and a few other Families, notably Centrarchidae, Percidae and Cichlidae) Covering Literature up to 1938, and, as far as It was Available to the Author, from 1939 to 1944." British Museum of Natural History, London (formally published in 1966).

O'Connell, C. P. (1955). The gas bladder and its relation to the inner ear in *Sardinops caerulea* and *Engraulis mordax*. *Fish. Bull. U.S.* (104) **56**, 505–533.

Ogilby, J. D. (1892). On some undescribed reptiles and fishes from Australia. *Rec. Aust. Mus.* **2**, 23–26.

Patterson, C. (1970). A clupeomorph fish from the Gault (Lower Cretaceous). *Zool. J. Linn. Soc.* **49**, 161–182.

Patterson, C. (1975). The distribution of Mesozoic freshwater fishes. *Mém. Mus. Natl. Hist Nat., Ser. A* **88**, 156–174.

Patterson, C. (1984). *Chanoides*, a marine Eocene otophysan fish (Teleostei: Ostariophysi). *J. Vertebr. Paleontol.* **4**, 430–456.

Patterson, C. (1994). Bony fishes. *In* "Major Features of Vertebrate Evolution" (R. Prothero and R. M. Schoch, eds.), Short Courses in Paleontol., No. 7. Paleontological Society, University of Tennessee, Knoxville.

Patterson, C., and Johnson, G. D. (1995). The intermuscular bones and ligaments of teleostean fishes. *Smithson. Contrib. Zool.* **559**, 1–85.

Patterson, C., and Rosen, D. E. (1977). Review of ichthyodectiform and other Mesozoic teleost fishes and the theory and practice of classifying fossils. *Bull. Am. Mus. Nat. Hist.* **158**, 81–172.

Poyato-Ariza, F. J. (1994). A new early Cretaceous gonorynchiform fish (Teleostei: Ostariophysi) from Las Hoyas (Cuenca, Spain). *Occas. Pap., Mus. Nat. Hist. Univ. Kans.* **164**, 1–37.

Regan, C. T. (1916). The British fishes of the subfamily Clupeinae and related species in other seas. *Ann. Mag. Nat. Hist.* [8] **18**, 1–19.

Regan, C. T. (1917a). A revision of the clupeid fishes of the genus *Pellonula* and of related genera in the rivers of Africa. *Ann. Mag. Nat. Hist.* [8] **19**, 198–207.

Regan, C. T. (1917b). A revision of the clupeoid fishes of the genera *Pomolobus, Brevoortia* and *Dorosoma* and their allies. *Ann. Mag. Nat. Hist.* [8] **19**, 297–316.

Regan, C. T. (1917c). A revision of the clupeid fishes of the genera *Sardinella, Harengula*, &c. *Ann. Mag. Nat. Hist.* [8] **19**, 377–395.

Regan, C. T. (1922). Clupeid genera *Clupeoides* and *Potamalosa*, and allied genera. *Ann. Mag. Nat. Hist.* [9] **10**, 587–590.

Roberts, T. (1969). Osteology and relationships of characoid fishes, particularly the genera *Hepsetus, Salminus, Hoplias, Ctenolucius*, and *Acestrorhynchus*. *Proc. Calif. Acad. Sci.* [4] **36**, 391–500.

Roberts, T. (1973). Interrelationships of ostariophysans. *Zool. J. Linn. Soc.* **53**, Suppl. 1, 373–395.

Rosen, D. E. (1973). Interrelationships of higher euteleostean fishes. *Zool. J. Linn. Soc.* **53**, Suppl. 1, 397–513.

Rosen, D. E. (1974). Phylogeny and zoogeography of salmoniform fishes and relationships of *Lepidogalaxias salamandroides*. *Bull. Am. Mus. Nat. Hist.* **153**, 265–326.

Rosen, D. E. (1982). Teleostean interrelationships, morphological function and evolutionary inference. *Am. Zool.* **22**, 261–273.

Rosen, D. E. (1985). An essay on euteleostean classification. *Am. Mus. Novit.* **2827**, 1–57.

Rosen, D. E., and Greenwood, P. H. (1970). Origin of the Weberian apparatus and the relationships of the ostariophysan and gonorynchiform fishes. *Am. Mus. Novit.* **2428**, 1–25.

Schaeffer, B. (1947). Cretaceous and Tertiary actinopterygian fishes from Brazil. *Bull. Am. Mus. Nat. Hist.* **89**, 1–39.

Smith, D. G. (1984). Elopiformes. Notacanthiformes and Anguilliformes: Relationships. *In* "Ontogeny and Systematics of Fishes"

(H. G. Moser, W. J. Richards, D. M. Cohen, M. P. Fahay, A. W. Kendall, Jr., and S. L. Richardson, eds.), Spec. Publ. No. 1, Suppl. to *Copeia*, pp. 94–102. American Society of Ichthyologists and Herpetologists, Lawrence, KS.

Stiassny, M. L. J. (1986). The limits and relationships of the acanthomorph fishes. *J. Zool.* (B) **1**, 411–460.

Svetovidov, A. N. (1952). "Fauna SSSR. Ryby. Tom II, Bip. 1. Seldevie (Clupeidae)," Zool. Inst. Akad. Nauk SSSR, New Ser. No. 48. Izd. Akad. Nauk SSSR, Moscow and Leningrad.

Svetovidov, A. N. (1963). "Fauna of USSR. Fishes. Vol. II, No. 1. Clupeidae," Zool. Inst. Acad. Sci. USSR, New Ser. No. 48. National Sciences Foundation, Washington, DC. (Israel Program for Scientific Translations, Jerusalem).

Taverne, L. (1973). La connexion otophysaire de *Gymnarchus* (Mormyriformes) et de *Papyrocranus* (Ostéoglossiformes) et la parenté des ostéoglossomorphes et des clupéomorphes. Etablissement d'une nouvelle systématique des poissons téléostéens. *Rev. Zool. Bot. Afr.* **87**, 391–401.

Wenz, S. (1984). *Rubiesichthys gregalis* n.g. n.sp., Pisces Gonorhynchiformes, du Crétacé inférieur du Montsech (Province de Lérida, Espagne). *Bull. Mus. Natl. Hist. Nat.* **6**, 275–285.

Whitehead, P. J. P. (1962). Abdominal scutes in the round herrings (Dussumieriidae). *Nature* (*London*) **195**, 511–512.

Whitehead, P. J. P. (1963a). A contribution to the classification of clupeoid fishes. *Ann. Mag. Nat. Hist.* [13] **5**, 737–750.

Whitehead, P. J. P. (1963b). A revision of the Recent round herrings (Pisces: Dussumieriidae). *Bull. Br. Mus.* (*Nat. Hist.*), *Zool.* **10**, 307–380.

Whitehead, P. J. P. (1985a). King herring: His place among the clupeoids. *Can. J. Fish. Aquat. Sci.* **42**, 3–20.

Whitehead, P. J. P. (1985b). "Clupeoid fishes of the World (Suborder Clupeoidei): An annotated and Illustrated Catalogue of the Herrings, Sardines, Pilchards, Sprats, Shads, Anchovies and Wolf Herrings. Part 1. Chirocentridae, Clupeidae and Pristigasteridae," FAO Fish. Synop. No. 125, Vol. 7, Part 1. Food and Agricultural Organization of the United Nations, Rome.

Whitehead, P. J. P., Nelson, G. J., and Wongratana, T. (1988). "Clupeoid Fishes of the World (Suborder Clupeoidei). Part 2. Engraulididae," FAO Fish. Synop. No. 125, Vol. 7, Part 2. Food and Agricultural Organization of the United Nations, Rome.

Wohlfart, T. (1936). Das Ohrlabyrinth der Sardine (*Clupea pilchardus* Walb.) und seine Beziehungen zur Schwimmblase und Seitenlinie. *Z. Morphol. Öekol. Tiere* **31**, 371–410.

Wongratana, T. (1980). Systematics of clupeoid fishes of the indo-Pacific region. Ph.D. Thesis, University of London.

Wongratana, T. (1983). Diagnosis of 24 new species and proposal of new name for a species of Indo-Pacific clupeoid fishes. *Jpn. J. Ichthyol.* **29**, 385–407.

Wongratana, T. (1987a). Four new species of clupeoid fishes (Clupeidae and Engraulidae) from Australian waters. *Proc. Biol. Soc. Wash.* **100**, 104–111.

Wongratana, T. (1987b). Two new species of anchovies of the genus *Stolephorus* (Engraulidae), with a key to species of *Engraulis, Encrasicholina*, and *Stolephorus*. *Am. Mus. Novit.* **2876**, 1–8.

Woodward, A. S. (1892). Doubly-armoured herrings. *Ann. Mag. Nat. Hist.* [6] **10**, 412–413.

Woodward, A. S. (1895). "Catalogue of the Fossil Fishes in the British Museum (Natural History), Cromwell Road, S.W. Part III. Containing the Actinopterygian Teleostomi of the Orders Chondrostei (concluded), Protospondyli, Aetheospondyli, and Isospondyli (in part)." British Museum (Natural History), London.

Woodward, A. S. (1901). "Catalogue of the Fossil Fishes in the British Museum (Natural History), Cromwell Road, S.W. Part

IV. Containing the Actinopterygian Teleostomi of the Suborders Isospondyli (in part), Ostariophysi, Apodes, Percesoces, Hemibranchii, Acanthopterygii, and Anacanthini." British Museum (Natural History), London.

Woodward, A. S. (1902–1912). "The Fossil Fishes of the English Chalk." Palaeontographical Society, London.

Woodward, A. S. (1942). The beginning of the teleostean fishes. *Ann. Mag. Nat. Hist.* [11] **9,** 902–912.

Interrelationships of Ostariophysan Fishes (Teleostei)

SARA V. FINK
Department of English
Michigan State University
East Lansing, Michigan 48824

WILLIAM L. FINK
Museum of Zoology
and
Department of Biology
The University of Michigan
Ann Arbor, Michigan 48109

I. Introduction

Ostariophysans make up nearly 75% of the freshwater fishes of the world and are thus an important part of the Recent fauna. Traditionally the group was diagnosed by the presence of the Weberian apparatus, a complex of bones and ligaments connecting the gasbladder and ear. In 1970, Rosen and Greenwood placed the gonorynchiforms (milkfishes and their relatives), which lack an otophysic connection, as the group Anotophysi within an enlarged Ostariophysi, renaming ostariophysans with an otophysic connection Otophysi. Fink and Fink (1981) provided the first phylogenetic analysis of interrelationships within the Ostariophysi and arrived at the following classification within Otophysi: Cypriniphysi (including Cypriniformes, the minnows and carps) is the sister-group of Characiphysi. Characiphysi includes the Characiformes (characins, or tetras) and its sister-group Siluriformes (gymnotoid electric fishes and catfishes). The evidence we cited was based on Recent members of the groups in question.

Here we review evidence on interrelationships among the five major ostariophysan lineages. We re-examine many features presented in our earlier study (Fink and Fink, 1981), present evidence gathered by ourselves and others since that publication, and comment on the relationships of some fossil taxa that have been proposed to be relevant to the interrelationships of the Recent taxa. Some characters and several fossils discussed by Fink *et al.* (1984) are not revisited here.

Reconsideration of the data relevant to ostariophysan interrelationships results in some alteration in the support for various nodes but does not challenge our earlier hypothesis. The most novel part of that hypothesis, the relationship proposed between catfishes and electric knifefishes, has received significant corroboration from studies of neuroanatomy and electroreception. We summarize that information and provide brief summaries also of recent literature relevant to interrelationships within major ostariophysan subgroups. We also provide some new data on gonorynchiforms.

II. Materials and Methods

Rather than including a lengthy list of materials that we have used over the years, we provide museum numbers and other information in the text where appropriate. Some of the larval material has disintegrated. In particular, the Michigan *Ictalurus* material illustrated in Fig. 4 is no longer extant and was not cataloged.

As we are sampling a tiny fraction of the total diversity of a large clade, we do not think it fruitful to

Copyright © 1996 by Academic Press, Inc.
All rights of reproduction in any form reserved.

consider numerical phylogenetic approaches, and we have not presented a data matrix. We cannot construct a matrix at this point which samples the morphological and species diversity of ostariophysans broadly and evenly enough to satisfy ourselves that any measures of tree quality are meaningful. Such measures are most useful in evaluating competing hypotheses, and no serious alternatives to the one accepted here, well-supported with data, have been suggested. We have sampled primarily species that appear to be morphologically or phylogenetically primitive in each group (Fink and Fink, 1981), though we have attempted to check further into phylogenetically derived groups to assure ourselves that we are dealing with a feature where it originally occurred as a novelty. Investigations by others of intragroup phylogenies have facilitated our ability to do this. As the number and quality of intragroup phylogenies increases, refinement of the character descriptions and knowledge of their distributions will also increase. As in Fink and Fink (1981), characters are listed for ease of description and comparison; features which may be associated phylogenetically or ontogenetically but are separated for descriptive purposes are noted in the text as appropriate.

The history of the terms "mesopterygoid" and "entopterygoid" or "endopterygoid" has been reviewed by Arratia (1992), and in accordance with her findings we use endoptyerygoid, a form of the term which has priority. We prefer endopterygoid over entopterygoid because in writing and in speech it bears less resemblance to the term ectopterygoid. The history of the term "Baudelot's ligament" is discussed by Patterson and Johnson (1995); despite its apparently serendipitous history, as it has become a well-known term, we continue to use it (as do Patterson and Johnson).

There is much work in progress on several groups of ostariophysan fishes by graduate students and in dissertations. We have refrained from citing unpublished work, except when there is no other source of pertinent information and then only when the work is available from UMI dissertation services.

Museum acronyms include AMNH (American Museum of Natural History, New York), BMNH (Natural History Museum, London), EY (Hebrew University of Jerusalem, Israel), FMNH (Field Museum of Natural History, Chicago), RGMC (Musée Royal de l'Afrique Centrale, Tervuren), MCZ (Museum of Comparative Zoology, Harvard University), UMMZ (University of Michigan Museum of Zoology), UMMP (University of Michigan Museum of Paleontology), and USNM (National Museum of Natural History, Washington, DC).

III. Classification

In our revision of ostariophysans in 1981, we proposed a classification that emphasized grouping and placed the electric knifefishes and catfishes into a group, Siluriformes. We thought that to be the most conservative approach, partly because it preserved the long-standing rank names Siluroidei and Gymnotoidei. However, it appears that even after nearly 3 decades of phylogenetics in ichthyology, the use of ranked names with their emphasis on "morphological gaps" still plays a role in how and which classifications are adopted. We think that explains the widespread use of the separate ordinal ranking of electric knifefishes (e.g., Carr and Maler, 1986; Lundberg et al., 1987; Nelson, 1994; Triques, 1993). Nevertheless, since the ordinal names of electric fishes and catfishes have become widespread in the literature, obscuring their relationship, we propose siluriphysi to name the clade consisting of the knifefishes and catfishes. This name is consistent in form with higher taxonomic names used by both Rosen and Greenwood (1970) and Fink and Fink (1981). For those interested in conveying morphology in names, it is intended to refer to the specializations shared by electric knifefishes and catfishes in the Weberian apparatus. Below, we use ordinal endings for the knifefish and catfish groups.

We are using names associated with taxonomic ranks, primarily as a matter of tradition in spite of their potential for misuse. Ranks can perpetuate the discredited notion that there is something special about groups more highly ranked than others, misleading the naive and reinforcing unrealistic notions about evolutionary processes. Ranks also imply an equivalence, sometimes incorrectly, between groups of equal rank. The following is a phylogenetic classification of major clades of Ostariophysi, with selected fossil taxa included. This classification is a summary of the phylogenetic hypothesis presented in Fig. 1.

Ostariophysi
  Anotophysi
    plesion †*Aethalionopsis*
    Gonorynchiformes
      incertae sedis: †*Dastilbe*, †*Gordichthys*,
        †*Rubiesichthys*, †*Tharrhias*
      Chanoidei
        Chanidae
          Chanos
      Gonorynchoidei
        Gonorynchidae
          plesion †Charitosominae
            †*Charitosomus, sedis mutabilis*

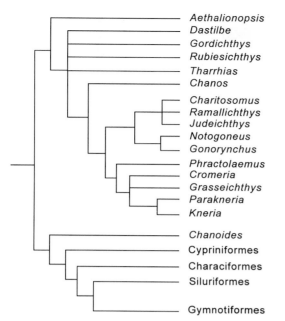

FIGURE 1 Interrelationships of major clades of ostariophysan teleosts.

†*Judeichthys, sedis mutabilis*
†*Ramallichthys, sedis mutabilis*
Gonorynchinae
  †*Notogoneus*
  Gonorynchus
Phractolaemidae
  Phractolaemus
Kneriidae
  Cromeria, sedis mutabilis
  Grasseichthys, sedis mutabilis
  Kneria + Parakneria, sedis mutabilis
Otophysi
  plesion †*Chanoides*
  Cypriniphysi
    Cypriniformes
  Characiphysi
    Characiformes
  Siluriphysi
    Siluriformes
    Gymnotiformes

IV. Review of Major Ostariophysan Lineages

Ostariophysi

Little new research has been published that deals with this most inclusive taxonomic level. Roberts (1982) provided a feature that may prove diagnostic of the group, keratinous unicellular projections of the

epidermis (unculi). These structures are more widespread among ostariophysan taxa than are keratinous tubercles. Roberts' survey found unculi in the gonorynchiform *Chanos*, as well as in *Kneria* and *Phractolaemus*, in all families of cypriniforms, in the characiform *Characidium*, and in a variety of catfishes. He did not examine gymnotiforms but found none in percids and noted that the paired fins of salmonids and osmeroids lacked "adhesive pads and therefore presumably are non-unculiferous" (Roberts, 1982, p. 75). According to Arratia and Huaquín (1995) unculi, like keratinous tubercles, have not been recorded from gymnotiforms.

Gonorynchiformes

Substantial new information is available on gonorynchiform morphology and relationships, for both Recent and fossil taxa, since 1981. We also have some new information to add, which came out of a study of the position of †*Ramallichthys orientalis* Gayet (1982a) from the Cenomanian of Israel, originally described as an ostariophysan and as a possible "pre-cypriniform." †*Ramallichthys* is the subject of a monograph by Gayet (1986a), in which this view is maintained on the basis of three cypriniform synapomorphies allegedly present in †*Ramallichthys*. More recently, in an analysis of gonorynchiform interrelationships by the same author, †*Ramallichthys* seems to be unequivocally placed within the gonorynchoids (Gayet, 1993a, fig. 1, though it is inconsistently placed just outside the "Gonorynchoidei" in the figure including stratigraphy, fig. 10). This placement is dependent, however, on gonorynchiforms being related to a non-otophysan group, a proposition we reject. According to Gayet (1993a, p. 188), if gonorynchiforms are ostariophysans, the presence of three cypriniform synapomorphies in †*Ramallichthys* reinstates the problem of the "position phylogénétique de ce taxon." According to Gayet (1993a, p. 188), Grande (1996) reported on the position of †*Ramallichthys* in a comparative study of some fossil gonorynchiforms and was unable to confirm the presence of the three alleged cypriniform characters in the specimens examined by Gayet (1986a) or in any other specimens. We concur with Grande's assessment, as discussed under the relevant characters (1, kinethmoid; 4, pre-ethmoids; 21, a dorsomedial process on the palatine articulating with the mesethmoid). Gayet (1986a, p. 72) cited two other otophysan and cypriniform features supposedly present in †*Ramallichthys*: a tendency to modify the neural elements of the anterior vertebrae and a tendency to form an articulation between the palatine and endopterygoid; but she described nothing in the anterior vertebrae or suspensorium of †*Ramallichthys* that is not present also in gonorynchoids, and she stated that neither an endopterygoid facet nor a palatine articular condyle can be observed in †*Ramallichthys* (Gayet, 1986a, p. 29). We conclude that the three supposed cypriniform features are insufficiently demonstrated in †*Ramallichthys*, as are any additional otophysan or cypriniform features, so that there are no incongruent apomorphies if the fish is placed in the Gonorynchidae. Even if †*Ramallichthys* did possess morphology more closely resembling the cypriniform synapomorphies than that figured by Gayet (1986a), it possesses such a preponderance of gonorynchiform, gonorynchoid, and gonorynchid apomorphies, as detailed in Grande (1996) and amplified in this chapter, that its relationships are hardly problematic. Indeed Grande (1996) concluded that the slight differences between †*Ramallichthys*, †*Judeichthys*, and one †*Charitosomus* species, †*C. hakelensis* (= †*Hakeliosomus* of Gayet, 1993b), are such that the three taxa should not only be placed within the same genus but might represent a single species. Much of this similarity is evident in the descriptions and illustrations of the three taxa by Gayet (1985b, 1986a, 1993b).

A number of other fossil taxa have been considered close relatives of the Recent *Chanos*, including †*Aethalionopsis* (Taverne, 1981), †*Dastilbe*, †*Gordichthys* (Poyato-Ariza, 1994), †*Parachanos* [considered a synonym of a monotypic †*Dastilbe* by Taverne (1981, p. 974), a decision questioned by Blum (1991a, p. 275), who provided evidence of two species of †*Dastilbe*], †*Rubiesichthys* (Wenz, 1984), and †*Tharrhias* (a review of fossil taxa historically allied with the Chanidae is provided by Patterson, 1984a). In most cases, however, the primary reason for such placement was morphological similarity to *Chanos*, and no compelling evidence of gonorynchiform characters has been presented. This point was made for †*Tharrhias* by Patterson (1984a), but it holds true for †*Aethalionopsis*, †*Dastilbe*, and †*Parachanos* as well.

Patterson (1984a) opened the issue of reevaluating the placement of these taxa. Utilizing the diagnostic characters of Fink and Fink (1981), he identified six ostariophysan and eight gonorynchiform characters in †*Tharrhias* (Patterson, 1984a, p. 135). Poyato-Ariza (1994, p. 33) similarly identified five ostariophysan and five gonorynchiform synapomorphies in †*Gordichthys* (which alone among known gonorynchiforms possesses vomerine and parasphenoid teeth). Wenz (1984, p. 283) identified three gonorynchiform characteristics in †*Rubiesichthys*. Blum (1991a) provided a cladistic analysis of these fossil taxa (except the more recently described †*Gordichthys*). Blum (1991a, p. 283) also noted some features shared by Recent gonorynchiforms that are absent in these fossil taxa (his ac-

count holds for †*Gordichthys* as well): a junction between the supraorbital laterosensory canal and the supratemporal commissure, further reduction in the parietals, greater elongation of the suspensorium (this appears to be inaccurate for *Kneria* and *Parakneria*, and possibly for †*Ramallichthys* and †*Judeichthys*), reduction in the number of infraorbitals [number unknown in †*Rubiesichthys* (Wenz, 1984) and increased again in *Kneria* and *Parakneria* according to Grande (1994, p. 13)], and fusions in the caudal skeleton. Blum also placed †*Aethalionopsis* as more primitive than the others based on its larger parietals and possession of three epurals (rather than two). There is still no diagnosis of the Chanidae and hence no clear synapomorphies placing any of these taxa in that family; Blum's characters plus the apparent absence of epicentral bones in all these taxa (see character 1, below) indicate that all of them are stem-group gonorynchiforms rather than close relatives of *Chanos*, lying outside of the group of extant gonorynchiforms and fossil gonorynchids (i.e., crown-group gonorynchiforms). We provide information concerning some of these stem-group gonorynchiforms for some characters, both in the discussion of gonorynchiforms below, and in Section V (review of the data of Fink and Fink, 1981). In our classification we list †*Aethalionopsis* as an anotophysan plesion on the basis of Blum's evidence and the rest of these fossil taxa as Gonorynchiformes *incertae sedis*.

Following is an account of synapomorphies and interrelationships of gonorynchiforms at various taxonomic levels, gathered in part from the literature and in part from our own research.

Gonorynchiform Synapomorphies

To the gonorynchiform synapomorphies presented in Section V (characters 6, 10, 29, 46, 49, 65, and possibly 38; 14 is discarded), we can add the following features:

1. In their survey of teleostean intermuscular bones and ligaments, Patterson and Johnson (1995) concluded that the presence of epicentral bones is a gonorynchiform synapomorphy. The epicentral series includes the bones often termed "cranial ribs" or "cephalic ribs" in gonorynchiforms that are illustrated, usually identified as epipleurals, in *Chanos* (Rosen and Greenwood, 1970, fig. 3), *Gonorynchus* (Patterson and Johnson, 1995, figs. 3, 5), *Phractolaemus* (Thys van den Audenaerde, 1961, fig. 13), *Kneria* (Grande, 1994, fig. 18; Lenglet, 1974, figs. 17–19, showing sexual dimorphism), and *Parakneria* (Lenglet, 1974, fig. 17). *Phractolaemus* has only one epicentral, usually termed a cephalic rib, and no other intermuscular bones. *Cromeria* appears to be variable; Grande (1994, p. 11) stated

it has one cephalic rib, but we observed specimens (BMNH 1969.11.14.124) with two such epicentrals (*Cromeria* also has epineurals and epipleurals). *Grasseichthys* has no intermuscular bones (Grande, 1994, p. 8; Patterson and Johnson, 1995, p. 23; also personal observation). Epicentrals are present in fossil gonorynchid genera taxa, including †*Ramallichthys* (Gayet, 1986a, p. 40), †*Judeichthys* (Gayet, 1985b, p. 74), †*Charitosomus hakelensis*, †*C. lineolatus*, †*C. major* (Gayet, 1993b, pp. 40, 64, 71), †*Charitopsis spinosus* (BMNH P.63170), and †*Notogoneus* (UMMP V56936). In stem-group gonorynchiform fossils, epicentrals appear to be absent, though epineurals and epipleurals have been noted for most (†*Aethalionopsis*, Taverne, 1981, p. 968; †*Gordichthys*, Poyato-Ariza, 1994, p. 19; epineurals in †*Rubiesichthys*, Wenz, 1984, p. 279; condition not noted for †*Tharrhias* or †*Dastilbe* by Blum, 1991a,b; but epineurals and epipleurals observed for †*Tharrhias*, BMNH P.62748, MCZ 11980).

Gayet (1986a, p. 65), in questioning gonorynchiform monophyly, cited the presence of epicentrals in gonorynchoids, implying their absence in *Chanos*. Epicentrals are present in *Chanos*, however, as shown in the illustration of a 25 mm specimen in Rosen and Greenwood (1970, fig. 3; distally bifid epicentrals identified as epipleurals). (Gayet also contrasted the "simples, non bifurqués" epineurals and epipleurals of *Chanos* with bifurcated epineurals and epipleurals in gonorynchoids, but *Chanos* has proximally branched epineurals and epipleurals. Bifid epineurals are illustrated by Rosen and Greenwood, and in our larger specimens [USNM 199831] the branches on epineurals associated with predorsal vertebrae are multifurcate; some epipleurals in our specimens are proximally bifurcate.)

Patterson and Johnson (1995, p. 15) note that epicentral ligaments are the primitive teleostean condition, and that epicentral bones, a derived condition, also occur in *Notopterus*, *Megalops*, clupeomorphs, gymnotiforms, *Thymallus*, *Parasudis*, *Alepisaurus*, and *Omosudis*.

2. Gonorynchiforms lack Baudelot's ligament (Patterson and Johnson, 1995, p. 19). The ligament is missing, among the non-acanthomorph taxa that they surveyed, only in gonorynchiforms, *Notacanthus*, *Hypentelium*, and *Gigantura*.

3. In gonorynchiforms, the anterior pleural rib, on the third vertebra, is distinctly larger than the next few ribs. The rib has been illustrated or observed to occur in *Chanos* (Rosen and Greenwood, 1970, fig. 3; Fink and Fink, 1981, fig. 6), †*Charitosomus hakelensis* (Gayet, 1986a, fig. 32), †*Charitosomus lineolatus* (Gayet, 1993b, fig. 31), *Cromeria* (Grande, 1994, p. 11), †*Dastilbe* (Blum, 1991a, p. 274), *Gonorynchus* (UMMZ 188723), †*Gordichthys* (Poyato-Ariza, 1994, p. 19), *Grasseichthys* (BMNH 1984.9.12.62), †*Judeichthys* (Gayet, 1985b, p. 74), *Kneria* (UMMZ 190116), †*Notogoneus* (UMMP V56851), *Parakneria* (Grande, 1994, p. 15), *Phractolaemus* (UMMZ 195066), †*Ramallichthys* (Gayet, 1986a, fig. 31), †*Rubiesichthys* (Wenz, 1984, p. 289), and †*Tharrhias* (Patterson, 1984a, fig. 3; Blum, 1991b, p. 286). In *Phractolaemus*, a number of more posterior ribs are also somewhat enlarged, and the first rib is only slightly larger than the others; in the kneriids, the first rib is again only slightly larger, a difference best seen in an anterior view. †*Aethalionopsis* presumably does not have this character, as Taverne (1981) makes no mention of it; this absence is consistent with Blum's (1991a) placement of †*Aethalionopsis* as outside of a group comprising the other known gonorynchiforms.

An alternative interpretation of the expanded third rib that should be taken into consideration in any future studies of fossil taxa is that it is a feature of gonorynchiforms plus otophysans, excluding †*Aethalionopsis*. This assumes that part of the tripus of otophysans is homologous with, and is an elaboration of, the proximal portion of the pleural rib. The morphology of †*Chanoides*, which has "extremely massive" anterior ribs that are broad proximally, and diagonally elongated parapophyses (Patterson, 1984b, p. 440) similar in configuration to those of *Chanos* (though more elongate), suggests that some enlargement of anterior ribs is an ostariophysan feature. In addition, the large articular head of what appears to be a rib-like transformator process of the tripus, present in new material of †*Chanoides* (Patterson, 1996; see character 85 in section VI below), suggests that the plesiomorph condition for otophysans was an enlarged rib on the third vertebra—though not necessarily any larger than other anterior ribs. This hypothesis requires that the body of this transformator process was markedly reduced in otophysans (it is slender in †*Chanoides*, though more robust than in extant otophysans). Differential enlargement of the anterior rib in particular may still be a gonorynchiform feature, and one might expect to find a number of enlarged anterior ribs in fossil gonorynchiforms, as indeed appears to be the case, for example, in †*Dastilbe* (Blum, 1991a, p. 266, bottom photograph), †*Gordichthys* (Poyato-Ariza, 1994, fig. 10), and †*Tharrhias* (Blum, 1991a, p. 265, left photograph).

4. In all extant gonorynchiforms observed or reported in the literature, the esophagus is unusually long, extending some distance into the visceral cavity, and the stomach is subdivided into a slender mucous portion of about the same diameter as the esophagus and a larger muscular portion. The mucous portion turns anteriad and widens into the muscular portion.

This morphology is present in *Chanos, Gonorynchus* (see Monod, 1963, figs. 76–81; mucous stomach not identified), *Phractolaemus* (see Thys van den Audenaerde, 1961, pp. 115–16 and fig. 5), and the kneriids *Kneria* and *Parakneria* (Lenglet, 1973, p. 254 and fig. 3). The visceral anatomy of *Cromeria* has not been described. The length of the esophagus has not been described for *Grasseichthys*, but a photograph of the abdominal cavity (Géry, 1965, fig. 9) shows a long, gourd-shaped muscular stomach with the intestine coming from its craniad end, suggesting that a long esophagus-mucous stomach structure probably enters the muscular stomach caudally. Although digestive system anatomy is not well-documented in most teleosts, the esophagus usually is fairly short and the stomach, if subdivided, has the muscular portion next to the esophagus and any mucous portion (pyloric section) next to the small intestine (see, e.g., Bertin, 1958).

5. In gonorynchiforms, the interior structure of the esophagus includes spiral folds bearing secondary folds and posteriorly directed papillae; the epithelium contains numerous large secretory cells. The gross structure and some information about histology has been provided for *Chanos* by Chandy (1956), for *Phractolaemus* in juveniles and adults by Thys van den Audenaerde (1961, pp. 112–15, fig. 4 and pls. 4, 5), and for *Kneria* and *Parakneria* by Lenglet (1973, pp. 253, 264). In *Phractolaemus*, Thys van den Audenaerde described a thin "chitinous" layer in alevins, which disappears in the adults, and "pseudo-cartilage" in the anterior secondary folds in adults. Since chitin is extremely rare in vertebrates, keratin may be the substance in question. No information is available in the literature on *Cromeria* or *Grasseichthys*, but cleared and stained *Cromeria* (BMNH 1969.11.14.124) show a spiral arrangement of villi stained with alcian blue, suggesting the presence of some form of these esophageal structures. We have not found any reports on the esophageal structure of *Gonorynchus*, and in the specimens examined by us the folds are primarily longitudinal, with only a slight and irregular spiral twisting (visible in our cleared and stained specimen, UMMZ 188723, 70 mm SL). Numerous villi are present, and the tissue appears highly glandular in our large alcohol specimen (FMNH 103977, 280 mm SL). The esophagus is especially elongate in *Gonorynchus*, extending nearly to the muscular stomach. Just before the turn into the stomach, the passage narrows internally, in a way comparable to *Phractolaemus* as figured by Thys van den Audenaerde (1961, pl. 5, fig. c), and a short mucous stomach, with villi like those in the esophagus, extends around the curve and empties into the muscular stomach. The characters listed below that unite *Gonorynchus* with *Phractolaemus* and kneriids suggest that *Gonorynchus* has reduced the spiraling of the esophageal folds in connection with lengthening of the esophagus and shortening of the mucous stomach. We have found no records for any other teleosts of such spiral folding, papillae, or villi in the esophagus.

6. In gonorynchiforms, there is an anteroventrally elongate second uroneural, a feature questioned by Fink *et al.* (1984) in their discussion of †*Ramallichthys*, but which we now accept. A disproportionately elongate second uroneural, whose anterior tip approaches the anterior tip of hypural 3, occurs in *Chanos* (Fink and Fink, 1981, fig. 23A); in the Cretaceous gonorynchoids †*Charitosomus hakelensis* (Patterson, 1970a, fig. 45; Gayet, 1993b, fig. 17), †*Judeichthys* (Gayet, 1985b, pl. 2, although the ventral extension appears less marked), and †*Ramallichthys* (Gayet, 1986a, fig. 37); and in the Eocene †*Notogoneus* (UMMP V56936), though apparently not in the Cretaceous †*Charitopsis* (Gayet, 1993b, fig. 42). The second uroneural is also especially broad in †*Charitosomus hakelensis*, †*Ramallichthys*, and †*Judeichthys*. The second uroneural is lost in all living gonorynchoids (Rosen and Greenwood, 1970, figs. 9–11). In other ostariophysans with a second uroneural, and in other primitive teleosts, the second uroneural is slender and shorter, extending no farther anteroventrally than about the anterior tip of hypural 4 (e.g., Fink and Fink, 1981, figs. 23B, C).

Gayet (1993b, p. 177) proposed that the premaxilla placed lateral to the maxilla is a gonorynchiform feature. The position of the premaxilla external to the anteromedial process or processes of the maxilla is the general teleostean condition, present in a wide array of lower and higher teleosts. Nor are the premaxillae displaced laterally, as they approach the midline in all gonorynchiform taxa. We find no support for this character.

Gonorynchoid Synapomorphies

1. In gonorynchoids there is a linear tubular supratemporal transmitting the lateral line canal to the pterotic, and the canal bifurcates within the pterotic so that the occipital commissural canal enters the parietal from the pterotic instead of from a dorsal branch of the supratemporal as it does in *Chanos*, otophysans, and generalized teleosts. The supratemporal canal is also carried by the supraoccipital, whereas in *Chanos*, generalized cypriniforms and otophysans, and most clupeomorphs, the occipital commissural canal is carried by the parietals alone. In generalized teleosts, the occipital commissural canal is carried by independent supratemporal ossicles (a survey of the supratemporal canal in teleosts was given by Arratia and Gayet, 1995,

p. 483). Passage of the canal through the supraoccipital occurs as an independently derived feature in some Cretaceous clupeomorphs (e.g., †*Spratticeps*, Patterson, 1970b), in cobitoids (Sawada, 1982, p. 86), and in some gobionines (Ramaswami, 1955a, p. 154).

Gayet illustrated the tubular supratemporal (1986a, figs. 7, 40A) and the branching within the pterotic canal (1986a, fig. 1) for †*Ramallichthys*. She illustrated this same pattern in the Cretaceous gonorynchoids †*Judeichthys* (1985b, fig. 1) and †*Charitosomus hakelensis* (1986a, fig. 2; 1993b, figs. 1, 2), and in the Recent gonorynchoids *Kneria* and *Phractolaemus* (1986a, fig. 4). [Gayet's figure of *Kneria* is after Lenglet's drawing (1974, fig. 3, Gayet indicated fig. 58 in error), which shows no parietals; bones also missing in Lenglet's *Parakneria* (1974, fig. 5), but in our *Kneria* the parietals are present, though small; the tubular part of the parietal of *Kneria* is shown in Greenwood *et al.*, 1966, fig. 6.] We assume this pattern of the occipital commissure involves fusion between a part of the supratemporal and the pterotic. In *Gonorynchus* the same pterotic-parietal canal is present, but Ridewood (1905, p. 363) noted that ''the tubular scales of the transverse commissure are readily removable from the parietal and supraoccipital bones,'' and Monod (1963, p. 257) said much the same. In our small *Gonorynchus* (UMMZ 188723, 70 mm SL), the canal-bearing ossicles are fused with the pterotic and at least with the lateral part of the parietal. *Grasseichthys* apparently has no sensory canals on the cranium (none were mentioned or illustrated in Grande, 1994); d'Aubenton (1961) described *Cromeria* as having two supratemporals and no parietal, while Grande (1994, p. 11) described the parietals as ''hardly noticeable'' and did not identify them in her Figure 10, although she indicated a supratemporal commissure that extends from the pterotic completely across the supraoccipital.

Gayet (1986a, fig. 3) illustrated this gonorynchoid occipital commissure pattern in a cypriniform, the homalopterid (*sensu* Sawada, 1982) *Beaufortia*, but her figure after Ramaswami (1952c, fig. 4) is in error. In copying Ramaswami's figure, Gayet included his posttemporal and supratemporal in the pterotic and turned the anterior semicircular canal into a sensory canal running from the parietal into the autosphenotic, which she wrongly labelled dermosphenotic. However, a gonorynchoid pattern of occipital commissure does occur in one homalopterid cypriniform, *Homaloptera zollingeri*, illustrated by Ramaswami (1952b, fig. 2; the pattern differs from that in gonorynchoids because there is no supratemporal in *H. zollingeri*). Ramaswami noted that the supratemporal is present in other species of *Homaloptera*, invoking fusion or loss of the bone to explain the modified pattern in

this one species. We do not regard this instance of homoplasy between a homalopterid and gonorynchoids as invalidating the gonorynchoid synapomorphy.

2. In gonorynchoids, one or more of the anterior supraneurals are expanded and spatulate. In †*Ramallichthys* (Gayet, 1986a, fig. 31), the anterior supraneural over the third neural arch is expanded into a long, triangular structure, with the supraneurals over the second and fourth neural arches aligned along the borders of the expanded supraneural between them. In †*Charitosomus hakelensis*, Gayet (1986a, fig. 31; 1993b: fig. 14) illustrated a specimen which shows virtually the same arrangement as †*Ramallichthys*, except that the enlarged supraneural lies over the second neural arch, not the third. In †*Judeichthys* the first supraneurals overlying neural arches 2–4, are somewhat enlarged relative to their successors, and the first is the largest. Additional examples of enlarged, spatulate supraneurals in fossil gonorynchids are illustrated in Gayet (1993b, figs. 15, 31). In †*Notogoneus*, Perkins (1970, fig. 8) did not record any enlarged anterior supraneurals, but UMMP V56851 has enlarged, spatulate supraneurals contacting their neighbors from vertebra 8 forwards. *Gonorynchus* shows virtually the same arrangement. In Monod's (1963, figs. 53, 54) large specimen of *Gonorynchus* the foremost of three enlarged supraneurals is over neural arch 3, whereas in the fish illustrated by Greenwood *et al.* (1966, fig. 8) it is over neural arch 2, as in our specimens.

The African gonorynchoids all have supraneurals reduced to one or none, but their shape and position, when present, suggest that the spatulate supraneurals present in the previously mentioned gonorynchids are a gonorynchoid, rather than a gonorynchid, synapomorphy. *Phractolaemus* (Thys van den Audenaerde, 1961, fig. 12), *Kneria*, and *Parakneria* (Lenglet, 1974, fig. 17) have a single, relatively wide supraneural, a roughly circular or rectangular bone, over the third neural arch, the position of the enlarged supraneural in †*Ramallichthys*. *Cromeria* has no supraneurals (Grande, 1994, fig. 5), nor does *Grasseichthys* (BMNH 1984.9.12.62) [we think that the plate-like structures identified in *Grasseichthys* by Grande (1994, fig. 9) as supraneurals are expanded neural arches, comparable to those of *Phractolaemus*]. The anterior supraneurals are not modified in *Chanos* (Fink and Fink, 1981, fig. 6), †*Tharrhias* (Patterson, 1984a, fig. 3), or other basal euteleosts.

3. In gonorynchoids, the parapophyses are fused to the centra. In the literature this is sometimes described as having the ribs articulating directly with the vertebral centra (e.g., Lenglet, 1974, p. 78; Thys van den Audenaerde, 1961, p. 138). Our material of

Gonorynchus (UMMZ 188723) indicates that vertebrae incorporate parapophyses by ontogenetic fusion. The vertebrae of all fossil gonorynchid taxa look like those of *Gonorynchus*, with variation chiefly in the relative size and position of the fused parapophyses (e.g., Gayet, 1986a, fig. 31 of †*Ramallichthys*; Gayet, 1985b, p. 74 of †*Judeichthys*; Gayet, 1993b, figs. 15, 20 of †*Charitosomus hakelensis* and †*C. formosus*). The same morphology occurs in †*Notogoneus* (UMMP V56851, V56936), and Monod (1963, p. 268, figs. 53–55) and Greenwood *et al.* (1966, fig. 8) illustrated the same condition in *Gonorynchus*. In *Chanos* the anterior parapophyses are autogenous (Rosen and Greenwood, 1970, fig. 3; Fink and Fink, 1981, fig. 6).

4. In gonorynchoids, the neural spine on neural arch 1 is greatly reduced or absent (Fink and Fink, 1981, p. 304).

5. In gonorynchoids, the ectopterygoid is reduced anteriorly (Fink and Fink, 1981, p. 304). Such reduction together with mobility of the palatine was suggested by Fink and Fink (1981, p. 304) to be a synapomorphy of gonorynchoids. In *Gonorynchus*, however, although the ectopterygoid is reduced anteriorly, it overlaps the palatine, and the joint does not appear to be mobile (based on our juvenile specimen, UMMZ 188723, 70 mm SL). In *Phractolaemus*, the bones overlap from lateral view (Thys van den Audenaerde, 1961, fig. 17), but they are separated from one another by connective tissue and can move with respect to one another (based on our adult specimen, UMMZ 195066). *Gonorynchus* and *Phractolaemus* have an anteriorly truncated ectopterygoid, however, and mobility may be a character of *Phractolaemus* plus kneriids, in which the ectopterygoid is either truncated anteriorly (*Kneria* and *Parakneria*) or absent (*Grasseichthys*; Grande, 1994). [We are unsure whether *Cromeria* has an ectopterygoid; neither d'Aubenton (1961) nor Grande (1994) illustrated or described such a bone, but Grande used absence of an ectopterygoid to diagnose *Grasseichthys* and said that all other gonorynchiforms have one.] Gayet (1986a, pp. 51, 52) incorrectly claimed that a reduced ectopterygoid like that in *Kneria* occurs in the fossil "chanoids" †*Aethalionopsis* and †*Parachanos* (Taverne, 1981, figs. 4 and 10). In Taverne's drawings of those fishes it is the anterior part of the endopterygoid that is reduced since that bone does not contact the autopalatine, whereas the ectopterygoid–palatine contact is normal and unmodified (see character 22). Gayet also stated that †*Ramallichthys* and †*Judeichthys* do not have a mobile palatine, although they show reduction in the ectopterygoid anteriorly, as do the various fossil gonorynchoids she described in her monograph (1993b) on fossil "Gono-

rynchoidei" (= gonorynchids, herein). Anterior reduction of the palatine is also a synapomorphy of Cypriniformes (see character 25).

6. In gonorynchoids, there is no A_ω intramandibular portion of the A_1 adductor musculature (Howes, 1985b). An A_ω is present in relatively primitive members of the four otophysan lineages [e.g., *Opsariichthys* (Takahasi, 1925, pl. 3, fig. 46); *Zacco*, UMMZ 202259; *Xenocharax* (Vari, 1979, fig. 40); *Diplomystes*, UMMZ 212700; *Sternopygus*, UMMZ 228961], and its presence is plesiomorphic for teleosts (Winterbottom, 1974). Howes (1985b, p. 299) noted that the muscle is absent in "various otophysan taxa."

7. In gonorynchoids, the A_1 section of the adductor mandibulae muscle is subdivided into an inner and an outer slip, with the outer slip inserting on the maxilla (Howes, 1985b). Howes described this feature in *Gonorynchus* (1985b, fig. 5), *Phractolaemus* (1985b, figs. 18, 19), *Kneria, Parakneria* (1985b, figs. 11, 12), and *Cromeria* (1985b, fig. 16). In *Gonorynchus*, the inner slip goes to "the thick connective tissue surrounding the coronoid process" of the lower jaw (Howes, 1985b, p. 281), whereas in the African forms it goes to the maxilla. Howes was unable to discern any adductor subdivisions in the diminutive *Grasseichthys*. He also did not suggest this as a gonorynchoid character in his Discussion. In otophysans and other primitive teleosts, the adductor mandibulae does not have inner and outer slips.

Gonorynchid Synapomorphies

1. In gonorynchids, there is a small patch of robust teeth on the posterior part of the endopterygoid and on the second basibranchial (Grande, 1996), as in *Gonorynchus* alone among Recent ostariophysans, and as in the Cretaceous gonorynchids †*Charitosomus* (Woodward, 1896), †*Judeichthys* (Gayet, 1985b, p. 72), and †*Ramallichthys* (Gayet, 1986a, pl. 1, fig. 2, text-fig. 25). The Eocene–Oligocene gonorynchid †*Notogoneus*, sister-group to *Gonorynchus* (Grande, 1996), lacks these teeth (Perkins, 1970; Grande, 1996, also personal observation). †*Judeichthys* is known from one specimen from the same beds in Israel as †*Ramallichthys*; Gayet (1985b, p. 66) erected a monospecific family Judeichthyidae for it. She distinguished Judeichthyidae from Gonorynchidae principally by "presence of teeth on the fifth pharyngeal and basihyal and absence of teeth on the endopterygoid." The three tooth patches preserved in the single specimen of †*Judeichthys* (Gayet, 1985b, pl. 1, fig. 2, text-fig. 3) are in the same position as those in †*Ramallichthys* (Gayet, 1986a, pl. 1, fig. 2, text-fig. 25): one is at the posterodorsal corner of the endopterygoid (the same position

of the endopterygoid teeth in *Gonorynchus*), one lies below it at the middle of the metapterygoid–quadrate suture, and one lies more anteriorly at about the middle of the vertical portion of the dentary–angular suture. In both †*Judeichthys* and †*Ramallichthys* Gayet sought to argue that whereas the anterior teeth were borne on the second basibranchial (basihyal is substituted in the diagnosis of Judeichthyidae, previously cited, but is not mentioned elsewhere in the text) as in *Gonorynchus*, the posterior teeth belong not to the endopterygoids, but instead the upper patch relates to the fourth pharyngobranchial (1986a, p. 31) or to the fifth (1985b, pp. 66, 73; 1986a, p. 56), and the lower to a posterior basibranchial, possibly the fourth. Gayet's arguments on these points rely on her opinions on the way in which the fossils have been crushed, which in her view should superimpose left and right endopterygoid tooth patches in fishes crushed laterally. We regard those arguments as insufficient to override the expectation that the dentition of fossil gonorynchoids (and Gayet acknowledges that †*Judeichthys* is one) is most reasonably interpreted by comparison with *Gonorynchus*. Gayet (1993a, character 14) has more recently accepted the position of the upper toothplates on the endopterygoids and that of the lower on the basibranchial (1993b, p. 99).

2. In gonorynchids, with the exception of *Gonorynchus*, the procurrent caudal rays are strongly asymmetrical, with the dorsal rays extending farther anteriorly than the ventral rays; the dorsal rays terminate anteriorly a distance of two or more neural spines forward of the hemal spines reached by the ventral rays. This asymmetry occurs in the Cretaceous gonorynchids †*Ramallichthys* (Gayet, 1986a, fig. 37), †*Judeichthys* (Gayet, 1985b, fig. 8), and †*Charitosomus formosus* (Gayet, 1993b, pl. 4, fig. 3, text-fig. 21), though it is not as marked in †*C. hakelensis* Davis [Taverne (1976, fig. 21) illustrated a specimen assigned to that species in which the procurrent rays extend forward to NPU5 dorsally and NPU4 ventrally]. The Eocene †*Notogoneus osculus* Cope shows the same asymmetry as †*Ramallichthys*. Such asymmetry in the procurrent caudal rays is clearly derived, and among the wide range of teleosts surveyed by Fujita (1990) it occurs only in stomiiforms (Fujita, 1990, figs. 73, 76, 80, 81, 83–87, 92). The opposite form of asymmetry, with the lower procurrent rays extending farther forward than the upper, is more common; it was used, for example, by Markle (1989, p. 81) as a synapomorphy of euclichthyids and morids among gadiforms. The asymmetry of †*Ramallichthys*, †*Judeichthys*, some species of †*Charitosomus*, and †*Notogoneus* might be synapomorphous for an extinct group of gonorynchids that excludes

Gonorynchus, but that possibility is contradicted by several derived features shared by †*Notogoneus* and *Gonorynchus* but absent in other fossil gonorynchids [e.g., ctenoid scales, extension of the squamation to completely cover the head (scales are reported to be absent in †*Ramallichthys* and †*Judeichthys*), configuration of the maxilla and premaxilla, a peculiar lacrimal with a keel near its ventral edge, and a splint-like metapterygoid (Grande, 1996)]. Grande also noted the presence of a slender, elongate ethmoid region, with a thin, flat mesethmoid, but this morphology appears to be shared also by †*Charitopsis* (Gayet, 1993b, pl. 7, fig. 4, and text-fig. 38).

3. In gonorynchids, hypurals 1 and 2 are fused to each other and to the compound ural centrum (Grande, 1996). Although Gayet described and illustrated the hypurals as separate from each other and hypural 1 as autogenous, Grande (1996) stated that these elements are fused in all fossil gonorynchid material she examined and that apparent autogeny is due to postmortem breakage.

Relationships among the African Freshwater Gonorynchiforms

This group is well-established as monophyletic, based on both osteology (Grande, 1994; Lenglet, 1974) and myology (Howes, 1985b). Howes reported on the cranial muscles of extant gonorynchiforms and corroborated the relationship among the kneriids *Kneria*, *Parakneria*, and *Cromeria*. He also corroborated their relationship to *Phractolaemus* but found evidence linking the kneriid *Grasseichthys* both to the other kneriids and to *Phractolaemus*. Howes left the relationship of these freshwater forms to *Gonorynchus* and *Chanos* unresolved. More recently, Grande's study (1994) on interrelationships among kneriids, based on osteology, corroborated monophyly of the Kneriidae but did not resolve whether *Cromeria* is the sister-taxon to *Grasseichthys* or to the *Kneria–Parakneria* clade. Lenglet (1973) provided evidence from visceral anatomy relating *Kneria* and *Parakneria* to *Phractolaemus*, particularly presence of a partial compartmentalization of the gasbladder (see discussion of character 54) and in the structure of the esophagus and mucous stomach (discussed previously; see Gonorynchiform Synapomorphies). Lenglet (1973) did not have material of *Cromeria* or *Grasseichthys* and noted that *Grasseichthys* has a short, straight intestine (based on Géry, 1965, fig. 9) in contrast to *Phractolaemus*, *Kneria*, and *Parakneria* (she did not note the similar structure of the stomach, also discussed previously). The monophyly of the Kneriidae is well-corroborated osteologically, however, and since elongation and coiling of the intestine occurs during juvenile growth, the absence of these features in *Grasseichthys* may be associated with paedomorphosis.

Cypriniformes

While no robust overview of cypriniform interrelationships has been published since 1981, several workers have examined major clades within the group. Wu *et al.* (1981) presented an ostensibly cladistic analysis of the Cyprinoidei (= Cypriniformes herein) but used only characiforms to represent outgroup conditions and recognized ancestral higher taxa; reworking of their data may prove useful. The same outgroup was used in the work of Chen *et al.* (1984) on the Cyprinidae. Their data were reanalyzed by Cavender and Coburn (1992) in a phylogenetic analysis of the Cyprinidae; this parsimony analysis did not duplicate the tree of Chen *et al.*, finding instead a less-resolved solution.

Cavender and Coburn (1992) provided the first published phylogenetic diagnosis of the Cyprinidae. They presented evidence of two major lineages of cyprinids, which they recognized as subfamilies; the Cyprininae containing three lineages named informally as barbins, cyprinins, and labeonins; and the Leuciscinae containing eight lineages in a fully resolved cladogram: tincins, rasborins, gobionins, acheilognathins, xenocyprins, cultrins, leuciscins, and phoxinins. Some of their subgroups coincide with some of those of Chen *et al.* Cavender and Coburn noted that some of these subgroups have considerably more support than others. Coburn and Cavender (1992) presented a hypothesis of relationships among phoxinins, including all North American cyprinids (except *Notemigonus*, which is a leuciscin), and a limited number of Eurasian members of the group.

Sawada (1982) examined the Cobitoidea in a cladistic study and provided evidence of catostomids and *Gyrinocheilus* as successive outgroups, but his assumptions about irreversibility of character evolution cause us to view his conclusions with caution. Smith (1992) published a chapter on the phylogenetics of Catostomidae.

Siebert (1987), in an unpublished dissertation, proposed that the Cypriniformes consist of two clades, Cyprinidae and Cobitidoidea, the latter containing all non-cyprinid cypriniformes; he also provided evidence for the monophyly of several cypriniform families.

Characiphysi

Little work has been done testing this node of the phylogeny. Striedter (1992), based on a small sample of taxa, suggested that the pretectum of the diencephalon has a cell group, termed the pretectal lateral line nucleus (PLL) in *Colossoma* and the nucleus electrosensorius (nE) in siluriphysans. In cypriniforms and other outgroups there is no such nucleus. Arratia (1992, p. 107) stated that the hyomandibula articulating with the neurocranium via a single articular facet is a characiphysan synapomorphy based on the presence of a double articular facet in gonorynchiforms and cypriniforms. Arratia listed only *Chanos* in her Table 1, but a double facet is illustrated for kneriids in Grande (1994) and Lenglet (1974); *Gonorynchus* has two heads, though these are not completely distinct (Monod, 1963, pp. 260–261), and †*Ramallichthys* appears to be similar (Gayet, 1986a, p. 31) (*Phractolaemus*, UMMZ 195066, has a single head). †*Gordichthys* appears to have two heads also (Poyato-Ariza, 1994, p. 16). A single articular facet is primitive for teleosts, however, and is also present in the otophysan plesion †*Chanoides* (Patterson, 1984b, p. 435). If the single articular facet in †*Chanoides* is representative of the plesiomorphic otophysan condition, a double articular facet was derived independently in gonorynchiforms and cypriniforms, and the single facet in characiphysans is primitive.

Characiformes

Little work has yet been published on characiform higher taxa. Vari (1983) provided a well-documented hypothesis that the characiform families Prochilodontidae and Curimatidae together are the sister-group of a lineage comprising Anostomidae and Chilodontidae. At this writing there are several authors using various molecular features to address characiform interrelationships, but these have not yet been published. Buckup (1993), in a study of characidiins, included a preliminary cladogram for a wide range of characiform taxa, based on an as yet unpublished study that focused on the search for outgroups to characidiins. He provided evidence that the Characidiinae are related to the Crenuchinae (comprising *Crenuchus* and *Poecilocharax*) and placed these together in the Crenuchidae. Vari (1995) addressed the relationships among lebiasinids, hepsetids, erythrinids, and ctenoluciids.

Siluriphysi

Siluriphysan monophyly has been addressed in studies of the electrosensory system, especially aspects of neural anatomy that are a part of that system. The empirical basis for many of these generalizations is limited to a small number of species by the necessity of detailed cytological and hodological studies. An extreme example is the study of the connections of the lateral preglomerular nucleus (Striedter, 1992), which is limited to a single species from each of the four otophysan terminal taxa. Despite this limitation, these studies have resulted in a substantial body of evidence supporting the hypothesis of a sister-group relationship between catfishes and electric knife-fishes. The following abstract of features was put together in concert with J. S. Albert, and some of them are from work in progress by him and M. J. Lannoo:

1. In siluriphysans, there are low-frequency, ampullary-shaped electroreceptor organs tuned to low-frequency (less than 30 Hz) ambient electric fields (Zakon, 1986). There are no electroreceptor organs in pertinent outgroups; within neopterygians, electroreception has evolved independently in osteoglossomorphs (as have structures similar to those described under 2, 6, and 7). Arratia and Huaquín (1995) summarized the distribution of ampullary organs in siluriforms, noting their presence in diplomystids and trichomycterids.

2. In siluriphysans, there are electrosensory afferents in the lateral line nerve or nerves. This feature was character 120 of Fink and Fink (1981). In siluriforms, the afferents are found in all six lateral line nerves (Tong and Finger, 1983); in gymnotiforms they are exclusively in the anteroventral lateral line nerve (Northcutt and Vischer, 1988). There are no electrosensory afferents in pertinent outgroup taxa.

3. In siluriphysans, the preotic lateral line nerve ganglia fuse during ontogeny so that the five anterior lateral line nerves enter the brain in a single bundle (Northcutt, 1992). In non-siluriphysans the ganglia are spatially discrete.

4. In siluriphysans, there is a laminated electrosensory lateral line lobe (ELL). The properties of the ELL were summarized by Carr *et al.* (1981), Heiligenberg and Dye (1982), Finger and Tong (1984), Finger (1986), and Carr and Maler (1986). An ELL also occurs in electroreceptive osteoglossomorphs, but it is different in morphology and not as highly laminated. The siluriphysan ELL is a medullary structure consisting of seven layers. The molecular layer of descending parasagittal fibers at the pial surface is composed of two sets of fibers: the dorsal molecular layer arising from the posterior eminentia granularis of the cerebellum and the ventral molecular layer composed of bilateral projections from a midbrain nucleus. Below these fibers is a layer of efferent cells whose axonal projections to the lateral lemniscus are bundled into the plexiform layer beneath them. Siluriphysans possess two populations of pyramidal cells recognized by the presence or absence of basilar dendrites. Two populations of granule cells reside in the layer deep to the plexiform layer. Situated in a deep neuropile layer of primary afferents are large cells, with dendritic extensions into the lower plexiform layer. The ventral lamina, the deep fiber layer, is the site of primary afferent inputs.

J. S. Albert and M. J. Lannoo (personal communication) also have noted other features of the siluriphysan ELL that extend beyond those cytological details. The electrotopic representation of the body surface in the ELL is maintained along both the longitudinal and dorsoventral axes (Carr *et al.*, 1981; Carr and Maler, 1986; New and Singh, 1994). In addition, the pyramidal cells in the siluriform ELL and in the medial segment of the gymnotiform ELL share the expression of the monoclonal antibody Zebrin II (Lannoo *et al.*, 1991, 1992).

5. In siluriphysans, the anterior lobe of the corpus cerebellum is large, extending anterior to the rhombomesencephalic isthmus, at least to the midlength of the optic tectum (J. S. Albert and M. J. Lannoo, personal communication). In outgroups, this lobe is smaller and does not extend as far forward.

6. In siluriphysans, there is a posterior eminentia granularis (EGp) with associated pathways. The siluriphysan EGp is a portion of the cerebellar molecular layer that receives direct lateral line efferents (exclusively electrosensory in gymnotiforms; Bass, 1982; Tong and Finger, 1983; Sas and Maler, 1987). These include descending toral efferents to the lateral inferior olive, ascending olivary efferents to the posterior eminentia granularis, and projection of cerebellar granule cells via the brachium conjunctivum to the torus semicircularis (Carr *et al.*, 1982). In outgroups, there is no such EGp with associated pathways.

7. In siluriphysans, there is a dorsal (or lateral) and a ventral (or medial) nucleus representing each of the nuclei of the lateral lemniscus. The dorsal or lateral member of each pair is electrosensory; the ventral or medial member is mechanosensory (Striedter, 1991). In outgroups, there is no electrosensory component to the lateral lemniscus (and thus no nuclei).

8. In siluriphysans, the torus semicircularis (TS) is large, almost entirely filling the mesencephalic ventricle; its lateral nucleus (TSl) is subdivided into mechanosensory and electrosensory subnuclei (Striedter, 1992). In outgroups, the TS is smaller and the TSl is not subdivided.

9. In siluriphysans, projections from the anterior diencephalic nucleus to the lateral nucleus preglomerulosus (PG1) are not present (Striedter, 1991). In other ostariophysans, such projections are present.

10. In siluriphysans, projections from PGl to the centroposterior nucleus (CP) are not present (Striedter, 1991). In other ostariophysans, such projections are present.

11. In siluriphysans, the nucleus electrosensorius (nE) of the synencephalon completes a feedback loop from TSl to PGl (Striedter, 1990, fig. 9a; 1991, fig. 14).

In outgroups, there is no structure with these connections.

12. In siluriphysans, the posterior nucleus of telencephalic area Dorsalis (Dp) is large, and the posterior pole of the nucleus is larger in cross-sectional area than the anterior pole (J. S. Albert and M. J. Lannoo, personal communication). In plesiomorphic members of the outgroup taxa this nucleus is small.

13. In siluriphysans, projections from division 2 of the medial nucleus of telencephalic area dorsalis (Dm) to PGl are lacking (Striedter, 1992). Such projections are present in the outgroups examined.

14. In siluriphysans, there is a lateral division of the central nucleus of telencephalic area Dorsalis (Dcl) (Striedter, 1992). In outgroups, there is no such division.

15. In siluriphysans, the posterior portion of the central nucleus of telencephalic area Dorsalis (Dcp) is large. This portion of the nucleus is small in outgroups (Striedter, 1992).

16. In siluriphysans, there is a lack of contact between the anterior cartilage of the autopalatine and the neurocranium (Arratia, 1992, p. 107).

Triques (1993, p. 125) proposed that two features of the urohyal, shortness in the longitudinal axis and lateral expansion of the anterior head of the urohyal, are siluriphysan synapomorphies. Arratia and Schultze (1990), however, described the ontogeny of the expanded urohyal head in catfishes as containing unique features not shared with any other teleost, including gymnotiforms, and neither an expanded head nor short length appears in the single illustration of a gymnotiform urohyal in Kusaka (1974, p. 84), though they do appear in Kusaka's illustrations of catfishes (1974, pp. 92–96). These characters need further study.

Siluriformes

Considerable work has been done on morphology and interrelationships of catfish higher taxa since 1981. Among these publications are those of Arratia (1987, 1992), Bornbusch (1991), Howes (1983a,b, 1985a), de Pinna (1989a,b, 1993), de Pinna and Vari (1995), Mo (1991; see also de Pinna and Ferraris, 1992), and Schaefer (1987, 1990). None of these works deals with siluriform monophyly directly, but some, such as Arratia's (1987, 1992) have inferences about character interpretation and polarity. Arratia in particular has added greatly to knowledge of primitive catfish morphology, as well as having provided an extensive analysis of catfish characters in her (1992) publication, which included a phylogeny of a limited number of catfish taxa. We discuss some areas of disagreement

on character interpretations in Section V below. The unpublished work of de Pinna (1993) stands as the most comprehensive examination of siluriform interrelationships.

Gymnotiformes

A number of publications about this group have appeared since the mid-1980s. At the time of our previous paper, the major work on gymnotiforms was that of Ellis (1913). The years following our publication have seen the discovery of many new gymnotiforms, including a large number of river-dwelling species (Lundberg et al., 1987). Neither these newly discovered Recent species nor the first discovery of fossil gymnotiform material, from the Upper Miocene of Bolivia (†Ellisella, Gayet and Meunier, 1991), results in any major changes in gymnotiform classification. Ostensibly phylogenetic research on interrelationships among gymnotiforms includes recent publications by Alves-Gomes et al. (1995) using mitochondrial DNA together with some information from morphology and electrophysiology, by Triques (1993) using morphology, and by Gayet et al. (1994) using morphology and some data on patterns of regeneration of skeletal structures. Albert (1996) noted that these studies were inadequate in one way or another; his own work includes neuroanatomical characters as well as morphological and electrogenic characters, and he reanalyzed the mitochondrial DNA data, obtaining results different from those preferred by Alves-Gomes et al. His conclusions regarding gymnotiform interrelationships are summarized below, as well as in Figure 2 of Albert and Fink (1996).

These studies differ most strongly in which taxa are placed at the base of the phylogeny. Using a differential weighting scheme for the molecular data in combination with nonmolecular characters, Alves-Gomes et al. (1995, fig. 10) placed Sternopygus at the base of the tree, apart from the other sternopygids, with the remaining sternopygids placed with the Apteronotidae. The position of the Gymnotus–Electrophorus clade was unresolved. The molecular analyses cast doubt on the monophyly of the Hypopomidae. Triques (1993) and Gayet et al. (1994) placed the Apteronotidae as the sister-group to other gymnotiforms, and the Sternopygidae as sister-group to the remaining taxa (comprising a hypopomid–rhamphichthyid clade and a Gymnotus–Electrophorus clade). Albert's results (1996) place the Gymnotus–Electrophorus clade at the base of the tree, with an apteronotid–sternopygid group as sister to a rhamphichthyid–hypopomid group (and all current families monophyletic). This basal position for the Gymno-

tus–Electrophorus clade is also supported by the molecular data. As noted by Alves-Gomes *et al.* (1995, p. 305), a consensus of the nine most parsimonious trees places the *Gymnotus–Electrophorus* clade as the sister-group to the other taxa, results confirmed in Albert's reanalysis of the mitochondrial data.

†*Ellisella* was considered by Gayet *et al.* to be the sister-group to non-apteronotid gymnotiforms but by Fink *et al.* (1997) to be a sternopygid, most closely resembling the extant *Distocyclus goajira*.

V. Review of the Evidence of Fink and Fink (1981)

In this section, we reevaluate the data presented in our 1981 paper based on comments and criticisms made in the literature and our own reworking of some characters. Those characters not discussed are simply listed in brief form. The bulk of the comments in the literature are those of Gayet (particularly 1986a), a large part of which is a point-by-point criticism of many of the characters used in Fink and Fink (1981). Each character discussed is identified by number, a brief identification of the character, and the taxon it was used to diagnose, as given in the 1981 paper; reinterpretations of generality of characters are noted in the text. We use our newly proposed names for catfish and electric fish higher taxa rather than those of Fink and Fink (1981).

Neurocranium

1. Kinethmoid; Cypriniformes. An anterior median ossification that has been termed a kinethmoid has been described in two taxa outside of the Cypriniformes, †*Chanoides* and †*Ramallichthys*. The bone in †*Chanoides*, described and discussed by Patterson (1984b, pp. 436, 452), appears to have differed from the kinethmoid of cypriniforms in orientation, articulation, presence of cartilage caps on either end of the bone, absence of relief associated with ligamentous attachment, and (by inference) in function. A structure in †*Ramallichthys* termed a kinethmoid was discussed and illustrated by Gayet (1986a, pl. 1, fig. 3; text-figs. 8, 9), but Gayet (1993b, p. 93) raised the possibility of that structure being the anterior part of the vomer. Grande (1996), in her study of fossil and Recent gonorynchid taxa, identified the "kinethmoid" illustrated in the holotype (EY 386) as part of the vomer (her fig. 12A), and what Gayet labeled as vomer Grande identifies as a process extending from the left maxilla. Grande (1996) also found no trace of a kinethmoid in EY 381 (or of a pre-ethmoid or dorsomedial

palatine process; see characters 4 and 21), a specimen which shows the vomer in ventral aspect. The examination of the holotype by one of us (S.V.F.) corroborates Grande's interpretation.

2. Mesethmoid articulates anterior to vomer; Characiphysi. The characiphysan arrangement is said to occur in two fossil genera, †*Lusitanichthys* and †*Salminops* (Gayet, 1986a, p. 47). The only illustration of the vomer in †*Lusitanichthys* (Gayet, 1985c, fig. 17; 1986a, fig. 44) shows the bone disarticulated so that it is separated from the mesethmoid by the antorbital, palatine, and premaxilla; in such conditions, the form of the articulation between vomer and mesethmoid must be an inference, not an observation. Similarly, in †*Salminops* the vomer in the single specimen is illustrated in a very vague way as separated from the preserved part of the mesethmoid by a considerable space (Gayet, 1985c, p. 95, fig. 2); expectation, not observation, must also contribute here.

3. Mesethmoid with premaxillary articular processes; Characiphysi.

4. "Pre-ethmoid"; Cypriniformes. Following Fink *et al.*'s (1984, p. 1040) rebuttal of Gayet's earlier comments on the pre-ethmoid of cypriniforms, Gayet (1986a, p. 49) found that there is a real difference between the pre-ethmoid of cypriniforms and the ethmopalatine of characiforms and that †*Ramallichthys* possesses the first, and †*Lusitanichthys* the second. The "pre-ethmoid" of †*Ramallichthys* is illustrated in the same two specimens as the "kinethmoid." In the holotype it appears to be part of what Gayet labeled as vomer (fig. 8), as Gayet said (1986a, p. 21); but Grande (1996, fig. 12A) identified Gayet's vomer as part of the left premaxilla. In specimen EY 381, Gayet identified part of the mesethmoid as pre-ethmoid (1986a, p. 21); as in the case of the "kinethmoid" of †*Ramallichthys*, Grande (1996) could not find any structures resembling a cypriniform pre-ethmoid on this specimen.

The "ethmopalatine" bone of †*Lusitanichthys* (Gayet, 1985c, figs. 17, 19; 1986a, fig. 44) is illustrated in the appropriate position in two specimens, but there is evidently no means of determining whether the bone is homologous with the characiform ethmopalatine cartilage; with any of the similar bones or cartilages in various cypriniforms (e.g., Patterson, 1984b, fig. 10); with either the "kinethmoid," "ethmopalatine," or extramaxilla ossifications of †*Chanoides*; or with similar kinds of elements in various other lower teleosts, including *Chanos* (Fink and Fink, 1981, fig. 3A), esocoids, and mormyrids and notopterids (Patterson and Rosen, 1977, p. 98). Whatever the homology among these structures in various lower teleostean taxa (and the homology has been questioned, justifiably, by Howes, 1985b, p. 279), the condition of the "pre-ethmoid" in cypriniforms differs in morphology and topography from any other known elements.

5. Compressed dorsal portion of mesethmoid; Siluriphysi. Although many catfishes have a broader mesethmoid, this condition is found in the fossil †Hypsidoris as well as in diplomystids and is hypothesized to be the plesiomorph condition within siluriforms.

6. Structures of interorbital septum reduced; Gonorynchiformes. Gayet (1986a, p. 49) stated that †Judeichthys has a well-developed interorbital septum, though it lacks an orbitosphenoid, but later (1993a, p. 170 and fig. 1) accepted this character as a gonorynchiform synapomorphy and stated that it is present in all fossil gonorynchiforms she examined, as far as can be determined (whether the pterosphenoids are reduced could not be determined). The "interorbital septum" of †Judeichthys (Gayet, 1985b, p. 70, pl. 1, fig. 2, text-fig. 3) appears to us to be a sclerotic ossification.

7. Absence of basisphenoid; Ostariophysi. Gayet (1986a, p. 49) disputed this character by reporting a personal communication from Chardon in which a "bélophragme" (pedicel of basisphenoid) was reported in *Chanos* and in which Kindred's (1919) opinion that the parasphenoid of siluriforms incorporates parts of the basisphenoid was supported by Chardon's observations on a 21 mm *Clarias*. We have examined counterstained series of *Clarias* ranging beyond 21 mm and have seen no evidence that cartilage is involved in the development of the parasphenoid. Without such evidence, there can be no grounds for assuming fusion between the dermal parasphenoid and the endoskeletal basisphenoid (the paucity of evidence for this fusion was noted in Patterson, 1977, p. 96, footnote). The parasphenoid laminae that extend dorsal to the trabeculum communis discussed by Fink and Fink (1981, p. 313) appear in the series of *Clarias* and *Silurus* to develop as extensions of the parasphenoid rather than as separate ossifications that later fuse with the parasphenoid.

8. Position of sacculi and lagenae more posterior and nearer to midline; Ostariophysi (Rosen and Greenwood, 1970).

9. Auditory foramen in prootic; Characiformes. Gayet (1986a, p. 50) raised questions about the distribution of this feature. She repeated Roberts's (1973) remark that some cypriniforms have the foramen, reported that she has observed the foramen to be missing in the characiform *Triportheus*, and said that Vari (1979) failed to report the foramen in the characiform families Citharinidae and Distichodontidae. Our observations of several cyprinids and catostomids suggest that the enlarged vagus foramen, sometimes incorporating the glossopharyngeal foramen (and extending forward into the prootic in *Carpiodes*), was mistaken by Roberts for the auditory foramen. Examination of several dried skeletons of *Triportheus* confirms the presence of the foramen. Vari (1979) illustrated the braincases only in lateral view, and, as Weitzman showed (1962, figs. 3, 4), the auditory foramen is usually not visible in lateral view. The foramen is present and large in the distichodontids *Distichodus* (BMNH 1901.12.21.35, 1903.7.28.143) and *Xenocharax* (BMNH 1899.9.26.126) and present but small in *Ichthyoborus* (BMNH 1907.12.2.3739, 1912.4.1.59), where it lies in the anterodorsal roof of a shallow groove running from the vagus foramen to the opening for the jugular vein. This groove is deeper in citharinids and in distichodontids such as *Phago* and *Eugnatichthys*, and the foramen, if present, is concealed in the roofed anterior part of the groove. The foramen is also present, though small, in *Characidium* and its relatives. In sum, our observations confirm the validity of this feature as a synapomorphy of the Characiformes.

10. Reduced parietals; Gonorynchiformes. Gayet (1986a, p. 50) cited Taverne's (1981) observation that the early Cretaceous †*Aethalionopsis* has unreduced parietals. In fact, Taverne described †*Aethalionopsis* as "lateroparietal" (parietals not meeting in the midline; fig. 3), although the parietals are illustrated as reaching much farther forward than other fossil and Recent gonorynchiforms. If †*Aethalionopsis* is a stem-group gonorynchiform as Patterson (1984a, p. 136) suggested, its lack of certain apomorphies of crown-group gonorynchiforms is to be expected. We note that Taverne (1981, p. 968) reported epineurals and epipleurals in †*Aethalionopsis* but made no mention of epicentral intermuscular bones, the presence of which is a synapomorphy of (crown-group) gonorynchiforms (Patterson and Johnson, 1995, p. 23).

11. Parietals not present as separate ossifications; Siluriphysi. Bamford (1948, pp. 373–374) provided evidence of the fusion of four small dermal parietal ossifications (two bilateral pairs) to the endochondral supraoccipital (which also begins as a bilateral pair) during ontogeny in *Galeichthys* (= *Arius*). Arratia *et al.* (1978, p. 164); and Arratia and Menu-Marque (1981, p. 100) noted similar fusion during ontogeny in some trichomycterids, and Alexander (1965, fig. 13) illustrated what may be a parietal incompletely fused to the supraoccipital on one side of an asymmetrical specimen of *Schilbe*. A separate parietal on one side only in a single juvenile specimen has been reported for *Diplomystes camposensis* (Arratia, 1987, pp. 48, 92) and *Rhamdia sapo* (Arratia and Gayet, 1995, p. 482), and Arratia and Menu Marque (1984, p. 496) reported

that small parietals are "distinguishable" in most, though not all, adults of *Trichomycterus roigi*. This rare occurrence of a unilateral separate parietal is consistent with the hypothesis that the absence of separate parietals generally in adult catfishes is due to early ontogenetic fusion between the parietals and the supraoccipital. Arratia and Gayet (1995, pp. 497, 499) added two additional observations that support the hypothesis of such fusion as a synapomorphy of catfishes. First is the presence in diplomystids of supraoccipital in the floor of the rudimentary posttemporal fossa, where the parietal usually occurs in other teleosts. Second, the presumed anterior pitline overlies the supraoccipital in some catfishes, whereas it overlies the parietal in other ostariophysans and teleosts. The presence of distinguishable parietals in some adults of *T. roigi* is presumably paedomorphic and does not invalidate the siluriform synapomorphy.

According to Lenglet (1974) and Grande (1994), parietals are also absent as separate ossifications in some *Kneria* and *Parakneria*, although Lenglet illustrated what appear to be parietal canals, perhaps fused to the pterotic, in *Kneria wittei* (1974, figs. 1, 3) and *Parakneria thysi* (1974, fig. 5). Specimens we examined of *Kneria wittei* (UMMZ 190116, 3 ex. 40.5–53.0 mm SL; BMNH 1976.10.20:142–160, 1 ex. 44.0 mm SL; RGMC 01-P-512-826 A, 1 ex. 21.6 mm SL) all have separate parietals, though these are little more than sensory canal ossicles.

12. Mediodorsal opening into posttemporal fossa; Characiformes. Fink and Fink (1981, p. 313) noted that this opening is absent in citharinids and gasteropelecids, presumably secondarily; the opening is present in distichodontids, the sister-taxon of citharinids. Although the relationships of gasteropelecids are not yet known, they are neither at the base of the characiform tree nor a relative of distichodontids and citharinids, and so they must be related to characiforms which possess the mediodorsal opening. Patterson (1984b, pp. 433–434) described a similar but nonhomologous opening in †*Chanoides* and *Chanos* and commented on the characiform-type opening present in certain clupeoids (he mentions *Clupea*). The opening in †*Chanoides* is located posterior and ventral to the posterior semicircular canal, whereas that in characiforms is anterior to the posterior semicircular canal and bordered by the epioccipital, supraoccipital, and parietal bones. Our observations on *Chanos* differ from Patterson's, in that our specimens do not have an opening like that of †*Chanoides*, although in very small specimens (less than about 40 mm SL) the epioccipital, supraoccipital, and parietal bones do not join together, leaving a gap in approximately the position

of the characiform opening. This gap is closed in larger specimens. However, a small characiform-like opening is present in our three cleared and stained specimens of *Kneria wittei* (UMMZ 190116), though these fish have no lateral opening (they do have a large posterior opening in the position Patterson described in †*Chanoides*). A similar small opening was observed on one side of another *Kneria wittei* specimen (BMNH 1976.10.20:142–160), but on the other side a tiny plate occluded most of the opening. Since neither Lenglet (1974) nor Grande (1994) mentioned or illustrated such an opening, its presence is probably variable within the genus. We have observed very small openings between the epioccipitals and parietals in cleared and stained or dry skeletal material of a few clupeoids, including *Dorosoma* (UMMZ 230784, UMMZ 194304) and *Brevoortia smithi* (UMMZ 179140), but not in *Brevoortia patronus* (UMMZ 179139-S) or material of *Alosa*, *Harengula*, *Opisthonema*, or *Pellona*. Gayet (1985c, p. 115) reported a dorsal opening into the posttemporal fossa as one of the two characiform characters of †*Lusitanichthys*, but the presence of similar openings in some clupeoids indicates that the opening in †*Lusitanichthys* need not have the significance that Gayet attributed to it. Gayet also described the fossa as a shallow depression (1981, p. 176), whereas that of characiforms is deep.

13. Absence of intercalar; Siluriphysi.

14. Expanded posterior occipital margin; Gonorynchiformes. Since 1981, we have examined counterstained specimens of *Gonorynchus*, *Phractolaemus*, *Kneria*, *Parakneria*, and *Cromeria* and stained material of *Grasseichthys*. These observations indicate that *Phractolaemus* and the kneriids (*sensu* Grande, 1994) examined do not have "a prominent posterodorsal cartilage margin" (Fink and Fink, 1981, p. 313), but they do have a tight, mostly bony joint between the first neural arch and the produced posterior margin of the neurocranium— the exoccipital in *Phractolaemus* and the exoccipital and supraoccipital in *Kneria*, *Parakneria*, *Cromeria*, and *Grasseichthys*. This joint is treated as part of the expansion of the first neural arch (character 65). The distinctive characteristics of this joint appear to be due to derived features of the first neural arch, rather than to the shape of the occipital margin, and hence are discussed with character 65. *Gonorynchus* also does not have a prominent occipital margin. Based on this survey we now omit this character.

15. Lagenar capsule; Characiformes. Gayet reported this character in the single fossil specimen of †*Salminops* (1985c, fig. 2). She maintained that destruction of the operculum near the rear of its skull was caused by well-developed capsules. But a much more

general explanation is that the occipital condyle caused this. The region labeled as the capsule is where the condyle would be expected.

Orbital Region

16. Eye reduced relative to circumorbital series; Siluriphysi. This should be modified to indicate that the character pertains to the adult condition. Some adult catfishes, such as the silurid *Kryptopterus*, have large eyes that closely approach the infraorbital canals; current hypotheses of catfish interrelationships indicate that such enlargement is secondary (de Pinna, 1993).

17. Sclerotic bones absent; Siluriphysi.

18. Infraorbital series reduced to little more than canal-bearing bones; Siluriphysi. Plate-like development in some advanced loricarioids is inferred to be secondary.

19. Supraorbital absent; Siluriphysi. Arratia (1987, p. 96) said that a supraorbital bone occurs in trichomycterids (illustrated in Arratia et al., 1978, fig. 15) but later (Arratia, 1992, p. 97) characterized the supraorbital in trichomycterids as an ossification of a ligament, not homologous with the dermal supraorbital in other ostariophysan groups. We agree that this element is neomorphic.

Suspensorium

20. Dermopalatine absent; Ostariophysi. The toothplates in some catfishes associated or fused with the autopalatine (see, e.g., Arratia, 1992, p. 114) appear to be accessory vomerine toothplates (Grande, 1987) and are assumed to be neomorphic and not homologous with the primitive teleostean dermopalatine. The toothplates in this region in a few characiforms are also interpreted as neomorphic (Fink and Fink, 1981).

21. Dorsomedial process of palatine contacting mesethmoid; Cypriniformes. Gayet (1986a, pp. 29, 51, fig. 9) claimed such a process is present in †*Ramallichthys* and illustrated it in one specimen, EY 381. In this illustration the process appears to be in a different position from that in cypriniforms. Grande (1996), however, found no evidence of this process, or of any other alleged cypriniform features of †*Ramallichthys*.

22. Palatine articulation in facet of endopterygoid; Cypriniformes. Gayet (1986a, p. 51) noted a specialized joint between the palatine and the ectopterygoid in certain gonorynchiforms. This is true in *Kneria* and *Parakneria*, which have a facet on the ectopterygoid articulating with the palatine, but we do not accept the relevance of this condition to cypriniform monophyly because the ectopterygoid, not the endopterygoid, is involved. Gayet also asserted, citing Taverne (1981),

that the Cretaceous anotophysans †*Aethalionopsis* and †*Parachanos* have a specialized articulation. But Taverne did not describe one, and his drawings of †*Aethalionopsis* and †*Parachanos* are lateral views showing a rod-like autopalatine lying dorsolateral to the tapering anterior part of the ectopterygoid, the usual teleost condition resembling that in *Chanos* (Fink and Fink, 1981, fig. 9). Gayet additionally claimed, on the basis of Howes (1978, fig. 9), that *Luciobrama* shows a primitive pattern in having an oblique contact between the anterior part of the endopterygoid and the palatine and therefore contradicts the cypriniform character as stated by Fink and Fink. Examination of skeletal material (BMNH 1889.6.24:48) shows that *Luciobrama* has the cypriniform endopterygoid facet, although it is unusually elongate.

23. Palatine ossification absent; Gymnotiformes. Arratia (1992, p. 31, fig. 12B, D) reported a differentiated autopalatine in *Hypopomus*, but, as she noted (p. 89), this results from "chondroidal osteogenesis" and is not homologous with the autopalatine of other ostariophysans. Specimens of *Hypopomus* that we examined show some slight ossification inside the palatal cartilage, as was observed also by Triques (1993). This character should be rephrased as autopalatine bone absent.

24. Palatine separate from the rest of the palatoquadrate arch, and with extension posterior to the lateral ethmoid that serves as the insertion site of the extensor tentaculi; Siluriformes. The development of the siluriform palatoquadrate arch in two separate parts was summarized in Arratia and Schultze (1991, pp. 14–17). Near absence of the posterior process of the palatine, as a secondary reduction, characterizes silurids (Bornbusch, 1991).

25. Anteriorly reduced ectopterygoid; Cypriniformes. Based on Weitzman (1962, fig. 10) and Gosline (1973, p. 768), Gayet (1986a, p. 52) claimed that some characiforms, in particular *Brycon*, have an anteriorly reduced ectopterygoid and a mobile ectopterygoid and endopterygoid. Our observations of *Brycon* skeletal material (UMMZ 205828, 2 ex., 209 and 210 mm SL) show that there is substantial overlap between the palatine and ectopterygoid and that both are immobile with respect to the rest of the suspensorium. Hemiodontids, parodontids, prochilodontids, and the derived anostomid *Gnathodolus* are the characiforms that have a mobile ectopterygoid, but in all these the ectopterygoid is mobile relative to the quadrate rather than to the palatine (Vari, 1983, p. 30).

Gayet (1986a, p. 52; 1993a, p. 181) also claimed that Fink and Fink (1981, fig. 8) misrepresented the structure of the palate in *Chanos*, but the difference between her interpretation and that figure is caused only by inclusion of cartilage in the latter. Substantial misrepresentation of these structures is to be found in Gayet's work (1985b, fig. 4), an anatomical impossibility mixing medial and lateral views. When corrected (1986a, fig. 21), the medial view is still misinterpreted as lateral.

Gayet (1986a) suggested that anterior reduction of the ectopterygoid in some gonorynchiforms invalidates that feature as a cypriniform synapomorphy, but we do not regard homoplasy as automatically invalidating proposed synapomorphies. The condition of the ectopterygoid in gonorynchiforms is discussed in the section on gonorynchoid synapomorphies.

26. Ectopterygoid greatly reduced or absent; Siluriphysi. All gymnotiforms lack an ectopterygoid, but a small, ovoid bone articulating with the ventral surface of the posterior part of the palatine occurs in members of the primitive catfish family Diplomystidae. Arratia (1987, p. 98) noted that the presence of this bone is variable and questioned its identification as an ectopterygoid but later (Arratia, 1992, p. 132) concurred with this identification. A more loosely articulated, elongate and often crescent-shaped bone in a similar position in some catfishes was inferred by Arratia (1992, pp. 84, 114, 132) to be neomorphic based in part on a parsimony argument, although it is also described as a "calcification" of a ligament, i.e., a sesamoid bone. This assessment must be tested by further work on siluriform interrelationships.

27. Endopterygoid reduced and not contacting quadrate, metapterygoid, or hyomandibula; Siluriformes. The identity and homology of various bones of the catfish suspensorium have been the subject of dispute in the literature of catfish morphology. This subject has been treated recently by Arratia, and we concur with her assessment of the homology of the endopterygoid across various siluriform taxa (1992, p. 84).

28. Endopterygoid with vertical strut; Gymnotiformes. The presence of this strut, whether large or small, in sternopygids, *Gymnotus*, most apteronotids, and most rhamphichthyoids led us to hypothesize that this strut was a gymnotiform synapomorphy. The feature is shown to be ambiguous at the base of Albert's (1996) cladogram due to its absence in *Electrophorus*. Should ontogenetic information show the presence of the strut in small *Electrophorus*, it can stand as a feature of the group which has been lost several times. Gayet *et al.* (1994, p. 295) referred to this feature as the ossification of a metapterygoid ligament, presumably a *lapsus* for mesopterygoid. They also mentioned attachment to the lateral ethmoid, although whether this was intended to refer to gymnotiforms or just an analogous structure (of the ectopterygoid)

in carangids is not clear. As they noted, there is a ligament generally in other ostariophysans between the endopterygoid and the lateral ethmoid (usually the posteroventral part), and the gymnotiform strut may represent an ossification of this ligament. If so, its attachment site differs from the primitive condition since in the gymnotiforms we have observed, it attaches to either the orbitosphenoid or the ventral surface of the frontal, well posterior to the lateral ethmoid (which is generally reduced in gymnotiforms and sometimes absent).

29. Anteriorly elongate suspensorium; Gonorynchiformes. Gayet (1986a, p. 53) accepted this character in Chanoidei but wrote that among gonorynchoids it is hardly developed in †*Judeichthys*, seems to have disappeared in *Gonorynchus*, is arguable in kneriids, and is absent in †*Ramallichthys*. We have attempted to quantify the character by measuring two distances, from the middle of the quadrate condyle to the anteroventral margin of the hyomandibula, and from the middle of the hyomandibular dorsal head to about the middle of its ventral margin. The ratios between the first and second measurements are as follows: *Chanos*, 1.08; †*Judeichthys*, 0.93; *Gonorynchus*, 1.35; *Kneria*, 0.93; *Parakneria*, 0.90; †*Ramallichthys*, 0.96; and *Phractolaemus ansorgei*, 1.91.

In contrast the ratio in most outgroup taxa measured ranged from 0.81 in *Xenocharax* (Fink and Fink, 1981, fig. 10) to 0.57 in †*Chanoides* (Patterson 1984b, fig. 2).

The character shows considerable homoplasy. Within otophysans, exceptions included several species of *Gila* (e.g., *G. robusta*, 1.0), *Mylocheilus* (0.92), *Notemigonus* (0.96), *Botia modesta* (1.2), *Nannostomus harrisoni* (1.85; Weitzman, 1964, fig. 71), and *Sternopygus* (1.04). Examination of studies of Sawada (1982) and Mago-Leccia (1976) shows that elongation is in fact common though by no means universal in cobitoids and gymnotiforms, but it is rare in characiforms and siluriforms. It is also rare in most non-ostariophysan lower teleosts, the exceptions being in osmeroids and esocoids.

30. Axe-shaped endochondral portion of metapterygoid; Otophysi. This metapterygoid shape character used by Fink and Fink as an otophysan synapomorphy deserves further consideration. The character description proposed was

the endochondral portion of the metapterygoid is an axe-shaped bone, either double-headed (most cypriniforms and characiforms) or single-headed, with the posterior half of the bone absent (siluriforms [= Siluriphysi herein], see character 31). The posterior border of the metapterygoid (siluriforms) or homologous ventral border (all other otophysans) is bony rather than cartilaginous (Figs. 9–12).

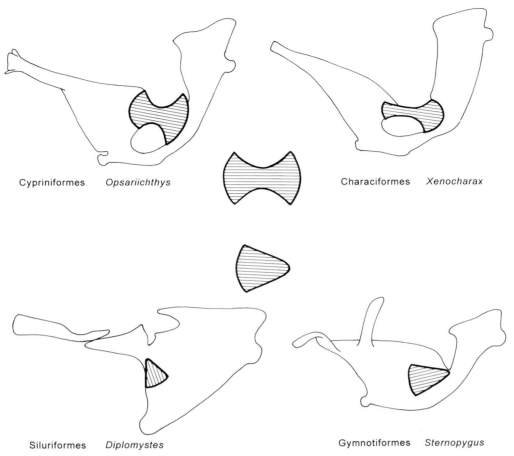

Cypriniformes   *Opsariichthys*   Characiformes   *Xenocharax*

Siluriformes   *Diplomystes*   Gymnotiformes   *Sternopygus*

FIGURE 2   Schematic diagram of the endochondral portion of the metapterygoid (hatched) in relation to the rest of the suspensorium within the four otophysan lineages, illustrating the double-headed axe-shape postulated by Fink and Fink (1981) to be the general otophysan condition (upper), and the single-headed axe-shape postulated to be shared by siluriphysans (lower). A counterclockwise rotation of the single-headed metapterygoid was inferred to have taken place in the evolutionary transformation from a general siluriphysan condition, positioned as in *Sternopygus*, to the derived siluriform condition, positioned as in *Diplomystes*.

The shape of the character is summarized in Fig. 2, showing the double-headed condition of cypriniforms and characiforms, inferred to be the primitive otophysan condition, and the single-headed condition, with loss of the posterior head, shared by siluriphysans. In siluriforms an evolutionary transformation involving counterclockwise rotation was inferred to have occurred. Which part of the bone was endochondral and which part lamellar outgrowth was inferred from inscriptions on the bone in alizarin-stained specimens and location of the cartilaginous margins.

Ontogenetic observations provide little further information concerning the proposed homology of the "axe-shape" since the development of the metapterygoid in each of the four otophysan lineages has unique characteristics. In cypriniforms, the metapterygoid region of the palatoquadrate cartilage, before ossifica-

tion, is simple and elongate (Fig. 3A, left). It extends dorsally to about the level of the interhyal articulation with the cartilage plate of the future hyomandibular and symplectic bones. The posterodorsal end of the palatoquadrate then broadens, lengthens, and develops a small process at its tip, giving a slightly bifurcate shape that remains after ossification (Fig. 3A, right). The emarginate ventral border of the endochondral portion of the bone develops by ontogenetic regression of cartilage and replacement by lamellar bone (Fig. 3). We have no developmental examples of cypriniforms with a metapterygoid–quadrate fenestra; presumably in these no lamellar bone forms. Dorsal expansion of the ventral lamellar plate then ceases and ossification of the cartilage core occurs, resulting in the axe-shaped portion visible in adults. In *Cyprinus* and catostomids, the ventral emargination of the carti-

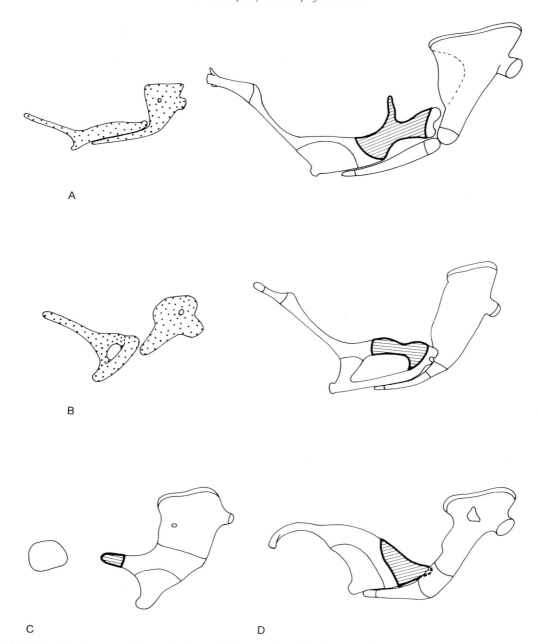

FIGURE 3 Development of the metapterygoid in examples of the four otophysan lineages. (A) Cypriniformes: *Zacco platypus*, UMMZ 231058, before ossification (left) and after (right); (B) Characiformes: *Hyphessobrycon callistus*, UMMZ 211676, before ossification (left) and after (right); (C) Siluriformes: *Siluris glanis*, UMMZ 213368, 9 days (separate anterior element is palatine cartilage); (D) Gymnotiformes: *Apteronotus leptorhynchus*, UMMZ 231059, 19 days. The endochondral part of the metapterygoid ossification is hatched.

lage and replacement by lamellar bone is more posterior, smaller, and occurs much later in development than in the other cypriniforms observed; the resulting shape of the endochondral portion of the metapterygoid cannot really be described as "axe-shaped." Whether this variation in catostomids and *Cyprinus* is inferred to be a derived condition within cypriniforms

is a question that can only be answered with further information about cypriniform interrelationships.

In characiforms, in the smallest cleared and stained specimens with an identifiable palatoquadrate, a large fenestra is present (Fig. 3B, left). The shape of the posterodorsal margin of the palatoquadrate cartilage plate is blunter than in cypriniforms but also simple, and the cartilage extends dorsally to about the same point. As in cypriniforms, the posterodorsal end of the palatoquadrate cartilage broadens, lengthens, and develops a small process at its tip (Fig. 3B, right). The double-headed axe-shape emerges with growth and ossification of the portion dorsal to the fenestra, rather than by regression of cartilage. Arratia (1992, pp. 87, 94) noted the bifurcation present in characiforms and cypriniforms as a possible synapomorphy they share. A similar bifurcation is present in *Brevoortia*.

The double-headed axe-shape in adult characiforms and many cypriniforms thus develops in distinct ways. In the cypriniforms examined, the ventral margin of the axe-shape borders an area that results from cartilage regression during ontogeny, while in characiforms it borders a foramen that appears to exist from the earliest stage of chondrification. Since these taxa shared a common otophysan ancestor, the development of the metapterygoid once took only a single form, but the ontogeny of present taxa does not provide a clue as to what that form was and whether it involved a double-headed axe-shape. Ontogeny thus provides no more information than does the distinctive adult morphology.

Ontogenetic information does, however, suggest a means of transformation from what we have termed the double-headed axe-shape to the single-headed shape of siluriphysans. The development in the two siluriphysan groups can be described as follows:

In gymnotiforms, based on specimens of *Apteronotus leptorhynchus* and *Eigenmannia* sp., the posterodorsal corner of the palatoquadrate cartilage is triangular in shape (Fig. 3D). With enlargement of the suspensorium, the region of the metapterygoid near the quadrate broadens, resulting in a triangular shape to the endochondral portion. The posterior part of the cartilage does not broaden during growth, so that a triangular perichondral metapterygoid ossification develops, sandwiching the cartilage posterodorsal to the quadrate; lamellar outgrowths develop dorsally.

The development of the catfish suspensorium is in many respects as unique as the adult morphology. Two cartilage elements develop: the palatine anteriorly and a separate element comprising hyomandibula, quadrate, and metapterygoid (in contrast to the usual teleostean pattern of a hyomandibula-symplectic element and a metapterygoid-quadrate-palatine element; this pattern is described in more detail by Arratia, 1992). No symplectic element ever appears, and no early fusion of the hyomandibula and quadrate portions has been observed. In the earliest stages of its development, the metapterygoid ossification is more nearly rectangular than triangular (Fig. 3C; also, e.g., Arratia, 1987, fig. 25A of *Ictalurus*). With enlargement of the suspensorium, however, the pos-

teroventral region of the metapterygoid broadens, resulting in an often elongate triangular shape to the endochondral portion (e.g., Arratia, 1987, figs. 25B and fig. 32A-B of some ictalurids, figs. 16A-C of juvenile diplomystids, and fig. 32A of *Heptapterus*) that generally shortens and broadens during growth. Lamellar outgrowths from the endochondral portion appear in the oldest of our specimens of *Silurus* (30 days). The hypothesized rotation of the metapterygoid from the posterodorsal to the anterodorsal region of the cartilage has not been observed. As in gymnotiforms (Fig. 3D), however, the tip of the metapterygoid cartilage does not broaden, and lamellar outgrowths develop slightly later. (In our *Clarias* developmental series, the metapterygoid is a small triangular ossification throughout the size range we have, and there are no lamellar outgrowths.)

The absence of broadening in the tip of the metapterygoid distal to the quadrate in catfishes and electric fishes suggests that the similarity in shape of the siluriphysan endochondral metapterygoid may be due to developmental truncation relative to other otophysans. The resulting form should probably not be equated with just the anterior portion of the axe-shape in characiforms and cypriniforms; the posterior half is not reduced or lost but rather does not expand as it does in the more plesiomorph taxa.

Some homoplasy in the metapterygoid character not noted by Fink and Fink (1981) should be mentioned. A small posterior emargination in the cartilage and replacement by lamellar bone occurs in esocoids and some osmeroids. Additionally, the gonorynchoids examined have a narrow, roughly triangular metapterygoid with a ventral bony border, positioned approximately posterior to the quadrate. This is visible in adult *Kneria* and *Phractolaemus*. In larval *Gonorynchus*, the cartilage triangle posterior to the quadrate never ossifies but shrinks in size. Simultaneously a long, thin, anteromedially directed cartilage process extending from the posterior tip of the triangle separates and ossifies, resulting in the bizarre splint-like metapterygoid remnant of the adult.

Gayet (1986a) questioned Fink and Fink's hypothesis (1981, p. 320) that the metapterygoid-quadrate fenestra is an otophysan feature. Our hypothesis was based principally on the supposition that cypriniforms that have a fenestra are phylogenetically primitive members of their lineages. Fink *et al.* (1984, p. 1040) pointed out that this parsimony assessment, as well as that of Gayet (1982b, p. 44), was merely guesswork pending further information on phylogenetic hypotheses of groups of cypriniforms. Sawada (1982, pp. 185–186) proposed that the fenestra is derived for two subgroups of his Cobitinae, but his methods assume irreversibility of characters and his conclusions merit reexamination. Sawada also apparently overlooked the presence of a fenestra in at least one homalopterid (Ramaswami, 1952b, fig. 5c). [The citation of Ramaswami, 1952c, by Fink and Fink (1981, p. 320) concerning a fenestra was an error.]

31. Endochondral portion of metapterygoid roughly triangular; Siluriphysi. As noted with character 30, ontogenetic observations provide no clue as to the hypothesized rotation of the metapterygoid in catfishes. Nevertheless, the similarity in the shape of the endochondral portion of the metapterygoid during development, detailed in the preceding paragraphs, provides some corroboration for the homology of shape postulated to be a synapomorphy of the Siluriphysi.

32. Metapterygoid anterodorsal to the quadrate; Siluriformes. Although there has been controversy about the identity of the metapterygoid element in siluriforms, in part due to its unusual position (e.g., Howes, 1985a; Howes and Teugels, 1989; Arratia *et al.*, 1978; Arratia and Menu Marque, 1984), the developmental information provided by our material provides no evidence of any compound origin for the metapterygoid, or for the hyomandibula. Arratia (1990, 1992, p. 74) reached the same conclusion.

33. Symplectic and posterior process of quadrate absent; Siluriformes.

34. Preopercle and interopercle shortened along an antero–posterior axis; Siluriformes.

35. Subopercle absent; Siluriformes.

36. Opercle approximately triangular; Siluriphysi. Some catfishes (e.g., *Corydoras* and its relatives) have an ovoid opercle that resembles the outgroup condition, but the distribution of such variation indicates it is secondarily derived, and the more triangular shape is primitive within catfishes. Arratia (1987, p. 102) stated that the opercle of trichomycterids is not approximately triangular, but we disagree; discounting the toothed projection of the opercle, the shape is distinctly triangular in small juveniles (e.g., *Trichomycterus areolatus*, UMMZ 212766, 10.8 mm SL), though the modification in shape that occurs during growth and with the addition of more teeth can rapidly make the basic triangular form highly attenuated (*T. areolatus*, UMMZ 212769, 13.1 mm SL). A clearly triangular form is present in some adult trichomycterids also, again discounting the toothed process and also the anteromedial process that articulates in a fossa on the interopercle (e.g., *T. rivulatus*, UMMZ 66324, 54 mm SL).

Jaws

37. Dorsomedial extension of premaxilla; Cypriniformes. Gayet (1986a, p. 53) allowed that the alternative state (dorsolateral extension) occurs in some characiforms but questioned its presence in *Xenocharax* and *Rhoadsia* (Fink and Fink, 1981, fig. 3C, D) and in the majority of siluriforms; among cypriniforms, she denied the presence of the character in *Catostomus, Labeo*, and *Garra* (Ramaswami, 1955b). Examination of skeletal and cleared and stained material shows well-developed but narrow processes in *Catostomus; Garra* has reduced processes, but the premaxilla still extends farthest dorsally along the midline. *Labeo* has lost the dorsomedial processes altogether, a derived feature which fails to invalidate the cypriniform synapomorphy. *Xenocharax* and *Rhoadsia* have smaller dorsal processes than many characids, but the dorsalmost extent is nevertheless lateral rather than along the midline. Dorsal processes are absent or nearly so in most catfishes and gymnotiforms, a possible synapomorphy of Siluriphysi.

38. Premaxilla a thin, flat bone; Gonorynchiformes. Gayet commented (1986a, p. 54) that in *Gonorynchus* the premaxilla is not "a very thin flat bone," the character given by Fink and Fink. This is true, and the oversight in Fink and Fink was due to lack of suitable material. The premaxilla of *Gonorynchus* has two robust articular thickenings, one through which the two premaxillae articulate across the midline via thick connective tissue and one that articulates with the maxilla (see, e.g., Monod, 1963, figs. 10–11). With the exception of these articular thickenings, however, the premaxilla is quite thin. The fossil gonorynchid †*Notogoneus* has very similar morphology (Grande, 1996, character 6, in part). According to Gayet (1993b, p. 95), the premaxilla in †*Ramallichthys*, †*Judeichthys*, and †*Charitosomus* is thicker but also has articular processes. The articular processes seem to be a synapomorphy of gonorynchids (though perhaps absent in †*Charitopsis spinosus*; Gayet, 1993b, p. 76). Whether the thin, plate-like condition is a gonorynchiform synapomorphy, reversed to some degree within gonorynchids, is unclear. The question may be addressed by examination of the gonorynchiform stem-group taxa. Reports in the literature that †*Tharrhias*, †*Dastilbe*, †*Rubiesichthys*, and †*Gordichthys* all have a thin, broad, plate-like premaxilla (e.g., Poyato-Ariza, 1994, pp. 29–30) suggest that the condition in gonorynchids is a secondary modification. [Poyato-Ariza included †*Aethalionopsis* in his characterization, but Taverne's description (1981, p. 974) was an inference, as the premaxilla was missing or fragmentary in the material he examined (p. 964).]

39. Absence of contact between maxilla and mesethmoid; Characiformes. Gayet (1986a, p. 55) commented that kneriids and the fossil "chanoid" †*Aethalionopsis* also lack such contact. This is true of kneriids, *Phractolaemus*, and the other African gonorynchoids

(Howes, 1985b), but again we do not regard homoplasy between characiforms and certain gonorynchoids as invalidating the character, which in this case is quite different in detail in the two groups. In †*Aethalionopsis* the maxilla is complete in only one specimen according to Taverne (1981, p. 964) and its head "devait s'appuyer contre la region mésethmoïdienne" [i.e., "would have contacted the mesethmoid region"].

40. Maxillary barbels; Siluriformes and some cypriniforms.

41. Absence of supramaxillae; Ostariophysi. Gayet (1986a, p. 66) questions whether the supramaxilla reported in a few characiforms is primitive or neomorphic and whether, following the report of a supramaxilla in †*Chanoides*, Fink and Fink (1981) still consider lack of supramaxillae synapomorphous at any level. A partial answer to the first question was given by Vari (1983, p. 10); based on the relationships of known supramaxilla-bearing characiforms (i.e., on parsimony), the bone is considered neomorphic. A proposal that the bone is primitively retained would require an unparsimonious argument. In answer to the second question, lack of supramaxillae still appears to characterize gonorynchiforms and crown-group otophysans, though †*Chanoides* renders it homoplastic, as Fink *et al.* (1984, p. 1039) acknowledged.

42. Absence of jaw teeth; Gonorynchiformes and Cypriniformes. The "teeth" reported by Monod (1963, p. 259) but not observed by Grande (1996) were observed by us in the old and rather poorly preserved material of *Gonorynchus* examined for our 1981 paper (MCZ 8441). The structures were clearly not bony. Since keratinous structures occur in gonorynchiforms, this possibility might be investigated.

43. Replacement teeth for outer-row dentary teeth and some premaxillary teeth formed in crypts; Characiformes.

44. Multicuspid teeth; Characiformes. Gayet echoed (1986a, p. 56) Roberts's (1967, p. 231) claim that multicuspid teeth in characins are derived by fusion of unicuspid teeth and that unicuspid teeth are therefore primitive for characiforms. We think that Gayet and Roberts are mistaken in equating the enameloid caps of the cusps, which form separately, with separate teeth. This recalls the discarded theories of Gervais (1854) and others that mammalian molars develop by fusion of separate teeth, because the enamel germs of the cusps form independently. A decision on whether multicuspid teeth are synapomorphous for all or only for some characiforms depends on a theory of relationships within the group, such as that summarized by Fink and Fink (1981, p. 306) as justifying their inference about the generality of multicuspid

teeth in the group. Neither Roberts nor Gayet presented any reasoned alternative.

45. Ligament between anterior portion of maxilla (near palatine articulation) and coronoid region of lower jaw; Siluriphysi. As noted by Arratia (1992, p. 91), diplomystids have cartilage in this region, to which the ligament attaches. This is true for most catfishes. In gymnotiforms we have examined the attachment is to the coronoid portion of the anguloarticular, but detailed examination of these fishes is needed.

Gill Arches

46. Epibranchial organs; Gonorynchiformes. The homology of epibranchial organs among gonorynchiforms has been questioned by some authors (e.g., Gayet, 1986a, p. 56; 1993a, p. 172) since epibranchial organs occur elsewhere among ostariophysans and among teleosts and since the epibranchial organs of some gonorynchiforms differ in many details. That epibranchial organs have evolved a number of times in lower teleosts means that evidence of homology should distinguish both aspects of morphology shared among the taxa in question and aspects of morphology differing from other taxa. The epibranchial organs of *Chanos* and *Gonorynchus* have differing complex features of skeletal support structures and soft anatomy that are unique to each taxon. Soft structure modifications shared by these taxa and by *Phractolaemus* and kneriids, which have simpler epibranchial organs, include chiefly the presence of a dorsal diverticulum with some associated gill rakers. Such characteristics are common among teleostean epibranchial organs and hence can provide only weak evidence of homology. However, the diverticula are positioned more laterally in gonorynchiforms than is usual in other taxa. Further, in most taxa epibranchial 4 is enlarged (see Nelson, 1967, fig. 1c-h, fig. 2a-j, fig. 3a-j; Greenwood and Rosen, 1971, figs. 1–7) and provides some of the skeletal support for the anterior part of the diverticulum; epibranchial 5 is often enlarged, but its ventral portion is generally closely applied to epibranchial 4. Gonorynchiforms are distinctive in that epibranchial 4 is not enlarged, and skeletal support of the anterior part of the pouch is provided almost entirely by epibranchial 5, which is at least slightly enlarged and projects dorsolaterally from its ventral articulation with ceratobranchial 4. In *Chanos* and *Gonorynchus* epibranchial 5 is quite large and its dorsal part curves medially, approaching (*Chanos*, Nelson, 1967, fig. 1j) or fusing to (*Gonorynchus*) the cartilage of the posterior process of epibranchial 4 (Monod, 1963, fig. 40, as "cartilage semilunaire"). In *Phractolae-*

mus, epibranchial 5 is large and projects dorsolaterally (drawn as the elongate conical tip of ceratobranchial 4 in Thys van den Audenaerde, 1961, fig. 21; in our adult specimen, UMMZ 195066, ceratobranchial 5 has a considerably shorter cartilage tip than that illustrated by Thys van den Audenaerde). In our specimens of *Kneria wittei*, epibranchial 5 projects laterally, and (*contra* Howes, 1985b, p. 300) lies within the wall of the diverticulum, ventral to some of the anterior row of gill rakers. Howes (1985b, p. 300) reported that *Parakneria* and *Cromeria* resemble *Kneria* and that *Grasseichthys* has a small diverticulum with a few gill-rakers. He also reported that *Grasseichthys* has a cartilaginous epibranchial 5 posterodorsal to epibranchial 4, an orientation that resembles that of the other kneriids.

Gonorynchus also shares this morphology, but this can be determined only through examination of larvae. The long, flat cartilage in adults that extends from the dorsal tip of ceratobranchial 5, sometimes identified as epibranchial 5 (e.g., Howes, 1985b, fig. 9; Monod, 1963, fig. 40), is in fact a neomorphic feature not present in other extant gonorynchiforms. It is not present in any of our larval specimens (AMNH 55560), and therefore it must develop later in ontogeny. It lies within and supports the posterior row of gill-rakers and the posterior wall of the diverticulum. Examination of seven larval specimens (AMNH 55560, 12.1–21.5 mm SL) shows that epibranchial 5 develops in the usual position for lower teleosts, articulating ventrally with ceratobranchial 4 (not ceratobranchial 5); as in other gonorynchiforms, it supports the anterior wall of the diverticulum. The dorsal tip of epibranchial 5 fuses with the large posterior process of epibranchial 4 (the two are in contact but separate in a specimen 18.5 mm SL and fused in a specimen 21.5 mm SL). [In the place where Howes's (1985b) fig. 9 identifies epibranchial 4, our *Gonorynchus* specimen has both epibranchials 4 and epibranchial 5, not in contact except at their dorsal fusion and ventral articulation, as illustrated in Monod (1963, fig. 40) (epibranchial 5 identified as "cartilage semilunaire").]

In summary, the epibranchial organ morphology hypothesized here as synapomorphous for gonorynchiforms consists of a slightly to moderately recurved dorsal diverticulum, more laterally situated than most other epibranchial organs, with the anterior row of gill-rakers supported largely by a somewhat enlarged epibranchial 5 directed posterodorsally from its ventral articulation with ceratobranchial 4.

47. Teeth absent from pharyngobranchials 2 and 3, and from basihyal; Gonorynchiforms, Cypriniformes, Siluriphysi.

48. Two posterior pharyngobranchial toothplates absent; Gonorynchiformes and Cypriniformes.

49. Teeth absent from ceratobranchial 5; Gonorynchiformes. Such absence also occurs in various characiform taxa.

50. Toothplate associated with basibranchials 1–3 absent; Cypriniformes and Siluriphysi. This toothplate is also absent in non-gonorynchid gonorynchiforms and in †*Notogoneus* (Grande, 1996).

51. Only one pharyngobranchial toothplate present; Siluriphysi.

52. Modified shape of ceratobranchial 5; Cypriniformes. Under this heading, Gayet (1986a, p. 56) gave her interpretation of the endopterygoid teeth in †*Judeichthys* and †*Ramallichthys* as belonging to the fifth pharyngobranchials. We note here only that teleosts have no fifth pharyngobranchial. The identity of the bone to which the teeth are attached is discussed previously; see Gonorynchid Synapomorphies.

53. Teeth on ceratobranchial 5 ankylosed to bone; Cypriniformes.

Gasbladder

54. Gasbladder with anterior and posterior chambers; Ostariophysi. This feature is very widespread within ostariophysans, although a few taxa have lost a gasbladder (e.g., *Gonorynchus*, the troglodytic catfish *Trogloglanis*). Additional posterior elaborations or an additional chamber occur in some taxa [e.g., malapterurids and at least some pangasiid catfishes (Howes, 1985a, p. 57; Bridge and Haddon, 1893, pl. 19, figs. 91 and 93; pl. 17, fig. 71); some serrasalmins (Fink and Machado-Allison, 1992, fig. 10)].

Among gonorynchiforms, *Chanos* is the only taxon to possess a clearly two-chambered gasbladder. *Phractolaemus* (Thys van den Audenaerde, 1961, fig. 2, pl. 8A, B) has a well-vascularized, elongate gasbladder with many partial partitions; the bladder presumably serves a respiratory function. The gasbladder of *Kneria* is also elongate and well-vascularized with many asymmetrical partial partitions; that of *Parakneria* is smaller and highly variable, but it also has partitions (Lenglet, 1973, p. 257, fig. 4; 261, fig. 6). The slight constrictions in the bladders of these two taxa correspond to internal partitions, and these slight constrictions—of which there are many along the length of the bladder—bear no compelling similarity to the single distinct constriction separating the anterior and posterior chambers in *Chanos* or most otophysan taxa. *Cromeria* and *Grasseichthys* have elongate, slender bladders (Swinnerton, 1901, p. 444; Géry, 1965, p. 387), but we have found no further anatomical information. The close resemblance between the gasbladder in

Chanos and otophysans, in this feature as well as characters 55–57, leads us to suggest that the absence of the constriction separating the gasbladder into two distinct chambers in gonorynchoids represents a secondary loss. The alternative, that this feature was independently acquired in *Chanos* and in otophysans, is equally parsimonious, but this interpretation is not supported by details of similarity nor by the mosaic patterns of distribution of those other features.

The morphology of the gasbladder in catfishes requires special comment. We previously suggested (Fink and Fink, 1981, p. 324) that the posterior otophysan chamber is lost in most catfishes. Gayet (1986a, p. 57) cited an *in litt.* communication from Chardon that questions this identification of the gasbladder in catfishes with the anterior chamber of other otophysans. We agree with these criticisms. In non-siluriform ostariophysans, the pneumatic duct enters the gasbladder between the anterior and posterior chambers (Rosen and Greenwood, 1970, figs. 7, 8). If one uses the entrance into the bladder of the pneumatic duct as a landmark in siluriforms, it is clear that the posterior chamber is usually present, even though an external constriction between the chambers is absent (e.g., Chardon, 1968, figs. 12, 13 of *Silurodes hypophthalmus*; fig. 20, *Diplomystes*; figs. 37, 40 of *Ictalurus nebulosus* [as *Ameiurus*]). Arratia (1987, p. 45) stated that the posterior chamber is not present in *D. camposensis*, but the morphology she described and illustrated in her Figure 20 is the same as that described by Chardon.

Thus, though the posterior portion of the gasbladder appears to be much reduced or lost in some catfishes with encapsulated gasbladders, the general siluriform condition appears to be not absence of the posterior chamber but absence of the constriction separating the anterior and posterior chambers. Instead, a partial transverse septum separates the anterior and posterior parts of the bladder.

55. Anterior chamber of gasbladder partially or completely covered by a shiny peritoneal tunic; Ostariophysi. This feature was discussed in a short note by Chardon and Vandewalle (1989), in which they questioned the homology of the layers of the gasbladder in gonorynchiforms and otophysans. In particular they argued that Rosen and Greenwood (1970) were in error when they identified the outer coating of the gasbladder in *Chanos* and otophysans as a "peritoneal tunic" and interpreted those authors as implying an origin in somatopleure for the structure. Chardon and Vandewalle maintained that the tunic is extra-peritoneal, in part because they claim to have found kidney tissue anteriorly within it. We are unable to confirm the presence of kidney within the covering in any of

the specimens we examined, including freshly thawed, unfixed market specimens of *Chanos*. In all specimens examined, the anterior wall of the bladder chamber is smoothly continuous. Kidney tissue is present anterior to the tunic in the spaces between the gasbladder and the transverse septum.

Unfortunately, the use of nonequivalent schematic diagrams, and lack of any histological figures of all relevant taxa, make evaluation of Chardon and Vandewalle's contribution difficult. This problem is increased when the discussion is ambiguous regarding just what the features are. For example, Chardon and Vandewalle (1989, p. 267) maintained that "non-otophysine fishes" have in the gasbladder a fusion of endoderm + splanchnopleure and that *Chanos* has "splanchno and somatopleura separated." *Chanos* is a non-otophysan, of course, and these seem to be two ways of saying the same thing. In any case, there is no contradiction, since in the figures both *Salmo* and *Chanos* are labeled as having endoderm fused with splanchnopleure, and together these are separated from the somatopleure. What appears to be the real issue is whether in otophysans there is a unique fusion of somatopleure and splanchopleure. That conclusion is apparently based on a histological section of an 18.5 mm larval *Barbus* in which only two tissue layers are visible in the gasbladder. The authors conclude that the somatopleure and splanchnopleure are fused, whereas the endoderm is free. This may be so, and further observations are needed.

Chanos has a well-developed gasbladder and silvery tunic; *Phractolaemus* (UMMZ 195066) has both gasbladder and tunic; *Gonorynchus* (FMNH 103977) lacks a gasbladder but has a silvery sheet in that area; *Kneria wittei* (UMMZ 190116) has a well-developed gasbladder but, based on our observations, lacks a silvery tunic (though Lenglet, 1973, p. 257, states that the wall of the gasbladder is silvery when the fish is preserved in formalin; she describes the gasbladder of *Parakneria*, in contrast, as usually fibrous and opaque). This distribution suggests that a silvery tunic has been lost within gonorynchoids.

Our examination of outgroup taxa suggests that a silvery peritoneum associated with the dorsal body wall and the gasbladder is also present in some clupeomorphs. In those examined (*Brevoortia*, *Chirocentrus*, *Dorosoma*, and *Lile*) there is at least some guanine lining the peritoneum dorsal to or around the gasbladder. In *Pellona* (UMMZ 207387), there is a silvery sheath around the gasbladder and broad fans of silvery connective tissue extending from the ribs to around the bladder. A silvery area associated with the gasbladder may be a synapomorphous feature shared by ostariophysans and clupeomorphs.

56. Peritoneal tunic of anterior chamber of gasbladder attached to anterior two pleural ribs; Ostariophysi. In *Gonorynchus*, although the gasbladder is lost, the silvery peritoneal tunic still attaches to the anterior pleural ribs.

57. Dorsal mesentery suspending gasbladder thickened; Ostariophysi. This thickening is prominent in *Chanos*, present in *Phractolaemus* (UMMZ 195066), and occurs in *Gonorynchus* (FMNH 109377) even though the gasbladder is absent. We have not observed or seen descriptions of any such thickening in other gonorynchiforms.

Anterior Vertebrae

58. Absence of supraneural anterior to first neural arch; Ostariophysi. Gayet (1986a, p. 57, fig. 32) reported that this supraneural is present in a specimen of †*Charitosomus hakelensis* and in *Diplomystes* (according to a personal communication from Chardon) and may be present in †*Ramallichthys* (1986a, fig. 31). In †*Charitosomus* and †*Ramallichthys* this feature obviously depends on interpretation of single fossils in which the structures are partially preserved or visible (dotted areas in Gayet, 1986a, figs. 31, 32). Grande (1996) could find no such structure in the specimen of †*Ramallichthys* illustrated by Gayet and reported that neural arch 1 is in fact incomplete. Grande also stated that no gonorynchiforms have such a supraneural, and so presumably she did not find one in the specimens of †*Charitosomus hakelensis* she examined.

In *Diplomystes*, the structure referred to by Chardon is presumably the median cartilage that caps the claustra in that genus (Chardon, 1968, pp. 33, 34, figs. 17, 18). Arratia (1987, p. 30) interpreted this cartilage as "probably" the second supraneural, but she also says that her observations of the Weberian apparatus of *Diplomystes* "do not show any significant differences" from those of Fink and Fink (1981). Fink and Fink (1981, p. 327) described the otophysan claustrum as "an ossification on the anteroventral border of the chondral block that roofs the neural canal anteriorly," and there seems no particular reason to regard this cartilage as developed from a supraneural that is absent in other ostariophysans rather than from neural arch material.

59. Absence of second supraneurals; Otophysi. Patterson (1984b, p. 445) pointed out that †*Chanoides* has both the two expanded supraneurals present in cypriniforms and a small supraneural associated with the neural spine of the fourth arch. He also cited a supraneural associated with the fourth neural spine in several characiforms not observed by Fink and Fink (1981). He concluded that the two expanded supra-

neurals of †*Chanoides* and cypriniforms must be those of the second and third neural arches, and the one in characiforms must be that of the third, a conclusion agreed to by Fink *et al.* (1984, p. 1037). This means that the second supraneural is present in otophysans primitively, and character 59 of Fink and Fink (1981) is wrong. See also character 61.

60. Ventral expansion of anterior one or two supraneurals to form synchondral joint with neural arches 3–4; Otophysi.

61. Absence of third supraneural; Characiphysi. In view of the discussion under character 59, this must be corrected. Characiphysans lack the second supraneural rather than the third.

Gayet argued (1986a, p. 58) that the single enlarged supraneural of characiphysans could correspond to the two enlarged supraneurals of †*Chanoides* and perhaps sometimes the following one, fused together, and that the larger of the two in cypriniforms could be the third and fourth fused together. In support of these proposals she cited enlarged supraneurals, especially in characiforms, claimed that in cypriniforms the posterior enlarged supraneural is situated in the position of the third and fourth, and suggested that the close approximation of the third and fourth in †*Ramallichthys* supports this hypothesis. Since †*Ramallichthys* is a gonorynchid rather than a cypriniform, the form of its supraneurals cannot be relevant to the condition in cypriniforms. In many cypriniforms the largest supraneural does extend back to the margin of the fourth neural spine, yet the portion which extends back appears to be simply lamellar outgrowth from the third supraneural. If it did include a fused fourth supraneural, one would expect both ontogenetic evidence of fusion and the presence of a separate fourth supraneural in those taxa in which the third does not extend near the fourth neural spine. Neither has yet been observed; without such evidence, we see no reason to postulate fusion. We have examined several ontogenetic series of characiforms, from taxa both with and without a fourth supraneural, and no fusion of elements occurs. No second supraneural ever appears, and the fourth either appears and remains separate or never appears. Further, absence of the fourth in characiforms is not associated with an expansion of the enlarged, third supraneural.

In the siluriform *Diplomystes*, Arratia (1987, fig. 9) illustrated a small ossification posterior to the large supraneural in one specimen and concluded, presumably because of this individual, that the enlarged supraneural is the "result of the early fusion of a large supraneural 3 and a small supraneural 4." But so far as we are aware, no full ontogenetic series of *Diplomystes* is yet available, and wherever ontogenetic information is available in siluriforms (Bamford, 1948, on *Galeichthys*; our larval and juvenile specimens of *Pylodictis*, *Plotosus*, and *Silurus*), there is no indication of ontogenetic fusion between supraneural ossifications. Arratia (1992, p. 125) identified the presence of a compound supraneural or two supraneural ossifications as a synapomorphy of *D. chilensis* and *D. viedmensis*, but Azpelicueta (1994) did not describe or illustrate (in her fig. 2) the supraneural of the latter taxon, or taxa that were formerly considered subspecies of *D. viedmensis*, as compound, though she examined cleared and stained specimens as small as 30 mm SL. We conclude, as did Fink *et al.* (1984, p. 1037), that as yet there is no evidence of fusion between supraneural ossifications in otophysans, and, so far as we are aware, none is known from other teleosts.

62. Anterior inclination and cranial articulation of enlarged supraneural; Characiphysi. Gayet (1986a) claimed that this feature is not observable in siluriforms. Many siluriforms have lost the supraneural ossification, but in those which have it, it clearly articulates with the neurocranium, as illustrated by Fink and Fink (1981, fig. 17). Arratia (1992, p. 125) stated that only diplomystids among catfish have an ossified supraneural. This may be true for adults, but the largest of our series of *Silurus* (23 days) and specimens of small juvenile ictalurids and a plotosid have a small, perichondrally ossified supraneural (*Pylodictis*, UMMZ 150867; *Plotosus*, UMMZ 155792). That the catfish supraneural, when it occurs, is tilted anteriorly and is homologous with the third, rather than the second, supraneural of cypriniforms is more apparent in juveniles than in adults; see, e.g., the supraneural cartilage in larval *Silurus* and *Ictalurus* (Fig. 4).

63. Roof over neural canal formed by anterior neural arches; Ostariophysi. In the African freshwater gonorynchiforms, the solidity of this roof is much reduced, although the anterior neural arches are still considerably enlarged.

64. Absence of unattached neural arch anterior to that of first vertebra; Ostariophysi. As discussed by Patterson and Johnson (1995, pp. 16–17), an accessory neural arch has a mosaic distribution within elopocephalan teleosts, and assessments of where it has been gained and/or lost remain unclear. Whether or not loss of such an arch can be interpreted as a character of ostariophysans will depend upon parsimony analyses within other lower teleostean groups and where ostariophysans fit among those groups. In 1981, we included *Polypterus* and *Amia* as outgroup taxa showing the unattached neural arch (Fink and Fink, 1981, p. 326), but the accessory neural arches in these non-teleostean taxa are associated with vertebrae fused into the occiput, a rare phenomenon in

teleosts (Patterson and Johnson report such fusion only in *Heterotis* among the taxa they examined).

65. First neural arch especially enlarged and contacting the occipital margin in an extensive, tight joint; Gonorynchiformes. Examination of cleared and stained material of additional gonorynchiforms shows that this character needs clarification and shows homoplasy in *Gonorynchus*. The enlargement of the arch is obvious in *Chanos*. Contrary to Gayet's comment (1993a, p. 169), Rosen and Greenwood (1970, fig. 3) did illustrate this enlargement, although the dorsal part of the arch is labeled supradorsal, an identification we do not follow for reasons described earlier (Fink and Fink, 1981, p. 325). In *Phractolaemus*, *Cromeria*, and *Grasseichthys*, the first neural arch is reduced in the longitudinal axis (e.g., Thys van den Audenaerde, 1961, figs. 12, 13; Grande, 1994, figs. 6, 9) but is still enlarged dorsally, as it is also in *Chanos* (Fink and Fink, 1981, fig. 6).

The first neural arch in *Gonorynchus*, though slightly enlarged, does not contact the occiput in adults, although a short anterodorsal process of the cartilaginous neural arch contacts the exoccipital margin slightly during development (observed in our largest larval specimen, AMNH 55560, 21.5 mm SL). This anterodorsal projection apparently regresses later since we observe neither it nor contact of the arch with the exoccipital in our juvenile *Gonorynchus* specimen (UMMZ 188723, 70 mm SL). But information from fossils indicates that loss of an anterodorsal process and a tight joint in adults may have occurred in *Gonorynchus* since specimens of its sister-taxon, the Eocene gonorynchid †*Notogoneus* (BMNH P.63350 and UMMP V56851), show a large first neural arch bearing an anterodorsal process that forms an extensive joint with the exoccipital. The presence of an anterodorsal process of a similar position and orientation in *Phractolaemus* and the kneriids suggests that such a process is a gonorynchoid feature secondarily lost in *Gonorynchus*. This hypothesis of reduced contact in *Gonorynchus* can be tested further when appropriate material of other fossil gonorynchids (*sensu* Grande, 1996) is studied.

Gayet (1986a, p. 59) argued that the scaphium, the transformed first neural arch of otophysans, seems identical to the enlarged first neural arch of gonorynchiforms, but the scaphium articulates to the centrum via a reduced pedicel, while the base of the first neural arch in gonorynchiforms is generally large and robust. In addition, even if one takes into account the claustrum as part of a subdivided first neural arch in otophysans, the first neural arch of otophysans is smaller than the largest of the other anterior arches, whether the largest is the third, the fourth, or (as in catfishes)

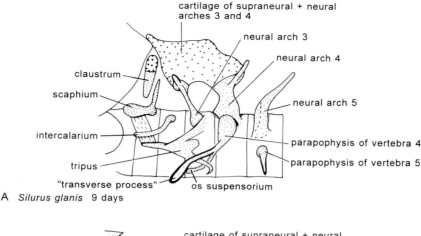

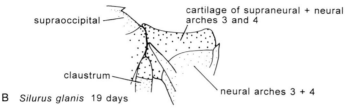

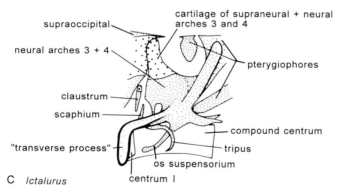

FIGURE 4   Early ontogenetic stages of the Weberian region in two siluriforms. (A) and (B) *Silurus glanis* (UMMZ 223314); (C) *Ictalurus* (uncataloged).

the third and fourth fused together. Gayet also argued that the first neural arch of *Gonorynchus* and that of †*Tharrhias* (Patterson, 1984a, fig. 3) are not enlarged or developed anteriorly. As described immediately above, we think the morphology of that structure in *Gonorynchus* is due to secondary reduction. If †*Tharrhias* is a stem-group gonorynchiform as Patterson (1984a) proposed, one would not expect it to show all the apomorphies of Recent members of the group (the anterior arch appears slightly enlarged longitudinally, but not dorsally, in Patterson, 1984a, fig. 3).

Gayet (1986a, p. 51) also reported a personal communication from Chardon that the character may be present in certain catfishes. The cartilage of the claus-

trum, which in many catfishes remains separate from the cartilage block over the anterior vertebral region, articulates tightly with the back of the neurocranium in many catfishes, but in the specimens we examined it does not extend as far dorsally as the margin of fused neural arches 3 + 4. We regard the condition in some catfishes, in which the claustrum is lost and the anterior portion of arches 3 + 4 articulates tightly with the neurocranium, as nonhomologous with the gonorynchiform condition because it involves a different neural arch.

66. Scaphium and claustrum present; Otophysi. Gayet (1986a, p. 59) referred to this as "transformation of the anterior vertebral elements into a Weberian

apparatus" and grouped it with characters 69 (intercalarium), 85 (tripus), and 88 (os suspensorium) as a single character. This strategy reduces the number of characters accounted for in a parsimony analysis, making it easier to "overthrow" the monophyly of the Otophysi by contradictory characters from fossil material. Gayet based her lumping of the characters on the presumption that all the elements of the Weberian apparatus are constant within the Otophysi. However, treating all these features as one requires the assumption that the Weberian apparatus appeared fully developed at a single step, something we are not prepared to assume. Indeed, Gayet's assumption is that "Weberian apparatus" is something other than a name applied to a complex of modifications, each of which could have occurred at different times. We know of no argument based on developmental biology or distribution of other characters that requires a saltatory origin of these various features. Perhaps an argument could be mustered that all these elements are necessary for the apparatus to function, but the lack of a claustrum in gymnotiforms and the lack of both claustrum and intercalarium in some siluriforms (Chardon, 1968) seem to refute such a claim, and it is denied by Gayet and Chardon's (1987) own attempt to erect a series of stages in the development of the Weberian apparatus, each illustrated by a particular fossil. This issue is also addressed in the Discussion below.

The scaphium of †*Chanoides* is slightly more primitive than that in extant otophysans in bearing foramina for dorsal and ventral nerve roots (Patterson, 1984b, p. 447).

67. Claustrum ossification absent; Gymnotiformes.

68. Anterior extension of scaphium; Characiphysi. Gayet (1986a, p. 59) questioned this feature because it is present in cypriniforms with a strongly foreshortened first centrum and because in some adults of *Opsariichthys* the first two centra are more foreshortened than in the specimen illustrated by Fink and Fink (1981, fig. 14). Both these observations are correct. The proportion of the scaphium extending anterior to centrum 1 needs to be examined in a thorough survey of cypriniform taxa to see if the outgroup condition for characiphysans can be determined and then to see if the degree of anterior extension in characiphysans is greater than in that outgroup condition.

69. Second neural arch reduced and modified into intercalarium; Otophysi. †*Chanoides* shows a more plesiomorphic morphology of the intercalarium than any extant otophysans: it is relatively large, perforated by foramina for dorsal and ventral roots of spinal nerves, and the manubrium fuses to the neural arch portion relatively late (Patterson, 1984b, p. 447).

70. Articular process of intercalarium absent; Siluriphysi.

71. Transverse process of third neural arch; Characiformes. Gayet (1986a, p. 59) accepted this character and claimed that it is present in †*Salminops*. In that unique fossil (Gayet, 1985c, fig. 8) the third neural arch is disarticulated from its centrum. Since both centrum and arch are disturbed, we do not see how the original orientation of the arch can be interpreted unequivocally or why the structure labeled as the transverse process should not be the base of the arch, or "the small prezygapophysis in this position" in other primitive teleosts (Fink and Fink, 1981, p. 329).

72. Anterior margin of third neural arch relatively close to neurocranium; Characiphysi. Fink and Fink (1981, p. 329) and Gayet (1986a, p. 59) noted that this character is associated with the foreshortening of the anterior centra.

73. Vertical anterior margin of dorsal part of third neural arch; Characiphysi. Fink and Fink (1981, p. 329) allowed that this was small or absent in characiforms; Gayet added that it is not present in all catfishes, referring to Chardon (1968). While the reduction of the third neural arch in most characiforms and the modification of the fused neural arches in many siluriforms means that this morphology is not observed in many characiphysans, its presence in the phylogenetically plesiomorph members of both those constituent lineages indicates that it may nevertheless be a characiphysan synapomorphy. The issue of variation or absence of a diagnostic character within subgroups is treated in the Discussion below.

74. Third neural arch with an anteroventral process that articulates with or is fused to a prominence on the second centrum; Siluriphysi. Examination of much more material of early ontogenetic stages of siluriforms indicates that this character diagnoses Gymnotiformes rather than Siluriphysi. The third neural arch does not overlap the second centrum in *Silurus* or in additional material of *Ictalurus* (Fig. 4).

75. Anterodorsal orientation of remnant of third neural spine; Characiphysi. Fink and Fink (1981, p. 330) noted that since this neural spine is absent in cypriniforms, the orientation might be an otophysan character. In the ontogenetic series of cypriniforms we have examined that have the spine, however, the tilt of the spine is dorsal or posterodorsal, never anterior (*Tribolodon hakonensis*, UMMZ 212151, 9.1 mm TL–14.7 mm SL). In series of characiforms, gymnotiforms, and siluriforms (Fig. 4), by contrast, the dorsal spine of the arch is tilted anteriorly. In †*Chanoides* two specimens show the third neural arch (Patterson, 1984b, figs 13, 14), and there appears to be no neural spine component, as in cypriniforms and various

adult characiforms cited by Fink and Fink (1981) and Gayet (1986a). Gayet identified the characiphysan character in †*Salminops*, but as pointed out under number 71 above, the original orientation of the third neural arch and spine in the unique specimen is a matter of interpretation rather than observation.

76. Third and fourth neural arches fused together and to the complex centrum; Siluriformes.

77. Fusion between fifth neural arch and centrum; Characiphysi. Gayet (1986a, p. 60) reported that this character is not universal in characiforms since the fifth neural arch is autogenous in the Oligocene characiform †*Eurocharax* (Gayet, 1985c, fig. 9). To us, the identification of the centra in the single specimen examined of that fish showing the Weberian apparatus seems questionable. If the fish is a characiform, as it appears to be, the tripus should be fused to its centrum, yet both it and a structure interpreted as the third neural arch are illustrated as having drifted well away from the third centrum and as overlying a centrum identified as the first. Gayet also said that the supposed Cretaceous characiform †*Salminops* does not have the fifth neural arch fused to the centrum. Since we do not agree that †*Salminops* is a characiform, we do not think it relevant.

78. Anterior vertebral centra foreshortened; Otophysi. As described by Fink and Fink, this character referred to some foreshortening of the first four centra, relative to more posterior centra. Gayet (1986a) considered this feature to have an irregular distribution. In *Chanos* and *Gonorynchus*, she saw reduction only in the first centrum, and in *Parakneria* and *Kneria*, Gayet reported only slight reduction of the first two centra. She lists several characiforms figured by Roberts (1969) as having significant reduction only in the first centrum. Her accounts of *Chanos*, *Gonorynchus*, *Parakneria*, and *Kneria* are correct, as they have the expected primitive state of non-otophysans. Her description of the character as illustrated by Roberts is incorrect, as can be confirmed by measuring the drawings which include the fifth centrum (the drawings of *Hepsetus* and *Hoplias* include only the anterior four centra, so no comparison with more posterior centra can be made). Gayet reported that in the cypriniforms *Opsariichthys*, *Tinca*, and *Barbus* only the first two centra are foreshortened.

Our examination of numerous skeletons shows that the situation is more complex than thought by Fink and Fink. Within otophysans, there is foreshortening of the anterior centra, especially the first, but the pattern is not always limited to the anterior four centra, and in some cases one or more of the anterior four centra may be larger than other anterior centra. For example, in material of the characiform *Hoplias*

(UMMZ 207341), centrum 3 (C3) is larger than C1, C2, C4, or C5, and C1, C2, and C4 are smaller than C5 or more posterior centra. In *Hoplerythrinus* (UMMZ 207298), C2 is about equal in length to C5; C1, C3 and C4 are smaller than C2; and C1–C5 are smaller than the more posterior centra. In large *Salminus* (UMMZ 207986), C2 is nearly twice as long as C1; C3 is slightly larger than C2; C4 is larger than C3; C5 is longest; C6 is about equal to C4; C7–C9 are smaller than C6; and C10 and remaining centra are slightly larger than C6. In the cypriniform *Opsariichthys* (UMMZ 187604), C1 and C2 are greatly foreshortened; C3 and C4 are larger, about equal in length to centrum 8 and successive centra; while C5–C7 are foreshortened relative to C3 and C4. In the cyprinid *Nocomis* (UMMZ 173092), C1 and C2 are much foreshortened; C3 is shorter than C4, C5 is near the size of C3; C6 is slightly smaller than C4; and C7 and subsequent centra are all longer.

Examination of specimens of outgroup taxa among salmonids (*sensu* Fink, 1984), *Esox*, *Umbra*, primitive stomiiforms (see Fink and Weitzman, 1982), osmeroids, *Kneria*, and *Gonorynchus* failed to find patterns of foreshortening like those in otophysans. In these outgroup taxa, the first and sometimes second centra are foreshortened, but the remaining centra are longer and approximately equal in size.

79. Further foreshortening of the first three centra; Characiphysi. Gayet (1986a) asserted that the character is not verifiable in siluriforms since the second to fourth centra are fused in all known fossils. But Fink and Fink (1981, p. 331) referred to juvenile *Ictalurus*, and we have also examined ontogenetic series of *Silurus*, in which the centra are unfused and the character is verifiable (e.g., Fig. 4A).

80. Centra 2–4 fused into "complex centrum"; Siluriformes.

81. Fusion of first two parapophyses with their centra; Otophysi. Gayet (1986a, p. 60) referred to homoplasy of this character in gonorynchoids and to an autogenous first parapophysis in †*Lusitanichthys* (Gayet, 1985c, fig. 22; the text of that paper, p. 113, says that the parapophyses are also apparently autogenous on the second centrum; this is not repeated in the 1986a paper). We do not regard the condition in †*Lusitanichthys*, if confirmed, as invalidating the otophysan character since we see the fish as *incertae sedis*, as it seems also to Gayet (1985c, 1986a, p. 74). Fusion of parapophyses with all centra is universal in gonorynchoids, so far as we know, and may be a character relating that group and otophysans. In most ontogenetic series of cypriniforms and characiforms, the processes develop as long outgrowths of the centra, apparently without any cartilage precursors such as precede the development of the other parapophy-

ses. In a small series of the cyprinid *Tribolodon*, however, an 11.1 mm SL specimen (UMMZ 212151) has cartilage clearly visible inside the ossified exterior of the process (which is fused with the centrum), indicating homology with a parapophysis.

82. Absence of parapophysis (= lateral process in cypriniforms) on first centrum; Characiphysi. Gayet (1986a, p. 60) repeated the observation of Fink and Fink (1981, p. 335) that in the cypriniforms *Cyprinus*, some catostomids, and some homalopterids the lateral process of the first centrum is absent, thus implying that homoplasy between characiphysans and certain cypriniforms renders the character invalid. Within cypriniforms, Sawada (1982, p. 138) reported the lateral process of the first centrum as general in cobitoids; it is also present in *Gyrinocheilus* (Ramaswami, 1952a, p. 134). Among catostomids, *Cycleptus*, traditionally regarded as a morphologically primitive American genus, has a well-developed lateral process, as does *Myxocyprinus* (Wu *et al.*, 1981, fig. 1a). Although we still lack a robust cladogram of cypriniforms, there is no indication that those few cypriniforms lacking a process on the first centrum are anything but derived, and thus Gayet's implication of homoplasy is irrelevant.

83. Absence of lateral process (= parapophysis) on vertebral centrum 2; Siluriformes.

84. Elongate lateral process of centrum 2 extending well into the body musculature; Cypriniformes. The contrasting state in other otophysans was characterized by Fink and Fink as "short." Revising this to "short, occasionally extending slightly lateral to the tripus" might help correct Gayet's (1986a, p. 60) reference to Robert's (1969, figs. 39, 47) drawings of the characiforms *Acestrorhynchus* and *Hepsetus* as showing that feature. These illustrations of juveniles show the lateralmost extent we were able to find in non-cypriniforms, and in adults of those species the lateral process is proportionally shorter. Some cypriniforms have relatively short lateral processes, which we interpret as a secondary condition (see, e.g., Howes, 1978, fig. 39B,C).

85. Tripus; Otophysi. A disarticulated transformator process of the tripus has been identified by Patterson (1996) in new, much larger material of †*Chanoides*. These slender bones with large articular heads strongly resemble short ribs and corroborate the identity of the transformator process of extant otophysans as a modified pleural rib.

86. Parapophysis portion of tripus attached to centrum by a lamella; Characiphysi. This characteristic appears to be present in †*Chanoides* as well (Patterson, 1984b). Its presence in that taxon renders the identified polarity of the character problematic: If such fu-

sion is primitive for the Otophysi, its absence would diagnose Cypriniformes; if lack of fusion between parapophysis and centrum, as occurs on centrum 5 in non-siluriform otophysans, is primitive for the Otophysi, then the character has appeared independently in †*Chanoides* and the Characiphysi. We know of no evidence that allows us to choose between these alternatives.

87. Transformator processes of the tripus separated posteriorly by the width of the complex centrum; Siluriformes.

88. Os suspensorium; Otophysi. This feature, like the transformator process of the tripus, is present in †*Chanoides*, based on newly prepared material (Patterson, 1996).

89. "Transverse process" of fourth centrum with ovoid, anterolateral face approaching the suspensorium of the pectoral girdle (Fink and Fink, 1981, figs. 17, 18); Siluriphysi.

90. "Transverse process" of the fourth centrum expanded broadly, articulating with the suspensorium of the pectoral girdle (Fink and Fink, 1981, figs. 17, 19C); Siluriformes.

91. "Transverse process" of the fourth centrum fused to complex centrum; Siluriformes.

92. Os suspensorium with an elongate anterior horizontal process closely applied to ventrolateral surface of vertebral centra; Siluriphysi.

93. Os suspensorium consisting of anterior horizontal processes only; Siluriformes.

94. All pleural rib elements projecting from the centra at an angle close to the horizontal (Fink and Fink, 1981, figs. 17, 18); Siluriphysi.

Pectoral Girdle

95. Suspensorium of pectoral girdle consisting of single ossified element comprising the supracleithrum, ossified Baudelot's ligament, and perhaps the posttemporal; Siluriformes. Arratia and Gayet (1995, p. 502) stated that developmental studies are unable to distinguish between the identity of this bone as either supracleithrum or fused posttemporo-supracleithrum. Arratia (1987, pp. 105–106) questioned the identity of the horizontal limb of this element with Baudelot's ligament, stating that it is ossified even early in ontogeny in diplomystids and ictalurids. However, our three specimens of *Pylodictis* (UMMZ 150867, ca. 15 mm SL) show ossification of this compound structure extending only about halfway along the ligament, and our series of *Clarias* and *Silurus* show development of the ligament from completely unossified to ossified along its entire length except where it attaches to the neurocranium. Even in the

largest of the *Silurus* (31.7 mm SL), the ossification occurs *within* the ligament—i.e., ligamentous tissue surrounds the slender ossification within it. Arratia and Gayet (1995) confirmed the hypothesis of Fink and Fink (1981) that the small plate-like bone at the posterodorsal margin of the neurocranium in siluriforms is the extrascapular.

96. Reduction of number of postcleithra; to one (Cypriniformes) or none (Gonorynchiformes and Siluriphysi). Gayet (1986a, p. 61) remarked that gonorynchiforms always must have had a reduced number of postcleithra, citing Taverne (1981, p. 967) as stating that early Cretaceous †*Aethalionopsis* lacks them. However, Taverne wrote "On ne trouve pas de postcleithrum", a statement that could imply absence or failure of preservation, not "l'absence de postcleithrum" as misquoted by Gayet (1993b, p. 101). Gayet (1993b, p. 101) similarly interpreted no mention of postcleithra in the description of †*Dastilbe elongatus* by Silva Santos (1947) to mean absence. Taverne wrote (1981, p. 973) that †*Aethalionopsis* is "presqu'identique" to †*Tharrhias*, in which there are two postcleithra (Patterson, 1984a, p. 135), structures that would be difficult to observe except in acid-prepared material (not available for †*Aethalionopsis* or †*Dastilbe*). Gayet (1986a, p. 61) reported three postcleithra in †*Lusitanichthys* and one, a broad rod, in †*Ramallichthys* and †*Judeichthys*. We have been unable to determine whether or not †*Notogoneus* has any postcleithra, but Gayet reported that the other fossil gonorynchiforms she examined (1993b) lack them. Poyato-Ariza (1994, p. 20) reports at least two in †*Gordichthys*.

Among these various fossil taxa that have been included in the Gonorynchiformes, however, only the gonorynchids are members of the crown-group Gonorynchiformes as recognized here, and among those only †*Ramallichthys* and †*Judeichthys* are reported to have a postcleithrum. If they do indeed have such a bone, either the element has reappeared in them, or absence of postcleithra has evolved more than once within crown-group gonorynchiforms.

97. Attachment of Baudelot's ligament to cranium; Characiphysi. Based on homoplasy noted by Fink and Fink in some cypriniforms lacking the first parapophysis, Gayet (1986a) suggested that cranial attachment is related to loss of the first parapophysis and is not a characiphysan synapomorphy. Although attachment to the cranium may occur without absence of the first parapophysis, as in *Gastromyzon borneensis*, the correlation in at least five independent cases [characiphysans, *Cyprinus*, a species of *Nemacheilus*, a species of *Homaloptera*, and some catostomids (Fink and Fink, 1981, pp. 335, 352)] between loss of the first parapophysis and attachment of the ligament to the cra-

nium indicates that indeed cranial attachment should not be construed as evidence independent of loss of the parapophysis. However, this is not a feature with rampant homoplasy. Patterson and Johnson (1995) found Baudelot's ligament, when present (absent in Gonorynchiformes; see Review of Major Ostariophysan Lineages, above), attached to the first centrum (or first plus one or more posterior centra) in all non-acanthomorph taxa they sampled except for the osteoglossomorphs *Heterotis* and *Osteoglossum*, the cobitid *Acanthopsoides*, and *Galaxias*. In these taxa the ligament has a double attachment to both basioccipital and the first centrum.

98. Baudelot's ligament robust and bifurcated distally; Siluriphysi. This feature was questioned by Arratia (1987, p. 106), who considered the large ossified extension of the posttemporal–supracleithrum to the skull base as a neomorph in catfishes. Apparently that decision was based in part on her not finding specimens which displayed the ontogeny of the structure. We have examined ontogenetic series of small catfishes and confirm that the structure in question is Baudelot's ligament. Specimens of *Clarias* (UMMZ 223314) examined of up to 16.6 mm show the entire ligament as unossified from its connection at the posterior skull base to its junction with the posttemporal–supracleithrum. A series of *Silurus* (UMMZ 213368) shows the ligament unossified in individuals up to about 20 mm and partially ossified in specimens of about 32 mm. Further, on methodological grounds, we see no reason to postulate evolution of a new structure with the exact locality and spatial extent of an old one, inasmuch as one then is encumbered by the need for further explanation of what happened to the old structure.

99. Anterior and posterior parts of bifurcate Baudelot's ligament attached to cleithrum; Gymnotiformes.

Pectoral Fins

100. More posterior pectoral-fin rays offset posteriorly from anterior ray; Siluriformes.

101. Flanges on dorsal and ventral halves of pectoral-fin rays about equal in size; Siluriphysi.

Pelvic Girdle and Fins

102. Bifurcated pelvic bone; Otophysi. This feature is present in cypriniforms, siluriforms, and some characiforms (gymnotiforms have no pelvic girdle). Gayet (1986a) followed Roberts (1973, p. 386) in assuming these characiforms are specialized, but evidence presented by Fink and Fink (1981, p. 306) indicates most of them (the Distichodontidae) form part of the primi-

tive sister-group to all other characiforms. Additional homoplasy overlooked by Fink and Fink does occur in this feature, however; citharinids do not have a bifurcate pelvic bone (*contra* Vari, 1979, p. 311). This information results in this character being ambiguous: a bifurcate bone could be an otophysan feature lost independently in citharinids and in other, non-distichodontid characiforms; or, a bifurcate bone could have evolved independently in cypriniforms, siluriforms, and distichodontids (in either case the gymnotiform condition does not affect the accounting). A characiform thought by Roberts to be primitive, but by Fink and Fink (1981), Buckup (1993), and Vari (1995) to be specialized, also shows bifurcation early in ontogeny (*Hepsetus*; Fink and Fink, 1981, p. 335), but the bone in a series of 25 specimens of *Hyphessobrycon callistus* (UMMZ 211676) shows no bifurcation.

Gayet also stated that *Alosa* has a bifurcate pelvic girdle. *Alosa pseudoharengus* specimens examined by us had a single anterior process. Presence of a bifurcate girdle in some *Alosa* species would not affect our conclusion, however, a point taken up in our discussion of homoplasy below.

103. Pelvic girdle and fin absent; Gymnotiformes.

Dorsal and Anal Fins and Fin Supports

104. Dorsal fin absent; Gymnotiformes.

105. Anal fin extending anteriorly to pectoral fin region; Gymnotiformes.

106. Middle radial ossification absent from all dorsal and anal fin pterygiophores; Siluriphysi.

107. Anal-fin rays articulating directly with proximal radials, and distal radials reduced; Gymnotiformes.

Caudal Fin and Fin Supports

108. Principal caudal fin–ray count 9/9 or fewer; Siluriphysi. Among gymnotiforms, this character can be observed only in apteronotids, since the caudal fin is absent in all other gymnotiforms.

109. Caudal support skeleton consolidated into a single element, and caudal fin greatly reduced in size or absent; Gymnotiformes.

110. Compound terminal vertebral centrum; Otophysi. Gayet (1986a) questioned why Fink and Fink considered this to be a synapomorphy of otophysans, given the homoplasy they documented in the feature (fusion in two lineages of gonorynchiforms; fusion in otophysans, with subsequent modification in many otophysans). As discussed below, homoplastic synapomorphies are evidence that must be accounted for.

In this case the evidence supports the Otophysi and one or more groups within gonorynchiforms.

111. Hemal spines of PU3 and anterior vertebrae fused with their respective centra; Ostariophysi. Gayet (1986a) maintained that the hemal spines of †*Ramallichthys* and †*Judeichthys* are autogenous in that region, and her reconstructions show them to be so, if one interprets the solid lines as joint lines. However, the structures she identified as hemapophyses and posterior zygapophyses in †*Ramallichthys* and †*Charitosomus formosus* (1986a, figs. 34–35; 1993b, fig. 20) are also separated by a solid line from the vertebral centrum, though use of these terms presumes that the structures are fused with the centrum. The observations of one of us (S.V.F.) on the holotype of †*Ramallichthys* do not support the interpretation of autogenous hemal spines, however, and T. Grande (personal communication) states they are fused in both †*Ramallichthys* and †*Judeichthys*.

Gayet (1986a) also noted that †*Lusitanichthys* has the elements unfused, in agreement with her original description (Gayet, 1981) and contradicting her subsequent reconstruction (1985c, fig. 26) of the fossil as a characiform. Her reinterpretation provides evidence placing the fossils outside the Ostariophysi as currently delimited.

†*Salminops* (Gayet, 1985c) was described as having the "last 3 hemal spines fused" and we take that to mean, at least, that HPU3 is fused with its centrum. The hemal spines of the fourth–sixth preural centra were stated to be unobservable, and that of the seventh is autogenous. The photograph of the caudal skeleton (Gayet, 1985c, fig. 3) shows the hemal spine of PU3 displaced posteriorly from where it is reconstructed to be, and one interpretation is that the bone was autogenous and moved posteriorly during decomposition. The other hemal elements are not interpretable from the photograph. In any case, the autogenous condition of PU7 provides evidence that the fossil lies outside the Ostariophysi.

112. Fusion of hemal spine of PU2, parhypural, and hypural 1 to centra; Characiphysi. This character was incorrectly stated in Fink and Fink (1981); the error was pointed out by M. Gayet (personal communication) but was again misstated by Fink *et al.* (1984). We hope it is finally stated correctly, as follows: In characiphysans, the hemal spine of PU2 fuses with its centrum, and the parhypural and hypural 1 fuse with the compound terminal centrum, as illustrated in Fink and Fink (1981, fig. 23). In characiforms, the condition of hypural 1 is transitory during ontogeny. Presence of the fusion early in ontogeny was noted for several characiforms by Fink and Fink, and our series of *Hyphessobrycon callistus* (UMMZ 211676) and

Serrasalmus (UMMZ 211677) also show this. Description and figures for a number of additional characiforms are presented in Oldani (1977, p. 138; 1979a, p. 55; 1979b, p. 67; 1983, p. 99).

113. Hypural 1 separated from compound centrum by hiatus in adults; Characiformes. Gayet (1986a) notes that there is homoplasy in this feature, as pointed out by Fink and Fink. The feature is found in the gonorynchiforms *Kneria*, *Parakneria*, and *Phractolaemus* and is apparently variable in *Cromeria* (Monod, 1968, fig. 267 shows a small hiatus; Rosen and Greenwood, 1970, fig. 11B do not) and *Grasseichthys* (Monod, 1968, figs. 268–271; Géry, 1965, fig. 7). In †*Chanoides*, there is a small hiatus (Patterson, 1984b, fig. 17).

In *Chanos*, H1 articulates proximally with the parhypural (Fink and Fink, 1981, fig. 23A; Rosen and Greenwood, 1970, fig. 9B). In the fossils, Gayet (1986a) stated that H1 articulates autogenously with the terminal centrum in †*Lusitanichthys* (evidence against chaphysan relationships), and in †*Salminops* it is "disconnected." The photograph of †*Salminops* (Gayet, 1985c, fig. 3) shows H1 to be somewhat displaced posterodorsally.

114. Hypural 2 fused with compound centrum; Otophysi. Again, Gayet's (1986a) comments are about homoplasy, already noted by Fink and Fink (1981). The character remains an otophysan synapomorphy and probably also a synapomorphy somewhere within gonorynchoids.

115. Two or fewer epurals; Gonorynchiformes, Cypriniformes, and Siluriphysi. This feature was considered highly homoplastic by Fink and Fink, yet it appears to have value in diagnosing groups within gonorynchiforms, for example. We are not sure why Gayet (1986a) mentioned the caudal skeleton of *Barbus* illustrated by Rosen and Greenwood (1970) as an exception to the feature except perhaps to show that some *Barbus* have two epurals and some have one.

Fin Spines

116. Dorsal- and pectoral-fin spines; Siluriformes. Absence of fin spines, whether dorsal or pectoral, appears to be secondary wherever it occurs among known siluriform taxa.

Miscellaneous

117. Alarm substance; Ostariophysi. Gayet (1986a) specifically noted that gymnotiforms lack alarm substance. In fact, both the alarm substance and alarm reaction are missing in that group, where communication is based on electrical signals (Hopkins, 1983).

Fink and Fink pointed this out with additional information and cited a comprehensive review on the subject. For example, many predaceous ostariophysans have alarm substance but lack the reaction. Gayet seemed to include this character, which could not be assessed in the fossils, to point out another case of homoplasy. Arratia and Huaquín (1995) summarized the distribution of alarm substance and fright reaction in catfishes.

118. Keratinized nuptial tubercles; Ostariophysi. Gayet (1986a) listed the distribution of these features as described by Fink and Fink—some gonorynchiforms, many cypriniforms, a few characiforms, a few siluriforms (in which they are not "nuptial"), and no gymnotiforms. Gayet apparently viewed as contradictory the conclusion that keratinized tubercles are an ostariophysan synapomorphy, though they are present chiefly in gonorynchiforms and cypriniforms. Multicellular tubercles with a keratinized cap, in which the cells have pyknotic nuclei or no nuclei, occur only in ostariophysans: kneriid and phractolaemid gonorynchiforms, all families of cypriniforms, and parodontid characiforms (tubercles with no or slight keratinization, and no cap, also occur in some lebiasinids, a few characids, and at least some mochokid catfishes) (Wiley and Collette, 1970; Collette, 1977). Outside the ostariophysans, keratinized tubercles are known only in some salmonoids, osmeroids, and percids; and usually in these groups only the surface cells have the keratin; tubercles in the salmonoid *Prosopium* have a thicker keratinized layer but the nuclei are not obliterated (Wiley and Collette, 1970).

More widely distributed in ostariophysans than keratinous multicellular tubercles, and apparently unique to them, are unicellular keratinous structures ("unculi" of Roberts, 1982); see Review of Major Ostariophysan Lineages.

119. Electroreception; Siluriphysi. Fink and Fink (1981) greatly simplified the description of this extremely complex character, and much has been done on siluriphysan electroreception biology in the interim; see Review of Major Ostariophysan Lineages.

120. Anterior lateral line nerve with recurrent branch innervating electroreceptors of the trunk; Siluriphysi. This feature was modified by Fink *et al.* (1984) and is summarized and updated in Review of Major Ostariophysan Lineages.

121. Electrogenesis; Gymnotiformes. An extensive literature on electrogenesis in gymnotiforms exists (Bass, 1986). Recent studies with catfishes have shown that several synodids (Hagedorn *et al.*, 1990) and *Clarias* (Baron *et al.*, 1994) are also electrogenic. In synodontids, at least, the electric organ is probably modified gasbladder sonic muscle.

122. Anus at least as far anterior as pectoral-fin insertion; Gymnotiformes. Gayet *et al.* (1994) argued that this feature and elongation of the anal fin are not independent characters. However, the anal fin does not extend forward to the anus in most gymnotiforms, indicating that other factors are involved in location of the anus.

123. Olfactory tracts elongate and lobes "near nasal rosette"; Otophysi. Based on examination of more characiform material, this cannot be considered an unambiguous otophysan feature. In most teleosts, including gonorynchiforms, the olfactory bulb is adjacent to the telencephalon and apart from the lateral ethmoid so that the olfactory tracts from the telencephalon to the olfactory bulb are short, and the olfactory nerve from the bulb to the foramen in the lateral ethmoid is relatively long. In cypriniforms and siluriforms, the bulb is situated near the lateral ethmoid and apart from the telencephalon so that the tracts from the telencephalon to the bulbs are relatively long, and the nerve to the nasal rosette is short. In gymnotiforms, the bulbs are adjacent to the telencephalon and apart from the lateral ethmoids. Characiforms are variable: in citharinids and distichodontids, which appear to form the sister-group to all other characiforms (Fink and Fink, 1981; Buckup, 1991), the condition resembles that in cypriniforms and siluriforms; in most other characiforms the condition resembles that in gymnotiforms. The key taxa are the parodontids and the four-family group comprising Anostomidae, Chilodontidae, Curimatidae, and Prochilodontidae; based on data surveyed by Buckup (1991), either the parodontids or the two lineages together comprise the sister-group to the remaining characiforms. In parodontids, the olfactory bulb is greatly enlarged (nearly as long [5.5 mm] as the optic lobe [6.0 mm] in a specimen of *Parodon pongoense*, CAS 53733, though not as broad). As a result, the bulb is both adjacent to the telencephalon *and* adjacent to the lateral ethmoid so that parodontids cannot be placed decisively in either category. [We cannot corroborate Vari's report (1979, p. 323) that parodontids have "a forward location of the bulb."] Among specimens examined of these four families, only a prochilodontid had the bulb adjacent to the lateral ethmoid rather than the telencephalon (*Prochilodus nigricans*, CAS 59303); *Curimata boulengeri* (CAS 69563), *Leporinus fasciatus* (CAS 11628 [IUM 12183]), and *Chilodus punctatus* (CAS 59141) all had the bulb adjacent to the telencephalon rather than to the lateral ethmoid. If an additional basal characiform clade (e.g., parodontids) had the bulb anteriorly situated, the simplest accounting would place the forward position as an otophysan feature, with reversals to the outgroup condition in gymnotiforms and in a sub-group of characiforms (three changes), as opposed to a forward position having evolved independently in cypriniforms, siluriforms, the citharinid plus distichodontid clade, and parodontids (four changes). The ambiguous condition in parodontids renders the entire character ambiguous, however, with five changes required whether the anterior position is considered an otophysan feature or an independently evolved feature in each lineage. Vari (1979, p. 323) stated some elongation in olfactory tracts was present also in the characiforms *Salminus* and in a subgroup of African characids; these instances would add two steps to either hypothesis of transformation.

124. Scales absent; Siluriformes. On the basis of morphological differences from scales and taxonomic distribution, the bony plates present in some loricarioids are inferred to be derived within catfishes.

125. No adipose fin; Gonorynchiformes, Cypriniformes, and Gymnotiformes. The fin is widely distributed in basal euteleosts, characiforms, and siluriforms.

In papers on the fossil gymnotiform †*Ellisella* (Gayet and Meunier, 1991; Gayet *et al.*, 1994), a region along the dorsal part of the body in one of the specimens was interpreted as an adipose fin. However, Fink *et al.* (1997) present an argument that the adipose fin reported in †*Ellisella* is a taphonomic artifact, resulting in part from compressing the body outline of a three-dimensional animal into two dimensions. An additional issue is the proposed homology between the dorsal filament of apteronotids and the adipose fin of other otophysans. This filament, a cutaneous connective-tissue structure along a posterior portion of the body, could conceivably represent the remnant of an adipose fin since it resembles in structure and appearance the base of the adipose fin in other taxa. More detailed comparative information on structure and development of the apteronotid filament and of the adipose fin in other taxa would be desirable to address this question. If the filament does represent a reduced adipose fin, then reduction would characterize gymnotiforms, and absence would characterize non-apteronotid gymnotiforms. However, on Albert's cladogram (1997), the dorsal filament is interpreted as a new feature.

The issue of adipose fin absence in gonorynchiforms and cypriniforms is raised in Gayet's extended treatment of †*Ramallichthys* (1986a). In contrast to her usual position on the reduced value of homoplastic characters, she considered this loss to be a feature of gonorynchiforms and/or cypriniforms (presumably either two independent synapomorphies or a synapomorphy uniting the two groups). Gayet also stated with confidence that both †*Ramallichthys* and †*Jude-*

ichthys lack an adipose fin (though she was unable to tell for †*Lusitanichthys* and †*Salminops*). Our only comment is that the photographs of the former two fossils contain no apparent soft tissues, and we cannot see how the presence or absence of an adipose fin could ever be ascertained. Grande (1996) drew the same conclusion, doubting whether soft tissues and scales would be preserved at least in the matrix around the single specimen of †*Judeichthys*. Our expectation, however, since †*Ramallichthys* and †*Judeichthys* are gonorynchids, is that they indeed lacked an adipose fin.

126. Sinus impar; Otophysi. Although not visible in any of her fossils, Gayet (1986a) comments on this feature to claim that, because the sinus impar functions as a part of the Weberian apparatus, it is not an independent synapomorphy. In a subsequent paper (Gayet, 1986b, p. 406), she went further, claiming that various aspects of the Weberian apparatus that support the clade Otophysi actually amount to a single character: "the primitive presence of a Weberian apparatus identical to that of the living forms, hence only one character." We deal with such arguments below in the Discussion.

127. *Adductor mandibulae* with superficial ventral division; Ostariophysi. Gayet (1986a) inferred that †*Ramallichthys* had such a muscle based on an insertion scar. Gosline (1989) has recently commented on the morphology of this muscle. Howes posed three questions on this feature: first, is it an ostariophysan character (Howes, 1985b, p. 276); second, is insertion of the muscle on the maxilla the primitive form of the muscle in ostariophysans (Howes, 1985b, p. 277); and third, what is the primitive state of the feature in siluriphysans (Howes, 1983a, p. 15)? All of these questions can be answered with a parsimony argument. First, a superficial division of the adductor occurs in ostariophysans among basal euteleosts and in eurypterygians (Fink and Weitzman, 1982), indicating that the muscle has evolved twice. In addition, there is a positional argument suggesting that the superficial division in eurypterygians is different from that of ostariophysans: In ostariophysans the muscle is ventrolateral, and in eurypterygians it is dorsolateral. We chose (Fink and Fink, 1981) to term the muscle in ostariophysans the superficial ventral division of the adductor to differentiate it from the A_1 of eurypterygians. On the second matter, as Howes (1985b) noted, the lateral division attaches to the maxilla in gonorynchiforms and cypriniforms so that on our cladogram, maxillary insertion is the ostariophysan feature. The condition of the muscle in characiphysans is more complicated, and the diagnostic state depends on the state in primitive characiformes and siluriphysans. In most distichodontids, such as *Xenocharax*, the lateral

division attaches to the maxilla (in some derived members of that group, it attaches to the mandible; Vari, 1979). In some genera thought to be primitive by Buckup (1993), such as *Parodon* (UMMZ 208048), the muscle inserts on the maxilla, whereas in others, such as *Chilodus* (UMMZ 204210), the muscle attaches to the mandible and perhaps somewhat on the primordial ligament. In characiforms thought traditionally to be primitive, but considered by Buckup to be phylogenetically derived, such as *Hoplias* (UMMZ 205987) and *Lebiasina* (UMMZ 185314), the muscle inserts on the mandible or, in *Hepsetus* (UMMZ 189121), on the mandible and the primordial ligament. Third, the question of the primitive attachment in siluriphysans also depends upon a parsimony argument, but here the distribution is also uncertain. In gymnotiforms the adductor musculature has a variety of patterns (Aguilera, 1986), sometimes without clear subdivisions (*Gymnotus* and *Electrophorus*), sometimes with a ventral superficial division (the apteronotids *Sternarchorhamphus* and *Sternarchorhynchus*), sometimes with ventral and dorsal parts of the superficial division (an apteronotid identified as *Apteronotus bonapartii*), and sometimes with a dorsal superficial division to the maxilla (e.g., the hypopomid *Steatogenys*), the lacrimal (e.g., *Sternopygus*), or both the maxilla and the mandible (the apteronotid *Adontosternarchus*). Which one of these patterns is primitive for gymnotiforms cannot be determined with available information. In primitive catfishes (e.g., *Diplomystes*) described by Howes (1983a, p. 12), the adductor muscle is not well-differentiated and inserts entirely or almost entirely on the mandible. New information about both gymnotiform and catfish morphology has shown the situation to be more complicated than originally stated by Fink and Fink (1981), and their suggestion that a mandibular insertion was shared by these two groups appears not to be true. The general condition in catfishes, undifferentiated adductor musculature with mandibular insertion, is similar to the outgroup state but emerges as secondary in the context of the rest of the data presented here.

VI. Discussion

In some of the literature we have reviewed, there are statements that some particular feature cannot be diagnostic of a clade because it occurs in other taxa. A variation of this is that a character modified from its primitive state or reversed within a group cannot be used as a synapomorphy of that group. We reject the argument that homoplastic features cannot be synapomorphous, i.e., cannot constitute evidence concerning relationships. Evidence cannot simply be discarded. Which features show homoplasy can be discovered only in the context of a cladogram, and then one finds that the homoplastic feature is two or more features, each of which diagnoses a clade. That some features may be so homoplastic as to be of little apparent value in bolstering a phylogenetic hypothesis may be true, but this is chiefly important when we evaluate alternative hypotheses. In the context of the arguments under discussion, the accusations of homoplasy are aimed at discrediting features not cladograms. In response to the second argument, we would maintain that characters which transform within a clade are no less valid as evidence than those that remain stable during the evolution of a group. To use a well-worn but still useful example, tetrapod limbs are not invalidated as diagnostic of tetrapods when we find a tetrapod without limbs or one with wings.

Homoplastic features may cause difficulty in reconstructing character evolution when two or more alternative interpretations of the evolution of a feature are equally parsimonious on a given tree. There are instances of such features in the characters discussed above, and the reader may wish to carefully consider their distributions. We have not always described the effects of alternative optimizations.

Independence of features is an assumption in phylogenetic analysis, and such independence has been questioned by Gayet regarding the Weberian apparatus. Gayet (1985a, 1986b) argued that our support for Otophysi comes down to one character, a Weberian apparatus, and that this character may well have evolved more than once. We do indeed argue, on the basis of parsimony (Fink et al., 1984), that most of the complex of features known as the Weberian apparatus evolved only once, but we certainly would not argue that it evolved *all* at once. Indeed, as we note at various points in the discussion of characters, †*Chanoides* has more plesiomorphic characteristics of the Weberian apparatus in some respects than any extant otophysan. Examples include the shape of the intercalarium and its separate manubrium and the form of the newly discovered transformator process of the tripus, more clearly rib-like than that in any extant otophysan (Patterson, 1996). And of course some aspects of morphology that are usually thought of as part of the Weberian apparatus (e.g., neural arches forming a roof over the neural canal, attachment of the gasbladder dorsally by strong connective tissues, a silvery tunic, and attachment of anterior ribs to the tunic), are hypothesized to have evolved before the otophysan features connected with sound transduction.

The value of fossils has been a point of contention in recent papers on ostariophysan relationships. We

think that fossil organisms should be treated as Recent ones in phylogenetic analysis and that they should be considered in the context of living members of their groups. In earlier papers, we (Fink et al., 1984) as well as Patterson (1984b) decried the common expectation that fossils would cause Recent groups to be rendered polyphyletic and the features that diagnose them to be homoplastic. Gayet's body of work on these fishes is a typical example of such expectations. Gayet seems to have abandoned some of her more radical theories about the evolution of the group, although she still appears to believe that the Weberian apparatus has evolved several times (especially Gayet, 1986c, 1993a; Gayet and Chardon, 1987). Much of the argument about these matters centers on the fossil taxon †*Ramallichthys*, which has variously been used to question gonorynchiform, cypriniform, and, most recently, ostariophysan monophyly (Gayet, 1993a). However, Gayet may be coming toward agreement with us and Grande (1996) that †*Ramallichthys* is a gonorynchoid (Gayet, 1993a, figs. 1, 10; see our discussion under gonorynchiformes). But more importantly, with Grande's (1996) work we find that, after more than a decade of argumentation and after a dozen publications by Gayet that include fanciful reconstructions with numerous errors of morphology and in which are proposed improbable biogeographic scenarios, massive morphological convergence, multiple origins of the Weberian apparatus based on "intermediate" fossils, and multiple origins of ostariophysan higher taxa, the relevant fossils (†*Charitosomus hakelensis*, †*Judeichthys*, and †*Ramallichthys*) may in fact be members of the same species, or at the least, members of the same gonorynchid genus.

Much progress has been made on ostariophysan interrelationships since the first *Interrelationships of Fishes*, and the pace of work is accelerating as phylogenetic principles are applied to more of the included groups. The next volume with this title will include citations to many works with detailed phylogenetic hypotheses. As these efforts proceed, the time for a more complete synthesis than we have attempted will be due, biogeographic patterns will be clearer, and more thorough analyses of character evolution based on more comprehensive sampling of ostariophysan subgroups will be possible.

VII. Summary

Interrelationships of the major clades of ostariophysan teleosts are examined. We summarize work on this group subsequent to our 1981 paper, review the data we previously presented in the light of new evidence, and comment on the relationships of certain extinct forms. The Gonorynchiformes is monophyletic and is the sister-group to the remaining Ostariophysi, termed the Otophysi. Cypriniformes is the sister-group to other otophysans; Characiformes is the sister-group of the catfishes plus Neotropical electric fishes. A new term is proposed for the latter group, Siluriphysi, comprising the Siluriformes and Gymnotiformes. Placement of †*Chanoides* as the sister-group to extant otophysans is supported. Evidence indicates that the extinct taxon †*Ramallichthys* is a member of the Gonorynchoidei.

Acknowledgments

A number of people have provided invaluable assistance toward the completion of this study. Mario de Pinna, Richard Vari, and Colin Patterson provided very useful reviews of the manuscript. James Albert shared his expertise on gymnotiforms and siluriphysan electrophysiology and neuroanatomy; Paulo Buckup, Tom Di-Benedetto, and Stanley Weitzman provided information on characiforms; Lee Fuiman provided us with some ontogenetic material of otophysan taxa and was instrumental in helping us to obtain additional material, including the *Silurus* series kindly provided by a hatchery in Ahrensburg, Germany. Frank Kirschbaum very kindly provided specimens of early ontogenetic stages of two gymnotiform species. Terry Grande made available both cleared and stained and fossil gonorynchiform material, including the holotype of †*Ramallichthys*, and Lance Grande provided facilities for examination of that material at the Field Museum in Chicago. Gloria Arratia, Stanley Blum, Mario de Pinna, Carl Ferraris, John Maisey, Scott Schaefer, Darrell Siebert, and Jerry Smith have at various times provided helpful information on specimens and discussions of characters. Will Fink aided in some of the data gathering. Last, but capping the list, without Colin Patterson's prompting us to respond to criticisms of our data back in 1988, the paper would not even have been initiated. At one time he was a coauthor, and much of the new data and analysis of gonorynchoid interrelationships is his work as much as ours.

References

Aguilera, O. (1986). The striated muscles in the gymnotiformes (Teleostei-Ostariophysi): Facial musculature. *Acta Biol. Venez.* **12**(2), 13–23.

Albert, J. S. (1996). A phylogenetic analysis of the American knife-fishes (Teleostei: Gymnotiformes). *Misc. Publ. Mus. Zool., Univ. Mich.* (submitted).

Albert, J. S., and Fink, W. L. (1996). *Sternopygus xingu*, a new species of electric fish from Brazil (Teleostei: Gymnotoidei), with comments on the phylogenetic position of *Sternopygus*. *Copeia*, pp. 85–102.

Alexander, R. McN. (1965). Structure and function in the catfish. *J. Zool.* **148**, 88–152.

Alves-Gomes, J. A., Ortí, G., Haygood, M., Heiligenberg, W., and Meyer, A. (1995). Phylogenetic analysis of the South American electric fishes (Order Gymnotiformes) and the evolution of their electrogenic system: A synthesis based on morphology, electrophysiology, and mitochondrial sequence data. *Mol. Biol. Evol.* **12**, 298–318.

Arratia, G. (1987). Description of the primitive family Diplomystidae (Siluriformes, Teleostei, Pisces): Morphology, taxonomy and phylogenetic implications. *Bonner Zool. Monogr.* **24**, 1–120.

Arratia, G. (1990). Development and diversity of the suspensorium of the trichomycterids and comparison with loricarioids (Teleostei: Siluriformes). *J. Morphol.* **205**, 193–218.

Arratia, G. (1992). Development and variation of the suspensorium of primitive catfishes (Teleostei: Ostariophysi) and their phylogenetic relationships. *Bonner Zool. Monogr.* **32**, 1–149.

Arratia, G., and Gayet, M. (1995). Sensory canals and related bones of Tertiary siluriform crania from Bolivia and North America and comparison with Recent forms. *J. Vertebr. Paleontol.* **15**, 482–505.

Arratia, G., and Huaquín, L. (1995). Morphology of the lateral line system and of the skin of diplomystid and certain primitive loricarioid catfishes and systematic and ecological considerations. *Bonner Zool. Monogr.* **36**, 1–110.

Arratia, G., and Menu-Marque, S. (1981). Revision of the freshwater catfishes of the genus *Hatcheria* (Siluriformes, Trichomycteridae) with commentaries on ecology and biogeography. *Zool. Anz.*, Jena **207**, 88–111.

Arratia, G., and Menu Marque, S. (1984). New catfishes of the genus *Trichomycterus* from the high Andes of South America (Pisces, Siluriformes) with remarks on distribution and ecology. *Zool. Jahrb., Abt. Syst. (Oekol.), Geogr. Biol.* **111**, 493–520.

Arratia, G., and Schultze, H.-P. (1990). The urohyal: Development and homology within osteichthyans. *J. Morphol.* **203**, 247–282.

Arratia, G., and Schultze, H.-P. (1991). Palatoquadrate and its ossifications: Development and homology within osteichthyans. *J. Morphol.* **208**, 1–81.

Arratia, G., Chang, A., Menu-Marque, S., and Rojas, G. (1978). About *Bullockia* gen. nov., *Trichomycterus mendozensis* n. sp. and revision of the family Trichomycteridae (Pisces, Siluriformes). *Stud. Neotrop. Fauna Environ.* **13**, 157–194.

Azpelicueta, M. (1994). Three East-Andean species of *Diplomystes* (Siluriformes: Diplomystidae). *Ichthyol. Explor. Freshwaters* **5**, 223–240.

Bamford, T. W. (1948). Cranial development of *Galeichthys felis*. *Proc. Zool. Soc. London* **118**, 364–391.

Baron, V. D., Orlov, A. A., and Golubtsov, A. S. (1994). African *Clarias* catfish elicits long-lasting weak electric pulses. *Experientia* **50**, 644–647.

Bass, A. (1982). Evolution of vestibulolateralis lobe of the cerebellum in electroreceptive and nonelectroreceptive teleosts. *J. Morphol.* **174**, 335–348.

Bass, A. (1986). Electric organs revisited. *In* "Electroreception" (T. H. Bullock and W. Heiligenberg, eds.), pp. 13–70. Wiley (Interscience), New York.

Bertin, L. (1958). Appareil digestif. *In* "Traité de Zoologie" (P-P. Grassé, ed.), pp. 1248–1302. Masson, Paris.

Blum, S. (1991a). *Dastilbe* Jordan, 1910. *In* "Santana Fossils: An Illustrated Atlas" (J. Maisey, ed.), pp. 274–285. TFH Publications, Neptune, NJ.

Blum, S. (1991b). *Tharrhias* Jordan and Branner, 1908. *In* "Santana Fossils: An Illustrated Atlas" (J. Maisey, ed.), pp. 286–296. TFH Publications, Neptune City, NJ.

Bornbusch, A. H. (1991). Monophyly of the catfish family Siluridae (Teleostei: Siluriformes), with a critique of previous hypotheses of the family's relationships. *Zool. J. Linn. Soc.* **101**, 105–120.

Bridge, T. W., and Haddon, A. C. (1893). The air-bladder and the Weberian ossicles in siluroid fishes. *Philos. Trans. R. Soc. London* **184**, 65–333.

Buckup, P. A. (1991). The Characidiinae: A phylogenetic study of the South American darters and their relationships with other characiform fishes. Unpublished Ph.D. Dissertation, University of Michigan, Ann Arbor.

Buckup, P. A. (1993). The monophyly of the Characidiinae, a Neotropical group of characiform fishes (Teleostei: Ostariophysi). *Zool. J. Linn. Soc.* **108**, 225–245.

Carr, C. E., and Maler, L. (1986). Electroreception in gymnotiform fish: Central anatomy and physiology. In "Electroreception" (T. H. Bullock and W. Heiligenberg, eds.), pp. 319–374. Wiley (Interscience), New York.

Carr, C. E., Maler, L., Heiligenberg, W. H., and Sas, E. (1981). Laminar organization of afferent and efferent systems of the torus semicircularis of gymnotiform fish: Morphological substrates for parallel processing in the electrosensory system. *J. Comp. Neurol.* **203**, 649–670.

Carr, C. E., Maler, L., and Sas, E. (1982). Peripheral organization and central projections of the electrosensory nerves in gymnotid fishes. *J. Comp. Neurol.* **211**, 139–153.

Cavender, T., and Coburn, M. (1992). Phylogenetic relationships of North American Cyprinidae. In "Systematics, Historical Ecology, and North American Freshwater Fishes" (R. Mayden, ed.), pp. 293–327. Stanford University Press, Stanford, CA.

Chandy, M. (1956). On the oesophagus of the mild-fish *Chanos chanos* (Forskal). *J. Zool. Soc. India* **8**, 79–84.

Chardon, M. (1968). Anatomie comparée de l'appareil de Weber et des structures connexes chez les Siluriformes. *Ann.—Mus. R. Afr. Cent., Tervuren, Belg. (Sér. 8), Sci. Zool.* **169**, 1–277.

Chardon, M., and Vandewalle, P. (1989). About identification of swimbladder sheets in Ostariophysi. *Fortschr. Zool.* **35**, 264–268.

Chen, X.-L., Yue, P.-Q., and Lin, R.-D. (1984). Major groups within the family Cyprinidae and their phylogenetic relationships. *Acta Zootaxon. Sin.* **9**, 424–440.

Coburn, M., and Cavender, T. (1992). Interrelationships of North American cyprinid fishes. In "Systematics, Historical Ecology, and North American Freshwater Fishes" (R. Mayden, ed.), pp. 328–373. Stanford University Press, Stanford, CA.

Collette, B. B. (1977). Epidermal breeding tubercles and bony contact organs in fishes. *Symp. Zool. Soc. London* **39**, 225–268.

Da Silva Santos, R. (1947). Uma redescrição de *Dastilbe elongatus*, com algumas considerações sobre o gênero *Dastilbe*. *Minist. Agric. Div. Geol. Minerol. Notas Prelim. Estud.* **42**, 1–47.

d'Aubenton, F. (1961). Morphologie du crâne de *Cromeria nilotica occidentalis* Daget 1954. *Bull. Inst. Fondam. Afr. Noire, Ser. A* **23**, 134–163.

de Pinna, M. C. C. (1989a). A new sarcoglanidine catfish, phylogeny of its subfamily, and an appraisal of the phyletic status of the Trichomycterinae (Teleostei, Trichomycteridae). *Am. Mus. Novit.* **2950**, 1–39.

de Pinna, M. C. C. (1989b). Redescription of *Glanapteryx anguilla*, with notes on the phylogeny of Glanapteryginae (Siluriformes, Trichomycteridae). *Proc. Acad. Natl. Sci. Philadelphia* **141**, 361–374.

de Pinna, M. C. C. (1993). Higher-level phylogeny of Siluriformes, with a new classification of the order (Teleostei, Ostariophysi). Unpublished Ph.D. Dissertation, City University of New York.

de Pinna, M. C. C., and Ferraris, C. (1992). Review of: "Anatomy, relationships and systematics of the Bagridae (Teleostei: Siluroidei) with a hypothesis of siluroid phylogeny" by T. Mo. 1991. *Copeia* pp. 1132–1134.

de Pinna, M. C. C., and Vari, R. P. (1995). Monophyly and phylogenetic diagnosis of the family Cetopsidae, with synonymization of the Helogenidae (Teleostei: Siluriformes). *Smithson. Contrib. Zool.* **571**, 1–26.

Ellis. M. M. (1913). The gymnotid eels of tropical America. *Mem. Carnegie Mus.* **6**, 109–204.

Finger, T. E., and Tong, S. L. (1984). Central organization of eighth nerve and mechanosensory lateral line systems in the brainstem of ictalurid catfish. *J. Comp. Neurol.* **229**, 129–151.

Finger, T. E. (1986). Electroreception in catfish. In "Electroreception" (T. H. Bullock and W. Heiligenberg, eds.), pp. 287–317. Wiley (Interscience), New York.

Fink, S. V., and Fink, W. L. (1981). Interrelationships of the ostariophysan fishes. *Zool. J. Linn. Soc.* **72**, 297–353.

Fink, S. V., Greenwood, P. H., and Fink, W. L. (1984). A critique of recent work on fossil ostariophysan fishes. *Copeia*, pp. 1033–1041.

Fink, S. V., Albert, J. S., and Fink, W. L. (1997). Phylogenetic relationships of the fossil knifefish *Ellisella kirschbaumi* (Ostariophysi; Gymnotoidei). *J. Vertebr. Paleontol.* (in press).

Fink, W. L. (1984). Salmoniformes: Introduction. In "Ontogeny and Systematics of Fishes" (H. G. Moser, W. J. Richards, D. M. Cohen, M. P. Fahay, A. W. Kendall, Jr., and S. L. Richardson, eds.), Spec. Publ. No. 1, Suppl. to *Copeia*, p. 139. American Society of Ichthyologists and Herpetologists, Lawrence, KS.

Fink, W. L., and Machado-Allison, A. (1992). Three new species of piranhas from Brazil and Venezuela (Teleostei: Characiformes). *Ichthyol. Explor. Freshwaters* **3**, 55–71.

Fink, W. L., and Weitzman, S. H. (1982). Relationships of the stomiiform fishes (Teleostei), with a description of *Diplophos*. *Bull. Comp. Zool.* **150**, 31–92.

Fujita, K. (1990). In "The Caudal Skeleton of Teleostean Fishes." Tokai University Press, Japan.

Gayet, M. (1981). Contribution à l'étude anatomique et systématique de l'ichthyofaune cénomanienne du Portugal. Deuxième partie: Les ostariophysaires. *Comun. Serv. Geol. Port.* **67**, 173–190.

Gayet, M. (1982a). Cypriniforme ou Gonorhynchiforme? *Ramallichthys*, nouveau genre du Cénomanien inférieur de Ramallah (Monts de Judée). *C. R. Seances Acad. Sci., Sér. 2* **295**, 405–407.

Gayet, M. (1982b). Considération sur la phylogénie et la paleobiogéographie des Ostariophysaires. *Geobios, Mem. Spec.* **6**, 39–52.

Gayet, M. (1985a). Rôle de l'évolution de l'appareil de Weber dans la phylogénie des Ostariophysi, suggéré par un nouveau Characiforme du Cénomanien supérieur marin du Portugal. *C. R. Seances Acad. Sci., Sér. 2* **300**, 895–898.

Gayet, M. (1985b). Gonorhynchiforme nouveau du Cénomanien inférieur marin de Ramallah (Monts de Judée): *Judeichthys haasi* nov. gen. nov. sp. (Teleostei, Ostariophysi, Judeichthyidae nov. fam.). *Bull. Mus. Natl. Hist. Nat., Sect. C [4]* **7**, 65–85.

Gayet, M. (1985c). Contribution à l'étude anatomique et systématique de l'ichthyofaune Cénomanienne du Portugal. Troisième partie: Complément a l'étude des Ostariophysaires. *Comun. Serv. Geol. Port.* **71**, 91–118.

Gayet, M. (1986a). *Ramallichthys* Gayet du Cénomanien inférieur marin de Ramallah (Judée), une introduction aux relations phylogénétiques des Ostariophysi. *Mém. Mus. Natl. Hist. Nat., Sér. C (Paris)* **51**, 1–81.

Gayet, M. (1986b). About ostariophysan fishes: A reply to S. V. Fink, P. H. Greenwood and W. L. Fink's criticisms. *Bull. Mus. Natl. Hist. Nat., Sect. C [4]* **8**, 393–409.

Gayet, M. (1986c). Problème de l'origine des osselets de Weber. *Océanis* **12**, 357–366.

Gayet, M. (1993a). Rélations phylogénétiques des Gonorhynchiformes (Ostariophysi). *Belg. J. Zool.* **123**, 165–192.

Gayet, M. (1993b). Gonorhynchoidei du Crétacé Supérieur marin du Liban et rélations phylogénétiques des Charitosomidae nov. fam. *Doc. Lab. Géol. Fac. Sci. Lyon* **126**, 1–131.

Gayet, M., and Chardon, M. (1987). Possible otophysic connections in some fossil and living ostariophysan fishes. *Proc. Congr. Eur. Ichthyol., 5th, Stockholm, 1985*, pp. 31–42.

Gayet, M., and Meunier, F. J. (1991). Première découverte de Gymnotiformes fossiles (Miocène supérieur de Bolivie). *C. R. Seances Acad. Sci., Sér. 2* **313**, 471–476.

Gayet, M., Meunier, F. J., and Kirschbaum, F. (1994). *Ellisella kirschbaumi* Gayet and Meunier, 1991, gymnotiforme fossile de Bolivie et ses relations phylogénétiques au sein des formes actuelles. *Cybium* **18**, 273–306.

Gervais P. (1854). "Histoire naturelle des mammifères." L. Curmer, Paris.

Géry, J. (1965). Poissons du Bassin de l'Ivindo. II. Clupeiformes. *Biol. Gabonica* **1**, 385–393.

Gosline, W. A. (1973). Considerations regarding the phylogeny of cypriniform fishes, with special reference to structures associated with feeding. *Copeia*, pp. 761–776.

Gosline, W. A. (1989). Two patterns of differentiation in the jaw musculature of teleostean fishes. *J. Zool.* **218**, 649–661.

Grande, L. (1987). Redescription of †*Hypsidoris farsonensis* (Teleostei: Siluriformes) with a reassessment of its phylogenetic relationships. *J. Vertebr. Paleontol.* **7**, 24–54.

Grande, T. (1994). Phylogeny and paedomorphosis in an African family of freshwater fishes (Gonorynchiformes: Kneriidae). *Fieldiana Zool.* [N.S.] **78**, 1–20.

Grande, T. (1996). The interrelationships of fossil and Recent gonorynchid fishes with comments on two Cretaceous taxa from Israel. *In* "Mesozoic Fishes: Systematics and Paleoecology" (G. Arratia and G. Viohl, eds.), pp. 296–315. Dr. Friedrich Pfeil, Munich.

Greenwood, P. H., and Rosen, D. E. (1971). Notes on the structure and relationships of the alepocephaloid fishes. *Am. Mus. Novit.* **2473**, 1–41.

Greenwood, P. H., Rosen, D. E., Weitzman, S. H., and Myers, G. S. (1966). Phyletic studies of teleostean fishes, with a provisional classification of living forms. *Bull. Am. Mus. Nat. Hist.* **141**, 339–456.

Hagedorn, M., Womble, M., and Finger, T. E. (1990). Synodontid catfish: A new group of weakly electric fish: Behavior and anatomy. *Brain, Behav. Evol.* **35**, 268–277.

Heiligenberg, W. F., and Dye, J. (1982). Labeling of electroreceptor afferents in a gymnotoid fish by intracellular injection of HRP: The mystery of multiple maps. *J. Comp. Physiol.* **148**, 287–296.

Hopkins, C. D. (1983). Functions and mechanisms in electroreception. *In* "Fish Neurobiology" (R. G. Northcutt and R. E. Davis, eds.), pp. 215–259. University of Michigan Press, Ann Arbor.

Howes, G. J. (1978). The anatomy and relationships of the cyprinid fish *Luciobrama macrocephalus* (Lacépède). *Bull. Br. Mus. (Nat. Hist.), Zool.* **34**, 1–64.

Howes, G. J. (1983a). Problems in catfish anatomy and phylogeny exemplified by the Neotropical Hypophthalmidae (Teleostei: Siluroidei). *Bull. Br. Mus. (Nat. Hist.), Zool.* **45**, 1–39.

Howes, G. J. (1983b). The cranial muscles of loricarioid catfishes, their homologies and value as taxonomic characters (Teleostei: Siluroidei). *Bull. Br. Mus. (Nat. Hist.), Zool.* **45**, 309–345.

Howes, G. J. (1985a). The phylogenetic relationships of the electric catfish family Malapteruridae (Teleostei: Siluroidei). *J. Nat. Hist.* **19**, 37–67.

Howes, G. J. (1985b). Cranial muscles of gonorynchiform fishes, with comments on generic relationships. *Bull. Br. Mus. (Nat. Hist.), Zool.* **49**, 273–303.

Howes, G. J., and Teugels, G. G. (1989). Observations on the ontogeny and homology of the pterygoid bones in *Corydoras paleatus* and some other catfishes. *J. Zool.* **219**, 441–456.

Kindred, J. E. (1919). The skull of *Ameiurus*. *Ill. Biol. Monogr.* **5**, 3–120.

Kusaka, T. (1974). "The Urohyal of Fishes." University of Tokyo Press, Japan.

Lannoo, M. J., Brochu, G., Maler, L., and Hawks, R. (1991). Zebrin II immunoreactivity in the rat and the weakly electric teleost *Eigenmannia* (Gymnotiformes) reveals three modes of Purkinje cell development. *J. Comp. Neurol.* **309**, 1–19.

Lannoo, M. J., Maler, L., and Hawks, R. (1992). Zebrin II distinguishes the ampullary organ receptive map from the tuberous organ receptive maps during development in the teleost electrosensory lateral line lobe. *Brain Res.* **586**, 176–180.

Lenglet, G. (1973). Contribution à l'étude de l'anatomie viscérale des Kneriidae. *Ann. Soc. R. Zool. Belg.* **103**, 239–270.

Lenglet, G. (1974). Contribution à l'étude ostéologique des Kneriidae. *Ann. Soc. R. Zool. Belg.* **104**, 51–103.

Lundberg, J. G., Lewis, W. M., Saunders, J. F., and Mago-Leccia, F. (1987). A major food web component in the Orinoco River channel: Evidence from planktivorous electric fishes. *Science* **237**, 81–83.

Mago-Leccia, F. (1976). Los peces gymnotiformes de Venezuela: Un estudio preliminar para la revision del grupo en la America del Sur. Unpublished Ph.D. Dissertation, Universidad Central de Venezuela.

Markle, D. F. (1989). Aspects of character homology and phylogeny of the Gadiformes. *Sci. Ser. Nat. Hist. Mus.* **32**, 59–88.

Mo, T. (1991). "Anatomy, Relationships and Systematics of the Bagridae (Teleostei: Siluroidei)—With a Hypothesis of Siluroid Phylogeny," *Theses Zool. No. 17.* Koeltz Scientific Books, Koenigstein.

Monod, T. (1963). Sur quelques points de l'anatomie de *Gonorhynchus gonorhynchus* (Linné 1766). *Mem. Inst. Fr. Afr. Noire* **68**, 255–310.

Monod, T. (1968). Le complexe urophore des poissons téléostéens. *Mem. Inst. Fr. Afr. Noire, Dakar* **81**, 1–705.

Nelson, G. J. (1967). Epibranchial organs in lower teleostean fishes. *J. Zool.* **153**, 71–89.

Nelson, J. S. (1994). "Fishes of the World," 3rd ed. Wiley, New York.

New, J. G., and Singh, S. (1994). Central topography of anterior lateral line nerve projections in the channel catfish, *Ictalurus punctatus*. *Brain, Behav. Evol.* **43**, 34–50.

Northcutt, R. G. (1992). The phylogeny of octavolateralis ontogenies: A reaffirmation of Garstang's phylogenetic hypothesis. *In* "The Evolutionary Biology of Hearing" (D. B. Webster, R. R. Fay, and A. N. Popper, eds), pp. 21–48. Springer-Verlag, New York.

Northcutt, R. G., and Vischer, H. A. (1988). *Eigenmannia* possesses autapomorphic rami of the anterior lateral line nerves. *Soc. Neurosci. Abstr.* **14**, 54.

Oldani, N. O. (1977). Identificación y morfología de larvas, juveniles y adultos de *Apareiodon affinis* (Steindachner) (Pisces, Parodontidae). *Physis (Buenos Aires), Secc. B* **37**, 133–140.

Oldani, N. O. (1979a). Identificación y morfología de larvas, juveniles y adultos de *Thoracocharax stellatus* (Kner, 1860) (Pisces, Gasteropelecidae). *Rev. Asoc. Cienc. Nat. Litoral* **10**, 49–60.

Oldani, N. O. (1979b). Identificación y morfología de larvas, juveniles y adultos de *Triportheus paranensis* (Günther, 1874) (Pisces, Characidae). *Rev. Asoc. Cienc. Nat. Litoral* **10**, 61–71.

Oldani, N. O. (1983). Identificación y morfología de larvas, juveniles y adultos de *Mylossoma paraguayensis* Norman, 1929 (Pisces, Characidae). *Stud. Neotrop. Fauna Environ.* **18**, 89–100.

Patterson, C. (1970a). Two upper Cretaceous salmoniform fishes from the Lebanon. *Bull. Br. Mus. (Nat. Hist.), Geol.* **19**, 207–296.

Patterson, C. (1970b). A clupeomorph fish from the Gault (Lower Cretaceous). *Zool. J. Linn. Soc.* **49**, 161–182.

Patterson, C. (1977). Cartilage bones, dermal bones and membrane bones, or the exoskeleton versus the endoskeleton. *In* "Problems in Vertebrate Evolution" (S. M. Andrews, R. S. Miles, and A. D. Walker, eds), pp. 77–121. Academic Press, London.

Patterson, C. (1984a). Family Chanidae and other teleostean fishes as living fossils. In "Living Fossils" (N. Eldridge and S. M. Stanley, eds.), pp. 132–139. Springer-Verlag, New York.

Patterson, C. (1984b). *Chanoides*, a marine Eocene otophysan fish (Teleostei: Ostariophysi). *J. Vertebr. Paleontol.* **4**, 430–456.

Patterson, C. (1996). In preparation.

Patterson, C., and Johnson, G. D. (1995). The intermuscular bones and ligaments of teleostean fishes. *Smithson. Contrib. Zool.* **599**, 1–83.

Patterson, C., and Rosen, D. E. (1977). Review of the ichthyodectiform and other Mesozoic teleost fishes and the theory and practice of classifying fossils. *Bull. Am. Mus. Nat. Hist.* **158**, 81–172.

Perkins, P. L. (1970). *Notogoneus osculus* Cope, an Eocene fish from Wyoming (Gonorynchiformes, Gonorynchidae). *Postilla* **147**, 1–18.

Poyato-Ariza, J. (1994). A new Early Cretaceous Gonorynchiform fish (Teleostei: Ostariophysi) from Las Hoyas (Cuenca, Spain). *Occas. Pap. Mus. Nat. Hist., Univ. Kans.* **164**, 1–37.

Ramaswami, L. S. (1952a). Skeleton of cyprinoid fishes in relation to phylogenetic studies. I. The systematic position of the genus *Gyrinocheilus* Vaillant. *Proc. Natl. Inst. Sci. India* **18**, 125–140.

Ramaswami, L. S. (1952b). Skeleton of cyprinoid fishes in relation to phylogenetic studies. III. The skull and other skeletal structures of homalopterid fishes. *Proc. Natl. Inst. Sci. India* **18**, 495–517.

Ramaswami, L. S. (1952c). Skeleton of cyprinoid fishes in relation to phylogenetic studies. IV. The skull and skeletal structures of gastromyzonid fishes. *Proc. Natl. Inst. Sci. India* **18**, 519–538.

Ramaswami, L. S. (1955a). Skeleton of cyprinoid fishes in relation to phylogenetic studies. 6. The skull and Weberian apparatus in the subfamily Gobioninae (Cyprinidae). *Acta Zool. (Stockholm)* **36**, 127–158.

Ramaswami, L. S. (1955b). Skeleton of cyprinoid fishes in relation to phylogenetic studies. 7. The skull and Weberian apparatus of Cyprininae (Cyprinidae). *Acta Zool. (Stockholm)* **36**, 199–242.

Ridewood, W. G. (1905). On the skull of *Gonorhynchus greyi*. *Ann. Mag. Nat. Hist.* [7] **15**, 361–372.

Roberts, T. R. (1967). Tooth formation and replacement in characoid fishes. *Stanford Ichthyol. Bull.* **8**, 231–249.

Roberts, T. R. (1969). Osteology and relationships of characoid fishes, particularly the genera *Hepsetus, Salminus, Hoplias, Ctenolucius,* and *Acestrorhynchus*. *Proc. Calif. Acad. Sci.* **36**, 391–500.

Roberts, T. R. (1973). Interrelationships of ostariophysans. In "Interrelationships of Fishes" (P. H. Greenwood, R. S. Miles, and C. Patterson, eds), pp. 373–395. Academic Press, London.

Roberts, T. R. (1982). Unculi (horny projections arising from single cells), an adaptive feature of the epidermis of ostariophysan fishes. *Zool. Scr.* **11**, 55–76.

Rosen, D. E., and Greenwood, P. H. (1970). Origin of the Weberian apparatus and the relationships of the ostariophysan and gonorynchiform fishes. *Am. Mus. Novit.* **2428**, 1–25.

Sas, E., and Maler, L. (1987). The organization of afferent input to the caudal lobe of the cerebellum of the gymnotid fish *Apteronotus leptorhynchus*. *Anat. Embryol.* **177**, 55–79.

Sawada, Y. (1982). Phylogeny and zoogeography of the superfamily Cobitoidea (Cyprinoidei, Cypriniformes). *Mem. Fac. Fish., Hokkaido Univ.* **28**, 63–223.

Schaefer, S. A. (1987). Osteology of *Hypostomus plecostomus* (Linnaeus), with a phylogenetic analysis of the loricariid subfamilies (Pisces: Siluroidei). *Contrib. Sci., Nat. Hist. Mus. Los Angeles Cty.* **394**, 1–31.

Schaefer, S. A., (1990). Anatomy and relationships of the scoloplacid catfishes. *Proc. Acad. Natl. Sci. Philadelphia* **142**, 167–210.

Siebert, D. J. (1987). Interrelationships among families of the order Cypriniformes (Teleostei). Unpublished Ph.D. Dissertation, City University of New York.

Smith, G. R. (1992). Phylogeny and biogeography of the Catostomidae, freshwater fishes of North American and Asia. In "Systematics, Historical Ecology, and North American Freshwater Fishes" (R. Mayden, ed.), pp. 778–826. Stanford University Press, Stanford, CA.

Striedter, G. F. (1990). The diencephalon of the channel catfish *Ictalurus punctatus*. I. Nuclear organization. *Brain, Behav. Evol.* **36**, 329–354.

Striedter, G. F. (1991). Auditory, electrosensory, and mechanosensory lateral line pathways through the forebrain in channel catfishes. *J. Comp. Neurol.* **312**, 311–332.

Striedter, G. F. (1992). Phylogenetic changes in the connections of the Lateral Preglomerular Nucleus in ostariophysan teleosts: A pluralistic view of brain evolution. *Brain, Behav. Evol.* **39**, 329–357.

Swinnerton, H. H. (1903). The osteology of *Cromeria nilotica* and *Galaxias attenuatus*. *Zool. Jahrb. Abt. Anat. Ontog.* **18**, 58–70.

Takahasi, N. (1925). On the homology of the cranial muscles of the cypriniform fishes. *J. Morphol.* **40**, 1–109.

Taverne, L. (1976). Les téléostéens fossiles du Crétacé moyen de Kipala (Kwango, Zaire). *Ann. Mus. R. Afr. Cent., Tervuren, Belg. (Sér. 8), Sci. Geol.* **79**, 1–50.

Taverne, L. (1981). Ostéologie et position systématique d'*Aethalionopsis robustus* (Pisces, Teleostei) du Crétacé inférieur de Bernissart (Belgique) et considérations sur les affinités des Gonorhynchiformes. *Bull. Cl. Sci., Acad. R. Belg.* [5] **67**, 958–982.

Thys van den Audenaerde, D. F. E. (1961). L'anatomie de *Phractolaemus ansorgei* Blgr. et la position systématique des Phractolaemidae. *Ann. Mus. R. Afr. Cent., Tervuren, Belg., (Sér. 8), Sci. Zool.* **103**, 100–167.

Tong, S. L., and Finger, T. E. (1983). Central organization of the electrosensory lateral line system in bullhead catfish, *Ictalurus nebulosus*. *J. Comp. Neurol.* **217**, 1–16.

Triques, M. L. (1993). Filogenia dos gêneros de Gymnotiformes (Actinopterygii, Ostariophysi), com base em characters esqueléticos. *Comun. Mus. Ciênc. PUCRS, Ser. Zool., Porto Alegre* **6**, 85–130.

Vari, R. P. (1979). Anatomy, relationships and classification of the families Citharinidae and Distichodontidae (Pisces, Characoidea). *Bull. Br. Mus. (Nat. Hist.), Zool.* **36**, 261–344.

Vari, R. P. (1983). Phylogenetic relationships of the families Curimatidae, Prochilodontidae, Anostomidae and Chilodontidae (Pisces: Characiformes). *Smithson. Contrib. Zool.* **378**, 1–60.

Vari, R. P. (1995). The Neotropical fish family Ctenoluciidae (Teleostei: Ostariophysi: Characiformes): Supra and intrafamilial phylogenetic relationships, with a revisionary study. *Smithson. Contrib. Zool.* **564**, 1–97.

Weitzman, S. H. (1962). The osteology of *Brycon meeki*, a generalized characid fish, with an osteological definition of the family. *Stanford Ichthyol. Bull.* **8**, 1–77.

Weitzman, S. H. (1964). Osteology and relationships of South American characid fishes of subfamilies Lebiasininae and Erythrininae with special reference to subtribe Nannostomina. *Proc. U.S. Natl. Mus.* **116**, 127–170.

Wenz, S. (1984). *Rubiesichthys gregalis* n. g., n. sp., Pisces Gonorynchiformes, du Crétacé inférieur du Montsech (Province de Lérida, Espagne). *Bull. Mus. Natl. Hist. Nat., Sect. C* [4] **3**, 275–285.

Wiley, M. L., and Collette, B. B. (1970). Breeding tubercles and contact organs in fishes: Their occurrence, structure, and significance. *Bull. Am. Mus. Nat. Hist.* **143**, 143–216.

Winterbottom, R. (1974). A descriptive synonymy of the striated muscles of the teleostei. *Proc. Acad. Natl. Sci. Philadelphia* **125**, 225–317.

Woodward, A. S. (1896). Extinct fishes of the teleostean family Gonorhynchidae. *Proc. Zool. Soc. London* **2**, 500–504.

Wu, X., Chen, Y., Chen, X., and Chen, J. (1981). A taxonomical system and phylogenetic relationship of the families of the suborder Cyprinoidei (Pisces). *Sci. Sin.* **24**, 563–572 (Chinese with English summary).

Zakon, H. H. (1986). The electroreceptive periphery. *In* "Electroreception" (T. H. Bullock and W. Heiligenberg, eds.), pp. 103–156. Wiley (Interscience), New York.

Note added in proof. Taverne (1995) provides evidence of the presence of a Weberian apparatus in material of *Clupavus maroccanus*; previously, this structure was reported on the basis of a single imprint (Gayet, 1981, p. 185). Taverne discusses and illustrates one specimen in which bony elements are also preserved, though, as he says, rather poorly. Although details of the structures are unclear and may remain so because of the state of preservation, it appears that this specimen may be an otophysan. Taverne's "parsimony" analysis places the taxon as a characiphysan based on the supposed presence of a single large supraneural (fragmented and only partially preserved) adjacent to the margin of the cranium, although in his assessment *C. maroccanus* lacks four otophysan characters of Fink and Fink (1981) (30, 78, 102, 110) and one ostariophysan character (41). Our reassessment of these characters would still count 30, 41, 78, and 110 as diagnostic of otophysans; therefore, parsimony analysis would place *C. maroccanus* as an otophysan plesion along with *Chanoides*. The supposed characiphysan morphology of the supraneural and its relation to the cranial margin seem to us sufficiently unclear in the illustrated specimen that we would reserve judgment on its presence until better material becomes available. The shape of the exoccipital in Taverne's figure (1995, fig. 3) resembles no otophysan that we know of, in part due to its rounded and projecting shape and in part due to the identified location of the vagal foramen, which is situated well posterior to its usual location.

Poyato-Ariza (1996) presented a detailed examination and a phylogenetic analysis of the taxa we have considered to be primitive anotophysans. He provides photographs of two acid-prepared specimens of *Tharrhias* (1996, fig. 14) that show articulation between the exoccipital and the first neural arch, corroborating the presence of this gonorynchiform character in that taxon. He presents addi-

tional characters for the Gonorynchiformes and concludes that *Aethalionopsis, Tharrhias, Dastilbe, Parachanos, Rubiesichthys,* and *Gordichthys* are all members, with the extant *Chanos,* of the Chanidae. His Chanidae is supported by an impressive number of characters (14) and provides considerable evidence that must be taken into account in future work on gonorynchiforms. We think that several of these are polarized incorrectly, including his (45), neural arches anterior to the dorsal fin autogenous; (62), epural number greater than 1; and (24), symplectic extending anteriorly to near the articular condyle of the quadrate. The same is true of (46), parapophyses of vertebrae 1 and 2 autogenous (unknown for several taxa), which is used to diagnose his Chaninae. All of these are generalized conditions that occur widely in lower teleosts. Others of his chanid characters appear to be present in other gonorynchiforms, although additional modified morphology makes any assessment problematic. These include his (21), anterior notch on the dentary, which is present and elongate in our *Kneria* and *Gonorynchus* (and present also, as he notes, in some other lower teleosts); and (18), greatly elevated coronoid process, which is present in fossil gonorynchoids as well as extant gonorynchoids, including perhaps *Phractolaemus*. In the latter taxon, since a symphysis occurs on the dentary just anterior to the small articular, one interpretation of the mandible is that it is mostly coronoid process. Various features of the suspensorium, such as (23), position of the quadrate-mandibular articulation anterior to a vertical line through the front of the orbit with associated modifications of other elements, and (30), enlarged opercle, provide more persuasive evidence supporting his Chanidae. In short, we interpret some of these characters as uninformative because they are plesiomorphic or more widespread and therefore possibly diagnostic of Gonorynchiformes, but Poyato-Ariza has also analyzed into discrete features some of the morphology that accounts for the strong chanid gestalt, previously perceived but not well accounted for, of many of these specimens. We await further analysis of Poyato-Ariza's evidence in the context of more material of extant gonorynchoids and assessment of the characters in fossil gonorynchids.

Taverne, L. (1995). Description de l'appareil de Weber du téléostéen crétacé marin *Clupavus maroccanus* et ses implications phylogénétiques. *Belg. J. Zool.* **125**, 267–282.

Poyato-Ariza, F. J. (1996). A revision of the otariophysan fish family Chanidae, with special reference to the Mesozoic forms. *Palaeo Ichthyologica* **6**, 5–52.

12

Relationships of Lower Euteleostean Fishes

G. DAVID JOHNSON
National Museum of Natural History
Smithsonian Institution
Washington, D.C.

COLIN PATTERSON
Natural History Museum
London, England

We all make mistakes; then we're sorry.

Popular song

I. Introduction

In the first *Interrelationships of Fishes* lower euteleosts, or "protacanthopterygians" as they were then called, were omitted, with only a comment in the Preface citing Weitzman (1967, on osmeroids and stomiatoids), McDowall (1969, on osmeroids and galaxioids), Rosen and Greenwood (1970, on gonorynchiforms and ostariophysans), Greenwood and Rosen (1971, on argentinoids and alepocephaloids), and Nelson (1970b, on salangids and argentinids; 1972, on esocoids and galaxioids).

Ten years later, in *Ontogeny and Systematics of Fishes*, Fink (1984a) summarized the history of protacanthopterygians as "erosion" and "attrition, most notably at the hands of Rosen (1973)" [in the first *Interrelationships of Fishes*]. Fink then saw the problems as these: (1) What are the relationships of the Esocoidei? (2) What are the relationships of the Ostariophysi? Do these fishes lie above or below the Esocoidei? (3) What is the pattern of relationships among the traditional "salmoniform" taxa, exclusive of Esocoidei and Ostariophysi? (4) What are the relationships of and within the Argentinoidei (sensu Greenwood and Rosen, 1971, i.e., argentinoids plus alepocephaloids)? (5)

What are the relationships of and within the Osmeroidei? (6) What are the relationships of and within Salmonidae? (7) Where does *Lepidogalaxias* belong? (8) What are the relationships within stomiiform fishes? (9) What of the Myctophoidei, as recognized by Greenwood *et al.* (1966, i.e., Aulopiformes and Myctophiformes in current terminology)? In that agenda, items (8) and (9) are treated elsewhere in this volume and do not concern us, but items (1) through (7) do.

Some classifications and/or cladograms of lower euteleosts, dating back to the first application of cladistic method, are summarized in Fig. 1. As is obvious from incongruence between all the patterns in Fig. 1, there has been protracted argument on how lower euteleostean groups are interrelated, how they are related to neoteleosts (stomiiforms and eurypterygians, Johnson, 1992), and what group is basal to other euteleosts. The most substantial treatment of these problems is in Begle's (1991, 1992) cladistic analyses of Osmeroidei (1991) and Argentinoidei (1992) (Fig. 1G). Begle's two papers resulted in the cladogram in Fig. 2, in which the terminals are the genera or higher taxa sampled in his matrix. His classification (Begle, 1991, fig.1; 1992, table 2) to family level, without ranks above the superfamily and sequenced according to the conventions of Wiley (1981), was as follows:

Euteleostei
  Esocae
    Ostariophysi *sedis mutabilis*

251

Copyright © 1996 by Academic Press, Inc.
All rights of reproduction in any form reserved.

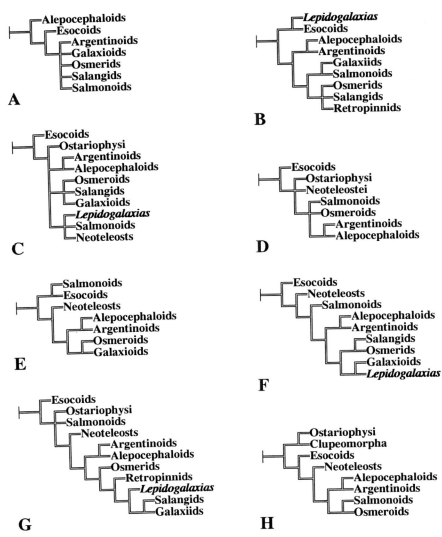

FIGURE 1 Cladograms or classifications of lower euteleostean fishes summarized as branching diagrams. Common names (e.g., alepocephaloids and esocoids) are used instead of formal taxon names to emphasize the equivalence of groups that may be given different ranks, and so names with different terminations, in classifications. (A) G. J. Nelson (1970b). (B) Rosen (1974). (C) Fink (1984b). (D) Sanford (1990). (E) Williams (1987; summarized in J. S. Nelson, 1994, p. 175). (F) J. S. Nelson (1994). (G) Begle (1991, 1992). (H) Patterson (1994).

Salmonoidei *sedis mutabilis*
(Neoteleostei + Osmerae) *sedis mutabilis*
  Neoteleostei
  Osmerae
    Argentinoidei
      Argentinoidea
        Argentinidae
        Microstomatidae
        Bathylagidae
        Opisthoproctidae
      Alepocephaloidea

Alepocephalidae (including Bathylaconidae, Bathyprionidae, Leptochilichthyidae, Platytroctidae)
Osmeroidei
  Osmeroidea
    Osmeridae (including Plecoglossidae)
  Galaxioidea
    Retropinnidae (including Prototroctidae)
    Lepidogalaxiidae

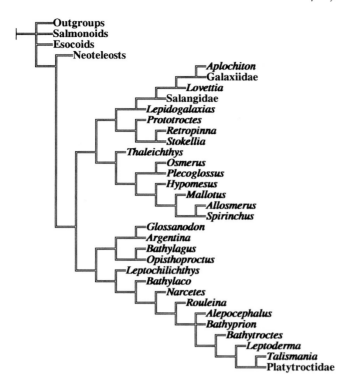

FIGURE 2 Result of Begle's (1991, 1992) cladistic analyses of a sample of lower euteleosts.

Salangidae (including Sundasalangidae)
Galaxiidae (including Aplochitonidae)

The original plan for this volume was that Douglas P. Begle should write about lower euteleosts since he had covered the ground in his two papers (Begle, 1991, 1992). However, in 1993 Begle left ichthyology to move into computer programming. To fill the gap, we agreed to take on lower euteleosts, believing that we had a new angle on them from our work on intermusculars (Patterson and Johnson, 1995). In particular, the discovery of cartilaginous epicentrals in salmonoids and osmeroids, together with the absence of ossified epipleurals in both, implied that they might be sister-groups (Patterson and Johnson, 1995, p. 26, fig. 9). In arguing for that proposal, we criticized the characters supporting some of the higher-level relationships in Begle's cladogram, but we anticipated that his conclusions at lower levels (e.g., within osmerids or within galaxioids) would withstand criticism. However, when we began checking Begle's character descriptions and coding, we found many errors. Because Begle's work resulted in the only fully resolved phylogeny of lower euteleosts, and because the characters that he used included the great majority of those previously used in systematic studies of those fishes,

we decided that we must check each character against specimens. At the same time, we found that our own survey of lower euteleostean intermusculars (Patterson and Johnson, 1995) was deficient both in sampling and in the quality of our material, casting doubt on the pairing of salmonoids and osmeroids that we proposed. Given this disquiet about both recent sources of published information, Begle's and our own, we were thrown into a detailed survey of lower euteleosts, limited only by time and by the availability of specimens. On sampling, we limited ourselves principally to the 33 taxa in Begle's sample (Fig. 2), but we excluded the alepocephaloid *Bathyprion* (unavailable) and added more argentinoids (bathylagids, opisthoproctids, and microstomatids, which Begle did not sample) and more platytroctid alepocephaloids. On characters, we concentrated on the 108 used by Begle (1992), the majority of them taken from the literature [e.g., nos. 1–28, 54, 55, from Fink (1984b); nos. 30–38, 51, 63, from Howes and Sanford (1987b)], because we were driven to discover every detail of his miscoding or misconduct. But during our survey almost 100 other characters came to light.

In presenting our data, we originally had a section criticizing our own work on the intermusculars, a longer critique of Begle's 108 characters, and a section describing new characters. Referees and others pointed out that this sequence gave too much attention to our critique of Begle's work and artificially separated characters into three different sets. We have therefore organized the primary data into a single survey of characters structured by anatomical region. We then discuss in sequence the outgroup and ingroup relationships of osmerids [Osmeridae and Salangidae of Begle (1991)], osmeroids [= Osmerae of Begle (1991)], argentinoids and alepocephaloids [= Argentinoidei of Greenwood and Rosen (1971)], salmonoids, esocoids, and euteleosteans as a whole. We do not include the "salangid" *Sundasalanx*, only member of the Sundasalangidae (Roberts, 1984), in our study because it is not euteleostean (Darrell Siebert, 1996, and personal communication).

We dedicate this work to two absent friends, Humphry Greenwood (1927–1995), senior editor of the first *Interrelationships*, and Donn Rosen (1929–1986), the length of whose chapter in that book was justified by the information it contained. May we be so excused.

II. Character Survey

The following survey of 200 characters is organized by anatomical region or system, beginning with the braincase and ending with reproductive and soft anatomical features. One purpose of this survey is to check and correct the character coding in Begle (1991, 1992), who based his cladistic analyses on matrices of 84 (1991) or 108 (1992) characters, each coded in 20 (1991) or 33 (1992) terminal taxa. We take Begle's more complete, 1992 matrix as definitive but mention significant differences in character description or coding between the data sets in Begle (1991) and (1992). Twenty-six of the 33 terminals in Begle's 1992 paper are genera, and 7 are collective at higher levels: "outgroups" (a hypothetical ancestor comprising a row of zeros), salmonoids, esocoids, neoteleosts, Platytroctidae, Salangidae, and Galaxiidae. Begle organized his character sequence partly by source [nos. 1–28 were from Fink (1984b), and nos. 29–38 from Howes and Sanford (1987b)] and partly according to his two publications [nos. 85–108 are only in Begle (1992), not Begle (1991)]. Where a character in our survey is in Begle (1992), we give his number (Table 1) and his character description and coding in square brackets, followed by comments on that coding. Our own character coding is in Appendices 1–4. This survey is based on cleared-and-stained specimens, but we include references to published illustrations, where they exist, so that workers without ready access to specimens may check our observations.

Our characters are distributed as follows:

A. Braincase: numbers 1–26
B. Suspensorium: numbers 27–48
C. Jaws: numbers 49–64
D. Circumorbital bones: numbers 65–68
E. Hyoid bar, branchiostegals, and operculum: numbers 69–79
F. Gill arches: numbers 80–112
G. Axial skeleton (including caudal skeleton and fin): numbers 113–147
H. Pectoral girdle and fin: numbers 148–159
I. Pelvic girdle and fin: numbers 160–165
J. Median fins: numbers 166–170
K. Squamation: numbers 171–173
L. Sensory canals and bones carrying them: numbers 174–181
M. Reproductive structures: numbers 182–187
N. Other soft anatomical features: numbers 188–198
O. Life cycle: numbers 199 and 200

A. Braincase

1. Dermethmoid. Paired dermal ethmoid bones (proethmoids; lateral dermethmoids of Fink and Weitzman, 1982, p. 36) occur in esocoids, some stomiiforms, and all osmerids except *Hypomesus*, *Plecoglossus*, and *Mallotus*, which, like salangids, have a median dermethmoid (we observed paired proethmoids in one specimen of *Hypomesus transpacificus*). As argued by Patterson (1975) and Fink and Weitzman (1982), paired dermethmoids are a derived feature. The proethmoids of esocoids are elongate, unlike the small, platelike bones in osmerids. Other osmeroids (retropinnids, *Aplochiton*, *Lovettia*, *Lepidogalaxias*, and galaxiids) lack the dermethmoid (no. 2 below).

2. Ethmoid endoskeleton. The ethmoid endoskeleton in elopocephalans and clupeocephalans is primitively well ossified, with paired lateral ethmoids, a median mesethmoid that incorporates the supraethmoid and rostrodermethmoid, and a median ventral ethmoid that is usually fused with the vomer (Patterson, 1975). In lower euteleosts the most completely ossified ethmoid endoskeleton is in argentinids (Kobyliansky, 1990, fig. 10); elsewhere, ethmoid ossification is generally reduced. Apart from lateral ethmoids, there is no endoskeletal ethmoid ossification in opisthoproctids, retropinnids (McDowall, 1969), salangids (Roberts, 1984), and most salmonids (Stearley and Smith, 1993). Coregonids and osmerids are unusual in having the supraethmoid and dermethmoid separate, not fused. That condition has been reported elsewhere only in a Cretaceous euteleost of unknown relationships (Patterson, 1970, fig. 2) and in some stomiiforms (Weitzman, 1967; Fink and Weitzman, 1982); in other stomiiforms the two components are thought to be fused (Fink, 1985, p. 13). In esocoids the proethmoids fuse with paired endoskeletal ossifications (Jollie, 1975; Wilson and Veilleux, 1982; Reist, 1987). In *Aplochiton*, *Lovettia*, *Lepidogalaxias*, and galaxiids there is an endoskeletal mesethmoid that appears to lack any dermethmoid component.

The ethmoid endoskeleton in *Mallotus* differs from that of all other osmerids except salangids in two features: it is elongate and unossified except for lateral ethmoids, which appear between about 60 and 100 mm SL, much later than the ethmoid endoskeleton in other osmerids where it is short and always includes at least two perichondral ethmoid bones in addition to the lateral ethmoids. The elongate ethmoid endoskeleton of salangids is unossified.

3. [Begle's 1: Median posterior shaft of vomer present (0) or absent (1). State (1) coded in all Osmeroidei except *Lepidogalaxias*.] Shaft wrongly coded as absent in *Plecoglossus* (Chapman, 1941a; Klyukanov, 1975), galaxiids, and *Lovettia* (McDowall, 1969, 1984). In Begle (1991) the character was different, with (0) for a short shaft and (1) for a long one; coding was wrong in *Lovettia* (1) and *Aplochiton* (0) but correct in galaxiids (0). In *Aplochiton* and salangids the vomer is absent; they represent a third state (2). A vomerine shaft is

present in our *Hypomesus olidus*, and one was illustrated as present in *H. japonicus* and *H. nipponensis* (but absent in *H. olidus*) by Klyukanov (1977, fig. 4).

4. [Begle's 16: Vomerine teeth present (0) or as fangs on head of bone (1) or absent (2). State (1) coded in *Osmerus*, state (2) in galaxiids, *Lovettia*, *Aplochiton*, salangids, *Alepocephalus*, platytroctids, *Leptoderma*, *Bathyprion*, and *Bathytroctes*.] Vomerine teeth occur in all platytroctids (Sazonov, 1986; Matsui and Rosenblatt, 1987) but are absent in *Rouleina* (Markle, 1976, 1978). Absence of the vomer in *Aplochiton* and salangids (no. 3) means that they should be coded as (?). In Begle (1991) the character was two-state, presence (0) vs absence (1), with *Spirinchus* wrongly coded (1); inexplicably, *Osmerus* was coded (2) and salangids, *Aplochiton*, *Lovettia*, and galaxiids were coded (3).

5. [Begle's 103: Vomer ends anteriorly at margin of ethmoid cartilage (0) or extends beyond it (1). State (1) coded in all sampled argentinoids: *Argentina*, *Bathylagus*, *Opisthoproctus*, and *Glossanodon*.] *Aplochiton* and salangids, which have no vomer, accordingly coded (0), not (?).

6. [Begle's 58: Orbitosphenoid present (0) or reduced or absent (1). State (1) coded in all osmeroids and in *Alepocephalus*, *Bathylagus*, *Opisthoproctus*, *Bathyprion*, *Narcetes*, and *Rouleina*, with platytroctids polymorphic.] We commented (Patterson and Johnson, 1995, p. 26). The bone is wrongly coded as present in esocoids and as absent or reduced in *Alepocephalus*, *Narcetes*, *Rouleina*, and *Bathylagus* (Chapman, 1943; Kobyliansky, 1986). It is absent in our small *Leptochilichthys*. The orbitosphenoid is present in microstomatids (*Microstoma* and *Nansenia*; Chapman, 1948). In Begle (1991) the character was presence vs absence of the bone.

7. [Begle's 63: Pterosphenoids not reduced, meeting in midline (0) or reduced and widely separated, not meeting in midline (1). State (1) coded in all Osmeroidei and all sampled alepocephaloids.] The character is from Howes and Sanford (1987b, p. 27). We noted (Patterson and Johnson, 1995, p. 27) that the character is ambiguous because widely separated pterosphenoids occur in the most primitive teleosts, such as Jurassic pholidophorids and leptolepids, *Hiodon*, and *Elops* (Howes and Sanford, 1987b, also reported the condition in esocoids, alepocephaloids, and gonorynchiforms). In any case, the pterosphenoids are widely separated in salmonoids, esocoids, *Argentina*, and *Glossanodon*, and in all of these Begle coded them as in contact medially. As presented and coded by Begle, the character is worthless and we discard it.

8. [Begle's 71: Pterosphenoid without (0) or with small ventral flange midway along its length (1). State (1) coded in all Osmeroidei except salangids and *Retro-*

pinna.] The character is not so simple. Among osmerids, *Hypomesus* and *Mallotus* have a ventral flange or process from the anterior part of the pterosphenoid and the pterosphenoid–prootic junction is entirely in cartilage bone. In *Osmerus* there is a process about midway along the bone that is directed posteroventrally, toward a similar anterodorsal process from the prootic. *Spirinchus* resembles *Osmerus*, but the pterosphenoid and prootic processes meet and interdigitate, forming the ventral margin of a foramen between the two bones so that their junction is partly in membrane bone. *Allosmerus* and *Plecoglossus* have both a ventral process from the middle of the bone and a posteroventral process that interdigitates with an anterodorsal process from the prootic, as in *Spirinchus*. In *Thaleichthys* the pterosphenoids are much modified and almost contact the parasphenoid (Klyukanov, 1970), but as in *Allosmerus*, *Plecoglossus*, and *Spirinchus* a posteroventral process from the pterosphenoid interdigitates with an anterodorsal process from the prootic. Since the process on the anterior part of the pterosphenoid (as in *Hypomesus* or *Mallotus*) and the posteroventral process toward the prootic (as in *Osmerus* or *Spirinchus*) coexist in *Allosmerus* and *Plecoglossus*, the two processes cannot be homologous.

In retropinnids, a broad ventral process (Howes and Sanford, 1987b, fig. 2B) is present in *Prototroctes* and *Stokellia*. In galaxiids the pterosphenoid (like many other structures) is remarkably variable but shares one feature with that of retropinnids and *Aplochiton*: there is a broad and shallow medial arm, ossified perichondrally around the epiphyseal bar and applied to the underside of the frontal. In retropinnids and some galaxiids, the pterosphenoids meet in the midline by means of this epiphyseal arm (cf. preceding character as presented by Begle). Some galaxiids retain a slender anterior, perichondrally and endochondrally ossified arm of the pterosphenoid (e.g., *G. maculatus*, *G. fontanus*, *G. zebratus*, and *Neochanna*), but in others and in *Aplochiton* the anterior arm is ossified only in membrane bone, and in others it is missing entirely so that the pterosphenoid is L-shaped in ventral view, with only the posterior and medial (epiphyseal) arms (e.g., *Paragalaxias dissimilis*, *Galaxias fasciatus*, and *G. platei*). A ventral flange or process (always in membrane bone when present) is variable in galaxiids, even within species and from one side to another in individuals (*G. platei*, dried skeleton); absent in *Paragalaxias dissimilis* and some *Galaxias fasciatus*; and present in *Aplochiton*, *G. fontanus*, *G. maculatus*, *G. brevipinnis*, and some *G. fasciatus*. In *Lepidogalaxias* the pterosphenoid, placed far laterally, lacks the epiphyseal arm (and epiphyseal cartilage), has the anterior part ossified only in membrane bone, and has a minute ventral process from the center of the orbital margin.

Salangids (Roberts, 1984) and *Lovettia* lack a pterosphenoid and so must be coded as (?) for any feature concerning the bone.

Begle's original character 71 should be discarded; it comprises four separate two-state characters: (A) Pterosphenoid present or absent (in salangids and *Lovettia*); (B) pterosphenoid unmodified or with extensive epiphysial arm; (C) pterosphenoid with or without ventral process or flange from anterior half of margin; and (D) pterosphenoid with or without posteroventral process towards prootic. We have not checked all of the non-osmeroids for comparable structures.

9. [Begle's 72: Anterior margin of prootic rounded and smooth (0) or notched, with a small dorsal projection (1). State (1) coded in *Allosmerus* and *Spirinchus*.] We take the dorsal projection to be that which is directed towards or meets a process from the pterosphenoid (no. 8 above). We have also observed this in *Osmerus* (*O. eperlanus* and *O. mordax*), *Thaleichthys* (where the anterior margin of the prootic is almost horizontal, through reduction in the myodome and widening of the otic region), and *Plecoglossus*. See also no. 10, below, in *Allosmerus, Plecoglossus, Spirinchus,* and *Thaleichthys.*

10. [Begle's 73: Prootic/pterosphenoid contact at dorsal margin of prootic (0) or more medial, by interdigitation of prootic and pterosphenoid (1). State (1) coded in *Osmerus* and *Plecoglossus*.] The medial contact is between the membrane bone processes described in nos. 8 and 9 above. It also occurs in *Allosmerus, Spirinchus,* and *Thaleichthys* but is absent in *Osmerus.* The character is best treated as a further state (2) of no. 9.

11. Interorbital septum. In the osmerids *Allosmerus, Osmerus, Spirinchus,* and *Thaleichthys* there is a cartilaginous interorbital septum in the anterior part of the orbit (Klyukanov, 1970, figs. 1–4; 1975, fig. 8). Klyukanov (1975, p. 13), accepting Weitzman's (1967, p. 533) belief that the orbitosphenoid was primitively absent, regarded the extensive interorbital cartilage in these osmerids (and in some salmonines) as primitive. But presence of the orbitosphenoid is undoubtedly primitive in teleosts (Patterson, 1975, p. 427), and we regard the osmerid cartilage as derived. There is no comparable structure in other osmeroids, or in coregonids, argentinoids, or esocoids. Among alepocephaloids, there is a comparable cartilaginous septum in a few derived forms (e.g., *Rinoctes* and *Photostylus*).

12. [Begle's 57: Basisphenoid present (0) or reduced or absent (1). State (1) coded in all Osmeroidei

and all sampled argentinoids and alepocephaloids except *Argentina, Talismania, Leptoderma,* and *Bathylaco,* with platytroctids polymorphic.] In Begle (1991) the character was presence vs absence of the bone. We commented (Patterson and Johnson, 1995, p. 26) that the basisphenoid is wrongly coded as absent in *Lepidogalaxias,* where the bone is large, and in *Alepocephalus, Bathytroctes, Narcetes, Rouleina,* and *Glossanodon* (Begle's text correctly says that the bone is present in the latter). The basisphenoid is well-developed in some bathylagids (Kobyliansky, 1986), in most platytroctids (Sazonov, 1986), and in the microstomatids *Microstoma* and *Nansenia* (Chapman, 1948). The basisphenoid is present in esocids but absent in umbrids. It is absent in all opisthoproctids.

13. Myodome. The posterior myodome is primitively large in teleosts (Patterson, 1975, p. 543), and an extensive myodome occurs among lower euteleosts in argentinids, alepocephaloids, *Esox,* salmonoids, osmerids, retropinnids, and *Aplochiton.* Among osmeroids the myodome is absent in salangids, *Lepidogalaxias, Lovettia,* and some galaxiids. In most of the galaxiids we examined (*Neochanna, Nesogalaxias, Paragalaxias, Galaxias fontanus,* and *G. occidentalis*) there is a myodome with a wide but very shallow orbital opening, whereas in others (*G. zebratus* and *Galaxiella*) it is absent. *Galaxias paucispondylus* shows an intermediate condition, with the prootic bridge developed only at the extreme posterior part of the prootic, so that the myodome is no more than a shallow pit.

14. Buccohypophyseal canal. The buccohypophyseal canal through the parasphenoid is normally closed during ontogeny in teleosts; a patent canal has been recorded only in Mesozoic forms, in *Elops, Megalops* (Holstvoogd, 1965), and perhaps in a few clupeomorphs (Patterson, 1975, p. 530). There is an obvious median buccohypophyseal canal through the ossification center of the parasphenoid in *Aplochiton, Lovettia,* and in almost all galaxiids (e.g., *G. occidentalis, G. zebratus, G. paucispondylus, Galaxiella,* and *Neochanna*). We have not seen a buccohypophyseal canal elsewhere in osmeroids (it is not there in *Lepidogalaxias*).

15. Basipterygoid process and efferent pseudobranchial artery. Primitively in teleosts, the parasphenoid has a basipterygoid process, penetrated a foramen for the efferent pseudobranchial artery which lies anterior to the internal carotid foramen (Patterson, 1975, pp. 529, 532). In Recent teleosts, a basipterygoid process is recorded only in osteoglossomorphs (osteoglossoids and some mormyroids), except for Gosline's (1969, p. 196) report of "a pair of knob-like projections from the parasphenoid" in the alepocephaloid *Searsia koefoedi,* leading him to

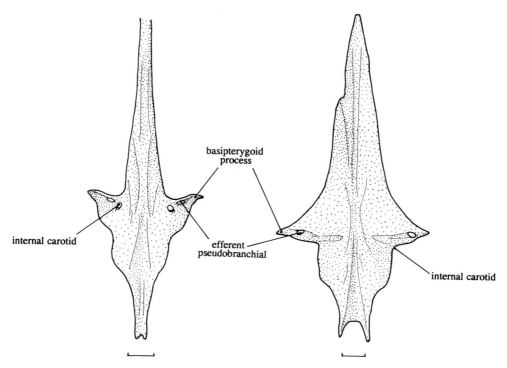

FIGURE 3 Basipterygoid process in alepocephaloids. Parasphenoid, in ventral view, of *Searsioides multispinus* (left, SIO 77-21, 113 mm SL) and *Bathylaco nigricans* (right, SIO 64-15, 86 mm SL). Scale bars: 1 mm. The internal carotid does not pass through the parasphenoid in *Bathylaco* but through the notch indicated.

conclude that it "appears to have a basipterygoid process." Gosline was right: *Searsia* and all other examined platytroctids have a basipterygoid process, penetrated in the usual way by a foramen for the efferent pseudobranchial artery (Fig. 3; also visible in Sazonov, 1986, figs. 10B, 11A). The only other Recent euteleost in which we have found a similar process is *Bathylaco* (Fig. 3). The efferent pseudobranchial foramen persists in various lower euteleosts (e.g, *Argentina*, coregonids, and *Salmo*; Patterson, 1975, p. 532) but is absent in osmerids and retropinnids, where the artery passes lateral to the parasphenoid, as in the majority of teleosts. However, retropinnids show a unique condition: in *Retropinna*, *Prototroctes*, and *Stokellia* there is a conspicuous earlike, paired cartilage projecting lateral to the parasphenoid, just in front of the opening of the myodome, and the cartilage is perforated by a foramen for the efferent pseudobranchial artery. In *Aplochiton* there is no such cartilage, but instead the parasphenoid contains a pair of foramina for the efferent pseudobranchials. There is no efferent pseudobranchial foramen in *Lovettia* or *Lepidogalaxias*, but galaxiids commonly have a foramen (e.g., *G.maculatus* and *G.zebratus*) or notch (e.g., *G.fontanus*, *G.platei*, and *Paragalaxias*) for the artery in the parasphenoid.

16. Otic bulla. In osmerids, there is variation in the form of the otic bulla (lateral wall of the saccular recess). *Thaleichthys* and *Spirinchus* have an inflated otic bulla with a large area of cartilage in its wall (Klyukanov, 1970, figs. 1, 2; 1975, fig. 7). In *Allosmerus*, *Mallotus*, and *Osmerus* the bulla is less inflated and the area of cartilage is smaller (Klyukanov, 1970, figs. 3, 4, 6), whereas in *Hypomesus* and *Plecoglossus* the bulla is uninflated (Klyukanov, 1970, fig. 5; 1975, figs. 5, 7). Klyukanov (1975, p. 13) considered the inflated bulla of *Thaleichthys* and *Spirinchus* to be primitive, but, as he correctly noted, the bulla is not inflated in outgroups, and when an inflated bulla of this type occurs elsewhere in teleosts (e.g., trachichthyid beryciforms) it can be shown to be derived.

17. [Begle's 74: Sphenotic spine blunt (0) or rodlike (1). State (1) coded in galaxiids, *Allosmerus*, *Hypomesus*, *Spirinchus*, and *Mallotus*, with salangids coded as (?) because the sphenotic is not ossified.] Within osmerids, we see no difference between the sphenotic spine (postorbital process) of *Osmerus* and that of *Hypomesus* or *Mallotus*. The spine is sharply pointed in the *Allos-*

merus, *Spirinchus*, and *Thaleichthys* that we studied. The character was discarded.

18. Lateral extent of frontal. In teleosts the supraorbital sensory canal primitively runs longitudinally approximately along the center of each frontal, and the frontal has a laminar portion, lateral to the sensory canal, that roofs the orbit. In *Lepidogalaxias* and *Lovettia* this lateral laminar portion is absent, and the sensory canal runs along the margin of the bone. Salangids have the frontal so weakly ossified (Roberts, 1984, figs. 3–6) that they cannot be assessed for this feature.

19. Extent of parietals. The parietals in teleosts primitively overlie the supraoccipital medially, meeting in the midline, and are overlapped by the frontals anteriorly so that their exposed surface is relatively small. In osmeroids there are three derived states. First, in all osmerids except *Hypomesus* and *Plecoglossus* the parietals are partially (*Mallotus* and *Osmerus*) or completely (*Allosmerus*, *Spirinchus*, and *Thaleichthys*) separated by the supraoccipital (they are also partially separated in *Lovettia*). Second, in all southern osmeroids (retropinnids, *Aplochiton*, *Lovettia*, *Lepidogalaxias*, and galaxiids) the parietals are not overlapped by the frontals but suture with them and extend forward to or beyond the postorbital process. Third, parietals are absent in salangids (Roberts, 1984).

In other lower euteleosts, the parietals are completely separated by the supraoccipital in all alepocephaloids (Greenwood and Rosen, 1971, p. 33) and esocoids, and in bathylagids and opisthoproctids among argentinoids. Among salmonoids the parietals are partially separated in *Thymallus*, *Stenodus*, and some *Coregonus* and are completely separated in all salmonines (Sanford, 1987, 1990).

20. Parietals and occipital (supratemporal) commissure. Primitively in teleosts, the parietals do not carry the occipital sensory canal, which lies in the extrascapulars. Among lower euteleosts, the parietal carries the medial part of the occipital commissural sensory canal, presumably through fusion with a medial extrascapular, in all argentinoids except bathylagids, where the commissure is secondarily absent. In argentinids the parietal carries the commissure across its posterior margin, the primitive position, but in microstomatids the lateral extrascapular is a long tube or gutter, directed forward over the autopterotic (the dermopterotic is absent, no. 21 below), and the occipital commissure passes across the anterior margin of the parietal. In opisthoproctids the parietal is rather short rostrocaudally, and the commissure passes across the middle of the bone.

21. Dermopterotic. The dermopterotic is primitively fused to the autopterotic in Recent teleosts. Gos-

line (1969, p. 197) reported that the dermopterotic is independent in *Alepocephalus rostratus*, and this is true also of our *A.agassizi*, *A.bairdi*, and *A.tenebrosus*, and of *Leptochilichthys* and *Talismania aphos*. In other alepocephalids we find the dermopterotic and autopterotic fused in *Bajacalifornia*, *Bathylaco*, *Bathytroctes*, *Binghamichthys*, *Narcetes*, *Rinoctes*, *Rouleina*, and *Talismania oregoni* and the dermopterotic absent in *Leptoderma* and *Photostylus*. In platytroctids the same three conditions occur: the dermopterotic and autopterotic are fused in *Paraholtbyrnia*; the dermopterotic is free in *Mentodus*, *Mirorictus*, *Pellisolus*, *Searsia*, and *Searsioides*; and the dermopterotic is absent in *Holtbyrnia*, *Platytroctes*, and *Sagamichthys*.

The relation between the temporal sensory canal and dermopterotic also varies in alepocephaloids. The primitive pattern, with the canal penetrating the bone from end to end, occurs in some platytroctids (*Paraholtbyrnia*; Sazonov, 1986, fig. 11) and in the alepocephalids *Bathylaco*, *Narcetes*, and *Talismania*. Derived conditions include a short enclosed canal in the dermopterotic (*Bajacalifornia* and *Binghamichthys*); the canal running superficial to the bone (*Bathytroctes*, *Rinoctes* and *Rouleina*); the canal running through one or more free ossicles (some platytroctids; e.g., Sazonov, 1986, fig. 10); the canal running superficial to a free bone (some platytroctids, *Alepocephalus*, and *Leptochilichthys*); and no dermopterotic (see above).

Among argentinoids, the dermopterotic is absent in microstomatids, some bathylagids (Kobyliansky, 1986, figs. 2–5), and apparently in all opisthoproctids except *Bathylychnops*.

22. Temporal fontanelles. The osmerids *Spirinchus* and *Thaleichthys* differ from other osmerids by eliminating the temporal fontanelles in the chondrocranial roof during ontogeny (Klyukanov, 1970, fig. 9). Klyukanov (1970) took this condition to be primitive, but comparison with outgroups (e.g., retropinnids and salmonoids) shows that it is derived. Closure of the fontanelles occurs as a derived state in a subgroup of *Oncorhynchus* (Stearley and Smith, 1993, character 1, where *Thaleichthys* is wrongly coded).

23. Posttemporal fossa. Primitively in teleosts the posttemporal fossa is extensive and is roofed by the dermopterotic and parietal. Among lower euteleosts, this primitive type of posttemporal fossa persists only in argentinids and esocids. There is an unroofed posttemporal fossa in other argentinoids (many of which have lost the dermopterotic, no. 21 above) and in salmonoids, osmeroids, and umbrids. Gosline (1969) reported that the alepocephaloids *Alepocephalus*, *Bathyprion*, and *Xenodermichthys* have no posttemporal fossa but that there is a roofed fossa in *Bathylaco*. We

could find no posttemporal fossa in *Bathylaco* and believe that alepocephaloids are characterized by loss of the fossa. They exhibit two different states. In *Bathylaco*, *Bathytroctes*, *Narcetes*, *Rinoctes*, and *Talismania antillarum*, the posterior margins of the parietal and dermopterotic form a straight transverse line, as in taxa with a roofed posttemporal fossa such as *Elops* and argentinids. In other alepocephaloids the margins of the parietal and dermopterotic, if a contact between them exists, form a ''V'' open posteriorly, as in taxa with an unroofed posttemporal fossa; we take the second condition to be derived and regard alepocephaloids that lack the dermopterotic (no. 21 above) as a special case of it.

24. [Begle's 25: Basioccipital without (0) or with (1) a caudally projecting peg on either side of the first vertebra, coded as present in galaxiids.] The peg is more restricted; when present, it carries Baudelot's ligament, which is double in some galaxiids, with one originating on the basioccipital and one on V1. In our cleared-and-stained material, pegs are present only in *Galaxias fasciatus* (McDowall, 1969, fig. 2; Begle's source). They are absent in *G. fontanus*, *G. zebratus*, *G. maculatus*, *G. occidentalis*, *G. paucispondylus*, *Galaxiella*, *Neochanna apoda*, and *Paragalaxias dissimilis*, although they may develop in larger individuals of species in which the ligament is double; pegs are absent in dried skeletons of *G. maculatus* and *G. platei* but present in *G. fasciatus*. Variability means that the character must be entered ''B'' in the matrix (Begle's coding for polymorphic taxa), making it empty if polymorphism for (0) and (1) is treated as (0). Character was discarded.

25. [Begle's 48: Occiput greatly depressed (1) only in salangids.] No comment.

26. [Begle's 55: Occipital condyle formed only by basioccipital (0) or tripartite, with exoccipital condyles (1). State (1) coded in neoteleosts and *Lepidogalaxias*, with salmonoids coded as (?).] Begle's query for salmonoids evidently refers to the primitive state in salmonoids, which might be state (0), as coded in *Coregonus* and *Stenodus* by Stearley and Smith (1993), or state (1), as in salmonines. Stearley and Smith (1993, p. 19) listed the tripartite condyle as a character of Salmonidae (including coregonines) and coded it as present in the coregonine *Prosopium* and in *Thymallus*. In our cleared-and-stained *Prosopium williamsoni* and *Thymallus thymallus* the condyle is not tripartite but is as illustrated by Rosen (1985, fig. 7B). According to Stearley and Smith's (1993) cladogram, the neoteleostean type of condyle (e.g., Rosen, 1985, fig. 3) is therefore synapomorphous for Salmoninae. There is also a tripartite condyle in opisthoproctids (seen in *Opisthoproc-*

tus, *Bathylychnops*, *Dolichopteryx*, *Macropinna*, and *Rhynchohyalus*).

B. Suspensorium

27. [Begle's 35: Palatine without (0) or with (1) distinctive dumbbell shape. State (1) coded in all osmerids, including *Plecoglossus*.] Begle (1991) credited the character to Chapman (1941b) and McAllister (1963), both of whom described the palatine as longer and more slender in *Hypomesus* and (in particular) *Mallotus*. Howes and Sanford (1987a, p. 167) correctly reported the derived state (their ''diabolo-shaped'') as absent in *Hypomesus* and *Mallotus* but in their second (1987b) paper used it as a character of Osmeridae, secondarily modified in those two genera. In *Hypomesus* the autopalatine is much shorter than in *Mallotus* and (particularly in *H. olidus* and *H. pretiosus*) differs from that of other osmerids in being shallower posteriorly rather than shorter. We find it impossible to discriminate the osmerid dumbbell or diabolo condition from that in, for example, the alepocephaloids *Bathytroctes*, *Rouleina*, and many platytroctids (Sazonov, 1986, figs. 4, 5).

28. Fusion between autopalatine and dermopalatine. The autopalatine and dermopalatine are separate in retropinnids but fused in *Lepidogalaxias* and all osmerids except *Hypomesus* and *Osmerus*. In *Plecoglossus*, Howes and Sanford (1987a, p. 146) reported fusion late in ontogeny (at ca. 100 mm), but the bones are separate in our specimens of ca. 150 mm. The two bones are separate in *Alepocephalus rostratus* (Gosline, 1969), *A. agassizi*, and *Leptoderma* but fused in our *A. tenebrosus* and in other alepocephalids, platytroctids, argentinoids (Kobyliansky, 1990, fig. 6), salmonoids, and esocoids. The dermopalatine is absent in galaxiids, *Aplochiton*, and *Lovettia* (Figs. 4B–4D), and the autopalatine is absent in salangids.

29. [Begle's 10: Palatine teeth present (0) or absent (1). State (1) coded in salangids, galaxiids, retropinnids, *Aplochiton*, *Lovettia*, and the alepocephaloids *Leptoderma* and *Rouleina*.] In retropinnids, this character depends on interpreting the single toothed bone that extends from beneath the autopalatine to the quadrate (Fig. 4A). McDowall (1969) and Williams (1987) interpreted the bone as fusion between a toothed palatine and a toothless ectopterygoid, but Begle evidently interpreted it as a toothed ectopterygoid (see no. 32). In one cleared-and-stained *Retropinna* (BMNH 1964.4.30.19) there are two separate bones on the right side, corresponding in position to a long toothed palatine and a short toothless ectopterygoid resembling that bone in *Lovettia* or *Lepidogalaxias* (Figs. 4D and

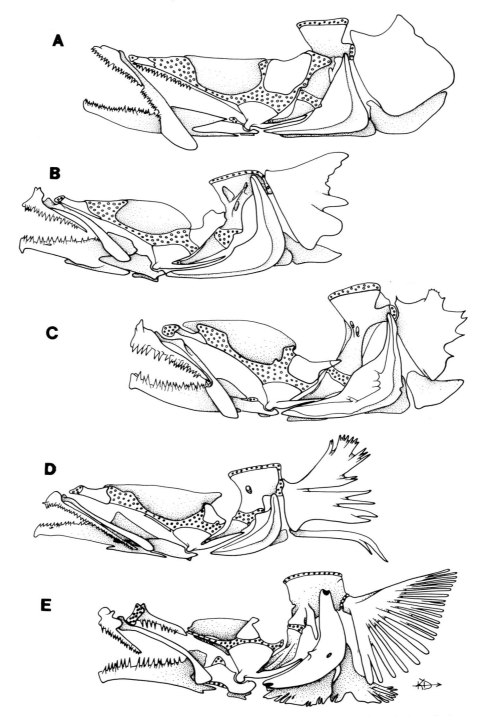

FIGURE 4 Suspensorium in galaxioids. Left jaws, palate, and operculum, in lateral view, of (A) *Retropinna retropinna*, MCZ 58015, 71 mm SL; (B) *Galaxias occidentalis*, AMNH 31478, 46 mm SL; (C) *Aplochiton zebra*, AMNH 31048, 77 mm SL; (D) *Lovettia sealei*, BMNH 1937.8.22.1, 41 mm SL; (E) *Lepidogalaxias salamandroides*, USNM 339265, 44 mm SL.

4E) but with a more extensive contact with the quadrate. On that basis, we agree with McDowall that retropinnids have a toothed palatine, usually fused with a toothless ectopterygoid. There is a similar prob-

lem in salangids, where there is a single elongate toothed bone, identified as "= ectopterygoid?" by Roberts (1984). In salangids the bone is well forward on the palatoquadrate and in *Protosalanx* is separated from the quadrate by the endopterygoid; we interpret it as the dermopalatine.

In bathylagids and microstomatids the palatine teeth are in a single row, differing from the patch of teeth that occurs in other argentinoids (argentinids and opisthoproctids; Kobyliansky, 1990, fig. 6).

30. [Begle's 70: Palatine contacting maxilla by a small knob if at all (0) or with lateral knob overlying maxilla (1). State (1) coded in galaxiids, *Lovettia*, *Aplochiton*, salangids, and *Lepidogalaxias*.] In galaxiids, *Lovettia*, and *Aplochiton* the lateral knob is the anterior end of the palatine cartilage, and the condition is the same as that in retropinnids (Figs. 4A–4D). In *Lepidogalaxias* (Fig. 4E) there is a distinctive cartilaginous lateral process that curves ventrally distally. In salangids the palatine cartilage is unmodified: in *Protosalanx* and *Salangichthys* the dermopalatine extends to the pointed tip of the cartilage, which merely ends in an oblique junction with the ethmoid; in *Neosalanx* the anterior end of the palatine cartilage is widened but does not overlie the maxilla. The derived state is autapomorphous for *Lepidogalaxias*.

31. [Begle's 3: Endopterygoid teeth broadly distributed over oral surface of bone (0) or a narrow band of larger teeth along medial margin (1) or teeth absent (2). State (1) coded in all Osmeroidei except *Lepidogalaxias* and salangids, which have state (2). State (2) also in all sampled argentinoids and alepocephaloids, which were wrongly coded (0) in Begle, 1991.] Esocoids and salmonoids, which lack endopterygoid teeth, are both wrongly coded (0) (Stearley and Smith, 1993, also wrongly code *Novumbra* as having endopterygoid teeth). Endopterygoid teeth, coded by Begle as absent in all Alepocephaloidea, are present in almost all platytroctids (Matsui and Rosenblatt, 1987, table 1), in pattern (1) (Sazonov, 1986, figs. 4 and 5).

Within osmeroids, there is variation in state (1). The endopterygoid teeth are in a single row in *Aplochiton*, *Lovettia*, galaxiids, and the osmerids *Allosmerus*, *Mallotus*, *Osmerus*, and *Thaleichthys*; a single row with 2 or 3 teeth lateral to the posterior end of the row in *Spirinchus*; a single row with 8 to 10 teeth lateral to its posterior end in *Plecoglossus*; and a single row with an extensive patch of teeth scattered lateral to the posterior end of the row in retropinnids and *Hypomesus*.

32. [Begle's 11: Ectopterygoid present (0) or absent (1). State (1) coded in galaxiids, *Aplochiton*, *Lovettia*, and salangids.] The ectopterygoid is wrongly coded as absent in *Lovettia* (Fig. 4D; McDowall, 1969). In our opinion (no. 29) retropinnids generally lack an independent ectopterygoid but have it fused with the dermopalatine.

33. [Begle's 23: Ectopterygoid posterior to autopalatine (0) or ventral to it (1). State (1) coded in retropinnids, with queries for galaxiids, *Lovettia*, *Aplochiton*, and salangids.] See no. 29 above; the bone is the dermopalatine and the character is therefore discarded.

34. [Begle's 32: Ectopterygoid with dorsal rim unmodified (0) or with a horizontal flange directed laterally (1). State (1) coded in *Plecoglossus*.] Howes and Sanford (1987a,b) found the derived state in *Hypomesus* and *Osmerus* as well as *Plecoglossus*. We agree with Begle (also Wilson and Williams, 1991, fig. 11) that no obvious flange exists in *Hypomesus* and *Osmerus*.

35. Ectopterygoid teeth. The ectopterygoid is primitively toothed in teleosts, as it is in neoteleosts. In lower euteleosts, the only records of ectopterygoid teeth are in alepocephaloids. In platytroctids, Sazonov (1986) reported ectopterygoid teeth in some individuals of *Sagamichthys* and in the larger species of *Holtbyrnia*, and Matsui and Rosenblatt (1987, p. 131) also found ectopterygoid teeth "variably present in larger individuals" of *Holtbyrnia* and *Sagamichthys* (we have seen such teeth in *H. innesi*). In alepocephalids, Nielsen and Larsen (1968) reported several rows of ectopterygoid teeth in 3 out of 10 specimens of *Bathylaco*, with a few teeth on one side in a 4th specimen, and we observed a single row of teeth (on both ectopterygoids) in a BMNH dried skeleton of *Alepocephalus rostratus*.

36. [Begle's 38: Metapterygoid without lateral shelf (0) or with short lateral shelf (1) or with prominent diagonal shelf (2). State (1) coded in *Mallotus* and *Plecoglossus* (where the shelf is described as horizontal) and state (2) in the other five osmerid genera.] *Mallotus* is wrongly coded, having state (2), so that (1) is autapomorphous for *Plecoglossus*. The shelf is very weak in *Hypomesus pretiosus*. Wilson and Williams (1991) cited "lateral ridge of metapterygoid" as a character of all osmerids.

37. [Begle's 89: Metapterygoid large and broad (0) or reduced and rodlike (1). State (1) coded in *Argentina*, *Bathylagus*, *Glossanodon*, and *Opisthoproctus*.] The metapterygoid is absent in *Bathylagus s.s.* and several other bathylagid genera (Kobyliansky, 1986) and is "small and insignificant" in microstomatids (Chapman, 1948, p. 10). In *Leptoderma*, *Aplochiton*, *Lovettia*, *Lepidogalaxias*, and most galaxiids the metapterygoid is also reduced and comparable in size and shape to the symplectic (Figs. 4B–4E). In *Lovettia* and *Lepidogalaxias* it is less than half the size of the symplectic and fails to contact the hyomandibular (Fig. 4; McDowall, 1969, fig. 3D; Roberts, 1984, fig. 22); they are recoded (1).

38. Metapterygoid position. Wilson and Williams (1991, figs. 11 and 12) published a cladistic analysis of osmerids in which *Hypomesus* is the sister of the other six Recent genera and is distinguished from them by one character, position of the metapterygoid: the metapterygoid is posterodorsal to the quadrate in *Hypomesus* but dorsal to it in *Allosmerus, Mallotus, Osmerus, Plecoglossus, Spirinchus,* and *Thaleichthys.* We agree with Wilson and Williams that appropriate outgroups show the same condition as *Hypomesus* [e.g., esocoids; argentinoids; coregonids and *Thymallus,* though not all salmonids (Stearley and Smith, 1993, fig. 7); and retropinnids (Fig. 4)]. In alepocephaloids both conditions occur among platytroctids (Sazonov, 1986, figs. 4 and 5) and among alepocephalids. In salangids the metapterygoid and quadrate are well ossified only in *Protosalanx* (Roberts, 1984, fig. 9), where the condition is as in *Hypomesus* and outgroups.

39. [Begle's 90: Metapterygoid without medial shelf (0) or with shelf at midpoint of bone (1). State (1) coded in all alepocephaloids except *Leptochilichthys, Bathylaco, Narcetes,* and *Rouleina.*] The shelf is the narrow, oblique one shown in most of the platytroctids in Sazonov's illustrations (1986, figs. 4 and 5). It is well-developed in our *Leptochilichthys* and *Rouleina* but absent in *Leptoderma,* where the metapterygoid is reduced with forward displacement of the quadrate (no. 37). Otherwise, we found the structure in our alepocephaloid material to agree with Begle's coding (we were unable to check *Bathyprion*). However, late in our work (July 1995) we saw Williams's (1987) dissertation; he recorded a medial shelf "of some form or another" in all examined alepocephaloids. Williams's sample included all the genera in Begle's except *Leptochilichthys* and *Bathyprion.* On rechecking our material, we found no shelf in *Leptoderma* but a vestige in *Bathylaco* and *Narcetes.* With those problems in interpretation, *Bathylaco* and *Narcetes* would best be coded as (?). Williams (1987) showed that a medial shelf on the metapterygoid like that in alepocephaloids also occurs in all osmerids (including *Plecoglossus*); we confirmed his observations.

40, 41. [Begle's 60, 61: Quadrate (60) and metapterygoid (61) without (0) or with (1) linear ridges, sometimes ramifying. State (1) coded for the quadrate in *Allosmerus, Spirinchus,* and *Thaleichthys* and for the metapterygoid in *Allosmerus* and *Spirinchus.*] The ridges are struts of membrane bone, and are commonly developed elsewhere, for example, on the quadrate in *Bathylagus* and other bathylagids (Kobyliansky, 1986) and on both quadrate and metapterygoid in *Alepocephalus* and platytroctids (Sazonov, 1986), on the quadrate in many salmonoids (e.g., *Oncorhynchus, Salvelinus,* and *Thymallus*), and on the metapterygoid

in some of those. In our osmerid material, struts occur on the quadrate in *Allosmerus, Mallotus, Thaleichthys,* and *Spirinchus,* where they are particularly strongly developed, and on the metapterygoid in *Allosmerus, Spirinchus,* and *Thaleichthys.* Polarity is questionable, and the character may be size related.

42. [Begle's 88: Ventral arm of symplectic short, less than half the length of the dorsal arm (0) or longer than the dorsal arm (1). State (1) coded in *Argentina, Glossanodon,* and *Opisthoproctus,* though the text says that it also occurs in *Bathylagus.*] The character expresses the forwardly displaced jaw articulation of argentinoids. The symplectic is primitively a straight bone (e.g., Patterson, 1973, figs. 7, 23, and 26), but in many teleosts it develops a more or less pronounced flexure near its midpoint. In bathylagids (Kobyliansky, 1990, figs. 6–8), *Opisthoproctus* (Trewavas, 1933, fig. 7), and microstomatids (Chapman, 1948, fig. 4) the two arms of the symplectic are about equal in length, as they are in osmerids (Weitzman, 1967, fig. 3), platytroctids (Sazonov, 1986, fig. 5), and umbrids (Wilson and Veilleux, 1982, fig. 7), for example. As presented by Begle, the primitive state is wrongly described, and the derived state occurs only in argentinids.

43. [Begle's 46: Hyomandibula not fused (0) or fused (1) to palatopterygoid. State (1) coded as autapomorphic for salangids.] Roberts (1984), who reported the salangid condition as unique, was unaware that the derived state is also reported in early ontogeny of *Clupea* (Norman, 1926; also *Alosa,* Shardo, 1995), *Sebastes* (Mackintosh, 1923), *Ictalurus* (Kindred, 1919), and *Heterotis* (Daget and d'Aubenton, 1957). Salangids maintain into adulthood a condition that is widespread, though not universal, in teleost embryos.

44. [Begle's 26: Hyomandibular without (0) or with (1) lateral spur at or below the level of the opercular process, projecting caudally to contact the preopercle. State (1) coded in galaxiids, *Aplochiton,* and all osmerid genera except *Thaleichthys.*] In Begle (1991) an undescribed state (2) was entered for the osmerid genera. Fink (1984b) used a lateral hyomandibular spur to characterize galaxiids (excluding aplochitonids), but Begle's wording of this character follows Howes and Sanford (1987b, p. 21). *Lepidogalaxias* (Fig. 4E) and esocoids (Wilson and Veilleux, 1982, fig. 7; Howes and Sanford, 1987b, fig. 5) are wrongly coded as lacking the spur. In osmerids, Wilson and Williams (1991, fig. 11) discriminated a vertical strut (= spur) in outgroups *Hypomesus, Mallotus, Plecoglossus,* and *Thaleichthys* from a lateral strut in *Osmerus, Allosmerus,* and *Spirinchus.*

45. [Begle's 98: Hyomandibular with lateral ridge short, less than half the length of the bone (0) or

longer, sometimes occupying the entire length of the hyomandibular shaft (1), or absent (2). State (1) coded in all osmerids and state (2) in retropinnids.] It is true that retropinnids, like salangids and *Lovettia*, which were both coded (0) by Begle, lack a lateral ridge on the hyomandibular; instead, as in *Aplochiton*, the metapterygoid sends a distinctive cartilaginous process across the hyomandibular towards the preopercle (Fig. 4). But osmerids (coded 1) do not have a long lateral ridge or crest but a short one (no. 44 above), whereas salmonoids, argentinids, alepocephalids (all coded 0), and outgroups such as *Elops* and *Chanos* all have a long one. We are unable to make sense of the character and so we discard it.

46. [Begle's 99: Hyomandibular with lateral ridge contacting (0) or failing to contact preopercle (1). State (1) coded in *Stokellia*, *Allosmerus*, *Thaleichthys*, *Leptoderma*, *Bathyprion*, *Bathytroctes*, *Bathylaco*, *Narcetes*, and *Rouleina*.] The character repeats no. 44 above and so is redundant in part. The coding is also confused and contradictory, for in the paragraph on the character Begle wrote "ridge contacts the preopercle in *Leptoderma*, *Bathyprion*, [etc.]," describing state (0), *not* the (1) coded for those genera. In the preceding character (no. 45), he (correctly) coded retropinnids as lacking a lateral ridge on the hyomandibular, but here he coded *Prototroctes* and *Retropinna* as having a ridge that contacts the preopercle. And under character 44 above he coded a "lateral spur . . . projecting caudally to contact the preopercle" as a *derived* feature present in all osmerid genera (including *Allosmerus*, here said to lack the contact) except *Thaleichthys*. With corrected coding the character is redundant except for the potential information on alepocephaloids, which is significant in Begle's cladogram as the only feature distinguishing *Leptochilichthys* (state 0) from the rest of the group (state 1, with reversal to 0 in *Alepocephalus* and platytroctids + *Talismania*). However, the hyomandibular crest unquestionably contacts the preopercle in our *Narcetes* and *Bathytroctes*, and in *Leptoderma* and *Bathylaco* the loss of contact has entirely different causes—forward inclination of the hyomandibular in the short-jawed *Leptoderma* and backward inclination in the long-jawed *Bathylaco*. Character was discarded.

47. [Begle's 100: Hyomandibular with anterior laminar extension (0) or with laminar bone reduced or absent (1). State (1) coded in neoteleosts, all Osmeroidei except *Prototroctes* and *Plecoglossus*, and all argentinoids and alepocephaloids except *Leptochilichthys* and *Narcetes*.] We commented (Patterson and Johnson, 1995, p. 27) at a time when we accepted Begle's accounts of characters and their distribution. We see no difference between the anterior part of the hyomandibular in *Prototroctes* (coded 1) and *Retropinna* (coded 0; Fig. 4A) or between it in *Narcetes* (coded 1) and

Bathytroctes or *Rouleina* (both coded 0). Laminar bone is certainly not absent on the hyomandibular of neoteleosts (e.g., Johnson *et al.*, 1996, figs. 6, 26–29). Character was discarded.

48. [Begle's 105: Opercular process of hyomandibular dorsally located and straight (0) or curved ventrally (1) or located at or below the midpoint of the bone (2). State (1) coded in platytroctids, *Leptoderma*, *Leptochilichthys*, and *Bathytroctes*; state (2) said to occur in *Bathyprion*, but it is coded (0).] State (2) occurs in *Bathylaco* (Markle, 1976, fig. 9). Figures 4 and 5 in Sazonov (1986) show the hyomandibular in 10 genera of platytroctids, and both states (0) and (1) occur. The opercular process is identical in our *Bathytroctes* (coded 1) and *Rouleina* (coded 0) and is similar but longer in *Narcetes* (coded 0). Character was discarded.

C. Jaws

49. [Begle's 29: Premaxilla with articular process not tightly adhering to maxillary head (0) or syndesmotically attached to it (1). Character from Howes and Sanford (1987a), with state (1) coded only in *Plecoglossus* and *Prototroctes*.] In this and the following two characters, *Opisthoproctus*, which has no premaxilla, is wrongly coded (0) rather than (?) by Begle.

50. [Begle's 66: Ascending process of premaxilla knoblike (0) or sharply triangular (1). State (1) coded in galaxiids and *Aplochiton*.] See no. 49.

51. [Begle's 83: Premaxilla without (0) or with (1) alveolar (postmaxillary) process extending beneath maxilla. State (1) coded in all osmeroids, all argentinoids and alepocephaloids, and neoteleosts, with state (0) in outgroups, salmonoids, and esocoids.] Begle credited the character to Rosen (1985), who regarded the "serial alignment" (p. 37) of the premaxilla and maxilla in salmonids as primitive. We commented (Patterson and Johnson, 1995, p. 27) and regard the character as wrongly coded in outgroups, esocoids, and salmonoids, where it is apomorphic for salmonines (there is an alveolar process in coregonids and *Thymallus*); with corrected coding [state (0) only in some salmonoids] the character is empty and is discarded. In Begle (1991) character 83 was different, referring to a long alveolar process (greater than half the length of the maxilla) coded as autapomorphic for *Stokellia*, but this is also present in *Prototroctes*, *Aplochiton*, *Lovettia*, and some galaxiids (Fig. 4; McDowall, 1969, fig. 7).

52. Premaxillary teeth. The premaxilla is primitively toothed. Among lower euteleosts it is toothless only in argentinoids and the alepocephalid *Leptochilichthys*. There is no premaxilla in opisthoproctids.

53. Premaxilla–maxilla contact. Primitively in teleosts the distal end of the alveolar process of the pre-

maxilla is free or attached to the maxilla by loose connective tissue. In bathylagids and microstomatids there is a distinctive articulation between the tip of the premaxilla and a notch or facet on the maxilla (Kobyliansky, 1990, fig. 4; note that there is no premaxilla in opisthoproctids, no. 52 above).

54. [Begle's 30: Maxilla and palatine without (0) or with (1) head-to-head articulation. Character from Howes and Sanford (1987a) with state (1) coded only in *Retropinna* and *Stokellia*.] No comment.

55. [Begle's 67: Maxilla more or less straight (0) or curved dorsally (1) in lateral profile. State (1) coded in galaxiids and *Aplochiton*.] We believe that this character refers to the configurations of retropinnid, aplochitonid, and galaxiid maxillae shown by McDowall (1969, fig. 7) and is based on a misreading of that figure (taking 7H, 7C, and 7F for 7F, 7G and 7H). In fact, the curvature of the maxilla in *Aplochiton* is most closely matched in McDowall's illustration by *Stokellia*. Character was discarded.

56. [Begle's 95: Teeth on maxilla present (0) or absent (1). State (1) coded in argentinoids and in *Alepocephalus*, *Leptoderma*, and *Leptochilichthys*.] The maxilla is also toothless in esocoids, *Prototroctes*, *Stokellia*, *Aplochiton*, *Lovettia*, galaxiids, and *Lepidogalaxias*, all wrongly coded (0).

57. Supramaxillae. Primitively in teleosts there are two supramaxillae. Among lower euteleosts, two supramaxillae occur in platytroctids (Sazonov, 1986, fig. 3) and the alepocephalids *Alepocephalus*, *Bajacalifornia*, *Bathytroctes*, *Binghamichthys*, *Narcetes*, *Rinoctes*, some *Rouleina* (Markle, 1976, p. 27), and *Talismania* among those we examined. There is one supramaxilla in esocoids, salmonoids, osmerids (lost during ontogeny in *Plecoglossus*; Howes and Sanford 1987a), salangids, and the alepocephaloids *Bathylaco*, some *Bathyprion* (Markle, 1976, p. 27), *Leptochilichthys*, *Leptoderma*, *Photostylus*, some *Rouleina*, and *Xenodermichthys* among those we examined. There are no supramaxillae in southern osmeroids (= galaxioids: retropinnids, *Aplochiton*, *Lovettia*, *Lepidogalaxias*, and galaxiids) or in argentinoids.

58. Shape and dentition of dentary. There is significant variation in the dentary dentition in osmerids. In *Hypomesus* (Patterson, 1970, fig. 37; Klyukanov, 1970, fig. 7) teeth are small and confined to the anterior quarter of the jaw, and there is a high coronoid process with a concave anterior margin. In juvenile *Plecoglossus* (Howes and Sanford, 1987a, fig. 4) the coronoid process is similar, and in *Plecoglossus* and *Mallotus* the teeth are also small and are confined to the anterior half of the jaw. In other osmerids (*Allosmerus*, *Osmerus*, *Spirinchus*, and *Thaleichthys*; Klyukanov, 1970, fig. 7; 1975, fig. 9) and

in salangids (Roberts, 1984) and southern osmeroids (McDowall, 1969) teeth are larger and occupy more than half the length of the jaw. In outgroups, Stearley and Smith (1993, character 65) assessed the *Hypomesus* pattern, which occurs also in coregonids, as primitive relative to the *Spirinchus* pattern, which occurs also in salmonids.

Among argentinoids, bathylagids and microstomatids share a derived pattern of the dentary, with a long single row of incisorlike teeth (Kobyliansky, 1990, fig. 5).

59. Meckelian fossa. The opening of the Meckelian fossa on the inner face of the dentary in osmerids was discussed by Klyukanov (1975, p. 7) and Howes and Sanford (1987a, p. 157). The opening is very small and placed in the anterior third of the bone in *Hypomesus* (Patterson, 1970, fig. 37) and juvenile *Plecoglossus* (Howes and Sanford, 1987a, fig. 4), similarly placed but slightly larger in *Osmerus* and salangids, larger and beneath the middle of the tooth row in *Mallotus*, and large and beneath the rear of the tooth row in *Allosmerus*, *Spirinchus*, and *Thaleichthys* (Klyukanov, 1975, fig. 9). Howes and Sanford (1987a) noted that the fossa is also small and anteriorly placed in retropinnids and *Aplochiton*. It is similar in coregonids and in *Glossanodon* (Kobyliansky, 1990, fig. 5). Weitzman (1967, p. 529) called the recess "small (or practically nonexistent) and far anterior" in salmonids and esocoids.

60. [Begle's 31: Paired postsymphysial cartilages absent (0) or present (1) at dentary symphysis. Character from Howes and Sanford (1987a), with state (1) coded only in *Osmerus* and *Plecoglossus*.] Howes and Sanford (1987a) found a median postsymphysial bone in one specimen of *O. mordax*; like them, we have seen it only in that specimen and not in other specimens of that species or of *O. eperlanus*. There is no demonstrable shared feature and the character becomes autapomorphous for *Plecoglossus*.

61. [Begle's 37: Dentary without (0) or with (1) medial tusk-like process at symphysis. Character said to be from Howes and Sanford (1987a,b), with state (1) coded in *Osmerus* and *Plecoglossus*.] Howes and Sanford (1987a) did not report a medial tusk-like process in *Osmerus*, and their "symphysial dentary process" (1987b, p. 24) said to be shared by *Osmerus* and *Plecoglossus* probably refers to the "postsymphysial notch" (Howes and Sanford, 1987a, p. 157) present in the margin of the dentary in an *Osmerus mordax* and in *Plecoglossus* of 40 to 60 mm SL. We have found no such notch in other specimens of *O. mordax*, or in *O. eperlanus*, and there is again no demonstrable shared feature so that the character is autapomorphous for *Plecoglossus*.

62. [Begle's 2: Articular fused with angular (0) or absent or greatly reduced, appearing late in ontogeny (1). State (1) coded in all Osmeroidei.] The character is from Fink (1984b), who gave no source or discussion. Howes and Sanford (1987b, p. 26) interpreted it as late ossification of the articular in all osmeroids and reported an osmeroid-like condition in *Argentina sphyraena*. The articular of *Argentina* (Kobyliansky, 1990, fig. 5) does not appear to differ from that of other argentinoids, and we accept the character as true of osmeroids, but it also occurs in esocoids (e.g., Nelson, 1973, fig. 6H and 6K; Jollie, 1975, p. 76).

63. [Begle's 102: Mouth terminal and large (0) or very small (1). State (1) coded in *Argentina, Bathylagus, Opisthoproctus* and *Glossanodon*.] No comment.

64. Tooth attachment. Fink (1981) identified four modes of tooth attachment in teleosts in a survey of the jaws and pharyngeals of a wide range of species. He distinguished hinged (depressible) teeth as type 4 and found them to be restricted to neoteleosts and the pharyngeals of *Esox*. Our observations on esocoids differ from Fink's (1981) in two ways. First, he assessed only the pharyngeal teeth of *Esox* as type 4, but we find the jaw teeth also to be type 4. Second, among umbrids according to Fink, *Dallia* has type 2 teeth in the jaws and pharyngeals, and *Umbra* has type 1 in both; we found the reverse, type 1 in *Dallia* and type 2 in *Umbra* (also in *Novumbra*). The same patterns as in umbrids are found in some lower neoteleosts (for example, type 1 in *Gonostoma* and type 2 in *Synodus*; Fink, 1981).

D. Circumorbital Bones

65, 66. Antorbital (65) and supraorbital (66). These two bones are primitively both present in teleosts and in lower euteleosts (e.g., platytroctids, argentinids, salmonoids, and osmerids). The antorbital is absent in esocoids, salangids, retropinnids, some galaxiids, and the alepocephalids *Leptoderma* and *Rouleina* (Markle, 1976). From our specimens and the literature, we could not determine whether opisthoproctids have an antorbital or not. The supraorbital is absent in umbrids, opisthoproctids, retropinnids, *Aplochiton, Lovettia, Lepidogalaxias*, some galaxiids, and the alepocephalids *Alepocephalus, Bajacalifornia, Bathyprion, Leptoderma, Narcetes, Rouleina,* and *Xenodermichthys* (Markle, 1976).

In bathylagids and microstomatids the supraorbital is long and there is an extensive contact between it and the dermosphenotic above the posterior part of the orbit (Kobyliansky, 1986, fig. 1; 1990, fig. 3).

67. Number of infraorbitals. Primitively in teleosts there are seven canal-bearing infraorbitals, including

the antorbital and lachrymal anteriorly and the dermosphenotic posteriorly (Nelson, 1969b). The osmerid *Spirinchus* has an extra infraorbital, with five between the lachrymal and dermosphenotic (Weitzman, 1967). We also found an extra infraorbital on one side of a specimen of *Thaleichthys*. Salangids have one infraorbital (behind the eye) or none (Roberts, 1984). *Aplochiton, Lovettia* and galaxiids have two infraorbitals, the lachrymal and the first infraorbital behind it (McDowall, 1969); *Lepidogalaxias* has none.

68. Lachrymal. Primitively, the lachrymal is attached to the lateral ethmoid by loose connective tissue. In *Aplochiton, Lovettia*, and galaxiids the lachrymal articulates with the lateral ethmoid by a cartilage-covered condyle. The lachrymal is primitively larger than the succeeding infraorbitals. In bathylagids, microstomatids (Kobyliansky, 1990, fig. 3), and opisthoproctids the first infraorbital is larger than the lachrymal.

E. Ventral Part of Hyoid Arch, Branchiostegals, and Operculum

69. [Begle's 20: Ventral border of ceratohyal straight with branchiostegals along most of its length (0), or deeply concave anteriorly with branchiostegals restricted to area behind the concavity (1), or with rectangular notch (2). State (1) coded in *Prototroctes, Retropinna,* and *Stokellia*, and state (2) in *Talismania* and platytroctids.] State (1) also occurs in *Aplochiton* (Chapman, 1944, fig. 5), and in all four genera (*Aplochiton* and retropinnids) cartilage extends along the ventral border of the deep posterior part (fig. 5A). Begle's state (2) refers not to a notch in the ventral border but to an excavation in the *dorsal* border in some alepocephaloids. It is a different character, part of a different transformation series (no. 70 below). In any event, our *Talismania* (*T.antillarum, T.aphos,* and *T.oregoni*) all have a rectangular distal ceratohyal with no dorsal notch and no fenestra, making state 2 autapomorphic (and polymorphic, no. 70) in platytroctids in Begle's sample.

70. Dorsal margin of ceratohyal. Begle (1992, character 20) mentioned a notch in the dorsal border of the ceratohyal (no. 69) in platytroctids and the alepocephalid *Talismania* and coded it as a derived feature. Sazonov (1986, figs. 6–8) illustrated the notch in several genera among platytroctids but showed that others have a fenestra in the bone (Fig. 6D), as does the alepocephalid *Rinoctes* (Markle and Merrett, 1980). The notch is clearly homologous with the fenestra (fig. 6). Elsewhere in teleosts, a ceratohyal fenestra ("beryciform foramen" of McAllister, 1968) is a primitive feature (Rosen and Patterson, 1969, p. 408). This

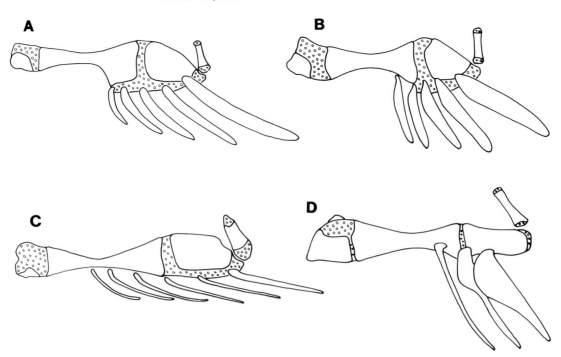

FIGURE 5 Hyoid in galaxioids. Left hyoid bar and branchiostegals, in lateral view, of (A) *Retropinna retropinna*; (B) *Galaxias occidentalis*; (C) *Lovettia sealei*; (D) *Lepidogalaxias salamandroides*. Same specimens as Fig. 3.

primitive, fenestrate type of ceratohyal can be discriminated by the presence of cartilage along the entire dorsal margin, whereas when the fenestra is lost, the dorsal margin is closed by perichondral bone (Fig. 6). The primitive cartilaginous dorsal margin of the ceratohyal is not always associated with an obvious fenestra in the bone, for example, in *Elops*, *Megalops*, *Etrumeus*, or *Clupea* (Fig. 6A), and among platytroctids Sazonov (1986, fig. 6) illustrated ceratohyals with a cartilaginous dorsal margin and no fenestra in *Pectinantus* and *Platytroctegen*. Our *Platytroctes* also have a cartilaginous dorsal margin and no fenestra (cf. Sazonov, 1986, fig. 6A). Beyond alepocephaloids, the only other Recent lower euteleosts with a cartilaginous dorsal margin to the ceratohyal are coregonids, where there is also a fenestra (Fig. 6B; Patterson, 1970, fig. 28; Stearley and Smith, 1993, character 90, with the fenestrate condition wrongly coded as derived). The only report of a ceratohyal fenestra in alepocephalids is in *Rinoctes* (Markle and Merrett, 1980; our cleared-and-stained *Rinoctes* lacks the hyoid and gill arches, but an alcohol specimen shows a fenestra). There is a complete cartilage-covered upper margin on the ceratohyal in *Bathylaco*, *Bathytroctes*, *Narcetes*, *Rinoctes*, *Talismania antillarum* and *T. oregoni*, and *Rouleina attrita* and *R.maderensis*, whereas *R.squamilateratus* and *Bajacalifornia* retain the dorsal notch with cartilage cover-

ing the margin of the bone in front of and behind the notch. Other alepocephalids we have seen (including *Talismania aphos*; cf. *T.antillarum* and *T.oregoni* above) have a waisted ceratohyal with the dorsal margin closed by perichondral bone (Fig. 6F). That type of ceratohyal occurs in all esocoids, argentinoids, salmonids (Fig. 6C), and osmeroids. The primitive type of ceratohyal, with a cartilaginous dorsal margin, persists in various neoteleosts (e.g., neoscopelids, many beryciforms, zeiforms, and some percoids).

71. [Begle's 62: Proximal ceratohyal ("epihyal") without (0) or with (1) midlateral foramen. State (1) coded in all osmerids except *Thaleichthys*, although the text implies that the latter should also be coded (1).] The "foramen" is the termination of the groove for the hyoidean artery; it is also present (for example) in *Pantodon*, *Elops*, *Megalops*, *Clupea* (Fig. 6A), and salmonoids (Fig. 6B and 6C) and is primitive for teleosts. We regard state (0) as autapomorphic for *Thaleichthys* (within osmerids) and discard the character.

72. [Begle's 68: Proximal ceratohyal ("epihyal") more than half the length of distal ceratohyal (0) or short, much less than half the length of the distal ceratohyal (1). State (1) coded in galaxiids and *Aplochiton*.] In galaxiids (e.g., Fig. 5B) the proximal ceratohyal is about half the length of the distal, or slightly more. The character is true only of *Aplochiton*.

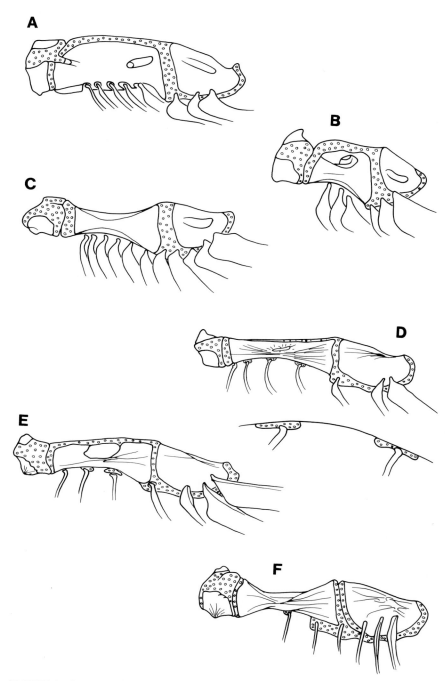

FIGURE 6 Ceratohyals and branchiostegal cartilages. Left hyoid bar, in lateral view, of (A) the clupeid *Clupea harengus*, BMNH 1932.2.15.1, 63 mm SL; (B) the coregonid *Prosopium williamsoni*, BMNH 1892.12.30.340, 81 mm SL; (C) the salmonid *Thymallus thymallus*, BMNH 1970.10.14.3, 70 mm SL; (D) and (E) the platytroctid alepocephaloids (D) *Paraholtbyrnia cyanocephala*, SIO 77-53, 125 mm SL [with (below) a close-up of two branchiostegal cartilages] and (E) *Searsia koefoedi*, SIO 77-38, 115 mm SL; (F) *Alepocephalus tenebrosus*, SIO uncat., 135 mm SL. A, B, and D show the primitive teleostean condition, with the upper border of the distal ceratohyal covered by cartilage and with a foramen (A) or fenestra (B and D) associated with the passage of the hyoidean artery. E shows a more derived condition, in which the fenestra is converted into a notch in the dorsal margin of the bone, though here the notch is still bridged by a strand of cartilage. C and F show the derived condition, typical of most teleosts, in which the distal ceratohyal is a waisted bone with its dorsal margin closed in perichondral bone. Branchiostegal cartilages, found in primitive alepocephaloids, are shown in D–F, with F showing a variant in which the posterior branchiostegal cartilage is continuous with that between the distal and proximal ceratohyals.

73. [Begle's 86: Interhyal elongate and rodlike (0) or short and dumbbell-shaped. State (1) coded in *Argentina*, *Bathylagus*, *Opisthoproctus*, and *Glossanodon*.] The character is related to the small mouth and forwardly inclined suspensorium in argentinoids. As described, it is true of argentinids, opisthoproctids, and microstomatids (the last not checked by Begle) but not true of *Bathylagus* (e.g., Kobyliansky, 1986, fig. 10), where the interhyal is as long as the epihyal and no shorter or more waisted than that, for example, of platytroctid alepocephaloids (Sazonov, 1986, fig. 6), *Umbra* (Wilson and Veilleux, 1982, fig. 8) or some galaxioids (Fig. 5). The interhyal is very short, though not always strongly waisted, in primitive salmonoids (e.g., coregonids, *Thymallus*, and *Brachymystax*).

74. [Begle's 87: Small cartilages connecting branchiostegals with hyoid arch absent (0) or present (1). State (1) coded only in *Talismania*.] These cartilages (Fig. 6) are widespread in platytroctids: we have seen them in *Holtbyrnia*, *Mentodus*, *Mirorictus*, *Paraholtbyrnia*, *Pellisolus*, *Sagamichthys*, and *Searsia*, but not in *Platytroctes* or *Searsioides*; they are illustrated by Sazonov (1986, figs. 6–8) in all platytroctid genera except *Platytroctes*, including juvenile *Searsioides*. They also occur among alepocephalids in *Bajacalifornia*, *Bathylaco*, *Bathytroctes*, *Leptochilichthys*, and *Rinoctes* and in the osmerids *Mallotus* (on the posterior three of four or five branchiostegals on the distal ceratohyal) and *Hypomesus olidus* (on the second or on the first and second of four branchiostegals on the distal ceratohyal). In *Alepocephalus tenebrosus* (Fig. 6F) a strand of cartilage extends forward along the ventral margin of the ceratohyal as far as the base of the second (of three) branchiostegals on the bone, and there is a separate cartilage at the base of the most anterior branchiostegal. We have not found branchiostegal cartilages in other alepocephaloids we checked (*Leptoderma*, *Narcetes*, *Photostylus*, *Rouleina*, *Alepocephalus agassizi*, and *Talismania oregoni*).

75. Branchiostegal attachment. In teleosts generally and in lower euteleosts the branchiostegals are normally differentiated into an anterior series inserting on the ventral or internal face of the hyoid bar (ventral or internal branchiostegals of McAllister, 1968) and a posterior series inserting on the external face of the bar (external branchiostegals of McAllister, 1968). These two series are recognizable in all lower euteleosteans except bathylagids and opisthoproctids, where the branchiostegals are all external.

76. [Begle's 82: Opercle extending dorsally above its articulation with the hyomandibular (0) or not (1). State (1) coded in galaxiids, *Lovettia*, *Aplochiton*, retropinnids, salangids, *Lepidogalaxias*, and all alepocephaloids.] State (1) is not true of *Bathylaco* (Fig. 7J and no. 78 below). The dorsal part of the opercle is also reduced in esocoids (e.g., Wilson, 1984, fig. 9).

77. [Begle's 65: Anterodorsal border of opercle horizontal and without spine (0) or with notch and spine (1) or with deep, narrow notch (2). State (1) coded in *Plecoglossus*, *Hypomesus*, *Osmerus*, and *Thaleichthys*; state (2) in *Allosmerus*, *Mallotus*, and *Spirinchus*.] The anterodorsal border of the opercle is characteristically emarginate in most osmerids (Fig. 7; Klyukanov, 1970, fig. 10; 1975, fig. 10). In our material of the seven genera (Fig. 7) it is unreasonable to regard *Allosmerus*, *Mallotus*, and *Spirinchus* as having a different condition from the other genera, though if one worked only from Klyukanov's somewhat schematic drawings, that conclusion might be possible. In a blind test where one of us asked the other to sort eight stained opercles, the groups found were *Allosmerus* with *Mallotus*; *Spirinchus* with *Hypomesus pretiosus*, *Osmerus*, *Thaleichthys*, and *Plecoglossus*, with the last most divergent; and *H. olidus* (which has an excavation like an inverted keyhole, Fig. 7A) on its own. Other groupings might be detected (Fig. 7), but we see Begle's state (2) only in *Allosmerus* and *Mallotus*. See also no. 78 below.

78. [Begle's 94: Dilatator spine on dorsal margin of opercle absent (0), present (1), or a large spinelike process extending dorsally above opercle (2).] State (1) coded in all alepocephaloids except *Bathylaco*, which has state (2). The character is redundant since it repeats nos. 76 and 77 above (Begle's 65 and 82). When the dorsal part of the opercle is emarginate or reduced the dilatator operculi muscle has to insert somewhere, and there is no difference between the large spine in front of the emargination in osmerids (Fig. 7A–H), *Dolicholagus* (Begle's *Bathylagus*, his 1991, fig. 7) and *Bathylaco* (Fig. 7J), or between the small spine (sometimes merely the dorsal margin of the articular facet, e.g., *Talismania*, Fig. 7K; *Lovettia*, Fig. 4D; *Lepidogalaxias*, Fig. 4E; and *Esox*) in galaxioids, salangids, and other alepocephaloids. If this character were retained, state (2) should be coded in osmerids (duplicating Begle's 65, our 77) and state (1) in galaxioids (duplicating Begle's 82, our 76). We have added *Bathylaco* to character 77 (Begle's 65, state 1) and discarded this character.

79. [Begle's 17: Posterior border of suspensorium rounded/smooth (0) or incised/emarginate (1). State (1) coded in galaxiids, *Aplochiton*, *Lovettia*, and *Lepidogalaxias*.] The character refers to the serrated or fimbriate margin of the opercle in those fishes, which is particularly deeply incised in *Lovettia* and *Lepidogalaxias* (Fig. 4D and 4E). The margin of the opercle is also

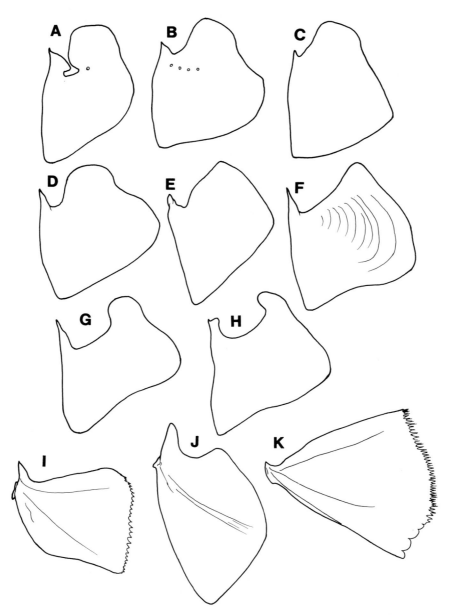

FIGURE 7 Opercles of osmerids and alepocephaloids. Left opercle, in lateral view, of (A) *Hypomesus olidus*, HSU 86-33, 70 mm SL; (B) *H.pretiosus*, HSU 81-87, 55 mm SL; (C) *Plecoglossus altivelis*, BMNH 1984.12.6.16, 63 mm SL; (D) *Osmerus mordax*, HSU 85-46, 75 mm SL; (E) *Spirinchus thaleichthys*, USNM 105639, 110 mm SL; (F) *Thaleichthys pacificus*, USMN 342051, 140 mm SL; (G) *Mallotus villosus*, AMNH 26286, 137 mm SL; (H) *Allosmerus elongatus*, USNM 342050, 90 mm SL; (I) *Pellisolus eubranchus*, SIO 60-287, 95 mm SL; (J) *Bathylaco nigricans*, SIO 64-15, 86 mm SL; (K) *Talismania aphos*, SIO 72-144, 105 mm SL.

fimbriate or incised in most bathylagids (Kobyliansky, 1986, figs. 6–9), and less strongly so in many alepocephaloids, including some platytroctids (Sazonov, 1986, fig. 4), *Alepocephalus, Bathytroctes, Leptoderma, Leptochilichthys,* and *Narcetes* and to some extent *Rouleina* and *Talismania* in Begle's sample. Rather than ex-

tend state (1) to all those fishes (in many of which the condition could not reasonably be distinguished from that in *Aplochiton* or some galaxiids), we restrict the derived state to the extremely deep incisions seen in *Lovettia* (Fig. 4D; McDowall, 1969, fig. 3D) and *Lepidogalaxias* (Fig. 4E; Roberts, 1984, fig. 22).

F. Gill Arches

80. [Begle's 106: Basihyal with scattered teeth (0), marginal fangs (1), small teeth on terminus (2), terminal fangs (3), or toothless (4). State (1) coded in salmonoids, galaxiids, *Aplochiton*, *Lovettia*, retropinnids, and all osmerids except *Hypomesus* and *Mallotus*; state (2) in *Glossanodon*; state (3) in *Argentina*; and state (4) in salangids, *Lepidogalaxias*, and all sampled Alepocephaloidea except *Leptochilichthys* (coded ?; the basihyal is absent).] Salangids are wrongly coded since basihyal teeth are present in *Protosalanx* (state 0) and *Salanx reevesi* (a median row) (Nelson, 1970b; Roberts, 1984). Platytroctids are wrongly coded since almost all have basihyal teeth (Matsui and Rosenblatt, 1987, table 1; Begle illustrated them in *Sagamichthys*, 1991, fig. 5), either in a median row or pattern (1). In salmonoids, state (1) is synapomorphous for salmonines (Stearley and Smith, 1993, character 85) since coregonids and *Thymallus* have state (0) (Norden, 1961, pl. 6). *Glossanodon* is polymorphic for states (2) and (4) (Cohen, 1958). Argentinids are the only argentinoids with basihyal teeth (Kobyliansky, 1990, p. 159, fig. 7; Greenwood and Rosen (1971, fig. 19) illustrated a toothed basihyal in the microstomatid *Nansenia* that differs greatly from the toothless bone in our specimens and in Kobyliansky's).

81. Basibranchial dentition. Primitively in teleosts an elongate toothplate covers basibranchials 1–3 (Nelson, 1969a, p. 494; Rosen, 1974, figs. 2–4). Derived states in lower euteleosts are (1) loss of teeth, leaving a toothless dermal bone (Stearley and Smith, 1993, character 89); and (2) loss of both teeth and the dermal bone (Stearley and Smith, 1993, character 88). State (1) occurs in all argentinoids; in all alepocephaloids except platytroctids, where basibranchial teeth are present in most genera (Matsui and Rosenblatt, 1987, table 1), and *Rinoctes* (basibranchial teeth recorded by Markle and Merrett, 1980; confirmed in two USNM specimens); among osmeroids in *Aplochiton*, *Lovettia*, *Lepidogalaxias*, and galaxiids (Rosen, 1974, fig. 5); and among salmonoids in our *Coregonus cylindraceum* (contra Stearley and Smith, 1993), *Prosopium*, *Thymallus* (contra Stearley and Smith, 1993), and many salmonines. State (2) occurs in salangids (Roberts, 1984). A third derived state is fragmentation ("secondary subdivision," Nelson, 1969a, p. 497) of the basibranchial toothplate, which characterizes esocoids (Rosen, 1974, fig. 1). In salmonids the toothless dermal plate is occasionally divided (*Salvelinus fontinalis* divided in Norden, 1961, pl. 6B, but not in Rosen, 1974, fig. 3B and 3C; and *Oncorhynchus keta*, Rosen, 1974, fig. 3E).

82. Toothplate of fourth basibranchial (Bb4). Presence of this toothplate is the single character cited to distinguish esocoids from all other euteleosts (Fink and Weitzman, 1982; Fink, 1984b; Begle, 1992). Patterson and Johnson (1995, p. 25) reviewed the distribution of a Bb4 toothplate in lower teleosts and concluded that the character is questionable; they failed to mention that the toothplate also occurs in percopsiforms (*Amblyopsis*, Rosen, 1962, fig. 13; *Aphredoderus*, Nelson, 1969a, pl. 92, fig. 2; and *Percopsis*, Rosen and Patterson, 1969, pl. 65, fig. 1) and was illustrated by Kobyliansky in the argentinoids *Bathylagichthys* (1986, fig. 11) and *Nansenia* (1990, fig. 8; personal observation), so destroying its validity in distinguishing esocoids.

83. [Begle's 64: First basibranchial unmodified (0) or with ventral cartilaginous vane (1). State (1) coded in retropinnids, osmerids, and all sampled argentinoids and alepocephaloids.] We commented (Patterson and Johnson, 1995, p. 26), noting that Begle's "unmodified" state occurs in argentinids and *Opisthoproctus* (it also occurs in bathylagids). Weitzman's (1974, fig. 75) comparison of the basibranchials in *Spirinchus* and three stomiiforms shows that the latter also have Begle's "derived" state, as do *Elops* (Fig. 8A; Nelson, 1968a, fig. 1) and many clupeoids (e.g., *Chirocentrus*, Fig. 8B; Nelson, 1970a, fig. 5). We see no difference between the configuration in *Argentina* (Fig. 8C) or *Retropinna* (Fig. 8D), both coded (1) by Begle, and in some salmonoids (Figs. 8E–8G), coded (0); *Thymallus* (Fig. 8G) has the "derived" state in much the same form as osmerids (Fig. 8H and 8I) and alepocephaloids (Fig. 8J). If retained, the character would have to be coded (0) in argentinoids and retropinnids but (1) in basal neoteleosts (stomiiforms) and some salmonoids and outgroups. We regard the character as so problematic and subjective that we discard it.

84. [Begle's 97: Basibranchials without (0) or with (1) a narrow median dorsal ridge "separating right and left portions of branchial basket." State (1) coded in all sampled alepocephaloids.] The character originated with Greenwood and Rosen (1971, p. 8) and Rosen (1974, p. 274). Markle (1976, p. 82) wrote, "the sharp medial basibranchial ridge is virtually absent in every alepocephalid [i.e., excluding platytroctids]." *Leptochilichthys* has an extraordinarily deep median keel extending from end to end of the copula and presumably representing a modified toothplate, but we agree with Markle that there is nothing notable in other alepocephalids. In platytroctids, where the basibranchial is normally toothed, the teeth are often arranged longitudinally along a median ridge or crest (Matsui and Rosenblatt, 1987, p. 14). State (1) therefore occurs only in some platytroctids and *Leptochilichthys*.

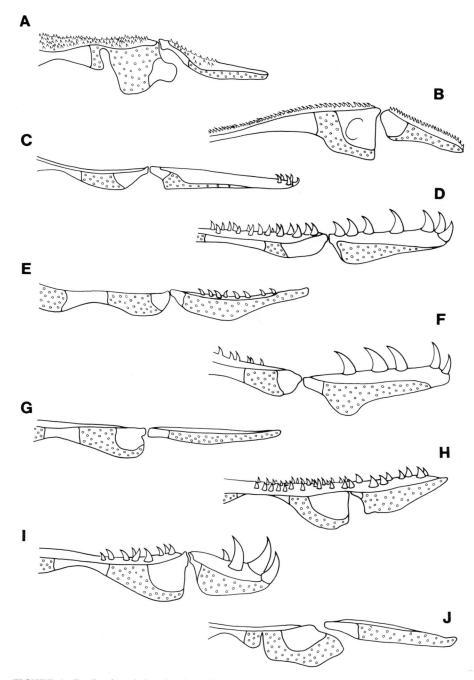

FIGURE 8 Basihyal and first basibranchial, in right lateral view, of (A) *Elops hawaiensis*, BMNH 1962.4.3.1, 69 mm SL; (B) *Chirocentrus dorab*, BMNH 1966.11.16.5, 147 mm SL; (C) *Argentina sphyraena*, USNM 238015, 127 mm SL; (D) *Retropinna retropinna*, BMNH 1964.4.30.19, 83 mm SL; (E) *Stenodus leucichthys*, BMNH 1985.7.16.22, 63 mm SL; (F) *Oncorhynchus clarki*, BMNH 1957.2.20.3, 90 mm SL; (G) *Thymallus thymallus*, BMNH 1970.10.14.3, 71 mm SL; (H) *Hypomesus transpacificus*, BMNH 1984.6.28.11, 65 mm SL; (I) *Osmerus eperlanus*, BMNH 1971.2.16.303, 63 mm SL; (J) *Alepocephalus tenebrosus*, SIO uncat., 135 mm SL.

85. [Begle's 49: Fourth hypobranchial absent (0) or present (1). State (1) coded only in salangids.] The character is from Roberts (1984). Like fusion of the hyomandibula and quadrate (no. 43), this is a larval character persisting into maturity only in salangids. We have observed separate fourth hypobranchials in larval *Coregonus, Salmo, Hypomesus, Osmerus,* and *Galaxias.*

86. [Begle's 79: Fourth ceratobranchial (Cb4) unmodified (0) or much wider in dorsal view than ceratobranchials 1–3, sometimes with distal end expanded (1). State (1) coded in all sampled argentinoids and alepocephaloids.] Begle (1991, fig. 6) illustrated Cb4 broad throughout its length in *Bathylagus* and CB4 expanded distally in the platytroctid *Sagamichthys.* Kobyliansky (1990, fig. 8) illustrated the broad Cb4 in other bathylagids but also showed that it does not occur in argentinids (*Argentina* and *Glossanodon*) or microstomatids. Among alepocephalids, *Alepocephalus, Bathytroctes,* and *Leptoderma* have the platytroctid condition, but other alepocephaloids do not. There appear to be two different states here, recoded (1) for the bathylagid configuration and (2) for the platytroctid.

87. [Begle's 80: Fourth gill arch unmodified (0) or with fleshy membrane along joint of ceratobranchial and epibranchial, partitioning the crumenal organ from the orobranchial chamber (1). State (1) coded in all sampled argentinoids and alepocephaloids.] The membrane is a continuation of esophageal tissue and spans the gap between the fourth arch and the fifth that is created by the accessory cartilage (no. 90, below). No comment.

88. Dentition of fifth ceratobranchial (Cb5). Among lower euteleosts, Cb5 is toothless in bathylagids, microstomatids (Kobyliansky, 1990), opisthoproctids, and in the alepocephalid *Photostylus.* In alepocephaloids other than *Photostylus,* the dentition on Cb5 varies. In *Bajacalifornia* and *Narcetes* there is a longitudinal band of teeth, a pattern that we take to be primitive by comparison with outgroups (e.g., Nelson, 1969a, pl. 85). One derived state is a single longitudinal row of teeth, seen in *Bathylaco.* An alternative derived state is a single marginal row of teeth on a fanlike medial expansion of the bone, seen in platytroctids, *Alepocephalus, Bathytroctes, Leptochilichthys, Leptoderma, Rinoctes, Rouleina,* and *Talismania.*

89. [Begle's 69: Fifth ceratobranchial without (0) or with bony anterior laminar extension, close to medial margin (1). State (1) coded in galaxiids and *Lepidogalaxias.*] In *Lepidogalaxias,* there is a distinctive anteriorly directed process from the anterolateral margin of Cb5, presumably serially homologous with successively larger ventrally directed processes on Cb1–4. In galax-

iids, we have seen nothing similar in any of the species examined and the character becomes autapomorphous.

90. [Begle's 78: Accessory cartilage between fifth ceratobranchial and epibranchial absent (0) or present (1). State (1) coded in all sampled argentinoids and alepocephaloids except *Bathyprion* and *Bathylaco.*] The character is from Greenwood and Rosen (1971), and Markle (1976) reported the cartilage absent in *Bathyprion* and *Bathylaco. Opisthoproctus* was wrongly coded (1) (Greenwood and Rosen, 1971, fig. 6), despite comments in Begle's text (1992, p. 356) on alternative evaluations for "absence of an accessory cartilage in opisthoproctids." Like Greenwood and Rosen (1971), we found no accessory cartilage in the opisthoproctids *Opisthoproctus* and *Rhynchohyalus,* or in *Macropinna,* but found it present in *Bathylychnops* and on one side in our *Dolichopteryx.* It is absent in the alepocephalid *Photostylus.*

91. [Begle's 39: Uncinate process on first epibranchial present (0) or absent (1). State (1) coded in galaxiids, *Lovettia,* and *Aplochiton.*] Absent also in *Lepidogalaxias.* In this and the succeeding characters referring to uncinate processes (nos. 92–94, 103–107) salangids should be coded as (?) since there is no ossification and hence no way of recognizing an uncinate process, which can be discriminated from an extension of the cartilaginous head of the structure only once it ossifies.

92. [Begle's 40: Uncinate process on second epibranchial present (0) or absent (1). State (1) coded in galaxiids, *Lovettia,* and *Aplochiton.*] It is absent also in all alepocephaloids except *Bathylaco* and *Bathytroctes* (we could not check *Bathyprion* or *Rinoctes*) and in most bathylagids (Kobyliansky, 1986, fig. 12A; 1990, fig. 9I,K; personal observation), but present in *Melanolagus* (Kobyliansky, 1986, fig. 12C).

93. [Begle's 41: Uncinate process on third epibranchial present (0) or absent (1, only in *Lepidogalaxias*).] No comment.

94. [Begle's 91: Uncinate process on fourth epibranchial (Eb4) present (0) or absent (1). State (1) coded in all sampled argentinoids and alepocephaloids except *Argentina, Glossanodon,* and some platytroctids.] The platytroctid feature may refer to Greenwood and Rosen's (1971) figure of *Searsia koefoedi* or to Sazonov's (1986, p. 70) report of a well-differentiated process in *Barbantus;* we have not been able to study *Barbantus* but have found no process in two cleared-and-stained *S. koefoedi* (probably those used by Matsui and Rosenblatt, 1987, p. 23, who also failed to find it). It might become distinct from the cartilaginous posterodorsal margin in large specimens. Greenwood and Rosen (1971, fig. 4) illustrated an uncinate process

on Eb4 in *Glossanodon pygmaeus*; it is lacking in our *G.polli* and *G.struhsakeri* and in Kobyliansky's (1990, fig. 9) *G.danieli*. The process is absent in *Lepidogalaxias*, *Hypomesus*, *Mallotus*, salmonoids, galaxiids, *Aplochiton*, *Lovettia*, and in our retropinnids (though shown in *Stokellia* by Rosen, 1974, fig. 16E); Begle wrongly coded it as present in all of these. As we argue below, Begle's outgroups and neoteleosts should also be coded for absence.

Rosen (1974, p. 278) gave a careful account of fourth epibranchial configuration in lower teleosts. He noted that the uncinate process is absent in osteoglossomorphs, clupeomorphs, *Chanos*, the ostariophysans he sampled, and *Elops* but is present in other elopomorphs. He inferred that the process might be a derived feature shared by elopomorphs and euteleosts, independently lost in *Elops* and in various euteleostean lineages, but wrote (p. 279) "It would be instructive to attempt a study of this bone in some of the fishes that appear to be primitive sister groups to some or all living forms, for example, in ichthyodectids and in some of the fishes called leptolepids." We checked Eb4 of Jurassic pholidophorids (*Pholidophorus bechei*, *P.germanicus*, and *P.macrocephalus*) and leptolepids (*Tharsis dubius* and the "Callovian *Leptolepis*" of Patterson, 1975), and it was checked in the Cretaceous ichthyodectiform *Cladocyclus* by Patterson and Rosen (1977, p. 103). In all, there is no uncinate process, as in osteoglossomorphs, *Elops*, *Chanos*, and clupeomorphs [Begle's outgroups should therefore be coded for absence]. The uncinate process of Eb4 is a derived feature, independently developed in non-elopid elopomorphs and at some level or levels within euteleosts.

Homology of the elopomorph process is established by comparison between *Elops* and *Megalops*. In *Elops*, Nelson (1968b, fig. 6) illustrated an "interarcual cartilage" connecting the tip of the uncinate process of Eb3 with the head of Eb4 (there is no uncinate process); we confirmed his observations and found a corresponding but smaller interarcual cartilage in *Megalops*, where it lies between the articulating tips of the uncinate processes of Eb3 and Eb4. The elopomorph uncinate process therefore develops from the head (anterior end) of Eb4, as one would infer from its shape. Observations of small *Esox* show that the uncinate process develops in the same way, by separation from the head of Eb4. We believe that the uncinate process of Eb4 in osmerids has a different origin; it, and Begle's coding of this character, is discussed with the following character.

95. [Begle's 50: Levator process on fourth epibranchial wide, with width at distal margin up to half the length of the underlying epibranchial (0) or very narrow, narrower than the width of the epibranchial

(1). State (1) coded in *Allosmerus*, *Hypomesus*, *Spirinchus*, *Mallotus*, *Argentina*, *Bathylagus*, *Glossanodon*, and *Opisthoproctus* (the latter said to have (0) in Begle's text).] The term levator process applied to the fourth epibranchial (Eb4) seems to originate with Rosen (1974, p. 278), who first differentiated "a distinct process" from the "elevation" found, for example, in *Elops*, *Hiodon*, and *Thymallus*, where there is continuity between the cartilaginous margin of the elevation and the posterior articular surface of the bone (Nelson, 1967, fig. 1). Later in the same paper, Rosen (1974, p. 279) equated the "reflected dorsal section" or elevation of Eb4 in *Elops* and *Megalops* with the levator process, but this loses a real distinction, and we prefer to restrict the definition to a distinct and separate process on which the fourth external levator muscle inserts. So defined, there is no levator process in osteoglossomorphs, most elopomorphs, and clupeomorphs (Nelson, 1967, figs. 2 and 3), or in salmonoids (Rosen, 1974, fig. 9), alepocephaloids (Greenwood and Rosen, 1971, figs. 1–3), opisthoproctids (Greenwood and Rosen, 1971, fig. 6; Stein and Bond, 1985, fig. 5), retropinnids (Rosen, 1974, fig. 16), galaxiids (Rosen, 1974, fig. 10), or *Lepidogalaxias* (Rosen, 1974, fig. 14). All those fishes also lack an uncinate process on Eb4, except for non-elopid elopomorphs, where the process segments from the anterior tip of Eb4 (no. 94, above).

Among esocoids, esocids have lost the fourth external levator (Holstvoogd, 1965, also personal observation) and there is no levator process or posterior elevation on Eb4, but there is an uncinate process (no. 95, above). In umbrids there is a fourth levator muscle in *Umbra* and *Dallia*, but none in *Novumbra* (personal observation). Rosen (1974, fig. 8) reported a "relic" levator process in one specimen of *Umbra*; in our material, the muscle inserts on the posterior elevation and there is no levator process. Other umbrids (Rosen, 1974, figs. 8 and 15) lack the posterior elevation (in *Dallia* the fourth levator inserts on the broadened posterior part of Eb4) but have an uncinate process, presumably developed from the front end of Eb4, like that of *Esox* (above).

A distinct levator process on Eb4 is present in *Albula* and *Halosaurus* among elopomorphs; in argentinid, microstomatid, and bathylagid argentinoids (Greenwood and Rosen, 1971, figs. 4–6; Kobyliansky, 1986, fig. 12; 1990, fig. 9); in all osmerids (Rosen, 1974, fig. 16), in some clupeomorphs (Nelson, 1967, figs. 2 and 3); in *Chanos* and some ostariophysans (Nelson, 1967, fig. 1; Fink and Fink, 1981, fig. 13); in most lower neoteleosts (consistently in stomiiforms and myctophiforms, variable in aulopiforms); and among examined acanthomorphs in *Polymixia* and percopsiforms (Pat-

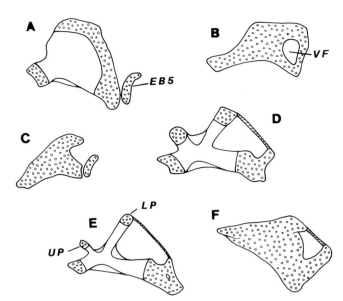

FIGURE 9 Epibranchial configuration in osmeroids. Fourth and fifth epibranchials, medial view of right side, in (A) *Prototroctes maraena*, BMNH 1984.10.3.1, 74 mm SL; (B) *Lovettia sealei*, BMNH 1937.8.22.1, 41 mm SL; (C) *Osmerus* sp., HSU uncat., 29 mm larva; (D) *Osmerus eperlanus*, BMNH 1971.2.16.303, 63 mm SL; (E) *Thaleichthys pacificus*, USNM 342051, 136 mm SL; (F) *Salangichthys microdon*, BMNH 1996.2.6.1, 83 mm SL. Abbreviations: EB5, fifth epibranchial; LP, levator process of fourth epibranchial; UP, uncinate process of fourth epibranchial; VF, vascular foramen for efferent artery, assumed to be enclosed by distal and proximal fusion between fourth and fifth epibranchials.

terson and Rosen, 1989, fig. 13). In all those fishes (except some clupeomorphs), the levator process meets Begle's description—its tip is narrower than the body of Eb4. In Rosen's (1974, fig. 16) illustrations of the process in osmerids, the narrowest is in *Plecoglossus*, coded (0) by Begle. In our material, *Hypomesus olidus* and *H.transpacificus* agree with Rosen's *H.olidus* (1974, fig. 16D), but in *H.pretiosus* the dorsal and ventral tips of the posterior processes are in contact and are fused in one specimen. In *Thaleichthys*, the only osmerid genus not illustrated by Rosen (*Osmerus* is in Greenwood and Rosen, 1971, fig. 7), Fig. 9E shows that Eb4 agrees exactly with that of *Spirinchus* (Rosen, 1974, fig. 16A). Salangids are a special case because Eb4 does not ossify so that (as with uncinate processes, no. 91 above) it is not easy to discriminate a distinct levator process from an extension of the cartilaginous posterior margin of Eb4. Rosen's (1974, fig. 16G) illustration of Eb4 in the salangid *Salangichthys microdon* is very different from our material of that species (Fig. 9F), in which Eb4 resembles Rosen's illustration of *Mallotus* (1974, fig. 16C) except for lack of ossification. In *Protosalanx* Eb4 is like our *Salangichthys* (Fig. 9F), whereas in *Salanx* it is like Rosen's (1974,

fig. 16G) ''*Salangichthys*,'' and in *Neosalanx* it is similar but the upper arm of Eb4 is shorter. Because Eb4 in salangids is so similar in shape to that of osmerids, we regard the levator process as present in salangids. Begle's data are recoded with (1) (presence of levator process) for neoteleosts, osmerids, and salangids, (0) (absence) for *Opisthoproctus*; and (2) for the condition in esocoids.

Among the lower euteleosts with a levator process on Eb4, there is also an uncinate process, articulating with that of Eb3, in *Argentina* and some *Glossanodon* (Greenwood and Rosen, 1971, fig. 4), the bathylagid *Bathylagichthys* (Kobyliansky, 1986, fig. 12), and the osmerids *Osmerus*, *Plecoglossus*, *Spirinchus*, and *Thaleichthys* (Fig. 9; Rosen, 1974, fig. 16). In *Allosmerus*, Rosen (1974, fig. 16B) illustrated a specimen in which the cartilaginous head of the levator process is wide and partially separated into what he labelled as uncinate and levator processes; in our *Allosmerus*, three out of six specimens show uncinate and levator processes separated by bone. This, and the configuration in other osmerids, indicates that the osmerid uncinate process of Eb4, wherever it occurs, is segmented from the levator process. It is therefore nonhomologous with the uncinate process in elopomorphs and *Esox*, which is segmented from the anterior tip of Eb4. We lack developmental information on argentinoids but guess from the configuration of Eb4 that the uncinate process, where it occurs, also segments from the anterior tip of Eb4.

The levator process on Eb4 is directed posteriorly and has no articulation or connection with the uncinate process of Eb3. This is so in ostariophysans (e.g., Rosen, 1973, fig. 3; Fink and Fink, 1981, fig. 13), argentinoids (Kobyliansky, 1990, fig. 9), the osmerids *Hypomesus* and *Mallotus* (Rosen, 1974, fig. 16), stomiiforms (Fink and Weitzman, 1982, fig. 11), and various aulopiforms and myctophiforms (Baldwin and Johnson, in this volume; Stiassny, in this volume). In *Aulopus* (Rosen, 1973, fig. 1) and *Parasudis* there is an uncinate process on Eb4 that approaches but does not directly articulate with that of Eb3. In acanthomorphs (e.g., Rosen, 1973, figs. 82–101; Rosen and Patterson, 1990, figs. 30–50) Eb4 has an uncinate process that articulates with and is tightly bound to the uncinate process of Eb3. In *Polymixia* and percopsiforms there is also a levator process on which the fourth levator muscle inserts; in *Aulopus*, *Parasudis*, and other acanthomorphs the muscle inserts posterior to the uncinate process.

To sum up on the distribution of levator and uncinate processes of Eb4, the levator process is a derived feature developed, presumably independently, within elopomorphs, clupeomorphs, ostariophysans, argentinoids, osmerids, and neoteleosts. Loss of a separate levator process appears to be a synapomorphy of Acanthopterygii, although it may occasionally occur secondarily within percomorphs (e.g., *Pomadasys* and *Lobotes*, Rosen and Patterson, 1990, figs. 7C and 42A). An uncinate process segments off from the levator process in some osmerids. An uncinate process characterizes acanthomorphs. Begle's character 91 (our 94, above), conflated two different kinds of uncinate process: the esocoid or argentinoid type, segmented from the head of Eb4; and the osmerid type, segmented from the levator process. It should be coded with (0) for absence of the uncinate process, (1) for the esocoid/argentinoid type [with (B) for polymorphism in *Glossanodon*], and (2) for the osmerid [with (B) for polymorphism in *Allosmerus*].

96. [Begle's 92: Fifth epibranchial (Eb5) less than half the length of the fourth (0) or almost as long as the fourth (1). State (1) coded in *Bathylagus* and *Opisthoproctus*.] Opisthoproctids (Greenwood and Rosen, 1971, fig. 6; Stein and Bond, 1985, fig. 5) show an elongate Eb5, as does the microstomatid *Nansenia* (Greenwood and Rosen, 1971, fig. 5B; Kobyliansky, 1990, fig. 9; in *Microstoma* the cartilage is partially fused to Eb4). In bathylagids (Greenwood and Rosen, 1971, fig. 6B; Kobyliansky, 1986, fig. 12; 1990, fig. 9) Eb5 varies in length, and in *Bathylagichthys* and *Leuroglossus* it is proportionally no longer than in argentinids. As described the character is autapomorphous for *Opisthoproctus*; if differently phrased it might be a feature of argentinoids but would merely reflect another aspect of the crumenal organ, already coded as two characters (nos. 87 and 90).

97. [Begle's 59: Fifth epibranchial separate (0) or fused to fourth, forming circular foramen for efferent artery (1). State (1) coded in galaxiids, *Lovettia*, *Aplochiton*, and salangids.] There are two different patterns, illustrated in Rosen (1974, figs. 10 and 16), Roberts (1984, figs. 14–17) and Fig. 9. In the galaxiid/aplochitonid pattern, an enclosed vascular foramen, as in *Aplochiton* (Rosen, 1974, fig. 10), *Lovettia* (Fig. 9B), and most galaxiids, is apparently formed by ontogenetic fusion between Eb4 and Eb5, with fusion initiated at the upper ends of the structures (as in Rosen, 1974, fig. 10A, *Galaxias divergens*, and on one side of our *G.fontanus*). In salangids (Fig. 9F), as in adult osmerids (Fig. 9D and E), there is a distinct levator process on Eb4, and a vascular notch is formed by approximation of the levator process and the upper end of Eb5, which fuses ontogenetically to Eb4 at its lower end. These states should be distinguished [(1) for the galaxiid/aplochitonid condition, (2) for the osmerid/salangid condition]. Among osmerids, Eb5 fused to Eb4 at its

lower end [state (2)] is shown in *Thaleichthys* and *Osmerus* in Fig. 9 and illustrated by Rosen (1974, fig. 16) in *Allosmerus*, *Hypomesus*, *Mallotus*, and *Spirinchus*. There is a free Eb5 in *Plecoglossus* (Rosen, 1974, fig. 16), but that shown in *Osmerus eperlanus* by Greenwood and Rosen (1971, fig. 7) does not exist in our material except in larvae (Fig. 9C); adult *O.eperlanus* and *O.mordax* both show the condition in Fig. 9D. Among retropinnids, Rosen (1974) illustrated Eb5 fused at its lower end in *Retropinna* and a free Eb5 in *Stokellia*; our material agrees and shows a free Eb5 in *Prototroctes* (Fig. 9A). *Lepidogalaxias* lacks Eb5 and a vascular foramen and so should be coded as (?). In esocoids, there is no Eb5 or vascular foramen in esocids, *Dallia*, and *Novumbra* (Rosen, 1974, fig. 8), but in *Umbra* Rosen illustrated a free Eb5 enclosing a vascular notch, and our specimens confirm the condition. Esocoids should therefore be coded (B), indicating polymorphism, or (?).

98 and 99. Suprapharyngobranchials. Nelson (1968b) reported a suprapharyngobranchial on the first gill arch (Spb1, no. 98) in *Alepocephalus* and wrote that a second suprapharyngobranchial (Spb2, no. 99) occurs among teleosts only in Elopidae. Spb2 is also present in *Megalops*. Markle and Merrett (1980) reported Spb2 in *Rinoctes* (we disregard their third suprapharyngobranchial; it lies in the wrong position, cf. Nelson, 1968b, p. 137). We find Spb1 and Spb2 in several platytroctids (*Holtbyrnia*, *Mirorictus*, *Paraholtbyrnia*, and *Pellisolus*) and in the alepocephalids *Narcetes*, *Rouleina*, and *Talismania*. Spb1 only is present in the platytroctids *Platytroctes*, *Searsia*, and *Searsioides* and in the alepocephalids *Alepocephalus*, *Bajacalifornia*, and *Bathylaco*. *Bathytroctes*, *Leptochilichthys*, *Leptoderma*, and *Photostylus* have no suprapharyngobranchials, as do argentinoids and other lower eutelosts.

100. Pharyngobranchial 1 (Pb1). Pb1 is primitively ossified in teleosts. It is cartilaginous in bathylagids and absent in microstomatids (Kobyliansky, 1990) and *Lepidogalaxias*. In esocoids, Pb1 is conical and its tip is closed in bone, not covered by cartilage, a derived condition that we have found elsewhere only in *Glossanodon*, where the bone is more elongate.

101. Articulation between Pb1 and epibranchials. In osmeroids the pattern varies. We take the primitive condition to be Pb1 articulating with the anterior tip of Eb1, as in *Lepisosteus*, *Amia*, *Hiodon*, *Elops*, argentinids, etc. (Nelson, 1968b). Among osmeroids, this is found in northern taxa—osmerids—and salangids. In southern osmeroids (retropinnids, *Aplochiton*, *Lovettia*, *Lepidogalaxias*, and galaxiids) Pb1 articulates with the lateral surface of Eb1, except in *Lepidogalaxias*, where Pb1 is absent. Among salmonoids, coregonids and *Thymallus* have a broad-based Pb1, articulating

with the tips of both Eb1 and Pb2. In salmonines Pb1 articulates with the lateral surface of Eb1 (Rosen, 1974, fig. 11), as in southern osmeroids. In all examined alepocephaloids except *Leptoderma* and *Photostylus* Pb1 is extremely broad-based and articulates with both Eb1 and Pb2. In *Photostylus* Pb1 is small and articulates only with Eb1; *Leptoderma* lacks Pb1. In esocids Pb1 articulates with both Eb1 and Pb2, whereas in umbrids it usually articulates only with the tip of Pb2; in one *Umbra* its base is broad and also articulates with Eb1.

102. Toothplate of pharyngobranchial 2 (Pb2). Pb2 primitively carries a toothplate in teleosts. Among lower euteleosts the toothplate is absent in esocoids, all argentinoids (Argentinidae, Bathylagidae, Microstomatidae and Opisthoproctidae); the alepocephalids *Alepocephalus*, *Bajacalifornia*, *Bathylaco*, *Bathyprion* (Markle, 1976), *Bathytroctes*, *Binghamichthys*, *Leptoderma*, *Narcetes*, *Photostylus*, and *Rouleina*; and all salmonoids except *Coregonus* and *Stenodus*.

103. [Begle's 43: Uncinate process on second pharyngobranchial present (0) or absent (1). State (1) coded in galaxiids, *Lepidogalaxias*, *Aplochiton*, retropinnids, and salangids.] It is also absent in *Lovettia*.

104. [Begle's 45). Uncinate process on second pharyngobranchial directed laterally or caudally (0) or directed anteriorly (1). State (1) coded in all osmerids.] Taxa lacking the process (no. 103 above) are all wrongly coded (0) rather than (?). As with no. 107, below, we found that distinguishing the two states is often subjective. In our judgment, among Begle's sample the process is also anterolaterally directed (as in osmerids) in salmonoids, as it is in *Argentina*, *Glossanodon*, and *Bathylagus* (e.g., Kobyliansky, 1990, fig. 9). It is anterolaterally directed in *Elops* and microstomatids but laterally directed in *Megalops*, esocoids, and opisthoproctids. In alepocephaloids, the process is directed dorsally rather than laterally; in some alepocephaloids its orientation is anterodorsal and in others dorsolateral. With these problems in evaluating the character, we discard it.

105. Toothplate of pharyngobranchial 3 (Pb3). Pb3 primitively carries a toothplate in teleosts. Among sampled lower euteleosts, the toothplate is absent in all argentinoids (Argentinidae, Bathylagidae, Microstomatidae, and Opisthoproctidae) and in the alepocephalids *Alepocephalus tenebrosus* and *Photostylus*.

106. [Begle's 42: Uncinate process on third pharyngobranchial present (0) or absent (1, only in *Lepidogalaxias*).] It is absent also in our *Retropinna*, *Lovettia*, and some galaxiids (e.g., *Galaxias maculatus*, *G. fontanus*, and *G. zebratus*).

107. [Begle's 44: Uncinate process of third pharyngobranchial not extending over second epibranchial

(0) or extending well over body of that bone (1). State (1) coded in all osmerids except *Hypomesus*.] As coded by Begle, the derived state distinguishes all osmerids except *Hypomesus* from other osmeroids and outgroups. We could not confirm the implied difference in length or orientation of the uncinate process of PB3 in direct comparisons of three species of *Hypomesus* with other osmerid genera, and although the uncinate process is longer in osmerids than in retropinnids, it is as long and has the same orientation over Eb2 in *Argentina* and various alepocephaloids, for example. Taxa lacking the process (no. 106) were wrongly coded (0) rather than (?) by Begle. Character was discarded.

108. [Begle's 47: Third pharyngobranchial with narrow anterior extension, reaching first pharyngobranchial (0), or without anterior extension (1). State (1) coded in galaxiids, *Lepidogalaxias*, *Lovettia*, retropinnids, and salangids.] By "reaching" Pb1 we assume that Begle meant Pb3 extends forward almost to the anterior tip of Pb2, as illustrated by Rosen (1974, figs. 11 and 12) in *Salmo* and *Hypomesus*. That condition occurs in all osmerids and in *Aplochiton*, argentinids, and alepocephaloids. Begle's state (1), "without anterior extension," is true, for example, of salangids (Roberts, 1984, figs. 14–16) and *Lepidogalaxias* (Rosen, 1974, fig. 15), though the configuration is very different in the two. It is not true of retropinnids or some galaxiids (e.g., Rosen, 1974, figs. 11C and 12A), where Pb3 has an anterior extension alongside about two-thirds of Pb2. We code the retropinnid state as "intermediate" (1) and recode Begle's state (1) as (2). Some galaxiids (e.g., *Galaxias fasciatus*) have state (0). *Opisthoproctus* has the intermediate state (1), as do other opisthoproctids, microstomatids, and some bathylagids. In esocoids the tip of Pb3 is broad and spatulate (Rosen, 1974, fig. 15); we assess *Esox* as having state (2) and *Umbra* state (0).

109. Upper pharyngeal toothplates (UP4 and UP5). Primitively in teleosts there are two principal upper pharyngeal toothplates (Nelson, 1969a; Johnson, 1992). UP4 is absent and UP5 is minute in bathylagids, microstomatids, and opisthoproctids (UP5 is absent in *Macropinna* and *Opisthoproctus*), and UP4 is absent in the alepocephalid *Photostylus* and in all salmonoids (see below). Identification of the single toothplate in most osmeroids (as either UP4 or UP5) is problematic (see below). In *Lovettia* both UP4 and UP5 are absent, and in *Aplochiton* and galaxiids the toothplate is fragmented, into four small toothplates in *Aplochiton* and into two to four in galaxiids (e.g., Rosen, 1974, figs. 11B and 11C; there is one toothplate in our *G.paucispondylus*).

Esocoids are unique among lower euteleosts in having lost UP5 but retained a large UP4, a condition that

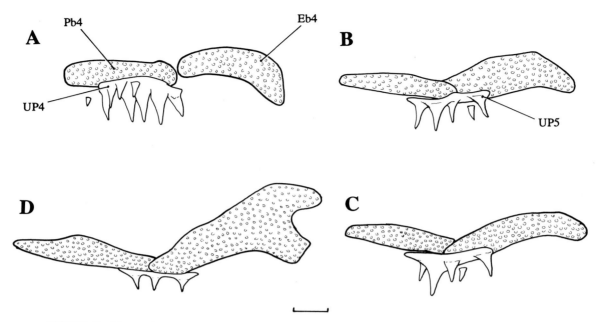

FIGURE 10 Upper pharyngeal toothplate (UP4 or UP5) and fourth pharyngobranchial (Pb4) and epibranchial (Eb4) in larvae. (A) *Esox americanus*, ROM 24422, 15 mm NL, lateral view of left side. (B) *Coregonus clupeaformis*, ROM 68375, 17 mm SL, same view as A. (C) as B, medial view of right side. (D) *Hypomesus* sp., USNM 340198, 33 mm SL, same view as C. Scale bar: 0.1 mm

we previously believed to be unique to ctenosquamates (Johnson, 1992). We base the identification of the toothplate on larval *Esox americanus* (Fig. 10A), where it develops in close association with the ventral surface of Pb4 rather than the tip of Eb4 (cf. Johnson, 1992, fig. 8). In salmonoids, larvae of *Coregonus* (Figs. 10B–10C) indicate by the same criterion that the single toothplate is UP5. In osmeroids the situation is ambiguous. Larval *Hypomesus* (Fig. 10D) suggest that the single toothplate is UP5 because it originates at the tip of Eb4, rather than farther forward, where UP4 originates in *Esox* and in euteleosts with both UP4 and UP5 (Johnson, 1992, fig. 8A). In our larval *Osmerus* there is a single tooth beneath the junction of Pb4 and Eb4, implying UP5, as in *Hypomesus*. In our larval *Mallotus*, there are three separate toothplates, with two beneath Pb4 and one, the smallest, beneath Eb4. In adult *Mallotus* there is usually a single toothplate, but occasionally there are two, one beneath Pb4 and one beneath the junction of Pb4 and Eb4. In our larval galaxiids the single tooth is beneath Pb4, implying UP4, but in adult galaxiids there are two or more toothplates, variable in position. Larval retropinnids might help to resolve the situation, but our assessment is that primitively there is a single toothplate in osmeroids, that it is PB5, and that the ambiguities described above are due to secondary subdivision of the toothplate.

110. [Begle's 54: Retractor dorsalis absent (0) or present (1). State (1) coded in neoteleosts and *Lepidogalaxias*.] Johnson (1992, p. 11) commented further on this.

111. [Begle's 81: Gill rakers on fourth and fifth arches unmodified (0) or expanded/elongate (1). State (1) coded in all sampled argentinoids and alepocephaloids.] We believe that Begle's coding of this character is based on misreading of Greenwood and Rosen (1971) rather than attributes of specimens. In discriminating the crumenal organ of alepocephaloids from that of argentinoids, Greenwood and Rosen (1971, pp. 9, 14) described the gill rakers in the former as broad-based and toothed and in the latter as long, slender, and toothless. However, these are general descriptions of all the gill rakers in alepocephaloids and argentinoids, not just those on the fourth and fifth arches. In argentinids and most alepocephalids, the gill rakers of the fourth and fifth arches are in no way differentiated from those of more anterior arches. In bathylagids, microstomatids, and opisthoproctids the rakers of the fourth and fifth arches are much longer than those on the second and third, and resemble those on the first arch. In some alepocephaloids the rakers on the upper part of the fourth and fifth arches may be slightly larger or more close-packed than those on the lower part of the arches, but there is no general character matching Begle's description.

The derived state, differentiation of the fourth and fifth arch rakers, is restricted to *Bathylagus* and *Opisthoproctus* in Begle's sample.

112. [Begle's 101: Gill rakers toothless (0) or with a series of marginal teeth, sometimes with one or a few fanglike teeth distally (1). State (1) coded in all sampled alepocephaloids.] Polarity is reversed because gill rakers are primitively toothed (they are modified toothplates), as they are in outgroups, esocoids, salmonoids, and primitive neoteleosts. The gill rakers are toothless in *Alepocephalus tenebrosus* and *Photostylus* among the alepocephaloids we have sampled, in argentinoids, and in all osmeroids except *Retropinna* (McDowall, 1969, p. 802). Coding should be (0) for toothed rakers and (1) for toothless.

G. Axial Skeleton (Including Caudal Skeleton and Fin)

113. Accessory neural arch. Distribution of the accessory neural arch (ANA) was reviewed by Patterson and Johnson (1995, p. 17). Among lower euteleosts, they recorded ANA in alepocephaloids, salmonoids, northern osmeroids (including salangids), and *Esox*, whereas ANA is absent in argentinoids, southern osmeroids (retropinnids and other galaxioids), and umbrids. With more extensive sampling, we now find that ANA occurs in all osmerid and salangid genera and in *Aplochiton* alone among southern osmeroids; in platytroctid alepocephaloids; and among alepocephalids in *Alepocephalus, Bathylaco, Bathytroctes,* and *Talismania* (where it may carry an epineural); but it is absent in *Bajacalifornia, Leptochilichthys, Leptoderma, Photostylus, Rinoctes,* and *Rouleina.* Among osmerids, *Spirinchus* is notable in having ANA reduced to a minute nubbin, whereas it is large in all others. Previously (Patterson and Johnson, 1995), we were unwilling to decide whether ANA is synapomorphous at some level and has been repeatedly lost or has arisen repeatedly and is nonhomologous from group to group. With more extensive sampling in euteleosts, we are now convinced that ANA was primitively present in that group and has been repeatedly lost.

114. Occipital gap. In osmeroids and outgroups such as salmonoids the articulation between the occipital condyle and V1 is normally close, but in *Lepidogalaxias* and *Lovettia* there is a distinct gap between the two. A gap also occurs in many alepocephaloids.

115. Baudelot's ligament. In teleosts Baudelot's ligament primitively originates on V1 (Patterson and Johnson, 1995, p. 19), as it does in all examined lower euteleosts except southern osmeroids. In retropinnids, *Aplochiton* and *Lovettia* Baudelot's ligament originates on the occiput (the entry of V1 for *Stokellia* in

Patterson and Johnson, 1995, table 4, is an error). In *Lepidogalaxias* the ligament originates on V1. In galaxiids it originates only on V1 or has a double origin, on the occiput and V1.

116. Fusion of neural arches to centrum. Teleostean neural arches are primitively autogenous and remain so in most lower euteleosts. Within osmeroids, all neural arches anterior to the dorsal fin are fused to the centra in *Aplochiton, Lovettia, Lepidogalaxias,* and galaxiids. They are also all fused in umbrids (Wilson and Veilleux, 1982).

117. Epineural fusion or autogeny. Epineural bones are primitively fused to neural arches in teleosts (Patterson and Johnson, 1995, p. 11). In lower euteleosts, all but the last two or three epineurals are fused in alepocephaloids (except in *Leptoderma,* where only the first two of about 25 are fused), and more than half of them are fused in argentinids (Patterson and Johnson, 1995, table 4). In salmonoids, the epineural of V1 may be fused (in some coregonid specimens; fused on one side in the *Thymallus* recorded by Patterson and Johnson, 1995, table 4), but the epineurals are generally all autogenous, as they are in all osmeroids. In esocoids the first two to five epineurals are fused in *Esox* (Patterson and Johnson, 1995, fig. 8), and none is fused in umbrids.

118. Epineural origin. Primitively, epineurals originate on the neural arch. In a few lower (nonacanthomorph) teleosts the point of origin of some epineurals descends on to the centrum, as in some osteoglossomorphs, *Esox,* and some aulopiforms (Patterson and Johnson, 1995, p. 12). Extensive sampling of osmerids shows that in all genera (including *Plecoglossus*) except *Hypomesus* epineural origin is displaced onto the anterodorsal part of the centrum from about V10 back to about V30; we also observed this displacement in *Hypomesus olidus* and *H. pretiosus* but not in *H. transpacificus,* where the epineurals remain on the neural arch. In salangids, where the epineural ligaments are unossified, their origin is displaced ventrally in the osmerid pattern in *Neosalanx* and *Protosalanx;* in *Salanx* the cartilage of the neural arches and parapophyses surrounds the centra so that the condition could not occur. In *Salangichthys* we could not clearly see the point of origin of the epineural ligaments. In *Stokellia* the origin of the epineural ligaments (there are no bones) is also displaced on to the centrum from about V8 to V20.

119. Epineural descent. Patterson and Johnson (1995, p. 27) suggested a synapomorphy of Argentinoidei: in the argentinoid and alepocephaloid genera then sampled, the tips of the first three (*Glossanodon, Leptoderma,* and *Searsia*) or four (*Argentina*) epineurals are displaced ventrally relative to their successors, in

the same way as is the distal part of the first epineural in the lampridiform *Velifer* (Johnson and Patterson, 1993, fig. 1). We have now checked other genera and found the first three epineurals descended distally in the alepocephaloids *Bathylaco, Bathytroctes, Mirorictus,* and *Searsioides;* the first four in the argentinoids *Bathylagus* and *Nansenia* and the alepocephaloids *Leptochilichthys* and *Pellisolus;* and among other alepocephaloids about the first 11 epineurals descended in *Rinoctes,* only the first two in *Bajacalifornia, Rouleina, Alepocephalus rostratus,* and *Talismania antillarum,* only the first in *Alepocephalus tenebrosus* and *Talismania aphos,* and none in *Xenodermichthys.* Our specimen of *Platytroctes* is poor, but Matsui and Rosenblatt's drawing (1987, fig. 6) indicates that only the first epineural is descended. The opisthoproctids *Opisthoproctus, Dolichopteryx,* and *Rhynchohyalus* have no epineural on V1 or V2 and none descended thereafter; *Macropinna* has no epineural on V1 and none descended; but *Bathylychnops* has no epineural on V1 and those on V2 and 3 descended. *Microstoma* has extremely long epineurals with none descended, a situation we take to be derived because the microstomatid *Nansenia* has four descended epineurals. There are no ossified epineurals in *Photostylus,* and our specimen of *Narcetes* is too poor to check. Summarizing, descent of the first two to four epineurals seems to stand up as a character of Argentinoidei (argentinoids and alepocephaloids). Genera or species lacking the character (only one descended in *Platytroctes, Talismania aphos,* and *Alepocephalus tenebrosus,* and none in most opisthoproctids, *Microstoma,* and *Xenodermichthys*) are all indicated, by other evidence (see below), as derived members of their subgroups.

120. Epineural ossification. Epineurals are primitively ossified in teleosts. Patterson and Johnson (1995, p. 12) found epineural ligaments but no bones in the retropinnids *Retropinna* and *Stokellia,* the salangid *Salangichthys,* and the umbrids *Dallia* and *Novumbra,* whereas in *Lepidogalaxias* there are no epineural bones or ligaments. We have checked other salangids and galaxioids and found epineural ligaments but no bones in *Prototroctes, Aplochiton,* and *Lovettia* but bones in all galaxiids. In *Novumbra* the epineurals ossify in larger specimens.

121. Cartilaginous and bony epicentrals. The epicentral series of intermusculars is primitively represented by ligaments in teleosts (Patterson and Johnson, 1995, p. 15). Epicentral bones develop in those ligaments in several taxa (e.g., *Megalops,* clupeomorphs, gonorynchiforms, and gymnotoids), and we found cartilages in the distal part of the anterior epicentral ligaments in salmonoids, osmeroids, the stomiiform *Maurolicus,* and the acanthomorph *Polymixia;*

epicentral cartilages also occur among aulopiforms in the three genera of Evermannellidae (Baldwin and Johnson, in this volume). Regarding the epicentral cartilages in *Maurolicus* and *Polymixia* as autapomorphous, we (Patterson and Johnson, 1995, p. 26) argued that epicentral cartilages support a grouping of Osmeroidei and Salmonoidei. As with osmerid epipleurals (no. 122, below), our sampling of Argentinoidei was deficient, as was the quality of our material. We have now found cartilage rods in the epicentrals of all argentinid, bathylagid, and opisthoproctid genera in which our material is sufficiently well prepared to show them (*Argentina,* the bathylagids *Bathylagus* and *Leuroglossus,* the microstomatid *Nansenia,* and the opisthoproctids *Bathylychnops, Dolichopteryx,* and *Macropinna*). Among alepocephaloids, cartilaginous epicentrals are absent in all examined platytroctids (we have good material of seven genera) and absent in *Alepocephalus, Leptoderma, Photostylus,* and *Talismania* but present in *Bathylaco* (in the epicentrals of V1–3 only) and *Rinoctes* (in an occipital epicentral and those of V1–5). We do not yet have good enough material to check for them in other alepocephaloid genera.

Among Osmeroidei, in addition to the records in Patterson and Johnson (1995), we have now seen cartilaginous epicentrals in all osmerid genera and in *Aplochiton* and have confirmed that the epicentral cartilages ossify in *Lepidogalaxias* by study of a size range of double-stained specimens. There are no cartilaginous epicentrals in *Lovettia* or in our material of the four genera of salangids; in all those fishes, and in *Aplochiton,* there are no ossified intermusculars. Cartilage rods in the epicentrals therefore characterize osmeroids (all osmerids and retropinnids, *Aplochiton, Lepidogalaxias,* and many galaxiids), salmonoids (Patterson and Johnson, 1995, p. 14), argentinoids, and a minority of sampled alepocephaloids. Given that argentinoids and alepocephaloids are sister-groups (Argentinoidei of Greenwood and Rosen, 1971), the most economical interpretation is that cartilaginous epicentrals characterize a group comprising Argentinoidei, Osmeroidei, and Salmonoidei and are secondarily absent in most alepocephaloids (as they are in salangids, *Lovettia,* and some diminutive galaxiids among osmeroids).

122. Epipleural bones. Ossified epipleurals are primitively present in elopocephalans (Patterson and Johnson, 1995, p. 13). We argued (Patterson and Johnson, 1995, p. 26) that the absence of epipleural bones in salmonoids and osmeroids is a derived feature indicating that they are sister-groups. Our interpretation of osmerids was deficient both in sampling and in literature search.

We recorded the intermusculars of three osmerids, *Hypomesus transpacificus*, *Osmerus mordax*, and *Plecoglossus* (Patterson and Johnson, 1995, table 3), and found no epipleural bones in them. We did see epipleural bones in *Spirinchus* (Patterson and Johnson, 1995, p. 26), but because that genus occupies a derived position in Begle's (1991) cladogram of osmerids, we regarded its epipleurals as secondary (reversal) and argued that osmerids primitively lack epipleural bones. But Klyukanov (1975) described epipleurals in *Thaleichthys* (from ca. V20 back into the caudal region), *Spirinchus* (on 7–13 abdominal vertebrae and "possibly" 1–2 caudal vertebrae), *Allosmerus* (on 7–9 abdominal vertebrae) and *Mallotus* (on 25–27 abdominal and 7–10 caudal vertebrae). Wilson and Williams (1991, fig. 11) recorded epipleural bones in all osmerid genera except *Plecoglossus*. We have now examined all osmerid genera and agree with Klyukanov that epipleural bones occur in *Thaleichthys* (on about the last 20 abdominal but not on the first caudal vertebra in fishes ca. 60–150 mm SL), *Spirinchus* (absent, or ossified only on 2 or 3 posterior abdominal vertebrae in fishes ca. 45 mm SL; present on the last 7–12 abdominal vertebrae in fishes ca. 100 mm SL), *Allosmerus* (on 5–12 posterior abdominal vertebrae and sometimes on the first 1 or 2 caudals in fishes ca. 80–100 mm SL) and *Mallotus* (absent in a 55 mm specimen, ossified on ca. 20 posterior abdominal vertebrae and 2 caudal vertebrae in a 60 mm specimen, present on ca. 25 abdominal and 5 caudal vertebrae in fishes ca. 140 mm SL). In agreement with Klyukanov (1975), and in contrast to Wilson and Williams (1991), we have found no epipleurals in *Osmerus* (*O.mordax* and *O.eperlanus*) or *Hypomesus* (*H.olidus*, *H.pretiosus*, and *H.transpacificus*). A decision on whether epipleural bones are primitive or derived for osmerids depends on relationships within the group (Section IVB below), but the pattern of epipleurals in osmerids is distinctive. Primitively in elopocephalans the epipleural bones occur on roughly equal numbers of abdominal and caudal vertebrae (Patterson and Johnson, 1995, tables 3 and 4), ossifying rostrally and caudally from a focus around the first caudal vertebra. When epipleural bones are present in osmerids, they are generally confined to the posterior abdominal vertebrae (see above).

In galaxiids, we (Patterson and Johnson, 1995, p. 26, table 4) recorded epipleural bones in *Galaxias zebratus* and noted McDowall's (1969, 1978) records of them in *Galaxias* and *Paragalaxias*. We have now seen a larger sample of galaxiids. Epipleurals are present in all our cleared-and-stained *Galaxias* species and also in our *Galaxiella*, *Nesogalaxias*, *Paragalaxias*, and in rudimentary form (small nubbins) in some *Neochanna* but not in others.

123. Epipleural extent. Primitively, epipleural bones are absent on the anterior vertebrae in teleosts, but they extend unusually far forward in two groups, aulopiforms and alepocephaloids (Patterson and Johnson, 1995, p. 14). We (Patterson and Johnson, 1995, p. 27) proposed that monophyly of alepocephaloids is corroborated by the fact that the ossified epipleural series extends forward to V3 in *Bathytroctes*, *Leptochilichthys*, *Rinoctes*, *Searsia*, and *Talismania*. Matsui and Rosenblatt (1987, fig. 6) illustrated that condition in *Mirorictus*, and we have now also observed it in *Alepocephalus tenebrosus*, *Bathylaco*, *Narcetes*, *Rouleina*, and the platytroctids *Paraholtbyrnia*, *Pellisolus*, and *Searsioides*. The epipleurals extend to V2 in our *Bajacalifornia* (in which the first rib is on V1) and *Alepocephalus agassizi* (first rib on V2), to V4 in *Bathyprion* (Markle, 1976) and *Alepocephalus agassizi*, to V6 in *Holtbyrnia* and *Sagamichthys*, to V7 in *Platytroctes*, to V9 in *Leptoderma* and *Rinoctes* [in contrast to the report of V3 in *Rinoctes* by Patterson and Johnson (1995); Markle and Merrett (1980), reported epipleurals to V7], and to V11 in *Xenodermichthys*; in *Photostylus* there are no ossified epipleurals. Gosline (1969, fig. 11) illustrated an epipleural on V2 in *Alepocephalus rostratus*, and Markle (1976) reported epipleurals to V2 in *Bathylaco*, *Asquamiceps*, and *Talismania oregoni*, V4 in *Bathytroctes*; V6 in *Alepocephalus bicolor*; V8 in *Ericara*; V11 in *Leptoderma*; and V12 in *Conocara* and *Xenodermichthys*. We take conditions in *Rinoctes* (V7–9), *Leptoderma* (V9–11), *Ericara* (V8), *Conocara* and *Xenodermichthys* (V12), and *Photostylus* (none) to be derived since in our view (Section VIII) all that we have examined are derived alepocephaloids.

124. Proximal forking of epineurals and epipleurals. Patterson and Johnson (1995, p. 28) proposed that monophyly of salmoniforms (argentinoids, alepocephaloids, salmonoids, and osmeroids) might be corroborated by absence of proximal forking of the epineural and epipleural bones. Proximal forking of the epineurals and epipleurals is general in elopomorphs, clupeomorphs, ostariophysans, and myctophiforms. It does not occur in stomiiforms; in esocoids proximal forking occurs only in the epineurals (esocids) or in no more than three or four bones in each series (*Umbra*); and in aulopiforms the bones are forked only in chlorophthalmids and some paralepids (Baldwin and Johnson, in this volume). The lack of proximal forking of intermusculars in stomiiforms, the basal neoteleostean group, weakens it as a salmoniform character, unless (as proposed in Section X) esocoids belong between salmoniforms and neoteleosts.

125. Supraneural pattern. There are two basic patterns of supraneural development in teleosts. In the first, exemplified in Fig. 11A by larval *Clupea*, the

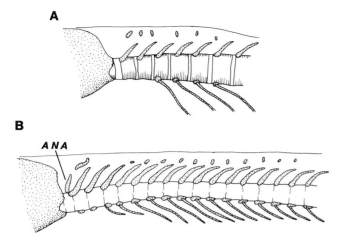

FIGURE 11 Patterns of supraneural development in larvae. Anterior vertebrae and supraneurals in (A) larval *Clupea harengus*, BMNH 1996.2.6.4, 20 mm SL; (B) larval *Salmo salar*, BMNH 1932.11.13.3, 19 mm SL. Cartilage stippled; chordacentra in A hatched. A shows the primitive pattern 1; B shows the euteleostean pattern 2. The accessory neural arch (ANA) is present in both *Clupea* and *Salmo*, but in *Clupea* (or in this lot of cleared-and-stained larvae) it develops late; it is present and already ossifed only in the largest specimen in the lot, 35 mm SL.

supraneurals develop caudally in series from the first, which lies anterior to the neural spine of V1. In adults resulting from this mode of development (Figs. 12A and 12B) there is normally no differentiation between the first and second supraneurals, and they are separated by the neural spine of V1. We will call this pattern 1. In the second (pattern 2), exemplified in Fig. 11B by larval *Salmo*, the first supraneural (anterior to the neural spine of V1) develops independently, and the remainder differentiate in rostral and caudal gradients from a focus roughly midway between the occiput and dorsal fin origin. In adults resulting from this mode of development (Figs. 12C, 12D, and 12H) the first supraneural is usually differentiated from the second (the first is substantially larger or thicker), and the two are separated by two (or more) neural spines. We have observed or inferred pattern 1, development in rostrocaudal series, in *Amia*, *Hiodon*, *Elops* (Fig. 12A), *Megalops*, *Albula*, *Denticeps*, several clupeoids (Fig. 11A), *Umbra* (Wilson and Veilleux, 1982, fig. 12E), microstomatids, bathylagids (Fig. 12B), and the alepocephalids *Bajacalifornia*, *Bathytroctes*, *Binghamichthys*, *Narcetes*, *Rinoctes*, and *Talismania antillarum*. We have observed or inferred pattern 2, differentiation of the first supraneural and a gap between it and the second, in *Esox*, salmonoids (Figs. 11B and 12D), osmeroids, argentinids (Fig. 12C), opisthoproctids, the stomiiforms *Diplophos* (Fink and Weitzman, 1982, fig. 5) and *Pollichthys*, *Aulopus*, the neoscopelid *Scopelen-*

gys, and the alepocephalids *Alepocephalus agassizi*, *Bathylaco*, *Talismania oregoni*, and *T.aphos*. Taxa with only one supraneural (over V1), such as *Chirocentrus* and the alepocephalids *Alepocephalus tenebrosus* and *Leptochilichthys*, cannot be assigned to pattern 1 or 2.

Two variants of pattern 2 occur in osmeroids: the first (pattern 2A) in osmerids and salangids and the second (pattern 2B) in galaxioids (retropinnids, *Aplochiton*, *Lovettia*, *Lepidogalaxias*, and galaxiids). In pattern 2A (Figs. 12E and 12F), the first supraneural is associated with the first two neural arches; this pattern also occurs in argentinids (*Argentina* and *Glossanodon*, Fig. 12C). In the galaxioid pattern, 2B, the first supraneural develops in continuity with the neural arch (Fig. 12I), and in the adult (Figs. 12G and 12H) V1 has no neural spine, and the supraneural is closely articulated or continuous by cartilage with the neural arch.

A different variant of pattern 2 occurs among esocoids in *Dallia*, *Novumbra*, and the Paleocene *Esox tiemani* (Wilson and Veilleux, 1982; Wilson, 1984): the supraneurals develop rostrally and caudally from a focus midway between the occiput and dorsal fin, as usual, but there is no anterior supraneural.

Ostariophysans show a third pattern (pattern 3) of supraneural development. Fink and Fink (1981, p. 324) identified absence of the supraneural anterior to the neural arch of V1 as an ostariophysan synapomorphy. *Chanos* (Rosen and Greenwood, 1970, fig. 3; Fink and Fink, 1981, fig. 6) and the early Cretaceous *Tharrhias* (Patterson, 1984b, fig. 3) lack that first supraneural but otherwise show a caudal gradient of supraneurals from above V2. In gonorynchid (Patterson and Johnson, 1995, fig. 5; Gayet, 1993, figs. 8 and 9) and kneriid (Lenglet, 1974, fig. 17; Grande, 1994, fig. 9) gonorynchiforms, the supraneural anterior to the neural spine of V2 is either absent or is smaller than its successors, and the supraneurals evidently develop rostrally and caudally from over V3. Otophysans (e.g., Fink and Fink, 1981, figs. 14–18; Patterson, 1984a, figs. 14 and 16) show the same pattern, with the supraneural over V2 either absent (characiphysans) or smaller than its successor (cypriniforms and the Eocene *Chanoides*; Patterson, 1984a, p. 445).

A fourth pattern of supraneural development (pattern 4) occurs in platytroctid alepocephaloids and in *Rouleina* and *Xenodermichthys*—the supraneurals are evenly spaced, with one every two or three vertebrae (Matsui and Rosenblatt, 1987, fig. 6).

A synapomorphy of eurypterygians (aulopiforms, myctophiforms and acanthomorphs), additional to the three listed by Johnson (1992), is that they have no more than three supraneurals. Exceptions to that statement, such as four supraneurals in the neoscope-

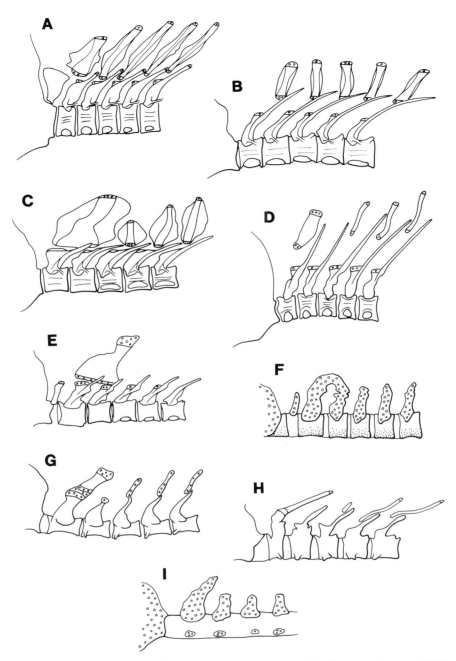

FIGURE 12 Anterior vertebrae and supraneurals in (A) *Elops hawaiensis*, BMNH 1962.4.3.1, 69 mm SL; (B) *Bathylagoides* sp., USNM 234768, 43 mm SL; (C) *Glossanodon polli*, USNM 203236, 101 mm SL; (D) *Oncorhynchus clarki*, BMNH 1957.2.20.3, 90 mm SL; (E) *Mallotus villosus*, AMNH 26286, 137 mm SL; (F) *Protosalanx chinensis*, HSU 85-38, 65 mm SL; (G) *Lovettia sealei*, BMNH 1937.8.22.1, 41 mm SL; (H) *Lepidogalaxias salamandroides*, USNM 339265, 44 mm SL; (I) Galaxiidae indet. larva, USNM 340197, 25 mm SL.

lids *Neoscopelus* and *Solivomer*, five in the stephanoberyciforms *Rondeletia* (Moore, 1993) and *Hispidoberyx*, eight in *Barbourisia* (Moore, 1993), and four or more in a few percomorphs (Johnson, 1984; Smith-Vaniz, 1984; Mabee, 1988), are all most economically explained by secondary increase, as demonstrated in centrarchids by Mabee (1988). We interpret supraneural development in lower eurypterygians (aulopiforms and myctophiforms) as pattern 2 and in acanthomorphs (Mabee, 1988) as a modified pattern 1.

Summarizing, pattern 1 supraneurals occur in osteoglossomorphs, elopomorphs, and clupeomorphs, and pattern 2 occurs in basal euteleosts (argentinids, some alepocephalids, salmonoids, osmeroids and stomiiforms) and lower neoteleosts, with the majority of that group (acanthomorphs) having a modifed pattern 1. Esocoids exhibit both pattern 1 (*Umbra*) and 2 (*Esox*).

126. Number of supraneurals. Primitively in teleosts the supraneurals are numerous, with one above each vertebra from the occiput back to the dorsal fin origin. That pattern persists in most lower euteleosts (argentinids, bathylagids, microstomatids, opisthoproctids, salmonoids, umbrids, and in the Paleocene *Esox tiemani* among esocids; Patterson and Johnson, 1995, p. 24). Among osmeroids, the primitive pattern occurs in most osmerids (see below) and in *Aplochiton, Lovettia, Lepidogalaxias,* and galaxiids. Retropinnids have a reduced number of supraneurals, covering half or fewer of the vertebrae between the occiput and dorsal origin. Salangids have a single supraneural, over V1 and V2 (Fig. 12F). Among osmerids, *Hypomesus, Osmerus, Plecoglossus, Spirinchus,* and *Thaleichthys* show the primitive pattern, with supraneurals extending almost to the dorsal origin (D). *Allosmerus* has about 16 supraneurals, with D at about V28, and *Mallotus* (Fig. 12E) has 1–8 supraneurals, with the same dorsal origin. Among alepocephaloids, the number of supraneurals is reduced in platytroctids, *Rouleina,* and *Xenodermichthys,* which have one supraneural to every two or three vertebrae (no. 125 above) and a total of about 5 supraneurals. In other sampled alepocephalids the most complete series of supraneurals is in *Bathylaco,* with 13 supraneurals and D at V19. *Bajacalifornia* also has 13 supraneurals with D at V21, *Bathytroctes* has 11 with D at V18, and *Talismania antillarum* has 11 with D at V19. Other alepocephalids have fewer than 10 supraneurals: *Talismania aphos* 9, D at V20; *Rinoctes,* 8, D at V20; *Narcetes,* 7, D at V23; *Alepocephalus agassizi,* 2; *A. tenebrosus* and *Leptochilichthys,* 1; and *A. bairdii, Leptoderma,* and *Photostylus,* none.

127. Laminar supraneurals. Supraneurals are primitively rodlike cartilages sheathed by a tube of perichondral bone. In *Argentina* and *Glossanodon* the anterior supraneurals are expanded rostrocaudally by sheets of membrane bone (Fig. 12C). The character is trivial but is one of the few indicators of argentinid monophyly.

128. Condition of last few neural and haemal spines. Gosline (1960, p. 332), in grouping salmonoids, osmeroids, and argentinoids (his Salmonoidei, which did not include alepocephaloids), wrote

the caudal skeletons of adult members . . . are, with the exception of those of the Salangidae and of the neotenic aplochitonid *Lovettia,* the most easily recognizable. . . . the last few preterminal vertebrae have neural and haemal spines with flattened, anteroposteriorly oriented blades. These together make up a flange or keel running above and below the posteriormost portion of the vertebral column.

Greenwood and Rosen (1971) illustrated these blades (their "preural flanges") in several argentinoids and in the alepocephaloid *Searsia,* and Markle (1976), Kobyliansky (1986, 1990), Sazonov (1986), and Fujita (1990) illustrated them in several other alepocephaloids and in argentinoids. In all these alepocephaloids and argentinoids, the preural flanges, when present, are confined to the proximal part of the anterior margin of the neural and haemal spine in a configuration that seems no different from that in (for example) *Chanos,* cyprinids, and chlorophthalmids (Fig. 13A; Fujita, 1990, figs. 30, 36, 41, and 98–100). But in salmonoids and osmeroids the last few neural and haemal arches generally have the structure described by Gosline (1960): they have laminar bone on both the anterior and posterior margins and tend to contact their neighbors in the midline, forming the "keel" above and below the column that Gosline described (Figs. 13C, 13E and 13F). This condition is illustrated by Rosen (1974, figs. 18 and 25–27) in several galaxiids, *Plecoglossus, Retropinna, Stokellia, Aplochiton, Lovettia* and several salmonoids, and by Fujita (1990, figs. 61–63 and 66–71) in *Hypomesus, Plecoglossus,* and several salmonoids. Other salmonoids showing the condition are illustrated in Shaposhnikova (1968a,b), Nybelin (1971, pl. 6) and Arratia and Schultze (1992). Among osmeroids, the condition does not occur (the neural and haemal spines are slender distally) in *Allosmerus, Osmerus* (Fujita, 1990, fig. 60), *Spirinchus* (Weitzman, 1967, fig. 4), *Thaleichthys* (Chapman, 1941b, fig. 15), salangids (Roberts, 1984, fig. 21; Fujita, 1990, fig. 64), *Lepidogalaxias* (Rosen, 1974, fig. 24) and some galaxiids (e.g., *Galaxias maculatus,* Fujita, 1990, fig. 65). In *Mallotus* the neural and haemal spines are normally slender distally, but they become laminar in large specimens.

129. Condition of NPU2. The condition of the neural spine of PU2 (NPU2) in lower euteleosts has been debated, particularly by Rosen (1973, pp. 422–432), who took the "broad half-spine" found in *Elops, Pterothrissus,* osmerids, and *Aulopus,* for example, to be derived and to characterize Eurypterygii or (Rosen, 1985, p. 52) eurypterygians plus osmeroids. Johnson (1992, p. 12) reviewed subsequent evaluations.

Fujita (1990, table 2) recorded a short NPU2 in *Elops, Pterothrissus, Hypomesus, Plecoglossus, Coregonus, Thymallus,* and *Diplophos* among the pre-eurypterygian teleosts he sampled. It also occurs in all other osmer-

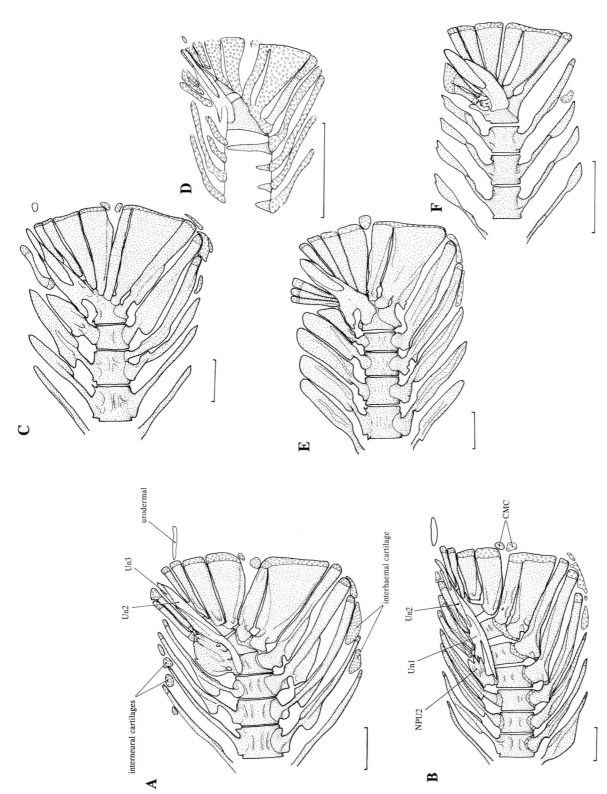

FIGURE 13 Caudal skeleton in (A) *Argentina silus*, BMNH 1967.3.5.2, 78 mm SL; (B) *Coregonus lavaretus*, BMNH 1996.2.6.20, 73 mm SL; (C) *Hypomesus transpacificus*, BMNH 1984.6.28.11, 65 mm SL; (D) *Hypomesus* sp., larva, BMNH 1937.8.22.1, 41 mm SL. Abbreviations as in text; scale bars = 1 mm. A shows the normal position of Un2 in teleosts, posteroventral to Un1, whereas B–F show the derived condition characteristic of salmonoids and osmeroids, where Un2 is anterodorsal to Un1. A shows the derived, long NPU2 characteristic of argentinoids (and many other groups) whereas B–E show the primitive, short NPU2. C and E–F show the laminar distal parts of the last few neural and haemal spines characteristic of salmonoids and osmeroids; these do not develop in *Coregonus lavaretus* but are conspicuous in other species (e.g., Nybelin, 1971, pl. 6; *C. nasus* and *C. albula*) and in *Prosopium*.

ids, all retropinnids, and commonly in alepocephaloids (Fig. 13; Greenwood and Rosen, 1971, fig. 11, *Searsia*; Markle, 1976, fig. 13A, *Bathylaco*; Sazonov, 1986, figs. 17–19, *Platytroctes, Barbantus, Mirorictus, Persparsia, Holtbyrnia,* and *Sagamichthys*; Matsui and Rosenblatt, 1987, fig. 7, *Paraholtbyrnia*; and in our *Pellisolus* as well as several of the mentioned genera).

Among the Jurassic teleosts studied by Arratia (1991, character 19), NPU2 is coded as equal in length to NPU3 only in the ichthyodectiform *Allothrissops* and in the *incertae sedis* taxa *Daitingichthys, Orthogonicleithrus, Pachythrissops,* and *Leptolepis talbragarensis* (it is variable in the last and in ichthyodectiforms; Patterson and Rosen, 1977, p. 110, fig. 46). Among generalized elopomorphs, NPU2 is long in *Megalops* and *Albula* but short in the Jurassic *Anaethalion* and the Cretaceous megalopid *Sedenhorstia* (Goody, 1969).

In salmonoids, Stearley and Smith (1993, character 97), in agreement with Arratia and Schultze (1992), referred to the short neural spine in coregonids and *Thymallus* as a neural arch with a detached neural spine. *Thymallus* and coregonids may give that impression, because in some specimens NPU2 is tipped with cartilage (Arratia and Schultze, 1992, figs. 21 and 22) and appears to be associated with the tip of the first epural. Occasional specimens have two neural arches on PU2, apparently associated distally with tips of the first and second epurals, whereas in other coregonids the tip of NPU2 is without cartilage and lies in front of the tip of the epural (Fig. 13B; Nybelin, 1971, pl. 6).

In the most primitive teleosts, such as the early Jurassic pholidophorids and *Pholidolepis* (Patterson, 1968), NPU2 is shorter than NPU3 but like NPU3 was ossified in cartilage and tipped with cartilage in life. The same is true of Jurassic "leptolepids" such as *Leptolepis coryphaenoides, Todiltia, Ascalabos,* and *Tharsis,* and of the Jurassic ichthyodectiforms *Allothrissops* and *Thrissops* (Patterson, 1968; Patterson and Rosen, 1977; Arratia, 1991). A bladelike or leaflike NPU2 tipped with membrane bone, like that of *Elops,* first appears in late Jurassic teleosts such as *Anaethalion* and *Luisichthys* (Patterson and Rosen, 1977; Arratia, 1991). It occurs commonly in Lower Cretaceous teleosts, for example in *Crossognathus* and pachyrhizodontids (Forey, 1977; Taverne, 1989), *Scombroclupeoides* (Patterson and Rosen, 1977, fig. 47) and *"Leptolepis" neocomiensis* (Patterson, 1970, fig. 48). This leaflike NPU2 is derived relative to the cartilage-tipped NPU2 of pholidophorids and true leptolepids, but in our view its distribution clearly indicates that it is primitive relative to a long NPU2.

130. [Begle's 5: Rudimentary neural arches in caudal skeleton independent (0) or fused to centrum followed in some cases by fusion to first uroneural (1) or fused to first uroneural followed in some cases by fusion to centrum (2). State (0) coded in outgroups, salmonoids, neoteleosts, *Lepidogalaxias,* and esocoids; state (1) in all argentinoids, all alepocephaloids except platytroctids, *Talismania, Bathylaco,* and *Leptoderma,* all southern osmeroids and *Thaleichthys;* state (2) in Osmeridae (minus *Thaleichthys*) and salangids. Platytroctids and *Talismania* are coded as polymorphic for (0) and (1); *Bathylaco* and *Leptoderma* are coded (?).] Begle's text implies that the (1) entered for *Thaleichthys* is an error. We commented on this character (Patterson and Johnson, 1995, p. 27). It is taken from Fink (1984b), and the "followed in some cases" in the description of states (1) and (2) refers to ontogenetic information on neural arches over PU1 in *Osmerus* and galaxiids presented by Fink and Weitzman (1982, p. 83). In Begle (1991) the character was two-state and referred to the first uroneural (Un1) as well as caudal neural arches; the derived state was coded only in osmerids and salangids, as in Fink (1984b). It is easy enough to determine whether Un1 is free or fused to a centrum, but determining the ontogenetic history of rudimentary neural arches from adult specimens is problematic. The effect of Begle's (1991) coding was to distinguish osmerids plus salangids from the remaining taxa in his sample, whereas the effect of his 1992 coding was also to distinguish all argentinoids and osmeroids (except *Lepidogalaxias*) from outgroups.

This part of the caudal skeleton of argentinoids and osmeroids has been discussed often and at length over the past quarter-century (Patterson, 1970; Greenwood and Rosen, 1971; Rosen, 1974; Markle, 1976; Fink and Weitzman, 1982) without any real agreement on its significance. Begle's (1992) coding of (1) for most alepocephaloids is not supported by the descriptions and illustrations in Greenwood and Rosen (1971), Markle (1976) and Sazonov (1986). In alepocephaloids, the neural arches of PU1 and U1, when present, usually remain autogenous, in state (0). In argentinoids (Fig. 13A; Patterson, 1970; Greenwood and Rosen, 1971; Kobyliansky, 1986, 1990; Fujita, 1990) the first uroneural is free (*Argentina* and opisthoproctids) or fused to PU1, and the caudal neural arches are generally well-developed and fused to the centra (except in opisthoproctids), justifying Begle's (1) in all except opisthoproctids. In retropinnids (Fig. 13E; McDowall, 1969; Rosen, 1974) the first uroneural is free, and there are no caudal neural arches; they should be coded (0).

131. [Begle's 77: PU1 bearing one (1) or more (0) rudimentary neural arches. State (1) coded in esocoids and *Lepidogalaxias.*] Begle (1991) referred this character to Rosen (1974) and Fink and Weitzman (1982); we cannot understand it. The primitive condition in tele-

osts is to have one arch per centrum, and to have one arch, rudimentary or not, over PU1 can only be primitive. Working through Fujita's (1990) figures, which are arranged systematically, the first fish one encounters with *no* (rather than one) neural arch over PU1 is *Esox* (Fujita, 1990, fig. 54), and the first fish in which the configuration might imply more than one arch is *Argentina*. Rosen's (1974) extensive sample of variant caudal skeletons in lower euteleosts shows more than one arch [Begle's (1)] in some galaxiids and in *Aplochiton* and an arch that bifurcates distally in some *Dallia*. Our *Lepidogalaxias*, like Rosen's (1974, fig. 24), all have a single arch with a complete neural spine on PU1. Character was discarded.

132. [Begle's 108: U1 free (0) or fused to PU1 (1). State (1) coded in *Bathylagus* and *Opisthoproctus*.] State (1) does not occur in *Opisthoproctus* (Greenwood and Rosen, 1971, fig. 15; checked in two further specimens, including that used by Begle). Begle's text assigns the fusion to argentinids, though it is not coded in them. Argentinids are variable, with fusion in most species (or individuals) of *Argentina* (fused in Gosline, 1960, fig. 10; Rosen and Patterson, 1969, fig. 71; Patterson, 1970, fig. 38; Greenwood and Rosen, 1971, fig. 12; unfused in Fig. 13A and Fujita, 1990, fig. 56) and *Glossanodon* (fused in Greenwood and Rosen, 1971, fig. 12; Fujita, 1990, fig. 57; unfused in Kobyliansky, 1990, fig. 14). State (1) occurs in microstomatids (Kobyliansky, 1990, fig. 14) and among Begle's sample in all neoteleosts, all Osmeroidei except *Lepidogalaxias*, several platytroctids (Sazonov, 1986, figs. 17–19), and *Leptoderma* (Greenwood and Rosen, 1971, fig. 9; personal observation). They are wrongly coded.

133. [Begle's 18: Number of hypurals six (0) or five (1). State (1) coded in galaxiids, *Aplochiton*, *Lovettia*, and *Lepidogalaxias*.] *Leptoderma* also has five (or fewer) hypurals (Greenwood and Rosen, 1971; Markle, 1976; personal observation) and is miscoded. There are no more than five hypurals in the umbrids *Dallia* and *Umbra* (Rosen, 1974; Wilson and Veilleux, 1982). Fusion between hypurals occurs in some galaxiids (Rosen, 1974, fig. 18) and salangids (Roberts, 1984, p. 201). In retropinnids (*Prototroctes*, *Retropinna*, *Stokellia*) hypural 1 is fused to the parhypural (Fig. 13E; McDowall, 1969, fig. 4; Rosen, 1974, fig. 27), an unusual condition that we have not observed elsewhere in lower euteleosts (cf. Fujita, 1990, table 2.1).

134. Number of epurals. Three epurals, primitive for Recent teleosts, occur in most lower euteleosts. There are no more than two epurals in argentinids; platytroctids and all alepocephalids except *Bathylaco* and *Narcetes* (Markle, 1976; Sazonov, 1986); a derived subgroup of salmonids (Stearley and Smith, 1993,

character 98; reversal to three inferred in *Oncorhynchus*); *Hypomesus*, *Mallotus*, and *Plecoglossus* among osmerids; *Aplochiton* and *Lovettia* among galaxioids; and umbrids among esocoids.

There is one epural in the alepocephalid *Leptochilichthys*, opisthoproctids, and galaxiids, and none in the alepocephalid *Leptoderma* or in *Lepidogalaxias*.

135. [Begle's 76: Uroneurals more than one (0) or one (1). State (1) coded in esocoids and *Lepidogalaxias*.] There is one uroneural in *Leptoderma* (Markle, 1976; personal observation), whereas our *Lepidogalaxias* consistently have two uroneurals.

136. Uroneural 1 (Un1) autogenous or fused. Begle (1991, character 5) correctly coded the fusion of Un1 with the compound centrum in all osmerids and salangids. Begle (1992) altered the character to a description of ontogenetic fusions between ural neural arches and other structures (no. 130, above). Fusion of Un1 also occurs in *Glossanodon*, bathylagids, and microstomatids (though it is drawn as if separate in *Microstoma* and *Nansenia* by Greenwood and Rosen, 1971, fig. 13; cf. Kobyliansky, 1990, fig. 14).

137. Membranous outgrowth of Un1. Begle (1992, fig. 9) cited a "membranous outgrowth of first uroneural" as a synapomorphy uniting all euteleosts (esocoids, ostariophysans, salmonoids, his Osmerae, and neoteleosts). The character is problematic and has involved unsubstantiated speculation on ontogenetic processes (e.g., Greenwood and Rosen, 1971; Rosen, 1974). The structure in question is the "stegural" of Monod (1968, pp. 62, 594), for which the type locality is *Salmo* and other salmonids. Ontogeny is now well known in salmonoids (Arratia and Schultze, 1992), and contrary to earlier suggestions there is no indication that the structure is compound in origin. Stearley and Smith (1993, character 96) considered the large, fan-shaped membranous outgrowth in salmonines to be derived relative to the smaller outgrowth in *Thymallus* and coregonids (Fig. 13B; Norden, 1961, pls. 14 and 15; Arratia and Schultze, 1992, figs. 21 and 25). The form of the membranous outgrowth, if present, can be assessed accurately only when Un1 is autogenous. In alepocephaloids, where Un1 is always free, it has a substantial anterodorsal outgrowth in platytroctids (Sazonov, 1986, figs. 17–19) and in most alepocephalids (Markle, 1976; outgrowth absent in *Alepocephalus*, *Asquamiceps*, *Bathytroctes*, *Conocara*, *Ericara*, *Leptoderma*, and *Talismania*). In argentinoids, Un1 is usually fused to the underlying compound centrum, but when it is free (*Argentina*, Fig. 13A; Patterson, 1970, fig. 38; Fujita, 1990, fig. 57; opisthoproctids; a 22 mm bathylagid larva) there is no conspicuous membranous outgrowth. In osmeroids, Un1 is fused to the compound centrum in osmerids and salangids, but

the configuration is very like that in retropinnids (Figs. 13C and 13E; Rosen, 1974, figs. 26 and 27), where Un1 is free and has a membranous outgrowth like that in salmonoids, and larval osmerids (Fig. 13D) confirm the condition. In other southern osmeroids (*Aplochiton, Lovettia, Lepidogalaxias,* and galaxiids) the membranous outgrowth is small or absent (Fig. 13F; Rosen, 1974, figs. 18, 19, and 24; in our larval galaxiids, ca. 25 mm SL, there is no conspicuous outgrowth). In esocoids (Rosen, 1974, figs. 20–23; Wilson and Veilleux, 1982, figs. 13 and 14; Fujita, 1990, figs. 54 and 55) the membranous outgrowth of Un1 is small but distinct, except in *Umbra krameri* and some *Dallia*, where it is absent. The membranous outgrowth on Un1 is generally well developed in neoteleosts (e.g., Fujita, 1990, figs. 72, 84, 96–102, 115–117, 175–178). In ostariophysans Un1 is free only in early Cretaceous stem gonorynchiforms such as *Dastilbe* (Blum, 1991, p. 279), *Tharrhias* (Patterson, 1984b, fig. 2), *Gordichthys,* and *Rubiesichthys* (Poyato-Ariza, 1994, figs. 16, 17). Although some specimens of *Gordichthys* have been interpreted as showing a membranous outgrowth (Poyato-Ariza, 1994, p. 28), the structure in question is no more distinctive than the low ridge developed in several Jurassic ''leptolepids'' (Arratia, 1991, p. 297) or in large *Elops* (Schultze and Arratia, 1988, p. 289). In clupeomorphs Un1 is free only in *Denticeps* (Greenwood, 1968, fig. 29; Monod, 1968, fig. 263; Grande, 1985, fig. 7) and in Cretaceous stem clupeomorphs (Grande, 1982, 1985). As in stem gonorynchiforms (above) there may be a low ridge on the anterodorsal surface of Un1, but there is no distinctive membranous outgrowth.

Summing up, an anterodorsal membranous outgrowth on Un1 characterizes euteleosts if ostariophysans are excluded and if absence of the outgrowth in argentinoids is considered secondary.

138. Extent of Un1. Primitively in Recent teleosts Un1 extends forward to PU2 [e.g., *Hiodon, Elops, Megalops,* many alepocephaloids (Markle, 1976; Sazonov, 1986), and salmonoids (Fig. 13B; Fujita, 1990)]. When Un1 is fused to the the compound centrum (no. 136 above), it never extends beyond PU1. It also never extends beyond PU1 in esocoids or in neoteleosts (Fujita, 1990).

139. Position of Un2. Greenwood and Rosen (1971, p. 25, fig. 16) mentioned one feature that they took to be evidence of relationship between osmeroids and salmonoids, the position of the second uroneural (Un2), as first described in salmonoids by Cavender (1970). Primitively, Un2 in teleosts is elongate, with a slender, tapering anterior portion that lies posteroventral to Un1 (e.g., Fig. 13A and Fujita, 1990, figs. 6–8, 15–30, 47–48, and 56–58, of elopomorphs, clu-

peomorphs, *Chanos,* characins, and argentinoids). In all salmonoids and osmeroids Un2 is rather broad and lies anterodorsal to Un1 (e.g., Figs. 13B–13F; Rosen, 1974, figs. 18, 19, and 25–27; Fujita, 1990, figs. 60–71; Arratia and Schultze, 1992). The character is not perfectly clean, since Fujita's illustrations of cyprinoids (1990, figs. 35–45) and of most aulopiforms (1990, figs. 100 and 103–114) show a configuration like that in osmeroids and salmonoids.

140. Third uroneural (Un3). Un3 is primitively present in lower euteleosts, as it is in salmonoids and osmerids. Un3 is absent in our salangid material (illustrated in *Salangichthys ishikawae* by Fujita, 1990, fig. 64, but unusual in form) and in southern osmeroids (retropinnids, *Aplochiton, Lovettia, Lepidogalaxias,* and galaxiids). Among argentinoids Un3 is present in argentinids, microstomatids, and opisthoproctids, and in *Bathylagichthys* and *Lipolagus* among bathylagids (Greenwood and Rosen, 1971; Kobyliansky, 1986, 1990). Among alepocephaloids, Un3 is present in most platytroctids (Sazonov, 1986) and most alepocephalids (Markle, 1976). It is absent in the alepocephalids *Leptoderma, Photostylus,* and *Rouleina* (Greenwood and Rosen, 1971; Markle, 1976). Esocoids lack Un3, as do all neoteleosts.

141. Interneural and interhaemal cartilages. Kobyliansky (1986, fig. 19) and Fujita (1990, figs. 58 and 59) illustrated the extensive series of large interneural and interhaemal cartilages in bathylagids: they may extend forward to about PU12, and always extend at least to PU5. The microstomatid *Nansenia* also has conspicuous interneural and interhaemal cartilages (Kobyliansky, 1990, fig. 14), extending forward to about PU7; there are fewer in *Microstoma.* The only other teleosts with comparable development of the interneural and interhaemal cartilages are myctophids and some stomiiforms (Fujita, 1990, table 3). In argentinids the interneural cartilage series sometimes extends to PU5 (Fig. 13A); in opisthoproctids interneural and interhaemal cartilages do not extend forward beyond PU4.

142. Caudal median cartilages (CMCs, Fujita, 1990). CMCs (Figs. 13 and 14) are two cartilages lying in the gap between the distal parts of hypurals 2 and 3, one at the posteroventral corner of H3 and one at the posterodorsal corner of H2. They occur only in euteleosts (Fujita, 1990, table 3), where their presence is unambiguously primitive for argentinoids, osmeroids, salmonoids and neoteleosts. There is a single CMC, rather than two, in *Mallotus,* retropinnids (*Prototroctes, Retropinna,* and *Stokellia*), some salmonids (Fujita, 1990, figs. 67–69), some *Salangichthys* (Fujita, 1990, fig. 64, *S.ishikawae;* there are two in our *S.microdon*), and some *Salanx* (our *S.prognathus;* there are two

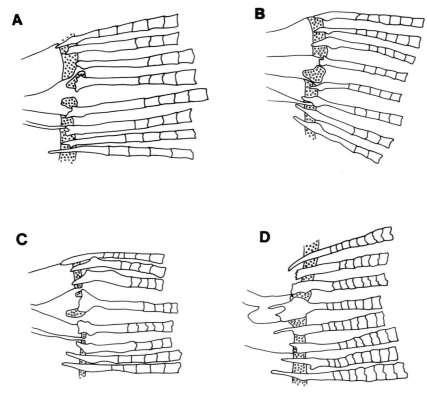

FIGURE 14 Caudal median cartilages (CMCs) in argentinoids. The figures show the distal parts of hypurals 1–3, the CMCs, and the bases of the central caudal fin rays. (A) *Osmerus mordax*, BMNH 1984.11.29.11, 85 mm SL, showing the configuration normal in lower euteleosts, with the CMCs each carrying the innermost ray of the upper and lower caudal lobes. (B) *Pseudobathylagus milleri*, SIO 80-258, 155 mm SL, with a single CMC lying between the innermost caudal rays. (C) *Argentina sphyraena*, USNM 238015, 127 mm SL, and (D) *Bathylaco nigricans*, USNM 206693, 285 mm SL, show the condition characteristic of primitive argentinoids and alepocephaloids, with the CMCs carrying only the lowermost ray of the upper caudal lobe. We interpret the bathylagid condition (B) as derived relative to the condition in C and D.

in our *S.ariakensis* and in Roberts's *S.cuvieri*, 1984, fig. 21). CMCs are secondarily absent in acanthomorphs, and primarily absent in osteoglossomorphs, elopomorphs, clupeomorphs, and ostariophysans. Among lower euteleosts, they are absent in *Aplochiton*, *Lovettia*, *Lepidogalaxias*, galaxiids, the alepocephalids *Leptoderma* and *Photostylus*, and in esocoids. Their absence in esocoids is problematic. Patterson and Johnson (1995, p. 25) used CMCs as a character to distinguish esocoids from other euteleosts (excluding ostariophysans). However, we now infer that CMCs have been lost in at least six euteleostean lineages: southern osmeroids (except retropinnids); some alepocephalids; stomiid stomiiforms (Fujita, 1990, figs. 84 and 85); synodontoids and a subgroup of ipnopids among aulopiforms (Baldwin and Johnson, this volume); and acanthomorphs (Johnson and Patterson, 1993). Whether the absence of CMCs in esocoids is primary

or secondary must be resolved by congruence with other characters.

143. CMCs and caudal finrays. Primitively, the upper CMC supports the lowermost ray of the upper caudal lobe, and the lower CMC supports the uppermost ray of the lower lobe (Fig. 14A; Fujita, 1990, figs. 60–63 and 73–78). In argentinids (*Argentina* and *Glossanodon*), microstomatids (*Microstoma* and *Nansenia*), the opisthoproctid *Bathylychnops*, platytroctid alepocephaloids (10 genera examined), and the alepocephalids *Bathylaco* and *Narcetes*, the CMCs together support the lowermost ray of the upper caudal lobe (Figs. 14C and 14D). This condition was illustrated in argentinids by Fujita (1990, figs. 56 and 57) and was reported in *Bathylaco*, *Narcetes*, and the platytroctid *Mentodus* by Markle (1976). Its distribution implies that the derived condition, with one finray on both CMCs, is synapomorphous for Argentinoidei (argen-

tinoids + alepocephaloids). If so, three reversals to the primitive pattern must be accepted: in bathylagids (Fig. 14B), opisthoproctids (except *Bathylychnops*), and alepocephalids (all except *Bathylaco* and *Narcetes*).

144. Caudal scutes. Fujita (1990, table 2) recorded upper and lower caudal scutes in *Elops*, a few clupeids, *Chanos, Alestes, Argentina, Glossanodon, Osmerus, Hypomesus*, and various aulopiforms. Their distribution correlates well with that of the urodermal (no. 146, below); the only substantial differences are that scutes occur in a few clupeomorphs and ostariophysans, where no urodermal is recorded, and scutes are not recorded in coregonids or myctophiforms, where there is a urodermal. It can be difficult to discriminate a small caudal scute (unpaired) from a procurrent ray (paired) [for example, we believe that *Plecoglossus* has caudal scutes, whereas Fujita (1990, fig. 63) took the structures to be procurrent rays].

145. [Begle's 9: Principal caudal rays 10/9 (0), 9/9 (1), or 8/8 (2). State (1) coded in retropinnids, state (2) in galaxiids, *Lovettia, Aplochiton*, and *Lepidogalaxias*.] In Begle (1991) the character was two-state, with (1) for 9/9 or less, and an undescribed state (2) was coded for galaxiids, *Lovettia, Aplochiton*, and *Lepidogalaxias*. Also (Begle, 1991, p. 53) retropinnids were wrongly diagnosed by "Principal caudal rays 8/8" rather than 9/9.

146. Urodermal. Fujita (1990, table 2) recorded a urodermal in *Elops, Argentina* (Fig. 13A), *Glossanodon, Osmerus, Hypomesus* (Fig. 13C), *Plecoglossus, Coregonus* (Fig. 13B), a few aulopiforms, and several myctophiforms. There is also a urodermal in microstomatids and opisthoproctids (Greenwood and Rosen, 1971, figs. 13 and 15; Kobyliansky, 1990, fig. 14), in the bathylagid *Bathylagichthys* (Kobyliansky, 1986, fig. 19), in the coregonids *Prosopium* (Arratia and Schultze, 1992) and *Stenodus*, and in all other osmerids (*Allosmerus, Mallotus, Spirinchus*, and *Thaleichthys*). No urodermal is recorded in Recent osteoglossomorphs, non-elopid elopomorphs, clupeomorphs, ostariophysans, alepocephaloids, salmonids, galaxioids, salangids, esocoids, stomiiforms, or acanthomorphs. In some Jurassic teleosts there are two urodermals (e.g., *Allothrissops, Ascalabos, Tharsis*, and various "*Leptolepis*" spp.; Patterson and Rosen, 1977; Arratia, 1991), whereas others have one (e.g., *Anaethalion* and *Leptolepides*). Arratia and Schultze (1992, p. 247) proposed that the urodermal is "independently acquired" in coregonids, osmerids and/or argentinids, and myctophiforms, but to us its distribution indicates that it is homologous throughout Recent teleosts and has been lost repeatedly.

147. [Begle's 27: Caudal fin margin incised/deeply forked (0) or rounded/emarginate (1). State (1) coded

in *Lepidogalaxias* and reported in some galaxiids.] State (1) also occurs in umbrids.

H. Pectoral Girdle and Fin

148. [Begle's 12: Extrascapular present (0), attached to pterotic (1), or absent (2). State (1) coded in *Argentina* and *Glossanodon*, and state (2) in galaxiids, *Aplochiton, Lovettia, Lepidogalaxias*, and salangids. In Begle (1991) the character was two-state, with (1) for absence.] State (1), "extrascapular is attached to the pterotic in the Argentinidae (Chapman, 1942a)" (Begle, 1992, p. 361), is evidently taken from Ahlstrom *et al.* (1984, tables 42 and 43), who characterized argentinids in that way, probably based on Cohen (1958, p. 101; 1964, p. 3). The character is nonexistent and comes from a misunderstanding of Chapman (1942a), who described the dermal portion of the pterotic in *Argentina* as the "supratemporal." McDowall (1969, p. 819) incorrectly reported that extrascapulars are absent in *Plecoglossus, Spirinchus*, and *Mallotus*; in all there are several ossicles, as in other osmerids (no. 179, below). Among alepocephaloids, extrascapulars are absent in *Leptoderma*, wrongly coded by Begle, and in *Photostylus, Rinoctes, Rouleina squamilaterata*, and *Platytroctes* in our sample. Among esocoids there is an extrascapular in *Esox* and *Novumbra* but none in *Kenoza, Dallia*, or *Umbra*. Begle's state (1) should be deleted and his (2) changed to (1).

149. Internal limb of posttemporal. The internal limb of the posttemporal is primitively ossified in teleosts and attaches to the intercalar. The internal limb is unossified in all southern osmeroids (retropinnids, *Aplochiton, Lovettia, Lepidogalaxias*, and galaxiids). The internal limb is typically ossified in esocoids, salmonoids, alepocephaloids, and argentinoids except opisthoproctids.

150. [Begle's 13: Cleithrum with ventral process descending to meet coracoid just in front of its articulation with scapula (0) or without such a process (1). State (1) coded in galaxiids, *Aplochiton, Lovettia, Lepidogalaxias*, and salangids.] The character is from McDowall (1969). The ventral process is also absent in *Stokellia*, esocoids, *Bathylagus, Opisthoproctus*, and all examined alepocephaloids except *Bathylaco* and some platytroctids; these are all wrongly coded (0). In argentinids, *Bathylaco*, platytroctids, salmonoids, and outgroups such as elopiforms, *Clupea, Chanos, Triplophos*, and *Aulopus* the process is present but different in form from that in osmerids and *Retropinna*. In the latter, the process is narrow-based and straight-sided, extending down to interdigitate with the margin of the coracoid (in *Prototroctes* the process is very small but does interdigitate with the coracoid). In ar-

gentinids, *Bathylaco*, platytroctids, salmonoids, and outgroups the process is long-based and triangular, and the coracoid lies against it. We therefore believe that there are three states of the character: a long-based triangular process (0), a narrow, columnar process (1), and absence (2).

151. Postcleithra. Primitively in teleosts there are three principal postcleithra (Pcl 1–3 of Gottfried, 1989). One or more extra bones, lateral to Pcl 2 and 3, have been reported in *Elops* and most salmonoids [Arratia and Schultze (1987) illustrated two in *Oncorhynchus gairdneri*, and Sanford (1987) recorded one as present in all salmonoids except *Prosopium* and *Thymallus*, with four on one side in an *O.gorbuscha*]; these extra bones are neglected here. Osteoglossomorphs have only one (or no) postcleithrum, positionally homologous with Pcl 1. Generalized clupeomorphs and characiform ostariophysans have three, as do many aulopiforms (Baldwin and Johnson, in this volume). Ctenosquamates have only two, positionally homologous with Pcl 2 and 3 (Gottfried, 1989). Postcleithra are apparently absent in stomiiforms. In lower euteleosts, Pcl 1–3 are present in salmonoids, argentinids, microstomatids (Kobyliansky, 1990, figs. 11 and 12), and the bathylagid *Bathylagichthys* (Kobyliansky, 1986, fig. 18). Other bathylagids have Pcl 2 and 3, Pcl 2 only, or none (Kobyliansky, 1986). Opisthoproctids lack postcleithra. Among alepocephaloids there is a single postcleithrum, positionally homologous with Pcl 1, in most platytroctids (Sazonov, 1986, figs. 12–15). In the alepocephalids *Bathylaco*, *Bathytroctes*, and *Narcetes* there is a single postcleithrum (Markle, 1976), placed lower than that in platytroctids and which might therefore be Pcl 2. Other alepocephalids lack postcleithra. Osmerids, salangids, retropinnids, *Aplochiton*, and *Lepidogalaxias* lack postcleithra [the "postcleithrum" illustrated in *Salanx* by Roberts (1984, fig. 18) is a cartilage]. In galaxiids, McDowall (1969) illustrated one rodlike postcleithrum (corresponding to Pcl 3) in *Galaxias fasciatus* and reported no postcleithra in several other species. We find a rodlike Pcl 3, sometimes merely a sliver of bone dissociated from the cleithrum, in *Galaxias fasciatus*, *G.maculatus*, *G.occidentalis*, *G.platei*, *G.zebratus*, *Galaxiella*, and *Paragalaxias* and nothing in *Galaxias fontanus*, *G.paucispondylus*, *Neochanna* and *Nesogalaxias*. There is a rodlike postcleithrum, dissociated from the cleithrum, in *Lovettia*. Esocoids have one postcleithrum, positionally homologous with Pcl 3.

In summary, among noneurypterygian euteleosts, absence of postcleithra characterizes stomiiforms, opisthoproctids, and osmeroids, with Pcl 3 reacquired in *Lovettia* and some galaxiids; absence of Pcl 3 charac-

terizes alepocephaloids; and absence of Pcl 1 and 2 characterizes esocoids.

152. [Begle's 7: Mesocoracoid present (0) or absent (1). State (1) coded in galaxiids, *Aplochiton*, *Lovettia*, retropinnids, *Lepidogalaxias*, salangids, esocoids, *Bathylagus*, *Leptoderma*, *Opisthoproctus*, and *Bathyprion*.] The bathylagids sampled by Begle (his "*Bathylagus*") were *Bathylagoides wesethi* and *Dolicholagus longirostris*, both of which lack the mesocoracoid; it is present in *Bathylagichthys* and *Melanolagus* (Kobyliansky, 1986).

153. [Begle's 93: Coracoid with ventral process short, not extending below ventral margin of pectoral girdle (0), or narrowly elongate, extending below pectoral girdle (1). State (1) coded in *Alepocephalus*, *Talismania*, platytroctids, *Leptoderma*, *Bathyprion*, and *Rouleina*.] The "ventral process" (e.g., Markle and Merrett, 1980, fig. 5; Markle and Krefft, 1985, fig. 3) is the postcoracoid process, which is long in all larval teleosts; retention of the process, as in many alepocephaloids and in salangids (Fig. 15F; Roberts, 1984, fig. 18) and *Lovettia* (Fig. 15I), is a paedomorphic feature. Platytroctids (Sazonov, 1986, figs. 12–15) have a short postcoracoid process, comparable to that illustrated by Markle (1976, figs. 22 and 23) in *Narcetes* or *Bathylaco*, both coded (0) by Begle.

154. Number of pectoral radials. Primitively there are four pectoral radials in teleosts, and this number is remarkably constant. In alepocephalids, there are two radials in *Bathylaco*, two in *Bathyprion*, three or four in *Leptoderma*, three in *Photostylus*, three or four in *Rinoctes*, two or three in *Rouleina*, and three in *Xenodermichthys* (Markle, 1976; Markle and Merrett, 1980; personal observation). Among bathylagids, there are three radials in *Dolicholagus* and *Lipolagus* (Kobyliansky, 1986, figs. 14 and 15).

155. Form of first pectoral radial. In most osmerids and in retropinnids the first pectoral radial is modified in comparison with outgroups (Figs. 15A and 15B): it is enlarged, thickened, and embraces the scapula proximally (Figs. 15C and 15D). The first pectoral radial is unmodified in *Mallotus*, salangids, *Aplochiton*, *Lovettia*, *Lepidogalaxias*, and galaxiids (Fig. 15).

156. Proximal articulation of third pectoral radial. In salangids, *Lovettia*, and *Lepidogalaxias* the third pectoral radial tapers proximally and fails to reach the scapulocoracoid (Figs. 15I and 15J).

157. Proximal articulation of fourth pectoral radial. In the osmerids *Allosmerus*, *Osmerus*, *Plecoglossus*, *Thaleichthys*, and in large *Mallotus* (Fig. 15G), the fourth pectoral radial tapers proximally and fails to articulate with the scapulocoracoid. In *Hypomesus* and *Spirinchus* (Fig. 15C), as in outgroups (Figs. 15A, 15B,

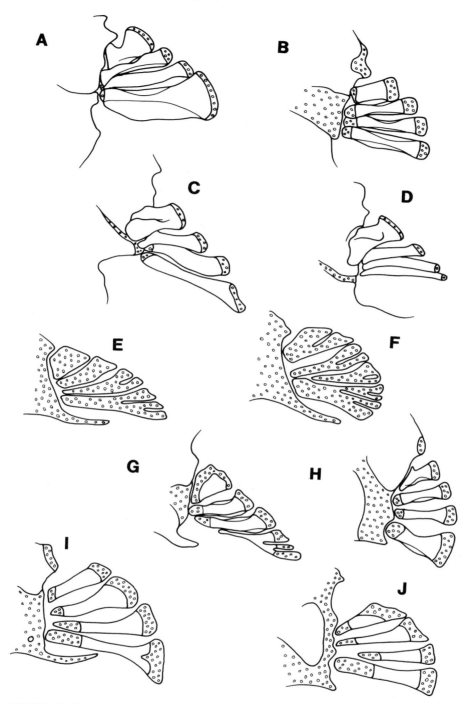

FIGURE 15 Pectoral radials of left side, lateral view, in (A) *Argentina sphyraena*, USNM 238015, 127 mm SL; (B) *Salmo trutta*, BMNH 1983.10.17.5, 74 mm SL; (C) *Spirinchus starksi*, USNM 342052, 95 mm SL; (D) *Retropinna retropinna*, BMNH 1964.4.30.19, 83 mm SL; (E) *Mallotus villosus*, HSU 89-282, 61 mm SL; (F) *Salangichthys microdon*, BMNH 1996.2.6.1, 81 mm SL; (G) *Mallotus villosus*, AMNH 26286, 137 mm SL; (H) *Galaxias occidentalis*, AMNH 31478, 79 mm SL; (I) *Lovettia sealei*, BMNH 1937.8.22.1, 41 mm SL; (J) *Lepidogalaxias salamandroides*, USNM 339265, 44 mm SL.

and 15D), the radial articulates with the scapulocoracoid.

158. Form of fourth pectoral radial. In *Mallotus* and salangids the fourth pectoral radial is multifid distally (Figs. 15E–15G), with at least three branches in *Mallotus* and often more in salangids. Sazonov (1986, fig. 12) illustrated a similar modification in the alepocephaloid *Platytroctes*. In salangids the distal parts of all the pectoral radials are subdivided (Fig. 15F; Roberts, 1984, fig. 18). In adult *Mallotus* the lowermost radial is trifid or quadrifid in all our specimens, and in small fishes the third radial is also bifid (Figs. 15E and 15G).

159. [Begle's 107: Pectoral fin small, develops late in ontogeny (0) or large and develops early in ontogeny (1). State (1) coded in *Argentina*, *Bathylagus*, and *Glossanodon*.] The character is taken from Ahlstrom *et al.* (1984). Begle's text allocates the derived state to all argentinoids (i.e., including *Opisthoproctus*). In either case Begle's coding appears to be based on misreading of Ahlstrom *et al.* (1984, p. 161, tables 42 and 43), who credited the derived state to Microstomatidae and Opisthoproctidae. So coded, the character becomes autapomorphous for *Opisthoproctus* in Begle's sample.

I. Pelvic Girdle and Fin

160. [Begle's 14: Posterior pubic symphysis present (0) or absent (1). State (1) coded in galaxiids, *Aplochiton*, *Lovettia*, *Lepidogalaxias*, and salangids.] The character is from Fink (1984b). The symphysis is also absent in esocoids (Wilson and Veilleux, 1982, fig. 11), *Opisthoproctus* (and other opisthoproctids), and in Begle's sample of bathylagids (though not in *Bathylagichthys*, Kobyliansky, 1986, fig. 13). In all osmerids except *Hypomesus*, *Osmerus*, and *Plecoglossus*, the symphysis is narrow, ligamentous, and very weak (Fig. 16D).

161. [Begle's 34: Pelvic bone without (0) or with (1) ventral condyle articulating with the first three or four hemitrichia. State (1) coded in all Osmeroidei except *Lepidogalaxias*.] The character originated with McDowall (1969, table 2) and was discussed by Howes and Sanford (1987b, p. 27). The osmerid condition is shown in Figs. 16C and 16D. There is no trace of this condyle in galaxiids, *Aplochiton*, or *Lovettia*, as Figs. 16F and 16G and McDowall's illustration (1969, fig. 6) show. We agree with Fink's (1984b, fig. 107) assessment that the condyle occurs only in osmerids, some salangids, and retropinnids.

162. Form of pelvic articular surface. In the osmerids *Hypomesus*, *Osmerus* and *Plecoglossus* the articular surface of the pelvic girdle is short and almost transverse, as in retropinnids and salmonoids (Fig. 16; Klyukanov, 1975, fig. 12). In other osmerid genera and in salangids (Figs. 16D and 16E) the girdle has a long cartilaginous medial border that does not project medially beyond the level of the ossified margin in front of it, and the articular surface of the girdle is elongate and oblique, lying at about 45^0 to the long axis of the fish [Weitzman (1967, fig. 5) and Klyukanov (1975, fig. 12) show the oblique articular surface in *Spirinchus* and *Thaleichthys* but do not indicate the extent of the cartilaginous symphyseal area].

163. Medial membrane bone extension of pelvic girdle. In *Lovettia* and *Lepidogalaxias* (Figs. 16G and 16H) the pelvic bone has a medial membrane bone lamina with a fimbriate margin. Though trivial, we have not observed this configuration elsewhere in osmeroids.

164. Pelvic radials. Primitively in teleosts there are three pelvic radials (Johnson, 1992), as in alepocephaloids (Sazonov, 1986, fig. 16), argentinids, salmonoids and osmeroids (Fig. 16). Esocoids are unique among lower euteleosts in having no pelvic radials (Johnson, 1992). The number of pelvic radials varies within alepocephalids. Most have three, but *Leptochilichthys* and *Photostylus* have two, and *Leptoderma* has one.

165. [Begle's 85: Pelvic splint present (0) or absent (1). State (1) coded in *Lovettia*, salangids, *Bathylagus*, *Talismania*, platytroctids, *Leptoderma*, *Opisthoproctus*, and *Rouleina*, with *Glossanodon* polymorphic.] *Lepidogalaxias* is wrongly coded (0), whereas *Lovettia* and *Bathylagus* (Kobyliansky, 1986, fig. 13) are wrongly coded (1). Platytroctids are polymorphic; there is a splint in *Matsuichthys* (Sazonov, 1992, p. 28), and we have observed one in *Pellisolus*, though not in several other platytroctid genera.

J. Median Fins

166. [Begle's 8: Dorsal fin anterior (0) or posterior (1). State (1) coded in all galaxioids, salangids, *Alepocephalus*, *Talismania* and *Leptoderma*.] However the character is defined, esocoids are mistakenly coded (0) [coded (1) in Begle (1991)]. We have assumed that "posterior" means originating behind the pelvics since that is the only definition that would differentiate *Aplochiton* or *Prototroctes* from *Mallotus* or *Allosmerus*. By that definition, all alepocephaloids have state (1).

167. Middle radials in dorsal and anal fin. Esocoids (esocids and umbrids) share a derived condition: in the dorsal and anal fins of *Esox* (Bridge, 1896) and *Umbra* (Wilson and Veilleux, 1982) ossified middle radials develop only on the central pterygiophores (none ossify in *Dallia* or *Novumbra*), not through to

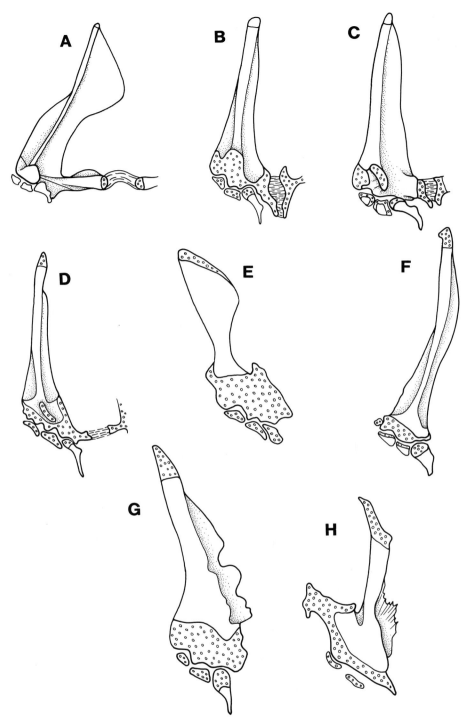

FIGURE 16 Right pelvic girdle, in ventral view, in an argentinid (A), a coregonid (B) and osmeroids (C–H). (A) *Argentina georgei*, USNM 187834, 125 mm SL; (B) *Coregonus lavaretus*, BMNH 1996.2.6.20, 73 mm SL; (C) *Hypomesus transpacificus*, BMNH 1984.6.28.11, 65 mm SL; (D) *Mallotus villosus*, AMNH 26286, 137 mm SL; (E) *Salangichthys microdon*, BMNH 1996.2.6.1, 83 mm SL; (F) *Galaxias fontanus*, BMNH 1983.6.21.1, 72 mm SL; (G) *Lovettia sealei*, BMNH 1937.8.22.1, 41 mm SL; (H) *Lepidogalaxias salamandroides*, USNM 339265, 44 mm SL.

the terminal pterygiophores as they do in other lower euteleosts. Thus esocoids maintain into maturity a state that is transient in the ontogeny of outgroups.

168. Fusion between posterior dorsal pterygiophores. All osmerids (including *Plecoglossus*) and salangids share a derived fusion between the distal parts of the posterior dorsal fin pterygiophores (Fig. 17). This also occurs in some specimens of *Lovettia* (three of four cleared-and-stained). Fusion between dorsal pterygiophores is reported elsewhere only in the aulopiform *Uncisudis* and in veliferid acanthomorphs (Baldwin and Johnson, this volume).

169. [Begle's 28: Adipose fin present (0) or absent (1). State (1) coded in galaxiids and *Lepidogalaxias*.] Although Begle's text (correctly) reports absence in esocoids and all alepocephaloids, presence of the adipose fin is shown as one of three characters uniting esocoids with other euteleosts (Begle, 1992, fig. 9).

170. [Begle's 36: Adipose cartilage absent (0), present (1), or present and pear-shaped (2). State (1) coded in salangids, *Allosmerus*, *Mallotus*, and *Spirinchus*, state (2) in *Hypomesus*, *Osmerus*, *Plecoglossus*, and *Thaleichthys*.] The character originated with Matsuoka and Iwai (1983) and was discussed by Howes and Sanford (1987b). Matsuoka and Iwai described state (2) in *Hypomesus*, *Osmerus*, *Plecoglossus*, and *Thaleichthys* and state (1) (a horizontal, fenestrated plate, arched dorsally in transverse section) in the salangid *Salangichthys* and the osmerid *Spirinchus*. According to our observations, Begle wrongly coded *Allosmerus*, which has state (2), and salangids, which exhibit both state (1) (*Salanx* and *Salangichthys*) and state (2) (*Protosalanx* and *Neosalanx*); and *Spirinchus* was miscoded by Begle and misidentified by Matsuoka and Iwai since our material of *S.starksi*, *S.dilatus*, and *S.lanceolatus* (the species named by Matsuoka and Iwai) shows state (2), and the specimen illustrated by Matsuoka and Iwai (1983, fig. 2b) is clearly *Mallotus*. Begle's state (2) therefore occurs in all osmerids except *Mallotus* and all salangids except *Salanx* and *Salangichthys*.

K. Squamation

171. [Begle's 15: Scales present (0) or absent (1). State (1) coded in galaxiids, *Aplochiton*, *Lovettia*, salangids, and *Leptoderma*.] Scales are present in mature male salangids, in a row above the anal fin (Roberts, 1984, p. 182), so they either represent a third state (2) or should be coded (0). Among alepocephalids, scales are also absent in *Mirognathus*, *Photostylus*, *Rinoctes*, and *Xenodermichthys* and are confined to lateral line ossicles in *Rouleina* (Markle and Merrett, 1980).

172. Radii on scales. Sanford (1987, 1990; fig. 1D) proposed absence of radii on the scales as a character

relating salmonoids to osmeroids and argentinoids in the taxon Salmonae. While radii are apparently absent on the scales of all salmonoids and osmeroids (Kobayasi, 1955; personal observation), we have observed well-developed radii on scales of the alepocephaloids *Bajacalifornia*, *Bathytroctes*, *Narcetes*, and *Talismania*, and they are illustrated in *Bathylaco* by Nielsen and Larsen (1970, fig. 3) (alepocephaloids usually lose all or most of their scales during capture). In argentinoids, radii are absent in *Argentina* (Cohen, 1964, fig. 4; Roberts, 1993, fig. 8; personal observation on *A.silus* and *A.sialis*), but we have observed them in the microstomatid *Microstoma*, and they occur in *Pseudobathylagus* (*P.milleri*; R. H. Rosenblatt, personal communication). We could find no scales in available specimens of *Glossanodon* or in other bathylagids and opisthoproctids. The presence of radii in alepocephaloids, bathylagids, and microstomatids (as in generalized elopomorphs and in clupeomorphs, ostariophysans, esocoids, and neoteleosts) indicates that absence of radii characterizes salmonoids and osmeroids.

173. Scaling of cheek and operculum. A synapomorphy of esocoids (esocids and umbrids) is that the cheek and operculum are scaled. This derived condition is otherwise found only in eurypterygians (aulopiforms and ctenosquamates), derived elopomorphs (halosaurs, notacanths, and scaled eels), and apparently in one alepocephalid, *Bathytroctes squamosus* Alcock (type-species of *Lepogenys* Parr, 1951).

L. Sensory Canals and Associated Bones

174. Postorbital contact between supraorbital and infraorbital canals. Primitively in teleosts, the supraorbital and infraorbital sensory canals are independent: the supraorbital canal runs straight back through the frontal to the parietal, and the infraorbital canal turns laterally from the dermopterotic into the dermosphenotic (e.g., Patterson, 1975, figs. 145 and 147). This pattern occurs sporadically in Recent lower teleosts (e.g., *Hiodon*, *Chanos*, many cypriniforms, and some characiforms; Gosline, 1965; Nelson, 1972), and in all alepocephaloids according to Gosline (1969, p. 191; also Matsui and Rosenblatt, 1987, p. 18, on platytroctids; but see *Leptoderma*, Greenwood and Rosen, 1971, fig. 23). Two derived states may be recognized. In the first, there is a postorbital junction between the supraorbital and infraorbital canals, but the primitive posterior (parietal) branch of the supraorbital canal is retained, as in *Elops*, *Megalops*, *Argentina*, *Glossanodon*, *Hypomesus*, and *Osmerus* (Nelson, 1972, figs. 13, 14, and 16; Patterson, 1970, fig. 29; Greenwood and Rosen, 1971, fig. 21). In the second, there is a postorbital junction between the canals, and the parietal branch

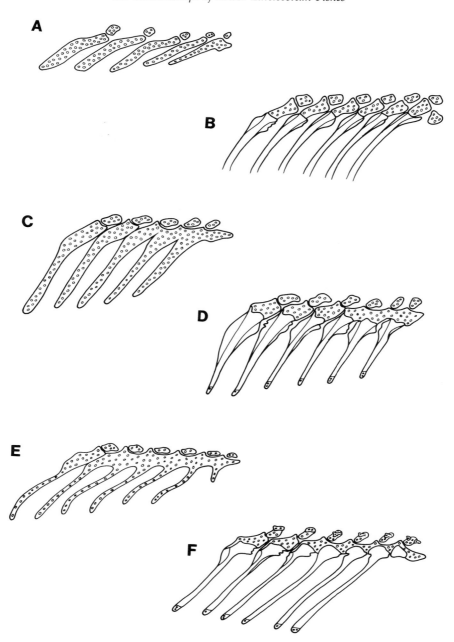

FIGURE 17 Posterior pterygiophores of dorsal fin in (A) Galaxiidae indet. larva, USNM 340197, 26 mm SL; (B) *Coregonus cylindraceus*, BMNH 1963.1.7.10, 49 mm SL; (C) *Osmerus* sp. larva, HSU uncat., 34 mm SL; (D) *Hypomesus olidus*, HSU 86-33, 46 mm SL; (E) *Salangichthys microdon*, BMNH 1996.2.6.1, 83 mm SL; (F) *Mallotus villosus*, AMNH 26286, 137 mm SL. A and B show the primitive teleostean condition, with separate posterior pterygiophores, both in larvae and adults. C–E show the derived osmerid condition, with fusion between the distal parts of the posterior pterygiophores, both in larvae and adults.

is absent, as in esocoids, *Thymallus*, and *Synodus* (Nelson, 1972, figs. 1–4, 7, 8, 14, and 16). In salmonoids, whereas *Salmo*, *Oncorhynchus*, and *Parahucho* have the same derived condition as *Thymallus* (Cavender, 1970, figs. 3 and 4), coregonids and primitive salmonines such as *Brachymystax* and *Hucho* seem to retain the parietal branch but have it displaced laterally in the frontal so that it is directed towards the lateral margin of the parietal (e.g., Berg, 1940, figs. 120 and 122; Shaposhnikova, 1968a, figs. 11–26; Cavender, 1970, fig. 2). In neoteleosts the postorbital junction between the supraorbital and infraorbital canals seems always to occur; the parietal branch persists, for example, in myctophids, *Polymixia*, and many beryciforms.

175. [Begle's 6: Infraorbital sensory canal turning upward behind the eye (0) or deflected posteroventrally (1). State (1) coded in galaxiids, *Aplochiton, Lovettia*, retropinnids, salangids, and *Lepidogalaxias*.] Salangids have state (0) and are wrongly coded (Nelson, 1972, fig. 15B). *Lepidogalaxias* is also wrongly coded since it lacks the infraorbital canal (Nelson, 1972, p. 35). Following Nelson's (1972) and Rosen's (1974, p. 272) comments, we recognize four states: infraorbital canal uninterrupted and unmodified (0); infraorbital canal interrupted, with the anterior portion running posteroventrally to cross the preopercular canal (1), in retropinnids; postorbital infraorbital bones and sensory canal absent (2), in galaxiids, *Aplochiton*, and *Lovettia*; and infraorbital bones and canal absent (3), in *Lepidogalaxias*.

176. [Begle's 19: Infraorbital canal extending to preopercle (1) or not (0). State (1) coded in retropinnids.] If character 94 is coded as suggested above, this character is redundant.

177. [Begle's 52: Supraorbital and preopercular canals each with five or more pores (0) or with three (1). State (1) coded in retropinnids.] The character is from Nelson (1972, p. 38). *Lepidogalaxias* also has three preopercular pores and lacks the supraorbital canal (Nelson, 1972, p. 35).

178. [Begle's 56: Temporal sensory canal present (0) or absent (1). State (1) coded in galaxiids, *Aplochiton, Lovettia*, salangids, and *Lepidogalaxias*.] The character is from Nelson (1972). Salangids have state (0) (Nelson, 1972, fig. 15B) and are wrongly coded.

179. Condition of extrascapular. The teleostean extrascapular is primitively a triangular bone (often with a posterior notch or some other indication of development from two primordia; Ridewood, 1904), carrying a triradiate sensory canal which is received from the posttemporal posteriorly, passes to the pterotic anteriorly, and communicates with its fellow in the midline in the occipital or supratemporal commissure (e.g., *Hiodon, Elops*, and *Megalops*). In several teleostean groups the medial portion of the occipital commissure is carried by the parietal, presumably by fusion of a medial extrascapular, as in argentinoids (no. 20, above), where only the lateral extrascapular occurs. Extrascapulars are absent in several alepocephaloids, osmeroids, and esocoids (no. 145, above). In other lower euteleosts the extrascapular is commonly fragmented into two or more ossicles, often no more than bony tubes surrounding parts of the sensory canal. In coregonids and *Thymallus* there is a tabular (Stearley

and Smith, 1993, character 84), triradiate lateral extrascapular, as in argentinids, *Bathylaco*, and some platytroctids (e.g., *Paraholtbyrnia*). In retropinnids there is a tubular, triradiate lateral extrascapular, as in salmonines, some platytroctids (e.g., *Searsia*), and some alepocephalids (e.g., *Alepocephalus bairdii, Binghamichthys*). In all osmerids the lateral extrascapular is fragmented into three (or more) separate ossicles, as it is in several platytroctids (Sazonov, 1986, fig. 3) and alepocephalids (e.g., *Alepocephalus agassizi, Narcetes, Rouleina maderensis*, and *Talismania*). In bathylagids (Kobyliansky, 1986, fig. 1), microstomatids (Kobyliansky, 1990, fig. 3), and most opisthoproctids (Chapman, 1942b, fig. 2, "ST"; Trewavas, 1933, fig. 5, "doc") there are one or more elongate, tubular ossicles extending anterodorsally above the pterotic, towards the rear end of the frontal. To determine whether these bones are extrascapulars [Kobyliansky's (1986, 1990) interpretation] or detached parts of the pterotic requires study of innervation. The opisthoproctid *Bathylychnops* has extremely wide or inflated sensory canals; the extrascapular comprises two cavernous ossicles, one gutterlike and one basically triradiate, from which the temporal canal passes forward to the dermopterotic and the occipital commissure passes medially to the parietal.

180. Sensory canal in posttemporal. Primitively in teleosts, the lateral part of the posttemporal is penetrated by the lateral line, passing forward from the supracleithrum to the extrascapular. This pattern persists in many generalized lower euteleosts: in salmonoids; *Allosmerus, Hypomesus, Osmerus*, and *Plecoglossus* among osmerids; *Holtbyrnia* (Sazonov, 1986, fig. 3), *Paraholtbyrnia, Pellisolus*, and *Searsia* among platytroctids; *Bathylaco* among alepocephalids; and *Argentina, Glossanodon, Bathylagichthys* (Kobyliansky, 1986, fig. 18), *Microstoma, Nansenia*, and the opisthoproctid *Bathylychnops* among argentinoids. Two derived states are recognizable. In the first, there is a separate sensory canal ossicle superficial to the posttemporal, as in *Mallotus, Spirinchus*, and *Thaleichthys* among osmerids; several platytroctids (e.g., *Mentodus, Mirorictus*, and *Searsioides*) and alepocephalids (e.g., *Talismania oregoni, Bathytroctes*, and *Narcetes*); all bathylagids except *Bathylagichthys* (Kobyliansky, 1986, figs. 14–17); and the opisthoproctids *Opisthoproctus* and *Rhynchohyalus*. In the second pattern, there is no canal in the posttemporal and no superficial ossicle, as in esocoids (Jollie, 1975); salangids, retropinnids, *Aplochiton, Lovettia, Lepidogalaxias*, and galaxiids among osmeroids; and *Leptoderma, Platytroctes, Photostylus*, and *Rouleina* among alepocephaloids.

181. Sensory canal in supracleithrum. As with the posttemporal (no. 180), the supracleithrum of teleosts

is primitively penetrated by the lateral line, and there are two derived states, a separate sensory canal ossicle superficial to the bone and no superficial ossicle or canal in the bone. There is general correlation between the condition of the supracleithrum and posttemporal, but some taxa retain the canal through the supracleithrum while having a separate ossicle or nothing over the posttemporal [*Esox*, the osmerids *Mallotus* and *Spirinchus*, the alepocephalids *Bathytroctes*, *Narcetes*, and *Talismania*, the bathylagids *Bathylagus*, *Leuroglossus*, and *Pseudobathylagus* (Kobyliansky, 1986, figs. 15 and 16), and *Opisthoproctus*]. There is an ossicle over the supracleithrum in the osmerid *Thaleichthys*.

M. Reproductive Structures

182. [Begle's 4: Egg not adhesive (0) or surrounded by adhesive "anchor membrane" (1). State (1) coded in all osmerids.] Salangid eggs have "instead of an anchor membrane, an anchoring structure that is composed of various types of filaments that turn out and onto the substrate (Wakiya and Takahashi, 1913 [1937 intended])" (Hearne, 1984, p. 155). According to Begle's (1992) description ("rupturing of an outer 'chorion' . . . producing a structure which adheres to the underlying substrate") the salangid structure qualifies as state (1), and the illustrations of osmerid and salangid eggs in Wakiya and Takahashi (1937, pl. 21), Korovina (1977), and especially Chyung (1961, pl. 51, figs. 242 and 243), indicate that the comparison is sensible.

183. [Begle's 22: Left and right ovaries present (0) or right ovary only (1). State (1) coded in retropinnids.] The character was credited by Begle (1991) to McDowall (1969), who wrote that *Retropinna* and *Stokellia* have only the *left* ovary and that *Prototroctes* has both. The derived state (left ovary only) was later reported in *Prototroctes* (McDowall, 1976, 1984); it also occurs in *Plecoglossus* (Chapman, 1941a).

184. [Begle's 104: Membranous ovarian tunic absent (0) or present (1). State (1) coded in *Talismania*, *Bathyprion*, and *Bathylaco*.] The character is from Markle (1976). According to Markle and Merrett (1980, p. 228), *Bathylaco* has the exposed type of ovary that they consider primitive, whereas *Leptoderma* has the enclosed type. Begle's coding is reversed for those two genera. *Bathyprion* has a type of enclosed ovary different from that in *Leptoderma*, *Rinoctes*, *Talismania*, etc.: it hangs free in the coelom rather than being enclosed by fusion between the ovarian tunic and peritoneum (Markle and Merrett, 1980, fig. 1).

185. Sperm structure. Patterson and Johnson (1995) noted a possible salmoniform character in Jamieson's (1991) records of a single annular mito-

chondrion in the sperm of alepocephaloids, salmonids, and *Galaxias*. Mattei (1991) also commented on sperm ultrastructure in alepocephaloids and salmonids, citing three characters that seem to relate the two groups, though an annular mitochondrion does not occur in all sampled alepocephaloids or salmonids.

186. [Begle's 33: Nuptial tubercles present (0) or absent (1). State (1) coded in galaxiids, *Aplochiton*, salangids, *Prototroctes*, *Lepidogalaxias*, and all argentinoids and alepocephaloids.] The character is from Wiley and Collette (1970; also Collette, 1977). Presence of nuptial tubercles is one of three characters used by Begle (1992, fig. 9) to unite esocoids with other euteleosts; esocoids lack nuptial tubercles and are wrongly coded (0). Salangids are wrongly coded (1) (Roberts, 1984, p. 183). Among Neoteleostei, tubercles are recorded only in Percidae and *Gadus* (Vladykov *et al.*, 1985), so the group should be coded (1). In salmonoids, tubercles occur only in the three coregonine genera and in *Salvelinus namaycush* (Stearley and Smith, 1993), so that the primitive state for the group is ambiguous. Polarity of the character is also questionable; Begle entered (0) for outgroups, but the only outgroup taxa with tubercles are ostariophysans. *Lovettia* was coded as lacking tubercles in Begle (1991) and as having them in Begle (1992); the latter seems correct (McDowall, 1971).

187. [Begle's 53: Anal fin rays and scales unmodified in males (0), scales anterior to anal fin greatly enlarged and anal fin skeleton modified in males (1), or both preanal scales and anal fin skeleton greatly modified (2). State (1) coded in *Mallotus* and salangids; state (2) in *Lepidogalaxias*.] In Begle (1991) the character was two-state, with an undescribed state (2) entered for *Lepidogalaxias*. No comment.

N. Other Soft Anatomical Features

188. [Begle's 51: Nasal lamellae arranged in a rosette (0) or parallel and longitudinal (1). State (1) coded in galaxiids, *Aplochiton*, *Lovettia*, retropinnids, and *Lepidogalaxias*.] The character is from Howes and Sanford (1987b). No comment.

189. Iris. In *Argentina* and *Glossanodon* there is a crescent of white tissue above the iris (Cohen, 1964). The character is trivial but like no. 127 is one of the few indicators of argentinid monophyly.

190. Swimbladder. Ahlstrom *et al.* (1984) cited a unique synapomorphy of Argentinoidea: the swimbladder (when present; it is absent in bathylagids and some opisthoproctids) is served by microrete mirabilia. All Alepocephaloidea are characterized by ab-

sence of the swimbladder (Greenwood and Rosen, 1971; Markle, 1976).

191. [Begle's 75: Pyloric caeca present (0) or absent (1). State (1) coded in *Lepidogalaxias* and esocoids.] Caeca are also absent in salangids (Roberts, 1984), *Allosmerus* (McAllister, 1963), *Lovettia* (McDowall, 1971), retropinnids (McDowall, 1976, 1979), and among galaxiids in many species of *Galaxias* and in *Galaxiella* and *Paragalaxias* (McDowall and Frankenberg, 1981, figs. 45 and 46).

192. Bone cells. The bones of teleosts are primitively cellular. In lower euteleosts, acellular bone is recorded in two groups, osmeroids and esocoids. In osmeroids acellular bone is reported in *Hypomesus*, *Osmerus*, *Thaleichthys*, and *Galaxias* (Kölliker, 1859; Moss, 1961, 1965). In esocoids it is reported in *Esox*, *Umbra*, and *Dallia* (Kölliker, 1859; Moss, 1961, 1965). Acellular bone has also been recorded in other lower teleosts by Moss (1961, 1965): in both species of *Hiodon* (Kölliker, 1859, found cellular bone, if his "*Hyodon claudulus*" is the same as *Hiodon clodalus* Lesueur, = *H. tergisus*); in the siluroid *Ictalurus* (Kölliker, 1859, also found acellular bone in *Trichomycterus*); and in the eel *Gymnothorax* (Kölliker, 1859, also found acellular bone in *Conger*, *Ophisurus* and *Nettastoma*). Thus bone cells have certainly been lost more than once in teleostean history.

193. [Begle's 96: Saclike shoulder organ absent (0) or present (1, platytroctids only).] No comment.

194. [Begle's 21: Horny midventral abdominal keel absent (0) or present (1). State (1) coded in retropinnids.] The character is from McDowall (1969). There is also a "ventral abdominal keel" (McDowall, 1978, p. 118) in various galaxiids, sometimes in males only. McDowall (1969, p. 810) wrote that this is "not comparable with the keel of retropinnids" but did not say why it is not homologous.

195. Peduncular flanges. In many galaxiids there are prominent fleshy or adipose "caudal peduncle flanges" (McDowall, 1970, p. 354; 1971, p. 39, etc.) extending forward from the caudal fin along the dorsal and ventral margins of the caudal peduncle. These structures are also present in *Lepidogalaxias*.

196. Photophores. In lower euteleosts, photophores occur in most platytroctids (Matsui and Rosenblatt, 1987, table 2; absent in *Barbantus*, *Mirorictus*, *Pellisolus*, *Platytroctegen*, *Tragularius*, and some *Holtbyrnia*), a few alepocephalids (*Anomalopterichthys*, *Rouleina*, *Xenodermichthys*, and *Photostylus*; Markle, 1976), and some opisthoproctids (Cohen, 1964).

197. [Begle's 24: Distinctive cucumber odor of fresh specimens absent (0) or present (1). State (1) coded in retropinnids, *Osmerus* and *Thaleichthys*.] Occurs also in *Argentina* (e.g., Yarrell, 1838; Smitt, 1895, p. 919)

and *Mallotus* (Smitt, 1895, p. 879, "stale cucumbers"), and the responsible substance has been isolated in *Hypomesus* (McDowall *et al.*, 1993). The odor also characterizes an Australasian chlorophthalmid, *Chlorophthalmus nigripinnis*, the "Cucumber fish" (Gomon et al., 1994, p. 268). Berra *et al.* (1982), noting that H. B. Bigelow and W. C. Schroeder were unable to detect the odor in *Osmerus*, tested six people with a *Prototroctes* and found that only three could detect the cucumber. One wonders how many qualified sniffers have put fresh *Glossanodon*, opisthoproctids, and alepocephaloids to the test, and what weight to attach to the lack of positive reports in other osmerids and in salangids.

198. [Begle's 84: Midlateral band of silver pigment absent (0) or present (1, *Allosmerus* only).] No comment.

O. Life Cycle

199. Anadromy and diadromy. Anadromy is a life history including a spawning migration from the sea into rivers; diadromy is a more general term, including anadromy, catadromy (spawning migration from rivers to the sea), and amphidromy (migration from sea to freshwater or vice versa, with no relation to reproduction). McDowall (1988, 1993) reviewed the distribution of the two life history patterns in actinopterygians. Among actinopterygians, anadromy occurs in some sturgeons, some alosine clupeids, some percoids, etc. (review in McDowall, 1988), but only in osmeroids and salmonoids can it be considered primitive for suprafamilial taxa. McDowall's (1993) survey of diadromy, the more general phenomenon, shows that osmeroids, salmonoids, and gobioids are the only actinopterygian higher taxa in which diadromy can be considered primitive. Among osmeroids, anadromy occurs in all osmerids except *Allosmerus*, *Hypomesus pretiosus*, *Mallotus*, and *Spirinchus starksi* (all marine), and *Plecoglossus* (amphidromous, larvae move into the sea and over winter before a return migration); in salangids (perhaps not in all; McAllister, 1988, p. 57); in retropinnids except *Prototroctes* (amphidromous, as *Plecoglossus*); and in *Lovettia*. *Aplochiton* is poorly known and may be anadromous or amphidromous. Among salmonoids anadromy occurs in coregonids (*Coregonus* and *Stenodus*) and in many salmonines and is therefore assessed as primitive for the group. *Galaxias* includes one catadromous species (*Galaxias maculatus*) and several amphidromous species; other galaxiids are freshwater, as is *Lepidogalaxias*. Anadromy and diadromy do not occur elsewhere in lower euteleosts.

200. Heterochrony. Paedomorphosis or neoteny is evident in salangids and to some extent in *Mallotus*

and *Lovettia* among osmeroids. The opposite phenomenon, peramorphosis or acceleration, is evident in *Lepidogalaxias*. Gosline (1960, pp. 345, 351) referred to *Lovettia* as "neotenic," citing its "definitely larval appearance," "membranous" median area of the skull roof, and absence of flanges or lamellae on the distal parts of the last few neural and haemal spines (in fact, those flanges are present, Fig. 13F). In the *Lovettia* we studied [40–45 mm SL, average adult size (Blackburn, 1950)] the braincase is largely cartilaginous, the pectoral endoskeleton is unossified in some specimens, there is a long postpectoral process (Fig. 15I), and the supraneurals are unossified. We agree with Gosline's assessment. In salangids, neoteny is documented by Roberts (1984). Early life-history stages of *Mallotus, Hypomesus, Osmerus, Plecoglossus,* and salangids are shown in Okiyama (1988, pp. 66–73). By about 30 mm SL, *Hypomesus, Osmerus,* and *Plecoglossus* look much like adults, whereas at 40 mm SL *Mallotus* still looks like a larva, and resembles salangids much more than it resembles osmerids of comparable size. Many details of skeletal development in *Mallotus* could be cited to back up the claim of neoteny; the resemblance in the pectoral endoskeleton between a 60 mm *Mallotus* and a salangid (Figs. 15G and 15H) will serve as an example. *Lepidogalaxias* exhibits peramorphosis, a unique acceleration of skeletal development, so that at an adult size of less than 60 mm SL it ossifies structures that remain cartilaginous or fail to develop in any of its much larger relatives. Examples of those structures include a septal bone and basisphenoid, ossified in the cartilage flooring the orbit and both otherwise absent in all osmeroids; the ossified epicentral cartilages, unknown elsewhere in teleosts; and the scales, absent in all galaxiids and in *Aplochiton* and *Lovettia*. Comparison between preflexion larvae of *Lepidogalaxias* (7 mm NL) and *Galaxiella* (8 mm NL) shows, for example, that in *Lepidogalaxias* all the gill arch elements are ossified, with well-developed teeth on CB5 and UP5, whereas in *Galaxiella* no gill arch elements are ossified and there are no pharyngeal teeth. In the vertebral column, however, *Lepidogalaxias* lags behind *Galaxiella*. By 13.5 mm SL, *Lepidogalaxias* has the head skeleton ossified essentially as in the adult, but centra are still undifferentiated; *Galaxiella* already has the anterior centra differentiated at 8 mm NL. This may explain the separation of PU1 and U1 in *Lepidogalaxias*, whereas the two are fused in all other osmeroids.

P. Molecular Sequence Data

Lower euteleosts are still very poorly sampled for molecular sequences. There are, so far as we know, no sequence data from any alepocephaloid, argentinoid or osmeroid, and the only comparative data are for a few homologous sequences from salmonoids and *Esox*.

Lê *et al.* (1989) presented a partial sequence (ca. 300 nucleotides) of large subunit ribosomal RNA (rRNA) in *Esox*, six other teleosts, and a range of outgroup taxa, but since their teleost sample included only *Clupea* and five acanthomorphs in addition to *Esox*, the results are not helpful (*Esox* as sister-group of acanthomorphs). In subsequent work by Lê (1991; Lê *et al.,* 1993) with a more extensive sample of teleosts, including *Salmo*, and longer 28S rRNA sequences, *Esox* was omitted. Müller-Schmidt *et al.* (1993) analyzed the amino-acid sequence of ependymins of *Esox, Oncorhynchus* (two paralogous sequences), *Clupea*, and the cyprinids *Carassius* (two paralogous sequences), and *Brachydanio*. Their work, particularly their evidence for a clupeomorph/ostariophysan clade, is discussed by Lecointre and Nelson (this volume). With no outgroup, they were restricted to phenetic methods. Nevertheless, because their alignment includes two paralogous pairs of sequences (*Oncorhynchus* and *Carassius*) and one from *Salmo salar* (not included in their analysis), it is possible to draw some inferences. If one assumes that divergence between the paralogous *Oncorhynchus* sequences dates from the tetraploidization of the salmonoid genome, the mid-Eocene salmonine *Eosalmo* places that event at least 60 MYA. The *Salmo* sequence differs from its orthologue in *Oncorhynchus* by only three residues (compared with 21 differences between the *Oncorhynchus* paralogues); *Oncorhynchus* and *Salmo* have been distinct for at least 6 MY (Stearley and Smith, 1993). The corrected DNA divergence between *Esox* and the *Oncorhynchus* paralogues (ca. 15.5%) is about 1.6 times as great as that between the *Oncorhynchus* paralogues (9.5%), placing the esocoid/salmonoid divergence at or before roughly 1.6 × 60 or 95 MYA, in the Cenomanian or earlier.

Bernardi *et al.* (1993) analysed the amino-acid sequence of growth hormone (ca. 180 residues) in 25 species of teleosts including four salmonines and *Esox*, with a shark and a sturgeon as outgroups. In their parsimony trees, *Esox* is twinned with the four salmonines in the sister clade to Percomorpha (13 perciforms and a pleuronectiform), with bootstrap support of 69% for the *Esox*/salmonine clade, as against 62% for salmonines. Scanning the alignment, the support for an *Esox*/salmonine clade looks remarkably strong (including several shared residues omitted in the phylogenetic analysis of Bernardi *et al.*), and there is no indication that *Esox* belongs either below salmonines (as in Figs. 1C, 1D, 1F, and 1G) or above them (as the sister of percomorphs in the sample of Bernardi *et al.*).

Complete small subunit rRNA sequences are now available from a range of teleosts, including *Salmo* and *Esox* (Littlewood *et al.*, 1996). The position of *Esox* is labile in parsimony trees generated from those data: it may appear as the sister of (i) *Chanos* + *Clupea* (the only ostariophysan and clupeomorph in the sample); (ii) as the sister of *Salmo* + Acanthomorpha (the only other euteleosts in the sample; nine acanthomorphs sequenced); or (iii) as the sister of *Salmo*. Position (i) is favored by the most conservative alignment, omitting portions where there is any doubt, and may be an effect of "long-branch attraction," since *Esox* is a comparatively long branch terminal, and the *Chanos* + *Clupea* branch is very long. Position (ii) is favored by a less restrictive alignment with all sites equally weighted, and position (iii) by the same alignment with transversions weighted twice as heavily as transitions, which is normally a sound strategy. None of the three solutions has significant bootstrap support.

Phillips *et al.* (1995) reported data from restriction maps (not sequences) of nuclear ribosomal DNA in *Osmerus* and a range of salmonoids. Their data indicate that *Hucho* is paraphyletic, with *H.hucho* closely related to *Brachymystax* and *H.perryi* (genus *Parahucho*) related to other salmonines. Their matrix of percent sequence divergence is interesting in showing the same distance (1.97%) between *Osmerus* and *Coregonus* as between *Coregonus* and *Thymallus* or *Brachymystax* and *H.hucho*, implying close relationship between osmerids and salmonoids.

To sum up, the molecular evidence concerns only salmonoids and esocoids. Growth hormone (rather strongly) and small subunit rRNA (very weakly) support salmonoid/esocoid relationship, for which there is, to our knowledge, no morphological support whatever. Restriction maps of rDNA suggest osmeroid/salmonoid relationship.

III. Discussion of Begle's (1991, 1992) Analyses

Begle's (1992) data matrix contained 108 characters, coded in 33 taxa, of which 26 are genera and 7 are collective at higher levels: outgroups (a hypothetical ancestor comprising a row of zeros), salmonoids, esocoids, neoteleosts, Platytroctidae, Salangidae, and Galaxiidae.

Eleven characters (his nos. 25, 27, 32, 41, 42, 46, 48, 49, 84, 87, and 96) are autapomorphic in Begle's matrix, but after checking and recoding, 4 of those (his nos. 27, 42, 46, and 87) turn out not to be so, whereas a further 7 characters are autapomorphic (his nos. 31, 37, 68, 69, 70, 92, and 107).

As shown in Section II, according to our reading of specimens and the literature, Begle's (1992) matrix contains errors of fact (not interpretation), ranging from minor to gross, in 89 (82%) of his 108 characters. In addition, in our opinion there are errors of interpretation in a further 3 characters (Begle's 10, 23, and 74), leaving 16 characters or 15% that we accept as coded in Begle's publications. Of those 16 characters, 6 (Begle's nos. 32, 46, 48, 49, 84, and 96) are autapomorphic, leaving only 10 that can be used to group taxa. And those 10 are all lifted intact from previous cladistic analyses (Begle's nos. 7, 9, 19, 21, and 54 from Fink, 1984b; nos. 30 and 51 from Howes and Sanford, 1987a,b; no. 53 from Rosen, 1974; and nos. 80 and 102 from Greenwood and Rosen, 1971).

Begle wrote (1991, p. 36; 1992, p. 351) "Every specimen was examined for every character, to check the observed variation against that described in the literature." Apart from the question of claiming to check cucumber odor, ovarian membranes, pyloric caeca, egg structure, etc. in cleared-and-stained specimens, the claim is demonstrably untrue since in a few genera we studied the same specimens (e.g., *Bathylaco*, USNM 206693; *Bathytroctes*, USNM 215493; *Lepto-*

TABLE 1   Key Relating the Character Numbers in This Paper to Those in Begle (1992)

```
                                                                              111111111
                     11111111112222222222333333333344444444445555555555666666666677777777778888888888999999999900000000 0
           123456789012345678901234567890123456789012345678901234567890123456789012345678901234567890123456789012345678
```

```
     111111  1111   11 11 1  11     11 1     1111 1  1111 1           111   1 11     1 1    1 1  11
     63837564234567  7376983924464563862763999000040289878 1271  9447 8755783    1193398817596774339957598 44416  84853
     32120526592800149369433744799404617016123637438555877068267011737052908907151067161853427946386 3456723548092
```

Note. The upper numbers (1–108, in numerical order) are Begle's, and the lower numbers (in random order) show where those 108 are covered in our survey of 200 characters.

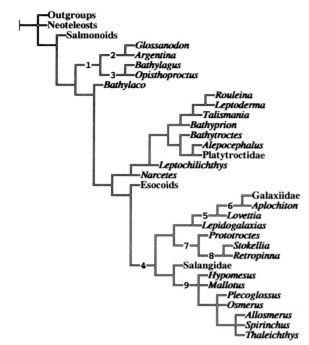

FIGURE 18 Result of parsimony analysis of Begle's (1992) data after correction and recoding. Strict consensus of three equally parsimonious trees, length 231 steps, C.I. 0.42, R.I. 0.73, found by Hennig86 (Farris, 1988). Numbers on branches indicate clades common to this tree and Begle's (Fig. 2).

derma, USNM 215604; *Narcetes*, VIMS 2120; and *Opisthoproctus*, MCZ 61958) and found them insufficiently dissected or incapable of yielding information supposedly checked in them.

Following the general acceptance of cladistic method, it is now fashionable in systematic ichthyology to present only brief descriptions of structure and to concentrate on lists of characters, a matrix, and the results from parsimony analyses. We believe that this fashion merely replaces one black box (evolutionary systematics) by another (the matrix). It is unproductive to divert the systematist's effort from what is primary—studying the fishes with as much care as possible—to what is secondary (and futile if the primary work is not done properly)—manipulating the matrix.

When we ran Begle's (1992) matrix, as published, on Hennig86 (Farris, 1988) we obtained the same result as Begle (Fig. 2).

When we ran the corrected matrix (to be published elsewhere) on Hennig86, not surprisingly, we found a very different result. There are three equally parsimonious trees with the strict consensus shown in Fig. 18. Comparison of Figs. 2 and 18 shows that of the 28 nodes in Fig. 2, only nine (less than one-third) are reproduced in Fig. 18 (numbered 1–9): those linking

the four argentinoids, *Argentina* with *Glossanodon*, and *Bathylagus* with *Opisthoproctus*; that linking all osmeroids; that linking the seven osmerid genera; those linking the three retropinnids and *Retropinna* with *Stokellia*; and those linking *Aplochiton* with galaxiids and *Lovettia* with those two.

The most striking difference between Figs. 2 and 18 is in the position of esocoids, which are in the basal polychotomy in Fig. 2 and are the sister group of osmeroids in Fig. 18. Unfortunately, this does not help to resolve the true position of esocoids but is more a reflection of the miscoding of esocoids in Begle's matrix (incorrect in 18 characters, fewer than in several other taxa, but effective because the number of non-zero entries increases from 4 to 17) and of the quality of Begle's data. The characters that place esocoids with osmeroids in Fig. 18 are absence of the orbitosphenoid, mesocoracoid and pubic symphysis, and reduction of the articular and the dorsal portion of the opercle; they are not convincing evidence that the two groups are immediately related.

Another major difference between Figs. 2 and 18 is in the position of salangids, which move from within galaxioids to the osmerids. This is a reflection of gross miscoding of salangids by Begle (incorrect in 27 characters, more than any other taxon) and approximates what we believe to be the true position of salangids (below).

IV. Monophyly and Interrelationships of Osmeridae

Osmerids are included with other osmeroids in Appendix 1, a matrix of 112 characters compiled from the survey in Section II by abstracting characters with potential to group two or more osmeroids (i.e., excluding autapomorphies of terminals). Analyzed by Hennig86 and Clados, Appendix 1 generates the trees in Fig. 19. Rather than work through every character at each node in Fig. 19, in this section we evaluate previous work on osmerid interrelationships and comment on the characters that justify our main conclusions. Unless there is an explicit reference to Section II, character numbers in this section are those in Fig. 19 and Appendix 1, where each is tied to the main survey in Section II.

A. Relationships of Salangidae

Salangids have generally been associated with Osmeridae (review in Roberts, 1984) but have not previously been included within that group. The result of our analysis (Fig. 19) is that salangids are nested well within osmerids, as the sister group of *Mallotus*.

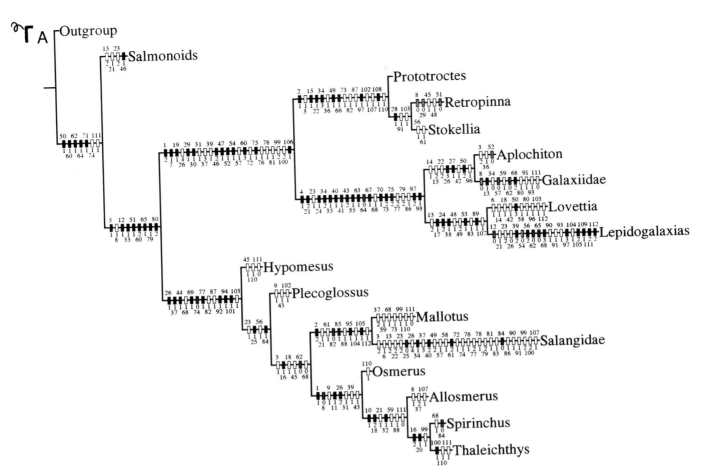

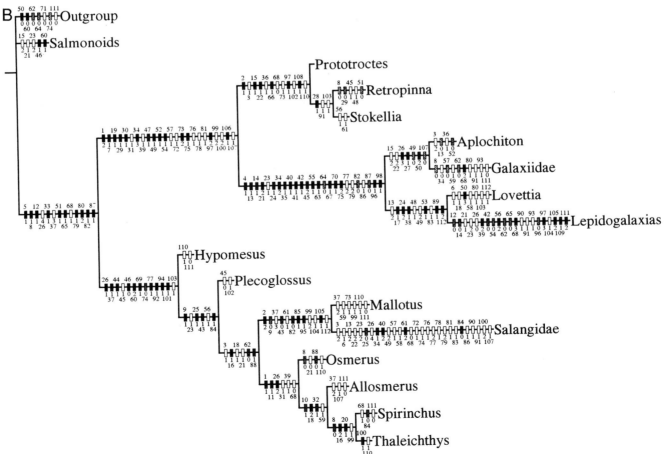

Unambiguous synapomorphies supporting that relationship are those at the *Mallotus*–salangid node under the alternative optimizations of the two trees in Fig. 19: the elongate, unossified ethmoid (2); fourth pectoral radial branched distally (85, Fig. 15; in young *Mallotus*, Fig. 15E, the third radial is also branched distally, as in salangids); adipose cartilage a fenestrate plate (95, only in *Salangichthys* and *Salanx* among salangids); and modification of the anal fin skeleton in males (105). In addition, *Mallotus* and salangids share reduction in the number of supraneurals (61, 1–8 in *Mallotus*, one in salangids; homoplasy with *Stokellia*, where the condition is clearly nonhomologous); reversal to the unmodified condition of the first pectoral radial (82, homoplasy with non-retropinnid southern osmeroids); modified scales on the anal fin in males (104, homoplasy with *Lepidogalaxias*); and retardation of skeletal ontogeny (112, homoplasy with *Lovettia*). Further, the adipose fin of *Mallotus* differs from that of all other osmerids in being long-based (McAllister, 1963; Matsuoka and Iwai, 1983, fig. 1, where "*Spirinchus*" = *Mallotus*; see character 170, section II) and resembles the adipose in salangids; and *Mallotus* has 17–22 pectoral fin rays, more than any other osmerid but comparable with the counts in primitive salangids (22–27 in *Protosalanx* and 17–28 in *Salangichthys*).

B. Relationships within Osmeridae

Seven published osmerid phylogenies are summarized in Fig. 20. In our experience, osmerids are unique in the disparity of opinion on their interrelationships. Components common to more than one phylogeny (numbered in Fig. 20) comprise a *Hypomesus* + *Mallotus* clade in Figs. 20A–20D, an *Allosmerus* + *Osmerus* clade in Figs. 20A and 20D, an *Osmerus* + *Allosmerus* + *Spirinchus* clade in Figs. 20C and 20D, an *Osmerus* + *Plecoglossus* clade in Figs. 20E and 20G, and a clade of five genera (excluding *Thaleichthys*) in Figs. 20A, 20C, and 20D. No other groups are shared by any two of the seven interpretations in Fig. 20. There is also notable disagreement on which osmerid genus is the most basal, though four of the seven schemes place *Thaleichthys* at the base of the tree.

Osmerid interrelationships are clearly a difficult problem. Although we do not regard our result (Fig.

19) as the final solution, we are confident about two aspects of it: that *Thaleichthys* and *Spirinchus* are derived osmerids, not basal as Weitzman (1967) and others (Fig. 20) have argued; and that *Hypomesus* is the basal genus (cf. Fig. 20F).

Primitive characters distinguishing *Hypomesus* from all other osmerids (Fig. 19) include the following: endopterygoid teeth in a patch posteriorly, rather than a single row (23); position of metapterygoid (25); all epineurals originating on the neural arch (56, *H. transpacificus* only); and unmodified fourth pectoral radial (84). Primitive characters distinguishing *Hypomesus* and *Plecoglossus* from all other osmerids include the following: presence of vomerine shaft (3, absent in some *Hypomesus*); uninflated otic bulla (16); parietals fully in contact (18); and distal keels on last few neural and haemal arches (62). Primitive characters distinguishing *Hypomesus*, *Plecoglossus*, and *Mallotus* plus salangids from other osmerids include the following: unpaired dermethmoid (1); absence of cartilaginous interorbital septum (11); hyomandibular crest not a triangular spur (26); and form of the lower jaw (31) and basihyal dentition (39).

If *Thaleichthys* and/or *Spirinchus* were basal osmerids (Figs. 20A, 20C, 20D, and 20G; Weitzman, 1967), all the characters cited in the previous paragraph must be wrong or misinterpreted. In Fig. 19, *Spirinchus* and *Thaleichthys* are paired as the most derived osmerids by three characters: globose otic bulla (16), closure of fontanelles in braincase roof (20), and separate sensory canal ossicle over posttemporal (99, also in *Mallotus*). *Allosmerus* is grouped with *Spirinchus* and *Thaleichthys* by a futher four characters: contact in membrane bone between pterosphenoid and prootic (10), complete separation of parietals (18), large Meckelian fossa (32), and presence of epipleural bones (59, also in *Mallotus*).

Thaleichthys has the posterior myodome reduced and the orbitotemporal and otic regions of the braincase wider and shallower than other osmerids, with the pterosphenoids displaced ventrally so that they almost contact the parasphenoid anteriorly (Klyukanov, 1975, fig. 8A). The deep myodome of other osmerids is shared with outgroup taxa such as retropinnids, salmonoids, and argentinids. In *Thaleichthys* there is a separate canal-bearing ossicle over the su-

FIGURE 19 Cladograms of osmeroids based on the data in Appendix 1 analyzed by Hennig86 (Farris, 1988); trees produced with Clados (Nixon, 1992). The single shortest tree is 259 steps long, C.I. 0.59, R.I. 0.77. The two trees show alternative optimizations of characters for which there are equally parsimonious solutions, with A favoring reversals and B favoring forward changes. Character numbers are printed alternately above and below hashmarks, with character states immediately below hashmarks. Black hashmarks indicate uncontradicted synapomorphies, white hashmarks indicate homoplastic forward changes, and grey hashmarks indicate homoplastic reversals. Placing *Osmerus* as the sister of *Mallotus* + salangids and the three terminal osmerid genera increases tree length by three steps, or 1.2%.

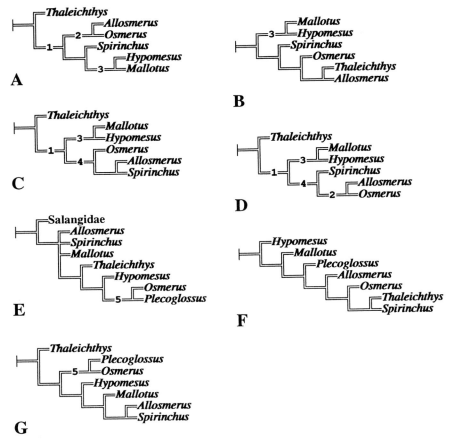

FIGURE 20 Alternative patterns of osmerid relationships. (A) Chapman (1941b). (B) McAllister (1963). (C) McAllister (1966). (D) Klyukanov (1977). (E) Howes and Sanford (1987b). (F) Wilson and Williams (1991). (G) Begle (1991). Numbers on branches denote components common to more than one tree.

pracleithrum, whereas in other osmerids the lateral line penetrates the bone (except in salangids, where the canal is superficial).

Spirinchus has the intercalar small, failing to extend forward toward the prootic as it does in other osmerids; the accessory neural arch (ANA) reduced to a tiny nubbin, whereas it is large in all other osmerids; and seven infraorbitals, whereas other osmerids and outgroups have six (we also observed seven infraorbitals on one side of a specimen of *Thaleichthys*).

While we have confidence in the distal (*Allosmerus*, *Spirinchus*, and *Thaleichthys*) and proximal (*Hypomesus* and *Plecoglossus*) parts of the osmerid tree in Fig. 19, we believe that the central part (*Osmerus* and *Mallotus* + salangids) is less secure. As noted in the legend to Fig. 19, exchanging the positions of *Osmerus* and *Mallotus* plus salangids increases tree length by only three steps, or 1.2%.

Hypomesus, the basal osmerid genus, contains four species. Our examination of three of them (we lacked *H.nipponensis*) yielded suggestions that the genus may

be paraphyletic because a primitive state occurs only in one or two of the three species. The characters concerned are vomerine shaft present only in *H.olidus*; palatine most slender in *H.olidus* and *H.pretiosus*; basihyal toothplate least modified in *H.pretiosus*; and all epineurals on the neural arch in *H.transpacificus*. One character that might be derived and is shared by the three species that we studied is the form of the fourth epibranchial (Rosen, 1974, fig. 16). Two epurals also occur in all *Hypomesus*, but the same is true of *Plecoglossus* and *Mallotus*, and in Fig. 19 reversal to three epurals characterizes a derived group of osmerids (character 68).

C. Monophyly of Osmeridae

With the data in Appendix 1, unequivocal synapomorphies of Osmeridae are those that appear at the osmerid node in Fig. 19 under both optimizations: a short hyomandibular crest (26); opercle with a dorsal notch (37); levator process on Eb4 (44); fusion between Un1 and PU1 (69); fragmentation of extrascapular into several ossicles (77, extrascapular absent in salangids); confluence of posterior dorsal pterygiophores (92); adipose cartilage (94); and adhesive membrane of egg (101). Pattern 2A supraneurals (60, state 2) characterize osmerids but appear at the osmerid node only in Fig. 19B. In Fig. 19A the character is at the osmeroid node, with pattern 2B (60, state 3) correctly characterizing galaxioids, whereas in Fig. 19B pattern 2B appears nowhere on the tree. We cannot explain this foible of the Clados program.

Begle (1991, p. 51) listed nine characters of Osmeridae in his sense (including *Plecoglossus* but excluding Salangidae). Three of them (nos. 69, 94, and 101) also occur in salangids and so characterize Osmeridae as we interpret the group. Two of the characters occur in all osmerids (sensu Begle) but not in salangids: metapterygoid shelf (no. 36 in section II; the metapterygoid is unossified in salangids except in *Protosalanx* and some *Salangichthys*) and notch in dorsal margin of opercle (37; salangids have the dorsal part of the opercle reduced). The remaining four characters comprise one that also occurs in salangids but is primitive at this level (103, nuptial tubercles), two that do not exist (orientation of uncinate processes on Pb2 and Pb3, nos. 104 and 107 in section II), and one that is questionable (dumbbell-shaped autopalatine, no. 27 in section II).

V. Monophyly and Interrelationships of Osmeroidei

Appendix 1 is a matrix of 112 characters in osmeroids that generates the trees in Fig. 19. As in the previous section, rather than work through every character at each node in Fig. 19, we here evaluate previous work on osmeroid interrelationships and describe the characters that justify our main conclusions. Unless there is an explicit reference to Section II, character numbers in this section are those in Fig. 19 and Appendix 1, where each is tied to the main survey in section II.

A. Relationships of Lepidogalaxias

The phylogenetic position of *Lepidogalaxias* has long been a problem. Begle (1991) reviewed that problem and placed *Lepidogalaxias* as the sister-group of salangids, *Aplochiton*, *Lovettia*, and galaxiids (Fig. 2). That conclusion was facilitated by gross miscoding of salangids (incorrect in 27 characters). In our view, the key to understanding *Lepidogalaxias* is its unique acceleration of skeletal development, detailed under no. 200 in Section II. Searching for the sister group of *Lepidogalaxias*, we tried hard to place it within Galaxiidae,

for example, as related to the diminutive *Galaxiella* (McDowall, 1978; McDowall and Frankenberg, 1981), particularly to *G. munda* or *G. nigrostriata*, with which it is sympatric. *Galaxiella* shares with *Lepidogalaxias* extremely well-developed caudal peduncle flanges, a rounded caudal fin, sexual dimorphism, and aestivation (McDowall and Pusey, 1983), but we were unable to find any skeletal features that support such a relationship. Instead, our analysis (Fig. 19) places *Lepidogalaxias* as sister to the Tasmanian *Lovettia*, a fish comparable in size to *Lepidogalaxias* (up to 70 mm SL). Although the two look very different (*Lovettia* looks like a smelt; *Lepidogalaxias* looks like a galaxiid), they share several features: absence of myodome (13, homoplasy with salangids); frontals lacking the posterolateral laminar portion that roofs the orbit (17) so that the pterosphenoid of *Lepidogalaxias* (a bone lacking in *Lovettia*) appears almost as a roofing bone; metapterygoid greatly reduced and removed from hyomandibular (24, state 2); opercle with deeply incised margin (38, Fig. 4; the subopercle is also deeply incised in *Lepidogalaxias*, whereas it is rodlike in *Lovettia*, with a single distal incision in some specimens); uncinate process of Pb3 absent (48, homoplasy with some galaxiids and *Retropinna*); Pb3 without anterior extension (49, state 2, homoplasy with salangids); gap between occiput and V1 (53); pectoral radials (Fig. 15I and 15J) with an identical configuration, the third tapering proximally and failing to contact the girdle (83, homoplasy with salangids); and a medial, membrane bone lamina on the pelvic girdle (89, Figs. 16G and 16H). *Lepidogalaxias* and *Lovettia* are also the only galaxioids to retain a separate ectopterygoid and agree in the shape of the hyoid bar (Fig. 5), which is shallow, without the abrupt deepening of the proximal ceratohyal seen in other galaxioids. Of course there are many differences between *Lepidogalaxias* and *Lovettia*, most of them due to autapomorphies of *Lepidogalaxias* such as the basisphenoid, vomerine and palatine teeth, epicentral bones, separate PU1 and U1, and scales; *Lovettia* agrees with *Aplochiton* and galaxiids in lacking these features.

B. Relationships of Aplochiton

If *Lepidogalaxias* and *Lovettia* are sister-groups, *Aplochiton* is the sister of Galaxiidae (Fig. 19), in agreement with Begle (1991). Begle cited four characters supporting this relationship; we accept two of them, the spurlike posterodorsal crest on the hyomandibular (26, state 3; crest very small in *Aplochiton*) and the form of the premaxillary ascending process (27). The only additional features we have noted that support immediate relationship between *Aplochiton* and galaxiids are passage of the efferent pseudobranchial artery through a notch or foramen in the parasphenoid (15, state 2), absence of the ectopterygoid (22, state 2, homoplasy with salangids), and fragmentation of the upper pharyngeal toothplate (50, state 2).

C. Relationships within Galaxioidea

Within Galaxioidea, Retropinnidae (*Retropinna*, *Prototroctes*, and *Stokellia*) are shown to be monophyletic by the unossified ethmoid (2); earlike sphenoid cartilage (15, state 1); fusion of ectopterygoid and dermopalatine (22); absence of both supraorbital and antorbital (34, state 3); fusion between first hypural and parhypural (66); single CMC (73, homoplasy with *Mallotus*); infraorbital canal extending to preopercle (97, state 1); right ovary absent (102, homoplasy with *Plecoglossus*); and horny abdominal keel (108). Begle (1991, p. 53) listed the last three of these, together with three pores in the supraorbital and preopercular sensory canals (177, section II), and three other characters that we reject.

Within retropinnids, our data, like Begle's (1991), indicate pairing of *Retropinna* and *Stokellia*. The three characters favoring that group are maxillo–palatine articulation (28), dorsal fin position (91), and absence of nuptial tubercles (103). However, this pattern of relationships requires independent loss of maxillary and gill-raker dentition in *Prototroctes* and *Stokellia* (or reacquisition of both in *Retropinna*).

Retropinnidae are the sister-group of Galaxiidae s.l. (*Aplochiton*, *Lovettia*, *Lepidogalaxias*, and galaxiids), a remarkably well characterized group (Fig. 19) shown to be monophyletic by: toothless vomer (4); buccohypophysial canal in parasphenoid (14, closed in *Lepidogalaxias*); absence of dermopalatine (21, state 2, reversed in *Lepidogalaxias*); reduced metapterygoid (24); lachrymal with cartilage-covered condyle meeting lateral ethmoid (35, no lachrymal in *Lepidogalaxias*); absence of basibranchial teeth (40) and of uncinate process on Eb1 (41) and Eb2 (42, reversed in *Lepidogalaxias*); Eb5 fused to Eb4 to form an enclosed vascular foramen (45, state 2, absent in *Lepidogalaxias*); all anterior neural arches fused to centrum (55); full-length NPU2 (63); well-developed NPU1 (64); five hypurals (67); loss of membranous outgrowth of Un1 (70) and CMCs (73); 16 principal caudal rays or fewer (75, state 2); and absence of extrascapular (77, state 2, homoplasy with salangids), ventral process of cleithrum (79, state 2, homoplasy with salangids), pubic symphysis (86, homoplasy with salangids), ventral pelvic condyle (87), scales (96, reversed in *Lepidogalaxias*), and posterior part of infraorbital (97, state 2) and temporal (98) sensory canals.

Retropinnidae and Galaxiidae s.l. together make up the Galaxioidea. Begle's (1991) Galaxioidea differed from ours in including Salangidae. He listed nine galaxioid characters, of which we accept four: absence of uncinate process on Pb2 (47), 18 or fewer principal caudal fin rays (75, state 1), absence of mesocoracoid (81, homoplasy with salangids), and nasal lamellae parallel (106). Other galaxioid characters include dermethmoid absent (1); pterosphenoid with extensive medial epiphysial arm (7); parietals extending forward to postorbital process (19); no supramaxilla (30); Pb1 articulating with lateral surface of Eb2 (46, state 2); Baudelot's ligament on occiput (54, reversed in *Lepidogalaxias*; double ligament in some galaxiids); no ossified epineurals (57, reversed in galaxiids and homoplastic with salangids); pattern 2B supraneurals (60, state 3); and absence of third uroneural (72, homoplasy with salangids), caudal scutes (74, homoplasy with salangids and salmonoids), urodermal (76, homoplasy with salangids), internal limb of posttemporal (78, homoplasy with salangids), and sensory canal in posttemporal and supracleithrum (99, 100, further homoplasies with salangids). Parsimony also resolves absence of maxillary teeth and ANA as galaxioid characters, with reversal of the first in *Retropinna* (Fig. 19, no. 29) and of the second in *Aplochiton* (Fig. 19, no. 52), but we know of no other well-attested instance of reacquisition of maxillary teeth or ANA.

D. Monophyly of Osmeroidei

Within Osmeroidei, Osmeridae (including Plecoglossidae and Salangidae; = Osmeroidea) are the sister-group of Galaxioidea. Begle (1992, p. 353) proposed eight osmeroid characters: vomer with a short shaft (no. 3 in Appendix 1 and Fig. 19; character coded by Begle as shaft absent), loss of orbitosphenoid (5), ventral process on pterosphenoid (8 and 9), reduced articular (33), ventral condyle on the pelvic girdle (87), reduced pterosphenoid (no. 7 in Section II), modified endopterygoid dentition (23), and marginal basihyal teeth (39). The last three were listed by Begle as "additional characters" whose occurrence at the osmeroid node varies with optimization. Absence of the orbitosphenoid, reduced articular, and ventral pelvic condyle withstand criticism, except that the first two characters also occur in esocoids, and the third appears at the osmeroid node only under one optimization (Fig. 19B). The vomerine shaft and basihyal dentition are ambiguous because of variation within osmerids and galaxioids. The pterosphenoid process and reduced pterosphenoid are wrongly interpreted, and the endopterygoid dentition fails because the osmeroid pattern also occurs in platytroctid alepocephaloids. Absence of the basisphenoid (12), a columnar coracoid process of the cleithrum (79), and fusion of PU1 and U1 (65) are additional osmeroid characters, the first miscoded by Begle and so placed as a character of osmeroids + argentinoids, the second miscoded as primitive, and the third miscoded and placed as a character only of a subgroup of Argentinoidea. The coracoid process is lost in Galaxiidae s.l., and the other two characters are reversed in *Lepidogalaxias*.

Further osmeroid characters include enlarged and modified first pectoral radial (82; reversed in *Mallotus*, salangids, and Galaxiidae s.l.), and postcleithra absent (80, state 2; Pcl 3 reappears in *Lovettia* and some galaxiids). Parsimony also resolves toothless gill rakers (51) as an osmeroid character, with reversal in *Retropinna*. We know of no other well-attested instance of reacquisition of gill-raker dentition.

The cucumber odor in osmeroids remains problematic. Under both optimizations in Fig. 19 the character (110) is resolved as independently acquired in retropinnids, *Hypomesus*, *Mallotus*, and *Spirinchus*. It occurs also in *Argentina* and a chlorophthalmid (no. 197, Section II); further sampling is needed, preferably using the methods of Berra *et al.* (1982) and McDowall *et al.* (1993).

Comparison of Fig. 19 with Begle's cladogram of osmeroids (Fig. 2) shows only three internal nodes (out of 13) in common, the pairing of *Aplochiton* and galaxiids, the three retropinnids, and the pairing of *Retropinna* and *Stokellia*.

E. Classification

The relationships indicated in Fig. 19 can be expressed in the following sequenced classification:

Suborder Osmeroidei
  Superfamily Osmeroidea
    Family Osmeridae
      Subfamily Hypomesinae
        Hypomesus
      Subfamily Plecoglossinae
        Plecoglossus
      Subfamily Osmerinae
        Tribe Salangini
          Mallotus
          Protosalanx
          Salangichthys
          Salanx
          Neosalanx
        Tribe Osmerini
          Osmerus
          Allosmerus
          Spirinchus
          Thaleichthys
  Superfamily Galaxioidea
    Family Retropinnidae
      Prototroctes
      Retropinna
      Stokellia
    Family Galaxiidae
      Subfamily Lepidogalaxiinae
        Lovettia
        Lepidogalaxias
      Subfamily Galaxiinae
        Tribe Aplochitonini
          Aplochiton
        Tribe Galaxiini
          Brachygalaxias
          Galaxias
          Galaxiella
          Neochanna
          Nesogalaxias
          Paragalaxias

VI. Monophyly of Argentinoidei (sensu Rosen and Greenwood, 1971; Argentinoidea + Alepocephaloidea)

Begle (1992, p. 355), like Greenwood and Rosen (1971), found that the only unambiguous characters of Argentinoidei are those drawn from the crumenal organ. Begle offered five additional characters whose occurrence at the argentinoid node varied with optimization: reduction in endopterygoid teeth and loss of nuptial tubercles, uncinate process on Eb4, maxillary teeth, and basihyal teeth. Those characters are discussed above (nos. 31, 56, 80, 94, and 186 in Section II), and none can be used to characterize Argentinoidei. The survey in Section II yields two further argentinoid synapomorphies. The first is ventral displacement or descent of the distal parts of the first two to four epineurals (no. 119 in Section II). As concluded in our discussion of the character, argentinoid genera or species in which it is lacking (only one descended in *Platytroctes*, *Talismania aphos*, and *Alepocephalus tenebrosus* and none in most opisthoproctids, *Microstoma*, and *Xenodermichthys*; no ossified epineurals in *Photostylus*) are all indicated, by other evidence (see below), as derived members of their subgroups. The second new argentinoid synapomorphy is no. 143 in Section II, support of the lowermost fin ray of the upper caudal lobe by both caudal median cartilages (CMCs, Figs. 14C and 14D). This occurs in argentinids (*Argentina* and *Glossanodon*), microstomatids (*Microstoma* and *Nansenia*), the opisthoproctid *Bathy-*

lychnops, platytroctids (10 genera examined) and the alepocephalids *Bathylaco* and *Narcetes*. It has not been reported elsewhere. In argentinoids, three reversals to the primitive pattern must be accepted: in bathylagids (Fig. 14B), opisthoproctids (except *Bathylychnops*), and alepocephalids.

In summary, Argentinoidei are characterized by three features: crumenal organ, descent of the first two to four epineurals, and CMCs together supporting the lowermost ray of the upper caudal lobe.

VII. Monophyly and Interrelationships of Argentinoidea

Begle (1992, p. 353) proposed six unambiguous characters of Argentinoidea: narrow levator process on Eb4; short, dumbbell-shaped interhyal; reduced metapterygoid; small, terminal mouth; anterior expansion of vomer; and fusion between PU1 and U1 (he omitted the elongate symplectic, his character 88, which he also found only in Argentinoidea). The characters are evaluated above (nos. 5, 37, 63, 73, 95, and 132 in Section II). Those that withstand criticism (nos. 5, 37, and 63) are related to the small mouth and forwardly displaced jaw articulation. Begle also discussed other characters proposed for Argentinoidea by Greenwood and Rosen (1971) and Ahlstrom *et al.* (1984). He accepted two of Greenwood and Rosen's characters, premaxilla freed from its ethmoid articulation and large supraneural laminae over PU1 and U1; and three characters of Ahlstrom *et al.*, development of dorsal and anal fins within the larval finfold, pustules on the chorion, and micro-rete mirabilia in the swimbladder.

Within Argentinoidea, Begle (1992) summarized divergent recent opinions on relationships, reanalyzed data of Ahlstrom *et al.* (1984), and settled on the pattern [Argentinidae [Microstomatidae [Bathylagidae, Opisthoproctidae]]]. However, Kobyliansky (1990) argued in detail that Microstomatidae and Bathylagidae should be combined as subfamilies of Microstomatidae (the older name); Kobyliansky did not discuss opisthoproctids, but his opinion that bathylagids and microstomatids are sister-groups coincides with Greenwood and Rosen (1971) and differs from Begle (1992).

Kobyliansky (1986, 1990) showed that *Bathylagichthys* is more primitive than the other seven bathylagid genera that he recognized in retaining a sensory canal ossicle fused to the posttemporal, an opercle without marginal serrations, a toothplate on Bb4, an uncinate process on Eb4, three postcleithra (vs two or fewer), and a urodermal. In microstomatids, we agree with

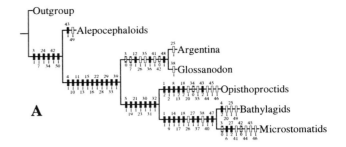

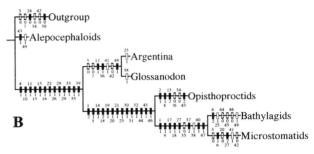

FIGURE 21 Cladograms of Argentinoidea based on the data in Appendix 2 analyzed by Hennig86 (Farris, 1988); trees produced with Clados (Nixon, 1992). Conventions as in Fig. 19. The single shortest tree is 74 steps long, C.I. 0.75, R.I. 0.70. The two trees show alternative optimizations, with A favoring reversals and B favoring forward changes.

Kobyliansky (1990, p. 174) that *Nansenia* is more primitive than the other two genera (*Microstoma* and *Xenophthalmichthys*, distinguished by "little but the tubular eyes" in the latter; Cohen, 1964, p. 23). *Nansenia* differs from *Microstoma*, and resembles primitive bathylagids, in having a toothplate on Bb4, a free Eb5, descended anterior epineurals, an adipose fin, and numerous interneural and interhaemal cartilages. We studied five of the six genera of opisthoproctids (*Bathylychnops*, *Dolichopteryx*, *Macropinna*, *Opisthoproctus*, and *Rhynchohyalus*). Among those five, *Bathylychnops* differs from the other four and resembles other primitive argentinoids in having a dermopterotic, a sensory canal in the posttemporal, an accessory cartilage between Cb5 and Eb5 (present also on one side in our *Dolichopteryx*), anterior epineurals descended distally, and CMCs together supporting a single finray.

Appendix 2 is a matrix of 50 characters in argentinoid subgroups, alepocephaloids, and an outgroup. Analyzed with Hennig86 (Farris, 1988) and Clados (Nixon, 1992) that matrix gives the trees in Fig. 21. The pattern resolved, [Argentinidae [Opisthoproctidae [Bathylagidae, Microstomatidae]]], is congruent with the conclusions of Greenwood and Rosen (1971) and Kobyliansky (1990), and it contradicts those of Ahlstrom *et al.* (1984) and Begle (1992). A sequenced classification of the group might be (genera in alphabetical order within family-group taxa):

Superfamily Argentinoidea
  Family Argentinidae
    Argentina
    Glossanodon
  Family Opisthoproctidae
    Bathylychnops
    Dolichopteryx
    Macropinna
    Opisthoproctus
    Rhynchohyalus
    Winteria
  Family Microstomatidae
    Subfamily Bathylaginae
      Tribe Bathylagichthyini
        Bathylagichthys
      Tribe Bathylagini
        Lipolagus
        Melanolagus
        Dolicholagus
        Leuroglossus
        Bathylagoides
        Pseudobathylagus
        Bathylagus
    Subfamily Microstomatinae
      Tribe Nanseniini
        Nansenia
      Tribe Microstomatini
        Microstoma
        Xenophthalmichthys

Characters of the nonterminal groups (based on interpretation of Fig. 21; character numbers from Appendix 2 in brackets) are as follows. Argentinoidea: Parietals carrying medial part of occipital commissural sensory canal (4), endopterygoid teeth absent (10), metapterygoid reduced (11), premaxilla and maxilla toothless (13 and 15), no supramaxillae (16), basibranchials 1–3 toothless (22), Pb2 and Pb3 toothless (28 and 29), ANA absent (33), and Un1 without membranous anterodorsal outgrowth (39). Argentinidae: Ventral arm of symplectic shorter than dorsal arm (12), anterior supraneurals expanded rostrocaudally (36) and white crescent above iris (48). Opisthoproctidae + Microstomatidae: occipital commissure anteriorly placed on parietal (5), lachrymal smaller than succeeding infraorbitals (19), basihyal and Cb5 toothless (21 and 23), UP4 absent (30), UP5 minute (31), and gill rakers of arches four and five differentiated from those of arches two and three (32). Microstomatidae: Mesethmoid with separate laminar dorsal and ventral ethmoids (1), palatine teeth in a single row (9), premaxilla articulating with maxilla posteriorly (14), bladelike

dentary teeth (17), supraorbital and dermosphenotic in contact above orbit (18), Pb1 unossified or absent (27), pattern 1 supraneurals (35), PU1 and U1 fused (37), numerous caudal interneural and interhaemal cartilages (40), and extrascapular ossicles present above pterotic (47) (see also Kobyliansky, 1990, table 1).

VIII. Monophyly and Interrelationships of Alepocephaloidea

Begle (1992) proposed four alepocephaloid synapomorphies and two additional characters that might diagnose the group: dorsally reduced opercle, dilatator spine on opercle, middorsal ridge on basibranchials, and toothed gill rakers, with reduced pterosphenoid and toothless maxilla as additional characters. These are discussed above (Section II, our nos. 7, 56, 76, 78, 84, and 112). The dorsally reduced opercle remains uncontradicted (except perhaps in *Bathylaco*, Fig. 7J) but also occurs in all galaxioids and in salangids and esocoids; the dilatator spine is merely another way of describing it. The middorsal ridge on the basibranchials exists only in some platytroctids (toothed) and *Leptochilichthys* (very high and toothless); toothed gill rakers are primitive; and reduced pterosphenoids and a toothless maxilla occur only in a few derived alepocephaloids. The configuration of the opercle is the only character to withstand criticism, a conclusion also reached by Matsui and Rosenblatt (1987, p. 23). Thus alepocephaloid monophyly is unsupported by Begle's data, as is evident from Fig. 18.

Greenwood and Rosen (1971) suggested a few more alepocephaloid characters: separation of parietals, and absence of adipose fin, swimbladder, and urodermal. Patterson and Johnson (1995, p. 27) proposed that monophyly of alepocephaloids is corroborated by the fact that the ossified epipleural series extends unusually far forward, to about V3. More extensive sampling (detailed under no. 123 in Section II) confirms the character, with partial reversal in a few derived genera (epipleurals to V9 in *Rinoctes*, V9–11 in *Leptoderma*, V8 in *Ericara*, V12 in *Conocara* and *Xenodermichthys*, and no epipleurals in *Photostylus*). Reviewing Begle's (1992) data has produced one further plausible alepocephaloid character, branchiostegal cartilages connecting the branchiostegals with the ceratohyal (Section II, no. 74). Elsewhere, we found these only in two osmerids, *Mallotus* and *Hypomesus olidus*, whereas they are widespread in alepocephaloids.

Appendix 3 is a matrix of 59 characters of platytroctids and selected alepocephalid genera (principally those sampled by Begle, 1992), together with argentinids and a hypothetical outgroup to root the tree. Analyzed by Hennig86, that matrix generates 14 shortest trees (length 170 steps, C.I. 0.51, R.I. 0.55) with a strict consensus giving no resolution beyond a monophyletic Alepocephaloidea; platytroctids or *Bathylaco* as the sister of the remaining alepocephaloids; and the terminal grouping [*Rouleina* [*Leptoderma, Photostylus*]]. The 14 trees fall into two sets: 12 in which *Bathylaco* is the basal alepocephaloid taxon and two in which platytroctids are basal. We reject the first set because they imply reacquisition of teeth on the endopterygoid, basihyal, basibranchial, and Pb2 in platytroctids (or independent loss of all in *Bathylaco*); the pattern is favored by parsimony only because the immediate outgroup, argentinids, also lacks teeth on the endopterygoid, basibranchial, and Pb2. In the second set of trees (platytroctids basal) the remaining alepocephaloids are grouped by loss of basihyal teeth (no. 21 in Appendix 3, redeveloped or retained in one lot of *Bajacalifornia*). Beyond platytroctids and *Bathylaco*, the 14 trees place the genera *Alepocephalus, Bajacalifornia, Bathyprion, Bathytroctes, Leptochilichthys, Narcetes,* and *Rinoctes* in a variety of patterns in relation to each other and to the terminal [*Rouleina* [*Leptoderma, Photostylus*]] group.

In order to eliminate the set of trees in which *Bathylaco* is the basal alepocephaloid, we weighted character 21 (basihyal teeth) at 2. With that weighting, the data in Appendix 3 give two trees. One is shown in Fig. 22; the second tree differs from it in pairing *Bajacalifornia* and *Narcetes* as the sister-group of the eight genera beyond them in Fig. 22 and placing those genera in the pattern [*Talismania* [*Rinoctes* [*Leptochilichthys* [*Bathyprion* [*Alepocephalus* [*Rouleina* [*Leptoderma, Photostylus*]]]]]]].

Based on Appendix 3 and Fig. 22, characters of Alepocephaloidea are parietals separated by supraoccipital (no. 5), posttemporal fossa absent (8), branchiostegal cartilages present (18), dorsal part of opercle reduced (19), epipleural bones extending forward to about V3 (32), urodermal absent (43), no more than one postcleithrum (45), and absence of adipose fin (52) and swimbladder (58).

Platytroctids are the sister-group of other alepocephaloids, as indicated by their possession of teeth on the endopterygoid (11), basihyal (21), and basibranchial (22). Platytroctids were originally held to be monophyletic because of one derived character, the shoulder organ (Parr, 1951; Markle, 1976). Matsui and Rosenblatt (1987, p. 23) proposed two further characters: the subcutaneous canal system [which we have not tried to evaluate, but Sazonov (1986, 1992) found it to be restricted to his subfamily Searsiinae] and

supraneurals spaced over every second or third vertebra (pattern 4, no. 125, Section II); the same pattern occurs in *Rouleina* and *Xenodermichthys*, where we infer that it is independently derived (33, state 2, Fig. 22). Matsui and Rosenblatt suggested three other platytroctid characters in the caudal skeleton: PU1–3 only half the length of more anterior centra (e.g., PU5); U2 overlapped laterally by the extended bases of hypurals 3 and 4; and first epural much longer than the second. We found that none of these characterizes platytroctids. The length of PU1–3 fails among platytroctids in *Mirorictus, Mentodus, Sagamichthys,* and some *Searsia koefoedi* in our material, and in *Pectinantus* and *Persparsia* as illustrated by Sazonov (1986, figs. 16 and 18), whereas among our alepocephalid material, *Bathylaco* has the "platytroctid" condition. U2 is fully embraced by the bases of hypurals 3 and 4 in several of our platytroctids, but in others (e.g., *Holtbyrnia* and *Pellisolus*) the hypural bases cover less than half the centrum. In alepocephalids, *Bathylaco* has U2 fully embraced by the hypural bases, and in *Talismania aphos* they cover more than half of U2. The ratio between the lengths of the second and first epurals (E2/E1) seems variable within species. In our platytroctids the ratio is >3/4 in *Pellisolus*, and in Sazonov's (1986) illustrations it is >3/4 in *Pellisolus, Holtbyrnia, Maulisia, Persparsia,* and *Platytroctegen*. In alepocephalids the ratio is <3/4 in *Bajacalifornia* and *Rouleina*.

Beyond platytroctids, *Bathylaco* is the sister-group of other alepocephalids, which are united by loss of the basipterygoid process (4), fewer supraneurals (34), a long NPU2 (37), cleithrum without a ventral process (44), reduced extrascapular (54), and absence of sensory canal in posttemporal (55). In Fig. 22, and in the alternative tree, *Bathytroctes* is placed as the sister of the remaining alepocephalids. The characters of *Bathytroctes* tending to favor this are presence of the supraorbital (16), uncinate process on Eb2 (25, otherwise present only in *Bathylaco*), ANA (29), numerous supraneurals (34), and postcleithrum (54). In some other respects, *Bathytroctes* is derived (e.g., characters 7, 18, 26, 27, and 35), and we think it possible that *Narcetes* (for which information is lacking in several characters) is the sister of the remaining alepocephalids, as suggested by its three epurals (38), argentinoid–platytroctid pattern of CMCs and fin rays (42), and other characters in which it seems more primitive than *Bathytroctes*.

We have little confidence in the characters in Appendix 3, too many of which are merely loss of primitive characters, and as the low C.I. and R.I. indicate, they are riddled with homoplasy. There is also much missing data in Appendix 3. Hence we have no confidence in the tree in Fig. 22 as an estimate of

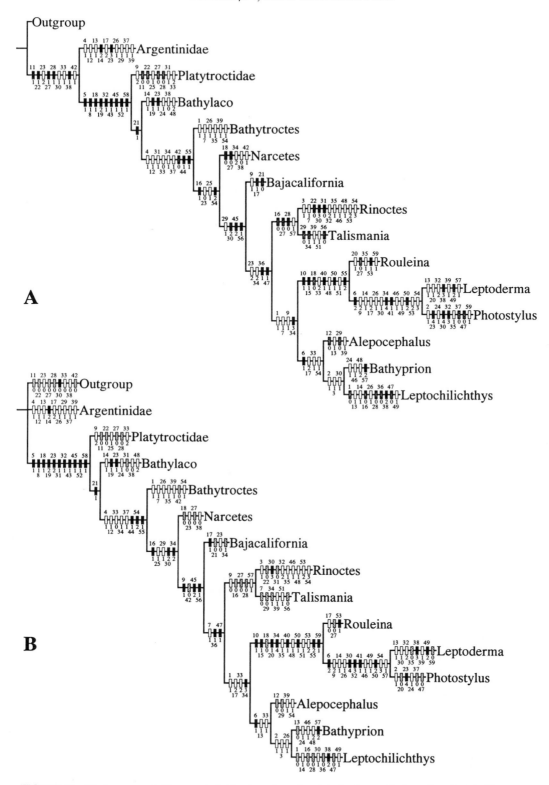

FIGURE 22 Cladograms of Alepocephaloidea based on the data in Appendix 3 analyzed with Hennig86 (Farris, 1988); trees produced with Clados (Nixon, 1992). Conventions as in Fig. 19; weighting as described in text. The two trees show alternative optimizations, with A favoring reversal and B favoring forward changes.

alepocephaloid interrelationships and will not detail characters at the nodes beyond *Bathytroctes*. Alepocephaloids are badly in need of study with better and more comprehensive material than we were able to assemble. Even with our limited material, it seems clear that some genera (*Alepocephalus* and *Talismania*) are nonmonophyletic (e.g., comments in Section II under characters 21, 23, 70, 74, 112, 119, 123, 125, and 179).

In Begle's (1992) cladogram of alepocephaloids (Fig. 2), the only other cladistic analysis of the group, *Leptochilichthys* is at the base, and in the terminal dichotomy platytroctids are twinned with *Talismania*. Platytroctids were misplaced by a combination of miscoding (incorrect in 18 characters) and misinterpreting primitive characters as derived (e.g., the two characters twinning platytroctids and *Talismania* are the condition of the dorsal border of the ceratohyal, and contact between the lateral crest of the hyomandibular and the preopercle). Other nodes in Begle's cladogram involve similar miscoding and/or misinterpretation (e.g., reacquisition of orbitosphenoid and basisphenoid).

On the basis of Fig. 22, a sequenced classification of alepocephaloids might be as follows:

Superfamily Alepocephaloidea
    Family Platytroctidae, ca. 14 genera
    Family Bathylaconidae, *Bathylaco*, *Herwigia*
    Family Alepocephalidae, ca. 23 genera

IX. Monophyly and Relationships of Salmonoidei

The interrelationships of salmonoids seem now firmly established (Sanford, 1987, 1990; Stearley and Smith, 1993): Coregonidae (*Coregonus*, *Prosopium*, and *Stenodus*) are the sister-group of Salmonidae, which comprise Thymallinae (*Thymallus*) and Salmoninae (eight Recent genera in Stearley and Smith's scheme; Phillips *et al.*, 1995, add *Parahucho*). Our character survey (Section II) provides a few characters, additional to those listed by Sanford (1990) and/or Stearley and Smith (1993), which reinforce or clarify the monophyly of Salmonidae and Salmoninae. Additional salmonid characters are a distal ceratohyal with no fenestra and closed dorsally by perichondral bone (no. 70, Section II), toothless Pb2 (no. 102), and no urodermal (no. 146). Additional salmonine characters are a tripartite occipital condyle (no. 26; in agreement with Sanford, 1990) and a long NPU2 (no. 129; in agreement with Stearley and Smith, 1993, but reversing their polarity).

Salmonoid monophyly is not in question, although among the salmonoid characters listed by Sanford (1990) and Stearley and Smith (1993), the most striking is still the tetraploid karyotype. The outgroup relationships of salmonoids are unsettled (Fig. 1). The sister-group relationship proposed between salmonoids and galaxiids (Rosen, 1974; Fig. 1B) and between salmonoids and neoteleosts (Fink, 1984b; Fig. 1C) can be dismissed as without support from plausible shared derived characters. Williams (1987; also Nelson, 1994, p. 175; Fig. 1E) proposed a sister-group relationship between salmonoids and esocoids on the basis of two characters of the suspensorium and jaw musculature. The first is an anteroventral wing of the hyomandibular that overlaps the medial face of the metapterygoid. Williams recorded this structure in all esocoids and coregonids but found it to be absent in most salmonids. Sanford (1987, 1990) found the anteroventral process only in *Coregonus* and *Prosopium* and used it as a synapomorphy of those genera. Williams's second character is an adductor mandibulae that inserts directly on the lower jaw and has no ligamentous connection with the maxilla. He recorded that condition in all esocoids, all salmonids, and *Coregonus*, but because *Prosopium* and *Stenodus* have the condition he took to be primitive, he regarded the character as questionable. Our survey has not yielded any characters indicative of immediate relationship between esocoids and salmonoids (Section II and Appendix 4). Relationships between salmonoids and all (Fig. 1F) or some (Figs. 1A, 1D, and 1G) argentinoids + osmeroids (= Osmerae of Begle) remain to be evaluated.

Sanford (1987, 1990; Fig. 1D) proposed two characters relating salmonoids to osmeroids and argentinoids: absence of radii on the scales and well-developed teeth on the margin of the basihyal, a character first proposed and discussed by Nelson (1970b; Fig. 1A). The two characters are discussed above (Section II, nos. 80 and 172). Radii are apparently absent on the scales of all salmonoids and osmeroids but occur in alepocephaloids (e.g., *Bajacalifornia*, *Bathylaco*, *Bathytroctes*, *Narcetes*, and *Talismania*) and argentinoids (e.g., *Microstoma* and *Pseudobathylagus*). Marginal basihyal teeth, Sanford's (1987, 1990) second character, do not occur in Argentinoidea, where the teeth are terminal in *Argentina* and some *Glossanodon*, but are otherwise absent. In alepocephaloids, marginal basihyal teeth occur in the platytroctids *Barbantus*, *Platytroctes*, *Sagamichthys*, *Searsia*, and *Searsioides* (Matsui and Rosenblatt, 1987, table 1), but others have a median row of teeth. Among osmeroids, primitive osmerids (*Hypomesus* and *Plecoglossus*) and *Prototroctes* have the marginal basihyal teeth no better developed than in umbrids (Rosen, 1974; figs. 1, 2, and 4), and the same is true of primitive salmonoids (coregonids and *Thymallus*).

Begle (1991, 1992; Fig. 2) placed argentinoids + osmeroids (his Osmerae) as the sister-group of Neoteleostei, with salmonoids more remote. He offered two characters relating Osmerae and neoteleosts (1992, p. 354): extended alveolar process of premaxilla and reduction of laminar bone on anterior margin of hyomandibular. Both characters (discussed above, Section II, nos. 47 and 51) are wrongly interpreted. Begle (1992, p. 354) offered seven characters supporting monophyly of his Osmerae: reduction or loss of basisphenoid (no. 12 in Section II); ventral vane on Bb1 (no. 83); fusion of rudimentary neural arches to centra in caudal skeleton (no. 130); reduction of pterosphenoid (no. 7); loss of endopterygoid teeth (no. 31); form of basihyal teeth (no. 80); and loss of nuptial tubercles (no. 186) (Begle cited the last four as "additional characters" whose occurrence at the osmeriform node varied with optimization). The seven characters are discussed in Section II. All are shown to involve errors of coding and/or interpretation.

Patterson and Johnson (1995) described two characters relating salmonoids to osmeroids: presence of cartilaginous epicentrals (Section II, no. 121), and absence of epipleural bones (no. 122). They suggested one character relating Argentinoidei to osmeroids + salmonoids as Salmoniformes, lack of proximal forking of intermuscular bones (no. 124), and noted a further possible salmoniform character in Jamieson's (1991) records of a single annular mitochondrion in the sperm of alepocephaloids, salmonids, and *Galaxias* (no. 185). Cartilaginous epicentrals are now found also to occur widely in Argentinoidea and in a few alepocephaloids. The group they characterize, then, is not osmeroids + salmonoids but those plus argentinoids and alepocephaloids. Epipleural bones occur in most osmerid and galaxiid genera, but their pattern in osmerids and their distribution in osmerids and galaxioids imply that they are secondarily derived within each group (character 59, Fig. 19) so that absence of the bones as a salmonoid + osmeroid character is not contradicted. Proximal forking of the epineurals and epipleurals is general in elopomorphs, clupeomorphs, ostariophysans, and myctophiforms. In stomiiforms it does not occur; in esocoids proximal forking occurs only in the epineurals (esocids) or in no more than three or four bones in each series (*Umbra*); and in aulopiforms the bones are forked only in chlorophthalmids and some paralepids (Baldwin and Johnson, in this volume). The lack of proximal forking of intermusculars in stomiiforms, the basal neoteleos-

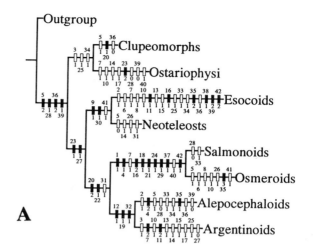

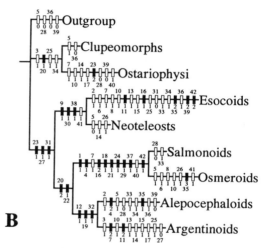

FIGURE 23 Cladograms of the major groups of Clupeocephala based on the data in Appendix 4 analyzed with Hennig86 (Farris, 1988); trees produced with Clados (Nixon, 1992). Conventions as in Fig. 19. The single shortest tree is 89 steps long, C.I. 0.60, R.I. 0.46. The two trees show alternative optimizations, with A favoring reversal and B favoring forward changes.

tean group, weakens it as a salmoniform character, unless (as suggested below) esocoids belong between salmoniforms and neoteleosts.

Appendix 4 is a matrix of 42 characters, abstracted from Section II, in salmonoids, the other major euteleostean groups, clupeomorphs, ostariophysans, and a hypothetical outgroup to root the tree. Analysed by Hennig86, that matrix produces one shortest tree, shown in Fig. 23. Salmonoids fall out as the sister group of osmeroids. Characters of that grouping are separate dermethmoid and supraethmoid (no. 1 in Appendix 4 and Fig. 23); open posttemporal fossa (no. 4); a single supramaxilla (7, homoplasy in ostariophysans and esocoids); UP4 absent (16, state 2); epineurals fused to neural arches on no more than V1 (18); no epipleural bones (21); last few neural and haemal spines keellike distally (24); Un2 anterodorsal to Un1 (29); scales without radii (37); nuptial tubercles (40, homoplasy with ostariophysans); and diadromy (42). With these 11 characters, the grouping of salmonoids and osmeroids has stronger support than any other in Fig. 23. It is perhaps surprising that the group has not been recognized before (Fig. 1).

In Fig. 23, salmonoids + osmeroids are grouped with alepocephaloids + argentinoids by two characters: cartilaginous epicentrals (no. 20) and absence of forking in epineurals and epipleurals (22, entered in Appendix 4 as a query for neoteleosts but resolved by parsimony as general there). Those two characters may seem far from convincing, but we know of no valid characters to support any other pattern of relationships for these groups.

X. Monophyly and Relationships of Esociformes

Esociforms or esocoids comprise only Esocidae (*Esox*, five Recent species) and Umbridae (*Dallia*, *Novumbra*, *Umbra*, and five Recent species). Monophyly of the group was first proposed in a cladistic framework by Nelson (1972) on the basis of modifications of the cephalic sensory-canal system. Among the esocoid characters given by Rosen (1974), the only striking ones are the paired, elongate proethmoids and modification of Eb4 by loss of the levator process or elevation (questionable in *Umbra* and also absent in *Lepidogalaxias*). Our character survey (Section II) yields six further characters supporting esocoid monophyly. First, the basibranchial toothplate is fragmented into two (no. 81, Section II). Second, Pb1 is conical and its tip is closed in bone (no. 100, Section II). Third, the single upper pharyngeal toothplate is UP4, not UP5 (no. 109, Section II). Fourth, there is only a single postcleithrum, homologous with postcleithrum 3 of other teleosts (no. 151, Section II). To our knowledge, the only other teleosts with postcleithrum 3 alone are some galaxioids, the aulopiform *Bathypterois*, and some derived acanthomorphs. Fifth, in the dorsal and anal fins of *Esox* and *Umbra* ossified middle radials develop only on the central pterygiophores (or on none of them in *Dallia* and *Novumbra*) (no. 167, Section II). And sixth, the cheek and operculum are scaled (no. 173, Section II), a derived feature otherwise found only in eurypterygians (aulopiforms and ctenosquamates), derived elopomorphs, and in one alepocephalid.

Within esociforms, monophyly of esocids and umbrids was justified by Nelson (1972). Relationships

within umbrids were discussed in detail by Wilson and Veilleux (1982), who agreed with Nelson (1972) that *Novumbra* is the sister of *Dallia* and *Umbra*. Reist (1987) conducted a phenetic analysis of morphometric characters in umbrids, criticized some of the characters used by Wilson and Veilleux, and concluded that *Novumbra* and *Umbra* are more closely related to each other than to *Dallia*. His argument is not convincing, and we accept Nelson's and Wilson and Veilleux's evidence that *Dallia* and *Umbra* are sisters. If so, one equivocal character of esocoids can be resolved. Supraneural pattern (no. 125 in Section II) in esocoids is equivocal because Recent esocids have pattern 2, *Umbra* has pattern 1, and *Dallia*, *Novumbra*, and the Paleocene *Esox tiemani* have a variant of pattern 2 in which the first supraneural is absent. Parsimony resolves pattern 2 or a modification of it as primitive for the group.

Esociform relationships are problematic. Patterson and Johnson (1995, p. 25) reviewed the recent consensus that esocoids are the sister-group of all other euteleosts, including ostariophysans. Begle (1992, fig. 9) summarized that hypothesis with a cladogram in which esocoids are linked with other euteleosts by three characters and are distinguished from them by one. Only one of the three euteleostean characters occurs in esocoids. The three are the adipose fin (absent in all esocoids), nuptial tubercles (absent in all esocoids), and membranous outgrowth of first uroneural. The single character cited to distinguish all other euteleosts from esocoids is the toothplate of Bb4 (no. 82, Section II), present also in some characiforms, bathylagids, microstomatids, and percopsiforms.

Johnson and Patterson (1993, p. 600) distinguished esocoids from other euteleosts (excluding ostariophysans) by the absence in esocoids of median caudal cartilages (CMC of Fujita, 1990). However, as discussed under no. 142 in Section II, we now infer that CMCs have been lost in at least six euteleostean lineages so that whether the absence of CMCs in esocoids is primary or secondary must be resolved by congruence with other characters.

Esocoids are included with clupeomorphs, ostariophysans, and the other major euteleostean groups in the matrix of 42 characters in Appendix 4 and in the trees generated from that matrix in Fig. 23. Esocoids are resolved as the sister group of neoteleosts. The unambiguous characters supporting that relationship are type 4 tooth attachment (no. 9 in Appendix 4) and acellular skeleton (no. 41, homoplasy with osmeroids). Parenti (1986) first used those characters to propose relationship between esocoids and neoteleosts. Less trustworthy characters (to us) are absence of Un3 (no. 30, homoplasy with derived members of several other groups, e.g., galaxioids and salangids) and scaling of cheek and operculum (no. 38), which may be interpreted equally parsimoniously as a character of esocoids and neoteleosts which is lost in stomiiforms, or as homoplastic in esocoids and neoteleosts. Presence of UP4 only (no. 16) is shared by esocoids and ctenosquamates.

Although the evidence relating esocoids to neoteleosts may seem anything but compelling, the alternative solutions have even less in their favor. The current consensus, that esocoids are the sister of all other euteleosts, with (Figs. 1C, 1D, and 1G) or without (Figs. 1F and 1H) ostariophysans, boils down to alternative evaluations of the adipose fin and CMCs. With the data in Appendix 4, the tree in Fig. 1F is 102 steps long, much less parsimonious than the shortest tree (length increased by 13 steps or 15%), whereas the tree in Fig. 1H is 93 steps long, an increase of four steps or 4.5%. We have found no new characters discriminating esocoids from all other euteleosts. The second suggestion (Williams, 1987; Fig. 1E) is that esocoids are the sister of salmonoids. With the data in Appendix 4, the tree in Fig. 1E is 103 steps long, 14 steps or 16% longer than the shortest. We have found no new characters indicating that esocoids are related to salmonoids.

If the pattern shown in Fig. 23 is correct, with esocoids the sister of neoteleosts, parsimony resolves one more of the problematic characters in esocoids. Absence of the adipose fin (no. 36 in Fig. 23) is secondary.

As for the molecular evidence on esocoids (Section IIP), sampling is grossly deficient in all four molecules so far sequenced in *Esox*, but growth hormone (rather strongly) and small subunit rRNA (weakly) both support a salmonoid–esocoid relationship for which there is, to our knowledge, no morphological support.

XI. Monophyly and Relationships of Euteleostei

The euteleost problem is discussed by Lecointre and Nelson (in this volume), who point out difficulties with each of the "characters" previously invoked to define Euteleostei: adipose fin (absent in esocoids and alepocephaloids); nuptial tubercles (absent in esocoids, argentinoids, alepocephaloids, and most neoteleosts); and stegural (questionable in argentinoids and, in their view, esocoids). We accept the molecular evidence (discussed by Lecointre and Nelson) that Clupeomorpha and Ostariophysi are sister-groups. This simplifies the euteleost problem by eliminating ostariophysans, but contributes to the question of how Euteleostei (minus Ostariophysi) might be de-

fined only in a negative way, by indicating that there are teleostean higher taxa (e.g., Clupeomorpha + Ostariophysi) without demonstrable morphological characters.

Appendix 4 and Fig. 23 suggest three characters that distinguish Euteleostei (minus Ostariophysi). The first is pattern 2 supraneurals (no. 23 in Appendix 4). Pattern 2 does not occur in osteoglossomorphs, elopomorphs, clupeomorphs, or ostariophysans; the first three have pattern 1, and the last has a pattern of its own which we call pattern 3. Pattern 2 occurs in all salmonoids and osmeroids, and in basal argentinoids and neoteleosts. Among alepocephaloids it occurs only in the primitive *Bathylaco* and the derived *Alepocephalus*, with platytroctids having the autapomorphous pattern 4. Parsimony resolves pattern 2 as primitive for salmoniforms, but the alepocephaloids are problematic.

The second euteleostean character is the stegural, Un1 with a membranous anterodorsal outgrowth (no. 27 in Appendix 4). This occurs in all euteleostean groups except argentinoids, where parsimony resolves absence as secondary. The third euteleostean character is caudal median cartilages (CMCs, no. 31 in Appendix 4). Their absence in esocoids is problematic, and, as Fig. 23 shows, it is equally parsimonious to treat them as a euteleostean character lost in esocoids or as independently acquired in salmoniforms and neoteleosts.

As for molecular evidence, the ostariophysan plus clupeomorph grouping has very strong support in trees based on partial sequences of large subunit ribosomal RNA (rRNA; Lê *et al.*, 1993) and complete sequences of small subunit rRNA (T. Littlewood, C. Patterson, and A. B. Smith, 1996). In those molecular trees Euteleostei (minus Ostariophysi) are only weakly distinct (bootstrap <50% for both large and small subunit samples), but in the small subunit alignment and on the present (admittedly rudimentary) sample euteleosts are clearly distinguished by the molecular synapomorphy shown in Fig. 24.

Our current estimate of higher level euteleostean relationships (Fig. 23) is summarized in the following classification, which introduces some new rankings and one new name and adapts two old names to new uses.

Clupeocephala
    Otocephala (new)
        Clupeomorpha
        Ostariophysi
    Euteleostei
        Protacanthopterygii
            Order Argentiniformes (new)

Branchiostoma	GUUUUC****G***GAAC
Lampetra	GUUUUC****G***GAAC
Squalus	GUUUUC****G***GAAC
Raja	GUUUUC****G***GAAC
Latimeria	GUUUUC****G***GAAC
Homo	GUUUUC****G***GAAC
Polyodon	GUUUCC****U***GAAC
Lepisosteus	GUUUCC****G***GAAC
Amia	GUUUCC****U***GAAC
Hiodon	GUUUCC****C***GAAC
Elops	GUUUCC****C***GAAC
Megalops	GUUUCC****G***GAAC
Albula	GUUUCC****U***GAAC
Ophichthus	GUUUCC****C***GAAC
Clupea	GUUUCC****G***GAAC
Chanos	GUUUCC****G***GAAC
Esox	GUUUUC***CCC**GAAC
Salmo	GUUUUU**CUUCU*GAAC
Lampris	GUUUUC*UUCUCU*GAAC
Polymixia	GUUUUC*UUCUCU*GAAC
Holocentrus	GUUUUC**UCUCU*GAAC
Lophius	GUUUUCUCUUCUCUGAAC
Fundulus	GUUUUU**CUCCU*GAAC
Sebastolobus	GUUUUC**UUUCU*GAAC
Solea	GUUUUC*UCUCUCUGAAC
Mola	GUUUUC*UUCUCU*GAAC

FIGURE 24 A molecular synapomorphy of Euteleostei (minus Ostariophysi). Positions 924–942 in an alignment of small subunit ribosomal RNA from a sample of craniates. However the nucleotides in bold type are aligned, insertions are synapomorphous for euteleosts.

Suborder Argentinoidei
    Suborder Alepocephaloidei
    Order Salmoniformes
        Suborder Salmonoidei
        Suborder Osmeroidei
Neognathi
    Order Esociformes
    Neoteleostei

In the first *Interrelationships* the Protacanthopterygii of Greenwood *et al.* (1966) were omitted, as we noted in the first sentence of this paper. We find it necessary to revive Protacanthopterygii, in something close to its original meaning, because of problems in ranking within "salmoniforms." To express the relationships in Fig. 23, either the current Osmeroidei and Salmonoidei must be downgraded, or the current Alepocephaloidea and Argentinoidea must be upgraded. We chose the latter because lowering the rank of Osmeroidei would raise problems in expressing well-resolved relationships within that group using conventional ranks, whereas raising the rank of Alepocephaloidea leaves room to express relationships within a diverse group when they are resolved. The name Neognathi was introduced by Rosen (1973) in the first *Interrelationships* for Euteleostei minus Ostariophysi. That group is now equivalent to Euteleostei. Previously (Patterson and Johnson, 1995, fig. 9) we proposed using Neognathi for Euteleostei minus Esociformes, a group that no longer exists. We now adapt Neognathi for esocoids and neoteleosts, where it is appropriate for fishes characterized by the derived type 4 tooth attachment.

XII. Conclusions

We began this paper with an agenda provided by Fink (1984a). We repeat his seven questions here, with an answer to each. (1) What are the relationships of the Esocoidei? This is still the hardest question to answer, but the solution now current, that esocoids are the sister group of all other euteleosts, has no support, and we propose, like Parenti (1986), that esocoids are the sister-group of neoteleosts. The two groups share type 4 tooth attachment and an acellular skeleton. (2) What are the relationships of the Ostariophysi? Do these fishes lie above or below the Esocoidei? Ostariophysans are the sister-group of clupeomorphs (Lecointre and Nelson, this volume) and esocoids belong above them. (3) What is the pattern of relationships among the traditional 'salmoniform' taxa, exclusive of the Esocoidei and Ostariophysi? See Figs. 19, 21, 22, and 23. (4) What are the relationships of and within the Argentinoidei (sensu Greenwood and Rosen, 1971, i.e., argentinoids plus alepocephaloids)? Argentinoidei, which we rank as Argentiniformes, are the sister-group of Salmoniformes, Salmonoidei + Osmeroidei. See Fig. 21 for relationships within argentinoids. Figure 22 shows our attempt at analyzing relationships within alepocephaloids, but we lacked material and opportunity for a proper study of the group. (5) What are the relationships of and within the Osmeroidei? Osmeroids are the sister-group of Salmonoidei. See Fig. 19 for relationships within osmeroids. (6) What are the relationships of and within the Salmonidae? See (5), Sanford (1990), Stearley and Smith (1993) and Section IX above. (7) Where does *Lepidogalaxias* belong? It is a galaxioid, the sister of *Lovettia*.

Much of this paper, and much of our effort in working towards it, have been effectively wasted in criticizing Begle's (1991, 1992) work. The criticism was necessary only because the work is published, in respected journals and so will be regarded by other biologists as reliable. As we have shown, it is not. In the present context—a volume on the state of the art in fish systematics—we cannot leave the matter without a final comment. The disregard for truth we found in Begle's two papers is ultimately the responsibility of the author, but also indicates in this instance a failure of the system of checks and balances that maintains standards in our science. Surely, at some stage between Begle's writing his papers and their publication, some ichthyologist might have cast an eye over the work and noticed absurdities like crediting *Esox* with an adipose fin, an anteriorly placed dorsal fin, nuptial tubercles, an orbitosphenoid, and endopterygoid and maxillary teeth. To discover that *Esox* lacks some of these features would not need Edward Phelps Allis; a few words with an angler or someone who had glanced at a print of pike in a pub should be enough.

XIII. Summary

Review and reanalysis of published data, together with new characters, suggest the following pattern of relationships among lower euteleosts: [[[[Platytrocti-dae, Alepocephalidae] [Argentinidae [Opisthoprocti-dae [Microstomatinae, Bathylaginae]]]] [[Coregonidae [Thymallinae, Salmoninae]] [Osmeridae (including Salangidae, Plecoglossidae) [Retropinnidae [[*Lovettia, Lepidogalaxias*] [*Aplochiton*, Galaxiidae]]]]]] [Esoci-formes, Neoteleostei]]. That is, salmonoids and os-meroids are sister-groups; together they are the sister-group of alepocephaloids plus argentinoids; and esocoids are the sister-group of neoteleosts. We group salmonoids and osmeroids as Salmoniformes; alepo-cephaloids and argentinoids as Argentiniformes; Argentiniformes plus Salmoniformes as Protacan-thopterygii; Esociformes plus Neoteleostei as Neog-nathi; and Protacanthopterygii plus Neognathi as Euteleostei. Within Alepocephaloidei, we lacked ma-terial for a proper study of the group. However, platy-troctids are not a derived subgroup within Alep-ocephalidae but are the sister-group of all other alepocephaloids, and bathylaconids are the sister of alepocephalids. Monophyly of Alepocephaloidei is supported by branchiostegal cartilages and epipleural bones that extend unusually far forward. Within Ar-gentinoidei, bathylagids (including microstomatids) are the sister of opisthoproctids, and argentinids are the sister of those two combined. Monophyly of Ar-gentinoidei is supported by about a dozen characters, mostly related to the small mouth and reduced denti-tion. Monophyly of Argentiniformes is supported by the crumenal organ, ventral displacement of the first three or four epineural bones, and support of a single caudal fin ray by the caudal median cartilages. Salan-gids are osmerids, not galaxioids. Within Osmeridae, *Hypomesus* is the basal genus, *Spirinchus* and *Thaleich-thys* are derived, not primitive, and salangids are the sister group of *Mallotus*. Within Galaxioidei, *Lepidoga-laxias* is the sister of the Tasmanian *Lovettia*, and *Aplochiton* is the sister of galaxiids. The grouping of salmonoids and osmeroids (Salmoniformes), al-though not previously proposed, is among the most strongly supported in lower euteleosts, with 11 char-acters, including separate dermethmoid and su-praethmoid, absence of epipleural bones, features of the caudal skeleton, scales without radii, nuptial tu-bercles, and anadromy. Monophyly of Protacanthopt-erygii is supported by epicentral cartilages and lack of proximal forking in the intermuscular bones. We find no morphological evidence to support the current view that Esociformes (Esocidae and Umbridae) are the most primitive euteleosts; a sister-group relation-ship between esociforms and Neoteleostei is sup-ported by type 4 tooth attachment and an acellular skeleton. Monophyly of Euteleostei is supported by pattern of supraneural development and presence of a stegural and caudal median cartilages in the cau-dal skeleton.

Acknowledgments

For the gift or loan of specimens, we are grateful to Tim Berra, Bill Fink, Norma Feinberg, Ron Fritzsche, Howard Gill, Tony Gill, Karsten Hartel, Doug Markle, John Olney, Ted Pietsch, Dick Rosen-blatt, Darrell Siebert, Melanie Stiassny, and Rick Winterbottom. For advice and assistance, we thank Carole Baldwin, Ed Brothers, Tim Littlewood, Doug Markle, and Dick Rosenblatt. We are most grateful to Robert Williams, Edmonton, Alberta, for the gift of a copy of his 1987 Ph.D. thesis in July 1995; because our work was virtually finished by then, we regret not giving it the attention it deserves in our paper. For thoughtful and welcome criticism of a draft, we are indebted to Lynne Parenti and particularly to Randy Mooi. Our research was supported by the Smithsonian Scholarly Studies Program.

I (GDJ) began collaborating with Colin Patterson in 1991, and since that time I have spent hundreds of hours working next to him at the microscope, usually surrounded by precariously stacked boxes of glycerin. Many of those hours produced exciting discover-ies; even more were spent painfully struggling to find characters, confronting our previous errors, or wallowing in what seemed like hopeless homoplasy. We have smiled and often laughed through most of those hours and they remain memorable among all those I have spent looking at fishes over the past 25 years. Working with Colin, there is obviously much to admire and benefit from—his deep knowledge of the fishes and the literature, his facility with words on the page, the intellect, focus and seemingly boundless energy he brings to the work and, of course, an unwavering deter-mination to get everything right.

References

Ahlstrom, E. H., Moser, H. G., and Cohen, D. M. (1984). Argenti-noidei: Development and relationships. *In* "Ontogeny and Sys-tematics of Fishes" (H. G. Moser, W. J. Richards, D. M. Cohen, M. P. Fahay, A. W. Kendall, Jr., and S. L. Richardson, eds.), Spec. Publ. No. 1, pp. 155–169. American Society of Ichthyolo-gists and Herpetologists, Lawrence, KS.

Arratia, G. (1991). The caudal skeleton of Jurassic teleosts; a phylo-genetic analysis. *In* "Early Vertebrates and Related Problems in Evolutionary Biology" (M.-M. Chang, Y.-H. Liu, and G.-R. Zhang, eds.), pp. 249–340. Science Press, Beijing.

Arratia, G., and Schultze, H.-P. (1987). A new halecostome fish

(Actinopterygii, Osteichthyes) from the late Jurassic of Chile and its relationships. *Dakoterra* **3**, 1–13.

Arratia, G., and Schultze, H.-P. (1992). Reevaluation of the caudal skeleton of certain actinopterygian fishes: III. Salmonidae. Homologization of caudal skeleton structures. *J. Morphol.* **214**, 187–249.

Begle, D. P. (1991). Relationships of the osmeroid fishes and the use of reductive characters in phylogenetic analysis. *Syst. Zool.* **40**, 33–53.

Begle, D. P. (1992). Monophyly and relationships of the argentinoid fishes. *Copeia*, pp. 350–366.

Berg, L. S. (1940). Classification of fishes, both Recent and fossil. *Tr. Zool. Inst. Leningr.* **5**, 87–517.

Bernardi, G., D'Onofrio, G., Caccio, S., and Bernardi, G. (1993). Molecular phylogeny of bony fishes, based on the amino acid sequence of the growth hormone. *J. Mol. Evol.* **37**, 644–649.

Berra, T. M., Smith, J. F., and Morrison, J. D. (1982). Probable identification of the cucumber odor of the Australian grayling *Prototroctes maraena*. *Trans. Am. Fish. Soc.* **111**, 78–82.

Blackburn, M. (1950). The Tasmanian whitebait, *Lovettia seali* (Johnston), and the whitebait fishery. *Aust. J. Mar. Freshwater Res.* **1**, 155–198.

Blum, S. (1991). *Dastilbe* Jordan 1910. *In* "Santana Fossils: An Illustrated Atlas" (J. G. Maisey, ed.), pp. 274–283. TFH Publications, Neptune City, NJ.

Bridge, T. W. (1896). The mesial fins of ganoids and teleosts. *J. Linn. Soc. London, Zool.* **25**, 530–602.

Cavender, T. M. (1970). A comparison of coregonines and other salmonids with the earliest known teleostean fishes. *In* "Biology of Coregonid Fishes" (C. C. Lindsey and C. S. Woods, eds.), pp. 1–32. University of Manitoba Press, Winnipeg.

Chapman, W. McL. (1941a). The osteology and relationships of the isospondylous fish, *Plecoglossus altivelis* Temminck and Schlegel. *J. Morphol.* **68**, 425–455.

Chapman, W. McL. (1941b). The osteology and relationships of the osmerid fishes. *J. Morphol.* **69**, 279–301.

Chapman, W. McL. (1942a). The osteology and relationships of the Argentinidae, a family of oceanic fishes. *J. Wash. Acad. Sci.* **32**, 104–117.

Chapman, W. McL. (1942b). The osteology and relationships of the bathypelagic fish *Macropinna microstoma*, with notes on its visceral anatomy. *Ann. Mag. Nat. Hist.* [11] **9**, 272–304.

Chapman, W. McL. (1943). The osteology and relationships of the bathypelagic fishes of the genus *Bathylagus* Günther with notes on the systematic position of *Leuroglossus stilbius* Gilbert and *Therobromus callorhinus* Lucas. *J. Wash. Acad. Sci.* **33**, 147–160.

Chapman, W. McL. (1944). On the osteology and relationships of the South American fish, *Aplochiton zebra* Jenyns. *J. Morphol.* **75**, 149–165.

Chapman, W. McL. (1948). The osteology and relationships of the Microstomatidae, a family of oceanic fishes. *Proc. Calif. Acad. Sci.* [4] **26**, 1–22.

Chyung, M. K. (1961). "Illustrated Encyclopedia. The Fauna of Korea (2). Fishes." Ministry of Education, Seoul.

Cohen, D. M. (1958). A revision of the fishes of the subfamily Argentininae. *Bull. Fla. State Mus. Biol. Sci.* **3**, 93–172.

Cohen, D. M. (1964). Suborder Argentinoidea. *Mem. Sears. Found. Mar. Res.* **1**(4), 1–70.

Collette, B. B. (1977). Epidermal breeding tubercles and bony contact organs in fishes. *Symp. Zool. Soc. Lond.* **39**, 225–268.

Daget, J., and d'Aubenton, F. (1957). Développement et morphologie du crâne d'*Heterotis niloticus* Ehr. *Bull. Inst. Fr. Afr. Noire,* A **19**, 881–936.

Farris, J. S. (1988). "Hennig86, Version 1.5," Program and documentation. Farris, Port Jefferson, NY.

Fink, S. V., and Fink, W. L. (1981). Interrelationships of the ostariophysan fishes. *Zool. J. Linn. Soc.* **72**, 297–353.

Fink, W. L. (1981). Ontogeny and phylogeny of tooth attachment modes in actinopterygian fishes. *J. Morphol.* **167**, 167–184.

Fink, W. L. (1984a). Salmoniforms: Introduction. *In* "Ontogeny and Systematics of Fishes" (H. G. Moser, W. J. Richards, D. M. Cohen, M. P. Fahay, A. W. Kendall, Jr., and S. L. Richardson, eds.), Spec. Publ. No. 1, p. 139. American Society of Ichthyologists and Herpetologists, Lawrence, KS.

Fink, W. L. (1984b). Basal euteleosts: Relationships. *In* "Ontogeny and Systematics of Fishes" (H. G. Moser, W. J. Richards, D. M. Cohen, M. P. Fahay, A. W. Kendall, Jr., and S. L. Richardson, eds.), Spec. Publ. No. 1, pp. 202–206. American Society of Ichthyologists and Herpetologists, Lawrence, KS.

Fink, W. L. (1985). Phylogenetic interrelationships of the stomiid fishes (Teleostei: Stomiiformes). *Misc. Publ. Mus. Zool., Univ. Mich.* **171**, 1–127.

Fink, W. L., and Weitzman, S. H. (1982). Relationships of the stomiiform fishes (Teleostei), with a description of *Diplophos*. *Bull. Mus. Comp. Zool.* **150**, 31–93.

Forey, P. L. (1977). The osteology of *Notelops* Woodward, *Rhacolepis* Agassiz and *Pachyrhizodus* Dixon (Pisces: Teleostei). *Bull. Br. Mus. (Nat. Hist.), Geol.* **28**, 123–204.

Fujita, K. (1990). "The Caudal Skeleton of Teleostean Fishes." Tokai University Press, Tokyo.

Gayet, M. (1993). Relations phylogénétiques des Gonorhynchiformes (Ostariophysi). *Belg. J. Zool.* **123**, 165–192.

Gomon, M. F., Glover, J. C. M., and Kuiter, R. H. (1994). "The Fishes of Australia's South Coast." State Print, Adelaide.

Goody, P. C. (1969). *Sedenhorstia dayi* (Hay), a new elopoid from the Cenomanian of Hajula in the Lebanon. *Am. Mus. Novit.* **2358**, 1–23.

Gosline, W. A. (1960). Contributions toward a classification of modern isospondylous fishes. *Bull. Br. Mus. (Nat. Hist.), Zool.* **6**, 325–365.

Gosline, W. A. (1965). Teleostean phylogeny. *Copeia*, pp. 186–194.

Gosline, W. A. (1969). The morphology and systematic position of the alepocephaloid fishes. *Bull. Br. Mus. (Nat. Hist.), Zool.* **18**, 183–218.

Gottfried, M. D. (1989). Homology and terminology of higher teleost postcleithral elements. *Trans. San Diego Soc. Nat. Hist.* **21**, 283–290.

Grande, L. (1982). A revision of the fossil genus *Diplomystus*, with comments on the interrelationships of clupeomorph fishes. *Am. Mus. Novit.* **2728**, 1–34.

Grande, L. (1985). Recent and fossil clupeomorph fishes with materials for revision of the subgroups of clupeoids. *Bull. Am. Mus. Nat. Hist.* **181**, 231–372.

Grande, T. (1994). Phylogeny and paedomorphosis in an African family of freshwater fishes (Gonorynchiformes: Kneriidae). *Fieldiana, Zool.* [N.S.] **78**, 1–20.

Greenwood, P. H. (1968). The osteology and relationships of the Denticipitidae, a family of clupeomorph fishes. *Bull. Br. Mus. (Nat. Hist.), Zool.* **16**, 213–273.

Greenwood, P. H., and Rosen, D. E. (1971). Notes on the structure and relationships of the alepocephaloid fishes. *Am. Mus. Novit.* **2473**, 1–41.

Greenwood, P. H., Rosen, D. E., Weitzman, S. H., and Myers, G. S. (1966). Phyletic studies of teleostean fishes, with a provisional classification of living forms. *Bull. Am. Mus. Nat. Hist.* **131**, 339–456.

Hearne, M. E. (1984). Osmeridae: Development and relationships. *In* "Ontogeny and Systematics of Fishes" (H. G. Moser, W. J. Richards, D. M. Cohen, M. P. Fahay, A. W. Kendall, Jr., and

S. L. Richardson, eds.), Spec. Publ. No. 1, pp. 153–155. American Society of Ichthyologists and Herpetologists, Lawrence, KS.

Holstvoogd, C. (1965). The pharyngeal bones and muscles in Teleostei, a taxonomic study. *Proc. K. Ned. Akad. Wet. Ser. C* **68**, 209–218.

Howes, G. J., and Sanford, C. P. J. (1987a). Oral ontogeny of the ayu, *Plecoglossus altivelis* and comparisons with the jaws of other salmoniform fishes. *Zool. J. Linn. Soc.* **89**, 133–169.

Howes, G. J., and Sanford, C. P. J. (1987b). The phylogenetic position of the Plecoglossidae (Teleostei, Salmoniformes), with comments on the Osmeridae and Osmeroidei. *Proc. Congr. Eur. Ichthyol., 5th, Stockholm, 1985*, pp. 17–30.

Jamieson, B. G. M. (1991). "Fish Evolution and Systematics: Evidence from Spermatozoa." Cambridge University Press, Cambridge, UK.

Johnson, G. D. (1984). Percoidei: Development and relationships. *In* "Ontogeny and Systematics of Fishes" (H. G. Moser, W. J. Richards, D. M. Cohen, M. P. Fahay, A. W. Kendall, Jr., and S. L. Richardson, Jr., eds.), Spec. Publ. No. 1, pp. 464–498. American Society of Ichthyologists and Herpetologists, Lawrence, KS.

Johnson, G. D. (1992). Monophyly of the euteleostean clades—Neoteleostei, Eurypterygii, and Ctenosquamata. *Copeia*, pp. 8–25.

Johnson, G. D., and Patterson, C. (1993). Percomorph phylogeny: A survey and a new proposal. *Bull. Mar. Sci.* **52**, 554–626.

Johnson, G. D., Baldwin, C. C., Okiyama, M., and Tominaga, Y. (1996). Osteology and relationships of *Pseudotrichonotus altivelis* (Teleostei: Aulopiformes: Pseudotrichonotidae). *Ichthyol. Res.* **43**, 17–45.

Jollie, M. (1975). Development of the head skeleton and pectoral girdle in *Esox. J. Morphol.* **147**, 61–88.

Kindred, J. E. (1919). The skull of *Amiurus. Ill. Biol. Monogr.* **5**, 1–121.

Klyukanov, V. A. (1970). Classification of smelts (Osmeridae) with respect to peculiarities of skeleton structure in the genus *Thaleichthys. Zool. Zh.* **49**, 399–417 (in Russian).

Klyukanov, V. A. (1975). The systematic position of the Osmeridae in the order Salmoniformes. *J. Ichthyol.* **15**, 1–17.

Klyukanov, V. A. (1977). Origin, evolution and distribution of Osmeridae. *In* "Principles of the Classification and Phylogeny of Salmonoid Fishes" (O. A. Skarlato, ed.), pp. 13–27 (in Russian). Zool. Inst., Akad. Nauk SSSR, Leningrad.

Kobayasi, H. (1955). Comparative studies of the scales in Japanese freshwater fishes, with special reference to phylogeny and evolution. [IV] Particular lepidology of freshwater fishes. I. Suborder Isospondyli. *Jpn. J. Ichthyol.* **4**, 64–75.

Kobyliansky, S. H. (1986). Materials for a revision of the family Bathylagidae (Teleostei, Salmoniformes). *Tr. P.P. Shirshov Inst. Oceanol.* **121a**, 6–50 (in Russian).

Kobyliansky, S. H. (1990). Taxonomic status of microstomatid fishes and problems of classification of suborder Argentinoidei (Salmoniformes, Teleostei). *Tr. P.P. Shirshov Inst. Oceanol.* **125**, 148–177 (in Russian).

Kölliker, A. (1859). Ueber verschiedene Typen in der mikroskopischen Structur des Skelettes der Knochenfische. *Verh. Phys.-Med. Ges. Würzburg* **9**, 3–17.

Korovina, V. M. (1977). Embryological material showing phylogenetic relationship between Osmeridae and Plecoglossidae. *In* "Principles of the Classification and Phylogeny of Salmonoid Fishes" (O. A. Skarlato, ed.), pp. 5–13 (in Russian). Zool. Inst., Akad. Nauk SSSR, Leningrad.

Lê, H. L. V. (1991). Evolution de l'ARN ribosomique 28S: Utilisation pour l'étude de la phylogénie des vértébrés et de l'horloge moléculaire. Doctoral Thesis, Université de Paris-Sud.

Lê, H. L. V., Perasso, R., and Billard, R. (1989). Phylogénie molécu-
laire préliminaire des 'poissons' basée sur l'analyse de séquences d'ARN ribosomique 28S. *C. R. Seances Acad. Sci., Paris Ser. 3* **309**, 493–498.

Lê, H. L. V., Lecointre, G., and Perasso, R. (1993). A 28S rRNA-based phylogeny of the gnathostomes: First steps in the analysis of conflict and congruence with morphologically based cladograms. *Mol. Phylogenet. Evol.* **2**, 31–51.

Lenglet, G. (1974). Contribution à l'étude ostéologique des Kneriidae. *Ann. Soc. R. Zool. Belg.* **104**, 51–103.

Leviton, A. E., Gibbs, R. H., Jr., Heal, E., and Dawson, C. E. (1985). Standards in herpetology and ichthyology: Part I. Standard symbolic codes for institutional resource collections in herpetology and ichthyology. *Copeia*, pp. 802–832.

Littlewood, T., Patterson, C., and Smith, A. B. (1996). In preparation.

Mabee, P. M. (1988). Supraneural and predorsal bones in fishes: Development and homologies. *Copeia*, pp. 827–838.

Mackintosh, N. A. (1923). The chondrocranium of the teleostean fish *Sebastes marinus. Proc. Zool. Soc. London*, pp. 501–513.

Markle, D. F. (1976). Preliminary studies on the systematics of deep-sea Alepocephaloidea (Pisces: Salmoniformes). Unpublished Ph.D. Dissertation, College of William and Mary, Gloucester Point, VA.

Markle, D. F. (1978). Taxonomy and distribution of *Rouleina attrita* and *Rouleina maderensis* (Pisces: Alepocephalidae). *Fish. Bull.* **76**, 79–87.

Markle, D. F., and Krefft, G. (1985). A new species and review of *Bajacalifornia* (Pisces: Alepocephalidae) with comments on the hook jaw of *Narcetes stomias. Copeia*, pp. 345–356.

Markle, D. F., and Merrett, N. R. (1980). The abyssal alepocephalid, *Rinoctes nasutus* (Pisces: Salmoniformes), a redescription and an evaluation of its systematic position. *J. Zool.* **190**, 225–239.

Matsui, T., and Rosenblatt, R. H. (1987). Review of the deep-sea fish family Platytroctidae (Pisces: Salmoniformes). *Bull. Scripps Inst. Oceanogr.* **26**, 1–159.

Matsuoka, M., and Iwai, T. (1983). Adipose fin cartilage found in some teleostean fishes. *Jpn. J. Ichthyol.* **30**, 37–46.

Mattei, X. (1991). Spermatozoon ultrastructure and its systematic implications in fishes. *Can. J. Zool.* **69**, 3038–3055.

McAllister, D. E. (1963). A revision of the smelt family, Osmeridae. *Bull.—Natl. Mus. Can.* **191**, 1–53.

McAllister, D. E. (1966). Numerical taxonomy and the smelt family, Osmeridae. *Can. Field Nat.* **80**, 227–238.

McAllister, D. E. (1968). Evolution of branchiostegals and classification of teleostome fishes. *Bull.—Natl. Mus. Can.* **221**, 1–239.

McDowall, R. M. (1969). Relationships of galaxioid fishes with a further discussion of salmoniform classification. *Copeia*, pp. 796–824.

McDowall, R. M. (1970). The galaxiid fishes of New Zealand. *Bull. Mus. Comp. Zool.* **139**, 341–432.

McDowall, R. M. (1971). Fishes of the family Aplochitonidae. *J. R. Soc. N. Z.* **1**, 31–52.

McDowall, R. M. (1976). Fishes of the family Prototroctidae (Salmoniformes). *Aust. J. Mar. Freshwater Res.* **27**, 641–659.

McDowall, R. M. (1978). A new genus and species of galaxiid fish from Australia (Salmoniformes: Galaxiidae). *J. R. Soc. N. Z.* **8**, 115–124.

McDowall, R. M. (1979). Fishes of the family Retropinnidae (Pisces: Salmoniformes)—a taxonomic revision and synopsis. *J. R. Soc. N. Z.* **9**, 85–121.

McDowall, R. M. (1984). Southern hemisphere freshwater salmoniforms: Development and relationships. *In* "Ontogeny and Systematics of Fishes" (H. G. Moser, W. J. Richards, D. M. Cohen, M. P. Fahay, A. W. Kendall, Jr., and S. L. Richardson, eds.),

Spec. Publ. No. 1, pp. 150–153. American Society of Ichthyologists and Herpetologists, Lawrence, KS.

McDowall, R. M. (1988). "Diadromy in Fishes. Migrations between Freshwater and Marine Environments." Croom Helm, London.

McDowall, R. M. (1993). A recent marine ancestry for diadromous fishes? Sometimes yes, but mostly no! *Environ. Biol. Fishes* **37**, 329–335.

McDowall, R. M., and Frankenberg, R. S. (1981). The galaxiid fishes of Australia. *Rec. Aust. Mus.* **33**, 443–605.

McDowall, R. M., and Pusey, B. J. (1983). *Lepidogalaxias salamandroides* Mees—a redescription, with natural history notes. *Rec. West. Aust. Mus.* **11**, 11–23.

McDowall, R. M., Clark, B. M., Wright, G. J., and Northcote, T. G. (1993). Trans-2-cis-6-Nonadienal: The cause of cucumber odor in osmerid and retropinnid smelts. *Trans. Am. Fish. Soc.* **122**, 144–147.

Monod, T. (1968). Le complexe urophore des poissons téléostéens. *Mém. Inst. Fondam. Afr. Noire* **81**, 1–705.

Moore, J. A. (1993). The phylogeny of the Trachichthyiformes (Teleostei: Percomorpha). *Bull. Mar. Sci.* **52**, 114–136.

Moss, M. L. (1961). Studies of the acellular bone of teleost fish. I. Morphological and systematic variations. *Acta Anat.* **46**, 343–362.

Moss, M. L. (1965). Studies of the acellular bone of teleost fish. V. Histology and mineral homeostasis of fresh-water species. *Acta Anat.* **60**, 262–276.

Müller-Schmidt, A., Ganss, B., Gorr, T., and Hoffmann, W. (1993). Molecular analysis of ependymins from the cerebrospinal fluid of the orders Clupeiformes and Salmoniformes: No indication for the existence of an euteleost infradivision. *J. Mol. Evol.* **36**, 578–585.

Nelson, G. J. (1967). Epibranchial organs in lower teleostean fishes. *J. Zool.* **153**, 71–89.

Nelson, G. J. (1968a). Gill arches of teleostean fishes of the division Osteoglossomorpha. *J. Linn. Soc. London, Zool.* **47**, 261–277.

Nelson, G. J. (1968b). Gill arch structure in *Acanthodes*. In "Current Problems of Lower Vertebrate Phylogeny" (T. Ørvig, ed.), Nobel Symp. 4, pp. 129–143. Almqvist & Wiksell, Stockholm.

Nelson, G. J. (1969a). Gill arches and the phylogeny of fishes, with notes on the classification of vertebrates. *Bull. Am. Mus. Nat. Hist.* **141**, 475–552.

Nelson, G. J. (1969b). Infraorbital bones and their bearing on the phylogeny and geography of osteoglossomorph fishes. *Am. Mus. Novit.* **2394**, 1–37.

Nelson, G. J. (1970a). The hyobranchial apparatus of teleostean fishes of the families Engraulidae and Chirocentridae. *Am. Mus. Novit.* **2410**, 1–30.

Nelson, G. J. (1970b). Gill arches of some teleostean fishes of the families Salangidae and Argentinidae. *Jpn. J. Ichthyol.* **17**, 61–66.

Nelson, G. J. (1972). Cephalic sensory canals, pitlines, and the classification of esocoid fishes, with notes on galaxiids and other teleosts. *Am. Mus. Novit.* **2492**, 1–49.

Nelson, G. J. (1973). Relationships of clupeomorphs, with remarks on the structure of the lower jaw in fishes. In "Interrelationships of Fishes" (P. H. Greenwood, R. S. Miles, and C. Patterson, eds.), pp. 333–349. Academic Press, London.

Nelson, J. S. (1994). "Fishes of the World," 3rd ed. Wiley, New York.

Nielsen, J. G., and Larsen, V. (1968). Synopsis of the Bathylaconidae (Pisces, Isospondyli) with a new eastern Pacific species. *Galathea Rep.* **9**, 221–238.

Nielsen, J. G., and Larsen, V. (1970). Ergebnisse der Forschungsreisen des FFS "Walter Herwig" nach Südamerika. XIII. Notes on the Bathylaconidae (Pisces, Isospondyli) with a new species from the Atlantic Ocean. *Arch. Fischereiwiss.* **21**, 28–39.

Nixon, K. C. (1992). "Clados, Version 1.2," Program and documentation. Nixon, Trumansburg, NY.

Norden, C. R. (1961). Comparative osteology of representative salmonid fishes, with particular reference to the Grayling (*Thymallus arcticus*) and its phylogeny. *J. Fish. Res. Board Can.* **18**, 679–791.

Norman, J. R. (1926). The development of the chondrocranium of the eel (*Anguilla vulgaris*), with observations on the comparative morphology and development of the chondrocranium in bony fishes. *Philos. Trans. R. Soc. London, Ser. B* **214**, 369–464.

Nybelin, O. (1971). On the caudal skeleton in *Elops* with remarks on other teleostean fishes. *Acta R. Soc. Sci. Litt. Gothoburg. Zool.* **7**, 1–52.

Okiyama, M., ed. (1988). "An Atlas of the Early Stage Fishes in Japan." Tokai University Press, Tokyo.

Parenti, L. R. (1986). The phylogenetic significance of bone types in euteleost fishes. *Zool. J. Linn. Soc.* **87**, 37–51.

Parr, A. E. (1951). Preliminary revision of the Alepocephalidae, with introduction of a new family, Searsidae. *Am. Mus. Novit.* **1531**, 1–21.

Patterson, C. (1968). The caudal skeleton in Lower Liassic pholidophorid fishes. *Bull. Br. Mus. (Nat. Hist.), Geol.* **16**, 201–239.

Patterson, C. (1970). Two Upper Cretaceous salmoniform fishes from the Lebanon. *Bull. Br. Mus. (Nat. Hist.), Geol.* **19**, 205–296.

Patterson, C. (1973). Interrelationships of holosteans. In "Interrelationships of Fishes" (P. H. Greenwood, R. S. Miles, and C. Patterson, eds.), pp. 233–305. Academic Press, London.

Patterson, C. (1975). The braincase of pholidophorid and leptolepid fishes, with a review of the actinopterygian braincase. *Philos. Trans. R. Soc. London, Ser. B* **269**, 275–579.

Patterson, C. (1984a). *Chanoides*, a marine Eocene otophysan fish (Teleostei: Ostariophysi). *J. Vertebr. Paleontol.* **4**, 430–456.

Patterson, C. (1984b). Family Chanidae and other teleostean fishes as living fossils. In "Living Fossils" (N. Eldredge and S. M. Stanley, eds.), pp. 132–139. Springer-Verlag, New York.

Patterson, C. (1994). Bony fishes. In "Major Features of Vertebrate Evolution" (D. R. Prothero and R. M. Schoch, eds.), Short Courses Paleontol. No. 7, pp. 57–84. Paleontological Society, University of Tennessee, Knoxville.

Patterson, C., and Johnson, G. D. (1995). The intermuscular bones and ligaments of teleostean fishes. *Smithson. Contrib. Zool.* **559**, 1–85.

Patterson, C., and Rosen, D. E. (1977). Review of ichthyodectiform and other Mesozoic teleost fishes and the theory and practice of classifying fossils. *Bull. Am. Mus. Nat. Hist.* **158**, 81–172.

Patterson, C., and Rosen, D. E. (1989). The Paracanthopterygii revisited: Order and disorder. *Sci. Ser., Nat. Hist. Mus. L.A. Cty.* **32**, 5–36.

Phillips, R. B., Oakley, T. H., and Davis, E. L. (1995). Evidence supporting the paraphyly of *Hucho* (Salmonidae) based on ribosomal DNA restriction maps. *J. Fish Biol.* **47**, 956–961.

Poyato-Ariza, F. J. (1994). A new early Cretaceous gonorynchiform fish (Teleostei: Ostariophysi) from Las Hoyas (Cuenca, Spain). *Occas. Pap. Mus. Nat. Hist., Univ. Kans.* **164**, 1–37.

Reist, J. D. (1987). Comparative morphometry and phenetics of the genera of esocoid fishes (Salmoniformes). *Zool. J. Linn. Soc.* **89**, 275–294.

Ridewood, W. G. (1904). On the cranial osteology of the fishes of the families Elopidae and Albulidae, with remarks on the morphology of the skull in lower teleostean fishes generally. *Proc. Zool. Soc. London* pt. 2, 35–81.

Roberts, C. D. (1993). Comparative morphology of spined scales and their phylogenetic significance in the Teleostei. *Bull. Mar. Sci.* **52**, 60–113.

Roberts, T. R. (1984). Skeletal anatomy and classification of the neotenic Asian salmoniform superfamily Salangoidea (icefishes or noodlefishes). *Proc. Calif. Acad. Sci.* **43**, 179–220.

Rosen, D. E. (1962). Comments on the relationships of the north American cave fishes of the family Amblyopsidae. *Am. Mus. Novit.* **2109**, 1–35.

Rosen, D. E. (1973). Interrelationships of higher euteleostean fishes. *In* "Interrelationships of Fishes" (P. H. Greenwood, R. S. Miles, and C. Patterson, eds.), pp. 397–513. Academic Press, London.

Rosen, D. E. (1974). Phylogeny and zoogeography of salmoniform fishes and relationships of *Lepidogalaxias salamandroides*. *Bull. Am. Mus. Nat. Hist.* **153**, 265–326.

Rosen, D. E. (1985). An essay on euteleostean classification. *Am. Mus. Novit.* **2827**, 1–57.

Rosen, D. E., and Greenwood, P. H. (1970). Origins of the Weberian apparatus and the relationships of the ostariophysan and gonorynchiform fishes. *Am. Mus. Novit.* **2428**, 1–25.

Rosen, D. E., and Patterson, C. (1969). The structure and relationships of the paracanthopterygian fishes. *Bull. Am. Mus. Nat. Hist.* **141**, 357–474.

Rosen, D. E., and Patterson, C. (1990). On Müller's and Cuvier's concepts of pharyngognath and labyrinth fishes and the classification of percoid fishes, with an atlas of percomorph dorsal gill arches. *Am. Mus. Novit.* **2983**, 1–57.

Sanford, C. P. J. (1987). The phylogenetic relationships of the salmonoid fishes. Ph.D. Thesis, University of London.

Sanford, C. P. J. (1990). The phylogenetic relationships of salmonoid fishes. *Bull. Br. Mus. (Nat. Hist.), Zool.* **56**, 145–153.

Sazonov, Y. I. (1986). Morphology and classification of the fishes of the family Platytroctidae (Salmoniformes, Alepocephaloidei). *Tr. P.P. Shirshov Inst. Oceanol.* **121a**, 51–96 (in Russian).

Sazonov, Y. I. (1992). *Matsuichthys* gen. novum, a new genus of the fish family Platytroctidae (Salmoniformes) with notes on the classification of the subfamily Platytroctinae. *J. Ichthyol.* **32**, 4, 26–37.

Schultze, H.-P., and Arratia, G. (1988). Reevaluation of the caudal skeleton of some actinopterygian fishes: II. *Hiodon, Elops,* and *Albula. J. Morphol.* **195**, 257–303.

Shaposhnikova, G. K. (1968a). Comparative morphology of the whitefishes (Coregonidae) from the USSR. *Tr. Zool. Inst. Leningr.* **46**, 207–256 (in Russian).

Shaposhnikova, G. K. (1968b). A comparative morphological study of Taimen (*Hucho* Günther) and Lenok (*Brachmystax* Günther). *Probl. Ichthyol.* **8**, 351–370.

Shardo, J. D. (1995). Comparative embryology of teleostean fishes. I. Development and staging of the American Shad, *Alosa sapidissima. J. Morphol.* **225**, 125–167.

Siebert, D. (1996). In preparation.

Smith-Vaniz, W. F. (1984). Carangidae: Relationships. *In* "Ontogeny and Systematics of Fishes" (H. G. Moser, W. J. Richards, D. M. Cohen, M. P. Fahay, A. W. Kendall, Jr., and S. L. Richardson, eds.), Spec. Publ. No. 1, pp. 522–530. American Society of Ichthyologists and Herpetologists, Lawrence, KS.

Smitt, F. A. (1895). "A History of Scandinavian Fishes" (B. Fries, C. U. Ekström, and C. Sundervall), 2nd ed., Part 2, revised and completed by F. A. Smitt, pp. 567–1240. Norstedt, Stockholm.

Stearley, R. F., and Smith, G. R. (1993). Phylogeny of the Pacific trouts and salmons (*Oncorhynchus*) and genera of the family Salmonidae. *Trans. Am. Fish. Soc.* **122**, 1–33.

Stein, D. L., and Bond, C. E. (1985). Observations on the morphology, ecology, and behaviour of *Bathylychnops exilis* Cohen. *J. Fish Biol.* **27**, 215–228.

Taverne, L. (1989). *Crossognathus* Pictet, 1858 du Crétacé inférieur de l'Europe et systématique, paléozoogéographie et biologie des

Crossognathiformes nov. ord. (Téléostéens) du Crétacé et du Tertiaire. *Palaeontographica, Abt. A* **207**, 79–105.

Trewavas, E. (1933). On the structure of two oceanic fishes, *Cyema atrum* Günther and *Opisthoproctus soleatus* Vaillant. *Proc. Zool. Soc. London,* pp. 601–614.

Vladykov, V. D., Renaud, C. B., and Laframboise, S. (1985). Breeding tubercles in three species of *Gadus* (cods). *Can. J. Fish. Aquat. Sci.* **42**, 608–615.

Wakiya, Y., and Takahashi, N. (1937). Study of fishes of the family Salangidae. *J. Coll. Agric. Tokyo Imp. Univ.* **14**, 267–295.

Weitzman, S. H. (1967). The origin of the stomiatoid fishes with comments on the classification of salmoniform fishes. *Copeia,* pp. 507–540.

Weitzman, S. H. (1974). Osteology and evolutionary relationships of the Sternoptychidae, with a new classification of stomiatoid families. *Bull. Am. Mus. Nat. Hist.* **153**, 327–478.

Wiley, E. O. (1981). "Phylogenetics." Wiley, New York.

Wiley, M., and Collette, B. B. (1970). Breeding tubercles and contact organs in fishes: Their occurrence, structure, and significance. *Bull. Am. Mus. Nat. Hist.* **143**, 143–216.

Williams, R. R. G. (1987). The phylogenetic relationships of the Salmoniform fishes based on the suspensorium and its muscles. Unpublished Ph.D. Dissertation, University of Alberta, Edmonton.

Wilson, M. V. H. (1984). Osteology of the Paleocene teleost *Esox tiemani. Palaeontology* **27**, 597–608.

Wilson, M. V. H., and Veilleux, P. (1982). Comparative osteology and relationships of the Umbridae (Pisces: Salmoniformes). *Zool. J. Linn. Soc.* **76**, 321–353.

Wilson, M. V. H., and Williams, R. R. G. (1991). New Paleocene genus and species of smelt (Teleostei: Osmeridae) from freshwater deposits of the Paskapoo Formation, Alberta, Canada, and comments on osmerid phylogeny. *J. Vertebr. Paleontol.* **11**, 434–451.

Yarrell, W. (1938). On a new species of smelt from the Isle of Bute (*Osmerus hebredicus*). *Br. Assoc. Adv. Sci., Rep.* **2**, 108–109.

Material Examined

List of specimens examined, in systematic sequence. Institutional abbreviations follow Leviton *et al.* (1985). All specimens are cleared and stained except those with the suffix (d), = dried skeleton; or (f), = fossil; or (s), = spirit specimen.

Outgroups

Amia calva L., BMNH 1996.2.6.14-15

Pholidophorus bechei Agassiz, BMNH P.64021 (f)

P. germanicus Quenstedt, BMNH P.3704 (f)

P. macrocephalus Agassiz, BMNH P.52518 (f)

Tharsis dubius (Blainville), BMNH P.12070 (stomach contents of *Pholidophorus macrocephalus*) (f)

"Callovian *Leptolepis*" of Patterson (1975), BMNH P.64022 (f)

Hiodon alosoides (Rafinesque), AMNH 23754SW (2), BMNH 1980.7.7.6

H. tergisus Lesueur, USNM 167970 (2)

Pantodon buchholzi Peters, USNM 336676

Elops hawaiensis Regan, BMNH 1962.4.3.1

E. machnata Forsskål, BMNH 1962.8.28.1

E. saurus L., USNM 272928

Megalops atlanticus Cuvier and Valenciennes, USNM 132933

M. cyprinoides (Broussonet), BMNH 1855.9.19.832

Albula vulpes (L.), USNM 128509, USNM 128391 (4), USNM 128393 (3)

Halosaurus guentheri Goode and Bean, USNM 319535

Denticeps clupeoides Clausen, BMNH 1969.4.28.1

Chirocentrus dorab (Forsskål), BMNH 1966.11.16.5–6

Clupea harengus L., BMNH 1932.2.15.1, BMNH 1970.2.17.22 (2)

larval *Clupea harengus*, BMNH 1996.2.6.4.4-12

Etrumeus teres (De Kay), USNM 188950

Pellona flavipinnis (Valenciennes), USNM 229344

Chanos chanos (Forsskål), BMNH 1996.2.6.16-19

Salmonoids

Coregonidae

Coregonus cylindraceus (Pallas), BMNH 1963.1.7.10 (3)

C. lavaretus (L.), BMNH 1996.2.6.20-21

larval *C. clupeaformis* (Mitchill), ROM 68375 (10)

Stenodus leucichthys (Güldenstadt), BMNH 1985.7.16.22 (5)

Prosopium williamsoni (Girard), BMNH 1892.12.30.340

Salmonidae

Thymallus thymallus (L.), BMNH 1970.10.14.3, BMNH 1986.5.20.259 (4)

Brachymystax lenok (Pallas), BMNH 1974.8.6.1

Hucho hucho (L.), BMNH 1985.1.25.1 (2)

Oncorhynchus clarki (Richardson), BMNH 1957.2.20.3

O. kisutch (Walbaum), BMNH 1979.7.18.1

Salmo gairdneri Richardson, BMNH 1985.12.20.1

S. salar L., BMNH 1996.2.6.22-24

larval *S. salar*, BMNH 1932.11.13.3 (16)

S. trutta L., BMNH 1981.9.22.81, BMNH 1983.10.17.5 (3), BMNH 1996.2.6.25-26

Salvelinus alpinus (L.), BMNH 1957.9.20.1-3

S. fontinalis (Mitchill), USNM 272669

Osmeroids

Osmeridae

Allosmerus elongatus (Ayres), USNM 342050 (4), UMMZ 93883 (2)

Hypomesus olidus (Pallas), HSU 86-33 (3)

H. pretiosus (Girard), HSU 81-187 (3)

H. transpacificus McAllister, BMNH 1984.6.28.11 (2)

larval *Hypomesus* sp., USNM 340198 (2)

Mallotus villosus (Müller), AMNH 26286 (2), HSU 89-282 (2), BMNH 1970.11.17.27 (2), BMNH 1970.11.17.18, BMNH 1970.11.17.55-56, USNM 306413 (3), USNM 130301 (3)

Neosalanx brevirostris (Pellegrin), HSU 85-38 (3), UMMZ 180147

Osmerus sp., larvae, HSU uncat.

O. eperlanus (L.), BMNH 1971.2.16.303

O. mordax (Mitchill), HSU 85-46, BMNH 1984.11.29.11, BMNH 1984.11.2.5

Plecoglossus altivelis Temminck & Schlegel, BMNH 1984.12.6.12-16, HSU 93-059 (2)

Protosalanx chinensis (Basilewsky), HSU 85-38 (3)

Salangichthys microdon Bleeker, BMNH 1996.2.6.1-3

Salanx (Salanx) ariakensis (Kishinouye), UMMZ 180137

S. (Hemisalanx) prognathus (Regan), UMMZ 180152 (2)

Spirinchus lanceolatus (Hikita), USNM 085563 (s)

S. starksi (Fisk), USNM 342052 (4)

S. thaleichthys (Ayres), USNM 104689 (2), USNM 104690 (6), USNM 105639

Thaleichthys pacificus (Richardson), USNM 342051 (2), UMMZ 129011, USNM 188123

Retropinnidae

Retropinna retropinna (Richardson), MCZ 58015, BMNH 1964.4.30.19, BMNH uncat. (d)

Prototroctes maraena Günther, BMNH 1984.10.3.1, UMMZ 212764 (2)

Prototroctes oxyrhynchus Günther, BMNH 1873.12.13.69 (d)

Stokellia anisodon (Stokell), BMNH 1984.64.30.11 (2)

Galaxiidae

Aplochiton taeniatus Jenyns, HSU 81-192 (10)

Aplochiton zebra Jenyns, AMNH 31048, HSU 81-192 (2), MCZ 46272 (2), BMNH 1868.6.22.9 (d)

Lovettia sealii (Johnston), BMNH 1937.8.22.1 (4)

Lepidogalaxias salamandroides Mees, USNM 339265 (3), larvae, USNM 342027 (3), BMNH uncat. (4)

Galaxiidae indet. larvae, USNM 340197 (12)

Galaxias brevipinnis Günther, BMNH uncat. (d)

G. fasciatus Gray, BMNH 1965.12.16.1 (2), BMNH 1853.2.14.8 (d), BMNH 1843.3.7.4 (d)

G. fontanus Fulton, BMNH 1983.6.21.1 (2)

G. maculatus (Jenyns), BMNH 1971.11.15.35, BMNH 1894.4.13.60 (d), BMNH 1896.6.17.79 (d)

G. occidentalis Ogilby, AMNH 31478 (3)

G. paucispondylus Stokell, AMNH 30889SW (2)

G. platei Steindachner, BMNH 1894.4.13.50 (d)

G. vulgaris Stokell, USNM 203883 (10)

G. zebratus (Castelnau), BMNH 1975.12.29.544 (2)

Galaxiella munda McDowall, AMNH 48833

G. nigrostriata (Shipway), larvae, USNM 342026 (4)
Nesogalaxias neocaledonicus Weber & de Beaufort, AMNH 31036SW, USNM 203885
Paragalaxias dissimilis (Regan), BMNH 1976.8.13.1
Neochanna apoda Günther, AMNH 30135, BMNH 1872.1.23.5 (d)

Argentinoids
Argentinidae

Argentina georgei Cohen and Atsaides, USNM 187834
A. silus (Ascanius), BMNH 1967.3.5.2 (2)
A. sphyraena L., USNM 238015, BMNH 1970.2.17.87 (2)
A. striata Goode and Bean, USNM 272945 (2)
Glossanodon polli Cohen, USNM 203236 (3)
G. struhsakeri Cohen, USNM 36618

Opisthoproctidae

Bathylychnops exilis Cohen, OS 012209
Dolichopteryx longipes (Vaillant), SIO 51-85
Macropinna microstoma Chapman, USNM 220876
Opisthoproctus soleatus Vaillant, AMNH 29688SW, BMNH 1933.5.23.1, MCZ 61958, SIO uncat.
Rhynchohyalus natalensis (Gilchrist and von Bonde), AMNH 29689SW

Bathylagidae

Bathylagoides sp., USNM 234768 (2)
B. wesethi (Bolin), USNM 339262 (3)
Bathylagus bericoides (Borodin), USNM 199825 (2)
indet. bathylagid larva, USNM 332419
Leuroglossus stilbius Gilbert, USNM 327747 (10)
Pseudobathylagus milleri (Jordan and Gilbert), SIO 80-258
Microstoma microstoma (Risso), AMNH 291684SW, BMNH 1888.11.29.67 (d)
Nansenia oblita (Facciola), AMNH 29685SW
Nansenia sp., USNM 203439

Alepocephaloids
Platytroctidae

Holtbyrnia latifrons Sazonov, SIO 71-112
H. innesi (Fowler), USNM 326306 (s)
Mirorictus taningi Parr, SIO 82-85, SIO 66-20
Mentodus rostratus (Günther), USNM 215612
Paraholtbyrnia cyanocephala Krefft, SIO 77-53
Pellisolus eubranchus Matsui and Rosenblatt, SIO 60-287

Platytroctes apus Günther, USNM 201650
Sagamichthys abei Parr, SIO 66-488
Searsia koefoedi Parr, SIO 77-38, SIO 77-53
Searsioides multispinus Sazonov, SIO 77-21

Alepocephalidae

Alepocephalus agassizi Goode and Bean, USNM 215572
A. bairdii Goode and Bean, BMNH 1996.2.6.27
A. rostratus Risso, BMNH 1886.8.4.7 (d), BMNH uncat. (d)
A. tenebrosus Gilbert, SIO uncat.
Bajacalifornia burragei Townsend and Nichols, LACM 9714-19, SIO 69-489
Bathylaco nigricans Goode and Bean, SIO 91-19, SIO 64-15, USNM 206693
Bathytroctes (*Grimatroctes*) sp., USNM 339266
B. microlepis Günther, USNM 215493
Binghamichthys sp., USNM 339263
Leptochilichthys agassizi Garman, USNM 200518
Leptoderma macrops Vaillant, USNM 215604, USNM 215605 (2)
Narcetes stomias (Gilbert), VIMS 2120
Photostylus pycnopterus Beebe, USNM 215656
Rinoctes nasutus (Koefoed), USNM 268400, USNM 189010 (s), USNM 215517 (s)
Rouleina attrita (Vaillant), USNM 215480
R. maderensis Maul, USNM 215473
R. squamilaterata (Alcock), USNM 137752, USNM 307293
Talismania antillarum (Goode and Bean), USNM 215556
T. aphos (Bussing), SIO 72-144
T. oregoni Parr, USNM 304453
Xenodermichthys copei (Gill), USNM 215524

Esociformes
Esocidae

Esox lucius L., ROM 598CS (2)
E. americanus Gmelin, BMNH 1982.11.10.16 (3)
larval *E. americanus*, ROM 24422 (9)

Umbridae

Dallia pectoralis Bean, AMNH 38034SW, BMNH 1984.6.26.2
Novumbra hubbsi Schultz, AMNH 45019 (2), AMNH 30883SW (4)
Umbra krameri Walbaum, BMNH 1979.7.23.1 (2), USNM 205523
U. limi (Kirtland), USNM 179712 (15)
U. pygmaea (DeKay), BMNH 1966.10.14.5, BMNH 1996.2.6.28, AMNH 33406 (3)

Neoteleosts

Stomiiformes

Diplophos taenia Günther, USNM 206614
Triplophos hemingi (McArdle), USNM 199832
Pollichthys mauli (Poll), BMNH 1984.1.1.13

Aulopiformes

Aulopus filamentosus Cloquet, USNM 292105
A. japonicus Günther, AMNH 28635SW

Myctophiformes

Neoscopelus macrolepidotus Johnson, USNM 188056
 (2), USNM 317160
Notoscopelus resplendens (Richardson), AMNH 25928
Scopelengys tristis Alcock, AMNH 97466
Solivomer arenidens Miller, USNM 29507

Acanthomorpha

Polymixia lowei Günther, USNM 308378

Appendix 1

Matrix of 112 Morphological and Other Characters of Osmeroids, with Salmonoids and a Hypothetical Outgroup

Character

```
                                                                                         1111111111 1111
          1111111111222222222233333333334444444444555555555566666666667777777777888888888899999999990000000000 1111
          1234567890123456789012345678901234567890123456789012345678901234567890123456789012345678901234567890 123456789012

Outgroup       0000000000000000000000000000000000000000?0000000000000000000000000000000000000000000000000000000000000000000
Salmonoids     00000000000000200000102000000000000000000000100000100101000000000000000000000000?00000000000000000000000000000000
Hypomesus      00B010010001000000000001000000010000011000B0002010110011000010101000010120100000001B0
Plecoglossus   000010011001000000000010110000010000010000011000001100001000201011001101000001210100001110000000010
Mallotus       021010100010001010100101100000010002000011000011000011000101210010120001011100010101110000101
Salangidae     022?11??2?0120??0?20?220040000B011003002???1110??2110000111022001100010110101212210111011100022101110100011
Allosmerus     101010011110000120010012000110002010000B1100001100010012000100120101001100110001010000100000
Osmerus        1010100101100010010000101200000100010001100011000011000011000010120010100010010000000110
Spirinchus     1010100111100020010101200001100010100012010010101100011000010120001100001010B00000B0
Thaleichthys   10101001111000202011010120001100010100012011101010010030011100011111101100?0102011001110110
Retropinna     21100100010010000400101013013010000021011101010030101110000111111011110112110000201001110110
Prototroctes   211010100010010000400110130130100000210111101100311100001111011110112110000201002001110110
Stokellia      21101011000100010001000400111110130130100000210111101100311100011111100111101112110000201002011001110110
Aplochiton     20221011000012000102221103101110121130111002210021001100301101010111012121221100000000?2122000001000010
Lovettia       2001112???012100111020040011101210311111002211231111101030110101011121212121100100010B00?1212200100110011
Lepidogalaxias 200110100002000101022000011012203121000?711211110121003001000130111212122101001110110?0312200121101022
Galaxiidae     20011010000112000102211030101110102210301011101001300B101012B21101100013BB21101100012121121100001000010?121220000011B0B0B0
```

Note. "B" Indicates Polymorphism for States (0) and (1)

Characters

In the following list, an asterisk after a number indicates a multistate character treated as unordered in the parsimony analysis. Numbers in square brackets after character numbers refer to the numbered sequence in Section II.

1* [1]. Dermethmoid median (0), paired (1) or absent (2).

2* [2]. Ethmoid endoskeleton short with one or more perichondral ossifications anterior to the lateral ethmoids (0), short and unossified (1), or long and unossified (2).

3* [3]. Vomer with (0) or without (1) shaft, or vomer absent (2).

4 [4]. Vomer toothed (0) or toothless (1) [vomer absent = (?)].

5 [6]. Orbitosphenoid present (0) or absent (1).

6 [8]. Pterosphenoid present (0) or absent (1).

7 [8]. Pterosphenoid unmodified (0) or with extensive medial epiphysial arm (1).

8 [8]. Pterosphenoid with (1) or without (0) ventral process or flange from anterior half of ventral margin.

9 [8]. Pterosphenoid with (1) or without (0) posteroventral membrane–bone process towards anterodorsal process from prootic.

10 [9]. Contact (1) or lack of it (0) between anterodorsal process of prootic and pterosphenoid.

11 [11]. Cartilaginous interorbital septum present in anterior part of orbit (1) or absent (0).

12 [12]. Basisphenoid present (0) or absent (1).

13 [13]. Posterior myodome deep and extending into basioccipital (0), small and shallow (1), or absent (2).

14 [14]. Buccohypophysial canal in parasphenoid absent (0) or present (1).

15* [15]. Efferent pseudobranchial artery passing lateral to parasphenoid and sphenoid endoskeleton (0), through an earlike cartilage projecting beyond the parasphenoid (1), or though a notch or canal in the parasphenoid (2).

16 [16]. Otic bulla not inflated and with little or no cartilage in its wall (0), somewhat inflated (1), or globose (2).

17 [18]. Frontals with (0) or without (1) laminar lateral part roofing orbit.

18 [19]. Parietals in contact medially (0), or partially (1) or completely (2) separated by supraoccipital. [There is no supraoccipital in salangids, coded (?); *Lovettia* has parietals separate, but supraoccipital does not extend between them].

19 [19]. Parietals overlapped anteriorly by frontals, so that their exposed area is relatively small (0), or sutured with frontals and extending forwards to or beyond the postorbital process (1).

20 [22]. Fontanelles in cartilaginous roof of otic region remaining open (0) or closed during ontogeny (1).

21* [28]. Dermopalatine and autopalatines separate (0) or fused (1), or dermopalatine absent (2) [unossified autopalatine coded ?; *Plecoglossus* coded (0) despite Howes and Sanford's (1987a) report of fusion].

22* [29, 32]. Ectopterygoid present (0), fused with palatine (1), or absent (2).

23* [31]. Endopterygoid teeth concentrated along dorsal margin of bone, with a patch of teeth posteriorly (0), or in a single row (1) or absent (2).

24 [37]. Metapterygoid large (0), comparable in size to symplectic (1), or reduced, less than half as large as symplectic and not contacting hyomandibular or symplectic (2).

25 [38]. Anterior margin of metapterygoid above (0) or anterior (1) to quadrate.

26* [44, 45]. Hyomandibular with a vertically elongate lateral crest (0), a short vertical crest fitting against the propercular (1), a triangular spur (2), an obliquely orientated spurlike crest (3), or no preopercular crest (4).

27 [50]. Ascending process of premaxilla knoblike (0) or sharply triangular (1).

28 [54]. Maxilla and palatine with (1) or without (0) head-to-head articulation.

29 [56]. Maxilla toothed (0) or toothless (1).

30 [57]. Supramaxilla present (0) or absent (1).

31 [58]. Dentary with toothed margin occupying less than half of length of lower jaw (0) or more than half (1) [*Plecoglossus* scored as juvenile (Howes and Sanford, 1987a, fig. 4)].

32 [59]. Meckelian fossa small and anteriorly placed (0) or large and opening beneath the hind end of the dentary tooth row (1).

33 [62]. Anguloarticular with substantial endoskeletal (articular) component (0) or with articular component small (1).

34* [65, 66]. Supraorbital and antorbital present (0), supraorbital only (1), antorbital only (2), or neither (3). Galaxiids exhibit all four conditions and are coded (0).

35 [68]. Lachrymal with (1) or without (0) cartilage-covered condyle at its articulation with lateral ethmoid [lachrymal absent in *Lepidogalaxias*; coded (?)].

36 [69]. Branchiostegals extend forward to ossified ventral border of distal ceratohyal (0) or are restricted to cartilage-covered margin of the deep posterior part of the hyoid bar (1) (Fig. 4).

37* [76, 77]. Dorsal margin of opercular entire and unmodified (0), with an anterodorsal notch (1), with notch and a tongue-like process behind it (2), or not extending above articulation with hyomandibular (3).

38 [79]. Posterior margin of opercular and subopercular deeply incised (1) or not (0).

39* [80]. Basihyal with scattered teeth (0), marginal fangs (1), or toothless (2).

40 [81]. Basibranchial teeth present (0), toothless dermal plate (1), or dermal plate absent (2).

41 [91]. Uncinate process on first epibranchial present (0) or absent (1).

42 [92]. Uncinate process on second epibranchial present (0) or absent (1).

43 [94]. Uncinate process on fourth epibranchial absent (0) or present (1).

44 [95]. Distinct levator process on fourth epibranchial absent (0) or present (1).

45* [97]. Epibranchial 5 free (0) or fused to Eb4 at its lower end (1) or at both ends (2) [Eb5 absent in *Lepidogalaxias*; coded (?)].

46* [101]. Base of first pharyngobranchial articulating with anterior tip of Eb1 (0), with both Eb1 and Eb2 (1), or with lateral surface of Eb1 (2) [Pb1 absent in *Lepidogalaxias*; coded (?)].

47 [103]. Uncinate process on second pharyngobranchial present (0) or absent (1).

48 [106]. Uncinate process on third pharyngobranchial present (0) or absent (1).

49 [108]. Third pharyngobranchial with narrow anterior extension reaching Pb1 or tip of Pb2 (0), without anterior extension (2), or intermediate (1).

50* [109]. Fourth and fifth upper pharyngeal toothplates (UP4, UP5) distinct and separate (0), a single toothplate (1), two to four separate toothplates (bearing no relation to the original two) (2), or toothless (3).

51 [112]. Gill rakers toothed (0) or toothless (1).

52 [113]. Accessory neural arch (ANA) present (0) or absent (1).

53 [114]. Unossified gap between occipital condyle and first centrum absent (0) or present (1).

54 [115]. Baudelot's ligament originating on first vertebra (0) or on occiput (1) [in galaxiids the ligament may show state (1) or may originate on both structures; they are coded (1)].

55 [116]. Neural arches of some vertebrae anterior to dorsal fin autogenous (0) or all fused to centrum (1).

56* [118]. Epineural bones and/or ligaments originate on neural arch (0), on centrum on several anterior vertebrae (1), or absent (2).

57 [120]. Epineural bones present (0) or absent (1).

58 [121]. Cartilage rods in epicentral ligaments present (0) or absent (1).

59 [122]. Epipleural bones absent (0) or present (1).

60* [125]. Supraneurals develop in pattern 1 (0), pattern 2 (1), pattern 2A (2), or pattern 2B (3).

61 [126]. Supraneurals numerous, ca. 15 or more (0), fewer than ten (1), or one (2).

62 [128]. Median keels of laminar bone absent (0) or present (1) on distal parts of last few neural and haemal spines.

63 [129]. NPU2 leaflike and about half the length of NPU3 (0) or similar in form to NPU3 (1).

64 [130]. NPU1 leaflike (0) or rudimentary or absent (1).

65 [132]. U1 free (0) or fused to PU1 (1).

66 [133]. Parhypural and hypural 1 separate (0) or fused (1).

67 [133]. Hypurals six (0) or five (1).

68* [134]. Epurals three (0), two (1), one (2) or none (3) [polymorphisms coded as greatest number observed].

69 [136]. First uroneural free (0) or fused to PU1 (1).

70 [137]. Membranous outgrowth of Un1 present (0) or absent (1).

71 [139]. Second uroneural slender and posteroventral to first (0) or broad and lateral or dorsolateral to first (1).

72. [140]. Number of uroneurals three (0) or two (1).

73* [142]. Upper and lower caudal median cartilages (CMCs) present (0), a single cartilage (1), or absent (2).

74 [144]. Upper and lower caudal scutes present (0) or absent (1).

75 [145]. Principal caudal rays 10/9 (0), 9/9 (1), or 8/8 (2).

76 [146]. Urodermal present (0) or absent (1).

77* [148, 179]. Extrascapular single or double (0), several canal-bearing ossicles (1), or absent (2).

78 [149]. Posttemporal with lower (intercalar) limb ossified (0) or not (1).

79* [150]. Cleithrum with long-based triangular ventral process towards coracoid (0), with narrow columnar process (1), or with no process (2).

80* [151]. Number of postcleithra three (0), one (1), or zero (2).

81 [152]. Mesocoracoid present (0) or absent (1).

82 [155]. First pectoral radial unmodified (0), or enlarged and embracing scapula (1) (Fig. 12C, D).

83 [156]. Third pectoral radial unmodified (0) or tapering proximally and failing to reach scapulocoracoid (1).

84 [157]. Fourth pectoral radial articulating with glenoid (0) or tapering proximally and failing to articulate with glenoid (1).

85 [158]. Fourth pectoral radial single (0) or multifid distally (1).

86 [160]. Posterior pubic symphysis present (0) or absent (1).

87 [161]. Pelvic girdle with (1) or without (0) ventral condyle.

88 [162]. Articular surface for pelvic fin short and transverse (0) or elongate and oblique (1).

89 [163]. Shaft of pelvic girdle with (1) or without (0) laminar medial membrane–bone expansion.

90 [165]. Pelvic splint present (0) or absent (1).

91 [166]. Dorsal fin placed above or close behind pelvics (0), or above or close to anal (1).

92 [168]. Posterior dorsal pterygiophores confluent (1) or nor (0).

93 [169]. Adipose fin present (0) or absent (1).

94 [170]. Adipose cartilage present (1) or absent (0).

95 [170]. Adipose cartilage beanlike (0) or a transversely arched, fenestrate plate (1) [taxa lacking the cartilage coded (?)].

96 [171]. Scales present (0) or absent (1).

97* [175, 176]. Infraorbital sensory canal uninterrupted and unmodified (0), interrupted, with the anterior portion running posteroventrally to cross the preopercular canal (1), postorbital infraorbital bones and sensory canal absent (2), or all infraorbital bones and canal absent (3).

98 [178]. Temporal sensory canal present (0) or absent (1).

99* [180]. Posttemporal penetrated by lateral line (0), with a separate canal-bearing ossicle (1), or with no relation to sensory canal (2).

100* [181]. Supracleithrum penetrated by lateral line (0), with a separate canal-bearing ossicle (1), or with no relation to sensory canal (2).

101 [182]. Egg with (1) or without (0) adhesive membrane or filaments.

102 [183]. Both ovaries present in females (0) or left only (1).

103 [186]. Nuptial tubercles absent (0) or present (1).

104 [187]. Scales on anal fin base of mature males unmodified (0) or enlarged (1).

105* [187]. Anal fin skeleton unmodified in mature males (0), anterior anal endoskeleton and central fin-rays modified (1), or entire anal fin skeleton greatly modified (2).

106 [188]. Nasal lamellae in a rosette (0) or parallel and longitudinal (1).

107 [191]. Pyloric caeca present (0) or absent (1).

108 [194]. Horny midventral abdominal keel absent (0) or present (1).

109 [195]. Prominent adipose "caudal peduncle flanges" absent (0) or present (1).

110 [197]. Cucumber odor absent (0) or present (1).

111* [199]. Life cycle entirely marine (0), diadromous (anadromous or amphidromous) (1), or entirely freshwater (2).

112* [200]. Skeletal ontogeny unmodified (0), retarded relative to sexual maturity (1), or accelerated (2).

Appendix 2

Matrix of 50 Morphological Characters in Subgroups of Argentinoids, Alepocephaloids, and an Outgroup.

	Character
	1111111111222222222233333333334444444444 5
	12345678901234567890123456789012345678901234567890
Outgroup	00000000000000000000000000000000000000000000000000
Alepocephaloids	0010001000000000000000001000B000001B00000B11000001?
Argentina	00010000011110110000010111011001111B0101000000101
Glossanodon	00010000011110110000010101B101110001111B1101000000101
Bathylagids	1012?0101110111111111111111111111001111010000101?
Microstomatids	1001111011101111110111101211111111001111001111001
Opisthoproctids	2111101101102?11021111110001111110100010012111000B1

Note. "B" Indicates Polymorphism for States (0) and (1).

Characters

In the following list, an asterisk after a number indicates a multistate character treated as unordered in the parsimony analysis. Numbers in square brackets after character numbers refer to the numbered sequence in Section II. Bathylagids are coded from *Bathylagichthys* where information is available (Kobyliansky, 1986, 1990), microstomatids are coded from *Nansenia*, and opisthoproctids from *Bathylychnops*.

1* [2]. Mesethmoid fully ossified (0), with separate laminar dorsal and ventral ethmoids (Kobyliansky, 1990, fig. 10) (1), or unossified (2).

2 [6]. Orbitosphenoid present (0) or absent (1).

3 [19]. Parietals in contact medially (0) or separated by supraoccipital (1).

4* [20]. Parietal with no direct relation to occipital commissural sensory canal (0), or carrying canal (presumably through fusion with a medial extrascapular) (1), or occipital commissure absent (2).

5 [20]. Occipital commissural canal crosses posterior margin of parietal (0) or lies anteriorly on the bone (1).

6 [21]. Dermopterotic present (0) or absent (1).

7 [23]. Posttemporal fossa roofed (0) or open (1).

8 [26]. Occipital condyle formed only by basioccipital (0) or tripartite, with exoccipital condyles (1).

9 [29]. Palatine teeth in a patch (0) or a single row (1).

10 [31]. Endopterygoid teeth present (0) or absent (1).

11 [37]. Metapterygoid large (0) or reduced (1).

12 [42]. Ventral arm of symplectic equal in length to dorsal arm (0) or shorter (1).

13* [52]. Premaxilla toothed (0), toothless (1), or absent (2).

14 [53]. Premaxilla articulating with maxilla posteriorly (1) or not (0).

15 [56]. Maxilla toothed (0) or toothless (1).

16 [57]. Two supramaxillae (0) or none (1).

17 [58]. Dentary with scattered small teeth (0) or with a long, single row of bladelike teeth (1).

18* [66]. Supraorbital and dermosphenotic in contact (1) above orbit or not (0), or supraorbital absent (2).

19 [68]. Lachrymal (foremost infraorbital) larger (0) or smaller (1) than succeeding infraorbital.

20 [75]. Anterior branchiostegals carried on medial or ventral margin of ceratohyal (0) or all branchiostegals on external face of ceratohyal (1).

21 [80]. Basihyal teeth present (0) or absent (1).

22 [81]. Basibranchials 1–3 with (0) or without (1) teeth.

23 [88]. Cb5 toothed (0) or toothless (1).

24 [90]. Accessory cartilage at tip of Cb5 absent (0) or present (1).

25 [94]. Uncinate process on Eb4 absent (0) or present (1).

26 [95]. Eb4 with (1) or without (0) distinct levator process.

27 [100]. Pb1 ossified (0), cartilaginous (1), or absent (2).

28 [102]. Pb2 toothed (0) or toothless (1).

29 [105]. Pb3 toothed (0) or toothless (1).

30 [109]. UP4 present (0) or absent (1).

31 [109]. UP5 normal (0) or minute (1).

32 [111]. Gill rakers on fourth and fifth arches similar to those on third (0) or much longer (1).

33 [113]. ANA present (0) or absent (1).

34 [119]. First two to four epineurals unmodified (0) or descended distally (1).

35 [125]. Supraneurals develop in pattern 1 (0) or pattern 2 (1).

36 [127]. Anterior supraneurals rodlike (0) or rostrocaudally expanded (1).

37 [132]. PU1 and U1 separate (0) or fused (1).

38 [136]. Un1 fused to PU1 (1) or free (0).

39 [137]. Membranous outgrowth of Un1 present (0) or absent (1).

40 [141]. Caudal interneural and interhaemal cartilages three or fewer (0) or four or more (1).

41 [143]. CMCs each support a fin ray (0), or together support lowermost ray of upper caudal lobe (1).

42 [144]. Caudal scutes present (0) or absent (1).

43 [151]. Postcleithra three (0), one (1) or none (2).

44 [152]. Mesocoracoid present (0) or absent (1).

45 [159]. Pectoral fin develops small and late (0) or early and large (1).

46 [165]. Pelvic splint present (0) or absent (1).

47 [179]. Lateral extrascapular single, triradiate and lying behind pterotic (0) or represented by one or more tubular ossicles extending forward above pterotic (1).

48 [189]. Eye with (1) or without (0) a crescent of white tissue above the iris.

49 [190]. Swimbladder present (0) or absent (1) (absent in *Bathylychnops*; present in *Opisthoproctus*).

50 [190]. Swimbladder with (1) or without (1) microrete mirabilia [swimbladder absent coded (?)].

Appendix 3

Matrix of 59 Morphological Characters in Subgroups of Alepocephaloidea, Argentinidae, and a Hypothetical Outgroup

	Character
	111111111122222222223333333333444444444455555 5555
	12345678901234567890123456789012345678901234567890123456789
Outgroup	00000000000000000000000000000000000000000000000000000000000
Argentinidae	000100000011120020000130011111001000111001000000000000000000
Platytroctidae	00001BB1200B00000120002010B001112?00B100011B100000B10BBB01B
Alepocephalus	10011111101B100121201120101102111301111000112010000010111010
Bajacalifornia	000110011011000111 20B100101112?1010011000011200000010211010
Bathylaco	00001001001B01000110111100110101100000000110100200010000010
Bathyprion	11111??1?0110B01??2011?1???1???1??011100001?211200010???210
Bathytroctes	100110110011000001201120011101?10110111000111000000101 10010
Leptochilichthys	01111111?0111100212011201110 11?1?3001200001200010010?11010
Leptoderma	10011211211111112021112011111112?40113111?11211122112321110
Narcetes	000110010011000100201100100 1???10200100001111000000010210010
Photostylus	110112112111011120201141111111413?41101011?11210112112321?11
Rinoctes	00111011001100000120102 0?0001302021110000112111000123 11110
Rouleina	1001101111110B11B0211120100112?12?11110100112011011112 21011
Talismania	00011B01B01100000120112010000211B1011110001120100011021 0110

Note. "B" Indicates polymorphism for States (0) and (1).

Characters

In the following list, an asterisk after a number indicates a multistate character treated as unordered in the parsimony analysis. Numbers in square brackets after character numbers refer to the numbered sequence in Section II.

1 [4]. Vomerine teeth present (0) or absent (1).

2 [6]. Orbitosphenoid present (0) or absent (1).

3 [12]. Basisphenoid present (0) or absent (1).

4 [15]. Basipterygoid process present (0) or absent (1).

5 [19]. Parietals in contact (0) or separated by supraoccipital (1).

6* [21] Dermopterotic and autopterotic fused (0), separate (1), or dermal component absent (2).

7 [21]. Temporal sensory canal enclosed in pterotic (0) or not (1).

8 [23]. Posttemporal fossa extensive and roofed (0) or absent (1).

9* [23]. Posterior margins of parietal and dermopterotic continuous and transverse (0), forming a "V" open posteriorly (1), or no contact between the bones (2, including dermopterotic absent).

10 [29]. Palatine teeth present (0) or absent (1).

11 [31]. Endopterygoid teeth present (0) or absent (1).

12 [35]. Ectopterygoid teeth present (0) or absent (1).

13 [56]. Maxillary teeth present (0) or absent (1).

14* [57]. Supramaxillae two (0), one (1) or none (2).

15 [65]. Antorbital present (0) or absent (1) (information from Markle, 1976).

16 [66]. Supraorbital present (0) or absent (1) (information from Markle, 1976).

17* [70]. Ceratohyal with dorsal margin covered by cartilage (0), with cartilage cover interrupted by a notch (1), or covered by perichondral bone (2) (Fig. 16).

18 [74]. Branchiostegal cartilages absent (0) or present (1).

19* [76, 78]. Dorsal part of opercular normal (unreduced) (0), with large dilatator spine (1), or with small spine (2).

20. Subopercular normal (0) or dagger-like (1) (information from Markle, 1976).

21 [80]. Basihyal teeth present (0) or absent (1).

22 [81]. Basibranchial teeth present (0) or absent (1).

23* [88]. CB5 dentition a longitudinal band of teeth (0), a longitudinal single row (1), a single marginal row on a fanlike medial expansion (2), a cluster at the anterior end of the bone (3), or absent (4).

24 [90]. Accessory cartilage above Cb5 present (0) or absent (1).

25 [92]. Uncinate process on Eb2 present (0) or absent (1).

26 [98]. Suprapharyngobranchial 1 present (0) or absent (1).

27 [99]. Suprapharyngobranchial 2 present (0) or absent (1).

28 [102]. Pb2 toothed (0) or toothless (1).

29 [113]. ANA present (0) or absent (1).

30* [119]. Distal parts of epineurals displaced ventrally on no vertebrae (0), on first three or four (1), on first one or two (2), or on more than four (3), or epineurals absent (4).

31 [121]. Epicentral cartilages present (0) or absent (1).

32* [123]. Epipleural bones not extending forward beyond about V20 (0), extending forward to V2–4 (1), to about V9 (V6–11) (2), or absent (3).

33* [125]. Supraneurals develop in pattern 1 (0), pattern 2 (1) or pattern 4 (2) [taxa with one supraneural or none coded (?)].

34 [126]. Number of supraneurals more than 70% of number of predorsal vertebrae (0), 50–70% (1), 30–50% (2), or only one or two supraneurals (3), or none (4) [taxa with pattern 4 supraneurals coded (?)].

35. Number of abdominal vertebrae greater than number of caudal vertebrae by six or more (0), or approximately equal (± 5) (1) (Markle, 1976).

36. Laminar bone developed on last few neural and haemal arches (0) or not (1) [information from Markle and Merrett, 1980].

37 [129]. NPU2 about half as long as NPU3 (0) or as long as NPU3 (1).

38* [134]. Number of epurals three (0), two (1), one (2), or zero (3).

39 [137]. Membranous anterodorsal outgrowth on Un1 present (0) or absent (1).

40 [140]. Un3 present (0) or absent (1).

41 [142]. CMCs present (0) or absent (1).

42 [143]. Each CMC supports a fin ray (0) or both support lowermost ray of upper lobe (1) (Fig. 14) [CMCs absent coded (?)].

43 [146]. Urodermal present (0) or absent (1).

44 [150]. Cleithrum with (0) or without (1) ventromedial process meeting coracoid.

45 [151]. Postcleithra three (0), one (1), or none (2).

46 [152]. Mesocoracoid present (0) or absent (1).

47 [153]. Postcoracoid process short (0) or long (1).

48* [154]. Pectoral radials four (0), three (1), or two (2).

49 [164]. Pelvic radials three (0), two (1), or one (2).

50. Pelvic rays 8 or 9 (0), 7 (1), or 6 or fewer (2).

51 [165]. Pelvic splint present (0) or absent (1).

52 [169]. Adipose fin present (0) or absent (1).

53* [171]. Scales present (0), restricted to lateral line (1), or absent (2).

54* [179]. Extrascapular a substantial triradiate bone (0), a triradiate tubular ossicle (1), two or more tubular ossicles (2), or absent (3).

55* [180]. Posttemporal containing sensory canal (0), canal passes through separate ossicle or ossicles (1), or no ossicle (2).

56 [181]. Supracleithrum containing sensory canal (0) or not (1).

57* [184]. Ovary exposed to coelom laterally (0), ovary enclosed by fusion of ovarian tunic with peritoneum (1), or ovary enclosed by tunic but hanging free in coelom (2).

58 [190]. Swimbladder present (0) or absent (1).

59 [196]. Photophores absent (0) or present (1).

Appendix 4

Matrix of 42 Characters in Clupeomorphs, the Major Groups of Euteleosts, and a Hypothetical Outgroup

	Characters
	1111111111222222222233333333334444
	1234567890123456789012345678901234567890123456789012
Outgroup	0000000000000000000000000000000000000000000000
Clupeomorphs	0010100000000000001000010010000010000011000
Ostariophysi	0010201001000100100?0020100000000102000100
Esocoids	0100211111010110000001010011?1001121012012
Alepocephaloids	0102100000010000001201?0001000111111000000
Argentinoids	0010202001211110101201101001001100020001000
Salmonoids	1001201000000002010211110010101010021011011
Osmeroids	1001111101000?020102111101111010000321011111
Neoteleosts	00000000100001000000?1001110110000020?1010

Note. The characters are abstracted from the survey in Section II. Polymorphism within terminals is not coded. Instead, the primitive condition is entered if it occurs in basal members of a group. Queries are entered where the condition in basal members is questionable, as specified in the character descriptions, or where the character is inapplicable.

Characters

In the following list, an asterisk after a number indicates a multistate character treated as unordered in the parsimony analysis. Numbers in square brackets after character numbers refer to the numbered sequence in Section II.

1 [2]. Dermethmoid and supraethmoid fused (0) or separate (1).

2 [19]. Parietals in contact in midline (0) or separated by supraoccipital (1).

3 [20]. Parietals carrying the occipital commissural sensory canal (1) or not (0).

4* [23]. Posttemporal fossa roofed (0), open (1), or absent (2). [Clupeomorphs coded from Cretaceous ellimmichthyids.]

5* [31]. Endopterygoid teeth cover medial surface of bone (0), are restricted to a row along the medial margin (1), or absent (2). [Clupeomorphs coded from Cretaceous ellimmichthyids.]

6 [45]. Lateral crest of hyomandibular long (0) or short (1).

7* [57]. Two supramaxillae (0), one (1), or none (2).

8 [62]. Articular well-developed (0) or reduced (1).

9 [64]. Tooth attachment type 1–3 (0) or type 4 (1).

10 [70]. Dorsal margin of distal ceratohyal cartilaginous (0) or closed in perichondral bone (1).

11* [81]. Basibranchial toothplate covers basibranchials 1–3 (0), fragmented into two (1), or toothless (2).

12 [90]. Accessory cartilage between Cb5 and Eb5 absent (0) or present (1).

13 [94]. Uncinate process on Eb4 absent (0) or present (1). [Neoteleosts coded (?) because process is absent in stomiiforms].

14 [95]. Levator process on Eb4 absent (0) or present (1). [Osmeroids coded (?) because primitive condition might be (0), as in galaxioids, or (1) as in osmerids.]

15 [102]. Toothplate of Pb2 present (0) or absent (1).

16* [109]. UP4 and UP5 present (0), UP4 only (1), or UP5 only (2).

17 [113]. ANA present (0) or absent (1).

18 [117]. Epineural bones fused to neural arches on several (five or more) anterior neural arches (0) or fused on no more than one (1). [Ostariophysans coded from the Eocene *Chanoides*.]

19 [119]. Distal parts of first two to four epineurals in series with their successors (0) or descended (1).

20* [121]. Epicentrals ligamentous (0), ossified (1), or with cartilage rods distally (2). [Ostariophysans coded (?) because epicentral bones are present in gonorynchiforms, absent in otophysans.]

21 [122]. Epipleurals ossified (0) or ligamentous (1).

22 [124]. Epineurals and epipleurals (when present) forked proximally (0) or not (1). [Neoteleosts coded (?) because primitive state is questionable.]

23* [125]. Supraneurals develop in pattern 1 (0), pattern 2 (1), or pattern 3 (2). [Alepocephaloids exhibit patterns 1, 2, and 4 and are coded (?).]

24 [128]. Last few neural and haemal spines slender distally (0) or expanded rostrocaudally, forming a keel above and below the vertebrae (1).

25 [129]. NPU2 leaflike, about half as long as NPU3 (0), or as long as NPU3 (1).

26 [132]. PU1 and U1 separate (0) or fused (1). [Ostariophysans coded from Cretaceous gonorynchiforms.]

27 [137]. Membranous anterodorsal outgrowth of Un1 absent (0) or present (1).

28 [138]. Un1 extends forwards to PU2 (0) or only to PU1 (1). [Ostariophysans coded from Cretaceous gonorynchiforms.]

29 [139]. Un2 slender and posteroventral to Un1 (0) or broad and anterodorsal to Un1 (1). [Inapplicable in esocoids; coded (?).]

30 [140]. Un3 present (0) or absent (1).

31 [142]. CMCs absent (0) or present (1).

32 [143]. CMCs each support a finray (0) or together support the lowermost ray of the upper caudal lobe (1) [absence of CMCs coded (0)].

33 [144]. Caudal scutes present (0) or absent (1).

34 [146]. Urodermal present (0) or absent (1).

35* [151]. Postcleithra 1–3 present (0), only Pcl 1 (1), only Pcl 3 (2), or none (3).

36* [166, 169]. Adipose fin absent, dorsal fin over pelvics (0); adipose fin absent, dorsal fin posterior (1); or adipose fin present (2).

37 [172]. Scales with (0) or without (1) radii.

38 [173]. Cheek and operculum naked (0) or scaled (1). [Neoteleosts coded (?) because state (1) occurs in eurypterygians but not in stomiiforms.]

39* [74]. Supraorbital sensory canal without postorbital junction with infraorbital canal and with parietal branch (0), with both postorbital junction and parietal branch (1), or with postorbital junction and no parietal branch (2).

40 [186]. Nuptial tubercles absent (0) or present (1).

41 [192]. Skeleton cellular (0) or acellular (1).

42* [199]. Marine (0), diadromous (1), or freshwater (2). [Ostariophysans coded from primitive gonorynchiforms and the Eocene *Chanoides*.]

13

Interrelationships of Stomiiform Fishes

ANTONY S. HAROLD
Department of Ichthyology and Herpetology
Royal Ontario Museum
Toronto, Ontario, Canada M5S 2C6

STANLEY H. WEITZMAN
Division of Fishes
National Museum of Natural History
Smithsonian Institution
Washington, D.C. 20560

I. Introduction

The order Stomiiformes is a morphologically diverse group of deep-water oceanic fishes. They vary widely in body form, from elongate and eel-like as in some stomiids like *Idiacanthus,* to the highly compressed deep-bodied hatchetfishes such as *Sternoptyx.* Most stomiiforms are mesopelagic, living down to about 1000 meters, with some species below that level (e.g., *Bathophilus* and some *Cyclothone* species). However, in keeping with the great variety of body form and structures associated with feeding and luminescence, stomiiforms occupy many oceanic habitats. Some taxa (e.g., *Polyipnus* and *Stomias*) exhibit a much higher degree of endemism than would be expected for pelagic fishes.

Rosen (1973) separated the stomiiforms (= Stomiatoidei) from other protacanthopterygian groups, placing them as the sister-group of all other neoteleosts (Eurypterygii). Derived characters shared by stomiiforms and other neoteleosts include Type 4 tooth attachment and the retractor dorsalis muscle (see Johnson, 1992, p. 11).

Like many organisms existing at these near lightless levels, stomiiforms possess specialized organs, photophores, for the production of light. The structure of stomiiform photophores is unique and has been used, among other characters, to diagnose the order (Fink and Weitzman, 1982). The position and

patterns, the varied histological nature, and the developmental patterns of the photophores within the order have also been applied to systematic problems, especially by Weitzman (1974, phylogeny reproduced here as Fig. 1), who used them to diagnose the families Sternoptychidae, Gonostomatidae, and Photichthyidae.

Some of the differences in phylogenetic hypotheses of Ahlstrom (1974) and Weitzman (1974) did not result from character choices or anatomical interpretation but rather from disparate phylogenetic philosophies (see discussion by Ahlstrom *et al.,* 1984). Although there was some agreement in groupings (e.g., all sternoptychids have type Alpha photophores that increase in number by a budding process), other elements of Ahlstrom's (1974) scheme were not based on shared–derived characters and are not recognized here.

Among the most problematic genera are *Diplophos, Manducus,* and *Triplophos.* These taxa could not be placed by Ahlstrom *et al.* (1984) or Fink (1984) with the other gonostomatids (e.g., *Gonostoma* and *Bonapartia*) to which they were previously assigned. *Manducus* has in the past been treated as a synonym of *Diplophos* (e.g., Fink and Weitzman, 1982), but Weitzman (in Ahlstrom *et al.,* 1984) later proposed that such a generic concept is unlikely to be supported by any derived characters (but see Harold, in press). We follow the treatment of Ahlstrom *et al.* (1984) in recognizing

Copyright © 1996 by Academic Press, Inc.
All rights of reproduction in any form reserved.

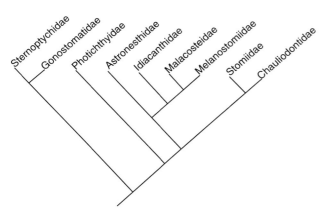

FIGURE 1 Cladogram showing interrelationships of stomiiform families, according to Weitzman (1974).

Diplophos and *Manducus*, pending further analysis. Whether the species of these two genera comprise a single genus or not, they appear to exhibit a greater number of plesiomorphies than any other stomiiform taxon. *Triplophos* was also discussed by Fink (1984) and Ahlstrom *et al.* (1984) and was thought possibly to be closer to the photichthyan taxa than to the gonostomatids or sternoptychids.

Until some of these problems are resolved a well-corroborated phylogeny and classification for the order will remain problematic. However, the majority of stomiiform species can be placed into one of the three families diagnosed by derived characters. The major problems remaining are *Diplophos*, *Manducus*, *Triplophos*, and the seven "photichthyid" genera (see below). In this paper we review the diagnoses and intrarelationships of the Sternoptychidae, Gonostomatidae, and Stomiidae (*sensu* Fink, 1985) and discuss a series of characters that bear on the phylogenetic relationships of the above problematic taxa. A parsimonious reconstruction is presented as an aid to examining the distributions of characters that are relevant to these taxa.

II. Materials and Methods

Phylogenetic analyses were performed using PAUP version 3.1.1 (Swofford, 1993) and Hennig86 version 1.5 (Farris, 1988), with all multistate characters run unordered. MacClade version 3.1 (Maddison and Maddison, 1992) was used to check character coding and to trace character evolution. A listing of material examined is given in Appendix 3.

Diplophos and *Manducus* usually have the plesiomorphic stomiiform condition, and in such cases polarity is unambiguous. Where outgroup conditions are inconsistent *Diplophos* and *Manducus* often behave as a functional outgroup to the other stomiiforms.

We provide two formal analyses here: (1) an assessment of Weitzman's (1974) sternoptychid phylogeny by combining the characters published in the 1974 paper with additional new characters into a single matrix (character state matrix and brief character descriptions are given in Appendix 1) and (2) an analysis of the interrelationships of *Diplophos*, *Manducus*, *Triplophos*, Sternoptychidae, Gonostomatidae, Stomiidae, and the "photichthyid" genera, separately, *Ichthyococcus*, *Photichthys*, *Pollichthys*, *Polymetme*, *Vinciguerria*, *Woodsia*, and *Yarrella* (character state matrix, brief character descriptions, and an apomorphy list are given in Appendix 2). In the list of derived characters in Appendix 2 an asterisk (*) indicates homoplastic states, and Roman numerals indicate equally parsimonious optimizations of the same character. Character numbers in the text refer to state 1 in the descriptions except for reversals and multistate characters in which cases the character number is followed by a state number in parentheses.

Terminology used for anatomical structures is based mainly on Weitzman (1974) and Fink and Weitzman (1982) for osteology, Weitzman (1986) and Harold (1994) for photophores, and Winterbottom (1974) for myology.

We list here all stomiiform families and included genera (* = not monophyletic based on evidence presented by Harold (in press); ** = not monophyletic based on evidence presented herein):

Gonostomatidae: *Bonapartia* Goode and Bean, 1896; *Cyclothone* Goode and Bean, 1883; *Gonostoma* Rafinesque, 1810*; *Margrethia* Jespersen and Tåning, 1919
"Photichthyidae"**: *Ichthyococcus* Bonaparte, 1840; *Photichthys* Hutton, 1872; *Pollichthys* Grey, 1959; *Polymetme* McCulloch, 1926; *Vinciguerria* Jordan and Evermann, 1896; *Woodsia* Grey, 1959; *Yarrella* Goode and Bean, 1896
Sternoptychidae: *Araiophos* Grey, 1961; *Argyripnus* Gilbert and Cramer, 1897; *Argyropelecus* Cocco, 1829; *Danaphos* Bruun, 1931; *Maurolicus* Cocco, 1838; *Polyipnus* Günther, 1887; *Sonoda* Grey, 1959; *Sternoptyx* Hermann, 1781; *Thorophos* Bruun, 1931; *Valenciennellus* Jordan and Evermann, 1896
Stomiidae: *Aristomias* Zugmayer, 1913; *Astronesthes* Richardson, 1845; *Bathophilus* Giglioli, 1882; *Borostomias* Regan, 1908; *Chauliodus* Bloch and Schneider, 1801; *Chirostomias* Regan and Trewavas, 1930; *Echiostoma* Lowe, 1843; *Eustomias* Vaillant, 1888; *Flagellostomias* Parr, 1927; *Grammatostomias* Goode and Bean, 1896; *Heterophotus* Regan and

Trewavas, 1929; *Idiacanthus* Peters, 1877; *Leptostomias* Gilbert, 1905; *Malacosteus* Ayres, 1848; *Melanostomias* Brauer, 1902; *Neonesthes* Regan and Trewavas, 1929; *Odontostomias* Norman, 1930; *Opostomias* Günther, 1887; *Pachystomias* Günther, 1887; *Photonectes* Günther, 1887; *Photostomias* Collett, 1889; *Rhadinesthes* Regan and Trewavas, 1929; *Stomias* Cuvier, 1816; *Tactostoma* Bolin, 1939; *Thysanactis* Regan and Trewavas, 1929; *Trigonolampa* Regan and Trewavas, 1930

Stomiiformes (unplaced): *Diplophos* Günther, 1873; *Manducus* Goode and Bean, 1896; *Triplophos* Brauer, 1902

III. Basal Branching Order of Major Stomiiform Clades

A. Monophyly of Stomiiformes

Rosen (1973) indicated that stomiiforms are united by the presence of a modified alignment of the second and third pharyngobranchials and a specialized double articular surface of the second epibranchial. Diagnosis of the order was later extended by Fink and Weitzman (1982) who proposed eight shared–derived characters: (a) unique photophore structure, (b) Type 3 tooth attachment, (c) sudivision of the medial section of the adductor mandibulae (A1β of Rosen, 1973) into two sections, one inserting dorsally directly onto the maxilla and the other onto the primordial ligament, (d) unique ethmoid–contralateral premaxilla ligament-crossing pattern, (e) a single, broad termination of the second epibranchial articulating with the second and third pharyngobranchials (after Rosen, 1973, p. 441), (f) posterior branchiostegal rays greatly enlarged, (g) some branchiostegals articulating with the ventral hypohyals, and (h) rete mirabila located at the posterior of the gas bladder (suggested by Marshall, 1960).

Fink and Weitzman's (1982) characters appear to be diagnostic of the order (we have not surveyed [h]), with one possible exception. The broad dorsal termination of the second epibranchial described in (e) above is very similar to the condition in myctophiforms (compare Paxton, 1972, fig. 9 with Fink and Weitzman, 1982, fig. 11; also see Baldwin and Johnson, this volume; Stiassny, this volume), casting some doubt on its validity as a stomiiform synapomorphy.

In their characterization of the Stomiiformes Lauder and Liem (1983, p. 144) suggested that the retractor dorsalis muscle may be specialized beyond its derived neoteleostean configuration. However, in terms of diagnosing the order, the latter character adds little support since modification of the retractor insertion is probably part of the same functional complex as are the pharyngobranchial–epibranchial modifications described by Rosen (1973).

B. Relationships of *Diplophos* and *Manducus*

Diplophos, *Manducus*, and *Triplophos* (Fig. 2) are elongate forms traditionally placed in the Gonostomatidae. A major advance in our understanding of stomiiform intrarelationships was the proposal by Fink and Weitzman (1982) and Fink (1984) that *Diplophos* (*sensu lato*, containing *Manducus* and implied monophyly of a group containing the species of both genera) was the likely sister-group of all other members of the order (Fig. 3). This hypothesis was based primarily on the occurrence of a modified, edentate basihyal in most stomiiforms, which in *Diplophos* consists of a large, toothed basihyal bone, such as is present in outgroup taxa (e.g., some aulopiforms, osmeroids, and salmonoids). Hence *Diplophos* and *Manducus* are together regarded as basal, and characterstates exhibited by these genera that are shared with outgroups are clearly plesiomorphic for Stomiiformes.

Ahlstrom *et al.* (1984) and Smith *et al.* (1991) did not support a broad *Diplophos* concept and recognized the genus *Manducus*. In the former work it was suggested, on the basis of modifications of the pectoral-fin radials, that *D. taenia* and *D. rebainsi* were possibly more closely related to the photichthyans than were *M. maderensis* and *M. greyi*. The condition in *Diplophos* (*sensu stricto*) in which four bony radials are present with the basal cartilages of the third and fourth fused is not clearly homologous with the situation in photichthyans which lack any vestige of the fourth radial. Harold (in press) reported that *D. rebainsi*, unlike *D. taenia* and the two *Manducus* species, shares with all other stomiiforms a derived, fully-developed neural spine on the second preural centrum. A reduced neural spine of the second preural centrum (NPU2) (as suggested by Ahlstrom *et al.*, 1984) may also be derived for the clade.

Monophyly of a clade comprising all stomiiforms minus *Diplophos* and *Manducus* (Fig. 4, clade B) was not examined until recently (Harold, in press). The existence of such a clade is supported by such characters as the following: vertically oriented extensor proprius muscle originating on connective tissue associated with pleural ribs (character 153); a full neural spine associated with the second preural centrum (168) (possibly convergent with a similar condition in *Diplophos rebainsi*); first pharyngobranchial ventrally

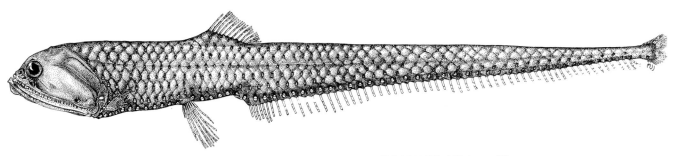

FIGURE 2   *Triplophos hemingi* (McArdle), USNM 203384, 147.4 mm SL.

bifurcate (181); and basihyal vertically oriented, cylindrical, and edentate (183), although a similar arrangement occurs in some aulopiforms (see Baldwin and Johnson, this volume).

IV. Infraorders Gonostomata and Photichthya

A. Monophyly of Gonostomata

The relationships of the generalized, non-barbeled stomiiforms were considered in detail by Weitzman (1974) who proposed a division of the order into two infraorders (Fig. 1), the Gonostomata (Sternoptychidae and Gonostomatidae) and the Photichthya (Photichthyidae, Astronesthidae, Idiacanthidae, Malacosteidae, Melanostomiidae, Stomiidae, and Chauliodontidae). These clades were based mainly on number of pectoral fin radials: four in Gonostomata and three in Photichthya. The presence of four radials at this level is evidently plesiomorphic, but in spite of this problem in the diagnosis of Gonostomata, the group as an entity appears to survive if *Diplophos*, *Manducus*, and *Triplophos* are excluded (Fig. 3; Fink, 1984; Ahlstrom *et al.*, 1984; Harold, in press). Some characters have been discussed pertaining to the relationships of the monotypic genus *Triplophos*, in partic-

ular its possible affinity to the Photichthya (Fink, 1984; Ahlstrom *et al.*, 1984; Harold, in press). Although *Triplophos* is not a photichthyan we discuss its relationships in the context of that group with which it shares certain derived characters.

A restricted Gonostomata, consisting of the Sternoptychidae and the four surviving gonostomatid genera (*Bonapartia*, *Cyclothone*, *Gonostoma*, and *Margrethia*), was proposed as a monophyletic group by Ahlstrom *et al.* (1984) based on the presence of a protracted metamorphosis, especially that of the photophores. Monophyly of the group was corroborated by Harold (in press) by the presence of an A2 section of the adductor mandibulae which is divided into separate dorsal and ventral muscles (character 151, Appendix 2).

B. Monophyly of Photichthya

The infraorder Photichthya was established by Weitzman (1974) for a numerically and morphologically diverse group of barbeled stomiiforms, the Stomiidae (*sensu* Fink, 1985) and seven genera of more generalized forms, the Photichthyidae (Weitzman,

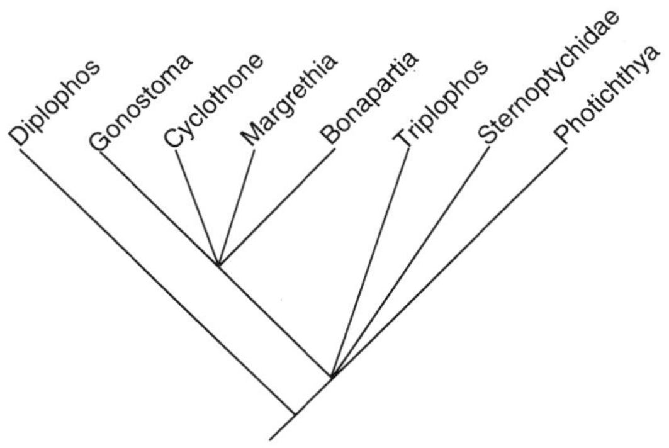

FIGURE 3   Cladogram showing interrelationships of stomiiform subgroups, according to Fink (1984).

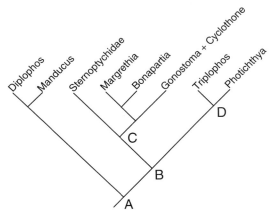

FIGURE 4   Cladogram showing interrelationships of stomiiform subgroups hypothesized by Harold (in press), with a possible resolution of *Triplophos* as discussed herein.

1974), previously ascribed to the Gonostomatidae (e.g., Grey, 1964). Two derived features were proposed as synapomorphies of all photichthyans: presence of three pectoral-fin radials (character 161) and presence of type Gamma photophores (167). Despite the occurrence of reversals to four radials in some stomiids, the number of these elements appears to be a valid character at this level. Fink (1985, p. 74, char. 265) argued that the presence of four pectoral radials in the stomiids *Chirostomias*, *Heterophotus*, and the clade comprising *Leptostomias* and *Thysanactis* was probably atavistic. Surveys of photophore histology are needed for further evaluation of the second character, our present knowledge being based, for the most part, on the work of Bassot (1966), which included only a small subset of stomiiform genera. Observations for many photichthyans, in particular, are needed.

V. Sternoptychidae

A. Monophyly of Sternoptychidae

The Sternoptychidae as recognized here was first proposed and diagnosed by Weitzman (1974, pp. 446–448). We have added more characters (for a total of 150) to our analysis of the Sternoptychidae but still agree with Weitzman's assessment regarding the monophyly of the family and most of his suggested generic phylogeny (Weitzman, 1974, fig. 113). We found *Thorophos* to be the sister-genus to the remaining sternoptychid genera (e.g., *Sternoptyx*) rather than the sister-genus to *Araiophos* as hypothesized by Weitzman (1974).

We used three outgroup taxa in our analysis of the family; *Diplophos*, *Manducus*, and the Gonostomatidae (*sensu stricto*). The analysis resulted in a singlemost parsimonious tree of length 264 (CI=0.66, RI=0.72) (Fig. 5A; see Appendix 1 for character state descriptions). An additional tree, at one step longer, was found (Fig. 5B). Since there is little stability in nodes C, D, and E we have used both trees in character optimization. The difference between these trees, discussed below, concerns the hypothesized relationships of *Maurolicus* and its phylogenetic relationships among other sternoptychid genera. The relationships of *Maurolicus* were also not fully resolved by Weitzman (1974). Because of space constraints characters are generally referred to only by their number in Appendix 1 in the sections below that document support for monophyly of the terminals and suprageneric clades depicted in Figure 5.

Clade A; Sternoptychidae: We found 22 unequivocal synapomorphies diagnosing the family in each of

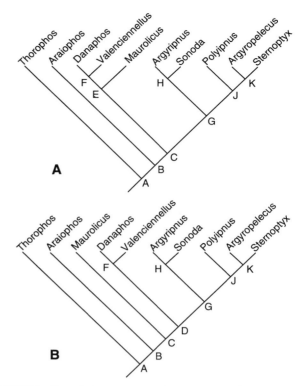

FIGURE 5 Two cladograms proposing interrelationships of sternoptychid genera, as analyzed herein. (A) the most parsimonious solution with *Maurolicus* as the sister-genus of *Danaphos* and *Valenciennellus* and (B) one transition step longer with *Maurolicus* as the sister-genus of *Danaphos* through *Sternoptyx*. Tree statistics: (A) TL=264, CI=0.66, RI=0.72; (B) TL=265, CI=0.66, RI=0.72.

the two trees (see Fig. 5). Among unique, unreversed characters diagnosing the Sternoptychidae are parietals separated by supraoccipital (character 12) and photophores occurring in clusters as a result of development by budding (118) (Table 1). Another 10 synapomorphies that show no homoplastic conditions are characters 38, 51, 66 (with three states), 79, 86 (with three states), 87, 88, 89 (with two states), 108, and 119. An additional 10 unambiguous synapomorphies, 39, 40, 47, 65, 78, 81, 94, 107, 109, and 144 corroborate the family's monophyly, but each has one or more homoplasies.

B. Interrelationships of Sternoptychid Genera

In our two cladograms *Thorophos* is hypothesized as the sister-group to the remaining sternoptychid genera (*Araiophos* through *Sternoptyx* in Figs. 5A and 5B) and *Maurolicus* is hypothesized as having two alternative relationships, one in each of the two trees.

Thorophos: This is the sister-genus to all other sternoptychid genera and is diagnosed by the following

TABLE 1 Occurrence and Meristic Values of Serial Photophores in Stomiiform Genera

	OA	IP	IV	VAV	AC	Sources[a]
Stomiiformes						
Diplophos	60–77	P	31–51	13–17	32–51	3, 15, 20, 27, 32
Manducus	45–48	P	29–33	12–14	28–39	3, 15, 32
Triplophos	50–56	P	24–30	5–7	35–41	3, 15, 32
Gonostomatidae						
Bonapartia	0	A	14–16	5–6	18–21	3, 32
Cyclothone[b]	6–10	A	12–14	4–5	12–16	3, 32
Gonostoma	11–21	A	11–16	3–10	15–23	1, 3, 15, 17, 25, 32
Margrethia	0	A	13–15	4	17	3, 32
"Photichthyidae"						
Ichthyococcus	23–29 or 32–35	P	25–28	9–14	12–14	3, 31
Photichthys	33–34	P	10 + 14–15 = 24–25	15–17	16–18	3
Pollichthys	18–21	P	21–23	7–9	18–21	3
Polymetme	17–18	P	19–21	7–8	21–25	3, 29, 31
Vinciguerria	21–25	P	21–24	7–11	12–15	3, 31
Woodsia	31–33	P	25	11–12 or 14–15	12–14	3, 19, 31
Yarrella	52–53	P	23–25	9–12	20–28	3, 31
Sternoptychidae						
Araiophos	0	P	(2) + (3) + 3–4 + (2) = 10–11	(3–5)	6–8	1, 2, 3, 34
Argyripnus	(6–7)	P	(6) + (10)	(18–32)[c]	(4–5) + (12–18) = 35–51	3, 28
Argyropelecus	7	P	(6) + (12)	(4)	(6) + (4)	3, 4, 16
Danaphos	6	P	(3) + (4) + 11 = 18	(5)	22–26	1, 3, 15, 34
Maurolicus	8–10 or (2) + 7	P	(6) + (11–13) = 18–19	(6–7)	1 + (12–18) + (6–10) = 19–29	1, 3, 30
Polyipnus	6 + 1[d]	P	(6) + (10)	(5)	0 or 3 + (4–14) + (4) = 8–21	3, 4, 16
Sonoda	7	P	(6) + (10)	(7–8)	(16–21) + (19–24) or (5–6) + (5–6) + (5–6) = 36–43 or 15–18	3
Sternoptyx	(3)	P	(5) + (10)	(3)	1 + (3) + (4)	3, 4, 33, 34
Thorophos	1 or (2) + 5	P	17	(5)	13–15	3, 34
Valenciennellus	(2) + 3	P	(3) + (4) + (16–17) = 23–24	(4–5)	3–6 or 9–17	1, 3, 15

six unambiguous synapomorphies for its two species: characters 77, 91, 106, 112, 113, and 136. All features listed above that diagnose *Thorophos* show one or a few homoplastic occurrences among other sternoptychid genera and/or generic groups.

Clade B; *Araiophos, Maurolicus, Danaphos, Valenciennellus, Argyripnus, Sonoda, Polyipnus, Argyropelecus,* and *Sternoptyx*: This new resolution, with *Araiophos* as the sister-group of all other sternoptychids with the exception of *Thorophos*, is the only major topological difference from Weitzman's (1974) hypothesis. Eight unequivocal synapomorphies support monophyly of clade B in the cladogram that indicates *Maurolicus* is the sister-genus of the genera *Danaphos* and *Valenciennellus* (Fig. 5A): no contact between parietal and pterotic (character 10), symplectic with both ends equal in size (43 [1]), interopercle much longer than subopercle (56), posterior portion of infraorbital series absent

(134), and 26, 37 (1), 97 (0), and 110. Seven characters (110 is equivocal in this tree) support clade B in the cladogram that indicates that *Maurolicus* is the sister-taxon to all remaining sternoptychids (Fig. 5B).

Araiophos: This genus contains one known species, *Araiophos eastropas* Ahlstrom and Moser, a diminutive pelagic fish showing many paedomorphic features. Small body size and reduced ossification of certain skeletal elements were not utilized as characters but can be inferred *a posteriori* to be derived. According to our analysis the genus can be diagnosed by the following unequivocal synapomorphies: character 25 (2), 89 (2), 104 (3), 133 (1) (also shared with *Thorophos nexilis*, but not with *T. euryops*), and 135 (1) (shared with *Danaphos*, and the three deep-bodied sternoptychids *Polyipnus, Argyropelecus,* and *Sternoptyx*).

Clade C; *Maurolicus* plus all remaining genera: Weitzman (1974, fig. 113) was uncertain of the posi-

TABLE 1 (*continued*)

	OA	IP	IV	VAV	AC	Sources[a]
Stomiidae						
Aristomias	28–38	P	8 + 14–17 or 19 = 22–25 or 27	15–18	9–12	5, 18, 23
Astronesthes	11–50	P	4–12 + 5–23 = 9–35	7–28	6–15	6, 10, 13, 14, 18
Bathophilus	18–27 or 33	P	4–6 + 12–18 = 16–24	11–13 or 17	5–9	11, 18, 24
Borostomias	37–54	P	10–13 + 20–31 = 30–44	15–25	9–15	6, 10, 18
Chauliodus	39–50	P	8–11 + 17–23 = 25–34	22–30	8–13	9, 18, 21
Chirostomias	39 or 42–45	P	8–9 + 25–28 = 33–37	16 or 19–20	9–10	18, 24
Echiostoma	37–49	P	10 + 24–28 = 34–38	14–19	9–13	11, 18, 24
Eustomias	36–59	P	7–9 + 24–36 = 31–45	11–21	15–25	11, 18, 24
Flagellostomias	42–49	P	8–10 + 31–34 = 39–44	14–16	15–18	11, 18, 24
Grammatostomias	34–40	P	6–7 + 15–18 = 23–25	19–22	10–13	18, 24
Heterophotus	49–56	P	10–11 + 32–35 = 42–46	13–14	12–15	18
Idiacanthus	52–61	P	31–36	15–18	13–18	7, 18
Leptostomias	59–72	P	10–11 + 39–48 = 49–59	20–24	11–14	11, 18, 24
Malacosteus	7–15	P	IC 12–22			5, 12, 18, 23
Melanostomias	33–43	P	10–11 + 23–30 = 33–41	12–16	8–11	11, 18, 24
Neonesthes	26–36	P	9–12 + 14–17 = 23–29	16–21	13–18	6, 18
Odontostomias	44–50	P	44–47	13–15	12–13	24, 26
Opostomias	44–53	P	8–9 + 27–28 = 35–37	24–26	13–16	11, 18
Pachystomias	29–34	P	8–9 + 14–18 = 22–27	13–14	8–9	5, 11, 18, 24
Photonectes	28–41 or 41–53	P	8–11 + 19–24 or 34–38 = 42–49 or 53–62	11–18	9–13	11, 18, 24
Photostomias	32–39	P	7 + 13–16 = 20–23	21–25	12–15	5, 18, 23
Rhadinesthes	49–51	P	6 or 10 + 25–26 = 31–32 or 35–36	20–23	16	6, 18, 24
Stomias	36–67 or 137–153	P	9–13 + 32–51 or 80–86 = 41–64 or 89–99	5–16 or 58–67	14–22	8, 18, 22
Tactostoma	43 + 18	P	8 + 46	19	12	18
Thysanactis	44–48	P	20 + 31–32 = 51–52	14–16	11–12	18, 24
Trigonolampa	43–53	P	10–12 + 21–26 = 31–38	20–24	10–14	11, 18, 24

Note. Photophore terminology defined in Weitzman (1986). A, absent; P, present; parentheses indicate photophores in common organ or cluster, formed by budding process.

[a]Ahlstrom, 1974 (1); Ahlstrom and Moser, 1969 (2); Ahlstrom *et al.*, 1984 (3); Baird, 1971 (4); Fink, 1985 (5); Gibbs, 1964a (6), Gibbs 1964b (7); Gibbs, 1986a (8), Gibbs, 1986b (9), Gibbs, 1986c (10), Gibbs 1986d (11); Goodyear and Gibbs, 1986 (12); Gibbs and McKinney, 1988 (13); Gibbs and Weitzman, 1965 (14); Grey, 1964 (15); Harold, 1994 (16); Harold (in press) (17); Kawaguchi and Moser, 1984 (18); Krefft, 1973 (19); Krefft and Parin, 1972 (20); Morrow, 1964a (21), Morrow, 1964b (22), Morrow 1964c (23); Morrow and Gibbs, 1964 (24); Mukhacheva, 1972 (25); Norman, 1930 (26); Ozawa *et al.*, 1990 (27); Parin, 1992 (28); Parin and Borodulina, 1990 (29); Parin and Kobyliansky, 1993 (30); Schaefer *et al.*, 1986a (31), Schaefer *et al.*, 1986b (32); Schultz, 1964 (33); Weitzman, 1974 (34).

[b]Photophores absent in *Cyclothone obscura* Brauer, 1902.

[c]Photophores listed as VAV are equivalent to VAV + ACA.

[d]L photophore of *Polyipnus* species listed as posterior OA.

tion of *Maurolicus* in his cladogram but placed it questionably as the sister-genus to the remaining genera from *Danaphos* through *Sternoptyx* (our clade D, Fig. 5B). The clade is supported in our analysis by the following six unequivocal synapomorphies: characters 45, 46, 77 (also occurs in *Thorophos*), 104 (2), 121, and 122.

Maurolicus: One of our present cladograms (Fig. 5B) places *Maurolicus* as the sister-genus to the remaining sternoptychid genera from *Danaphos* through *Sternoptyx* while the other, the slightly more parsimonious solution (Fig. 5A), places it as the sister-genus to *Da-*

naphos and *Valenciennellus*; see alternate clade E below. Unequivocal synapomorphies for *Maurolicus* supported by both sets of relationships are as follows: characters 4 (also found in *Argyripnus* and *Sonoda*), 71 (also found in *Polyipnus*), 80 (also found in *Polyipnus*), 86 (2), 93 (2), 101 (also found in *Sternoptyx*), and 111 (also found in clade J; *Polyipnus*, *Argyropelecus*, and *Sternoptyx*).

Clade D (Fig. 5B only); *Danaphos*, *Valenciennellus*, *Argyripnus*, *Sonoda*, *Polyipnus*, *Argyropelecus*, and *Sternoptyx*: This clade is supported by one of the cladograms (Fig. 5B), and in terms of composition it is equivalent to clade C of Fig. 5A without *Maurolicus*. There are five putative synapomorphies for clade D: characters 75, 84, 116, 120 (with two derived states), and 144.

Clade E (Fig. 5A only); *Maurolicus* plus *Danaphos* and *Valenciennellus*: This clade can be diagnosed by the following unequivocal synapomorphies: 9, 22 (also found in *Thorophos*), 35, 41 (2), 70 (also found in *Polyipnus*), and 96 (1).

Clade F; *Danaphos* and *Valenciennellus*: This clade is recognized in both of our cladograms; one in which it is the sister-group to *Maurolicus* (Fig. 5A) and the other in which it is the sister-group to clade G (*Argyripnus* through *Sternoptyx*) (Fig. 5B). In both cases it is diagnosed by the following synapomorphies: 1, 3, 5, 7, 21, 42, 48 (1), 58, 82, 90, 94 (0), 97 (0), 110 (0), 116, and 145. The presence of a ligament between the interopercle and the epihyal, character 48 (1), is similar to the condition in ctenosquamates (Stiassny, this volume; character 5) but is unlikely, based on parsimony, to be homologous. An additional, homoplastic character, 83 (1), is optimized as an unequivocal synapomorphy for *Danaphos* and *Valenciennellus* only in the cladogram in which *Maurolicus* is hypothesized to be their sister-group (Fig. 5A).

Danaphos: *Danaphos* is diagnosed by the following unequivocal, albeit homoplastic, synapomorphies: character 17 (also found in *Argyropelecus*), 101 (also found in *Sternoptyx*), and 135 (also found in *Araiophos* and clade J).

Valenciennellus: *Valenciennellus* is diagnosed by the following unequivocal synapomorphies: character 23 (also found in *Sonoda* and clade J), 89 (2) (also found in *Araiophos* and *Sternoptyx*), and 123 (also found in *Thorophos, Araiophos* and *Sonoda*).

Clade G; *Argryipnus, Sonoda, Polyipnus, Argyropelecus*, and *Sternoptyx*: This clade is diagnosed by the following unequivocal synapomorphies: 85, 112 (also occurs in *Thorophos*), 124, 139 (lost in *Sternoptyx*), and 150 (lost in *Argyropelecus*).

Clade H; *Argyripnus* plus *Sonoda*: The following unequivocal synapomorphies diagnose this clade: characters 4 (also shared by *Maurolicus*), 18 (1), 37 (2), 57 (1), 95, 115, 125, 126, and 131.

Argyripnus: This genus can be diagnosed by the following unequivocal synapomorphies: characters 83 (2) and 86 (3) (but also found in *Polyipnus* and *Argyropelecus*).

Sonoda: This genus can be diagnosed by the following synapomorphies: characters 23 (also found in *Valenciennellus, Polyipnus*, and *Argyropelecus*), 24, 26 (2), 123 (also found in *Thorophos, Araiophos*, and *Valenciennellus*), and 145 (also present in *Danaphos* and *Valenciennellus*).

Clade J; *Polyipnus, Argyropelecus*, and *Sternoptyx*:

Monophyly of this clade of deep-bodied sternoptychids was recently corroborated by Harold (1993) and is here diagnosed by the following 26 unequivocal synapomorphies: 6, 13, 20, 26 (3), 30, 43 (2), 50, 52, 53 (1), 54, 55 (1), 59, 60, 66 (2), 69 (1), 74, 76, 96 (2), 98, 99, 102, 127–130, and 138. An additional seven, homoplastic characters optimize as synapomorphies of clade J: 40 (0), 65 (0), 78 (0), 105, 111, 135, and 136.

Polyipnus: We corroborate Harold's (1994) conclusion that *Polyipnus* is monophyletic. Characters 25 (1), 93 (1), 137, and 142 are uniquely derived and diagnose this genus. Five other characters with various homoplasies add further support to monophyly of *Polyipnus*: 70, 71, 80 (1), 91, 104 (1), and 144 (0).

Clade K; *Argyropelecus* plus *Sternoptyx*: The following 18 nonhomoplastic synapomorphies diagnose clade K: characters 14, 29, 31, 41 (1), 49, 57 (2), 61, 77 (1), 103, 114 (2), 117, 137, 140, 141, and 146–149. Seven other, homoplastic characters further corroborate monophyly of clade K: characters 2, 8, 39 (0), 44 (0), 47 (0), 94 (0), and 113.

Argyropelecus: According to Harold (1993) this genus is monophyletic. Among the characters we have analyzed here we have found three that are homoplastic which corroborate monophyly: characters 17, 24, and 115.

Sternoptyx: This genus of very deep-bodied sternoptychids can be diagnosed by the following 13 consistent characters: 11, 15, 18, 32, 34, 48 (2), 53 (2), 55 (2), 66 (3), 69 (2), 72, 92, and 120. Six homoplastic characters were optimized as derived for *Sternoptyx*, further supporting a hypothesis of monophyly: characters 56 (0), 63, 69 (2), 89 (2), 101, and 139 (0).

VI. Gonostomatidae

A. Monophyly of Gonostomatidae

The family Gonostomatidae was restricted to the genera *Bonapartia, Cyclothone, Gonostoma*, and *Margrethia* (Fig. 6) by Fink (1984) and Ahlstrom *et al.* (1984). Characters were not described by these authors but a list of synapomorphies is provided by Harold (in press). These include (a) loss of IP photophores (see Table 1), (b) specialized jaw dentition, (c) a short premaxilla, and (d) PB2 configuration.

B. Intrarelationships of the Gonostomatidae

Fink's (1984) cladogram shows four gonostomatid genera as a clade but their relationships are unresolved. Harold (in press) provided a list of derived characters relevant to relationships within this re-

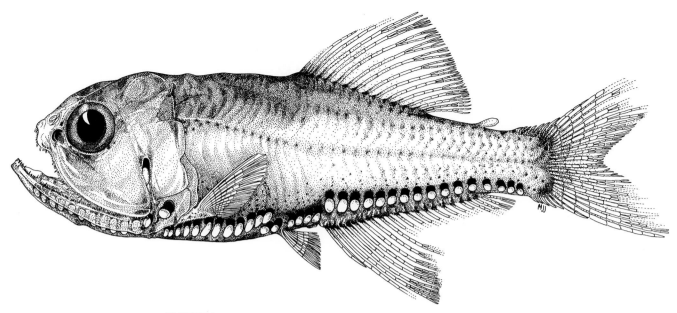

FIGURE 6 *Margrethia obtusirostra* Jespersen and Tåning, USNM 203295.

stricted Gonostomatidae (Fig. 7). *Margrethia* and *Bonapartia* are sister-groups in the figured set of relationships, a hypothesis supported by a laterally-displaced A2α muscle insertion and elongate anterior rays of the anal fin.

The *Bonapartia–Margrethia* clade is the sister-group of the remaining gonostomatids, presently ascribed to the genera *Gonostoma* and *Cyclothone*. Placement of *Cyclothone* has for some time been problematic, owing mainly to the reductive states of much of its morphology. Ahlstrom *et al.* (1984) argued that much of the confusing morphology arose through paedomorphosis (especially the presence of the generalized, rapid photophore development which otherwise characterizes *Diplophos*, *Manducus*, *Triplophos*, and the Photich-

thya) and that *Cyclothone* is a gonostomatid based on other, derived features.

Gonostoma and *Cyclothone* are most closely related, but it appears that some *Gonostoma* species, especially *G. bathyphilum* (Fig. 7), likely share a more recent common ancestor with the species of *Cyclothone* than do other members of *Gonostoma* (Harold, in press). Evidence for monophyly of the *Gonostoma–Cyclothone* clade consists, for example, of (a) lack of an adductor arcus palatini muscle (W. L. Fink, unpublished), (b) elongate vertebral centra, (c) division of the ventral section of the A2 muscle found in other gonostomatans (i.e., A2β) into two separate muscles so that A2 is represented by a total of three separate muscles, and (d) an elongate, anterodorsally flexed dorsal ramus of the hyomandibula. In this reconstruction *Cyclothone* is a clade nested within *Gonostoma* which renders the latter genus paraphyletic. Further corroboration of this set of phylogenetic relationships is needed, however, before taking the action of synonymizing the genus *Cyclothone* with *Gonostoma*.

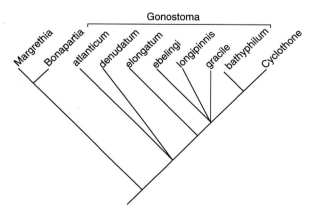

FIGURE 7 Cladogram showing interrelationships of gonostomatids, reproduced from Harold (in press).

VII. Photichthya

A. *Relationships of "Photichthyid" Genera and the Stomiidae*

Within the infraorder Photichthya are two family-level groups, the "Photichthyidae" (lightfishes) and the Stomiidae (barbeled stomiiforms). Stomiids are distinctive members of this group sharing the derived

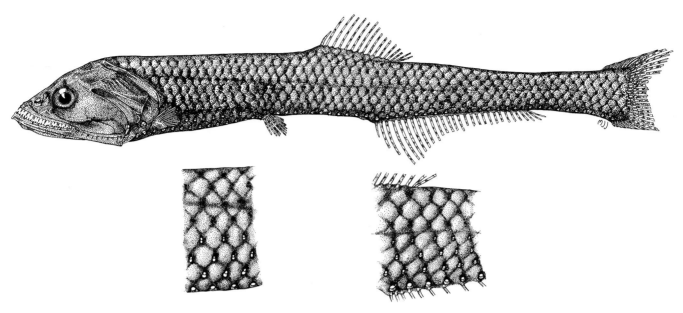

FIGURE 8 *Yarrella blackfordi* Goode and Bean, USNM 186280, 149 mm SL.

luminous chin barbel (Fink, 1984, 1985), among other features. On the other hand, the seven genera of the "Photichthyidae" appear to share no uniquely derived characters.

"Photichthyids" have moderately elongate, compressed body morphology and two roughly parallel, ventrolateral rows of photophores (OA and IV) (Figs. 8–11). *Ichthyococcus* has many uniquely derived characters, including highly reduced premaxillae and fusion of the maxilla and the anterior supramaxilla into a beak-like structure, but these characters serve only to diagnose that genus. In habitat "photichthyids" vary from mesopelagic (e.g., *Vinciguerria*) to benthopelagic (*Polymetme*).

Weitzman (1974) tentatively diagnosed the superfamily Photichthyoidea (containing the single family Photichthyidae) based on the presence of serial photophores with a duct and lumen (character 201). The proposed photophore character was thought to be derived and therefore available as evidence of monophyly. Fink's (1985) studies of "photichthyids" suggested that the family was probably not monophyletic. Without an alternative the family continues to be recognized in its original formulation (Eschmeyer, 1990; Nelson, 1994).

The polarity of the supporting photophore character is ambiguous since somewhat similar structure is present in the type Beta photophores of gonostomatids (see Herring and Morin, 1978, pp. 306–307, fig. 9.7). The lack of a duct or lumen in stomiid type Gamma photophores which was thought to be a primitive feature of that group by Weitzman (1974, p. 472) may alternatively be a synapomorphy.

The examination of "photichthyids" by Fink (1984, 1985) as stomiid outgroups revealed that *Woodsia* and *Ichthyococcus* may share derived characters with the stomiids that are not present in the other members of their family. We have extended Fink's examination of these taxa by including other characters and the remaining stomiiform taxa.

Our most parsimonious solution (Fig. 12) supports the notion that the "photichthyids" do not constitute

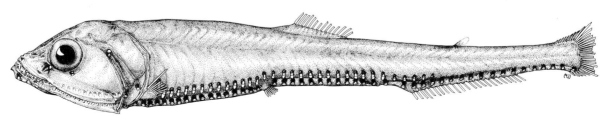

FIGURE 9 *Photichthys argenteus* Hutton, USNM 203387, 55.7 mm SL.

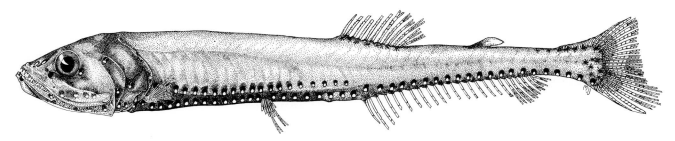

FIGURE 10 *Pollichthys mauli* (Poll), USNM 203408, 47.0 mm SL.

a natural group and that *Woodsia* is the most likely sister-group of the Stomiidae. Among the "photichthyids" the likely basal taxa are *Yarrella* and *Polymetme*. These two genera lack the following derived features that are shared by *Vinciguerria, Pollichthys, Photichthys, Ichthyococcus, Woodsia*, and the Stomiidae (clade G, Fig. 12): character 155, presence of a posterior ORB photophore, ventral or posterior to the eye (we propose that the ORB 2 photophore of "photichthyids" is the homologue of the postorbital photophore [PO of Fink, 1985, and others] of stomiids, based on topographical position and parsimony), and character 156, a radiating pattern of photocytes (A cells) within the body of the photophores (e.g., *Chauliodus*, Herring and Morin, 1978, pp. 306–307, fig. 9.7.5). This last character appears to occur through convergence in the *Margrethia–Bonapartia* clade (Harold, in press) in the Gonostomatidae and in *Diplophos rebainsi*.

There is some evidence that *Polymetme* is more closely related to the taxa here included in clade G than is *Yarrella*, but this is a problem in need of much further study. The one nonhomoplastic character in support of this clade F is character 179, the dorsal uncinate process of the second pharyngobranchial forming an acute angle, as opposed to an obtuse or vertical angle, with the anterior ramus of the bone. Within clade G the two genera of relatively diminutive species, *Vinciguerria* and *Pollichthys*, are sister-groups

(clade L), their monophyly supported, in particular, by character 173, an anteroventrally elongate hyomandibular spine that is bound by a ligament to the lateral surface of the mesopterygoid, and character 184, presence of a tooth plate adhering to the medial surface of the pharyngobranchial of the second gill arch. Features related to shared size reduction in *Vinciguerria* and *Pollichthys* have not been investigated but may be a source of possibly corroborative data.

The remaining photichthyan taxa, *Photichthys, Ichthyococcus, Woodsia*, and the Stomiidae are represented by clade H in our diagram. This clade is supported by such characters as anal fin located well posterior to the dorsal fin (character 159) (although a specialized condition occurs in the majority of stomiid genera in which both the dorsal and the anal fins are immediately anterior to the caudal fin) and mesopterygoid reduction (172).

Within clade H, *Woodsia* is proposed to be the sister-group of the Stomiidae. *Woodsia* and the stomiids share the reduction or lack of gill rakers in adults (character 162). This character was used as a synapomorphy of the stomiid genera by Fink (1985, p. 45, character 103). In *Woodsia* species gill rakers only develop near the ceratobranchial–epibranchial joint, elsewhere on the arch (including the entire hypobranchial) rakers are represented only by small tooth plates (*c.f., Astronesthes* species). This suggests that

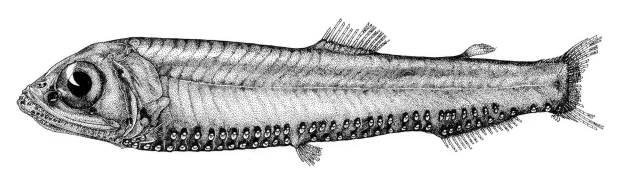

FIGURE 11 *Woodsia nonsuchae* (Beebe), SIO 67–67-10, 114.5 mm SL.

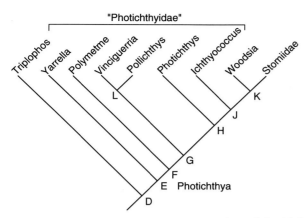

FIGURE 12 Cladogram showing interrelationships of photichthyans and *Triplophos*, as analyzed herein. Tree statistics: TL=83, CI=0.69, RI=0.82.

gill raker reduction appeared as an evolutionary novelty in the common ancestor of *Woodsia* and the Stomiidae.

Another character relating to clade K, discussed by Fink (1985, p. 45, character 104), is the occurrence of relatively broad basibranchials, especially through the body of cartilage between basibranchials 2 and 3 (character 197). We also find an unusual configuration of the basibranchial 3–4 joint in *Woodsia* shared with the basal stomiid genus *Neonesthes* which is probably related to the previous character.

Fink (1985, p. 59, character 211) suggested that the relatively high number of epineurals fused to their respective neural arches in *Ichthyococcus* and the stomiids may be a synapomorphy for those taxa (character 46). If this character is a synapomorphy at the stated level, then it must be lost in *Woodsia*. In terms of parsimony, however, it is equally likely that the high number of fused epineurals in *Ichthyococcus* is convergent with the similar condition in the stomiids.

B. Interrelationships of Triplophos

The genus *Triplophos* remains a problem. *Triplophos* lacks all the Gonostomatid specializations cited here. Possible derived characters shared with the "photichthyid" *Yarrella* were discussed by Harold (in press), suggesting that *Triplophos* is more closely related to the Photichthya than to the Gonostomata. These characters are character 191, reduction of lateral and medial plates of the anterior process of the pelvic girdle, and character 192, the modified arrangement of ural centra. Interpretation of these characters, though they appear to be derived for stomiiforms, is ambiguous. They may be synapomorphies for *Triplophos* plus the

Photichthya that reverse in *Polymetme* plus the remaining photichthyans (minus *Yarrella*), or they may be independently derived in *Triplophos* and in *Yarrella*.

Other characters suggesting a possible sister-group relationship between *Triplophos* and the Photichthya include a complete outer row of dentary teeth (character 175), apparently restricted among photichthyans to the basal genera *Yarrella*, *Polymetme*, and *Pollichthys*. *Triplophos* also shares with most photichthyans character 190(2), presence of two epurals. Three epurals are present in *Diplophos*, *Manducus*, and the four gonostomatid genera, but their number is generally two or three among examined outgroups (e.g., selected osmerids, salmonids, aulopiforms, and myctophiforms). Presence of two epurals is the more likely derived condition for stomiiforms based on functional outgroup comparison to *Diplophos* and *Manducus*.

C. Intrarelationships of the Stomiidae

The family Stomiidae (= superfamily Stomiatioidea of Weitzman, 1974) was diagnosed by Fink (1984, p. 183) based on a series of synapomorphies. Included among these are presence of a mental barbel, divided geniohyoideus muscle, a portion of the adductor mandibulae inserting on the postorbital (PO) photophore, and the lack of gill rakers in adults (but see section VII. A. above). The set of generic relationships proposed by Fink (1984, 1985, p. 11, figs. 1–6), based on a strict consensus of six equally parsimonious cladograms, is repeated here (Fig. 13). We have not reanalyzed stomiid intrarelationships, although the amount of homoplasy in the analysis (which may not be excessive for a study dealing with 26 terminal taxa), and our hypothesis that *Woodsia* is resolved as the sister-group of the family (Fig. 12), suggests that this is an area in need of further study.

VIII. Conclusions

Research into the phylogenetic relationships of the Stomiiformes over the last 20 years has allowed for diagnoses of the order and several major subgroups. Rosen's (1973) work, which proposed recognition of a group at the ordinal level for the first time (Stomiatiformes), placed these fishes into a hierarchical context that remains largely intact. Weitzman (1974) and Fink and Weitzman (1982) started the process of analyzing relationships within the order, one of their most significant conclusions being that *Diplophos* and/or *Manducus* is the sister-group to all other stomiiforms.

Major stomiiform clades have been delimited and are quite well supported: Sternoptychidae (Weitz-

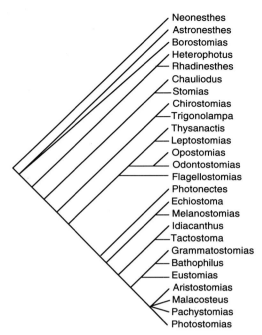

Neonesthes
Astronesthes
Borostomias
Heterophotus
Rhadinesthes
Chauliodus
Stomias
Chirostomias
Trigonolampa
Thysanactis
Leptostomias
Opostomias
Odontostomias
Flagellostomias
Photonectes
Echiostoma
Melanostomias
Idiacanthus
Tactostoma
Grammatostomias
Bathophilus
Eustomias
Aristostomias
Malacosteus
Pachystomias
Photostomias

FIGURE 13 Cladogram showing interrelationships of stomiid genera, reproduced from Fink (1984).

man, 1974), Stomiidae (Fink, 1985), and Gonostomatidae (Harold, in press). Based on our own set of anatomical characters we propose a set of relationships among these families and the remaining, hitherto unanalyzed, "photichthyid" genera. Our analysis, though far from complete, suggests that the "Photichthyidae," as currently formulated, do not constitute a diagnosable group (Fig. 12). Ambiguity remains in the deciphering of the interrelationships of the "photichthyid" and stomiid genera, but clearly *Woodsia* and *Ichthyococcus* are more closely related to the stomiids than any other stomiiform genera. Further studies of photophore histology, and an expanded survey of myology, hold promise for the resolution of some remaining problems.

IX. Summary

The order Stomiiformes is morphologically a highly diverse group of oceanic fishes comprising several well-circumscribed clades: Gonostomatidae in a restricted sense (*Bonapartia, Cyclothone, Gonostoma,* and *Margrethia* only), Sternoptychidae, and Stomiidae. Intrarelationships and interrelationships of these three families, *Diplophos, Manducus, Triplophos,* and the problematic "Photichthyidae," are based on literature survey as well as on new character data and phylogenetic analyses. *Diplophos* and *Manducus* together form

the sister-group of all other stomiiforms. Of the remaining taxa the Gonostomatidae and Sternoptychidae are sister-groups comprising the clade Gonostomata (infraorder), a clade which is the sister-group of the infraorder Photichthya. The latter is a morphologically diverse clade consisting of the forms bearing a chin barbel, the Stomiidae, and what appears to be a paraphyletic group of non-barbeled forms that is presently referred to the "Photichthyidae." Of the "photichthyid" genera *Woodsia* is the probable sister-group of the Stomiidae.

Acknowledgments

It is with great pleasure that we dedicate this chapter to Colin Patterson. We thank W. L. Fink, R. Winterbottom, and M. Weitzman for making available various unpublished notes and illustrations on the morphology of stomiiforms. For her assistance with PAUP and MacClade software we thank A. Weitzman. The manuscript was improved through reviews by W. L. Fink and C. Patterson. We also wish to thank the following museum staff who made material available for study: D. Catania (CAS), K. E. Hartel (MCZ), E. Holm (ROM), T. Iwamoto (CAS), S. L. Jewett and L. Palmer (USNM), R. Lavenberg (LACM), M. McGrouther (AMS), N. V. Parin (IOAN), A. Post (IHS), J. Randall (BPBM), W. J. Richards (Rosenstiel School of Marine and Atmospheric Science, University of Miami), A. Suzumoto (BPBM), and R. Winterbottom (ROM). Financial support was gratefully received by A.S.H. in the form of a Natural Sciences and Engineering Research Council of Canada Postdoctoral Fellowship and a Tilton Postdoctoral Fellowship from the California Academy of Sciences, San Francisco. The manuscript was prepared with the aid of staff and resources of the Department of Ichthyology and Herpetology, Royal Ontario Museum, Toronto, Canada.

References

Ahlstrom, E. H. (1974). The diverse patterns of metamorphosis in gonostomatid fishes—an aid to classification. *In* "The Early Life History of Fish" (J. H. S. Blaxter, ed.), pp. 659–674. Springer-Verlag, Berlin.

Ahlstrom, E. H., and Moser, H. G. (1969). A new gonostomatid from the tropical eastern Pacific. *Copeia* **1969**, 493–500.

Ahlstrom, E. H., Richards, W. J., and Weitzman, S. H. (1984). Families Gonostomatidae, Sternoptychidae, and associated stomiiform groups: Development and relationships. *In* "Ontogeny and Systematics of Fishes" (H. G. Moser, W. J. Richards, D. M. Cohen, M. P. Fahay, A. W. Kendall, Jr., and S. L. Richardson, eds.), Spec. Publ. No. 1, pp. 184–198. American Society of Ichthyologists and Herpetologists, Lawrence, KS.

Baird, R. C. (1971). The systematics, distribution and zoogeography of the marine hatchetfishes (family Sternoptychidae). *Bull. Mus. Comp. Zool.* **142**, 1–128.

Bassot, J.-M. (1966). On the comparative morphology of some luminous organs. *In* "Bioluminescence in Progress" (F. H. Johnson and Y. Haneda, eds.), pp. 557–610. Princeton University Press, Princeton, NJ.

Eschmeyer, W. N. (1990). "Catalog of the Genera of Recent Fishes." California Academy of Sciences, San Francisco.

Farris, J. S. (1988). "Hennig86, Version 1.5," Program and documentation, Farris, Port Jefferson Station, NY.

Fink, W. L. (1984). Stomiiforms: Relationships. *In* "Ontogeny and Systematics of Fishes" (H. G. Moser, W. J. Richards, D. M. Cohen, M. P. Fahay, A. W. Kendall, Jr., and S. L. Richardson, eds.), Spec. Publ. No. 1, pp. 181–184. American Society of Ichthyologists and Herpetologists, Lawrence, KS.

Fink, W. L. (1985). Phylogenetic interrelationships of the stomiid fishes (Teleostei: Stomiiformes). *Misc. Publ. Mus. Zool., Univ. Mich.* **171**, 1–127.

Fink, W. L., and Weitzman, S. H. (1982). Relationships of the stomiiform fishes (Teleostei), with a description of *Diplophos*. *Bull. Mus. Comp. Zool.* **150(2)**, 31–93.

Gibbs, R. H., Jr. (1964a). Family Astronesthidae. *In* "Fishes of the Western North Atlantic" (H. B. Bigelow, D. M. Cohen, M. M. Dick, R. H. Gibbs, Jr., M. Grey, J. E. Morrow, Jr., L. P. Schultz, and V. Walters, eds.), Part 4, Sears Found. Mar. Res., Mem. No. 1, pp. 311–350. Yale University, New Haven, CT.

Gibbs, R. H., Jr. (1964b). Family Idiacanthidae. *In* "Fishes of the Western North Atlantic." (H. B. Bigelow, D. M. Cohen, M. M. Dick, R. H. Gibbs, Jr., M. Grey, J. E. Morrow, Jr., L. P. Schultz, and V. Walters, eds.), Part 4, Sears Found. Mar. Res., Mem. No. 1, pp. 512–522. Yale University, New Haven, CT.

Gibbs, R. H., Jr. (1986a). Family No. 67: Stomiidae. *In* "Smith's Sea Fishes" (M. M. Smith and P. C. Heemstra, eds.), pp. 229–230. Macmillan South Africa, Johannesburg.

Gibbs, R. H., Jr. (1986b). Family No. 68: Chauliodontidae. *In* "Smith's Sea Fishes" (M. M. Smith and P. C. Heemstra, eds.), p. 230. Macmillan South Africa, Johannesburg.

Gibbs, R. H., Jr. (1986c). Family No. 69: Astronesthidae. *In* "Smith's Sea Fishes" (M. M. Smith and P. C. Heemstra, eds.), pp. 231–234. Macmillan South Africa, Johannesburg.

Gibbs, R. H., Jr. (1986d). Family No. 72: Melanostomiidae. *In* "Smith's Sea Fishes" (M. M. Smith and P. C. Heemstra, eds.), pp. 236–243. Macmillan South Africa, Johannesburg.

Gibbs, R. H., Jr., and McKinney, J. F. (1988). High-count species of the stomiid fish genus *Astronesthes* from the southern Subtropical Convergence Region: Two new species and redescription of *Cryptostomias* (= *Astronesthes*) *psychrolutes*. *Smithson. Contrib. Zool.* **460**, 1–25.

Gibbs, R. H., Jr., and Weitzman, S. H. (1965). *Cryptostomias psychrolutes*, a new genus and species of astronesthid fish from the southwestern Pacific Ocean. *Vidensk. Medd. Dansk Naturh. Foren.* **128**, 265–271.

Goodyear, R. H., and Gibbs, R. H., Jr. (1986). Ergebnisse der forschungsreisen des FFS "Walther Herwig" nach Südamerika. X. Systematics and zoogeography of stomiatoid fishes of the *Astronesthes cyaneus* species group (family Astronesthidae), with descriptions of three new species. *Arch. Fischereiwiss.* **20**, 107–131.

Grey, M. (1964). Family Gonostomatidae. *In* "Fishes of the Western North Atlantic" (H. B. Bigelow, D. M. Cohen, M. M. Dick, R. H. Gibbs, Jr., M. Grey, J. E. Morrow, Jr., L. P. Schultz, and V. Walters, eds.), Part 4, Sears Found. Mar. Res., Mem. No. 1, pp. 78–240. Yale University, New Haven, CT.

Harold, A. S. (1993). Phylogenetic relationships of the sternoptychid *Argyropelecus* (Teleostei: Stomiiformes). *Copeia* **1993**, 123–133.

Harold, A. S. (1994). A taxonomic revision of the sternoptychid genus *Polyipnus* (Teleostei: Stomiiformes), with an analysis of phylogenetic relationships. *Bull. Mar. Sci.* **54**, 428–534.

Harold, A. S. Phylogenetic relationships of the Gonostomatidae (Teleostei: Stomiiformes). *Bull. Mar. Sci.* (in press).

Herring, P. J., and Morin, J. G. (1978). Bioluminescence in fishes. *In* "Bioluminescence in Action" (P. J. Herring, ed.), pp. 273–329. Academic Press, London.

Johnson, G.D. (1992). Monophyly of the euteleostean clades—

Neoteleostei, Eurypterygii, and Ctenosquamata. *Copeia* **1993**, 8–25.

Kawaguchi, K., and Moser, H. G. (1984). Stomiatoidea: Development. *In* "Ontogeny and Systematics of Fishes" (H. G. Moser, W. J. Richards, D. M. Cohen, M. P. Fahay, A. W. Kendall, Jr., and S. L. Richardson, eds.), Spec. Publ. No. 1, pp. 169–181. American Society of Ichthyologists and Herpetologists, Lawrence, KS.

Krefft, G. (1973). Ergebnisse der forschungsreisen des FFS "Walther Herwig" nach Südamerika. XXVIII. *Woodsia meyerwaardeni* spec. nov., ein neuer Gonostomatidae aus dem Südatlantik. *Arch. Fischereiwiss.* **24**, 129–139.

Krefft, G., and Parin, N. V. (1972). Ergebnisse der forschungsreisen des FFS "Walther Herwig" nach Südamerika. XXV. *Diplophos rebainsi* n. sp. (Osteichthyes, Stomiatoidei, Gonostomatidae), a new gonostomatid fish from southern seas. *Arch. Fischereiwiss.* **22**, 94–100.

Lauder, G. V., and Liem, K. F. (1983). The evolution and interrelationships of actinopterygian fishes. *Bull. Mus. Comp. Zool.* **150**, 195–197.

Maddison, W. P., and Maddison, D. R. (1992). "MacClade: Analysis of Phylogeny and Character Evolution. Version 3.04." Sinauer Assoc., Sunderland, MA.

Marshall, N. B. (1960). Swimbladder structure of deep-sea fishes in relation to their systematics and biology. *'Discovery' Rep.* **31**, 1–121.

Morrow, J. E., Jr. (1964a). Family Chauliodontidae. *In* "Fishes of the Western North Atlantic" (H. B. Bigelow, D. M. Cohen, M. M. Dick, R. H. Gibbs, Jr., M. Grey, J. E. Morrow, Jr., L. P. Schultz, and V. Walters, eds.), Part 4, Sears Found. Mar. Res., Mem. No. 1, pp. 274–289. Yale University, New Haven, CT.

Morrow, J. E., Jr. (1964b). Family Stomiatidae. *In* "Fishes of the Western North Atlantic" (H. B. Bigelow, D. M. Cohen, M. M. Dick, R. H. Gibbs, Jr., M. Grey, J. E. Morrow, Jr., L. P. Schultz, and V. Walters, eds.), Part 4, Sears Found. Mar. Res., Mem. No. 1, pp. 290–310. Yale University, New Haven, CT.

Morrow, J. E., Jr. (1964c). Family Malacosteidae. *In* "Fishes of the Western North Atlantic" (H. B. Bigelow, D. M. Cohen, M. M. Dick, R. H. Gibbs, Jr., M. Grey, J. E. Morrow, Jr., L. P. Schultz, and V. Walters, eds.), Part 4, Sears Found. Mar. Res., Mem. No. 1, pp. 523–549. Yale University, New Haven, CT.

Morrow, J. E., Jr., and Gibbs, R. H., Jr. (1964). Family Melanostomiatidae. *In* "Fishes of the Western North Atlantic" (H. B. Bigelow, D. M. Cohen, M. M. Dick, R. H. Gibbs, Jr., M. Grey, J. E. Morrow, Jr., L. P. Schultz, and V. Walters, eds.), Part 4, Sears Found. Mar. Res., Mem. No. 1, pp. 351–511. Yale University, New Haven, CT.

Mukhacheva, V. A. (1972). Materialy po sistematike, rasprostraneniiu i biologii vidov roda *Gonostoma* (Pisces, Gonostomatidae). *Tr. Inst. Okeanol. im. P. P. Shirshova, Akad. Nauk SSSR* **93**, 205–249.

Nelson, J. S. (1994). "Fishes of the World." Wiley, New York.

Norman, J. R. (1930). Oceanic fishes and flatfishes collected in 1925–1927. *'Discovery' Rep.* **2**, 261–370.

Ozawa, T., Oda, K., and Ida, T. (1990). Systematics and distribution of the *Diplophos taenia* Species Complex (Gonostomatidae), with a description of a new species. *Jpn. J. Ichthyol.* **37**, 98–115.

Parin, N. V. (1992). *Argyripnus electronus*, a new sternoptychid fish from the Sala y Gomez Ridge. *Jpn. J. Ichthyol.* **39**, 135–137.

Parin, N. V., and Borodulina, O. D. (1990). Survey of the genus *Polymetme* (Photichthyidae) with a description of two new species. *Vopr. Ikhtiol.* **30**, 733–743.

Parin, N. V., and Kobyliansky, S. G. (1993). Review of the genus *Maurolicus* (Sternoptychidae, Stomiiformes), with re-establishing validity of five species considered junior synonyms of *M. muelleri* and descriptions of nine new species. *Tr. Inst. Okeanol. im. P.P. Shirshova, Akad. Nauk SSSR* **128**, 69–107.

Paxton, J. R. (1972). Osteology and relationships of the lanternfishes (family Myctophidae). *Natl. Hist. Mus. Los Ang. Cty. Sci. Bull.* **13**, 1–81.

Rosen, D. E. (1973). Interrelationships of higher euteleostean fishes. *Zool. J. Linn. Soc.* **53**, Suppl. 1, 397–513.

Schaefer, S., Johnson, R. K., and Badcock, J. (1986a). Family No. 73: Photichthyidae. *In* "Smith's Sea Fishes" (M. M. Smith and P. C. Heemstra, eds.), pp. 243–247. Macmillan South Africa, Johannesburg.

Schaefer, S., Johnson, R. K., and Badcock, J. (1986b). Family No. 74: Gonostomatidae. *In* "Smith's Sea Fishes" (M. M. Smith and P. C. Heemstra, eds.), pp. 247–253. Macmillan South Africa, Johannesburg.

Schultz, L. P. (1964). Family Sternoptychidae. *In* "Fishes of the Western North Atlantic" (H. B. Bigelow, D. M. Cohen, M. M. Dick, R. H. Gibbs, Jr., M. Grey, J. E. Morrow, Jr., L. P. Schultz, and V. Walters, eds.), Part 4, Sears Found. Mar. Res., Mem. No. 1, pp. 241–273. Yale University, New Haven, CT.

Smith, D. G., Hartel, K. E., and Craddock, J. E. (1991). Larval development, relationships, and distribution of *Manducus maderensis*, with comments on the transformation of *M. greyae* (Pisces, Stomiiformes). *Breviora* **49**, 1–17.

Swofford, D. L. (1993). "PAUP: Phylogenetic Analysis Using Parsimony, Version 3.1.1." Smithsonian Institution, Washington, DC.

Weitzman, S. H. (1974). Osteology and evolutionary relationships of the Sternoptychidae, with a new classification of stomiatoid families. *Bull. Am. Mus. Nat. Hist.* **53**, 327–478.

Weitzman, S. H. (1986). Order Stomiiformes: Introduction. *In* "Smith's Sea Fishes" (M. M. Smith and P. C. Heemstra, eds.), pp. 227–229. Macmillan South Africa, Johannesburg.

Winterbottom, R. (1974). A descriptive synonymy of the striated muscles of the Teleostei. *Proc. Acad. Natl. Sci. Philadelphia* **125(12)**, 225–317.

Appendix 1 Character Data for Sternoptychidae

Character State Matrix (characters 1–150)

Triplophos

0000000000 0000000000 0?00001100 0000000100 0000000000 0000000000 0000001000
0001000000 0000000000 0000000000 0001110000 0000000000 0000000000 0000000000
0000000000

Diplophos

0100000100 0000000000 0?00000100 0000000000 0000000000 0000000000 0000001000
0000000000 0000000000 0000000000 0000000000 0000000000 0000000000 0000000000
0000000000

Gonostomatidae

0000000000 0000000000 0?00000100 0000000000 0000000000 0000000000 0000001000
0000000000 0000000000 0000001000 0000000000 0000000000 00?0000000 0000000000
0000000000

Thorophos euryops

0000000000 0100000000 0100000000 0000000111 0001001000 1000000000 0010111000
0000002110 1000011110 1001001001 0001111110 0110000110 0010000000 0000010000
0000000000

Thorophos nexilis

0000000000 0100000000 0100000000 0000000111 0001001000 1000000000 0010111000
0000002110 1000011110 1001001001 0001111110 0110000110 0010000000 0010010000
0000000000

Araiophos

0000000001 0100000000 0000210000 0000011111 0011001000 1000010000 0010111000
0000000110 1000011120 000100000? 0003101111 ??????01?0 0010000000 0111100000
?000000000

Maurolicus

0001000011 0100000000 0100010100 0000111110 2011111000 1000010000 0000111001
1000002111 1000021110 0021011000 0002001111 1000000110 1100000000 0001000000
0000000000

Danaphos

1010101010 0100001000 1100000100 0000100111 2111111100 1000010100 0000111001
0010102110 1111011111 0000010001 1002011110 0001010111 1100000000 0101100000
0001100000

Valenciennellus

1010101010 0100000100 1110000100 0000100111 2111111100 1000010100 0000111001
0010102110 1111011121 0000010001 0002011110 0001010111 1110000000 0101000000
0001100000

Argyripnus

0001000001 0100010100 0000010100 0000002111 0001111000 1000001000 0001111000
0010000110 1021131110 0001101000 0002010101 0101110111 1101110000 1101000010
0000000001

Sonoda

0001000001 0100010100 0011020100 0000002111 0011111 0?0 1000011000 0001111000
0010102110 1???111110 0001101000 0002001101 01011101?1 1111110000 1101?00010
0001100001

Polyipnus

0000010001 0110010211 0010130101 0011001110 0021111001 1111110011 0101021111
1001112011 0001131110 1011021111 0101101101 1100010111 1101001111 0101110110
0111000001

Argyropelecus

0100010101 0111001211 0011031111 10?0001100 1020110011 1111112011 1000021010
0001011010 1011131110 0000021111 0112110111 1112111111 1101001111 0101111111
1110011110

Sternoptyx

0100010101 1111100301 ?0??????11 1112001100 1020110211 1121202011 1110031120
0111111010 0???111120 0100021111 111?111111 1112011112 1101001111 0101111101
1001011111

Character List

1. Interfrontal joint. 0, separate; 1, fused anteriorly.
2. Dorsal frontal surface. 0, smooth; 1, pitted.
3. Longitudinal frontal fossa. 0, separate; 1, joined anteriorly.
4. Longitudinal frontal fossa. 0, shallow; 1, deep.
5. Frontal crest. 0, present; 1, absent anteriorly.
6. Frontal crest. 0, low or absent; 1, prominent.
7. Interorbital space. 0, broad; 1, narrow.
8. Parietal surface. 0, smooth; 1, pitted.
9. Parietal–sphenotic relationship. 0, no contact; 1, contact.
10. Parietal–pterotic relationship. 0, contact; 1, no contact.
11. Parietal–intercalar relationship. 0, no contact; 1, contact.
12. Parietals. 0, not separated by supraoccipital; 1, separated.
13. Parietal crest. 0, absent; 1, present.
14. Parasphenoid shape. 0, straight to slightly convex; 1, strongly convex.
15. Parasphenoid glossopharyngeal tunnel. 0, absent; 1, present.
16. Parasphenoid lateral wing. 0, moderately well-developed; 1, posterior process and wings form a cap.
17. Basisphenoid. 0, present; 1, absent.
18. Posterior myodome. 0, large; 1, small; 2, moderate size; 3, dorsally elongate.

19. Posterior myodome. 0, horizontal; 1, vertical.
20. Sphenotic size. 0, small; 1, moderate to large.
21. Posttemporal fossa bounded by frontal. 0, yes; 1, no.
22. Neural arch attachment. 0, some free; 1, all fused.
23. Posttemporal fossa bounded by sphenotic. 0, none; 1, small to large amount.
24. Posttemporal fossa bounded by pterotic. 0, large amount; 1, moderate amount.
25. Posttemporal fossa bounded by epioccipital. 0, large amount; 1, moderate amount; 2, small amount.
26. Posttemporal fossa bounded by parietal. 0, large amount; 1, moderate amount; 2, small amount; 3, reduced and specialized.
27. Posttemporal fossa bounded by intercalar. 0, none; 1, small amount.
28. Posttemporal fossa bounded by exoccipital. 0, none; 1, varying amounts.
29. Otic bullae. 0, well-developed, large; 1, not visible.
30. Exoccipital pedicles. 0, moderately well-developed; 1, enlarged.
31. Basioccipital. 0, as deep or deeper than exoccipital, in posterior view; 1, not as deep as exoccipital.
32. Exoccipital plates dorsal to foramen magnum. 0, in contact; 1, no contact.
33. Centrum-like face of basioccipital. 0, moderate size; 1, shallow or small.
34. Epioccipital process. 0, small to prominent; 1, prominent posterodorsal process; 2, slender spine-like process.
35. Palatine teeth. 0, one row present; 1, absent.
36. Palatine posterior process. 0, present; 1, absent.
37. Palatine shape. 0, double-headed; 1, posterior head lost; 2, cartilaginous bar with tooth plate ventrally.
38. Mesopterygoid teeth. 0, present; 1, absent.
39. Mesopterygoid shape. 0, short; 1, long.
40. Mesopterygoid fenestra. 0, absent; 1, present.
41. Ectopterygoid shape. 0, broad, elongate, moderate size; 1, slender, elongate; 2, triangular to quadrangular, large.
42. Symplectic shape, overall. 0, elongate slender; 1, very elongate.
43. Symplectic shape, terminations. 0, dorsal end enlarged, not club-shaped; 1, both ends equal; 2, club-shaped.
44. Articulation of dorsal border of quadrate. 0, with metapterygoid; 1, ectopterygoid and/or mesopterygoid plus metapterygoid.

45. Position of lower jaw adductor pocket. 0, posterior 1/3 or 1/4 of mandible; 1, middle or anterior of mandible.
46. Mandible shape. 0, elongate, posterior portion 2–3 times depth of anterior; 1, deep, 3–4 times depth of anterior.
47. Mandible tooth line and coronoid platform. 0, platform absent; 1, platform present.
48. Mandibulohyoid ligament system. 0, mandible to epihyal and to interopercle; 1, separate to interopercle and epihyal to interopercle; 2, no fibers to interopercle.
49. Antorbital. 0, present; 1, absent.
50. Supraorbital. 0, present; 1, absent.
51. Infraorbital series. 0, six bones, excluding antorbital; 1, reduced to four or fewer.
52. Opercle shape, notch in dorsal border. 0, present but altered or reduced relative to outgroups; 1, modified.
53. Opercular spine. 0, present; 1, developed into a lateral ridge; 2, strong lateral spiny process.
54. Opercle shape. 0, roughly rectangular or quadrangular; 1, elongate rectangular.
55. Subopercle shape. 0, rectangular to half rectangular; 1, triangular; 2, dorsoventrally elongate.
56. Relative size of interopercle and subopercle. 0, interopercle length about equal to that of subopercle; 1, interopercle much longer than subopercle.
57. Interopercle shape. 0, short dorsal process; 1, elongate dorsal process; 2, highly elongate, narrow dorsal process.
58. Subopercle ossification. 0, complete; 1, incomplete.
59. Preopercular angle. 0, gradual, up to right angle; 1, abrupt right angle.
60. Preopercular spines. 0, absent; 1, present.
61. Relative size of preopercular limbs. 0, dorsal and ventral limbs about equal; 1, dorsal limb much longer than ventral.
62. Interpremaxillary ligament. 0, strong; 1, weak.
63. Premaxillary–proethmoid (rostrodermethmoid lateral process) crossed ligament. 0, present; 1, absent.
64. Premaxillary–proethmoid (rostrodermethmoid lateral process) uncrossed ligament. 0, present; 1, absent.
65. Palatopremaxillary ligament. 0, separate maxillary head–palatine and maxillary–premaxillary ligament; 1, continuous palatopremaxillary ligament; ?, absent.
66. Palatomaxillary ligament. 0, short; 1, long; 2, moderate length; 3, very short.
67. Maxillary–proethmoid (rostrodermethmoid lateral process) ligament. 0, present; absent.

68. Suspensory palatine ligament. 0, long; 1, short.
69. Premaxillary ascending process. 0, short to moderately long; 1, elongate; 2, almost none.
70. Maxillary angle. 0, not angulate; 1, angulate.
71. Maxillary width. 0, slender; 1, posteriorly expanded.
72. Maxillary toothed border. 0, convex; 1, concave.
73. Anterior supramaxilla. 0, present; 1, absent.
74. Hyomandibular length. 0, moderately long, about half of cranial length; 1, very long, about three-quarters of cranial length.
75. Hyomandibular spine. 0, present; 1, absent.
76. Posterior ceratohyal length. 0, short, less than length of anterior ceratohyal; 1, elongate, greater than half of anterior ceratohyal length.
77. Anterior ceratohyal shape. 0, moderately constricted in middle; 1, greatly constricted; 2, not greatly constricted.
78. Largest end of anterior ceratohyal. 0, posterior or both ends about equal; 1, anterior.
79. Total number of branchiostegal rays. 0, 12 to 22, rarely 11 in *Ichthyococcus*; 1, 10 or fewer.
80. Urohyal shape. 0, incised posterior margin; 1, not incised.
81. Basihyal. 0, present; 1, absent.
82. Supraethmoid relative position to frontals. 0, ventral; 1, dorsal; ?, supraethmoid absent.
83. Length of posterior supraethmoid process. 0, long; 1, process absent; 2, short; ?, supraethmoid absent.
84. Relative size of supraethmoid. 0, large; 1, small to moderate; ?, supraethmoid absent.
85. Proethmoids (lateral processes of rostro-dermethmoid). 0, present; 1, absent.
86. Capsular ethmoids. 0, present, well-developed; 1, absent; 2, fused together; 3, fused to supraethmoid.
87. Ventral ethmoid. 0, present; 1, absent.
88. Myodome bone. 0, well-developed, separate; 1, absent.
89. Lateral ethmoid. 0, well-developed; 1, small to moderate size; 2, absent.
90. Lateral vomerine teeth. 0, present; 1, absent.
91. Median vomerine teeth. 0, absent; 1, present.
92. Vomer. 0, present; 1, absent.
93. Vomer, anterodorsal extension. 0, not reaching supraethmoid; 1, reaching supraethmoid; 2, special process dorsal to supraethmoid; 2, extends dorsally, ventral to supraethmoid.
94. Ethmoid cornu. 0, absent or weakly developed; 1, moderately to well-developed.
95. Ethmoid prenasal process. 0, absent; 1, present.
96. Ethmoid cartilage. 0, broad; 1, narrow; 2, broad, modified.
97. Rib-bearing vertebrae. 0, first rib associated with vertebra 2; 1, first rib associated with vertebra 3.
98. Enlarged ribs. 0, absent; 1, present.
99. Ribs directly supporting pelvic girdle. 0, absent; 1, present.
100. Epipleurals. 0, present; 1, absent.
101. Epineurals. 0, present; 1, absent.
102. Expanded neural and haemal spines. 0, absent; 1, present.
103. Specialized supraneurals (dorsal blade). 0, absent; 1, present.
104. Epurals. 0, three separate elements; 1, two separate elements; 2, one element; 3, epurals absent; ?, fused to uroneural
105. Caudal radials, other than epurals. 0, present; 1, absent.
106. Uroneurals. 0, two present; 1, one present (second uroneural absent).
107. Parhypural. 0, free from preural centrum 1 and hypural 1; 1, fused to preural centrum 1 and/or hypural 1.
108. Hypurals 1 and 2. 0, autogenous; 1, fused.
109. Hypurals 3, 4, and 5. 0, autogenous; 1, fused (3–5 or 3–6 fused).
110. Ural centrum 2. 0, not fused to PU1 + U1; 1, fused to PU1 + U1.
111. Sagitta: postcaudal trough. 0, present; 1, absent.
112. Sagitta: crista superior. 0, present; 1, absent.
113. Sagitta: crista inferior. 0, present; 1, absent.
114. Sagitta: rostrum. 0, well-developed, prominent; 1, very short; 2, absent or low eminence.
115. Sagitta: lateral surface. 0, convex; 1, flat.
116. Sagitta: lateral profile. 0, longer than deep (height about 1.3 to 2.0 times in length; 1, deeper than long (height about 0.4 to 0.9 times in length).
117. Sagitta: length relative to cranial length. 0, large (4.7 to 7.0 times in length of cranium); 1, small (15 to 50 times in length of cranium).
118. Photophore development. 0, *in situ* formation through white phase; 1, budding (photophores in clusters).
119. Adipose fin shape. 0, short-based; 1, long-based.
120. Number of pelvic radials. 0, three; 1, six; 2, one.
121. Body shape. 0, highly elongate with shallow head; 1, deep body and head.
122. Photophores ventrally on caudal peduncle. 0, singly or in clusters of 2; 1, clusters of 4 or more.

123. Position of anal-fin origin. 0, posterior to dorsal-fin origin; 1, anterior to dorsal-fin origin. Highly variable in the Gonostomatidae, hence coded "?" for that outgroup.
124. Anal-fin hiatus. 0, absent; 1, present.
125. Attachment of pterygiophores immediately anterior and posterior to anal-fin hiatus. 0, nonligamentous; 1, ligamentous.
126. Anterior portion of pelvic girdle ischial process. 0, present; 1, absent.
127. Pelvic girdle orientation. 0, horizontal; 1, approximately vertical.
128. Abdominal keel-like structure. 0, absent; 1, present.
129. Body depth. 0, 3.7 to 7.7 percent of standard length; 1, 0.8 to 2.0 percent of standard length.
130. Iliac spines. 0, absent; 1, present.
131. Photophores: posterior inferior OP size. 0, about equal to other OP; 1, greatly enlarged.
132. Photophores: SO. 0, present; 1, absent.
133. Photophores: OA. 0, more than 1; none or 1.
134. Posterior infraorbitals. 0, posterior infraorbitals, behind eye, present; 1, posterior part of series, those posterior to eye, absent (equivalent to infraorbitals 5 and 6 and possibly 4).
135. Number of anterior infraorbitals. 0, three or four present; 1, entire series represented by two anterior elements (probably equivalent to numbers 1 and 2 in other taxa).
136. NPU2 shape. 0, narrow; 1, broad and flat.
137. Palatopremaxillary ligament. 0, single slip; 1, ligament originating on palatine and subdivided into branches to premaxilla, maxilla, and supraethmoid.
138. Posttemporal. 0, short and weak, not well-ossified; 1, elongate and strong, well-ossified.
139. Cleithrum ventral lateral wing. 0, no posterior notch; 1, posterior notch through which fin rays pass.
140. Cleithrum shape. 0, ventral anterior portion smoothly curved; 1, ventral anterior portion highly angular.
141. Pectoral radial articulation. 0, radial II articulating with scapula and coracoid; 1, radial II articulating only with scapula.
142. Photophores: L (lateral). 0, absent; 1, present.
143. Posttemporal and supracleithrum relationship. 0, free; 1, fused.
144. Distal pterygiophore perichondral ossifications. 0, present; 1, absent.
145. Urohyal size. 0, moderate to large; 1, small.
146. Number of hypobranchial 1 gill rakers. 0, more than three; 1, three or fewer.
147. Hypobranchial 1 middorsal tabular process. 0, absent; 1, present.

148. Pubic process relationship to posterior pleural rib. 0, not parallel or bound together; 1, shaft of pubic process tightly bound and parallel to distal end of last pleural rib.
149. Hypobranchial 1 shape. 0, approximately straight; 1, curved dorsally in an arc.
150. Photophores: PV number. 0, more than 10; 1, 10.

Appendix 2 Character Data for Photichthya and Other Stomiiforms

Character State Matrix (first character at left is 151)

Manducus maderensis

0100000000 0000000000 0000000000 0000000010 0010100000 0

Diplophos taenia

0100000000 1000000000 0000000001 0000000010 0010100000 0

Diplophos rebainsi

0100010000 0000000100 0000000001 0000000010 0010100000 0

Triplophos hemingi

0210000100 0000000101 0000100001 1010010102 1100000000 0

Thorophos

1211001100 0000002100 0000000000 1010100001 0020000101 0

Araiophos

1211001100 0000002100 0000000000 1010100001 0020000111 0

Clade C sternoptychids

1211001100 0000002100 0000000000 1010100001 0020000101 0

Gonostomatidae

1211011100 0000000100 0000000000 1010000000 0010000010 0

Polymetme

0210001100 2000001101 1000100010 1010000002 0000000001 1

Yarrella

0210000100 2000001111 1000100000 1010010002 1100000001 1

Pollichthys

0210111100 2000001101 0010100010 1011001102 0000000001 1

Vinciguerria

0210111100 2000001111 0010000010 1011101102 0000000001 1

Ichthyococcus

0210111111 2011111100 0100010110 1010010002 0040010001 1

Photichthys

0210111111 2000001100 0101011010 1110000002 0001000001 1

Woodsia

0210111111 2111111100 0101011110 1110000002 0001001001 1

Stomiidae

0210111111 2111111100 0101011110 1110000002 0031011001 0

Outgroup

000?????00 000000?000 0000000000 0000000000 000?0000?0 ?

Character List

151. Lateral adductor mandibulae (A2). 0, single, undivided; 1, subdivided into A2α (dorsal) and A2β (ventral) sections.
152. Median adductor mandibulae (A1β). 0, single, undivided; 1, divided into two large muscles to primordial ligament and maxilla; 2, reduction of one or both A1β sections.
153. Extensor proprius pelvicus muscle. 0, approximately horizontally oriented, overlying the adductor superficialis; 1, vertically oriented, markedly diverged from adductor superficialis and originating in connective tissue associated with one or two pleural ribs.
154. Metamorphosis of photophores. 0, rapid; 1, protracted.
155. Posterior ORB photophore (= PO or postorbital of stomiids). 0, absent; 1, present.
156. Photophore A cell configuration. 0, irregular; 1, radiating.
157. Accessory photophore rows (e.g., LLP photophores). 0, present; 1, absent.
158. Row of small photophores on lower jaw. 0, present; 1, absent.
159. Position of anal fin origin. 0, below or anterior to dorsal fin; 1, posterior to dorsal fin.
160. Posterior four branchiostegal rays. 0, bases separated by space; 1, bases crowded together with at least the posterior two in, or nearly in, contact.
161. Number of pectoral fin radials. 0, four; 1, three.
162. Gill raker development. 0, presence of full gill rakers along entire branchial arch; 1, gill rakers absent or restricted to near the ceratobranchial–epibranchial joint in adults.
163. Pectoral fin rays of larvae. 0, short; 1, elongate. (after Ahlstrom *et al.*, 1984).
164. Gut of larvae. 0, contained by body wall; 1, trailing. (after Ahlstrom *et al.*, 1984).
165. Pigmentation pattern of larvae. 0, profuse pigmentation not present below lateral midline; 1, area below lateral midline profusely pigmented. (after Ahlstrom *et al.*, 1984).
166. Cross-sectional shape of larvae. 0, ovate or elliptical; 1, circular. (after Ahlstrom *et al.*, 1984).
167. Photophore type. 0, type Beta; 1, type Gamma; 2, type Alpha.
168. NPU2 spine. 0, short, roughly half the length of the spine immediately anterior; 1, fully-developed spine.
169. Palatovomerine ligament. 0, absent; 1, present.
170. Anterior palatomaxillary ligament configuration. 0, lies ventral to lateral process of rostrodermethmoid; 1, looped over dorsal surface of lateral process of rostrodermethmoid.
171. Contralateral and ipsilateral branches of premaxillary–rostrodermethmoid ligament. 0, separate; 1, fused into a continuous sheet of connective tissue.
172. Mesopterygoid. 0, well-developed; 1, reduced or absent.
173. Hyomandibular spine. 0, short or moderately elongate and not contacting mesopterygoid; 1, elongate and bound to the lateral surface of the mesopterygoid by a ligament.
174. Palatine shape. 0, posterior (mesopterygoid) articular process similar in size to anterior (maxillary) process; 1, posterior process reduced and much smaller than anterior process.
175. Anterior or lateral dentary tooth row. 0, short with only a few teeth near symphysis; 1, full row, lining outer margin of dentary.
176. Premaxillary symphysis shape. 0, ascending processes concave medially resulting in a median recess; 1, ascending processes with straight medial surfaces and in contact along their lengths.
177. Size of medial jaw teeth. 0, various sizes; 1, mainly large, without small teeth interspersed.
178. Third pharyngobranchial tooth plate (UP3). 0, highly dentigerous; 1, teeth reduced to about four or fewer or absent altogether.
179. Second pharyngobranchial dorsal uncinate process. 0, nearly vertical to axis of horizontal shaft of bone; 1, forming an anterior, acute angle of between about 60 and 80 degrees.
180. First epibranchial uncinate process. 0, distally cartilaginous, without a medial bony flange; 1, medial bony flange present, extending distally from cartilaginous process.
181. First pharyngobranchial shaft. 0, cylindrical; 1, ventrally bifurcate, with the lateral limb continuous with a strong ligament to the anterolateral surface of the first epibranchial.

182. Number of posterior ceratohyal branchiostegal rays. 0, more than six; 1, six or fewer.
183. Basihyal configuration. 0, flattened, horizontal and dentigerous; 1, cylindrical, vertical, and edentate.
184. Second pharyngobranchial tooth plates. 0, absent or, if present, adhering loosely; 1, fused or adhering tightly.
185. Hypurals three and four. 0, separate elements; 1, fused into a plate.
186. Second basibranchial superficial tooth plates. 0, present; 1, absent.
187. Third basibranchial tooth plates. 0, lateral to basibranchial and superficial to tendon from third hypobranchial; 1, paired tooth plates closely adherent on dorsal surface.
188. Fourth basibranchial tooth plates. 0, absent; 1, present.
189. Nasal size. 0, small, not extending anterior to nasal capsule; 1, large, extending anteriorly dorsal to rostrodermethmoid lateral process.
190. Number of epurals. 0, three; 1, one; 2, two.
191. Lateral and medial plate-like processes of anterior or pubic process of the pelvic bone. 0, present; 1, absent.
192. Configuration of ural centrum PU2 + U1. 0, short, equidimensional; 1, elongate cylindrical.
193. Parapophysis of first vertebra. 0, parapophysis of first vertebra larger than that of second but not elongate; 1, parapophysis one elongate and continuous with Baudelot's ligament; 2, parapophyses one and two subequal and short; 3, parapophyses one and two subequal, short, and the first associated with completely ossified Baudelot's ligament; 4, first parapophysis highly reduced.
194. Number of branchiostegal (BR) photophores. 0, fewer than 14; 1, 14 or more.
195. Pleural rib of third vertebra. 0, present; 1, absent.
196. Epineurals fused to neural arches. 0, less than half the body length; 1, more than half the body length. (after Fink, 1985, p. 59).
197. Basibranchials. 0, narrow; 1, broad. (after Fink, 1985, p. 45).
198. Hypurals three and four. 0, autogenous; 1, fused.
199. OA photophores. 0, present; 1, absent.
200. Ossified accessory neural arch. 0, present; 1, absent.
201. Serial photophore duct and lumen (after Weitzman, 1974). 0, absent; 1, present.

Apomorphy List (Figs. 4 and 12)

Diplophos plus *Manducus*: 189, 195
Clade A, order Stomiiformes: 152(1)
Clade B: 152(2), 153, 158, 181, 183, 168*
Clade C, infraorder Gonostomata: 151, 154, 199*II
Clade D: 190(2), 170*, 175*, 186*I, 191*I, 192*I
Clade E, infraorder Photichthya: 161(2), 167(1), 171*I, 200*, 201*
Clade F: 179, 157*, 186*, 191*I, 191*II
Clade G: 155, 156*, 171*I, 175*II
Clade H: 159, 160, 172, 176, 170*, 174*I, 175*I, 177*I, 182*I, 194*I
Clade J: 163, 164, 165, 166, 178, 196*I
Clade K: 162, 197, 174*II, 177*II, 182*II, 194*II
Clade L: 173, 184, 187, 188*
Diplophos: 168* (*D. rebainsi* only), 180*
Manducus: none
Sternoptychidae: 190(1), 198, 199*I, 200*
Gonostomatidae: 199*I
Triplophos: 180*, 186*II, 188*, 191*II, 192*II
Yarrella: 169*, 171*II, 186*II, 191*II, 192*II
Polymetme: 171*II
Vinciguerria: 169*, 175*I, 185*
Pollichthys: 175*II
Photichthys: 174*II, 177*II, 182*II, 194*II
Ichthyococcus: 193(4), 174*I, 177*I, 182*I, 186*II, 194*I
Woodsia: 196*I
Stomiidae: 193(3), 196*II, 201*

Appendix 3

The stomiiform taxa examined in this study are the same as those listed in Harold (in press).

The following outgroups were examined. Argentinidae: *Argentina sphyraena* (USNM 238016), *A. striata* (USNM 188224), and *Glossanodon struhsakeri* (USNM 36418). Aulopidae: *Aulopus filamentosus* (USNM 225043). Chlorophthalmidae: *Chlorophthalmus agassizi* (USNM 159377) and *Parasudis truculentus* (USNM 159407). Galaxiidae: *Galaxias maculatus* (USNM 203872) and *G. vulgaris* (USNM 203886). Myctophidae: *Diaphus theta* (ROM 344CS), *Lampanyctus* sp. (ROM 586CS), and *Myctophum affine* (USNM 317161). Neoscopelidae: *Neoscopelus macrolepidotus* (USNM 317160). Osmeridae: *Mallotus villosus* (USNM 306411), *Spirinchus dilatus* (USNM 104689), and *Thaleichthys pacificus* (CAS 15378). Salmonidae: *Coregonus artedi* (ROM 1022CS), *C. clupeaformis* (ROM 1028CS), *C. nasus* (ROM 1012CS), *Oncorhynchus gorbuscha* (ROM 1005CS), *O. mykiss* (ROM 534CS), *Prosopium cylindraceum* (ROM 0028CS), *Salmo trutta* (ROM 1013CS), *Salvelinus fontinalis* (ROM 0998CS), *Stenodus leucichthys* (ROM 1016CS), and *Thymallus arcticus* (ROM 0039CS). Synodontidae: *Synodus variegatus* (USNM 217675).

Interrelationships of Aulopiformes

CAROLE C. BALDWIN *and*
G. DAVID JOHNSON

Department of Vertebrate Zoology
National Museum of Natural History
Smithsonian Institution
Washington, D.C. 20560

I. Introduction

In 1973, Rosen erected the order Aulopiformes for all non-ctenosquamate eurypterygians, that is, the Iniomi of Gosline *et al.* (1966) minus the Myctophiformes (Myctophidae and Neoscopelidae). Rosen's aulopiforms included 15 families (Alepisauridae, Anotopteridae, Aulopidae, Bathysauridae, Bathypteroidae, Chlorophthalmidae, Evermannellidae, Giganturidae, Harpadontidae, Ipnopidae, Omosudidae, Paralepididae, Scopelarchidae, Scopelosauridae, and Synodontidae) and 17 fossil genera, a morphologically diverse group of benthic and pelagic fishes that range in habitat from estuaries to the abyss.

Rosen (1973) diagnosed the Aulopiformes by the presence of an elongate uncinate process on the second epibranchial (EB2) bridging the gap between a posterolaterally displaced second pharyngobranchial (PB2) and the third pharyngobranchial (PB3). He noted that paralepidid fishes lack this distinctive configuration of EB2 and thus questioned their placement in the order. Subsequently, R. K. Johnson (1982) recognized that certain paralepidids (*Paralepis*, and *Notolepis*) have an enlarged EB2 uncinate process but questioned Rosen's use of this feature to diagnose aulopiform because he believed the same condition occurs in neoscopelids. Instead he suggested that the modification is a primitive iniome condition and that the small EB2 uncinate process of myctophids is secondarily derived. R. K. Johnson (1982) resurrected a more traditional view of iniome

relationships in which Rosen's (1973) aulopiforms and myctophiforms are united in the order Myctophiformes.

In 1985, Rosen altered his concept of a monophyletic Aulopiformes, noting that *Aulopus* shares several derived features with ctenosquamates, most notably a median rostral cartilage. Hartel and Stiassny (1986) considered a true median rostral cartilage a character of acanthomorphs and concluded that the morphology of the rostral cartilage is highly variable below that level. Nevertheless, Stiassny (1986) supported Rosen's (1985) view of a paraphyletic Aulopiformes, proposing that *Chlorophthalmus, Parasudis* and *Aulopus* form the sister group of ctenosquamates based on an elevated, reoriented cranial condyle on the maxilla and concurrent exposure of a "maxillary saddle" for reception of the palatine prong. G.D. Johnson (1992) discussed the shortcomings of Rosen's (1985) analysis and observed that neither Rosen nor Stiassny (1986) mentioned the distinctive gill-arch configuration originally described by Rosen (1973) as unique to aulopiforms. He added an additional gill-arch character to Rosen's (1973) complex, the absence of a cartilaginous condyle on PB3 for articulation of EB2, and concluded that a suite of gill-arch modifications constitutes a complex specialization supporting the monophyly of Rosen's (1973) Aulopiformes. In addition, Johnson (1992) offered further evidence (absence of the fifth upper pharyngeal toothplate and associated third internal levator muscle) for the monophyly of Rosen's (1973) Ctenosquamata, which include myctophids, neoscopelids and acanthomorphs, but not aulopi-

Copyright © 1996 by Academic Press, Inc.
All rights of reproduction in any form reserved.

Gosline et al. (1966):
Order Iniomi
  Myctophoidea
    Aulopidae
    Bathysauridae
    Synodontidae
    Harpadontidae
    Bathypteroidae
    Ipnopidae
    Chlorophthalmidae
    Notosudidae (=Scopelosauridae)
    Myctophidae
    Neoscopelidae
  Alepisauroidea
    Paralepididae
    Omosudidae
    Alepisauridae
    Anotopteridae
    Evermannellidae
    Scopelarchidae

Rosen (1973):
Order Aulopiformes, new name
  Suborder Aulopoidei, new name
    Aulopidae
    Bathysauridae
    Bathypteroidae
    Ipnopidae
    Chlorophthalmidae
    Notosudidae (=Scopelosauridae)
  Suborder Alepisauroidei
    [15 fossil genera]
    Superfamily Synodontoidea, new usage
      [2 fossil genera]
      Synodontidae
      Harpadontidae
      Giganturidae (? + Rosauridae)
    Superfamily Alepisauroidea
      Paralepididae
      Omosudidae
      Alepisauridae
      Anotopteridae
      Evermannellidae
      Scopelarchidae

Sulak (1977):
Benthic Myctophiformes:
  Aulopidae
    Aulopus (including *Hime, Latropiscus*)
  Synodontidae
    Subfamily Harpadontinae
      Harpadon (incl. *Peltharpadon*)
      Saurida
    Subfamily Bathysaurinae
      Bathysaurus (incl. *Macristium*)
    Subfamily Synodontinae
      Synodus (incl. *Xystodus*)
      Trachinocephalus
  Chlorophthalmidae
    Subfamily Chlorophthalminae
      Chlorophthalmus
      Parasudis
      Bathysauropsis (incl. *Bathysaurops*)
    Subfamily Ipnopinae
      Tribe Ipnopini
        Ipnops (incl. *Ipnoceps*)
      Tribe Bathypteroini
        Bathypterois (incl. *Benthosaurus*)
      Tribe Bathymicropini
        Bathymicrops
        Bathytyphlops (incl. *Macristiella*)

R. K. Johnson (1982):
Myctophiformes:
  Aulopoids
    Aulopidae
  Myctophoids + Chlorophthalmoids
    Myctophoids
      Myctophidae
      Neoscopelidae
    Chlorophthalmoids
      Notosudidae
      Scopelarchidae
      Chlorophthalmidae
      Ipnopidae
  Synodontoids + Alepisauroids
    Synodontoids
      Synodontidae
      Harpadontidae
      Bathysauridae
    Alepisauroids
      Paralepididae
      Anotopteridae
      Evermannellidae
      Omosudidae
      Alepisauridae

FIGURE 1   Four previously hypothesized classifications of aulopiform or myctophiform fishes.

forms (see also Stiassny, this volume). Johnson *et al.* (1996) argued that *Aulopus* is not closely related to ctenosquamates but is the cladistically primitive member of their Synodontoidei, a lineage that also includes *Pseudotrichonotus, Synodus, Trachinocephalus, Saurida,* and *Harpadon.* Finally, Patterson and Johnson (1995) provided corroborative evidence from the intermuscular bones and ligaments for Rosen's (1973) Aulopiformes, in the extension of the epipleural series anteriorly to the first or second vertebra.

Various schemes of relationships among iniomous fishes have accompanied confusion about the recognition of a monophyletic Aulopiformes (Fig. 1). Gosline

et al. (1966) recognized two "suborders": myctophoids (Aulopidae, Bathysauridae, Synodontidae, Harpadontidae, Bathypteroidae, Ipnopidae, Chlorophthalmidae, Notosudidae [= scopelosaurids of Marshall, 1966—see Paxton, 1972; Bertelsen *et al.,* 1976], Myctophidae, and Neoscopelidae); and alepisauroids (Paralepididae, Omosudidae, Alepisauridae, Anotopteridae, Evermannellidae, and Scopelarchidae). Rosen (1973) added synodontids and harpadontids (his synodontoids) and 17 fossil genera to the Alepisauroidei, described a new suborder, the Aulopoidei, for Aulopidae, Bathysauridae, Bathypteroidae, Ipnopidae, Chlorophthalmidae, and Notosudidae and, as noted, restricted the Myctophiformes to myctophids and neoscopelids.

Sulak (1977) examined aspects of the osteology of the benthic "myctophiforms" and envisioned them forming two divergent lineages exhibiting progressively greater differentiation from the basal aulopid body plan, an expanded Synodontidae that included bathysaurids, synodontids, and harpadontids, and an expanded Chlorophthalmidae for chlorophthalmids (including *Bathysauropsis*) and ipnopids (including bathypteroids).

To examine a previously proposed relationship between the Evermannellidae and Scopelarchidae (e.g., Marshall, 1955; Gosline *et al.*, 1966), R. K. Johnson (1982) studied the distribution of selected characters among iniomes. He did not present a formal classification but described three perceived iniomous clades. One comprised only aulopids, a second was equivalent to Rosen's (1973) alepisauroids minus scopelarchids, and the third included myctophids, neoscopelids, chlorophthalmids, ipnopids, notosudids, and scopelarchids. R. K. Johnson's (1982) phylogeny corroborated Sulak's (1977) placement of bathysaurids in the synodontid + harpadontid lineage, but he noted that only two clades resulting from his analysis, the myctophoids (Myctophidae, and Neoscopelidae) and the alepisauroids (Paralepididae, Anotopteridae, Evermannellidae, Omosudidae, and Alepisauridae) were well supported.

Okiyama (1984b) examined R. K. Johnson's (1982) hypothesis in light of evidence from aulopiform larvae. He did not produce an independent hypothesis of relationships but noted that his data offer little support for a notosudid + scopelarchid + chlorophthalmid + ipnopid lineage; rather, in his similarity matrix, scopelarchids share the most derived features (two) with evermannellids. Larval morphology also does not support a close association between bathysaurids and the synodontid + harpadontid lineage, but, as Okiyama (1984a) noted, *Bathysaurus* larvae are highly specialized.

To demonstrate the potential systematic value of the intermuscular ligaments and bones in teleostean fishes, Patterson and Johnson (1995) investigated aulopiform interrelationships based on this skeletal system. Their data provided support for a monophyletic Synodontoidei (*sensu* Johnson *et al.*, 1996) and a sister-group relationship between evermannellids and scopelarchids. Novel relationships depicted in their strict consensus of 24 equally parsimonious trees include the following: a clade comprising all aulopiform taxa except ipnopids (represented by *Bathypterois* in their analysis) and *Parasudis*; sister-group relationships between *Chlorophthalmus* and synodontoids,

notosudids and the evermannellid–scopelarchid lineage, and bathysaurids and giganturids; and a paraphyletic Paralepididae, with *Paralepis* forming the sister group of a monophyletic clade comprising *Omosudis* and *Alepisaurus*. Patterson and Johnson (1995) noted that the paraphyly of the Paralepididae suggested by their data may be artificial, a result of the greatly reduced number of intermuscular elements in *Macroparalepis*.

No other comprehensive studies of aulopiform relationships have been undertaken, and thus considerable conflict about the evolutionary history of aulopiform fishes existed when we initiated this study, the goal of which was to hypothesize a phylogeny of extant aulopiform genera based on cladistic analysis of a wide range of morphological data. R. K. Johnsons's (1982) cladistic analysis of iniome relationships used commonality rather than outgroup comparison to assess character polarity, and we thus found that many of his polarity decisions were reversed in our analysis. Patterson and Johnson's (1995) phylogeny is of limited value because it was constructed on the basis of a single complex. Despite their shortcomings these publications, as well as those of Rosen (1973), Sulak (1977) and Okiyama (1984b), proved useful in this study, and we derived many informative characters from them.

II. Methods

Osteological abbreviations are listed in Appendix 1, and a full list of materials examined is given in Appendix 2. Terminology for bones of the pelvic girdle follows Stiassny and Moore (1992), and that for the intermuscular bones and ligaments follows Patterson and Johnson (1995). In all line drawings, scale bars represent 1 mm, and open circles indicate cartilage.

A. Data Analysis

Character data were analyzed using heuristic methods in Swofford's (1991) PAUP Version 3.0, and character distributions were explored using MacClade Version 3.04 of Maddison and Maddison (1992). Ctenosquamates, represented by the cladistically primitive Myctophidae, Neoscopelidae, *Metavelifer*, and *Polymixia* (Stiassny, 1986; G. D. Johnson, 1992; Johnson and Patterson, 1993), were considered the first outgroup, and stomiiforms, represented by the cladistically primitive *Diplophos* (Fink and Weitzman, 1982), the second. The analysis included all aulopiform genera except the notosudid *Luciosudis*; the recently described ipnopid, *Discoverichthys* (Merrett and

Nielsen, 1987); and the paralepidids *Dolichosudis*, *Magnisudis*, and *Notolepis*.

All characters were weighted equally, and all multistate characters were treated as unordered unless otherwise noted. Steps in the transformation of a single character are denoted by subscripts following the character number (e.g., 1_2 is state 2 of character 1). Many characters have more than one equally parsimonious reconstruction, and we optimized ambiguous characters on the tree using ACCTRAN, a method that favors reversals over parallel acquisitions when the choice is equally parsimonious (Farris, 1970; Swofford and Maddison, 1987). Ambiguous character states resolved using ACCTRAN are denoted in Discussion (Section VI) with an asterisk, e.g., $(34_1{}^*)$.

Character data also were analyzed using Hennig86 (Farris, 1988) and the results exported to Clados Version 1.2 (Nixon, 1992) for construction of a tree on which characters and states are indicated (Fig. 6). There are some discrepancies in the distribution of character states between PAUP–MacClade and Hennig86–Clados, primarily because (1) for ambiguous characters optimized with e.g., ACCTRAN, MacClade recognizes that ambiguity may still exist at certain nodes, whereas Clados forces a resolution at all nodes; and (2) PAUP–MacClade allows polymorphisms in terminal taxa, whereas Hennig86–Clados does not. Character states on the tree (Fig. 6) that appear as synapomorphies in Clados but not MacClade are marked with a large dot; they are not discussed in the text, which is based on the PAUP–MacClade results.

B. Taxonomy

Parin and Kotlyar (1989) resurrected the aulopid genus *Hime* Starks (type species *A. japonicus* Günther) for Pacific aulopids based on a difference in the length of the dorsal-fin base between Atlantic and Pacific species but used length of the anal-fin base as a taxonomic feature within *Hime*. We find the evidence for generic distinction unconvincing and thus follow Mead (1966a) in recognizing a single genus, *Aulopus*, for all aulopid species.

We place *Harpadon* and *Saurida* in the Synodontidae as did Sulak (1977), *Omosudis* in the Alepisauridae, and *Anotopterus* in the Paralepididae (see Discussion). "*Scopelarchoides*" herein refers to *S. signifer* which, according to R. K. Johnson (1974a), may be an incorrect generic assignment for that species. He hypothesized that *S. nicholsi* (the type species of *Scopelarchoides*) and *S. danae* are more closely related to *Scopelarchus* than to other species of *Scopelarchoides* but retained *Scopelarchoides* for *S. signifer* pending further investigation.

Early in our study it became apparent that *Bathysauropsis gigas* (Kamohara) is not closely related to *B. gracilis* Regan and *B. malayanus* (Fowler). *Bathysauropsis gracilis* is the type species of *Bathysauropsis* Regan, and thus all reference to *Bathysauropsis* is to *B. gracilis* and *B. malayanus*. A new genus, *Bathysauroides*, is erected for *Bathysauropsis gigas*.

III. *Bathysauroides* Gen. Nov.

Diagnosis—An aulopiform distinguished from all other genera by the following combination of characters: a low number of caudal vertebrae (5–7, or ca. 11–15% of total vertebrae in *Bathysauroides gigas*), slightly elliptical eyes with an anterior aphakic space and gill rakers present as toothplates.

Type species—*Bathysaurops gigas* Kamohara 1952.

Etymology—From the Greek *bathys*, deep, and *sauros*, lizard, in reference to the deep habitat and superficial resemblance to lizardfishes.

Gender—Masculine.

Justification—Our hypothesis of cladistic relationships among aulopiform genera (Fig. 2) is best reflected by removing *Bathysauropsis gigas* from *Bathysauropsis* Regan and placing it in a distinct genus. In addition to the diagnostic characters listed above, *Bathysauroides gigas* can be distinguished from its former congeners based on the following features identified in this study or taken from the original description of *Bathysaurops gigas* (Kamohara, 1952): palatine with more prominent teeth than premaxilla; epipleurals extending anteriorly to the 1st vertebra (vs 2nd); epineurals on about the 3rd through 17th vertebrae originating on centrum (vs neural arch); 16–17 pectoral-fin rays (vs 22–24); basihyal with two rows of large teeth (vs no basihyal teeth); pectoral fin extending to vertical through middle of dorsal-fin base (vs beyond base of dorsal); anus much closer to pelvic fins than to anal fin (vs closer to anal fin); and adipose fin inserting above anterior part of anal-fin base (vs well behind anal base)

IV. Monophyly of Aulopiformes

We agree with Rosen (1973) that a lateral displacement of the proximal end of PB2 and a concomitant elongation of the uncinate process on EB2 to bridge the large gap between EB2 and PB3 are derived for aulopiforms (Character 1, Fig. 3). We disagree with R. K. Johnson's (1982) assessment of an elongate EB2 uncinate process as a primitive iniome condition because the first and second outgroups for iniomes are

FIGURE 2 Proposed phylogenetic relationships among aulopiform genera based on strict consensus of nine equally parsimonious trees (length = 364, CI = 0.55, RI = 0.80).

acanthomorphs and stomiiforms, neither of which has an EB2 uncinate process. We also disagree with his interpretation of the unbranched anterior portion of the EB2 of *Neoscopelus* as an elongate uncinate process. There is nothing in the EB2 morphology of *Neoscopelus* (Rosen, 1973, fig. 71) to suggest that it is configured differently from that of myctophids and stomiiforms—that is, the cartilaginous tip is somewhat expanded such that it articulates with both PB2 and PB3 (Rosen, 1973, figs. 18–22 and 69–70). Furthermore, like those two groups, the EB2 of *Neoscopelus* articulates with a cartilaginous condyle on PB3,

the absence of which is another aulopiform synapomorphy (Character 2; Johnson, 1992).

Rosen (1973, figs. 14–16) questioned an aulopiform affinity for paralepidids because of (1) the primitive, salmoniform-like appearance of the dorsal gill arches of juvenile *Paralepis speciosa* and (2) the absence in adult *Paralepis* and *Lestrolepis* of the long EB2 uncinate process and laterally displaced PB2 characteristic of other aulopiforms. The first is invalid because Rosen's (1973, fig. 16) "juvenile *Paralepis*" is not a paralepidid. We examined the specimen upon which his description and illustration were based (AMNH 17232) and

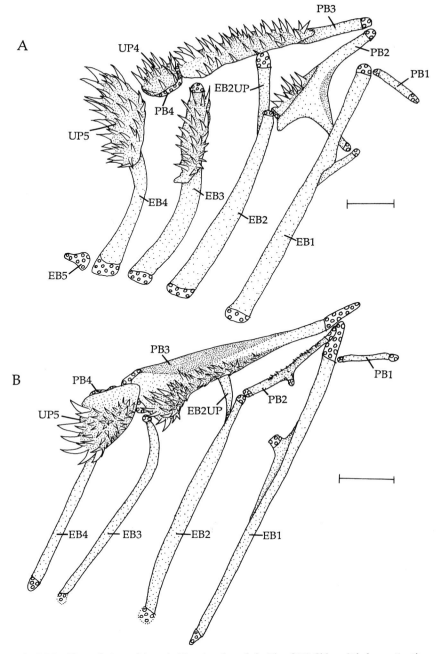

FIGURE 3 Ventral view of dorsal gill arches from left side of (A) *Chlorophthalmus atlanticus,* USNM 339774 and (B) *Synodus variegatus,* USNM 339776.

concluded on the basis of meristic and other features that it is an argentinoid, probably *Bathylagus.* Several features characteristic of bathylagid (and not aulopiform) gill arches are evident in Rosen's fig. 16: PB2 is broad anteriorly rather than tapered, UP4 is absent, UP5 (labelled UP4 or UP5 by Rosen) is extremely reduced to a small ovoid plate, and there is a long levator process on EB4. Note also that the muscle labelled "RAB" by Rosen inserts on EB4 rather than the pha-

ryngobranchials, indicating that it is the oesophageal sphincter, not the retractor dorsalis. Rosen's (1973) "juvenile *Paralepis*" also has an uncinate process on PB3 for articulation with the uncinate process of EB2, a primitive teleostean feature lacking in aulopiforms (Johnson, 1992).

Rosen's (1973) second claim, that paralepidids lack a laterally displaced PB2 and concomitant elongation of the EB2 uncinate process is not true of *Paralepis,*

Arctozenus, Anotopterus, or *Sudis.* In those taxa, as in other aulopiforms, the uncinate process of EB2 (which is cartilaginous in *Paralepis* and *Arctozenus*) spans the gap between PB3 and the posterolaterally displaced PB2. In other paralepidids examined (*Macroparalepis, Uncisudis, Lestidium, Lestidiops, Stemonosudis,* and *Lestrolepis*), the uncinate process of EB2 is parallel and closely applied to the main arm of EB2, which undoubtedly explains why Rosen overlooked it. The configuration of the dorsal gill arches of those paralepidids involves several diagnostic modifications that we discuss in more detail in a later section (see character 9).

Additional evidence corroborating the monophyly of Rosen's (1973) Aulopiformes is found in the pattern of the intermuscular bones (Patterson and Johnson, 1995). The group is uniquely characterized by having attached epipleural bones extending forward to at least the second, and frequently the first, vertebra (character 54). Epipleurals are most commonly restricted to midbody as they are in stomiiforms, myctophiforms, and *Polymixia* (the only acanthomorph with epipleural bones). Our analysis also indicates that another feature of the intermusculars, the displacement of one or more of the anterior epipleurals dorsally into the horizontal septum (character 55), a feature used by Patterson and Johnson (1995) to indicate relationships within the Aulopiformes, is best interpreted as a synapomorphy of the group.

Another aulopiform character is their lack of a swimbladder (character 112; see Marshall, 1954, 1960; Marshall and Staiger, 1975). Many deep-sea fishes lack a swimbladder, but the presence of a swimbladder primitively in stomiiforms (including *Diplophos*) and ctenosquamates (most myctophids and neoscopelids, lampridiforms, and polymixiids)—see Marshall (1960), Woods and Sonoda (1973)—suggests that loss of the swimbladder in aulopiforms is independent of losses in other teleosts. R. K. Johnson (1982) hypothesized three losses of the swimbladder among iniomes: in aulopids, in the chlorophthalmoid lineage of his myctophoid + chlorophthalmoid clade, and in the ancestor of his alepisauroid + synodontoid lineage. Rosen's (1973) hypothesis of a monophyletic Aulopiformes requires a single loss in the ancestral aulopiform.

We agree with R. K. Johnson's suggestion that peritoneal pigment in larvae may be diagnostic of Rosen's (1973) aulopiforms (character 116). Larvae of *Diplophos,* myctophiforms, and primitive acanthomorphs lack peritoneal pigment, as do several aulopiforms (notosudids, some ipnopids, and the scopelarchid *Benthalbella*), presumably secondarily. Larvae of *Bathysauropsis* and *Bathysauroides* are unknown.

Finally, we have found new evidence for aulopiform monophyly in the morphology of the pelvic girdle. Primitively in euteleosts, the pelvic plates often approach one another or abut medially in the region of the medial processes (Stiassny and Moore, 1992), as in *Diplophos* and myctophiforms (Fig. 4A), but the medial processes are never fused. Uniquely in aulopiforms, the medial processes of the pelvic girdle are long broad plates that are joined medially by cartilage (character 87, Figs. 4B–4D, and 5).

Stiassny (1986) rejected Rosen's concept of Aulopiformes, arguing that three genera of that group (*Aulopus, Chlorophthalmus,* and *Parasudis*) form the sister group of ctenosquamates based on a particular type of association between the maxilla and the palatine (her fig. 5). This single feature (character 44) does not outweigh the branchial, intermuscular, swimbladder, larval pigmentation, and pelvic girdle evidence that unites aulopiforms. Furthermore, placement by Johnson *et al.* (1996) of the Aulopidae as the sister-group of other synodontoids and our placement of the Chlorophthalmidae as the sister group of other chlorophthalmoids are in direct conflict with Stiassny's (1986) hypothesis.

V. Character Analysis

Our hypothesis of the relationships among aulopiform genera (Fig. 2) was derived from the data matrix in Table 1. The tree represents a strict consensus of nine fully resolved trees (each 364 steps in length, CI=0.55, RI=0.80 in the PAUP analysis), all of the ambiguity occurring within the Paralepididae and Scopelarchidae. The Hennig86 analysis yielded the same trees, although there were small differences in tree statistics.

Based on our analysis, we divide aulopiform genera into four clades: Synodontoidei (Aulopidae, Pseudotrichonotidae, and Synodontidae), Chlorophthalmoidei (Chlorophthalmidae, *Bathysauropsis,* Notosudidae, and Ipnopidae), Alepisauroidei (Alepisauridae, Paralepididae, Evermannellidae, and Scopelarchidae), and Giganturoidei (*Bathysauroides,* Bathysauridae, and Giganturidae). In the following comparison of phylogenetically informative characters among aulopiforms, character numbers refer to those in the matrix (Table 1) and on the Clados-derived tree (Fig. 6).

A. Gill Arches

1. Second Epibranchial Uncinate Process (Fig. 3)—As discussed above (in Monophyly of Aulopiformes) the presence of an uncinate process on EB2 articulating with PB3 characterizes all aulopiform except *Bathypterois* and some paralepidids. In *Bathypterois* (Fig. 7B),

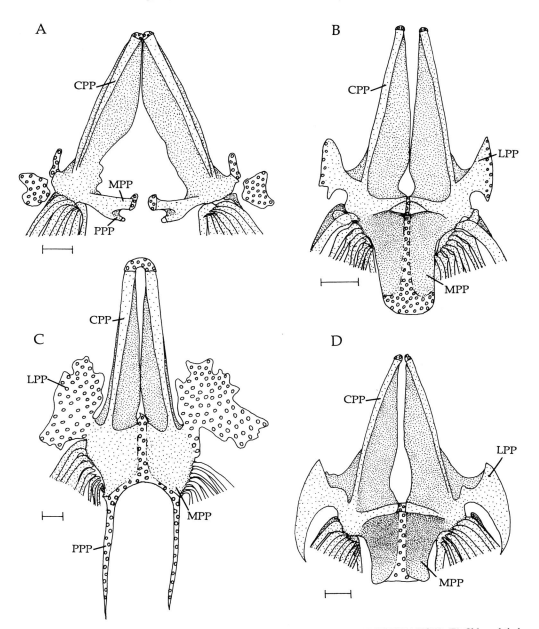

FIGURE 4 Ventral view of pelvic girdle of (A) *Myctophum obtusirostre*, AMNH 29140SW, (B) *Chlorophthalmus agassizi*, USNM 159385, (C) *Bathypterois pectinatus*, FMNH 88982, and (D) *Scopelosaurus hoedti*, USNM 264256.

the EB2 uncinate process falls well short of PB3, but PB2 is posterolaterally displaced as it is in other aulopiforms. In certain paralepidids (*Macroparalepis, Uncisudis, Lestidium, Lestidiops, Stemonosudis, and Lestrolepis* [Fig. 8B]), PB2 is reoriented and the resulting configuration of EB2 and its uncinate process is very different from that of other aulopiforms. We describe this condition more fully in Character 9 below and, to avoid duplicating what we interpret as a unique specializa-

tion of paralepidids, we do not assign a different state to that condition here. Other features clearly place *Bathypterois* and all paralepidids deep within the Aulopiformes, and thus the variation in the EB2 uncinate process in those taxa is derived relative to the primitive aulopiform condition.

(1_0) = EB2 uncinate process absent
(1_1) = EB2 uncinate process present and enlarged;

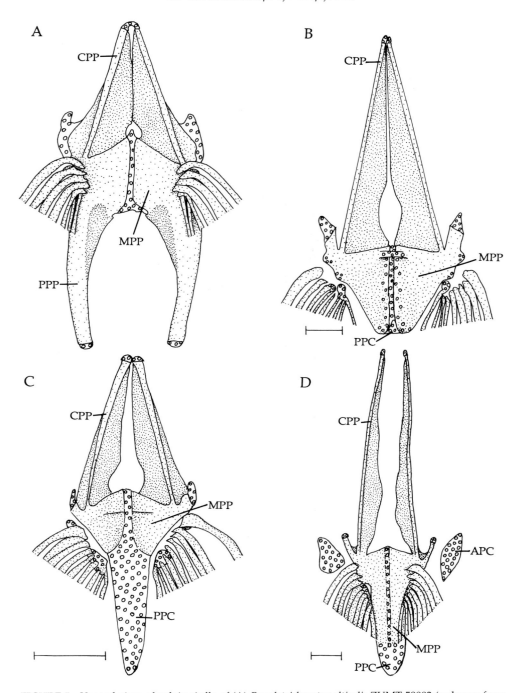

FIGURE 5 Ventral view of pelvic girdle of (A) *Pseudotrichonotus altivelis* ZUMT 59882 (redrawn from Johnson *et al.*, 1996), (B) *Scopelarchoides signifer*, USNM 274385, (C) *Evermannella indica*, USNM 235141, and (D) *Lestrolepis intermedia*, USNM 290253. The dorsally projecting autogenous pelvic cartilages in *Evermannella* are not shown because they are obscured by the pelvic girdle; in *Lestrolepis*, these cartilages have been manually displaced from their dorsally directed orientation for illustration.

PB2 displaced posterolaterally (except in some paralepidids)

(1_2) = EB2 uncinate process present, but not enlarged; PB2 displaced posterolaterally

2. *Cartilaginous Condyle on Dorsal Surface of Third Pharyngobranchial*—Aulopiforms lack a condyle on PB3 articulating with EB2 (Johnson, 1992). This condyle is a primitive euteleostean condition and is pres-

TABLE 1 Matrix for 118 Characters in 37 Aulopiform Genera, the Stomiiform *Diplophos*, the Myctophiforms Myctophidae (*Myctophum*, *Lampanyctus*) and *Neoscopelus*, and the acanthomorphs *Metavelifer* and *Polymixia*.

	1–5	6–10	11–15	16–20	21–25	26–30	31–35	36–40	41–45	46–50	51–55	56–60	61–65	66–70	71–75	76–80	81–85	86–90	91–95	96–100	101–105	106–110	111–115	116–118
Diplophos	00100	01000	00010	00000	00000	00000	00110	00000	00000	00000	00000	00000	00020	00011	00100	00000	00000	00000	00011	20100	00000	0001?	00000	000
Myctophidae	00000	00000	00000	01000	?1100	00000	00200	00000	00020	00000	00000	00000	00100	00010	02000	0000?	00000	00000	10000	00000	00000	00000	0?0?0	000
Neoscopelus	00000	00000	00000	01000	01000	00000	00000	00000	00020	00000	00000	00000	00100	00010	02000	00021	00000	00000	00000	00000	00000	00000	0?000	000
Metavelifer	0000?	00000	00000	00000	01000	10001	00000	00000	00010	50000	0003?	?????	00000	00210	02012	00030	000?0	00000	00000	10100	00000	00?1?	0?000	000
Polymixia	00000	00000	00000	01000	01000	00001	00000	00001	00020	00000	00000	00000	10000	00010	02000	00030	00000	00000	00000	00000	00000	00110	00001	000
Aulopus	11000	00000	00020	02000	01000	00120	00000	00000	00020	00000	00011	00000	01000	00011	00020	0103?	20000	01100	00000	30000	00000	00000	01000	100
Pseudotrichonotus	11100	01000	00020	00000	01000	00100	00000	00010	00010	00000	00011	10000	01000	00110	00120	01030	21200	01100	00000	00000	00000	00010	01000	200
Synodus	11111	00000	00000	22000	00001	00100	00000	01111	10000	00000	10011	10000	01000	00001	10220	01030	21200	01100	00000	00000	00000	00000	01000	200
Trachinocephalus	11111	00000	00010	22000	00001	10120	00000	01111	10000	00000	10011	10000	01000	00001	10220	01030	21200	01100	00000	00000	00000	00000	01000	200
Harpadon	11101	00001	11010	22100	00001	10120	01100	01100	10000	00000	20011	00000	00030	00001	00320	11030	01100	01100	00000	20000	00000	00000	01000	200
Saurida	11101	00000	11010	22100	00001	10120	01100	01100	10000	00000	10011	00000	00030	00001	00120	11030	20100	01100	00000	00000	00000	00000	01000	200
Bathypterois	21000	21000	00020	00000	01100	00010	00010	00000	01001	00000	01110	00000	00000	00210	01310	00020	00000	01011	00000	10000	00000	?2101	01100	?00
Bathymicrops	11100	20000	00020	00010	11110	00000	00010	00000	01101	50000	01110	00000	00000	00220	01530	10130	04200	01071	00000	20010	00000	?2211	01100	000
Bathytyphlops	11002	21000	00020	00000	11110	00000	00010	00000	01101	30000	01111	00000	00000	00120	01310	00030	00100	01011	00000	00010	00000	?2111	01100	100
Ipnops	11000	21000	00020	00010	01100	00000	00010	00000	01001	30000	0112?	?0000	01000	00220	02310	00130	02100	01011	00000	20000	00000	?5111	01100	000
Scopelosaurus	11000	20000	00010	00001	11100	00100	00100	00100	00001	10000	00011	01000	10000	00210	01010	00020	00000	01021	00000	00000	00000	11101	01010	000
Ahliesaurus	11000	20000	00010	00001	11100	00000	00100	00100	00001	10000	00011	01000	10000	00210	01010	00020	00000	01021	00000	00000	00000	11?01	01010	000
Chlorophthalmus	11000	10000	00020	00000	00001	00001	00000	00000	00021	00000	00011	00000	01100	00120	01010	0000?	00000	01021	00000	30000	00001	10001	01000	100
Parasudis	11000	10000	00020	00000	01000	00001	00000	00021	00021	00000	00010	00000	00100	10210	01010	00000	00000	01021	00000	00000	00001	10001	01000	101
Bathysauropsis	11000	10000	00020	00000	01100	0000?	00010	00000	00001	00000	00011	00110	00100	00010	01010	00000	00000	0102?	10010	30000	00000	11001	01???	??0
Omosudis	11001	01101	11002	00100	00000	0000?	?0000	20000	00000	20000	00021	00110	00071	10210	01421	00070	01010	01020	10010	00100	12100	00001	11001	101

Taxon																							
Alepisaurus	11001	01001	11002	00100	00000	00001	00000	20000	20000	00021	00100	00021	10310	01421	00000	01010	01020	10010	20100	12100	00001	11001	101
Coccorella	11001	01101	11003	10000	10000	00001	00020	20000	20000	00021	00000	00021	31120	01010	00000	00000	01020	10110	30101	11000	03001	01020	100
Odontostomops	11001	01100	11002	10000	10020	01001	00020	20000	20000	00021	00700	00021	31120	01010	00201	00201	11020	10110	30101	11000	00001	01020	100
Evermannella	11001	01100	11002	10000	10020	01001	00020	20000	20000	00021	00000	00021	31120	01010	00201	00201	11020	10110	30101	11000	03001	01020	100
Scopelarchus	11001	01000	11002	00000	00001	00001	21000	20000	20000	00021	00000	00000	01020	01340	00000	00000	01021	01010	00101	10000	03001	01020	100
Scopelarchoides	11001	01000	11001	00000	00001	00001	21000	20000	20000	00021	00000	00000	01020	01040	00000	00000	21020	01010	00101	10000	03001	01020	100
Benthalbella	11001	00000	11001	00000	00001	00001	21000	20000	20000	00027	00000	00070	07020	01040	00070	00000	01020	01010	00101	10000	03001	01020	000
Rosenblattichthys	11001	01000	11001	00000	00001	00001	21000	20000	20000	00027	00070	01040	07020	01040	00000	00070	01020	01010	00101	10000	03001	01020	100
Paralepis	11001	01000	11002	00002	01000	21000	21111	20000	00021	00120	01020	01421	20110	01421	00000	00000	01020	10010	30100	10000	00001	01000	100
Arctozenus	11001	01000	11002	00002	01000	21000	21111	20000	00021	00020	00020	01021	20110	01021	02000	00000	01020	10010	00100	10000	00001	01000	100
Lestrolepis	11001	01010	27002	00002	01000	21000	21111	20000	00021	00001	00001	01121	20120	01121	00700	00020	01020	10011	01100	11010	00001	01000	101
Lestidium	11001	01010	27002	00002	01000	21000	21111	20000	00021	00001	00001	01021	20110	01021	00000	00010	01020	10011	01100	11010	00001	01000	100
Stemonosudis	11001	01010	27002	00002	01000	21000	21111	20000	00021	00001	00001	01121	20110	01121	00000	00020	01021	10011	01100	11010	00001	01000	101
Uncisudis	11001	01010	27002	00002	01000	21000	21111	20000	00022	?0???	?0010	?1722	??110	?1722	?0010	00020	01020	10011	01100	11010	00001	01000	100
Macroparalepis	11001	01010	27002	00002	01000	21000	21111	20000	00021	00001	00010	01221	20110	01221	00000	00000	01020	10011	01100	11010	00001	01000	100
Lestidiops	11001	01010	27002	00002	01000	21000	21111	20000	00007?	00?00	00700	01120	??110	01120	00700	00010	01020	10011	00100	11010	00001	01000	101
Sudis	11001	01000	27002	00002	20000	20000	?1117	20000	00021	?7000	?7000	01122	20110	01122	?0000	?0000	01020	10011	?0100	11010	00001	01101	100
Anotopterus	11101	01001	27002	00002	00000	00000	01111	20000	10021	00002	00002	01371	20710	01371	00007	0₃20?	01020	0007?	11100	01200	00001	01000	?00
Bathysauroides	11001	00000	01110	01000	00000	01000	00000	20000	00021	11000	11000	01010	00000	01010	20000	00001	01020	00000	30000	00000	11001	01???	?00
Bathysaurus	11001	00000	01110	01101	00000	01101	00000	10000	30021	20040	20040	00310	20000	00310	00001	00001	01020	00000	30000	00000	010₁0¹	01100	110
Gigantura	1100?	0?000	01???	?00??	?0?00	07000	60007	?0000	20021	20300	20300	0053?	20700	0053?	20220	10200	07020	00010	20200	02000	04001	01000	110

Note. "?" indicates a missing data point. Polymorphisms are represented by a superscript over a subscript, e.g., $\frac{0}{1}$.

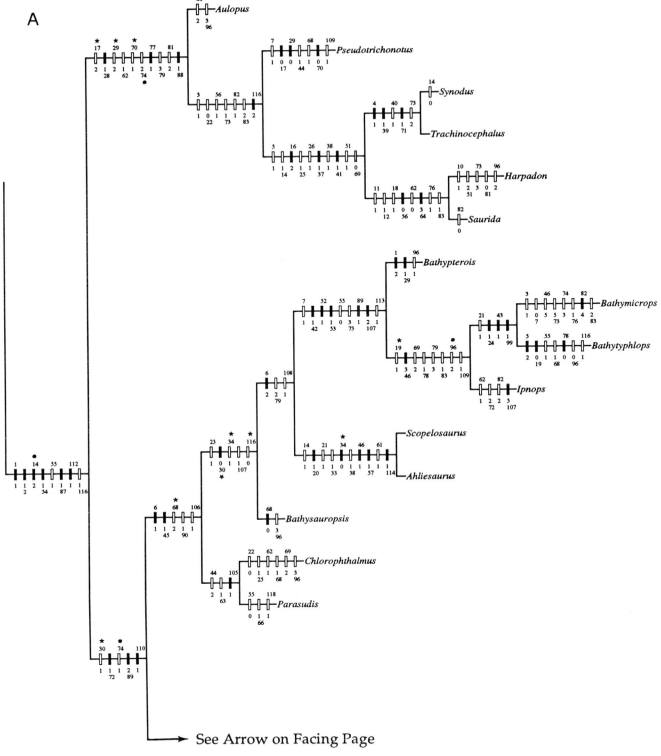

FIGURE 6 Distribution of derived character states among aulopiform genera. (A) Synodontoidei and Chlorophthalmoidei; (B) Alepisauroidei and Giganturoidei. Black bars denote non-homoplasious forward changes; open bars indicate homoplasious forward changes. Character state numbers are directly below bars; character numbers are alternated above and below bars. Ambiguous character states resolved using ACCTRAN are marked with an asterisk. Derived character states that appear in the Hennig86–Clados analysis but not in PAUP–MacClade are marked with a black dot.

B

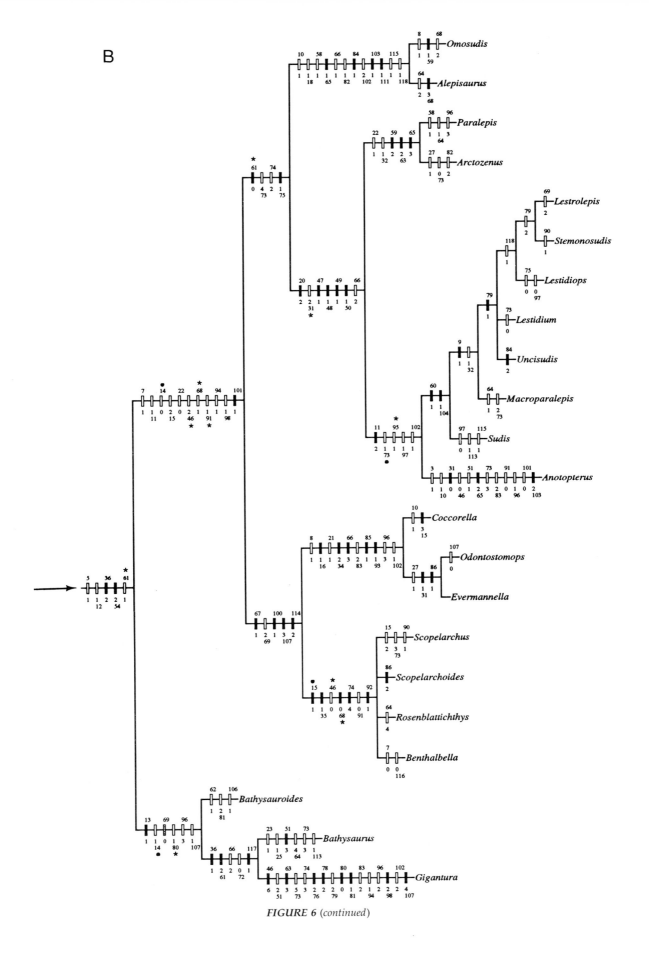

FIGURE 6 (continued)

ent in *Diplophos*, myctophiforms, and ctenosquamates in general.

(2_0) = PB3 with cartilaginous condyle articulating with EB2

(2_1) = PB3 without cartilaginous condyle articulating with EB2

3. Fourth Pharyngobranchial Toothplate (Fig. 3)—Johnson et al. (1996) described the distribution of the fourth pharyngobranchial toothplate (UP4) among aulopiforms and outgroups and concluded that although polarity for Aulopiformes is equivocal (absent in *Diplophos*, present in ctenosquamates), loss of UP4 is a synapomorphy of Pseudotrichonotidae, Synodontidae, and Harpadontidae. Our analysis corroborates this hypothesis. Independent losses of UP4 occur in *Bathymicrops* and *Anotopterus*.

(3_0) = UP4 present

(3_1) = UP4 absent

4. Articulation of First Pharyngobranchial (Fig. 3B)—In *Synodus* and *Trachinocephalus* the first pharyngobranchial (PB1) articulates with the proximal base of the elongate cartilaginous tip of EB1. In all other aulopiforms and outgroups, PB1 (if present) articulates at the distal end of the cartilaginous tip of EB1.

(4_0) = PB1 articulates at distal tip of EB1

(4_1) = PB1 articulates at proximal base of cartilaginous tip of EB1

5. Gill Rakers or Toothplates—R. K. Johnson (1982) hypothesized independent replacement of gill rakers by toothplates in scopelarchids and the ancestor of his synodontoids + alepisauroids. Our analysis suggests that scopelarchids are closely related to alepisauroids but synodontids are not, and thus the presence of toothplates is both a synapomorphy of alepisauroids plus giganturoids and of synodontids. Gill rakers are lathlike in most chlorophthalmoids, pseudotrichonotids, aulopids and the outgroups. *Bathytyphlops* has all rakers present as toothplates except for a single elongate raker on EB1. *Metavelifer* has normal rakers on the first arch but reduced rakers on the others. Gill rakers and toothplates are lacking in *Gigantura*.

(5_0) = Gill rakers long, lathlike

(5_1) = Gill rakers present as toothplates

(5_2) = Single elongate gill raker on EB1

6. Second Pharyngobranchial with Extra Uncinate Process (Figs. 7, and 8)—The typical aulopiform PB2 is tipped with cartilage at the proximal and distal ends and has an uncinate process for articulation with the uncinate process of EB1 (Fig. 3, and 8B). Ipnopids (Fig. 7B) and notosudids (Fig. 7A) have an extra PB2 uncinate process proximally that extends along the lateral surface of the distal portion of EB2. It is best developed in *Bathypterois*. In other chlorophthalmoids, there is no extra uncinate process, but the proximal cartilaginous head of PB2 is expanded laterally so that it extends slightly along the lateral aspect of EB2 (Fig. 8A). This appears to be intermediate between the condition in ipnopids and notosudids, in which there is a clear separation of the proximal cartilaginous head of PB2 into two cartilage-tipped processes, and that of other aulopiforms and the outgroups in which the proximal head of PB2 is not expanded or divided and contacts EB2 squarely. An expanded proximal PB2 base is a synapomorphy of chlorophthalmoids, the extra uncinate process appearing in the ancestor of ipnopids and notosudids.

(6_0) = PB2 without extra uncinate process

(6_1) = PB2 without extra uncinate process but with an expanded proximal base

(6_2) = PB2 with extra uncinate process

7. Second Pharyngobranchial Toothplate (Fig. 3)—A toothplate fused to PB2 (UP2) is present in synodontoids (except *Pseudotrichonotus*), chlorophthalmids, *Bathysauropsis*, *Bathymicrops*, notosudids, *Bathysauroides*, and *Bathysaurus*. It is lacking in other aulopiforms, including all alepisauroids except the scopelarchid *Benthalbella*, which has a very small toothplate proximally. R. K. Johnson (1982) noted that *Scopelarchoides signifer* also has UP2, but he did not illustrate it as such (R. K. Johnson, 1974a, fig. 10), and UP2 is lacking in our specimen of *S. signifer*. Presence of UP2 in ctenosquamates and absence in primitive stomiiforms suggests that polarity for aulopiforms is equivocal. Nevertheless, all reconstructions of aulopiform phylogeny produced in this study indicate that UP2 is primitively present in aulopiforms, and thus its absence in alepisauroids is derived. UP2 is lost in the ancestor of the Ipnopidae (R. K. Johnson, 1982) and independently in *Pseudotrichonotus*. It is reacquired in *Bathymicrops* and *Benthalbella*. PB2 is lacking in adults of the highly specialized *Gigantura*.

(7_0) = UP2 present

(7_1) = UP2 absent

8. Second Pharyngobranchial Uncinate Process (Fig. 9B)—Evermannellids and *Omosudis* have a long, laterally directed uncinate process on PB2 that articulates with a laterally directed uncinate process on the first epibranchial (EB1). The EB1 uncinate process is more medially directed in other aulopiforms (e.g., Figs. 3, 7, and 8) and the outgroups, and thus a much shorter PB2 uncinate process bridges the gap between the two bones. The EB1 uncinate process is also medially

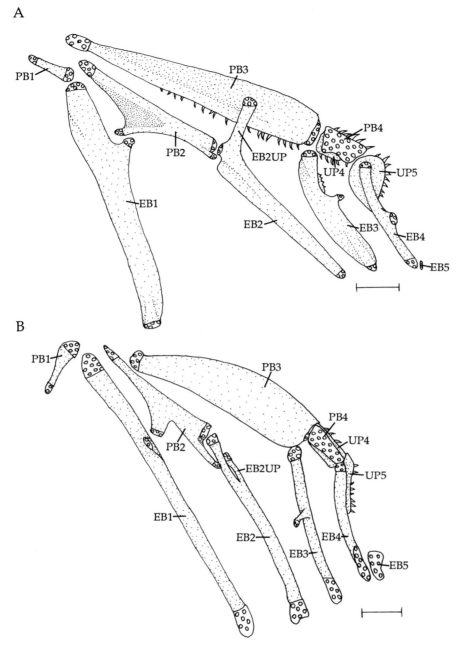

FIGURE 7 Dorsal view of dorsal gill arches from left side of (A) *Scopelosaurus hoedti*, USNM 264256 and (B) *Bathypterois pectinatus*, FMNH 88982.

directed and the uncinate process of PB2 is short in *Alepisaurus*, paralepidids, and scopelarchids, and thus it is most parsimonious to hypothesize independent origins of the long uncinate process in *Omosudis* and the ancestral evermannellid.

(8_0) = PB2 with short uncinate process
(8_1) = PB2 with long uncinate process

    9. Uncinate Process of Second Epibranchial Adjacent to

Second Epibranchial (Fig. 8B)—The dorsal gill arches of the paralepidid genera *Macroparalepis, Uncisudis, Lestidium, Lestrolepis, Stemonosudis,* and *Lestidiops* ("*Macroparalepis* and above" for short) are distinctive among aulopiforms in that both EB2 and its uncinate process terminate distally at or near PB3. This arrangement is the result of a shift in the position of PB2, which primitively runs anteromedial to posterolateral in aulopiforms and outgroups but extends almost an-

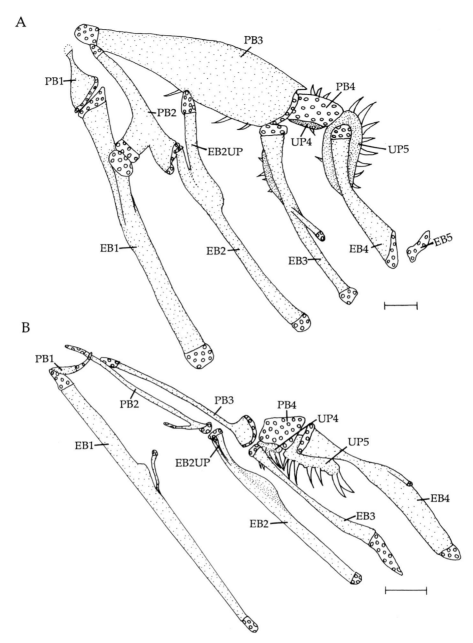

FIGURE 8 Dorsal view of dorsal gill arches from left side of (A) *Bathysauropsis gracilis*, AMS IA6934 and (B) *Lestrolepis intermedia*, USNM 290253.

terior to posterior in *Macroparalepis* and above, lying directly or nearly so against PB3. This modification brings the proximal end of PB2 close to PB3, and thus EB2 approaches PB3 at its articulation with PB2. The main arm of the second epibranchial and its uncinate process therefore are adjacent as they approach PB3 and PB2, respectively. The uncinate process of EB2 is in such close proximity to EB2 in some paralepidids

as to be easily overlooked (e.g., Rosen, 1973, fig. 15), and in *Lestrolepis*, it is a small strut that falls well short of PB3. Those conditions are not found elsewhere among aulopiforms or outgroups.

(9_0) = EB2 uncinate process diverges from EB2 as it approaches PB3; PB2 oriented anteromedial to posterolateral

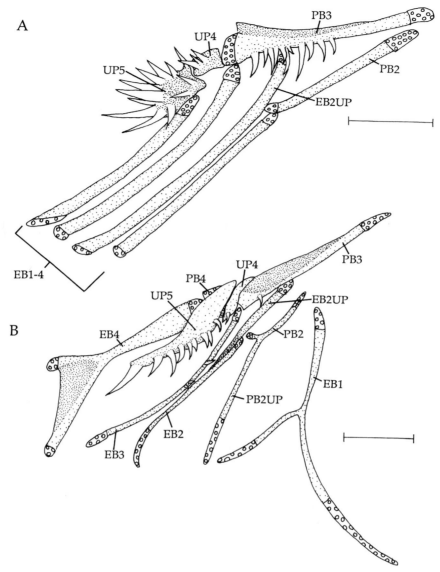

FIGURE 9 Ventral view of dorsal gill arches from left side of (A) *Scopelarchus analis*, USNM 234988 and (B) *Coccorella atlantica*, USNM 235189.

(9_1) = EB2 uncinate process adjacent to EB2 as both approach PB3; PB2 oriented anterior to posterior

10. *Third Pharyngobranchial Produced (Fig. 9B)*—In *Harpadon, Omosudis, Alepisaurus, Anotopterus*, and *Coccorella*, PB3 extends anteriorly well beyond the distal ends of EB1 and PB2. In other aulopiforms and outgroups, PB3 extends to or falls slightly short of the terminal points of those bones. Parsimony indicates convergence in the evolution of a produced PB3—in the ancestor of alepisaurids and independently in *Harpadon, Anotopterus*, and *Coccorella*.

(10_0) = PB3 not extending anteriorly beyond the tips of EB1 and PB2
(10_1) = PB3 extending anteriorly beyond the tips of EB1 and PB2

11. *Distribution of PB3 Teeth (Fig. 9).*—Alepisauroids share a reduced third pharyngobranchial toothplate (UP3), the teeth being reduced in number and restricted to the lateral edge of the ventral surface of PB3. A further reduction occurs in the Paralepididae (including *Anotopterus*), where UP3 is lacking in all members of the family examined except *Paralepis* and

Arctozenus. Primitively in aulopiforms (including *Bathysaurus* and larval *Gigantura*) and outgroups, UP3 is not confined to the lateral edge and sometimes covers most of PB3 (Fig. 3). UP3 is independently reduced in *Harpadon* and *Saurida*, where it resembles that of alepisauroids in being restricted to the lateral edge of PB3.

(11_0) = UP3 covering large area of ventral surface of PB3

(11_1) = UP3 restricted to lateral edge of ventral surface of PB3

(11_2) = UP3 absent

12. Size of PB3 Teeth—Giganturoids share with alepisauroids very large PB3 teeth. We did not measure teeth and thus our analysis of the character is not quantitative, but compare the size of PB3 teeth in *Aulopus* (Rosen, 1973, fig. 4) with that of *Alepisaurus* (Rosen, 1973, fig. 9) and *Gigantura* (Rosen, 1973, fig. 17). *Harpadon* and *Saurida* again are homoplastic in exhibiting large PB3 teeth.

(12_0) = PB3 teeth small

(12_1) = PB3 teeth large

13. First Pharyngobranchial—A suspensory pharyngobranchial is absent in *Pseudotrichonotus, Omosudis, Alepisaurus, Anotopterus, Coccorella, Scopelarchus,* and some *Scopelarchoides,* e.g., *S. danae* and *S. nicholsi* (see R. K. Johnson, 1974a). It is reduced to a small cartilage in *Odontostomops, Evermannella* and *Scopelarchoides signifer.* It is thus a variable feature within the Aulopiformes, especially in alepisauroids among which relationships are uncertain, and we are unable to explain convincingly its pattern of evolution within the group.

In *Bathysauroides* and *Bathysaurus,* the suspensory pharyngobranchial is an unusually long bone, roughly one-third the size of the first epibranchial. PB1 is relatively smaller in other aulopiforms and outgroups, approximately one-fifth the size of the first epibranchial or smaller (as in *Chlorophthalmus, Synodus, Scopelosaurus, Bathypterois, Bathysauropsis,* and *Lestrolepis,* Figs. 3, 7, and 8). Although PB1 is large in larval *Gigantura* (see Fig. 19) it is absent in adults, and we conservatively code this character as "missing data" in *Gigantura.* Regardless, a long PB1 emerges as a synapomorphy of giganturoids.

(13_0) = PB1 normal, reduced, or absent

(13_1) = PB1 very long

14. Fifth Epibranchial (Figs. 3, 7, and 8)—A fifth epibranchial (EB5—see Bertmar, 1959; Nelson, 1967) is lacking in all ctenosquamates, and present as a tiny cartilage in *Diplophos.* It is present and large in *Aulopus, Pseudotrichonotus,* chlorophthalmids, *Bathysaur-*

opsis, and ipnopids, and reduced or absent in other synodontoids and chlorophthalmoids. Alepisauroids lack EB5, and giganturoids have it in the form of a small cartilaginous element (except the highly modified *Gigantura* in which it is absent). There are many possible reconstructions of the evolution of this feature, but in all of them a small cartilaginous EB5 is a synapomorphy of notosudids.

(14_0) = EB5 absent

(14_1) = EB5 present as a small cartilage

(14_2) = EB5 present as a large cartilage

15. Dentition of Fifth Ceratobranchial (Fig. 10)—Primitively in aulopiforms and outgroups, small teeth cover the medially expanded anterior surface of CB5 (Fig. 10B). In most alepisauroids (Fig. 10A), teeth are restricted to the medial edge of CB5, although in the scopelarchid genera *Scopelarchoides, Benthalbella,* and *Rosenblattichthys,* most teeth are medial, but one or two are scattered across the center of the dorsal surface. We tentatively interpret the reduction of the tooth-bearing area of CB5 as a synapomorphy of alepisauroids, with one or more modifications in scopelarchids. CB5 teeth are lacking in *Coccorella* (R. K. Johnson, 1982).

(15_0) = CB5 with teeth scattered all over anterodorsal surface

(15_1) = CB5 with most teeth restricted to medial edge of anterodorsal surface

(15_2) = CB5 with all teeth restricted to medial edge of anterodorsal surface

(15_3) = CB5 without teeth

16. Shape of Fifth Ceratobranchial—In aulopiforms and outgroups, CB5 is primitively rod-shaped posteriorly, with a medially expanded tooth-bearing surface anteriorly (Fig. 10B). Synodontids have a somewhat V-shaped CB5 in which the medial expansion is robust (Fig. 10C). Evermannellids also have a somewhat V-shaped CB5, the medial expansion of which is very slender (R. K. Johnson, 1982, figs. 15D, and 15H).

(16_0) = CB5 not V-shaped

(16_1) = CB5 V-shaped, the medial limb slender

(16_2) = CB5 V-shaped, the medial limb robust

17. Gap Between the Fourth Basibranchial Cartilage and Fifth Ceratobranchials (Fig. 10)—In *Diplophos* and most aulopiforms, BB4 extends posteriorly beyond its articulation with the CB4s to articulate with the proximal tips of the CB5s (Fig. 10A). In myctophids, neoscopelids, and *Polymixia,* BB4 is reduced in length, not extending beyond the bases of the CB4s; the CB5s closely approach but do not articulate with BB4. The main body of BB4 is also reduced in *Aulopus* and synodon-

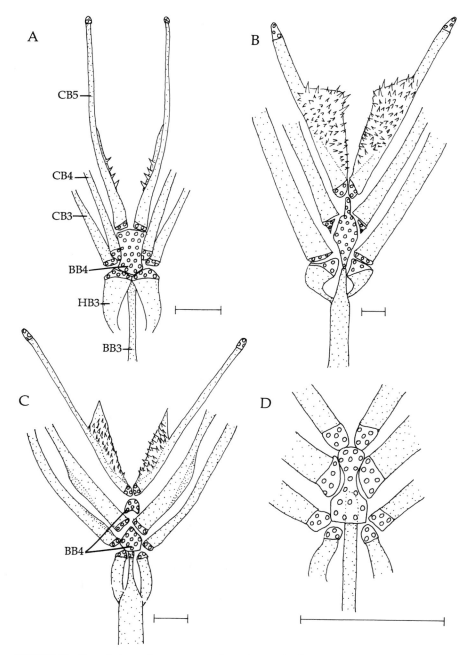

FIGURE 10 Dorsal view of posterior portion of ventral gill arches, posterior toward top of page. (A) *Lestrolepis* sp., USNM 307290, (B) *Aulopus filamentosus*, USNM 292105, (C) *Synodus variegatus*, USNM 339776, and (D) a larval *Synodus variegatus*, USNM 339775.

tids, but in those taxa the CB5s are well separated from the main body of BB4 by a tail of cartilage (aulopids) or one or two small separate cartilages (synodontids) that extend posteriorly from the main body of BB4 (Figs. 10B and 10C). We interpret the conditions in synodontoids and primitive ctenosquamates as distinct.

The absence of the derived synodontoid condition in pseudotrichonotids renders a hypothesis of independent acquisition of the gap in aulopids and the ancestor of synodontids as parsimonious as a single origin in the synodontoid ancestor with reversal in *Pseudotrichonotus*. However, there is some indication that the different configurations of BB4 and CB5s within synodontoids are homologous: in synodontids, the small posterior cartilages that are separate from BB4 in adults (Fig. 10C) are continuous with that element in larvae (Fig. 10D), a condition very similar to that of *Aulopus* (Fig. 10B) in which a cartilaginous tail extends posteriorly from BB4.

(17_0) = No gap between CB5s and BB4 cartilage
(17_1) = Gap between CB5s and BB4 cartilage, CB5s not articulating with reduced BB4
(17_2) = CB5s separated from main body of BB4 by tail or small nubbins of cartilage extending posteriorly from BB4.

18. Third Basibranchial Extends beyond Fourth Basibranchial Cartilage—In *Harpadon*, *Saurida*, and alepisaurids, BB3 extends beneath BB4, terminating posteriorly beyond the posterior end of BB4. In other aulopiforms and outgroups, BB3 terminates beneath the anterior end of BB4 (Fig. 10) or slightly more posteriorly in *Macroparalepis* and *Lestrolepis*.

(18_0) = BB3 terminates beneath the anterior end of BB4 cartilage
(18_1) = BB3 terminates beyond the posterior end of BB4 cartilage

19. Fourth Basibranchial Ossified—BB4 is ossified in only two aulopiforms, the ipnopids *Bathymicrops* and *Ipnops*. Because of the proposed sister-group relationship between *Bathymicrops* and *Bathytyphlops*, which has a cartilaginous BB4, we hypothesize ossification of BB4 in the ancestor of *Bathytyphlops* + *Bathymicrops* + *Ipnops* with reversal in *Bathytyphlops*. Alternatively, BB4 ossified independently in *Bathymicrops* and *Ipnops*.

(19_0) = BB4 cartilaginous
(19_1) = BB4 ossified

20. Elongate First Basibranchial—BB1 is typically very small in aulopiforms and outgroups. In notosudids and most paralepidids (including *Anotopterus*), BB1 is elongate, such that the first gill arch is widely separated from the hyoid arch. The condition in notosudids and paralepidids is different, however, in that BB1 is a long ossified element in the former and mostly cartilaginous in the latter. The basibranchials are of approximately equal length in *Paralepis*, but as in other paralepidids, BB1 comprises a short ossified segment anteriorly followed by a long posterior cartilage.

(20_0) = BB1 not elongate
(20_1) = BB1 elongate, ossified
(20_2) = BB1 usually elongate, comprising a short ossified anterior segment followed by a long posterior cartilage

21. Elongate Second Basibranchial—The first and second gill arches are widely separated by an elongate BB2 in notosudids, *Bathytyphlops*, *Bathymicrops*, and evermannellids. Stiassny (this volume) considered an elongate BB2 a synapomorphy of lampanyctine myctophids, but elongate basibranchials are lacking in other outgroups.

(21_0) = BB2 not elongate
(21_1) = BB2 elongate

22. Gillrakers or Toothplates on Third Hypobranchials—Gillrakers are primitively present on HB3 in aulopiforms and ctenosquamates. Loss of gillrakers on HB3 occurred three times within aulopiforms: once in the ancestor of *Pseudotrichonotus* + synodontids, in *Chlorophthalmus*, and again in the ancestral alepisauroid. The reappearance of HB3 gillrakers is a synapomorphy of *Paralepis* and *Arctozenus*.

(22_0) = Gillrakers or toothplates present on HB3
(22_1) = Gillrakers or toothplates lacking on HB3

23. Gillrakers or Toothplates on Basibranchial(s)—Myctophids have gillrakers extending onto BB2 and BB3, but other outgroups and most aulopiforms lack gillrakers on the basibranchials. *Bathysauropsis*, notosudids, and all ipnopids have gillrakers or toothplates at least on BB2 and sometimes on BB1 and BB3. In those chlorophthalmoids, the basibranchials are deeper than in most other aulopiforms, creating a surface for attachment of the rakers. *Bathysaurus* is the only other aulopiform with gillrakers (present as toothplates in that genus) on BB2. In some paralepidids, the gill filaments and toothplates extend alongside of but do not articulate with BB1 and BB2, which are very thin bones with little surface area laterally for the attachment of rakers.

(23_0) = Gillrakers or toothplates lacking on basibranchials
(23_1) = Gillrakers or toothplates on BB2, sometimes BB1 and BB3

24. Ligament between First Hypobranchial and Ventral Hypohyal—In most aulopiforms and all outgroups, a ligament connects HB1 to the hyoid arch, usually the hypohyal but sometimes the anterior ceratohyal (e.g., *Synodus*). *Bathymicrops* and *Bathytyphlops* are unique among aulopiforms in having this ligament ossified. In *Bathymicrops*, the dorsal and ventral hypohyals are not separate from one another or from the ceratohyal, but the ossified ligament articulates with the hyoid near its junction with the branchial skeleton.

(24_0) = Ligament from HB1 to ventral hypohyal not ossified

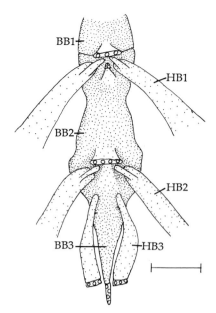

FIGURE 11 Ventral view of midportion of ventral gill arches of *Synodus variegatus*, USNM 339776, hyoid arch removed.

(24_1) = Ligament from HB1 to ventral hypohyal ossified

25. *First Hypobranchial with Ventrally Directed Process (Fig. 11)*—A small process projects ventrally from HB1 in synodontids, *Chlorophthalmus*, and *Bathysaurus*. Ventral hypobranchial processes on HB1 are lacking in other aulopiform and outgroup taxa.

(25_0) = HB1 without ventrally directed processes
(25_1) = HB1 with a ventrally directed process

26. *Second Hypobranchial with Ventrally Directed Process (Fig. 11)*—Synodontids also have a small process on HB2, a feature that occurs elsewhere among aulopiforms and outgroups only in *Metavelifer*.

(26_0) = HB2 without ventrally directed process
(26_1) = HB2 with ventrally directed process

27. *Third Hypobranchials Fused Ventrally*—In most aulopiforms and all outgroups, the third hypobranchials are variously separated widely from one another or bound closely together ventrally. The third hypobranchials are fused ventrally only in the evermannellids *Evermannella* and *Odontostomops* and the paralepidid *Arctozenus*.

(27_0) = Third hypobranchials not fused ventrally
(27_1) = Third hypobranchials fused ventrally

B. Hyoid Arch

28. *Ventral Ceratohyal Cartilage (Fig. 12)*—An autogenous cartilage extending along part of the ventral margin of the anterior ceratohyal (Fig. 12D) is a derived feature of synodontoids (Johnson *et al.*, 1996). An autogenous cartilage on the ventral surface of the anterior ceratohyal is lacking in other aulopiform and outgroups (Figs. 12A–12C).

(28_0) = Anterior ceratohyal without autogenous ventral cartilage
(28_1) = Anterior ceratohyal with autogenous cartilage along ventral margin

29. *Number of Branchiostegals on the Posterior Ceratohyal*—McAllister (1968) noted that aulopids, synodontids, and harpadontids differ from other "myctophiforms" in having numerous branchiostegals. Johnson *et al.* (1996) considered six or more branchiostegals on the posterior ceratohyal as a synapomorphy of synodontoids. We concur, as four or fewer is a primitive feature for aulopiforms. *Pseudotrichonotus* has only two, and thus an increase in the number of branchiostegals on the posterior ceratohyal could have been independently acquired in aulopids and in the ancestor of synodontids. Five branchiostegals on the posterior ceratohyal is an autapomorphy of *Bathypterois*.

(29_0) = Four or fewer branchiostegals on posterior ceratohyal
(29_1) = Five branchiostegals on posterior ceratohyal
(29_2) = Six or more branchiostegals on posterior ceratohyal

30. *Number of Branchiostegals on the Anterior Ceratohyal (Fig. 12)*—Synodontoids, most chlorophthalmoids, and most outgroups have five or more branchiostegals on the anterior ceratohyal (Figs. 12B and 12D). *Metavelifer*, *Polymixia*, chlorophthalmids, alepisauroids (Figs. 12A and 12C) and giganturoids have four or fewer (four in all taxa except *Polymixia*, *Coccorella*, and *Alepisaurus* which have three). A reduced number of branchiostegals on the anterior ceratohyal is interpreted in our analysis as a synapomorphy of chlorophthalmoids + alepisauroids + giganturoids with reversal in the ancestor of *Bathysauropsis* + notosudids + ipnopids; it could have evolved independently in the ancestors of the Chlorophthalmidae and Alepisauroidei + Giganturoidei.

(30_0) = Five or more branchiostegals on anterior ceratohyal
(30_1) = Four or fewer branchiostegals on anterior ceratohyal

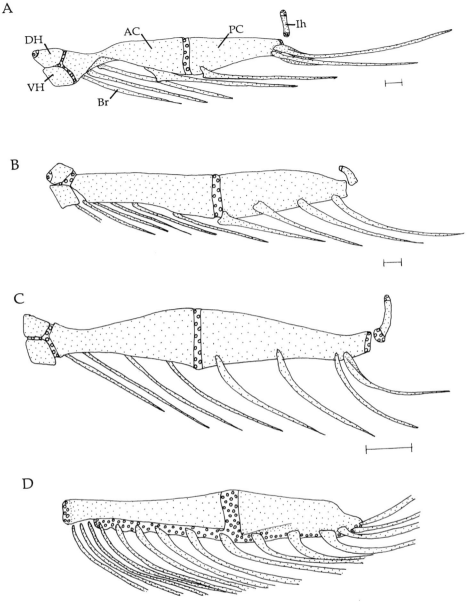

FIGURE 12 Lateral view of hyoid arch from left side of (A) *Stemonosudis rothschildi*, AMS I.22826001, (B) *Scopelosaurus hoedti*, USNM 264256, (C) *Evermannella indica*, USNM 235141, and (D) *Synodus variegatus*, USNM 31518.

31. *Proximity of Posteriormost Two Branchiostegals (Fig. 12)*—In all paralepidids except *Anotopterus*, the posteriormost two branchiostegals are inserted very close to one another on the posteroventral corner of the posterior ceratohyal (Fig. 12A). Branchiostegals on the posterior ceratohyal in other aulopiforms and the outgroups are well spaced (Figs. 12B and 12D) except in *Evermannella* and *Odontostomops*, in which the posteriormost two are also close but articulate more anteriorly on the posterior ceratohyal than in paralepidids (Fig. 12C).

(31_0) = All branchiostegals on posterior ceratohyal evenly spaced

(31_1) = Two posteriormost branchiostegals close, inserting on ventral margin of posterior ceratohyal

(31_2) = Two posteriormost branchiostegals close, inserting on posteroventral corner of posterior ceratohyal

32. *3 + 1 Arrangement of Branchiostegals on the Anterior Ceratohyal (Fig. 12A)*—In most paralepidids, the ventral margin of the anterior ceratohyal is deeply

indented. Three of the four anterior ceratohyal bran-chiostegals are inserted close to one another on the anterior side of the indentation, and the fourth is inserted posterior to the indentation. McAllister (1968) indicated that the unusual spacing of branchio-stegals may be diagnostic of the Paralepididae, but in *Anotopterus* and *Sudis* (and other aulopiforms with four anterior ceratohyal branchiostegals), the ele-ments are more evenly spaced, although there may be a well-defined indentation in the bone. Parsimony suggests the 3+1 pattern evolved twice: once in the ancestor of *Arctozenus* and *Paralepis* and again in the ancestor of *Macroparalepis*, *Uncisudis*, *Lestidium*, *Lestid-iops*, *Stemonosudis*, and *Lestrolepis*.

(32_0) = Branchiostegals on anterior ceratohyal roughly evenly spaced

(32_1) = Branchiostegals on anterior ceratohyal ar-ranged in "3+1" pattern

33. Hypohyal Branchiostegals (Fig. 12B)—Scopelo-saurus and *Ahliesaurus* have the anteriormost branchio-stegal inserting on the ventral hypohyal. In other aulopiforms and most outgroups, all branchiostegals insert on the ceratohyals. *Diplophos* also has a branchiostegal on the ventral hypohyal, and the myc-tophid *Lampanyctus* has three branchiostegals on the very elongate ventral hypohyal.

(33_0) = No branchiostegals on ventral hypohyal

(33_1) = Anteriormost branchiostegal on ventral hy-pohyal

(33_2) = Anteriormost three branchiostegals on ven-tral hypohyal

*34. Basihyal Morphology—*The morphology of the basihyal is variable among aulopiform and out-groups, but typically it is horizontal, has a triangular ossification anteriorly (in dorsal view), and is covered by an edentate or strongly toothed dermal plate. A small, obliquely aligned basihyal occurs in ipnopids and *Bathysauropsis* (Hartel and Stiassny, 1986, fig. 7). Notosudids share with most aulopiforms and out-groups a horizontal basihyal, a condition interpreted here as a reversal. An obliquely aligned basihyal is independently present in *Diplophos*.

A more extreme form of the ipnopid condition is found in evermannellids, where the basihyal lies at 90° to the first basibranchial (R. K. Johnson, 1982, fig. 13B). Our phylogeny suggests that the configurations of the basihyal in the two groups are unrelated.

(34_0) = Basihyal oriented horizontally

(34_1) = Basihyal oriented obliquely

(34_2) = Basihyal oriented at 90° angle to BB1

*35. Basihyal Teeth—*Basihyal teeth are variously present or absent among aulopiforms and outgroups.

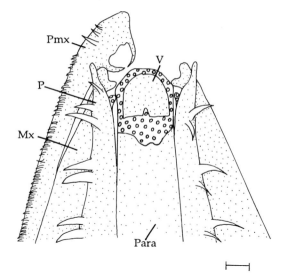

FIGURE 13 Ventral view of palate of *Lestrolepis intermedia*, USNM 290253; premaxilla removed from left side.

They are present in scopelarchids as large, posteriorly curved structures that form a single row down the center of the bone (R. K. Johnson, 1974a, fig. 9).

(35_0) = Basihyal teeth absent or unmodified if present

(35_1) = Basihyal teeth present as large, posteriorly curved structures

C. Jaws, Suspensorium, and Circumorbitals

*36. Dominant Tooth-bearing Bone (Fig. 13)—*In alepi-sauroids and *Bathysauroides*, the palatine is the domi-nant tooth-bearing bone of the upper "jaws." In most of those taxa, the premaxilla also bears teeth, but they are considerably smaller than the palatine teeth. *Ba-thysaurus* has premaxillary and palatine teeth about equally developed. The large teeth in *Gigantura* are on the premaxilla, but the dermopalatine never develops, and the autopalatine is lost ontogenetically. In all other aulopiforms and outgroups, the palatine may bear teeth, but the premaxilla (and maxilla in stomii-forms) is the dominant tooth-bearing bone of the up-per jaw.

(36_0) = Premaxilla (or premaxilla and maxilla) is the dominant tooth-bearing bone of upper jaw

(36_1) = Premaxilla and palatine are the dominant tooth-bearing bones of upper jaw

(36_2) = Palatine is the dominant tooth-bearing bone of "upper jaw"

*37. Quadrate with Produced Anterior Limb (Fig. 14)—*Johnson *et al.* (1996) described a series of suspensorial

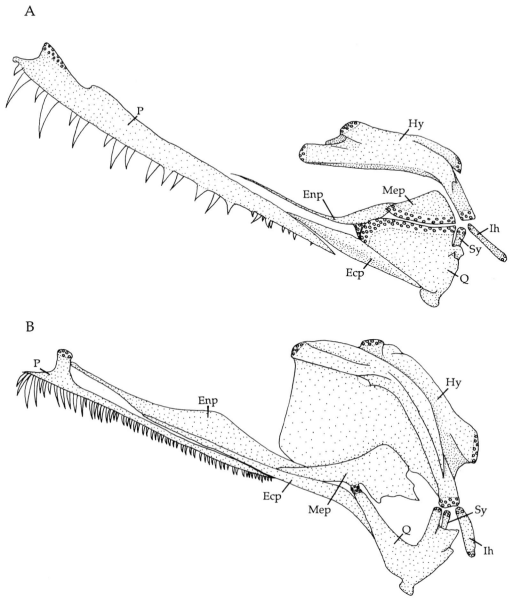

FIGURE 14 Lateral view of suspensorium from left side of (A) *Bathysaurus mollis*, VIMS 6107 and (B) *Synodus variegatus*, USNM 315318.

modifications that support the monophyly of syno-dontids. Among these is the loss of the typical fan-shaped quadrate due to the presence of a produced anterior limb (Fig. 14B).

(37_0) = Quadrate fan-shaped
(37_1) = Quadrate with produced anterior limb

38. Quadrate with Two Distinct Cartilaginous Heads (Fig. 14)—Synodontids have two distinct cartilaginous condyles on the quadrate, one anteriorly and one pos-

teriorly, both anterior to the symplectic incisure (Fig. 14B). In most other aulopiforms and outgroups, a single large cartilage borders the dorsal and anterodorsal margins of the fan-shaped quadrate (Fig. 14A). Separation of the quadrate cartilage into two discrete condyles occurs independently in notosudids.

(38_0) = Quadrate with single large cartilage on dorsal border
(38_1) = Quadrate cartilage separated into two condyles

39. Large Concavity in Dorsal Margin of Quadrate (Fig. 14)—Among those aulopiforms having the quadrate cartilage separated into two condyles, *Synodus* and *Trachinocephalus* are unique in having a large concavity between them (Fig. 14B).

(39_0) = No concavity in quadrate (excluding symplectic incisure)
(39_1) = Concavity between anterior and posterior cartilaginous condyles

40. Posterior Cartilaginous Condyle of Quadrate Articulates with Hyomandibular (Fig. 14)—In *Synodus* and *Trachinocephalus*, the posterior cartilaginous condyle of the quadrate articulates with the ventral cartilaginous condyle of the hyomandibular (Fig. 14B). In other aulopiforms and outgroups, the posterior portion of the single quadrate cartilage or the posterior of the separate cartilaginous condyles articulates with a ventral cartilaginous condyle on the metapterygoid (e.g., as in *Bathysaurus*, Fig. 14A). In synodontids, the metapterygoid is displaced anteriorly, well forward of the posterior limb of the quadrate, and lacks a ventral cartilaginous condyle.

(40_0) = Posterior portion of quadrate articulates dorsally with metapterygoid
(40_1) = Posterior cartilaginous condyle of quadrate articulates dorsally with hyomandibular.

41. Metapterygoid Produced Anteriorly (Fig. 14)— In synodontids, the metapterygoid has an anterior extension that overlies the posterior part of the ectopterygoid (Fig. 14B). In other aulopiforms and outgroups, the metapterygoid overlies the quadrate and does not extend anteriorly over the ectopterygoid (Fig. 14A).

(41_0) = Metapterygoid overlies quadrate
(41_1) = Metapterygoid extends anteriorly over posterior portion of ectopterygoid

42. Metapteryoid Free of Hyomandibular—In ipnopids, the posterior end of the metapterygoid overlies the hyomandibular, but the two bones are not tightly bound together and do not articulate with one another through cartilage. In *Bathypterois*, the suspensorium is even less articulated, as the hyomandibular is also free from the symplectic. In other aulopiforms and outgroups, the metapterygoid may overlie the hyomandibular, but it is always tightly bound to it— through a cartilage process on the posterior margin of the metapterygoid, connective tissue, or sometimes a bony strut extending from the dorsal aspect of the metapterygoid posteriorly to the hyomandibular shaft.

(42_0) = Metapterygoid bound to hyomandibular
(42_1) = Metapterygoid free from hyomandibular

43. Hyomandibular and Opercle Oriented Horizontally—In *Bathytyphlops* and *Bathymicrops*, the hyomandibular is rotated so that it lies almost parallel to the long axis of the body. The opercle is similarly rotated and lies directly above the hyomandibular (Sulak, 1977, fig. 11). Accompanying those changes are an elongation of the ectopterygoid and reduction of the endopterygoid (Sulak, 1977, fig. 11). In other aulopiforms and outgroups, the orientation of the hyomandibular ranges from slightly oblique, as in *Chlorophthalmus* (Sulak, 1977, fig. 7A), to oblique (ca. 45°), as in *Harpadon* (Sulak, 1977, fig. 6C; Johnson *et al.*, 1996, fig. 27). In none of those taxa does the hyomandibular approach a horizontal orientation, and the opercle is never rotated dorsally to lie above it.

(43_0) = Hyomandibular oriented vertically or subvertically, opercle posterior to suspensorium
(43_1) = Hyomandibular oriented ca. horizontally, opercle rotated dorsally to lie above hyomandibular

44. Maxillary Saddle (Fig. 15)—Regan (1911) first noted that iniomous fishes usually differ from isospondylous fishes in having a process on the palatine that projects outward and upward and articulates with a depression (maxillary saddle) in the proximal end of the maxilla. Gosline *et al.* (1966) referred to the palatine projection as a "palatine prong" and noted that its presence in only some iniomes renders it of questionable value in distinguishing iniomous fishes from those of the Isospondyli. Stiassny (1986) found that, among Rosen's (1973) aulopiforms, only *Aulopus*, *Chlorophthalmus*, and *Parasudis* have a palatine prong system and that the presence of that feature as well as a deeply folded articular head on the maxilla (Rosen, 1973, p. 505) unites those genera with the Ctenosquamata.

We concur with Stiassny (1986) that a palatine prong system is unique among aulopiforms to aulopids and chlorophthalmids, but our analysis supports the monophyly of Rosen's (1973) aulopiforms, and we therefore disagree with Stiassny's interpretation of this feature as evidence of a paraphyletic Aulopiformes. It is most parsimonious to hypothesize independent evolution of the palatine prong system in ctenosquamates, aulopids, and chlorophthalmids, but we note that only one additional step is required for evolution of the system at the level of Eurypterygii with reversals within the Aulopiformes. *Pseudotrichonotus* is unique among aulopiforms in having a well-developed maxillary saddle for attachment of the palato-maxillary ligament; it lacks a palatine prong

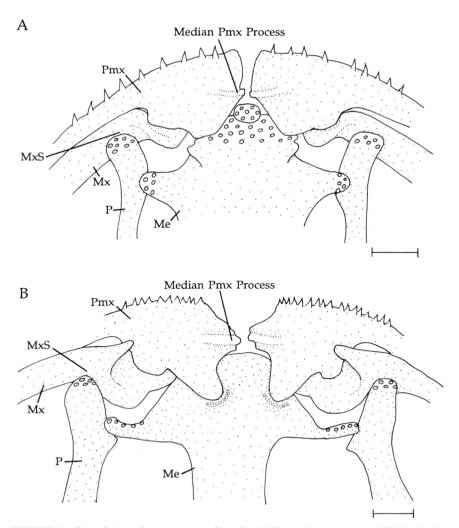

FIGURE 15 Dorsal view of snout region of head in (A) *Parasudis truculentus*, USNM 159850 and (B) *Chlorophthalmus atlanticus*, USNM 339774.

(Johnson *et al.*, 1996). As noted by Olney *et al.* (1993), *Metavelifer* has lost the connection between the palatine and maxilla, a modification that allows greater protrusibility of the upper jaw.

(44_0) = Palatine prong and maxillary saddle absent
(44_1) = Palatine prong absent, maxillary saddle present
(44_2) = Palatine prong and maxillary saddle present

45. Dorsomedially Directed Premaxillary Process (Fig. 15)—In chlorophthalmoids, a dorso-medially directed process arises from near the medial edge of each premaxilla and is ligamentously bound to its contralateral member. We have not observed this process in other aulopiforms or outgroups, which typically have a smooth medial premaxillary margin (as in *Lestrolepis*, Fig. 13).

(45_0) = Premaxilla without dorso-medially directed process on medial edge
(45_1) = Premaxilla with dorso-medially directed process on medial edge

46. Number of Infraorbitals—Aulopiforms primitively have six infraorbitals, as do stomiiforms, myctophiforms, and *Polymixia*. *Metavelifer* has three, as does *Bathymicrops*. *Ipnops* and *Bathytyphlops* have five infraorbitals, a synapomorphy of those genera and *Bathymicrops* in our analysis, with further reduction in the latter.

R. K. Johnson (1982) noted that seven infraorbitals is a synapomorphy of notosudids, a hypothesis corroborated by our analysis. Alepisaurids, most paralepidids, and evermannellids have eight infraorbitals, an ambiguous character state interpreted here as a synapomorphy of alepisauroids, with reversal to six in scopelarchids and *Anotopterus*.

(46_0) = Six infraorbitals
(46_1) = Seven infraorbitals
(46_2) = Eight infraorbitals
(46_3) = Five infraorbitals
(46_4) = Three infraorbitals
(46_5) = No infraorbitals

47. Long Snout—Paralepidids have a long snout ranging from approximately 50% of the head length in *Paralepis*, *Arctozenus*, *Macroparalepis*, *Lestidium*, *Lestidiops*, and *Lestrolepis* to well over 50% head length in *Anotopterus*, *Sudis*, *Uncisudis*, and *Stemonosudis*. In all other aulopiforms and outgroups, the snout is considerably less than 50% head length.

(47_0) = Snout length much < 50% head length
(47_1) = Snout length > 50% head length

48. Premaxillary Fenestra (Fig. 13)—R. K. Johnson (1982) considered the presence of a fenestrated premaxilla as a synapomorphy of paralepidids and *Anotopterus*, noting that although Rosen (1973) viewed the feature as diagnostic of his alepisauroids, it does not appear to be present in alepisaurids, evermannellids, and other iniome fishes. We concur and note that all paralepidids examined have at least a partial fenestra in the anterior end of the premaxilla. The fenestra is usually complete in larger specimens.

(48_0) = No premaxillary fenestra
(48_1) = Anterior premaxilla with fenestra

49. Palatine Articulates with Premaxilla (Fig. 13)—In paralepidids, the palatine terminates anteriorly in a long process that articulates by connective tissue with the medial surface of the premaxilla, just posterior to the premaxillary fenestra. Various associations between the palatine and maxilla (when present) exist among aulopiforms and outgroups, but the palatine typically terminates near the point where it articulates with the maxilla and does not extend anteriorly to meet the premaxilla.

The palatine also articulates with the premaxilla in *Harpadon*, but this condition differs from that in paralepidids in that the premaxilla replaces the maxilla as the site of articulation with the palatine and ethmoid because the maxilla is present only as a small remnant.

(49_0) = Palatine without process for articulation with premaxilla
(49_1) = Palatine with long process for articulation with premaxilla

50. Lacrimal Oriented Horizontally on Snout—The lacrimal typically borders the orbit anteriorly. In para-lepidids, the second infraorbital is in the position normally occupied by the lacrimal, and the lacrimal is located horizontally on the snout, well rostral to the orbit. Our identification of this bone as the lacrimal rather than the antorbital is based on its relatively large size and association with the upper jaw; as is typical of the lacrimal, this bone extends along a portion of the upper border of the maxilla. In our small specimens of *Sudis*, the position of the lacrimal cannot be determined.

(50_0) = Lacrimal bordering orbit anteriorly
(50_1) = Lacrimal anterior to orbit, oriented horizontally.

51. Maxilla Reduced—Johnson *et al.* (1996) discussed the problems associated with Sulak's (1977) interpretation of a reduced maxilla in *Harpadon* and *Saurida*, especially his assessment of a division of the maxilla into anterior and posterior elements. They concluded that a reduced maxilla is a synapomorphy of synodontids and harpadontids and suggested that the reduced maxilla in *Bathysaurus*, which exists as a remnant anteriorly, is not homologous with the posterior remnant in *Harpadon*. The condition in *Bathysaurus*, however, may be homologous with the posterior maxillary remnant of *Gigantura*, as our phylogeny supports Patterson and Johnson's (1995) hypothesis of a sister-group relationship between *Bathysaurus* and *Gigantura* (see character 117 below and Discussion); nevertheless, we conservatively consider anterior and posterior maxillary remnants as distinct.

The maxilla of most alepisauroids is slender relative to that of other aulopiforms, but usually retains the same shape in that the posterior end is expanded. We do not recognize the alepisauroid condition as a separate state except in *Anotopterus*, in which the maxilla is a very slender, strut-like bone with no posterior expansion.

(51_0) = Maxilla well developed with posterior end expanded
(51_1) = Maxilla intact but slender, posterior end not expanded
(51_2) = Maxilla present as posterior remnant
(51_3) = Maxilla present as anterior remnant

D. Cranium

We did not examine the cranial morphology of aulopiforms in detail but describe certain aspects of the ipnopid cranium below.

52. Frontal Expanded Laterally over Orbit—The skull of ipnopids is dorsoventrally compressed, and the frontals extend laterally over the eyes in *Bathypterois*,

Bathymicrops, and *Bathytyphlops* and beneath the large photosensitive plates in *Ipnops*. In other aulopiforms and outgroups, the frontals lie completely between the orbits.

(52_0) = Frontal not expanded laterally
(52_1) = Frontal expanded laterally

53. *Sphenotic Process*—The sphenotic is modified in all ipnopids such that a process of the bone extends anteriorly beneath the greatly expanded frontal. In its most extreme form, the process extends forward to the lateral ethmoid, as in *Bathytyphlops*. It extends about halfway to the lateral ethmoid in *Bathymicrops*, and only a little forward in *Ipnops*. It is least developed but present as a small rounded extension beneath the frontal in *Bathypterois*. In other aulopiforms and outgroups, the sphenotic abuts the frontal but has no anteriorly directed process.

(53_0) = Sphenotic without anterior process
(53_1) = Sphenotic with anterior process

E. *Intermuscular Bones and Ligaments*

Unless otherwise noted, all characters described in this section are from Patterson and Johnson (1995). Our survey of aulopiform intermusculars is more extensive than that of Patterson and Johnson, but it is still incomplete (Table 1). Further investigation is needed.

54. *Epipleurals Extend Anteriorly to First or Second Vertebra*—In synodontoids and chlorophthalmoids, epipleurals extend anteriorly to the second vertebra (V2); in alepisauroids and giganturoids, they extend to V1. When present, epipleurals begin on V3 in the outgroups. *Metavelifer* and all acanthomorphs except *Polymixia* lack epipleurals, a condition we code as a separate state for this character but as "missing data" for other epipleural characters in this section to avoid erroneously inflating tree length and modifying other tree statistics. Patterson and Johnson (1995) hypothesized that the anterior extension of epipleurals to V2 is a synapomorphy of Aulopiformes, but their consensus tree indicates that the extension of epipleurals to V1 (from V2) evolved independently in the ancestor of the Evermannellidae + Scopelarchidae and the remaining alepisauroids. Our phylogeny suggests that the extension of epipleurals to V1 occurred once, in the ancestor of alepisauroids + giganturoids.

(54_0) = Epipleurals originate on V3
(54_1) = Epipleurals originate on V2
(54_2) = Epipleurals originate on V1
(54_3) = Epipleurals absent

55. *One or More Epipleurals Displaced Dorsally Into Horizontal Septum*—The presence of one or more anterior epipleurals displaced dorsally into the horizontal septum is a synapomorphy of aulopiforms. *Aulopus* and *Gigantura* have a single epipleural in the horizontal septum, but most other aulopiforms have more than one displaced. *Bathypterois*, *Bathymicrops*, and *Parasudis* have all epipleurals beneath the horizontal septum, and those taxa fall outside of a clade comprising the remaining aulopiforms in Patterson and Johnson's (1995) tree constructed solely on the basis of intermuscular characters. It is more parsimonious to interpret the absence of dorsally displaced epipleurals in *Bathypterois*, *Bathymicrops*, and *Parasudis* as reversals. Our small cleared and stained specimen of *Ipnops* lacks ossified epipleurals anteriorly, and we were not able to conclusively identify ligamentous epipleurals anteriorly in that specimen.

(55_0) = All epipleurals beneath the horizontal septum
(55_1) = One or more epipleurals displaced dorsally into horizontal septum

56. *Abrupt Transition of Epipleurals in and beneath the Horizontal Septum*—Two states of the derived condition of dorsally displaced anterior epipleurals occur in aulopiforms. In one, the transition between epipleurals in and beneath the horizontal septum is abrupt, such that the last posterolaterally directed epipleural in the horizontal septum is followed immediately by a ventrolaterally directed epipleural that is completely below the horizontal septum. This occurs only in *Pseudotrichonotus*, *Synodus*, and *Trachinocephalus*. In other aulopiforms, including *Harpadon* and *Saurida*, the transition is gradual, occurring over a series of vertebrae.

We initially coded this character as having three states (no epipleurals in the horizontal septum, an abrupt transition of epipleurals in and beneath the horizontal septum, and a gradual transition of those epipleurals) to determine the primitive aulopiform state. Parsimony indicates that a gradual transition is primitive for aulopiforms and a synapomorphy of the order. However, it does not seem valid to consider both the presence of one or more dorsally displaced epipleurals (55_1) and a gradual transition of epipleurals in and beneath the horizontal septum as synapomorphies of aulopiforms because the latter is a state of the former. Accordingly, for this character, we group the most common outgroup condition (no dorsally displaced epipleurals) and the primitive ingroup condition (gradual transition of epipleurals) as a single state. This allows us to recognize the abrupt transition of epipleurals in some synodontoids as a derived feature without creating two synapomorphies of aulopiforms where only one is warranted.

(56_0) = No epipleurals displaced dorsally into the horizontal septum or the transition between epipleurals in and beneath the horizontal septum is gradual.

(56_1) = Abrupt transition between epipleurals in and beneath the horizontal septum

57. One or More Epipleurals Forked Distally—In the region where the epipleurals leave the horizontal septum in notosudids, around V19 (*Ahliesaurus*) or V20–V24 (*Scopelosaurus*), the epipleurals are bifurcate distally. In other aulopiforms and outgroups, epipleurals are not forked distally at the transition in and beneath the horizontal septum (or none are in the septum).

(57_0) = Epipleurals not forked distally

(57_1) = Epipleurals forked distally at transition of epipleurals in and beneath the horizontal septum

58. Epipleural on First and Second Vertebrae Fused to Centrum—In *Omosudis*, *Alepisaurus*, and *Paralepis*, the epipleurals on V1 and V2 are fused to the centrum. Those epipleurals are free in other paralepidids, aulopiforms, and outgroups. Fusion of the epipleurals in *Paralepis* is independent of that in the alepisaurid lineage.

(58_0) = Epipleurals on V1 and V2 autogenous

(58_1) = Epipleurals on V1 and V2 fused to centrum

59. Epipleurals Not Attached to Axial Skeleton—Most epipleurals are not attached to the axial skeleton in *Omosudis*, *Paralepis*, and *Arctozenus*, but most or all are attached in *Alepisaurus*, other aulopiforms, and outgroups. As with the epineurals (see character 63 below), *Paralepis* and *Arctozenus* have the anterior epipleurals forked anteriorly, and the branch that attaches the bone to the axial skeleton disappears posteriorly leaving a large series of unattached epipleurals.

(59_0) = Most or all epipleurals attached to axial skeleton

(59_1) = Most epipleurals not attached to axial skeleton

(59_2) = Most epipleurals are free dorsal branches

60. Reduced Number of Epipleurals—Most aulopiforms have a long series of epipleurals that begin on V1 or V2. Most outgroups also have a well-developed series of epipleurals, although they begin more posteriorly than in aulopiforms (see character 54). In the paralepidids *Lestrolepis*, *Macroparalepis*, and *Sudis*, the epipleural series is confined to the first five or fewer vertebrae. Epipleurals are not evident in our small specimens of *Uncisudis*, *Stemonosudis*, and *Lestidiops*.

(60_0) = Long series of epipleurals

(60_1) = Epipleural series not extending posteriorly beyond V5

61. Origin of Epineurals—In *Scopelosaurus* and *Ahliesaurus*, anterior epineurals originate on the neural arch. The origin of subsequent epineurals descends to the centrum or parapophysis, and then it reascends in posterior epineurals to the neural arch. A similar configuration of ventrally displaced epineurals occurs in evermannellids, scopelarchids, and *Bathysauroides*. In those taxa, the origin of epineurals always returns to the neural arch posteriorly, and usually less than half of the epineurals originate on the centrum. In *Bathysaurus*, the first five epineurals originate on the neural arch, and the origin of the rest descends to the centrum. In *Gigantura*, all epineurals originate on the centrum. In other aulopiforms, *Diplophos*, myctophiforms, and *Metavelifer*, all epineurals originate on the neural arch. In *Polymixia*, epineurals on V3–10 originate on the centrum, those more anterior and posterior originate on the neural arch or spine. The origin of some of the central epineurals on the centrum (with reascension posteriorly) and the origin of most or all epineurals on the centrum (without reascension posteriorly) are derived conditions within the Aulopiformes.

(61_0) = All epineurals originate on neural arch

(61_1) = Some epineurals originate on the centrum or parapophysis; these flanked anteriorly and posteriorly by epineurals originating on the neural arch

(61_2) = Most or all epineurals originate on centrum; epineurals not reascending to neural arch posteriorly

62. First One to Three Epineurals with Distal End Displaced Ventrally—In some aulopiforms, the first one to three epineurals are turned downward such that they extend lower than their successors. The distal end of the epineurals on V1–V3 is so modified in *Aulopus*, *Pseudotrichonotus* (V1–V2), *Synodus* (V1–V2), *Trachinocephalus* (V1), *Chlorophthalmus* (V1–V2), *Ipnops* (V1), and *Bathysauroides* (V1–V2). Patterson and Johnson (1995) suggested that *Chlorophthalmus* may be the sister group of synodontoids based on this feature, but our analysis indicates that having the distal end of the anteriormost one or more epineurals turned downward evolved independently in *Chlorophthalmus*, *Bathysauroides*, and the ancestral synodontoid. The condition is reversed in *Harpadon* and *Saurida*.

(62_0) = Distal end of epineurals not displaced ventrally

(62_1) = Distal end of first one to three epineurals displaced ventrally

63. *Some Epineurals and Epipleurals Forked Proximally*—Beginning on about V12 or V15, the epineurals in *Chlorophthalmus* and *Parasudis* are forked proximally. Posteriorly, the dorsomedial branch, which attaches the epineural to the axial skeleton, disappears, leaving a short series of unattached epineurals. A similar condition occurs primitively in myctophiforms. Patterson and Johnson (1995) did not identify forked epineurals in *Paralepis*, but our examination of additional specimens of that genus indicates that the anteriormost five or six epineurals are forked proximally, the dorsal branch disappearing posteriorly, leaving a long series of unattached epineurals (see character 65). A nearly identical pattern characterizes *Arctozenus*. A unique branching of the epineurals characterizes *Gigantura* (Patterson and Johnson, 1995).

Proximal branching of epipleurals occurs in the same pattern as that of the epineurals among aulopiforms, and we group the branching of the two series of bones as a single character.

(63_0) = No epineurals (or epipleurals) forked proximally
(63_1) = Epineurals (and epipleurals) from about V12–V15 to near end of series forked proximally
(63_2) = Epineurals (and epipleurals) on about V1–V5 forked proximally
(63_3) = "*Gigantura*" pattern of branching

64. *Epineurals Fused to Neural Arch*—Epineurals are fused to the neural arch on V1–V10 in *Harpadon* and *Saurida*. Epineurals typically are not fused to the axial skeleton in aulopiforms and outgroups, although they are fused to the neural arch on V1–V5 in *Diplophos* and *Alepisaurus* and on V1 in *Paralepis* and *Macroparalepis*; most are fused to the centrum in *Rosenblattichthys* and *Bathysaurus*. Fusion of epineurals to the axial skeleton has thus evolved several times within aulopiforms, and our analysis suggests that this condition is phylogenetically significant only as a synapomorphy of *Harpadon* and *Saurida*.

(64_0) = Epineurals not fused to axial skeleton
(64_1) = Epineural fused to neural arch on V1
(64_2) = Epineurals fused to neural arch on V1–V5
(64_3) = Epineurals fused to neural arch on V1–V10
(64_4) = Most epineurals fused to centrum

65. *Epineurals Attached to Axial Skeleton*—In *Alepisaurus* and *Omosudis*, most epineurals are not attached to the axial skeleton, and in *Anotopterus*, all are unattached. *Paralepis* and *Arctozenus* have the anterior epineurals forked and attached to the axial skeleton by the dorsal branch of the fork; on about V5 or V6, only the ventral branch remains, and the epineurals are thus unattached posteriorly. In other aulopiforms and outgroups, all or most epineurals are attached.

(65_0) = Most or all epineurals attached to axial skeleton
(65_1) = Most epineurals unattached
(65_2) = All epineurals unattached
(65_3) = Unattached epineurals represent only free ventral branches of forked epineurals

66. *Epicentrals*—Paralepidids, *Bathysaurus*, and *Gigantura* lack epicentrals. *Omosudis* and *Alepisaurus* have them ossified and beginning on V3. *Parasudis* has all epicentrals ossified and beginning on V1. All other aulopiforms and outgroups have ligamentous epicentrals, except the anterior epicentrals are cartilaginous in evermannellids (see next character), and the ligamentous epicentrals of *Polymixia* contain a cartilaginous rod distally (Patterson and Johnson, 1995). It is equally parsimonious to consider ligamentous epicentrals, ossified epicentrals, or no epicentrals as the ancestral condition for the clade comprising alepisaurids and paralepidids, but it seems unlikely that ligamentous epicentrals transformed into ossified ones and then were lost or that ligamentous epicentrals were lost and then regained as ossified epicentrals. There is evidence from *Parasudis* that ligamentous epicentrals ossify and from giganturoids that ligamentous epicentrals are lost (the ancestral chlorophthalmoid and giganturoid intermuscular systems are characterized by ligamentous epicentrals), and thus we believe it most likely that the ossification of ligamentous epicentrals is a synapomorphy of alepisaurids, and loss of ligamentous epicentrals is a derived feature of paralepidids. To reflect this, we partially ordered this character such that a single step is required to lose or ossify ligamentous epicentrals, but two steps are required to lose ossified epicentrals or gain ossified epicentrals when none existed ancestrally. Considering this character entirely unordered does not change the phylogeny but eliminates a synapomorphy of alepisaurids and one of paralepidids.

(66_0) = Epicentrals ligamentous
(66_1) = Epicentrals ossified
(66_2) = Epicentrals absent
(66_3) = Epicentrals cartilaginous anteriorly, ligamentous posteriorly

67. *Anterior Epicentrals Closely Applied to Distal End of Epipleurals*—Evermannellids are unique among aulopiforms in having the anterior epicentrals present as small rods of cartilage closely applied to the distal ends of the epipleurals. This is unusual because epi-

centrals are almost always attached to the centrum or parapophyses. A similar condition occurs in scopelarchids except that the anterior epicentrals are in ligament.

(67₀) = All epicentrals attached to centrum or parapophyses
(67₁) = Anterior epicentrals attached to distal end of epipleurals

F. Postcranial Axial Skeleton

68. Number of Supraneurals—Presence of three supraneurals preceding the dorsal fin is a synapomorphy of eurypterygians (Johnson and Patterson, this volume), but many aulopiforms have two or fewer, and numerous reconstructions of the reductions are possible. We interpret a single supraneural as a synapomorphy of chlorophthalmoids with reversals in *Chlorophthalmus*, *Bathysauropsis*, and *Bathytyphlops*. Presence of two supraneurals is a synapomorphy of alepisauroids, with further reduction to one (*Omosudis*) or none (*Alepisaurus*) in alepisaurids (or in their common ancestor) and reversal to three in the ancestral scopelarchid.

(68₀) = Three or more supraneurals
(68₁) = Two supraneurals
(68₂) = One supraneural
(68₃) = No supraneurals

69. Number of Caudal Vertebrae—Aulopiforms, stomiiforms, and ctenosquamates primitively have about half (40–60%) of the vertebrae as caudal vertebrae. A reduction in the number of caudal vertebrae occurs independently in the synodontid–harpadontid clade (17–19%) and in giganturoids (11–24%). Scopelarchids and evermannellids have 62–70% caudal vertebrae, a condition that we interpret as synapomorphic for those families. A large number of caudal vertebrae occur independently in *Chlorophthalmus* (62%), *Bathymicrops* (68%), and *Arctozenus* (70%). It seems reasonable that both the very low and very high numbers of caudal vertebrae were derived from the primitive aulopiform condition of about 50% (coded as 69₁), and thus we consider the three states to form an ordered transformation series (69₀ ↔ 69₁ ↔ 69₂).

(69₀) = < 25% caudal vertebrae
(69₁) = 40–60% caudal vertebrae
(69₂) = > 60% caudal vertebrae

70. Accessory Neural Arch—An accessory neural arch on V1 is present in *Diplophos*, *Aulopus*, and synodontids. It is absent in all ctenosquamates, *Pseudotricho-*

notus, chlorophthalmoids, alepisauroids, and giganturoids. Polarity of this character for aulopiforms is equivocal, but in our analysis, an accessory neural arch is a synapomorphy of synodontoids.

(70₀) = Accessory neural arch absent
(70₁) = Accessory neural arch present

71. First Neural Arch with Brush-like Growth—There is a unique brush-like posterodorsal outgrowth of bone on the first neural arch of *Synodus* and *Trachinocephalus* (Patterson and Johnson, 1995).

(71₀) = No brush-like growth on first neural arch
(71₁) = Brush-like growth on first neural arch

72. Number of Open Neural Arches—In chlorophthalmoids (except *Ipnops*), alepisauroids, and *Bathysauroides*, the neural arch on V1 and sometimes V2–V4 is open dorsally. In ctenosquamates, all neural arches are closed dorsally (see also Stiassny, this volume), whereas many are open in synodontoids, *Bathysaurus*, *Gigantura*, and *Diplophos*. The latter is the primitive aulopiform condition, and thus a reduced number of open neural arches is a synapomorphy of the Chlorophthalmoidei + Alepisauroidei + Giganturoidei. Having many open neural arches is a reversal uniting giganturids and bathysaurids.

(72₀) = Many neural arches open dorsally
(72₁) = Neural arches open on V1 and sometimes V2–V4
(72₂) = All neural arches closed dorsally

73. Origin of First Rib—The origin of the first rib varies among aulopiforms from V1 to V5. The first rib originates on V3 primitively in aulopiforms, but its origin changes within all aulopiform suborders. Nearly 75 reconstructions of this character are possible in aulopiforms, the only hypothesis of relationship common to all of them being that a more posterior origin (V4) of the first rib is a synapomorphy of Pseudotrichonotidae + Synodontidae, with the origin shifting to V5 in the ancestor of *Synodus* and *Trachinocephalus* and to V2 in *Harpadon*. Our analysis also suggests that the origin of the first rib moved anteriorly from V3 to V2 in the ancestral ipnopid and from V3 to V1 in the ancestor of the alepisaurid + paralepidid clade.

(73₀) = First rib originates on V3
(73₁) = First rib originates on V4
(73₂) = First rib originates on V5
(73₃) = First rib originates on V2
(73₄) = First rib originates on V1
(73₅) = Ribs absent

74. Ossification of Ribs—In synodontoids, alepisaurids, and paralepidids, all ribs ossify in membrane bone. In most scopelarchids, ribs are ligamentous, but in *Scopelarchoides signifer*, most ribs ossify in membrane bone, and only the last two are ligamentous. In all other aulopiforms except *Bathymicrops* and *Gigantura*, which lack ribs, at least some ribs ossify in membrane bone. In the outgroups, all ribs ossify in cartilage. Having any or all ribs ossify in membrane bone is derived for aulopiforms, but the distribution of the two states is such that the ancestral aulopiform condition could be either. However, ossification of only some ribs in membrane bone is primitive for the clade comprising chlorophthalmoids, alepisauroids, and giganturoids, and thus having all ribs ossify in membrane bone in paralepidids and alepisaurids is derived for that group.

(74_0) = All ribs ossify in cartilage
(74_1) = Some ribs ossify in membrane bone
(74_2) = All ribs ossify in membrane bone
(74_3) = Ribs absent
(74_4) = Some or all ribs ligamentous

75. *Origin of Baudelot's Ligament*—Baudelot's ligament originates on more than one vertebra in most paralepidids (V1 and V2) and alepisaurids (V2–V4). In all other aulopiforms and outgroups, Baudelot's ligament originates on V1 (V1 and the occiput in *Metavelifer*).

(75_0) = Baudelot's ligament originates on V1
(75_1) = Baudelot's ligament originates on more than one vertebra
(75_2) = Baudelot's ligament originates on V1 and the occiput

76. *Ossification of Baudelot's Ligament*—Baudelot's ligament is ossified in *Harpadon* and *Saurida*, a derived condition that occurs independently in *Bathymicrops*. Baudelot's ligament is lacking in *Gigantura*.

(76_0) = Baudelot's ligament is ligamentous
(76_1) = Baudelot's ligament is ossified
(76_2) = Baudelot's ligament is absent

G. Caudal Skeleton

77. *Modified Proximal Segmentation of Caudal-fin Rays*—Johnson et al. (1996, Figs. 20, 23, and 26) described a peculiar proximal segmentation of most principal caudal rays in synodontoids in which a small proximal section is separated from the remainder of the ray by a distinctive joint. The ends of the hemitrichs that meet at this joint are round, whereas those meeting at joints of the normal segmentation of caudal rays are laterally compressed and curved.

(77_0) = Proximal portion of principal caudal-fin rays not modified
(77_1) = Proximal portion of most principal caudal rays with modified segment

78. *Segmentation Begins on Distal Half of Each Caudal Ray*—In most aulopiforms and outgroups, segmentation of caudal rays begins on the proximal half of each ray, sometimes very close to the attachment of the rays to the caudal skeleton. In *Ipnops* and *Bathymicrops*, segmentation of caudal rays begins much farther posteriorly, on the distal half of each ray. Our analysis suggests evolution of posteriorly displaced segmentation in the ancestor of *Ipnops*, *Bathymicrops*, and *Bathytyphlops*, with reversal in the last. *Gigantura* lacks segmentation of caudal-fin rays.

(78_0) = Segmentation begins on proximal half of each caudal ray
(78_1) = Segmentation begins on distal half of each caudal ray
(78_2) = Caudal rays not segmented

79. *Median Caudal Cartilages*—A pair of autogenous median caudal cartilages ("CMCs" of Fujita, 1990) is present primitively in aulopiforms and outgroups except acanthomorphs which have none. CMCs are also absent in synodontoids, *Bathymicrops*, *Bathytyphlops*, and *Ipnops*. The dorsal CMC is absent in *Neoscopelus*, notosudids, *Bathypterois*, *Lestrolepis*, and *Stemonosudis* and reduced in size in *Uncisudis*, *Lestidium*, and *Lestidiops*. *Gigantura* has a single median CMC.

(79_0) = Two CMCs, about equal in size
(79_1) = Two CMCs, the dorsal one minute
(79_2) = One CMC
(79_3) = No CMCs

80. *Urodermal*—Fujita (1990) noted that a small ossified urodermal occurs near the proximal end of a caudal-fin ray of the dorsal caudal-fin lobe in some myctophids, *Neoscopelus*, one species of *Aulopus*, one *Chlorophthalmus*, and *Bathysaurus*. A urodermal is lacking in all other aulopiforms and other outgroups examined except *Bathysauroides*.

(80_0) = No urodermal
(80_1) = Small urodermal in upper caudal lobe

81. *Expanded Neural and Haemal Spines on Posterior Vertebrae*—Synodontoids (except *Harpadon*) have broad laminar expansions on the last three to six preural vertebrae (Johnson et al., 1996, figs. 16, 20, and 21). Neural and haemal spines on PU2 and PU3 are expanded in *Bathysauroides* and on PU2 in *Gigantura*.

(81_0) = Posterior neural and haemal spine not expanded

(81_1) = Neural and haemal spines of PU2 expanded
(81_2) = Neural and haemal spines of PU2 and PU3 (to PU6 in some) expanded

82. Number of Hypurals—Presence of six hypurals is primitive for aulopiforms and outgroups. Loss of the sixth hypural occurs in the ancestor of the synodontoid clade Pseudotrichonotidae + Synodontidae (Johnson *et al.*, 1996) and independently in the ancestral alepisaurid. Five hypurals also characterize *Arctozenus*, but the reduction is the result of fusion of the first and second hypurals (or failure of the two bones to differentiate). Other reductions in number of hypurals occur in *Anotopterus* (four or five; one and two sometimes fused, sixth lost) and *Bathymicrops* (two plates in the young specimen we examined, one comprising the parhypural and first two hypurals fused distally, and the other hypurals 3–5 fused distally—the distal portions of the plates are cartilaginous, and further differentiation of hypurals may accompany their ossification).

(82_0) = Six hypurals
(82_1) = Five hypurals; the sixth lost or fused
(82_2) = Five hypurals; the first and second not differentiated
(82_3) = Four hypurals; the first and second not differentiated, the sixth lost or fused
(82_4) = Two hypurals

83. Number of Epurals—The presence of three epurals is primitive for aulopiforms and the four aulopiform suborders, but the number is reduced within each. In synodontoids, a single epural is a synapomorphy of pseudotrichonotids and synodontids, with reversal to two in the ancestral harpadontid. Within the Chlorophthalmoidei, two epurals is a derived feature of *Ipnops*, *Bathytyphlops*, and *Bathymicrops*, with further reduction to one in the last. Among alepisauroids, evermannellids share a single epural, a reduction independently derived in *Anotopterus* and *Gigantura*. Adults of several other aulopiforms, including *Parasudis*, *Omosudis*, *Alepisaurus*, and some paralepidids, also have only two epurals, but one of them is split, suggesting that it may represent partial fusion of two epural bones. Where available, ontogenetic evidence supports this hypothesis, and we do not recognize this condition as distinct from that of three epurals here (but see character 118 below). Accordingly, although R. K. Johnson (1982) interpreted the reduction of epurals to one or two as a synapomorphy of evermannellids, omosudids, and alepisaurids, we disagree.

(83_0) = Adults with two or three epurals; if two, one split

(83_1) = Adults with two epurals, neither split
(83_2) = Adults with one epural

H. Median Fins

84. Fusion of Adjacent Pterygiophores (Figs. 16A, and 16B)—In *Omosudis* (Fig. 16A), the posterior portion of the proximal–middle element of the penultimate anal-fin pterygiophore is fused to the anterior aspect of the same element of the ultimate pterygiophore. The nine posteriormost pterygiophores are fused in this manner in *Alepisaurus* (Fig. 16B). The only other aulopiform examined with fused pterygiophores is the paralepidid, *Uncisudis*, which has most of the dorsal-fin pterygiophores fused. Among the outgroups, pterygiophores are fused only in *Metavelifer*. The three aulopiforms in which we observed fused pterygiophores are young specimens, and the fused cartilaginous pterygiophores may separate upon ossification. Nevertheless, the cartilaginous pterygiophores of no other young aulopiform specimens examined are fused.

(84_0) = No fusion of pterygiophores of dorsal or anal fin
(84_1) = Adjacent posterior anal-fin pterygiophores fused
(84_2) = Adjacent dorsal-fin pterygiophores fused

85. Pterygiophores of Dorsal Fin Triangular Proximally (Fig. 16C)—The proximal end of each dorsal-fin pterygiophore in all evermannellid genera is roughly triangular, the result of an expansion of the small flanges that flank the central axis. No other aulopiforms or outgroups have the proximal ends of the dorsal-fin pterygiophores triangular.

(85_0) = Pterygiophores of dorsal fin not triangular proximally
(85_1) = Pterygiophores of dorsal fin triangular proximally

86. Pterygiophores of Anal Fin Triangular Proximally—*Evermannella* and *Odontostomops* have anterior pterygiophores of the anal fin that are triangular proximally. The anal-fin pterygiophores are not modified in *Coccorella* or in other aulopiforms and outgroups except *Scopelarchoides*, in which the posterior pterygiophores of the anal fin are broadened proximally.

(86_0) = Pterygiophores of anal fin not triangular proximally
(86_1) = Anterior pterygiophores of anal fin triangular proximally
(86_2) = Posterior pterygiophores of anal fin triangular proximally

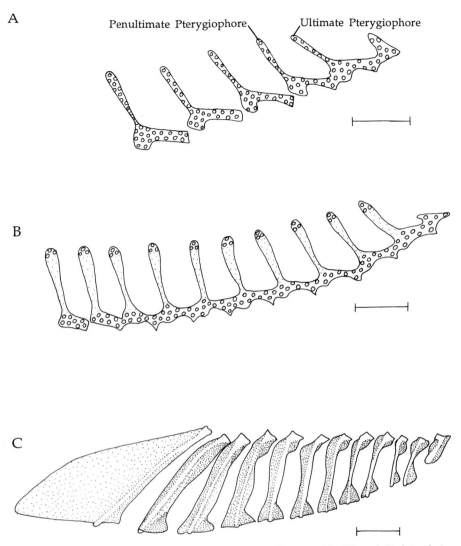

FIGURE 16   Proximal–middle pterygiophores of three alepisauroids: (A) and (B), lateral view of posterior anal-fin pterygiophores from left side of *Omosudis lowei*, USNM 219982 and *Alepisaurus* sp., MCZ 60345, respectively; (C) lateral view of dorsal-fin pterygiophores from left side of *Evermannella indica*, USNM 235141.

I. Pelvic and Pectoral Girdles and Fins

87. Medial Processes of Pelvic Girdle Joined Medially by Cartilage (Figs. 4 and 5)—As already noted, aulopiforms have very elongate medial pelvic processes that are joined by cartilage. The medial processes are typically much smaller and do not articulate with one another in stomiiforms and myctophiforms and, although they overlap in some acanthomorphs (Stiassny and Moore, 1992), they are never fused medially as in aulopiforms.

(87_0) = Medial processes not joined medially
(87_1) = Medial processes joined medially by cartilage

88. Posterior Processes of Pelvic Girdle Elongate and Widely Separated (Fig. 5A)—Posterior pelvic processes are lacking in most aulopiforms (character 89), and they are small, usually slender processes of various shape and orientation in the outgroups. In all synodontoids, the posterior processes are very well developed, widely separated, elongate structures that give the pelvic girdle a "bowed" appearance.

(88_0) Posterior pelvic processes small (or absent)
(88_1) Posterior pelvic processes elongate, widely separated

89. Posterior Processes of Pelvic Girdle Absent (Figs. 4B, 4D, and 5B–5D)—Posterior pelvic processes are primitively present in aulopiforms. In chlorophthalmids, *Bathysauropsis*, notosudids, alepisauroids, and giganturoids, the posterior pelvic processes are absent. In *Bathypterois*, the cartilage joining the medial processes divides posteriorly and forms two slender, widely separated cartilages (Fig. 4C). A similar condition occurs in *Ipnops* and *Bathytyphlops*, but the posterior cartilaginous processes are very short. In some alepisauroids, the cartilage between the medial processes continues posteriorly as a single, median cartilage, and the posterior tip of this cartilage is sometimes bifurcate. We interpret this condition as a terminal bifurcation of a median cartilage, in contrast to the formation of short cartilaginous posterior processes in some ipnopids. It is sometimes difficult to distinguish the two conditions, especially in, e.g., *Bathytyphlops* where the cartilaginous posterior processes are not as widely separated as in *Bathypterois* and *Ipnops*.

It is most parsimonious to hypothesize loss of the posterior processes in the ancestor of chlorophthalmoids + alepisauroids + giganturoids with evolution of cartilaginous posterior processes in the ancestral ipnopid. The small specimens of *Bathymicrops* that we examined appear to lack posterior processes, but investigation of larger material is needed. In young *Paralepis*, the lateral edges of the median cartilaginous plates ossify first, creating the impression of well-separated, ossified posterior processes, but these are not homologous with the posterior processes of synodontoids.

(89_0) = Ossified posterior processes of pelvic girdle present
(89_1) = Posterior processes are cartilaginous
(89_2) = Posterior processes of pelvic girdle absent

90. Lateral Pelvic Processes (Figs. 4B–4D)—Where the central process bends laterally and terminates, it is capped by a very large cartilaginous process in chlorophthalmoids. In *Chlorophthalmus*, *Parasudis*, and *Scopelosaurus*, the process is partially or entirely ossified, but it is cartilaginous in our small specimens of other chlorophthalmoids. All aulopiforms and outgroups examined have a lateral pelvic-fin process, but it is typically only a small nubbin of cartilage capping the tip of the central process. In young specimens of some alepisauroids, a large cartilage also caps the central process, but in adults, only a small lateral cartilage is present, along with an autogenous cartilage that apparently is pinched off of the large cartilage (character 91). The retention of a large cartilaginous or ossified lateral pelvic process is a synapomorphy of chlorophthalmoids. A similar cartilage is present

in *Scopelarchus analis*, an acquisition independent of that in chlorophthalmoids.

(90_0) = Lateral pelvic processes small
(90_1) = Lateral pelvic processes large, sometimes ossifying in adults

91. Autogenous Pelvic Cartilages (Fig. 5D)—Paralepidids (except *Anotopterus*), *Alepisaurus*, and evermannellids have a well-developed cartilage that extends dorsally into the body musculature from the region where the lateral pelvic-fin rays articulate with the girdle. In some young specimens, this cartilage is attached by a small rod of cartilage to the cartilage capping the central process, suggesting that it originates as part of the lateral cartilage. A similar cartilage is present in myctophids (Fig. 4A) but lacking in other aulopiforms and outgroups.

(91_0) = Autogenous pelvic cartilages absent
(91_1) = Autogenous pelvic cartilages present

92. Ventrally Directed Posterior Cartilage of the Pelvic Fin (Fig. 5B)—In scopelarchids, the cartilage joining the medial processes continues posteriorly beyond the posterior tips of the medial processes as a broad cartilaginous plate. It narrows posteriorly then abruptly curves ventrally, terminating as a small, ventrally directed process that is bound by connective tissue to the abdominal cavity wall (R. K. Johnson, 1974a). In other alepisauroids, the median cartilage may extend posteriorly beyond the medial processes, but it never deviates from the horizontal.

(92_0) = Cartilage between medial processes, if present, not terminating in ventrally directed process
(92_1) = Cartilage between medial processes terminating in ventrally directed process

93. Posterior Pelvic Cartilage Elongate (Fig. 5C)—In evermannellids, the cartilage joining the medial pelvic processes also extends posteriorly as a broad plate, but it is uniquely elongate in this family, extending well beyond the posterior tips of the medial processes, reaching up to two-thirds the length of the bony girdle (R. K. Johnson, 1982).

(93_0) = Cartilage extending posteriorly from between medial processes, if present, not elongate
(93_1) = Cartilage extending posteriorly from between medial processes elongate

94. Position of Pectoral and Pelvic Fins—In alepisauroids, the pectoral fins are positioned low on the body (closer to the ventral midline than to the lateral midline), and the pelvics are abdominal. These are primitive teleostean and neoteleostean features, but they are derived within aulopiforms, which primitively have high-set pectorals and subthoracic pelvics as in synodontoids, chlorophthalmoids, *Bathysaurus*, and *Bathysauroides*. As noted by Rosen (1973), a more dorsal placement of the pectoral fins and an anterior shift in the pelvic fins appear to be synapomorphies of aulopiforms plus ctenosquamates, i.e., the Eurypterygii.

(94_0) = Pectoral fins set high on body, pelvics subthoracic

(94_1) = Pectoral fins set low on body, pelvics abdominal

95. Relative Position of Abdominal Pelvic Fins—Primitively in alepisauroids, the abdominal pelvic fins are inserted beneath or behind a vertical through the origin of the dorsal fin. In *Sudis, Macroparalepis, Uncisudis, Lestidiops, Stemonosudis,* and *Lestrolepis* the dorsal fin originates more posteriorly than in most other alepisauroids (except in *Anotopterus* in which it is lacking), and the abdominal pelvic fins insert anterior to a vertical through the origin of the dorsal. Pelvic fins are absent in juvenile and adult *Gigantura*; in larvae, they are abdominal and insert beneath the origin of the dorsal fin.

(95_0) = Pelvic fins subthoracic or, if abdominal, inserting beneath or behind a vertical through the origin of the dorsal fin

(95_1) = Pelvic fins abdominal, inserting anterior to vertical through dorsal fin

96. Number of Postcleithra—Gottfried (1989) considered the presence of two postcleithra (the second and third of primitive teleosts) as a synapomorphy of ctenosquamates and noted that although the number of postcleithra varies among aulopiforms, the presence of three in basal taxa such as *Aulopus* indicates that three is primitive for aulopiforms. However, most synodontoids, chlorophthalmoids, and alepisauroids have two or fewer postcleithra, and our analysis suggests the primitive number for the order is two. Loss of the dorsal postcleithrum may be a synapomorphy of eurypterygians, not ctenosquamates, as proposed by Stiassny (this volume). Further study of the homology of postcleithral elements among aulopiforms and ctenosquamates is needed, but Gottfried (1989) noted the two postcleithra of *Synodus* and *Trachinocephalus* appear to be the same two (the second and third) that characterize ctenosquamates.

Of phylogenetic significance within aulopiforms is the presence of three postcleithra in evermannellids (a synapomorphy of the three included genera), *Bathysauroides,* and *Bathysaurus. Gigantura* lacks postcleithra. The number of postcleithra is reduced in most ipnopids (one in *Bathypterois* and none in *Bathymicrops*

and *Ipnops*), but the primitive state for the family is ambiguous.

(96_0) = Two postcleithra

(96_1) = One postcleithrum

(96_2) = Postcleithra absent

(96_3) = Three postcleithra

97. Cleithrum with Strut Extending to Dorsal Postcleithrum (Fig. 17)—In certain paralepidids, there is a distinctive projection extending from the cleithrum to the dorsal postcleithrum. It is very narrow where it arises from the cleithrum and then broadens posteriorly at or near its contact with the postcleithrum, and it is often closely applied to the lateral surface of the scapula. This strut occurs among aulopiforms only in *Anotopterus, Macroparalepis, Uncisudis, Lestidium, Stemonosudis,* and *Lestrolepis* and is absent in the outgroups, but some aulopiforms have a small, blunt posterior cleithral projection in the same region.

(97_0) = Cleithrum with small rounded posterior projection or projection absent

(97_1) = Cleithrum with strut extending posteriorly to postcleithrum

98. Orientation of Pectoral-Fin Base—The pectoral-fin base is oriented more horizontally than vertically in alepisauroids, *Diplophos,* and *Metavelifer,* and more vertically in other aulopiforms, myctophiforms, and *Polymixia.* The latter is primitive for aulopiforms. Parr (1928) noted that scopelarchids and evermannellids differ markedly in insertion and development of pectoral fins, but our observations suggest that although the insertion of the pectorals in scopelarchids (and paralepidids) is not as low on the body as in evermannellids and alepisaurids, in all of those taxa the base of the fin is more horizontal than in cladistically primitive aulopiforms. In preserved specimens, this reorientation of the pectoral-fin base is easily identified because rather than lying flat against the body the fin projects ventrolaterally. In the reoriented position, the fin movement is more up and down than front and back as in other aulopiforms.

The very high-set pectoral fins of *Gigantura,* which also have a nearly horizontal base, are autapomorphic.

(98_0) = Pectoral-fin base more vertical than horizontal

(98_1) = Pectoral-fin base more horizontal than vertical, inserted on the ventrolateral surface of the body

(98_2) = Pectoral-fin base horizontal, inserted on dorsolateral surface of body

99. Greatly Elongated Supracleithrum—*Bathytyphlops* and *Bathymicrops* have a very long supracleithrum,

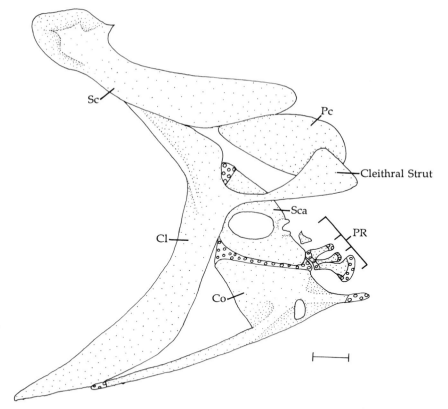

FIGURE 17 Lateral view of pectoral girdle from left side of *Stemonosudis rothschildi*, AMS I. 22826001.

equal to or longer than the cleithrum (Sulak, 1977, fig. 10). In other aulopiforms and outgroups, the supracleithrum is shorter than the cleithrum. Merrett and Nielsen (1987) noted that the supracleithrum in *Discoverichthys praecox* is about equal in length to the cleithrum, suggesting a possible relationship with *Bathymicrops* and *Bathytyphlops*.

(99_0) = Supracleithrum shorter than cleithrum
(99_1) = Supracleithrum equal to or longer than cleithrum

100. Ventral Limb of Posttemporal Not Ossified—R. K. Johnson (1982) noted that scopelarchids and evermannellids are unique among iniomes in having an unforked posttemporal. In those families, an ossified dorsal limb articulates with the epiotic, but there is no ossified ventral limb. Instead, a ligament connects the main body of the posttemporal to the intercalar. A forked posttemporal in which both the dorsal and ventral limbs are ossified is present in other aulopiforms and outgroups.

(100_0) = Posttemporal forked, both branches ossified

(100_1) = Posttemporal unforked, the ventral branch ligamentous

J. External Morphology

101. Margin of Anal Fin Indented—A derived feature of the Alepisauroidei is the shape of the anal fin, the margin of which is deeply indented near the anterior end. In other aulopiforms and outgroups, the margin of the anal fin may be straight, slightly convex, or concave, but it is usually not deeply indented. Some *Polymixia* (e.g., *P. nobilis*) have the anal fin indented similar to that of alepisauroids. Absence of an indented anal fin in *Anotopterus* is a reversal.

(101_0) = Margin of anal fin not indented
(101_1) = Margin of anal fin indented

102. Scales—Ossified scales on the body and lateral line are primitive for aulopiforms. R. K. Johnson (1982) noted that ossified scales or scale-like structures are absent in three families, Evermannellidae, Omosudidae, and Alepisauridae. However, our investigation indicates that ossified lateral-line structures are present in all evermannellids. Body scales are absent in evermannellids, as they are in alepisaurids and all paralepidids examined except *Paralepis* and *Arctozenus*. Only alepisaurids and giganturids lack ossified body and lateral line scales.

(102_0) = Body and lateral-line scales present and ossified

(102_1) = Body scales absent, lateral-line scales or structures at least partially ossified

(102_2) = Body and lateral-line scales or structures absent

103. Fleshy Mid-lateral Keel—R. K. Johnson (1982) considered the presence of a fleshy, mid-lateral keel on each side of the posterior section of the body as a derived feature of alepisaurids (the keel is restricted to the caudal peduncle in *Omosudis* and covers the posterior one-third to one-half of the body in *Alepisaurus*). We agree and note that a fleshy midlateral keel does not occur elsewhere among aulopiforms or outgroups except in *Anotopterus*, which has a pair of fleshy keels on each side of the caudal peduncle (Rofen, 1966c, fig. 182).

(103_0) = Fleshy mid-lateral keel absent

(103_1) = Single fleshy mid-lateral keel on posterior portion of body

(103_2) = Pair of fleshy mid-lateral keels on caudal peduncle

104. Body Transparent, Glassy in Life—Rofen (1966b, p. 210) noted that *Sudis* and all other paralepidids except *Paralepis* and *Arctozenus* are "transparent or nearly so, glassy in life, the surface of the skin iridescent in a kaleidoscope of colors." Our review of the literature indicates that other paralepidids, aulopiforms, and outgroups may be iridescent, but if so they are silvery and not transparent. We have not examined any living or fresh specimens of Paralepididae, but we tentatively consider the glassy appearance described by Rofen (1966a) as a synapomorphy of *Sudis, Macroparalepis, Uncisudis, Lestidium, Lestidiops, Stemonosudis,* and *Lestrolepis*.

(104_0) = Appearance in life not transparent or glassy

(104_1) = Appearance in life transparent, glassy

105. Scale Pockets in Continuous Flap of Skin—Hartel and Stiassny (1986) hypothesized a sister-group relationship between *Parasudis* and *Chlorophthalmus*, citing as evidence the presence of scale pockets in a continuous flap of skin. The skin flap is pigmented distally, and thus the overall appearance of pigmentation in those genera is a zig-zag or herringbone pattern (Hartel and Stiassny, 1986; Mead, 1966d). Other au-

lopiforms, including other chlorophthalmoids, do not have scales implanted in pockets along a continuous flap of pigmented skin.

(105_0) = Scale pockets not in continuous flap of skin

(105_1) = Scale pockets in a continuous flap of marginally pigmented skin

106. Elliptical or Keyhole Aphakic Space—Mead (1966d) noted that *Chlorophthalmus* and *Parasudis* have a keyhole-shaped pupil, created by a conspicuous aphakic (i.e., lensless) space anteriorly. Marshall (1966) and Bertelsen *et al.* (1976) described a similar, but elliptical, lensless space in notosudids, and we have observed the same condition in *Bathysauropsis malayanus, B. gracilis,* and *Bathysauroides.* An aphakic space is lacking in other aulopiforms and outgroups. If the two forms of aphakic space are considered as separate states, the character is ambiguous, and neither state is phylogenetically informative. If we accept the two conditions as primary homologues, the presence of an aphakic space is a synapomorphy of chlorophthalmoids. We code this character as "missing" in ipnopids, which have greatly reduced or modified eyes. The aphakic space of *Bathysauroides* is best interpreted as independently derived; a modification of the iris of that species (incomplete or divided anteriorly at least in subadults) may be further evidence that the eye morphology of *Bathysauroides* is unique.

(106_0) = No aphakic space

(106_1) = Elliptical or keyhole shaped aphakic space

107. Eye Morphology—A laterally directed round eye characterizes synodontoids, chlorophthalmids, alepisaurids, paralepidids, and *Odontostomops*. Within the Chlorophthalmoidei, there is a trend toward reduction in eye size, from slightly flattened or elliptical in *Bathysauropsis* and notosudids, to minute in most ipnopids. *Ipnops* lacks recognizable eyes but has broad, lensless light-sensitive organs on the surface of the head (see Mead, 1966c). It is most parsimonious to hypothesize a reduction in eye size in the ancestor of *Bathysauropsis*, notosudids, and ipnopids with further reduction in the last. An elliptical eye also characterizes *Bathysauroides* and *Bathysaurus*. Giganturids are unique among aulopiform in having anteriorly directed telescopic eyes.

Scopelarchids and most evermannellids have dorsally directed semitubular or tubular eyes. The laterally directed round eyes in *Odontostomops*, which is the sister group of *Evermannella*, are best interpreted as a reversal. Lending support to the interpretation of tubular eyes as a synapomorphy of the Evermannellidae and Scopelarchidae is the fact that larvae of both families have dorsoventrally elongate eyes (R. K.

Johnson, 1974a, 1982), implying a similar ontogeny of the adult condition (Character 114).

(107_0) = Eyes laterally directed, round
(107_1) = Eyes slightly flattened to elliptical
(107_2) = Eyes minute or absent
(107_3) = Eyes dorsally directed, semitubular or tubular
(107_4) = Eyes anteriorly directed, telescopic
(107_5) = Eyes are broad, lensless plates on dorsal surface of head

108. Gular Fold—Mead (1966b) noted that *Bathypterois* has a thick gular fold that covers the ventral surface of the branchiostegal membranes where they overlap anteriorly. Hartel and Stiassny (1986) noted that a well-developed gular fold is characteristic of all ipnopids as well as *Bathysauropsis*, and they considered this feature as further evidence that *Bathysauropsis* is an ipnopid.

We examined the gular region of all aulopiforms and found that the thickness of the gular fold varies with size. Nevertheless, the gular fold of ipnopids is different from the typical aulopiform condition in that the posterior edge of the fold is crescent-shaped and is not tightly bound to the branchiostegal membranes except along the lateral edges. In most other aulopiforms, the posterior margin of the gular fold is tent-shaped and tightly bound to the branchiostegal membranes. Thus in ipnopids, a probe inserted beneath the fold can be extended to near the symphysis of the dentary bones, whereas in other aulopiforms and most outgroups, extension of a probe beneath the fold anteriorly is impossible because of the attachment of the fold to the branchiostegal membranes.

Notosudids, but not *Bathysauropsis*, also have a crescent-shaped gular fold that is loosely bound to the branchiostegal membranes, a derived feature that we consider further evidence of a sister-group relationship between notosudids and ipnopids. A crescent-shaped gular fold is independently derived in *Polymixia*.

(108_0) = Gular fold tent-shaped
(108_1) = Gular fold crescent-shaped

109. Adipose Fin—Presence of a dorsal adipose fin is primitive for aulopiforms, although several outgroups (*Diplophos* and acanthomorphs) lack an adipose fin. Among aulopiforms, an adipose fin is lacking in *Pseudotrichonotus*, *Bathysaurus*, and the ipnopids *Bathymicrops*, *Bathytyphlops*, and *Ipnops*.

(109_0) = Adipose fin present
(109_1) = Adipose fin absent

K. Internal Soft Anatomy

110. Mode of Reproduction—R. K. Johnson (1982) hypothesized that synchronous hermaphroditism evolved three times among iniome fishes—once in the ancestor of his chlorophthalmid + ipnopid + notosudid + scopelarchid lineage, once in bathysaurids, and again in the ancestor of his alepisauroid clade. Our phylogeny suggests that all of those taxa form a monophyletic group, and thus we hypothesize a single origin of hermaphroditism, in the ancestor of our chlorophthalmoid + alepisauroid + giganturoid lineage. Synodontoids, myctophiforms, and *Polymixia* have separate sexes (see R. K. Johnson, 1982, for references) and, although the mode of reproduction in many stomiiforms is unknown, gonochorism also appears to be the primitive aulopiform strategy.

(110_0) = Separate sexes
(110_1) = Synchronous hermaphrodites

111. Thin-Walled, Heavily Pigmented Stomach—R. K. Johnson (1982) considered the presence of a highly distensible black stomach as a derived feature of alepisaurids. He noted that other iniomes examined by him have a heavily muscularized, unpigmented stomach.

(111_0) = Stomach not highly distensible, with thick unpigmented walls
(111_1) = Stomach highly distensible, with thin heavily pigmented walls

112. Swimbladder—Aulopiforms lack a swimbladder, but ctenosquamates primitively have one (absent in some myctophiforms) as do most stomiiforms (e.g., gonostomatids, sternoptychids, photichthyids, some astronesthids, and stomiids) (Marshall, 1954, 1960; Marshall and Staiger, 1975; R. K. Johnson, 1982).

(112_0) = Swimbladder present
(112_1) = Swimbladder absent

L. Larval Morphology

113. Enlarged Pectoral Fins—Okiyama (1984b) noted that ipnopid larvae share the derived condition of greatly enlarged, fanlike pectoral fins. Larvae of *Sudis hyalina* also have elaborate pectorals (Okiyama, 1984a, fig. 113F), and larvae of *Bathysaurus* have all fins except the caudal greatly enlarged (Okiyama, 1984a, fig. 111C). The pectoral fins of larval *Rosenblattichthys* are well developed relative to other scopelarchids (see R. K. Johnson, 1984a, fig. 127A,B) but not nearly as much as in ipnopids. Okiyama (1984b, p. 256) indicated in his character matrix that alepisaurids have elongate pectoral fins, but the illustrations of *A. brevirostris* and *A. ferox* (Okiyama, 1984a, fig. 112A,B) do not reflect this condition. Pectoral fins are enlarged in certain myctophiforms (e.g., some *Lampanyctus*, Moser *et al.*, 1984, Fig. 124F) but not in other outgroups.

(113_0) = Pectoral fins not enlarged in larvae
(113_1) = Pectoral fins enlarged in larvae

114. Elongate Eyes—The eyes are dorsoventrally elongate in larval scopelarchids and evermannellids. R. K. Johnson (1984b) noted that the eyes are not elongate in larvae of *Odontostomops*, and thus evermannellids and scopelarchids may have independently acquired them. We have not examined larval *Odontostomops*, but in illustrations of *O. normalops* (R. K. Johnson, 1982, Figs. 5D and 6D) the eye appears to be slightly wider than in other evermannellids, but it is dorsoventrally elongate rather than round.

Notosudid larvae also have narrow eyes, but they differ from evermannellid eyes in being elongate in the anteroposterior plane (Bertelsen *et al.*, 1976; Okiyama, 1984a, fig. 111A). Some myctophids have dorsoventrally elongate eyes, but round eyes are primitive for aulopiforms.

(114_0) = Eyes in larvae round
(114_1) = Eyes in larvae elongate; the horizontal axis longer than the vertical
(114_2) = Eyes in larvae elongate; the vertical axis longer than the horizontal

115. Head Spination—Head spines are uncommon in larvae of non-acanthomorph teleosts, but serrate cranial ridges and preopercular spines are present in a strikingly similar configuration in *Alepisaurus ferox* and *Omosudis* (Okiyama, 1984a, figs. 112B, 112E, and 112F). Larvae of *A. brevirostris* apparently lack head spines (Rofen, 1966b, fig. 171; Okiyama, 1984a, fig. 112A), and thus ornamentation in the two genera could be nonhomologous. However, the presence of two nearly identical patterns of head spines among a group of teleosts that are not known for elaborate head ornamentation leads us to believe that the conditions in *A. ferox* and *Omosudis* are homologous.

The paralepidid *Sudis* also has head ornamentation, in the form of serrate cranial ridges and a large, strongly serrate spine at the angle of the preopercle. Other paralepidids lack head spines, and it is thus most parsimonious to hypothesize independent acquisition of head ornamentation in *Sudis* and the Alepisauridae.

(115_0) = Head spines lacking in larvae
(115_1) = Head spines present in larvae

116. Peritoneal Pigment—As noted, R. K. Johnson (1982) suggested that peritoneal pigment in larvae may be diagnostic of Rosen's (1973) Aulopiformes, a notion supported in our analysis, despite the absence of peritoneal pigment in larvae of some chlorophthalmoids (notosudids, *Ipnops*, *Bathymicrops*, and some *Bathypterois*). Okiyama (1984b) and R. K. Johnson (1982) described several states of this character: a single, unpaired peritoneal pigment "section"; multiple, unpaired pigment sections; a single unpaired section changing ontogenetically to several unpaired sections; multiple paired pigment spots; and absence of peritoneal pigment. Johnson *et al.* (1996) considered the presence of paired peritoneal pigment spots in larvae and juveniles a synapomorphy of *Pseudotrichonotus* and synodontids. These spots are retained in the abdominal wall of adults as tiny dense discs of pigment. Our investigation suggests that the presence of one or more unpaired peritoneal pigment sections is primitive for aulopiforms, and thus we concur with Johnson *et al.* (1996) that the presence of paired peritoneal pigment sections in some larval synodontoids is derived.

(116_0) = Peritoneal pigment absent in larvae
(116_1) = Single or multiple unpaired peritoneal pigment sections in larvae
(116_2) = Multiple paired peritoneal pigment sections in larvae

117. Ontogenetic Reduction of Large Maxilla (Fig. 18)—Adults of *Gigantura* have only a small maxillary remnant posteriorly, but in larval giganturids ("*Rosaura*") the maxilla is a very large, leaf-shaped bone that tapers abruptly anteriorly near its articulation with the premaxilla (Fig. 18B). Rosen (1971) discussed the relationships of Regan's (1903) Macristiidae, a "myctophoid" family described on the basis of a single specimen of *Macristium chavesi* that is now lost. He described a new *Macristium*-like larval fish (the "Chain" larva) and concluded that it is probably the young stage of *Bathysaurus*, a notion corroborated by R. K. Johnson (1974b). In his paper, Rosen (1971) illustrated a lateral view of the skull bones of the "Chain larva" (Fig. 18A). Adult *Bathysaurus* have only a small, anterior remnant of the maxilla (e.g., Sulak, 1977, fig. 5A), but Rosen's illustration shows a very large maxilla in the larva that bears a striking resemblance to that of larval *Gigantura*. It is large and leaf-shaped and tapers abruptly anteriorly (Fig. 18).

Dramatic ontogenetic reduction of a large maxilla is thus shared by *Bathysaurus* and *Gigantura*, and we have not observed it elsewhere in the Aulopiformes, including synodontids in which the maxilla is reduced in adults (see e.g., Okiyama, 1984a, figs. 111D–111G). Larval *Bathysauroides* are undescribed, but adults have a well-developed maxilla; we thus predict that the

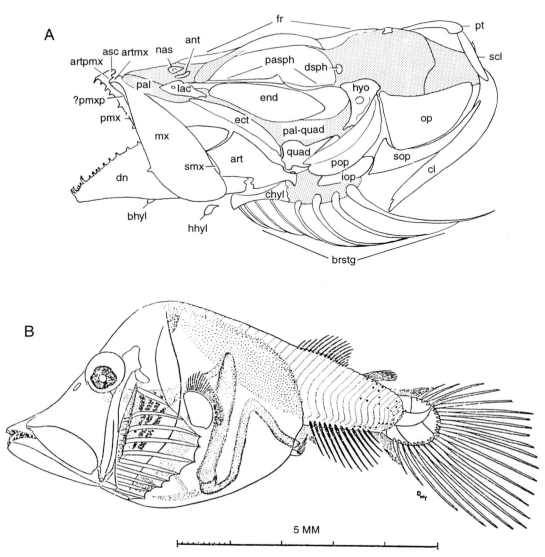

FIGURE 18 Larvae of (A) *Bathysaurus* (from Rosen, 1971, fig. 5, depicting the syncranium of the "Chain" larva) and (B) *Gigantura* (from Tucker, 1954, fig. 1).

maxilla in larval *Bathysauroides* is not enlarged or reduced ontogenetically.

(117_0) = Maxilla not enlarged in larva, not greatly reduced ontogenetically
(117_1) = Maxilla enlarged in larva, greatly reduced ontogenetically

118. Ontogenetic Fusion of Epurals—Adult *Parasudis* have two epurals, but the anterior is split distally. Larval *Parasudis* have three epurals, suggesting the adult condition is the result of partial ontogenetic fusion of the first and second epurals. Partial fusion of two epurals also apparently occurs in *Omosudis*, *Alepisaurus*, *Lestrolepis*, *Lestidiops*, and *Stemonosudis*,

which have two epurals in adults, one of which is split proximally. As in larval *Parasudis*, larval *Stemonosudis* have three cartilaginous epurals. We have not examined this feature in larvae of other paralepidids and alepisaurids listed above, but it is reasonable to assume that the divided epurals in adults of those taxa are also the result of ontogenetic fusion. Our analysis indicates that such ontogenetic fusion occurred three times within aulopiforms: in *Parasudis*, in the ancestral alepisaurid, and in the ancestor of the paralepidid clade comprising *Lestidiops*, *Stemonosudis*, and *Lestrolepis*.

(118_0) = No ontogenetic fusion of epurals
(118_1) = Partial ontogenetic fusion of two epurals

VI. Discussion

Monophyly of Rosen's (1973) Aulopiformes is supported by seven derived features. Four characters were previously recognized as synapomorphies of the order: (1_1) an enlarged EB2 uncinate process (Rosen, 1973); (2_1) absence of a cartilaginous condyle on PB3 for articulation of EB2 (Johnson, 1992); (54_1) anterior extension of the epipleural series to at least V2 (Patterson and Johnson, 1995); and (116_1) peritoneal pigment in larvae (R. K. Johnson, 1982). Two previously described characters were not recognized as aulopiform synapomorphies: (55_1) displacement of one or more of the anterior epipleurals dorsally into the horizontal septum (Patterson and Johnson, 1995) and (112_1) absence of a swimbladder (e.g., R. K. Johnson, 1982). We identified a seventh diagnostic feature of aulopiforms, (87_1) fusion of the medial processes of the pelvic girdle. Additionally, although not recognized formally in our analysis, a benthic existence may be a synapomorphy of aulopiforms. Because stomiiforms and primitive ctenosquamates are pelagic (polymixiids and *Metavelifer* are benthopelagic), and adults of most aulopiforms are benthic, a transition from a pelagic to a benthic environment may have characterized the ancestral aulopiform. Several aulopiforms have reinvaded the pelagic realm.

Aulopiform genera comprise four major clades that we designate the suborders Synodontoidei, Chlorophthalmoidei, Alepisauroidei, and Giganturoidei. Below we summarize the evidence supporting the monophyly of those clades and relationships among them. Within suborders, we emphasize characters supporting newly proposed clades as well as those previously undescribed or recognized as synapomorphies at different taxonomic levels.

Limits and relationships of the Synodontoidei of Johnson *et al.* (1996) are well supported in this study, with each clade being diagnosed by five or more unambiguous derived features (Fig. 6). *Aulopus* is cladistically the most primitive synodontoid, a hypothesis that conflicts with previous proposals (e.g., Rosen, 1985; Hartel and Stiassny, 1986) in which aulopids and sometimes chlorophthalmids were considered more closely related to ctenosquamates than to other aulopiforms. Synodontoids share eight derived features (Fig. 6), including two not recognized by Johnson *et al.* (1996): (17_2^*) gap between BB4 and CB5s and (88_1) elongate, widely separated posterior pelvic processes. Most of the homoplasy within the group occurs in the highly modified *Pseudotrichonotus* and the secondarily free-swimming *Harpadon*.

Sulak (1977) considered *Bathysaurus* as a subfamily of his expanded Synodontidae, but our data reject that notion. *Bathysaurus* lacks all synapomorphies of synodontoids and the clade comprising *Pseudotrichonotus* + synodontids and has only 2 of the 10 derived features uniting synodontids (Fig. 6): (5_1) gill rakers reduced to toothplates and (69_0) reduced number of caudal vertebrae.

Rosen (1973) argued that synodontids and harpadontids are closely related to alepisauroids and included the superfamily Synodontoidea in his suborder Alepisauroidei. He appears to have based this on three characters, a single upper pharyngeal toothplate (UP4 or UP5), enlarged orobranchial teeth, and gill rakers present as toothplates. Johnson (1992) noted that all alepisauroids except *Anotopterus* have both UP4 and UP5. UP4 is absent (3_1) only in *Pseudotrichonotus* and synodontids (Fig. 3B), a derived feature of that clade. Enlarged orobranchial teeth also fails as a synapomorphy of synodontids and alepisauroids because the enlarged teeth are premaxillary in synodontids and their relatives, whereas in alepisauroids, premaxillary teeth are often minute, and the enlarged teeth are on the palatine (36_2). Rosen's third character, (5_1) gill rakers present as toothplates, is apparently independently derived in synodontoids and alepisauroids.

The remaining aulopiforms—chlorophthalmoids, alepisauroids, and giganturoids—form a novel clade diagnosed on the basis of four derived features: (30_1^*) anterior ceratohyal bearing four or fewer branchiostegals, (72_1) neural arches open dorsally only on the anteriormost four or fewer vertebrae, (89_2) ossified posterior pelvic processes absent, and (110_1) sexual reproduction by synchronous hermaphroditism. Most of these fishes inhabit depths of 1000 to 6000 m, and the evolution of synchronous hermaphroditism may have contributed to their successful radiation into the deep. Synodontoids have separate sexes and are primarily shallow-water fishes.

The Chlorophthalmoidei include the Chlorophthalmidae, *Bathysauropsis* (c.f. *B. gracilis* and *B. malayanus*), Notosudidae, and Ipnopidae. Monophyly of chlorophthalmoids is supported by the following: (6_1) proximal end of PB2 expanded laterally; (45_1) medial edge of premaxilla with a dorsomedially directed process; (68_2^*) one supraneural; (90_1) central process of pelvic girdle capped laterally by a very large winglike process, ossified in some taxa; and (106_1) pupil elliptical or keyhole-shaped, with a prominent aphakic space anteriorly (except in ipnopids where eyes are minute or greatly modified).

The Chlorophthalmidae (*Chlorophthalmus* and *Parasudis*) share three previously described derived features (44_2, 63_1, and 105_1) relating to squamation, intermuscular, and jaw morphology (Hartel and Stiassny, 1986; Stiassny, 1986; Patterson and Johnson, 1995). Ipnopids have a small, obliquely aligned basihyal, and its presence in *Bathysauropsis gracilis* led Hartel and Stiassny (1986, fig. 7) to reassign *Bathysauropsis* to the Ipnopidae (from the Chlorophthalmidae). Our phylogeny indicates that *Bathysauropsis* is the sister group of ipnopids + notosudids, and thus we interpret (34_1^*) an obliquely aligned basihyal as a synapomorphy of *Bathysauropsis*, notosudids, and ipnopids, with reversal in notosudids. The *Bathysauropsis* clade also shares (23_1) gill rakers extending onto lateral surfaces of deep basibranchials, (30_0^*) five or more branchiostegals on anterior ceratohyal, and (107_1) reduced or modified eyes relative to the very large, round eyes of chlorophthalmids.

A sister-group relationship between notosudids and ipnopids has not been proposed previously. Bertelsen *et al.* (1976) suggested that notosudids are most closely related to chlorophthalmids, R. K. Johnson (1982) placed notosudids as the sister group of his scopelarchid + chlorophthalmid + ipnopid clade, and Patterson and Johnson (1995) considered notosudids as the sister group of the Scopelarchidae + Evermannellidae. R. K. Johnson (1982) based his hypothesis on two derived features, absence of a swimbladder and presence of synchronous hermaphroditism, but we consider those features as synapomorphies of aulopiforms and the chlorophthalmoid + alepisauroid + giganturoid clade, respectively. Patterson and Johnson (1995) cited the origin of epineurals on the centrum or parapophysis on about vertebrae 5–15 as evidence for their placement of notosudids, but our analysis suggests independent evolution of ventrally displaced epineurals in notosudids and alepisauroids. Our hypothesis of a sister-group relationship between notosudids and ipnopids is supported by (6_2) an unusual modification of PB2 in which the proximal end has an extra uncinate process, (79_2) absence of at least one CMC, and (108_1) a thick, crescent-shaped gular fold.

The notosudid genera *Scopelosaurus* and *Ahliesaurus* share 10 derived features (Fig. 6), including (20_1) elongate BB1, (33_1) anteriormost branchiostegal on ventral hypohyal, (38_1) quadrate with two cartilaginous heads, and (57_1) epipleurals forked distally at transition of epipleurals in and beneath horizontal septum. Although we did not examine the monotypic *Luciosudis*, information from Bertelsen *et al.* (1976) suggests that *L. normani* has at least two synapomorphies of *Scopelosaurus* and *Ahliesaurus*, (46_1) seven infraorbitals and (114_1) horizontally elongate eyes in larvae. We conclude that the Notosudidae are monophyletic, but

further study is needed to elucidate relationships among the three genera.

Ipnopids (*Bathypterois*, *Bathymicrops*, *Bathytyphlops*, and *Ipnops*) share nine derived features (Fig. 6), including (113_1) an enlarged pectoral fin in larvae, a condition that occurs elsewhere among aulopiforms and the outgroups only in *Sudis*, *Rosenblattichthys*, and *Bathysaurus*. Ipnopids also share the following: (7_1) UP2 usually absent, (42_1) metapterygoid free from hyomandibular, (52_1) frontal expanded laterally over orbit, (53_1) sphenotic with an anteriorly directed process extending beneath frontal, (73_3) ribs, when present, beginning on V2, and (89_1) posterior processes of pelvic girdle cartilaginous; most have (107_2) minute eyes. Some of these features are reversed in *Bathymicrops*, which lacks ribs, has UP2, and apparently lacks posterior pelvic processes. Nevertheless, our analysis places *Bathymicrops* as the sister group of *Bathytyphlops*, as proposed by Sulak (1977). The two share (43_1) a horizontally oriented hyomandibular and opercle, (99_1) a long supracleithrum, (21_1) an elongate BB2, and (24_1) ossification of the ligament between HB1 and the hyoid.

Bathypterois, formerly placed in a separate family (Bathypteroidae; see, e.g., Mead (1966b)), is the sister group of the other ipnopid genera, which are united on the basis of several, mostly reductive, derived features: (46_3) five (or fewer) infraorbitals, (79_3) loss of CMCs, (83_2) two (or one) epurals, and (109_1) absence of an adipose fin. They also share (69_2) a high percentage of caudal vertebrae. *Ipnops* and *Bathymicrops* have (19_1) ossified BB4 and (78_1) segmentation of caudal rays beginning on distal half of each ray, additional features treated as synapomorphies of *Ipnops* + *Bathymicrops* + *Bathytyphlops* in our analysis, with reversal in *Bathytyphlops*.

We did not examine the single known specimen of the ipnopid *Discoverichthys praecox*, but we used data from Merrett and Nielsen (1987) to explore its relationships. Although the configuration of the gill arches, pelvic girdle, and intermuscular are unknown, *Discoverichthys* lacks a swimbladder and is hermaphroditic, suggesting that it belongs in the chlorophthalmoid + alepisauroid + giganturoid clade of aulopiforms. Because the premaxilla is the dominant tooth-bearing bone of the upper jaw, and the gillrakers are lathlike, *Discoverichthys* is best placed in the chlorophthalmoid lineage. It has the well-developed gular fold of notosudids, *Bathysauropsis*, and ipnopids, the small oblique basihyal of *Bathysauropsis* and ipnopids, the minute eye of most ipnopids, and, like *Bathymicrops*, *Bathytyphlops*, and *Ipnops*, it lacks an adipose fin. *Discoverichthys* does not have the opercle and hyomandibular reoriented as in *Bathymicrops* and *Bathytyphlops*, nor does it share with those genera a greatly elongated supracleithrum. We tentatively conclude that *Discoverichthys* is most closely related to the clade comprising *Bathymicrops*, *Bathytyphlops*, and *Ipnops*, but it does not appear to belong to the *Bathymicrops* + *Bathytyphlops* group.

Alepisauroids and giganturoids form another new clade in our tree and share several derived features, most notably the following: (5_1) gill rakers present as toothplates, (36_2) palatine the dominant tooth-bearing bone of the "upper jaw", (54_2) epipleurals extending to V1, and (61_1*) origin of some (or all) epineurals on centrum. Adults of *Gigantura* lack a dermopalatine and most elements of the branchial skeleton but share with alepisauroids, *Bathysaurus*, and *Bathysauroides* the anterior extension of epipleurals to V1 (see discussion of *Gigantura* below).

Our Alepisauroidei comprise the Alepisauridae (including *Omosudis*), Paralepididae (including *Anotopterus*), Evermannellidae, and Scopelarchidae. Rosen's (1973) alepisauroids were characterized by gill-arch morphology, especially attenuation of epibranchial and pharyngobranchial elements, absence of UP2, UP5, and a toothplate on EB3 (ET3), and large pharyngobranchial teeth. UP4 and UP5 are present in most alepisauroids, and large pharyngobranchial teeth also characterize giganturoids. Aulopiforms vary considerably in length of epibranchial and pharyngobranchial elements and the presence of ET3, and neither convincingly diagnoses alepisauroids. However, alepisauroids do have distinctive gill arches, characterized in part by (7_1) absence of UP2. Other diagnostic features of alepisauroid gill arches include: (11_1) teeth on UP3 (when present) restricted to lateral edge, (15_2) teeth on CB5 restricted to medial edge, and (22_0) gillrakers (present as toothplates) not extending onto HB3. Alepisauroids also share the following: (46_2*) eight infraorbitals, (68_1*) two supraneurals, (91_1*) autogenous lateral pelvic cartilages, (94_1) abdominal pelvic fins, (98_1) a nearly horizontal (or more horizontal than vertical) pectoral-fin base, and (101_1) an indented anal fin. Furthermore, the pelagic lifestyle of alepisauroids may represent a single evolutionary transition from the benthic existence of primitive aulopiforms.

We agree with R. K. Johnson (1982) that *Omosudis* and *Alepisaurus* are sister taxa. They share 12 unambiguous derived characters, including features of the gill arches, intermuscular system, caudal skeleton, external morphology, internal soft anatomy, and head spination in larvae (Fig. 6), the following several of which are previously unrecognized alepisaurid synapomorphies: (10_1) PB3 extending anteriorly beyond EB1 and PB2, (18_1) BB3 extending beneath BB4, (58_1) epipleurals on V1 and V2 fused to centrum, (65_1) most epineur-

als unattached, and (84_1) adjacent posterior anal-fin pterygiophores fused. The close relationship between *Omosudis* and *Alepisaurus* is best represented by referring *Omosudis* to the Alepisauridae.

Patterson and Johnson (1995) hypothesized a sister-group relationship between the *Omosudis* + *Alepisaurus* clade and paralepidids. They based this on three derived features: (74_2) all ribs ossified in membrane bone, (76_1) Baudelot's ligament originating on more than one vertebra, and epineurals on the first five or fewer vertebrae fused to the neural arch. Examination of additional taxa indicates that epineurals are free from the axial skeleton except in the two genera, *Paralepis* and *Macroparalepis*, examined by Patterson and Johnson (1995). An additional but ambiguous synapomorphy of alepisaurids and paralepidids is (73_4) ribs originating on V1. Further study of this group is clearly needed.

We concur with R. K. Johnson (1982) that paralepidids and *Anotopterus* form a monophyletic lineage. In addition to his character, (48_1) a fenestrate premaxilla, they share (20_2) an elongate BB1, (47_1) a prolonged snout, (49_1) an anterior extension of the palatine to meet the premaxilla, (50_1) a long horizontally oriented lacrimal on the elongate snout, and (66_2) absence of epicentrals. Relationships among the speciose paralepidids are poorly understood, and we have contributed little toward their resolution. Our preliminary data do not corroborate all aspects of the classifications of Rofen (1966a) and Post (1987), wherein *Sudis* is given subfamilial or familial status, respectively, and the remaining genera are divided between two tribes or subfamilies. Post (1987) included *Arctozenus*, *Magnisudis*, *Notolepis*, and *Paralepis* in his subfamily Paralepidinae based on apparently primitive aulopiform features (e.g., cycloid body scales, no luminous organs, and no ventral adipose fin). We examined two genera of Post's Paralepidinae, *Paralepis* and *Arctozenus*, and found that they share three intermuscular characters (59_2, 63_2, and 65_2) as well as (22_1) gill rakers (present as toothplates) on HB3 and (32_1) branchiostegals on anterior ceratohyal in 3+1 pattern. They lack the diagnostic features of the lineage comprising *Anotopterus* and all other paralepidid genera, including *Sudis*: (11_2) UP3 absent, (97_1) cleithral strut present; and (102_1) body scales absent but ossified lateral-line scales present. A toothplate fused to PB3 is a conservative feature among euteleosts, and its absence is strong evidence of the phylogenetic integrity of this paralepidid group. Placement of *Anotopterus* as the sister group of one paralepidid clade requires its inclusion in the Paralepididae.

Sudis shares with *Lestidiops*, *Lestidium*, *Lestrolepis*, *Macroparalepis*, *Stemonosudis*, and *Uncisudis* (60_1) a reduced number of epipleurals and (104_1) a transparent, "glassy" body. A close association between the main branch of EB2 and its uncinate process (9_1) and (32_1) a 3 + 1 pattern of branchiostegals on the anterior ceratohyal unite all of those genera, excluding *Sudis*, as a monophyletic assemblage. *Uncisudis, Lestidium, Lestidiops, Stemonosudis,* and *Lestrolepis* have (79_1) the dorsal CMC reduced to a tiny nubbin (or absent). *Lestidiops, Stemonosudis,* and *Lestrolepis* exhibit (118_1) partial ontogenetic fusion of two epurals. Finally, *Lestrolepis* and *Stemonosudis* share (79_2) absence of the dorsal CMC. No further resolution of relationships among paralepidid genera is evident from our data, and further study is needed.

We agree with R. K. Johnson (1982) that *Coccorella, Evermannella,* and *Odontostomops* constitute a monophyletic Evermannellidae but diagnose the family based on 10 additional derived features (Fig. 6). Most striking among these are (34_2) basihyal oriented at about a 90° angle to first basibranchial, (66_3) anterior epicentrals cartilaginous, (85_1) pterygiophores of dorsal fin triangular proximally, and (93_1) a long tail of cartilage extending posteriorly from the pelvic girdle. Our data do not corroborate R. K. Johnson's (1982) hypothesis of a sister-group relationship between *Coccorella* and *Evermannella*. Rather, three derived features indicate that *Evermannella* and *Odontostomops* are sister taxa: (27_1) third hypobranchials fused ventrally, (31_1) posteriormost two branchiostegals close, and (86_1) proximal ends of anal-fin pterygiophores expanded.

The Scopelarchidae are monophyletic, as proposed by R. K. Johnson (1974a, 1982), the four genera (*Benthalbella, Scopelarchus, Scopelarchoides,* and *Rosenblattichthys*) sharing reversals of several derived alepisauroid conditions $(46_0^*, 68_0^*,$ and $91_0)$ as well as three novel derived features: (35_1) large, posteriorly curved basihyal teeth; (74_4) some or all ribs in ligament; and (92_1) a median cartilage extending posteriorly from the pelvic girdle that bends down to terminate as a small, ventrally directed process. Our data do not elucidate relationships within the Scopelarchidae.

Although scopelarchids were traditionally placed near evermannellids (e.g., Marshall, 1955; Gosline *et al.,* 1966), R. K. Johnson (1982) suggested that resemblances between the two families may be superficial. Five unambiguous synapomorphies support a sister-group relationship between the Scopelarchidae and Evermannellidae: (67_1) attachment of anterior epicentrals to distal ends of epipleurals, (69_2) high percentage (>60%) of caudal vertebrae, (100_1) unossified ventral posttemporal limb, (114_2) dorsoventrally elongate eyes in larvae, and (107_3) dorsally directed, semitubular or tubular eyes in adults. Eyes are lateral and not

tubular in *Odontostomops,* and R. K. Johnson (1982) hypothesized that tubular eyes are a synapomorphy of *Coccorella* and *Evermannella*. Our hypothesis of a sister-group relationship between *Evermannella* and *Odontostomops* indicates that the absence of tubular eyes in *Odontostomops* is best interpreted as a reversal of the primitive evermannellid + scopelarchid condition.

R. K. Johnson (1982) hypothesized that evermannellids, not paralepidids, are the sister group of the alepisaurid clade and that scopelarchids are part of a clade comprising notosudids, chlorophthalmids, and ipnopids. His arrangement of evermannellids and alepisaurids is five steps longer than ours, and inclusion of scopelarchids in our Chlorophthalmoidei requires at least 18 additional steps. Patterson and Johnson's (1995) placement of the Evermannellidae + Scopelarchidae clade as the sister group of notosudids, which was based on a single feature of the intermusculars, is 19 steps longer than our hypothesis.

Our giganturoids include *Bathysauroides, Bathysaurus,* and *Gigantura,* but historically relationships of these fishes have been perceived differently: *Bathysauroides* (along with *Bathysauropsis gracilis* and *B. malayanus*) was considered a chlorophthalmid (Sulak, 1977) or ipnopid (Hartel and Stiassny, 1986); *Bathysaurus* was considered a synodontid by Sulak (1977) and a close relative of aulopids and chlorophthalmids by Rosen (1973); and *Gigantura,* which has only sometimes been included in the aulopiforms (see discussion below), was considered closely related to synodontids by Rosen (1973). Support for the Giganturoidei is not strong because most derived features shared by *Bathysauroides* and *Bathysaurus* are absent in the highly modified *Gigantura,* but our analysis suggests they are united on the basis of five derived features: (13_1) elongate PB1 (PB1 absent in adult *Gigantura*); (69_0) reduced number (<25%) of caudal vertebrae; (80_1^*) small urodermal in upper caudal lobe (absent in *Gigantura*); (96_3) three postcleithra (none in *Gigantura*); and (107_1) elliptical eyes (eyes greatly modified in *Gigantura*).

Gigantura has usually been placed in a separate order (e.g., Regan, 1925; Berg, 1940; Walters, 1961). Regan (1925) suggested that giganturids might be related to synodontids, and Rosen (1973) concluded that giganturids are alepisauroid aulopiforms, most closely related to synodontids and harpadontids. Rosen's hypothesis was not based on explicit evidence, and, as he noted, the gill arches of adult *Gigantura* are much reduced and do not exhibit the distinctive EB2 uncinate process diagnostic of aulopiforms. The gill arches of larval *Gigantura,* however, are more complete, and our examination of them indicates the pres-

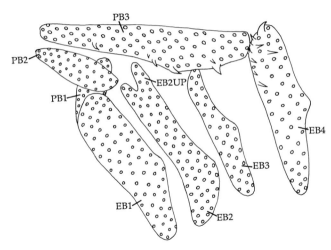

FIGURE 19 Ventral view of dorsal gill arches from right side of larval *Gigantura chuni*, MCZ 60324.

ence of the characteristic EB2 uncinate process (Fig. 19). Furthermore, Patterson and Johnson (1995) noted that intermuscular data, particularly $(54_{1,2})$ the anterior extension of epipleurals, support inclusion of giganturids in the Aulopiformes. *Gigantura* also has three additional aulopiform synapomorphies: (55_1) first epipleural in horizontal septum (Patterson and Johnson, 1995), (112_1) swimbladder absent, and (116_1) peritoneal pigment in larvae. We believe the evidence convincingly places the bizarre giganturids in the Aulopiformes.

Giganturids are aligned with chlorophthalmoids, alepisauroids, and other giganturoids based on (110_1) reproduction by synchronous hermaphroditism (Johnson and Bertelsen, 1991), and they share with alepisauroids and other giganturoids (54_2) anterior extension of epipleurals to V1 and (12_1) large pharyngobranchial teeth.

Patterson and Johnson (1995) suggested a sister-group relationship between *Bathysaurus* and *Gigantura* based on two derived features: (69_0) reduction in number of caudal vertebrae and (61_2^*) origin of most or all epineurals on centra rather than neural arches. A reduced number of caudal vertebrae is a synapomorphy of giganturoids, and the latter character is ambiguous (it could be a synapomorphy of giganturoids with reversal in *Bathysauroides gigas*), but our analysis supports Patterson and Johnson's (1995) interpretation. *Bathysaurus* and *Gigantura* also share (66_2) epicentral series absent (this occurs elsewhere among aulopiforms only in paralepidids); (72_0) most neural arches open dorsally (a reversal of the primitive chlorophthalmoid + alepisauroid + giganturoid condition); and (117_1) maxilla reduced ontogenetically from a very

large broad bone in larvae to a small anterior (*Bathysaurus*) or posterior (*Gigantura*) remnant in adults.

In summary, Aulopiformes are monophyletic and comprise four monophyletic suborders. Our suborder Synodontoidei is the same as that of Johnson *et al.* (1996). Our suborder Chlorophthalmoidei is similar to R. K. Johnson's (1982) chlorophthalmoid clade except that we exclude the Scopelarchidae. Our suborder Alepisauroidei comprises the same recent genera as Rosen's (1973) superfamily Alepisauroidea, and our suborder Giganturoidei combines the new genus *Bathysauroides* with the giganturid–bathysaurid lineage proposed by Patterson and Johnson (1995). Among the most significant aspects of our phylogeny are the following: *Aulopus* is a synodontoid and thus not closely related to ctenosquamates. Synodontoids are not alepisauroids but the primitive sister group of all other aulopiforms. *Bathysaurus* is not a synodontoid but a giganturoid. *Bathysauropsis* is polyphyletic, *B. gracilis* and *B. malayanus* being more closely related to notosudids and ipnopids than to *B. gigas*. *Bathysaurops gigas* Kamohara (= *Bathysauropsis gigas*) is the type species of a new genus, *Bathysauroides*, which is related to bathysaurids and giganturids. *Omosudis* is reassigned to the Alepisauridae, and *Anotopterus* is reassigned to the Paralepididae. Scopelarchids are alepisauroids and the sister group of evermannellids. And finally, *Gigantura* is an aulopiform and may be the sister group of *Bathysaurus*. Further study is needed to elucidate relationships within the Paralepididae and Scopelarchidae and to test all poorly supported relationships hypothesized herein. We have examined certain aspects of aulopiform morphology in detail, but there is much yet to be studied; we view this work as a foundation for further study of this diverse order of fishes.

VII. CLASSIFICATION

As diagnosed here, the extant aulopiforms comprise 43 genera. *Bathysauropsis* and *Bathysauroides* have no familial assignment in our phylogeny, but we assign the remaining 41 genera to 12 families. A new classification of aulopiform genera reflecting phylogenetic relationships as perceived herein follows (suborders are listed in phyletic sequence):

Order Aulopiformes
   Suborder Synodontoidei
      Family Aulopidae (*Aulopus*)
      Family Pseudotrichonotidae (*Pseudotrichonotus*)
      Family Synodontidae (*Harpadon, Saurida, Synodus, Trachinocephalus*)

Suborder Chlorophthalmoidei
    Family Chlorophthalmidae (*Chlorophthalmus, Parasudis*)
    Bathysauropsis (*B. gracilis, B. malayanus*)
    Family Notosudidae (*Ahliesaurus, Luciosudis, Scopelosaurus*)
    Family Ipnopidae (*Bathymicrops, Bathypterois, Bathytyphlops, Discoverichthys, Ipnops*)
Suborder Alepisauroidei
    Family Alepisauridae (*Alepisaurus, Omosudis*)
    Family Paralepididae (*Anotopterus, Arctozenus, Dolichosudis, Lestidiops, Lestidium, Lestrolepis, Macroparalepis, Magnisudis, Notolepis, Paralepis, Stemonosudis, Sudis, Uncisudis*)
    Family Evermannellidae (*Coccorella, Evermannella, Odontostomops*)
    Family Scopelarchidae (*Benthalbella, Rosenblattichthys, Scopelarchoides, Scopelarchus*)
Suborder Giganturoidei
    Bathysauroides gigas (new genus)
    Family Bathysauridae (*Bathysaurus*)
    Family Giganturidae (*Gigantura*)

VIII. Summary

Relationships among aulopiform genera are investigated based on cladistic analysis of 118 morphological characters. Monophyly of Rosen's (1973) Aulopiformes, which he diagnosed on the basis of unique modifications in the dorsal gill arches, is corroborated by features of the intermuscular system, internal soft anatomy, and larval pigmentation as well as new evidence from the morphology of the pelvic girdle. Our analysis suggests four aulopiform clades, listed below in phyletic sequence: (1) Synodontoidei (Aulopidae, Pseudotrichonotidae, and Synodontidae—including *Harpadon* and *Saurida*); (2) Chlorophthalmoidei (Chlorophthalmidae, *Bathysauropsis*, Notosudidae, and Ipnopidae); (3) Alepisauroidei (Alepisauridae—including *Omosudis*, Evermannellidae, Scopelarchidae, and Paralepididae—including *Anotopterus*); and (4) Giganturoidei (Bathysauridae, Giganturidae, and *Bathysauroides*, a new genus erected for *Bathysauropsis gigas* [Kamohara]).

Acknowledgments

In addition to acknowledging Colin's remarkable contributions to science, we remember here his personal side, which has enriched both our lives. One of us (GDJ) has had the good fortune to work in close collaboration with Colin over the past several years, during which time we have become close friends. We have shared discoveries about everything from fishes to music, lots of laughs, and an occasional pint or two. Cheers, Colin—here's to many more years of the same.

For reviewing the manuscript, we thank A. G. Harold, R. K. Johnson, C. Patterson, R. H. Rosenblatt, and M. L. J. Stiassny. For loans or gifts of specimens, we thank B. Chernoff, M. F. Gomon, K. E. Hartel, T. Iwamoto, R. K. Johnson, J. A. Musick, M. Okiyama, T. Orrell, C. Patterson, J. R. Paxton, R. H. Rosenblatt, M. L. J. Stiassny, and K. J. Sulak. For use of and assistance with a MacIntosh computer and help with preparation of the final character matrix, we are grateful to M. Lang.

References

Berg, L. S. (1940). Classification of fishes both recent and fossil. *Trav. Inst. Zool. Acad. Sci. URSS* **5**(2), 346–517.

Bertelsen, E., Krefft, G., and Marshall, N. B. (1976). The fishes of the family Notosudidae. *Dana Rep.* **86,** 1–114.

Bertmar, G. (1959). On the ontogeny of the chondral skull in Characidae, with a discussion of the chondrocranial base and the visceral chondrocranium in fishes. *Acta Zool. (Stockholm)* **42,** 151–162.

Farris, J. S. (1970). Methods of computing Wagner trees. *Syst. Zool.* **19,** 83–92.

Farris, J. S. (1988). "Hennig86, Version 1.5," Program and documentation. Farris, Port Jefferson Station, N.Y.

Fink, W. L., and Weitzman, S. H. (1982). Relationships of the stomiiform fishes (Teleostei), with a description of *Diplophos*. *Bull. Mus. Comp. Zool.* **150**(2), 31–93.

Fujita, K. (1990). "The Caudal Skeleton of Teleostean Fishes." Tokai University Press, Tokyo.

Gosline, W. A., Marshall, N. B., and Mead, G. W. (1966). Order Iniomi. Characters and synopsis of families. *In* "Fishes of the Western North Atlantic," Sears Found. Mar. Res., Mem. No. 1, Part 5, pp. 1–18. Yale University, New Haven, CT.

Gottfried, M. D. (1989). Homology and terminology of higher teleost postcleithral elements. *Trans. San Diego Soc. Nat. Hist.* **21**(18), 283–290.

Hartel, K. E., and Stiassny, M. L. J. (1986). The identification of larval *Parasudis* (Teleostei, Chlorophthalmidae); with notes on the anatomy and relationships of aulopiform fishes. *Breviora* **487,** 1–23.

Johnson, G. D. (1992). Monophyly of the euteleostean clades—Neoteleostei, Eurypterygii, and Ctenosquamata. *Copeia,* **1992,** 8–25.

Johnson, G. D., and Patterson, C. (1993). Percomorph phylogeny: A survey of acanthomorphs and a new proposal. *Bull. Mar. Sci.* **52**(1), 554–626.

Johnson, G. D., Baldwin, C. C., Okiyama, M., and Tominaga, Y. (1996). Osteology and relationships of *Pseudotrichonotus altivelis* (Teleostei: Aulopiformes: Pseudotrichonotidae). *Ichthyol. Res.* **43,** 17–45.

Johnson, R. K. (1974a). A revision of the alepisauroid family Scopelarchidae (Pisces, Myctophiformes). *Fieldiana, Zool.* **66,** 1–249.

Johnson, R. K. (1974b). A *Macristium* larva from the Gulf of Mexico with additional evidence for the synonymy of *Macristium* with *Bathysaurus* (Myctophiformes: Bathysauridae). *Copeia,* **1974,** 973–977.

Johnson, R. K. (1982). Fishes of the families Evermannellidae and Scopelarchidae: Systematics, morphology, interrelationships, and zoogeography. *Fieldiana, Zool.* [N.S.] **12,** 1–252.

Johnson, R. K. (1984a). Scopelarchidae: Development and relationships. *In* "Ontogeny and Systematics of Fishes" (H. G. Moser, W. J. Richards, D. M. Cohen, M. P. Fahay, A. W. Kendall, Jr., and S. L. Richardson, eds.), Spec. Publ. No. 1, pp. 245–250.

American Society of Ichthyologists and Herpetologists, Lawrence, KS.

Johnson, R. K. (1984b). Evermannellidae: development and relationships. In "Ontogeny and Systematics of Fishes" (H. G. Moser, W. J. Richards, D. M. Cohen, M. P. Fahay, A. W. Kendall, Jr., and S. L. Richardson, eds.), Spec. Publ. No. 1, pp. 250–254. American Society of Ichthyologists and Herpetologists, Lawrence, KS.

Johnson, R. K. and Bertelsen, E. (1991). The fishes of the family Giganturidae: Systematics, development, distribution and aspects of biology. Dana Rep. 91, 1–45.

Kamohara, T. (1952). Revised descriptions of the offshore bottom-fishes of Prov. Tosa, Shikoku, Japan. Rep. Kochi Univ., Nat. Sci. No. 3, 1–122.

Leviton, A. E., Gibbs, R. H., Jr., Heal, E., and Dawson, C. E. (1985). Standards in herpetology and ichthyology: Part I. Standard symbolic codes for institutional resource collections in herpetology and ichthyology. Copeia, pp. 802–832.

Maddison, W. P., and Maddison, D. R. (1992). "MacClade: Analysis of Phylogeny and Character Evolution. Version 3.04." Sinauer Assoc., Sunderland, MA.

Marshall, N. B. (1954). "Aspects of Deep Sea Biology." Philosophical Library, New York.

Marshall, N. B. (1955). Alepisauroid fishes. 'Discovery' Rep. 27, 303–336.

Marshall, N. B. (1960). Swimbladder structure of deep-sea fishes in relation to their systematics and biology. 'Discovery' Rep. 31, 1–122.

Marshall, N. B. (1966). Family Scopelosauridae. In "Fishes of the Western North Atlantic," Sears Found. Mar. Res., Mem. No. 1, Part 5, pp. 194–203. Yale University, New Haven, CT.

Marshall, N. B., and Staiger, J. C. (1975). Aspects of the structure, relationships, and biology of the deep-sea fish Ipnops murrayi Gunther (Family Bathypteroidae). Bull. Mar. Sci. 25, 101–111.

McAllister, D. E. (1968). Evolution of branchiostegals and classification of teleostome fishes. Bull.—Natl. Mus. Can. 221, 1–237.

Mead, G. W. (1966a). Family Aulopidae. In "Fishes of the Western North Atlantic," Sears Found. Mar. Res., Mem. No. 1, Part 5, pp. 19–29. Yale University, New Haven, CT

Mead, G. W. (1966b). Family Bathypteroidae. In "Fishes of the Western North Atlantic," Sears Found. Mar. Res., Mem. No. 1, Part 5, pp. 114–146. Yale University, New Haven, CT.

Mead, G. W. (1966c). Family Ipnopidae. In "Fishes of the Western North Atlantic," Sears Found. Mar. Res., Mem. No. 1, Part 5, pp. 147–161. Yale University, New Haven, CT.

Mead, G. W. (1966d). Family Chlorophthalmidae. In "Fishes of the Western North Atlantic," Sears Found. Mar. Res., Mem. No. 1, Part 5, pp. 162–189. Yale University, New Haven, CT.

Merrett, N. R., and Nielsen, J. G. (1987). A new genus and species of the family Ipnopidae (Pisces, Teleostei) from the eastern North Atlantic, with notes on its ecology. J. Fish Biol. 31, 451–464.

Moser, H. G., Ahlstrom, E. H., and Paxton, J. R. (1984). Myctophidae: Development. In "Ontogeny and Systematics of Fishes" (H. G. Moser, W. J. Richards, D. M. Cohen, M. P. Fahay, A. W. Kendall, Jr., and S. L. Richardson, eds.), Spec. Publ. No. 1, pp. 218–239. American Society of Ichthyologists and Herpetologists, Lawrence, KS.

Nelson, G. J. (1967). Epibranchial organs in lower teleostean fishes. J. Zool. 153, 71–89.

Nixon, K. C. (1992). "Clados Version 1.2." Cornell University, Ithaca, NY.

Okiyama, M. (1984a). Myctophiformes: Development. In "Ontogeny and Systematics of Fishes" (H. G. Moser, W. J. Richards, D. M. Cohen, M. P. Fahay, A. W. Kendall, Jr., and S. L. Richard-

son, eds.), Spec. Publ. No. 1, pp. 206–218. American Society of Ichthyologists and Herpetologists, Lawrence, KS.

Okiyama, M. (1984b). Myctophiforms: relationships. In "Ontogeny and Systematics of Fishes" (H. G. Moser, W. J. Richards, D. M. Cohen, M. P. Fahay, A. W. Kendall, Jr., and S. L. Richardson, eds.), Spec. Publ. No. 1, pp. 254–259. American Society of Ichthyologists and Herpetologists, Lawrence, KS.

Olney, J. E., Johnson, G. D., and Baldwin, C.C. (1993). Phylogeny of lampridiform fishes. Bull. Mar. Sci. 52(1), 137–169.

Parin, N. V., and Kotlyar, A. N. (1989). A new aulopodid species, Hime microps, from the eastern South Pacific, with comments on geographic variations of H. japonica. Jpn. J. Ichthyol. 35, 407–413.

Parr, A. E. (1928). Deepsea fishes of the Order Iniomi from the waters around the Bahama and Bermuda Islands. Bull. Bingham Oceanogr. Coll. 3(3), 1–193.

Patterson, C., and Johnson, G. D. (1995). The intermuscular bones and ligaments of teleostean fishes. Smithson. Contrib. Zool. 559, 1–83.

Paxton, J. R. (1972). Osteology and relationships of the lanternfishes (Family Myctophidae). Bull. Nat. Hist. Mus. L.A. Cty., Sci. 13, 1–81.

Post, A. 1987. Results of the research cruises of FRV "Walther Herwig" to South America. LXVII. Revision of the subfamily Paralepidinae (Pisces, Aulopiformes, Alepisauroidei, Paralepididae). I. Taxonomy, morphology and geographical distribution. Arch. Fischereiwiss. 38, 75–131.

Regan, C. T. (1903). On a collection of fishes from the Azores. Ann. Mag. Nat. Hist. [7] 12, 344–348.

Regan, C. T. (1911). The anatomy and classification of the teleostean fishes of the Order Iniomi. Ann. Mag. Nat. Hist. [8] 7, 120–133.

Regan, C. T. (1925). The fishes of the genus Gigantura, A. Brauer; based on specimens collected in the Atlantic by the "Dana" expeditions, 1920–22. Ann. Mag. Nat. Hist. [9] 15, 53–59.

Rofen, R. R. (1966a). Family Paralepididae. In "Fishes of the Western North Atlantic," Sears Found. Mar. Res., Mem. No. 1, Part 5, pp. 205–461. Yale University, New Haven, CT

Rofen, R. R. (1966b). Family Omosudidae. In "Fishes of the Western North Atlantic," Sears Found. Mar. Res., Mem. No. 1, Part 5, pp. 464–481. Yale University, New Haven, CT

Rofen, R. R. (1966c). Family Anotopteridae. In "Fishes of the Western North Atlantic," Sears Found. Mar. Res., Mem. No. 1, Part 5, pp. 498–510. Yale University, New Haven, CT.

Rosen, D. E. (1971). The macristiidae, a ctenothrissiform family based on juvenile and larval scopelomorph fishes. Am. Mus. Novit. 2452, 1–22.

Rosen, D. E. (1973). Interrelationships of higher euteleosteans. In "Interrelationships of Fishes" (P. H. Greenwood, R. S. Miles, and C. Patterson, eds.), pp. 397–513. Academic Press, London.

Rosen, D. E. (1985). An essay on euteleostean classification. Am. Mus. Novit. 2827, 1–57.

Stiassny, M. L. J. (1986). The limits and relationships of the acanthomorph teleosts. J. Zool. (B) 1, 411–460.

Stiassny, M. L. J., and Moore, J. A. (1992). A review of the pelvic girdle of acanthomorph fishes, with comments on hypotheses of acanthomorph intrarelationships. Zool. J. Linn. Soc. 104, 209–242.

Sulak, K. J. (1977). The systematics and biology of Bathypterois (Pisces: Chlorophthalmidae) with a revised classification of benthic myctophiform fishes. Galathea Rep. 14, 49–108

Swofford, D. L. (1991). "PAUP: Phylogenetic Analysis Using Parsimony, Version 3.0s." Computer program distributed by the Illinois Natural History Survey, Champaign.

Swofford, D. L., and Maddison, W. P. (1987). Reconstructing ancestral character states under Wagner parsimony. Math. Biosci. 87, 199–229.

Tucker, D. W. (1954). Report on the fishes collected by S. Y. "Rosaura" in the North and Central Atlantic, 1937–38. Part IV. Families Carcharhinidae, Torpedinidae, Rosauridae (nov.), Salmonidae, Alepocephalidae, Searsidae, Clupeidae. *Bull. Br. Mus. (Nat. Hist.)* **2**(6), 163–214.

Walters, V. (1961). A contribution to the biology of the Giganturidae, with description of a new genus and species. *Bull. Mus. Comp. Zool.* **125**, 297–319.

Woods, L. P., and Sonoda, P. M. (1973). Order Berycomorphi (Beryciformes). *In* "Fishes of the Western North Atlantic," Sears Found. Mar. Res., Mem. No. 1, Part 6, pp. 263–396. Yale University, New Haven, CT.

Appendix 1

Abbreviations Used in Text Figures

AC	Anterior Ceratohyal
APC	Autogenous Pelvic Cartilage
BBn	nth Basibranchial
Br	Branchiostegal
CBn	nth Ceratobranchial
Cl	Cleithrum
Co	Coracoid
CPP	Central Pelvic Process
DH	Dorsal Hypohyal
Ecp	Ectopterygoid
EBn	nth Epibranchial
Enp	Endopterygoid
HBn	nth Hypobranchial
Hy	Hyomandibular
Ih	Interhyal
LPP	Lateral Pelvic Process
Me	Mesethmoid
Mep	Metapterygoid
MPP	Medial Pelvic Process
Mx	Maxilla
MxS	Maxillary Saddle
P	Palatine
Para	Parasphenoid
PBn	nth Pharyngobranchial
Pc	Postcleithrum
PC	Posterior Ceratohyal
Pmx	Premaxilla
PPC	Posterior Pelvic Cartilage
PPP	Posterior Pelvic Process
PR	Pectoral-fin Radial
Q	Quadrate
Sca	Scapula
Sc	Supracleithrum
Sy	Symplectic
UP	Uncinate Process
UPn	nth Upper Pharyngeal Toothplate
V	Vomer
VH	Ventral Hypohyal

Appendix 2

Material Examined

Our analysis included examination of representatives of more than 40 neoteleostean genera listed below using institutional abbreviations specified by Leviton *et al.* (1985). Whole and cleared and stained specimens or parts of specimens (e.g., gill arches and paired fins) dissected from very large specimens were examined for most taxa. Cleared and stained lots are indicated by "cs."

Aulopiformes—*Ahliesaurus berryi*: USNM 240503, 240505 (cs). *Alepisaurus brevirostris* USNM 200817 (gill arches, pelvic fin cs), 201275. *Alepisaurus* sp.: MCZ 60345 (cs). *Anotopterus pharao*: CAS 164180 (cs); SIO 5553 (cs); USNM 140825 (cs), 201286, 221035, 221035 (cs), 206844; SIO 62-775 (cs). *Arctozenus rissoi* USNM 302410 (1 cs), 283485 (cs). *Aulopus filamentosus*: USNM 292105 (cs), 301018. *Aulopus japonicus*: AMNH 28635SW (cs); FMNH 71831 (cs). *Aulopus* sp.: AMNH 28635 (cs). *Bathymicrops regis*: BMNH 1989.7.25.56.61 (cs). *Bathypterois longipes* USNM 35635. *Bathypterois pectinatus*: FMNH 88982 (cs). *Bathypterois* sp. MCZ 40567 (cs). *Bathypterois viridensis* USNM 117215. *Bathysauropsis gracilis* AMS IA6934 (cs): NMV A6932. *Bathysauropsis malayanus* USNM 098888 (holotype of *Bathysaurops malayanus*). *Bathysaurus ferox* AMS I.29591001; MCZ 62409 (cs); USNM 316825. *Bathysaurus mollis*: VIMS 6107 (cs). *Bathysauroides gigas*: AMS I, 22822001 (cs); NMV A5770, A4438, A4440 (cs). *Bathytyphlops marionae* USNM 336666 (cs), 336713 (formerly VIMS 06104); 341861 (gill arches cs). *Benthalbella dentata*: SIO 63-379 (cs). *Benthalbella elongata* USNM 207279. *Benthalbella macropinna* USC E1671. *Chlorophthalmus agassizi*: AMNH 40829SW (cs); USNM 159385 (cs), 302386. *Chlorophthalmus atlanticus* USNM 339774 (1 cs). *Coccorella atlantica*: USNM 235170, 235189 (cs), 235199 (cs). *Evermannella balbo* USNM 301265. *Evermannella indica*: U.H. 71-3-9 (cs); USNM 235141. *Gigantura chuni* AMNH 55345SW (cs); MCZ 60324 (cs). *Gigantura indica* MCZ 54133 (cs): SIO 76-9; USNM 215407. *Harpadon nehereus*: AMNH 17563 (cs); FMNH 179018 (cs); USNM 308838. *Harpadon squamosus*: FMNH 80823 (cs). *Ipnops agassizi*: USNM 54618 (gill arches cs). *Ipnops meadi* SIO 61-175 (cs). *Ipnops murrayi* USNM 101371, 336711 (formerly VIMS 6736), 336712 (formerly VIMS 6737). *Lestidiops affinis* MCZ 60632 (cs). *Lestidiops* sp.: USNM 307290 (cs). *Lestidium atlanticum*: USNM 201183 (cs), uncat. (cs). *Lestidium* sp.: USNM 341877 (1 cs). *Lestrolepis intermedia* USNM 290253 (2 cs). *Lestrolepis* sp. USNM 307290 (1 cs). *Macroparalepis affine*: USNM 302410 (cs); 201184 (cs). *Macroparalepis* sp.: FMNH

49988 (cs); USNM 201186 (cs). *Odontostomops normalops*: USNM 235029 (cs), 274377 (1 cs). *Omosudis lowei* USNM 219982 (cs), 206838, 287310. *Paralepis brevirostris*: USNM 196109 (cs). *Paralepis coregonoides*: USNM 196098, 290253 (cs). *Parasudis truculentus*: FMNH 67150 (cs); MCZ 62398 (cs); USNM 159096 (1 cs), 159407 (cs), 159850 (cs). *Pseudotrichonotus altivelis*: USNM 280366 (cs); ZUMT 55678 (cs), 59882 (cs). *Rosenblattichthys hubbsi* MCZ 52821 (cs). *Saurida brasiliensis*: USNM 185852 (cs); 187994 (cs). *Saurida gracilis*: USNM 256409 (cs). *Saurida normani*: USNM 341878 (cs). *Saurida parri*: USNM 193763 (cs), 340398 *Saurida undosquamous*: USNM 325180 (cs). *Scopelarchus analis*: MCZ 62599 (cs); USNM 234988 (cs). *Scopelarchoides nicholsi*: USNM 201154 (cs), 207295. *Scopelarchoides signifer*: USNM 274385 (cs). *Scopelosaurus argenteus* MCZ 63321 (cs), 62105 (cs), 62405 (cs). *Scopelosaurus fedorovi* SIO 60-251 (cs). *Scopelosaurus hoedti*: USNM 264256 (2 cs). *Syno*dontidae: USNM 309851 (cs). *Stemonosudis rothschildi* AMS I. 22826001 (cs). *Stemonosudis* sp. USNM 330273 (cs). *Sudis atrox* MCZ 60336 (cs); USNM 330285 (cs). *Sudis hyalina* USNM 340399 *Synodus jenkensi*: USNM 321745 (1 cs). *Synodus synodus*: USNM 318960 (1 cs). *Synodus variegatus*: USNM 140825 (cs); 315318 (cs). *Trachinocephalus myops*: FMNH 45392 (cs); MCZ 62106 (cs); USNM 305292, 185861 (cs); 339775 (larva, cs); 339776 (cs). *Uncisudis advena* MCZ 68531 (cs). Stomiiformes—*Diplophos taenia*: MCZ 55469 (cs); USNM 206614 (cs), 274404. Myctophiformes—*Lampanyctus cuprarius* USNM 300490 (cs). *Myctophum obtusirostre*: AMNH 29140SW (cs). *Neoscopelus macrolepidotus*: USNM 188056 (cs); 317160 (cs). *Neoscopelus* sp. USNM 159417 (cs). *Notoscopelus resplendens*: AMNH 25928SW (cs). Lampridiformes—*Metavelifer*: BPBM 23953 (cs). Polymixiiformes—*Polymixia lowei* USNM 137750, 185204 (cs), 308378 (cs).

15

Basal Ctenosquamate Relationships and the Interrelationships of the Myctophiform (Scopelomorph) Fishes

MELANIE L. J. STIASSNY

Department of Herpetology and Ichthyology
American Museum of Natural History
New York, New York 10024

I. Introduction

In his seminal paper for the first *Interrelationships of Fishes*, Rosen (1973) presented what was to become a powerfully influential view of relationships among the huge radiation of euteleostean fishes. Of all of the innovations presented in that work, it is perhaps his recognition of a monophyletic Neoteleostei resolved into four main components (the Stenopterygii [= Stomiiformes[1]], Cyclosquamata [= Aulopiformes], Scopelomorpha [= Myctophiformes], and Acanthomorpha) that represents the contribution of most enduring value (Fink and Weitzman, 1982; Lauder, 1983; Lauder and Liem, 1983; Stiassny, 1986; Stiassny and Moore, 1992; G. D. Johnson, 1992; Olney *et al.*, 1993; Patterson and Johnson, 1995).

In arguing for the recognition of a monophyletic Ctenosquamata (Scopelomorpha + Acanthomorpha), and in the belief that there is "abundant evidence that the relationships of the Myctophidae and Neoscopelidae lie with the paracanthopterygians and acanthop-

terygians, and not with the aulopiforms as has long been supposed," Rosen (1973, p. 452) dismembered the traditional concept of the iniomous fishes (see, e.g., Regan, 1911; Gosline *et al.*, 1966), and in its place proposed the Myctophiformes [Myctophidae + Neoscopelidae]) as the sister-group of the Acanthomorpha. While the morphological evidence presented by Rosen (1973) in support of this alignment was far from convincing and his conclusions were not universally accepted (e.g., R. K. Johnson, 1974, 1982; Okiyama, 1984b; Rosen, 1985; Hartel and Stiassny, 1986), a critical review by Johnson (1992) did provide one compelling character in support of ctenosquamate monophyly. However, support for the phylogenetic integrity of a taxonomic entity of this size and complexity (minimally some 15,000 species presenting some of the most challenging outstanding problems in teleostean systematics), resting upon a single, albeit apparently "uniquely derived and unreversed," character necessarily gives some cause for concern. Nelson (1994) was quite correct in noting that further testing of this hypothesis is clearly desirable. It is in this context that the initial phase of the present study was undertaken to determine whether there are additional derived morphological features which bear upon Rosen's (1973) and Johnson's (1992) concept of a monophyletic euteleostean ctenosquamate clade.

---

[1] Following Steyskal (1980), Fink and Weitzman (1982) amend the spelling of the group name Stomiatiformes, as used by Rosen (1973), to Stomiiformes. Nelson's (1994) important summary deviates from Rosen's (1973) scheme and follows Olney *et al.* (1993) in including the ateleopodiforms in his superorder Stenopterygii.

Copyright © 1996 by Academic Press, Inc.
All rights of reproduction in any form reserved.

The question of myctophiform monophyly is, in many respects, less controversial, and a close association between the neoscopelid and myctophid fishes has long been referenced in the taxonomic literature (e.g., Goode and Bean, 1896; Regan, 1911; Fowler, 1925; Gregory and Conrad, 1936). Nonetheless, despite this long history of association there have been remarkably few formulations of character data that may be critically interpreted as lending evidence for the monophyly of the group. Stiassny (1986, p. 423), in an analysis of the limits and relationships of Rosen's (1973) Acanthomorpha, briefly reviewed the issue of myctophiform monophyly but concluded that the "... question of myctophiform monophyly remains open." A similar sentiment is mirrored by Johnson (1992, p. 20) who stated that "the question of the monophyly of the Myctophiformes remains somewhat problematic." Once again, recognition of this unsatisfactory state of affairs has prompted the second part of this study which is aimed at investigating the status of Rosen's (1973) concept of a monophyletic Myctophiformes.

Perhaps not surprisingly in view of the foregoing, the intrarelationships of myctophiform fishes are also poorly understood. This is particularly true with respect to an understanding of the relationships of the three neoscopelid genera (*Neoscopelus*, *Scopelengys*, and *Solivomer*), both to one another and to the myctophid radiation itself. To put it another way, the assumption of neoscopelid monophyly that is implicit in much of the contemporary literature has never been satisfactorily justified, and further investigation is clearly desirable. Thus the final component of the present study is an analysis of myctophiform relationships incorporating new data from an anatomical investigation of the previously poorly documented neoscopelid myctophiforms. Although a detailed analysis is beyond the scope of the present investigation, some theories of relationship among the species-rich myctophid clade are also reviewed.

II. Methods and Materials

In seeking to recognize morphological features bearing on the question of ctenosquamate and myctophiform monophyly, and to delimit subclades, standard phylogenetic methods have been employed. For selection of appropriate outgroup materials, emphasis has been placed upon representation from stomiiform, cyclosquamate (aulopiform), and basal acanthomorph taxa. Within the non-holacanthopterygian Eurypterygii I have attempted to include representation from most major lineages. And from within these

lineages selection is concentrated, where feasible, upon the more morphologically generalized members of basal clades (Stomiiformes: Fink, 1984; Fink and Weitzman, 1982; Aulopiformes: R. K. Johnson, 1982; G. D. Johnson *et al.*, 1996; Baldwin and Johnson, this volume; Myctophiformes: Stiassny, 1986; Paxton *et al.*, 1984; Acanthomorphs: Rosen, 1985; Stiassny, 1986; Stiassny and Moore, 1992; Olney *et al.*, 1993; Moore, 1993; Johnson and Patterson, 1993; Patterson and Johnson, 1995).

Osteological specimens are either cleared and stained for bone and/or cartilage (c&s), dried skeletal preparations (skel), or radiographs (rad) of alcoholic specimens (alc). Nomenclature for the bony condyles and ligaments of the skull and upper jaws follows Stiassny (1986), and that of the musculature follows Winterbottom (1974). A list of abbreviations used in the text figures may be found in Appendix 1, and a listing of materials examined is given in Appendix 2.

III. Character Survey

A. Ctenosquamate Monophyly and Relationships

In reviewing the evidence supporting a monophyletic origin for Rosen's three euteleostean levels, Neoteleostei, Eurypterygii, and Ctenosquamata, Johnson (1992) provided a useful critique of much of the pertinent anatomical literature. Regarding ctenosquamate monophyly, he reviewed the putative synapomorphies listed by Rosen (1973, p. 506) and found that they are either not restricted to ctenosquamates, are extremely variable even among basal ctenosquamates, or both. With some reservation Johnson (1992) rejected Rosen's character evidence for a monophyletic Ctenosquamata but provided a description of a novel restructuring of the dorsal gill arches that he believed uniquely characterizes ctenosquamate fishes.

(1) Absence of the fifth upper pharyngeal tooth-plates and the associated third levatores interni muscles (Figs. 1A and 1B; G. D. Johnson, 1992, pp. 21–23)

The present study confirms Johnson's assertion that the dorsal gill arches of ctenosquamates exhibit a reductive restructuring. Johnson (1992) was correct in noting that *ctenosquamates are derived in relation to non-ctenosquamate euteleosts in lacking fifth upper toothplates (UP5) and in the possession of a reduced component of two pairs of levatores interni (LI1 and LI2) muscles.*

An additional derived feature of this system found in the Myctophidae is the development of a novel

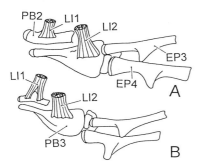

FIGURE 1 Isolated upper pharyngeal apparatus (in dorsal view) of (A) *Scopelengys* (AMNH 49534) and (B) *Electrona* (MCZ 116337).

partition of the first levator internus into two distinct muscle heads. These muscle divisions insert on either side of a prominent ascending process on the second pharyngobranchial. Johnson (1992, fig. 7b) illustrated this condition in a representative lampanyctine (*Diaphus*), and the condition typical for myctophines is illustrated here with reference to *Electrona* (Fig. 1B). *A subdivision of the first levator internus muscle is here interpreted as a synapomorphy of the Myctophidae.*

(2) Two posterior ceratohyal branchiostegal rays (Figs. 2 and 3)

In an extensive review of the evolution of branchiostegal rays in teleostomes, McAllister (1968) recognized

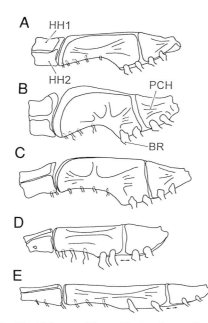

FIGURE 2 Hyoid bar and branchiostegal rays (in lateral view) of (A) *Neoscopelus* (AMNH 49533), (B) *Scopelengys* (AMNH 97466), (C) *Solivomer* (USNM 135928), (D) *Electrona* (MCZ 116337), and (E) *Lampanyctus* (AMNH 97539).

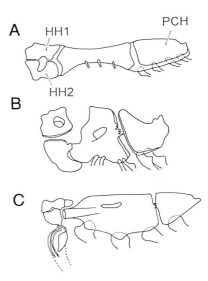

FIGURE 3 Hyoid bar and branchiostegal rays (in lateral view) of (A) *Chlorophthalmus* (AMNH 70606), (B) *Velifer* (RUSI 13821), and (C) *Polymixia* (AMNH 86102).

among "myctophiform" (= iniomous) fishes several "basic types of branchiostegal arrangement." He noted that uniquely among inioms, the Myctophidae and Neoscopelidae have two branchiostegal rays articulating with the posterior ceratohyal (= epihyal) (e.g., Fig. 2; Paxton, 1972, fig. 9a); in aulopiforms (and other basal euteleosts) the number of posterior ceratohyal branchiostegals ranges from three to nine (e.g., Fig. 3A; McAllister, 1968, plate 13; R. K. Johnson, 1974, fig. 8f; 1982, figs. 13a and 13b). Further review of this character system indicates that while it is certainly the case that myctophiforms invariably have only two branchiostegal rays articulating with the posterior ceratohyal, such a condition is not restricted to them. In fact a reduction to two posterior ceratohyal branchiostegals is standard in acanthomorph lineages (see, e.g., Fig. 3B; McAllister, 1968, plates 12–16). *Polymixia* (Fig. 3C) is noteworthy in that only a single branchiostegal ray articulates with the posterior ceratohyal, and such a reduction is interpreted here as an autapomorphy of that taxon. In view of the foregoing it is suggested here that *the presence of two (or fewer) branchiostegal rays on the posterior ceratohyal is a synapomorphy of ctenosquamate fishes.*

Two additional synapomorphies of the myctophid hyoid are recognized here. The first is *a marked reduction of the dorsal hypohyal (HH1)* (see, e.g., Figs. 2D and 2E; according to Paxton, 1972, this element is entirely lost in some myctophid lineages), and the second is *the presence of a distinctive gap between the posterior ceratohyal branchiostegals and the first two anterior ceratohyal branchiostegals* (indicated by arrows in Figs. 2D and 2E). This

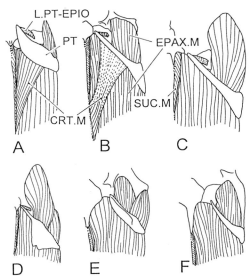

FIGURE 4 Dorsal neurocranial musculature of (A) *Diplophos* (AMNH 84352), (B) *Parasudis* (AMNH 84635), (C) *Solivomer* (AMNH 36364), (D) *Lobianchia* (AMNH 29619), (E) *Melamphaes* (AMNH 29756), and (F) *Percopsis* (AMNH 52491).

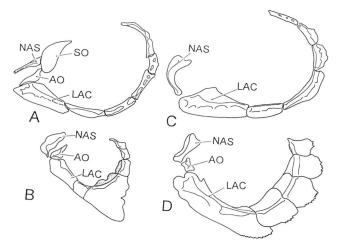

FIGURE 5 Left infraorbital series of (A) *Aulopus* (AMNH 28635), (B) *Lampanyctus* (AMNH 97539), (C) *Velifer* (USNM 13821), and (D) *Polymixia* (AMNH 86102).

feature is most striking in the myctophine myctophids and, while still discernible, is somewhat less evident in the more derived lampanyctine genera which tend to have markedly elongate hyoid bars.

(3) Loss of the craniotemporalis musculature (Figs. 4C–4F)

Stiassny (1986) suggested that the presence of a pair of muscle slips (the craniotemporalis muscles) passing anterolaterally from behind the head in the midline and inserting onto the posterodorsal margin of the dorsal limb of the posttemporal (e.g., Figs. 4A and 4B) may represent a teleostean synapomorphy. In her discussion of this character she noted that the cranio-temporalis musculature is invariably absent in acanthomorphs but did not comment on its absence in myctophids, although her illustrations (Stiassny, 1986, figs. 25a and 25c) indicated this to be the case, at least in *Electrona* and *Diaphus*. Further review of the distribution of this muscle slip confirms that it is present in all non-ctenosquamate euteleostean lineages (see, e.g., Fig. 4A), albeit reduced to a muscularly invested connective tissue sheet in most cyclosquamates (e.g., Fig. 4B). Extended review also confirms that the craniotemporalis musculature is entirely lacking in all myctophiforms (e.g., Figs. 4C and 4D) as well as in all acanthomorphs (Stiassny, 1986, fig. 25b; figs. 4E and 4F). Interestingly, in both basal acanthomorph clades (*Polymixia* and representative lampridiforms, *Velifer* and *Metavelifer*), epaxial devel-

opment is extensive and the posttemporal is entirely overlain by the epaxial muscle block; however, no craniotemporalis musculature underlies it.

In view of the distribution recognized here, it is suggested that *the loss of the craniotemporalis musculature is an additional synapomorphy of the Ctenosquamata.*

(4) Loss of the supraorbital bones of the circumorbital series (Figs. 5 and 6)

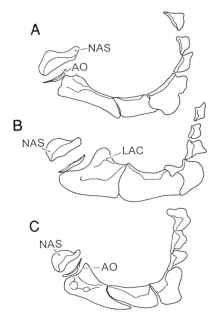

FIGURE 6 Left infraorbital series of (A) *Neoscopelus* (AMNH 28142), (B) *Scopelengys* (AMNH 49533), and (C) *Solivomer* (USNM 135928).

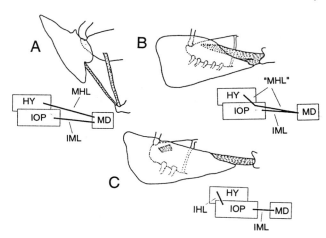

FIGURE 7 Mandibular linkage system (in lateral view) of (A) *Diplophos* (AMNH 84352), (B) *Aulopus* (AMNH 28635), and (C) *Solivomer* (USNM 135928). Schematic representations below each figure.

A single supraorbital is present in the anterodorsal margin of the orbit of most neoteleostean fishes (e.g., Fig. 5A; Fink and Weitzman, 1982, fig. 8). In a discussion of the distribution of supraorbital bones in "iniomis" R. K. Johnson (1982, p. 94) cited the absence of these elements in myctophids and neoscopelids as a derived character state uniting the two families. While I am able to confirm that supraorbitals are indeed lacking in neoscopelids (Fig. 6) and myctophids (e.g., Fig. 5B; see also Jollie, 1954), this feature is not unique to scopelomorphs but also characterizes the acanthomorph radiation (e.g., Figs. 5C and 5D). While Johnson (1982) noted that supraorbitals have been lost in the evermannellid genus *Odontostomops*, as well as in certain other aulopiform lineages, the presence of supraorbitals in most basal aulopiform lineages suggests that these losses are best interpreted as having occurred independently of that in ctenosquamates. *The absence of supraorbital bones in the circumorbital series of all scopelomorphs and all acanthomorphs is interpreted here as a ctenosquamate synapomorphy.*

(5) Presence of a discrete interoperculohyoid ligament (Fig. 7C)

Johnson (1992) suggested that the presence of an interoperculohyoid ligament, a character first described by Lauder (1982, 1983), may be a synapomorphy of eurypterygian fishes. However, he qualified the observation with the proviso that neither he nor Lauder had undertaken an exhaustive study of this character. My own observations of this ligament system indicate that the presence of a discrete interoperculohyoid ligament is found only in ctenosquamate fishes.

Primitively in teleosts the interoperculum and the hyoid bar are each connected with the mandible via separate ligaments: the elongate and often cordlike mandibulohyoid ligament (MHL) and the interoperculomandibular ligament (IML) (Fig. 7A; Lauder, 1982, fig. 1b). At the level of the Eurypterygii, Lauder (1982, 1983) correctly notes that a large component of the mandibulohyoid ligament inserts onto the interoperculum to form a novel functional linkage between the hyoid bar, the interoperculum, and the mandible (Lauder, 1982, fig. 1c). However, in all cyclosquamates the main mass of the mandibulohyoid ligament maintains its original connection with the mandible (e.g., Fig. 7B). This is not the case in scopelomorphs or in most acanthomorph lineages where direct structural contact between the hyoid and the mandible is lost (e.g., Fig. 7C).

In ctenosquamates the entire mandibulohyoid ligament (now = interoperculohyoid ligament) passes from the lateral face of the epihyal to insert on the dorsomedial surface of the interoperculum at a point distant from the interoperculomandibular ligament (IHL; Fig. 7C). Lauder (1982, 1983) was correct in noting the *functional* innovation of a linkage between the hyoid and interoperculum at the level of the eurypterygian fishes (mediated by partial migration of the mandibulohyoid ligament onto the interoperculum). However, it is only at the level of the Ctenosquamata that a full separation of this ligament from the mandible occurs. In other words, *the presence of a true interoperculohyoid ligament (a fully repositioned mandibulohyoid ligament), discrete and separated by a distinct gap from the origin of the interoperculomandibular ligament, is interpreted here as a synapomorphy of ctenosquamate fishes.*

(6) Neural arches of the first vertebra fused into single unit (Figs. 8B–8D)

Typically among euteleosts the first neural arch is composed of bilateral elements which often meet in the dorsal midline but are not fused into a single element (e.g., Fig. 8A). In contrast to that plesiomorphic configuration, the first neural arch of neoscopelids (Fig. 8B), myctophids (Fig. 8C), and acanthomorphs (e.g., Fig. 8D) is fused mediodorsally to form a single structural unit. *The presence of a single, medially fused neural arch on the first vertebral centrum is interpreted as a synapomorphy of the Ctenosquamata.*

Possible Additional Ctenosquamate Characters

(a) Gottfried (1989) provided a useful review of the distribution of postcleithral elements among actinopterygian fishes. Within the Teleostei he recognized a reductional series and argued that the presence of only two postcleithral pairs (pcl 2 and pcl 3, i.e., the ab-

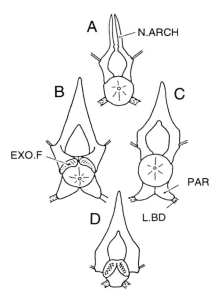

FIGURE 8 First vertebra (in anterior view) of (A) *Aulopus* (AMNH 92091), (B) *Neoscopelus* (AMNH 27405), (C) *Hygophum* (AMNH 29491), and (D) *Polymixia* (AMNH 88707).

sence of pcl 1) in ctenosquamates is a synapomorphy of that clade. I have not undertaken an exhaustive review of this character system, and while I have found no evidence countering Gottfried's interpretation, Baldwin and Johnson (this volume) interpret the presence of two postcleithra as the basal aulopiform condition (secondarily reversed to three among various aulopiform subclades). In view of this alternative interpretation I include this character here in a listing of "possible additional ctenosquamate characters," with the proviso that it may in fact be a eurypterygian character, secondarily reversed in certain aulopiform subclades.

(b) Another character recognized by Lauder (1982, 1983) as a ctenosquamate synapomorphy is a reorientation of the rectus communis muscle. Primitively in neoteleosts the rectus communis originates on the fifth ceratobranchial and passes rostrad to insert musculously onto the descending process of the third hypobranchial (e.g., Fig. 9A). A similar configuration is found in both *Scopelengys* and *Solivomer* (Figs. 9C and 9D), whereas *Neoscopelus* exhibits a condition found in all myctophids in which a portion of the rectus communis inserts onto a tendinous fascia developed on the dorsal surface of the sternohyoideus musculature (Fig. 9E). The condition in *Polymixia* is essentially similar, with a shared insertion of the rectus communis onto the third hypobranchial and the sternohyoideus (Fig. 9B). It is my observation that it is only in lampridiforms and holacanthopterygians that the

rectus communis (now = pharyngohyoideus) has lost its plesiomorphic association with the third hypobranchial and inserts fully onto the urohyal.

(c) As one of the characters linking the myctophoids (neoscopelids + myctophids) with the acanthomorphs, Rosen (1973) observed that myctophoid swimbladders and gas-glands are of the advanced "acanthopterygian type." And while Johnson (1974) has questioned the "advanced" nature of the swimbladder of *Neoscopelus*, my own observations of the swimbladders of *Neoscopelus* (and *Solivomer*) accord with those of Marshall (1960) and indicate that this system may yield additional evidence of ctenosquamate monophyly. More detailed study of the structure and development of the myctophiform swimbladder, particularly in relation to the developmental origin of the gas-gland, is clearly desirable (Marshall, 1960).

(d) Patterson and Johnson (1995) noted that an accessory neural arch, the presence of which is widespread among neoteleosts (Rosen, 1985), is consistently absent in ctenosquamates. My observations confirm this assertion, however, Patterson and Johnson (1995) were also correct in their observation that an accessory neural arch is mosaically distributed among stomiiforms and, while present in most cyclosquamate lineages, its occurrence is sporadic in that clade also. Such a mosaic distribution outside of the Ctenosquamata renders interpretation of this character problematic.

(e) Patterson and Johnson (1995, p. 31) also noted that there is a derived absence of fusion between anterior epineurals and neural arches in all ctenosqua-

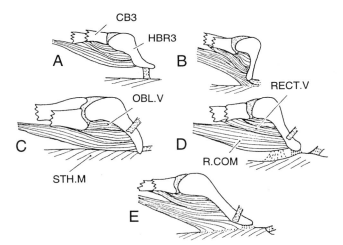

FIGURE 9 Muscular associations of the third hypobranchial element (in lateral view) of (A) *Aulopus* (AMNH 28635), (B) *Polymixia* (AMNH 70602), (C) *Solivomer* (AMNH 36364), (D) *Scopelengys* (AMNH 12841), and (E) *Neoscopelus* (AMNH 84363).

mates. Once again, this derived feature occurs mosaically in basal neoteleosts and eurypterygians, and interpretation at the level of the Ctenosquamata is therefore problematic.

B. Myctophiform Monophyly and Relationships

As noted in the Introduction, a close association between the neoscopelid and myctophid fishes has long been referenced in the taxonomic literature. Yet, despite this association there have been remarkably few attempts to critically appraise the question of myctophiform monophyly (Stiassny, 1986). In fact, Rosen (1985), in a troublingly inconsistent paper, challenged the notion of a monophyletic Myctophiformes and, in its place, posited a sister-group relationship between the myctophid lineage and the Acanthomorpha, while placing the neoscopelids in an unresolved polytomy with some aulopiform lineages. Although Johnson (1992) effectively discounted most of the character data that Rosen (1985) had mustered as evidence for myctophiform polyphyly, the fact remains that myctophids *do* exhibit an interesting array of derived morphological features, many of which mirror in intriguing ways the conditions found in certain acanthomorph, and even percomorph, lineages but that are not found in neoscopelids or basal acanthomorphs. Such an apparent "prefiguring" is seen in the caudal fin of myctophids. For example, in myctophids the first uroneural is fused to the compound preural1 +ural1 centrum, the second ural centrum is not present as an independent element, and the parhypural is fused with the first hypural plate (see, e.g., Paxton, 1972, fig. 12). This is not the case in most aulopiforms, in all neoscopelids, or in basal acanthomorph lineages (see, e.g., Rosen and Patterson, 1969, figs. 2, 3, 16, and 73), but it is true for many holacanthopterygian lineages. Similarly, the presence of a subocular shelf, the loss of supramaxillae, the loss of a posterior process on the maxilla, and the fusion of the anterior neural arches with their associated vertebral centra are all "advanced" features present in myctophids and many holacanthopterygians but not in neoscopelids or other basal ctenosquamates. No doubt it was this homoplasic "prefiguring" in myctophids of advanced percomorph features that so impressed Rosen in his later paper. However, critical review reveals that the weight of evidence argues in favor of myctophiform monophyly as was originally conceived by Rosen in 1973 and supported by Stiassny (1986). The morphological evidence for that association is reviewed briefly here.

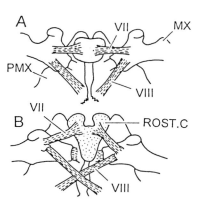

FIGURE 10 Ligamentous attachment of the maxillae (in anteromedial view) in (A) *Aulopus* (AMNH 28635) and (B) *Neoscopelus* (AMNH 27405).

(7) Presence of a median dorsal keel on the mesethmoid (Stiassny, 1986, fig. 19; Figs. 15C and 15D)

Regan (1911) was among the first to address the issue of myctophiform monophyly in a comparative context, and he provided a short diagnosis for the Myctophidae (including the then single known neoscopelid, *Neoscopelus macrolepidotus*). He claimed that myctophids were distinguished from other inioms by two features. The first was the presence of an extension of the parasphenoid contacting the frontals between the lateral ethmoids. Although this feature has been cited in the subsequent literature (e.g., Gregory and Conrad, 1936; Miller, 1947) I find nothing resembling it in any myctophiform taxon. On the contrary, I find myctophiforms have an unremarkable parasphenoid–frontal configuration and therefore must discount Regan's first character as an error. However, Regan's (1911) second character, the presence of a well-developed median dorsal keel or process on the ethmoid (= mesethmoid), does appear to characterize the myctophiforms (Figs. 15C and 15D; Stiassny, 1986, fig. 19). Among outgroups a similar dorsal mesethmoid process is present only in certain melamphaid genera (Stiassny, 1986; Moore, 1993). Based upon the limited distribution observed here, I interpret *the presence of a well-developed median dorsal keel on the mesethmoid as a synapomorphy of the Myctophiformes* and concur with Moore's (1993) interpretation that a median dorsal ethmoid process is a synapomorphy of the Melamphaidae but with an independent derivation from that of the myctophiforms.

(8) The median maxillo–premaxillary ligaments (VIII) insert onto contralateral buccal elements (Stiassny, 1986, fig. 9a; Fig. 10B)

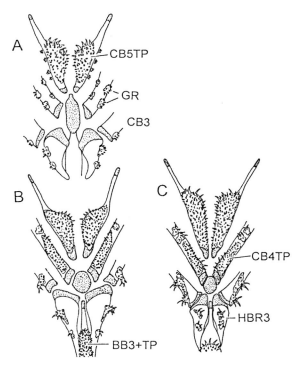

FIGURE 11 Lower pharyngeal jaw apparatus (in dorsal view) in (A) *Aulopus* (AMNH 28635), (B) *Solivomer* (USNM 135419), and (C) *Lampanyctus* (AMNH 97539).

In a review of the buccal jaw ligaments of teleosts, Stiassny (1986) noted the presence of a derived configuration of the median maxillo–premaxillary ligaments (VIII) in myctophiform fishes. Typically among neoteleosts, a well-developed maxillo–premaxillary ligament passes from the medial face of the maxilla to insert on the medial face of the premaxillary ascending process of the same side of the head (see, e.g., Fig. 10A). In a derived configuration unique to the myctophids and neoscopelids the maxillo–premaxillary ligaments are crossed and each ligament passes from the medial face of the maxilla of one side to insert on the contralateral premaxilla (e.g., Fig. 10B). *The presence of crossed median maxillo–premaxillary ligaments (VIII) is interpreted here as a synapomorphy of the Myctophiformes.*

(9) Presence of a large toothplate fused to the proximal face of the fourth ceratobranchial (Fig. 11)

As can be seen in the accompanying figures, neoscopelid (e.g., Fig. 11B) and myctophid (e.g., Fig. 11C) fishes are characterized by the presence of well-developed toothplates fused to the dorsal face of the fourth ceratobranchials. Most outgroup representatives examined in this study lack fourth ceratobranchial toothplates (e.g., Fig. 11A), and while free ceratobranchial

toothplates are present in many acanthomorphs (e.g., *Polymixia* and *Hoplostethus*), in none of the outgroups examined are toothplates fused to the underlying fourth ceratobranchial elements. In view of the limited distribution of this feature I interpret *the presence of a toothplate fused to the proximal end of each fourth ceratobranchial to be a synapomorphy of the Myctophiformes.* Among myctophines fused ceratobranchial toothplates have apparently been lost in *Electrona, Benthosoma,* and *Centrobranchus* but are present in other Myctophini and Gonichthyini, as well as in all lampanyctines and neoscopelids examined.

(10) The first levator externus muscle reduced or absent (Figs. 12B and 12C)

The standard teleostean complement of levatores externi is four pairs connecting the skull to the epibranchial series of the pharyngeal jaw apparatus. While the third levator externus is sometimes missing in certain percomorph lineages (Winterbottom, 1974), as far as I am able to ascertain, a reduction of the first levator externus to a thin muscle slip, or its replacement by a ligament, is a specialization found only in myctophiform fishes. In neoscopelids the first levator is reduced to an extremely thin muscle slip that originates medial to the remaining levatores (e.g., Fig. 12B), while in myctophids the muscle is reduced further and replaced by a ligament that connects the skull to the first epibranchial (e.g., Fig. 12C). The typical outgroup condition is represented by *Aulopus* (Fig. 12A) in which the robust first levator externus originates lateral to the remaining levatores. *The reduction of the first levator externus muscle to a thin slip (neoscopelids) or ligament (myctophids) is interpreted here as an additional synapomorphy of myctophiform fishes.*

(11) Enlarged conelike parapophyses on the first vertebral centrum meeting in the ventral mid-

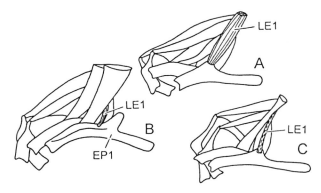

FIGURE 12 First epibranchial and associated muscles (in lateral view) in (A) *Aulopus* (AMNH 28635), (B) *Solivomer* (AMNH 36364), and (C) *Myctophum* (AMNH 23659).

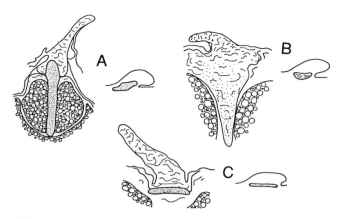

FIGURE 13 Adipose fin (in sagittal section) of (A) *Gonichthys*, (B) *Neoscopelus*, and (C) *Salangichthys* (redrawn after Matsuoka and Iwai, 1982).

line (Stiassny, 1986; Figs. 8B and 8C; Figs. 14C and 14D)

Typically among neoteleosts Baudelot's ligament originates from a small parapophysis fused to the first vertebral centrum (e.g., Figs. 8A, 14A and 14B). Rarely, the parapophysis is autogenous, as in *Parasudis* (Stiassny, 1986, figs. 8e and 8f) and *Chlorophthalmus*. However in myctophiforms the parapophyses are autogenous, enlarged, and cone-shaped and meet in the ventral midline of the first centrum (e.g., Figs. 8B and 8C, and 14D). Similar structures are not present in any outgroup taxon and *the presence of large, cone-like, ventrally projecting autogenous parapophyses that meet in the ventral midline is interpreted as an additional synapomorphy of the Myctophiformes.*

(12) Adipose fin support ventrally inserted into the supracarinalis posterior muscle mass (Figs. 13A and 13B)

In a review of adipose fin cartilages in teleostean fishes Matsuoka and Iwai (1982) noted that in their ventral insertion into the underlying muscle layer (= the supracarinalis posterior muscle) myctophid and neoscopelid adipose fin support structures are unique among teleosts (Figs. 13A and 13B). A similar muscular intrusion is not present in the "salmoniform," cypriniform, and siluriform outgroups they examined (e.g., Fig. 13C), in which the supporting cartilages are situated superficial to the body musculature (Matsuoka and Iwai, 1982, fig 3), or a cartilage is lacking as in the aulopiforms examined. In myctophids the adipose fin support lies along the base of the fin (Fig. 13A) and is composed of hyaline cartilage. The fin support of neoscopelids does not stain with alcian blue but is a chondroid tissue and contains networks of fine fibers extending the length of the fin

base (Fig. 13B). *The presence of adipose fin supports (either of hyaline cartilage or chondroid tissue) penetrating the supracarinalis posterior muscle mass is interpreted here as a synapomorphy of the Myctophiformes.*

The intrarelationships of myctophiform fishes are poorly understood, and this is particularly true with respect to the relationships of the three neoscopelid genera (*Neoscopelus, Scopelengys,* and *Solivomer*). Smith (1949) was the first to elevate Fowler's (1925) subfamilial division of the Myctophidae to familial rank, and although he provided no justification for this action his classification of myctophiforms into the Myctophidae and Neoscopelidae has been followed by subsequent authors. Despite the apparent consensus regarding the placement of *Neoscopelus, Solivomer,* and *Scopelengys* into the family Neoscopelidae, the phylogenetic justification for this taxonomic alignment has rarely been considered, and with few exceptions the character evidence supporting a monophyletic Neoscopelidae has not been reviewed critically. In the course of the present study such a review has been undertaken and the results indicate that there is support for the concept of a monophyletic Neoscopelidae. However, despite much effort, the morphological evidence that I have been able to muster is not overwhelming and probably is a good reflection of the essentially generalized nature of the extant members of this clade.

Neoscopelidae

R. K. Johnson (1982), in a wide-ranging review of the fishes of the families Evermannellidae and Scopelarchidae, undertook an investigation of the intrarelationships of "iniomous fishes" and summarized his results in the form of a branching diagram (Johnson, 1982, fig. 20). As evidence for the monophyly of the Neoscopelidae he referenced three characters. The first is "sclerotic bones absent," and while I agree that neoscopelids lack sclerotic ossification, I can find no support for this being a derived character at the level of the Neoscopelidae. For example, Johnson (1982) himself noted that among other "iniom" lineages, sclerotic ossifications are also absent in paralepidids, anotopterids, evermannellids, omosudids, and alepisaurids, and I find that sclerotic ossification is also lacking in myctophids. In sum, I find no support for this character at the level of the Neoscopelidae.

Johnson's second character refers to vertebral number, and according to his estimate neoscopelids are characterized by a reduced number of (28–34) vertebrae. In addition to problems with the "spanning" technique used to code continuous meristic variation into discrete states, Johnson had no specimens of *Soli-*

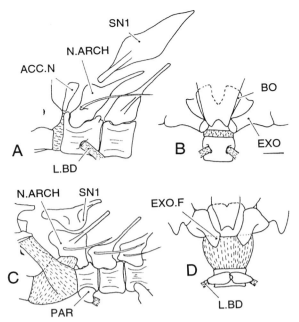

FIGURE 14 Cervical region of (A) *Aulopus* (AMNH 28635, in lateral view), (B) *Aulopus* (AMNH 28635, in ventral view), (C) *Solivomer* (USNM 135419, in lateral view), and (D) *Solivomer* (USNM 135419, in ventral view).

vomer at his disposal as *Solivomer* has a vertebral range of 39–41. Okiyama (1984a) provided a summary of vertebral number among aulopiform and myctophiform lineages, and it is evident that Johnson's vertebral codings cannot be maintained.

The third feature, cited by Johnson (1982, p. 88) as a "gap in ossification between the skull and first centrum with reduction in size of first centrum relative to succeeding centra," is part of a complex series of modifications of the occipito–cervical region that characterizes the Neoscopelidae and this complex is considered in further detail here.

(13) Presence of an elaborate and extensive cervical gap (Figs. 14C and 14D, and 16A–16C)

The nature of the occipital joint with the first vertebra in teleost fishes has attracted much attention in the recent literature (Rosen, 1985). And, as noted above, Johnson (1982) believed that the presence of a cervical gap characterizes neoscopelid fishes, however, he noted that a "similar gap" is also present in scopelarchids, chlorophthalmids, and ipnopids (see also Gosline *et al.*, 1966). In a more extensive review of the occipital region, Rosen (1985) argued that presence of a cervical gap characterizes most primitive neoteleosts. While I agree with Rosen's assessment that the presence of a cervical gap is a primitive feature of non-

acanthomorph neoteleosts (e.g., Figs. 14A and 14B), there is considerable variation in the extent and configuration of the cervical gap among these lineages. Based on a survey of this variation I conclude that neoscopelids are unique among neoteleosts in the possession of an extensive and uniquely elaborated cervical gap. The neoscopelid gap is characterized by the development of greatly enlarged exoccipital facets (e.g., Fig. 14D), the presence of prominent corresponding "articular" facets on the first (autogenous) neural arch (Fig. 8B), and the presence of complex encasing sheets of connective tissue surrounding the expanded, rubbery notochordal tube (Fig. 14B). I have not found a morphological configuration similar to this in any of the outgroup taxa and interpret *the presence of an extensive cervical gap spanned by connective tissue sheets, greatly enlarged exoccipital facets, and prominent facets on the neural arch of the first vertebra to be a complex synapomorphy of the Neoscopelidae.*

(14) Trilobate rostral cartilage (Stiassny, 1986; Figs. 15A and 15B, and 10B)

Hartel and Stiassny (1986) pointed out that the homologies of the rostral "cartilages" of non-acanthomorph neoteleosts are poorly established and that there is considerable diversity in the composition and associations of this structure throughout the non-acanthomorph Neoteleostei (see also Rosen, 1985). While Hartel and Stiassny (1986) concentrated their review on variation in aulopiform rostral morphology, their observation is equally true with regard to myctophi-

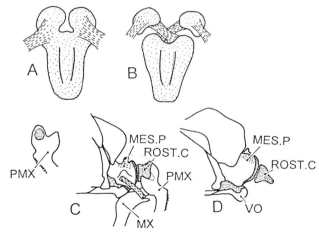

FIGURE 15 Isolated rostral cartilage of (A) *Solivomer* (USNM 135928) and (B) *Scopelengys* (AMNH 97466) and rostrum (in lateral view) of (C) *Hygophum* (AMNH 25019) and (D) *Lobianchia* (AMNH 29460).

forms. However, as can be seen from the following review, in this clade two basic patterns are found.

Neoscopelidae: Rosen (1985) and Stiassny (1986) noted that in neoscopelids a single, median rostral cartilage is ligamentously associated with the buccal jaws and that the cartilage has a characteristic trilobate structure (e.g., Fig.10B). Further review of neoscopelid rostral cartilages reveals that the rostral cartilage of *Scopelengys* (Fig.15B) differs from that of *Neoscopelus* and *Solivomer* (Fig. 15A) in that the lateral lobes are ligamentously bound to the median body of the cartilage rather than having a fully chondrified connection (see also Rosen, 1985, figs. 42a and 42b). Despite this variation, the median cartilage of all neoscopelids is clearly trilobate in form. A similarly trilobate rostral cartilage is not found in any of the outgroup taxa, and while the question of the "correspondence" of this structure with the single median cartilage of acanthomorphs remains to be addressed, *presence of a trilobate median cartilage ligamentously attached to the maxillae and premaxillae is a synapomorphy of the Neoscopelidae.*

Myctophidae: Among myctophids, and in contrast with the neoscopelid condition, a cylindrical median rostral cartilage is bound strongly to the tip of the mesethmoid (e.g., Figs. 15C and 15D; see also Jollie, 1954, fig. 15). The median cartilage is loosely attached to the maxillae and premaxillae by ligaments but usually remains attached to the mesethmoid when the buccal jaws are dissected away. Within the Myctophidae there is some variation in rostral morphology. For example, in most myctophine[2] myctophids the median cartilage is truncate and almost square in outline (Fig. 15C). In addition to the medial rostral cartilage, most myctophines have a pair of small disk-shaped lateral cartilages firmly bound to the medial face of the premaxillary ascending processes; however, these lateral cartilages have no structural association with the median cartilage. It is unclear whether the myctophine lateral cartilages are homologous with the lateral lobes of the neoscopelid rostral cartilage, although the condition in *Scopelengys* is suggestive of a possible transformation. Lampanyctines lack any trace of lateral premaxillary cartilages, and the median cartilage is more elongate than the myctophine cartilage, and it is somewhat nipple-shaped (e.g., Fig. 15D). Despite these modifications, *the presence of a cylindrical median rostral cartilage, firmly bound to the*

---

[2] Interestingly, in the Gonichthyini a small median cartilage is disassociated from the mesethmoid and lies between the ascending processes of the premaxillae. Lateral premaxillary cartilages are lacking in the Gonichthyini.

mesethmoid and only loosely attached to the buccal jaws, is interpreted here as a synapomorphy of the Myctophidae.

Additional Neoscopelid Characters Cited

Stiassny (1986) cited the following two characters as evidence of neoscopelid monophyly: the presence of "an enlarged bony protuberance on the median process of the maxilla" and a "broad strap-like" maxillo–rostroid ligament (VII) that "bifurcates on contacting the rostral cartilage." Further comparative review reveals that both of these features are variously developed among neoscopelids and show considerable intraspecific variation. Therefore these characters have been discounted in the present study.

While an explicitly phylogenetic hypothesis of the intrarelationships of the three neoscopelid genera has never previously been proposed, Nafpaktitis (1977), following Fraser-Brunner (1949), summarized the opinion of most authors with his assessment that *Neoscopelus* represents the most "generalized" member of the family; *Scopelengys*, with its weakly ossified skeleton, small eyes, flabby musculature, lack of photophores, and lack of swimbladder, represents a specialized but degenerate "offshoot"; and that *Solivomer* may be viewed as an "intermediate" between the two. Whether or not the standard scenario in which the sluggish, energy-deprived, deepwater *Scopelengys* is derived from an active benthopelagic and luminescent *Neoscopelus*-like ancestor (Nafpaktitis, 1977) is reasonable is, at this point, moot. Nonetheless, the fact that *Scopelengys* is so weakly ossified and degenerate makes an assessment of the precise intrarelationships of these three genera problematic. Despite the ever present possibility that apparently derived features shared by *Neoscopelus* and *Solivomer* may in fact be *secondarily* absent in *Scopelengys* as a result of reductive processes, I argue for a sister-group relationship between the former two genera, and the character data in support of this claim are presented below:

(15) Four supraneurals anterior to the dorsal fin (Figs. 16A and 16B)

Within the Neoteleostei there is a trend toward consolidation and reduction in number of supraneural elements. The presence of three (or fewer) supraneurals characterizes the majority of eurypterygian lineages (Patterson and Rosen, 1989; personal observation). However, *Solivomer* (Fig. 16A) and *Neoscopelus* (Fig. 16B) are unique among the taxa examined in this study in the possession of four, rather than three, supraneural elements. *Scopelengys* (Fig. 16C) and all myctophids (e.g., Fig. 16D) display the plesiomorphic

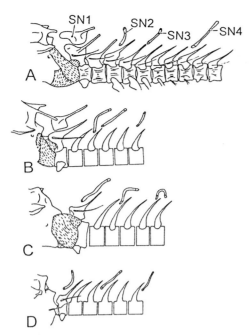

FIGURE 16 Posterior neurocranium and anterior vertebrae (in lateral view) of (A) *Solivomer* (USNM 135419) and schematic representation of (B) *Neoscopelus* (AMNH 28142), (C) *Scopelengys* (AMNH 49534), and (D) *Benthosoma* (AMNH29526).

eurypterygian condition of three supraneurals. Accordingly *the presence of four supraneurals is interpreted here as a synapomorphy of* Solivomer + Neoscopelus.

(16) Hatchet-shaped first supraneural (Figs. 16A and 16B)

The first supraneural in both *Solivomer* and *Neoscopelus* is of a characteristic hatchet-shape (Figs. 16A and 16B). While G. D. Johnson *et al.* (1996) argued that the presence of elongate and expanded supraneurals is a synapomorphy of synodontoid aulopiforms (e.g., Fig. 14A), I find the short hatchet-shaped first supraneurals of *Solivomer* and *Neoscopelus* to represent an alternative condition. As *Scopelengys* and all myctophids (Figs. 16C and 16D) share the plesiomorphic condition of an unelaborated, tubelike first supraneural, *the presence of a hatchet-shaped first supraneural is interpreted here as a synapomorphy of* Solivomer + Neoscopelus.

(17) Gill rakers modified into elongate toothplates arrayed along the first epibranchial (Figs. 17A and 17B)

Typically, among neoteleosts the first epibranchial element does not bear toothplates, although a few gill rakers are present in some taxa (e.g., *Scopelengys*, Fig. 17C). *Neoscopelus* (Fig. 17A) and *Solivomer* (Fig. 17B)

differ in that both have the distal outer gill rakers modified into elongate toothplates. Interestingly, in myctophids superficially similar toothplates are present along the distal limb of the first epibranchials (Fig. 17D); however, these toothplates are clearly derived from the novel medial toothplate row (page 417) rather than being modified outer gill rakers as in *Solivomer* and *Neoscopelus*. *The presence of gill rakers modified into elongate toothplates arrayed along the distal limb of the first epibranchial is interpreted here as a synapomorphy of* Solivomer + Neoscopelus.

Myctophidae

There can be little doubt of monophyly of the Myctophidae as it is currently recognized (Paxton, 1972, 1979; Moser *et al.*, 1984), and in the course of this study a number of novel myctophid synapomorphies have been identified (see below). The widespread notion of myctophids as being "of an unremarkable fishy shape" but strikingly characterized by "numerous light organs" (Hulley, 1995) hardly does justice to this morphologically specialized and diverse lineage.[3] It can be argued that preoccupation with photophores and photophore patterns within the family has diverted attention away from the probably equally informative and certainly extraordinarily rich variety of morphological innovation exhibited by this interesting clade.

Found in the upper 1000 meters of all oceans of the world, the Myctophidae are currently comprised of some 230–250 species classified into 30–35 genera (Paxton, 1979; Eschmeyer, 1990; Nelson, 1994). Evidence for monophyly of this assemblage is manifold. In fact, I find the clade to be bristling with synapomorphies in just about every anatomical system examined. This is an enviable situation for any morphologi-

---

[3] For an extensive discussion of the phylogenetic significance and biological function of myctophid photophores see, e.g., Paxton (1972), Moser and Åhlström (1974), and Nafpaktitis *et al.* (1977). It is of some interest, in the context of the present study, that among neoscopelids only *Neoscopelus* has photophores. Given the present proposal of neoscopelid monophyly and of a sister-group relationship of *Neoscopelus* and *Solivomer*, an interpretation of secondary loss of photophores in *Solivomer* and *Scopelengys* is equally parsimonious with the alternative interpretation of an independent acquisition of photophores in *Neoscopelus*. R. K. Johnson (1982, pp. 91–92) and Paxton *et al.* (1984, p. 241) consider the question of homology of neoscopelid and myctophid photophores at some length and both note the reported morphological differences between the photophores of *Neoscopelus* and those of myctophids. At present, however, insufficient morphological and developmental data on the photophores of *Neoscopelus* are available, and I can only reiterate the belief expressed by Paxton *et al.* (1984) that further study of the ultrastructure of *Neoscopelus* photophores would be of particular value.

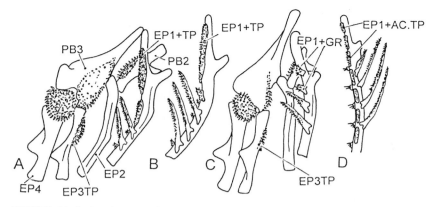

FIGURE 17 Isolated upper pharyngeal jaw elements (in ventral view) of (A) *Neoscopelus* (AMNH 28142), (B) *Solivomer* (USNM 135419), (C) *Scopelengys* (AMNH 49534), and (D) *Lampanyctus* (AMNH 97539).

cal systematist to encounter, but unfortunately the space limitations imposed by a review of this nature necessitate brevity. Therefore, before briefly considering the intrarelationships of the family I will describe a selection of characters that I believe lend strong support for the hypothesized monophyly of the myctophid clade.

Already the following myctophid synapomorphies have been identified:

(18) Subdivision of the first levator internus muscle into two heads, each inserting independently onto the second pharyngobranchial element (Fig. 1B; see also page 407)

(19) Reduction of the dorsal hypohyal element (HH1) (Figs. 2D and 2E; see also page 407)

(20) Presence of a distinct gap between the epihyal branchiostegals and the first two ceratohyal branchiostegals (Figs. 2D and 2E; see also page 407)

(21) Presence of a cylindrical median rostral cartilage firmly bound to the mesethmoid and only loosely attached to the buccal jaws (Fig. 15D,E; see also page 415)

Additionally, the following selection of characters is presented in corroboration of myctophid monophyly:

(22) Medial row of serially arranged toothplates arrayed along the length of the first branchial arch (Figs. 17D, 18A, and 18C)

Typically among teleosts an inner and an outer row of denticulate gill rakers are arrayed along the length of the first ceratobranchial (and often also the hypobranchial and epibranchial elements) (e.g., Fig. 18B). While in some outgroup taxa, supernumerary toothplates are irregularly interspersed between the serially

arranged gill rakers; uniquely in myctophids the first branchial arch carries a medial row of serially arranged oblong toothplates (e.g., Figs. 17D and 18C). With the exception of the two genera of Gonichthyini, *Gonichthys* and *Centrobranchus*, a medial row of toothplates has been located on the first branchial arch of all the myctophid taxa examined. Similar toothplates are not found in neoscopelids nor in any other outgroup taxon examined. Therefore, based upon this limited distribution, *the presence of a row of serially arranged, oblong, median toothplates, arrayed along the length of the first branchial arch, is interpreted here as a synapomorphy of the Myctophidae.*

(23) Maxillary control mediated via an $A1_\beta$ muscle component of the adductor mandibulae complex (Fig. 19B)

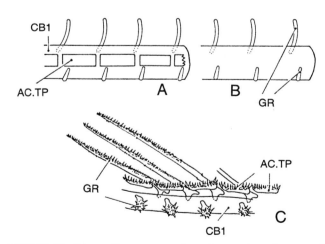

FIGURE 18 Schematic representation of a section of outer gill arch, gill rakers, and associated toothplates of (A) a myctophid and (B) a neoscopelid; section of outer gill arch of (C) *Hygophum* (AMNH 29491).

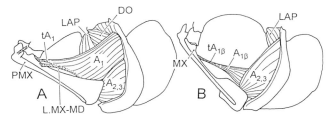

FIGURE 19 Adductor mandibulae muscle complex (in lateral view) of (A) *Scopelengys* (AMNH 12841) and (B) *Protomyctophum* (MCZ 102601).

Among neoteleosts, the maxilla is controlled either by an external A1 division of the adductor mandibulae, which inserts along the maxillo–mandibular ligament (ligamentum primordium) and attaches anteriorly on the maxilla via a separate tendon (tA1), or by an internal $A1_\beta$ muscle system. The $A1_\beta$ muscle, which originates medial to the A2 section, inserts onto the maxilla near the maxillary fulcrum via a $tA1_\beta$ tendon and has no connection or association with the maxillo–mandibular ligament.

An A1 system of maxillary control is widespread among teleosts (Rosen and Patterson, 1969; Winterbottom, 1974; Fink and Weitzman, 1982), while an $A1_\beta$ system has a more limited distribution among neoteleosts (Rosen, 1973). Fink and Weitzman (1982) reviewed maxillary control mechanisms in neoteleosts and discussed the phylogenetic significance of the occurrence of an $A1_\beta$ muscle system. They concluded that an $A1_\beta$ system has arisen independently a number of times within the Neoteleostei and cited as one example the presence of an $A1_\beta$ system in myctophids (Fig. 19B; see also Jollie, 1954, fig. 25; Rosen and Patterson, 1969, plate 55; Rosen, 1973, fig. 40; Winterbottom, 1974, fig. 4), noting the absence of $A1_\beta$ (and the presence of an A1 system) in neoscopelids (Fig. 19A; see also Winterbottom, 1974, fig. 3).

While there is considerable variation within the Myctophidae in the degree of subdivision and complexity of the A2 component of the adductor mandibulae, as well as in the degree of muscular encroachment of $A1_\beta$ onto the maxilla, a simple $A1_\beta$ muscle is invariably present. As an $A1_\beta$ is absent in neoscopelid and aulopiform outgroups (see also Rosen, 1973), I concur with Fink and Weitzman (1982) in their assessment of the derived nature of the $A1_\beta$ system of myctophids and suggest here that *the presence of a simple $A1_\beta$ system of maxillary control is a synapomorphy of the Myctophidae.* Interestingly, among basal acanthomorphs a complex $A1_\beta$ system is found in *Polymixia* (Rosen and Patterson, 1969, plate 53; Rosen, 1973, fig. 41), and certain paracanthopterygians also have an $A1_\beta$ system of maxillary control (Rosen and Patterson, 1969; Rosen,

1973), although in both of these cases the morphological details of the $A1_\beta$ systems differ from that found in myctophids.

(24) Bony connection between the descending process of the third hypobranchial element with the urohyal (Figs. 20C–20G)

Lauder (1983) identified what he believed to be a specialization of myctophiform fishes in which the

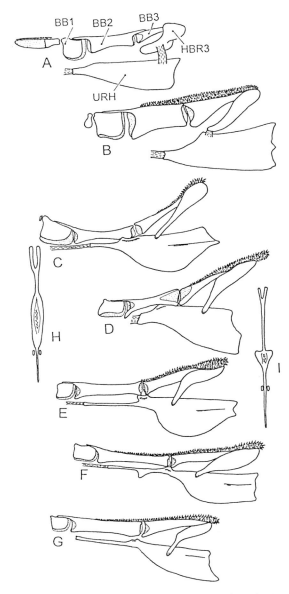

FIGURE 20 Ventral branchial apparatus (in lateral view) in (A) *Aulopus* (AMNH 28635), (B) *Solivomer* (USMN 135419), (C) *Benthosoma* (AMNH 29526), (D) *Gonichthys* (AMNH 1915), (E) *Lampanyctus* (AMNH 97539), (F) *Ceratoscopelus* (AMNH 29473), and (G) *Notolychnus* (AMNH 29479). In ventral view, (H) *Benthosoma* (AMNH 29526), and (I) *Lampanyctus* (AMNH 97539).

descending processes of the third hypobranchial extend laterally to form a rigid bony connection with the urohyal. Stiassny (1986) noted that while Lauder was correct in highlighting the peculiarity of the myctophid urohyal–hypobranchial 3 connection, he had incorrectly attributed this derived condition to neoscopelids (e.g., Fig. 20B). While all myctophids are characterized by the presence of a bony overlap of the elongate hypobranchial 3 elements that are firmly bound to either side of the median urohyal (see, e.g., Figs. 20C–20G), this condition is unique to that clade and does not occur in neoscopelids (e.g., Fig. 20B) nor in any non-myctophiform outgroups (e.g., Fig. 20A). Therefore, I concur with my initial interpretation (Stiassny, 1986) that *the development of a rigid bony connection between the third hypobranchial and the median urohyal is a synapomorphy of the Myctophidae.*

(25) Bony articulation between the urohyal and the second, or second and third, basibranchial element (Figs. 20C–20G)

Stiassny (1986, fig. 7) illustrated, but did not comment upon, a novel bony articulation between the urohyal and the basibranchial series in *Myctophum.* Further review of this character reveals that a similar articulation is present in most myctophid taxa (e.g., Figs. 20C–20F), although in a few taxa (e.g., *Notolychnus*) bone-to-bone contact is lacking, but an articular facet is retained on the urohyal (Fig. 20G). A similar articulation is lacking in neoscopelids (e.g., Fig. 20B) and in non-myctophiform outgroups (e.g., Fig. 20A). Once again, this distribution suggests that *the presence of an articulation between the basibranchial series and the urohyal element, mediated via an articulation facet developed on the dorsal face of the urohyal, is a synapomorphy of the Myctophidae.*

Although a detailed analysis of myctophid intrarelationships is beyond the scope of the present investigation, I comment briefly on a chapter by Paxton *et al.* (1984) as that study is the most recent and comprehensive analysis of relationships within the myctophid clade. Paxton *et al.* is an abbreviated synthesis and reanalysis of data initially presented in a prior series of papers (Paxton, 1972; Moser and Ahlstrom, 1970, 1972, 1974; Moser *et al.,* 1984). The authors acknowledged that their previous interpretation of myctophid intrarelationships was not strictly phylogenetic, being instead based on notions of overall similarity among the various taxa. As part of a more phylogenetically rigorous treatment Paxton *et al.* (1984) arrayed some 25 osteological, 17 larval, and 17 photophore characters culled from their previous studies, among representatives of 35 myctophid gen-

era. The characters were coded in a data matrix with polarity assessment determined on the basis of a number of discussed criteria, and a phylogenetic diagram representing myctophid intrarelationships was also presented (Fig. 21A). However, the phylogenetic diagram of Paxton *et al.* was apparently not derived from the matrix, but rather the authors constructed a phylogenetic tree based on their "knowledge of the family" and subsequently "used the apomorphic states of the 59 characters to define the various branching points" (1984, p. 243). When the Paxton *et al.* matrix (1984, pp. 242–243) is subjected to parsimony analysis using PAUP version 3.1.1 (Swofford, 1993) 16 equally parsimonious trees result; the strict consensus of these is presented here as Figure 21B. As can be seen by a comparison of the matrix-derived consensus tree (Fig. 21B) with the Paxton *et al.* tree (Fig. 21A) there is some discrepancy in the resultant hypothesis of myctophid intrarelationships, most evident in the alternate placement of the odd diminutive genus, *Notolychnus* (see also Fraser-Brunner, 1949; Moser and Åhlström, 1970, 1972, 1974; Paxton, 1972, for a discussion of the problematic *Notolychnus*).

While I do not intend to dwell at length on the question of myctophid intrarelationships, nonetheless in the course of this study I have identified some additional characters that help to elucidate the phylogenetic placement of *Notolychnus* at the subfamilial level, and these are outlined briefly below:

(26) Loss of fused third epibranchial toothplate (EP3TP; Figs. 22A–22C)

R. K. Johnson (1982) noted that myctophids lack third epibranchial toothplates (EP3TP), a character state that he considered derived among "inioms." While I agree that the loss of a toothplate *fused* to the ventral face of the third epibranchial element is a derived character at the level of the Ctenosquamata, I find that the loss of fusion of this toothplate is restricted among myctophids to the myctophine genera (e.g., Figs. 22A–22C). In these taxa toothplates, or gill rakers, are present but never fused to the epibranchial element. In contrast, all of the lampanyctine genera examined retain a toothplate fused to the ventral face of the third epibranchial (e.g., Figs. 22E–22G). Similarly, an EP3TP is present in the eurypterygian outgroup taxa examined (e.g., Fig. 22D). Based on this limited distribution, *the absence of a toothplate fused to the ventral face of the third epibranchial is interpreted as a derived character state.*

(27) Elongation of the second basibranchial element (Figs. 20E–20G)

In most taxa examined the second basibranchial is a moderate sized element that is approximately twice

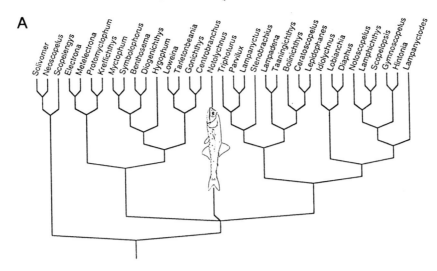

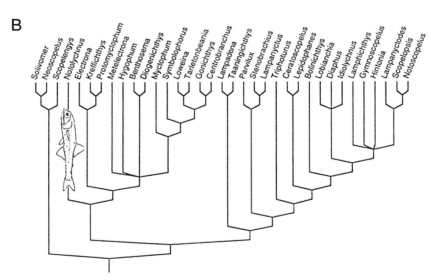

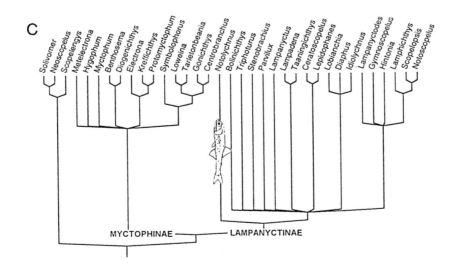

the length of the first basibranchial (e.g., Figs. 20A–20D). However, this is not the case in the lampanyctine myctophids examined, and invariably in these taxa the second basibranchial is a markedly elongate element that is minimally three, and often four, times the length of the first basibranchial. A similar elongation of the second basibranchial has not been found in myctophines or outgroup taxa and *the presence of an elongate second basibranchial is interpreted as a derived character state.*

(28) Urohyal with elongate anterior process and reduced articulation facet (Figs. 20E–20G)

Primitively among eurypterygians the urohyal is a robust element with a short bifurcating anterior process (e.g., Figs. 20A–20B). In myctophids, with the exception of the Gonichthyini (e.g., Fig. 20D), the urohyal is a thin, blade-like element, and in the lampanyctine genera examined the anterior process is markedly elongate and distinctly demarcated from the body of the urohyal blade (e.g., Figs. 20E–20G). Additionally, in lampanyctines the articulation facet on the dorsal surface of the urohyal is short (e.g., Fig. 20I). In contrast in the myctophine genera the anterior processes are less elongate and are formed as a tapering continuation of the urohyal blade (e.g., Figs. 20C and 20D), and the articulation facet is elongate (e.g., Fig. 20H). *The presence of an elongate, strongly demarcated anterior process and short articulation facet is tentatively interpreted here as a derived character state.*

(29) Metapterygoid strut projects dorsally from the main body of the bone (Paxton, 1972, fig. 6B)

Paxton (1972) identified two conditions of a bony strut developed on the posterodorsal margin of the metapterygoid in myctophid fishes. This metapterygoid strut either projects dorsally and extends well above the dorsal margin of the bone (Paxton, 1972, fig. 6B) or it is restricted to, or lies ventral to, the dorsal margin of the bone (Paxton, 1972, fig. 6A). Paxton (1972, table 3) lists the distribution of these two character states among 32 myctophid genera. In assessing the polarity of this character, while some variation exists, it is evident that among neoscopelids and other outgroups the metapterygoid strut most

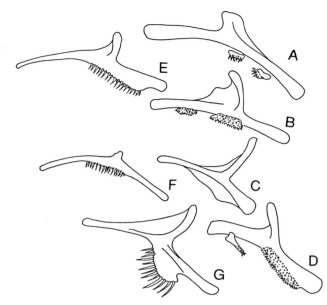

FIGURE 22 Isolated third epibranchials (in lateral view) of (A) *Myctophum* (AMNH 23659), (B) *Protomyctophum* (MCZ 102601), (C) *Gonichthys* (AMNH 1915), (D) *Lampanyctus* (AMNH 97539), (E) *Notolychnus* (AMNH 29479), (F) *Solivomer* (USNM 135419), and (G) *Diaphus* (AMNH 29449).

closely approximates the second condition, i.e., the strut is situated ventral to the dorsal border of the bone and I therefore interpret *the presence of a dorsally projecting metapterygoid strut as the derived character state.*

When these four characters are coded (Table 1) and added to the Paxton *et al.* (1984, pp. 242–243) data matrix, and the augmented matrix is subjected to parsimony analysis using PAUP, 264 equally parsimonious trees (length 148) result[4]; the strict consensus of

---

[4] The sharp increase from 16 to 264 equally parsimonious trees and the spurious realignment of *Lampichthys* and *Lampanyctodes* in Fig. 21C are most likely a result of numerous question marks in the data matrix (Table 1).

FIGURE 21 Alternate schemes of myctophid intrarelationship. (A) Phylogenetic diagram of the Myctophidae (after Paxton *et al.*, 1984). When redrawn using MacClade Version 3.0 (Maddison and Maddison, 1992) a length of 152 is calculated. (B) Strict consensus of 16 equally parsimonious trees (lengths 140) derived from the Paxton *et al.* (1984) matrix as computed by PAUP version 3.1.1 (heuristic search, stepwise addition, with all multistate characters unordered). (C) Estimate of relationships of the family Myctophidae as derived from the present study. This tree is the strict consensus of 264 equally parsimonious trees (lengths 148) derived using PAUP version 3.1.1 (heuristic search, stepwise addition with all multistate characters unordered). The matrix used is that of Paxton *et al.* (1984) with the addition of four characters (Table 1).

TABLE 1　Additional Data Added to the Paxton et al. (1984, pp. 242–243) Matrix

Genus	EP3 + TP	Basibranchial 1	Urohyal shape	Metapterygoid strut
Kreffichthys	?	?	?	?
Protomyctophum	1	0	0	0
Electrona	1	0	0	0
Metelectrona	?	?	?	0
Benthosoma	1	0	0	0
Diogenichthys	1	0	0	0
Hygophum	1	0	0	0
Myctophum	1	0	0	0
Symbolophorus	?	?	?	0
Loweina	?	?	?	0
Tarltonbeania	?	?	?	0
Gonichthys	1	0	0	0
Centrobranchus	1	0	0	1
Notolychnus	0	1	1	1
Lobianchia	0	1	1	1
Diaphus	0	1	1	0 and 1
Idiolychnus	?	?	?	?
Lampanyctodes	?	?	?	1
Gymnoscopelus	0	1	1	1
Scopelopsis	?	?	?	1
Lamphichthys	?	?	?	1
Notoscopelus	0	1	1	1
Hintonia	?	?	?	1
Lampadena	0	1	1	1
Taaningichthys	?	?	?	1
Ceratoscopelus	0	1	1	1
Lepidophanes	0	1	1	1
Bolinichthys	?	?	?	1
Triphoturus	?	?	?	1
Stenobrachius	0	1	1	1
Parvilux	?	?	?	1
Lampanyctus	0	1	1	1
Solivomer	0	0	0	0
Neoscopelus	0	0	0	0
Scopelengys	0	0	0	0

these is presented here (Fig. 21C). Perhaps most relevant to the discussion at hand is the resolution of *Notolychnus* as the basal lampanyctine clade. In this scheme, the Myctophinae are monophyletic and uniquely diagnosed by the loss of toothplates fused to the third epibranchials, and a monophyletic Lampanyctinae (including *Notolychnus*) is uniquely diagnosed by the presence of an elongate second basibranchial and a specialized urohyal morphology.

Obviously, in such a cursory review I do not wish to convey the impression that the intrarelationships of the myctophid fishes are represented adequately by the diagram presented here as Fig. 21C, or that the Paxton et al. (1984) matrix is not without need of much critical review and revision. Rather, I add these data in the hope that the resultant analysis will stimulate further work on this interesting clade, and at the present time I can do little more than echo the sentiment of Paxton et al. (1984, p. 244) who ended their consideration of myctophid relationships in the hope that "new, less plastic characters and better definitions of polarity will help resolve the problems."

IV. Discussion

It has been my aim in this study to investigate the relationships of the Myctophiformes, an order of teleostean fishes originally conceived by Donn Rosen in his influential contribution to the first *Interrelationships* volume (Rosen, 1973). In many respects my study has done little more than provide additional support for Rosen's visionary contribution, a far from unhappy state of affairs.

For ease of reference, the main conclusions of this study are perhaps best summarized in the form of a phylogenetic scheme (Fig. 23). As can be seen from this figure, the placement of a monophyletic Myctophiformes as the sister taxon to the Acanthomorpha is strongly supported. Relationships within the Myctophiformes are somewhat less well-resolved; however, strong support for a monophyletic Myctophidae has been located in a wide array of character systems. As I have already intimated, there is considerable morphological variation evident among the members of this widespread oceanic clade and it is in some ways unfortunate that much prior work has concentrated on the disposition of photophores among the various taxa and, as a result, the true extent of anatomical variation has not been fully explored. The resolution of myctophid intrarelationships remains an outstanding problem in teleostean systematics and I hope that the present study may serve as a useful starting point for further work.

Another finding of some significance is the corroboration of the monophyly of the Neoscopelidae. Evidence for this association is, however, far from overwhelming, and I suggest that this is a reflection of the morphologically generalized nature of the living members of this clade. While this makes for a difficult task in resolving the intrarelationships of these fishes,

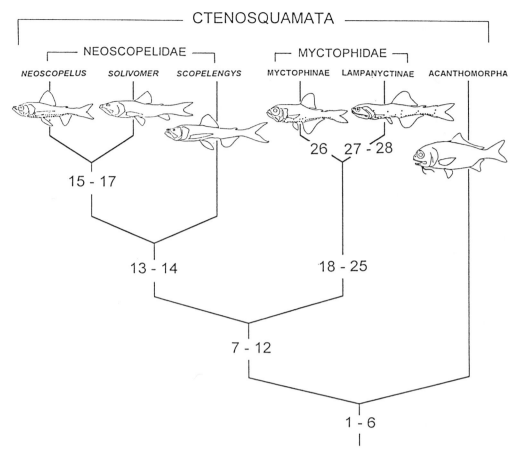

FIGURE 23 Scheme of ctenosquamate intrarelationships as supported by the present study. Numbers on stems correspond to characters discussed in the text.

it is of a certain utility when these taxa are selected as a source of outgroup data in ongoing analyses of the intrarelationships of the species-rich, and ever problematic, Acanthomorpha.

V. Summary

The anatomy and relationships of the Myctophiformes (Scopelomorpha), a grouping of euteleostean fishes originally assembled by Donn Rosen in his influential contribution to the first edition of this book (Rosen, 1973) are investigated. The placement of a monophyletic Myctophiformes as the sister taxon (and therefore the proximate outgroup) to the species-rich Acanthomorpha is strongly supported by data culled from a wide range of anatomical systems. While relationships within the Myctophiformes are less well-resolved, there is a considerable body of morphological evidence presented in support of a monophyletic Myctophidae. The more morphologically generalized Neoscopelidae are also found to be monophyletic, although data in support of this association are far from overwhelming.

While the present study has clarified the higher level relationships of the Myctophiformes, the resolution of myctophid intrarelationships remains an outstanding problem in teleostean systematics.

Acknowledgments

It is often hard to know where to begin, and where to end, when writing acknowledgments; there are always so many people to thank. But in this case at least the beginning is clear and I'd like to start by acknowledging the monumental contribution that Colin Patterson has made to our discipline. One of the real pleasures of working on this volume has been to see the unanimously enthusiastic response from authors when they learned that the volume is to be dedicated to Colin. It is very clear that his impeccable standards of intellectual rigor coupled with a prodigious output of high-quality empirical work are a continuing inspiration to us all.

Many thanks are due also to my fellow editors, Lynne Parenti and Dave Johnson; their good spirits and hard work have been much appreciated. Lynne and Dave, along with Colin Patterson,

Carole Baldwin, and John Paxton provided helpful commentary on an earlier version of this paper, and I am grateful to them all. I also thank Geoffrey Moser, who encouraged my interest in the myctophiforms and generously welcomed me into the exclusive "brotherhood" of myctophophiles.

For the loan of materials I am grateful to Karsten Hartel (MCZ) and Susan Jewett (USMN). My thanks also to Lita Elvers and Marcelo de Carvalho, who have both been a tremendous help to me throughout a protracted editorial process. Finally, I thank Amy Simmons for her ongoing encouragement and support.

References

Eschmeyer, W. N. (1990). "Catalog of the Genera of Recent Fishes." California Academy of Sciences, San Francisco.

Fink, W. L. (1984). Stomiiforms: Relationships. In "Ontogeny and Systematics of Fishes" (H. G. Moser, W. J. Richards, D., M. Cohen, M. P. Fahay, A. W. Kendall, Jr., and S. L. Richardson, eds.), Spec. Publ. No. 1, pp. 181–184. American Society of Ichthyologists and Herpetologists, Lawrence, KS.

Fink, W. L., and Weitzman, S. H. (1982). Relationships of the stomiiform fishes (Teleostei), with a description of *Diplophos*. *Bull. Mus. Comp. Zool.* **150**(2), 31–93.

Fowler, H. W. (1925). New taxonomic names of West African marine fishes. *Am. Mus. Novit.* **162**, 1–5.

Fraser-Brunner, A. (1949). A classification of the fishes of the family Myctophidae. *Proc. Zool. Soc. London* **118**, 1019–1106.

Goode, G. B., and Bean, T. H. (1896). Oceanic ichthyology, a treatise on the deep-sea and pelagic fishes of the world, based chiefly upon the collections made by the steamers Blake, Albatross, and Fish Hawk in the northeastern Atlantic with an atlas containing 417 figures. *Spec. Bull. U.S. Natl. Mus.* **2**, 1–1553.

Gosline, W. A., Mead, G. W., and Marshall, N. B. (1966). Order Iniomi. Characters and synopsis of families. In "Fishes of the Western North Atlantic," Sears Found. Mar. Res. Mem. No. 1, Part 5, pp. 1–18. Yale University, New Haven, CT.

Gottfried, M. D. (1989). Homology and terminology of higher teleost postcleithral elements. *Trans. San Diego Soc. Nat. Hist.* **21**(18), 283–290.

Gregory, W. K., and Conrad, G. M. (1936). Pictorial phylogenies of deep sea Isospondyli and Iniomi. *Copeia*, **1**, 21–36.

Hartel, K. E., and Stiassny, M. L. J. (1986). The identification of larval *Parasudis* (Teleostei, Chlorophthalmidae); with notes on the anatomy and relationships of aulopiform fishes. *Breviora* **487**, 1–23.

Hulley, P. A. (1995). Lanternfishes. In "Encyclopedia of Fishes" (J. R. Paxton and W. N. Eschmeyer, eds.), pp. 127–128. Academic Press, San Diego, CA.

Johnson, G. D. (1992). Monophyly of the euteleostean clades—Neoteleostei, Eurypterygii, and Ctenosquamata. *Copeia*, **1**, 8–25.

Johnson, G. D., and Patterson, C. (1993). Percomorph phylogeny: A survey of acanthomorphs and a new proposal. *Bull. Mar. Sci.* **52**(1), 554–626.

Johnson, G. D., Baldwin, C. C., Okiyama, M., and Tominaga, Y. (1996). Osteology and relationships of *Pseudotrachinotus altivelis* (Teleostei: Aulopiformes: Pseudotrichonotidae). *Ichthyol. Res.* **43**, 17–45.

Johnson, R. K. (1974). A revision of the alepisauroid family Scopelarchidae (Pisces, Myctophiformes). *Fieldiana, Zool.* [N.S.] **66**, 1–249.

Johnson, R. K. (1982). Fishes of the families Evermanellidae and Scopelarchidae: Systematics, morphology, interrelationships, and zoogeography. *Fieldiana, Zool.* [N.S.] **12**, 1–252.

Jollie, M. T. (1954). The general anatomy of *Lampanyctus leucopsarus* (Eigenmann and Eigenmann). Ph.D. Dissertation, Stanford University, Stanford, CA.

Lauder, G. V. (1982). Patterns of evolution in the feeding mechanism of actinopterygian fishes. *Am. Zool.* **22**, 275–285.

Lauder, G. V. (1983). Functional design and evolution of the pharyngeal jaw apparatus in euteleostean fishes. *Zool. J. Linn. Soc.* **77**, 1–38.

Lauder, G. V. and Liem, K. F. (1983). The evolution and interrelationships of the actinopterygian fishes. *Bull. Mus. Comp. Zool.* **150**, 95–197.

Maddison, W. P., and Maddison, D. R. (1992). "MacClade: Analysis of Phylogeny and Character Evolution. Version 3.04," Sinauer Assoc., Sunderland, MA.

Marshall, N. B. (1960). Swimbladder structure of deep-sea fishes in relation to their systematics and biology. *'Discovery' Rep.* **31**, 1–121.

Matsuoka, M., and Iwai, T. (1982). Adipose fin cartilage found in some teleostean fishes. *Jpn. J. Ichthyol.* **30**(1), 37–46.

McAllister, D. E. (1968). The evolution of branchiostegals and associated gular, opercular and hyoid bones and the classification of teleostome fishes, living and fossil. *Bull. Natl. Mus. Can.* **221**, 1–239.

Miller, R. R. (1947). A new genus and species of deep sea fish of the family Myctophidae from the Philippine Islands. *Proc. U.S. Natl. Mus.* **97**(3211), 81–90.

Moore, J. A. (1993). Phylogeny of the Trachichthyiformes (Teleostei: Percomorpha). *Bull. Mar. Sci.* **52**(1), 114–136.

Moser, H. G., and Åhlström, E. H. (1970). Development of laternfishes (family Myctophidae) in the California Current. Part I. Species with narrow-eyed larvae. *Bull. Los Angeles Cty. Mus. Nat. Hist.* **7**, 1–145.

Moser, H. G., and Åhlström, E. H. (1972). Development of the lanternfish, *Scopelopsis multipunctatus* Brauer 1906, with a discussion of its phylogenetic position in the family Myctophidae and its role in a proposed mechanism for the evolution of photophore patterns in lanternfishes. *Fish. Bull.* **70**, 541–564.

Moser, H. G., and Åhlström, E. H. (1974). Role of larval stages in systematic investigations of marine teleosts: The Myctophidae, a case study. *Fish. Bull.* **72**(2), 391–413.

Moser, H. G., Åhlström, E. H., and Paxton, J. R. (1984). Myctophidae: Development. In "Ontogeny and Systematics of Fishes" (H. G. Moser, W. J. Richards, D. M. Cohen, M. P. Fahay, A. W. Kendall, Jr., and S. L. Richardson, eds.), Spec. Publ. No. 1, pp. 218–239. American Society of Ichthyologists and Herpetologists, Lawrence, KS.

Nafpaktitis, B. G. (1977). Family Neoscopelidae. In "Fishes of the Western North Atlantic," Sears Found. Mar. Res., Mem. No. 1, Part 7, pp. 1–12. Yale University, New Haven, CT.

Nafpaktitis, B. G., Backus, R. H., Craddock, J. E., Haedrick, R. L., Robinson, B. H., and Karnella, C. (1977). Family Myctophidae. In "Fishes of the Western North Atlantic," Sears Found. Mar. Res., Mem. No. 1, Part 7, pp. 13–265. Yale University, New Haven, CT.

Nelson, J. S. (1994). "Fishes of the World," 3rd ed. Wiley, New York.

Okiyama, M. (1984a). Myctophiformes: Development. In "Ontogeny and Systematics of Fishes" (H. G. Moser, W. J. Richards, D. M. Cohen, M. P. Fahay, A. W. Kendall, Jr., and S. L. Richardson, eds.), Spec. Publ. No. 1, pp. 206–218. American Society of Ichthyologists and Herpetologists, Lawrence, KS.

Okiyama, M. (1984b). Myctophiformes: Relationships. In "Ontogeny and Systematics of Fishes" (H. G. Moser, W. J. Richards, D. M. Cohen, M. P. Fahay, A. W. Kendall, Jr., and S. L. Richardson, eds.), Spec. Publ. No. 1, pp. 254–259. American Society of Ichthyologists and Herpetologists, Lawrence, KS.

Olney, J. E., Johnson, G. D., and Baldwin, C. C. (1993). Phylogeny of lampridiform fishes. *Bull. Mar. Sci.* **52**(1), 137–169.

Patterson, C., and Johnson, G. D. (1995). The intermuscular bones and ligaments of teleostean fishes. *Smithson. Contrib. Zool.* **559**, 1–83.

Patterson, C., and Rosen, D. E. (1989). The Paracanthopterygii revisited: Order and disorder. *Sci. Ser., Nat. His. Mus. L.A. Cty.*, **32**, 5–36.

Paxton, J. R. (1972). Osteology and relationships of the lanternfishes (Family Myctophidae). *Bull. Nat. Hist. Mus. L.A. Cty.* **13**, 1–81.

Paxton, J. R. (1979). Nominal genera and species of lanternfishes (Family Myctophidae). *Contrib. Sci., Nat. Hist. Mus. L.A. Cty.* **322**, 1–28.

Paxton, J. R., Åhlström, E. H., and Moser, H. G. (1984). *In* "Ontogeny and Systematics of Fishes" (H. G. Moser, W. J. Richards, D. M. Cohen, M. P. Fahay, A. W. Kendall, Jr., and S. L. Richardson, eds.), Spec. Publ. No. 1, pp. 239–244. American Society of Ichthyologists and Herpetologists, Lawrence, KS.

Regan, C. T. (1911). The anatomy and classification of the teleostean fishes of the order Iniomi. *Ann. Mag. Nat. Hist.* [8] **7**(37), 120–123.

Rosen, D. E. (1973). Interrelationships of higher euteleostean fishes. *In* "Interrelationships of Fishes" (P. H. Greenwood, R. S. Miles, and C. Patterson, eds.), pp. 397–513. Academic Press, London.

Rosen, D. E. (1985). An essay on euteleostean classification. *Am. Mus. Novit.* **2827**, 1–57.

Rosen, D. E. and Patterson, C. (1969). The structure and relationships of the paracanthopterygian fishes. *Bull. Am. Mus. Nat. Hist.* **141**(3), 361–474.

Smith, J. L. B. (1949). "The Sea Fishes of South Africa." Central News Agency, Cape Town.

Steyskal, G. C. (1980). The grammar of family-group names as exemplified by those of fishes. *Proc. Biol. Soc. Wash.* **93**(1), 168–177.

Stiassny, M. L. J. (1986). The limits and relationships of the acanthomorph teleosts. *J. Zool. (B)* **1**, 411–460.

Stiassny, M. L. J., and Moore, J. A. (1992). A review of the pelvic girdle of acanthomorph fishes, with comments on hypotheses of acanthomorph intrarelationships. *Zool. J. Linn. Soc.* **104**, 209–242.

Swofford, D. L. (1993). "PAUP: Phylogenetic Analysis Using Parsimony, Version 3.1.1," Smithsonian Institution, Washington, DC.

Winterbottom, R. (1974). Descriptive synonomy of the striated muscles of the Teleostei. *Proc. Acad. Natl. Sci. Philadelphia* **125**(12), 225–317.

Appendix 1
Abbreviations Used in Text Figures

A_1; $A_{2,3}$; $A_{1\beta}$	Adductor mandibulae sections	L.PT–EPIO	Posttemporal–epiotic ligament
ACC.N	Accessory neural arch	MD	Mandible
AC.TP	Accessory toothplate	MES.P	Mesethmoid process
AO	Antorbital	MHL	Mandibulohyoid ligament
BB1–3	Basibranchial 1–3	MX	Maxilla
BO	Basioccipital	N.ARCH	Neural arch
BR	Branchiostegal rays	NAS	Nasal
CB1–5	Ceratobranchial 1–5	OBL.V	Obliquus ventralis muscle
CRT.M	Craniotemporalis muscle	PAR	Parapophysis
DO	Dilatator operculi muscle	PB2–3	Pharyngobranchial 2–3
EP1–4	Epibranchial 1–4	PMX	Premaxilla
EPAX.M	Epaxialis muscle	PCH	Posterior ceratohyal
EXO	Exoccipital	PT	Posttemporal
EXO.F	Exoccipital facet	R.COM	Rectus communis muscle
GR	Gill raker	RECT.V	Rectus ventralis muscle
HBR3	Hypobranchial 3	ROST.C	Rostral cartilage
HH1–2	Hypohyal 1–2	SN1–4	Supraneural 1–4
HY	Hyoid bar	STH.M	Sternohyoideus muscle
IML	Interoperculomandibular ligament	SO	Supraorbital
IOP	Interoperculum	SUC.M	Supracarinalis muscle
LAP	Levator arcuus palatini muscle	tA_1, $t A_{1\beta}$	Tendons of A_1 and $A_{1\beta}$
L.BD	Baudelot's ligament	TP	Toothplate
LE1	Levator externus 1	VII	Maxillo–rostroid ligament
LI1–2	Levatores interni 1–2	VIII	Median maxillo–premaxillary ligament
L.MX–MD	Maxillo–mandibular ligament	VO	Vomer

Appendix 2

The following is a list of materials examined from the collections of the American Museum of Natural History, New York (AMNH), the U.S. National Museum of Natural History, Washington (USNM), and the Museum of Comparative Zoology, Cambridge (MCZ). Material type is abbreviated in parenthesis.

Stomiiformes: *Diplophos maderensis* AMNH 84352 (alc, rad), *Maurolicus muelleri* AMNH 76762 (alc, c&s), *Sternoptyx diaphana* AMNH 29752 (c&s), *Gonostoma elongata* AMNH 29758 (c&s), *Gonostoma denudatum* AMNH 29836 (alc, c&s), *Astronesthes lucifer* AMNH 89876 (alc, rad); Aulopiformes: *Aulopus japonicus* AMNH 28635 (alc, c&s), *Aulopus purpurissatus* AMNH 92091 (skel), *Synodus poeyi* AMNH 82532 (c&s), *Saurida brasiliensis* AMNH 88664 (skel), *Harpodon sp.* AMNH 20585 (alc, rad), *Bathypterois bigelowi* AMNH 53087 (alc), *Chlorophthalmus agassizi* AMNH 76024 (alc), AMNH 70606 (c&s), *Chlorophthalmus nigripinnis* AMNH 95809 (skel), *Parasudis truculenta* AMNH 84635 (alc, rad), *Omosudis lowei* AMNH 29743 (alc), *Paralepis atlantica* AMNH 58079 (alc); Myctophiformes: *Neoscopelus macrolepidotus* AMNH 84363 (alc, rad), AMNH 28142 (alc, c&s), AMNH 49533 (alc, c&s), AMNH 27405 (c&s); *Scopelengys tristis* AMNH 12841 (alc, rad), AMNH 24143 (alc), AMNH 49534 (alc, rad, c&s), AMNH 97466 (c&s), USNM 201151 (alc, rad); *Solivomer arenidens* AMNH 36364 (alc, rad), USNM 29507 (c&s), USNM 35928 (alc, rad), USNM 135930 (alc, rad), USNM 135419 (c&s); *Electrona risso* MCZ 55468 (alc), MCZ 116336 (c&s), MCZ 116337 (c&s); *Protomyctophum arcticum* MCZ 102601 (alc, c&s); *Myctophum spinosum* AMNH 18152 (alc); *Myctophum obtusirostre* AMNH 23659 (alc, c&s); AMNH 25022 (c&s); *Myctophum punctatum* AMNH 75750 (c&s); *Benthosoma glaciale* AMNH 29526 (alc, c&s); *Hygophum benoiti* AMNH 14239 (alc, rad , *Hygophum hygomii* AMNH 29491 (alc, c&s); *Hygophum macrochir* AMNH 25019 (c&s); *Diogenichthys atlanticus* AMNH 29478 (alc, rad); *Centrobranchus nigroocelatus* AMNH 73692 (alc, rad, c&s); *Gonichthys tenuiculus* AMNH 1915 (alc, c&s); *Gymnoscopelus nicholsi* AMNH 1916 (alc, rad); *Notolychnus valdiviae* AMNH 29479 (alc, rad, c&s); *Diaphus mollis* AMNH 29449 (c&s); *Diaphus garmani* AMNH 22970 (c&s, rad); *Diaphus rafinesquii* AMNH 97537 (c&s, rad); *Lobianchia gemellarii* AMNH 29516 (alc, rad); *Lobianchia dofleini* AMNH 29619 (c&s); *Ceratoscopelus maderensis* AMNH 29460 (c&s); *Ceratoscopelus townsendi* AMNH 29473 (c&s); *Gymnoscopelus nicholsi* AMNH 1916 (alc, rad); *Lampanyctus crocodilus* AMNH 71957 (alc, rad); *Lampanyctus idostigma* AMNH 29748 (rad); *Lampanyctus nigrum* AMNH 97539 (c&s); *Lampanyctus regale* AMNH 12840 (rad); *Lampadena speculigera* AMNH 58088 (alc, rad); *Lepidophanes guentheri* AMNH 29507 (c&s); *Notoscopelus caudispinosus* AMNH 29530 (alc, rad); *Notoscopelus elongatus* AMNH 72024 (c&s); *Notoscopelus resplendens* AMNH 29528 (c&s); *Stenobrachius leucopsarus* AMNH 12839 (alc, rad); Acanthomorpha: *Velifer hypselopterus* AMNH 49575 (alc, rad), USNM 13821 (c&s); *Metavelifer multiradiatus* AMNH 91801 (skel); *Lampris guttatus* AMNH 56281 (skel); *Polymixia lowei* AMNH 88707 (skel), AMNH 86102 (c&s), AMNH 70602 (alc); *Percopsis omiscomaycus* AMNH 41145 (c&s), AMNH 52491 (alc); *Melamphaes cristiceps* AMNH 29756 (alc); *Zeus japonicus* AMNH 26924 (alc); *Beryx splendens* AMNH 95743 (skel); *Trachichthys australis* AMNH 48817 (alc).

16

Phylogenetic Significance of the Pectoral–Pelvic Fin Association in Acanthomorph Fishes: A Reassessment Using Comparative Neuroanatomy

LYNNE R. PARENTI
Division of Fishes
National Museum of Natural History
Smithsonian Institution
Washington, D.C. 20560

JIAKUN SONG
Division of Fishes
National Museum of Natural History
Smithsonian Institution
Washington, D.C. 20560
and
Department of Zoology
University of Maryland
College Park, Maryland 20742

I. Introduction

Features of the pectoral and pelvic fins are prominent in the characteristic "trends" for higher taxa listed by Greenwood *et al.* (1966, pp. 350–354) in their pre-cladistic summary of teleost phylogeny. For example, characters of their Division III, or Euteleostei, include anterior migration of the pelvic girdle and its linkage with the pectoral girdle, more dorsal placement of the pectoral-fin base, and reduction in number of pectoral radials and pelvic-fin rays. The Ostariophysi, the most primitive euteleosts in their analysis, were characterized by abdominal pelvic fins. The Protacanthopterygii were characterized by "occasional trends" for pelvic fins to advance anteriorly, pectoral fins to be elevated or positioned more dorsally along the body, and pelvics commonly with more than six rays. Paracanthopterygii were characterized by pelvic fins being thoracic, jugular, or mental ("in all but one species"), occasionally with up to 17 pelvic rays, and fin spines developed or not. Atherinomorpha were characterized by the pelvic girdle being abdominal, subabdominal, or thoracic; fin spines present or not; and four cuboidal pectoral radials recessed within an excavation in the scapulocoracoid margin. Acanthopterygii were characterized by pelvic fins, if present, being thoracic or jugular, with a pelvic fin typically consisting of a spine and five segmented rays (the I,5 pelvic-fin ray formula) except in "berycoids and a few other forms," and pectoral fins inserted high on the sides of the body.

Rosen (1973, p. 506) summarized "trends" in fin characters shared by paracanthopterygians and acanthopterygians as tendencies to develop a pelvic-fin spine, anal-fin spines, and more numerous dorsal-fin

427

Copyright © 1996 by Academic Press, Inc.
All rights of reproduction in any form reserved.

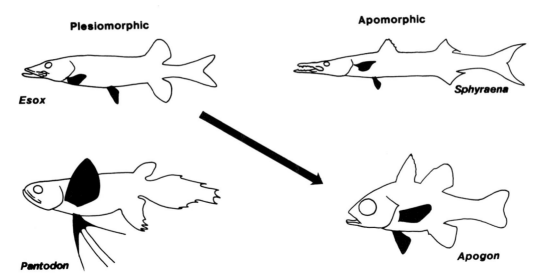

FIGURE 1 Diagrammatic representation of relative position of pectoral and pelvic fins, blackened, in relatively plesiomorphic and apomorphic teleost fishes. General transformation from plesiomorphic to apomorphic condition marked by arrow. *Pantodon* and *Sphyraena* have independently derived relative fin positions; their relative overall plesiomorphy or apomorphy is based on other characters. (Outlines modified from Greenwood *et al.*, 1966, and Nelson, 1994).

spines; to reduce the number of pelvic- and caudal-fin rays; to have the pectorals in a more dorsal position; and to have the pelvics more anterior, in a thoracic, jugular, or mental position. Nearly a decade later, Rosen (1982) depicted teleostean relationships in a cladogram (fig. 5) on which he also plotted (fig. 10) the states for several characters, including pectoral- and pelvic-fin position and fin spine occurrence. He (p. 266) stated in an explicitly phylogenetic context the long-held conclusion that "pectoral fins with a higher insertion on the side and pelvic fins with a more anterior insertion on the abdomen are taken to be derived" and that "fin spines primitively are absent." Rosen (1982) considered presence of pelvic-fin spines to be the most derived condition. Most "lower" teleosts, such as *Esox*, have the plesiomorphic condition of pectorals set low on the side of the body, pelvics abdominal, and fins without spines (Fig. 1). Most "higher" or "advanced" teleosts, such as *Apogon*, have the apomorphic condition of pectorals set high on the side of the body, pelvics thoracic or jugular, and dorsal-, anal-, and pelvic-fin spines present. The character of relative placement of the pectoral and pelvic fins is bimodal, however, in both lower and higher teleosts. Those lower teleosts, such as *Pantodon*, or higher teleosts, such as *Sphyraena*, that have the alternate condition (Fig. 1) are interpreted as secondarily derived.

Rosen's (1982, fig. 10) plot of pectoral and pelvic character states on a teleost cladogram illustrated the problem, apparent to Greenwood *et al.* (1966), Gosline (1971), and others, that there is no clear or abrupt transition from lower or "segmented-rayed" to higher or "spiny-finned" teleosts based on these fin characters. Atherinomorph taxa (atheriniforms of Rosen, 1982, fig. 10), in particular, display plesiomorphic and apomorphic states, precluding their classification as lower or higher teleosts using these fin characters (Gosline, 1971, fig. 28B). Resolution of pectoral- and pelvic-fin character homology is critical for testing monophyly and phylogenetic position of atherinomorph fishes comprising the orders Atheriniformes, Cyprinodontiformes, and Beloniformes (following Nelson, 1994). Parenti (1993) hypothesized that some atherinomorphs share derived characters with some paracanthopterygian fishes, including absence (inferred loss) of a ligamentous connection between the pelvic bones and the ventral postcleithrum. She concluded that a sister-group relationship between taxa in those two large groups warranted further investigation. In contrast to Parenti (1993, and also Rosen and Parenti, 1981), Stiassny (1990, 1993) postulated that atherinomorphs are most closely related to the mugiliform fishes.

Johnson and Patterson (1993) proposed a new higher taxon (Smegmamorpha) that includes atherinomorphs and mugiliforms, and other taxa but not paracanthopterygians. They diagnosed their Acanthopterygii, comprising Stephanoberyciformes, Zeiformes, Beryciformes, and Percomorpha (including

Smegmamorpha), by three pelvic-fin characters: presence of a spine, free radials reduced in size and/or number, and presence of an anteromedial process of the pelvic bone. These three character states are present in the atherinomorph order Atheriniformes but absent or variably present in the two other atherinomorph orders, Cyprinodontiformes and Beloniformes. To accept that atherinomorphs are percomorphs means to accept reversal or loss in Cyprinodontiformes and Beloniformes of the acanthopterygian pelvic characters of Johnson and Patterson (1993).

Recent cladistic proposals of acanthopterygian phylogeny have presented the most parsimonious interpretation of character distributions, but they have not identified unambiguous derived characters of pelvic-fin morphology or of the pectoral–pelvic association. Most notably, Stiassny and Moore (1992) provided a phylogenetic analysis of eight pelvic-fin characters of 13 representative acanthomorph genera, including the Madagascan atheriniform *Rheocles* as a representative atherinomorph and *Neoscopelus* as an outgroup. Their most highly-resolved phylogenetic reconstruction was a majority-rule consensus of 154 minimum length trees; *Rheocles* was included in a subgroup of acanthomorphs from which *Polymixia* and *Aphredoderus* (a representative paracanthopterygian) were excluded, although more conclusive statements about the phylogenetic position of *Rheocles*, hence atherinomorphs, were not supported by the pelvic-girdle characters surveyed. Fin characters have been troublesome in these phylogenetic studies because of their often ambiguous or incomplete description and the absence of assessment of character homology. Here, we investigate whether new evidence from neuroanatomy can help solve this systematic problem.

II. Nerves as Systematic Characters

Characters used traditionally in phylogenetic studies of fishes, including those cited above, are principally of external morphology, the skeletal system, bones and cartilage, and, to a lesser degree, striated muscles (see Winterbottom, 1974). Few systematic studies of vertebrates incorporate broad, comparative anatomical surveys of data from other characters such as nerves, reproductive anatomy, or hormonal systems (e.g., Freihofer, 1978; Jamieson, 1991).

Nerve patterns in the atherinomorph silverside genus *Menidia* were described in detail by Herrick (1899) nearly a century ago. Despite the apparent complexity and richness of this character system, nerve patterns have not been studied regularly and readily by systematic ichthyologists. Complexity of patterns and lack of an easy method to study them have precluded development of a comparative framework for interpretation of nerve patterns across broad taxonomic boundaries.

Freihofer's (1963, 1970, 1972, 1978) description of select portions of the nervous system that he used to interpret teleost phylogeny provided a limited, yet highly valuable, modern glimpse into phylogenetic patterns of neuroanatomical characters. He concentrated on patterns of the RLA (ramus lateralis accessorius) which consists of a sensory branch of the facial (VII) cranial nerve, the posterior lateral line nerve, and sensory branches of the vagal (X) cranial nerve. Unfortunately, his description of mixed, rather than individual, components from different cranial nerves prevented phylogenetic interpretation or homology resolution of his nerve patterns by subsequent workers.

During the past 20 years, cell, molecular, and developmental biologists have increased greatly our knowledge and appreciation of the formation and stability of nerve patterns. For example, during nerve system development, neural axons are guided by a particular protein family to their correct targets (e.g., a muscle group) and are repelled by another protein family diffused from an incorrect target (see review in Marx, 1995). Thus, a one-to-one relationship between a neuron and its target is established early in ontogeny. In addition to the individual neuron–target relationship, motor neurons innervate muscles only in a particular body segment that is designed by certain homeobox gene families (Hatta *et al.*, 1990; Holland *et al.*, 1991; Noden, 1991). Perhaps of more interest, during morphogenesis, temporal and spatial sequences of the nerve branching pattern are established by following differentiation of the musculature that they supply (Moody *et al.*, 1989). This implies that the hierarchical pattern of nerve branching may be used to infer the hierarchical pattern of muscle differentiation within a segment; in other words, tracing the nerve branch pattern in adults should reveal the ontogenetic origin of the muscles and provide a basis for interpreting muscle homology (Song, 1990; Song and Boord, 1993). Applying this principal, Song and Boord (1993) found that the peripheral branching pattern of the mandibular ramus of the trigeminal (V) cranial nerve and the organization of the trigeminal motor column in the brain are highly conserved in craniate phylogeny regardless of various modifications in mandibular muscle attachment and structure. Proximal, intermediate, and distal series of the mandibular nerve branches supply three major muscle groups and are in register with three neuronal populations of the trigeminal motor column. The adult branching pattern is established in response to mandibular muscle differentiation. Innervation of muscles of the mandibular segment of the head and location of motoneurons reflect their segmental origins and are reliable criteria for homologizing mandibular muscles among craniates.

Following the same principle, if the spinal-nerve innervation of fin muscles and body segments in later ontogenetic stages reflects their segmental origins, then innervation should help us homologize teleost pectoral- and pelvic-fin structures by identifying segmental order and/or the number of segments involved in fin formation. This is particularly relevant for pelvic-fin homology for, as Winterbottom (1974, p. 226) stated, "as the pelvic fin ... migrates through segments, it is composed of the mesenchymal elements of those segments. ... In the naming of the muscles to the pelvics ... morphologically similar muscles receive the same name regardless of whether the pelvic fins are jugular, thoracic or abdominal in position." Innervation of pelvic-fin muscles, then, should be a better indication of homology than fin position.

Peripheral innervation patterns are used to homologize target structures of nerves, as has been done for parts of the lateral line system and dermal bones (Song, 1989, 1993; Song and Northcutt, 1991a,b). We reassess the phylogenetic significance of the well-studied morphological and functional association between the pectoral and pelvic fins in acanthomorph fishes. Our goal is to generate interest in the nervous system in systematic ichthyology by demonstrating how to combine comparative anatomy of nerves with osteology and myology and, therefore, providing evidence to address additional questions of teleost phylogeny and character homology.

III. Notes on Materials and Methods

A. Materials

All material was triple-stained with Alizarin Red S, Alcian Blue, and Sudan Black B for demonstration of bone, cartilage, and nerves according to the method of Song and Parenti (1995), unless otherwise indicated, and is listed in Appendix 1. Institutional abbreviations are as in Leviton *et al.* (1985). The terms Paracanthoptergyii, Atherinomorpha, and Percomorpha are used in the sense of G. Nelson (1989, fig. 1); otherwise, higher classification follows J. S. Nelson (1994), with some modifications by Johnson and Patterson (1993). Monophyly of all higher taxa is not implied.

B. Phylogenetic Systematics

It is common in phylogenetic systematics to hypothesize relationships among higher taxa using the cladistically primitive member of each taxon to represent or to estimate ancestral character states of that taxon. For example, since Rosen and Parenti (1981) suggested that Old World silverside fishes such as the Madagascan *Bedotia* and *Rheocles* or the Australian–New Guinean *Melanotaenia* are relatively plesiomorphic taxa in a monophyletic Atherinomorpha, one of these Old World genera has been used to estimate the relationships of atherinomorphs to other acanthomorphs (e.g., Stiassny and Moore, 1992; Johnson and Patterson, 1993). Dyer's (1993) hypothesis that the New World Atherinopsidae is the most primitive taxon in a monophyletic Atheriniformes notwithstanding, we acknowledge that it is more fruitful for phylogenetic analysis of a higher taxon such as the Atherinomorpha to examine taxa in each of its three subgroups (atheriniforms, cyprinodontiforms, and beloniforms) rather than to examine a single, representative atherinomorph taxon for two reasons: (1) to gain a better representation of the range of diversity within atherinomorphs and (2) to gain an opportunity to test the monophyly of atherinomorphs in each analysis. For example, a phylogenetic analysis that includes an array of atherinomorphs, such as the silverside *Melanotaenia*, the killifish *Fundulus*, and the halfbeak *Hemirhamphodon*, that contradicted a monophyletic Atherinomorpha, on the other hand, might question the utility of that analysis in assessing relationships of other higher taxa. Therefore, we have surveyed an array of taxa in those higher groups in which we have a particular interest: Atherinomorpha, Paracanthopterygii, Percomorpha (*sensu* G. Nelson, 1989), and Smegmamorpha.

C. Innervation Patterns

1. Tracing Innervation Patterns

The peripheral innervation patterns of the pectoral- and pelvic-fin musculatures were examined by dissecting triple-stained specimens observed through a Zeiss zoom dissecting microscope with transmitted light. Most specimens were cut into two sagittal sections along a plane to one side of the vertebral column. To identify a particular spinal nerve course and to trace an individual nerve branch from the vertebra through which it exits to its innervation targets, it is necessary to dissect specimens at specific nerve conjunction points. At those points, using fine jeweller's forceps, the tip of a small scalpel blade, and microsurgical scissors, we separated nerve branches, or muscles, from each other, following the membranes (or myosepta) that were visible even though the muscle fibers became transparent during trypsin digestion. We followed the ventral ramus of the spinal nerve from the vertebral foramen from which it emerged to its branching point. Then we followed its branches, one-by-one, to their target muscle (or skin region) along the myoseptum between individual fin muscles or along the bone membrane and through foramina. While tracing individual nerve branch courses, specimens were kept in a 30 or 70% glycerin solution (with a 0.5% KOH base). Results were recorded by freehand or camera-lucida facilitated drawings and photomicrographs.

2. Terminology of Innervation Patterns

The pectoral–pelvic fin muscles and skin are innervated by spinal nerves. Because our goal was to identify the pattern of innervation of the pectoral- and pelvic-fin muscles (or skin) in a phylogenetic context, we first had to define a criterion to homologize the order or sequence of the spinal nerves.

Herrick's (1899) classic study of innervation in the silverside *Menidia* remains the only authority on teleost nerve patterns. He defined the first spinal nerve in *Menidia* as the nerve that emerges through the foramen on the occipital region of the cranium, mixed with "the spino-occipital nerve" (which consists of "the occipital nerve" and "the occipito-spinal nerve") and "the ramus of cervicalis nerve." In primitive neopterygians, such as the gar (*Lepisosteus platyrhinchus*), the roots of the first (or the hyobranchial and) three spinal nerves on the medulla region in the cranium are visible before they meet and emerge through a single foramen on the occipital region. In teleosts, in contrast, as Herrick (1899, p. 80) indicated, one potential problem is as follows: "Passing up the taxonomic series a progressively larger number of spinal segments became fused with the head and either wholly or partially degenerate." It is difficult, if not impossible, to determine in juvenile or adult teleosts how many spinal segments may have fused with the head ontogenetically and/or phylogenetically without detailed developmental study of each taxon or tracing of the peripheral branching pattern experimentally using neural tracers to reconstruct and analyze their segmental origin in the motor column and/or ganglionic organization from histological sections. Identification of a homologous first spinal nerve among neopterygian taxa is thus arduous. In order to conduct a phylogenetic analysis using preserved material in this study, we established a protocol to address this prob-

lem. We numbered spinal nerves based upon the order of the vertebra from which each exited instead of using Herrick's (1899) terminology. We defined the first spinal nerve (S1) as that one emerging from the first free vertebra, the second (S2) as that one emerging from the second vertebra, and so on. The first free vertebra and its spinal nerve foramen, as well as following vertebrae and their foramina, were identified in all triple-stained specimens. There is only one spinal nerve foramen on each free vertebra in teleosts, as far as we know, and the order and sequence of the spinal nerves are congruent with the order and sequence of the vertebrae through which they emerge. We called the group of nerves that emerge from the occipital foramen the spino-occipital nerve (SO) and recognized homology of its components in different taxa by analyzing its branching and innervation patterns. Then we identified a set of characters to describe the homologous innervation of pectoral- and pelvic-fin muscles, relative position of the fins, and pelvic fin migration in a broad array of teleost taxa. Muscle terminology follows Winterbottom (1974).

Terms for the ventral ramus of the spinal nerve used in this study are as follows:

SO Spino-occipital nerve (= "the first spinal nerve branch b+c and a part of the second spinal nerve" of Herrick, 1899): that group of nerves emerging through the foramen on the occipital region of the cranium. It branches into three main rami at the pectoral fin in most teleost taxa as follows:

SO_1 Ramus 1 of the SO (= "ramus cervicalis" of Herrick, 1899): innervates muscles of the anteroventral portion of the pectoral girdle, such as the sternohyoideus and pharyngoclavicularis, and skin of the ventral and lateral regions of the cleithrum in some teleosts.

SO_2 Ramus 2 of the SO: innervates the dorsomedial portion of the medial region of the pectoral fin, such as the region of the adductor superficialis muscle in some teleosts.

SO_3 Ramus 3 of the SO: emerges through the coracoid foramen and innervates the medial (and anteroventral) portion of the lateral region of the pectoral fin, such as the region of the abductor superficialis muscle in some teleosts.

SO' Spino-occipital "prime" nerve: emerges through the same foramen on the occipital region of the cranium just caudal to SO and is identified only in some teleosts; innervates the dorsomedial portion of the medial region of the pectoral fin, such as the region of the adductor profundus muscle, the region that is innervated by S1 in teleosts that lack SO'.

S1   The spinal nerve that emerges through the foramen on the first free vertebra (= "the second spinal nerve" of Herrick, 1899): innervates the dorsomedial portion of the medial region of the pectoral fin, such as the adductor profundus region in some teleosts; or branches dorsomedially and ventrolaterally to innervate the arrector dorsalis and the arrector ventralis muscles and the region parallel to the pectoral-fin rays, target areas that are innervated by S2 in some other teleosts.

S2   The spinal nerve that emerges through the foramen on the second free vertebra (= "the third spinal nerve" of Herrick, 1899) and divides into two or three branches: innervates the arrector dorsalis and the arrector ventralis muscles, the skin of the middle and ventral portion of the pectoral fin, and the ventral musculature and skin lying immediately posterior to the pectoral girdle; in some teleosts it innervates the anteroventral margin of the pelvic-fin muscle or the abdominal wall in those teleosts with a recognizable SO'.

S3   The spinal nerve that emerges through the foramen on the third free vertebra.

S4   The spinal nerve that emerges through the foramen on the fourth free vertebra.

This definition of spinal nerves allows us to address questions of serial homology of body segments by comparing the pattern of the nerve group that emerges from the cranium (SO and SO') and the spinal nerves that emerge from the anterior vertebrae (e.g., S1, S2, and S3) to the fin muscle groups in different teleost taxa. We expect to identify a segmentally homologous nerve-to-muscle group pattern. This pattern should reflect ontogenetic constraints during segmentation. Segmentally homologous nerves may be recognized by the same sequence and the same innervation but with a different initial nerve emerging from the spinal foramina in different neopterygian taxa, such as SO', S1, and S2 in some taxa and S1, S2, and S3 in others. Recognition of a pattern of these shifts in innervation may in the future allow us to address the question of ontogenetic and/or phylogenetic significance of proposed fusion of spinal segments to the head in neopterygians. This does not preclude our study, especially as we have no evidence to indicate that the first vertebra in acanthomorphs is not homologous, but instead reinforces the need for logical proposals of homologous nerves as a starting point.

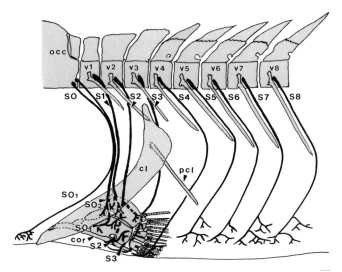

FIGURE 2   Esox americanus, USNM 237235; diagrammatic illustration of innervation pattern of pectoral fin by ventral components of spinal nerves; bone stippled, nerves black. Abbreviations: occ, occipital region of skull; v1 through v8, vertebrae 1 through 8; SO, spino-occipital nerve; SO_1, SO_2, and SO_3, branches of SO; S1 through S8, ventral components of spinal nerves exiting vertebrae 1 through 8; cor, coracoid; cl, cleithrum; pcl, postcleithrum.

The number of branches of each spinal nerve may vary from one fish taxon to another. For example, S2 may have three branches in one taxon and just one or two in other taxa. This variation is correlated with differences in the number or volume of individual muscles and morphology of the fins. Analyzing the variation in branching patterns of homologous spinal nerves in different taxa provides information on the functional and behavioral significance of fin characters in teleosts.

IV. Results

Esox (Fig. 1) has the plesiomorphic pectoral and pelvic fin association. Pectoral fins are set low on the body, distinctly separate from abdominal pelvic fins. Pectoral-fin muscles (Fig. 2) are innervated by spino-occipital nerves exiting the occipital region of the skull (SO), by the first and second spinal nerves (S1 and S2), and, as in some other plesiomorphic taxa, by the third and the fourth spinal nerves (S3 and S4) as well. Postcleithra lie in the body wall musculature and have no independent muscle innervation and hence no independent movement. Pelvic-fin muscles are innervated by motor components of spinal nerves from the muscle segments in which the fin lies or from nearby segments; for *Esox*, spinal nerves 11 through 17 (S11–S17) are involved.

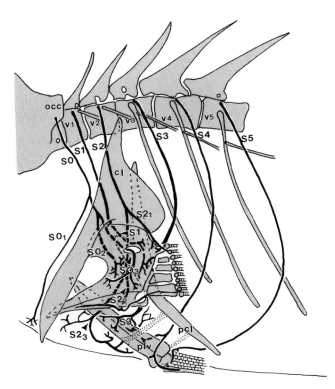

FIGURE 3 *Apogon maculatus*, USNM 306725; diagrammatic illustration of innervation pattern of pectoral and pelvic fins by ventral components of spinal nerves; bones (except pelvic rays) finely stippled, ligaments coarsely stippled; nerves black. Abbreviations: occ, occipital region of skull; v1 through v5, vertebrae 1 through 5; SO, spino-occipital nerve; SO_1, SO_2, and SO_3, branches of SO; S1 through S5, ventral components of spinal nerves exiting vertebrae 1 through 5; $S2_1$, $S2_2$, and $S2_3$, branches of S2; cor, coracoid; cl, cleithrum; pcl, postcleithrum; plv, pelvic bones.

Apogon (Figs. 1 and 3) has the apomorphic pectoral and pelvic fin association. Pelvic fins lie ventral to the pectoral fins. Pectoral-fin muscles are innervated by motor components of SO, S1, and S2, as they are in *Esox* and all other teleosts. Yet, branches of the motor components of S2 innervate not just the pectoral-fin muscles (through branches $S2_1$ and $S2_2$) but also pelvic-fin muscles (via $S2_3$). We interpret the phylogenetic significance of this character in the Discussion.

From initial investigation of a relatively plesiomorphic and a relatively apomorphic taxon representing the extremes of our phylogenetic transition series (Fig. 1), we observed that the pectoral innervation pattern is relatively stable; in the examination of all our material, we found no exception to the innervation pattern of the pectoral fin, which comprises minimally the motor components of SO, S1, and S2. The pelvic-fin muscle innervation pattern is variable, correlated with phylogenetic position, and may or may not reflect topographic position of the pelvic fins in adults.

Therefore, variability in pelvic-fin muscle innervation became the focus of our survey.

Characters recorded for 30 taxa are illustrated diagrammatically using *Pantodon* (Fig. 4) as an example: precaudal vertebrae (14), motor components of the spinal nerves that innervate pelvic-fin muscles (spinal nerves S3 through S5), and relative size and position of the pelvic girdle recorded as the vertebrae below which the girdle lies (vertebrae 4 through 6). The number of spinal motor nerves that innervate pelvic-fin muscles may be used to approximate the relative size of the pelvic girdles; the higher the number of nerves, the longer the pelvic girdle. These data are diagramed in Figs. 5 though 9.

Position of the pelvic fins relative to the vertebral column, recorded as the vertebra at the anterior extent of the pelvic bones, may be used to describe the ultimate position of pelvic fins in adults. Identification of nerves innervating pelvic-fin muscles and number of precaudal and caudal vertebrae are given for each of the 30 taxa in Table 1.

The osteoglossomorph *Pantodon*, the esociform *Esox*, and the ostariophysan *Cyclocheilichthys* are outgroup taxa for Acanthomorpha (Fig. 5). Pelvic-fin muscles are innervated by the motor component of S3 and/or one or more posterior spinal nerves in these genera. Position of pelvics may be relatively anterior, as in *Pantodon*; posterior, as in *Esox*; or midbody, as in *Cyclocheilichthys* (Fig. 5). Acanthomorph taxa, beginning with *Polymixia*, are listed in approximate phylogenetic order in Figs. 5 through 9. Pelvics may lie under the skull, as in *Porichthys*, which we recorded as opposite vertebra 0, or more posterior, as in *Parazen*, in which the pelvic bones lie opposite vertebrae 5 through 10 (Fig. 6). Despite variation in the ultimate position of the pelvic fins in adults, innervation of pelvic-fin muscles in a group of primitive acanthomorphs generally begins with spinal nerve S3 or S4, as in *Polymixia*, *Aphredoderus*, *Microgadus*, *Porichthys*, *Stephanoberyx*, and *Parazen*, or S5, as in *Percopsis* (Figs. 5 and 6).

Branches of the motor components of S2 innervate the pectoral-fin muscles (through branches $S2_1$ and $S2_2$) and also pelvic-fin muscles (via $S2_3$) without variation in a group of percomorphs (*sensu* G. Nelson, 1989; Figs. 6 and 8, in part, Fig. 9): *Holocentrus*, *Perca*, *Apogon*, *Elassoma*, *Agonostomus*, *Sphyraena*, *Bathygobius*, *Callogobius*, *Gunnellichthys*, *Sicyopterus*, *Sicyogaster*, and *Gobiesox*.

Pectoral-fin muscle innervation in atherinomorphs, illustrated for the silverside *Leuresthes* (Fig. 10), is as in other teleosts; branches of SO, S1 and S2 innervate pectoral-fin muscles. Pelvic-fin muscles are innervated by motor components of S3 in *Melanotaenia* (Fig.

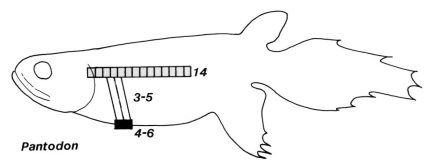

FIGURE 4 *Pantodon;* sketch illustrates number of precaudal vertebrae (14), ventral components of spinal nerves innervating pelvic-fin muscles (3 through 5), and relative size and position of pelvic bones (opposite vertebrae 4 through 6). These three characters are recorded for 30 taxa listed in approximate phylogenetic order in Figs. 5 through 9.

6) or S5 in *Bedotia* (Fig. 7), both atheriniforms, as in primitive acanthomorphs. In all other atherinomorphs surveyed, pelvic-fin muscles are innervated by motor components of more posterior spinal nerves, S6 or posterior (Fig. 7). Atheriniform and beloniform atherinomorphs also have a relatively high number of precaudal vertebrae (19 or more), which is higher than that in other primitive acanthomorphs examined, although a high number of precaudal vertebrae is not limited to atherinomorphs among acanthomorphs.

Posterior innervation of pelvic-fin muscles is characteristic for most atherinomorphs, even in those taxa in which pelvics are "thoracic" or "jugular" in adults, such as in the cyprinodontiform poeciliid *Tomeurus* (Fig. 11). In adult males of that genus, pelvics lie opposite vertebrae 1 and 2; in adult females, opposite vertebrae 4 and 5. In both sexes, pelvic-fin muscles are innervated by motor components of posterior nerves (here S6 and S7) as in other atherinomorphs.

Gobiesocids, or clingfishes, represented by *Sicyogaster* and *Gobiesox* (Fig. 9), have the complex innervation pattern: the motor component of spinal nerve S2 branches to innervate both the pectoral- and pelvic-fin muscles. Clingfishes have a complex, ventral sucking disc that is divided into an anterior and a posterior portion. The unambiguous pelvic-fin rays are anterior to the pectoral fin in adults (Fig. 12). Motor components of spinal nerves S2 and S3 that innervate the anterior pelvic-fin muscles in *Sicyogaster* and *Gobiesox* cross anterior to those sections of S1 and S2 that innervate the pectoral-fin muscles. Motor components of S4 and S5 innervate muscles that control movement of two ossified plates on each side (Fig. 12), one posteromedial and one posterolateral pair. In some gobiesocids, the posteromedial pair has what may be rudimentary fin rays; the posterolateral pair has a "fringed" posterior margin that does not appear to be composed entirely of fin rays (Fig. 12). These plates have long been interpreted as postcleithra; we use innervation patterns to reinterpret them as pelvic elements in the Discussion.

V. Discussion

Motor components of spinal nerves innervate pectoral- and pelvic-fin muscles in identifiable patterns that show promise for inferring phylogenetic relationships among teleost taxa and for reassessing character homology. Pectoral-fin muscles are innervated by motor components of spinal nerves SO, S1, and S2 in all teleosts surveyed. Innervation of the pelvic-fin muscles by motor components of spinal nerve 3 and/or one or more posterior spinal nerves is interpreted as the plesiomorphic condition in teleosts. Ultimate position of the pelvic fins in adults may be relatively anterior in these relatively plesiomorphic taxa, as in *Pantodon* or *Tomeurus*, or posterior, as in *Esox* or *Hemirhamphodon*; nonetheless, pelvic innervation is never anterior to S3.

Identification of the first spinal nerve that innervates pelvic-fin muscles is one way to approximate the original body segments of the pelvic fins. The course of the nerve branch, from the vertebral foramen to the pelvic-fin muscles, follows the topographic shift of the fin during ontogeny. Goodrich (1930) used the posterior innervation of the relatively anterior pelvic fins of gadid fishes to infer the anterior migration of pelvics during ontogeny. In the cyprinodontiform *Tomeurus*, pelvic fins are anterior in adult males and females. In both sexes, pelvic-fin muscles are inner-

TABLE 1 Spinal Nerve Innervation of Pelvic Fins

Taxon	2	3	4	5	6	7	8	9	10	11	12	13	14	15	16	17	18	Vertebrae
Pantodon		x	x	x														14+12
Esox										x	x	x	x	x	x	x		34+15
Cyclocheilichthys					x	x	x	x										16+14
Polymixia			x	x	x													13+17
Percopsis				x	x	x	x											17+18
Aphredoderus			x	x	x													13+15
Microgadus		x	x	x														18+35
Porichthys		x	x	x														11+30
Stephanoberyx			x	x	x													10+20
Parazen			x	x	x	x												12+22
Holocentrus	x	x	x	x														11+16
Leuresthes								x	x	x	x	x	x	x?				32+19
Odontesthes								x	x	x	x	x						21+28
Bedotia				x	x	x												19+18
Melanotaenia		x	x	x														20+14
Fundulus						x	x	x	x									14+19
Tomeurus ♂					x	x												16+21
Tomeurus ♀					x	x												13+25
Hemirhamphodon														x	x			25+14
Gasterosteus			x	x	x	x?												15+16
Perca	x	x	x	x	x	x												20+20
Apogon	x	x	x	x														10+15
Elassoma	x	x	x	x	x	x												12+17
Agonostomus	x	x	x															11+13
Sphyraena	x	x	x															11+13
Bathygobius	x	x	x															10+15
Callogobius	x	x	x															11+16
Gunnellichthys	x	x																26+32
Sicyopterus	x	x	x															10+16
Sicyogaster	x	x	x	x														14+19
Gobiesox	x	x	x	x														12+14

vated by S6 and S7; we infer that nerves exit the vertebrae and course anteriorly during ontogeny due to differential growth. Position of pelvic fins relative to the vertebral column to describe ultimate position of pelvic fins in adults is less ambiguous than terms such as abdominal, thoracic, or jugular. Ontogeny indicated by the innervation pattern allows use of the pelvic-fin position as a character in phylogenetic analysis; without this information, for example, "jugular" pelvics in two teleost taxa might erroneously be considered homologous.

We diagnose a large group of percomorph fishes by a complex character that we call "anchoring" of the pelvics to the pectoral fins (Table 2). Motor components of SO, S1, and S2 innervate pectoral-fin muscles; S2 branches to innervate pelvic- as well as pectoral-fin muscles. The pelvic fins may be anterior and closely associated with the pectoral fins, as in *Perca*, or relatively dissociated from the pectoral fins, as in *Sphyraena*, but are never in a posterior position. Sharing motor components of a single spinal nerve may facilitate coordinated movement of the pectoral and pelvic

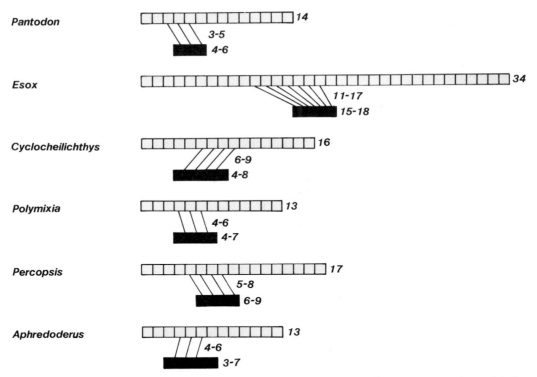

FIGURE 5 Diagram of precaudal vertebrae, ventral components of spinal nerves innervating pelvic-fin muscles, and relative size and position of pelvic bones recorded as range of vertebrae opposite which pelvic bones lie. See Fig. 4.

fins. Goodrich (1909) observed that in some teleosts with anterior pelvics, innervation of the pelvic fins may be shared with that of the pectoral fins, although this condition has been characterized as rare (see Winterbottom, 1974, p. 226).

Beryciformes *sensu stricto*, represented here by *Holocentrus*, were interpreted by Johnson and Patterson (1993, fig. 24) as primitive to their expanded Percomorpha, including Atherinomorpha. We place *Holocentrus* in that group of percomorph fishes that includes *Perca, Apogon, Elassoma, Agonostomus, Sphyraena*, gobioids, and gobiesocids (but excludes atherinomorphs) diagnosed by the apomorphic innervation pattern (Table 2). This innervation pattern is uncontradicted by the 12 representatives of that large group of percomorph fishes that we examined. These taxa are also diagnosed by medially sutured pelvic bones (Stiassny and Moore, 1992). Further study is needed to determine whether other Beryciformes of Johnson and Patterson (1993) that do not have medially sutured pelvic bones share the apomorphic innervation pattern of *Holocentrus* or the plesiomorphic pattern of stephanoberyciforms. Our data support the separation in a classification of *Holocentrus* from Stephanoberyciformes and Zeiformes as suggested

by Johnson and Patterson (1993) and, in part, by Moore (1993).

Monophyly of Johnson and Patterson's (1993) Smegmamorpha is contradicted by distribution of the apomorphic innervation pattern of pelvic-fin muscles (Table 2). Smegmamorphs *Elassoma* and *Agonostomus* have the apomorphic pattern, whereas atherinomorphs and *Gasterosteus* have the plesiomorphic pattern. *Gasterosteus* is not precluded from a close phylogenetic relationship with atherinomorphs. This character survey, however, is not meant as a rigorous test of smegmamorph monophyly as the remaining smegmamorph taxa, synbranchoids and mastacembeloids, lack pelvic fins.

Innervation patterns may be used to test or reinterpret character correlations. Pelvic-fin muscle innervation is inferred to have shifted within acanthomorphs from the plesiomorphic (S3, S4, or S5) to a more posterior innervation (S6 or posterior) in all atherinomorphs surveyed except *Melanotaenia* and *Bedotia*, atheriniforms which have the plesiomorphic condition. Atheriniform monophyly is, therefore, contradicted. Pelvic innervation is posterior even in atherinomorphs in which the pelvics are anterior in adults. Posterior pelvic innervation is not necessarily correlated with other

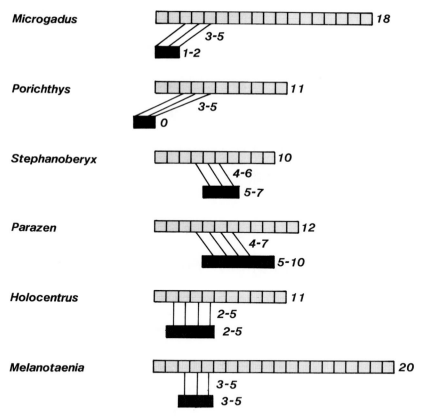

FIGURE 6 Diagram of characters as in Figs. 4 and 5. In *Porichthys*, pelvic bones lie under the skull.

derived pelvic characters that have been used to place atherinomorphs among acanthomorphs, such as the I,5 pelvic-fin ray formula. A pelvic spine and five segmented rays characterize atheriniforms (here *Melanotaenia, Leuresthes, Odontesthes,* and *Bedotia*). Cyprinodontiforms (*Fundulus* and *Tomeurus*) and beloniforms (*Hemirhamphodon*) have segmented-rayed pelvics with as many as eight rays (Parenti, 1993).

Atherinomorph monophyly is well-supported (Parenti, 1993). Posterior pelvic innervation in atherinomorphs may be correlated, in part, with the phylogenetic addition of precaudal vertebrae. An increase in precaudal vertebral number (19 or more) may be an additional diagnostic character of atherinomorphs that is modifiied in cyprinodontiforms (Fig. 7). Cyprinodontiforms that we examined (*Fundulus* and male and female *Tomeurus*; Table 1) have what we identified initially as a relatively low number (16 or fewer) of precaudal vertebrae (without hemal arches) and a relatively high number of caudal vertebrae (with hemal arches). The total number of vertebrae in these taxa (33–38) is nearly within the range of the other atherinomorphs (34–51) examined, however. Rosa-Molinar

et al. (1994) recently identified the "genital region" of the vertebral column in the cyprinodontiform poeciliid *Gambusia;* six vertebrae, 11 through 16, are markedly different from all other vertebrae. In both sexes, vertebrae 11 through 13 form hemal arches early in development, resorb them, and then form parapophyses that bear pleural ribs; vertebrae 14 through 16 bear hemal arches that increase in size during growth in males only. Thus, vertebrae may transform ontogenetically from what we identify as caudal to precaudal. Rosa-Molinar *et al.* (1994) proposed that this transformation is regulated by differential expression of control genes, such as homeobox genes. If such a transformation is characteristic of some group of cyprinodontiform fishes that includes the poeciliids *Gambusia, Tomeurus,* and the fundulid *Fundulus,* then our proposal that addition of precaudal vertebrae is diagnostic of atherinomorphs is not contradicted. Determining homology of individual vertebrae and identifying where precaudal vertebrae might have been added cannot be accomplished as part of this study and, we suspect, may require identification of genes that control duplication of segments. We would not

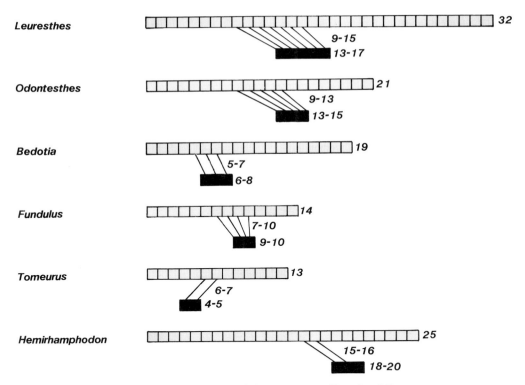

FIGURE 7 Diagram of characters as in Figs. 4 and 5.

have suspected that segments were added (or duplicated) in a taxon such as *Tomeurus*, for example, without the combination of information from ontogeny and innervation patterns, however.

For us, perhaps the greatest promise that innervation patterns hold is in the reinterpretation or clarification of homology. Innervation patterns of the pectoral- and pelvic-fin muscles suggest that homology of elements of the gobiesocid ventral sucking disc has been misinterpreted for over a century. Bones that support the posterior portion of the disc have been interpreted as postcleithra because in size, shape, position, and number they resemble postcleithra. Differential growth, however, may result in a bone in an adult lying in a position that confounds attempts to resolve homology. For example, a bone lying medial and ventral to the pectoral girdle in the atherinomorph phallostethid *Gulaphallus* was identified as a pelvic-fin spine that, through differential growth, had acquired the shape and position of a postlcleithrum (Parenti, 1986).

Günther (1861, p. 493) questioned previous interpretations of identification of the bones that support the gobiesocid sucking disc:

The pubic bones are united by suture, and form together a heart-shaped disc, the point of which is produced backwards. The

anterior portion of the disc is concave, with a bony longitudinal bridge and a feeble transverse ridge. The disc is fixed to the humeral bones by the convex portions of its anterior margin whilst the convex portions of the lateral margins serve as base for the ventral fins.* [referring to a footnote that reads] The structure of the ventral disc has not yet been correctly described. Stannius (1846) for instance (p. 91) denies the presence of the coracoid. The ventral fins have been taken for a detached portion of the pectorals, etc.

Guitel (1888) introduced his detailed anatomical study of gobiesocids with a recitation of various published opinions on the anatomy of the sucking disc. The majority of authors he cited concluded that the anterior portion of the disc was formed by the pectoral fins and the posterior portion by fusion of the pelvic-fin bones (Guitel, 1888, p. 12). Apparently, this was not interpreted as a statement on homology of the posteriormost, platelike bones that support the posterior portion of the disc because Guitel (1888) identified those as postcleithral elements, the "coracoïdien antérieur" and "coracoïdien postérieur." Subsequent studies followed Guitel (1888); for example, these bones were called post clavicles by Starks (1905), proximal and distal post cleithra by Briggs (1955), and dorsal and ventral postcleithra by Springer and Fraser (1976, fig. 9).

Reconsideration of homology of ventral sucking disc elements was discouraged possibly because the

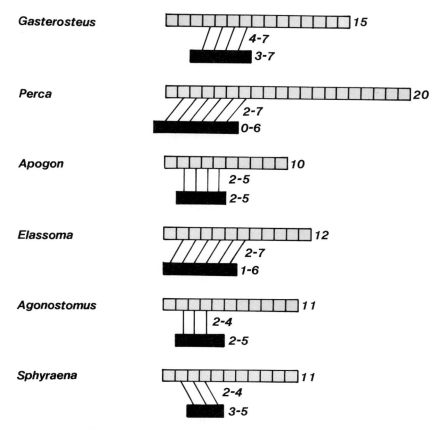

FIGURE 8 Diagram of characters as in Figs. 4 and 5.

disc is formed early in ontogeny; larval specimens of about 12 mm TL give no clues as to homology of the bones of the disc. Osteology of the disc of a 12-mm-TL specimen is nearly identical with that of an adult (Runyan, 1961, figs. 24b and 25b, respectively). Further, from our examination of larval specimens of the same size, we are unable to determine which of the bones is preformed in cartilage and which is formed of dermal bone and hence has no cartilaginous precursor. Smaller specimens may allow determination of homology of the bones. If innervation patterns are stable, however, they should "record" ontogenetic transposition of elements.

Presence of what we identify as rudimentary fin rays on the posteromedial bone in the posterior portion of the sucking disc of *Gobiesox* first led us to reconsider homology of these bones using triple-stained specimens. We interpret the bones as pelvic elements for at least three reasons. First, muscles that control movement of these bones are innervated by motor components of spinal nerves 4 and 5. In our understanding, this makes them pelvic bones, not pectoral bones. Second, and perhaps more important, if these are pectoral bones, it is unlikely that they are postclei-

thra because in most teleosts postcleithra lie in the body musculature and do not move independently of the rest of the pectoral girdle. Third, morphology of these bones is reminiscent of the enlarged posterior pelvic processes and articular surfaces of the pelvic bones of other percomorph fishes (e.g., Stiassny and Moore, 1992, fig. 12a, *Holocentrus*). The unambiguous pelvic bones of clingfishes, those with which the pelvic spine and rays articulate, are attenuate posteriorly. When placed together with what we interpret as the remaining pelvic pieces they mirror generalized percomorph pelvic-fin morphology and innervation. That is, if we were to bring all three bones identified as pelvic bones and the motor components of the nerves that innervate them together (Fig. 12), we would have a recognizably percomorph innervation pattern as in *Apogon* (Fig. 3). Spinal nerves S2, S3, S4, and S5 innervate the pelvic-fin muscles in both, S2 and S3 innervate the anterior pelvic-fin muscles, and S4 and S5 innervate the posterior pelvic-fin muscles.

Gobiesocids have the apomorphic innervation pattern and are, therefore, allied with percomorphs in our scheme (Table 2). We are unable to resolve further

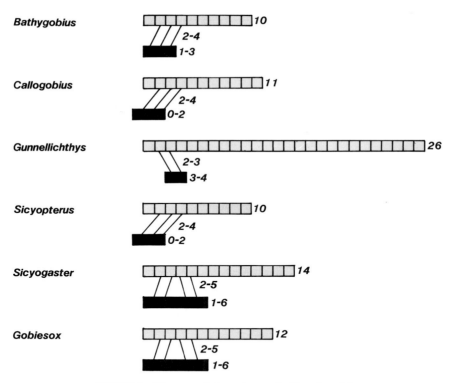

FIGURE 9   Diagram of characters as in Figs. 4 and 5.

TABLE 2   Thirty Taxa Surveyed and Listed by Plesiomorphic or Apomorphic Innervation Pattern

Plesiomorphic	Apomorphic
Outgroup taxa:	
Pantodon	
Cyclocheilichthys	
Esox	
Acanthomorphs:	
Polymixia	*Holocentrus*
Percopsis	*Perca*
Aphredoderus	*Apogon*
Microgadus	*Elassoma**
Porichthys	*Agonostomus**
	Sphyraena
Melanotaenia*	*Bathygobius*
Bedotia*	*Callogobius*
Leuresthes*	*Gunnellichthys*
Odontesthes*	*Sicyopterus*
Fundulus*	*Sicyogaster*
Tomeurus*	*Gobiesox*
Hemirhamphodon*	
Stephanoberyx	
Parazen	
*Gasterosteus**	

Note. Apomorphic pattern of "anchoring" is described in text. Smegmamorph taxa are starred; atherinomorph taxa are bold.

their proposed relationships with callionymoids or notothenioids viz. Gosline (1970). We note, however, that if the posteriorly attenuate anterior pelvic bones of gobiesocids are homologous with the posteriorly attenuate pelvic bones of the trachinioid (notothenioid of Gosline, 1970) *Cheimarrichthys*, these two taxa are more closely related to each other than either is to callionymoids, represented in our study by *Synchiropus*, which lack posteriorly attenuate pelvic bones.

VI. Summary

Identification of a segmentally homologous spinal nerve-to-pectoral- and pelvic-fin innervation pattern in teleosts that is constrained by ontogeny allowed us to reconsider the phylogenetic relationships among acanthomorphs. Transition from a relatively lower, or primitive, to a higher, or advanced, teleost may be recognized using pectoral- and pelvic-fin muscle innervation patterns. In primitive teleosts, here outgroup taxa and primitive acanthomorphs, pelvic-fin muscles are innervated by the motor component of S3 and/or one or more posterior spinal nerves. In advanced acanthomorphs, the pelvic fins are anchored to the pectoral fins; that is, motor components of SO, S1, and S2 innervate pectoral-fin muscles; S2 branches and also innervates pelvic-fin muscles. This innervation pattern is interpreted here as conservative

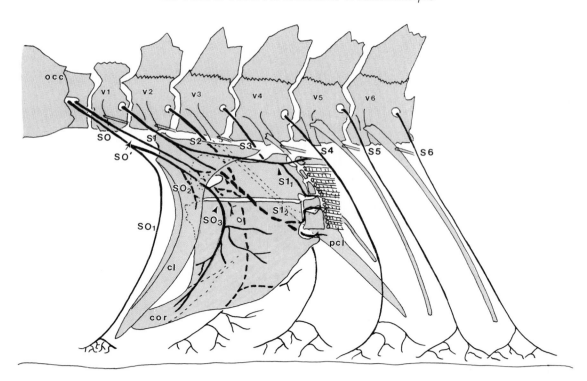

FIGURE 10 *Leuresthes sardina*, USNM 177811; diagrammatic illustration of innervation pattern of pectoral fin by ventral components of spinal nerves; bone stippled, nerves black. Abbreviations: occ, occipital region of skull; v1 through v6, vertebrae 1 through 6; SO, SO', spino-occipital nerves; SO_1, SO_2, and SO_3, branches of SO; S1 through S6, ventral components of spinal nerves exiting vertebrae 1 through 6; $S1_1$ and $S1_2$, branches of S1; cor, coracoid; cl, cleithrum; pcl, postcleithrum.

within taxa and, therefore, more useful for inferring phylogenetic relationships than are other morphological fin characters such as the I,5 pelvic-fin ray formula or relative position of the pectoral and pelvic fins.

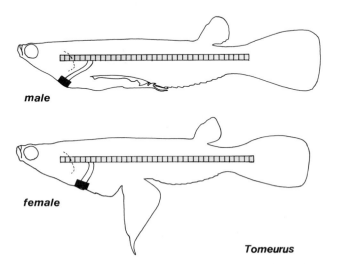

FIGURE 11 *Tomeurus gracilis*; sketch illustrates total vertebrae, ventral components of spinal nerves innervating pelvic-fin muscles (S6 and S7; see Fig. 7), and relative size and position of pelvic bones in adult males and females. Pelvic bones represented by black boxes. Outlines redrawn from Rosen and Bailey (1963, fig. 7).

Following this scheme, *Polymixia*, paracanthopterygians, atherinomorphs, gasterosteiforms, stephanoberyciforms, and zeiforms are primitive acanthomorphs, whereas remaining percomorphs are advanced acanthomorphs. Atherinomorph monophyly is supported by an apparent increase in the number of precaudal vertebrae and of one large subgroup (minus *Melanotaenia* and *Bedotia* here) by the posterior innervation of the pelvic-fin muscles. Although smegmamorph monophyly is not supported, a close relationship between atherinomorphs and gasterosteiforms, both proposed smegmamorph taxa, is not precluded by our findings.

We are optimistic about the utility of innervation patterns in further reinterpretation or clarification of homology within teleosts. Stability of innervation patterns, that is, maintenance of the one-to-one relationship between a neuron and its target, has contributed to reinterpretation of the bones supporting the posterior portion of the sucking disc of gobiesocids as pelvic-fin bones, not pectoral-fin bones. Nerve patterns provide a valuable test of homology and a source of new hypotheses of phylogeny, especially where other morphological characters have been transformed, through ontogeny, such that their identification is obscure.

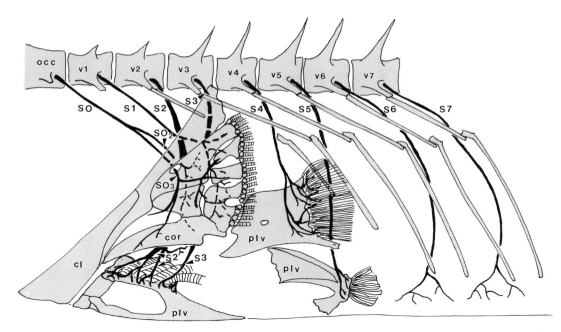

FIGURE 12 *Sicyogaster maeandricus,* USNM 82154; diagrammatic illustration of innervation pattern of pectoral and pelvic fins by ventral components of spinal nerves; bones (except pelvic rays) stippled, nerves black. Abbreviations: occ, occipital region of skull; v1 through v7, vertebrae 1 through 7; SO, spino-occipital nerve; SO_2 and SO_3, branches of SO; S1 through S7, ventral components of spinal nerves exiting vertebrae 1 through 7; cor, coracoid; cl, cleithrum; plv, pelvic bones. S2 and S3 innervate anterior portions, and S4 and S5 innervate posterior portions of pelvic musculature.

Acknowledgments

A. C. Gill (BMNH) started this project with us and contributed to our early discussions on the systematic significance of pelvic-fin characters. V. G. Springer (USNM) provided references and discussion, especially on gobiesocids. They, C. C. Baldwin (USNM), and G. D. Johnson (USNM) critiqued a draft of the manuscript. These comments and those of two reviewers are gratefully acknowledged. This study was supported, in part, by a Smithsonian Postdoctoral Fellowship to JS.

Loan of specimens and/or other information was provided by N. Feinberg and M. L. J. Stiassny (AMNH), R. Mooi (MPM), and E. Rosa-Molinar (University of Nebraska).

We dedicate this paper to Colin Patterson, whose enduring and valuable contributions serve as a model for systematic ichthyologists.

References

Briggs, J. C. (1955). "A Monograph of the Clingfishes (Order Xenopterygii)." Nat. Hist. Mus., Stanford University, Stanford, CA.

Dyer, B. (1993). A phylogenetic study of atheriniform fishes with a systematic revision of the South American silversides (Atherinomorpha, Atherinopsinae, Sorgentinini). Ph.D. Dissertation, University of Michigan, Ann Arbor.

Freihofer, W. C. (1963). Patterns of the ramus lateralis accessorius and their systematic significance in teleostean fishes. *Stanford Ichthyol. Bull.* **8**(2), 80–189.

Freihofer, W. C. (1966). The Sihler technique of staining nerves for systematic study especially of fishes. *Copeia*, p. 470–475.

Freihofer, W. C. (1970). Some nerve patterns and their systematic significance in paracanthopterygian, salmoniform, gobioid, and apogonid fishes. *Proc. Calif. Acad. Sci.* **38**(12), 215–264.

Freihofer, W. C. (1972). Trunk lateral line nerves, hyoid arch gill rakers, and olfactory bulb location in atheriniform, mugilid, and percoid fishes. *Occas. Pap. Calif. Acad. Sci.* **95**, 1–31.

Freihofer, W. C. (1978). Cranial nerves of a percoid fish, *Polycentrus schomburgkii* (family Nandidae), a contribution to the morphology and classification of the order Perciformes. *Occas. Pap. Calif. Acad. Sci.* **128**, 1–78.

Goodrich, E. S. (1909). "A Treatise on Zoology. Part IX. Vertebrata Craniata. (First fascicle: Cyclostomes and fishes)." 1964 Reprint, A. Asher and Co., Amsterdam.

Goodrich, E. S. (1930). "Studies on the Structure and Development of Vertebrates." 1958 Reprint, Dover, New York.

Gosline, W. A. (1970). A reinterpretation of the teleostean fish order Gobiesociformes. *Proc. Calif. Acad. Sci.* **38**(19), 363–382.

Gosline, W. A. (1971). "Functional Morphology and Classification of Teleostean Fishes." University of Hawaii Press, Honolulu.

Greenwood, P. H., Rosen, D. E., Weitzman, S. H., and Myers, G. S. (1966). Phyletic studies of teleostean fishes, with a provisional classification of living forms. *Bull. Am. Mus. Nat. Hist.* **131**, 339–456.

Guitel, F. (1888). Recherches sur les lepadogasters. *Arch. Zool. Exp. Gén.* [2] **6**, 423–627.

Günther, A. C. (1861). "Catalogue of the Acanthopterygian Fishes in the Collection of the British Museum," Vol. 3. British Museum (Natural History), London.

Hatta, K., Schilling, T. F., Brewiller, R. A., and Kimmel, C. B. (1990). Specification of jaw muscle identity in zebra fish: Correlation with engrailed homeo-protein expression. *Science* **250**, 802–805.

Herrick, C. J. (1899). "The Cranial and First Spinal Nerves of *Menidia*. A Contribution upon the Nerve Components of Bony Fishes." State Hospitals Press, Utica, NY.

Holland, N. D., Holland, L. Z., Davis, C. A., and Honma, Y. (1991). Expression domains of engrailed gene in lamprey embryos. *Am. Zool.* **31**, 46A (Abstr. 241).

Jamieson, B. G. M. (1991). "Fish Evolution and Systematics: Evidence from Spermatozoa." Cambridge University Press, Cambridge, UK.

Johnson, G. D., and Patterson, C. (1993). Percomorph phylogeny: A survey of acanthomorphs and a new proposal. *Bull. Mar. Sci.* **52**(1), 554–626.

Leviton, A. E., Gibbs, R. H., Jr., Heal, E., and Dawson, C. E. (1985). Standards in herpetology and ichthyology: Part I. Standard symbolic codes for institutional resources collections in herpetology and ichthyology. *Copeia*, pp. 802–832.

Marx, J. (1995). Helping neurons find their way. *Science* **268**, 971–973.

Moody, S. A., Quigg, M. S., and Frankfurter, A. (1989). Development of the peripheral trigeminal system in the chick revealed by an isotope-specific anti-beta-tubin monoclonal antibody. *J. Comp. Neurol.* **279**, 567–580.

Moore, J. A. (1993). Phylogeny of the Trachichthyiformes (Teleostei: Percomorpha). *Bull. Mar. Sci.* **52**(1), 114–136.

Nelson, G. (1989). Phylogeny of major fish groups. *In* "The Hierarchy of Life" (B. Fernholm, K. Bremer, and H. Jörnvall, eds.) pp. 325–336. Elsevier, Amsterdam.

Nelson, J. S. (1994). "Fishes of the World," 3rd ed. Wiley, New York.

Noden, D. M. (1991). Vertebrate craniofacial development: The relation between ontogenetic process and morphological outcome. *Brain, Behav. Evol.* **38**, 190–225.

Parenti, L. R. (1986). Homology of pelvic fin structures in female phallostethid fishes (Atherinomorpha, Phallostethidae). *Copeia*, pp. 305–310.

Parenti, L. R. (1993). Relationships of atherinomorph fishes (Teleostei). *Bull. Mar. Sci.* **52**(1), 170–196.

Rosa-Molinar, E., Hendricks, S. E., Rodriguez-Sierra, J. F., and Fritzsch, B. (1994). Development of the anal fin appendicular support in the western mosquitofish, *Gambusia affinis affinis* (Baird and Girard, 1854): A reinvestigation and reinterpretation. *Acta Anat.* **151**, 20–35.

Rosen, D. E. (1973). Interrelationships of higher euteleostean fishes. *In* "Interrelationships of fishes" (P. H. Greenwood, R. S. Miles, and C. Patterson, eds.), pp. 397–513. Academic Press, London.

Rosen, D. E. (1982). Teleostean interrelationships, morphological function and evolutionary inference. *Am. Zool.* **22**(2), 261–273.

Rosen, D. E., and Bailey, R. M. (1963). The poeciliid fishes (Cyprinodontiformes), their structure, zoogeography, and systematics. *Bull. Am. Mus. Nat. Hist.* **126**, 1–176.

Rosen, D. E., and Parenti, L. R. (1981). Relationships of *Oryzias*, and the groups of atherinomorph fishes. *Am. Mus. Novit.* **2719**, 1–25.

Runyan, S. (1961). Early development of the clingfish, *Gobiesox strumosus* Cope. *Chesapeake Sci.* **2**(3–4), 113–141.

Song, J. (1989). Lateral line innervation and dermal bone homologies in the Florida gar (*Lepisosteus platyrhinchus*) and other fishes. *Abstr. 69th Annu. Meet. Am. Soc. Ichthyol. Herpetol., 1989*, p. 147.

Song, J. (1990). Organization of the trigeminal motor nucleus of the clearnose skate, *Raja eglanteria, Am. Zool.* **30**(4), 121A, 684.

Song, J. (1993) Morphology, distribution and innervation of the lateral line receptors in gobies (Teleosts). *Abstr. Annu. Meet. Soc. Neurosci., 1993*, Vol. 19, Part 2, p. 1580.

Song, J., and Boord, R. L. (1993). Motor components of the trigeminal nerve and organization of the mandibular arch muscles in vertebrates. Phylogenetically conservative patterns and their ontogenetic basis. *Acta Anat.* **148**, 139–149.

Song, J., and Northcutt, R. G. (1991a). The morphology, distribution and innervation of the lateral line receptors of the Florida Gar, *Lepisosteus platyrhincus, J. Brain, Behav. Evol.* **37**(1), 10–37.

Song, J., and Northcutt, R. G. (1991b). The primary projections of the lateral line nerves of the Florida Gar, *Lepisosteus platyrhincus, J. Brain, Behav. Evol.* **37**(1), 38–63.

Song, J., and Parenti, L. R. (1995). Clearing and staining whole fish specimens for simultaneous demonstration of bone, cartilage, and nerves. *Copeia*, pp. 114–118.

Springer, V. G., and Fraser, T. H. (1976). Synonymy of the fish families Cheilobranchidae (= Alabetidae) and Gobiesocidae, with descriptions of two new species of *Alabes. Smithson. Contrib. Zool.* **234**, 1–23.

Stannius, F. H. (1846). "Zootomie der Fische." Berlin.

Starks, E. C. (1905). The osteology of *Caularchius maendricus* (Girard). *Biol. Bull. (Woods Hale, Mass.)* **9**(5), 292–303.

Stiassny, M. L. J. (1990). Notes on the anatomy and relationships of the bedotiid fishes of Madagascar, with a taxonomic revision of the genus *Rheocles* (Atherinomorpha: Bedotiidae). *Am. Mus. Novit.* **2979**, 1–33.

Stiassny, M. L. J. (1993). What are grey mullets? *Bull. Mar. Sci.* **52**(1), 197–219.

Stiassny, M. L. J., and Moore, J. A. (1992). A review of the pelvic girdle of acanthomorph fishes, with comments on hypotheses of acanthomorph relationships. *Zool. J. Linn. Soc.* **104**, 209–242.

Winterbottom, R. (1974). A descriptive synonymy of the striated muscles of the Teleostei. *Proc. Acad. Natl. Sci. Philadelphia* **125**(12), 225–317.

Appendix 1

Material Examined

Lepisosteus platyrhinchus, collection of Jiakun Song (head and anterior portion of the body cleared and stained with Sudan Black only); *Pantodon buchholzi*, USNM 336676 (3; 2 triple-stained, 1 cleared and stained for bone and cartilage only); *Cyclocheilichthys apogon*, USNM 330098 (1); *Ictalurus punctatus*, collection of Jiakun Song (2 cleared and stained with Sudan Black only); *Esox americanus*, USNM 237235 (1); *Polymixia lowei*, USNM 323212 (2); *Aphredoderus sayanus*, USNM 230945 (4); USNM 217374 (6 cleared and counterstained for bone and cartilage, 2 cleared and stained only for bone); *Percopsis omiscomaycus*, USNM 308217 (4); MPM 10732 (5); *Typhlichthys subterraneus*, USNM 175248 (1); *Microgadus tomcod*, AMNH 120384 (7); *Porichthys* sp., collection of Jiakun Song (1, cleared and stained with Sudan Black only); *Agonostomus* sp., USNM 318360 (5); *Agonostomus monticola*, USNM 329330 (2); *Odontesthes nigricans*, USNM 214436 (1); *Leuresthes sardina*, USNM 177811 (2); *Bedotia* sp., USNM 301519 (2); *Melanotaenia splendida*, USNM

308410 (2); *Fundulus heteroclitus*, USNM 326631 (5; 2 triple-stained, 2 cleared and stained with Sudan Black only, 1 cleared and stained for bone and nerves only); *Tomeurus gracilis*, USNM 225464 (2); *Hemirhamphodon kuekenthali*, USNM 329584 (2), USNM 330097 (4); *Stephanoberyx monae*, USNM 83892 (1); *Holocentrus ascensionis*, USNM 336715 (1); *Parazen pacificus*, USNM 187077 (1); *Gasterosteus aculeatus microcephalus*, USNM 75397 (2); *Elassoma evergladei*, USNM 336675 (6 cleared and stained for nerves by P. Mabee according to Sihler technique of Freihofer, 1966); *Apogon maculatus*, USNM 306725 (6 cleared and stained for nerves by P. Mabee according to Sihler technique of Freihofer, 1966); *Perca fluviatilis*, USNM 207079 (1); *Gobiesox nudus*, USNM 158677 (1); *Sicyogaster maeandricus*, USNM 82154 (1); Gobiesocidae, indet. larvae, USNM 338834 (5 cleared and stained solely with alcian blue); *Synchiropus splendidus*, USNM 270216 (1 cleared and stained solely with alizarin); *Cheimarrichthys fosteri*, USNM 214023 (1 cleared and stained solely with alizarin); *Bathygobius soporator*, USNM 326630 (2); *Callogobius depressus*, USNM 214163 (1); *Gunnellichthys monostigma*, USNM 20742 (1); *Sicyopterus micrurus*, USNM 313865 (1); *Sphyraena chinensis*, USNM 148033 (1).

Morphology, Characters, and the Interrelationships of Basal Sarcopterygians

RICHARD CLOUTIER

URA 1365 du CNRS
Université des Sciences et Technologies de Lille
Sciences de la Terre
59655 Villeneuve d'Ascq, France

PER ERIK AHLBERG

Department of Palaeontology
The Natural History Museum
Cromwell Road, London SW7 5BD, United Kingdom

I. Historical Background[1]

A new chapter in sarcopterygian systematics opened in 1970 when Schultze (1970) published a brief overview of sarcopterygian tooth structure which explicitly used the distribution of derived characters to delineate monophyletic groups and to argue for tetrapod monophyly. Three years later, Andrews (1973) produced the first attempt at an overall cladistic analysis of the sarcopterygian fishes. Although it retained certain characteristics of previous analyses, such as *a priori* separation of lungfishes and tetrapods from "crossopterygians" and a tendency to search for "key characters," it centered on a wide-ranging and thorough review of character state distributions and character polarities. Andrews (1973) concluded that character incongruence and thus evolutionary parallelism was rife among the "crossopterygians," but that the skull roof pattern might provide a reasonable guide to their relationships. On this basis she presented a fully resolved phylogeny (Fig. 1a) which divided the Crossopterygii into Binostia (actinistians and porolepiforms) and Quadrostia (osteolepiforms, rhizodonts, and onychodonts). However, she made no greater

claim for it than that, as it "seems to do least injustice to knowledge in its present state, it is proposed as a model for future discussion" (Andrews, 1973, p. 137). The rest of the decade saw gradual progress in sarcopterygian phylogenetics. Miles (1975) used parsimony arguments to place lungfishes and "crossopterygians" side by side as sister-groups in the clade Sarcopterygii. He included the Tetrapoda within Crossopterygii but did not consider the possibility that the latter group might be paraphyletic relative to the Dipnoi. However, in a subsequent paper (Miles, 1977) he placed the Actinistia as sister-group to the Dipnoi + Choanata. Schultze (1977) applied cladistic principles to the distribution of fin and limb characters among the Sarcopterygii.

At the 26th Symposium of Vertebrate Palaeontology and Comparative Anatomy held in 1978 at Reading, England, Gardiner presented a paper on sarcopterygian cladistic phylogeny which started a series of methodological debates. As a response to a comment by Parrington, Patterson confirmed—in support of Gardiner's arguments for a cladistic view of relationships—that a lungfish shares more characters with a cow than with a salmon. "The salmon, the lungfish, and the cow" soon became familiar as a classic example of a three-taxon statement. The debate was continued in the pages of *Nature* (Halstead, 1978; Halstead *et al.*, 1979; Gardiner *et al.*, 1979).

---

[1] All generic taxa referred in the text are extinct with the exception of *Polypterus*, *Latimeria*, *Protopterus*, *Lepidosiren*, and *Neoceratodus*.

Copyright © 1996 by Academic Press, Inc.
All rights of reproduction in any form reserved.

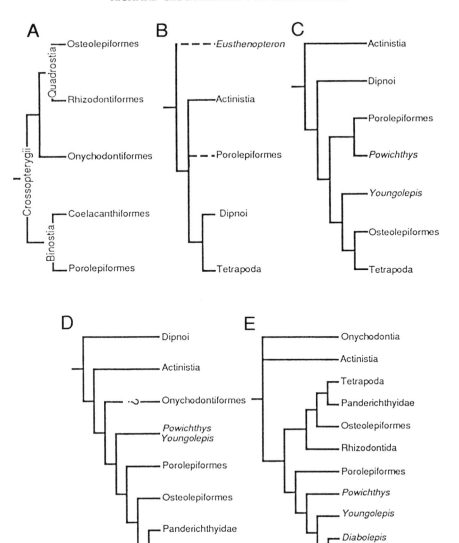

FIGURE 1 Previously published sarcopterygian phylogenies. (A) Andrews (1973); note that this phylogenetic hypothesis does not consider either the Dipnoi or the Tetrapoda. (B) Rosen *et al.* (1981); a pattern-cladistic approach with fossil taxa (dashed lines) inserted after the analysis of the Recent forms. *Eusthenopteron* occupies a basal position, far removed from the Tetrapoda. (C) Panchen and Smithson (1987); (D) Schultze (1987); (E) Ahlberg (1991b). Phylogenies (C)–(E) are based on fossil and Recent data. Osteolepiforms and "panderichthyids" (=elpistostegids; see text) consistently group with tetrapods, but otherwise there is little agreement between the three.

"The Terrestrial Environment and the Origin of Land Vertebrates" (edited by A. L. Panchen), published in 1980, gave a snapshot of the growing influence of cladistic methodology on sarcopterygian phylogenetics. With hindsight the two most significant papers are those of Patterson (1980) and Gardiner (1980). The former was a powerful critique of noncladistic approaches to the problem of tetrapod origins, while the latter presented a cladogram where lungfishes were

the living and fossil sister-group of tetrapods, and "crossopterygians" were paraphyletic. Forey (1980) used a pattern-cladistic approach to argue that *Latimeria* is a sarcopterygian, the sister-group of Dipnoi + Tetrapoda, rather than a chondrichthyan.

The points made by Patterson (1980), Gardiner (1980), and Forey (1980) were developed further by Rosen *et al.* (1981) in a provocative paper (Fig. 1b) which succeeded in starting a vigorous debate. The following years saw a surge of publications on this topic, many of them couched as rebuttals of Rosen *et al.* but nevertheless characterized by an explicitly cladistic approach (Holmes, 1985; Jarvik, 1980; Long, 1985, 1989; Maisey, 1986a; Panchen and Smithson, 1987; Schultze, 1981, 1987, 1991, 1994; Ahlberg, 1989, 1991b; Cloutier, 1990, 1991a,b; Chang, 1991a,b; Chang and Smith, 1992; Young *et al.*, 1992; Ahlberg and Milner, 1994). Much of the interest focused on the three-taxon problem of the living groups and on the position of the Osteolepiformes (Figs. 1c–1e). Forey (1987) and Forey *et al.* (1991) continued to build on the work of Rosen *et al.*

Alongside the systematic reviews came a sequence of descriptive works (Vorobyeva, 1977; Jarvik, 1980; Campbell and Barwick, 1982a,b, 1984, 1987, 1988; Chang, 1982, 1991b; Chang and Yu, 1984; Andrews, 1985; Schultze and Arsenault, 1985; Long, 1989; Ahlberg, 1989; Ahlberg *et al.*, 1994; Clack, 1989, 1994a,b; Coates and Clack, 1990, 1991; Vorobyeva and Schultze, 1991) which significantly expanded the data set available for phylogenetic analysis. The development of computer programs for phylogenetic inference ushered in the present stage of the debate, characterized by exhaustive analyses of large data sets (Cloutier, 1990, 1991a,b; Ahlberg, 1991b; Forey *et al.*, 1991; Lebedev and Coates, 1995; Schultze, 1994; Schultze and Marshall, 1993). On a parallel front, several workers have approached the three-taxon problem of lungfish, coelacanth, and tetrapod relationships from a molecular perspective (Meyer and Wilson, 1990; Gorr *et al.*, 1991; Stock and Swofford, 1991; Hedges *et al.*, 1993). At present the bulk of the molecular evidence favors a lungfish–tetrapod sister-group relationship (Meyer, 1995).

The last 20-odd years of research into sarcopterygian phylogeny have produced an enormously enlarged data base as well as substantial improvements in phylogenetic methodology and practice. No complete phylogenetic consensus has developed, but several significant and probably permanent changes of opinion have taken place:

1. There is almost universal agreement (but see Jarvik, 1980) about the status of the Sarcopterygii as a clade and about the characters which define the group (Rosen *et al.*, 1981; Schultze, 1987; Panchen and Smithson, 1987; Ahlberg, 1991b).

2. Placement by Rosen *et al.* (1981) of the Dipnoi as the living and fossil sister-group of the Tetrapoda has not generally found favor. Most paleontologists favor the "traditional" osteolepiform–tetrapod relationship, redefined in terms of shared derived characters (Schultze, 1987; Panchen and Smithson, 1987;

Long, 1989; Cloutier, 1990, 1991a; Ahlberg, 1991b; Vorobyeva and Schultze, 1991; Young *et al.*, 1992; Ahlberg and Milner, 1994). The majority of these authors remove the elpistostegids ("panderichthyids"; see below) from osteolepiforms and regard the osteolepiforms as the sister-group of the clade [elpistostegids + tetrapods]. Panchen and Smithson (1987), however, place the elpistostegids within the osteolepiforms. The main nonparticipants in this consensus are Chang (1991b), who regards tetrapods as the sister-group of all other sarcopterygians, and Forey *et al.* (1991), who continue to place the lungfishes as the living and fossil sister-group of tetrapods.

3. A number of authors have come to view the Porolepiformes and the Lower Devonian genera *Youngolepis* and *Powichthys* as immediate relatives of the Dipnoi (Maisey, 1986a; Ahlberg, 1989, 1991b; Cloutier, 1991b; Chang, 1991a,b; Chang and Smith, 1992). This is linked with the recognition of *Diabolepis* (Chang and Yu, 1984) as a primitive lungfish (Fig. 1e), although some authors (Forey *et al.*, 1991) accept the latter proposition without agreeing with the former. Another group of workers (Schultze, 1987; Panchen and Smithson, 1987; Long, 1989) place the porolepiforms and *Youngolepis* as separate plesions on the stem to [osteolepiforms + tetrapods] (Figs. 1c and 1d).

4. As regards the three-taxon problem of coelacanths, lungfishes, and tetrapods, most recent paleontological analyses favor a lungfish–tetrapod sister-group relationship (Maisey, 1986a; Panchen and Smithson, 1987; Cloutier, 1990, 1991a; Ahlberg, 1991b; Trueb and Cloutier, 1991; Forey *et al.*, 1991). However, there is also support for a Recent sister-group relationship between coelacanths and tetrapods (Schultze, 1987; Long, 1989; Vorobyeva and Schultze, 1991) or between coelacanths and lungfishes (Chang, 1991b).

While it is heartening to see emerging areas of agreement, the substantial remaining disputes cannot be ignored. These are largely due to differences in character interpretation and scope. Thus, the reassertion of the osteolepiform–tetrapod relationship is directly linked to the rejection of the interpretation of Rosen *et al.* (1981) of osteolepiform snout anatomy in favor of Jarvik's (1942, 1980) model. Many of the characters used by Maisey (1986a), Ahlberg (1989, 1991b), Chang (1991a,b), and others to link porolepiforms and lungfishes are ignored by Schultze (1987) and Panchen and Smithson (1987).

Our purpose in this chapter is to review the state of knowledge for all the major sarcopterygian groups and to present a phylogenetic analysis based on a large data set. Our respective contributions in part reflect our slightly different fields of research. Cloutier

provided the information about actinistians, onychodonts, and dipnoans, while Ahlberg was the principal
contributor on tetrapods, elpistostegids, porolepiforms, and *Youngolepis + Powichthys*. The character
matrix represents our shared knowledge and derives
from several sources (e.g., Ahlberg, 1991b); however,
the largest component was the previously unpublished character matrix from Cloutier's Ph.D. thesis
(1990). For this reason, and because he performed
the phylogenetic analysis and interpretation, Cloutier
takes the formal position as principal author. It was
a salutary experience to discover, during the prepara-
tory stages of this collaboration, that there were considerable and sometimes irreconcilable differences between our character tables. Some of these were due
to difficulties with poorly preserved specimens, but
in other cases the disagreement was purely one of
interpretation. This may be symptomatic of the debate
as a whole. The significance of these problems is considered in more detail in the Discussion.

II. The Principal Sarcopterygian Groups

The following section attempts to summarize most
of our knowledge of sarcopterygian morphology and
diversity (Fig. 2). The major omission is the tetrapod
crown group, which is not discussed in detail; its diversity is so great that any attempt at a systematic
overview would swamp the rest of the paper, and we
felt that a full account of the known stem tetrapod
genera was more important and timely. Detailed treat-
ments of crown tetrapod systematics can be found in
Carroll (1988) and elsewhere.

A. Actinistia

The Actinistia (=Coelacanthi, Coelacanthia, Coelacanthiformes, Coelacanthii, Coelacanthina) include
approximately 125 species belonging to 50 genera
(Cloutier and Forey, 1991). They range in time from
the Middle Devonian to Recent and reached their maximum of diversity during the Lower Triassic (Cloutier
and Forey, 1991). However, they are known from the
fossil record only from the Middle Devonian to the
Upper Cretaceous. The earliest actinistian known is
Euporosteus eifelianus from the Givetian of Germany
(Stensiö, 1937), whereas the youngest fossil species
is *Megalocoelacanthus dobiei* from the latest Campanian
to middle Maastrichtian of New Jersey (Schwimmer
et al., 1994). Ørvig (1986) identified a fragment of bone
possibly referrable to an actinistian, on histological
characteristics, from the Paleocene of southern Swe-

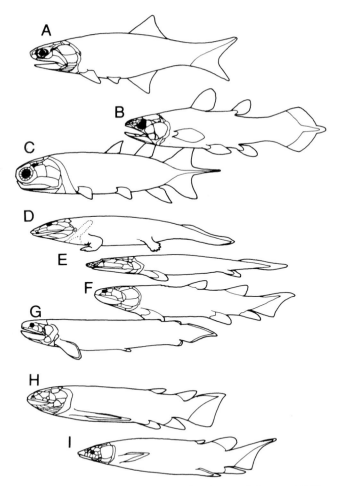

FIGURE 2   A representative range of early osteichthyans. (A) The
actinopterygian *Mimia toombsi* (after Gardiner, 1984). (B)–(I), Sarcopterygii; (B) the actinistian *Rhabdoderma elegans* (after Forey, 1981);
(C) the onychodont *Strunius walteri* (after Jessen, 1966); (D) the
tetrapod *Ichthyostega groenlandica* (modified from Jarvik, 1952, 1980);
(E) the elpistostegid *Panderichthys rhombolepis* (after Vorobyeva and
Schultze, 1991); (F) the osteolepiform *Osteolepis macrolepidotus* (after
Jarvik, 1948); (G) the rhizodont, ?*Strepsodus anculonamensis* (modi-
fied from Andrews, 1985); (H) the porolepiform *Glyptolepis paucidens*
(original); (I) the dipnoan *Dipterus valenciennesi* (after Ahlberg and
Trewin, 1995). Not to scale. Except for *Rhabdoderma* and *Strepsodus*,
which are of Carboniferous age, all these genera date from the
Middle-Upper Devonian (Eifelian–Famennian).

den. *Latimeria chalumnae* is the only living representative of this group and it has been found only in the
Comoros Archipelago, Mozambique Strait, and Cha-
lumna River (Republic of South Africa).

Although the monophyly of the Actinistia has
never been questioned, the diagnosis has been addressed repeatedly (Andrews, 1973; Forey, 1981, 1984;
Maisey, 1986a; Panchen and Smithson, 1987; Cloutier,
1991a,b, 1996a). The Actinistia is monophyletic based

on the following synapomorphies (Cloutier, 1996a): (1) absence of maxilla, (2) absence of surangular, (3) absence of branchiostegal rays, (4) absence of submandibulars, (5) presence of rostral organ, (6) numerous supraorbitals, (7) presence of extracleithrum (absent in the Diplocercidae), (8) triangular entopterygoid, (9) short dentary, (10) dorsal margin of angular elevated into a process, (11) coronoid IV oriented vertically, and (12) anterior position of anterior dorsal fin. Characters 2 and 10 are possibly dependent characters, characters 1 and 9 are homoplastic with respect to the Dipnoi, character 3 is homoplastic with respect to the Onychodontida, and character 4 is homoplastic with respect to the Tetrapoda.

Three additional characters are considered as potential synapomorphies for the actinistians (Cloutier, 1991a,b), but their condition remains unknown or unclear in *Miguashaia bureaui*: (13) presence of ventral process of lateral rostral, (14) tandem jaw articulation, and (15) posteriorly expanded U-shaped urohyal.

In addition to the characters listed above, Forey (1991) considered the following three characters as synapomorphies of the Actinistia: (1) second dorsal and anal fin lobes both contain a skeleton of several segments which resemble the endoskeletons of the paired fins, (2) head divided by a prominent intracranial joint in which the otico-occipital part of the neurocranium extends anteriorly to form a track-and-groove joint with the basisphenoid, and (3) small premaxilla. The first character is probably absent in *Miguashaia bureaui*; although the median fin endoskeletons are not preserved, the lack of a posterior dorsal fin lobe (and the very slight lobation of the anal fin) suggests that the endoskeletons were of a generalized sarcopterygian pattern (Ahlberg, 1992a); this character might be a synapomorphy of the clade [Actinistia except *Miguashaia*] (Cloutier, 1996a). The condition of the second character cannot be determined in *M. bureaui*; thus, it remains a potential synapomorphy of the group. The size of the premaxilla in *Miguashaia* is comparable to that of other sarcopterygians.

Schultze (1973) proposed *Miguashaia bureaui* (middle Frasnian, Escuminac Formation, Québec, Canada) to be the most primitive actinistian. According to Cloutier (1991a,b, 1996a), *Miguashaia* is an actinistian because it shares the aforementioned derived characters with other actinistians, and within this clade it is the sister-taxon of the rest of the group. The clade [Actinistia except *Miguashaia*] is characterized by the following characters: absence of intertemporal, lacrimal fused with jugal, otic canal passing in postparietal, unbranched distal ends of lepidotrichia, tail notochord straight and horizontal, epichordal and hypochordal lepidotrichia symmetrical, and presence of a supplementary caudal fin.

Interrelationships among actinistians have been investigated recently by Cloutier (1991a,b) and Forey (1991). Although there are minor disagreements between them (notably the position of the Coelacanthidae), both phylogenies are strongly asymmetrical and have good stratigraphic correlation. Most actinistian classifications fail to reflect the phylogeny of the group, but Schultze (1993) erected a classification which reflects Cloutier's (1991b) phylogeny.

Miguashaiidae Schultze 1993 is a monospecific family containing the middle Frasnian *Miguashaia bureaui* from the Escuminac Formation of Québec. The anatomy of *Miguashaia* is fairly well known (Cloutier, 1996a) with the exception of the neurocranium, gill arches, and axial skeleton. An isolated scale from the Givetian of Latvia is referred to the genus *Miguashaia* (Cloutier *et al.*, 1996). *Miguashaia* does not belong to the Coelacanthidae as suggested by Carroll (1988).

Diplocercidae Stensiö 1922 is a monogeneric family including fewer than 10 taxa of Devonian and Lower Carboniferous age. *Nesides* Stensiö has been synonymized with *Diplocercides* by Cloutier and Forey (1991). Traditionally the genera *Euporosteus* and *Chagrinia* have been incorporated within the Diplocercidae; however, based on their phylogenetic position (Cloutier, 1991a,b) they are removed from this family. *Diplocercides* is the only Devonian actinistian in which the neurocranium has been studied in detail (Jarvik, 1954, 1964; Bjerring, 1993).

Hadronectoridae Lund and Lund 1984 is a family originally created to include the three Namurian actinistians *Allenypterus*, *Hadronector*, and *Polyosteorhynchus* from the Heath Formation (Bear Gulch fauna), Montana. Lund and Lund (1985) provided detailed descriptions of the anatomy of these genera. Cloutier (1991a) excluded *Allenypterus* in order to keep the family monophyletic. The presence of a series of bifurcating supraorbital pores associated with the supraorbital sensory canal corroborates the monophyly of this clade (Cloutier, 1991a).

"Rhabdodermatidae" Berg 1958 is a paraphyletic family including mainly Carboniferous genera (e.g., *Rhabdoderma* and *Caridosuctor*). The Carboniferous *Rhabdoderma elegans* is the best known representative of this group (Forey, 1981).

Laugiidae Berg 1940 includes the genera *Laugia* and *Coccoderma* and ranges in time from the Lower Triassic to the Lower Cretaceous. In contrast to its classification in Schultze (1993), *Synaptotylus* is excluded from this family in order to respect the monophyly of the Laugiidae.

Whiteiidae Schultze 1993 is a family created to include the Triassic genus *Whiteia* from Madagascar and western Canada. A great deal of the anatomy of these forms remains to be redescribed, on the basis of available material. This might require a reidentification of the North American species (R. Cloutier, personal observation).

Coelacanthidae Agassiz 1843 is a family in which, to keep its monophyly, Cloutier (1991b) included all actinistians sharing the common ancestor of *Coelacanthus* and *Latimeria*. However, Schultze (1993) restricted the definition of the family to include *Coelacanthus*, *Axelia*, *Wimania*, and *Ticinepomis*. The phylogenetic position as well as the definition of the Coelacanthidae differ between Cloutier's (1991b) and Forey's (1991) hypotheses.

Mawsoniidae Schultze 1993 includes Triassic and Jurassic genera (*Alcoveria*, *Diplurus*, *Chinlea*, *Mawsonia*, and *Axelrodichthys*) forming a clade which is the sister-group to the Latimeriidae (Cloutier, 1991b; Forey, 1991). *Diplurus newarki* (Schaeffer, 1952) and *Axelrodichthys araripiensis* (Maisey, 1986b) are the best known representatives of this family.

Latimeriidae Berg 1940 is a family including the only living representative (*Latimeria chalumnae*) as well as a few Jurassic and Cretaceous genera (*Holophagus*, *Undina*, *Libys*, *Macropomoides*, and *Macropoma*). Lambers (1992) provided a revision of the genus *Libys*. The gigantic, recently described *Megacoelacanthus* probably belongs to this family, based on the characteristics of the pterygoid and basisphenoid, and not to the Coelacanthidae as suggested by Schwimmer *et al.* (1994).

Many authors have considered the actinistians as an evolutionarily conservative group (see Cloutier, 1991a,b) and *Latimeria chalumnae* as an example of a living fossil (Forey, 1984). As early as the Famennian, the actinistians had acquired their characteristic body shape. The anterior dorsal fin is located quite anteriorly compared to other sarcopterygians and is never lobated (Cloutier, 1996a). The anal and posterior dorsal fins are usually strongly lobed and contain endoskeletons and musculature which match those of the paired fins (Millot and Anthony, 1958) rather than the median fins of other sarcopterygians. Ahlberg (1992a) interpreted this pattern as evidence of a type of ho-

meotic transformation, with paired fin structures being expressed at the anal and posterior dorsal fin sites. Actinistians have a diphycercal caudal fin, symmetrical dorsoventrally and possessing a supplementary caudal lobe; the only exceptions are *Miguashaia*, which has a heterocercal tail (plesiomorphic within the Actinistia), and *Allenypterus* which has a modified tapering diphycercal tail.

Most of the synapomorphies diagnosing the clade concern the skull structure, primarily those parts related to the lower jaw and feeding mechanism. Lund and Lund (1985) and Lund *et al.* (1985) explained the conservatism of the jaw apparatus as a response to a specialized mechanism of suction feeding.

Latimeria is the only living sarcopterygian to possess an intracranial joint (Lauder, 1980). Through geological time, the ethmosphenoid has become relatively longer than the otico-occipital part of the neurocranium (Forey, 1991).

In addition to the mechanoreceptive lateral line system, actinistians have a unique electroreceptive organ located in the anterior part of the ethmosphenoid, the rostral organ (Northcutt and Bemis, 1993). In *Latimeria*, this organ is probably used to localize prey; its use seems to be linked with a unique headstand behavior (Fricke *et al.*, 1987).

Among piscine sarcopterygians, the Actinistia is the only group in which the mode of reproduction has been documented in the fossil record. Oviparity has been documented by Schultze (1985) in the Carboniferous *Rhabdoderma exiguum*. This contrasts with the Recent *Latimeria*, which is ovoviviparous.

B. Dipnoi

The lungfish clade, Dipnoi, is a universally recognized natural group whose record extends from the Lower Devonian (Pragian) to Recent (Schultze, 1992a). Approximately 280 species are divided into 64 genera, most represented only by tooth plates (ca. 125 species). The Dipnoi reached their maximum diversity during the Devonian (more than 85 species) and Triassic (more than 45 species). The 6 living species are classified into three genera: *Protopterus* (*P. dolloi*, *P. annectens*, *P. aethiopicus*, and *P. amphibius*) from tropical Africa (Greenwood, 1987), *Lepidosiren* (*L. paradoxa*) from South America, and *Neoceratodus* (*N. forsteri*) from Australia (Kemp, 1987). The oldest known members of the group include *Uranolophus wyomingensis* Denison 1968a (Beartooth Butte Formation, Wyoming), *Diabolepis speratus* (Chang and Yu, 1984) (Xitun Formation, Yunnan, China), and *Speonesydrion iani* Campbell and Barwick 1983 (Bloomfield Limestone, New South Wales, Australia) from the Pragian, and

Sorbitorhynchus deleaskitus Wang *et al.* 1990 (Dale Formation, Guangxi Province, China) and *Dipnorhynchus suessmilchi* (Etheridge, 1906) (New South Wales, Australia) from the Emsian. Interrelationships among dipnoans have only partially been assessed and remain highly debated (Miles, 1977; Marshall, 1987; Campbell and Barwick, 1990; Schultze *et al.*, 1993; Schultze and Marshall, 1993; Long, 1993).

The monophyly of the Dipnoi has never been challenged; however, since the discovery of *Diabolepis speratus* (Chang and Yu, 1984) the definition and diagnosis of the group have been debated (Maisey, 1986a; Campbell and Barwick, 1987; Panchen and Smithson, 1987; Schultze, 1987; Schultze and Campbell, 1987; Cloutier, 1990; Smith and Chang, 1990; Chang, 1991b). The Dipnoi are diagnosed by five uniquely shared derived characters (Cloutier, 1996b): the absence of marginal teeth on the lower jaw, the presence of tooth plates on the entopterygoids and prearticulars, a median B-bone located anteriorly or separating the parietals and postparietals, location of the anterior margin of the parietals well posterior to the orbits, and the presence of a labial cavity. Dipnoans are also characterized by the absence of vomerine fangs, maxilla, extratemporal, fossa autopalatina, and intracranial joint. Most of the characters diagnosing the Dipnoi reflect the peculiar nature of the dentition and cranial architecture.

Miles (1977) considered *Uranolophus wyomingensis* to be the most primitive dipnoan. Campbell and Barwick (1984) regarded *Speonesydrion iani* and *Dipnorhynchus suessmilchi* as the most primitive genera but considered these as belonging to another lineage than *Uranolophus*. Their arguments were primarily stratigraphical. Schultze and Marshall (1993) used *Dipnorhynchus suessmilchi* as the functional outgroup for their phylogenetic analysis of lungfishes. The most controversial species in this respect is *Diabolepis speratus*, which has been considered either as a primitive sarcopterygian related to *Powichthys* and Porolepiformes (Panchen and Smithson, 1987), a primitive dipnoan (Maisey, 1986a; Smith and Chang, 1990), the sister-group of all other dipnoans (Cloutier, 1990; Chang, 1991b), the sister-group of the Dipnoi (Chang and Yu, 1984; Ahlberg, 1991b), or a taxon of undetermined status (Campbell and Barwick, 1987; Schultze and Campbell, 1987). Campbell and Barwick (1990) ignored *Diabolepis* in their study on dipnoan phylogeny, whereas Schultze and Marshall (1993) discussed the significance of *Diabolepis* to basal dipnoan relationships without actually including it in the analysis. We regard *Diabolepis* as a dipnoan, the sister-group of all other Dipnoi. The clade [Dipnoi excepting *Diabolepis*] is supported by four synapomorphies: the palatal position of both anterior and posterior nares, the absence of premaxillae (though arguably homoplastic "premaxillae" have been identified in *Ganorhynchus*, *Orlovichthys*, and *Scaumenacia*; R. Cloutier, personal observation), the presence of C-bones, and the passage of the occipital sensory canal through both extrascapulars and postparietals.

Dipnoan classification is unsettled because there is no consensus concerning the phylogeny of the group. Campbell and Barwick (1983, 1987, 1990) divide the Dipnoi into three lineages on the basis of dentition (tooth-plated, dentine-plated, and denticulated), but this phylogeny is unparsimonious (Schultze and Marshall, 1993; Cloutier, 1996b) and has not found general favor. Schultze's (1993) classification is based on the strict consensus tree presented by Marshall (1987, fig. 5).

In order to provide a reasonably manageable taxonomic framework for the bewildering diversity of the lungfishes, we present here a list of families combining information from Miles (1977), Campbell and Barwick (1990), Long (1992), Schultze (1993), Schultze and Marshall (1993), and Cloutier (1996b). This should not be regarded as a definitive statement of dipnoan taxonomy; several of the families are certainly or probably nonmonophyletic and require reevaluation.

Diabolepididae Schultze 1993 has only one representative, the Early Devonian *Diabolepis speratus*, known exclusively from cranial material (Chang and Yu, 1984) and isolated dentitional elements (Smith and Chang, 1990). Its discovery had a crucial impact on our understanding of dipnoan–porolepiform relationships (Maisey, 1986a; Cloutier, 1990; Ahlberg, 1991b).

Uranolophidae Miles 1977 includes primitive Early Devonian dipnoans with a denticulated palate. As in *Diabolepis* and *Dipnorhynchus*, the B-bone separates the parietals but not the postparietals. *Uranolophus wyomingensis* is the only species belonging to this family. It is known from a single complete specimen and numerous skulls and lower jaws (Denison, 1968a,b; Campbell and Barwick, 1988).

Dipnorhynchidae Berg 1940 includes Early Devonian, dentine-plated dipnoans with a plesiomorphic skull roof pattern. The group comprises the genera *Dipnorhynchus* and *Speonesydrion*. Campbell and Barwick (1990) considered *Speonesydrion* as one of the basal members of their tooth-plated lineage. All species belonging to this family are known from partial skulls and lower jaws (Campbell and Barwick, 1982b, 1983, 1984). The neurocranium of *Dipnorhynchus suessmilchi* is the best known of any Early Devonian lungfish.

Chirodipteridae Campbell and Barwick 1990 is a Middle to Late Devonian family characterized by a peculiar type of dentition including dentine tuberosities arranged radially. The members are *Chirodipterus, Pillararhynchus, Gogodipterus,* and *Palaedaphus. Chirodipterus australis* (Gogo Formation, Lower Frasnian, Australia) is the best known representative (Miles, 1977).

Stomiahykidae Bernacsek 1977 includes Middle to Late Devonian genera (*Stomiahykus* and *Archaeonectes*) with a large tusklike tuberosity at the anterior end of the mesial row of the entopterygoid tooth plate (Campbell and Barwick, 1990). Long (1992) considered the Stomiahykidae to be closely related to the Chirodipteridae.

Dipteridae Owen 1846 is a paraphyletic group of Middle and Late Devonian forms with tooth plates, short posterior dorsal fin, and cosmine. *Dipterus valenciennesi* (Caithness Flagstone Group, Eifelian–Givetian, Scotland) is the best understood representative, being known from whole bodies with well-preserved heads (Forster-Cooper, 1937; White, 1965) and parts of the porolepiform-like postcranial endoskeleton (Ahlberg and Trewin, 1995). More than 20 species referred to *Dipterus* are only known from isolated tooth plates (Schultze, 1992b).

Rhynchodipteridae Berg 1940 includes Middle to Late Devonian, long-snouted dipnoans with denticulated palates. *Rhynchodipterus, Griphognathus,* and *Soederberghia* belong to this group. The tooth-plated, long-snouted genus *Rhinodipterus* is regarded by one of us (RC) as a rhynchodipterid, but some authors reject this view (Schultze, 1992a). The palatal condition of *Iowadipterus* remains unknown (Schultze, 1992a). The anatomy of *Griphognathus* has been described in detail (Miles, 1977; Campbell and Barwick, 1988).

Fleurantiidae Berg 1940 includes the latest Givetian to Famennian dipnoans characterized by the following cranial features (Cloutier, 1996b): (1) rostral part of the skull elongated, (2) wide mouth gape, (3) single median E-bone, (4) long bone L_1+L_2 extending medially to the M-bone, and (5) elongated entopterygoids bearing large conical teeth and numerous small denticles, both organized in radiating rows. Cloutier (1996b) discussed the interrelationships among fleurantiids (i.e., *Fleurantia, Jarvikia, Andreyevichthys,* and *Barwickia*). *Fleurantia* is the best known member of this family, although informative cranial material

of *Andreyevichthys* is under study (R. Cloutier, personal observation).

Phaneropleuridae Huxley 1861 includes Middle to Late Devonian lungfishes with enlarged bones B, C, and E, and a general absence of bone D (*Scaumenacia, Phaneropleuron,* and *Pentlandia*). The length of the posterior dorsal fin is more than one-quarter of the total length (Cloutier, 1996b). Schultze and Marshall (1993) considered this family to be paraphyletic. Phaneropleurids are closely related to fleurantiids.

Uronemidae Traquair 1890 is a monogeneric Carboniferous group, represented by *Ganopristodus* (=*Uronemus*), characterized by highly modified tooth plates with one long lingual tooth ridge and reduced lateral rows (Smith *et al.,* 1987) and a single bone replacing the intertemporal and supratemporal.

Sagenodontidae Romer 1966 is composed of Carboniferous dipnoans with large bone B, reduced parietals and E-bones, and bone L_1+L_2 in contact with bone B (*Sagenodus*). Tooth plates are formed by ridges rather than isolated teeth. The anal and posterior dorsal fins remain separated from the diphycercal caudal fin (Chorn and Schultze, 1989).

Ctenodontidae Woodward 1891 is composed of Carboniferous dipnoans with a single large bone replacing the intertemporal and bones L_1+L_2 (*Ctenodus, Tranodis,* and *Straitonia*). The skull roof pattern of *Ctenodus* and *Tranodis* is similar to that of the uronemids. The position of *Straitonia* is questionable because a single element is present between bones E and B as in *Sagenodus*.

Conchopomatidae Berg 1940 is a monogeneric (*Conchopoma*), Carboniferous to Permian family characterized by a denticulated parasphenoid with a rounded anterior margin, concave anterior margin of bone B, and a single median fin fringe. Schultze (1975) revised the genus *Conchopoma*.

Gnathorhizidae Miles 1977 is composed of Late Carboniferous to Early Triassic forms (*Palaeophichthys* and *Gnathorhiza*) in which the otic canal passes in the postparietal and with numerous cases of dermal bone reduction (large median bone E; single bone occupying space of bones 3, L_1, L_2, and M; and single bone in place of intertemporal, supratemporal, and tabular). In the lower jaw, the oral canal passes in the infradentaries, whereas the mandibular canal is in an open groove (Schultze and Marshall, 1993). Gnathorhizids have been considered as the sister-group of the Lepidosirenidae by Lund (1970) but not by other authors (Berman, 1968; Miles, 1977; Schultze and Marshall, 1993).

"Ceratodontidae" Gill 1872 is a paraphyletic group including Triassic to Tertiary species. The ceratodontid skull roof shows a reduced number of bones in the median and lateral series. More than 40 species of *Ceratodus* have been described, mostly from tooth plates. Ceratodontids may be paraphyletic with respect to the neoceratodontids. Schultze (1981) investigated the relationships among so-called ceratodontids (*Ptychoceratodus, Microceratodus, Arganodus, Tellerodus,* and *Paraceratodus*).

Neoceratodontidae Miles 1977 includes Triassic (*Epiceratodus*) to Recent (*Neoceratodus*) species with a reduced postparietal and a single bone occupying the space of the parietal and bones L_1, L_2, and M. Schultze (1981) suggested that *Asiatoceratodus* is closely related to *Ceratodus* owing to the presence in both genera of a single bone replacing bones A, B, and C.

Lepidosirenidae Bonaparte 1841 is an apomorphic family including two extant genera (*Protopterus* and *Lepidosiren*) and extinct representatives (e.g., the Cretaceous *Protopterus regulatus*). The skull is highly derived compared to most dipnoans (Miles, 1977; Schultze and Marshall, 1993). There are no cheek bones, nor skull roof bones lateral to the parietal and postparietal; the vomer is reduced to a patch of small, conical teeth; the nasal region is kinetic relative to the braincase; and the jaw adductor muscles attach above the skull roof.

Devonian dipnoan morphology is fairly well documented owing to the material from the Gogo Formation of Australia (e.g., *Chirodipterus australis* and *Griphognathus whitei*), Caithness Flagstones of Scotland (*Dipterus valenciennesi*), and Escuminac Formation of Québec (*Scaumenacia curta* and *Fleurantia denticulata*). However, the anatomy of late Paleozoic, Mesozoic, and Cenozoic species remains poorly understood because most of the species are known only from isolated elements.

The cranial anatomy of dipnoans is better understood than that of the postcranium but has generated much debate concerning the homology of the dermal bones. Forster-Cooper (1937) erected a neutral alphanumerical system of nomenclature for the skull roof (bones A–F, H–J, K–Q, and X–Z) and cheek (bones T, 1–11, and 13–14). Ahlberg (1991b) and Cloutier (1996b) agreed on the homologies of certain bones with that of other sarcopterygians: A=median extrascapular, X=intertemporal, Y_1=supratemporal,

Y_2=tabular, J=parietal, I=postparietal, 1=lacrimal, 4=postorbital, 8=squamosal, 9=preopercular, and 10=quadratojugal. In Recent forms the skull roof is greatly reduced compared to Paleozoic species, and the endocranium is cartilaginous rather than ossified. Dipnoans display a great deal of intraspecific variation, particularly in the skull roof pattern (Cloutier, 1996b) and probably the greatest amount of diversity in dermal skull roof pattern among sarcopterygians. The interpretation of the palatal and cheek bones differs greatly among authors; Rosen *et al.* (1981) argued for the presence of a maxilla, premaxilla, and choana in dipnoans based on their observation on *Griphognathus whitei*. Branchial and hyoid arches have been described by Miles (1977) for the Frasnian genera *Chirodipterus* and *Griphognathus*.

Dipnoans are characterized by peculiar types of dentition. Most of them lack a marginal dentition; the maxillae are absent and the premaxillae, when present, are greatly reduced and bear only a few teeth (e.g., *Scaumenacia, Andreyevichthys,* and *Ganorhynchus*). Paired prearticular and entopterygoid tooth plates constitute the primary dentitional apparatus of most species; such tooth plates are known from the Emsian to the Recent. Two other distinct types of dentition have been reported: dentine-plated (e.g., *Dipnorhynchus*) and denticulated types (e.g., *Uranolophus* and *Griphognathus*). Campbell and Barwick (1983, 1987, 1990) asserted that dipnoans are divided into two lineages characterized by their dentition—the tooth-plated and denticulated types. However, this hypothesis is unparsimonious (Schultze and Marshall, 1993; Cloutier, 1996b) based on the congruence of cranial and postcranial characters. In contrast to other gnathostomes, the teeth composing the plates are not shed during growth but rather added anteriorly and laterally. This mode of growth allows ontogenetic studies because an adult carries its own dental ontogenetic history (Cloutier *et al.*, 1993). Because of the constant wear on the crushing and/or shearing surfaces, dipnoans have a hypermineralized tissue infilling the numerous pulp cavities in the tooth plate, the petrodentine (Smith, 1984).

Several clearcut patterns, which might justify the term "trends," can be observed in the history of the Dipnoi. The earliest representatives of the group have, with the exception of *Diabolepis* (Chang and Yu, 1984), already acquired a wholly characteristic "lungfish head" featuring autostyly and a palatal bite. Their postcranial skeletons however seem hardly to be modified from the generalized sarcopterygian condition and bear a certain resemblance to those of porolepiforms (Denison, 1968a; Campbell and Barwick, 1988;

Ahlberg, 1989, 1991b, 1992b; Ahlberg and Trewin, 1995). During the Middle to Late Devonian, new dipnoan groups (Fleurantiidae, Phaneropleuridae) arise which have derived postcranial morphologies with long-based median fins (Ahlberg and Trewin, 1995; Cloutier, 1996b). All known Carboniferous and later lungfishes, other than *Sagenodus* (Chorn and Schultze, 1989), have very derived postcrania with diphycercal fin fringes rather than separate median fins (Ahlberg and Trewin, 1995).

In parallel with this morphological change there is a presumed environmental shift from open marine environments such as Taemas or Gogo to "Old Red Sandstone" facies and finally to apparently nonmarine and oxygen-poor environments like coal swamps (Campbell and Barwick, 1988). Skeletal structures associated with air-breathing (cranial ribs and long parasphenoid stalk) are absent in the primitive marine lungfishes but appear during the Middle Devonian and seem to define a clade within the Dipnoi (Long, 1993).

The dramatic changes in dipnoan median fin morphology during the Paleozoic were used by Dollo (1895) to infer the evolution of the group (as well as his principle of the irreversibility of evolution). Paedomorphosis has been suggested as a primary process in this group in relation to the fusion of the median fins, reduction of lepidotrichia, reduction of ossification (Bemis, 1984), and dentitional pattern (Cloutier *et al.*, 1993; Long, 1993). However, in the absence of a fully resolved phylogeny (Schultze and Marshall, 1993) these hypotheses cannot be fully evaluated (Ahlberg and Trewin, 1995). Interestingly, the mode of growth of lungfish tooth plates (see previous discussion) allows the possibility of observing developmental heterochrony in phylogeny. Westoll (1949) and Schaeffer (1952) compared the evolutionary rates of dipnoans with the bradytelic evolution of actinistians. However, rates of evolution have never been calculated in a phylogenetic perspective for the Dipnoi.

C. Tetrapoda

Although Jarvik (1942, 1972, 1980) and Bjerring (1989, 1991) continue to argue for tetrapod diphyly and a urodele–porolepiform relationship, the monophyletic status of the Tetrapoda is supported by a wealth of characters and is accepted by virtually all other workers (Schultze, 1970, 1981, 1987; Jurgens, 1973; Gaffney, 1979; Rosen *et al.*, 1981; Shubin and Alberch, 1986; Panchen and Smithson, 1987; Vorobyeva and Schultze, 1991). It is also supported by molecular evidence (Hedges *et al.*, 1993).

The tetrapods are defined here as a clade characterized by the possession of limbs with digits rather than paired fins, a pelvis with a sacrum, and zygapophyses; early members can also be recognized by a suite of derived jaw characters (Ahlberg, 1991a, 1995; Ahlberg *et al.*, 1994). This apomorphy-based definition encompasses the crown group and part of the stem group and would thus be seen as unsatisfactory according to the criteria of De Queiroz and Gauthier (1990). However, at present there is a sharp divide between early limbed vertebrates such as *Ichthyostega* and *Acanthostega*, whose membership in the tetrapod stem group can be taken as a well-founded starting assumption, and tetrapod-like "fishes" such as *Panderichthys* and *Elpistostege*, whose membership in the stem group needs to be tested. We therefore retain the traditional definition for the present. Note that a similar situation exists with respect to the Dipnoi.

The tetrapods have a fossil record reaching back into the Frasnian (Warren and Wakefield, 1972; Ahlberg and Milner, 1994; Ahlberg, 1995). The "traditional" early tetrapod groups, Labyrinthodontia and Lepospondyli, were exposed as nonnatural during the past decade (Smithson, 1985; Milner *et al.*, 1986; Panchen and Smithson, 1988). They have not been replaced by a new consensus. However, it is clear that all the Devonian genera, except perhaps *Tulerpeton* (Lebedev and Coates, 1995), fall outside the crown group. Within the crown group the Amniota and Lissamphibia are Recent sister-groups. The temnospondyls are members of the lissamphibian clade, while the anthracosaurs probably belong with the amniotes, but opinions differ as to the position of the loxommatids and the old "lepospondyl" groups (Milner *et al.*, 1986; Panchen and Smithson, 1988; Trueb and Cloutier, 1991; Carroll, 1992; Ahlberg and Milner, 1994; Lebedev and Coates, 1995).

The origin of the tetrapod crown group clearly antedates the first appearance of anthracosaurs and temnospondyls in the late Viséan (Ahlberg and Milner, 1994). The tentative assignment of the Russian Famennian tetrapod *Tulerpeton* to the amniote–anthracosaur clade (Lebedev and Coates, 1995) suggests an even earlier date for the split.

In the context of sarcopterygian interrelationships, the most interesting tetrapods are the Devonian genera. They are not generally placed in higher taxonomic categories, as their interrelationships are poorly resolved. We recognize eight genera:

Ichthyostega Säve-Söderbergh 1932 is represented by numerous specimens from the Upper Famennian of eastern Greenland. Most of the skeleton except the manus is known, but the braincase is peculiar and poorly understood (Jarvik, 1980). The pes has seven digits (Coates and Clack, 1990), the shoulder lacks a scapular blade (Jarvik, 1980), and the tail carries lepidotrichia (Jarvik, 1952). *Ichthyostega* is uniquely characterized by the possession of an unpaired median postparietal.

Acanthostega Jarvik 1952 occurs alongside *Ichthyostega* in the upper Famennian of eastern Greenland. Long known only from two incomplete skulls, it is described in full from new specimens collected in 1987 (Clack, 1988, 1989, 1994a,b; Coates, 1991; Coates and Clack, 1990, 1991). *Acanthostega*'s manus has eight digits (Coates and Clack, 1990), and the same may be true for the pes (M. I. Coates, personal communication). The proportions of the forelimb elements are markedly more fishlike than those of *Ichthyostega*, and the lepidotrichial tail fin is even larger (Coates, 1995). The scapulocoracoid is comparable to that of *Ichthyostega* (Coates and Clack, 1991; M. I. Coates, personal communication).

Tulerpeton Lebedev 1984 is known from a single incomplete body and a number of isolated bones, all from the upper Famennian Andreyevka-1 locality near Tula, central Russia (Alexeev *et al.*, 1994). *Tulerpeton* has a manus with six digits (Lebedev, 1984). In certain other respects it resembles post-Devonian tetrapods; the limb bones are slender and a scapular blade is present in the shoulder girdle (Lebedev, 1984; Lebedev and Coates, 1995). Associated bones from the site, which have not been formally attributed to *Tulerpeton*, show derived characters like open lateral line sulci (Lebedev and Clack, 1993) which are not present in the other Devonian tetrapods.

Ventastega Ahlberg *et al.* 1994 is described from cranial material collected at the upper Famennian localities of Pavãri and Ketleri in Latvia. Tetrapod clavicles, interclavicles, and ilia from these localities may also belong to this genus (Ahlberg *et al.*, 1994). *Ventastega* is the only upper Famennian tetrapod known to possess coronoid fangs. In other respects it broadly resembles *Ichthyostega* and *Acanthostega*.

Hynerpeton Daeschler *et al.* 1994 is a genus of middle or upper Famennian age, based on a scapulocoracoid+cleithrum from the Duncannon Member of the Catskill Formation, Pennsylvania. The shoulder girdle clearly belongs to a stem tetrapod and resembles that of *Ichthyostega* as well as the girdle fragments from Scat Craig.

Metaxygnathus Campbell and Bell 1977 is represented by a single lower jaw ramus from the Cloghnan Shale of New South Wales, Australia, probably of lower Famennian age (Campbell and Bell, 1977). The jaw carries coronoid fangs. The assignation of *Metaxygnathus* to the Tetrapoda has been disputed (Schultze and Arsenault, 1985; Schultze, 1987). However, its tetrapod nature is confirmed by a suite of derived characters which are shared with *Acanthostega*, *Ichthyostega*, and *Ventastega* but not with sarcopterygian fishes (Ahlberg *et al.*, 1994).

Elginerpeton Ahlberg 1995 is strictly speaking, known only from cranial remains, but it has been associated with postcranial tetrapod bones which probably also belong to it. This genus comes from the upper Frasnian of Scat Craig, Scotland and is thus together with *Obruchevichthys* (see following discussion) the earliest tetrapod known from skeletal remains. It has several autapomorphies including large size (skull length in excess of 40 cm; Ahlberg, 1995), triangular head shape with an acutely pointed snout, and, on the inner face of the mandible, a broad field of exposed meckelian bone ventral to the very narrow prearticular. The postcranial tetrapod bones from the site include an *Ichthyostega*-like tibia (Ahlberg, 1991a) and robust scapulocoracoids and ilia (Ahlberg, 1995).

Obruchevichthys Vorobyeva 1977 is only known from two incomplete lower jaws, one from the upper Frasnian of Latvia and one from an unknown locality in western Russia. It shares several derived characters with *Elginerpeton* and appears to be the sister-group of that genus (Ahlberg, 1995). It is likely that the *Elginerpeton–Obruchevichthys* clade (plesion Elginerpetontidae; Ahlberg, 1995) is the sister-group of all other Tetrapoda.

Space does not permit us to list the post-Devonian tetrapod groups.

Apart from the aforementioned genera, the Devonian tetrapod record includes some well-preserved upper Frasnian trackways from Genoa River, Victoria, Australia (Warren and Wakefield, 1972) and ?Famennian trackways (with more than 150 footprints) from Valentia Island, southwestern Ireland (Stössel, 1995). A supposed Lower Devonian trackway from Australia (Warren *et al.*, 1986) cannot be confidently identified as belonging to a tetrapod, while the isolated Devonian "tetrapod footprint" described from marine Brazilian deposits by Leonardi (1983) is probably a starfish trace fossil (Rocek and Rage, 1994).

Two other genera were described originally as Devonian tetrapods. *Elpistostege* Westoll 1938 has proved to be a tetrapod-like fish (see following discussion), while *Ichthyostegopsis* Säve-Söderbergh 1932 appears to be synonymous with *Ichthyostega*.

Ichthyostega, Acanthostega, Ventastega, Metaxygnathus, Hynerpeton, Elginerpeton, and *Obruchevichthys* retain primitive characters not seen in any later tetrapods. *Ichthyostega, Acanthostega,* and *Ventastega* share certain features such as a spade-shaped head, dorsally placed orbits, external nostril close to the jaw margin, reduction or loss of the lateral rostral bone, a closed palate with a mobile basal articulation, and entopterygoids which meet anteriorly in a midline point between the vomers (Jarvik, 1980; Ahlberg *et al.*, 1994; Clack, 1994a). The braincases of *Acanthostega* and *Ichthyostega* both show a basicranial fissure and some development of a cranial notochord.

Ichthyostega, Acanthostega, and *Tulerpeton* possess more than five digits (Coates and Clack, 1990; Lebedev, 1984). Their humeri, like those of later tetrapods, are structurally comparable to those of osteolepiforms and rhizodonts but have a distinctive L shape (Andrews and Westoll, 1970a,b; Rackoff, 1980; Panchen, 1985; Panchen and Smithson, 1987; Ahlberg, 1989). A late Frasnian humerus from Scat Craig, associated with *Elginerpeton* (Ahlberg, 1991a; Ahlberg and Milner, 1994), is morphologically intermediate between those of Famennian tetrapods and osteolepiforms. *Ichthyostega* and *Acanthostega* lack scapular blades but have well-developed cleithra. Their vertebral columns are broadly similar to that of *Eusthenopteron* but have weakly developed zygapophyses (Andrews and Westoll, 1970a; Jarvik, 1980; Coates, 1995). Both genera have ribcages; the ribs of *Ichthyostega* are extremely broad, overlapping structures (Jarvik, 1980).

D. Onychodontida

The Onychodontida (=Onychodontiformes, Struniiformes) is known only from three genera (*Grossius, Onychodus,* and *Strunius*) ranging from the Pragian (Zhu and Janvier, 1994) to the Famennian (Schultze, 1993). Zhu and Janvier (1994) described a lower jaw from the Posongchong Formation of China as the oldest known onychodontid. *Grossius* is known from a single three-dimensional skull from the Frasnian of Spain (Schultze, 1973). *Onychodus* is represented primarily by parasymphysial tooth spirals and isolated bones (Jessen, 1966) but also by well-preserved articulated material from the lower Frasnian Gogo Formation of Western Australia (Andrews, 1973; Long, 1991). Jessen (1966) described two species of *Strunius* (*S. walteri* and *S. rolandi*) from the Frasnian of Germany (Upper Plattenkalk, Bergisch Gladbach) which at the moment remain the best described members of

the group. Aquesbi (1988) described an onychodontid from Morocco represented by a single poorly preserved specimen. Material of *Onychodus* sp. from the Gogo Formation is being studied by S. M. Andrews (National Museums of Scotland, Edinburgh). The interrelationships among onychodonts have never been examined because of the lack of comparative material.

The monophyly of the Onychodontida has never been addressed in detail. However, the presence of spiral parasymphysial teeth located dorsal to the dentaries has been suggested as a synapomorphy of the group (Jessen, 1966; Schultze, 1969, 1973). Aquesbi (1988) mentioned that the Onychodontida is characterized by the following: (1) a double series of long sigmoid parasymphysial teeth with striated or crenulated enamel, (2) a large infradentary bordering the dentary ventrally, (3) the absence of an interclavicle, and (4) a reduced opercular series. However, the Gogo *Onychodus* material contradicts character 2 (P. E. Ahlberg, personal observation). Smith (1989) proposed the organization of the enamel crystallites of the teeth into fine ribs with a superficial chevron pattern as an onychodontid synapomorphy.

The skull combines characters similar to actinopterygians (e.g., well-developed dorsal process on the maxilla and large preoperculum oriented horizontally) and typical sarcopterygian features such as an intracranial joint. It seems likely that the "actinopterygian-like" characters are actually plesiomorphic osteichthyan traits.

The sole family, Onychodontidae Woodward 1891, is coextensive with the Onychodontida.

E. Porolepiformes

The Porolepiformes are an exclusively Devonian group. The earliest known representatives are several species of *Porolepis* from the Pragian (= Siegenian; Harland *et al.*, 1990) of the Rhineland and Spitsbergen (Schultze, 1993); the latest is *Holoptychius* sp. from the latest Famennian of central Russia (Alexeev *et al.*, 1994), eastern Greenland (Bendix-Almgreen, 1976), and elsewhere. Schultze (1993) claims a Tournaisian record for *Holoptychius* on the basis of its occurrence in the *Groenlandaspis* Series of eastern Greenland. However, the attribution of this Series to the Carboniferous is questionable. Porolepiforms are absent from the Tournaisian of central Russia (Alexeev *et al.*, 1994), North America, and Britain.

Jarvik (1942) was the first worker to recognize that *Porolepis*, then the only known member of the family Porolepididae, shares many characters with Holoptychiidae such as *Holoptychius* and *Glyptolepis*. He united the Porolepididae and Holoptychiidae in the order Porolepiformes Berg (1937). The monophyly of this group has been accepted by all subsequent authors except Maisey (1986a), who interpreted the porolepiforms as a paraphyletic assemblage of stem lungfishes. However, this interpretation was largely based on the characteristics of *Powichthys* and *Youngolepis*, which fall outside the Porolepiformes *sensu* Jarvik (1942).

We define the Porolepiformes as a clade characterized by the possession of dendrodont teeth (Schultze, 1969; Panchen and Smithson, 1987), subsquamosals (Cloutier, 1990; Ahlberg, 1991b; Cloutier and Schultze, 1996), and a unique skull roof pattern in which the intertemporal and supratemporal are absent and the postotic sensory canal passes through the growth center of the postparietal bone (Ahlberg, 1992c). The clade thus defined is equivalent to Porolepiformes of Jarvik (1942), although the diagnostic characters are different.

The porolepiforms are not a diverse group. We recognize eight genera, two in the Porolepididae and six in the Holoptychiidae:

"Porolepididae" Berg 1940 is a paraphyletic group defined by the possession of cosmine. The best known genus is *Porolepis* Woodward 1891, which ranges in age from Pragian to Givetian (Schultze, 1993). It is unclear whether *Porolepis* is a clade or simply a paraphyletic assemblage of primitive porolepiforms. *Heimenia* Ørvig 1969, also has cosmine, but the scale morphology is intermediate between those of *Porolepis* and the Holoptychiidae (Ørvig, 1969). Only scales and a single lower jaw of *Heimenia* (Jarvik, 1972, pl. 12-6) have been figured or described to date.

Holoptychiidae Owen 1860 is a clade defined by the possession of round scales, lack of cosmine, lack of median gular plate, and a relatively short ethmosphenoid cranial division (Ahlberg, 1992c). *Glyptolepis* Miller *ex* Agassiz 1841, ranges from the Eifelian to the early Frasnian (Lyarskaya, 1981). As traditionally defined, this genus may be paraphyletic with respect to other holoptychids (Ahlberg, 1992c). *Quebecius* Schultze 1973, from the middle Frasnian of Miguasha, Québec (Cloutier *et al.*, 1996), resembles *Glyptolepis* but is distinguished by a unique cheekplate pattern (Cloutier and Schultze, 1996). The early Frasnian genus *Laccognathus* Gross 1941, is defined by an autapomorphic dermal ornament composed of large tubercles with thick enamel (Ørvig, 1957) and by the possession of very large infradentary foramina (Gross, 1941; Ahlberg, 1992c). *Holoptychius* Agassiz *in* Murchison 1839, ranges in time from the middle Frasnian (Jarvik, 1972; Cloutier

et al., 1996) to the end of the Devonian (Alexeev *et al.*, 1994; see previous discussion). It has dermal ornament composed of laminar bone rather than dentine (Ørvig, 1957). *Duffichthys* Ahlberg 1992c, is represented by autapomorphic lower jaws from the upper Frasnian of Scat Craig, Scotland. The Middle Devonian genus *Hamodus* Obruchev 1933 is only known from isolated, very large dendrodont teeth with barbed tips. A further holoptychiid genus, *Paraglyptolepis* Vorobyeva 1987, has been described from the Givetian of Estonia. However, on basis of the available material it is not clear that this genus can be distinguished from *Glyptolepis*.

In the Early and Middle Devonian, porolepiforms tend to be the largest predators in the faunas where they occur (Ahlberg, 1992b). Maximum size for the group seems to be close to 2m. They seem to have been the first sarcopterygian group to evolve elaborate branched lateral line systems; their cranial sensory canals have numerous first- and second-order side branches, which cover almost the whole skull surface except the operculogular series. The gross morphology of the holoptychiids is extremely stereotyped (Ahlberg, 1992b).

On the whole, porolepiform cranial anatomy is fairly similar to that of osteolepiforms. This is particularly true for the intracranial joint, which runs through the profundus foramen in both groups. However, the ethmosphenoid braincase block and lower jaw are much closer to those of *Powichthys* and *Youngolepis* (Jessen, 1980; Chang, 1982, 1991a; Ahlberg, 1991b). The basibranchial skeleton lacks a sublingual rod, unlike that of osteolepiforms (Jarvik, 1972). The vertebral column is lungfish-like, as are the archipterygial pectoral fins (Ahlberg, 1989, 1991b). However, the pelvic fins have asymmetrical endoskeletons of a more generalized sarcopterygian pattern (Ahlberg, 1989). The spread of anatomical information among the Porolepiformes is patchy. The postcranial endoskeleton has only been described from *Glyptolepis*, although personal observation (by P.E. Ahlberg) of *Laccognathus* specimens in the care of Emilia Vorobyeva shows a very similar vertebral column and fin supports.

F. *Powichthys* and *Youngolepis*

These two genera from the Early Devonian (Lochkovian–Pragian) of Arctic Canada (*Powichthys*) and South China and Vietnam (*Youngolepis*) show affinities with both porolepiforms and lungfishes. They are known mostly from cranial remains, although the shoulder girdle of *Youngolepis* was described by Chang

(1991a) and a cleithrum associated with *Powichthys* was figured by Jessen (1980).

Powichthys Jessen 1975, from Prince of Wales Island in the Canadian Arctic, is the more porolepiform-like of the two and was referred to the Porolepiformes by its discoverer (Jessen, 1975, 1980). However, it lacks the derived porolepiform skull roof pattern and has polyplocodont rather than dendrodont tooth folding. *Powichthys* resembles the porolepiforms most closely in the structure of its ethmosphenoid. It has a pair of well-developed internasal pits between the vomers, and there is a large profundus foramen in the postnasal wall. However, the snout also contains rostral tubuli like those in lungfishes (Jessen, 1980). An opercular series associated with the genus (but not formally attributed to it; Jessen, 1980) appears to have contained a preoperculosubmandibular bone similar to that of porolepiforms. The lower jaw (again not formally attributed, but very probably belonging to *Powichthys*) has a porolepiform-like parasymphysial tooth plate attachment and three infradentary foramina similar to those of *Holoptychius* and *Laccognathus*. However, the immobilized intracranial joint lies at the level of the trigeminal and lateral ophthalmic nerves, as in coelacanths.

Youngolepis Chang 1982, from Yunnan, China (Chang, 1982, 1991b) and Vietnam (Tong-Dzuy Thanh and Janvier, 1990, 1994), has an extraordinary braincase which combines porolepiform- and lungfish-like features with apparent actinopterygian characteristics. The latter are most obvious around the posterior part of the braincase floor. There is a basicranial fissure rather than a fenestra. A "basicranial process" from the lateral commissure reaches forward toward a "processus descendens" from the sphenoid (Chang, 1982) just as in *Mimia* (Gardiner, 1984). The sphenoid is pierced by separate foramina for the carotid and efferent pseudobranchial arteries. Although these features are otherwise known only from actinopterygians, they are probably to be interpreted as plesiomorphic osteichthyan characters (Ahlberg, 1994). In *Youngolepis* their co-occurrence with an unconstricted cranial notochord, and an apparent remnant of the intracranial joint in the side wall of the braincase (Chang, 1982), raises the possibility that they are reversals from a fully developed intracranial joint.

Other parts of the anatomy of *Youngolepis* seem to show a mixture of porolepiform, lungfish, and general sarcopterygian characters. The lower jaw resembles that associated with *Powichthys* and has infradentary foramina, the snout contains rostral tubuli as in *Powichthys* and lungfishes, the cheekplate is broadly osteolepiform-like with extensive squamosal–

maxillary contact, and the shoulder girdle has a flattened but essentially tripodal scapulocoracoid.

G. Osteolepiformes

The Osteolepiformes (Berg, 1937; Jarvik, 1942) is by far the most diverse of the extinct sarcopterygian groups though much less diverse than the Actinistia, Dipnoi, or Tetrapoda. Approximately 60 species from some 25 genera have been described, ranging in age from Middle Devonian (Eifelian) to Lower Permian (Sakmarian). Among them is *Eusthenopteron foordi*, the most thoroughly studied fossil sarcopterygian and one of the best known of all fossil vertebrates (Whiteaves, 1883, 1889; Goodrich, 1902; Jarvik, 1937, 1942, 1944a,b, 1954, 1963, 1980; Andrews and Westoll, 1970a). Yet for all this our anatomical knowledge of the osteolepiforms remains patchy; *Eusthenopteron* stands out against a host of incomplete and poorly understood genera.

The overall impression given by the osteolepiforms is of a rather homogenous group of similar-looking fishes. However, this homogeneity does not necessarily imply monophyly; it is possible that the group Osteolepiformes is paraphyletic relative to the Rhizodontida, Elpistostegalia + Tetrapoda, or both.

The cranial anatomy is best illustrated by *Eusthenopteron*, although comparable information on the neurocranium is available from *Megalichthys* (Romer, 1937), *Gogonasus* (Long, 1988a), and *Medoevia* (Lebedev, 1995). The dermal skull bones are well known in the Scottish Middle Devonian genera *Osteolepis*, *Thursius*, and *Gyroptychius* (Jarvik, 1948). The braincase is divided by an intracranial joint running through the foramen for the profundus nerve. There is only one external nostril on each side of the head, but a large palatal opening surrounded by vomer, dermopalatine, maxilla, and premaxilla appears to have transmitted a choana (Jarvik, 1942; Panchen and Smithson, 1987). This interpretation was challenged by Rosen *et al.* (1981), who tried to show that the size of the opening had been exaggerated by Jarvik. However, new evidence from acid-prepared specimens (Long, 1988a; Lebedev, 1995; P. E. Ahlberg, personal observation) corroborates Jarvik's description and shows that the opening bears a very close resemblance to the choanae of Devonian tetrapods (Jarvik, 1980; Clack, 1994a). The otoccipital braincase block broadly resembles that of actinistians but retains lateral otic fissures which end in large vestibular fontanelles.

Dermal bone characteristics of the group include a large squamosal which separates the rather narrow preopercular from the maxilla. Cosmine is primitively present in osteolepiforms but has been lost in many genera. Vertebrae are either rhachitomous or ringcentra (Andrews and Westoll, 1970a,b). Ribs, if present, are short. The paired fin skeletons are short, uniserial metapterygia, and all fin radials are unjointed and unbranched (Andrews and Westoll, 1970a,b). The humerus is structurally very similar to that of basal tetrapods, although the actual shape is rather different (Ahlberg, 1991a,b). In the majority of osteolepiforms the paired and median fin bases carry enlarged scales, the so-called basal scutes.

The classification of the Osteolepiformes is in urgent need of revision. Schultze (1993) divides them into the following families:

"Osteolepididae" Cope 1889 is a paraphyletic group of primitive osteolepiforms. A typical representative is *Thursius* Sedgwick and Murchison 1828. Within this group, the megalichthyids can be recognized as a clade on the basis of several cranial characters (Young *et al.*, 1992). The osteolepidids range in time from Devonian (Eifelian) to Permian (Sakmarian).

Canowindridae Young *et al.* 1992 is a clade characterized by a posteriorly broad postparietal shield, lateral extrascapulars which almost meet in the midline anteriorly, and exclusion of the postorbital from the orbital margin. Three canowindrid genera are known: *Canowindra* Thomson 1973, *Beelarongia* Long 1987, and *Koharalepis* Young *et al.* 1992. At present the group appears restricted to the Upper Devonian of Australia and Antarctica.

Tristichopteridae Cope 1889 (=Eusthenopteridae Berg 1940) is a clade characterized by the presence of posttemporal bones between the operculars and lateral extrascapulars. Tristichopterids also lack cosmine, have thin round scales with a median ridge on the inner surface, and possess a characteristic triphycercal caudal fin. The earliest known tristichopterid is *Tristichopterus* (Andrews and Westoll, 1970b) from the upper Givetian John O'Groats Sandstone, Scotland, while the latest is *Eusthenodon* from the upper Famennian *Remigolepis* Series of eastern Greenland (Jarvik, 1952). *Marsdenichthys* from the Frasnian of Australia is held by Long (1985) to be the most primitive known member of the group. *Eusthenopteron* is the only tristichopterid to have been studied in great detail, and relationships within the group remain obscure.

Rhizodopsidae Berg 1940 is a group ranging from the Carboniferous to Permian. The best known

genus is *Rhizodopsis* (Andrews and Westoll, 1970b; Moy-Thomas and Miles, 1971).

H. Rhizodontida

The Rhizodontida (Andrews and Westoll, 1970b) are a group of Devonian and Carboniferous fishes chiefly remarkable for their great size. The Scottish Lower Carboniferous form *Rhizodus hibberti* seems to have reached a length of 7 m (Andrews, 1985). Most known rhizodont specimens consist of partly or wholly disarticulated material. Only one complete individual, a juvenile of ?*Strepsodus anculonamensis*, has been described; it has an elongate body with small median fins clustered near the diamond-shaped symmetrical tail (Andrews, 1985, fig. 2; this paper, Fig. 2g).

Despite our incomplete knowledge of rhizodont anatomy, the Rhizodontida can unambiguously be recognized as a clade. Synapomorphies of the group include robust lepidotrichia with extremely long unjointed proximal portions, and the presence on the cleithrum of a depressed posterior flange and an elaborate double overlap area for the clavicle (Andrews and Westoll, 1970b; Andrews, 1985; Long, 1989; Young *et al.*, 1992). The humerus has a bulbous head which forms a ball-and-socket joint with the round glenoid, and the radials of the pectoral fin skeleton are both jointed and branched (Andrews and Westoll, 1970b). Cosmine is always absent and the scales are round and thin.

The best understood part of rhizodont anatomy is the pectoral girdle. Parts of the dermal skull have been described from ?*Strepsodus* and from *Screbinodus* (Andrews, 1985) and *Barameda* (Long, 1989). The bone pattern is broadly similar to that of osteolepiforms, but *Barameda* (the most complete rhizodont) shows some unusual characters such as an extratemporal which contacts the supratemporal ("intertemporal" of Long, 1989) and a reduced postrostral mosaic. Long (1989) interpreted *Barameda* as having two external nostrils on each side. However, the reconstructed pattern conflicts with evidence from a detached *Strepsodus* premaxilla (BMNH P364(2); P. E. Ahlberg, personal observation) which shows a continuous overlap area for the lateral rostral in the region where Long placed the anterior nostril. Andrews (1985) reconstructed an osteolepiform-like arrangement of bones in the narial region but felt uncertain whether one or two nostrils were present. It is also unclear whether rhizodonts possess a choana. The condition of the nasal region is important, as the supposed possession of two external nostrils was one of the main features which led Long (1989), Ahlberg (1991b), and Young *et al.* (1992) to place rhizodonts below osteolepiforms in the tetrapod stem group.

A number of braincase fragments have been described from the Antarctic genus *Notorhizodon* (Young *et al.*, 1992). They include a well-developed intracranial joint and closely resemble the corresponding parts of *Eusthenopteron*; this also applies to the palatoquadrate of *Notorhizodon* (Young *et al.*, 1992). The lower jaw, known from *Notorhizodon* and in part also from *Barameda* (Long, 1989) and *Strepsodus* (Andrews, 1985), resembles those of osteolepiforms and porolepiforms. It has a very strongly developed symphysial fang pair on the dentary.

Probably the earliest known rhizodont is *Notorhizodon* from the "Middle–Late Devonian" Aztec Siltstone of Antarctica (Young *et al.*, 1992), while the last known representative is *Strepsodus* from the Westphalian Coal Measures of England (Schultze, 1993). The only analysis undertaken to date of rhizodont interrelationships was that of Young *et al.* (1992), which produced the following topology: [*Notorhizodon* + [*Barameda* + [*Screbinodus* + [*Rhizodus* + *Strepsodus*]]]]. This hypothesis is biogeographically interesting in that the two genera judged to be most primitive both derive from East Gondwana.

I. Elpistostegalia

This group is generally known as Panderichthyida Vorobyeva 1989, but Elpistostegalia Camp and Allison 1961 has priority (Schultze, 1996). The name of the single constituent family should likewise be Elpistostegidae Romer 1947, rather than Panderichthyidae Vorobyeva and Lyarskaya 1968. The group is of great phylogenetic interest because its members display a melange of tetrapod-like and osteolepiform-like characters. It is stratigraphically restricted and of low diversity; at present we recognize only two genera and three species.

Elpistostege Westoll 1938 is represented by the single species *Elpistostege watsoni* from the middle Frasnian of Miguasha, Québec, Canada. Two incomplete skulls and a section of vertebral column are known (Schultze and Arsenault, 1985) of this youngest elpistostegid.

Panderichthys rhombolepis Gross 1941 was originally described on the basis of incomplete lower jaws from the early Frasnian of Latvia. The discovery of several complete specimens at Lode quarry in Latvia (Lyarskaya and Mark-Kurik, 1972) resulted in a series of descriptive and interpretive papers (Vorobyeva, 1977, 1980, 1986, 1989; Vorobyeva and Tsessarskii, 1986; Vorobyeva and Schultze,

1991; Vorobyeva and Kuznetsov, 1992; Worobjewa, 1975) which have made this the best known elpistostegid.

Panderichthys stolbovi Vorobyeva 1960 is a slightly younger, though still early Frasnian, species from Russia that is known from a snout fragment and some incomplete lower jaws (Vorobyeva, 1960, 1962, 1971). *Panderichthys stolbovi* appears very similar to *P. rhombolepis*, but the two can be distinguished by their slightly different dermal ornament (P. E. Ahlberg, personal observation).

In addition to these *bona fide* elpistostegids, *Panderichthys bystrovi* Gross 1941 and *Obruchevichthys gracilis* Vorobyeva 1977 have also been attributed to the group. The material of "*Panderichthys*" *bystrovi* comes from the late Famennian locality of Ketleri in Latvia. The holotype, a mandibular fragment, needs to be redescribed; it certainly comes from a sarcopterygian fish but does not appear to belong to an elpistostegid (P. E. Ahlberg, personal observation). The maxilla and premaxilla attributed to *P. bystrovi* by Vorobyeva (1962) actually belong to a tetrapod, *Ventastega curonica* (Ahlberg *et al.*, 1994). *Obruchevichthys*, which is known only from two late Frasnian mandibular fragments, also appears to be a primitive tetrapod (Ahlberg, 1991a, 1995; Ahlberg *et al.*, 1994; see below). Genuine elpistostegids are thus at present restricted to the early and middle Frasnian.

Elpistostegid cranial anatomy resembles that of osteolepiforms in many respects. In *Panderichthys rhombolepis* and *P. stolbovi* (the two species where this region is known) the anterior end of the palate compares with that in *Eusthenopteron*: the vomers have well-developed posterior processes which suture to the sides of the parasphenoid and prevent the entopterygoids from meeting in the midline. This is quite different from the tetrapod pattern. The lower jaw likewise lacks obvious tetrapod characteristics (Ahlberg, 1991a) but resembles those of tristichopterids such as *Eusthenodon* (P. E. Ahlberg, personal observation) and *Platycephalichthys* (Vorobyeva, 1962) as well as that of the rhizodont *Notorhizodon* (Young *et al.*, 1992). The bone and sensory line pattern around the external nostril matches that of *Eusthenopteron* (Vorobyeva and Schultze, 1991; Jarvik, 1980).

The most obvious tetrapod-like structure in the elpistostegid skull is a pair of frontals anterior to the parietals (Westoll, 1938; Vorobyeva, 1977; Schultze and Arsenault, 1985; Vorobyeva and Schultze, 1991). It is worth noting in passing that the elpistostegid skull roof pattern furnishes powerful support for Westoll's (1938) terminology of dermal skull bones in osteichthyan fishes (Schultze and Arsenault, 1985;

Ahlberg, 1991b). As in tetrapods, but unlike osteolepiforms, the parietals and postparietals of elpistostegids are immovably sutured together. The intracranial joint must thus have been immobile and was possibly obliterated altogether.

Just as striking as these anatomical structures is the tetrapod-like morphology of the elpistostegid head. The skull is flattened and spade-shaped, the orbits are dorsal and crowned by bony "eyebrows," the interorbital skull roof is narrow and concave, and the external nostrils are almost marginal. These features must be related to the mode of life of the animals, and may indicate a shallow-water or marginal lifestyle (Ahlberg and Milner, 1994; Schultze, 1996).

A similar situation obtains with respect to the postcranial skeleton; the straight tail, lack of separate dorsal and anal fins, and probably dorsoventrally flattened body of *Panderichthys rhombolepis* can all be matched in *Ichthyostega* and *Acanthostega* and suggest that the elpistostegids may have had a capacity for terrestrial locomotion (Vorobyeva and Kuznetsov, 1992). The humerus and scapulocoracoid of *P. rhombolepis* combine tetrapod-like features with apparent autapomorphies (Vorobyeva and Schultze, 1991; Vorobyeva and Kuznetsov, 1992). Unfortunately these elements are unknown in the other two species. The vertebral column of *P. rhombolepis* has peculiar bladelike ribs which are sutured to the intercentra and neural arches. Pleurocentra are absent (Vorobyeva and Tsessarskii, 1986). Similar bladelike elements in a vertebral column attributed to *Elpistostege* are identified as neural arches by Schultze and Arsenault (1985), but these too appear to be ribs (R. Cloutier, personal observation).

The combination of characters seen in the Elpistostegalia raises important phylogenetic questions. Their synapomorphies with tetrapods are striking and suggest that the two are sister-groups (Vorobyeva and Schultze, 1991; Cloutier, 1990; Ahlberg, 1991b; Ahlberg and Milner, 1994; but see Panchen and Smithson, 1987). Many of their plesiomorphic characters are general osteichthyan traits and thus unproblematical, but others are shared specifically with tristichopterids and rhizodonts and could pose a challenge to osteolepiform monophyly (see preceeding sections). Most interesting of all are the characters that bear on the question of elpistostegid monophyly. Vorobyeva and Schultze (1991) propose five elpistostegid synapomorphies: (1) median rostral separated from premaxilla, (2) paired posterior postrostrals, (3) large median gular, (4) lateral recess in nasal capsule, and (5) subterminal mouth (=prominent snout). Character 1 is erroneous, as a comparable median rostral is developed in both osteolepiforms and basal tetrapods (Jarvik, 1980;

Ahlberg, 1995). Characters 2, 3, and 4 are all indeterminable in tetrapods (the postrostral mosaic and gular series have been lost altogether, while the nasal capsules are unossified and thus unknown in all fossil tetrapods) and thus not really testable. This seems to leave a prominent snout as the only elpistostegid synapomorphy.

Set against this character are a couple of features (intertemporal present in *Panderichthys rhombolepis* but absent in *Elpistostege* and known Devonian tetrapods; more tetrapod-like ornament in *Elpistostege* than in *P. rhombolepis*, P. E. Ahlberg, personal observation) which suggest the Elpistostegalia might be paraphyletic with respect to the tetrapods. At present our understanding of *Elpistostege* and *P. stolbovi* is too incomplete to allow the question of elpistostegid monophyly or paraphyly to be settled. However, the implications of the question are profound: if the elpistostegids are paraphyletic with respect to tetrapods, the many morphological details which give a common elpistostegid appearance to *Elpistostege* and *P. rhombolepis* (snout outline, shape and position of "eyebrows," etc.) will actually be attributes of the tetrapod stem lineage. A paraphyletic group Elpistostegalia would, in other words, provide much more detailed information about the earliest stages of tetrapod evolution than would an elpistostegid clade. The investigation of this area should be a priority for future sarcopterygian research programs.

III. The Character Set

The character set which we present is essentially a consensus list based on our earlier works (Ahlberg, 1989, 1991b; Cloutier, 1990), with the addition of a few characters from other sources (e.g., Chang and Smith, 1992). A total of 140 characters were combined (Appendix 1), and these include only those characters that can be recognized in early fossil sarcopterygians. The reasons for this approach are worth examining.

The debate over the relative merits of fossil and recent data goes back two decades (Løvtrup, 1977; Patterson and Rosen, 1977; Patterson, 1981, 1982a,b; Rosen *et al.*, 1981; Schoch, 1986; Doyle and Donoghue, 1986, 1987; Donoghue *et al.*, 1989; Forey, 1987; Panchen and Smithson, 1987; Schultze, 1987, 1994; Campbell and Barwick, 1987, 1988; Gauthier *et al.*, 1988; Huelsenbeck, 1991). It has often been clouded by conflation with the separate issue of whether fossils can be taken to represent actual ancestors of Recent forms. However, the view that fossil and recent organisms should all be treated as terminal taxa in a phylogenetic analysis has gradually gained near-universal acceptance.

There are two main schools of thought about the treatment of fossil data in a cladistic analysis. One argues that the cladogram should be constructed on the basis of character distributions among the living taxa and that fossils should only then be mapped onto the topology; the fossils are thus not allowed to modify the topology. This approach was applied by Patterson and Rosen (1977) to teleosts, and by Forey (1987)—with certain reservations—to sarcopterygians. The other school rejects this division and uses both fossil and Recent data in the initial analysis. Most paleontologists appear to fall into the latter school (Schultze, 1987; Panchen and Smithson, 1987; Vorobyeva and Schultze, 1991; Chang, 1991a,b).

We follow Gauthier *et al.* (1988) in rejecting *a priori* primacy for the characters of Recent taxa. In some cases, the character combinations displayed by sequences of plesions manifestly have the capacity to overturn phylogenetic judgements based on living taxa alone, and we can see no justification for artificially preventing this outcome.

While it is generally possible to get better and more detailed anatomical information from Recent taxa than from fossils, this is not equally true for all characters. Features of adult skeletal morphology are often just as well understood in well-preserved fossils as in Recent organisms. The real disadvantage of fossils lies in the complete absence of certain kinds of data such as physiology, development, and in most cases soft anatomy and gene sequences. We designate these as "neontological" characters and use the term "paleontological" for such characters as can be detected in both living and fossil organisms.

In practice, all character sets are affected by at least one of three types of problems which limit their usefulness in phylogenetic analysis. These are the following:

(1) Incomplete distribution: the characters are not known in all of the relevant taxa owing to anatomical incompleteness of the organisms.

(2) Poor understanding or definition of characters: the characters have limited "information content," making it difficult to distinguish homology from homoplasy.

(3) Low number of characters: the character set is too small to provide adequate support for all nodes.

Purely neontological data sets are usually not much affected by problem 3 but will suffer from 1 in direct proportion to the number of fossil taxa involved in the phylogenetic analysis. Paleontological data sets are often prone to problems 2 and 3. They are also affected by 1, in so far as fossil taxa are often incomplete. However, whereas "neontological" characters are typically known in all living taxa and unknown in all fossil ones, the gaps in a paleontological data set are determined by the preservation of different fossils and are likely to be more randomly distributed across the range of taxa.

The Sarcopterygii includes three crown groups, namely the Tetrapoda (amniotes + lissamphibians), the Dipnoi (*Neoceratodus, Protopterus,* and *Lepidosiren*), and the Actinistia (*Latimeria*). The tetrapod and dipnoan crown groups date back approximately to the basal Carboniferous and the Lower Triassic, respectively (Ahlberg and Milner, 1994; Schultze and Marshall, 1993), whereas the monospecific actinistian crown group has no fossil record at all. However, each of these crown groups is associated with a recognized stem group which reaches back to the Devonian; there is no disagreement that *Acanthostega* and *Ichthyostega* are stem tetrapods, *Dipnorhynchus* and *Uranolophus* are stem dipnoans, and *Diplocercides* and *Euporosteus* are stem actinistians. Alongside the long-lived clades we find exclusively Paleozoic groups such as onychodonts, osteolepiforms, elpistostegids, porolepiforms, and rhizodonts. Some of these appear to be clades, but others may be paraphyletic taxa (Rosen *et al.,* 1981; Young *et al.,* 1992). We also have the more isolated Early Devonian genera *Powichthys, Youngolepis,* and *Kenichthys* (Jessen, 1975, 1980; Chang, 1982, 1991a,b; Chang and Smith, 1992; Chang and Zhu, 1993).

The main phylogenetic uncertainties revolve around the relationships *between* the long-lived clades and various extinct groups; all the most debated phylogenetic nodes lie in or below the Devonian. Furthermore, outgroup-based phylogenetic analyses of the long-lived clades (Cloutier, 1991a,b; Forey, 1991; Schultze *et al.,* 1993; Ahlberg and Milner, 1994; Lebedev and Coates, 1995) indicate that the most plesiomorphic and phylogenetically basal members of each clade are Devonian fossil genera. Fossil taxa thus occupy crucial positions in the analysis, and it is clear that paleontological characters will be very important for sorting out their relationships. However, we have decided to go one step further in omitting neontological data from the analysis altogether.

The overall neontological data set divides naturally into morphological information and molecular data. Molecular phylogenetics is a relatively new field, but a number of workers have already tackled the three-taxon problem of lungfishes, actinistians and tetrapods (Meyer and Wilson, 1990, 1991; Gorr *et al.,* 1991; Stock *et al.,* 1991; Stock and Swofford, 1991; Sharp *et al.,* 1991; Normark *et al.,* 1991; Hedges *et al.,* 1993). No consensus view has yet emerged from this research, and there are significant methodological disagreements within the field, although the bulk of the molecular evidence seems to support a lungfish–tetrapod sister-group relationship (Meyer, 1995). As a detailed discussion of the role of molecular data is given by Marshall and Schultze (1992) and Schultze (1994), we will not give any further consideration to molecular data in this paper.

In recent years the most important investigations of neontological anatomy have been those of Fritzsch (1987, 1988, 1992), Trueb and Cloutier (1991), and Northcutt and Bemis (1993). Fritzsch described some possible coelacanth–tetrapod synapomorphies from the structure of the inner ear, as did Northcutt and Bemis (1993) who focused on neurological characters; Trueb and Cloutier favored the topology [Actinistia + [Dipnoi + Tetrapoda]]. Our main reason for not using these data is a wish to focus attention on the flood of new paleontological information which has become available during the past two decades (see Historical Background). A subsidiary consideration is the distribution of the data. Because neontological characters are unknown in all the fossil taxa and can only support a few of the many nodes in the phylogenetic reconstruction, the use of large numbers of such characters seems likely to affect the analysis in unpredictable ways.

Under the circumstances, we prefer to focus our present analysis entirely on the paleontological data set (Appendix 2). In effect, we want to see whether the "basal radiation" of sarcopterygians—i.e., the short-lived groups and the early representatives of the surviving clades—contains any obvious phylogenetic pattern. The results can then be compared with neontological (both molecular and morphological) and "total evidence" phylogenies for the Sarcopterygii in order to map out areas of agreement and disagreement.

IV. Discussion

The data matrix includes 140 characters and a total of 158 apomorphic character-states (Appendix 3). Appendix 1 provides the complete list of characters and their respective character-states. All characters were entered unordered and unweighted. All but one taxon (the actinopterygian *Polypterus*) are extinct ranging in time from the Lower Devonian to the Upper Carboniferous (Appendix 2). Most of the data matrix (Appendix 3) was coded based on our respective observation of original material with the exception of *Howqualepis* (Long, 1988b), *Speonesydrion* (Campbell and Barwick, 1983, 1984), *Dipnorhynchus* (Campbell and Barwick, 1982a,b), *Beelarongia* (Long, 1987; Young *et al.*, 1992), and *Barameda* (Long, 1989).

Fifty-four most parsimonious trees at 277 steps were found using the heuristic search (C.I. = 0.578; C.I. excluding uninformative characters = 0.572; R.I. = 0.818) using the 140 characters coded for 32 taxa (including five outgroup taxa). The tree was rooted on a monophyletic outgroup including *Polypterus*, *Cheirolepis*, *Mimia*, *Moythomasia*, and *Howqualepis*. The Adams and strict consensus trees show the same topology (Fig. 3). Four topological variants were found: (1) among dipnoans, (2) at the base of the Tetrapodomorpha, (3) among osteolepiforms, and (4) among tetrapods. As the characters selected for the analysis were chosen for their potential to resolve relationships between (rather than within) acknowledged clades, the topological variation at variants (1) and (4) can be disregarded as unimportant.

The monophyly of the Actinistia, Onychodontida, Dipnoiformes, Porolepiformes, Rhizodontida, Elpistostegalia, and Tetrapoda is corroborated (Table 1). However, the Osteolepiformes and Youngolepidida do not appear to be monophyletic groups. The monophyly of most clades is robust (72% for the Elpistostegalia and Rhizodontida to 98% for the Actinistia and Dipnoi, based on 100 bootstrap replicates); a monophyletic Osteolepiformes has been replicated only 34%.

The Actinistia is the sister-group of the remaining sarcopterygians (i.e., [Onychodontida + Rhipidistia]). Two clades constitute the Rhipidistia: (1) the Dipnomorpha including the Dipnoiformes and Porolepiformes and (2) the Tetrapodomorpha. In contrast to the conclusions of Chang and Smith (1992), *Powichthys* and *Youngolepis* do not form a monophyletic group but rather consecutive plesions in the stem group of the Dipnoi; this pattern had been postulated by Ahlberg (1991b). Thus we consider the Dipnoiformes to include *Powichthys*, *Youngolepis*, and Dipnoi. There is no evidence for the monophyly of the Youngolepididae (Gardiner, 1984), Youngolepiformes (Chang and Smith, 1992), and Youngolepidida (Young *et al.*, 1992). The dipnoan *Diabolepis* is considered to be the sister-group of the remaining Dipnoi as suggested by Cloutier (1990) and Chang (1991b). The Rhizodontida is the sister-group of the Osteolepidida as suggested by numerous authors (Cloutier, 1990; Ahlberg, 1991b; Vorobyeva and Schultze, 1991; Young *et al.*, 1992). The interrelationships within the Osteolepidida are as follow: ["Osteolepiformes" + [Elpistostegalia + Tetrapoda]]. As first suggested by Schultze and Arsenault (1985), the Elpistostegalia is the sister-group of the Tetrapoda. However, in terms of extant organisms, the Dipnoi is the Recent sister-group of the Tetrapoda as suggested by Rosen *et al.* (1981).

The distribution of characters is given for the major nodes concerning interrelationships among sarco-

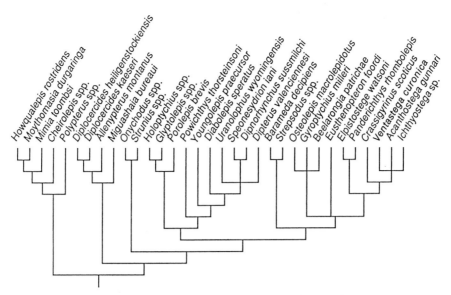

FIGURE 3 Interrelationships of 28 sarcopterygian taxa. Adams and strict consensus tree based on the 54 most parsimonious trees at 277 steps [C.I. = 0.578; R.I. = 0.818].

pterygian higher taxa (Fig. 4). Complete lists of character changes for these nodes are given in Table 1. Only the uniquely shared, derived characters common to the 54 most parsimonious trees are discussed in this section unless controversial anatomical structures are involved.

A. Sarcopterygii Romer 1955

Although unquestioned, the monophyly of the Sarcopterygii was supported by more than 30 characters (Table 1). Some of the synapomorphies represent the presence of new structures [tectals (char. 42), the

TABLE 1 Distribution of Characters Common to the 54 Most Parsimonious Trees for Major Sarcopterygian Clades

Taxa	Uniquely shared derived characters	Reversals	Homoplasies
Sarcopterygii	4, 18, 42, 49, 52, 54, 63(2), 88, 93(1), 105–106, 110, 120, 128		3, 28, 29(2), 34, 37, 40(2), 74, 81–85, 89, 94, 96, 112–113, 121, 124, 137
[Onychodontida + Rhipidistia]	93(2)		12, 35, 48, 64, 79, 92, 103, 139
Rhipidistia	56	62, 66, 118	14, 21, 29, 115–116
Dipnomorpha	2, 41, 43, 130	34, 74, 121, 135, 137	1, 31(1), 59, 78, 123
Dipnoiformes	10, 77, 100, 119(1)	48, 81, 86, 89–90, 139	87
[Youngolepis + Dipnoi]	17, 45, 108	59, 82–84, 102	
Tetrapodomorpha	44, 71, 119(2), 122	32, 79, 92	
Osteolepidida		113, 124	70
[Elpistostegalia + Tetrapoda]	15, 38, 67, 127	81–84, 139	20, 25, 47, 58, 135(2)
Actinistia	76, 95, 97, 138		19, 53, 58, 87, 91, 117, 129
Onychodontida	41(2)		11, 36, 57(2)
Porolepiformes	14(2), 51, 55, 63, 75, 101	29, 73	20, 36, 37(2), 129
Dipnoi	30, 65, 80	12, 14, 35, 79, 85, 92, 111	9, 19, 31(2), 47, 58, 91
Rhizodontida	114, 132(2)		
Elpistostegalia	33		
Tetrapoda	24, 39, 60–61, 68, 125	64	62, 66, 112(2)

Note. The category "uniquely shared derived characters" lists all characters with a C.I. equal to 1. The names of Taxa are those used in the text and in Fig. 4. See trees in Figs. 3 and 4. Appendix 1 provides the character descriptions.

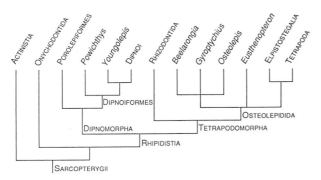

FIGURE 4 Interrelationships of sarcopterygian higher taxa. The topology corresponds to the phylogenetic tree illustrated in Fig. 3. See Table 1 and text for the distribution of characters.

squamosal (char. 54), the splenial (char. 93(1)), the jugal canal (char. 106), the humerus (char. 120), and basal plates in the dorsal fin supports (char. 128)]. In addition, structural and topographical changes occurred in the skull in comparison to basal actinopterygians: the premaxilla do not form part of the orbit (char. 18), more than four sclerotic plates compose the sclerotic ring (char. 49), the dermohyal is absent (char. 52), the hyomandibular has two proximal articular heads instead of one (char. 88), the preopercular canal does not end at the dorsal margin of the preopercular (char. 105), and the mandibular canal does not pass through the dentary (char. 110).

Characters 18, 52, 105, 106, 110, and 128 are complementary to the synapomorphies already identified for the Sarcopterygii (see Forey, 1980; Rosen *et al.*, 1981; Gardiner, 1984; Maisey, 1986a; Panchen and Smithson, 1987; Schultze, 1987; Cloutier, 1990; Ahlberg, 1991b). The presence of cosmine (char. 1) and submandibulars (char. 64; Gardiner, 1984; Schultze, 1987) are not sarcopterygian synapomorphies because actinistians lack these structures. Numerous branchial and hyoid characters used by previous authors (Forey, 1980; Maisey, 1986a; Panchen and Smithson, 1987) were not included in our analysis owing to the lack of information in the taxa analyzed.

B. [Onychodontida + Rhipidistia]

Of the nine characters supporting this clade, only one is not subsequently transformed. In addition to the splenial and angular, the infradentary series includes the postsplenial and surangular [char. 93(2)]. Based on this topology, the absence of surangular in actinistians (Cloutier, 1991a,b, 1996a) cannot be considered a synapomorphy of the group.

Of special interest at this node is the presence of character 48—palatal opening ("choana") surrounded

by the premaxilla, maxilla, dermopalatine, and vomer. The condition is known in *Eusthenopteron*, *Panderichthys*, tetrapods, and holoptychiids (P. E. Ahlberg, personal observation); the palatal opening is absent in dipnoans. Because the condition is unknown in onychodontids, it is possible that this character supports only the Rhipidistia. Nevertheless, the distribution of this character is contradictory to that hypothesized by various authors (Panchen and Smithson, 1987; Schultze, 1987, 1991). The interpretation of a palatal opening in porolepiforms agrees with the identification of a fenestra endochoanalis in *Glyptolepis* by Jarvik (1972) and Bjerring (1991). Thus, if one accepts this interpretation as suggested by the distribution of the characters, it follows that the posterior external nostril and the choana are nonhomologous since onychodonts and porolepiforms have two external nares. Schultze (1987, 1991) and Chang (1991b) argued that porolepiforms lack a true choana because the palatal part of the fenestra ventrolateralis is covered by the vomer (and possibly the dermopalatines). However, Section 62 of Jarvik's grinding series of *Glyptolepis groenlandica* (Jarvik, 1972: fig. 8C) shows on both sides of the snout a palatal opening which communicates unambiguously with the posterior external nostril and the nasal cavity. Furthermore, a specimen of *Holoptychius* from Dura Den (Scotland) prepared by one of us (P. E. Ahlberg) shows a small round opening at the junction of the maxilla, premaxilla, dermopalatine, and vomer, which seems to correspond precisely with the choana reconstructed by Jarvik (1972) and Bjerring (1991) from the grinding series of *Glyptolepis*. This opening is similar in size to the external nostrils. Although this topic could benefit from further study, the currently available evidence thus supports Jarvik's and Bjerring's interpretation.

C. Rhipidistia Cope 1871

Nine transformations corroborate the monophyly of the Rhipidistia (Table 1), of which only one is uniquely derived. In rhipidistians, the preopercular does not contact the maxilla (char. 56) because of a suture between the squamosal (or subsquamosals in porolepiforms) and the quadratojugal; in dipnoans in which the maxilla is absent, the preopercular does not reach the ventral margin of the cheek.

In his overview of sarcopterygian tooth structure, Schultze (1970) identified three types of dentine folding (plicidentine) in order (1) to characterize certain sarcopterygian clades, (2) to suggest a close relationship between Osteolepiformes and Tetrapoda, and (3) to demonstrate the monophyly of the Tetrapoda. Among the three types, polyplocodont plicidentine was said to be plesiomorphic, but no transformation series were inferred. Based on our tree, one has to consider that the dendrodont (in Porolepiformes) and eusthenodont types of folding (Schultze, 1970) evolved from a polyplocodont pattern. Thus the polyplocodont folding is not a synapomorphy of the clade [Rhizodontida + Osteolepidida] as suggested by Long (1989) and Young *et al.* (1992).

D. Dipnomorpha Ahlberg 1991b

The Dipnoi is closely related to the Porolepiformes (Maisey, 1986a; Ahlberg, 1989, 1991b; Cloutier, 1990, 1991a; Chang, 1991a,b; Chang and Smith, 1992) and not the sister-group of remaining sarcopterygians (Fig. 1d; Schultze 1987, 1994) nor that of the Tetrapoda (Fig. 1b; Rosen *et al.*, 1981; Gardiner, 1984). In our analysis, the clade [Dipnoiformes + Porolepiformes] is supported by the following four characters: mesh canals of the cosmine pore-canal system without horizontal partition (char. 2), median extrascapular overlapping the lateral extrascapulars [char. 41(1)], three or more tectals (char. 43), and presence of posterior branched radial complex associated with the posterior dorsal fin (char. 130).

Ahlberg (1991b) was the first to propose a large suite of characters (17) to support the monophyly of the clade [Porolepiformes + [*Powichthys* + [*Youngolepis* + [*Diabolepis* + Dipnoi]]]] (Fig. 1e). Of the characters used by Ahlberg (1991b), only 9 were included in our analysis (our characters 31, 34, 59, 71, 79, 100, 123, 130, and 137); some were combined (e.g., char. 137) or simply deleted owing to the lack of morphological information. Only character 130 is fully congruent with Ahlberg's hypothesis. Because our analysis was performed at a lower taxonomic level, some of the characters considered as uniquely shared derived are homoplastic with respect to various taxa (e.g., character 123 is also present in the rhizodontid *Barameda*).

The absence of a contact between the supraorbital and the parietal (char. 34) was considered by Ahlberg (1991b) as a dipnomorph uniquely shared derived character. Recent studies of the basal actinopterygian *Cheirolepis canadensis* by Arratia and Cloutier (1996) show that a supraorbital is present in this species and that there is no contact between it and the parietal. Thus the polarity of the character is different than that of Ahlberg (1991b) as well as its distribution.

The presence of preoperculosubmandibulars (char. 59) is frequently considered as a porolepiform synapomorphy (Jarvik, 1972; Vorobyeva and Schultze, 1991; Cloutier, 1990). Based on this analysis and Ahlberg (1991b), it is reinterpreted as a dipnomorph synapomorphy. It is likely that some of the 9-bones of dip-

noans are homologous to the preoperculosubmandibular found in porolepiforms and *Powichthys*.

E. Dipnoiformes Cloutier 1990

The Dipnoiformes is defined as the clade [*Powichthys* + [*Youngolepis* + Dipnoi]]. *Powichthys* and *Youngolepis* are consecutive plesions in the stem group of the Dipnoi; this pattern had been postulated by Ahlberg (1991b). The distribution of characters could be subject to further changes because all the basal taxa of this clade (*Powichthys*, *Youngolepis*, *Diabolepis*, *Dipnorhynchus*, and *Speonesydrion*) are only known from incomplete specimens (mainly partial skulls). Characters 10, 77, 100, and 119 are congruently distributed; there are an additional seven characters.

Chang and Smith (1992) and Chang and Zhu (1993) considered a broad marginal "tooth field" on the coronoids (char. 10) as a character shared by *Youngolepis* and *Powichthys* (also present in *Kenichthys*). Coronoids are absent in dipnoans.

The presence of rostral tubuli is shared by *Powichthys*, *Youngolepis*, and basal dipnoans (char. 77). The infraorbital canal follows the dorsal margin of the premaxilla (char. 100); this condition might be related to the condition of character 17 at the following node.

The proximal articular surface of the humerus is flat [char. 119(1)] rather than concave (plesiomorphic condition) or convex (Tetrapodomorpha synapomorphy). However, the distribution of character 119 could be an artifact of the selection of the taxa and the availability of information. Among basal dipnoiforms, the condition is only inferred in *Youngolepis* based on the condition of the glenoid fossa (Chang, 1991a). However, in advanced dipnoans the articular surface is concave (Schultze, 1987; Ahlberg, 1989).

Within the Dipnoiformes, the clade [*Youngolepis* + Dipnoi] is supported by characters 17, 45, and 108 and five reversals (Table 1). The position of the premaxilla (char. 17) in *Youngolepis* is interpreted as a transitional state between a plesiomorphic condition in which it forms the anterior part of the upper maxillary arcade and the derived dipnoan condition where the premaxilla is absent or reduced to a small dentigenous part. The posterior naris is still external but located very close to the jaw margin [char. 45(1)]; this condition precedes the dipnoan one in which the posterior naris occupies a palatal position. On the lower jaw, the middle pit line developed into an enclosed oral canal or a structure of intermediate morphology (char. 108).

F. Tetrapodomorpha Ahlberg 1991b

The Tetrapodomorpha includes the Rhizodontida, Elpistostegalia, Tetrapoda, and the so-called osteolepiforms. Significant modifications involve the anatomy of the nasal region and anterior palate as well as the pectoral appendage. The Tetrapodomorpha shares a single external naris which corresponds to the anterior naris (char 44); however, the condition is unknown in the Rhizodontida. The vomers articulate with each other medially (char. 71), and paired intervomerine pits are absent [char. 79(0)]. The anatomy of the humerus is modified: its proximal articular surface is convex [char. 119(2)] and the deltoid and supinator processes are present (char. 122).

Vorobyeva and Schultze (1991) refer to this clade as the Choanata; the main differences between their interpretation of the clade and ours concern the position of the Osteolepiformes and the distribution of some characters. They listed 21 characters of which 10 were used in our analysis (our characters 22–23, 41, 44, 48, 63, 66, 70, 119, and 140). The distribution of characters 44 and 119 are congruent in both hypotheses; however, Young *et al.* (1992) considered that the Rhizodontida lacks character 44.

The presence of a palatal opening surrounded by the maxilla, premaxilla, vomer, and dermopalatine (char. 48) is interpreted in our analysis as a synapomorphy of the Rhipidistia because of its presence in holoptychiids. The median extrascapular is overlapped by the lateral extrascapulars (char. 41) not only in members of this clade but also in actinistians. This relationship among the extrascapulars is considered to be plesiomorphic for the Sarcopterygii and is not diagnostic for the Tetrapodomorpha as suggested by Jarvik (1980), Vorobyeva and Schultze (1991), and Young *et al.* (1992).

Vorobyeva and Schultze (1991) define a character as "median gular always present" which they consider to be a synapomorphy of this clade; in our analysis, the presence of a median gular (char. 66) characterizes the rhipidistian node (and a large actinopterygian clade) and changes in holoptychiids and tetrapods. The polarity of character 66 is ambiguous because most basal actinopterygians possess a median gular with the exception of *Polypterus*. Based on our topology, the presence of a median gular is homoplastic with respect to actinopterygians and rhipidistians. Vorobyeva and Schultze (1991) also mention the presence of the posterior process on the vomer as a synapomorphy of this clade, although they note that the character is absent in some osteolepiforms. This character (char. 70) is in fact absent in the rhizodont *Barameda* (Long, 1989) as well as in the osteolepiform *Medoevia* (Lebedev, 1995) and *Gogonasus* (Long, 1988a), and the condition is unknown in *Strepsodus* and many osteolepiforms. The process is known to be present in tristichopterids, elpistostegids, and *Crassigyrinus*.

Other characters listed by Vorobyeva and Schultze (1991), not used in our analysis, deserve some comments. A separate median rostral is present in osteolepiforms, elpistostegids, and basal tetrapods (Jarvik, 1980; Ahlberg, 1995); thus a median rostral fused with the premaxilla is not a synapomorphy of this clade. A "long parasphenoid extending below oticooccipital region" is in fact restricted to crown-group Tetrapoda. The presence of seven submandibulars is also found in porolepiforms (Cloutier and Schultze, 1996).

G. Osteolepidida Boulenger 1901

The Osteolepidida are defined herein as the clade ["Osteolepiformes" + [Elpistostegalia + Tetrapoda]]. This "traditional" osteolepiform–tetrapod relationship is defined in terms of shared derived characters and congruence of characters (Schultze, 1987; Panchen and Smithson, 1987; Long, 1989; Cloutier, 1990, 1991b; Ahlberg, 1991b; Vorobyeva and Schultze, 1991; Young *et al.*, 1992; Ahlberg and Milner, 1994). This node is corroborated by character 70 and two reversals (chars. 113 and 124). The vomer has a distinctive posterior process (char. 70) which is lost in some basal tetrapods. (As mentioned above, this character is arguably primitively absent in some osteolepiforms and may thus in fact define a somewhat less inclusive clade.) The anocleithrum is exposed externally [char. 113(0)]; this character is considered as a reversal, although the plesiomorphic condition in actinopterygians deals with the postcleithrum. The mesomeres of the pectoral fin lack postaxial radials [char. 124(0)] in contrast to porolepiforms, dipnoans, and rhizodonts.

In contrast to the hypotheses of Janvier (1980), Long (1985), and Vorobyeva and Schultze (1991), the monophyly of the Osteolepiformes is not corroborated by our analysis; however, the paraphyly of the group is not demonstrated either. The monophyly of the osteolepiforms is jeopardized by the relative position of *Eusthenopteron*. Three equally parsimonious topologies have been obtained: *Eusthenopteron* is either the sister-group of (1) [Osteolepididae + Canowindridae], (2) [Elpistostegalia + Tetrapoda], or (3) [remaining osteolepiforms + [Elpistostegalia + Tetrapoda]]. In topologies 2 and 3, the Osteolepiformes is paraphyletic, whereas in topology 1 the monophyly is demonstrated. In topology 1, the presence of a large median postrostral (char. 23) and basal scutes on the fins (char. 131) would be considered as osteolepiform synapomorphies. A close relationship with the clade [Elpistostegalia + Tetrapoda] is optimized by character 140 (presence of well-ossified ribs) and the reversal of character 35 (absence of extratemporal). In terms of anatomy, the third topology is less robust because the clade [[Osteolepididae + Canowindridae] + [Elpistostegalia + Tetrapoda]] is supported by the reversal of two highly homoplastic characters: the presence of rhombic scales (char. 3) and the bilateral halves of the neural arch are separated (char. 136).

H. [Elpistostegalia + Tetrapoda]

The relationship between the Elpistostegalia and the Tetrapoda is well-corroborated. As mentioned by Schultze and Arsenault (1985) and Vorobyeva and Schultze (1991), the shape of the skull (char. 15) and the composition of the median series of skull roofing bones (char. 25) are diagnostic of this clade; the orbits are located dorsally, the interorbital distance is narrow and concave, and the skull as a whole is flattened. Paired frontals are present anterior to the parietals (char. 25); the superficially similar condition observed in *Polypterus* spp. is homoplastic with respect to this clade and these elements are not homologous to the frontals. The spiracle is present as a large, posteriorly open notch (char. 38) in the posterior part of the skull table enclosed between the tabular and the cheek bones. The unpaired fins (dorsal and anal fins) are lost (char. 127). The shape of the caudal fin is modified, the epichordal lepidotrichia being more developed than the hypochordal ones [char. 135(2)].

V. Conclusions

The Sarcopterygii represents a well-diagnosed clade composed of seven unambiguous subclades (i.e., Actinistia, Onychodontida, Dipnoiformes, Porolepiformes, Rhizodontida, Elpistostegalia, and Tetrapoda) and one questionable taxon (i.e., Osteolepiformes). A cladistic analysis of 140 osteological characters yielded the following phylogenetic conclusions (congruence with published analyses indicated by references):

1. The Actinistia is the sister-group of the remaining sarcopterygians (i.e., [Onychodontida + Rhipidistia]) (Panchen and Smithson, 1987).

2. The Onychodontida is the sister-group to the Rhipidistia.

3. The Dipnomorpha (including the Dipnoiformes and Porolepiformes) is the sister-group of the Tetrapodomorpha (Maisey, 1986a; Cloutier, 1990; Ahlberg, 1991b).

4. The Dipnoiformes (including *Powichthys*, *Youngolepis*, and Dipnoi) is the sister-group of the Porolepiformes (Maisey, 1986a; Cloutier, 1990, 1991a; Ahlberg, 1991b; Chang, 1991b; Chang and Smith, 1992).

5. The Rhizodontida is the sister-group of the Osteolepidida (Long, 1989; Cloutier, 1990; Ahlberg, 1991b; Vorobyeva and Schultze, 1991; Young *et al.*, 1992).

6. The interrelationships within the Osteolepidida are as follow: ["Osteolepiformes" + [Elpistostegalia + Tetrapoda]] (Schultze, 1987, 1991, 1994; Cloutier, 1990, 1991a; Ahlberg, 1991b; Young *et al.*, 1992).

7. The Elpistostegalia is the sister-group of the Tetrapoda (Schultze and Arsenault, 1985; Schultze, 1987, 1991, 1994; Cloutier, 1990; Ahlberg, 1991b; Vorobyeva and Schultze, 1991).

8. In terms of extant organisms, the Dipnoi is the Recent sister-group of the Tetrapoda (Forey, 1980, 1987; Gardiner, 1980, 1984; Rosen *et al.*, 1981; Maisey, 1986a; Cloutier, 1990, 1991a; Ahlberg, 1991b; Forey *et al.*, 1991; Trueb and Cloutier, 1991).

The classification proposed in this paper requires a reinterpretation of the diagnosis (characters) and definition (taxa) of already existing taxonomic categories. Available taxonomic names have been used and reassigned to clades that agree the closest to their original definition. We are not assigning Linnean rank to the taxonomic categories. Instead indentation signifies relative hierarchical rank. The classification of the Sarcopterygii based on this cladistic analysis is summarized as follows:

Sarcopterygii Romer 1955
    Actinistia Cope 1871
    [Onychodontida + Rhipidistia] clade
        Onychodontida[†] Andrews 1973
        Rhipidistia Cope 1887
            Dipnomorpha Ahlberg 1991b
                Porolepiformes[†] Jarvik 1942
                Dipnoiformes Cloutier 1990
            Tetrapodomorpha Ahlberg 1991b
                Rhizodontida[†] Andrews and Westoll 1970b
                Osteolepidida Boulenger 1901
                    "Osteolepiformes[†]" Berg 1937
                    [Elpistostegalia + Tetrapoda] clade
                        Elpistostegalia[†] Camp and Allison 1961
                        Tetrapoda Haworth 1825

VI. Summary

Sarcopterygians are classified into three extant groups (i.e., Actinistia, Dipnoiformes, and Tetrapoda) and five extinct Paleozoic taxa (i.e., Onychodontida, Porolepiformes, Rhizodontida, Osteolepiformes, and Elpistostegalia); sarcopterygian fishes (excluding tetrapods) account for approximately 500 species belonging to approximately 160 genera. The diagnosis, taxonomic content, stratigraphic range, evolutionary trends, and classification of the eight sarcopterygian higher clades are described. Fifty-four most parsimonious trees were found using 140 osteological characters (referred to as "the paleontological characters") coded for 27 sarcopterygian basal taxa. The character distribution is discussed for the most parsimonious sarcopterygian topology: [Actinistia + [[Porolepiformes + Dipnoiformes] + [Rhizodontida + ["osteolepiforms" + [Elpistostegalia + Tetrapoda]]]]]]. The controversial Devonian genera *Youngolepis* and *Powichthys* are included in the Dipnoiformes; the Dipnoiformes together with the Porolepiformes constitute the Dipnomorpha which is the sister-group of the Tetrapodomorpha. Although osteolepiforms are closely related to the clade [Elpistostegalia + Tetrapoda], their monophyly is not corroborated. The Elpistostegalia is the sister-group of the Tetrapoda, whereas dipnoans are the living sister-group to the tetrapods.

Acknowledgments

It is a pleasure to have this opportunity to acknowledge Colin Patterson's great contributions to vertebrate paleontology and systematics. He has played a major part—through his own work and through his influence on others—in placing the study of sarcopterygian interrelationships on a rigorous cladistic basis and thus shaping the field in which we work. On a personal level, we have both had the pleasure of working in the rigorous intellectual climate which he has encouraged and maintained at the Natural History Museum, London, and we owe him more pints of beer than we can readily remember. Thank you, Colin.

This paper draws on some 10 years' worth of accumulated work, and there are many people who have helped and influenced us in various ways during that time. We thank, in no particular order, Hans-Peter Schultze, Jenny Clack, Peter Forey, Brian Gardiner, Michael Coates, Erik Jarvik, Philippe Janvier, Mahala Andrews, Chang Mee-Mann, Gloria Arratia, Oleg Lebedev, John Long and Marius Arsenault, as well as many others whom we cannot list here.

References

Ahlberg, P. E. (1989). Paired fin skeletons and relationships of the fossil group Porolepiformes (Osteichthyes: Sarcopterygii). *Zool. J. Linn. Soc.* **96**, 119–166.

Ahlberg, P. E. (1991a). Tetrapod or near-tetrapod remains for the Upper Devonian of Scotland. *Nature (London)* **354**, 298–301.

Ahlberg, P. E. (1991b). A re-examination of sarcopterygian interrelationships, with special reference to the Porolepiformes. *Zool. J. Linn. Soc.* **103**, 241–287.

Ahlberg, P. E. (1992a). Coelacanth fins and evolution. *Nature (London)* **358**, 459.

Ahlberg, P. E. (1992b). The palaeoecology and evolutionary history of the porolepiform sarcopterygians. *In* "Fossil Fishes as Living Animals" (E. Mark-Kurik, ed.), pp. 71–90. Academy of Sciences of Estonia, Tallinn.

Ahlberg, P. E. (1992c). A new holoptychiid porolepiform fish from the Upper Frasnian of Elgin, Scotland. *Palaeontology* **35**, 813–828.

Ahlberg, P. E. (1994). The intracranial joint in vertebrate phylogeny. *J. Morphol.* **220** (ICVM-4 Abstr.), 319.

Ahlberg, P. E. (1995). *Elginerpeton pancheni* and the earliest tetrapod clade. *Nature (London)* **373**, 420–425.

Ahlberg, P. E., and Milner, A. R. (1994). The origin and early diversification of tetrapods. *Nature (London)* **368**, 507–514.

Ahlberg, P. E., and Trewin, N. H. (1995). The postcranial skeleton of the Middle Devonian lungfish *Dipterus valenciennesi*. *Trans. R. Soc. Edinburgh: Earth Sci.* **85**, 159–175.

Ahlberg, P. E., Luksevics, E., and Lebedev, O. (1994). The first tetrapod finds from the Devonian (Upper Famennian) of Latvia. *Philos. Trans. R. Soc. London, Ser. B* **343**, 303–328.

Alexeev, A. A., Lebedev, O. A., Barskov, I. S., Barskova, M. I., Kononova, L. I., and Chizova, V. A. (1994). On the stratigraphic position of the Famennian and Tournaisian fossil vertebrate beds in Andreyevka, Tula Region, Central Russia. *Proc. Geol. Assoc.* **105**, 41–52.

Andrews, S. M. (1973). Interrelationships of crossopterygians. *In* "Interrelationships of Fishes", (P. H. Greenwood, R. S. Miles, and C. Patterson, eds.), pp. 137–177. Academic Press, London.

Andrews, S. M. (1985). Rhizodont crossopterygian fish from the Dinantian of Foulden, Berwickshire, Scotland, with a re-evaluation of this group. *Trans. R. Soc. Edinburgh: Earth Sci.* **76**, 67–95.

Andrews, S. M., and Westoll, T. S. (1970a). The postcranial skeleton of *Eusthenopteron foordi* Whiteaves. *Trans. R. Soc. Edinburgh* **68**(9), 207–329.

Andrews, S. M., and Westoll, T. S. (1970b). The postcranial skeleton of rhipidistian fishes excluding *Eusthenopteron. Trans. R. Soc. Edinburgh* **68**(12), 391–489.

Aquesbi, N. (1988). Étude d'un Onychodontiforme (Osteichthyes, Sarcopterygii) du Dévonien moyen (Eifelien) du Maroc. *Bull. Mus. Natl. Hist. Nat.* [4] **10**, 181–196.

Arratia, G., and Cloutier, R., (1996). Reassessment of the morphology of *Cheirolepis canadensis* (Actinopterygii). *In* "Devonian Fishes and Plants from Miguasha, Quebec, Canada" (H.-P. Schultze and R. Cloutier, eds.), pp. 165–197. Verlag Dr. Fredrich Pfeil, Munich.

Bemis, W. E. (1984). Paedomorphosis and the evolution of the Dipnoi. *Paleobiology* **10**, 293–307.

Bendix-Almgreen, S. E. (1976). Palaeovertebrate faunas of Greenland. *In* "Geology of Greenland" (A. Escher and W. S. Watt, eds.), pp. 536–573. Geol. Surv. Greenland, Copenhagen.

Berg, L. S. (1937). A classification of fish-like vertebrates. *Bull. Acad. Sci. URSS*, pp. 1277–1280.

Berg, L. S. (1940). Classification of fishes, both recent and fossil. *Tr. Zool. Inst., Akad. Nauk SSSR* **5**, 1–517.

Berg, L. S. (1955). Classification of fish-like vertebrates, living and fossil. 2., corrected and enlarg. *Tr. Zool. Inst., Akad. Nauk SSSR* **20**, 1–286.

Berg, L. S. (1958). "System der rezenten und fossilen fischartigen und Fische." Deutcher Verlag der Wissenschaften. VEB, Berlin (translation by W. Gross of Berg, 1955).

Berman, D. S (1968). Lungfish from the Lueders Formation (Lower Permian, Texas) and the *Gnathorhiza*-lepidosirenid ancestry questioned. *J. Paleontol.* **42**, 827–835.

Bernacsek, G. M. (1977). A lungfish cranium from the Middle Devonian of the Yukon Territory, Canada. *Palaeontographica, Abt. A* **157**, 175–200.

Bjerring, H. C. (1989). Apertures of craniate olfactory organs. *Acta Zool. (Stockholm)* **70**, 71–85.

Bjerring, H. C. (1991). Some features of the olfactory organ in the Middle Devonian porolepiform *Glyptolepis groenlandica*. *Palaeontographica, Abt. A* **219**, 89–95.

Bjerring, H. C. (1993). Yet another interpretation of the coelacanthi-form basicranial muscle and its innervation. *Acta Zool. (Stockholm)* **74**, 289–299.

Camp, C. L., and Allison, H. J. (1961). Bibliography of fossil vertebrates 1949–1953. *Mem. Geol. Soc. Am.* **84**, 1–53.

Campbell, K. S. W., and Barwick, R. E. (1982a). A new species of the lungfish *Dipnorhynchus* from New South Wales. *Palaeontology* **25**, 509–527.

Campbell, K. S. W., and Barwick, R. E. (1982b). The neurocranium of the primitive dipnoan *Dipnorhynchus sussmilchi* (Etheridge). *J. Vertebr. Paleontol.* **2**, 286–327.

Campbell, K. S. W., and Barwick, R. E. (1983). Early evolution of dipnoan dentitions and a new genus *Speonesydrion*. *Mem. Assoc. Australas. Palaeontol.* **1**, 17–49.

Campbell, K. S. W., and Barwick, R. E. (1984). *Speonesydrion*, an Early Devonian dipnoan with primitive toothplates. *Palaeo Ichthyologica* **2**, 1–48.

Campbell, K. S. W., and Barwick, R. E. (1987). Paleozoic lungfishes—a review. *J. Morphol., Suppl.* **1**, 93–131.

Campbell, K. S. W., and Barwick, R. E. (1988). *Uranolophus*: A reappraisal of a primitive dipnoan. *Mem. Assoc. Australas. Palaeontol.* **7**, 87–144.

Campbell, K. S. W., and Barwick, R. E. (1990). Paleozoic dipnoan phylogeny: Functional complexes and evolution without parsimony. *Paleobiology* **16**, 143–169.

Campbell, K. S. W., and Bell, M. W. (1977). A primitive amphibian from the Late Devonian of New South Wales. *Alcheringa* **1**, 369–381.

Carroll, R. L. (1988). "Vertebrate Paleontology and Evolution." Freeman, New York.

Carroll, R. L. (1992). The primary radiation of terrestrial vertebrates. *Annu. Rev. Earth Planet. Sci.* **20**, 45–84.

Chang, M. M. (1982). The braincase of *Youngolepis*, a Lower Devonian crossopterygian from Yunnan, South-Western China. Ph.D. Thesis, University of Stockholm.

Chang, M. M. (1991a). Head exoskeleton and shoulder girdle of *Youngolepis*. *In* "Early Vertebrates and Related Problems of Evolutionary Biology" (M. M. Chang, Y. H. Liu, and G. R. Zhang, eds.), pp. 355–378. Science Press, Beijing.

Chang, M. M. (1991b). "Rhipidistians," dipnoans and tetrapods. *In* "Origins of the Higher Groups of Tetrapods: Controversy and Consensus" (H.-P. Schultze and L. Trueb, eds.), pp. 3–28. Cornell University Press (Comstock), Ithaca, NY.

Chang, M. M., and Smith, M. M. (1992). Is *Youngolepis* a porolepiform? *J. Vertebr. Paleontol.* **12**, 294–312.

Chang, M. M., and Yu, X. (1984). Structure and phylogenetic significance of *Diabolichthys speratus* gen. et sp. nov., a new dipnoan-like form from the Lower Devonian of Eastern Yunnan, China. *Proc. Linn. Soc. N. S. W.* **107**, 171–184.

Chang, M. M., and Zhu, M. (1993). A new Middle Devonian osteolepidid from Qujing, Yunnan. *Mem. Assoc. Australas. Palaeontol.* **15**, 183–198.

Chorn, J., and Schultze, H.-P. (1989). A complete specimen of *Sagenodus* (Dipnoi) from the Upper Pensylvanian of the Hamilton quarry, Kansas. *In* "Regional Geology and Paleontology of Upper Paleozoic Hamilton Quarry Area in Southeastern Kansas" (G. Mapes and R. Mapes, eds.), Kans. Geol. Surv. Guidebook Ser. 6, pp. 173–176.

Clack, J. A. (1988). New material of the early tetrapod *Acanthostega* from the Upper Devonian of East Greenland. *Palaeontology* **31**, 699–724.

Clack, J. A. (1989). Discovery of the earliest-known tetrapod stapes. *Nature (London)* **342**, 425–430.

Clack, J. A. (1994a). *Acanthostega gunnari*, a Devonian tetrapod from Greenland; the snout, palate and ventral parts of the braincase, with a discussion of their significance. *Medd. Grønl. Geosci.* **31**, 1–24.

Clack, J. A. (1994b). Earliest known tetrapod braincase and the evolution of the fenestra ovalis. *Nature (London)* **369**, 392–394.

Cloutier, R. (1990). Phylogenetic interrelationships of the actinistians (Osteichthyes: Sarcopterygii): Patterns, trends, and rates of evolution. Ph.D. Thesis, University of Kansas, Lawrence.

Cloutier, R. (1991a). Interrelationships of Palaeozoic actinistians: Patterns and trends. *In* "Early Vertebrates and Related Problems of Evolutionary Biology" (M. M. Chang, Y. H. Liu, and G. R. Zhang, eds.), pp. 379–428. Science Press, Beijing.

Cloutier, R. (1991b). Patterns, trends and rates of evolution within the Actinistia. *Environ. Biol. Fishes* **32**, 23–58.

Cloutier, R., (1996a). The primitive actinistian *Miguashaia bureaui* Schultze (Sarcopterygii). *In* "Devonian Fishes and Plants from Miguasha, Quebec, Canada" (H.-P. Schultze and R. Cloutier, eds.), pp. 227–247. Verlag Dr. Freidrich Pfeil, Munich.

Cloutier, R., (1996b). Dipnoi: Sarcopterygii). *In* "Devonian Fishes and Plants from Miguasha, Quebec, Canada" (H.-P. Schultze and R. Cloutier, eds.), pp. 198–226. Verlag Dr. Friedrich Pfeil, Munich.

Cloutier, R., and Forey, P. L. (1991). Diversity of extinct and living actinistian fishes (Sarcopterygii). *Environ. Biol. Fishes* **32**, 59–74.

Cloutier, R., and Schultze, H.-P. (1996). Porolepiforms fishes (Sarcopterygii). *In* "Devonian Fishes and Plants from Miguasha, Quebec, Canada" (H.-P. Schultze and R. Cloutier, eds.), pp. 248–270. Verlag Dr. Friedrich Pfeil, Munich.

Cloutier, R., Smith, M. M., and Krupina, N. I. (1993). Growth of the dental system of the Famennian dipnoan, *Andreyevichthys epitomus* from Russia: Morphometrics and morphogenesis of the entopterygoid tooth plate. *In* "The Gross Symposium. Scientific Sessions: Abstracts" (S. Turner, ed.). Universitè des Sciences et Technologies de Lille, Villeneuve d'Ascq, France.

Cloutier, R., Loboziak, S., Candilier, A.-M., and Blieck, A., (1996). Biostratigraphy of the Upper Devonian Escuminac Formation, eastern Québec, Canada: A comparative study based on miospores and fishes. *Rev. Palaeobot. Palynol* (in press).

Coates, M. I. (1991). New palaeontological contributions to limb ontogeny and phylogeny. *In* "Developmental Patterning of the Vertebrate Limb" (J. R. Hinchliffe, J. M. Hurle, and D. Summerbell, eds.), pp. 325–337. Plenum, New York.

Coates, M. I. (1995). The origin of vertebrate limbs. *In* "The Evolution of Developmental Mechanisms" (M. Akam, P. W. H. Holland, P. W. Ingham, and G. A. Wray, eds.), pp. 169–180. The Company of Biologists Limited, Cambridge, UK.

Coates, M. I., and Clack, J. A. (1990). Polydactyly in the earliest known tetrapod limbs. *Nature (London)* **347**, 66–69.

Coates, M. I., and Clack, J. A. (1991). Fish-like gills and breathing in the earliest known tetrapod. *Nature (London)* **352**, 234–235.

Cope, E. D. (1889). Synopsis of the families of Vertebrata. *Am. Nat.* **23**, 849–877.

Daeschler, E. B., Shubin, N. H., Thomson, K. S., and Amaral, W. W. (1994). A Devonian tetrapod from North America. *Science* **265**, 639–642.

Denison, R. H. (1968a). Early Devonian lungfishes from Wyoming, Utah and Idaho. *Fieldiana, Geol.* **17**, 353–413.

Denison, R. H. (1968b). The evolutionary significance of the earliest known lungfish, *Uranolophus*. *In* "Current Problems of Lower Vertebrate Phylogeny" (T. Ørvig, ed.), Nobel Symp. 4, pp. 247–258. Almqvist & Wiksell, Stockholm.

De Queiroz, K., and Gauthier, J. (1990). Phylogeny as a central principle in taxonomy: Phylogenetic definitions of taxon names. *Syst. Zool.* **39**, 307–322.

Dollo, L. (1895). Sur la phylogénie des Dipneustes. *Bull. Soc. Belge Géol. Paléontol. Hydrol.* **9**, 79–128.

Donoghue, M. J., Doyle, J. A., Gauthier, J., Kluge, A. G., and Rowe, T. (1989). The importance of fossils in phylogeny reconstruction. *Annu. Rev. Ecol. Syst.* **20**, 431–460.

Doyle, J. A., and Donoghue, M. J. (1986). Seed plant phylogeny and the origin of angiosperms: An experimental cladistic approach. *Bot. Rev.* **52**, 321–431.

Doyle, J. A., and Donoghue, M. J. (1987). The importance of fossils in elucidating seed plant phylogeny and macroevolution. *Rev. Palaeobot. Palynol.* **50**, 63–95.

Etheridge, R. (1906). The cranial buckler of a dipnoan fish, probably *Ganorhynchus*, from the Devonian beds of the Murrumbidgee River, New South Wales. *Rec. Aust. Mus.* **6**, 129–132.

Forey, P. L. (1980). *Latimeria*: A paradoxical fish. *Proc. R. Soc. London, Ser. B* **208**, 369–384.

Forey, P. L. (1981). The coelacanth *Rhabdoderma* in the Carboniferous of the British Isles. *Palaeontology* **24**, 203–229.

Forey, P. L. (1984). The coelacanth as a living fossil. *In* "Living fossils" (N. Eldredge and S. M. Stanley, eds.), pp. 166–169. Springer-Verlag, New York.

Forey, P. L. (1987). Relationships of lungfishes. *J. Morphol., Suppl.* **1**, 75–91.

Forey, P. L. (1991). *Latimeria chalumnae* and its pedigree. *Environ. Biol. Fishes* **32**, 75–97.

Forey, P. L., Gardiner, B. G., and Patterson, C. (1991). The lungfish, the coelacanth and the cow revisited. *In* "Origins of the Higher Groups of Tetrapods: Controversy and Consensus" (H.-P. Schultze and L. Trueb, eds.), pp. 145–172. Cornell University Press (Comstock), Ithaca, NY.

Forster-Cooper, C. (1937). The Middle Devonian fish fauna of Achanarras. *Trans. R. Soc. Edinburgh* **59**, 223–240.

Fricke, H., Reinicke, O., Hofer, H., and Nachtigall, W. (1987). Locomotion of the coelacanth *Latimeria chalumnae* in its natural environment. *Nature (London)* **329**, 331–333.

Fritzsch, B. (1987). The inner ear of the coelacanth fish *Latimeria* has tetrapod affinities. *Nature (London)* **327**, 153–154.

Fritzsch, B. (1988). Phylogenetic and ontogenetic origin of the dorsolateral auditory nucleus of anurans. *In* "The Evolution of the Amphibian Auditory System" (B. Fritzsch, M. Ryan, W. Wilsczynski, W. Hetherington, and T. Walkowiak, eds.), pp. 561–586. Wiley, New York.

Fritzsch, B. (1992). The water-to-land transition: Evolution of the tetrapod basilar papilla, middle ear, and auditory nuclei. *In* "The Evolutionary Biology of Hearing" (D. B. Webster, R. R. Fray, and A. N. Popper, eds.), pp. 351–375. Springer-Verlag, New York.

Gaffney, E. S. (1979). Tetrapod monophyly: A phylogenetic analysis. *Bull. Carnegie Mus. Nat. Hist.* **13**, 92–105.

Gardiner, B. G. (1980). Tetrapod ancestry: A reappraisal. *In* "The Terrestrial Environment and the Origin of Land Vertebrates" (A. L. Panchen, ed.), pp. 177–185. Academic Press, London.

Gardiner, B. G. (1984). The relationships of the palaeoniscid fishes, a review based on new specimens of *Mimia* and *Moythomasia* from the Upper Devonian of Western Australia. *Bull. Br. Mus. (Nat. Hist.), Geol.* **37**(4), 173–428.

Gardiner, B. G., Janvier, P., Patterson, C., Forey, P. L., Greenwood, P. H., Miles, R. S., and Jefferies, R. P. S. (1979). The salmon, the lungfish and the cow: A reply. *Nature (London)* **277**, 175–176.

Gauthier, J., Kluge, A. G., and Rowe, T. (1988). Amniote phylogeny and the importance of fossils. *Cladistics* **4**, 105–209.

Gill, T. (1872). Arrangement of the families of fishes. *Smithson. Misc. Collect.* **11**, I-XLVI, 1–49.

Goodrich, E. S. (1902). On the pelvic girdle and fin of *Eusthenopteron*. *Q. J. Microsc. Soc.* **45**, 311–324.

Gorr, T., Kleinschmidt, T., and Fricke, H. (1991). Close tetrapod relationships of the coelacanth *Latimeria* indicated by haemoglobin sequences. *Nature (London)* **351**, 394–395.

Greenwood, P. H. (1987). The natural history of African lungfishes. *J. Morphol., Suppl.* **1**, 163–179.

Gross, W. (1941). Über den Unterkiefer einiger devonischer Crossopterygier. *Abh. Preuss. Akad. Wiss., Mat.-Naturwiss. Kl.* pp. 1–51.

Halstead, L. B. (1978). The cladistic revolution—can it make the grade? *Nature (London)* **276**, 759.

Halstead, L. B., White, E. I., and MacIntyre, G. T. (1979). L. B. Halstead and colleagues reply *Nature (London)* **277**, 176.

Harland, W. B., Armstrong, R. L., Cox, A. V., Craig, L. E., Smith, A. G., and Smith, D. G. (1990). A Geologic Time Scale 1989. Cambridge University Press, Cambridge, UK.

Hedges, S. B., Hass, C. A., and Maxson, L. R. (1993). Relations of fish and tetrapods. *Nature (London)* **363**, 501–502.

Holmes, E. B. (1985). Are lungfishes the sister group of tetrapods? *Biol. J. Linn. Soc.* **25**, 379–397.

Huelsenbeck, J. P. (1991). When are fossils better than extant taxa in phylogenetic analyses? *Syst. Zool.* **40**, 458–469.

Huxley, T. H. (1861). Preliminary essay upon the systematic arrangement of the fishes of the Devonian epoch. *Mem. Geol. Surv. U.K., Figures and Descr. Br. Org. Remains* **10**, 1–40.

Janvier, P. (1980). Osteolepid remains from the Devonian of the Middle East, with particular reference to the endoskeletal shoulder girdle. *In* "The Terrestrial Environment and the Origin of Land Vertebrates" (A. L. Panchen, ed.) pp. 223–254. Academic Press, London.

Jarvik, E. (1937). On the species of *Eusthenopteron* found in Russia and the Baltic states. *Bull. Geol. Inst. Univ. Uppsala* **27**, 63–127.

Jarvik, E. (1942). On the structure of the snout of crossopterygians and lower gnathostomes in general. *Zool. Bidr. Uppsala* **21**, 235–675.

Jarvik, E. (1944a). On the dermal bones, sensory canals and pit-lines of the skull in *Eusthenopteron foordi* Whiteaves, with some remarks on *E. säve-söderberghi* Jarvik. *K. Sven. Vetenskapsakad. Handl.* [3] **21**(3), 1–48.

Jarvik, E. (1944b). On the exoskeletal shoulder-girdle of teleostomian fishes, with special reference to *Eusthenopteron foordi* Whiteaves. *K. Sven. Vetenskapsakad. Handl.* [3] **21**(7), 1–32.

Jarvik, E. (1948). On the morphology and taxonomy of the Middle Devonian osteolepid fishes of Scotland. *K. Sven. Vetenskapsakad. Handl.* [3] **25**, 1–301.

Jarvik, E. (1952). On the fish-like tail in the ichthyostegid stegocephalians. *Medd. Groenl.* **114**, 1–90.

Jarvik, E. (1954). On the visceral skeleton in *Eusthenopteron* with a discussion of the parasphenoid and palatoquadrate in fishes. *K. Sven. Vetenskapsakad. Handl.* [4] **5**, 1–104.

Jarvik, E. (1963). The composition of the intermandibular division of the head in fish and tetrapods and the diphyletic origin of the tetrapod tongue. *K. Sven. Vetenskapsakad. Handl.* [4] **9**, 1–74.

Jarvik, E. (1964). Specializations in early vertebrates. *Ann. Soc. R. Zool. Belg.* **94**, 11–95.

Jarvik, E. (1972). Middle and Upper Devonian Porolepiformes from East Greenland with special reference to *Glyptolepis groenlandica* n. sp. *Medd. Groenl.* **187**, 1–295.

Jarvik, E. (1980). "Basic Structure and Evolution of Vertebrates," Vol. 1. Academic Press, London.

Jessen, H. (1966). Die Crossopterygier des Oberen Plattenkalkes (Devon) der Bergisch-Gladbach-Paffrather Mulde (Rheinisches Schiefergebirge) unter Berücksichtigung von amerikanischem und europäischem *Onychodus*-material. *Ark. Zool.* [2] **18**, 305–389.

Jessen, H. (1975). A new choanate fish, *Powichthys thorsteinssoni* n.g. n.sp., from the early Lower Devonian of the Canadian Arctic Archipelago. *Colloq. Int. C. N. R. S.*, **218**, 213–222.

Jessen, H. (1980). Lower Devonian Porolepiformes from the Canadian Arctic with special reference to *Powichthys thorsteinssoni* Jessen. *Palaeontographica, Abt. A* **167**, 180–214.

Jurgens, J. D. (1973). The morphology of the nasal region of Amphibia and its bearing on the phylogeny of the group. *Ann. Univ. Stellenbosch, Ser. A* **46**, 3–136.

Kemp, A. (1987). The biology of the Australian lungfish, *Neoceratodus forsteri* (Krefft 1870). *J. Morphol., Suppl.* **1**, 181–198.

Lambers, P. (1992). On the ichthyofauna of the Solnhofen Lithographic Limestone (Upper Jurrasic, Germany). Ph.D. Dissertation, Rijksuniversiteit Groningen, Groningen, The Netherlands.

Lauder, G. V. (1980). On the evolution of jaw adductor musculature in primitive gnathostome fishes. *Breviora* **460**, 1–10.

Lebedev, O. A. (1984). The first find of a Devonian tetrapod in the U.S.S.R. *Dokl. Akad. Nauk SSSR* **278**, 1407–1473.

Lebedev, O. A. (1995). Morphology of a new osteolepidid fish from Russia. *Bull. Mus. Natl. Hist. Nat., Sect. C*, [4] **17**, 287–341.

Lebedev, O. A., and Clack, J. A. (1993). Devonian tetrapod remains from Andreyevka, Tula, Russia. *Palaeontology* **36**, 721–734.

Lebedev, O. A., and Coates, M. I. (1995). The postcranial skeleton of the Devonian tetrapod *Tulerpeton curtum* Lebedev. *Zool. J. Linn. Soc.* **114**, 307–348.

Leonardi, G. (1983). *Notopus petri* nov. gen., nov. sp.: Une empreinte d'amphibien du Dévonien au Paraná (Brésil). *Geobios* **16**, 233–239.

Long, J. A. (1985). The structure and relationships of a new osteolepiform fish from the Late Devonian of Victoria, Australia. *Alcheringa* **9**, 1–22.

Long, J. A. (1987). An unusual osteolepiform fish from the Late Devonian of Victoria, Australia. *Palaeontology* **30**, 839–852.

Long, J. A. (1988a). Late Devonian fishes from Gogo, Western Australia. *Natl. Geogr. Res.* **4**, 436–450.

Long, J. A. (1988b). New palaeoniscoid fishes from the Late Devonian and Early Carboniferous of Victoria. *Mem. Assoc. Australas. Palaeontol.* **7**, 1–64.

Long, J. A. (1989). A new rhizodontiform fish from the Early Carboniferous of Victoria, Australia, with remarks on the phylogenetic position of the group. *J. Vertebr. Paleontol.* **9**, 1–17.

Long, J. A. (1991). Arthrodire predation by *Onychodus* (Pisces, Crossopterygii) from the Late Devonian Gogo Formation, Western Australia. *Rec. West. Aust. Mus.* **15**, 479–481.

Long, J. A. (1992). Cranial anatomy of two new Late Devonian lungfishes (Pisces: Dipnoi) from Mt. Howitt, Victoria. *Rec. Aust. Mus.* **44**, 299–318.

Long, J. A. (1993). Cranial ribs and the origin of dipnoan air-breathing. *Mem. Assoc. Australas. Palaeontol.* **15**, 199–209.

Løvtrup, S. (1977). "The Phylogeny of Vertebrata." Wiley, London.

Lund, R. (1970). Fossil fishes from Southwestern Pennsylvania. Part I: Fishes from the Duquesne Limestones (Conemaugh, Pennsylvania). *Ann. Carnegie Mus.* **44**, 71–101.

Lund, R., and Lund, W. L. (1984). New genera of coelacanths from the Bear Gulch Limestone (Lower Carboniferous) of Montana (U.S.A.). *Geobios* **17**(2), 237–244.

Lund, R., and Lund, W. L. (1985). Coelacanths from the Bear Gulch Limestone (Namurian) of Montana and the evolution of the Coelacanthiformes. *Bull. Carnegie Mus. Nat. Hist.* **25**, 1–74.

Lund, W. L., Lund, R., and Klein, G. A. (1985). Coelacanth feeding mechanisms and ecology of Bear Gulch coelacanths. *C. R. Neuvi. Congr. Int. Stratigr. Géol. Carbonifère* **5**, 492–500.

Lyarskaya, L. A. (1981). Baltic Devonian Placodermi Asterolepididae. Zinatne, Riga.

Lyarskaya, L. A., and Mark-Kurik, E. (1972). Eine neue Fundstelle oberdevonscher Fische im Baltikum. *Neues Jahrb. Mineral., Geol. Palaeontol., Monatsh.* **7**, 407–414.

Maisey, J. G. (1986a). Heads and tails: A chordate phylogeny. *Cladistics* **2**, 201–256.

Maisey, J. G. (1986b). Coelacanth from the Lower Cretaceous of Brazil. *Am. Mus. Novit.* **2866**, 1–30.

Marshall, C. R. (1987). Lungfish: Phylogeny and parsimony. *J. Morphol., Suppl.* **1**, 151–162.

Marshall, C. R., and Schultze, H.-P. (1992). Relative importance of molecular, neontological, and paleontological data in understanding the biology of the vertebrate invasion of land. *J. Mol. Evol.* **35**, 93–101.

Meyer, A. (1995). Molecular evidence on the origin of tetrapods and the relationships of the coelacanth. *Trends Ecol. Evol.* **10**(3), 95–136.

Meyer, A., and Wilson, A. C. (1990). Origin of tetrapods inferred from their mitochondrial DNA affiliation to lungfish. *J. Mol. Evol.* **31**, 359–364.

Meyer, A., and Wilson, A. C. (1991). Coelacanth's relationships. *Nature (London)* **353**, 19.

Miles, R. S. (1975). The relationships of the Dipnoi. *Colloq. Int. C. N. R. S.* **218**, 133–148.

Miles, R. S. (1977). Dipnoan (lungfish) skulls and the relationships of the group: A study based on new species from the Devonian of Australia. *Zool. J. Linn. Soc.* **61**, 1–328.

Miller, H. (1841). "The Old Red Sandstone or New Walks in an Old Field." Johnstone, Edinburgh.

Milner, A. R., Smithson, T. R., Milner, A. C., Coates, M. I., and Rolfe, W. D. I. (1986). The search for early tetrapods. *Mod. Geol.* **10**, 1–28.

Millot, J., and Anthony, J. (1958). "Anatomie de *Latimeria chalumnae*," Tome I. Squelette et muscles. CNRS, Paris.

Moy-Thomas, J. A., and Miles, R. S. (1971). "Palaeozoic Fishes," 2nd ed. Chapman & Hall, London.

Murchison, R. I. (1839). "The Silurian System." John Murray, London.

Normark, B. B., McCune, A. R., and Harrison, R. G. (1991). Phylogenetic relationships of neopterygian fishes, inferred from mitochondrial DNA sequences. *Mol. Biol. Evol.* **8**, 819–834.

Northcutt, R. G., and Bemis, W. E. (1993). Cranial nerves of the coelacanth, *Latimeria chalumnae* (Osteichthyes: Sarcopterygii: Actinistia) and comparisons with other Craniata. *Brain, Behav. Evol.* **42**, Suppl. 1, 1–75.

Obruchev, D. (1933). Description of four new fish species from the Devonian of Leningrad Province. *Mater. Cent. Sci. Geol. Prospect. Inst., Palaeontol. Stratigr. Mag.* **1**, 12–14.

Ørvig, T. (1957). Remarks on the vertebrate fauna of the Lower Upper Devonian of Escuminac Bay, P.Q., Canada, with special reference to the porolepiform crossopterygians. *Ark. Zool.* [2] **10**, 367–426.

Ørvig, T. (1969). Vertebrates from the Wood Bay group and the position of the Emsian-Eifelian boundary in the Devonian of Vestspitsbergen. *Lethaia* **2**(4), 273–319.

Ørvig, T. (1986). A vertebrate bone from the Swedish Palaeocene. *Geol. Foeren. Stockholm Foerh.* **108**, 139–141.

Owen, R. (1846). "Lectures on the Comparative Anatomy and Physiology of the Vertebrate Animals," Part I. Longman, Brown, Green and Longman, London.

Owen, R. (1860). "Palaeontology, or a Systematic Summary of Extinct Animals and their Geological Relations." Black, Edinburgh.

Panchen, A. L. (1985). On the amphibian *Crassigyrinus scoticus* Watson from the Carboniferous of Scotland. *Philos. Trans. R. Soc. London Ser. B* **309**, 461–568.

Panchen, A. L., and Smithson, T. R. (1987). Character diagnosis, fossils, and the origin of tetrapods. *Biol. Rev. Cambridge Philos. Soc.* **62,** 341–438.

Panchen, A. L., and Smithson, T. R. (1988). The relationships of the earliest tetrapods. *In* "The Phylogeny and Classification of the Tetrapods" (M. J. Benton, ed.) Vol. 1, pp. 1–32. Oxford University Press (Clarendon), Oxford.

Patterson, C. (1980). Origin of tetrapods: historical introduction to the problem. *In* "The Terrestrial Environment and the Origin of Land Vertebrates" (A. L. Panchen, ed.), pp. 159–175. Academic Press, London.

Patterson, C. (1981). Significance of fossils in determining evolutionary relationships. *Annu. Rev. Ecol. Syst.* **12,** 195–223.

Patterson, C. (1982a). Morphological characters and homology. *Syst. Assoc. Spec. Vol.* **21,** 21–74.

Patterson, C. (1982b). Classes and cladists or individuals and evolution. *Syst. Zool.* **31,** 284–286.

Patterson, C., and Rosen, D. E. (1977). Review of ichthyodectiform and other Mesozoic teleost fishes and the theory and practice of classifying fossils. *Bull. Am. Mus. Nat. Hist.* **158,** 81–172.

Rackoff, J. S. (1980). The origin of the tetrapod limb and the ancestry of tetrapods. *Syst. Assoc. Spec. Vol.* **15,** 255–292.

Rocek, Z., and Rage, J. C. (1994). The presumed amphibian footprint *Notopus petri* from the Devonian: A probable starfish trace fossil. *Lethaia* **27,** 241–244.

Romer, A. S. (1937). The braincase of the Carboniferous crossopterygian *Megalichthys nitidus. Bull. Mus. Comp. Zool.* **82,** 1–73.

Romer, A. S. (1947). Review of the Labyrinthodontia. *Bull. Mus. Comp. Zool.* **99,** 7–368.

Romer, A. S. (1966). "Vertebrate Paleontology," 3rd ed. University of Chicago Press, Chicago.

Rosen, D. E., Forey, P. L., Gardiner, B. G., and Patterson, C. (1981). Lungfishes, tetrapods, paleontology, and plesiomorphy. *Bull. Am. Mus. Nat. Hist.* **167,** 159–276.

Säve-Söderbergh, G. (1932). Preliminary note on Devonian stegocephalians from East Greenland. *Medd. Groenl.* **94,** 1–107.

Schaeffer, B. (1952). Rates of evolution in the coelacanth and dipnoan fishes. *Evolution (Lawrence, Kans.)* **6,** 101–111.

Schoch, R. M. (1986). "Phylogeny Reconstruction in Paleontology." Van Nostrand-Reinhold, New York.

Schultze, H.-P. (1969). Die Faltenzähne der rhipidistiiden Crossopterygier, der Tetrapoden und der Actinopterygier-Gattung *Lepisosteus. Palaeontogr. Ital.* 65 [N. S. 35], 59–137.

Schultze, H.-P. (1970). Folded teeth and the monophyletic origin of tetrapods. *Am. Mus. Novi.* **2408,** 1–10.

Schultze, H.-P. (1973). Crossopterygier mit heterozerker Schwanzflosse aus dem Oberdevon Kanadas, nebst einer Beschreibungvon Onychodontida-Resten aus dem Mitteldevon Spaniens und aus dem Karbon der USA. *Palaeontographica, Abt. A* **143,** 188–208.

Schultze, H.-P. (1975). Die Lungenfisch-Gattung *Conchopoma* (Pisces, Dipnoi). *Senckenbergiana Lethaea* **56,** 191–231.

Schultze, H.-P. (1977). The origin of the tetrapod limb within the rhipidistian fishes. *In* "Major Patterns in Vertebrate Evolution" (M. K. Hecht, P. C. Goody, and B. M. Hecht, eds.), pp. 541–544. Plenum, New York.

Schultze, H.-P. (1981). Hennig und der Ursprung der Tetrapoda. *Palaeontol. Zh.* **55,** 71–86.

Schultze, H.-P. (1985). Reproduction and spawning sites of *Rhabdoderma* (Pisces, Osteichthyes, Actinistia) in Pennsylvanian deposits of Illinois, USA. *C. R. Neuv. Congr. Int. Stratigr. Géol. Carbonifère* **5,** 326–330.

Schultze, H.-P. (1987). Dipnoans as sarcopterygians. *J. Morphol. Suppl.* **1,** 39–74.

Schultze, H.-P. (1991). A comparison of controversial hypotheses on the origins of tetrapods. *In* "Origins of the Major Groups of Tetrapods: Controversies and Consensus" (H.-P. Schultze and L. Trueb, eds.), pp. 29–67. Cornell University Press (Comstock), Ithaca, NY.

Schultze, H.-P. (1992a). A new long-headed dipnoan (Osteichthyes) from the Middle Devonian of Iowa, USA. *J. Vertebr. Paleontol.* **12,** 42–58.

Schultze, H.-P. (1992b). Dipnoi. *In* "Fossilium catalogus I:Animalia" (F. Westphal, ser. ed.), *Pars 131.* Kugler Publications, Amsterdam and New York.

Schultze, H.-P. (1993). Osteichthyes: Sarcopterygii. *In* "The Fossil Record 2" (M. J. Benton, ed.), pp. 657–663. Chapman & Hall, London.

Schultze, H.-P. (1994). Comparison of hypotheses on the relationships of sarcopterygians. *Syst. Biol.* **43,** 155–173.

Schultze, H.-P., (1996). The elpistostegid fish *Elpistostege*, the closest the Miguasha fauna comes to a tetrapod. *In* "Devonian Fishes and Plants from Miguasha, Quebec, Canada" (H.-P. Schultze and R. Cloutier, eds.), pp. 316–327. Verlag Dr. Pfeil, München.

Schultze, H.-P., and Arsenault, M. (1985). The panderichthyid fish *Elpistostege*: A close relative of tetrapods? *Palaeontology* **28,** 293–309.

Schultze, H.-P., and Campbell, K. S. W. (1987). Characterization of the Dipnoi, a monophyletic group. *J. Morphol. Suppl.* **1,** 25–37.

Schultze, H.-P., and Marshall, C. R. (1993). Contrasting the use of functional complexes and isolated characters in lungfish evolution. *Mem. Assoc. Australas. Palaeontol.* **15,** 211–224.

Schultze, H.-P., Cloutier, R., and Marshall, C. R. (1993). Contrasting the use of functional complexes and isolated characters in lungfish evolution. *Assoc. Australas. Palaeontol., K. S. W. Campbell Symp., Abstr. Vol.*, p. 21.

Schwimmer, D. R., Stewart, J. D., and Williams, G. D. (1994). Giant fossil coelacanths of the late Cretaceous in the eastern United States. *Geology* **55,** 503–506.

Sedgwick, A., and Murchison, R. I. (1828). On the structure and relations of the deposits contained between the Primary Rocks and the Oolitic Series in the North of Scotland. *Trans. Geol. Soc. London* [2] **3,** 125–160.

Sharp, P. M., Lloyd, A. T., and Higgins, D. G. (1991). Coelacanth's relationships. *Nature (London)* **353,** 218–219.

Shubin, N. H., and Alberch, P. (1986). A morphogenetic approach to the origin and basic organization of the tetrapod limb. *Evol. Biol.* **20,** 319–387.

Smith, M. M. (1984). Petrodentine in extant and fossil dipnoan dentitions: Microstructure, histogenesis and growth. *Proc. Linn. Soc. N. S. W.* **107,** 367–407.

Smith, M. M. (1989). Distribution and variation in enamel structure in the oral teeth of sarcopterygians: its significance for the evolution of a protoprismatic enamel. *Hist. Biol.* **3,** 97–126.

Smith, M. M., and Chang, M. M. (1990). The dentition of *Diabolepis speratus* Chang & Yu, with further consideration of its relationships and the primitive dipnoan dentition. *J. Vertebr. Paleontol.* **10,** 420–433.

Smith, M. M., Smithson, T. R., and Campbell, K. S. W. (1987). The relationships of *Uronemus*: A Carboniferous dipnoan with highly modified tooth plates. *Philos. Trans. R. Soc. London* **317,** 299–327.

Smithson, T. R. (1985). On the relationships and morphology of the Carboniferous amphibian *Eoherpeton watsoni. Zool. J. Linn. Soc.* **85,** 317–410.

Stensiö, E. (1922). Über zwei Coelacanthiden aus dem Oberdevon von Wildungen. *Palaeontol. Zh.* **4,** 167–210.

Stensiö, E. (1937). On the Devonian coelacanthids of Germany with special reference to the dermal skeleton. *K. Sven. Vetenskapsakad. Handl.* **3**(16), 1–56.

Stock, D. W., and Swofford, D. L. (1991). Coelacanth relationships. *Nature (London)* **353**, 217–218.

Stock, D. W., Moberg, K. D., Maxson, L. R., and Whitt, G. S. (1991). A phylogenetic analysis of the 18S ribosomal RNA sequence of the coelacanth *Latimeria chalumnae*. *Environ. Biol. Fishes* **32**, 99–117.

Stössel, I. (1995). The discovery of a new Devonian tetrapod trackway in SW Ireland. *J. Geol. Soc., London* **152**, 407–413.

Thomson, K. S. (1973). Observations on a new rhipidistian fish from the Upper Devonian of Australia. *Palaeontographica, Abt. A* **143**, 209–220.

Tong-Dzuy Thanh, and Janvier, P. (1990). Les Vertébrés du Dévonien inférieur de Bac Bo oriental (Provinces de Bac Thaï et Lang Son, Viêt Nam). *Bull. Mus. Natl. Hist. Nat., Sect. C* [4] **12**(2), 143–223.

Tong-Dzuy Thanh, and Janvier, P. (1994). Early Devonian fishes from Trang Xa (Bac Thai, Vietnam) with remarks on the distribution of the vertebrates in the Song Cau Group. *J. Southeast Asian Earth Sci.* **10**, 235–243.

Traquair, R. H. (1890). List of the fossil Dipnoi and Ganoidei of Fife and the Lothians. *Proc. R. Soc. Edinburgh* **17**, 385–400.

Trueb, L, and Cloutier, R. (1991). A phylogenetic investigation of the inter- and intrarelationships of the Lissamphibia (Amphibia: Temnospondyli). *In* "Origins of the Higher Groups of Tetrapods: Controversy and Consensus" (H.-P. Schultze and L. Trueb, eds), pp. 223–313. Cornell University Press (Comstock), Ithaca, NY.

Vorobyeva, E. I. (1960). New data on the crossopterygian fish genus *Panderichthys* from the Devonian of the USSR. *Paleontol. Zh.* **1**, 87–96.

Vorobyeva, E. I. (1962). Rhizodont crossopterygians of the Main Devonian Field. *Tr. Paleontol. Inst.* **94**, 1–139.

Vorobyeva, E. I. (1971). The ethmoid region of *Panderichthys* and some problems of the cranial morphology of crossopterygians. *Tr. Paleontol. Inst.* **130**, 142–159.

Vorobyeva, E. I. (1977). Morphology and nature of evolution of crossopterygian fishes. *Tr. Paleontol. Instit.* **163**, 1–239.

Vorobyeva, E. I. (1980). Observations on two rhipidistian fishes from the Upper Devonian of Lode, Latvia. *Zool. J. Linn. Soc.* **70**, 191–201.

Vorobyeva, E. I. (1986). The current state of the problem of amphibian origin. *In* "Studies in Herpetology" (Z. Roček, ed.), pp. 25–28. Charles University, Prague.

Vorobyeva, E. I. (1987). Porolepid crossopterygian from the Middle Devonian of Estonia. *Paleontol. Zh.* **1987**(1), 76–85.

Vorobyeva, E. I. (1989). Panderichthyida—a new order of Paleozoic crossopterygian fishes (Rhipidistia). *Dokl. Akad. Nauk SSSR* **306**, 188–189.

Vorobyeva, E. I., and Kuznetsov, A. (1992). The locomotor apparatus of *Panderichthys rhombolepis* (Gross), a supplement to the problem of fish-tetrapod transition. *In* 'Fossil Fishes as Living Animals" (E. Mark-Kurik, ed.), pp. 131–140. Academy of Sciences of Estonia, Tallinn.

Vorobyeva, E. I., and Lyarskaya, L. A. (1968). Crossopterygian and dipnoan remains from the Amata beds of Latvia and their burial. *In* "Ocherki po filogenii i sistematike iskopaemykh ryb i beschelyustnykh" (D. V. Obruchev, ed.), pp. 71–86. Nauka, Moscow.

Vorobyeva, E. I., and Schultze, H.-P. (1991). Description and systematics of panderichthyid fishes with comments on their relationship to tetrapods. *In* "Origins of the Higher Groups of Tetrapods: Controversy and Consensus" (H.-P. Schultze and L.

Trueb, eds.), pp. 68–109. Cornell University Press (Comstock), Ithaca, NY.

Vorobyeva, E. I., and Tsessarskii, A. A. (1986). On origin of vertebrae in lower tetrapods. *Zh. Obshch. Biol.* **47**, 735–747.

Wang, S., Drapala, V., Barwick, R. E., and Campbell, K. S. W. (1990). A new Early Devonian lungfish, *Sorbitorhynchus deleaskitus* n. gen. et sp., from Guangxi, China. *Paleobiology* **16**, 168–169.

Warren, A., Jupp, R., and Bolton, B. R. (1986). Earliest tetrapod trackway. *Alcheringa* **10**, 183–186.

Warren, J. W., and Wakefield, N. A. (1972). Trackways of tetrapod vertebrates from the Upper Devonian of Victoria, Australia, *Nature (London)* **238**, 469–470.

Westoll, T. S. (1938). Ancestry of the tetrapods. *Nature (London)* **141**, 127.

Westoll, T. S. (1949). On the evolution of the Dipnoi. *In* "Genetics, Paleontology and Evolution" (G. L. Jepsen, G. G. Simpson, and E. Mayr, eds.), pp. 121–184. Princeton University Press, Princeton, NJ.

White, E. I. (1965). The head of *Dipterus valenciennesi*. *Bull. Br. Mus. (Nat. Hist.) Geol.* **11**, 3–45.

Whiteaves, J. F. (1883). On some remarkable fossil fishes from the Devonian rocks of Scaumenac Bay, P. Q., with descriptions of a new genus and three new species. *Can. Nat. Q. J. Sci. [N.S.]* **10**, 27–35.

Whiteaves, J. F. (1889). Illustrations of the fossil fishes of the Devonian rocks of Canada. Part II. *Trans. R. Soc. Can.* **6**, 77–96.

Woodward, A. S. (1891). Catalogue of the Fossil Fishes in the British Museum (Natural History), Part II. Br. Mus. (Nat. Hist.), London.

Worobjewa, E. (1975). Bemerkungen zu *Panderichthys rhombolepis* (Gross) aus Lode in Lettland (Gauja Schichten, Oberdevon). *Neues Jahrb. Mineral., Geol. Palaeontol., Monatsh.* **1975**, 407–414.

Young, G. C., Long, J. A., and Ritchie, A. (1992). Crossopterygian fishes from the Devonian of Antarctica: Systematics, relationships and biogeographic significance. *Rec. Austr. Mus., Suppl.* **14**, 1–77.

Zhu, M., and Janvier, P. (1994). Un Onychodontide (Vertebrata, Sarcopterygii) du Dévonien Inférieur de Chine. *C. R. Séances Acad. Sci., Ser. 2* **319**, 951–956.

Appendix 1: Character List

1. Cosmine. 0 = absent; 1 = present.
2. Mesh canals. 0 = pore cavity with horizontal partition; 1 = pore cavity without horizontal partition.
3. Condition of scales. 0 = rhombic scales; 1 = rounded scales.
4. Peg on rhombic scale. 0 = narrow; 1 = broad.
5. Boss on internal face of scale. 0 = absent; 1 = present.
6. Ganoine. 0 = absent; 1 = present.
7. Acrodin. 0 = absence of acrodin caps on teeth; 1 = presence of acrodin caps on teeth.
8. Marginal teeth on dentary. 0 = present; 1 = absent.
9. Dental plate. 0 = denticles on entopterygoid, or naked bone; 1 = tooth plate on entopterygoid; 2 = dentine plate on entopterygoid.

10. Dentition on coronoids. 0 = narrow marginal tooth row; 1 = broad marginal "tooth field."
11. Spiral parasymphysial teeth. 0 = absent; 1 = present.
12. Fang pairs in inner tooth arcade. 0 = absent; 1 = present.
13. Fang pair on anterior end of dentary. 0 = absent; 1 = present.
14. Plicidentine. 0 = absent; 1 = polyplocodont plicidentine; 2 = dendrodont plicidentine.
15. Skull shape. 0 = lateral orbits, interorbital skull roof wide and arched; 1 = dorsal orbits, interorbital skull roof narrow and flat or concave.
16. Premaxilla. 0 = present; 1 = absent.
17. Position of premaxilla. 0 = marginal; 1 = ventral part turned in.
18. Position of premaxilla. 0 = premaxilla forming part of orbit; 1 = premaxilla not forming part of orbit.
19. Maxilla. 0 = present; 1 = absent.
20. Posterodorsal process of maxilla. 0 = present; 1 = absent.
21. Shape of posterodorsal process of maxilla. 0 = smooth, convex posterodorsal margin; 1 = distinct posterodorsal angle.
22. Position of median rostral. 0 = rostral does not contribute to jaw margin; 1 = rostral contributes to jaw margin.
23. Postrostrals. 0 = postrostral mosaic of small variable bones; 1 = large median postrostral, with or without accessory bones.
24. Paired nasals meeting in midline of skull. 0 = absent; 1 = present.
25. Paired frontals. 0 = absent; 1 = present.
26. E-bone. 0 = absent; 1 = present.
27. C-bone. 0 = absent; 1 = present.
28. Supraorbitals. 0 = absent; 1 = present.
29. Number of supraorbitals (including the "posterior tectal" of Jarvik). 0 = one; 1 = two; 2 = more than two.
30. B-bone. 0 = absent; 1 = present.
31. Position of anterior margin of parietal. 0 = between or in front of orbits; 1 = slightly posterior to orbits; 2 = much posterior to orbits.
32. Pineal opening. 0 = open; 1 = closed.
33. Median supraorbital ridges ("eyebrows"). 0 = absent; 1 = present.
34. Parietal–supraorbital contact. 0 = absent; 1 = present.
35. Extratemporal. 0 = absent; 1 = present.
36. Intertemporal. 0 = present; 1 = absent.
37. Supratemporal series. 0 = single bone which contacts the extrascapular posteriorly and the intertemporal or dermosphenotic anteriorly; 1 = two bones (supratemporal and tabular) between extrascapular and intertemporal or postorbital; 2 = single bone (probably the tabular) in posterior position, bounded anteriorly by lateral extension of postparietal.
38. Spiracle. 0 = small hole on kinetic margin between skull roof and cheek; 1 = large, posteriorly open notch.
39. Extrascapulars. 0 = present; 1 = absent.
40. Number of extrascapulars. 0 = four; 1 = two; 2 = three; 3 = five.
41. Median extrascapular overlap. 0 = median extrascapular overlapped by lateral extrascapulars; 1 = median extrascapular overlaps the lateral extrascapulars; 2 = median extrascapular abuts the lateral extrascapulars.
42. Tectals. 0 = absent; 1 = present.
43. Number of tectals (not counting the "posterior tectal" of Jarvik; see char. 29). 0 = one; 1 = three or more.
44. Anterior and posterior nares. 0 = both present; 1 = only anterior naris present.
45. Position of posterior naris. 0 = external, far from jaw margin; 1 = external, close to jaw margin; 2 = palatal (palatal posterior naris of lungfishes deemed nonhomologous with tetrapod choana).
46. Position of posterior naris. 0 = associated with the orbit; 1 = not associated with the orbit.
47. Position of anterior naris. 0 = facial; 1 = marginal; 2 = palatal.
48. Palatal opening ("choana") surrounded by premaxilla, maxilla, dermopalatine, and vomer. 0 = absent; 1 = present.
49. Number of sclerotic plates. 0 = four or less; 1 = more than four.
50. Condition of lacrimal and jugal. 0 = separate bones; 1 = fused together.
51. Prespiracular. 0 = absent; 1 = present.
52. Dermohyal. 0 = present; 1 = absent.
53. Postspiracular. 0 = absent; 1 = present.
54. Squamosal and preopercular. 0 = one bone ("preopercular"); 1 = two separate bones.
55. Subsquamosals. 0 = absent; 1 = present.
56. Preopercular–maxillary contact. 0 = preopercular contacts maxilla (if maxilla absent, preopercular reaches ventral margin of cheek); 1 = preopercular does not contact maxilla (if maxilla absent, preopercular does not reach ventral margin of cheek).
57. Quadratojugal. 0 = present, small; 1 = present, large; 2 = absent.

58. Jugal–quadratojugal contact. 0 = absent; 1 = present.
59. Preoperculosubmandibular. 0 = absent; 1 = present.
60. Opercular. 0 = present; 1 = absent.
61. Subopercular. 0 = present; 1 = absent.
62. Branchiostegal rays. 0 = present; 1 = absent.
63. Number of branchiostegal rays per side. 0 = 10 or more; 1 = two to seven; 2 = one.
64. Submandibulars. 0 = absent; 1 = present.
65. Width of submandibulars. 0 = narrow; 1 = broad.
66. Median gular. 0 = present; 1 = absent.
67. Relative size of median gular. 0 = small; 1 = large.
68. Lateral gular. 0 = present; 1 = absent.
69. Size of lateral gular. 0 = lateral gular and branchiostegal rays of similar size; 1 = lateral gular covering approximately half the intermandibular space.
70. Posterior process of vomer. 0 = absent; 1 = present.
71. Articulation of vomer. 0 = vomers do not articulate with each other; 1 = vomers articulate with each other.
72. Articulation of pterygoid. 0 = pterygoids do not articulate with each other; 1 = pterygoids articulate with each other.
73. Articulation of parasphenoid. 0 = parasphenoid not sutured to vomer; 1 = parasphenoid sutured to vomer.
74. Denticulated spiracular groove on parasphenoid. 0 = present; 1 = absent.
75. Buccohypophysial foramen of parasphenoid. 0 = single; 1 = double.
76. Rostral organ. 0 = absent; 1 = present.
77. Rostral tubuli. 0 = absent; 1 = present.
78. Fossa autopalatina. 0 = absent; 1 = present.
79. Paired intervomerine pits. 0 = absent; 1 = present.
80. Labial cavity. 0 = absent; 1 = present.
81. Dermal joint between parietal and postparietal. 0 = absent; 1 = present.
82. Dorsal endoskeletal articulation between otoccipital and ethmosphenoid blocks of braincase. 0 = absent; 1 = present.
83. Ventral endoskeletal articulation between otoccipital and ethmosphenoid blocks of braincase. 0 = absent; 1 = present.
84. Basicranial fenestra with arcual plates. 0 = absent; 1 = present.
85. Unconstricted cranial notochord. 0 = absent; 1 = present.

86. Otico-sphenoid bridge. 0 = present; 1 = absent.
87. Position of intracranial joint relative to cranial nerves. 0 = joint passes through profundus foramen; 1 = joint passes through trigeminal foramen.
88. Condition of hyomandibular. 0 = hyomandibular with one proximal articular head; 1 = hyomandibular with two proximal articular heads.
89. Posttemporal fossae. 0 = absent; 1 = present.
90. Postorbital process on braincase (equivalent to character A3 of Chang and Smith, 1992). 0 = present; 1 = absent.
91. Dentary. 0 = long; 1 = short.
92. Anterior end of dentary. 0 = not modified; 1 = modified into support for parasymphysial tooth whorl.
93. Number of infradentaries. 0 = one; 1 = two; 2 = four.
94. Number of coronoids. 0 = four or more; 1 = three; 2 = two.
95. Condition of most posterior coronoid. 0 = not distinctly differentiated from other coronoids; 1 = well developed and oriented vertically.
96. Prearticular position. 0 = at posterior end of coronoid series, contacts dentary dorsally; 1 = ventral to the coronoid series, does not contact the dentary dorsally.
97. Articulation of symplectic with articular. 0 = absent; 1 = present.
98. Trajectory of supraorbital canal. 0 = canal passing between anterior and posterior nares; 1 = canal passing anterior to both nares.
99. Contact of supraorbital canal. 0 = supraorbital and infraorbital canals in contact rostrally; 1 = canals not in contact rostrally.
100. Relationship of infraorbital canal to premaxilla. 0 = infraorbital canal enters premaxilla; 1 = infraorbital canal follows dorsal margin of premaxilla.
101. Trajectory of otic canal. 0 = otic canal does not pass through growth center of postparietal; 1 = otic canal passes through growth center of postparietal.
102. Contact of otic canal. 0 = otic canal not joining supraorbital canal; 1 = otic canal joining supraorbital canal.
103. Position of anterior pit line. 0 = anterior pit line on postparietal; 1 = anterior pit line on parietal.
104. Position of posterior pit line. 0 = posterior pit line on posterior half of postparietal; 1 = posterior pit line on anterior half of postparietal.

105. Preopercular canal. 0 = preopercular canal ends at dorsal margin of preopercular; 1 = canal does not end at dorsal margin of preopercular.
106. Jugal canal. 0 = absent; 1 = present.
107. Position of infraorbital canal. 0 = ventral to anterior naris; 1 = dorsal to anterior naris.
108. Pit lines of lower jaw. 0 = middle pit line not developed into enclosed canal ("oral canal"); 1 = middle pit line developed into enclosed oral canal or intermediate morphology.
109. Pit line of lower jaw. 0 = anterior pit line not developed into enclosed canal; 1 = anterior pit line developed into enclosed canal linking oral and mandibular canals.
110. Trajectory of mandibular canal. 0 = mandibular canal passing through dentary; 1 = mandibular canal not passing through dentary.
111. Trajectory of mandibular canal. 0 = mandibular canal passing through most posterior infradentary; 1 = mandibular canal not passing through most posterior infradentary.
112. Anocleithrum. 0 = element developed as postcleithrum; 1 = element developed as anocleithrum *sensu stricto*; 2 = element absent.
113. Condition of anocleithrum/postcleithrum. 0 = exposed on surface; 1 = subdermal.
114. Depressed lamina of cleithrum. 0 = absent; 1 = present.
115. Dorsal end of cleithrum. 0 = pointed; 1 = broad and rounded.
116. Relationship of clavicle to cleithrum. 0 = ascending process of clavicle overlaps cleithrum laterally; 1 = ascending process of clavicle wraps round anterior edge of cleithrum, overlapping it both laterally and mesially.
117. Extracleithrum. 0 = absent; 1 = present.
118. Interclavicle. 0 = present; 1 = absent.
119. Proximal articular surface of humerus. 0 = concave; 1 = flat; 2 = convex.
120. Endoskeletal supports in pectoral fins. 0 = multiple elements articulating with girdle; 1 = single element ("humerus") articulating with girdle.
121. Entepicondylar foramen. 0 = absent; 1 = present.
122. Deltoid and supinator processes. 0 = absent; 1 = present.
123. Number of mesomeres in pectoral fin. 0 = three to five; 1 = seven or more.
124. Trifurcations in pectoral fin skeleton (i.e., mesomeres carrying both pre- and postaxial radials). 0 = absent; 1 = present.
125. Digits. 0 = absent; 1 = present.
126. Pelvis contacting vertebral column. 0 = no; 1 = yes.
127. Dorsal and anal fins. 0 = present; 1 = absent.
128. Basal plates in dorsal fin supports. 0 = absent; 1 = present.
129. Anterior dorsal fin support. 0 = separate radials and basal plate; 1 = single element.
130. Posterior branched radial complex in posterior dorsal fin. 0 = absent; 1 = present.
131. Basal scutes on fins. 0 = absent; 1 = present.
132. Relative length of proximal unsegmented part of lepidotrichium. 0 = much less than segmented part; 1 = similar to segmented part; 2 = much greater than segmented part.
133. Distal end of lepidotrichium. 0 = branched; 1 = single.
134. Epichordal lepidotrichia in tail. 0 = absent; 1 = present.
135. Relative size of epichordal and hypochordal lepidotrichia. 0 = epichordals less developed than hypochordals; 1 = epichordals and hypochordals equally developed; 2 = epichordals more developed than hypochordals.
136. Neural arches. 0 = bilateral halves of neural arch separated; 1 = halves fused.
137. Supraneural spines. 0 = present on thoracic and abdominal vertebrae; 1 = restricted to a few vertebrae at anterior end of column, or absent.
138. Condition of intercentra. 0 = ossified; 1 = not ossified.
139. Condition of pleurocentra. 0 = not ossified; 1 = ossified.
140. Ribs. 0 = absence of well-ossified ribs; 1 = presence of well-ossified ribs.

Appendix 2: List of Genera Used in the Phylogenetic Analysis

The genera are entered into the data set in alphabetical order so as to preclude any bias toward preconceived groupings during the phylogenetic analysis. The tetrapods, actinistians, and dipnoans included in the analysis are all stem-group members except *Crassigyrinus*, which is a probable crown tetrapod (Lebedev and Coates, 1995).

Acanthostega: Late Devonian (Famennian) tetrapod from eastern Greenland. Most of the skeleton is known.

Allenypterus: Early Carboniferous (Namurian) actinistian from Montana. Represented by complete, laterally compressed specimens.

Barameda: Early Carboniferous rhizodont from Victoria, Australia. Skull roof, palate, and some postcranial elements preserved as natural molds.

Beelarongia: Late Devonian (Frasnian) canowindrid osteolepiform from Victoria, Australia. Skull roof, cheek, shoulder girdle, and pectoral fin preserved as natural molds.

Cheirolepis: Middle to Late Devonian (Eifelian–Frasnian) actinopterygian known from Scotland and Québec. Represented by numerous complete, laterally compressed specimens.

Crassigyrinus: Early Carboniferous (Viséan–Serpukhovian) tetrapod from Scotland. Largely complete except for the tail.

Diabolepis: Early Devonian (Lochkovian) dipnoan from Yunnan, China. Only the skull roof, palate, and lower jaw have been described.

Diplocercides heiligenstockiensis: Late Devonian (Frasnian) actinistian from Bergisch-Gladbach, Germany. Skull roof, cheek, and part of the axial skeleton are known.

Diplocercides kaeseri: Late Devonian (Frasnian) actinistian from Hessen, Germany. The neurocranium has been described extensively.

Dipnorhynchus: a dipnoan. The best known species is *D. suessmilchii* from the Lower Devonian (Emsian) of New South Wales, Australia. Skull roof, braincase, palate, and lower jaw are known. Our coding also includes data from *D. kiandrensis* and *D. kurikae*.

Dipterus: Middle–Late Devonian (Eifelian–Frasnian) dipnoan from Scotland and Germany, represented by numerous complete bodies.

Elpistostege: Late Devonian (Frasnian) elpistostegid from Québec, Canada. Known from two incomplete skulls and a piece of vertebral column.

Eusthenopteron: widespread Late Devonian (Frasnian–Famennian) tristichopterid osteolepiform. The coding is based on *E. foordi* from Québec, which is represented by numerous complete bodies.

Glyptolepis: Middle–Late Devonian (Eifelian–Frasnian) holoptychiid porolepiform from Europe and Greenland. The whole body is known.

Gyroptychius: Middle Devonian (Eifelian–Givetian) "osteolepid" osteolepiform from Europe and Greenland. The whole body is known but only the dermal bones have been fully described.

Holoptychius: Late Devonian (Frasnian–Famennian) holoptychiid porolepiform of apparently worldwide distribution. Complete bodies, but endoskeleton rarely preserved.

Howqualepis: Late Devonian (Frasnian) actinopterygian from Victoria, Australia. Natural molds of complete bodies.

Ichthyostega: Late Devonian (Famennian) tetrapod from eastern Greenland. The dermal skull and most of the postcranium are known.

Miguashaia: Late Devonian (Frasnian) actinistian from Québec, Canada. Represented by complete but crushed bodies.

Mimia: Late Devonian (Frasnian) actinopterygian from Gogo, Western Australia. Whole body known in outstanding, three-dimensional detail.

Moythomasia: widespread Late Devonian (Frasnian) actinopterygian. Scored on basis of Gogo material comparable to that of *Mimia*.

Onychodus: widespread Late Devonian (Frasnian) onychodont. The best material comprises skull bones and some postcranial elements from Gogo.

Osteolepis: Middle Devonian (Eifelian–Givetian) "osteolepid" osteolepiform from Scotland. Many complete bodies, but little information about the internal skeleton.

Panderichthys: Late Devonian (Frasnian) elpistostegid from Latvia and Russia. *P. rhombolepis* is represented by complete bodies from Lode, Latvia.

Polypterus: a primitive Recent actinopterygian from equatorial Africa.

Porolepis: a "porolepid" porolepiform from the Lower Devonian (Pragian–Emsian) of Europe and Spitsbergen. The dermal skull, shoulder girdle, scales, and ethmosphenoid have been described.

Powichthys: porolepiform-like genus from the Lower Devonian (Lochkovian) of Arctic Canada. Braincase and skull roof known. Associated lower jaw, operculogular elements, and palatoquadrate probably also belong to genus.

Speonesydrion: Early Devonian (Siegenian) dipnoan from New South Wales, Australia. Only part of the skull has been described.

Strepsodus: Carboniferous (Dinantian–Westphalian) rhizodont from Europe and North America. Known from one complete but poorly preserved body and many isolated elements.

Strunius: a small onychodont from the Upper Devonian (Frasnian) of Germany and Latvia. Represented by complete but rather poorly preserved bodies.

Uranolophus: Early Devonian (Pragian) dipnoan from Wyoming. The skull roof, palate, lower jaw, and postcranial dermal skeleton are known.

Ventastega: Late Devonian (Famennian) tetrapod from Latvia. Lower jaw, palate, and cheekplate are known; associated clavicles, interclavicles, and ilia probably also belong to the genus.

Youngolepis: sarcopterygian genus from the Lower Devonian (Lochkovian) of Yunnan, China. Head and shoulder girdle are known.

Appendix 3: Data Set of 140 Characters for 28 Sarcopterygian Taxa

```
                       1111111111122222222223323333333334444444444555555555566666666667
              1234567890123456789012345678901234567890123456789012345678901234567890
Acanthostega       0L0L?000000111100101L0L1100110000101111LL101LL111001010110111L0L1L1L0
Allenypterus       0L1L00000000000011LL0000001200101011003?1?0010011011100110001L0L1L010
Barameda           0L1L10000001110001 0??L0000011000011010020 1????0??0010????????0?????00010
Beelarongia        1?0100002???????????0??001???0001?0000110?0002???10?0?001010?0?00?????????
Cheirolepis        0L0L010000000000P000000000010001000P0001L0L0000000000010002?000000L0000?
Crassigyrinus      0L0L0000000?111100101LLL1100110000100111LL101LL01?0010101110111L0L1L1L1
Diabolepis         11???0001L0000011???0??0001?12100?010?????011??????????????????????????0
Diplocercides kaeseri 0L1L00000000000011LL?0000012001010120020?1?0010011?1110011000 1L0L1L01?
D. heiligenstockiensis 0L1L0000000?00000011LL?0000012001010120020?1?0010?1101100110001L0L1L01?
Dipnorhynchus      1?0100L12L000?01LL1LLL000011?120000010 03?1102120??????????????????????????
Dipterus           111L00011L00000LL1LLL000111112100000 1002?1?02120100101011000021 1??01?
Elpistostege       0L0100002?0???100101L0001000110011011???101LL1??0??????????????1001??
Eusthenopteron     0L1L100000101000100101010000110000100100020101LL01100111011000021000011
Glyptolepis        0L1L0000001102000101L0000001000101112002111000101101101110100011 01L010
Gyroptychius       1L010000001010001000100000100001100011010020101LL0??0010101100000210000 11
Holoptychius       0L1L0000001102000101L0000001200101120021110001011011110100011 01L010
Howqualepis        0L00011000000000000001L00000L0000L000001L0L0000?0000000000000000L0000?
Ichthyostega       0L1L?000000111100101L0L1100110000101111LL101LL11?001010110111L0L1L1L0
Miguashaia         0L1L00000000000011LL0000001200101001002010?0010?100110011000 1L0L1L01?
Mimia              0L00011000000000000001L00000L0000L000000L0L00000000000000000000L00000
Moythomasia        0L00011000000000000001L00000L0000L000000L0L00000000000000000000L00000
Onychodus          0L1L0000001100000100001000000120010111002210001 0?100101002L0001L101L01L
Osteolepis         100100000010?00010010100000110001101002010 1GLL0110010101100000210000 1?
Panderichthys      0L0100000111100101L000100110001100110 20101LL11?0010101100002 1001011
Polypterus         0L000100000000000000LLLL01000L0010L00L000L0L0010 00LL0L00L1L0001L0L1L010
Porolepis          110100000001020001 01L0000001001100112002111000 10??01101111010001000010
Powichthys         1101000001?1010001???000000?01000?01002110001 11100?0?????????1000??0010
Speonesydrion      1?0?00012L000001LL???L00001??12000 001 0?????????????????????????000?11000??
Strepsodus         0L1L100000011100010 01?????00011 ??0???100??1????0?10???1?????????L10??01?
Strunius           0L1L0000001?000001000000000??001011?100211 00?1?00?20001?101L01?
Uranolophus        110100L10L000001LL1LLL000011?121000010 03??102120????????????????021100010
Ventastega         0L?L?0000001111001 01?0?????1?????0????1????1?1L?11?00??1011110?????????????0
Youngolepis        1101000001?1010011001 0?0000??0110?10100????0110??0??????10?10?0???1000010
```

```
                       1111111111111111111111111111111111111111111111111111111111111
              77777777788888888889999999999000000000011111111112222222222333333333334
              1234567890123456789012345678901234567890123456789012345678901234567890
Acanthostega       1101L0?0000000?LL11000210101 00??LL110??1117?01?002111?0111LLL0??1201011
Allenypterus       0011010???111??????????101?1 111??001L111?001111001?1?1????000 1?001111110 0
Barameda           101??0???1??????????????0??? ????????01??????????0110?2111 0??????100??????
Beelarongia        ?????????1?????????????0??? ??????????01???????01 10?2111 0??????100??????
Cheirolepis        ?0??0?????0?????L0??00000000 ?000?00000L000000000?0??L00?0?L?0000 0???0
Crassigyrinus      1011?0??0 00?000LL???00210101 0?????LL?????112L01?002111??7111LLL0LL??01001
Diabolepis         0010001?010?????????102LLL?10 ?0010???10 1??????????????????????????????
Diplocercides kaeseri ?0?1010000111111111101?1111 ??0101100 01 1??01001??????0??????011111?100
D. heiligenstockiensis ???1?10???111??????????101 ?1111??0???11?0011??0100 1??????0?0???001111?100
Dipnorhynchus      ?101L01LL100000LL?LL102LLL011 1L0010??11110??????????????????????????????
Dipterus           ?101001LL100000LL??L102LLL011 1L0110111101101101????11000101010?0001
Elpistostege       ????????0????????0?????????? ????????????????????????????????????????01 00?
Eusthenopteron     1011000000111110111100210101 0001101 10001100110021110000100100111 1011
Glyptolepis        0000100110111111101110121010 10011111100011110110100110001100001010010
Gyroptychius       101100??00111117????002?0?010 0011011000 11001?0?????00?01?10011010 1?
Holoptychius       00001001101111117?0121010100 111111000111101100?1?????0?0??0001 0?????
Howqualepis        ?0?000??00?0??0L0??00000100000 0000000000L000000??0??L00?0?L?000000?????
Ichthyostega       110100?00000001LL???002101000??LL110??112L0110021112?111LLL0??1211011
Miguashaia         ?0?0?????1????????101?11??? 001?011?001 1?0001 1?????0?01?7000100??100
Mimia              000000000000000L0000000000000000000000 0L0000000?000L00000L000000000 00
Moythomasia        000000000000000L0000000000000000000000 0L0000000?000L00000L000000000 00
Onychodus          L0L1000010111?110??0121010?? 001101107?101100001 0110??0?0??001 10?010
Osteolepis         ?0?1?00020111110 1??002?0??10 0011011 0001100110 0?????0?01?010010 01010
Panderichthys      1011L000007?????1?0 0210101 000 1LL1100011100110 0211?000?1LLL0001201001
Polypterus         L010L00000001L00100200010001 100000010001001 00 00?00000L000011 10??1
Porolepis          0000101101111110 1?01210101011 0?00011??????????????????????????????
Powichthys         00?000101001111101?0001210 1010101107?0001 1???????????????????????????
Speonesydrion      ?101001LL?000?0LL???102LLL?11L 00?????111 0????????????????????????????
Strepsodus         ?????????1??????????0?2 ?????01 ????????1111002111 0 1070 1020 11?????
Strunius           ?????????1??????????01?LLL??0 0110111001 1??0000 1?????0?0?7?0001 1??? 0
Uranolophus        L101L01LL10000?LL???102LLL??1L 0010 1??111?1101100?????0?0???000 ?01?000
Ventastega         110??0??00?????????00 2201?100 ??????1???11?????0?0?????1?????????????
Youngolepis        00100011100001 0110001210101010 0101101011??0?10?11?????????????????????
```

Note. 0 = plesiomorphic state; 1, 2, 3 = apomorphic states; ? = character not available; L = logical impossibility; P = polymorphic states (0 and 1).

Systematic Index

Taxa listed are at the generic level and higher. Page references in *italic* indicate illustrations.